COLLEGE MATHEMATICS

FOURTH EDITION

S. T. TAN

Stonehill College

BROOKS/COLE PUBLISHING COMPANY

An International Thomson Publishing Company

Pacific Grove • Albany • Belmont • Bonn • Boston • Cincinnati • Detroit • Johannesburg • London
Madrid • Melbourne • Mexico City • New York • Paris • Singapore • Tokyo • Toronto • Washington

To Pat, Bill, and Michael

Sponsoring Editor: *Margot Hanis*
Marketing Team: *Caroline Croley, Margaret Parks,*
  Debra Johnston
Editorial Assistant: *Kimberly Raborn*
Production Coordinator: *Marjorie Z. Sanders*
Production: *Cecile Joyner, The Cooper Company*
Manuscript Editor: *Carol Reitz*
Interior Design: *Julia Gecha*

Interior Illustration: *S T Associates*
Cover Design: *Lisa Henry*
Cover Illustration: *Judith L. Harkness*
Typesetting and Color Separation: *The PRD Group, Inc.*
Printing and Binding: *R. R. Donnelley & Sons Company/*
  Crawfordsville

Photo Credits: page xvi: (top) © Wolfgang Spunbarg/Photo Edit; (middle) Photo Disc, Inc.; (bottom) © Fotocronache Olympia/
Photo Edit. **xvii:** (top three photos) Photo Disk, Inc.; (bottom) © David Young-Wolff/Photo Edit. **xviii:** (top) Photo Disk,
Inc.; (middle) Elizabeth Zuckerman/Photo Edit; (bottom) Photo Disk, Inc. **2, 70, 172, 272, 322, 370, 460, 776, 928:** Photo Disc,
Inc. **210, 568:** © Tony Freeman/Photo Edit. **536:** © F. Ruggeri/The Image Bank. **682, 870:** © 1996 R. J. Western. **766:**
NASA/photo composite. **1026:** © Juergen Gebhardt/The Image Bank. **1076:** © Terje Rakke/The Image Bank.

For more information, contact:

BROOKS/COLE PUBLISHING COMPANY
511 Forest Lodge Road
Pacific Grove, CA 93950
USA

International Thomson Publishing Europe
Berkshire House 168-173
High Holborn
London WC1V 7AA
England

Thomas Nelson Australia
102 Dodds Street
South Melbourne, 3205
Victoria, Australia

Nelson Canada
1120 Birchmount Road
Scarborough, Ontario
Canada M1K 5G4

International Thomson Editores
Seneca 53
Col. Polanco
11560 México, D. F., México

International Thomson Publishing GmbH
Königswinterer Strasse 418
53227 Bonn
Germany

International Thomson Publishing Japan
Hirakawacho Kyowa Building, 3F
2-2-1 Hirakawacho
Chiyoda-ku, Tokyo 102
Japan

International Thomson Publishing Asia
221 Henderson Road
#05-10 Henderson Building
Singapore 0315

Printed in the United States of America

10 9 8 7 6 5 4 3 2 1

Library of Congress Cataloging-in-Publication Data

Tan, Soo Tang.
    College mathematics / S. T. Tan.—4th ed.
        p. cm.
    Includes index.
    ISBN 0-534-36121-8
    1. Mathematics. 2. Management—Mathematics. 3. Social sciences—
    Mathematics. I. Title.
QA7.2.T36 1999
510—dc21
                                        98-15592
                                           CIP

CONTENTS

* Sections marked with an asterisk are not prerequisites for later material.

PREFACE

College Mathematics, Fourth Edition, treats the standard topics in mathematics and calculus that are usually covered in a two-semester course for students in the managerial, life, and social sciences. The only prerequisite for understanding this book is a year of high school algebra. The author's objective in writing *College Mathematics* was to provide a textbook that is both readable by students and useful as a teaching tool for instructors. We hope that with this edition we have come one step closer to realizing our goal. This edition incorporates many suggestions by users and reviewers.

FEATURES

The following list includes some of the book's many important features:

- **Coverage of Topics** Since the book contains more than enough material for the usual two-semester or three-quarter course, the instructor may be flexible in choosing the topics most suitable for his or her course. The following chart on chapter dependency is provided to help the instructor design a course that is most suitable for the intended audience.

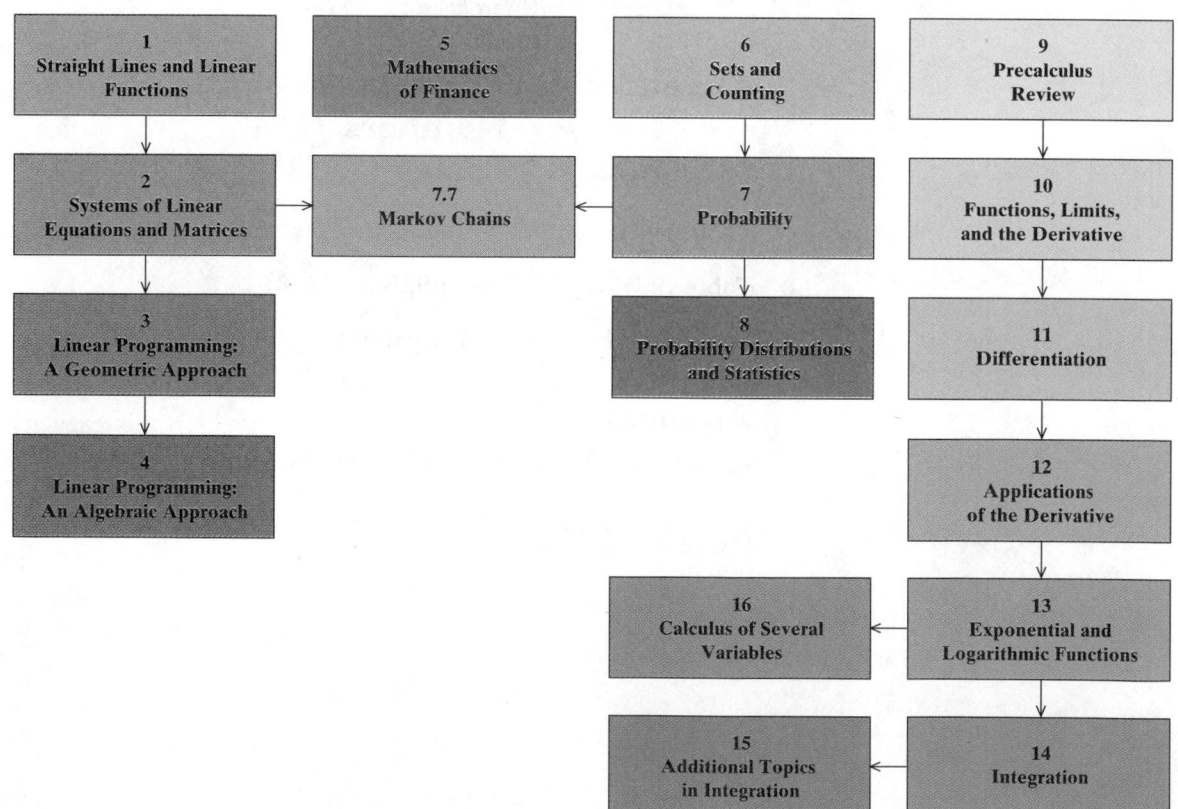

- **Approach** The problem-solving approach is stressed throughout the book. Numerous examples and solved problems are used to amplify each new concept or result in order to facilitate students' comprehension of the material. Figures are used extensively to help students visualize the concepts and ideas being presented.

- **Level of Presentation** Our approach is intuitive and we state the results informally. However, we have taken special care to ensure that this approach does not compromise the mathematical content and accuracy. Proofs of certain results are given, but they may be omitted if desired.

- **Applications** The text is application oriented. Many interesting, relevant, and up-to-date applications are drawn from the fields of business, economics, social and behavioral sciences, life sciences, physical sciences, and other fields of general interest. Some of these applications have their source in newspapers, weekly periodicals, and other magazines. Applications are found in the illustrative examples in the main body of the text as well as in the exercise sets.

- **Exercises** Each section of the text is accompanied by an extensive set of exercises containing an ample set of problems of a routine, computational nature that will help students master new techniques. The routine problems are followed by an extensive set of application-oriented problems that test students' mastery of the topics. Each chapter of the text also contains a set of review exercises. Answers to all odd-numbered exercises appear in the back of the book.

- **Portfolios** These interviews are designed to convey to the student the real-world experiences of professionals who have a background in mathematics and use it in their professions.

TECHNOLOGY

In this book our primary emphasis is on the use of the graphing utility, which is employed in two ways:

1. It is used to explore mathematical concepts and also to shed further light on selected worked examples in the text. In this capacity, it serves to augment the analytic solution obtained in the text with a graphical and/or numerical solution, thus adding to a greater understanding of the problem. Exercises in this category appear under the heading:

 EXPLORING WITH TECHNOLOGY

2. It is used to solve problems whose solutions require a prodigious amount of calculations. Indeed, it is the availability of the graphing utility that makes it possible to include the many real-life problems in this book. Such exercises are found at the end of many subsections in the book under the heading:

/ / / USING TECHNOLOGY

NEW IN THE 4TH EDITION

Exploring with Technology Questions

These optional questions appear throughout the main body of the text and serve to enhance the student's understanding of the concepts and theory presented. Complete solutions to these exercises are given in the *Complete Solutions Manual*.

Using Technology Subsections

These pages contain optional material and are placed at the end of the sections for which their use is appropriate. The subsections are written in the traditional example–exercise format with answers given at the back of the book. They may be used in the classroom if desired or as material for self-study by the student.

As many up-to-date and relevant applications have been introduced in these subsections, they provide students with an opportunity to interpret results in a real-life setting.

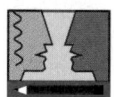

Group Discussion Questions

These are optional questions, appearing throughout the main body of the text, that can be discussed in class or assigned as homework. These questions generally require more thought and effort than the usual exercises. Complete solutions to these exercises are given in the *Complete Solutions Manual*.

Critical Thinking Questions

These are new exercises that go beyond the usual exercises involving computation and provide the student with a little more challenge. Many of these questions call for the interpretation of results or the theory presented in the section.

New Internet-Linked Supplements

Students and instructors will now have access to these additional materials at the Brooks/Cole World Wide Web site: http://www.brookscole.com/math/authors/tans

- Study hints, review material, and practice chapter tests for students
- Group projects and extended problems for each chapter
- Instructions, including keystrokes, for the procedures referenced in the text for specific calculators (TI-82, TI-85, and other popular models)
- Modified Using Technology sections for CAS systems, including the command statements for Mathematica, Maple, and other popular systems

Sources

We have included sources for those applications that are based on real-life data.

New Exercises

A wealth of innovative and timely exercises have been added to pique student interest. More exercises (both rote and applied) calling for graphical interpretations have also been added.

Helpful Changes

We have included sources for those applications that are based on real-life data.

The discussion of simple interest has been moved from Section 1.3 to Section 5.1 and is now covered along with compound interest.

A new real-life example—the motion of a magnetic levitation train—is used to motivate topics in limits, derivatives, and integration. The use of the motion of a familiar object in the different settings serves to enhance comprehension of the topics.

Chapter 10: A new illustrative example, "Spending by Businesses on Computer Security Equipment," has been added to the discussion of mathematical models in Section 10.3. The motion of a maglev is now used to motivate an intuitive discussion of the Intermediate Value Theorem.

Chapter 11: The use of the application "Estimating the Size of the Rings of Neptune" in Section 11.7, Differentials, is yet another example of the importance attached to our philosophy of motivating topics with real-life examples. A discussion of relative error has also been added to this section.

Chapter 12: The chapter has been reorganized and condensed. Sections 12.1 to 12.3 now cover the material previously covered in five sections. Topics are now organized into Applications of the First Derivative, Applications of the Second Derivative, and Curve Sketching.

Chapter 14: An early introduction to differential equations and a discussion of initial value problems have been added.

SUPPLEMENTS

- *Student's Solutions Manual,* available to both students and instructors, includes the solutions to odd-numbered exercises. ISBN 0-534-36122-6

- *Instructor's Complete Solutions Manual* includes solutions to all exercises. ISBN 0-534-36123-4

- *Test Bank with Chapter Tests,* free to adopters of the book, contains sample tests for each chapter. ISBN 0-534-36133-1

- *Thomson World Class Testing Tools* (Macintosh ISBN 0-534-36135-8; Windows ISBN 0-534-36125-0) This fully-integrated suite of test creation, delivery, and classroom management tools includes World Class Test, Test Online, and World Class Management software. World Class Testing Tools allows professors to deliver tests via print, floppy, hard drive, LAN, or Internet. With these tools, professors can create cross-platform exam files from publisher files or existing WESTest 3.2 test banks, create and edit questions, and provide their own feedback to objective test questions—enabling the system to work as a tutorial or an examination. In addition, professors can generate questions algorithmically, creating tests that include multiple-choice, true/false, and matching questions. Professors can also track the progress of an entire class or an individual student. Testing and tutorial results can be integrated into the class management tool, which offers scoring, gradebook, and reporting capabilities.

ACKNOWLEDGMENTS

I wish to express my personal appreciation to each of the following reviewers, whose many suggestions have helped make a much improved book.

Reviewers of the Fourth Edition:

Daniel D. Anderson
University of Iowa

Ronald Barnes
University of Houston-Downtown

Charles E. Cleaver
The Citadel

Michael W. Ecker
Pennsylvania State University, Wilkes-Barre Campus

Karen J. Hay
Phoenix College

Jeff Knisley
East Tennessee State University

Georgia B. Pyrros
University of Delaware

Nathan P. Ritchey
Youngstown State University

Arnold L. Schroeder
Long Beach City College

Barbara J. Shabell
California State Polytechnic University

Joseph F. Stokes
Western Kentucky University

Lowell Stultz
Texas Township Campus

Reviewers of the Previous Editions:

Wayne Andrepont
University of Southern Louisiana

Ghidewon Abay Asmerom
Virginia Commonwealth University

Wilson Banks
Illinois State University

Richard H. Bouldin
University of Georgia

Chris C. Braunschweiger
Marquette University

Ralph Bravaco
Stonehill College

David H. Carlson
Oregon State University

Lee Corbin
College of the Canyons

Duane E. Deal
Ball State University

Robert Eiken
Illinois Central College

Joseph W. Fitzpatrick
University of Texas at El Paso

Jimmy R. Hickey
Baylor University

Franz X. Hiergeist
West Virginia University

James R. McKinney
California State Polytechnic University at Pomona

Stephen J. Merrill
Marquette University

W. Kent Moore
Valdosta State College

Robert A. Moreland
Texas Tech University

Richard D. Porter
Northeastern University

Kimmo I. Rosenthal
Union College

Derald D. Rothman
Moorhead State University

Wesley J. Rozema
Northern Arizona University

Wesley Sanders
Sam Houston State University

James Wallen
Moorhead State University

Murray Wanger
Sacramento City College

Raymond Young
Embry-Riddle Aeronautical University

Thomas M. Zachariah
Loyola Marymount University

A special thanks also goes to Arthur J. Rosenthal, Salem State College, and Bruce R. Johnson, University of Victoria. My thanks also go to the editorial and production staffs of Brooks/Cole: Margot Hanis, Jennifer Wilkinson, Caroline Croley, Margaret Parks, Debra Johnston, Marjorie Sanders, and Vernon Boes for their thoughtful contributions and patient assistance and cooperation during the development and production of this book. Finally, I wish to thank Cecile Joyner of the Cooper Company and Carol Reitz, for doing an excellent job ensuring the accuracy and readability of this fourth edition, and Julie Gecha for her design of the interior of the book.

S. T. Tan

APPLICATIONS

*In **College Mathematics** we attempt to solve a wide variety of problems arising from many diverse fields of study. A small sample of the types of practical problems we will consider follows. In developing the tools for solving these and many other problems, we will delve into many areas of mathematics, including linear algebra, linear programming, probability, statistics, logic, and calculus.*

ALLOCATION OF FUNDS   Madison Finance has a total of $20 million earmarked for homeowner and auto loans. On the average, homeowner loans have a 10% annual rate of return, whereas auto loans yield a 12% annual rate of return. Management has also stipulated that the total amount of homeowner loans should be greater than or equal to four times the total amount of automobile loans. Determine the total amount of loans of each type Madison should extend to each category in order to maximize its returns.

AUTOMOBILE SAFETY   In an experiment conducted to study the effectiveness of an eye-level third brake light in the prevention of rear-end collisions, 250 of a state's 500 highway patrol cars were equipped with such lights. At the end of a 1-year trial period, the record revealed that of the cars equipped with a third brake light, 14 were involved in rear-end collisions. There were 22 such incidents involving cars not equipped with the accessory. Based on these data, what is the probability that a highway patrol car equipped with a third brake light will be rear-ended within a 1-year period?

SERUM CHOLESTEROL LEVELS   The serum cholesterol levels (in mg/dl) in a current Mediterranean population are found to be normally distributed with a mean of 160 and a standard deviation of 50. Scientists at the National Heart, Lung, and Blood Institute consider this pattern ideal for a minimal risk of heart attacks. Find the percentage of the population having blood cholesterol levels between 160 and 180 mg/dl.

POPULATION GROWTH   A study prepared for a Sunbelt town's Chamber of Commerce projected that the town's population in the next three years would grow according to the rule

$$P(x) = 50,000 + 30x^{3/2} + 20x$$

where $P(x)$ denotes the population x months from now. How fast will the population be increasing nine months from now? Sixteen months from now?

AIR POLLUTION   According to the South Coast Air Quality Management District, the level of nitrogen dioxide, a brown gas that impairs breathing, present in the atmosphere on a certain May day in downtown Los Angeles is approximated by

$$A(t) = 0.03t^3(t - 7)^4 + 60.2 \qquad (0 \le t \le 7)$$

where $A(t)$ is measured in pollutant standard index and t is measured in hours, with $t = 0$ corresponding to 7 A.M. How fast is the level of nitrogen dioxide increasing at 11 A.M.?

URBAN–SUBURBAN POPULATION FLOW
Because of the continued successful implementation of an urban renewal program, it is expected that each year 3% of the population currently residing in the city will move to the surburbs, and 6% of the population currently residing in the surburbs will move to the city. At present, 65% of the total population of the metropolitan area live in the city itself, whereas the remaining 35% live in the suburbs. If we assume that the total population of the metropolitan area remains constant, what will the distribution of the population be like one year from now?

LEARNING CURVES   The Eastman Optical Company produces a 35-mm single-lens reflex camera. Eastman's training department determined that after completing the basic training program, a new, previously inexperienced employee would be able to assemble $Q(t) = 50 - 30e^{-0.5t}$ model F cameras per day, t months after the employee began work on the assembly line. How many model F cameras can a new employee assemble per day after basic training? How many model F cameras can the average experienced employee assemble per day?

SUBWAY FARES A city's Metropolitan Transit Authority (MTA) operates a subway line for commuters from a certain suburb to the downtown metropolitan area. Currently, an average of 6000 passengers a day take the trains, paying a fare of $1.50 per ride. The Board of the MTA, contemplating raising the fare to $1.75 per ride in order to generate a larger revenue, engaged the services of a consulting firm. The firm's study revealed that for each 25-cent increase in fare, the ridership would be reduced by an average of 1000 passengers a day. The consulting firm recommended that MTA stick to the current fare of $1.50 per ride, which already yields a maximum revenue. Show that the consultants' recommendations were correct.

ADVERTISING EXPENDITURES The Ross-Simons Company has a monthly advertising budget of $60,000. Their marketing department estimates that if they spend x dollars on newspaper advertising and y dollars on television advertising, then the monthly sales will be given by $z = f(x, y) = 90x^{1/4}y^{3/4}$ dollars. Determine how much money Ross-Simons should spend on newspaper ads and on television ads per month to maximize its monthly sales.

OPTIMAL DRIVING SPEED A truck gets $400/x$ miles per gallon when driven at a constant speed of x miles per hour (between 50 and 70 miles per hour). If the price of fuel is $1 per gallon and the driver is paid $8 an hour, at what speed between 50 and 70 miles per hour is it most economical to drive?

COLLEGE
MATHEMATICS

This chapter introduces the Cartesian coordinate system, a system that allows us to represent points in the plane in terms of ordered pairs of real numbers. This, in turn, enables us to compute the distance between two points algebraically. We also study straight lines. *Linear functions*, whose graphs are straight lines, can be used to describe many relationships between two quantities. These relationships can be found in fields of study as diverse as business, economics, the social sciences, physics, and medicine. In addition, we see how some practical problems can be solved by finding the point(s) of intersection of two straight lines. Finally, we learn how to find an algebraic representation of the straight line that "best" fits a set of data points that are scattered about a straight line.

Which process should the company use? The Robertson Controls Company must decide between two manufacturing processes for its Model C electronic thermostats. In Example 4, page 47, you will see how to determine which process will be more profitable.

STRAIGHT LINES AND LINEAR FUNCTIONS

1.1 THE CARTESIAN COORDINATE SYSTEM

The Cartesian Coordinate System

The real number system is made up of the set of real numbers together with the usual operations of addition, subtraction, multiplication, and division. We assume that you are familiar with the rules governing these algebraic operations (see Appendix A).

Real numbers may be represented geometrically by points on a line. Such a line is called the **real number,** or **coordinate, line.** We can construct the real number line as follows: Arbitrarily select a point on a straight line to represent the number zero. This point is called the **origin.** If the line is horizontal, then choose a point at a convenient distance to the right of the origin to represent the number 1. This determines the scale for the number line. Each positive real number x lies x units to the right of zero, and each negative real number $-x$ lies x units to the left of zero.

In this manner, a one-to-one correspondence is set up between the set of real numbers and the set of points on the number line, with all the positive numbers lying to the right of the origin and all the negative numbers lying to the left of the origin (Figure 1.1).

Figure 1.1
The real number line.

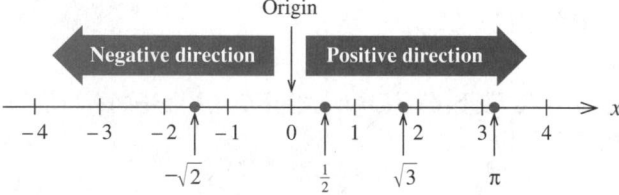

In a similar manner, we can represent points in a plane (a two-dimensional space) by using the **Cartesian coordinate system,** which we construct as follows: Take two perpendicular lines, one of which is normally chosen to be horizontal. These lines intersect at a point O, called the **origin** (Figure 1.2). The horizontal line is called the **x-axis,** and the vertical line is called the **y-axis.** A number scale is set up along the x-axis, with the positive numbers lying to the right of the origin and the negative numbers lying to the left of it. Similarly, a number scale is set up along the y-axis, with the positive numbers lying above the origin and the negative numbers lying below it.

Figure 1.2
The Cartesian coordinate system.

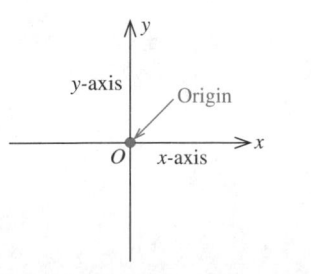

REMARK   The number scales on the two axes need not be the same. Indeed, in many applications different quantities are represented by x and y. For example, x may represent the number of typewriters sold and y the total revenue resulting from the sales. In such cases it is often desirable to choose different number scales to represent the different quantities. Note, however, that the zeros of both number scales coincide at the origin of the two-dimensional coordinate system.   ○ ○ ○

Figure 1.3

An ordered pair in the coordinate plane.

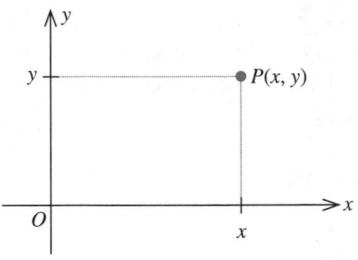

We can represent a point in the plane uniquely in this coordinate system by an **ordered pair** of numbers—that is, a pair (x, y), where x is the first number and y the second. To see this, let P be any point in the plane (Figure 1.3). Draw perpendiculars from P to the x-axis and y-axis, respectively. Then the number x is precisely the number that corresponds to the point on the x-axis at which the perpendicular through P hits the x-axis. Similarly, y is the number that corresponds to the point on the y-axis at which the perpendicular through P crosses the y-axis.

Conversely, given an ordered pair (x, y), with x as the first number and y the second, a point P in the plane is uniquely determined as follows: Locate the point on the x-axis represented by the number x and draw a line through that point parallel to the y-axis. Next, locate the point on the y-axis represented by the number y and draw a line through that point parallel to the x-axis. The point of intersection of these two lines is the point P (Figure 1.3).

In the ordered pair (x, y), x is called the **abscissa**, or **x-coordinate**, y is called the **ordinate**, or **y-coordinate**, and x and y together are referred to as the **coordinates** of the point P. The point P with x-coordinate equal to a and y-coordinate equal to b is often written $P(a, b)$.

The points $A(2, 3)$, $B(-2, 3)$, $C(-2, -3)$, $D(2, -3)$, $E(3, 2)$, $F(4, 0)$, and $G(0, -5)$ are plotted in Figure 1.4.

REMARK   In general, $(x, y) \neq (y, x)$. This is illustrated by the points A and E in Figure 1.4.   ○ ○ ○

The axes divide the plane into four quadrants. Quadrant I consists of the points P with coordinates x and y, denoted by $P(x, y)$, satisfying $x > 0$ and $y > 0$; Quadrant II, the points $P(x, y)$, where $x < 0$ and $y > 0$; Quadrant III, the points $P(x, y)$, where $x < 0$ and $y < 0$; and Quadrant IV, the points $P(x, y)$, where $x > 0$ and $y < 0$ (Figure 1.5).

Figure 1.4

Several points in the coordinate plane.

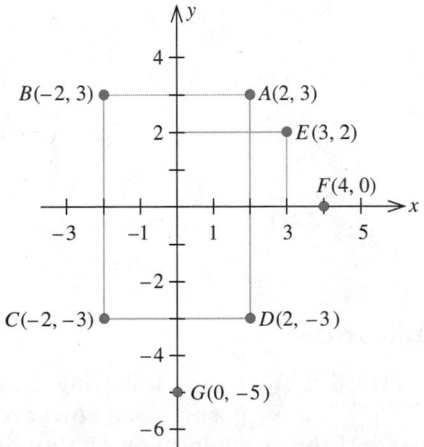

Figure 1.5

The four quadrants in the coordinate plane.

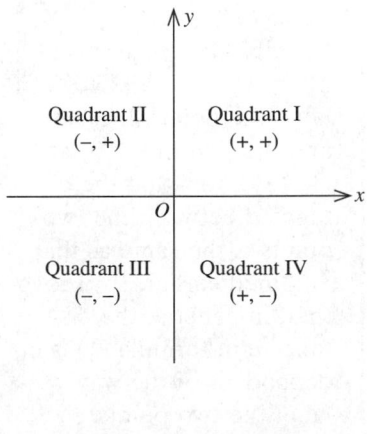

The Distance Formula

One immediate benefit that arises from using the Cartesian coordinate system is that the distance between any two points in the plane may be expressed solely in terms of the coordinates of the points. Suppose, for example, that (x_1, y_1) and (x_2, y_2) are any two points in the plane (Figure 1.6). Then the distance d between these two points is, by the Pythagorean Theorem,

$$d = \sqrt{(x_2 - x_1)^2 + (y_2 - y_1)^2}$$

For a proof of this result see exercise 36, page 10.

Figure 1.6
The distance between two points in the coordinate plane.

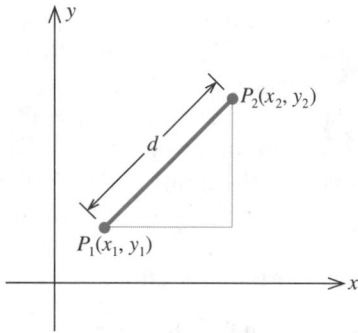

DISTANCE FORMULA   The distance d between two points $P_1(x_1, y_1)$ and $P_2(x_2, y_2)$ in the plane is given by

$$d = \sqrt{(x_2 - x_1)^2 + (y_2 - y_1)^2} \qquad \textbf{(1)}$$

EXAMPLE 1   Find the distance between the points $(-4, 3)$ and $(2, 6)$.

Solution   Let $P_1(-4, 3)$ and $P_2(2, 6)$ be points in the plane. Then we have

$$x_1 = -4, \; y_1 = 3, \; x_2 = 2, \text{ and } y_2 = 6$$

Using formula (1), we have

$$\begin{aligned}
d &= \sqrt{[2 - (-4)]^2 + (6 - 3)^2} \\
&= \sqrt{6^2 + 3^2} \\
&= \sqrt{45} \\
&= 3\sqrt{5}
\end{aligned}$$

Refer to Example 1
Suppose we label the point $(2, 6)$ as P_1 and the point $(-4, 3)$ as P_2. (a) Show that the distance d between the two points is the same as that obtained earlier. (b) Prove that, in general, the distance d in formula (1) is independent of the way we label the two points.

Application

EXAMPLE 2   In the following diagram (Figure 1.7), S represents the position of a power relay station located on a coastal high-way and M shows the location of a marine biology experimental station on a

Figure 1.7
The cable will connect the relay station S to the experimental station M.

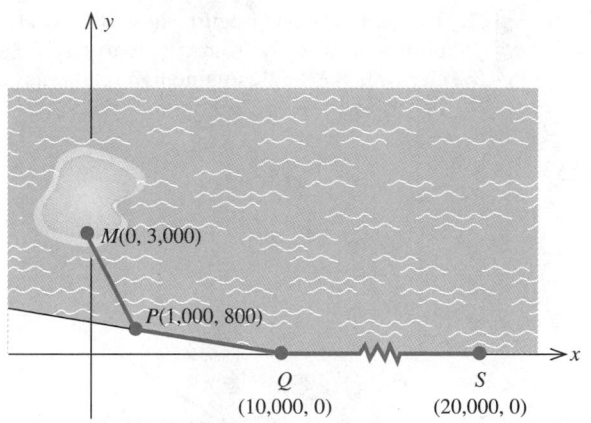

nearby island. A cable is to be laid connecting the relay station with the experimental station. If the cost of running the cable on land is $2 per running foot and the cost of running the cable under water is $6 per running foot, find the total cost for laying the cable.

Solution   The length of cable required on land is given by the distance from P to Q plus the distance from Q to S—that is,

$$\sqrt{(10,000 - 1000)^2 + (0 - 800)^2} + \sqrt{(20,000 - 10,000)^2 + (0 - 0)^2}$$
$$= \sqrt{9000^2 + 800^2} + 10,000$$
$$= \sqrt{81,640,000} + 10,000$$
$$\approx 19,035.49$$

or approximately 19,035.49 feet. Next, we see that the length of cable required underwater is given by

$$\sqrt{(0 - 1000)^2 + (3000 - 800)^2} = \sqrt{1000^2 + 2200^2}$$
$$= \sqrt{5,840,000}$$
$$\approx 2416.61$$

or approximately 2416.61 feet. Therefore, the total cost for laying the cable is

$$2(19,035.49) + 6(2416.61) = 52,570.64$$

or approximately $52,571.   ◐ ◐ ◐

SELF–CHECK EXERCISES 1.1

1. a. Plot the points $A(4, -2)$, $B(2, 3)$, and $C(-3, 1)$.
   b. Find the distance between the points A and B; between B and C; between A and C.
   c. Use the Pythagorean Theorem to show that the triangle with vertices A, B, and C is a right triangle.

2. The accompanying figure shows the location of cities A, B, and C. Suppose that a pilot wishes to fly from city A to city C but must make a mandatory stopover in city B. If the single-engine light plane has a range of 650 miles, can the pilot make the trip without refueling in city B?

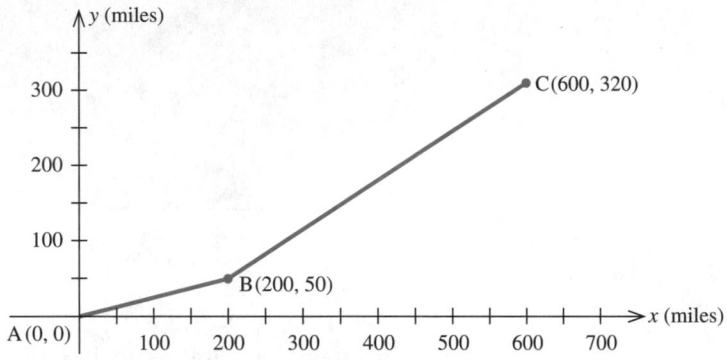

Solutions to Self-Check Exercises 1.1 can be found on page 11.

1.1 EXERCISES

In exercises 1–6, refer to the accompanying figure and determine the coordinates of the given point and the quadrant in which it is located.

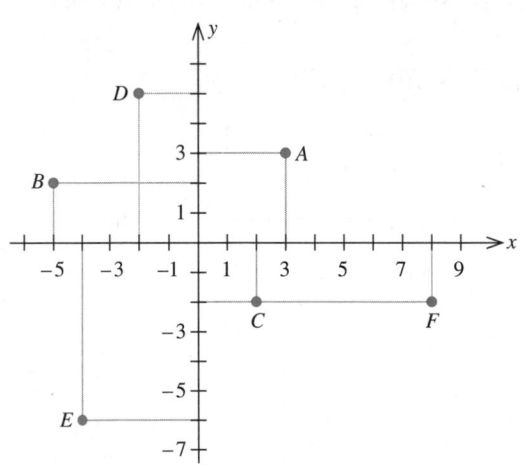

1. *A*   2. *B*   3. *C*

4. *D*   5. *E*   6. *F*

In exercises 7–12, refer to the accompanying figure.

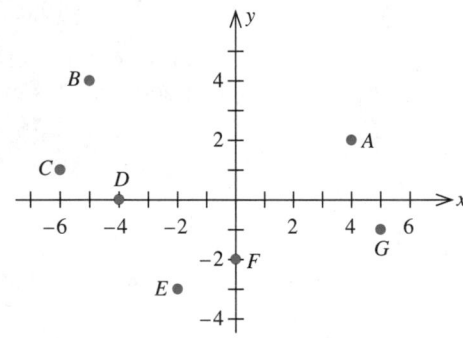

7. Which point has coordinates (4, 2)?

8. What are the coordinates of point *B*?

9. Which points have negative *y*-coordinates?

10. Which point has a negative *x*-coordinate and a negative *y*-coordinate?

11. Which point has an *x*-coordinate that is equal to zero?

12. Which point has a *y*-coordinate that is equal to zero?

In exercises 13–20, sketch a set of coordinate axes and plot the given point.

13. $(-2, 5)$ **14.** $(1, 3)$

15. $(3, -1)$ **16.** $(3, -4)$

17. $(8, -7/2)$ **18.** $(-5/2, 3/2)$

19. $(4.5, -4.5)$ **20.** $(1.2, -3.4)$

In exercises 21–24, find the distance between the given points.

21. $(1, 3)$ and $(4, 7)$ **22.** $(1, 0)$ and $(4, 4)$

23. $(-1, 3)$ and $(4, 9)$ **24.** $(-2, 1)$ and $(10, 6)$

25. Find the coordinates of the points that are 10 units away from the origin and have a *y*-coordinate equal to -6.

26. Find the coordinates of the points that are 5 units away from the origin and have an *x*-coordinate equal to 3.

27. Show that the points $(3, 4)$, $(-3, 7)$, $(-6, 1)$, and $(0, -2)$ form the vertices of a square.

28. Show that the triangle with vertices $(-5, 2)$, $(-2, 5)$, and $(5, -2)$ is a right triangle.

29. Distance Traveled   A grand tour of four cities begins at city A and makes successive stops at cities B, C, and D before returning to city A. If the cities are located as shown in the accompanying figure, find the total distance covered on the tour.

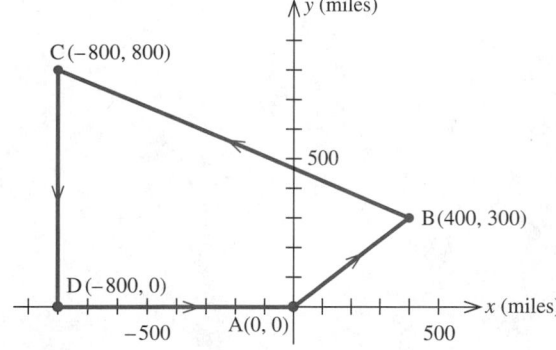

30. Delivery Charges   A furniture store offers free setup and delivery services to all points within a 25-mile radius of its warehouse distribution center. If you live 20 miles east and 14 miles south of the warehouse, will you incur a delivery charge? Justify your answer.

31. Travel Time   Towns A, B, C, and D are located as shown in the accompanying figure. Two highways link

town A to town D. Route 1 runs from town A to town D via town B, and route 2 runs from town A to town D via town C. If a salesman wishes to drive from town A to town D and traffic conditions are such that he could expect to average the same speed on either route, which highway should he take in order to arrive in the shortest time?

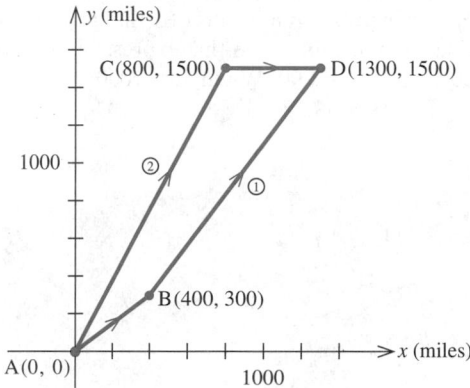

32. Minimizing Shipping Costs   Refer to the figure for exercise 31. Suppose that a fleet of 100 automobiles are to be shipped from an assembly plant in town A to town D. They may be shipped either by freight train along route 1 at a cost of 11 cents per mile per automobile or by truck along route 2 at a cost of $10\frac{1}{2}$ cents per mile per automobile. Which means of transportation minimizes the shipping cost? What is the net savings?

33. Consumer Decisions   Mr. Barclay wishes to determine which antenna he should purchase for his home. The TV store has supplied him with the following information:

Range in miles

VHF	UHF	Model	Price
30	20	A	$40.00
45	35	B	$50.00
60	40	C	$60.00
75	55	D	$70.00

Barclay wishes to receive channel 17 (VHF), which is located 25 miles east and 35 miles north of his home, and channel 38 (UHF), which is located 20 miles south and 32 miles west of his home. Which model will allow him to receive both channels at the least cost? (Assume that the terrain between Barclay's home and both broadcasting stations is flat.)

34. Calculating the Cost of Laying Cable   In the accompanying diagram, S represents the position of a power relay station located on a coastal highway, and M shows the location of a marine biology experimental station on a nearby island. A cable is to be laid connecting the relay station with the experimental station. If the cost of running the cable on land is \$2 per running foot and the cost of running cable underwater is \$6 per running foot, find an expression in terms of x that gives the total cost of laying the cable. Use this expression to find the total cost when $x = 900$. When $x = 1000$.

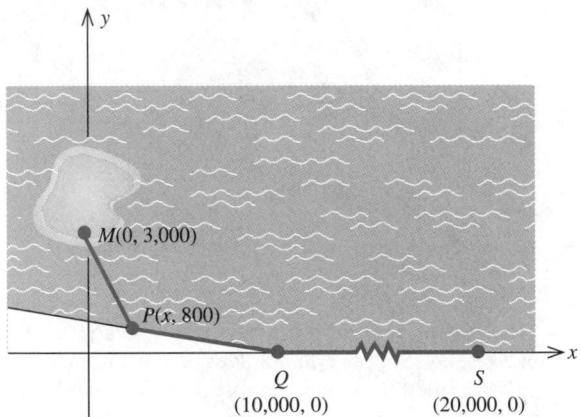

$M(0, 3{,}000)$

$P(x, 800)$

Q
$(10{,}000, 0)$

S
$(20{,}000, 0)$

35. In the Cartesian coordinate system, the two axes are perpendicular to each other. Consider a coordinate system in which the x- and y-axis are noncollinear (that is, the axes do not lie along a straight line) and are not perpendicular to each other (see the accompanying figure).

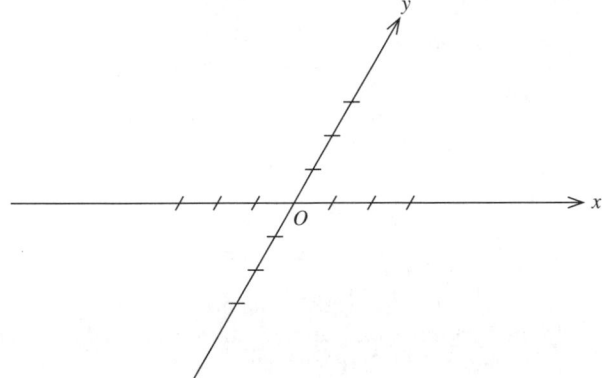

a. Describe how a point is represented in this coordinate system by an ordered pair (x, y) of real numbers. Conversely, show how an ordered pair (x, y) of real numbers uniquely determines a point in the plane.

b. Suppose you want to find a formula for the distance between two points, $P_1(x_1, y_1)$ and $P_2(x_2, y_2)$, in the plane. What advantage does the Cartesian coordinate system have over the coordinate system under consideration? Comment on your answer.

36. Let (x_1, y_1) and (x_2, y_2) be two points lying in the xy-plane. Show that the distance between the two points is given by

$$d = \sqrt{(x_2 - x_1)^2 + (y_2 - y_1)^2}$$

[*Hint:* Refer to the accompanying figure and use the Pythagorean Theorem.]

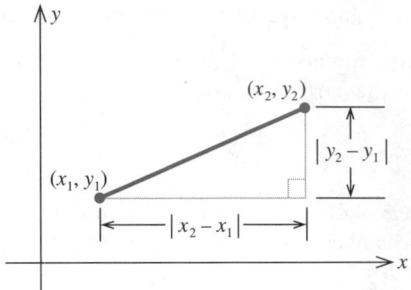

(x_2, y_2)

(x_1, y_1)

$|y_2 - y_1|$

$|x_2 - x_1|$

SOLUTIONS TO SELF–CHECK EXERCISES 1.1

1. a. The points are plotted in the accompanying figure.

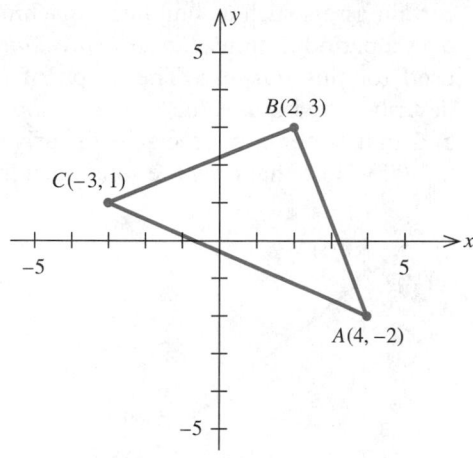

b. The distance between A and B is

$$d(A, B) = \sqrt{(2 - 4)^2 + [3 - (-2)]^2}$$
$$= \sqrt{(-2)^2 + 5^2} = \sqrt{4 + 25} = \sqrt{29}$$

The distance between B and C is

$$d(B, C) = \sqrt{(-3 - 2)^2 + (1 - 3)^2}$$
$$= \sqrt{(-5)^2 + (-2)^2} = \sqrt{25 + 4} = \sqrt{29}$$

The distance between A and C is

$$d(A, C) = \sqrt{(-3 - 4)^2 + [1 - (-2)]^2}$$
$$= \sqrt{(-7)^2 + 3^2} = \sqrt{49 + 9} = \sqrt{58}$$

c. We will show that

$$[d(A, C)]^2 = [d(A, B)]^2 + [d(B, C)]^2$$

From (b), we see that $[d(A, B)]^2 = 29$, $[d(B, C)]^2 = 29$, and $[d(A, C)]^2 = 58$, and the desired result follows.

2. The distance between city A and city B is

$$d(A, B) = \sqrt{200^2 + 50^2} \approx 206$$

or 206 miles. The distance between city B and city C is

$$d(B, C) = \sqrt{[600 - 200]^2 + [320 - 50]^2}$$
$$= \sqrt{400^2 + 270^2} \approx 483$$

or 483 miles. Therefore, the total distance the pilot would have to cover is 689 miles, so she must refuel in city B.

1.2 STRAIGHT LINES

In computing income tax, business firms are allowed by law to depreciate certain assets such as buildings, machines, furniture, automobiles, and so on, over a period of time. *Linear depreciation,* or the *straight-line method,* is often used for this purpose. The graph of the straight line shown in Figure 1.8 describes the book value V of a computer that has an initial value of $100,000 and that is being depreciated linearly over five years with a scrap value of $30,000. Note that only the solid portion of the straight line is of interest here.

Figure 1.8
Linear depreciation of an asset.

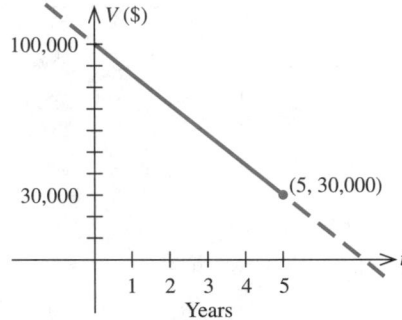

The book value of the computer at the end of year t, where t lies between 0 and 5, can be read directly from the graph. But there is one shortcoming in this approach: The result depends on how accurately you draw and read the graph. A better and more accurate method is based on finding an *algebraic* representation of the depreciation line. (We will continue our discussion of the linear depreciation problem in Section 1.3.)

In order to see how a straight line in the xy-plane may be described algebraically, we need to first recall certain properties of straight lines.

Figure 1.9
m is undefined.

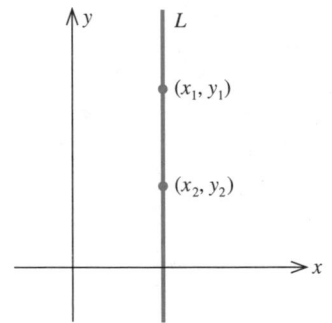

Slope of a Line

Let L denote the unique straight line that passes through the two distinct points (x_1, y_1) and (x_2, y_2). If $x_1 = x_2$, then L is a vertical line and the slope is undefined (Figure 1.9). If $x_1 \neq x_2$, we define the slope of L as follows.

SLOPE OF A NONVERTICAL LINE

If (x_1, y_1) and (x_2, y_2) are any two distinct points on a nonvertical line L, then the slope m of L is given by

$$m = \frac{\Delta y}{\Delta x} = \frac{y_2 - y_1}{x_2 - x_1} \qquad (2)$$

(Figure 1.10).   (*Continued on next page*)

Figure 1.10

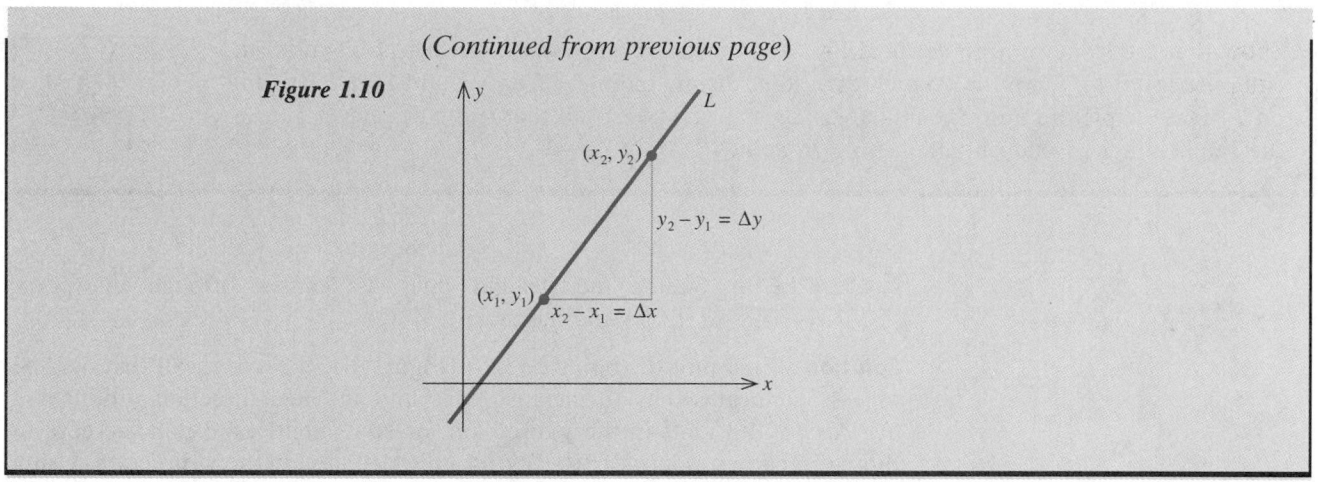

(*Continued from previous page*)

Since the slope of a nonvertical line is a ratio of two real numbers, the slope of a straight line is a constant whenever it is defined.

The number $\Delta y = y_2 - y_1$ (Δy is read "delta y") is a measure of the vertical change in y, and $\Delta x = x_2 - x_1$ is a measure of the horizontal change in x as shown in Figure 1.10. From this figure we can see that the slope m of a straight line L is a measure of the *rate of change of y with respect to x.*

Figure 1.11

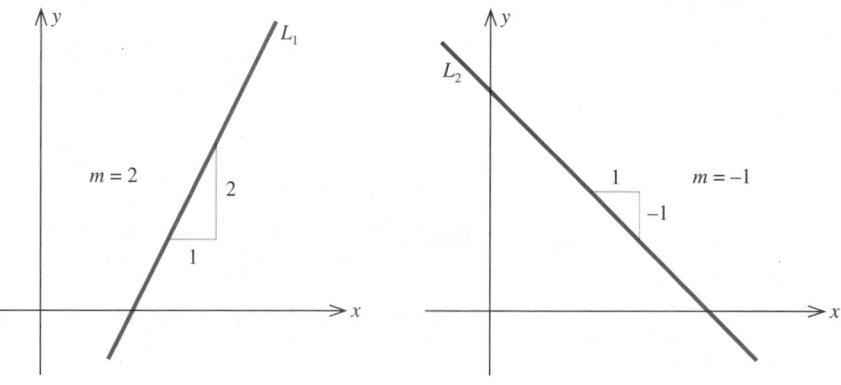

(a) The line rises ($m > 0$).   **(b)** The line falls ($m < 0$).

Figure 1.12
A family of straight lines.

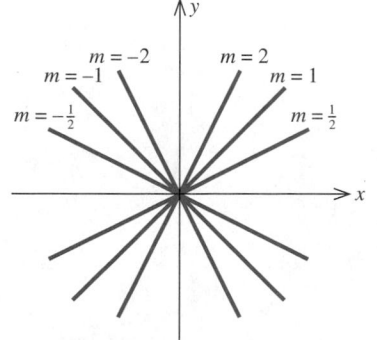

Furthermore, our earlier observation concerning the slope of a nonvertical straight line tells us that this rate of change is constant.

Figure 1.11a shows a straight line L_1 with slope 2. Observe that L_1 has the property that a unit increase in x results in a 2-unit increase in y. To see this, let $\Delta x = 1$ in equation (2) so that $m = \Delta y$. Since $m = 2$, we conclude that $\Delta y = 2$. Similarly, Figure 1.11b shows a line L_2 with slope -1. Observe that a straight line with positive slope slants upward from left to right (y increases as x increases), whereas a line with negative slope slants downward from left to right (y decreases as x increases). Finally, Figure 1.12 shows a family of straight lines passing through the origin with indicated slopes.

Show that the slope of a nonvertical line is independent of the two distinct points used to compute it. [*Hint:* Suppose we pick two other distinct points, $P_3(x_3, y_3)$ and $P_4(x_4, y_4)$ lying on L. Draw a picture and use similar triangles to demonstrate that using P_3 and P_4 gives the same values as that obtained using P_1 and P_2.]

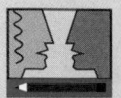

EXAMPLE 1   Sketch the straight line that passes through the point $(-2, 5)$ and has slope $-4/3$.

Solution   First plot the point $(-2, 5)$ (Figure 1.13). Next, recall that a slope of $-4/3$ indicates that an increase of 1 unit in the x-direction produces a *decrease* of 4/3 units in the y-direction, or equivalently, a 3-unit increase in the x-direction produces a $3(4/3)$, or 4-unit, decrease in the y-direction. Using this information, we plot the point $(1, 1)$ and draw the line through the two points.

Figure 1.13
L has slope $-4/3$ and passes through $(-2, 5)$.

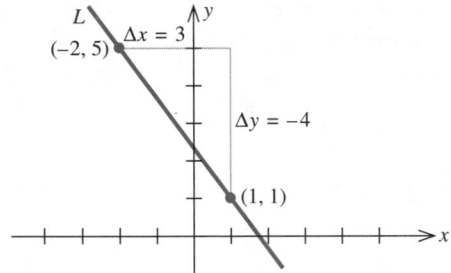

○ ○ ○

EXAMPLE 2   Find the slope m of the line that passes through the points $(-1, 1)$ and $(5, 3)$.

Solution   Choose (x_1, y_1) to be the point $(-1, 1)$ and (x_2, y_2) to be the point $(5, 3)$. Then, with $x_1 = -1$, $y_1 = 1$, $x_2 = 5$, and $y_2 = 3$, we find, using equation (2),

$$m = \frac{y_2 - y_1}{x_2 - x_1} = \frac{3 - 1}{5 - (-1)} = \frac{2}{6} = \frac{1}{3}$$

Figure 1.14
L passes through $(5, 3)$ and $(-1, 1)$.

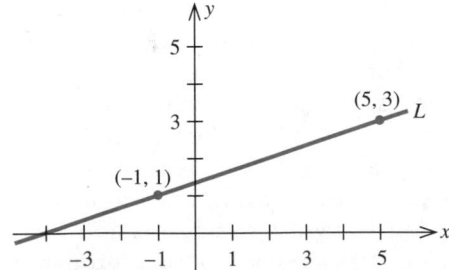

Figure 1.15
The slope of the horizontal line L is 0.

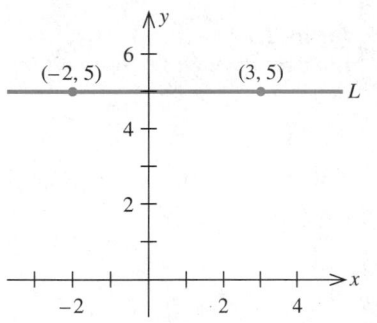

(Figure 1.14). Try to verify that the result obtained would have been the same had we chosen the point $(-1, 1)$ to be (x_2, y_2) and the point $(5, 3)$ to be (x_1, y_1).

EXAMPLE 3   Find the slope of the line that passes through the points $(-2, 5)$ and $(3, 5)$.

Solution   The slope of the required line is given by

$$m = \frac{5 - 5}{3 - (-2)} = \frac{0}{5} = 0$$

(Figure 1.15).

REMARK   In general, the slope of a horizontal line is zero.

We can use the slope of a straight line to determine whether a line is parallel to another line.

PARALLEL LINES

Two distinct lines are **parallel** if and only if their slopes are equal or their slopes are undefined.

Figure 1.16
L_1 and L_2 have the same slope and hence are parallel.

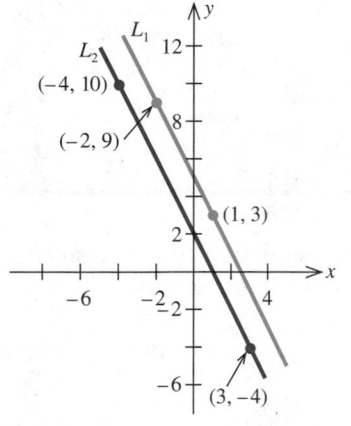

EXAMPLE 4   Let L_1 be a line that passes through the points $(-2, 9)$ and $(1, 3)$, and let L_2 be the line that passes through the points $(-4, 10)$ and $(3, -4)$. Determine whether L_1 and L_2 are parallel.

Solution   The slope m_1 of L_1 is given by

$$m_1 = \frac{3 - 9}{1 - (-2)} = -2$$

The slope m_2 of L_2 is given by

$$m_2 = \frac{-4 - 10}{3 - (-4)} = -2$$

Since $m_1 = m_2$, the lines L_1 and L_2 are in fact parallel (Figure 1.16).

Equations of Lines

We now show that every straight line lying in the xy-plane may be represented by an equation involving the variables x and y. One immediate benefit of this is that problems involving straight lines may be solved algebraically.

Let L be a straight line parallel to the y-axis (perpendicular to the x-axis) (Figure 1.17). Then L crosses the x-axis at some point $(a, 0)$ with the x-coordinate given by $x = a$, where a is some real number. Any other point on L has the form $(a, \bar{y})$, where $\bar{y}$ is an appropriate number. Therefore, the vertical line L is described by the sole condition

$$x = a$$

Figure 1.17
The vertical line x = a.

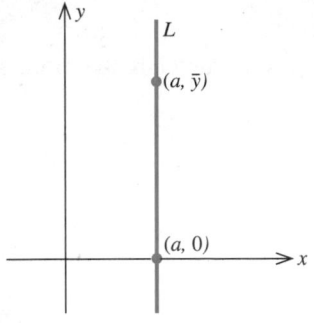

and this is accordingly an equation of L. For example, the equation $x = -2$ represents a vertical line 2 units to the left of the y-axis, and the equation $x = 3$ represents a vertical line 3 units to the right of the y-axis (Figure 1.18).

Figure 1.18
The vertical lines x = -2 and x = 3.

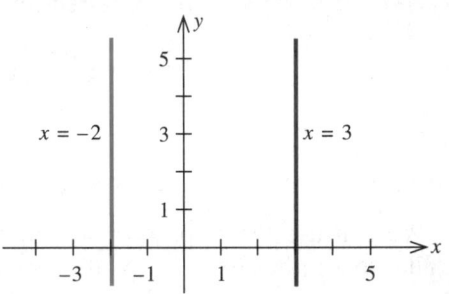

Figure 1.19
L passes through (x_1, y_1) and has slope m.

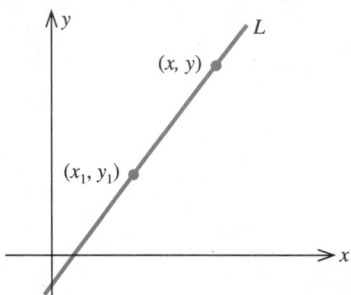

Next, suppose that L is a nonvertical line so that it has a well-defined slope m. Suppose (x_1, y_1) is a fixed point lying on L and (x, y) is a variable point on L distinct from (x_1, y_1) (Figure 1.19). Using equation (2) with the point $(x_2, y_2) = (x, y)$, we find that the slope of L is given by

$$m = \frac{y - y_1}{x - x_1}$$

Upon multiplying both sides of the equation by $x - x_1$, we obtain equation (3).

POINT-SLOPE FORM

The equation of the line that has slope m and passes through the point (x_1, y_1) is given by

$$y - y_1 = m(x - x_1) \tag{3}$$

Equation (3) is called the point-slope form of the equation of a line since it utilizes a given point (x_1, y_1) on a line and the slope m of the line.

Figure 1.20
L passes through (1, 3) and has slope 2.

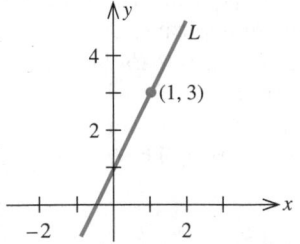

EXAMPLE 5   Find an equation of the line that passes through the point (1, 3) and has slope 2.

Solution   Using the point-slope form of the equation of a line with the point (1, 3) and $m = 2$, we obtain

$$y - 3 = 2(x - 1) \qquad [(y - y_1) = m(x - x_1)]$$

which, when simplified, becomes

$$2x - y + 1 = 0$$

(Figure 1.20).

EXAMPLE 6 Find an equation of the line that passes through the points $(-3, 2)$ and $(4, -1)$.

Solution The slope of the line is given by

$$m = \frac{-1 - 2}{4 - (-3)} = -\frac{3}{7}$$

Using the point-slope form of the equation of a line with the point $(4, -1)$ and the slope $m = -3/7$, we have

$$y + 1 = -\frac{3}{7}(x - 4) \qquad [(y - y_1) = m(x - x_1)]$$

$$7y + 7 = -3x + 12$$

or $\qquad 3x + 7y - 5 = 0$

(Figure 1.21).

Figure 1.21
L passes through $(-3, 2)$ and $(4, -1)$.

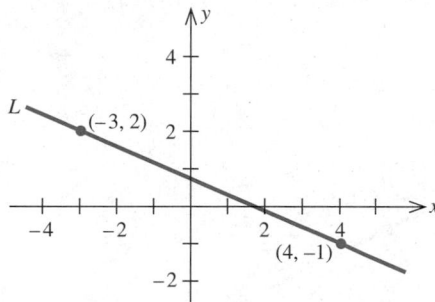

We can use the slope of a straight line to determine whether a line is perpendicular to another line.

PERPENDICULAR LINES	If L_1 and L_2 are two distinct nonvertical lines that have slopes m_1 and m_2, respectively, then L_1 is perpendicular to L_2 (written $L_1 \perp L_2$) if and only if $$m_1 = -\frac{1}{m_2}$$

If the line L_1 is vertical (so that its slope is undefined), then L_1 is perpendicular to another line, L_2, if and only if L_2 is horizontal (so that its slope is zero). For a proof of these results, see exercise 77, page 26.

EXAMPLE 7 Find an equation of the line that passes through the point $(3, 1)$ and is perpendicular to the line of Example 5.

Solution Since the slope of the line in Example 5 is 2, the slope of the required line is given by $m = -1/2$, the negative reciprocal of 2. Using the point-slope form of the equation of a line, we obtain

$$y - 1 = -\frac{1}{2}(x - 3) \qquad [(y - y_1) = m(x - x_1)]$$

$$2y - 2 = -x + 3$$

or $\qquad x + 2y - 5 = 0$

(Figure 1.22).

A straight line L that is neither horizontal nor vertical cuts the x-axis and the y-axis at, say, points $(a, 0)$ and $(0, b)$, respectively (Figure 1.23). The numbers a and b are called the **x-intercept** and **y-intercept,** respectively, of L.

Now, let L be a line with slope m and y-intercept b. Using equation (3), the point-slope form of the equation of a line, with the point given by $(0, b)$

Figure 1.22
L_2 is perpendicular to L_1 and passes through (3, 1).

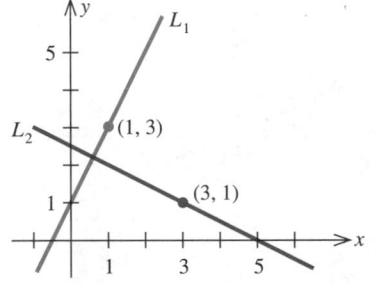

Figure 1.23
The line L has x-intercept a and y-intercept b.

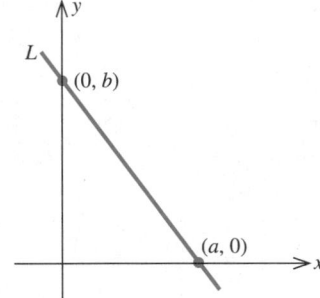

EXPLORING WITH TECHNOLOGY

1. Use a graphing utility to plot the straight lines L_1 and L_2 with equations $2x + y - 5 = 0$ and $41x + 20y - 11 = 0$ on the same set of axes using the standard viewing rectangle.
   a. Can you tell if the lines L_1 and L_2 are parallel to each other?
   b. Verify your observations by computing the slopes of L_1 and L_2 algebraically.

2. Use a graphing utility to plot the straight lines L_1 and L_2 with equations $x + 2y - 5 = 0$ and $5x - y + 5 = 0$ on the same set of axes using the standard viewing rectangle.
   a. Can you tell if the lines L_1 and L_2 are perpendicular to each other?
   b. Verify your observation by computing the slopes of L_1 and L_2 algebraically.

and slope m, we have

$$y - b = m(x - 0)$$

or $$y = mx + b$$

This is called the **slope-intercept form** of an equation of a line.

SLOPE-INTERCEPT FORM

The equation of the line that has slope m and intersects the y-axis at the point $(0, b)$ is given by

$$y = mx + b \qquad \textbf{(4)}$$

EXAMPLE 8 Find an equation of the line that has slope 3 and y-intercept -4.

Solution Using equation (4) with $m = 3$ and $b = -4$, we obtain the required equation:

$$y = 3x - 4$$

EXAMPLE 9 Determine the slope and y-intercept of the line whose equation is $3x - 4y = 8$.

Solution Rewrite the given equation in the slope-intercept form and obtain

$$y = \frac{3}{4}x - 2$$

Comparing this result with equation (4), we find $m = 3/4$ and $b = -2$, and we conclude that the slope and y-intercept of the given line are 3/4 and -2, respectively.

Exploring with Technology

1. Use a graphing utility to plot the straight lines with equations $y = -2x + 3$, $y = -x + 3$, $y = x + 3$, and $y = 2.5x + 3$ on the same set of axes using the standard viewing rectangle. What effect does changing the coefficient m of x in the equation $y = mx + b$ have on its graph?

2. Use a graphing utility to plot the straight lines with equations $y = 2x - 2$, $y = 2x - 1$, $y = 2x$, $y = 2x + 1$, and $y = 2x + 4$ on the same set of axes using the standard viewing rectangle. What effect does changing the constant b in the equation $y = mx + b$ have on its graph?

3. Describe in words the effect of changing both m and b in the equation $y = mx + b$.

Figure 1.24
Sales of a sporting goods store.

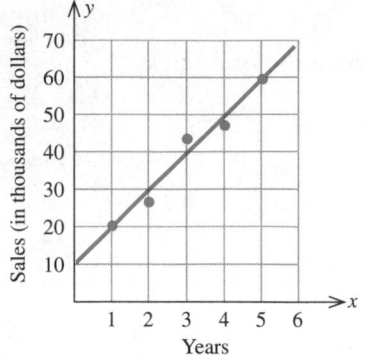

Applications

EXAMPLE 10   The sales manager of a local sporting goods store plotted sales versus time for the last five years and found the points to lie approximately along a straight line (Figure 1.24). By using the points corresponding to the first and fifth years, find an equation of the *trend line*. What sales figure can be predicted for the sixth year?

Solution   Using equation (2) with the points $(1, 20)$ and $(5, 60)$, we find that the slope of the required line is given by

$$m = \frac{60 - 20}{5 - 1} = 10$$

Next, using the point-slope form of the equation of a line with the point $(1, 20)$ and $m = 10$, we obtain

$$y - 20 = 10(x - 1)$$

or

$$y = 10x + 10$$

as the required equation.

The sales figure for the sixth year is obtained by letting $x = 6$ in the last equation, giving

$$y = 70$$

or $70,000.   ◦ ◦ ◦

EXAMPLE 11   Suppose that an art object purchased for $50,000 is expected to appreciate in value at a constant rate of $5,000 per year for the next five years. Use equation (4) to write an equation predicting the value of the art object in the next several years. What will its value be three years from the purchase date?

Solution   Let x denote the time (in years) that has elapsed since the purchase date, and let y denote the object's value (in dollars). Then $y = 50,000$ when $x = 0$. Furthermore, the slope of the required equation is given by $m = 5000$ since each unit increase in x (one year) implies an increase of 5000 units (dollars) in y. Using equation (4) with $m = 5000$ and $b = 50,000$, we obtain

$$y = 5000x + 50,000$$

Three years from the purchase date, the value of the object will be given by

$$y = 5000(3) + 50,000$$

or $65,000.   ◦ ◦ ◦

Refer to Example 11.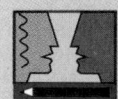
Can the equation predicting the value of the art object be used to predict long-term growth?

General Equation of a Line

We have considered several forms of the equation of a straight line in the plane. These different forms of the equation are equivalent to each other. In fact, each is a special case of the following equation.

**GENERAL FORM OF
A LINEAR EQUATION**

The equation

$$Ax + By + C = 0 \qquad\qquad (5)$$

where A, B, and C are constants and A and B are not both zero, is called the general form of a linear equation in the variables x and y.

We now state (without proof) an important result concerning the algebraic representation of straight lines in the plane.

An equation of a straight line is a linear equation; conversely, every linear equation represents a straight line.

This result justifies the use of the adjective *linear* in describing equation (5).

Figure 1.25
The straight line $3x - 4y = 12$.

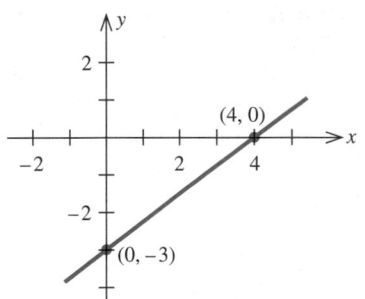

EXAMPLE 12 Sketch the straight line represented by the equation

$$3x - 4y - 12 = 0$$

Solution Since every straight line is uniquely determined by two distinct points, we need find only two such points through which the line passes in order to sketch it. For convenience, let us compute the points at which the line crosses the x- and y-axes. Setting $y = 0$, we find $x = 4$, so the line crosses the x-axis at the point $(4, 0)$. Setting $x = 0$ gives $y = -3$, so the line crosses the y-axis at the point $(0, -3)$. A sketch of the line appears in Figure 1.25.

Following is a summary of the common forms of the equations of straight lines discussed in this section.

**EQUATIONS OF
STRAIGHT LINES**

Vertical line: $x = a$

Horizontal line: $y = b$

Point-slope form: $y - y_1 = m(x - x_1)$

Slope-intercept form: $y = mx + b$

General form: $Ax + By + C = 0$

SELF-CHECK EXERCISES 1.2

1. Determine the number a so that the line passing through the points $(a, 2)$ and $(3, 6)$ is parallel to a line with slope 4.

2. Find an equation of the line that passes through the point $(3, -1)$ and is perpendicular to a line with slope $-1/2$.

3. Does the point $(3, -3)$ lie on the line with equation $2x - 3y - 12 = 0$? Sketch the graph of the line.

4. The percentage of people over 65 who have high school diplomas is summarized in the following table:

Year (x)	1960	1965	1970	1975	1980	1985	1990
Percent with Diplomas (y)	20	25	30	36	42	47	52

Source: U.S. Department of Commerce

a. Plot the percentage of people over 65 who have high school diplomas (y) versus the year (x).

b. Draw the straight line L through the points $(1960, 20)$ and $(1990, 52)$.

c. Find an equation of the line L.

d. Assume the trend continues and estimate the percentage of people over 65 who will have high school diplomas by the year 2000.

Solutions to Self-Check Exercises 1.2 can be found on page 26.

1.2 EXERCISES

In exercises 1–4, find the slope of the line shown in each figure.

1.

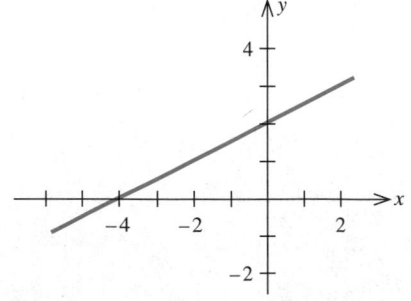

2.

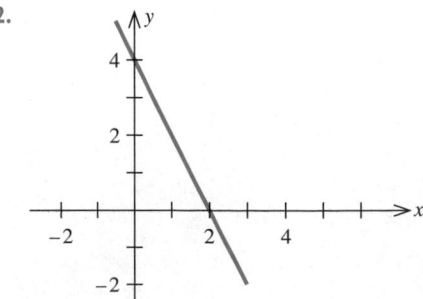

3.

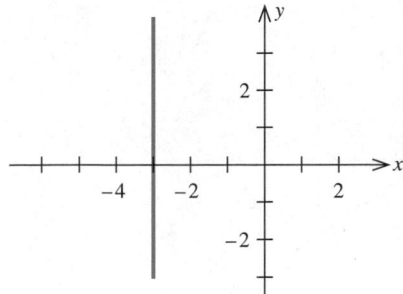

4.

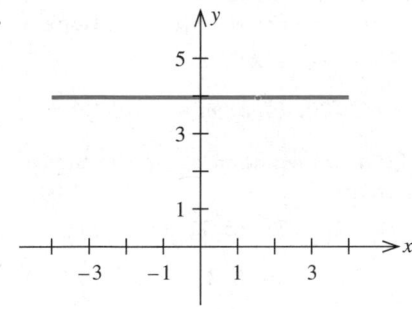

In exercises 5–10, find the slope of the line that passes through the given pair of points.

5. (4, 3) and (5, 8)   **6.** (4, 5) and (3, 8)

7. (−2, 3) and (4, 8)   **8.** (−2, −2) and (4, −4)

9. (a, b) and (c, d)

10. (−a + 1, b − 1) and (a + 1, −b)

11. Given the equation $y = 4x - 3$, answer the following questions.
a. If x increases by 1 unit, what is the corresponding change in y?
b. If x decreases by 2 units, what is the corresponding change in y?

12. Given the equation $2x + 3y = 4$, answer the following questions.
a. Is the slope of the line described by this equation positive or negative?
b. As x increases in value, does y increase or decrease?
c. If x decreases by 2 units, what is the corresponding change in y?

In exercises 13 and 14, determine whether the lines through the given pairs of points are parallel.

13. $A(1, -2)$, $B(-3, -10)$ and $C(1, 5)$, $D(-1, 1)$

14. $A(2, 3)$, $B(2, -2)$ and $C(-2, 4)$, $D(-2, 5)$

In exercises 15 and 16, determine whether the lines through the given pairs of points are perpendicular.

15. $A(-2, 5)$, $B(4, 2)$ and $C(-1, -2)$, $D(3, 6)$

16. $A(2, 0)$, $B(1, -2)$ and $C(4, 2)$, $D(-8, 4)$

17. If the line passing through the points $(1, a)$ and $(4, -2)$ is parallel to the line passing through the points $(2, 8)$ and $(-7, a + 4)$, what is the value of a?

18. If the line passing through the points $(a, 1)$ and $(5, 8)$ is parallel to the line passing through the points $(4, 9)$ and $(a + 2, 1)$, what is the value of a?

19. Find an equation of the horizontal line that passes through $(-4, -3)$.

20. Find an equation of the vertical line that passes through $(0, 5)$.

In exercises 21–26, match the statement with one of the graphs a–f.

21. The slope of the line is zero.

22. The slope of the line is undefined.

23. The slope of the line is positive and its y-intercept is positive.

24. The slope of the line is positive and its y-intercept is negative.

25. The slope of the line is negative and its x-intercept is negative.

26. The slope of the line is negative and its x-intercept is positive.

a.

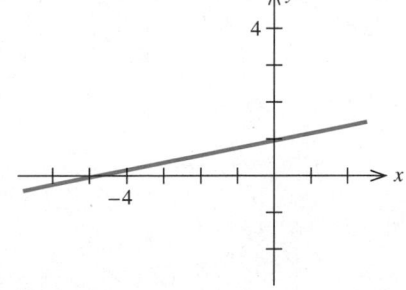

b.

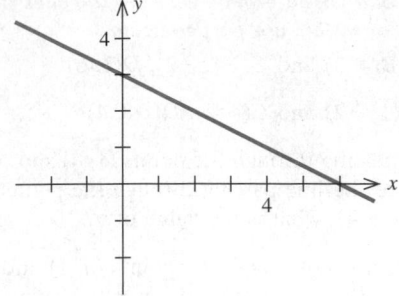

c.

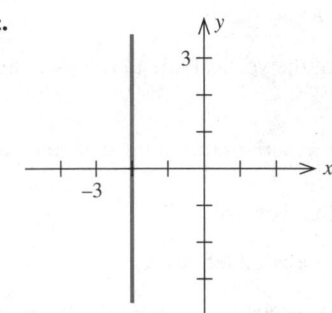

d.

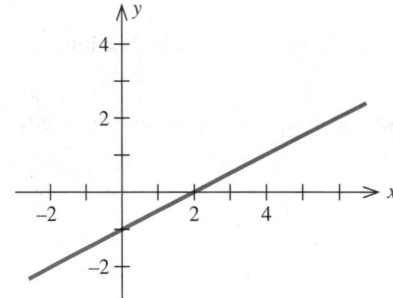

e.

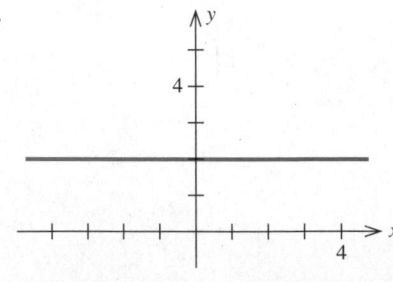

f.

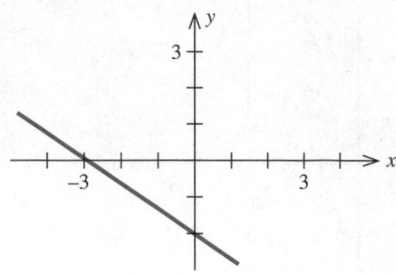

In exercises 27–30, find an equation of the line that passes through the given point and has the indicated slope m.

27. $(3, -4)$; $m = 2$ **28.** $(2, 4)$; $m = -1$

29. $(-3, 2)$; $m = 0$ **30.** $(1, 2)$; $m = -1/2$

In exercises 31–34, find an equation of the line that passes through the given points.

31. $(2, 4)$ and $(3, 7)$ **32.** $(2, 1)$ and $(2, 5)$

33. $(1, 2)$ and $(-3, -2)$ **34.** $(-1, -2)$ and $(3, -4)$

In exercises 35–38, find an equation of the line that has slope m and y-intercept b.

35. $m = 3$; $b = 4$ **36.** $m = -2$; $b = -1$

37. $m = 0$; $b = 5$ **38.** $m = -1/2$; $b = 3/4$

In exercises 39–44, write the given equation in the slope-intercept form and then find the slope and y-intercept of the corresponding line.

39. $x - 2y = 0$ **40.** $y - 2 = 0$

41. $2x - 3y - 9 = 0$ **42.** $3x - 4y + 8 = 0$

43. $2x + 4y = 14$ **44.** $5x + 8y - 24 = 0$

45. Find an equation of the line that passes through the point $(-2, 2)$ and is parallel to the line $2x - 4y - 8 = 0$.

46. Find an equation of the line that passes through the point $(2, 4)$ and is perpendicular to the line $3x + 4y - 22 = 0$.

In exercises 47–52, find an equation of the line that satisfies the given condition.

47. The line parallel to the *x*-axis and 6 units below it

48. The line passing through the origin and parallel to the line passing through the points $(2, 4)$ and $(4, 7)$

49. The line passing through the point (a, b) with slope equal to zero

50. The line passing through $(-3, 4)$ and parallel to the x-axis

51. The line passing through $(-5, -4)$ and parallel to the line passing through $(-3, 2)$ and $(6, 8)$

52. The line passing through (a, b) with undefined slope

53. Given that the point $P(-3, 5)$ lies on the line $kx + 3y + 9 = 0$, find k.

54. Given that the point $P(2, -3)$ lies on the line $-2x + ky + 10 = 0$, find k.

In exercises 55–60, sketch the straight line defined by the given linear equation by finding the x- and y-intercepts. [Hint: See Example 12.]

55. $3x - 2y + 6 = 0$ 56. $2x - 5y + 10 = 0$

57. $x + 2y - 4 = 0$ 58. $2x + 3y - 15 = 0$

59. $y + 5 = 0$ 60. $-2x - 8y + 24 = 0$

61. Show that an equation of a line through the points $(a, 0)$ and $(0, b)$ with $a \neq 0$ and $b \neq 0$ can be written in the form

$$\frac{x}{a} + \frac{y}{b} = 1$$

(Recall that the numbers a and b are the x- and y-intercepts, respectively, of the line. This form of an equation of a line is called the **intercept form.**)

In exercises 62–65, use the results of exercise 61 to find an equation of a line with the given x- and y-intercepts.

62. x-intercept 3; y-intercept 4

63. x-intercept -2; y-intercept -4

64. x-intercept $-1/2$; y-intercept $3/4$

65. x-intercept 4; y-intercept $-1/2$

In exercises 66 and 67, determine whether the given points lie on a straight line.

66. $A(-1, 7)$, $B(2, -2)$, and $C(5, -9)$

67. $A(-2, 1)$, $B(1, 7)$, and $C(4, 13)$

68. **Social Security Contributions** For wages less than the maximum taxable wage base, social security contributions by employees are 7.65% of the employee's wages.
    a. Find an equation that expresses the relationship between the wages earned (x) and the social security taxes paid (y) by an employee who earns less than the maximum taxable wage base.

b. For each additional dollar that an employee earns, by how much is his or her social security contribution increased? (Assume that the employee's wages are less than the maximum taxable wage base.)
c. What social security contributions will an employee who earns \$35,000 (which is less than the maximum taxable wage base) be required to make?

69. **College Admissions** Using data compiled by the Admissions Office at Faber University, college admissions officers estimate that 55% of the students who are offered admission to the freshman class at the university will actually enroll.
    a. Find an equation that expresses the relationship between the number of students who actually enroll (y) and the number of students who are offered admission to the university (x).
    b. If the desired freshman class size for the upcoming academic year is 1100 students, how many students should be admitted?

70. **Weight of Whales** The equation $W = 3.51L - 192$, expressing the relationship between the length L (in feet) and the expected weight W (in British tons) of adult blue whales, was adopted in the late 1960s by the International Whaling Commission.
    a. What is the expected weight of an 80-foot blue whale?
    b. Sketch the straight line that represents the equation.

71. **Cost of a Commodity** A manufacturer obtained the following data relating the cost y (in dollars) to the number of units (x) of a commodity produced:

Number of Units Produced (x)	0	20	40	60	80	100
Cost in Dollars (y)	200	208	222	230	242	250

a. Plot the cost (y) versus the quantity produced (x).
b. Draw the straight line through the points $(0, 200)$ and $(100, 250)$.
c. Derive an equation of the straight line of (b).
d. Taking this equation to be an approximation of the relationship between the cost and the level of production, estimate the cost of producing 54 units of the commodity.

72. **Ideal Heights and Weights for Women** The Venus Health Club for Women provides its members with the

following table, which gives the average desirable weight for women of a certain height:

Height x (in inches)	60	63	66	69	72
Weight y (in pounds)	108	118	129	140	152

a. Plot the weight (y) versus the height (x).
b. Draw a straight line L through the points corresponding to heights of 5 feet and 6 feet.
c. Derive an equation of the line L.
d. Using the equation of (c), estimate the average desirable weight for a woman who is 5 feet 5 inches tall.

73. Sales Growth   Metro Department Store's annual sales (in millions of dollars) during the past five years were:

Annual Sales (y)	5.8	6.2	7.2	8.4	9.0
Year (x)	1	2	3	4	5

a. Plot the annual sales (y) versus the year (x).
b. Draw a straight line L through the points corresponding to the first and fifth years.
c. Derive an equation of the line L.
d. Using the equation found in (c), estimate Metro's annual sales four years from now ($x = 9$).

74. Is there a difference between the statements "The slope of a straight line is zero" and "The slope of a straight line does not exist (is not defined)"? Explain your answer.

75. Consider the slope-intercept form of a straight line $y = mx + b$. Describe the family of straight lines obtained by keeping
a. the value of m fixed and allowing the value of b to vary
b. the value of b fixed and allowing the value of m to vary

76. Show that two distinct lines with equations $a_1x + b_1y + c_1 = 0$ and $a_2x + b_2y + c_2 = 0$, respectively, are parallel if and only if $a_1b_2 - b_1a_2 = 0$.
[*Hint:* Write each equation in the slope-intercept form and compare.]

77. Prove that if a line L_1 with slope m_1 is perpendicular to a line L_2 with slope m_2, then $m_1m_2 = -1$.
[*Hint:* Refer to the accompanying figure. Show that $m_1 = b$ and $m_2 = c$. Next, apply the Pythagorean Theorem and the distance formula to the triangles OAC, OCB, and OBA to show that $1 = -bc$.]

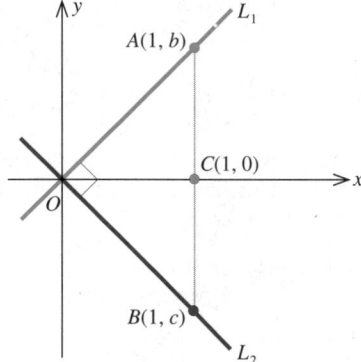

SOLUTIONS TO SELF-CHECK EXERCISES 1.2

1. The slope of the line that passes through the points (a, 2) and (3, 6) is

$$m = \frac{6 - 2}{3 - a}$$

$$= \frac{4}{3 - a}$$

Since this line is parallel to a line with slope 4, m must be equal to 4; that is,

$$\frac{4}{3 - a} = 4$$

or, upon multiplying both sides of the equation by $3 - a$,

$$4 = 4(3 - a)$$
$$4 = 12 - 4a$$
$$4a = 8$$
and $$a = 2$$

2. Since the required line L is perpendicular to a line with slope $-\frac{1}{2}$, the slope of L is

$$-\frac{1}{-\frac{1}{2}} = 2$$

Next, using the point-slope form of the equation of a line, we have

$$y - (-1) = 2(x - 3)$$
$$y + 1 = 2x - 6$$
or $$y = 2x - 7$$

3. Substituting $x = 3$ and $y = -3$ into the left-hand side of the given equation, we find

$$2(3) - 3(-3) - 12 = 3$$

which is not equal to zero (the right-hand side). Therefore, $(3, -3)$ does not lie on the line with equation $2x - 3y - 12 = 0$.

Setting $x = 0$, we find $y = -4$, the y-intercept. Next, setting $y = 0$ gives $x = 6$, the x-intercept. We now draw the line passing through the points $(0, -4)$ and $(6, 0)$, as shown.

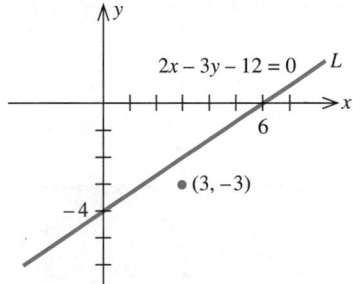

4. **a.** and **b.** See the accompanying figure.

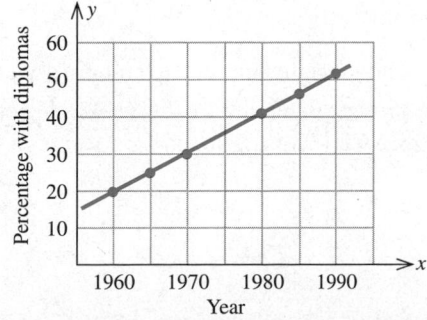

Using Technology

Corresponding technology sections for CAS may be found at the Brooks/Cole Web site at http://www.brookscole.com/math/authors/tans/

Graphing a Straight Line

The first step in plotting a straight line with a graphing utility is to select a suitable viewing rectangle. We usually do this by experimenting. For example, you might first plot the straight line using the **standard viewing rectangle** $[-10, 10]$ by $[-10, 10]$. If necessary, you then might adjust the viewing rectangle by enlarging it or reducing it to obtain a sufficiently complete view of the line or at least the portion of the line that is of interest.

EXAMPLE 1 Plot the straight line $2x + 3y - 6 = 0$ in the standard viewing rectangle.

Solution The straight line in the standard viewing rectangle is shown in Figure T1.

Figure T1
The straight line $2x + 3y - 6 = 0$ in the standard viewing rectangle.

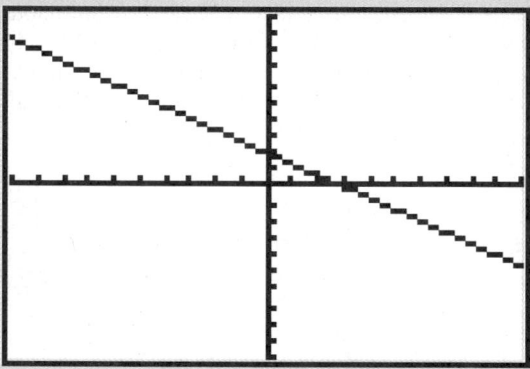

EXAMPLE 2 Plot the straight line $2x + 3y - 30 = 0$ in (a) the standard viewing rectangle and (b) the viewing rectangle $[-5, 20] \times [-5, 20]$.

Solution

a. The straight line in the standard viewing rectangle is shown in Figure T2a.

b. The straight line in the viewing rectangle $[-5, 20] \times [-5, 20]$ is shown in Figure T2b. This figure certainly gives a more complete view of the straight line.

Figure T2
(a) The graph of $2x + 3y - 30 = 0$ in the standard viewing rectangle.
(b) The graph of $2x + 3y - 30 = 0$ in the viewing rectangle $[-5, 20] \times [-5, 20]$.

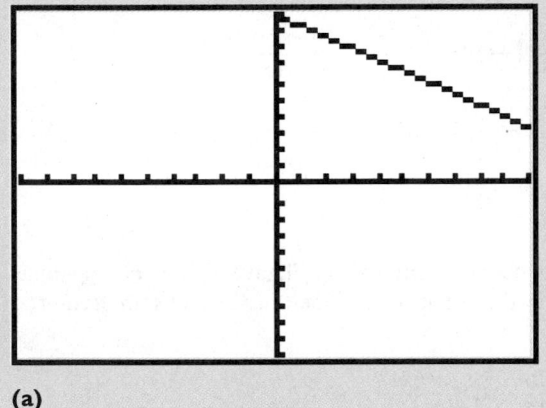

(a)

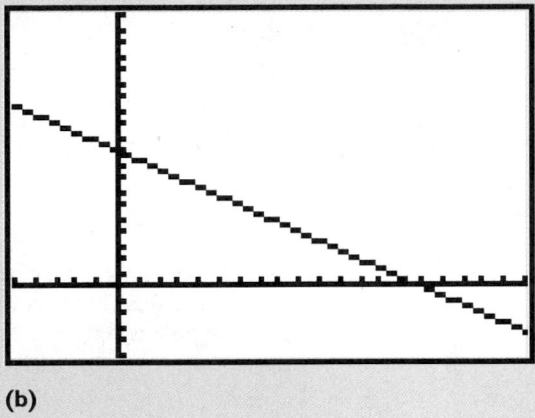

(b)

○ ○ ○

EXERCISES

In exercises 1–6, plot the straight line with the given equation in the standard viewing rectangle.

1. $3.2x + 2.1y - 6.72 = 0$

2. $2.3x - 4.1y - 9.43 = 0$

3. $1.6x + 5.1y = 8.16$ 4. $-3.2x + 2.1y = 6.72$

5. $2.8x = -1.6y + 4.48$ 6. $3.3y = 4.2x - 13.86$

In exercises 7–10, plot the straight line with the given equation in (a) the standard viewing rectangle and (b) the indicated viewing rectangle.

7. $12.1x + 4.1y - 49.61 = 0; [-10, 10] \times [-10, 20]$

8. $4.1x - 15.2y - 62.32 = 0; [-10, 20] \times [-10, 10]$

9. $20x + 16y = 300; [-10, 20] \times [-10, 30]$

10. $32.2x + 21y = 676.2; [-10, 30] \times [-10, 40]$

In exercises 11–18, plot the straight line with the given equation in an appropriate viewing window. Note: The answer is not unique.

11. $20x + 30y = 600$ 12. $30x - 20y = 600$

13. $22.4x + 16.1y - 352 = 0$

14. $18.2x - 15.1y = 274.8$ 15. $1.2x + 20y = 24$

16. $30x - 2.1y = 63$ 17. $-4x + 12y = 50$

18. $20x - 12.2y = 240$

c. The slope of L is

$$m = \frac{52 - 20}{1990 - 1960} = \frac{32}{30} = \frac{16}{15}$$

Using the point-slope form of the equation of a line with the point (1960, 20), we find

$$y - 20 = \frac{16}{15}(x - 1960) = \frac{16}{15}x - \frac{(16)(1960)}{15}$$

or

$$y = \frac{16}{15}x - \frac{6272}{3} + 20$$

$$= \frac{16}{15}x - \frac{6212}{3}$$

d. To estimate the percentage of people over 65 who will have high school diplomas by the year 2000, let $x = 2000$ in the equation obtained in (c). Thus, the required estimate is

$$y = \frac{16}{15}(2000) - \frac{6212}{3} \approx 62.67$$

or approximately 63%.

1.3 LINEAR FUNCTIONS AND MATHEMATICAL MODELS

Mathematical Models

Before we can apply mathematics to solving real-world problems, we need to formulate these problems in the language of mathematics. This process is referred to as **mathematical modeling.** A mathematical model may give a precise description of the problem under consideration, or, more likely than not, it may provide only an acceptable approximation of the problem. For example, the accumulated amount A at the end of t years when a sum of P dollars is deposited in a fixed bank account and earns interest at the rate of r percent per year compounded m times a year is given *exactly* by the formula (model)

$$A = P\left(1 + \frac{r}{m}\right)^{mt}$$

On the other hand, the size of a cancer tumor may be approximated by the volume of a sphere,

$$V = \frac{4}{3}\pi r^3$$

where r is the radius of the tumor in centimeters.

The many techniques used in constructing mathematical models of practical problems range from theoretical considerations of a problem on the one extreme to an interpretation of data associated with the problem on the other. The model giving the accumulated amount for a fixed bank account mentioned earlier may be derived theoretically (see Chapter 5). In Section 1.5 we will see how linear equations (models) can be constructed from a given set of data points. Also, in the ensuing chapters we will see how other mathematical models, including statistical and probability models, are used to describe and analyze real-world situations.

We will now look at an important way of describing the relationship between two quantities using the notion of a *function*.

Functions

A manufacturer would like to know how his company's profit is related to its production level; a biologist would like to know how the population of a certain culture of bacteria will change with time; a psychologist would like to know the relationship between the learning time of an individual and the length of a vocabulary list; and a chemist would like to know how the initial speed of a chemical reaction is related to the amount of substrate used. In each instance, we are concerned with the same question: How does one quantity depend on another? The relationship between two quantities is conveniently described in mathematics by using the concept of a function.

FUNCTION	A function f is a rule that assigns to each value of x one and only one value of y.

The number y is normally denoted by $f(x)$, read "f of x," emphasizing the dependency of y on x.

An example of a function may be drawn from the familiar relationship between the area of a circle and its radius. Letting x and y denote the radius and area of a circle, respectively, we have, from elementary geometry,

$$y = \pi x^2$$

This equation defines y as a function of x, since for each admissible value of x (a nonnegative number representing the radius of a certain circle), there corresponds precisely one number $y = \pi x^2$ giving the area of the circle. This *area function* may be written as

$$f(x) = \pi x^2 \qquad (6)$$

For example, to compute the area of a circle with a radius of 5 inches, we simply replace x in equation (6) by the number 5. Thus, the area of the circle is

$$f(5) = \pi 5^2 = 25\pi$$

or 25π square inches.

Suppose we are given the function $y = f(x)$.* The variable x is referred to as the **independent variable,** and the variable y is called the **dependent variable.** The set of all values that may be assumed by x is called the **domain** of the function f, and the set comprising all the values assumed by $y = f(x)$ as x takes on all possible values in its domain is called the **range** of the function f. For the area function (6), the domain of f is the set of all nonnegative numbers x, and the range of f is the set of all nonnegative numbers y.

We now focus our attention on an important class of functions known as linear functions. Recall that a linear equation in x and y has the form $Ax + By + C = 0$, where A, B, and C are constants and A and B are not both zero. If $B \neq 0$, the equation can always be solved for y in terms of x; in fact, as we saw in Section 1.2, the equation may be cast in the slope-intercept form:

$$y = mx + b \qquad (m, b \text{ constants}) \qquad \textbf{(7)}$$

Equation (7) defines y as a function of x. The domain and range of this function are the set of all real numbers. Furthermore, the graph of this function, as we saw in Section 1.2, is a straight line in the plane. For this reason, the function $f(x) = mx + b$ is called a *linear function*.

LINEAR FUNCTION	The function f defined by $$f(x) = mx + b$$ where m and b are constants, is called a **linear function.**

Linear functions play an important role in the quantitative analysis of business and economic problems. First, many problems arising in these and other fields are *linear* in nature or are *linear* in the intervals of interest and thus can be formulated in terms of linear functions. Second, because linear functions are relatively easy to work with, assumptions involving linearity are often made in the formulation of problems. In many cases these assumptions are justified, and acceptable mathematical models are obtained that approximate real-life situations.

In the rest of this section, we will look at several applications that can be modeled using linear functions.

Simple Depreciation

We first discussed linear depreciation in Section 1.2 as a real-world application of straight lines. The following example illustrates how to derive an equation describing the book value of an asset being depreciated linearly.

⊙ ⊙ ⊙

* It is customary to refer to a function f as $f(x)$.

EXAMPLE 1 A printing machine has an original value of $100,000 and is to be depreciated linearly over five years with a $30,000 scrap value. Find an expression giving the book value at the end of year t. What will the book value of the machine be at the end of the second year? What is the rate of depreciation of the printing machine?

Solution Let V denote the printing machine's book value at the end of the tth year. Since the depreciation is linear, V is a linear function of t. Equivalently, the graph of the function is a straight line. Now, in order to find an equation of the straight line, observe that $V = 100,000$ when $t = 0$; this tells us that the line passes through the point (0, 100,000). Similarly, the condition that $V = 30,000$ when $t = 5$ says that the line also passes through the point (5, 30,000). The slope of the line is given by

$$m = \frac{100,000 - 30,000}{0 - 5} = -\frac{70,000}{5} = -14,000$$

Using the point-slope form of the equation of a line with the point (0, 100,000) and the slope $m = -14,000$, we have

$$V - 100,000 = -14,000(t - 0)$$

or
$$V = -14,000t + 100,000$$

the required expression. The book value at the end of the second year is given by

$$V = -14,000(2) + 100,000 = 72,000$$

or $72,000. The rate of depreciation of the machine is given by the negative of the slope of the depreciation line. Since the slope of the line is $a = -14,000$, the rate of depreciation is $14,000 per year. The graph of $V = -14,000t + 100,000$ is sketched in Figure 1.26.

Figure 1.26
Linear depreciation of an asset.

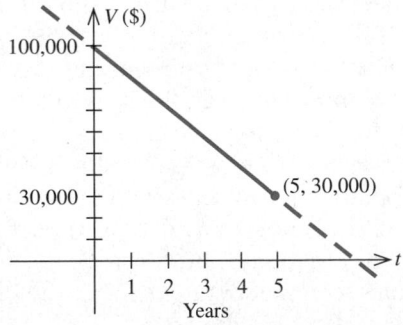

Linear Cost, Revenue, and Profit Functions

Whether a business is a sole proprietorship or a large corporation, the owner or chief executive must constantly keep track of operating costs, revenue resulting from the sale of products or services, and perhaps most important,

As a branch manager for a major Boston bank, Busa wears many hats in a typical day—supervisor, financial counselor, diplomat—juggling them all with equal dexterity. She is responsible for the successful operation of the branch, overseeing a staff of eight, and she must be familiar with the full range of Shawmut's financial products. She also must deal with irate customers as well as those who just want a little extra attention.

PORTFOLIO

Carol Busa

Title: Assistant Vice-President Financial Center Manager
Institution: A major Boston bank

"All banks provide more or less the same services. What distinguishes one bank from another is convenience," says Busa. An experienced banker in Boston's highly competitive financial district, Busa has learned that banks must deliver more than basic checking and savings accounts. Customers want a broad range of financial products and the knowledge and service to back them up. Busa exhorts her staff to work closely with customers to make banking as trouble-free as possible.

Busa deals with a wide assortment of customers, such as a junior executive looking for a car loan and a middle-aged mother seeking to tap her home equity for a child's college education.

Working with each customer, Busa examines several loan scenarios, entering different variables into the computer to determine the best financial mix for the customer's circumstance.

Although the computer frees Busa from doing manual calculations, she has to be able to explain how each variable affects the equation. For example, given the junior executive's modest salary, she might recommend a smaller car loan based on his ability to handle monthly payments. A $10,000 loan at 8% interest would require payments of $244.13 per month over a 4-year period, whereas an $8000 loan would reduce the monthly obligation to $195.30—a more manageable amount.

To finance tuition bills, Busa might suggest an equity line of credit versus a home equity loan. The line of credit would allow the mother to draw on the money as needed rather than borrowing a large sum at one time. The one disadvantage in borrowing from a line of credit is its variable interest rate. Customers are usually charged the current prime rate (the interest banks charge their best customers) plus 1.5%. As the prime fluctuates, it becomes impossible to forecast how much interest will accrue over a 20-year repayment cycle. The best Busa can do is calculate the accrued interest during the first one or two years, hoping the prime rate will remain constant for that long.

Forecasting interest payments is only part of Busa's job. She is often called on to help customers with more mundane tasks, such as balancing their checking accounts. "You'd be surprised how many people can't balance a checkbook," she says. Busa usually asks them to try one more time with an up-to-date computer statement. If they still can't balance, then she or her staff help them find the error.

If customers sometimes feel overwhelmed reconciling their accounts each month, they can take some comfort in the fact that Busa has to balance her branch bank accounts each day. Even more daunting is Shawmut's responsibility for balancing accounts in all 300-plus branches daily, exponentially increasing the chance of arithmetic error.

❍ ❍ ❍

the profits realized. Three functions provide management with a measure of these quantities: the total cost function, the revenue function, and the profit function.

COST, REVENUE, AND PROFIT FUNCTIONS

Let x denote the number of units of a product manufactured or sold. Then, the **total cost function** is

$$C(x) = \text{Total cost of manufacturing } x \text{ units of the product}$$

The **revenue function** is

$$R(x) = \text{Total revenue realized from the sale of } x \text{ units of the product}$$

The **profit function** is

$$P(x) = \text{Total profit realized from manufacturing and selling } x \text{ units of the product}$$

Generally speaking, the total cost, revenue, and profit functions associated with a company are more likely than not to be nonlinear (these functions are best studied using the tools of calculus). But *linear* cost, revenue, and profit functions do arise in practice, and we will consider such functions in this section. Before deriving explicit forms of these functions, we need to recall some common terminology.

The costs incurred in operating a business are usually classified into two categories. Costs that remain more or less constant regardless of the firm's activity level are called **fixed costs**. Examples of fixed costs are rental fees and executive salaries. Costs that vary with production or sales are called **variable costs**. Examples of variable costs are wages and costs for raw materials.

Suppose a firm has a fixed cost of F dollars, a production cost of c dollars per unit, and a selling price of s dollars per unit. Then the *cost function* $C(x)$, the *revenue function* $R(x)$, and the *profit function* $P(x)$ for the firm are given by

$$C(x) = cx + F$$
$$R(x) = sx$$

and

$$P(x) = R(x) - C(x) \qquad (\text{Revenue} - \text{cost})$$
$$= (s - c)x - F$$

where x denotes the number of units of the commodity produced and sold. The functions C, R, and P are linear functions of x.

EXAMPLE 2   Puritron, a manufacturer of water filters, has a monthly fixed cost of $20,000, a production cost of $20 per unit, and a selling price of $30 per unit. Find the cost function, the revenue function, and the profit function for Puritron.

Solution   Let x denote the number of units produced and sold. Then

$$C(x) = 20x + 20,000$$
$$R(x) = 30x$$

and

$$P(x) = R(x) - C(x)$$
$$= 30x - (20x + 20,000)$$
$$= 10x - 20,000$$

○ ○ ○

Linear Demand and Supply Curves

Figure 1.27
The graph of a linear demand function.

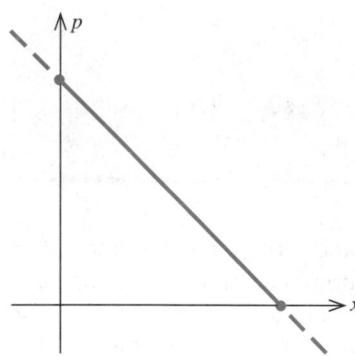

In a free market economy, consumer demand for a particular commodity depends on the commodity's unit price. A **demand equation** expresses this relationship between the unit price and the quantity demanded. The corresponding graph of the demand equation is called a **demand curve.** In general, the quantity demanded of a commodity decreases as its unit price increases, and vice versa. Accordingly, a **demand function,** defined by $p = f(x)$, where p measures the unit price and x measures the number of units of the commodity, is generally characterized as a decreasing function of x; that is, $p = f(x)$ decreases as x increases.

The simplest demand function is defined by a linear equation in x and p, where both x and p assume only nonnegative values. Its graph is a straight line having a negative slope. Thus, the demand curve in this case is that part of the graph of a straight line that lies in the first quadrant (Figure 1.27).

EXAMPLE 3   The quantity demanded of a certain model of clock is 48,000
···   units when the unit price is $8. At $12 per unit, the quantity demanded drops to 32,000 units. Find the demand equation, assuming that it is linear. What is the unit price corresponding to a quantity demanded of 40,000 units? What is the quantity demanded if the unit price is $14?

Solution   Let p denote the unit price of a clock (in dollars) and let x (in units of 1000) denote the quantity demanded when the unit price of the clocks is p. When $p = 8$, $x = 48$ and the point (48, 8) lies on the demand curve. Similarly, when $p = 12$, $x = 32$ and the point (32, 12) also lies on the demand curve. Since the demand equation is linear, its graph is a straight line. The slope of the required line is given by

$$m = \frac{12 - 8}{32 - 48} = \frac{4}{-16} = -\frac{1}{4}$$

so using the point-slope form of the equation of a line with the point (48, 8), we find that

$$p - 8 = -\frac{1}{4}(x - 48)$$

or

$$p = -\frac{1}{4}x + 20$$

Figure 1.28

The graph of the demand equation
$p = -\frac{1}{4}x + 20.$

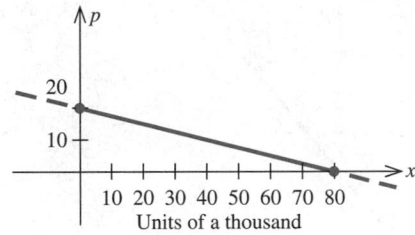

is the required equation. The demand curve is shown in Figure 1.28. If the quantity demanded is 40,000 units ($x = 40$), the demand equation yields

$$y = -\frac{1}{4}(40) + 20 = 10$$

and we see that the corresponding unit price is $10. Next, if the unit price is $14 ($p = 14$), the demand equation yields

$$14 = -\frac{1}{4}x + 20$$

$$\frac{1}{4}x = 6$$

or $x = 24$

so the quantity demanded will be 24,000 units.   ◦ ◦ ◦

Figure 1.29

The graph of a linear supply function.

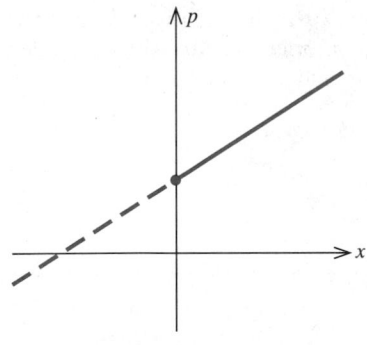

In a competitive market, a relationship also exists between the unit price of a commodity and its availability in the market. In general, an increase in a commodity's unit price will induce the manufacturer to increase the supply of that commodity. Conversely, a decrease in the unit price generally leads to a drop in the supply. An equation that expresses the relationship between the unit price and the quantity supplied is called a **supply equation,** and the corresponding graph is called a **supply curve.** A **supply function,** defined by $p = f(x)$, is generally characterized by an increasing function of x; that is, $p = f(x)$ increases as x increases.

As in the case of a demand equation, the simplest supply equation is a linear equation in p and x, where p and x have the same meaning as before but the line has a positive slope. The supply curve corresponding to a linear supply function is that part of the straight line that lies in the first quadrant (Figure 1.29).

EXAMPLE 4 The supply equation for a commodity is given by $4p -$
$5x = 120$, where p is measured in dollars and x is measured in units of 100.

a. Sketch the corresponding curve.

b. How many units will be marketed when the unit price is $55?

Figure 1.30
The graph of the supply equation
$4p - 5x = 120$.

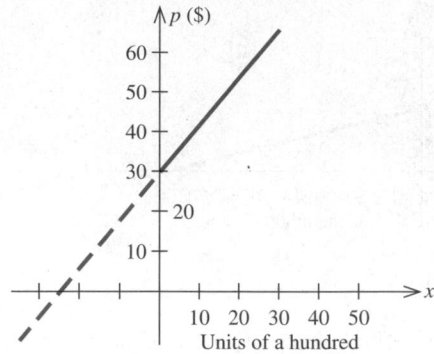

Solution **a.** Setting $x = 0$ and $p = 0$, we find that the p- and x-intercepts are 30 and -24, respectively. The supply curve is sketched in Figure 1.30.

b. Substituting $p = 55$ in the supply equation, we have $4(55) - 5x = 120$, or $x = 20$, so the amount marketed will be 2000 units.   ⚬ ⚬ ⚬

SELF–CHECK EXERCISES 1.3

1. A manufacturer has a monthly fixed cost of $60,000 and a production cost of $10 for each unit produced. The product sells for $15 per unit.
a. What is the cost function?
b. What is the revenue function?
c. What is the profit function?
d. Compute the profit (loss) corresponding to production levels of 10,000 and 14,000 units.

2. The quantity demanded for a certain make of 30″ × 52″ area rug is 500 when the unit price is $100. For each $20 decrease in the unit price, the quantity demanded increases by 500 units. Find the demand equation and sketch its graph.

Solutions to Self-Check Exercises 1.3 can be found on page 41.

1.3 EXERCISES

*In exercises 1–10, determine whether the equation defines
y as a linear function of x. If so, write it in the form y =
mx + b.*

1. $2x + 3y = 6$

2. $-2x + 4y = 7$

3. $x = 2y - 4$

4. $2x = 3y + 8$

5. $2x - 4y + 9 = 0$

6. $3x - 6y + 7 = 0$

7. $2x^2 - 8y + 4 = 0$

8. $3\sqrt{x} + 4y = 0$

9. $2x - 3y^2 + 8 = 0$

10. $2x + \sqrt{y} - 4 = 0$

11. A manufacturer has a monthly fixed cost of $40,000 and a production cost of $8 for each unit produced. The product sells for $12 per unit.
a. What is the cost function?

b. What is the revenue function?

c. What is the profit function?

d. Compute the profit (loss) corresponding to production levels of 8000 and 12,000 units.

12. A manufacturer has a monthly fixed cost of $100,000 and a production cost of $14 for each unit produced. The product sells for $20 per unit.

a. What is the cost function?

b. What is the revenue function?

c. What is the profit function?

d. Compute the profit (loss) corresponding to production levels of 12,000 and 20,000 units.

13. Find the constants m and b in the linear function $f(x) = mx + b$ so that $f(0) = 2$ and $f(3) = -1$.

14. Find the constants m and b in the linear function $f(x) = mx + b$ so that $f(2) = 4$ and the straight line represented by f has slope -1.

15. **Linear Depreciation** An office building worth $1,000,000 when completed in 1995 is being depreciated linearly over 50 years. What will the book value of the building be in 2000? In 2005? (Assume the scrap value is $0.)

16. **Linear Depreciation** An automobile purchased for use by the manager of a firm at a price of $14,000 is to be depreciated using the straight-line method over five years. What will the book of the automobile be at the end of three years? (Assume the scrap value is $0.)

17. **Consumption Functions** A certain economy's consumption function is given by the equation

$$C(x) = 0.75x + 6$$

where $C(x)$ is the personal consumption expenditure in billions of dollars and x is the disposable personal income in billions of dollars. Find $C(0)$, $C(50)$, $C(100)$.

18. **Sales Tax** In a certain state, the sales tax T on the amount of taxable goods is 6% of the value of the goods purchased (x), where both T and x are measured in dollars.

a. Express T as a function of x.

b. Find $T(200)$, $T(5.60)$.

19. **Social Security Benefits** Social security recipients receive an automatic cost-of-living adjustment (COLA) once each year. Their monthly benefit is increased by the same percentage that consumer prices have increased during the preceding year. Suppose that consumer prices have increased by 5.3% during the preceding year.

a. Express the adjusted monthly benefit of a social security recipient as a function of his or her current monthly benefit.

b. If Mr. Harrington's monthly social security benefit is now $620, what will his adjusted monthly benefit be?

20. **Profit Functions** Auto-Time, a manufacturer of 24-hour variable timers, has a monthly fixed cost of $48,000 and a production cost of $8 for each timer manufactured. The timers sell for $14 each.

a. What is the cost function?

b. What is the revenue function?

c. What is the profit function?

d. Compute the profit (loss) corresponding to production levels of 4000, 6000, and 10,000 timers, respectively.

21. **Profit Functions** The management of T.M.I., Inc., finds that the monthly fixed costs attributable to the company's blank tape division amount to $12,100. If the cost for producing each reel of tape is $.60 and each reel of tape sells for $1.15, find the company's cost function, revenue function, and profit function.

22. **Linear Depreciation** In 1995, the National Textile Company installed a new machine in one of its factories at a cost of $250,000. The machine is depreciated linearly over ten years with a scrap value of $10,000.

a. Find an expression for the machine's book value in the tth year of use $(0 \leq t \leq 10)$.

b. Sketch the graph of the function of (a).

c. Find the machine's book value in 1999.

d. Find the rate at which the machine is being depreciated.

23. **Linear Depreciation** A minicomputer purchased at a cost of $60,000 in 1994 has a scrap value of $12,000 at the end of four years. If the straight-line method of depreciation is used:

a. Find the rate of depreciation.

b. Find the linear equation expressing the minicomputer's book value at the end of t years.

c. Sketch the graph of the function of (b).

d. Find the minicomputer's book value at the end of the third year.

24. **Linear Depreciation** Suppose that an asset has an original value of $\$C$ and is depreciated linearly over N years with a scrap value of $\$S$. Show that the asset's book value at the end of the tth year is described by the function

$$V(t) = C - \left(\frac{C - S}{N}\right)t$$

[*Hint:* Find an equation of the straight line passing through the points $(0, C)$ and (N, S). (Why?)]

25. Rework exercise 15 using the formula derived in exercise 24.

26. Rework exercise 16 using the formula derived in exercise 24.

27. Drug Dosages   A method sometimes used by pediatricians to calculate the dosage of medicine for children is based on the child's surface area. If a denotes the adult dosage (in mg), and if S is the child's surface area (in square meters), then the child's dosage is given by

$$D(S) = \frac{Sa}{1.7}$$

a. Show that D is a linear function of S.

[*Hint:* Think of D as having the form $D(S) = mS + b$. What is the slope m and the y-intercept b?]

b. If the adult dose of a drug is 500 mg, how much should a child whose surface area is 0.4 square meter receive?

28. Drug Dosages   Cowling's Rule is a method for calculating pediatric drug dosages. If a denotes the adult dosage (in mg) and if t is the child's age (in years), then the child's dosage is given by

$$D(t) = \left(\frac{t+1}{24}\right) a$$

a. Show that D is a linear function of t.

[*Hint:* Think of $D(t)$ as having the form $D(t) = mt + b$. What is the slope m and the y-intercept b?]

b. If the adult dose of a drug is 500 mg, how much should a four-year-old child receive?

29. Celsius and Fahrenheit Temperatures   The relationship between temperature measured in the Celsius scale and the Fahrenheit scale is linear. The freezing point is 0°C and 32°F and the boiling point is 100°C and 212°F, respectively.

a. Find an equation giving the relationship between the temperature F measured in the Fahrenheit scale and the temperature C measured in the Celsius scale.

b. Find F as a function of C and use this formula to determine the temperature in Fahrenheit corresponding to a temperature of 20°C.

c. Find C as a function of F and use this formula to determine the temperature in Celsius corresponding to a temperature of 70°F.

30. Cricket Chirping and Temperature   Entomologists have discovered that there is a linear relationship between the number of chirps of crickets of a certain spe-

cies and the air temperature. When the temperature is 70°F, the crickets chirp at the rate of 120 times per minute, and when the temperature is 80°F, they chirp at the rate of 160 times per minute.

a. Find an equation giving the relationship between the air temperature T and the number of chirps per minute N of the crickets.

b. Find N as a function of T, and use this formula to determine the rate at which the crickets chirp when the temperature is 102°F.

31. Demand for Clock Radios   In the accompanying figure, L_1 is the demand curve for the model A clock radio manufactured by Ace Radio Inc., and L_2 is the demand curve for the model B clock radio.

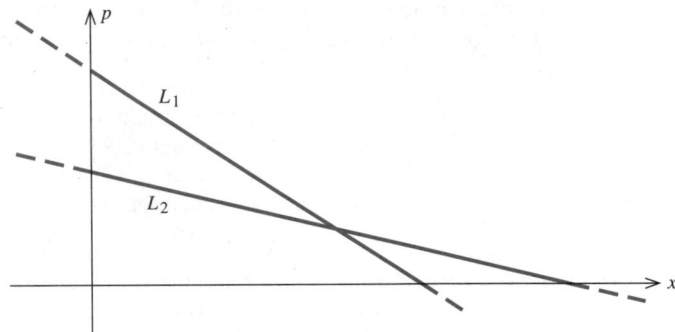

Which line has the greater slope? Interpret your results.

32. Supply of Clock Radios   In the accompanying figure, L_1 is the supply curve for the model A clock radio manufactured by Ace Radio Inc., and L_2 is the supply curve for the model B clock radio.

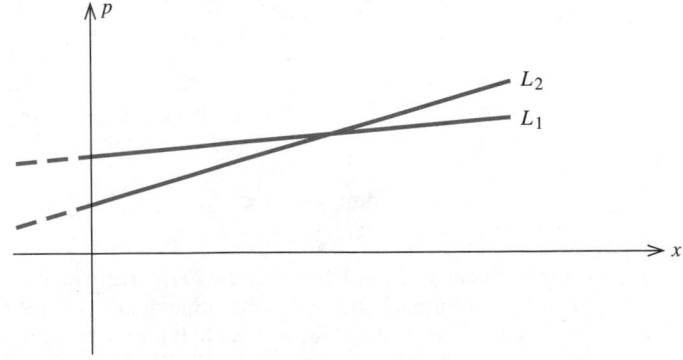

Which line has the greater slope? Interpret your results.

For each of the demand equations in exercises 33–36, where x represents the quantity demanded in units of 1000 and p is the unit price in dollars, (a) sketch the demand curve; (b) determine the quantity demanded corresponding to the given unit price p.

33. $2x + 3p - 18 = 0$; $p = 4$

34. $5p + 4x - 80 = 0$; $p = 10$

35. $p = -3x + 60$; $p = 30$

36. $p = -0.4x + 120$; $p = 80$

37. **Demand Functions**   At a unit price of $55, the quantity demanded of a certain commodity is 1000 units. At a unit price of $85, the demand drops to 600 units. Given that it is linear, find the demand equation. Above what price will there be no demand? What quantity would be demanded if the commodity were free?

38. **Demand Functions**   The quantity demanded for a certain commodity is 200 units when the unit price is set at $90. The quantity demanded is 1200 units when the unit price is $40. Find the demand equation and sketch its graph.

39. **Demand Functions**   Assume that a certain commodity's demand equation has the form $p = ax + b$, where x is the quantity demanded and p is the unit price in dollars. Suppose that the quantity demanded is 1000 units when the unit price is $9 and 6000 when the unit price is $4. What is the quantity demanded when the unit price is $7.50?

40. **Demand Functions**   The demand equation for the Sicard wristwatch is

$$p = -0.025x + 50$$

where x is the quantity demanded per week and p is the unit price in dollars. Sketch the graph of the demand equation. What is the highest price (theoretically) anyone would pay for the watch?

For each supply equation in exercises 41–44, where x is the quantity supplied in units of 1000 and p is the unit price in dollars, (a) sketch the supply curve; (b) determine the number of units of the commodity the supplier will make available in the market at the given unit price.

41. $3x - 4p + 24 = 0$; $p = 8$

42. $\frac{1}{2}x - \frac{2}{3}p + 12 = 0$; $p = 24$

43. $p = 2x + 10$; $p = 14$

44. $p = \frac{1}{2}x + 20$; $p = 28$

45. **Supply Functions**   Suppliers of 15-inch black-and-white television sets will make 10,000 available in the market if the unit price is $45. At a unit price of $50, 20,000 units will be made available. Assuming that the relationship between the unit price and the quantity supplied is linear, derive the supply equation. Sketch the supply curve and determine the quantity suppliers will make available when the unit price is $70.

46. **Supply Functions**   Producers will make 2000 refrigerators available when the unit price is $330. At a unit price of $390, 6000 refrigerators will be marketed. Find the equation relating the unit price of a refrigerator to the quantity supplied if the equation is known to be linear. How many refrigerators will be marketed when the unit price is $450? What is the lowest price at which a refrigerator will be marketed?

SOLUTIONS TO SELF-CHECK EXERCISES 1.3

1. Let x denote the number of units produced and sold. Then
   a. $C(x) = 10x + 60,000$
   b. $R(x) = 15x$
   c. $P(x) = R(x) - C(x) = 15x - (10x + 60,000)$
   $$= 5x - 60,000$$
   d. $P(10,000) = 5(10,000) - 60,000$
   $$= -10,000$$
   or a loss of $10,000.
   $P(14,000) = 5(14,000) - 60,000$
   $$= 10,000$$
   or a profit of $10,000.

Using Technology

Evaluating a Function

A graphing utility can be used to find the value of a function f at a given point with minimal effort. However, to find the value of y for a given value of x in a linear equation such as $Ax + By + C = 0$, the equation must first be cast in the slope-intercept form $y = mx + b$, thus revealing the desired rule $f(x) = mx + b$ for y as a function of x.

EXAMPLE I   Consider the equation $2x + 5y = 7$.

a. Plot the straight line with the given equation in the standard viewing rectangle.

b. Find the value of y when $x = 2$ and verify your result by direct computation.

c. Find the value of y when $x = 1.732$.

Solution   **a.** The straight line with equation $2x + 5y = 7$ or, equivalently, $y = -\frac{2}{5}x + \frac{7}{5}$ in the standard viewing rectangle is shown in Figure T1.

Figure T1
The straight line $2x + 5y = 7$ in the standard viewing rectangle.

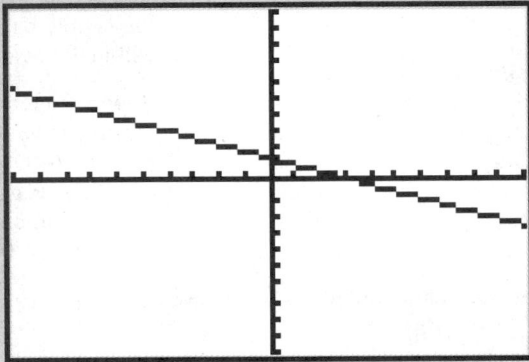

b. Using the evaluation function of the graphing utility and the value of 2 for x, we find $y = 0.6$. This result is verified by computing

$$y = -\frac{2}{5}(2) + \frac{7}{5} = -\frac{4}{5} + \frac{7}{5} = \frac{3}{5} = 0.6$$

when $x = 2$.

c. Once again using the evaluation function of the graphing utility, this time with the value 1.732 for x, we find $y = 0.7072$. The efficacy of the graphing utility is clearly demonstrated here!   ○ ○ ○

 When evaluating $f(x)$ at $x = a$, remember that the number a must lie between xMin and xMax.

EXAMPLE 2 According to Pacific Gas and Electric, the nation's largest utility company, the demand for electricity (in megawatts) in year t is approximately

$$D(t) = 295t + 328 \qquad (0 \le t \le 10)$$

with $t = 0$ corresponding to 1990.

a. Plot the graph of D in the viewing window $[0, 10] \times [0, 3500]$.

b. What will the demand for electricity be in 1999?

Source: Pacific Gas and Electric

Solution **a.** The graph of D is shown in Figure T2.

Figure T2
The graph of D in the viewing window $[0, 10] \times [0, 3500]$.

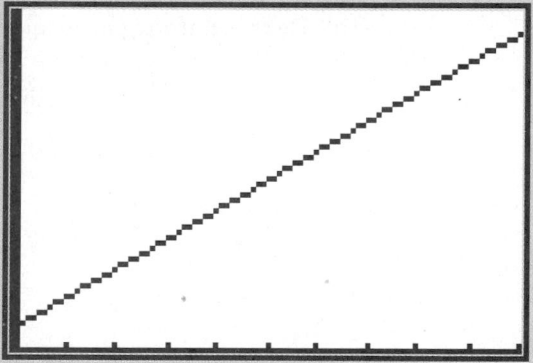

b. Evaluating the function at $x = 9$, we find $y = 2983$. Therefore, the demand for electricity in 1999 will be approximately 2983 megawatts. ○ ○ ○

EXERCISES

In exercises 1–8, use the evaluation function of your graphing calculator to find the value of y corresponding to the given value of x.

1. $3.1x + 2.4y - 12 = 0$; $x = 2.1$

2. $1.2x - 3.2y + 8.2 = 0$; $x = 1.2$

3. $2.8x + 4.2y = 16.3$; $x = 1.5$

4. $-1.8x + 3.2y - 6.3 = 0$; $x = -2.1$

5. $22.1x + 18.2y - 400 = 0$; $x = 12.1$

6. $17.1x - 24.31y - 512 = 0$; $x = -8.2$

7. $2.8x = 1.41y - 2.64$; $x = 0.3$

8. $0.8x = 3.2y - 4.3$; $x = -0.4$

2. Let p denote the price of a rug (in dollars) and let x denote the quantity of rugs demanded when the unit price is $\$p$. The condition that the quantity demanded is 500 when the unit price is $\$100$ tells us that the demand curve passes through the point (500, 100). Next, the condition that for each $\$20$ decrease in the unit price the quantity demanded increases by 500 tells us that the demand curve is linear and that its slope is given by $-20/500$, or $-1/25$. Therefore, letting $m = -1/25$ in the demand equation

$$p = mx + b$$

we find

$$p = -\frac{1}{25}x + b$$

To determine b, use the fact that the straight line passes through (500, 100) to obtain

$$100 = -\frac{1}{25}(500) + b$$

or $b = 120$. Therefore, the required equation is

$$p = -\frac{1}{25}x + 120$$

The demand curve is sketched in the accompanying figure.

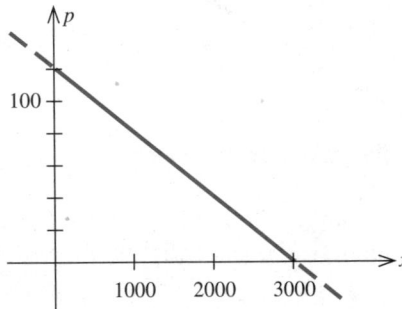

The graph of the demand curve
$p = -\frac{1}{25}x + 120.$

1.4 INTERSECTION OF STRAIGHT LINES

Finding the Point of Intersection

The solution of certain practical problems involves finding the point of intersection of two straight lines. To see how such a problem may be solved algebraically, suppose we are given two straight lines L_1 and L_2 with equations

$$y = m_1x + b_1 \qquad \text{and} \qquad y = m_2x + b_2$$

(where m_1, b_1, m_2, and b_2 are constants) that intersect at the point $P(x, y)$ (Figure 1.31).

Figure 1.31
*L_1 and L_2 intersect at the point
$P(x, y)$.*

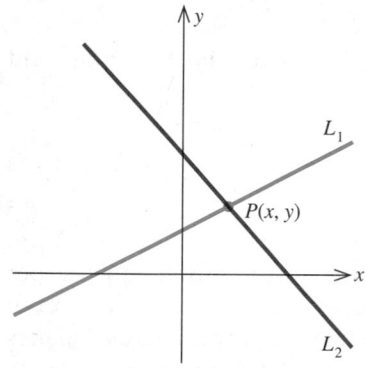

The point $P(x, y)$ lies on the line L_1 and so satisfies the equation $y = m_1 x + b_1$. It also lies on the line L_2 and so satisfies the equation $y = m_2 x + b_2$. Therefore, to find the point of intersection $P(x, y)$ of the lines L_1 and L_2, we solve the system comprising the two equations

$$y = m_1 x + b_1 \quad \text{and} \quad y = m_2 x + b_2$$

for x and y.

EXAMPLE I Find the point of intersection of the straight lines that have equations $y = x + 1$ and $y = -2x + 4$.

Solution We solve the given simultaneous equations. Substituting the value y as given in the first equation into the second, we obtain

$$x + 1 = -2x + 4$$
$$3x = 3$$

or

$$x = 1$$

Substituting this value of x into either one of the given equations yields $y = 2$. Therefore, the required point of intersection is $(1, 2)$ (Figure 1.32).

Figure 1.32
*The point of intersection of L_1 and
L_2 is $(1, 2)$.*

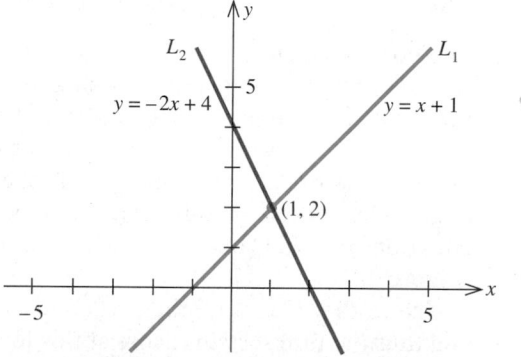

EXPLORING WITH TECHNOLOGY

I. Use a graphing utility to plot the straight lines L_1 and L_2 with equations $y = 3x - 2$ and $y = -2x + 3$, respectively, on the same set of axes using the standard viewing rectangle. Then use **TRACE** and **ZOOM** to find the point of intersection of L_1 and L_2. Repeat using the "intersection" function of your graphing utility.

2. Find the point of intersection of L_1 and L_2 algebraically.

3. Comment on the effectiveness of each method.

We now turn to some applications involving the intersections of pairs of straight lines.

Break-Even Analysis

Consider a firm with (linear) cost function $C(x)$, revenue function $R(x)$, and profit function $P(x)$ given by

$$C(x) = cx + F$$
$$R(x) = sx$$

and
$$P(x) = R(x) - C(x) = (s - c)x - F$$

where c denotes the unit cost of production, s denotes the selling price per unit, F denotes the fixed cost incurred by the firm, and x denotes the level of production and sales. The level of production at which the firm neither makes a profit nor sustains a loss is called the **break-even level of operation** and may be determined by solving the equations $p = C(x)$ and $p = R(x)$ simultaneously. For at this level of production, x_0, the profit is zero, so

$$P(x_0) = R(x_0) - C(x_0) = 0$$

and
$$R(x_0) = C(x_0)$$

The point $P_0(x_0, p_0)$, the solution of the simultaneous equations $p = R(x)$ and $p = C(x)$, is referred to as the **break-even point;** the number x_0 and the number p_0 are called the **break-even quantity** and the **break-even revenue,** respectively.

Geometrically, the break-even point $P_0(x_0, p_0)$ is just the point of intersection of the straight lines representing the cost and revenue functions, respectively. This follows because $P_0(x_0, p_0)$, being the solution of the simultaneous equations $p = R(x)$ and $p = C(x)$, must lie on both these lines simultaneously (Figure 1.33).

Note that if $x < x_0$, then $R(x) < C(x)$ so that $P(x) = R(x) - C(x) < 0$ and thus the firm sustains a loss at this level of production. On the other hand, if $x > x_0$, then $P(x) > 0$ and the firm operates at a profitable level.

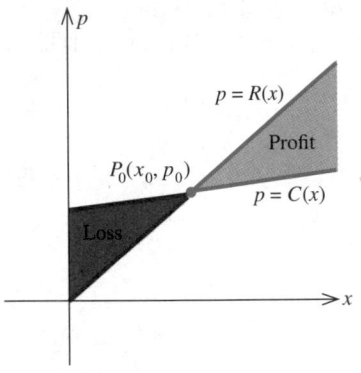

Figure 1.33
P_0 is the break-even point.

EXAMPLE 2   Prescott, Inc., manufactures its products at a cost of $4 per unit and sells them for $10 per unit. If the firm's fixed cost is $12,000 per month, determine the firm's break-even point.

Solution   The cost function C and the revenue function R are given by $C(x) = 4x + 12,000$ and $R(x) = 10x$, respectively (Figure 1.34).
Setting $R(x) = C(x)$, we obtain

$$10x = 4x + 12,000$$
$$6x = 12,000$$
$$x = 2000$$

Substituting this value of x into $R(x) = 10x$ gives

$$R(2000) = (10)(2000) = 20,000$$

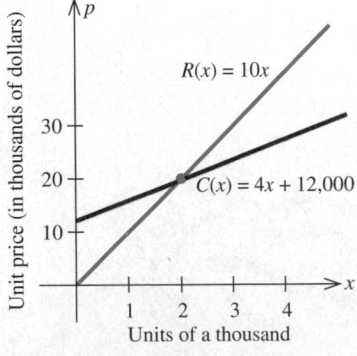

Figure 1.34
The point at which $R(x) = C(x)$ is the break-even point.

So, for a break-even operation, the firm should manufacture 2000 units of its product, resulting in a break-even revenue of $20,000.   ○ ○ ○

EXAMPLE 3   Using the data given in Example 2, answer the following questions:

a. What is the loss sustained by the firm if only 1500 units are produced and sold per month?

b. What is the profit if 3000 units are produced and sold per month?

c. How many units should the firm produce in order to realize a minimum monthly profit of $9000?

Solution   The profit function P is given by the rule

$$P(x) = R(x) - C(x)$$
$$= 10x - (4x + 12,000)$$
$$= 6x - 12,000$$

a. If 1500 units are produced and sold per month, we have

$$P(1500) = 6(1500) - 12,000 = -3000$$

so the firm will sustain a loss of $3000 per month.

b. If 3000 units are produced and sold per month, we have

$$P(3000) = 6(3000) - 12,000 = 6000$$

or a monthly profit of $6000.

c. Substituting 9000 for $P(x)$ in the equation $P(x) = 6x - 12,000$, we obtain

$$9000 = 6x - 12,000$$
$$6x = 21,000$$
$$x = 3500$$

Thus, the firm should produce at least 3500 units in order to realize a $9000 minimum monthly profit.   ○ ○ ○

EXAMPLE 4   The management of the Robertson Controls Company must decide between two manufacturing processes for its model C electronic thermostat. The monthly cost of the first process is given by $C_1(x) = 20x + 10,000$ dollars, where x is the number of thermostats produced; the monthly cost of the second process is given by $C_2(x) = 10x + 30,000$ dollars. If the projected sales are 800 thermostats at a unit price of $40, which process should management choose in order to maximize the company's profit?

Solution   The break-even level of operation using the first process is obtained by solving the equation

$$40x = 20x + 10,000$$

or $$20x = 10,000$$

Thus $\hspace{4cm} x = 500$

giving an output of 500 units. Next, we solve the equation

$$40x = 10x + 30,000$$

or $\hspace{4cm} 30x = 30,000$

Thus $\hspace{4cm} x = 1000$

giving an output of 1000 units for a break-even operation using the second process. Since the projected sales are 800 units, we conclude that management should choose the first process, which will give the firm a profit.   ◦ ◦ ◦

EXAMPLE 5   Referring to Example 4, decide which process Robertson's management should choose if the projected sales are (a) 1500 units and (b) 3000 units.

Solution   In both cases, the production is past the break-even level. Since the revenue is the same regardless of which process is employed, the decision will be based on how much each process costs.

a. If $x = 1500$, then

$$C_1(x) = (20)(1500) + 10,000 = 40,000$$
$$C_2(x) = (10)(1500) + 30,000 = 45,000$$

Hence, management should choose the first process.

b. If $x = 3000$, then

$$C_1(x) = (20)(3000) + 10,000 = 70,000$$
$$C_2(x) = (10)(3000) + 30,000 = 60,000$$

In this case, management should choose the second process.   ◦ ◦ ◦

EXPLORING WITH TECHNOLOGY

1. Use a graphing utility to plot the straight lines L_1 and L_2 with equations $y = 2x - 1$ and $y = 2.1x + 3$, respectively, on the same set of axes using the standard viewing rectangle. Do the lines appear to intersect?

2. Plot the straight lines L_1 and L_2 using the viewing rectangle $[-100, 100] \times [-100, 100]$. Do the lines appear to intersect? Can you find the point of intersection using **TRACE** and **ZOOM**? Using the "intersection" function of your graphing utility?

3. Find the point of intersection of L_1 and L_2 algebraically.

4. Comment on the effectiveness of the solution methods in parts 2 and 3.

◦   ◦   ◦

Market Equilibrium

Under pure competition, the price of a commodity eventually settles at a level dictated by the condition that the supply of the commodity be equal to the demand for it. If the price is too high, consumers will be more reluctant to buy, and if the price is too low, the supplier will be more reluctant to make the product available in the marketplace. **Market equilibrium** is said to prevail when the quantity produced is equal to the quantity demanded. The quantity produced at market equilibrium is called the **equilibrium quantity,** and the corresponding price is called the **equilibrium price.**

From a geometric point of view, market equilibrium corresponds to the point at which the demand curve and the supply curve intersect. In Figure 1.35, x_0 represents the equilibrium quantity and p_0 the equilibrium price. The point (x_0, p_0) lies on the supply curve and therefore satisfies the supply equation. At the same time, it also lies on the demand curve and therefore satisfies the demand equation. Thus, to find the point (x_0, p_0), and hence the equilibrium quantity and price, we solve the demand and supply equations simultaneously for x and p. For meaningful solutions, x and p must both be positive.

Figure 1.35
Market equilibrium is represented by the point (x_0, p_0).

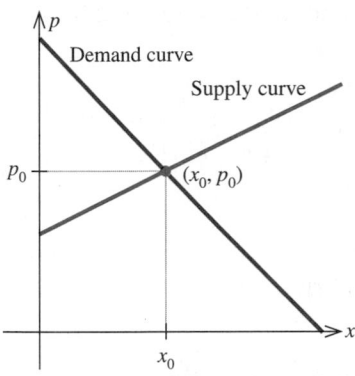

EXAMPLE 6 The management of the Thermo-Master Company, which manufactures an indoor-outdoor thermometer in its Mexico subsidiary, has determined that the demand equation for its product is

$$5x + 3p - 30 = 0$$

where p is the price of a thermometer in dollars and x is the quantity demanded in units of a thousand. The supply equation for these thermometers is

$$52x - 30p + 45 = 0$$

where x (measured in thousands) is the quantity Thermo-Master will make available in the market at p dollars each. Find the equilibrium quantity and price.

Solution   We need to solve the system of equations

$$5x + 3p - 30 = 0$$

and $$52x - 30p + 45 = 0$$

for x and p. Let us use the *method of substitution* to solve it. As the name suggests, this method calls for choosing one of the equations in the system, solving for one variable in terms of the other, and then substituting the resulting expression into the other equation. This gives an equation in one variable that can then be solved in the usual manner.

Let us solve the first equation for p in terms of x. Thus,

$$3p = -5x + 30$$

and $$p = -\frac{5}{3}x + 10$$

Next, we substitute this value of p into the second equation, obtaining

$$52x - 30\left(-\frac{5}{3}x + 10\right) + 45 = 0$$

$$52x + 50x - 300 + 45 = 0$$

$$102x - 255 = 0$$

or
$$x = \frac{255}{102} = \frac{5}{2}$$

The corresponding value of p is found by substituting this value of x into the equation for p obtained earlier. Thus

$$p = -\frac{5}{3}\left(\frac{5}{2}\right) + 10 = -\frac{25}{6} + 10$$

$$= \frac{35}{6} \approx 5.83$$

We conclude that the equilibrium quantity is 2500 units (remember that x is measured in units of a thousand) and the equilibrium price is $5.83 per thermometer.   ⊙ ⊙ ⊙

EXAMPLE 7   The quantity demanded of a certain model of videocassette recorder (VCR) is 8000 units when the unit price is $260. At a unit price of $200, the quantity demanded increases to 10,000 units. The manufacturer will not market any VCRs if the price is $100 or lower. However, for each $50 increase in the unit price above $100, the manufacturer will market an additional 1000 units. Both the demand and the supply equations are known to be linear.

a. Find the demand equation.

b. Find the supply equation.

c. Find the equilibrium quantity and price.

Solution   Let p denote the unit price in hundreds of dollars, and let x denote the number of units of VCRs in thousands.

a. Since the demand function is linear, the demand curve is a straight line passing through the points (8, 2.6) and (10, 2). Its slope is

$$m = \frac{2 - 2.6}{10 - 8} = -0.3$$

Using the point (10, 2) and the slope $m = -0.3$ in the point-slope form of the equation of a line, we see that the required demand equation is

$$p - 2 = -0.3(x - 10)$$

or $p = -0.3x + 5$ (Figure 1.36)

b. The supply curve is the straight line passing through the points (0, 1) and (1, 1.5). Its slope is

Figure 1.36
Market equilibrium occurs at the point (5, 3.5).

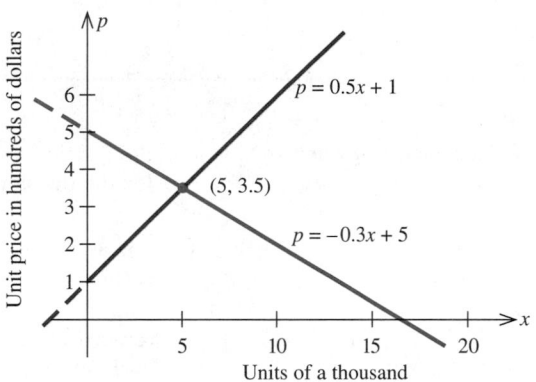

$$m = \frac{1.5 - 1}{1 - 0} = 0.5$$

Using the point $(0, 1)$ and the slope $m = 0.5$ in the point-slope form of the equation of a line, we see that the required supply equation is

$$p - 1 = 0.5(x - 0)$$

or $\qquad\qquad p = 0.5x + 1 \qquad$ (Figure 1.36)

c. To find the market equilibrium, we solve simultaneously the system comprising the demand and supply equations obtained in (a) and (b)—that is, the system

$$p = -0.3x + 5$$
$$p = 0.5x + 1$$

Subtracting the first equation from the second gives

$$0.8x - 4 = 0$$

and $x = 5$. Substituting this value of x in the second equation gives $p = 3.5$. Thus, the equilibrium quantity is 5000 units and the equilibrium price is \$350 (Figure 1.36). ◐ ◐ ◐

SELF-CHECK EXERCISES 1.4

1. Find the point of intersection of the straight lines with equations $2x + 3y = 6$ and $x - 3y = 4$.

2. There is no demand for a certain make of videocassette tape when the unit price is \$12. However, when the unit price is \$8, the quantity demanded is 8000 a week. The suppliers will not market any tapes if the unit price is \$2 or lower. At \$4 per tape, however, the manufacturer will make 5000 tapes available per week. Both the demand and supply functions are known to be linear.
   a. Find the demand equation.
   b. Find the supply equation.
   c. Find the equilibrium quantity and price.

Solutions to Self-Check Exercises 1.4 can be found on page 56.

1.4 EXERCISES

In exercises 1–6, find the point of intersection of each of the given pairs of straight lines.

1. $y = 3x + 4$
   $y = -2x + 14$

2. $y = -4x - 7$
   $-y = 5x + 10$

3. $2x - 3y = 6$
   $3x + 6y = 16$

4. $2x + 4y = 11$
   $-5x + 3y = 5$

5. $y = \frac{1}{4}x - 5$
   $2x - \frac{3}{2}y = 1$

6. $y = \frac{2}{3}x - 4$
   $x + 3y + 3 = 0$

In exercises 7–10, find the break-even point for the firm whose cost function C and revenue function R are given.

7. $C(x) = 5x + 10,000, \quad R(x) = 15x$

8. $C(x) = 15x + 12,000, \quad R(x) = 21x$

9. $C(x) = 0.2x + 120, \quad R(x) = 0.4x$

10. $C(x) = 150x + 20,000, \quad R(x) = 270x$

11. **Break-Even Analysis**   Auto-Time, a manufacturer of 24-hour variable timers, has a monthly fixed cost of $48,000 and a production cost of $8 for each timer manufactured. The units sell for $14 each.
    a. Sketch the graphs of the cost function and the revenue function, and hence find the break-even point graphically.
    b. Find the break-even point algebraically.
    c. Sketch the graph of the profit function.
    d. At what point does the graph of the profit function cross the *x*-axis? Interpret your result.

12. **Break-Even Analysis**   A division of Carter Enterprises produces "Personal Income Tax" diaries. Each diary sells for $8. The monthly fixed costs incurred by the division are $25,000, and the variable cost of producing each diary is $3.
    a. Find the break-even point for the division.
    b. What should the level of sales be in order for the division to realize a 15% profit over the cost of making the diaries?

13. **Break-Even Analysis**   A division of the Gibson Corporation manufactures bicycle pumps. Each pump sells for $9, and the variable cost of producing each unit is 40% of the selling price. The monthly fixed costs incurred by the division are $50,000. What is the break-even point for the division?

14. **Leasing**   The Ace Truck Leasing Company leases a certain size truck for $30 a day and 15 cents a mile, whereas the Acme Truck Leasing Company leases the same size truck for $25 a day and 20 cents a mile.
    a. Find the functions describing the daily cost of leasing from each company.
    b. Sketch the graphs of the two functions on the same set of axes.
    c. If a customer plans to drive at most 70 miles, which company should he rent a truck from for one day?

15. **Decision Analysis**   A product may be made using machine I or machine II. The manufacturer estimates that the monthly fixed costs of using machine I are $18,000, whereas the monthly fixed costs of using machine II are $15,000. The variable costs of manufacturing one unit of the product using machine I and machine II are $15 and $20, respectively. The product sells for $50 each.
    a. Find the cost functions associated with using each machine.
    b. Sketch the graphs of the cost functions of (a) and the revenue functions on the same set of axes.
    c. Which machine should management choose to maximize their profit if the projected sales are 450 units? 550 units? 650 units?
    d. What is the profit for each case in (c)?

16. The annual sales of the Crimson Drug Store are expected to be given by $S = 2.3 + 0.4t$ million dollars t years from now, whereas the annual sales of the Cambridge Drug Store are expected to be given by $S = 1.2 + 0.6t$ million dollars t years from now. When will the annual sales of the Cambridge Drug Store first surpass the annual sales of the Crimson Drug Store?

For each pair of supply and demand equations in exercises 17–20, where x represents the quantity demanded in units of 1000 and p is the unit price in dollars, find the equilibrium quantity and the equilibrium price.

17. $4x + 3p - 59 = 0 \quad$ and $\quad 5x - 6p + 14 = 0$

18. $2x + 7p - 56 = 0 \quad$ and $\quad 3x - 11p + 45 = 0$

19. $p = -2x + 22$ and $p = 3x + 12$

20. $p = -0.3x + 6$ and $p = 0.15x + 1.5$

21. Equilibrium Quantity and Price The quantity demanded of a certain brand of videocassette recorder (VCR) is 3000 per week when the unit price is $485. For each decrease in unit price of $20 below $485, the quantity demanded increases by 250 units. The suppliers will not market any VCRs if the unit price is $300 or lower. But at a unit price of $525, they are willing to make available 2500 units in the market. The supply equation is also known to be linear.
  a. Find the demand equation.
  b. Find the supply equation.
  c. Find the equilibrium quantity and price.

22. Equilibrium Quantity and Price The demand equation for the Miramar Heat Machine, a ceramic heater, is $x - 4p - 800 = 0$, where x is the quantity demanded per week and p is the wholesale unit price in dollars. The supply equation is $x - 20p + 1000 = 0$, where x is the quantity the supplier will make available in the market when the wholesale price is p dollars each. Find the equilibrium quantity and the equilibrium price for the Miramar heaters.

23. Equilibrium Quantity and Price The demand equation for the Schmidt-3000 fax machine is $3x + p - 1500 = 0$, where x is the quantity demanded per week and p is the unit price in dollars. The supply equation is $2x - 3p + 1200 = 0$, where x is the quantity the supplier will make available in the market when the unit price is p dollars. Find the equilibrium quantity and the equilibrium price for the fax machines.

24. Equilibrium Quantity and Price The quantity demanded per month of Russo Espresso Makers is 250 when the unit price is $140. The quantity demanded per month is 1000 when the unit price is $110. The suppliers will market 750 espresso makers if the unit price is $60 or lower. At a unit price of $80 they are willing to make available 2250 units in the market. Both the demand and supply equations are known to be linear.
  a. Find the demand equation.
  b. Find the supply equation.
  c. Find the equilibrium quantity and the equilibrium price.

25. Suppose the demand and supply equations for a certain commodity are given by $p = ax + b$ and $p = cx + d$,

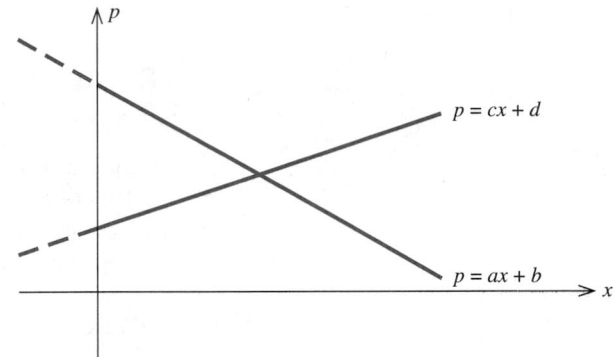

respectively, where $a < 0$, $c > 0$, and $b > d > 0$ (see the accompanying figure).
  a. Find the equilibrium quantity and equilibrium price in terms of a, b, c, and d.
  b. Use (a) to determine what happens to the market equilibrium if c is increased while a, b, and d remain fixed. Interpret your answer in economic terms.
  c. Use (a) to determine what happens to the market equilibrium if b is decreased while a, c, and d remain fixed. Interpret your answer in economic terms.

26. Suppose the cost function associated with a product is $C(x) = cx + F$ dollars and the revenue function is $R(x) = sx$, where c denotes the unit cost of production, s denotes the unit selling price, F denotes the fixed cost incurred by the firm, and x denotes the level of production and sales. Find the break-even quantity and the break-even revenue in terms of the constants c, s, and F and interpret your results in economic terms.

27. Let L_1 and L_2 be two nonvertical straight lines in the plane with equations $y = m_1x + b_1$ and $y = m_2x + b_2$, respectively. Find conditions on m_1, m_2, b_1, and b_2 so that (a) L_1 and L_2 do not intersect, (b) L_1 and L_2 intersect at one and only one point, and (c) L_1 and L_2 intersect at infinitely many points.

28. Find conditions on a_1, a_2, b_1, b_2, c_1, and c_2 so that the system of linear equations

$$a_1x + b_1y = c_1$$
$$a_2x + b_2y = c_2$$

has (a) no solution, (b) a unique solution, and (c) infinitely many solutions.

[*Hint:* Use the results of exercise 27.]

USING TECHNOLOGY

FINDING THE POINTS OF INTERSECTION OF TWO GRAPHS

A graphing utility can be used to find the point(s) of intersection of the graphs of two functions. Once again, it is important to remember that if the graphs are straight lines, the linear equations defining these lines must first be recast in the slope-intercept form.

EXAMPLE 1 Find the points of intersection of the straight lines with equations $2x + 3y = 6$ and $3x - 4y - 5 = 0$.

Solution Solving each equation for y in terms of x, we obtain

$$y = -\frac{2}{3}x + 2 \qquad \text{and} \qquad y = \frac{3}{4}x - \frac{5}{4}$$

as the respective equations in the slope-intercept form. The graphs of the two straight lines in the standard viewing rectangle are shown in Figure T1.

Figure T1
The straight lines $2x + 3y = 6$ and $3x - 4y - 5 = 0$.

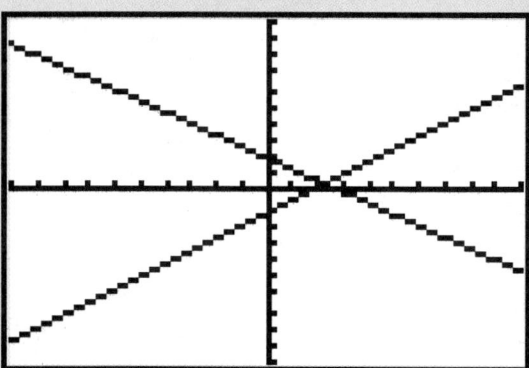

Then, using **TRACE** and **ZOOM** or the function for finding the point of intersection of two graphs, we find that the point of intersection, accurate to four decimal places, is (2.2941, 0.4706). ○ ○ ○

54

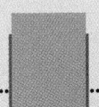

EXERCISES

In exercises 1–6, find the point of intersection of the pair of straight lines with the given equations. Express your answers accurate to four decimal places.

1. $y = 2x + 5$ and $y = -3x + 8$

2. $y = 1.2x + 6.2$ and $y = -4.3x + 9.1$

3. $2x - 5y = 7$ and $3x + 4y = 12$

4. $1.4x - 6.2y = 8.4$ and $4.1x + 7.3y = 14.4$

5. $2.1x = 5.1y + 71$ and $3.2x = 8.4y + 16.8$

6. $8.3x = 6.2y + 9.3$ and $12.4x = 12.3y + 24.6$

SOLUTIONS TO SELF-CHECK EXERCISES 1.4

1. The point of intersection of the two straight lines is found by solving the system of linear equations

$$2x + 3y = 6$$
$$x - 3y = 4$$

Solving the first equation for y in terms of x, we obtain

$$y = -\frac{2}{3}x + 2$$

Substituting this expression for y into the second equation, we obtain

$$x - 3\left(-\frac{2}{3}x + 2\right) = 4$$

$$x + 2x - 6 = 4$$

$$3x = 10$$

or $x = \frac{10}{3}$. Substituting this value of x into the expression for y obtained earlier, we find

$$y = -\frac{2}{3}\left(\frac{10}{3}\right) + 2 = -\frac{2}{9}$$

Therefore, the point of intersection is $\left(\frac{10}{3}, -\frac{2}{9}\right)$.

2. **a.** Let p denote the price per cassette and x the quantity demanded. The given conditions imply that $x = 0$ when $p = 12$ and $x = 8000$ when $p = 8$. Since the demand equation is linear, it has the form

$$p = mx + b$$

Now, the first condition implies that

$$12 = m(0) + b \quad \text{or} \quad b = 12$$

Therefore,

$$p = mx + 12$$

Using the second condition, we find

$$8 = 8000m + 12$$

or

$$m = -\frac{4}{8000} = -0.0005$$

Therefore, the required demand equation is

$$p = -0.0005x + 12$$

b. Let p denote the price per cassette and x the quantity made available at that price. Then, since the supply equation is linear, it also has the form

$$p = mx + b$$

The first condition implies that $x = 0$ when $p = 2$, so we have

$$2 = m(0) + b \qquad \text{or} \qquad b = 2$$

Therefore

$$p = mx + 2$$

Next, using the second condition, $x = 5000$ when $p = 4$, we find

$$4 = 5000m + 2$$

giving $m = 0.0004$. So the required supply equation is

$$p = 0.0004x + 2$$

c. The equilibrium quantity and price are found by solving the system of linear equations

$$p = -0.0005x + 12$$
$$p = 0.0004x + 2$$

Equating the two equations yields

$$-0.0005x + 12 = 0.0004x + 2$$
$$0.0009x = 10$$

or $x \approx 11{,}111$. Substituting this value of x into either equation in the system yields

$$p = 6.44$$

Therefore, the equilibrium quantity is 11,111 and the equilibrium price is $6.44.

1.5 THE METHOD OF LEAST SQUARES (OPTIONAL)

The Method of Least Squares

In Example 10, Section 1.2, we saw how a linear equation may be used to approximate the sales trend for a local sporting goods store. The *trend line*, as we saw, may be used to predict the store's future sales. Recall that we obtained the trend line in Example 10 by requiring that the line pass through two data points, the rationale being that such a line seems to *fit* the data reasonably well.

In this section we describe a general method known as the **method of least squares** for determining a straight line that, in some sense, best fits a set of data points when the points are scattered about a straight line. In order to illustrate the principle behind the method of least squares, suppose, for simplicity, that we are given five data points,

$$P_1(x_1, y_1), \quad P_2(x_2, y_2) \quad P_3(x_3, y_3), \quad P_4(x_4, y_4), \quad \text{and} \quad P_5(x_5, y_5)$$

describing the relationship between the two variables x and y. By plotting these data points, we obtain a graph called a **scatter diagram** (Figure 1.37).

Figure 1.37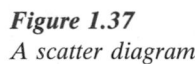
A scatter diagram.

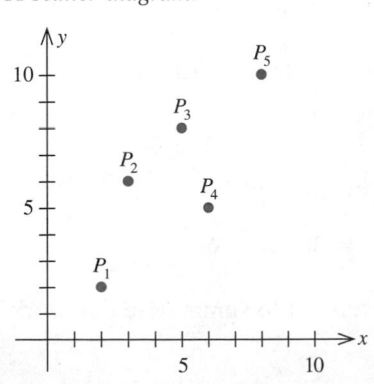

Figure 1.38
d_i is the distance between the straight line and a given data point.

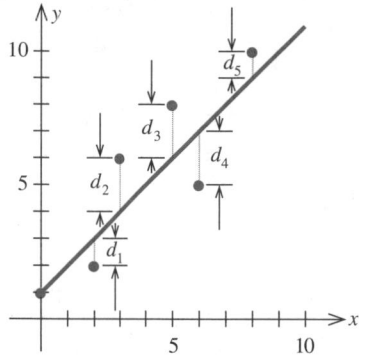

If we try to *fit* a straight line to these data points, the line will miss the first, second, third, fourth, and fifth data points by the amounts d_1, d_2, d_3, d_4, and d_5, respectively (Figure 1.38). We can think of the amounts d_1, d_2, ..., d_5 as the errors made when the values y_1, y_2, ..., y_5 are approximated by the corresponding values of y lying on the straight line L.

The **principle of least squares** states that the straight line L that fits the data points *best* is the one chosen by requiring that the sum of the squares of d_1, d_2, ..., d_5, that is,

$$d_1^2 + d_2^2 + d_3^2 + d_4^2 + d_5^2$$

be made as small as possible. In other words, the least-squares criterion calls for minimizing the sum of the squares of the errors. The line L obtained in this manner is called the **least-squares line,** or *regression line*.

The method for computing the least-squares line that best fits a set of data points is contained in the following result, which we state without proof.

THE METHOD OF LEAST SQUARES

Suppose we are given n data points

$$P_1(x_1, y_1), \quad P_2(x_2, y_2), \quad P_3(x_3, y_3), \ldots, P_n(x_n, y_n)$$

Then, the least-squares (regression) line for the data is given by the linear equation (function)

$$y = f(x) = mx + b$$

where the constants m and b satisfy the **normal equations**

$$nb + (x_1 + x_2 + \cdots + x_n)m = y_1 + y_2 + \cdots + y_n \tag{8}$$

and
$$(x_1 + x_2 + \cdots + x_n)b + (x_1^2 + x_2^2 + \cdots + x_n^2)m$$
$$= x_1 y_1 + x_2 y_2 + \cdots + x_n y_n \tag{9}$$

simultaneously.

EXAMPLE I   Find the least-squares line for the data

$$P_1(1, 1), \quad P_2(2, 3), \quad P_3(3, 4), \quad P_4(4, 3), \quad \text{and} \quad P_5(5, 6)$$

Solution   Here we have $n = 5$ and

$$x_1 = 1, \quad x_2 = 2, \quad x_3 = 3, \quad x_4 = 4, \quad x_5 = 5$$
$$y_1 = 1, \quad y_2 = 3, \quad y_3 = 4, \quad y_4 = 3, \quad y_5 = 6$$

In order to use equations (8) and (9), it is convenient to summarize our work in the form of a table:

	x	y	x^2	xy
	1	1	1	1
	2	3	4	6
	3	4	9	12
	4	3	16	12
	5	6	25	30
Sum	15	17	55	61

Using this table and (8) and (9), we obtain the normal equations

$$5b + 15m = 17 \tag{10}$$

and

$$15b + 55m = 61 \tag{11}$$

Solving equation (10) for b gives

$$b = -3m + \frac{17}{5} \tag{12}$$

which upon substitution into equation (11) gives

$$15\left(-3m + \frac{17}{5}\right) + 55m = 61$$

$$-45m + 51 + 55m = 61$$

$$10m = 10$$

or

$$m = 1$$

Substituting this value of m into equation (12) gives

$$b = -3 + \frac{17}{5}$$

or

$$b = \frac{2}{5} = 0.4$$

Therefore, the required least-squares line is

$$y = x + 0.4$$

The scatter diagram and the least-squares line are shown in Figure 1.39.

● ● ●

Application

EXAMPLE 2 The proprietor of the Leisure Travel Service compiled the following data relating the annual profit of the firm to its annual advertising expenditure (both measured in thousands of dollars):

Annual Advertising Expenditure (x)	12	14	17	21	26	30
Annual Profit (y)	60	70	90	100	100	120

Figure 1.39
The least-squares line $y = x + 0.4$ and the given data points.

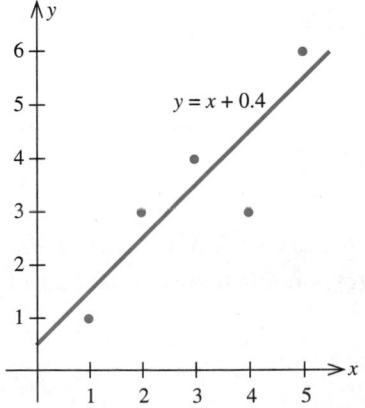

a. Determine the equation of the least-squares line for these data.

b. Draw a scatter diagram and the least-squares line for these data.

c. Use the result obtained in (a) to predict Leisure Travel's annual profit if the annual advertising budget is $20,000.

Solution

a. The calculations required for obtaining the normal equations are summarized in the accompanying table.

	x	y	x^2	xy
	12	60	144	720
	14	70	196	980
	17	90	289	1,530
	21	100	441	2,100
	26	100	676	2,600
	30	120	900	3,600
Sum	120	540	2,646	11,530

The normal equations are

$$6b + 120m = 540 \tag{13}$$

and

$$120b + 2646m = 11{,}530 \tag{14}$$

Solving equation (13) for b gives

$$b = -20m + 90 \tag{15}$$

which upon substitution into equation (14) gives

$$120(-20m + 90) + 2646m = 11{,}530$$
$$-2400m + 10{,}800 + 2646m = 11{,}530$$
$$246m = 730$$

or

$$m = 2.97$$

Substituting this value of m into equation (15) gives

$$b = -20(2.97) + 90$$

or

$$b = 30.6$$

Therefore, the required least-squares line is given by

$$y = f(x) = 2.97x + 30.6$$

b. The scatter diagram and the least-squares line are shown in Figure 1.40.

c. Leisure Travel's predicted annual profit corresponding to an annual budget of $20,000 is given by

$$f(20) = 2.97(20) + 30.6 = 90$$

or $90,000.

Figure 1.40
Profit versus advertising expenditure.

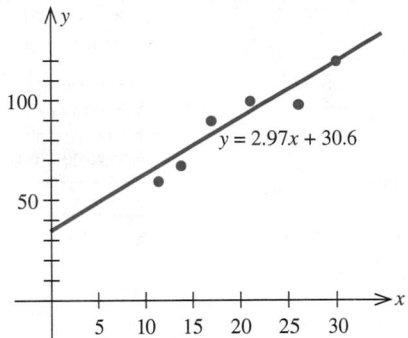

$y = 2.97x + 30.6$

SELF–CHECK EXERCISES 1.5

1. Find an equation of the least-squares line for the data

x	1	3	4	5	7
y	4	10	11	12	16

2. In a market research study for the Century Communications Company, the data were provided based on the projected monthly sales x (in thousands) of a videocassette version of a box-office-hit adventure movie with a proposed wholesale unit price of p dollars.

x	2.2	5.4	7.0	11.5	14.6
p	38	36	34.5	30	28.5

Find the demand equation if the demand curve is the least-squares line for these data.

Solutions to Self-Check Exercises 1.5 can be found on page 66.

1.5 EXERCISES

 A calculator is recommended for this exercise set.

In exercises 1–6, (a) find the equation of the least-squares line for the given data and (b) draw a scatter diagram for the given data and graph the least-squares line.

1.

x	1	2	3	4
y	4	6	8	11

2.

x	1	3	5	7	9
y	9	8	6	3	2

3.

x	1	2	3	4	4	6
y	4.5	5	3	2	3.5	1

4.

x	1	1	2	3	4	4	5
y	2	3	3	3.5	3.5	4	5

5. $P_1(1, 3)$, $P_2(2, 5)$, $P_3(3, 5)$, $P_4(4, 7)$, $P_5(5, 8)$

6. $P_1(1, 8)$, $P_2(2, 6)$, $P_3(5, 6)$, $P_4(7, 4)$, $P_5(10, 1)$

7. College Admissions The accompanying data were compiled by the admissions office at Faber College during the past five years. The data relate the number of college brochures and follow-up letters (x) sent to a preselected list of high school juniors who had taken the PSAT and the number of completed applications (y) received from these students (both measured in units of 1000).

x	4	4.5	5	5.5	6
y	0.5	0.6	0.8	0.9	1.2

a. Determine the equation of the least-squares line for these data.
b. Draw a scatter diagram and the least-squares line for these data.
c. Use the result obtained in (a) to predict the number of completed applications expected if 6400 brochures and follow-up letters are sent out during the next year.

8. Net Sales The management of Kaldor, Inc., a manufacturer of electric motors, submitted the accompanying data in its annual stockholders report. The table shows the net sales (in millions of dollars) during the five years that have elapsed since the new management team took over. (The first year the firm operated under the new management corresponds to the time period $x = 1$, and the four subsequent years correspond to $x = 2, 3, 4, 5$.)

Year (x)	1	2	3	4	5
Net Sales (y)	426	437	460	473	477

a. Determine the equation of the least-squares line for these data.
b. Draw a scatter diagram and the least-squares line for these data.
c. Use the result obtained in (a) to predict the net sales for the upcoming year.

9. SAT Verbal Scores The accompanying data were compiled by the superintendent of schools in a large metropolitan area. The table shows the average SAT verbal scores of high school seniors during the five years since the district implemented its "back-to-basics" program.

Year (x)	1	2	3
Average Score (y)	436	438	428

Year (x)	4	5
Average Score (y)	430	426

a. Determine the equation of the least-squares line for these data.
b. Draw a scatter diagram and the least-squares line for these data.
c. Use the result obtained in (a) to predict the average SAT verbal score of high school seniors two years from now ($x = 7$).

10. Auto Operating Costs The accompanying figures were compiled by Clarke, Kingsley, and Company, a consulting firm that specializes in auto operating costs, relating the annual mileage (in thousands of miles) that an average new compact car is driven to the cost per mile (in cents) of operating the car.

Annual Mileage (x)	5	10	15
Cost per Mile (y)	50.3	34.8	30.1

Annual Mileage (x)	20	25	30
Cost per Mile (y)	27.4	25.6	23.5

a. Determine an equation of the least-squares line for these data.
b. Draw a scatter diagram and the least-squares line for these data.
c. Use the result obtained in (a) to estimate the cost per mile of operating a new company car if it is driven 8000 miles during the first year of ownership.

11. Size of Average Farm The size of the average farm in the United States has been growing steadily over the years. The accompanying data, obtained from the U.S. Department of Agriculture, gives the size of the average farm y (in acres) from 1940 through 1991. (Here $x = 0$ corresponds to the beginning of the year 1940.)

Year (x)	0	10	20
Number of Acres (y)	168	213	297

Year (x)	30	40	51
Number of Acres (y)	374	427	467

a. Find the equation of the least-squares line for these data.
b. Use the result of (a) to estimate the size of the average farm in the year 2000.
Source: The World Almanac

12. Welfare Costs According to the Massachusetts Department of Welfare, the spending (in billions of dollars) by Medicaid, the national health-care plan for the poor, over the 5-year period from 1988 to 1992 is summarized in the accompanying table. (Here $x = 0$ represents the beginning of the year 1988.)

Year (x)	0	1	2
Expenditure (y)	1.550	1.662	1.786

Year (x)	3	4
Expenditure (y)	1.888	2.009

a. Find an equation of the least-squares line for these data.
b. Use the result of (a) to estimate Medicaid spending for the year 1996, assuming the trend continued.
Source: Massachusetts Department of Welfare

13. Mass Transit Subsidies The accompanying table gives the projected state subsidies (in millions of dollars) to the Massachusetts Bay Transit Authority (MBTA) over a 5-year period.

Year (x)	1	2	3	4	5
Subsidy (y)	20	24	26	28	32

a. Find an equation of the least-squares line for these data.
b. Use the result of (a) to estimate the state subsidy to the MBTA for the eighth year ($x = 8$).
Source: Massachusetts Bay Transit Authority

14. Social Security Wage Base The social security (FICA) wage base (in thousands of dollars) from 1993 to 1998 is given in the accompanying table.

Year	1993	1994	1995
Wage Base (y)	57.6	60.6	61.2

Year	1996	1997	1998
Wage Base (y)	62.7	65.4	68.4

a. Find an equation of the least-squares line for these data. (Let $x = 1$ represent the year 1993.)
b. Use the result of (a) to estimate the FICA wage base in the year 2002.
Source: The World Almanac

15. Production of All-Aluminum Cans Steel has been playing a decreasing role in the manufacture of beverage cans in the United States. According to the Can Manufacture Institute, the use of bimetallic cans has been dwindling while the use of all-aluminum cans has been growing steadily. The accompanying table gives the production of all-aluminum cans over the period from 1975 through 1989.

Year	1975	1977	1979	1981
No. of Cans (in billions)	16.7	26	33.3	48.3

Year	1983	1985	1987	1989
No. of Cans (in billions)	57	65.8	74.2	83.3

a. Find an equation of the least-squares line for these data. (Let $x = 1$ represent 1975.)
b. Use the result of (a) to estimate the number of cans produced in 1993, assuming the trend continued.
Source: Can Manufacturing Institute

16. Health-Care Spending The following data, compiled by the Organization for Economic Cooperation and Development (OECD) in 1990, give the per capita Gross Domestic Product (GDP) and the corresponding per capita spending on health care for selected countries.

Country	Turkey	Spain	Netherlands
GDP (per capita in thousands of dollars)	4.25	10	14
Health-Care Spending (per capita in dollars)	178	667	1194

USING TECHNOLOGY

FINDING AN EQUATION OF A LEAST-SQUARES LINE

A graphing utility is especially useful in calculating an equation of the least-squares line for a set of data. We simply enter the given data in the form of lists into the calculator and then use the linear regression function to obtain the coefficients of the required equation.

EXAMPLE 1   Find an equation of the least-squares line for the data

x	1.1	2.3	3.2	4.6	5.8	6.7	8
y	-5.8	-5.1	-4.8	-4.4	-3.7	-3.2	-2.5

Solution   First we enter the data as

$$x_1 = 1.1, y_1 = -5.8, x_2 = 2.3, y_2 = -5.1, x_3 = 3.2, y_3 = -4.8, x_4 = 4.6,$$

$$y_4 = -4.4, x_5 = 5.8, y_5 = -3.7, x_6 = 6.7, y_6 = -3.2, x_7 = 8, \text{ and } y_7 = -2.5$$

Then, using the linear regression function from the statistics menu, we find

$$a = -6.29996900666, \quad b = 0.460560979389, \quad \text{corr} = 0.994488871079, \quad n = 7$$

Therefore, an equation of the least-squares line ($y = a + bx$) is

$$y = -6.3 + 0.46x$$

The correlation coefficient of 0.99449 attests to the excellent fit of the regression line.   ◦ ◦ ◦

EXAMPLE 2   According to Pacific Gas and Electric, the nation's largest utility company, the demand for electricity from 1990 through the year 2000 is summarized in the following table:

t	0	2	4	6	8	10
y	333	917	1500	2117	2667	3292

Here $t = 0$ corresponds to 1990 and y gives the amount of electricity demanded in year t measured in megawatts. Find an equation of the least-squares line for these data.
Source: Pacific Gas and Electric

Solution   First we enter the data as

$$x_1 = 0, y_1 = 333, x_2 = 2, y_2 = 917, x_3 = 4, y_3 = 1500, x_4 = 6, y_4 = 2117,$$

$$x_5 = 8, y_5 = 2667, x_6 = 10, \text{ and } y_6 = 3292$$

Then, using the linear regression function from the statistics menu, we find

$$a = 328.476190476 \quad \text{and} \quad b = 295.171428571$$

Therefore, an equation of the least-squares line is

$$y = 328 + 295t$$

○ ○ ○

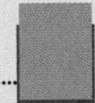

EXERCISES

In exercises 1–4, find an equation of the least-squares line for the data.

1.

x	2.1	3.4	4.7	5.6	6.8	7.2
y	8.8	12.1	14.8	16.9	19.8	21.1

2.

x	1.1	2.4	3.2	4.7	5.6	7.2
y	−0.5	1.2	2.4	4.4	5.7	8.1

3.

x	−2.1	−1.1	0.1	1.4	2.5	4.2	5.1
y	6.2	4.7	3.5	1.9	0.4	−1.4	−2.5

4.

x	−1.12	0.1	1.24	2.76	4.21	6.82
y	7.61	4.9	2.74	−0.47	−3.51	−8.94

5. **Waste Generation** According to data from the Council on Environmental Quality, the amount of waste (in millions of tons per year) generated in the United States from 1960 to 1990 was:

Year	1960	1965	1970	1975
Amount (y)	81	100	120	124

Year	1980	1985	1990
Amount (y)	140	152	164

a. Find an equation of the least-squares line for these data. (Let x be in units of 5, and let $x = 1$ represent 1960.)
b. Use the result of (a) to estimate the amount of waste generated in the year 2000, assuming the trend continues.
Source: Council on Environmental Quality

6. **Median Price of Homes** According to data from the Association of Realtors, the median price (in thousands of dollars) of existing homes in a certain metropolitan area from 1982 to 1992 was:

Year	1982	1983	1984	1985	1986	1987
Price (y)	66.4	69.8	72.8	76.0	79.6	83.1

Year	1988	1989	1990	1991	1992
Price (y)	86.3	89.5	92.3	96.0	99.5

a. Find an equation of the least-squares line for these data. (Let $x = 1$ represent 1982.)
b. Use the result of (a) to estimate the median price of a house in the year 1997, assuming the trend continued.
Source: Association of Realtors

65

Country	Sweden	Switzerland	Canada
GDP (per capita in thousands of dollars)	15.5	17.8	19.5
Health-Care Spending (per capita in dollars)	1500	1388	1640

a. Letting x denote a country's GDP (in thousands of dollars per capita) and y denote the per capita health-care spending (in dollars), find an equation of the least-squares line for these data giving the typical relationship between GDP and health-care spending for the selected countries.

b. The per capita GDP of the United States is $20,000. If the health-care spending of the United States were in line with that of these sample OECD countries, what would it be? [*Note:* The actual per capita health-care spending of the United States in 1990 was $2444.]

Source: Organization for Economic Cooperation and Development

SOLUTIONS TO SELF-CHECK EXERCISES 1.5

1. The calculations required for obtaining the normal equations may be summarized as follows:

	x	y	x^2	xy
	1	4	1	4
	3	10	9	30
	4	11	16	44
	5	12	25	60
	7	16	49	112
Sum	20	53	100	250

The normal equations are

$$5b + 20m = 53$$
$$20b + 100m = 250$$

Solving the first equation for b gives

$$b = -4m + \frac{53}{5}$$

which, upon substitution into the second equation, yields

$$20\left(-4m + \frac{53}{5}\right) + 100m = 250$$

$$-80m + 212 + 100m = 250$$

$$20m = 38$$

$$m = 1.9$$

Substituting this value of m into the expression for b found earlier, we find

$$b = -4(1.9) + \frac{53}{5}$$

or
$$b = 3$$

Therefore, the required least-squares line is

$$y = 1.9x + 3$$

2. The calculations required for obtaining the normal equations may be summarized as follows:

x	p	x^2	xp
2.2	38.0	4.84	83.6
5.4	36.0	29.16	194.4
7.0	34.5	49.00	241.5
11.5	30.0	132.25	345.0
14.6	28.5	213.16	416.1
Sum 40.7	167.0	428.41	1280.6

The normal equations are

$$5b + 40.7m = 167$$
$$40.7b + 428.41m = 1280.6$$

Solving this system of linear equations simultaneously, we find that

$$m = -0.81 \quad \text{and} \quad b = 39.99$$

Therefore, the required least-squares line is given by

$$p = f(x) = -0.81x + 39.99$$

which is the required demand equation provided $0 \leq x \leq 49.37$.

Group projects for each chapter can be found at the Brooks/Cole Web site at http://www.brookscole.com/math/authors/tans/

CHAPTER I SUMMARY OF PRINCIPAL FORMULAS AND TERMS

Formulas

1. Distance between two points $d = \sqrt{(x_2 - x_1)^2 + (y_2 - y_1)^2}$

2. Slope of a line $m = \dfrac{y_2 - y_1}{x_2 - x_1}$

3. Equation of a vertical line $x = a$

4. Equation of a horizontal line $y = b$

5. Point-slope form of the
   equation of a line $y - y_1 = m(x - x_1)$

6. Slope-intercept form of the
   equation of a line $y = mx + b$

7. General equation of a line $Ax + By + C = 0$

Terms

Cartesian coordinate system	Linear function
Ordered pair	Total cost function
Coordinates	Revenue function
Parallel lines	Profit function
Perpendicular lines	Demand function
Function	Supply function
Independent variable	Break-even point
Dependent variable	Market equilibrium
Domain	Equilibrium quantity
Range	Equilibrium price

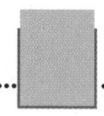

CHAPTER I REVIEW EXERCISES

*In exercises 1–4, find the distance between the two given
points.*

1. (2, 1) and (6, 4)   **2.** (9, 6) and (6, 2)

3. (−2, −3) and (1, −7)

4. $(\frac{1}{2}, \sqrt{3})$ and $(-\frac{1}{2}, 2\sqrt{3})$

*In exercises 5–10, find an equation of the line L that passes
through the point (−2, 4) and satisfies the given condition.*

5. L is a vertical line.   **6.** L is a horizontal line.

7. L passes through the point $(3, \frac{7}{2})$.

8. The x-intercept of L is 3.

9. L is parallel to the line $5x - 2y = 6$.

10. L is perpendicular to the line $4x + 3y = 6$.

11. Find an equation of the line with slope −1/2 and y-
   intercept −3.

12. Find the slope and y-intercept of the line with equation
   $3x - 5y = 6$.

13. Find an equation of the line passing through the point
   (2, 3) and parallel to the line with equation $3x + 4y - 8 = 0$.

14. Find an equation of the line passing through the point
   (−1, 3) and parallel to the line joining the points
   (−3, 4) and (2, 1).

15. Find an equation of the line passing through the point
   (−2, −4) that is perpendicular to the line with equation
   $2x - 3y - 24 = 0$.

*In exercises 16 and 17, sketch the graph of the given
equation.*

16. $3x - 4y = 24$ **17.** $-2x + 5y = 15$

18. Sales of a certain clock radio are approximated by the
   relationship

   $$S(x) = 6000x + 30{,}000 \qquad (0 \le x \le 5)$$

where $S(x)$ denotes the number of clock radios sold in year x ($x = 0$ corresponds to the year 1993). Find the number of clock radios expected to be sold in 1998.

19. A company's total sales (in millions of dollars) are approximately linear as a function of time (in years). Sales in 1989 were $2.4 million, whereas sales in 1994 amounted to $7.4 million.
   a. Find an equation giving the company's sales as a function of time.
   b. What were the sales in 1992?

20. Show that the triangle with vertices $A(1, 1)$, $B(5, 3)$, and $C(4, 5)$ is a right triangle.

21. Clark's Rule is a method for calculating pediatric drug dosages based on a child's weight. If a denotes the adult dosage (in mg) and if w is the child's weight (in lb), then the child's dosage is given by

$$D(w) = \frac{aw}{150}$$

   a. Show that D is a linear function of w.
   b. If the adult dose of a substance is 500 mg, how much should a child who weighs 35 lb receive?

22. An office building worth $6 million when it was completed in 1990 is being depreciated linearly over 30 years.
   a. What is the rate of depreciation?
   b. What will the book value of the building be in 2000?

23. In 1996, a manufacturer installed a new machine in her factory at a cost of $300,000. The machine is depreciated linearly over 12 years with a scrap value of $30,000.
   a. What is the rate of depreciation of the machine per year?
   b. Find an expression for the book value of the machine in year t ($0 \le t \le 12$).

24. A company has a fixed cost of $30,000 and a production cost of $6 for each unit it manufactures. A unit sells for $10.
   a. What is the cost function?
   b. What is the revenue function?
   c. What is the profit function?
   d. Compute the profit (loss) corresponding to production levels of 6000, 8000, and 12,000 units, respectively.

25. There is no demand for a certain commodity when the unit price is $200 or more, but for each $10 decrease in price below $200, the quantity demanded increases by 200 units. Find the demand equation and sketch its graph.

26. Bicycle suppliers will make 200 bicycles available in the market per month when the unit price is $50 and 2000 bicycles available per month when the unit price is $100. Find the supply equation if it is known to be linear.

In exercises 27 and 28, find the point of intersection of the lines with the given equations.

27. $3x + 4y = -6$ and $2x + 5y = -11$

28. $y = \frac{3}{4}x + 6$ and $3x - 2y + 3 = 0$

29. The cost function and the revenue function for a certain firm are given by $C(x) = 12x + 20,000$ and $R(x) = 20x$, respectively. Find the break-even point for the company.

30. Given the demand equation $3x + p - 40 = 0$ and the supply equation $2x - p + 10 = 0$, where p is the unit price in dollars and x represents the quantity in units of a thousand, determine the equilibrium quantity and the equilibrium price.

31. The accompanying data were compiled by the Admissions Office of Carter College during the past five years. The data relate the number of college brochures and follow-up letters (x) sent to a preselected list of high school juniors who took the PSAT and the number of completed applications (y) received from these students (both measured in thousands).

Number of Brochures Sent (x)	1.8	2	3.2
Number of Completed Applications (y)	0.4	0.5	0.7

Number of Brochures Sent (x)	4	4.8	
Number of Completed Applications (y)	1	1.3	

   a. Derive an equation of the straight line L that passes through the points $(2, 0.5)$ and $(4, 1)$.
   b. Use this equation to predict the number of completed applications that might be expected if 6400 brochures and follow-up letters are sent out during the next year.

Additional study hints and sample chapter tests for each chapter can be found at the Brooks/Cole Web site at http://www.brookscole.com/math/authors/tans/

The linear equations in two variables studied in Chapter 1 are readily extended to the case involving more than two variables. For example, a linear equation in three variables represents a plane in three-dimensional space. In this chapter we see how some real-world problems can be formulated in terms of systems of linear equations, and we also develop two methods for solving these equations.

In addition, we see how *matrices* (ordered rectangular arrays of numbers) can be used to write systems of linear equations in compact form. We then go on to consider some real-life applications of matrices. Finally, we show how matrices can be used to describe the Leontief input–output model—an important tool used by economists. For his work in formulating this model, Wassily Leontief was awarded the Nobel Prize in 1973.

How fast is the traffic moving? The flow of downtown traffic is controlled by traffic lights installed at each of the six intersections. One of the roads is to be resurfaced. In Example 5, page 105, you will see how the flow patterns must be altered in order to ensure a smooth flow of traffic even during rush hour.

2

Systems of Linear Equations and Matrices

71

2.1 SYSTEMS OF LINEAR EQUATIONS—INTRODUCTION

Systems of Equations

Recall that in Section 1.4 we had to solve two simultaneous linear equations in order to find the *break-even point* and the *equilibrium point*. These are two examples of situations in which solving a real-world problem calls for solving a system of linear equations in two or more variables. In this chapter we take up a more systematic study of such systems.

We begin by considering a system of two linear equations in two variables. Recall that such a system may be written in the general form

$$ax + by = h$$
$$cx + dy = k \tag{1}$$

where a, b, c, d, h, and k are real constants and neither a and b nor c and d are both zero.

Now let us study the nature of the solution of a system of equations in more detail. Recall that the graph of each equation in system (1) is a straight line in the plane, so that geometrically the solution to the system is the point(s) of intersection of the two straight lines L_1 and L_2, represented by the first and second equations of the system.

Figure 2.1 depicts each of the three cases that may occur. The two lines L_1 and L_2 may

a. intersect at exactly one point,

b. be parallel and coincident, or

c. be parallel and distinct.

Note that one and only one of these cases must occur. In the first case, the system has a unique solution corresponding to the single point of intersec-

Figure 2.1

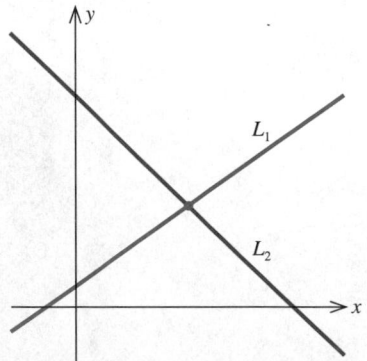

(a) Unique solution.

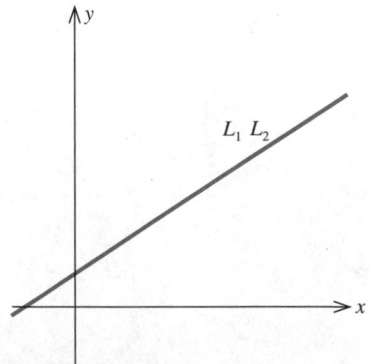

(b) Infinitely many solutions.

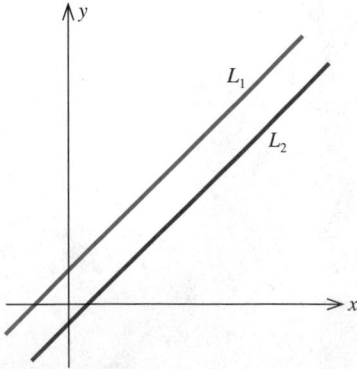

(c) No solution.

Generalize the discussion on page 72 to the case where there are three straight lines in the plane defined by three linear equations. What if there are n lines defined by n equations?

tion of the two lines. In the second case, the system has infinitely many solutions corresponding to the points lying simultaneously on both lines. Finally, in the third case, the system has no solution, since the two lines do not intersect.

Let us illustrate each of these possibilities by considering some specific examples.

1. *A system of equations with exactly one solution*

Consider the system

$$2x - y = 1$$
$$3x + 2y = 12$$

Solving the first equation for y in terms of x, we obtain the equation

$$y = 2x - 1$$

Substituting this expression for y into the second equation yields

$$3x + 2(2x - 1) = 12$$
$$3x + 4x - 2 = 12$$
$$7x = 14$$
$$x = 2$$

Finally, substituting this value of x into the expression for y obtained earlier gives

$$y = 2(2) - 1 = 3$$

Therefore, the unique solution of the system is given by $x = 2$ and $y = 3$. Geometrically, the two lines represented by the two linear equations that make up the system intersect at the point $(2, 3)$ (Figure 2.2).

Figure 2.2
A system of equations with one solution.

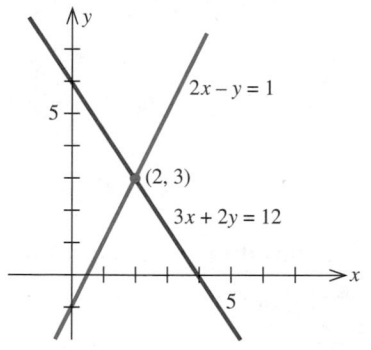

REMARK   We can check our result by substituting the values $x = 2$ and $y = 3$ into the equations. Thus

$$2(2) - (3) = 1 \qquad (\checkmark)$$
$$3(2) + 2(3) = 12 \qquad (\checkmark)$$

From the geometrical point of view, we have just verified that the point $(2, 3)$ lies on both lines. $\circ \; \circ \; \circ$

2. *A system of equations with infinitely many solutions*

Consider the system

$$2x - y = 1$$
$$6x - 3y = 3$$

Solving the first equation for y in terms of x, we obtain the equation

$$y = 2x - 1$$

Substituting this expression for y into the second equation gives

$$6x - 3(2x - 1) = 3$$
$$6x - 6x + 3 = 3$$

or $$0 = 0$$

Figure 2.3

A system of equations with infinitely many solutions; any point on the line is a solution.

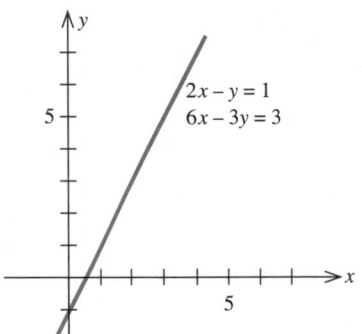

This result simply tells us that the second equation is equivalent to the first. (To see this, just multiply both sides of the first equation by 3.) Our computations have revealed that the system of two equations is equivalent to the single equation $2x - y = 1$. Thus, any ordered pair of numbers (x, y) satisfying the equation $2x - y = 1$ (or $y = 2x - 1$) constitutes a solution to the system.

In particular, by assigning the value t to x, where t is any real number, we find that $y = 2t - 1$, so the ordered pair $(t, 2t - 1)$ is a solution of the system. The variable t is called a **parameter.** For example, setting $t = 0$ gives the point $(0, -1)$ as a solution, and setting $t = 1$ gives the point $(1, 1)$ as another solution of the system. Since t represents any real number, there are infinitely many solutions to the system. Geometrically, the two equations in the system represent the same line, and all solutions of the system are points lying on the line (Figure 2.3). Such a system is said to be **dependent.**

3. *A system of equations that has no solution*

Consider the system

$$2x - y = 1$$
$$6x - 3y = 12$$

The first equation is equivalent to $y = 2x - 1$. Substituting this expression for y into the second equation gives

$$6x - 3(2x - 1) = 12$$
$$6x - 6x + 3 = 12$$

or $$0 = 9$$

which is clearly impossible. Thus, there is no solution to the system of equations. To interpret this situation geometrically, cast both equations in the slope-intercept form, obtaining

$$y = 2x - 1$$

and $$y = 2x - 4$$

1. Consider a system comprising two linear equations in two variables. Can the system have exactly two solutions? Exactly three solutions? Exactly a finite number of solutions?
2. Suppose that at least one of the equations in a system comprising two equations in two variables is nonlinear. Can the system have no solution? Exactly one solution? Exactly two solutions? Exactly a finite number of solutions? Infinitely many solutions? Illustrate each answer with a sketch.

Figure 2.4
A system of equations with no solution.

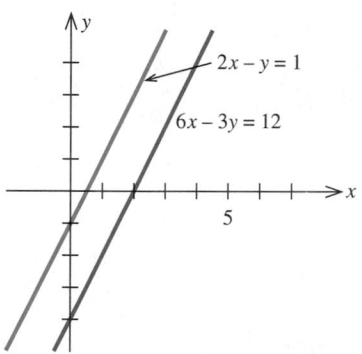

We see at once that the lines represented by these equations are parallel (each has slope 2) and distinct since the first has intercept -1 and the second has intercept -4 (Figure 2.4). Systems with no solutions, such as this one, are said to be **inconsistent.**

REMARK   We have used the method of substitution in solving each of these systems. If you are familiar with the method of elimination, you might want to re-solve each of these systems using this method. We will study the method of elimination in detail in Section 2.2.   ○ ○ ○

Application

In Section 1.4 we presented some real-world applications of systems involving two linear equations in two variables. Here is an example involving a system of three linear equations in three variables.

EXAMPLE 1   The Ace Novelty Company wishes to produce three types of souvenirs: types A, B, and C. To manufacture a type-A souvenir requires 2 minutes on machine I, 1 minute on machine II, and 2 minutes on machine III. A type-B souvenir requires 1 minute on machine I, 3 minutes on machine II, and 1 minute on machine III. A type-C souvenir requires 1 minute on machine I and 2 minutes each on machines II and III. There are 3 hours available on machine I, 5 hours available on machine II, and 4 hours available on machine III for processing the order. How many souvenirs of each type should Ace Novelty make in order to use all of the available time? Formulate, but do not solve, the problem.

Solution   The given information may be tabulated as follows:

	Type A	Type B	Type C	Time Available
Machine I	2	1	1	180
Machine II	1	3	2	300
Machine III	2	1	2	240

We have to determine the number of each of *three* types of souvenirs to be made. So, let x, y, and z denote the respective numbers of type-A, type-B, and type-C souvenirs to be made. The total amount of time that machine I is used is given by $2x + y + z$ minutes and must equal 180 minutes. This leads to the equation

$$2x + y + z = 180 \qquad \text{(Time spent on machine I)}$$

Similar considerations on the use of machines II and III lead to the following equations:

$$x + 3y + 2z = 300 \qquad \text{(Time spent on machine II)}$$
$$2x + y + 2z = 240 \qquad \text{(Time spent on machine III)}$$

Since the variables x, y, and z must satisfy simultaneously the three conditions represented by the three equations, the solution to the problem is found by solving the following system of linear equations:

$$2x + y + z = 180$$
$$x + 3y + 2z = 300$$
$$2x + y + 2z = 240 \qquad \qquad \circ \; \circ \; \circ$$

Solutions of Systems of Equations

We will complete the solution of the problem posed in Example 1 later on (page 91). For the moment, let us look at the geometrical interpretation of a system of linear equations, such as the system in Example 1, in order to gain some insight into the nature of the solution.

A linear system comprising three linear equations in three variables x, y, and z has the general form

$$a_1x + b_1y + c_1z = d_1$$
$$a_2x + b_2y + c_2z = d_2 \qquad \qquad \textbf{(2)}$$
$$a_3x + b_3y + c_3z = d_3$$

Just as a linear equation in two variables represents a straight line in the plane, it can be shown that a linear equation $ax + by + cz = d$ (a, b, and c not simultaneously equal to zero) in three variables represents a plane in three-dimensional space. Thus, each equation in system (2) represents a *plane* in three-dimensional space, and the *solution(s) of the system* is precisely the point(s) of intersection of the three planes defined by the three linear equations that make up the system. As before, the system has one and only one solution, infinitely many solutions, or no solution, depending on whether and how the planes intersect one another. Figure 2.5 illustrates each of these possibilities.

In Figure 2.5a, the three planes intersect at a point corresponding to the situation in which system (2) has a unique solution. Figure 2.5b depicts the situation in which there are infinitely many solutions to the system. Here, the three planes intersect along a line, and the solutions are represented by the infinitely many points lying on this line. In Figure 2.5c, the three planes are parallel and distinct, so there is no point in common to all three planes, and system (2) has no solution in this case.

Figure 2.5

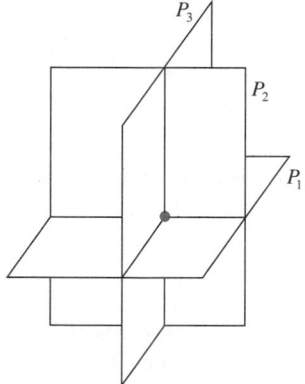

(a) A unique solution.

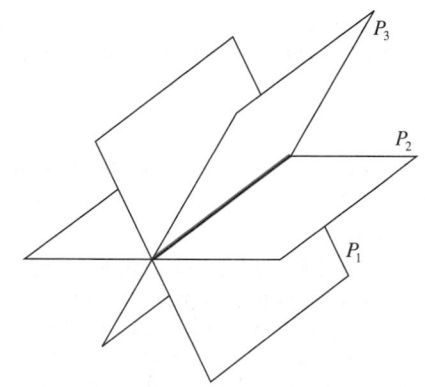

(b) Infinitely many solutions.

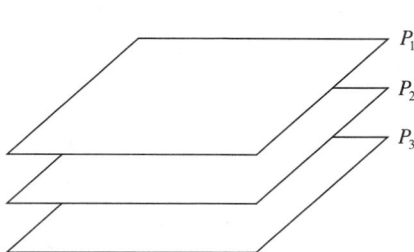

(c) No solution.

REMARK   The situations depicted in Figure 2.5 are by no means exhaustive. You may consider various other orientations of the three planes that would illustrate the three possible outcomes in solving a system of linear equations involving three variables.   ○ ○ ○

LINEAR EQUATIONS IN n VARIABLES

A linear equation in n variables $x_1, x_2, \ldots, x_n$ is one of the form

$$a_1x_1 + a_2x_2 + \cdots + a_nx_n = c$$

where $a_1, a_2, \ldots, a_n$ (not all zero) and c are constants.

For example, the equation

$$3x_1 + 2x_2 - 4x_3 + 6x_4 = 8$$

is a linear equation in the four variables x_1, x_2, x_3, and x_4.

When the number of variables involved in a linear equation exceeds three, we no longer have the geometrical interpretation we had for the lower-dimensional spaces. Nevertheless, the algebraic concepts of the lower-dimensional spaces generalize to higher dimensions. For this reason, a linear equation in n variables $a_1x_1 + a_2x_2 + \cdots + a_nx_n = c$, where $a_1, a_2, \ldots, a_n$ are not all zero, is referred to as an *n-dimensional hyperplane*. We may interpret the solution(s) to a system comprising a finite number of such linear equations to be the *point(s) of intersection* of the hyperplanes defined by the equations that make up the system. As in the case of systems involving two or three variables, it can be shown that only three possibilities exist regarding the nature of the solution of such a system: (a) a unique solution, (b) infinitely many solutions, and (c) no solution.

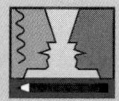

Refer to the Remark on this page.

Using the orientations of three planes, illustrate the outcomes in solving a system of three linear equations in three variables that result in no solution or infinitely many solutions.

SELF-CHECK EXERCISES 2.1

1. Determine whether the system of linear equations

$$2x - 3y = 12$$
$$x + 2y = 6$$

has (a) a unique solution, (b) infinitely many solutions, or (c) no solution. Find all solutions whenever they exist. Make a sketch of the set of lines described by the system.

2. A farmer has 200 acres of land suitable for cultivating crops A, B, and C. The cost per acre of cultivating crops A, B, and C is $40, $60, and $80, respectively. The farmer has $12,600 available for cultivation. Each acre of crop A requires 20 hours of labor, each acre of crop B requires 25 hours of labor, and each acre of crop C requires 40 hours of labor. The farmer has a maximum of 5950 hours of labor available. If he wishes to use all of his cultivatable land, the entire budget, and all the labor available, how many acres of each crop should he plant? Formulate, but do not solve the problem.

Solutions to Self-Check Exercises 2.1 can be found on page 80.

2.1 EXERCISES

In exercises 1–12, determine whether each system of linear equations has (a) one and only one solution, (b) infinitely many solutions, or (c) no solution. Find all solutions whenever they exist.

1. $x - 3y = -1$
  $4x + 3y = 11$

2. $2x - 4y = 5$
  $3x + 2y = 6$

3. $x + 4y = 7$
  $\frac{1}{2}x + 2y = 5$

4. $3x - 4y = 7$
  $9x - 12y = 14$

5. $x + 2y = 7$
  $2x - y = 4$

6. $\frac{3}{2}x - 2y = 4$
  $x + \frac{1}{3}y = 2$

7. $2x - 5y = 10$
  $6x - 15y = 30$

8. $5x - 6y = 8$
  $10x - 12y = 16$

9. $4x - 5y = 14$
  $2x + 3y = -4$

10. $\frac{5}{4}x - \frac{2}{3}y = 3$
  $\frac{1}{4}x + \frac{5}{3}y = 6$

11. $2x - 3y = 6$
  $6x - 9y = 12$

12. $\frac{2}{3}x + y = 5$
  $\frac{1}{2}x + \frac{3}{4}y = \frac{15}{4}$

13. Determine the value of k for which the system of linear equations

$$2x - y = 3$$
$$4x + ky = 4$$

has no solution.

14. Determine the value of k for which the system of linear equations

$$3x + 4y = 12$$
$$x + ky = 4$$

has infinitely many solutions. Then find all the solutions corresponding to this value of k.

In exercises 15–25, formulate, but do not solve, the problem. You will be asked to solve these problems in the next section.

15. Agriculture   The Johnson Farm has 500 acres of land allotted for cultivating corn and wheat. The cost of cultivating corn and wheat (including seeds and labor) is $42 and $30 per acre, respectively. Mr. Johnson has $18,600 available for cultivating these crops. If he wishes to use all the allotted land and his entire budget for cultivating these two crops, how many acres of each crop should he plant?

16. Investments   Michael has a total of $2000 on deposit with two savings institutions. One pays interest at the rate of 6% per year, whereas the other pays interest at the rate of 8% per year. If Michael earned a total of $144 in interest during a single year, how much does he have on deposit in each institution?

17. **Mixtures** The Coffee Shoppe sells a coffee blend made from two coffees, one costing $2.50/lb and the other costing $3/lb. If the blended coffee sells for $2.80/lb, find how much of each coffee is used to obtain the desired blend. (Assume the weight of the blended coffee is 100 lb.)

18. **Investments** Kelly has a total of $30,000 invested in two municipal bonds that have yields of 8% and 10% interest per year, respectively. If the interest Kelly receives from the bonds in a year is $2640, how much does she have invested in each bond?

19. **Ridership** The total number of passengers riding a certain city bus during the morning shift is 1000. If the child's fare is 25 cents, the adult fare is 75 cents, and the total revenue from the fares in the morning shift is $650, how many children and how many adults rode the bus during the morning shift?

20. **Real Estate** Cantwell Associates, a real estate developer, is planning to build a new apartment complex consisting of one-bedroom units and two- and three-bedroom townhouses. A total of 192 units is planned, and the number of family units (two- and three-bedroom townhouses) will equal the number of one-bedroom units. If the number of one-bedroom units will be three times the number of three-bedroom units, find how many units of each type will be in the complex.

21. **Investment Planning** The annual interest on Mr. Carrington's three investments amounted to $21,600: 6% on a savings account, 8% on mutual funds, and 12% on money-market certificates. If the amount of Carrington's investment in money-market certificates was twice the amount of his investment in the savings account, and the interest earned from his investment in money-market certificates was equal to the dividends he received from his investment in mutual funds, find how much money he placed in each type of investment.

22. **Box-Office Receipts** A theater has a seating capacity of 900 and charges $2 for children, $3 for students, and $4 for adults. At a certain screening with full attendance, there were half as many adults as children and students combined. The receipts totaled $2800. How many children attended the show?

23. **Management Decisions** The management of Hartman Rent-A-Car has allocated $1 million to buy a fleet of new automobiles consisting of compact, intermediate, and full-size cars. Compacts cost $8000 each, intermediate-size cars cost $12,000 each, and full-size cars cost $16,000 each. If Hartman purchases twice as many compacts as intermediate-size cars, and the total number of cars to be purchased is 100, determine how many cars of each type will be purchased. (Assume that the entire budget will be used.)

24. **Investment Clubs** The management of a private investment club has a fund of $200,000 earmarked for investment in stocks. To arrive at an acceptable overall level of risk, the stocks that management is considering have been classified into three categories: high-risk, medium-risk, and low-risk. Management estimates that high-risk stocks will have a rate of return of 15% per year; medium-risk stocks, 10% per year; and low-risk stocks, 6% per year. The investment in low-risk stocks is to be twice the sum of the investments in stocks of the other two categories. If the investment goal is to have an average rate of return of 9% per year on the total investment, determine how much the club should invest in each type of stock.

25. **Diet Planning** A dietician wishes to plan a meal around three foods. The percentage of the daily requirements of proteins, carbohydrates, and iron contained in each ounce of the three foods is summarized in the accompanying table. Determine how many ounces of each food the dietician should include in the meal to meet exactly the daily requirement of proteins, carbohydrates, and iron (100% of each).

	Food I	Food II	Food III
Percentage of Proteins	10	6	8
Percentage of Carbohydrates	10	12	6
Percentage of Iron	5	4	12

SOLUTIONS TO SELF–CHECK EXERCISES 2.1

1. Solving the first equation for y in terms of x, we obtain

$$y = \frac{2}{3}x - 4$$

Next, substituting this result into the second equation of the system, we find

$$x + 2\left(\frac{2}{3}x - 4\right) = 6$$

$$x + \frac{4}{3}x - 8 = 6$$

$$\frac{7}{3}x = 14$$

or
$$x = 6$$

Substituting this value of x into the expression for y obtained earlier, we have

$$y = \frac{2}{3}(6) - 4 = 0$$

Therefore, the system has the unique solution $x = 6$ and $y = 0$. Both lines are shown in the accompanying figure.

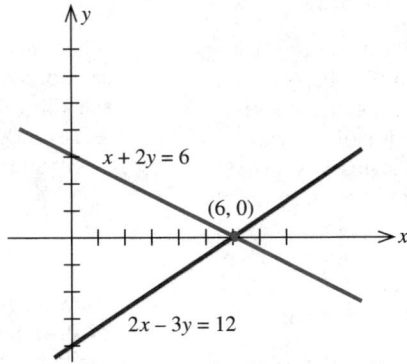

2. Let $x, y,$ and z denote the number of acres of crop A, crop B, and crop C, respectively, to be cultivated. Then, the condition that all the cultivatable land be used translates into the equation

$$x + y + z = 200$$

Next, the total cost incurred in cultivating all three crops is $40x + 60y + 80z$ dollars, and since the entire budget is to be expended, we have

$$40x + 60y + 80z = 12{,}600$$

Finally, the amount of labor required to cultivate all three crops is $20x + 25y + 40z$ hours, and since all the available labor is to be used, we have

$$20x + 25y + 40z = 5950$$

Thus, the solution is found by solving the following system of linear equations:

$$\begin{aligned} x + y + z &= 200 \\ 40x + 60y + 80z &= 12{,}600 \\ 20x + 25y + 40z &= 5{,}950 \end{aligned}$$

2.2 SOLVING SYSTEMS OF LINEAR EQUATIONS I

The Gauss-Jordan Method

The method of substitution used in Section 2.1 is well suited to solving a system of linear equations when the number of linear equations and variables is small. But for large systems, the steps involved in the procedure become difficult to manage.

A suitable technique for solving systems of linear equations of any size is the **Gauss-Jordan elimination method.** One advantage of this technique is its adaptability to the computer. This method involves a sequence of operations on a system of linear equations to obtain at each stage an **equivalent system**—that is, a system having the same solution as the original system. The reduction is complete when the original system has been transformed so that it is in a certain standard form from which the solution can be easily read.

The operations of the Gauss-Jordan elimination method are:

1. Interchange any two equations.

2. Replace an equation by a nonzero constant multiple of itself.

3. Replace an equation by the sum of that equation and a constant multiple of any other equation.

To illustrate the Gauss-Jordan elimination method for solving systems of linear equations, let us apply it to the solution of the following system:

$$\begin{aligned} 2x + 4y &= 8 \\ 3x - 2y &= 4 \end{aligned}$$

We begin by working with the first, or x-column. First, we transform the system into an equivalent system in which the coefficient of x in the first equation is 1:

$$\begin{aligned} 2x + 4y &= 8 \\ 3x - 2y &= 4 \end{aligned} \tag{3a}$$

$$\begin{aligned} x + 2y &= 4 \\ 3x - 2y &= 4 \end{aligned} \qquad \text{[Multiply the first equation in (3a)} \tag{3b}$$
by $\frac{1}{2}$ (operation 2).]

Next, we eliminate x from the second equation.

$$\begin{aligned} x + 2y &= 4 \\ -8y &= -8 \end{aligned}$$

[Replace second equation in (3b) by the sum of -3 times first equation + second equation (operation 3).]   **(3c)**

$$\begin{aligned} -3x - 6y &= -12 \\ \underline{3x - 2y = 4} \\ -8y = -8 \end{aligned}$$

Then we obtain the following equivalent system in which the coefficient of y in the second equation is 1.

$$\begin{aligned} x + 2y &= 4 \\ y &= 1 \end{aligned}$$

[Multiply second equation in (3c) by $-\frac{1}{8}$ (operation 2).]   **(3d)**

Next, we eliminate y in the first equation.

$$\begin{aligned} x &= 2 \\ y &= 1 \end{aligned}$$

[Replace first equation in (3d) by the sum of -2 times the second equation + the first equation (operation 3).]

$$\begin{aligned} x + 2y &= 4 \\ \underline{-2y = -2} \\ x = 2 \end{aligned}$$

This system is now in standard form and we can read off the solution to (3a) as $x = 2$ and $y = 1$. We can also express this solution as $(2, 1)$ and interpret it geometrically as the point of intersection of the two lines represented by the two linear equations that make up the given system of equations.

Let us consider another example involving a system of three linear equations and three variables.

EXAMPLE 1   Solve the following system of equations:

$$\begin{aligned} 2x + 4y + 6z &= 22 \\ 3x + 8y + 5z &= 27 \\ -x + y + 2z &= 2 \end{aligned}$$

Solution   First, we transform this system into an equivalent system in which the coefficient of x in the first equation is 1:

$$\begin{aligned} 2x + 4y + 6z &= 22 \\ 3x + 8y + 5z &= 27 \\ -x + y + 2z &= 2 \end{aligned}$$ **(4a)**

$$\begin{aligned} x + 2y + 3z &= 11 \\ 3x + 8y + 5z &= 27 \\ -x + y + 2z &= 2 \end{aligned}$$

[Multiply first equation in (4a) by $\frac{1}{2}$.]   **(4b)**

Next, we eliminate the variable x from all equations except the first:

$$x + 2y + 3z = 11$$
$$2y - 4z = -6$$
$$-x + y + 2z = 2$$

[Replace second equation in (4b) by the sum of -3 times first equation + second equation.]   **(4c)**

$$-3x - 6y - 9z = -33$$
$$\underline{3x + 8y + 5z = 27}$$
$$2y - 4z = -6$$

$$x + 2y + 3z = 11$$
$$2y - 4z = -6$$
$$3y + 5z = 13$$

[Replace the third equation in (4c) by the sum of first equation + third equation.]   **(4d)**

$$x + 2y + 3z = 11$$
$$\underline{-x + y + 2z = 2}$$
$$3y + 5z = 13$$

Then we transform system (4d) into yet another equivalent system, in which the coefficient of y in the second equation is 1:

$$x + 2y + 3z = 11$$
$$y - 2z = -3$$
$$3y + 5z = 13$$

[Multiply second equation in (4d) by $\frac{1}{2}$.]   **(4e)**

We now eliminate y from all equations except the second using operation 3 of the elimination method.

$$x + 7z = 17$$
$$y - 2z = -3$$
$$3y + 5z = 13$$

[Replace first equation in (4e) by the sum of first equation + (-2) times second equation.]   **(4f)**

$$x + 2y + 3z = 11$$
$$\underline{-2y + 4z = 6}$$
$$x + 7z = 17$$

$$x + 7z = 17$$
$$y - 2z = -3$$
$$11z = 22$$

[Replace third equation in (4f) by the sum of (-3) times second equation + third equation.]   **(4g)**

$$-3y + 6z = 9$$
$$\underline{3y + 5z = 13}$$
$$11z = 22$$

Finally, multiplying the third equation by 1/11 in (4g) leads to the system

$$x + 7z = 17$$
$$y - 2z = -3$$
$$z = 2$$

Eliminating z from all equations except the third (try it!) then leads to the system

$$\begin{aligned} x \qquad\quad &= 3 \\ y \quad &= 1 \\ z &= 2 \end{aligned} \qquad \textbf{(4h)}$$

In its final form, the solution to the given system of equations can be easily read off! We have $x = 3$, $y = 1$, and $z = 2$. Geometrically, the point $(3, 1, 2)$ lies in the intersection of the three planes described by the three equations comprising the given system.   ○ ○ ○

Augmented Matrices

Observe from the preceding example that in each step of the reduction process the variables x, y, and z play no significant role except as a reminder of the position of each coefficient in the system. With the aid of **matrices,** which are rectangular arrays of numbers, we can eliminate writing the variables at each step of the reduction and thus save ourselves a great deal of work. For example, the system

$$\begin{aligned} 2x + 4y + 6z &= 22 \\ 3x + 8y + 5z &= 27 \\ -x + y + 2z &= 2 \end{aligned} \qquad \textbf{(5)}$$

may be represented by the matrix

$$\left[\begin{array}{rrr|r} 2 & 4 & 6 & 22 \\ 3 & 8 & 5 & 27 \\ -1 & 1 & 2 & 2 \end{array} \right] \qquad \textbf{(6)}$$

The augmented matrix representing system (5).

The submatrix, consisting of the first three columns of matrix (6), is called the **coefficient matrix** of system (5). The matrix itself, (6), is referred to as the **augmented matrix** of system (5) since it is obtained by joining the matrix of coefficients to the column (matrix) of constants. The vertical line separates the column of constants from the matrix of coefficients.

The next example shows how much work you can save by using matrices instead of the standard representation of the systems of linear equations.

EXAMPLE 2   Write the augmented matrix corresponding to each equivalent system given in (4a) through (4h).

Solution   The required sequence of augmented matrices follows.

Equivalent System	**Augmented Matrix**

a. $\begin{aligned} 2x + 4y + 6z &= 22 \\ 3x + 8y + 5z &= 27 \\ -x + y + 2z &= 2 \end{aligned} \qquad \left[\begin{array}{rrr|r} 2 & 4 & 6 & 22 \\ 3 & 8 & 5 & 27 \\ -1 & 1 & 2 & 2 \end{array} \right] \qquad \textbf{(7a)}$

	Equivalent System	**Augmented Matrix**	

b.

$$\begin{aligned} x + 2y + 3z &= 11 \\ 3x + 8y + 5z &= 27 \\ -x + y + 2z &= 2 \end{aligned}$$

$$\left[\begin{array}{ccc|c} 1 & 2 & 3 & 11 \\ 3 & 8 & 5 & 27 \\ -1 & 1 & 2 & 2 \end{array}\right]$$

(7b)

c.

$$\begin{aligned} x + 2y + 3z &= 11 \\ 2y - 4z &= -6 \\ -x + y + 2z &= 2 \end{aligned}$$

$$\left[\begin{array}{ccc|c} 1 & 2 & 3 & 11 \\ 0 & 2 & -4 & -6 \\ -1 & 1 & 2 & 2 \end{array}\right]$$

(7c)

d.

$$\begin{aligned} x + 2y + 3z &= 11 \\ 2y - 4z &= -6 \\ 3y + 5z &= 13 \end{aligned}$$

$$\left[\begin{array}{ccc|c} 1 & 2 & 3 & 11 \\ 0 & 2 & -4 & -6 \\ 0 & 3 & 5 & 13 \end{array}\right]$$

(7d)

e.

$$\begin{aligned} x + 2y + 3z &= 11 \\ y - 2z &= -3 \\ 3y + 5z &= 13 \end{aligned}$$

$$\left[\begin{array}{ccc|c} 1 & 2 & 3 & 11 \\ 0 & 1 & -2 & -3 \\ 0 & 3 & 5 & 13 \end{array}\right]$$

(7e)

f.

$$\begin{aligned} x + 7z &= 17 \\ y - 2z &= -3 \\ 3y + 5z &= 13 \end{aligned}$$

$$\left[\begin{array}{ccc|c} 1 & 0 & 7 & 17 \\ 0 & 1 & -2 & -3 \\ 0 & 3 & 5 & 13 \end{array}\right]$$

(7f)

g.

$$\begin{aligned} x + 7z &= 17 \\ y - 2z &= -3 \\ 11z &= 22 \end{aligned}$$

$$\left[\begin{array}{ccc|c} 1 & 0 & 7 & 17 \\ 0 & 1 & -2 & -3 \\ 0 & 0 & 11 & 22 \end{array}\right]$$

(7g)

h.

$$\begin{aligned} x &= 3 \\ y &= 1 \\ z &= 2 \end{aligned}$$

$$\left[\begin{array}{ccc|c} 1 & 0 & 0 & 3 \\ 0 & 1 & 0 & 1 \\ 0 & 0 & 1 & 2 \end{array}\right]$$

(7h)   ○ ○ ○

The augmented matrix in (7h) is an example of a matrix in *row-reduced form*. In general, an augmented matrix with m rows and n columns (called an $m \times n$ matrix) is in row-reduced form if it satisfies the following conditions.

ROW-REDUCED FORM

1. Each row consisting entirely of zeros lies below any other row having nonzero entries.

2. The first nonzero entry in each row is 1 (called a **leading 1**).

3. In any two successive (nonzero) rows, the leading 1 in the lower row lies to the right of the leading 1 in the upper row.

4. If a column contains a leading 1, then the other entries in that column are zeros.

EXAMPLE 3 Determine which of the following matrices are in row-reduced form. If a matrix is not in row-reduced form, state which condition is violated.

a. $\begin{bmatrix} 1 & 0 & 0 & | & 0 \\ 0 & 1 & 0 & | & 0 \\ 0 & 0 & 1 & | & 3 \end{bmatrix}$ **b.** $\begin{bmatrix} 1 & 0 & 0 & | & 4 \\ 0 & 1 & 0 & | & 3 \\ 0 & 0 & 0 & | & 0 \end{bmatrix}$ **c.** $\begin{bmatrix} 1 & 2 & 0 & | & 0 \\ 0 & 0 & 1 & | & 0 \\ 0 & 0 & 0 & | & 1 \end{bmatrix}$

d. $\begin{bmatrix} 0 & 1 & 2 & | & -2 \\ 1 & 0 & 0 & | & 3 \\ 0 & 0 & 1 & | & 2 \end{bmatrix}$ **e.** $\begin{bmatrix} 1 & 2 & 0 & | & 0 \\ 0 & 0 & 1 & | & 3 \\ 0 & 0 & 2 & | & 1 \end{bmatrix}$ **f.** $\begin{bmatrix} 1 & 0 & | & 4 \\ 0 & 3 & | & 0 \\ 0 & 0 & | & 0 \end{bmatrix}$

g. $\begin{bmatrix} 0 & 0 & 0 & | & 0 \\ 1 & 0 & 0 & | & 3 \\ 0 & 1 & 0 & | & 2 \end{bmatrix}$

Solution The matrices in (a), (b), and (c) are in row-reduced form.

d. This matrix is not in row-reduced form. Conditions 3 and 4 are violated: The leading 1 in row 2 lies to the left of the leading 1 in row 1. Also, column 3 contains a leading 1 in row 3 and a nonzero element above it.

e. This matrix is not in row-reduced form. Conditions 2 and 4 are violated: The first nonzero entry in row 3 is a 2, not a 1. Also, column 3 contains a leading 1 and has a nonzero entry below it.

f. This matrix is not in row-reduced form. Condition 2 is violated: The first nonzero entry in row 2 is not a leading 1.

g. This matrix is not in row-reduced form. Condition 1 is violated: Row 1 consists of all zeros and does not lie below the other nonzero rows.

○ ○ ○

The foregoing discussion suggests the following adaptation of the Gauss-Jordan elimination method in solving systems of linear equations using matrices. First, the three operations on the equations of a system (see page 81) translate into the following row operations on the corresponding augmented matrices.

ROW OPERATIONS

1. Interchange any two rows.

2. Replace any row by a nonzero constant multiple of itself.

3. Replace any row by the sum of that row and a constant multiple of any other row.

We obtained the augmented matrices in Example 2 by using the same operations that we used on the equivalent system of equations in Example 1.

In order to help us describe the Gauss-Jordan elimination method using matrices, let us introduce some terminology. We begin by defining what is meant by a **unit column.**

UNIT COLUMN	A column in a coefficient matrix is in unit form if one of the entries in the column is a 1 and the other entries are zeros.

For example, in the coefficient matrix of (7d), only the first column is in unit form; in the coefficient matrix of (7h), all three columns are in unit form. Now, the sequence of row operations that transforms the augmented matrix (7a) into the equivalent matrix (7d) in which the first column

$$\begin{matrix} 2 \\ 3 \\ -1 \end{matrix}$$

of (7a) is transformed into the unit column

$$\begin{matrix} 1 \\ 0 \\ 0 \end{matrix}$$

is called *pivoting* the matrix about the element (number) 2. Similarly, we have pivoted about the element 2 in the second column of (7d), shown circled,

$$\begin{matrix} 2 \\ ②\\ 3 \end{matrix}$$

in order to obtain the augmented matrix (7g). Finally, pivoting about the element 11 in column 3 of (7g)

$$\begin{matrix} 7 \\ -2 \\ ⑪ \end{matrix}$$

leads to the augmented matrix (7h), in which all columns are in unit form. The element about which a matrix is pivoted is called the *pivot element.*

Before looking at the next example, let us introduce the following notation for the three types of row operations.

**NOTATION FOR
ROW OPERATIONS**

Letting R_i denote the ith row of a matrix, we write:

Operation 1: $R_i \leftrightarrow R_j$ to mean: Interchange row i with row j

Operation 2: cR_i to mean: Replace row i with c times row i

Operation 3: $R_i + aR_j$ to mean: Replace row i with the sum of row i and a times row j

EXAMPLE 4   Pivot the matrix

$$\begin{bmatrix} ③ & 5 & | & 9 \\ 2 & 3 & | & 5 \end{bmatrix}$$

about the circled element.

Solution   Using the notation just introduced, we obtain

$$\begin{bmatrix} 3 & 5 & | & 9 \\ 2 & 3 & | & 5 \end{bmatrix} \xrightarrow{\frac{1}{3}R_1} \begin{bmatrix} 1 & \frac{5}{3} & | & 3 \\ 2 & 3 & | & 5 \end{bmatrix} \xrightarrow{R_2 - 2R_1} \begin{bmatrix} 1 & \frac{5}{3} & | & 3 \\ 0 & -\frac{1}{3} & | & -1 \end{bmatrix}$$

The first column, which originally contained the entry 3, is now in unit form, with a 1 where the pivot element used to be, and we are done.

Alternate Solution   In the first solution, we used operation 2 to obtain a 1 where the pivot element was originally. Alternatively, we can use operation 3 as follows:

$$\begin{bmatrix} 3 & 5 & | & 9 \\ 2 & 3 & | & 5 \end{bmatrix} \xrightarrow{R_1 - R_2} \begin{bmatrix} 1 & 2 & | & 4 \\ 2 & 3 & | & 5 \end{bmatrix} \xrightarrow{R_2 - 2R_1} \begin{bmatrix} 1 & 2 & | & 4 \\ 0 & -1 & | & -3 \end{bmatrix}$$

⦾ ⦾ ⦾

REMARK   In Example 4, the two matrices

$$\begin{bmatrix} 1 & \frac{5}{3} & | & 3 \\ 0 & -\frac{1}{3} & | & -1 \end{bmatrix} \quad \text{and} \quad \begin{bmatrix} 1 & 2 & | & 4 \\ 0 & -1 & | & -3 \end{bmatrix}$$

look quite different, but they are in fact equivalent. You can verify this by observing that they represent the systems of equations

$$x + \frac{5}{3}y = 3 \qquad\qquad x + 2y = 4$$

and

$$-\frac{1}{3}y = -1 \qquad\qquad -y = -3$$

respectively, and both have the same solution: $x = -2$ and $y = 3$. Example 4 also shows that we can sometimes avoid working with fractions by using the appropriate row operation.

⦾ ⦾ ⦾

A summary of the Gauss-Jordan method follows.

THE GAUSS-JORDAN ELIMINATION METHOD	**1.** Write the augmented matrix corresponding to the linear system.
	2. Interchange rows (operation 1), if necessary, to obtain an augmented matrix in which the first entry in the first row is nonzero. Then pivot the matrix about this entry.
	3. Interchange the second row with any row below it, if necessary, to obtain an augmented matrix in which the second entry in the second row is nonzero. Pivot the matrix about this entry.
	4. Continue until the final matrix is in row-reduced form.

Before writing the augmented matrix, be sure to write all equations with the variables on the left and constant terms on the right of the equals sign. Also, make sure that the variables are in the same order in all equations.

EXAMPLE 5   Solve the system of linear equations given by

$$3x - 2y + 8z = 9$$
$$-2x + 2y + z = 3 \tag{8}$$
$$x + 2y - 3z = 8$$

Solution   Using the Gauss-Jordan elimination method, we obtain the following sequence of equivalent augmented matrices:

$$\begin{bmatrix} ③ & -2 & 8 & | & 9 \\ -2 & 2 & 1 & | & 3 \\ 1 & 2 & -3 & | & 8 \end{bmatrix} \xrightarrow{R_1 + R_2} \begin{bmatrix} 1 & 0 & 9 & | & 12 \\ -2 & 2 & 1 & | & 3 \\ 1 & 2 & -3 & | & 8 \end{bmatrix}$$

$$\xrightarrow[R_3 - R_1]{R_2 + 2R_1} \begin{bmatrix} 1 & 0 & 9 & | & 12 \\ 0 & 2 & 19 & | & 27 \\ 0 & 2 & -12 & | & -4 \end{bmatrix}$$

$$\xrightarrow{R_2 \leftrightarrow R_3} \begin{bmatrix} 1 & 0 & 9 & | & 12 \\ 0 & ② & -12 & | & -4 \\ 0 & 2 & 19 & | & 27 \end{bmatrix}$$

$$\xrightarrow{\frac{1}{2}R_2} \begin{bmatrix} 1 & 0 & 9 & | & 12 \\ 0 & 1 & -6 & | & -2 \\ 0 & 2 & 19 & | & 27 \end{bmatrix}$$

$$\xrightarrow{R_3 - 2R_2} \begin{bmatrix} 1 & 0 & 9 & | & 12 \\ 0 & 1 & -6 & | & -2 \\ 0 & 0 & ③\!\!① & | & 31 \end{bmatrix}$$

$$\xrightarrow{\frac{1}{31}R_3} \begin{bmatrix} 1 & 0 & 9 & | & 12 \\ 0 & 1 & -6 & | & -2 \\ 0 & 0 & 1 & | & 1 \end{bmatrix}$$

$$\xrightarrow[R_2 + 6R_3]{R_1 - 9R_3} \begin{bmatrix} 1 & 0 & 0 & | & 3 \\ 0 & 1 & 0 & | & 4 \\ 0 & 0 & 1 & | & 1 \end{bmatrix}$$

The solution to system (8) is given by $x = 3$, $y = 4$, and $z = 1$ and may be verified by substitution into system (8) as follows:

$$3(3) - 2(4) + 8(1) = 9 \qquad (\checkmark)$$
$$-2(3) + 2(4) + 1 \quad = 3 \qquad (\checkmark)$$
$$3 + 2(4) - 3(1) = 8 \qquad (\checkmark)$$

 When searching for an element to serve as a pivot, it is important to keep in mind that you may work only with the row containing the potential pivot or any row *below* it. In order to see what can go wrong if this caution is not heeded, consider the following augmented matrix for some linear system:

$$\begin{bmatrix} 1 & 1 & 2 & | & 3 \\ 0 & 0 & 3 & | & 1 \\ 0 & 2 & 1 & | & -2 \end{bmatrix}$$

Observe that column 1 is in unit form. The next step in the Gauss-Jordan elimination procedure calls for obtaining a nonzero element in the second position of the second row. If you use the first row (which is *above* the row under consideration) to help you obtain the pivot, you might proceed as follows:

$$\begin{bmatrix} 1 & 1 & 2 & | & 3 \\ 0 & 0 & 3 & | & 1 \\ 0 & 2 & 1 & | & -2 \end{bmatrix} \xrightarrow{R_2 \leftrightarrow R_1} \begin{bmatrix} 0 & 0 & 3 & | & 1 \\ 1 & 1 & 2 & | & 3 \\ 0 & 2 & 1 & | & -2 \end{bmatrix}$$

As you can see, not only have we obtained a nonzero element to serve as the next pivot, but it is already a 1, thus obviating the next step. This seems like a good move. But beware, we have undone some of our earlier work: The first column is no longer in the unit form where a 1 appears first. The correct move in this case is to interchange the second row with the third row.

The next example illustrates how to handle a situation in which the entry in the first row of the augmented matrix is zero.

1. Can the phrase "a nonzero constant multiple of itself" in a type 2 row operation be replaced by "any constant multiple of itself"? Explain.

2. Can a row of an augmented matrix be replaced by one obtained by adding a constant to every element in that row without changing the solution of the system of linear equations? Explain.

EXAMPLE 6   Solve the system of linear equations given by

$$
\begin{aligned}
2y + 3z &= 7 \\
3x + 6y - 12z &= -3 \\
5x - 2y + 2z &= -7
\end{aligned}
$$

Solution   Using the Gauss-Jordan elimination method, we obtain the following sequence of equivalent augmented matrices:

$$
\begin{bmatrix} 0 & 2 & 3 & | & 7 \\ 3 & 6 & -12 & | & -3 \\ 5 & -2 & 2 & | & -7 \end{bmatrix}
\xrightarrow{R_1 \leftrightarrow R_2}
\begin{bmatrix} ③ & 6 & -12 & | & -3 \\ 0 & 2 & 3 & | & 7 \\ 5 & -2 & 2 & | & -7 \end{bmatrix}
\xrightarrow{\frac{1}{3}R_1}
\begin{bmatrix} 1 & 2 & -4 & | & -1 \\ 0 & 2 & 3 & | & 7 \\ 5 & -2 & 2 & | & -7 \end{bmatrix}
\xrightarrow{R_3 - 5R_1}
$$

$$
\begin{bmatrix} 1 & 2 & -4 & | & -1 \\ 0 & ② & 3 & | & 7 \\ 0 & -12 & 22 & | & -2 \end{bmatrix}
\xrightarrow{\frac{1}{2}R_2}
\begin{bmatrix} 1 & 2 & -4 & | & -1 \\ 0 & 1 & \frac{3}{2} & | & \frac{7}{2} \\ 0 & -12 & 22 & | & -2 \end{bmatrix}
\xrightarrow[R_3 + 12R_2]{R_1 - 2R_2}
\begin{bmatrix} 1 & 0 & -7 & | & -8 \\ 0 & 1 & \frac{3}{2} & | & \frac{7}{2} \\ 0 & 0 & ㊵ & | & 40 \end{bmatrix}
\xrightarrow{\frac{1}{40}R_3}
$$

$$
\begin{bmatrix} 1 & 0 & -7 & | & -8 \\ 0 & 1 & \frac{3}{2} & | & \frac{7}{2} \\ 0 & 0 & 1 & | & 1 \end{bmatrix}
\xrightarrow[R_2 - \frac{3}{2}R_3]{R_1 + 7R_3}
\begin{bmatrix} 1 & 0 & 0 & | & -1 \\ 0 & 1 & 0 & | & 2 \\ 0 & 0 & 1 & | & 1 \end{bmatrix}
$$

The solution to the system is given by $x = -1$, $y = 2$, and $z = 1$, and may be verified by substitution into the system.   ○ ○ ○

Application

EXAMPLE 7   Complete the solution to Example 1 in Section 2.1, page 75.

Solution   To complete the solution of the problem posed in Example 1, recall that the mathematical formulation of the problem led to the following system of linear equations:

$$
\begin{aligned}
2x + y + z &= 180 \\
x + 3y + 2z &= 300 \\
2x + y + 2z &= 240
\end{aligned}
$$

where x, y, and z denote the respective numbers of type-A, type-B, and type-C souvenirs to be made.

Solving the foregoing system of linear equations by the Gauss-Jordan elimination method, we obtain the following sequence of equivalent augmented matrices:

$$\begin{bmatrix} ② & 1 & 1 & | & 180 \\ 1 & 3 & 2 & | & 300 \\ 2 & 1 & 2 & | & 240 \end{bmatrix} \xrightarrow{R_1 \leftrightarrow R_2} \begin{bmatrix} 1 & 3 & 2 & | & 300 \\ 2 & 1 & 1 & | & 180 \\ 2 & 1 & 2 & | & 240 \end{bmatrix}$$

$$\xrightarrow[R_3 - 2R_1]{R_2 - 2R_1} \begin{bmatrix} 1 & 3 & 2 & | & 300 \\ 0 & ⓪{-5} & -3 & | & -420 \\ 0 & -5 & -2 & | & -360 \end{bmatrix}$$

$$\xrightarrow{-\frac{1}{5}R_2} \begin{bmatrix} 1 & 3 & 2 & | & 300 \\ 0 & 1 & \frac{3}{5} & | & 84 \\ 0 & -5 & -2 & | & -360 \end{bmatrix}$$

$$\xrightarrow[R_3 + 5R_2]{R_1 - 3R_2} \begin{bmatrix} 1 & 0 & \frac{1}{5} & | & 48 \\ 0 & 1 & \frac{3}{5} & | & 84 \\ 0 & 0 & ① & | & 60 \end{bmatrix}$$

$$\xrightarrow[R_2 - \frac{3}{5}R_3]{R_1 - \frac{1}{5}R_3} \begin{bmatrix} 1 & 0 & 0 & | & 36 \\ 0 & 1 & 0 & | & 48 \\ 0 & 0 & 1 & | & 60 \end{bmatrix}$$

Thus, $x = 36$, $y = 48$, and $z = 60$; that is, Ace Novelty should make 36 type-A souvenirs, 48 type-B souvenirs, and 60 type-C souvenirs in order to use all available machine time.   ◦ ◦ ◦

SELF–CHECK EXERCISES 2.2

1. Solve the system of linear equations

$$2x + 3y + z = 6$$
$$x - 2y + 3z = -3$$
$$3x + 2y - 4z = 12$$

using the Gauss-Jordan elimination method.

2. A farmer has 200 acres of land suitable for cultivating crops A, B, and C. The cost per acre of cultivating crop A, crop B, and crop C is $40, $60, and $80, respectively. The farmer has $12,600 available for land cultivation. Each acre of crop A requires 20 hours of labor, each acre of crop B requires 25 hours of labor, and each acre of crop C requires 40 hours of labor. The farmer has a maximum of 5950 hours of labor available. If he wishes to use all of his cultivatable land, the entire budget, and all the labor available, how many acres of each crop should he plant?

Solutions to Self-Check Exercises 2.2 can be found on page 99.

2.2 EXERCISES

In exercises 1–4, write the augmented matrix corresponding to the given system of equations.

1. $2x - 3y = 7$
$\quad 3x + y = 4$

2. $3x + 7y - 8z = 5$
$\quad x \qquad + 3z = -2$
$\quad 4x - 3y \qquad = 7$

3. $\quad - y + 2z = 6$
$\quad 2x + 2y - 8z = 7$
$\quad 3y + 4z = 0$

4. $3x_1 + 2x_2 \qquad = 0$
$\quad x_1 - x_2 + 2x_3 = 4$
$\quad 2x_2 - 3x_3 = 5$

In exercises 5–8, write the system of equations corresponding to the given augmented matrix.

5. $\begin{bmatrix} 3 & 2 & | & -4 \\ 1 & -1 & | & 5 \end{bmatrix}$

6. $\begin{bmatrix} 0 & 3 & 2 & | & 4 \\ 1 & -1 & -2 & | & -3 \\ 4 & 0 & 3 & | & 2 \end{bmatrix}$

7. $\begin{bmatrix} 1 & 3 & 2 & | & 4 \\ 2 & 0 & 0 & | & 5 \\ 3 & -3 & 2 & | & 6 \end{bmatrix}$

8. $\begin{bmatrix} 2 & 3 & 1 & | & 6 \\ 4 & 3 & 2 & | & 5 \\ 0 & 0 & 0 & | & 0 \end{bmatrix}$

In exercises 9–18, indicate whether the matrix is in row-reduced form.

9. $\begin{bmatrix} 1 & 0 & | & 3 \\ 0 & 1 & | & -2 \end{bmatrix}$

10. $\begin{bmatrix} 1 & 1 & | & 3 \\ 0 & 0 & | & 0 \end{bmatrix}$

11. $\begin{bmatrix} 0 & 1 & | & 3 \\ 1 & 0 & | & 5 \end{bmatrix}$

12. $\begin{bmatrix} 0 & 1 & | & 3 \\ 0 & 0 & | & 5 \end{bmatrix}$

13. $\begin{bmatrix} 1 & 0 & 0 & | & 3 \\ 0 & 1 & 0 & | & 4 \\ 0 & 0 & 1 & | & 5 \end{bmatrix}$

14. $\begin{bmatrix} 1 & 0 & 0 & | & -1 \\ 0 & 1 & 0 & | & -2 \\ 0 & 0 & 2 & | & -3 \end{bmatrix}$

15. $\begin{bmatrix} 1 & 0 & 1 & | & 3 \\ 0 & 1 & 0 & | & 4 \\ 0 & 0 & -1 & | & 6 \end{bmatrix}$

16. $\begin{bmatrix} 1 & 0 & | & -10 \\ 0 & 1 & | & 2 \\ 0 & 0 & | & 0 \end{bmatrix}$

17. $\begin{bmatrix} 0 & 0 & 0 & | & 0 \\ 0 & 1 & 2 & | & 4 \\ 0 & 0 & 0 & | & 0 \end{bmatrix}$

18. $\begin{bmatrix} 1 & 0 & 0 & | & 3 \\ 0 & 1 & 0 & | & 6 \\ 0 & 0 & 0 & | & 4 \\ 0 & 0 & 1 & | & 5 \end{bmatrix}$

In exercises 19–26, pivot the given system about the circled element.

19. $\begin{bmatrix} ② & 4 & | & 8 \\ 3 & 1 & | & 2 \end{bmatrix}$

20. $\begin{bmatrix} 3 & 2 & | & 6 \\ ④ & 2 & | & 5 \end{bmatrix}$

21. $\begin{bmatrix} ⊖1 & 2 & | & 3 \\ 6 & 4 & | & 2 \end{bmatrix}$

22. $\begin{bmatrix} ① & 3 & | & 4 \\ 2 & 4 & | & 6 \end{bmatrix}$

23. $\begin{bmatrix} ② & 4 & 6 & | & 12 \\ 2 & 3 & 1 & | & 5 \\ 3 & -1 & 2 & | & 4 \end{bmatrix}$

24. $\begin{bmatrix} 1 & 3 & 2 & | & 4 \\ ② & 4 & 8 & | & 6 \\ -1 & 2 & 3 & | & 4 \end{bmatrix}$

25. $\begin{bmatrix} 0 & 1 & 3 & | & 4 \\ 2 & 4 & ① & | & 3 \\ 5 & 6 & 2 & | & -4 \end{bmatrix}$

26. $\begin{bmatrix} 1 & 2 & 3 & | & 5 \\ 0 & ⊖3 & 3 & | & 2 \\ 0 & 4 & -1 & | & 3 \end{bmatrix}$

In exercises 27–30, fill in the missing entries by performing the indicated row operations to obtain the row-reduced matrices.

27. $\begin{bmatrix} 3 & 9 & | & 6 \\ 2 & 1 & | & 4 \end{bmatrix} \xrightarrow{\frac{1}{3}R_1} \begin{bmatrix} \cdot & \cdot & | & \cdot \\ 2 & 1 & | & 4 \end{bmatrix} \xrightarrow{R_2 - 2R_1}$

$\begin{bmatrix} 1 & 3 & | & 2 \\ \cdot & \cdot & | & \cdot \end{bmatrix} \xrightarrow{-\frac{1}{5}R_2} \begin{bmatrix} 1 & 3 & | & 2 \\ \cdot & \cdot & | & \cdot \end{bmatrix} \xrightarrow{R_1 - 3R_2}$

$\begin{bmatrix} 1 & 0 & | & 2 \\ 0 & 1 & | & 0 \end{bmatrix}$

28. $\begin{bmatrix} 1 & 2 & | & 1 \\ 2 & 3 & | & -1 \end{bmatrix} \xrightarrow{R_2 - 2R_1} \begin{bmatrix} 1 & 2 & | & 1 \\ \cdot & \cdot & | & \cdot \end{bmatrix} \xrightarrow{-R_2}$

$\begin{bmatrix} 1 & 2 & | & 1 \\ \cdot & \cdot & | & \cdot \end{bmatrix} \xrightarrow{R_1 - 2R_2} \begin{bmatrix} 1 & 0 & | & -5 \\ 0 & 1 & | & 3 \end{bmatrix}$

29. $\begin{bmatrix} 1 & 3 & 1 & | & 3 \\ 3 & 8 & 3 & | & 7 \\ 2 & -3 & 1 & | & -10 \end{bmatrix} \xrightarrow[R_3 - 2R_1]{R_2 - 3R_1}$

$\begin{bmatrix} 1 & 3 & 1 & | & 3 \\ \cdot & \cdot & \cdot & | & \cdot \\ \cdot & \cdot & \cdot & | & \cdot \end{bmatrix} \xrightarrow{-R_2}$

$$\begin{bmatrix} 1 & 3 & 1 & | & 3 \\ \cdot & \cdot & \cdot & | & \cdot \\ 0 & -9 & -1 & | & -16 \end{bmatrix} \xrightarrow[R_3 + 9R_2]{R_1 - 3R_2}$$

$$\begin{bmatrix} \cdot & \cdot & \cdot & | & \cdot \\ 0 & 1 & 0 & | & 2 \\ \cdot & \cdot & \cdot & | & \cdot \end{bmatrix} \xrightarrow[-R_3]{R_1 + R_3} \begin{bmatrix} 1 & 0 & 0 & | & -1 \\ 0 & 1 & 0 & | & 2 \\ 0 & 0 & 1 & | & -2 \end{bmatrix}$$

30. $\begin{bmatrix} 0 & 1 & 3 & | & -4 \\ 1 & 2 & 1 & | & 7 \\ 1 & -2 & 0 & | & 1 \end{bmatrix} \xrightarrow{R_1 \leftrightarrow R_2}$

$$\begin{bmatrix} \cdot & \cdot & \cdot & | & \cdot \\ \cdot & \cdot & \cdot & | & \cdot \\ 1 & -2 & 0 & | & 1 \end{bmatrix} \xrightarrow{R_3 - R_1} \begin{bmatrix} 1 & 2 & 1 & | & 7 \\ 0 & 1 & 3 & | & -4 \\ \cdot & \cdot & \cdot & | & \cdot \end{bmatrix}$$

$$\xrightarrow[R_3 + 4R_2]{R_1 + \frac{1}{2}R_2} \begin{bmatrix} \cdot & \cdot & \cdot & | & \cdot \\ 0 & 1 & 3 & | & -4 \\ \cdot & \cdot & \cdot & | & \cdot \end{bmatrix} \xrightarrow{\frac{1}{11}R_3}$$

$$\begin{bmatrix} 1 & 0 & \frac{1}{2} & | & 4 \\ 0 & 1 & 3 & | & -4 \\ \cdot & \cdot & \cdot & | & \cdot \end{bmatrix} \xrightarrow[R_2 - 3R_3]{R_1 - \frac{1}{2}R_3} \begin{bmatrix} 1 & 0 & 0 & | & 5 \\ 0 & 1 & 0 & | & 2 \\ 0 & 0 & 1 & | & -2 \end{bmatrix}$$

31. Write a system of linear equations for the augmented matrix of exercise 27. Using the results of exercise 27, determine the solution of the system.

32. Repeat exercise 31 for the augmented matrix of exercise 28.

33. Repeat exercise 31 for the augmented matrix of exercise 29.

34. Repeat exercise 31 for the augmented matrix of exercise 30.

In exercises 35–50, solve the system of linear equations using the Gauss-Jordan elimination method.

35. $\begin{aligned} x - 2y &= 8 \\ 3x + 4y &= 4 \end{aligned}$

36. $\begin{aligned} 3x + y &= 1 \\ -7x - 2y &= -1 \end{aligned}$

37. $\begin{aligned} 2x - 3y &= -8 \\ 4x + y &= -2 \end{aligned}$

38. $\begin{aligned} 5x + 3y &= 9 \\ -2x + y &= -8 \end{aligned}$

39. $\begin{aligned} x + y + z &= 0 \\ 2x - y + z &= 1 \\ x + y - 2z &= 2 \end{aligned}$

40. $\begin{aligned} 2x + y - 2z &= 4 \\ x + 3y - z &= -3 \\ 3x + 4y - z &= 7 \end{aligned}$

41. $\begin{aligned} 2x + 2y + z &= 9 \\ x \quad\;\; + z &= 4 \\ 4y - 3z &= 17 \end{aligned}$

42. $\begin{aligned} 2x + 3y - 2z &= 10 \\ 3x - 2y + 2z &= 0 \\ 4x - y + 3z &= -1 \end{aligned}$

43. $\begin{aligned} -x_2 + x_3 &= 2 \\ 4x_1 - 3x_2 + 2x_3 &= 16 \\ 3x_1 + 2x_2 + x_3 &= 11 \end{aligned}$

44. $\begin{aligned} 2x + 4y - 6z &= 38 \\ x + 2y + 3z &= 7 \\ 3x - 4y + 4z &= -19 \end{aligned}$

45. $\begin{aligned} x_1 - 2x_2 + x_3 &= 6 \\ 2x_1 + x_2 - 3x_3 &= -3 \\ x_1 - 3x_2 + 3x_3 &= 10 \end{aligned}$

46. $\begin{aligned} 2x + 3y - 6z &= -11 \\ x - 2y + 3z &= 9 \\ 3x + y &= 7 \end{aligned}$

47. $\begin{aligned} 2x \quad\;\; + 3z &= -1 \\ 3x - 2y + z &= 9 \\ x + y + 4z &= 4 \end{aligned}$

48. $\begin{aligned} 2x_1 - x_2 + 3x_3 &= -4 \\ x_1 - 2x_2 + x_3 &= -1 \\ x_1 - 5x_2 + 2x_3 &= -3 \end{aligned}$

49. $\begin{aligned} x_1 - x_2 + 3x_3 &= 14 \\ x_1 + x_2 + x_3 &= 6 \\ -2x_1 - x_2 + x_3 &= -4 \end{aligned}$

50. $\begin{aligned} 2x_1 - x_2 - x_3 &= 0 \\ 3x_1 + 2x_2 + x_3 &= 7 \\ x_1 + 2x_2 + 2x_3 &= 5 \end{aligned}$

The problems in exercises 51–61 correspond to those in exercises 15–25, Section 2.1. Use the results of your previous work to help you solve these problems.

51. Agriculture The Johnson Farm has 500 acres of land allotted for cultivating corn and wheat. The cost of cultivating corn and wheat (including seeds and labor) is $42 and $30 per acre, respectively. Mr. Johnson has $18,600 available for cultivating these crops. If he wishes to use all the allotted land and his entire budget for cultivating these two crops, how many acres of each crop should he plant?

52. Investments Michael has a total of $2000 on deposit with two savings institutions. One pays interest at the rate of 6% per year, whereas the other pays interest at the rate of 8% per year. If Michael earned a total of $144 in interest during a single year, how much does he have on deposit in each institution?

53. Mixtures The Coffee Shoppe sells a coffee blend made from two coffees, one costing $2.50/lb and the other costing $3/lb. If the blended coffee sells for $2.80/

lb, find how much of each coffee is used to obtain the desired blend. (Assume the weight of the blended coffee is 100 lb.)

54. **Investments**   Kelly has a total of $30,000 invested in two municipal bonds that have yields of 8% and 10% interest per year, respectively. If the interest Kelly receives from the bonds in a year is $2640, how much does she have invested in each bond?

55. **Ridership**   The total number of passengers riding a certain city bus during the morning shift is 1000. If the child's fare is 25 cents, the adult fare is 75 cents, and the total revenue from the fares in the morning shift is $650, how many children and how many adults rode the bus during the morning shift?

56. **Real Estate**   Cantwell Associates, a real estate developer, is planning to build a new apartment complex consisting of one-bedroom units and two- and three-bedroom townhouses. A total of 192 units is planned, and the number of family units (two- and three-bedroom townhouses) will equal the number of one-bedroom units. If the number of one-bedroom units will be three times the number of three-bedroom units, find how many units of each type will be in the complex.

57. **Investment Planning**   The annual interest on Mr. Carrington's three investments amounted to $21,600: 6% on a savings account, 8% on mutual funds, and 12% on money-market certificates. If the amount of Carrington's investment in money-market certificates was twice the amount of his investment in the savings account, and the interest earned from his investment in money-market certificates was equal to the dividends he received from his investment in mutual funds, find how much money he placed in each type of investment.

58. **Box-Office Receipts**   A theater has a seating capacity of 900 and charges $2 for children, $3 for students, and $4 for adults. At a certain screening with full attendance there were half as many adults as children and students combined. The receipts totaled $2800. How many children attended the show?

59. **Management Decisions**   The management of Hartman Rent-A-Car has allocated $1 million to buy a fleet of new automobiles consisting of compact, intermediate, and full-size cars. Compacts cost $8000 each, intermediate-size cars cost $12,000 each, and full-size cars cost $16,000 each. If Hartman purchases twice as many compacts as intermediate-size cars, and the total number of cars to be purchased is 100, determine how many cars

of each type will be purchased. (Assume that the entire budget will be used.)

60. **Investment Clubs**   The management of a private investment club has a fund of $200,000 earmarked for investment in stocks. To arrive at an acceptable overall level of risk, the stocks that management is considering have been classified into three categories: high-risk, medium-risk, and low-risk. Management estimates that high-risk stocks will have a rate of return of 15% per year; medium-risk stocks, 10% per year; and low-risk stocks, 6% per year. The investment in low-risk stocks is to be twice the sum of the investments in stocks of the other two categories. If the investment goal is to have an average rate of return of 9% per year on the total investment, determine how much the club should invest in each type of stock.

61. **Diet Planning**   A dietician wishes to plan a meal around three foods. The percentage of the daily requirements of proteins, carbohydrates, and iron contained in each ounce of the three foods is summarized in the accompanying table.

	Food I	Food II	Food III
Percentage of Proteins	10	6	8
Percentage of Carbohydrates	10	12	6
Percentage of Iron	5	4	12

Determine how many ounces of each food the dietician should include in the meal to meet exactly the daily requirement of proteins, carbohydrates, and iron (100% of each).

62. **Investments**   Mr. and Mrs. Garcia have a total of $100,000 to be invested in stocks, bonds, and a money-market account. The stocks have a rate of return of 12% per year, while the bonds and the money-market account pay 8% and 4% per year, respectively. They have stipulated that the amount invested in the money-market account should be equal to the sum of 20% of the amount invested in stocks and 10% of the amount invested in bonds. How should the Garcias allocate their resources if they require an annual income of $10,000 from their investments?

63. **Box-Office Receipts**   For the opening night at the Opera House, a total of 1000 tickets were sold. Front

USING TECHNOLOGY

SOLVING SYSTEMS OF LINEAR EQUATIONS I

Solving a System of Linear Equations Using the Gauss-Jordan Method

The three matrix operations can be performed on a matrix using a graphing utility. The commands are summarized below:

Operation	Calculator Function		
	(TI-82)	(TI-85)	
$R_i \leftrightarrow R_j$	**RowSwap**(A, i, j)	**rSwap**(A, i, j)	or equivalent
cR_i	***Row**(c, A, i)	**multR**(c, A, i)	or equivalent
$R_i + aR_j$	***Row+**(a, A, j, i)	**mRAdd**(a, A, j, i)	or equivalent

When a row operation is performed on a matrix, the result is stored as an answer in the calculator. If another operation is performed on this matrix, then the matrix is erased. Should a mistake be made in the operation, then the previous matrix is lost. For this reason, you should store the results of each operation. We do this by pressing **STO,** followed by the name of a matrix, and then **ENTER.** We use this process in the following example.

EXAMPLE 1 Use a graphing calculator to solve the following system of
.... linear equations by the Gauss-Jordan method (see Example 5 in Section 2.2):

$$3x - 2y + 8z = 9$$
$$-2x + 2y + z = 3$$
$$x + 2y - 3z = 8$$

Solution Using the Gauss-Jordan method, we obtain the following sequence of equivalent matrices.

$$\begin{bmatrix} 3 & -2 & 8 & | & 9 \\ -2 & 2 & 1 & | & 3 \\ 1 & 2 & -3 & | & 8 \end{bmatrix} \xrightarrow{\text{*Row+}(1, A, 2, 1) \blacktriangleright B}$$

$$\begin{bmatrix} 1 & 0 & 9 & | & 12 \\ -2 & 2 & 1 & | & 3 \\ 1 & 2 & -3 & | & 8 \end{bmatrix} \xrightarrow{\text{*Row+}(2, B, 1, 2) \blacktriangleright C}$$

96

$$\begin{bmatrix} 1 & 0 & 9 & | & 12 \\ 0 & 2 & 19 & | & 27 \\ 1 & 2 & -3 & | & 8 \end{bmatrix} \xrightarrow{\textbf{*Row}+(-1, C, 1, 3) \blacktriangleright B}$$

$$\begin{bmatrix} 1 & 0 & 9 & | & 12 \\ 0 & 2 & 19 & | & 27 \\ 0 & 2 & -12 & | & -4 \end{bmatrix} \xrightarrow{\textbf{*Row}(\frac{1}{2}, B, 2) \blacktriangleright C}$$

$$\begin{bmatrix} 1 & 0 & 9 & | & 12 \\ 0 & 1 & 9.5 & | & 13.5 \\ 0 & 2 & -12 & | & -4 \end{bmatrix} \xrightarrow{\textbf{*Row}+(-2, C, 2, 3) \blacktriangleright B}$$

$$\begin{bmatrix} 1 & 0 & 9 & | & 12 \\ 0 & 1 & 9.5 & | & 13.5 \\ 0 & 0 & -31 & | & -31 \end{bmatrix} \xrightarrow{\textbf{*Row}(-\frac{1}{31}, B, 3) \blacktriangleright C}$$

$$\begin{bmatrix} 1 & 0 & 9 & | & 12 \\ 0 & 1 & 9.5 & | & 13.5 \\ 0 & 0 & 1 & | & 1 \end{bmatrix} \xrightarrow{\textbf{*Row}+(-9, C, 3, 1) \blacktriangleright B}$$

$$\begin{bmatrix} 1 & 0 & 0 & | & 3 \\ 0 & 1 & 9.5 & | & 13.5 \\ 0 & 0 & 1 & | & 1 \end{bmatrix} \xrightarrow{\textbf{*Row}+(-9.5, B, 3, 2) \blacktriangleright C} \begin{bmatrix} 1 & 0 & 0 & | & 3 \\ 0 & 1 & 0 & | & 4 \\ 0 & 0 & 1 & | & 1 \end{bmatrix}$$

The last matrix is in row-reduced form, and we see that the solution of the system is $x = 3$, $y = 4$, and $z = 1$. ○ ○ ○

Using rref (TI-85) to Solve a System of Linear Equations

The operation **rref** (or equivalent function in your calculator, if there is one) will transform an augmented matrix into one that is in row-reduced form. For example, using **rref,** we find

$$\begin{bmatrix} 3 & -2 & 8 & | & 9 \\ -2 & 2 & 1 & | & 3 \\ 1 & 2 & -3 & | & 8 \end{bmatrix} \xrightarrow{\textbf{rref}} \begin{bmatrix} 1 & 0 & 0 & | & 3 \\ 0 & 1 & 0 & | & 4 \\ 0 & 0 & 1 & | & 1 \end{bmatrix}$$

as obtained earlier!

Using SIMULT (TI-85) to Solve a System of Equations

The operation **SIMULT** (or equivalent operation on your calculator, if there is one) of a graphing utility can be used to solve a system of n linear equations in n variables, where n is an integer between 2 and 30.

EXAMPLE 2   Use the **SIMULT** operation to solve the system of Example 1.

Solution   Call for the **SIMULT** operation. Since the system under consideration has three equations in three variables, enter $n = 3$. Next, enter a1, 1 = 3, a1, 2 = -2, a1, 3 = 8, . . . , b1 = 9, a2, 1 = -2, . . . , b3 = 8. Select ⟨**SOLVE**⟩ and the display

$$x1 = 3$$
$$x2 = 4$$
$$x3 = 1$$

appears on the screen giving $x = 3$, $y = 4$, and $z = 1$ as the required solution.

○ ○ ○

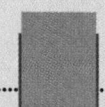

EXERCISES

In exercises 1–6, use a graphing utility to solve the system of equations (a) by the Gauss-Jordan method, (b) using the rref operation, and (c) using SIMULT.

1. $x_1 - 2x_2 + 2x_3 - 3x_4 = -7$
$3x_1 + 2x_2 - x_3 + 5x_4 = 22$
$2x_1 - 3x_2 + 4x_3 - x_4 = -3$
$3x_1 - 2x_2 - x_3 + 2x_4 = 12$

2. $2x_1 - x_2 + 3x_3 - 2x_4 = -2$
$x_1 - 2x_2 + x_3 - 3x_4 = 2$
$x_1 - 5x_2 + 2x_3 + 3x_4 = -6$
$-3x_1 + 3x_2 - 4x_3 - 4x_4 = 9$

3. $2x_1 + x_2 + 3x_3 - x_4 = 9$
$-x_1 - 2x_2 \quad\quad - 3x_4 = -1$
$x_1 \quad\quad - 3x_3 + x_4 = 10$
$x_1 - x_2 - x_3 - x_4 = 8$

4. $x_1 - 2x_2 - 2x_3 + x_4 = 1$
$2x_1 - x_2 + 2x_3 + 3x_4 = -2$
$-x_1 - 5x_2 + 7x_3 - 2x_4 = 3$
$3x_1 - 4x_2 + 3x_3 + 4x_4 = -4$

5. $2x_1 - 2x_2 + 3x_3 - x_4 + 2x_5 = 16$
$3x_1 + x_2 - 2x_3 + x_4 - 3x_5 = -11$
$x_1 + 3x_2 - 4x_3 + 3x_4 - x_5 = -13$
$2x_1 - x_2 + 3x_3 - 2x_4 + 2x_5 = 15$
$3x_1 + 4x_2 - 3x_3 + 5x_4 - x_5 = -10$

6. $2.1x_1 - 3.2x_2 + 6.4x_3 + 7x_4 - 3.2x_5 = 54.3$
$4.1x_1 + 2.2x_2 - 3.1x_3 - 4.2x_4 + 3.3x_5 = -20.81$
$3.4x_1 - 6.2x_2 + 4.7x_3 + 2.1x_4 - 5.3x_5 = 24.7$
$4.1x_1 + 7.3x_2 + 5.2x_3 + 6.1x_4 - 8.2x_5 = 29.25$
$2.8x_1 + 5.2x_2 + 3.1x_3 + 5.4x_4 + 3.8x_5 = 43.72$

orchestra seats cost $80 apiece, rear orchestra seats cost $60 apiece, and front balcony seats cost $50 apiece. The combined number of tickets sold for the front orchestra and rear orchestra exceeded twice the number of front balcony tickets sold by 400. The total receipts for the performance were $62,800. Determine how many tickets of each type were sold.

64. **Production Scheduling** A manufacturer of women's blouses makes three types of blouses: sleeveless, short-sleeve, and long-sleeve. The time required by each department to produce a dozen blouses of each type is shown in the accompanying table.

Department	Sleeveless	Short-sleeve	Long-sleeve
Cutting	9 min.	12 min.	15 min.
Sewing	22 min.	24 min.	28 min.
Packaging	6 min.	8 min.	8 min.

The cutting, sewing, and packaging departments have available a maximum of 80, 160, and 48 labor hours, respectively, per day. How many dozens of each type of blouse can be produced each day if the plant is operated at full capacity?

SOLUTIONS TO SELF-CHECK EXERCISES 2.2

1. We obtain the following sequence of equivalent augmented matrices:

$$\left[\begin{array}{ccc|c} 2 & 3 & 1 & 6 \\ 1 & -2 & 3 & -3 \\ 3 & 2 & -4 & 12 \end{array}\right] \xrightarrow{R_1 \leftrightarrow R_2} \left[\begin{array}{ccc|c} ① & -2 & 3 & -3 \\ 2 & 3 & 1 & 6 \\ 3 & 2 & -4 & 12 \end{array}\right] \xrightarrow[R_3-3R_1]{R_2-2R_1}$$

$$\left[\begin{array}{ccc|c} 1 & -2 & 3 & -3 \\ 0 & 7 & -5 & 12 \\ 0 & 8 & -13 & 21 \end{array}\right] \xrightarrow{R_2 \leftrightarrow R_3} \left[\begin{array}{ccc|c} 1 & -2 & 3 & -3 \\ 0 & ⑧ & -13 & 21 \\ 0 & 7 & -5 & 12 \end{array}\right] \xrightarrow{R_2-R_3}$$

$$\left[\begin{array}{ccc|c} 1 & -2 & 3 & -3 \\ 0 & 1 & -8 & 9 \\ 0 & 7 & -5 & 12 \end{array}\right] \xrightarrow[R_3-7R_2]{R_1+2R_2} \left[\begin{array}{ccc|c} 1 & 0 & -13 & 15 \\ 0 & 1 & -8 & 9 \\ 0 & 0 & 51 & -51 \end{array}\right] \xrightarrow{\frac{1}{51}R_3}$$

$$\left[\begin{array}{ccc|c} 1 & 0 & -13 & 15 \\ 0 & 1 & -8 & 9 \\ 0 & 0 & ① & -1 \end{array}\right] \xrightarrow[R_2+8R_3]{R_1+13R_3} \left[\begin{array}{ccc|c} 1 & 0 & 0 & 2 \\ 0 & 1 & 0 & 1 \\ 0 & 0 & 1 & -1 \end{array}\right]$$

The solution to the system is $x = 2$, $y = 1$, and $z = -1$.

2. Referring to the solution of exercise 2, Self-Check Exercises 2.1, we see that the problem reduces to solving the following system of linear equations:

$$\begin{aligned} x + y + z &= 200 \\ 40x + 60y + 80z &= 12{,}600 \\ 20x + 25y + 40z &= 5{,}950 \end{aligned}$$

Using the Gauss-Jordan elimination method, we have

$$
\begin{bmatrix}
1 & 1 & 1 & \bigm| & 200 \\
40 & 60 & 80 & \bigm| & 12600 \\
20 & 25 & 40 & \bigm| & 5950
\end{bmatrix}
\xrightarrow[R_3 - 20R_1]{R_2 - 40R_1}
$$

$$
\begin{bmatrix}
1 & 1 & 1 & \bigm| & 200 \\
0 & 20 & 40 & \bigm| & 4600 \\
0 & 5 & 20 & \bigm| & 1950
\end{bmatrix}
\xrightarrow{\frac{1}{20}R_2}
\begin{bmatrix}
1 & 1 & 1 & \bigm| & 200 \\
0 & 1 & 2 & \bigm| & 230 \\
0 & 5 & 20 & \bigm| & 1950
\end{bmatrix}
$$

$$
\xrightarrow[R_3 - 5R_2]{R_1 - R_2}
\begin{bmatrix}
1 & 0 & -1 & \bigm| & -30 \\
0 & 1 & 2 & \bigm| & 230 \\
0 & 0 & 10 & \bigm| & 800
\end{bmatrix}
\xrightarrow{\frac{1}{10}R_3}
$$

$$
\begin{bmatrix}
1 & 0 & -1 & \bigm| & -30 \\
0 & 1 & 2 & \bigm| & 230 \\
0 & 0 & 1 & \bigm| & 80
\end{bmatrix}
\xrightarrow[R_2 - 2R_3]{R_1 + R_3}
\begin{bmatrix}
1 & 0 & 0 & \bigm| & 50 \\
0 & 1 & 0 & \bigm| & 70 \\
0 & 0 & 1 & \bigm| & 80
\end{bmatrix}
$$

From the last augmented matrix in reduced form, we see that $x = 50$, $y = 70$, and $z = 80$. Therefore, the farmer should plant 50 acres of crop A, 70 acres of crop B, and 80 acres of crop C.

2.3 SOLVING SYSTEMS OF LINEAR EQUATIONS II

In this section we continue our study of systems of linear equations. More specifically, we look at systems that have infinitely many solutions and those that have no solution. We also study systems of linear equations in which the number of variables is not equal to the number of equations in the system.

Solution(s) of Linear Equations

Our first example illustrates the situation in which a system of linear equations has infinitely many solutions.

EXAMPLE 1   *A system of equations with an infinite number of solutions.*
Solve the system of linear equations given by

$$
\begin{aligned}
x + 2y - 3z &= -2 \\
3x - y - 2z &= 1 \\
2x + 3y - 5z &= -3
\end{aligned}
\tag{9}
$$

Solution   Using the Gauss-Jordan elimination method, we obtain the following sequence of equivalent augmented matrices:

$$\left[\begin{array}{ccc|c} ① & 2 & -3 & -2 \\ 3 & -1 & -2 & 1 \\ 2 & 3 & -5 & -3 \end{array}\right] \xrightarrow[R_3 - 2R_1]{R_2 - 3R_1} \left[\begin{array}{ccc|c} 1 & 2 & -3 & -2 \\ 0 & ⑦ & 7 & 7 \\ 0 & -1 & 1 & 1 \end{array}\right] \xrightarrow{-\frac{1}{7}R_2}$$

$$\left[\begin{array}{ccc|c} 1 & 2 & -3 & -2 \\ 0 & 1 & -1 & -1 \\ 0 & -1 & 1 & 1 \end{array}\right] \xrightarrow[R_3 + R_2]{R_1 - 2R_2} \left[\begin{array}{ccc|c} 1 & 0 & -1 & 0 \\ 0 & 1 & -1 & -1 \\ 0 & 0 & 0 & 0 \end{array}\right]$$

The last augmented matrix is in row-reduced form. Interpreting it as a system of linear equations gives

$$\begin{aligned} x - z &= 0 \\ y - z &= -1 \end{aligned}$$

a system of two equations in the three variables x, y, and z.

Let us now single out one variable, say z, and solve for x and y in terms of it. We obtain

$$\begin{aligned} x &= z \\ y &= z - 1 \end{aligned}$$

If we assign a particular value to z, say $z = 0$, we obtain $x = 0$ and $y = -1$, giving the solution $(0, -1, 0)$ to system (9). By setting $z = 1$, we obtain the solution $(1, 0, 1)$. In general, if we set $z = t$, where t represents some real number, we obtain a solution given by $(t, t - 1, t)$. Since the parameter t may be any real number, we see that the system (9) has infinitely many solutions. Geometrically, the solutions of the system (9) lie on the straight line in three-dimensional space given by the intersection of the three planes determined by the three equations in the system.   ⊙ ⊙ ⊙

REMARK   In Example 1 we chose the parameter to be z because it is more convenient to solve for x and y (both the x- and y-columns are in unit form) in terms of z.   ⊙ ⊙ ⊙

The next example shows what happens in the elimination procedure when the system does not have a solution.

EXAMPLE 2   *A system of equations that has no solution.* Solve the system of linear equations given by

$$\begin{aligned} x + y + z &= 1 \\ 3x - y - z &= 4 \\ x + 5y + 5z &= -1 \end{aligned} \qquad \textbf{(10)}$$

Solution   Using the Gauss-Jordan elimination method, we obtain the following sequence of equivalent augmented matrices:

$$
\left[\begin{array}{ccc|c} \textcircled{1} & 1 & 1 & 1 \\ 3 & -1 & -1 & 4 \\ 1 & 5 & 5 & -1 \end{array}\right]
\xrightarrow[R_3 - R_1]{R_2 - 3R_1}
\left[\begin{array}{ccc|c} 1 & 1 & 1 & 1 \\ 0 & -4 & -4 & 1 \\ 0 & 4 & 4 & -2 \end{array}\right]
$$

$$
\xrightarrow{R_3 + R_2}
\left[\begin{array}{ccc|c} 1 & 1 & 1 & 1 \\ 0 & -4 & -4 & 1 \\ 0 & 0 & 0 & -1 \end{array}\right]
$$

Observe that the third row in the last matrix reads $0x + 0y + 0z = -1$—that is, $0 = -1$! We conclude therefore that system (10) is inconsistent and has no solution. Geometrically, we have a situation in which two of the planes intersect in a straight line but the third plane is parallel to this line of intersection of the two planes and does not intersect it. Consequently, there is no point of intersection of the three planes.   ◦ ◦ ◦

Example 2 illustrates the following more general result of using the Gauss-Jordan elimination procedure.

SYSTEMS WITH NO SOLUTION	If there is a row in the augmented matrix containing all zeros to the left of the vertical line and a nonzero entry to the right of the line, then the system of equations has no solution.

It may have dawned on you that in all the previous examples we have dealt only with systems involving exactly the same number of linear equations as there are variables. However, systems in which the number of equations is different from the number of variables also occur in practice. Indeed, we will consider such systems in Examples 3 and 4.

The following theorem provides us with some preliminary information on a system of linear equations.

THEOREM I	**a.** If the number of equations is greater than or equal to the number of variables in a linear system, then one of the following is true:
	  i. The system has no solution
	  ii. The system has exactly one solution
	  iii. The system has infinitely many solutions.
	b. If there are fewer equations than variables in a linear system, then the system either has no solution or it has infinitely many solutions.

REMARK   Theorem 1 may be used to tell us, before we even begin to solve a problem, what the nature of the solution may be.   ○ ○ ○

Although we will not prove this theorem, you should recall that we have illustrated geometrically part (a) for the case in which there are exactly as many equations (three) as there are variables. To show the validity of part (b), let us once again consider the case in which a system has three variables. Now, if there is only one equation in the system, then it is clear that there are infinitely many solutions corresponding geometrically to all the points lying on the plane represented by the equation.

Next, if there are two equations in the system, then *only* the following possibilities exist:

1. The two planes are parallel and distinct.

2. The two planes intersect in a straight line.

3. The two planes are coincident (the two equations define the same plane) (Figure 2.6).

Figure 2.6

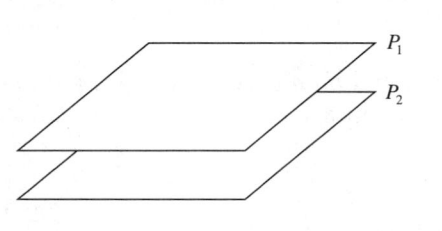
(a) No solution.

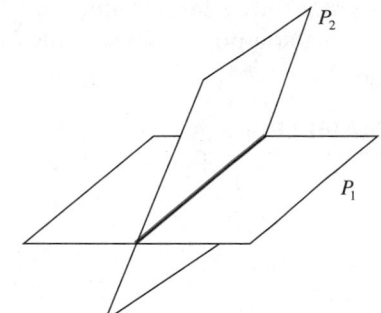
(b) Infinitely many solutions.

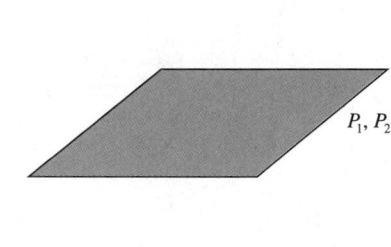
(c) Infinitely many solutions.

Thus, either there is no solution or there are infinitely many solutions corresponding to the points lying on a line of intersection of the two planes or on a single plane determined by the two equations. In the case where two planes intersect in a straight line, the solutions will involve one parameter, and in the case where the two planes are coincident, the solutions will involve two parameters.

Give a geometric interpretation of Theorem 1 for a linear system comprising equations involving two variables. Specifically, illustrate what can happen if there are three linear equations in the system (the case involving two linear equations has already been discussed in Section 2.1). What if there are four linear equations? What if there is only one linear equation in the system?

EXAMPLE 3   *A system with more equations than variables.* Solve the following system of linear equations:

$$x + 2y = 4$$
$$x - 2y = 0$$
$$4x + 3y = 12$$

Solution   We obtain the following sequence of equivalent augmented matrices:

$$\begin{bmatrix} \textcircled{1} & 2 & 4 \\ 1 & -2 & 0 \\ 4 & 3 & 12 \end{bmatrix} \xrightarrow[R_3 - 4R_1]{R_2 - R_1} \begin{bmatrix} 1 & 2 & 4 \\ 0 & \textcircled{-4} & -4 \\ 0 & -5 & -4 \end{bmatrix} \xrightarrow{-\frac{1}{4}R_2}$$

$$\begin{bmatrix} 1 & 2 & 4 \\ 0 & 1 & 1 \\ 0 & -5 & -4 \end{bmatrix} \xrightarrow[R_3 + 5R_2]{R_1 - 2R_2} \begin{bmatrix} 1 & 0 & 2 \\ 0 & 1 & 1 \\ 0 & 0 & 1 \end{bmatrix}$$

The last row of the row-reduced augmented matrix implies that $0 = 1$, which is impossible, so we conclude that the given system has no solution. Geometrically, the three lines defined by the three equations in the system do not intersect at a point. (To see this for yourself, draw the graphs of these equations.)   ○ ○ ○

EXAMPLE 4   *A system with more variables than equations.* Solve the following system of linear equations:

$$x + 2y - 3z + w = -2$$
$$3x - y - 2z - 4w = 1$$
$$2x + 3y - 5z + w = -3$$

Solution   First, observe that the given system comprises three equations in four variables, and so, by Theorem 1(b), either the system has no solution or it has infinitely many solutions. To solve it, we use the Gauss-Jordan method and obtain the following sequence of equivalent augmented matrices:

$$\begin{bmatrix} \textcircled{1} & 2 & -3 & 1 & -2 \\ 3 & -1 & -2 & -4 & 1 \\ 2 & 3 & -5 & 1 & -3 \end{bmatrix} \xrightarrow[R_3 - 2R_1]{R_2 - 3R_1} \begin{bmatrix} 1 & 2 & -3 & 1 & -2 \\ 0 & \textcircled{-7} & 7 & -7 & 7 \\ 0 & -1 & 1 & -1 & 1 \end{bmatrix} \xrightarrow{-\frac{1}{7}R_2}$$

$$\begin{bmatrix} 1 & 2 & -3 & 1 & -2 \\ 0 & 1 & -1 & 1 & -1 \\ 0 & -1 & 1 & -1 & 1 \end{bmatrix} \xrightarrow[R_3 + R_2]{R_1 - 2R_2} \begin{bmatrix} 1 & 0 & -1 & -1 & 0 \\ 0 & 1 & -1 & 1 & -1 \\ 0 & 0 & 0 & 0 & 0 \end{bmatrix}$$

The last augmented matrix is in row-reduced form. Observe that the given system is equivalent to the system

$$x - z - w = 0$$
$$y - z + w = -1$$

of two equations in four variables. Thus, we may solve for two of the variables in terms of the other two. Letting $z = s$ and $w = t$ (s, t, parameters), we find that

$$x = s + t$$
$$y = s - t - 1$$
$$z = s$$
$$w = t$$

The solutions may be written in the form $(s + t, s - t - 1, s, t)$, where s and t are any real numbers. Geometrically, the three equations in the system represent three hyperplanes in four-dimensional space (since there are four variables) and their "points" of intersection lie in a two-dimensional subspace of four-space (since there are two parameters).   ◦ ◦ ◦

REMARK   In Example 4 we assigned parameters to z and w rather than to x and y because x and y are readily solved in terms of z and w.   ◦ ◦ ◦

Application

The following example illustrates a situation in which a system of linear equations has infinitely many solutions.

EXAMPLE 5   Figure 2.7 shows the flow of downtown traffic in a certain city during the rush hours on a typical weekday. The arrows indicate the direction of traffic flow on each one-way road, and the average number of vehicles entering and leaving each intersection per hour appears beside each road. Fifth and Sixth Avenues can each handle up to 2000 vehicles per hour without causing congestion, whereas the maximum capacity of each of the two streets is 1000 vehicles per hour. The flow of traffic is controlled by traffic lights installed at each of the four intersections.

Figure 2.7

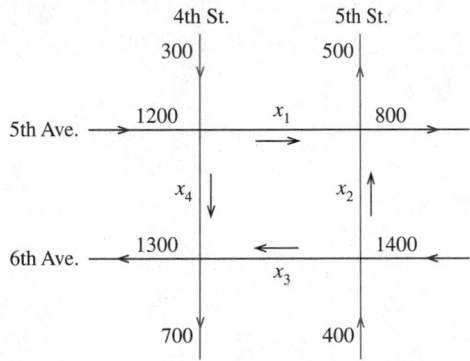

a. Write a general expression involving the rates of flow—x_1, x_2, x_3, x_4—and suggest two possible flow patterns that will ensure that there is no traffic congestion.

b. Suppose that the part of 4th Street between 5th Avenue and 6th Avenue is to be resurfaced and that traffic flow between the two junctions has to be slowed down to at most 300 vehicles per hour. Find two possible flow patterns that will result in a smooth flow of traffic.

Solution

a. To avoid congestion, all traffic entering an intersection must also leave that intersection. Applying this condition to each of the four intersections in a clockwise direction beginning with the 5th Avenue and 4th Street intersection, we obtain the following equations:

$$1500 = x_1 + x_4$$
$$1300 = x_1 + x_2$$
$$1800 = x_2 + x_3$$
$$2000 = x_3 + x_4$$

This system of four linear equations in the four variables x_1, x_2, x_3, x_4 may be rewritten in the more standard form

$$
\begin{aligned}
x_1 + x_4 &= 1500 \\
x_1 + x_2 &= 1300 \\
x_2 + x_3 &= 1800 \\
x_3 + x_4 &= 2000
\end{aligned}
$$

Using the Gauss-Jordan elimination method to solve the system, we obtain

$$
\begin{bmatrix}
1 & 0 & 0 & 1 & | & 1500 \\
1 & 1 & 0 & 0 & | & 1300 \\
0 & 1 & 1 & 0 & | & 1800 \\
0 & 0 & 1 & 1 & | & 2000
\end{bmatrix}
\xrightarrow{R_2 - R_1}
\begin{bmatrix}
1 & 0 & 0 & 1 & | & 1500 \\
0 & 1 & 0 & -1 & | & -200 \\
0 & 1 & 1 & 0 & | & 1800 \\
0 & 0 & 1 & 1 & | & 2000
\end{bmatrix}
$$

$$
\xrightarrow{R_3 - R_2}
\begin{bmatrix}
1 & 0 & 0 & 1 & | & 1500 \\
0 & 1 & 0 & -1 & | & -200 \\
0 & 0 & 1 & 1 & | & 2000 \\
0 & 0 & 1 & 1 & | & 2000
\end{bmatrix}
$$

$$
\xrightarrow{R_4 - R_3}
\begin{bmatrix}
1 & 0 & 0 & 1 & | & 1500 \\
0 & 1 & 0 & -1 & | & -200 \\
0 & 0 & 1 & 1 & | & 2000 \\
0 & 0 & 0 & 0 & | & 0
\end{bmatrix}
$$

The last augmented matrix is in row-reduced form and is equivalent to a system of three linear equations in the four variables x_1, x_2, x_3, x_4. Thus, we may express three of the variables—say, x_1, x_2, x_3—in terms of x_4. Setting

$x_4 = t$ (t, a parameter), we may write the infinitely many solutions of the system as

$$x_1 = 1500 - t$$
$$x_2 = -200 + t$$
$$x_3 = 2000 - t$$
$$x_4 = t$$

Observe that for a meaningful solution, $200 \le t \le 1000$, since x_1, x_2, x_3, and x_4 must all be nonnegative. For example, picking $t = 300$ gives the flow pattern

$$x_1 = 1200, \quad x_2 = 100, \quad x_3 = 1700, \quad x_4 = 300$$

Selecting $t = 500$ gives the flow pattern

$$x_1 = 1000, \quad x_2 = 300, \quad x_3 = 1500, \quad x_4 = 500$$

b. In this case, x_4 must not exceed 300. Again, using the results of (a), we find, upon setting $x_4 = t = 300$, the flow pattern

$$x_1 = 1200, \quad x_2 = 100, \quad x_3 = 1700, \quad x_4 = 300$$

obtained earlier. Picking $t = 250$ gives the flow pattern

$$x_1 = 1250, \quad x_2 = 50, \quad x_3 = 1750, \quad x_4 = 250 \qquad \circ \; \circ \; \circ$$

SELF-CHECK EXERCISES 2.3

1. The following augmented matrix in row-reduced form is equivalent to the augmented matrix of a certain system of linear equations. Use this result to solve the system of equations.

$$\begin{bmatrix} 1 & 0 & -1 & | & 3 \\ 0 & 1 & 5 & | & -2 \\ 0 & 0 & 0 & | & 0 \end{bmatrix}$$

2. Solve the system of linear equations

$$2x - 3y + z = 6$$
$$x + 2y + 4z = -4$$
$$x - 5y - 3z = 10$$

using the Gauss-Jordan elimination method.

3. Solve the system of linear equations

$$x - 2y + 3z = 9$$
$$2x + 3y - z = 4$$
$$x + 5y - 4z = 2$$

using the Gauss-Jordan elimination method.

Solutions to Self-Check Exercises 2.3 can be found on page 112.

2.3 EXERCISES

In exercises 1–12, given that the augmented matrix in row-reduced form is equivalent to the augmented matrix of a system of linear equations, (a) determine whether the system has a solution and (b) find the solution or solutions to the system, if they exist.

1.
$$\left[\begin{array}{ccc|c} 1 & 0 & 0 & 3 \\ 0 & 1 & 0 & -1 \\ 0 & 0 & 1 & 2 \end{array}\right]$$

2.
$$\left[\begin{array}{ccc|c} 1 & 0 & 0 & 3 \\ 0 & 1 & 0 & -2 \\ 0 & 0 & 1 & 1 \end{array}\right]$$

3.
$$\left[\begin{array}{cc|c} 1 & 0 & 2 \\ 0 & 1 & 4 \\ 0 & 0 & 0 \end{array}\right]$$

4.
$$\left[\begin{array}{ccc|c} 1 & 0 & 0 & 3 \\ 0 & 1 & 0 & 1 \\ 0 & 0 & 0 & 0 \end{array}\right]$$

5.
$$\left[\begin{array}{ccc|c} 1 & 0 & 1 & 4 \\ 0 & 1 & 0 & -2 \end{array}\right]$$

6.
$$\left[\begin{array}{cccc|c} 1 & 0 & 0 & 0 & 3 \\ 0 & 1 & 1 & 0 & -1 \\ 0 & 0 & 0 & 1 & 2 \end{array}\right]$$

7.
$$\left[\begin{array}{cccc|c} 1 & 0 & 0 & 0 & 2 \\ 0 & 1 & 0 & 0 & 1 \\ 0 & 0 & 1 & 0 & 3 \\ 0 & 0 & 0 & 0 & 1 \end{array}\right]$$

8.
$$\left[\begin{array}{ccc|c} 1 & 0 & 0 & 4 \\ 0 & 1 & 0 & -1 \\ 0 & 0 & 1 & 3 \\ 0 & 0 & 0 & 1 \end{array}\right]$$

9.
$$\left[\begin{array}{cccc|c} 1 & 0 & 0 & 0 & 2 \\ 0 & 1 & 0 & 0 & -1 \\ 0 & 0 & 1 & 1 & 2 \\ 0 & 0 & 0 & 0 & 0 \end{array}\right]$$

10.
$$\left[\begin{array}{cccc|c} 0 & 1 & 0 & 1 & 3 \\ 0 & 0 & 1 & -2 & 4 \\ 0 & 0 & 0 & 0 & 0 \\ 0 & 0 & 0 & 0 & 0 \end{array}\right]$$

11.
$$\left[\begin{array}{cccc|c} 1 & 0 & 3 & 0 & 2 \\ 0 & 1 & -1 & 0 & 1 \\ 0 & 0 & 0 & 0 & 0 \\ 0 & 0 & 0 & 0 & 0 \end{array}\right]$$

12.
$$\left[\begin{array}{cccc|c} 1 & 0 & 3 & -1 & 4 \\ 0 & 1 & -2 & 3 & 2 \\ 0 & 0 & 0 & 0 & 0 \\ 0 & 0 & 0 & 0 & 0 \end{array}\right]$$

In exercises 13–32, solve the system of linear equations using the Gauss-Jordan elimination method.

13. $2x - y = 3$
    $x + 2y = 4$
    $2x + 3y = 7$

14. $x + 2y = 3$
    $2x - 3y = -8$
    $x - 4y = -9$

15. $3x - 2y = -3$
    $2x + y = 3$
    $x - 2y = -5$

16. $2x + 3y = 2$
    $x + 3y = -2$
    $x - y = 3$

17. $3x - 2y = 5$
    $-x + 3y = -4$
    $2x - 4y = 6$

18. $4x + 6y = 8$
    $3x - 2y = -7$
    $x + 3y = 5$

19. $x - 2y = 2$
    $7x - 14y = 14$
    $3x - 6y = 6$

20. $x + 2y + z = -2$
    $-2x - 3y - z = 1$
    $2x + 4y + 2z = -4$

21. $3x + 2y = 4$
    $-\frac{3}{2}x - y = -2$
    $6x + 4y = 8$

22. $3y + 2z = 4$
    $2x - y - 3z = 3$
    $2x + 2y - z = 7$

23. $2x_1 - x_2 + x_3 = -4$
    $3x_1 - \frac{3}{2}x_2 + \frac{3}{2}x_3 = -6$
    $-6x_1 + 3x_2 - 3x_3 = 12$

24. $x + y - 2z = -3$
    $2x - y + 3z = 7$
    $x - 2y + 5z = 0$

25. $x - 2y + 3z = 4$
    $2x + 3y - z = 2$
    $x + 2y - 3z = -6$

26. $x_1 - 2x_2 + x_3 = -3$
    $2x_1 + x_2 - 2x_3 = 2$
    $x_1 + 3x_2 - 3x_3 = 5$

27. $4x + y - z = 4$
    $8x + 2y - 2z = 8$

28. $x_1 + 2x_2 + 4x_3 = 2$
    $x_1 + x_2 + 2x_3 = 1$

29. $2x + y - 3z = 1$
    $x - y + 2z = 1$
    $5x - 2y + 3z = 6$

30. $3x - 9y + 6z = -12$
    $x - 3y + 2z = -4$
    $2x - 6y + 4z = 8$

31. $x + 2y - z = -4$
    $2x + y + z = 7$
    $x + 3y + 2z = 7$
    $x - 3y + z = 9$

32. $3x - 2y + z = 4$
    $x + 3y - 4z = -3$
    $2x - 3y + 5z = 7$
    $x - 8y + 9z = 10$

33. Management Decisions The management of Hartman Rent-A-Car has allocated $840,000 to purchase 60 new automobiles to add to their existing fleet of rental cars. The company will choose from compact, mid-sized, and full-sized cars costing $10,000, $16,000, and $22,000 each, respectively. Find formulas giving the options available to the company. Give two specific options. (*Note:* Your answers will not be unique.)

34. Nutrition A dietician wishes to plan a meal around three foods. The meal is to include 8800 units of vitamin A, 3380 units of vitamin C, and 1020 units of calcium. The number of units of the vitamins and calcium in each ounce of the foods is summarized in the accompanying table.

	Food I	Food II	Food III
Vitamin A	400	1200	800
Vitamin C	110	570	340
Calcium	90	30	60

Determine the amount of each food the dietician should include in the meal in order to meet the vitamin and calcium requirements.

35. Nutrition Refer to exercise 34. In planning for another meal, the dietician changes the requirement of vitamin C to 2160 units instead of 3380 units. All other requirements remain the same. Show that such a meal cannot be planned around the same foods.

36. Investments Mr. and Mrs. Garcia have a total of $100,000 to be invested in stocks, bonds, and a money-market account. The stocks have a rate of return of 12% per year, while the bonds and the money-market account pay 8% and 4% per year, respectively. They have stipulated that the amount invested in stocks should be equal to the sum of the amount invested in bonds and three times the amount invested in the money-market account. How should the Garcias allocate their resources if they require an annual income of $10,000 from their investments?

37. Traffic Control The accompanying figure shows the flow of traffic near a city's Civic Center during the rush hours on a typical weekday. Each road can handle a maximum of 1000 cars per hour without causing congestion. The flow of traffic is controlled by traffic lights at each of the five intersections.
a. Set up a system of linear equations describing traffic flow.

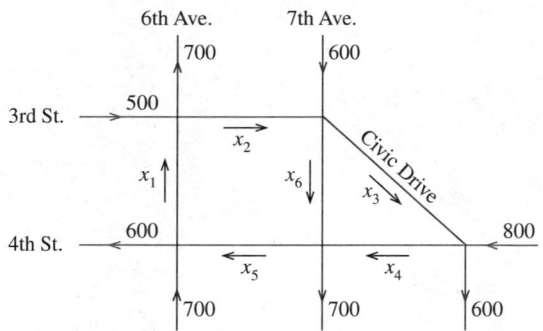

b. Solve the system devised in (a) and suggest two possible traffic-flow patterns that will ensure that there is no traffic congestion.
c. Suppose that 7th Avenue between 3rd and 4th Streets is soon to be closed for road repairs. Find one possible flow pattern that will result in a smooth flow of traffic.

38. Traffic Control The accompanying figure shows the flow of downtown traffic during the rush hours on a typical weekday. Each avenue can handle up to 1500 vehicles per hour without causing congestion, whereas the maximum capacity of each street is 1000 vehicles per hour. The flow of traffic is controlled by traffic lights at each of the six intersections.

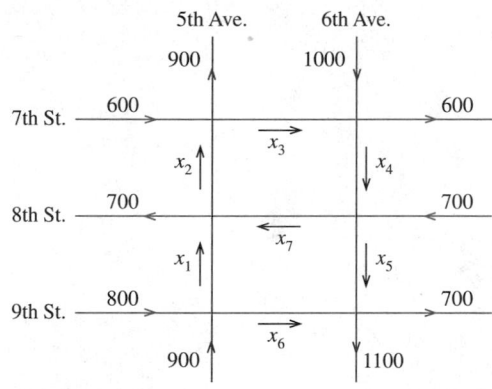

a. Set up a system of linear equations describing the traffic flow.
b. Solve the system devised in (a) and suggest two possible traffic-flow patterns that will ensure there is no traffic congestion.
c. Suppose that the traffic flow along 9th Street between 5th and 6th Avenues, x_6, is restricted due to sewer construction. What is the minimum permissible traffic flow along this road that will not result in traffic congestion?

USING TECHNOLOGY

SOLVING SYSTEMS OF LINEAR EQUATIONS II

We can use the row operations of a graphing utility to solve a system of m linear equations in n unknowns by the Gauss-Jordan method, as we did in the previous technology section. We can also use the **rref** or equivalent operation to obtain the row-reduced form without going through all the steps of the Gauss-Jordan method. The **SIMULT** function, however, cannot be used to solve a system where the number of equations and the number of variables are not the same.

EXAMPLE 1   Solve the system

$$
\begin{aligned}
x_1 - 2x_2 + 4x_3 &= 2 \\
2x_1 + x_2 - 2x_3 &= -1 \\
3x_1 - x_2 + 2x_3 &= 1 \\
2x_1 + 6x_2 - 12x_3 &= -6
\end{aligned}
$$

Solution   First we enter the augmented matrix A into the calculator as

$$
A = \left[\begin{array}{ccc|c}
1 & -2 & 4 & 2 \\
2 & 1 & -2 & -1 \\
3 & -1 & 2 & 1 \\
2 & 6 & -12 & -6
\end{array}\right]
$$

Then using the **rref** or equivalent operation, we obtain the equivalent matrix

$$
\left[\begin{array}{ccc|c}
1 & 0 & 0 & 0 \\
0 & 1 & -2 & -1 \\
0 & 0 & 0 & 0 \\
0 & 0 & 0 & 0
\end{array}\right]
$$

in reduced form. Thus, the given system is equivalent to

$$
\begin{aligned}
x_1 \quad\quad &= 0 \\
x_2 - 2x_3 &= -1
\end{aligned}
$$

Letting $x_3 = t$, where t is a parameter, we see that the solutions are $(0, 2t - 1, t)$. ◦ ◦ ◦

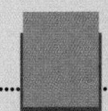 **EXERCISES**

In exercises 1–6, use a graphing utility to solve the system of equations using the rref or equivalent operation.

1.
$$2x_1 - x_2 - x_3 = 0$$
$$3x_1 - 2x_2 - x_3 = -1$$
$$-x_1 + 2x_2 - x_3 = 3$$
$$2x_2 - 2x_3 = 4$$

2.
$$3x_1 + x_2 - 4x_3 = 5$$
$$2x_1 - 3x_2 + 2x_3 = -4$$
$$-x_1 - 2x_2 + 4x_3 = 6$$
$$4x_1 + 3x_2 - 5x_3 = 9$$

3.
$$2x_1 + 3x_2 + 2x_3 + x_4 = -1$$
$$x_1 - x_2 + x_3 - 2x_4 = -8$$
$$5x_1 + 6x_2 - 2x_3 + 2x_4 = 11$$
$$x_1 + 3x_2 + 8x_3 + x_4 = -14$$

4.
$$x_1 - x_2 + 3x_3 - 6x_4 = 2$$
$$x_1 + x_2 + x_3 - 2x_4 = 2$$
$$-2x_1 - x_2 + x_3 + 2x_4 = 0$$

5.
$$x_1 + x_2 - x_3 - x_4 = -1$$
$$x_1 - x_2 + x_3 + 4x_4 = -6$$
$$3x_1 + x_2 - x_3 + 2x_4 = -4$$
$$5x_1 + x_2 - 3x_3 + x_4 = -9$$

6.
$$1.2x_1 - 2.3x_2 + 4.2x_3 + 5.4x_4 - 1.6x_5 = 4.2$$
$$2.3x_1 + 1.4x_2 - 3.1x_3 + 3.3x_4 - 2.4x_5 = 6.3$$
$$1.7x_1 + 2.6x_2 - 4.3x_3 + 7.2x_4 - 1.8x_5 = 7.8$$
$$2.6x_1 - 4.2x_2 + 8.3x_3 - 1.6x_4 + 2.5x_5 = 6.4$$

39. Determine the value of k so that the following system of linear equations has a solution and find the solution:

$$2x + 3y = 2$$
$$x + 4y = 6$$
$$5x + ky = 2$$

40. Determine the value of k so that the following system of linear equations has infinitely many solutions and find the solutions:

$$3x - 2y + 4z = 12$$
$$-9x + 6y - 12z = k$$

Solutions to Self-Check Exercises 2.3

1. Let x, y, and z denote the variables. Then the given row-reduced augmented matrix tells us that the system of linear equations is equivalent to the two equations

$$x \qquad - z = 3$$
$$y + 5z = -2$$

Letting $z = t$, where t is a parameter, we find the infinitely many solutions given by

$$x = \quad t + 3$$
$$y = -5t - 2$$
$$z = \quad t$$

2. We obtain the following sequence of equivalent augmented matrices:

$$\begin{bmatrix} 2 & -3 & 1 & | & 6 \\ 1 & 2 & 4 & | & -4 \\ 1 & -5 & -3 & | & 10 \end{bmatrix} \xrightarrow{R_1 \leftrightarrow R_2}$$

$$\begin{bmatrix} 1 & 2 & 4 & | & -4 \\ 2 & -3 & 1 & | & 6 \\ 1 & -5 & -3 & | & 10 \end{bmatrix} \xrightarrow[R_3 - R_1]{R_2 - 2R_1}$$

$$\begin{bmatrix} 1 & 2 & 4 & | & -4 \\ 0 & -7 & -7 & | & 14 \\ 0 & -7 & -7 & | & 14 \end{bmatrix} \xrightarrow{-\frac{1}{7}R_2}$$

$$\begin{bmatrix} 1 & 2 & 4 & | & -4 \\ 0 & 1 & 1 & | & -2 \\ 0 & -7 & -7 & | & 14 \end{bmatrix} \xrightarrow[R_3 + 7R_2]{R_1 - 2R_2} \begin{bmatrix} 1 & 0 & 2 & | & 0 \\ 0 & 1 & 1 & | & -2 \\ 0 & 0 & 0 & | & 0 \end{bmatrix}$$

The last augmented matrix, which is in row-reduced form, tells us that the given system of linear equations is equivalent to the following system of two equations:

$$x \qquad + 2z = \quad 0$$
$$y + \quad z = -2$$

Letting $z = t$, where t is a parameter, we see that the infinitely many solutions are given by

$$x = -2t$$
$$y = -t - 2$$
$$z = \quad t$$

3. We obtain the following sequence of equivalent augmented matrices:

$$\begin{bmatrix} ① & -2 & 3 & | & 9 \\ 2 & 3 & -1 & | & 4 \\ 1 & 5 & -4 & | & 2 \end{bmatrix} \xrightarrow[R_3 - R_1]{R_2 - 2R_1}$$

$$\begin{bmatrix} 1 & -2 & 3 & | & 9 \\ 0 & 7 & -7 & | & -14 \\ 0 & 7 & -7 & | & -7 \end{bmatrix} \xrightarrow{R_3 - R_2} \begin{bmatrix} 1 & -2 & 3 & | & 9 \\ 0 & 7 & -7 & | & -14 \\ 0 & 0 & 0 & | & 7 \end{bmatrix}$$

Since the last row of the final augmented matrix is equivalent to the equation $0 = 7$, a contradiction, we conclude that the given system has no solution.

2.4 MATRICES

Using Matrices to Represent Data

Many practical problems are solved by using arithmetic operations on the data associated with the problems. By properly organizing the data into *blocks* of numbers, we can then carry out these arithmetic operations in an orderly and efficient manner. In particular, this systematic approach enables us to use the computer to full advantage.

Let us begin by considering how the monthly output data of a manufacturer may be organized. The Acrosonic Company manufactures four different loudspeaker systems in three separate locations. The company's May output is summarized in Table 2.1.

Table 2.1

	Model A	Model B	Model C	Model D
Location I	320	280	460	280
Location II	480	360	580	0
Location III	540	420	200	880

Now, if we agree to preserve the relative location of each entry in Table 2.1, we can summarize the set of data further, as follows:

$$\begin{bmatrix} 320 & 280 & 460 & 280 \\ 480 & 360 & 580 & 0 \\ 540 & 420 & 200 & 880 \end{bmatrix}$$

A matrix summarizing the data in Table 2.1.

The array of numbers displayed here is an example of a *matrix*. Observe that the numbers in the first row give the output of models A, B, C, and D of Acrosonic loudspeaker systems manufactured in location I; similarly, the numbers in the second and third rows give the respective outputs of these loudspeaker systems in locations II and III. The numbers in each column of the matrix give the outputs of a particular model of loudspeaker system manufactured in each of the company's three manufacturing locations.

More generally, a matrix is an ordered rectangular array of real numbers. For example, each of the following arrays is a matrix:

$$A = \begin{bmatrix} 3 & 0 & -1 \\ 2 & 1 & 4 \end{bmatrix} \quad B = \begin{bmatrix} 3 & 2 \\ 0 & 1 \\ -1 & 4 \end{bmatrix} \quad C = \begin{bmatrix} 1 \\ 2 \\ 4 \\ 0 \end{bmatrix} \quad D = \begin{bmatrix} 1 & 3 & 0 & 1 \end{bmatrix}$$

The real numbers that make up the array are called the **entries,** or *elements,* of the matrix. The entries in a row in the array are referred to as a **row** of the matrix, whereas the entries in a column in the array are referred to as a **column** of the matrix. Matrix A, for example, has two rows and three columns, which may be identified as follows:

$$\begin{array}{cccc} & \text{Column 1} & \text{Column 2} & \text{Column 3} \\ \text{Row 1} & \begin{bmatrix} 3 & 0 & -1 \\ \text{Row 2} & 2 & 1 & 4 \end{bmatrix} \end{array}$$

A 2×3 matrix

The **size,** or *dimension,* of a matrix is described in terms of the number of rows and columns of the matrix. For example, matrix A has two rows and three columns and is said to have size 2 by 3, denoted 2×3. In general, a matrix having m rows and n columns is said to have size $m \times n$.

MATRIX

A **matrix** is an ordered rectangular array of numbers. A matrix with m rows and n columns has size $m \times n$. The entry in the ith row and jth column is denoted by a_{ij}.

A matrix of size $1 \times n$—a matrix having one row and n columns—is referred to as a **row matrix,** or *row vector,* of dimension n. For example, the matrix D is a row vector of dimension 4. Similarly, a matrix having m rows and one column is referred to as a **column matrix,** or *column vector,* of dimension m. The matrix C is a column vector of dimension 4. Finally, an $n \times n$ matrix—that is, a matrix having the same number of rows as columns—is called a **square matrix.** For example, the matrix

$$\begin{bmatrix} -3 & 8 & 6 \\ 2 & \frac{1}{4} & 4 \\ 1 & 3 & 2 \end{bmatrix}$$

A 3×3 square matrix

is a square matrix of size 3×3, or simply of size 3.

EXAMPLE 1   Consider the matrix

$$P = \begin{bmatrix} 320 & 280 & 460 & 280 \\ 480 & 360 & 580 & 0 \\ 540 & 420 & 200 & 880 \end{bmatrix}$$

representing the output of loudspeaker systems of the Acrosonic Company discussed earlier (see Table 2.1).

a. What is the size of the matrix P?

b. Find a_{24} (the entry in the second row and fourth column of the matrix P) and give an interpretation of this number.

c. Find the sum of the entries that make up the first row of P and interpret the result.

d. Find the sum of the entries that make up the fourth column of P and interpret the result.

Solution

a. The matrix P has three rows and four columns and hence has size 3×4.

b. The required entry lies in the second row and fourth column and is the number zero. This means that no model D loudspeaker system was manufactured in location II in May.

c. The required sum is given by

$$320 + 280 + 460 + 280 = 1340$$

which gives the total number of loudspeaker systems manufactured in location I in May as 1340 units.

d. The required sum is given by

$$280 + 0 + 880 = 1160$$

giving the output of model D loudspeaker systems in all locations of the company in May as 1160 units.   ๐ ๐ ๐

Equality of Matrices

Two matrices are said to be *equal* if they have the same size and their corresponding entries are equal. For example,

$$\begin{bmatrix} 2 & 3 & 1 \\ 4 & 6 & 2 \end{bmatrix} = \begin{bmatrix} (3-1) & 3 & 1 \\ 4 & (4+2) & 2 \end{bmatrix}$$

Also,

$$\begin{bmatrix} 1 & 3 & 5 \\ 2 & 4 & 3 \end{bmatrix} \neq \begin{bmatrix} 1 & 2 \\ 3 & 4 \\ 5 & 3 \end{bmatrix}$$

since the matrix on the left has size 2×3, whereas the matrix on the right has size 3×2, and

$$\begin{bmatrix} 2 & 3 \\ 4 & 6 \end{bmatrix} \neq \begin{bmatrix} 2 & 3 \\ 4 & 7 \end{bmatrix}$$

since the corresponding elements in the second row and second column of the two matrices are not equal.

EQUALITY OF MATRICES	Two matrices are equal if they have the same size and their corresponding entries are equal.

EXAMPLE 2   Solve the following matrix equation for x, y, and z:

$$\begin{bmatrix} 1 & x & 3 \\ 2 & y-1 & 2 \end{bmatrix} = \begin{bmatrix} 1 & 4 & z \\ 2 & 1 & 2 \end{bmatrix}$$

Solution   Since the corresponding elements of the two matrices must be equal, we find that $x = 4$, $z = 3$, and $y - 1 = 1$, or $y = 2$.   ◦ ◦ ◦

Addition and Subtraction

Two matrices A and B of the *same size* can be added or subtracted to produce a matrix of the same size. This is done by adding or subtracting the corresponding entries in the two matrices. For example,

$$\begin{bmatrix} 1 & 3 & 4 \\ -1 & 2 & 0 \end{bmatrix} + \begin{bmatrix} 1 & 4 & 3 \\ 6 & 1 & -2 \end{bmatrix} = \begin{bmatrix} 1+1 & 3+4 & 4+3 \\ -1+6 & 2+1 & 0+(-2) \end{bmatrix} = \begin{bmatrix} 2 & 7 & 7 \\ 5 & 3 & -2 \end{bmatrix}$$

Adding two matrices of the same size

and $$\begin{bmatrix} 1 & 2 \\ -1 & 3 \\ 4 & 0 \end{bmatrix} - \begin{bmatrix} 2 & -1 \\ 3 & 2 \\ -1 & 0 \end{bmatrix} = \begin{bmatrix} 1-2 & 2-(-1) \\ -1-3 & 3-2 \\ 4-(-1) & 0-0 \end{bmatrix} = \begin{bmatrix} -1 & 3 \\ -4 & 1 \\ 5 & 0 \end{bmatrix}$$

Subtracting two matrices of the same size

ADDITION AND SUBTRACTION OF MATRICES	If A and B are two matrices of the same size, **1.** the *sum $A + B$* is the matrix obtained by adding the corresponding entries in the two matrices. **2.** the *difference $A - B$* is the matrix obtained by subtracting the corresponding entries in B from A.

EXAMPLE 3   The total output of the Acrosonic Company for June is summarized in Table 2.2.

Table 2.2

	Model A	**Model B**	**Model C**	**Model D**
Location I	210	180	330	180
Location II	400	300	450	40
Location III	420	280	180	740

The output for May was given earlier in Table 2.1. Find the total output of the company for May and June.

Solution   As we saw earlier, the production matrix for the Acrosonic Company in May is given by

$$A = \begin{bmatrix} 320 & 280 & 460 & 280 \\ 480 & 360 & 580 & 0 \\ 540 & 420 & 200 & 880 \end{bmatrix}$$

Next, from Table 2.2, we see that the production matrix for June is given by

$$B = \begin{bmatrix} 210 & 180 & 330 & 180 \\ 400 & 300 & 450 & 40 \\ 420 & 280 & 180 & 740 \end{bmatrix}$$

Finally, the total output of the Acrosonic Company for May and June is given by the matrix

$$A + B = \begin{bmatrix} 320 & 280 & 460 & 280 \\ 480 & 360 & 580 & 0 \\ 540 & 420 & 200 & 880 \end{bmatrix} + \begin{bmatrix} 210 & 180 & 330 & 180 \\ 400 & 300 & 450 & 40 \\ 420 & 280 & 180 & 740 \end{bmatrix}$$

$$= \begin{bmatrix} 530 & 460 & 790 & 460 \\ 880 & 660 & 1030 & 40 \\ 960 & 700 & 380 & 1620 \end{bmatrix}$$

○ ○ ○

The following laws hold for matrix addition.

LAWS FOR MATRIX ADDITION	If A, B, and C are matrices of the same size, then
	1. $A + B = B + A$ (Commutative law)
	2. $(A + B) + C = A + (B + C)$ (Associative law)

The *commutative law* for matrix addition states that the order in which matrix addition is performed is immaterial. The *associative law* states that when adding three matrices together, we may first add A and B and then add the resulting sum to C. Equivalently, we can add A to the sum of B and C.

A *zero matrix* is one in which all entries are zero. The zero matrix O has the property that

$$A + O = O + A = A$$

for any matrix A having the same size as that of O. For example, the zero matrix of size 3×2 is

$$O = \begin{bmatrix} 0 & 0 \\ 0 & 0 \\ 0 & 0 \end{bmatrix}$$

If A is any 3×2 matrix, then

$$A + O = \begin{bmatrix} a_{11} & a_{12} \\ a_{21} & a_{22} \\ a_{31} & a_{32} \end{bmatrix} + \begin{bmatrix} 0 & 0 \\ 0 & 0 \\ 0 & 0 \end{bmatrix} = \begin{bmatrix} a_{11} & a_{12} \\ a_{21} & a_{22} \\ a_{31} & a_{32} \end{bmatrix} = A$$

where a_{ij} denotes the entry in the ith row and jth column of the matrix A.

The matrix obtained by interchanging the rows and columns of a given matrix A is called the *transpose* of A and is denoted A^T. For example, if

$$A = \begin{bmatrix} 1 & 2 & 3 \\ 4 & 5 & 6 \\ 7 & 8 & 9 \end{bmatrix}$$

then

$$A^T = \begin{bmatrix} 1 & 4 & 7 \\ 2 & 5 & 8 \\ 3 & 6 & 9 \end{bmatrix}$$

TRANSPOSE OF A MATRIX

If A is an $m \times n$ matrix with elements a_{ij}, then the **transpose** of A is the $n \times m$ matrix A^T with elements a_{ji}.

Scalar Multiplication

A matrix A may be multiplied by a real number, called a **scalar** in the context of matrix algebra. The **scalar product,** denoted by cA, is a matrix obtained by multiplying each entry of A by c. For example, the scalar product of the matrix

$$A = \begin{bmatrix} 3 & -1 & 2 \\ 0 & 1 & 4 \end{bmatrix}$$

and the scalar 3 is the matrix

$$3A = 3\begin{bmatrix} 3 & -1 & 2 \\ 0 & 1 & 4 \end{bmatrix} = \begin{bmatrix} 9 & -3 & 6 \\ 0 & 3 & 12 \end{bmatrix}$$

SCALAR PRODUCT

If A is a matrix and c is a real number, then the scalar product cA is the matrix obtained by multiplying each entry of A by c.

EXAMPLE 4   Given

$$A = \begin{bmatrix} 3 & 4 \\ -1 & 2 \end{bmatrix} \quad \text{and} \quad B = \begin{bmatrix} 3 & 2 \\ -1 & 2 \end{bmatrix}$$

find the matrix X satisfying the *matrix equation* $2X + B = 3A$.

Solution From the given equation $2X + B = 3A$, we find that

$$2X = 3A - B$$

$$= 3\begin{bmatrix} 3 & 4 \\ -1 & 2 \end{bmatrix} - \begin{bmatrix} 3 & 2 \\ -1 & 2 \end{bmatrix}$$

$$= \begin{bmatrix} 9 & 12 \\ -3 & 6 \end{bmatrix} - \begin{bmatrix} 3 & 2 \\ -1 & 2 \end{bmatrix} = \begin{bmatrix} 6 & 10 \\ -2 & 4 \end{bmatrix}$$

so

$$X = \frac{1}{2}\begin{bmatrix} 6 & 10 \\ -2 & 4 \end{bmatrix} = \begin{bmatrix} 3 & 5 \\ -1 & 2 \end{bmatrix}$$

 o o o

EXAMPLE 5 The management of the Acrosonic Company has decided to increase its July production of loudspeaker systems by 10% (over its June output). Find a matrix giving the targeted production for July.

Solution From the results of Example 3, we see that Acrosonic's total output for June may be represented by the matrix

$$B = \begin{bmatrix} 210 & 180 & 330 & 180 \\ 400 & 300 & 450 & 40 \\ 420 & 280 & 180 & 740 \end{bmatrix}$$

The required matrix is given by

$$(1.1)B = 1.1\begin{bmatrix} 210 & 180 & 330 & 180 \\ 400 & 300 & 450 & 40 \\ 420 & 280 & 180 & 740 \end{bmatrix}$$

$$= \begin{bmatrix} 231 & 198 & 363 & 198 \\ 440 & 330 & 495 & 44 \\ 462 & 308 & 198 & 814 \end{bmatrix}$$

and is interpreted in the usual manner.

o o o

SELF-CHECK EXERCISES 2.4

1. Perform the indicated operations:

$$\begin{bmatrix} 1 & 3 & 2 \\ -1 & 4 & 7 \end{bmatrix} - 3\begin{bmatrix} 2 & 1 & 0 \\ 1 & 3 & 4 \end{bmatrix}$$

2. Solve the following matrix equation for x, y, and z:

$$\begin{bmatrix} x & 3 \\ z & 2 \end{bmatrix} + \begin{bmatrix} 2 - y & z \\ 2 - z & -x \end{bmatrix} = \begin{bmatrix} 3 & 7 \\ 2 & 0 \end{bmatrix}$$

3. Jack owns two gas stations, one downtown and the other in the Wilshire district. Over two consecutive days his gas stations recorded gasoline sales represented by the following matrices:

	Regular-unleaded	Unleaded-plus	Super-unleaded
Downtown	1200	750	650
Wilshire	1100	850	600

$$A = \begin{bmatrix} 1200 & 750 & 650 \\ 1100 & 850 & 600 \end{bmatrix}$$

and

	Regular-unleaded	Unleaded-plus	Super-unleaded
Downtown	1250	825	550
Wilshire	1150	750	750

$$B = \begin{bmatrix} 1250 & 825 & 550 \\ 1150 & 750 & 750 \end{bmatrix}$$

Find a matrix representing the total sales of the two gas stations over the 2-day period.

Solutions to Self-Check Exercises 2.4 can be found on page 124.

2.4 EXERCISES

In exercises 1–6, refer to the following matrices:

$$A = \begin{bmatrix} 2 & -3 & 9 & -4 \\ -11 & 2 & 6 & 7 \\ 6 & 0 & 2 & 9 \\ 5 & 1 & 5 & -8 \end{bmatrix}$$

$$B = \begin{bmatrix} 3 & -1 & 2 \\ 0 & 1 & 4 \\ 3 & 2 & 1 \\ -1 & 0 & 8 \end{bmatrix}$$

$$C = \begin{bmatrix} 1 & 0 & 3 & 4 & 5 \end{bmatrix} \quad D = \begin{bmatrix} 1 \\ 3 \\ -2 \\ 0 \end{bmatrix}$$

1. What is the size of A? of B? of C? of D?

2. Find a_{14}, a_{21}, a_{31}, and a_{43}.

3. Find b_{13}, b_{31}, and b_{43}.

4. Identify the row matrix. What is its transpose?

5. Identify the column matrix. What is its transpose?

6. Identify the square matrix. What is its transpose?

In exercises 7–12, refer to the following matrices:

$$A = \begin{bmatrix} -1 & 2 \\ 3 & -2 \\ 4 & 0 \end{bmatrix} \quad B = \begin{bmatrix} 2 & 4 \\ 3 & 1 \\ -2 & 2 \end{bmatrix}$$

$$C = \begin{bmatrix} 3 & -1 & 0 \\ 2 & -2 & 3 \\ 4 & 6 & 2 \end{bmatrix} \quad D = \begin{bmatrix} 2 & -2 & 4 \\ 3 & 6 & 2 \\ -2 & 3 & 1 \end{bmatrix}$$

7. What is the size of A? of B? of C? of D?

8. Explain why the matrix $A + C$ does not exist.

9. Compute $A + B$.

10. Compute $2A - 3B$.

11. Compute $C - D$.

12. Compute $4D - 2C$.

In exercises 13–20, perform the indicated operations.

13. $\begin{bmatrix} 6 & 3 & 8 \\ 4 & 5 & 6 \end{bmatrix} - \begin{bmatrix} 3 & -2 & -1 \\ 0 & -5 & -7 \end{bmatrix}$

14. $\begin{bmatrix} 2 & -3 & 4 & -1 \\ 3 & 1 & 0 & 0 \end{bmatrix} + \begin{bmatrix} 4 & 3 & -2 & -4 \\ 6 & 2 & 0 & -3 \end{bmatrix}$

15. $\begin{bmatrix} 1 & 4 & -5 \\ 3 & -8 & 6 \end{bmatrix} + \begin{bmatrix} 4 & 0 & -2 \\ 3 & 6 & 5 \end{bmatrix}$

$\quad - \begin{bmatrix} 2 & 8 & 9 \\ -11 & 2 & -5 \end{bmatrix}$

16. $3\begin{bmatrix} 1 & 1 & -3 \\ 3 & 2 & 3 \\ 7 & -1 & 6 \end{bmatrix} + 4\begin{bmatrix} -2 & -1 & 8 \\ 4 & 2 & 2 \\ 3 & 6 & 3 \end{bmatrix}$

17. $\begin{bmatrix} 1.2 & 4.5 & -4.2 \\ 8.2 & 6.3 & -3.2 \end{bmatrix} - \begin{bmatrix} 3.1 & 1.5 & -3.6 \\ 2.2 & -3.3 & -4.4 \end{bmatrix}$

18. $\begin{bmatrix} 0.06 & 0.12 \\ 0.43 & 1.11 \\ 1.55 & -0.43 \end{bmatrix} - \begin{bmatrix} 0.77 & -0.75 \\ 0.22 & -0.65 \\ 1.09 & -0.57 \end{bmatrix}$

19. $\dfrac{1}{2}\begin{bmatrix} 1 & 0 & 0 & -4 \\ 3 & 0 & -1 & 6 \\ -2 & 1 & -4 & 2 \end{bmatrix}$

$\quad + \dfrac{4}{3}\begin{bmatrix} 3 & 0 & -1 & 4 \\ -2 & 1 & -6 & 2 \\ 8 & 2 & 0 & -2 \end{bmatrix}$

$\quad - \dfrac{1}{3}\begin{bmatrix} 3 & -9 & -1 & 0 \\ 6 & 2 & 0 & -6 \\ 0 & 1 & -3 & 1 \end{bmatrix}$

20. $0.5\begin{bmatrix} 1 & 3 & 5 \\ 5 & 2 & -1 \\ -2 & 0 & 1 \end{bmatrix} - 0.2\begin{bmatrix} 2 & 3 & 4 \\ -1 & 1 & -4 \\ 3 & 5 & -5 \end{bmatrix}$

$\quad + 0.6\begin{bmatrix} 3 & 4 & -1 \\ 4 & 5 & 1 \\ 1 & 0 & 0 \end{bmatrix}$

In exercises 21–24, solve for u, x, y, and z in the matrix equation.

21. $\begin{bmatrix} 2x-2 & 3 & 2 \\ 2 & 4 & y-2 \\ 2z & -3 & 2 \end{bmatrix} = \begin{bmatrix} 3 & u & 2 \\ 2 & 4 & 5 \\ 4 & -3 & 2 \end{bmatrix}$

22. $\begin{bmatrix} x & -2 \\ 3 & y \end{bmatrix} + \begin{bmatrix} -2 & z \\ -1 & 2 \end{bmatrix} = \begin{bmatrix} 4 & -2 \\ 2u & 4 \end{bmatrix}$

23. $\begin{bmatrix} 1 & x \\ 2y & -3 \end{bmatrix} - 4\begin{bmatrix} 2 & -2 \\ 0 & 3 \end{bmatrix} = \begin{bmatrix} 3z & 10 \\ 4 & -u \end{bmatrix}$

24. $\begin{bmatrix} 1 & 2 \\ 3 & 4 \\ x & -1 \end{bmatrix} - 3\begin{bmatrix} y-1 & 2 \\ 1 & 2 \\ 4 & 2z+1 \end{bmatrix} = 2\begin{bmatrix} -4 & -u \\ 0 & -1 \\ 4 & 4 \end{bmatrix}$

In exercises 25 and 26, let

$$A = \begin{bmatrix} 2 & -4 & 3 \\ 4 & 2 & 1 \end{bmatrix} \quad B = \begin{bmatrix} 4 & -3 & 2 \\ 1 & 0 & 4 \end{bmatrix}$$

and $\quad C = \begin{bmatrix} 1 & 0 & 2 \\ 3 & -2 & 1 \end{bmatrix}$

25. Verify by direct computation the validity of the commutative law for matrix addition.

26. Verify by direct computation the validity of the associative law for matrix addition.

In exercises 27–30, let

$$A = \begin{bmatrix} 3 & 1 \\ 2 & 4 \\ -4 & 0 \end{bmatrix} \quad and \quad B = \begin{bmatrix} 1 & 2 \\ -1 & 0 \\ 3 & 2 \end{bmatrix}$$

Verify each equation by direct computation.

27. $(3+5)A = 3A + 5A$

28. $2(4A) = (2 \cdot 4)A = 8A$

29. $4(A+B) = 4A + 4B$

30. $2(A-3B) = 2A - 6B$

In exercises 31–34, find the transpose of the given matrix.

31. $\begin{bmatrix} 3 & 2 & -1 & 5 \end{bmatrix}$

32. $\begin{bmatrix} 4 & 2 & 0 & -1 \\ 3 & 4 & -1 & 5 \end{bmatrix}$

33. $\begin{bmatrix} 1 & -1 & 2 \\ 3 & 4 & 2 \\ 0 & 1 & 0 \end{bmatrix}$

34. $\begin{bmatrix} 1 & 2 & 6 & 4 \\ 2 & 3 & 2 & 5 \\ 6 & 2 & 3 & 0 \\ 4 & 5 & 0 & 2 \end{bmatrix}$

35. **Cholesterol Levels** Mr. Cross, Mr. Jones, and Mr. Smith each suffer from coronary heart disease. As part of their treatment, they were put on special low-cholesterol diets: Mr. Cross on diet I, Mr. Jones on diet II, and Mr. Smith on diet III. Progressive records of each patient's cholesterol level were kept. At the beginning of the first, second, third, and fourth months the cholesterol levels of the three patients were:

Cross: 220, 215, 210, and 205

Jones: 220, 210, 200, and 195

Smith: 215, 205, 195, and 190

Represent this information in a 3×4 matrix.

Using Technology

MATRIX ADDITION AND SUBTRACTION, SCALAR MULTIPLICATION, AND THE TRANSPOSE OF A MATRIX

The graphing utility can be used to perform matrix addition, matrix subtraction, and scalar multiplication. It can also be used to find the transpose of a matrix.

EXAMPLE 1 Let

$$A = \begin{bmatrix} 1.2 & 3.1 \\ -2.1 & 4.2 \\ 3.1 & 4.8 \end{bmatrix} \quad \text{and} \quad B = \begin{bmatrix} 4.1 & 3.2 \\ 1.3 & 6.4 \\ 1.7 & 0.8 \end{bmatrix}$$

Find $(2.1A + 3.2B)^T$.

Solution We first enter the matrices A and B into the calculator. Then, using the matrix operations, we enter the expression $(2.1A + 3.2B)^T$ and obtain the desired matrix:

$$(2.1A + 3.2B)^T = \begin{bmatrix} 15.64 & -.25 & 11.95 \\ 16.75 & 29.3 & 12.64 \end{bmatrix}$$

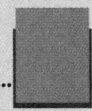 **EXERCISES**

In exercises 1–8, refer to the following matrices, and use a graphing utility to perform the indicated operations.

$$A = \begin{bmatrix} 1.2 & 3.1 & -5.4 & 2.7 \\ 4.1 & 3.2 & 4.2 & -3.1 \\ 1.7 & 2.8 & -5.2 & 8.4 \end{bmatrix}$$

$$B = \begin{bmatrix} 6.2 & -3.2 & 1.4 & -1.2 \\ 3.1 & 2.7 & -1.2 & 1.7 \\ 1.2 & -1.4 & -1.7 & 2.8 \end{bmatrix}$$

1. $12.5A$ **2.** $-8.4B$ **3.** $A - B$

4. $B - A$ **5.** $1.3A + 2.4B$ **6.** $2.1A - 1.7B$

7. $(A + B)^T$ **8.** $3A^T + 4B^T$

36. Bookstore Inventories The Campus Bookstore's inventory of books is:

Hardcover: Textbooks, 5280; Fiction, 1680; Nonfiction, 2320; Reference, 1890

Paperback: Fiction, 2810; Nonfiction, 1490; Reference, 2070; Textbooks, 1940

The College Bookstore's inventory of books is:

Hardcover: Textbooks, 6340; Fiction, 2220; Nonfiction, 1790; Reference, 1980

Paperback: Fiction, 3100; Nonfiction, 1720; Reference, 2710; Textbooks, 2050

a. Represent Campus's inventory as a matrix A.
b. Represent College's inventory as a matrix B.
c. Assume the two companies decide to merge, and write a matrix C that represents the total inventory of the newly amalgamated company.

37. Banking The numbers of three types of bank accounts on January 1 in the Central Bank and its branches are represented by matrix A.

$$A = \begin{array}{l} \text{Main office} \\ \text{West side branch} \\ \text{East side branch} \end{array} \begin{bmatrix} 2820 & 1470 & 1120 \\ 1030 & 520 & 480 \\ 1170 & 540 & 460 \end{bmatrix}$$

with columns labeled: Checking accounts, Savings accounts, Fixed deposit accounts

The number and types of accounts opened during the first quarter are represented by matrix B, and the number and types of accounts closed during the same period are represented by matrix C. Thus,

$$B = \begin{bmatrix} 260 & 120 & 110 \\ 140 & 60 & 50 \\ 120 & 70 & 50 \end{bmatrix} \quad \text{and} \quad C = \begin{bmatrix} 120 & 80 & 80 \\ 70 & 30 & 40 \\ 60 & 20 & 40 \end{bmatrix}$$

a. Find matrix D, which represents the number of each type of account at the end of the first quarter at each location.
b. Because a new manufacturing plant is opening in the immediate area, it is anticipated that there will be a 10% increase in the number of accounts at each location during the second quarter. Write a matrix E to reflect this anticipated increase.

SOLUTIONS TO SELF-CHECK EXERCISES 2.4

1. $\begin{bmatrix} 1 & 3 & 2 \\ -1 & 4 & 7 \end{bmatrix} - 3 \begin{bmatrix} 2 & 1 & 0 \\ 1 & 3 & 4 \end{bmatrix} = \begin{bmatrix} 1 & 3 & 2 \\ -1 & 4 & 7 \end{bmatrix} - \begin{bmatrix} 6 & 3 & 0 \\ 3 & 9 & 12 \end{bmatrix}$

$= \begin{bmatrix} -5 & 0 & 2 \\ -4 & -5 & -5 \end{bmatrix}$

2. We are given

$$\begin{bmatrix} x & 3 \\ z & 2 \end{bmatrix} + \begin{bmatrix} 2-y & z \\ 2-z & -x \end{bmatrix} = \begin{bmatrix} 3 & 7 \\ 2 & 0 \end{bmatrix}$$

Performing the indicated operation on the left-hand side, we obtain

$$\begin{bmatrix} 2+x-y & 3+z \\ 2 & 2-x \end{bmatrix} = \begin{bmatrix} 3 & 7 \\ 2 & 0 \end{bmatrix}$$

By the equality of matrices, we have

$$2 + x - y = 3$$
$$3 + z = 7$$
$$2 - x = 0$$

from which we deduce that $x = 2$, $y = 1$, and $z = 4$.

3. The required matrix is

$$A + B = \begin{bmatrix} 1200 & 750 & 650 \\ 1100 & 850 & 600 \end{bmatrix} + \begin{bmatrix} 1250 & 825 & 550 \\ 1150 & 750 & 750 \end{bmatrix}$$

$$= \begin{bmatrix} 2450 & 1575 & 1200 \\ 2250 & 1600 & 1350 \end{bmatrix}$$

2.5 MULTIPLICATION OF MATRICES

Matrix Product

In Section 2.4 we saw how matrices of the same size may be added or subtracted and how a matrix may be multiplied by a scalar (real number), an operation referred to as scalar multiplication. In this section we see how, with certain restrictions, one matrix may be multiplied by another matrix.

In order to define matrix multiplication, let us consider the following problem: On a certain day, Al's Service Station sold 1600 gallons of regular-unleaded, 1000 gallons of unleaded-plus, and 800 gallons of super-unleaded gasoline. If the price of gasoline on this day was $1.14 for regular-unleaded, $1.29 for unleaded-plus, and $1.39 for super-unleaded gasoline, find the total revenue realized by Al's for that day.

The day's sale of gasoline may be represented by the matrix

$$A = [1600 \quad 1000 \quad 800] \qquad \text{[Row matrix } (1 \times 3)]$$

Next, we let the unit selling price of regular-unleaded, unleaded-plus, and super-unleaded gasoline be the entries in the matrix

$$B = \begin{bmatrix} 1.14 \\ 1.29 \\ 1.39 \end{bmatrix} \qquad \text{[Column matrix } (3 \times 1)]$$

The first entry in matrix A gives the number of gallons of regular-unleaded gasoline sold, and the first entry in matrix B gives the selling price for each gallon of regular-unleaded gasoline, so their product $(1600)(1.14)$ gives the revenue realized from the sale of regular-unleaded gasoline for the day. A similar interpretation of the second and third entries in the two matrices suggests that we multiply the corresponding entries to obtain the respective revenues realized from the sale of regular-unleaded, unleaded-plus, and super-unleaded gasoline. Finally, the total revenue realized by Al's from the sale of gasoline is given by adding these products to obtain

$$(1600)(1.14) + (1000)(1.29) + (800)(1.39) = 4226$$

or $4226.

This example suggests that if we have a row matrix of size $1 \times n$,

$$A = [a_1 \quad a_2 \quad a_3 \cdots a_n]$$

and a column matrix of size $n \times 1$,

$$B = \begin{bmatrix} b_1 \\ b_2 \\ b_3 \\ \vdots \\ b_n \end{bmatrix}$$

we may define the *product* of A and B, written AB, by

$$AB = [a_1 \quad a_2 \quad a_3 \quad \cdots \quad a_n] \begin{bmatrix} b_1 \\ b_2 \\ b_3 \\ \vdots \\ b_n \end{bmatrix} = a_1b_1 + a_2b_2 + a_3b_3 + \cdots + a_nb_n \quad \textbf{(11)}$$

EXAMPLE 1   Let

$$A = [1, \quad -2, \quad 3, \quad 5] \qquad \text{and} \qquad B = \begin{bmatrix} 2 \\ 3 \\ 0 \\ -1 \end{bmatrix}$$

Then

$$AB = [1, \quad -2, \quad 3, \quad 5] \begin{bmatrix} 2 \\ 3 \\ 0 \\ -1 \end{bmatrix} = (1)(2) + (-2)(3) + (3)(0) + (5)(-1) = -9$$

○ ○ ○

EXAMPLE 2   Judy's stock holdings are given by the matrix

$$\begin{matrix} \text{GM} \quad \text{IBM} \quad \text{BAC} \\ A = [700 \quad 400 \quad 200] \end{matrix}$$

At the close of trading on a certain day, the prices (in dollars per share) of these stocks are

$$B = \begin{bmatrix} 50 \\ 120 \\ 42 \end{bmatrix} \begin{matrix} \text{GM} \\ \text{IBM} \\ \text{BAC} \end{matrix}$$

What is the total value of Judy's holdings as of that day?

Solution   Judy's holdings are worth

$$AB = [700 \quad 400 \quad 200] \begin{bmatrix} 50 \\ 120 \\ 42 \end{bmatrix} = (700)(50) + (400)(120) + (200)(42)$$

or $91,400.   ◦ ◦ ◦

Returning once again to the matrix product AB in equation (11), observe that the number of columns of the row matrix A is *equal* to the number of rows of the column matrix B. Observe further that the product matrix AB has size 1×1 (a real number may be thought of as a 1×1 matrix). Schematically,

Size of A Size of B

$(1 \times n)$ $(n \times 1)$

(1×1)

Size of AB

More generally, if A is a matrix of size $m \times n$ and B is a matrix of size $n \times p$ (the number of columns of A equals the numbers of rows of B), then the *matrix product* of A and B, AB, is defined and is a matrix of size $m \times p$. Schematically,

Size of A Size of B

$(m \times n)$ $(n \times p)$

$(m \times p)$

Size of AB

Next, let us illustrate the mechanics of matrix multiplication by computing the product of a 2×3 matrix A and a 3×4 matrix B. Suppose

$$A = \begin{bmatrix} a_{11} & a_{12} & a_{13} \\ a_{21} & a_{22} & a_{23} \end{bmatrix}$$

and

$$B = \begin{bmatrix} b_{11} & b_{12} & b_{13} & b_{14} \\ b_{21} & b_{22} & b_{23} & b_{24} \\ b_{31} & b_{32} & b_{33} & b_{34} \end{bmatrix}$$

From the schematic

same

Size of A (2×3) (3×4) Size of B

(2×4)

Size of AB

we see that the matrix product $C = AB$ is defined (the number of columns of A equals the number of rows of B) and has size 2×4. Thus,

$$C = \begin{bmatrix} c_{11} & c_{12} & c_{13} & c_{14} \\ c_{21} & c_{22} & c_{23} & c_{24} \end{bmatrix}$$

The entries of C are computed as follows: The entry c_{11} (the entry in the *first* row, *first* column of C) is the product of the row matrix comprising the entries from the *first* row of A and the column matrix comprising the *first* column of B. Thus,

$$c_{11} = \begin{bmatrix} a_{11} & a_{12} & a_{13} \end{bmatrix} \begin{bmatrix} b_{11} \\ b_{21} \\ b_{31} \end{bmatrix} = a_{11}b_{11} + a_{12}b_{21} + a_{13}b_{31}$$

The entry c_{12} (the entry in the *first* row, *second* column of C) is the product of the row matrix comprising the *first* row of A and the column matrix comprising the *second* column of B. Thus,

$$c_{12} = \begin{bmatrix} a_{11} & a_{12} & a_{13} \end{bmatrix} \begin{bmatrix} b_{12} \\ b_{22} \\ b_{32} \end{bmatrix} = a_{11}b_{12} + a_{12}b_{22} + a_{13}b_{32}$$

The other entries in C are computed in a similar manner.

EXAMPLE 3 Let

$$A = \begin{bmatrix} 3 & 1 & 4 \\ -1 & 2 & 3 \end{bmatrix} \quad \text{and} \quad B = \begin{bmatrix} 1 & 3 & -3 \\ 4 & -1 & 2 \\ 2 & 4 & 1 \end{bmatrix}$$

Compute AB.

Solution The size of matrix A is 2×3, and the size of matrix B is 3×3. Since the number of columns of matrix A is equal to the number of rows of matrix B, the matrix product $C = AB$ is defined. Furthermore, the size of matrix C is 2×3. Thus,

$$\begin{bmatrix} 3 & 1 & 4 \\ -1 & 2 & 3 \end{bmatrix} \begin{bmatrix} 1 & 3 & -3 \\ 4 & -1 & 2 \\ 2 & 4 & 1 \end{bmatrix} = \begin{bmatrix} c_{11} & c_{12} & c_{13} \\ c_{21} & c_{22} & c_{23} \end{bmatrix}$$

It remains now to determine the entries $c_{11}, c_{12}, c_{13}, c_{21}, c_{22}$, and c_{23}. We have

$$c_{11} = \begin{bmatrix} 3 & 1 & 4 \end{bmatrix} \begin{bmatrix} 1 \\ 4 \\ 2 \end{bmatrix} = (3)(1) + (1)(4) + (4)(2) = 15$$

$$c_{12} = \begin{bmatrix} 3 & 1 & 4 \end{bmatrix} \begin{bmatrix} 3 \\ -1 \\ 4 \end{bmatrix} = (3)(3) + (1)(-1) + (4)(4) = 24$$

$$c_{13} = \begin{bmatrix} 3 & 1 & 4 \end{bmatrix} \begin{bmatrix} -3 \\ 2 \\ 1 \end{bmatrix} = (3)(-3) + (1)(2) + (4)(1) = -3$$

$$c_{21} = \begin{bmatrix} -1 & 2 & 3 \end{bmatrix} \begin{bmatrix} 1 \\ 4 \\ 2 \end{bmatrix} = (-1)(1) + (2)(4) + (3)(2) = 13$$

$$c_{22} = \begin{bmatrix} -1 & 2 & 3 \end{bmatrix} \begin{bmatrix} 3 \\ -1 \\ 4 \end{bmatrix} = (-1)(3) + (2)(-1) + (3)(4) = 7$$

and $$c_{23} = \begin{bmatrix} -1 & 2 & 3 \end{bmatrix} \begin{bmatrix} -3 \\ 2 \\ 1 \end{bmatrix} = (-1)(-3) + (2)(2) + (3)(1) = 10$$

so the required product AB is given by

$$AB = \begin{bmatrix} 15 & 24 & -3 \\ 13 & 7 & 10 \end{bmatrix}$$

○ ○ ○

EXAMPLE 4   If

$$A = \begin{bmatrix} 3 & 2 & 1 \\ -1 & 2 & 3 \\ 3 & 1 & 4 \end{bmatrix} \quad \text{and} \quad B = \begin{bmatrix} 1 & 3 & 4 \\ 2 & 4 & 1 \\ -1 & 2 & 3 \end{bmatrix}$$

then

$$AB = \begin{bmatrix} 3 \cdot 1 + 2 \cdot 2 + 1 \cdot (-1) & 3 \cdot 3 + 2 \cdot 4 + 1 \cdot 2 & 3 \cdot 4 + 2 \cdot 1 + 1 \cdot 3 \\ (-1) \cdot 1 + 2 \cdot 2 + 3 \cdot (-1) & (-1) \cdot 3 + 2 \cdot 4 + 3 \cdot 2 & (-1) \cdot 4 + 2 \cdot 1 + 3 \cdot 3 \\ 3 \cdot 1 + 1 \cdot 2 + 4 \cdot (-1) & 3 \cdot 3 + 1 \cdot 4 + 4 \cdot 2 & 3 \cdot 4 + 1 \cdot 1 + 4 \cdot 3 \end{bmatrix}$$

$$= \begin{bmatrix} 6 & 19 & 17 \\ 0 & 11 & 7 \\ 1 & 21 & 25 \end{bmatrix}$$

$$BA = \begin{bmatrix} 1 \cdot 3 + 3 \cdot (-1) + 4 \cdot 3 & 1 \cdot 2 + 3 \cdot 2 + 4 \cdot 1 & 1 \cdot 1 + 3 \cdot 3 + 4 \cdot 4 \\ 2 \cdot 3 + 4 \cdot (-1) + 1 \cdot 3 & 2 \cdot 2 + 4 \cdot 2 + 1 \cdot 1 & 2 \cdot 1 + 4 \cdot 3 + 1 \cdot 4 \\ (-1) \cdot 3 + 2 \cdot (-1) + 3 \cdot 3 & (-1) \cdot 2 + 2 \cdot 2 + 3 \cdot 1 & (-1) \cdot 1 + 2 \cdot 3 + 3 \cdot 4 \end{bmatrix}$$

$$= \begin{bmatrix} 12 & 12 & 26 \\ 5 & 13 & 18 \\ 4 & 5 & 17 \end{bmatrix}$$

○ ○ ○

As the last example shows, in general, $AB \neq BA$ for any two square matrices A and B. However, the following laws are valid for matrix multiplication.

LAWS FOR MATRIX MULTIPLICATION

If the products and sums are defined for the matrices A, B, and C, then

1. $(AB)C = A(BC)$ (Associative law)

2. $A(B + C) = AB + AC$ (Distributive law)

The square matrix of size n having 1s along the main diagonal and zeros elsewhere is called the **identity matrix** of size n.

IDENTITY MATRIX

The identity matrix of size n is given by

$$I_n = \begin{bmatrix} 1 & 0 & . & . & . & 0 \\ 0 & 1 & . & . & . & 0 \\ . & . & & . & & . \\ . & . & & . & & . \\ . & . & & . & & . \\ 0 & 0 & . & . & . & 1 \end{bmatrix} \quad (n \text{ rows})$$

$(n \text{ columns})$

The identity matrix has the property that $I_nA = A$ for any $n \times r$ matrix A, and $BI_n = B$ for any $s \times n$ matrix B. In particular, if A is a square matrix of size n, then

$$I_nA = AI_n = A$$

EXAMPLE 5   Let

$$A = \begin{bmatrix} 1 & 3 & 1 \\ -4 & 3 & 2 \\ 1 & 0 & 1 \end{bmatrix}$$

Then

$$I_3A = \begin{bmatrix} 1 & 0 & 0 \\ 0 & 1 & 0 \\ 0 & 0 & 1 \end{bmatrix}\begin{bmatrix} 1 & 3 & 1 \\ -4 & 3 & 2 \\ 1 & 0 & 1 \end{bmatrix} = \begin{bmatrix} 1 & 3 & 1 \\ -4 & 3 & 2 \\ 1 & 0 & 1 \end{bmatrix} = A$$

$$AI_3 = \begin{bmatrix} 1 & 3 & 1 \\ -4 & 3 & 2 \\ 1 & 0 & 1 \end{bmatrix}\begin{bmatrix} 1 & 0 & 0 \\ 0 & 1 & 0 \\ 0 & 0 & 1 \end{bmatrix} = \begin{bmatrix} 1 & 3 & 1 \\ -4 & 3 & 2 \\ 1 & 0 & 1 \end{bmatrix} = A$$

so $I_3A = AI_3$, confirming our result for this special case.   ○ ○ ○

Application

EXAMPLE 6   The Ace Novelty Company received an order from Magic World Amusement Park for 900 "Giant Pandas," 1200 "Saint Bernards," and 2000 "Big Birds." Ace's management decided that 500 Giant Pandas, 800 Saint Bernards, and 1300 Big Birds could be manufactured in their Los Angeles plant, and the balance of the order could be filled by their Seattle plant. Each Panda requires 1.5 square yards of plush, 30 cubic feet of stuffing, and 5 pieces of trim; each Saint Bernard requires 2 square yards of plush, 35 cubic feet of stuffing, and 8 pieces of trim; and each Big Bird requires 2.5 square yards of plush, 25 cubic feet of stuffing, and 15 pieces of trim. The plush costs $4.50 per square yard, the stuffing costs 10 cents per cubic foot, and the trim costs 25 cents per unit.

a. Find how much of each type of material must be purchased for each plant.

b. What is the total cost of materials incurred by each plant and the total cost of materials incurred by Ace Novelty in filling the order?

Solution   The quantities of each type of stuffed animal to be produced at each plant location may be expressed as a 2×3 *production matrix P*. Thus,

$$P = \begin{array}{c} \text{L.A.} \\ \text{Seattle} \end{array} \begin{array}{ccc} \text{Pandas} & \text{St. Bernards} & \text{Birds} \\ \begin{bmatrix} 500 & 800 & 1300 \\ 400 & 400 & 700 \end{bmatrix} \end{array}$$

Similarly, we may represent the amount and type of material required to manufacture each type of animal by a 3×3 *activity matrix A*. Thus,

$$A = \begin{array}{c} \text{Pandas} \\ \text{St. Bernards} \\ \text{Birds} \end{array} \begin{array}{ccc} \text{Plush} & \text{Stuffing} & \text{Trim} \\ \begin{bmatrix} 1.5 & 30 & 5 \\ 2 & 35 & 8 \\ 2.5 & 25 & 15 \end{bmatrix} \end{array}$$

Finally, the unit cost for each type of material may be represented by the 3×1 *cost matrix C*:

$$C = \begin{array}{c} \text{Plush} \\ \text{Stuffing} \\ \text{Trim} \end{array} \begin{bmatrix} 4.50 \\ 0.10 \\ 0.25 \end{bmatrix}$$

a. The amount of each type of material required for each plant is given by the matrix *PA*. Thus,

$$PA = \begin{bmatrix} 500 & 800 & 1300 \\ 400 & 400 & 700 \end{bmatrix} \begin{bmatrix} 1.5 & 30 & 5 \\ 2 & 35 & 8 \\ 2.5 & 25 & 15 \end{bmatrix}$$

$$= \begin{array}{c} \text{L.A.} \\ \text{Seattle} \end{array} \begin{array}{ccc} \text{Plush} & \text{Stuffing} & \text{Trim} \\ \begin{bmatrix} 5600 & 75,500 & 28,400 \\ 3150 & 43,500 & 15,700 \end{bmatrix} \end{array}$$

b. The total cost of materials for each plant is given by the matrix PAC:

$$PAC = \begin{bmatrix} 5600 & 75{,}500 & 28{,}400 \\ 3150 & 43{,}500 & 15{,}700 \end{bmatrix} \begin{bmatrix} 4.50 \\ 0.10 \\ 0.25 \end{bmatrix}$$

$$= \begin{matrix} \text{L.A.} \\ \text{Seattle} \end{matrix} \begin{bmatrix} 39{,}850 \\ 22{,}450 \end{bmatrix}$$

or \$39,850 for the L.A. plant and \$22,450 for the Seattle plant. Thus, the total cost of materials incurred by Ace Novelty is \$62,300.   ◦ ◦ ◦

Matrix Representation

Example 7 shows how a system of linear equations may be written in a compact form with the help of matrices. (We will use this matrix equation representation in Section 2.6.)

EXAMPLE 7   Write the system of linear equations

$$\begin{aligned} 2x - 4y + z &= 6 \\ -3x + 6y - 5z &= -1 \\ x - 3y + 7z &= 0 \end{aligned}$$

in matrix form.

Solution   Let us write

$$A = \begin{bmatrix} 2 & -4 & 1 \\ -3 & 6 & -5 \\ 1 & -3 & 7 \end{bmatrix}, \quad X = \begin{bmatrix} x \\ y \\ z \end{bmatrix}, \quad \text{and} \quad B = \begin{bmatrix} 6 \\ -1 \\ 0 \end{bmatrix}$$

Note that A is just the 3×3 matrix of coefficients of the system, X is the 3×1 column matrix of unknowns (variables), and B is the 3×1 column matrix of constants. We now show that the required matrix representation of the system of linear equations is

$$AX = B$$

To see this, observe that

$$AX = \begin{bmatrix} 2 & -4 & 1 \\ -3 & 6 & -5 \\ 1 & -3 & 7 \end{bmatrix} \begin{bmatrix} x \\ y \\ z \end{bmatrix} = \begin{bmatrix} 2x - 4y + z \\ -3x + 6y - 5z \\ x - 3y + 7z \end{bmatrix}$$

Equating this 3×1 matrix with matrix B now gives

$$\begin{bmatrix} 2x - 4y + z \\ -3x + 6y - 5z \\ x - 3y + 7z \end{bmatrix} = \begin{bmatrix} 6 \\ -1 \\ 0 \end{bmatrix}$$

which, by matrix equality, is easily seen to be equivalent to the given system of linear equations.   ◦ ◦ ◦

SELF-CHECK EXERCISES 2.5

1. Compute

$$\begin{bmatrix} 1 & 3 & 0 \\ 2 & 4 & -1 \end{bmatrix} \begin{bmatrix} 3 & 1 & 4 \\ 2 & 0 & 3 \\ 1 & 2 & -1 \end{bmatrix}$$

2. Write the following system of linear equations in matrix form:

$$\begin{aligned} y - 2z &= 1 \\ 2x - y + 3z &= 0 \\ x \quad\ + 4z &= 7 \end{aligned}$$

3. On July 1, the stock holdings of Ash and Joan Robinson were given by the matrix

$$A = \begin{array}{c} \\ \text{Ash} \\ \text{Joan} \end{array} \begin{array}{cccc} \text{AT\&T} & \text{GTE} & \text{IBM} & \text{GM} \\ \begin{bmatrix} 2000 & 1000 & 500 & 5000 \\ 1000 & 500 & 2000 & 0 \end{bmatrix} \end{array}$$

and the closing prices of AT&T, GTE, IBM, and GM were $24, $47, $150, and $14 per share, respectively. Use matrix multiplication to determine the separate values of Ash's and Joan's stock holdings as of that date.

Solutions to Self-Check Exercises 2.5 can be found on page 139.

2.5 EXERCISES

In exercises 1–4, the sizes of matrices A and B are given. Find the size of AB and BA whenever they are defined.

1. *A* is of size 2 × 3 and *B* is of size 3 × 5

2. *A* is of size 3 × 4 and *B* is of size 4 × 3

3. *A* is of size 1 × 7 and *B* is of size 7 × 1

4. *A* is of size 4 × 4 and *B* is of size 4 × 4

5. Let *A* be a matrix of size *m* × *n* and *B* be a matrix of size *s* × *t*. Find conditions on *m*, *n*, *s*, and *t* so that both matrix products *AB* and *BA* are defined.

6. Find condition(s) on the size of a matrix *A* so that *A²* (that is, *AA*) is defined.

In exercises 7–24, compute the indicated products.

7. $\begin{bmatrix} 1 & 2 \\ 3 & 0 \end{bmatrix} \begin{bmatrix} 1 \\ -1 \end{bmatrix}$

8. $\begin{bmatrix} -1 & 3 \\ 5 & 0 \end{bmatrix} \begin{bmatrix} 7 \\ 2 \end{bmatrix}$

9. $\begin{bmatrix} 3 & 1 & 2 \\ -1 & 2 & 4 \end{bmatrix} \begin{bmatrix} 4 \\ 1 \\ -2 \end{bmatrix}$

10. $\begin{bmatrix} 3 & 2 & -1 \\ 4 & -1 & 0 \\ -5 & 2 & 1 \end{bmatrix} \begin{bmatrix} 3 \\ -2 \\ 0 \end{bmatrix}$

11. $\begin{bmatrix} -1 & 2 \\ 3 & 1 \end{bmatrix} \begin{bmatrix} 2 & 4 \\ 3 & 1 \end{bmatrix}$

12. $\begin{bmatrix} 1 & 3 \\ -1 & 2 \end{bmatrix} \begin{bmatrix} 1 & 3 & 0 \\ 3 & 0 & 2 \end{bmatrix}$

13. $\begin{bmatrix} 2 & 1 & 2 \\ 3 & 2 & 4 \end{bmatrix} \begin{bmatrix} -1 & 2 \\ 4 & 3 \\ 0 & 1 \end{bmatrix}$

14. $\begin{bmatrix} -1 & 2 \\ 4 & 3 \\ 0 & 1 \end{bmatrix} \begin{bmatrix} 2 & 1 & 2 \\ 3 & 2 & 4 \end{bmatrix}$

15. $\begin{bmatrix} 0.1 & 0.9 \\ 0.2 & 0.8 \end{bmatrix} \begin{bmatrix} 1.2 & 0.4 \\ 0.5 & 2.1 \end{bmatrix}$

16. $\begin{bmatrix} 1.2 & 0.3 \\ 0.4 & 0.5 \end{bmatrix} \begin{bmatrix} 0.2 & 0.6 \\ 0.4 & -0.5 \end{bmatrix}$

17. $\begin{bmatrix} 6 & -3 & 0 \\ -2 & 1 & -8 \\ 4 & -4 & 9 \end{bmatrix} \begin{bmatrix} 1 & 0 & 0 \\ 0 & 1 & 0 \\ 0 & 0 & 1 \end{bmatrix}$

18. $\begin{bmatrix} 2 & 4 \\ -1 & -5 \\ 3 & -1 \end{bmatrix} \begin{bmatrix} 2 & -2 & 4 \\ 1 & 3 & -1 \end{bmatrix}$

19. $\begin{bmatrix} 3 & 0 & -2 & 1 \\ 1 & 2 & 0 & -1 \end{bmatrix} \begin{bmatrix} 2 & 1 & -1 \\ -1 & 2 & 0 \\ 0 & 0 & 1 \\ -1 & -2 & 2 \end{bmatrix}$

20. $\begin{bmatrix} 2 & 1 & -3 & 0 \\ 4 & -2 & -1 & 1 \\ -1 & 2 & 0 & 1 \end{bmatrix} \begin{bmatrix} 2 & -1 \\ 1 & 4 \\ 3 & -3 \\ 0 & -5 \end{bmatrix}$

21. $4 \begin{bmatrix} 1 & -2 & 0 \\ 2 & -1 & 1 \\ 3 & 0 & -1 \end{bmatrix} \begin{bmatrix} 1 & 3 & 1 \\ 1 & 4 & 0 \\ 0 & 1 & -2 \end{bmatrix}$

22. $3 \begin{bmatrix} 2 & -1 & 0 \\ 2 & 1 & 2 \\ 1 & 0 & -1 \end{bmatrix} \begin{bmatrix} 2 & 3 & 1 \\ 3 & -3 & 0 \\ 0 & 1 & -1 \end{bmatrix}$

23. $\begin{bmatrix} 1 & 0 \\ 0 & 1 \end{bmatrix} \begin{bmatrix} 4 & -3 & 2 \\ 7 & 1 & -5 \end{bmatrix} \begin{bmatrix} 1 & 0 & 0 \\ 0 & 1 & 0 \\ 0 & 0 & 1 \end{bmatrix}$

24. $2 \begin{bmatrix} 3 & 2 & -1 \\ 0 & 1 & 3 \\ 2 & 0 & 3 \end{bmatrix} \begin{bmatrix} 1 & 0 & 0 \\ 0 & 1 & 0 \\ 0 & 0 & 1 \end{bmatrix} \begin{bmatrix} 1 & 2 & 0 \\ 0 & -1 & -2 \\ 1 & 3 & 1 \end{bmatrix}$

In exercises 25 and 26, let

$$A = \begin{bmatrix} 1 & 0 & -2 \\ 1 & -3 & 2 \\ -2 & 1 & 1 \end{bmatrix} \qquad B = \begin{bmatrix} 3 & 1 & 0 \\ 2 & 2 & 0 \\ 1 & -3 & -1 \end{bmatrix}$$

and

$$C = \begin{bmatrix} 2 & -1 & 0 \\ 1 & -1 & 2 \\ 3 & -2 & 1 \end{bmatrix}$$

25. Verify the validity of the associative law for matrix multiplication.

26. Verify the validity of the distributive law for matrix multiplication.

27. Let

$$A = \begin{bmatrix} 1 & 2 \\ 3 & 4 \end{bmatrix} \qquad \text{and} \qquad B = \begin{bmatrix} 2 & 1 \\ 4 & 3 \end{bmatrix}$$

Compute AB and BA and hence deduce that matrix multiplication is, in general, not commutative.

28. Let

$$A = \begin{bmatrix} 0 & 3 & 0 \\ 1 & 0 & 1 \\ 0 & 2 & 0 \end{bmatrix} \qquad B = \begin{bmatrix} 2 & 4 & 5 \\ 3 & -1 & -6 \\ 4 & 3 & 4 \end{bmatrix}$$

and

$$C = \begin{bmatrix} 4 & 5 & 6 \\ 3 & -1 & -6 \\ 2 & 2 & 3 \end{bmatrix}$$

a. Compute AB.
b. Compute AC.
c. Using the results of (a) and (b), conclude that $AB = AC$ does not imply that $B = C$.

29. Let

$$A = \begin{bmatrix} 3 & 0 \\ 8 & 0 \end{bmatrix} \qquad \text{and} \qquad B = \begin{bmatrix} 0 & 0 \\ 4 & 5 \end{bmatrix}$$

Show that $AB = O$, thereby demonstrating that for matrix multiplication the equation $AB = O$ does not imply that one or both of the matrices A and B must be the zero matrix.

30. Let

$$A = \begin{bmatrix} 2 & 2 \\ -2 & -2 \end{bmatrix}$$

Show that $A^2 = O$. Compare this with the equation $a^2 = 0$, where a is a real number.

31. Find the matrix A such that

$$A \begin{bmatrix} 1 & 0 \\ -1 & 3 \end{bmatrix} = \begin{bmatrix} -1 & -3 \\ 3 & 6 \end{bmatrix}$$

Hint: Let $A = \begin{bmatrix} a & b \\ c & d \end{bmatrix}$

32. Let

$$A = \begin{bmatrix} 3 & 1 \\ 0 & 2 \end{bmatrix} \qquad \text{and} \qquad B = \begin{bmatrix} 4 & -2 \\ 2 & 1 \end{bmatrix}$$

a. Compute $(A + B)^2$.
b. Compute $A^2 + 2AB + B^2$.
c. From the results of (a) and (b), show that in general $(A + B)^2 \neq A^2 + 2AB + B^2$.

33. Let

$$A = \begin{bmatrix} 2 & 4 \\ 5 & -6 \end{bmatrix} \quad \text{and} \quad B = \begin{bmatrix} 4 & 8 \\ -7 & 3 \end{bmatrix}$$

a. Find A^T and show that $(A^T)^T = A$.
b. Show that $(A + B)^T = A^T + B^T$.
c. Show that $(AB)^T = B^T A^T$.

34. Let

$$A = \begin{bmatrix} 1 & 3 \\ -2 & -1 \end{bmatrix} \quad \text{and} \quad B = \begin{bmatrix} 3 & -4 \\ 2 & -2 \end{bmatrix}$$

a. Find A^T and show that $(A^T)^T = A$.
b. Show that $(A + B)^T = A^T + B^T$.
c. Show that $(AB)^T = B^T A^T$.

In exercises 35–40, write the given system of linear equations in matrix form.

35. $2x - 3y = 7$
$3x - 4y = 8$

36. $2x \quad = 7$
$3x - 2y = 12$

37. $2x - 3y + 4z = 6$
$\quad 2y - 3z = 7$
$x - \quad y + 2z = 4$

38. $x - 2y + 3z = -1$
$3x + 4y - 2z = 1$
$2x - 3y + 7z = 6$

39. $-x_1 + x_2 + x_3 = 0$
$2x_1 - x_2 - x_3 = 2$
$-3x_1 + 2x_2 + 4x_3 = 4$

40. $3x_1 - 5x_2 + 4x_3 = 10$
$4x_1 + 2x_2 - 3x_3 = -12$
$-x_1 \quad + x_3 = -2$

41. Investments William and Michael's stock holdings are given by the matrix

$$A = \begin{matrix} \text{William} \\ \text{Michael} \end{matrix} \begin{bmatrix} \overset{\text{BAC}}{200} & \overset{\text{GM}}{300} & \overset{\text{IBM}}{100} & \overset{\text{TRW}}{200} \\ 100 & 200 & 400 & 0 \end{bmatrix}$$

At the close of trading on a certain day, the prices (in dollars per share) of the stocks are given by the matrix

$$B = \begin{matrix} \text{BAC} \\ \text{GM} \\ \text{IBM} \\ \text{TRW} \end{matrix} \begin{bmatrix} 54 \\ 48 \\ 98 \\ 82 \end{bmatrix}$$

a. Find AB.
b. Explain the meaning of the entries in the matrix AB.

42. Box-Office Receipts The Cinema Center comprises four theaters: Cinemas I, II, III, and IV. The admission price for one feature at the Center is $2 for children, $3 for students, and $4 for adults. The attendance for the Sunday matinee is given by the matrix

$$A = \begin{matrix} \text{Cinema I} \\ \text{Cinema II} \\ \text{Cinema III} \\ \text{Cinema IV} \end{matrix} \begin{bmatrix} \overset{\text{Children}}{225} & \overset{\text{Students}}{110} & \overset{\text{Adults}}{50} \\ 75 & 180 & 225 \\ 280 & 85 & 110 \\ 0 & 250 & 225 \end{bmatrix}$$

Write a column vector B representing the admission prices. Then compute AB, the column vector showing the gross receipts for each theater. Finally, find the total revenue collected at the Cinema Center for admission that Sunday afternoon.

43. Real Estate Bond Brothers, Inc., a real estate developer, builds houses in three states. The projected number of units of each model to be built in each state is given by the matrix

$$A = \begin{matrix} \text{N.Y.} \\ \text{Conn.} \\ \text{Mass.} \end{matrix} \begin{bmatrix} \overset{\text{I}}{60} & \overset{\text{II}}{80} & \overset{\text{III}}{120} & \overset{\text{IV}}{40} \\ 20 & 30 & 60 & 10 \\ 10 & 15 & 30 & 5 \end{bmatrix}$$

The profits to be realized are $20,000, $22,000, $25,000, and $30,000, respectively, for each Model I, II, III, and IV house sold.

a. Write a column matrix B representing the profit for each type of house.
b. Find the total profit Bond Brothers expects to earn in each state if all the houses are sold.

44. Politics: Voter Affiliation Matrix A gives the percentage of eligible voters in the city of Newton, classified according to party affiliation and age group.

$$A = \begin{matrix} \text{Under 30} \\ \text{30 to 50} \\ \text{Over 50} \end{matrix} \begin{bmatrix} \overset{\text{Dem.}}{0.50} & \overset{\text{Rep.}}{0.30} & \overset{\text{Ind.}}{0.20} \\ 0.45 & 0.40 & 0.15 \\ 0.40 & 0.50 & 0.10 \end{bmatrix}$$

The population of eligible voters in the city by age group is given by the matrix B:

$$\begin{matrix} \text{Under 30} & \text{30 to 50} & \text{Over 50} \end{matrix}$$
$$B = \begin{bmatrix} 30{,}000 & 40{,}000 & 20{,}000 \end{bmatrix}$$

Find a matrix giving the total number of eligible voters in the city who will vote Democrat, Republican, and Independent.

USING TECHNOLOGY

MATRIX MULTIPLICATION

The graphing utility can be used to perform matrix multiplication.

EXAMPLE 1 Let

$$A = \begin{bmatrix} 1.2 & 3.1 & -1.4 \\ 2.7 & 4.2 & 3.4 \end{bmatrix} \quad B = \begin{bmatrix} 0.8 & 1.2 & 3.7 \\ 6.2 & -0.4 & 3.3 \end{bmatrix} \quad \text{and}$$

$$C = \begin{bmatrix} 1.2 & 2.1 & 1.3 \\ 4.2 & -1.2 & 0.6 \\ 1.4 & 3.2 & 0.7 \end{bmatrix}$$

Find $(1.1A + 2.3B)C^T$.

Solution First we enter the matrices A, B, and C into the calculator. Then, using matrix operations, we enter the expression $(1.1A + 2.3B)C^T$. We obtain the desired matrix

$$(1.1A + 2.3B)C^T = \begin{bmatrix} 25.81 & 10.05 & 29.047 \\ 43.175 & 74.724 & 43.893 \end{bmatrix} \qquad \circ \ \circ \ \circ$$

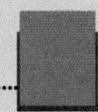
EXERCISES

In exercises 1–8, refer to the following matrices and use a graphing utility to perform the indicated operations. Express your answers accurate to two decimal places.

$$A = \begin{bmatrix} 1.2 & 3.1 & -1.2 & 4.3 \\ 7.2 & 6.3 & 1.8 & -2.1 \\ 0.8 & 3.2 & -1.3 & 2.8 \end{bmatrix}$$

$$B = \begin{bmatrix} 0.7 & 0.3 & 1.2 & -0.8 \\ 1.2 & 1.7 & 3.5 & 4.2 \\ -3.3 & -1.2 & 4.2 & 3.2 \end{bmatrix}$$

$$C = \begin{bmatrix} 0.8 & 7.1 & 6.2 \\ 3.3 & -1.2 & 4.8 \\ 1.3 & 2.8 & -1.5 \\ 2.1 & 3.2 & -8.4 \end{bmatrix}$$

1. AC
2. CB
3. $(A + B)C$
4. $(2A + 3B)C$
5. $(2A - 3.1B)C$
6. $C(2.1A + 3.2B)$
7. $(AC)^T$
8. $(CA)^T$

In exercises 9–12, refer to the following matrices and use a graphing utility to perform the indicated operations. Express your answers accurate to two decimal places.

$$A = \begin{bmatrix} 2 & 5 & -4 & 2 & 8 \\ 6 & 7 & 2 & 9 & 6 \\ 4 & 5 & 4 & 4 & 4 \\ 9 & 6 & 8 & 3 & 2 \end{bmatrix}$$

$$B = \begin{bmatrix} 2 & 6 & 7 & 5 \\ 3 & 4 & 6 & 2 \\ -5 & 8 & 4 & 3 \\ 8 & 6 & 9 & 5 \\ 4 & 7 & 8 & 8 \end{bmatrix}$$

$$C = \begin{bmatrix} 6.2 & 7.3 & -4.0 & 7.1 & 9.3 \\ 4.8 & 6.5 & 8.4 & -6.3 & 8.4 \\ 5.4 & 3.2 & 6.3 & 9.1 & -2.8 \\ 8.2 & 7.3 & 6.5 & 4.1 & 9.8 \\ 10.3 & 6.8 & 4.8 & -9.1 & 20.4 \end{bmatrix}$$

$$D = \begin{bmatrix} 4.6 & 3.9 & 8.4 & 6.1 & 9.8 \\ 2.4 & -6.8 & 7.9 & 11.4 & 2.9 \\ 7.1 & 9.4 & 6.3 & 5.7 & 4.2 \\ 3.4 & 6.1 & 5.3 & 8.4 & 6.3 \\ 7.1 & -4.2 & 3.9 & -6.4 & 7.1 \end{bmatrix}$$

9. Find AB and BA.
10. Find CD and DC. Is $CD = DC$?
11. Find $AC + AD$.
12. Find
    a. AC
    b. AD
    c. $A(C + D)$
    d. Is $A(C + D) = AC + AD$?

137

45. College Admissions A university admissions committee anticipates an enrollment of 8000 students in its freshman class next year. To satisfy admission quotas, incoming students have been categorized according to their sex and place of residence. The number of students in each category is given by the matrix

$$A = \begin{array}{c} \text{In-state} \\ \text{Out-of-state} \\ \text{Foreign} \end{array} \overset{\begin{array}{cc} \text{Male} & \text{Female} \end{array}}{\begin{bmatrix} 2700 & 3000 \\ 800 & 700 \\ 500 & 300 \end{bmatrix}}$$

By using data accumulated in previous years, the admissions committee has determined that these students will elect to enter the College of Letters and Science, the College of Fine Arts, the School of Business Administration, and the School of Engineering according to the percentages that appear in the following matrix:

$$B = \begin{array}{c} \text{Male} \\ \text{Female} \end{array} \overset{\begin{array}{cccc} \text{L. \& S.} & \text{Fine Arts} & \text{Bus. Ad.} & \text{Eng.} \end{array}}{\begin{bmatrix} 0.25 & 0.20 & 0.30 & 0.25 \\ 0.30 & 0.35 & 0.25 & 0.10 \end{bmatrix}}$$

Find the matrix AB that shows the number of in-state, out-of-state, and foreign students expected to enter each discipline.

46. Production Planning Refer to Example 6 in this section. Suppose that Ace Novelty received an order from another amusement park for 1200 Pink Panthers, 1800 Giant Pandas, and 1400 Big Birds. The quantity of each type of stuffed animal to be produced at each plant is shown in the following production matrix:

$$P = \begin{array}{c} \text{L.A.} \\ \text{Seattle} \end{array} \overset{\begin{array}{ccc} \text{Panthers} & \text{Pandas} & \text{Birds} \end{array}}{\begin{bmatrix} 700 & 1000 & 800 \\ 500 & 800 & 600 \end{bmatrix}}$$

Each Panther requires 1.3 square yards of plush, 20 cubic feet of stuffing, and 12 pieces of trim. If the materials required to produce the other two stuffed animals and the unit cost for each type of material are the same as those given in Example 6,
a. find how much of each type of material must be purchased for each plant.
b. find the total cost of materials that will be incurred at each plant.
c. find the total cost of materials incurred by Ace Novelty in filling the order.

47. Production Planning The total output of loudspeaker systems of the Acrosonic Company in their three production facilities for May and June is given by the matrices A and B, respectively, where

$$A = \begin{array}{c} \text{Location I} \\ \text{Location II} \\ \text{Location II} \end{array} \overset{\begin{array}{cccc} \text{Model} & \text{Model} & \text{Model} & \text{Model} \\ \text{A} & \text{B} & \text{C} & \text{D} \end{array}}{\begin{bmatrix} 320 & 280 & 460 & 280 \\ 480 & 360 & 580 & 0 \\ 540 & 420 & 200 & 880 \end{bmatrix}}$$

$$B = \begin{array}{c} \text{Location I} \\ \text{Location II} \\ \text{Location III} \end{array} \overset{\begin{array}{cccc} \text{Model} & \text{Model} & \text{Model} & \text{Model} \\ \text{A} & \text{B} & \text{C} & \text{D} \end{array}}{\begin{bmatrix} 210 & 180 & 330 & 180 \\ 400 & 300 & 450 & 40 \\ 420 & 280 & 180 & 740 \end{bmatrix}}$$

The unit production costs and selling prices for these loudspeakers are given by matrices C and D, respectively, where

$$C = \begin{array}{c} \text{Model A} \\ \text{Model B} \\ \text{Model C} \\ \text{Model D} \end{array} \begin{bmatrix} 120 \\ 180 \\ 260 \\ 500 \end{bmatrix} \quad \text{and} \quad D = \begin{array}{c} \text{Model A} \\ \text{Model B} \\ \text{Model C} \\ \text{Model D} \end{array} \begin{bmatrix} 160 \\ 250 \\ 350 \\ 700 \end{bmatrix}$$

Compute the following matrices and explain the meaning of the entries in each matrix.
a. AC **b.** AD **c.** BC **d.** BD
e. $(A + B)C$ **f.** $(A + B)D$
g. $A(D - C)$ **h.** $B(D - C)$
i. $(A + B)(D - C)$

48. Diet Planning A dietician plans a meal around three foods. The number of units of vitamin A, vitamin C, and calcium in each ounce of these foods is represented by the matrix M, where

$$M = \begin{array}{c} \text{Vitamin A} \\ \text{Vitamin C} \\ \text{Calcium} \end{array} \overset{\begin{array}{ccc} \text{Food I} & \text{Food II} & \text{Food III} \end{array}}{\begin{bmatrix} 400 & 1200 & 800 \\ 110 & 570 & 340 \\ 90 & 30 & 60 \end{bmatrix}}$$

The matrices A and B represent the amount of each food (in ounces) consumed by a girl at two different meals, where

$$A = \overset{\begin{array}{ccc} \text{Food I} & \text{Food II} & \text{Food III} \end{array}}{\begin{bmatrix} 7 & 1 & 6 \end{bmatrix}} \quad \text{and}$$

$$B = \overset{\begin{array}{ccc} \text{Food I} & \text{Food II} & \text{Food III} \end{array}}{\begin{bmatrix} 9 & 3 & 2 \end{bmatrix}}$$

Calculate the following matrices and explain the meaning of the entries in each matrix:
a. MA^T **b.** MB^T **c.** $M(A + B)^T$

SOLUTIONS TO SELF-CHECK EXERCISES 2.5

1. We compute

$$\begin{bmatrix} 1 & 3 & 0 \\ 2 & 4 & -1 \end{bmatrix} \begin{bmatrix} 3 & 1 & 4 \\ 2 & 0 & 3 \\ 1 & 2 & -1 \end{bmatrix}$$

$$= \begin{bmatrix} 1(3) + 3(2) + 0(1) & 1(1) + 3(0) + 0(2) & 1(4) + 3(3) + 0(-1) \\ 2(3) + 4(2) - 1(1) & 2(1) + 4(0) - 1(2) & 2(4) + 4(3) - 1(-1) \end{bmatrix}$$

$$= \begin{bmatrix} 9 & 1 & 13 \\ 13 & 0 & 21 \end{bmatrix}$$

2. Let

$$A = \begin{bmatrix} 0 & 1 & -2 \\ 2 & -1 & 3 \\ 1 & 0 & 4 \end{bmatrix} \qquad X = \begin{bmatrix} x \\ y \\ z \end{bmatrix} \qquad \text{and} \qquad B = \begin{bmatrix} 1 \\ 0 \\ 7 \end{bmatrix}$$

Then the given system may be written as the matrix equation

$$AX = B$$

3. Write

$$B = \begin{bmatrix} 24 \\ 47 \\ 150 \\ 14 \end{bmatrix} \begin{matrix} \text{AT\&T} \\ \text{GTE} \\ \text{IBM} \\ \text{GM} \end{matrix}$$

and compute

$$AB = \begin{matrix} \text{Ash} \\ \text{Joan} \end{matrix} \begin{bmatrix} 2000 & 1000 & 500 & 5000 \\ 1000 & 500 & 2000 & 0 \end{bmatrix} \begin{bmatrix} 24 \\ 47 \\ 150 \\ 14 \end{bmatrix}$$

$$= \begin{bmatrix} 240,000 \\ 347,500 \end{bmatrix} \begin{matrix} \text{Ash} \\ \text{Joan} \end{matrix}$$

We conclude that Ash's stock holdings were worth \$240,000 and Joan's stock holdings were worth \$347,500 on July 1.

2.6 THE INVERSE OF A SQUARE MATRIX

The Inverse of a Square Matrix

In this section, we discuss a procedure for finding the inverse of a matrix and show how the inverse can be used to help us solve a system of linear equations. The inverse of a matrix also plays a central role in the Leontief input–output model, which we will discuss in Section 2.7.

Recall that if a is a nonzero real number, then there exists a unique real number a^{-1} (that is, $1/a$) such that

$$a^{-1}a = \left(\frac{1}{a}\right)(a) = 1$$

The use of the inverse of a real number enables us to solve algebraic equations of the form

$$ax = b \tag{12}$$

For if $a \neq 0$, then $a^{-1} = 1/a$. Upon multiplying both sides of (12) by a^{-1}, we have

$$a^{-1}(ax) = a^{-1}b$$

or

$$\left(\frac{1}{a}\right)(ax) = \frac{1}{a}(b)$$

and

$$x = \frac{b}{a}$$

For example, since the inverse of 2 is $2^{-1} = 1/2$, we can solve the equation

$$2x = 5$$

by multiplying both sides of the equation by $2^{-1} = 1/2$, giving

$$2^{-1}(2x) = 2^{-1} \cdot 5$$

or

$$x = \frac{5}{2}$$

We can use a similar procedure to solve the matrix equation

$$AX = B$$

where A, X, and B are matrices of the proper sizes. In order to do this, we need the matrix equivalent of the inverse of a real number. Such a matrix, whenever it exists, is called the *inverse of a matrix*. More specifically, we have the following definition.

INVERSE OF A MATRIX

Let A be a square matrix of size n. A square matrix A^{-1} of size n such that

$$A^{-1}A = AA^{-1} = I_n$$

is called the **inverse** of A.

In defining the inverse of a matrix A, why is it necessary to require that A be a square matrix?

Let us show that the matrix

$$A = \begin{bmatrix} 1 & 2 \\ 3 & 4 \end{bmatrix}$$

has as its inverse

$$A^{-1} = \begin{bmatrix} -2 & 1 \\ \frac{3}{2} & -\frac{1}{2} \end{bmatrix}$$

Since

$$AA^{-1} = \begin{bmatrix} 1 & 2 \\ 3 & 4 \end{bmatrix} \begin{bmatrix} -2 & 1 \\ \frac{3}{2} & -\frac{1}{2} \end{bmatrix} = \begin{bmatrix} 1 & 0 \\ 0 & 1 \end{bmatrix}$$

and

$$A^{-1}A = \begin{bmatrix} -2 & 1 \\ \frac{3}{2} & -\frac{1}{2} \end{bmatrix} \begin{bmatrix} 1 & 2 \\ 3 & 4 \end{bmatrix} = \begin{bmatrix} 1 & 0 \\ 0 & 1 \end{bmatrix} = I$$

we see that A^{-1} is the inverse of A, as asserted.

Not every square matrix has an inverse. A matrix that does not have an inverse is called **singular.** An example of a singular matrix is given by

$$B = \begin{bmatrix} 0 & 1 \\ 0 & 0 \end{bmatrix}$$

If B had an inverse given by

$$B^{-1} = \begin{bmatrix} a & b \\ c & d \end{bmatrix}$$

where a, b, c, and d are some appropriate numbers, then, by the definition of an inverse, we would have $BB^{-1} = I$, that is,

$$\begin{bmatrix} 0 & 1 \\ 0 & 0 \end{bmatrix} \begin{bmatrix} a & b \\ c & d \end{bmatrix} = \begin{bmatrix} 1 & 0 \\ 0 & 1 \end{bmatrix}$$

or

$$\begin{bmatrix} c & d \\ 0 & 0 \end{bmatrix} = \begin{bmatrix} 1 & 0 \\ 0 & 1 \end{bmatrix}$$

which implies that $0 = 1$—an impossibility! This contradiction shows that B does not have an inverse.

A Method for Finding the Inverse of a Square Matrix

The methods of Section 2.5 can be used to find the inverse of a nonsingular matrix. In order to discover such an algorithm, let us find the inverse of the matrix A, given by

$$A = \begin{bmatrix} 1 & 2 \\ -1 & 3 \end{bmatrix}$$

Suppose A^{-1} exists and is given by

$$A^{-1} = \begin{bmatrix} a & b \\ c & d \end{bmatrix}$$

where a, b, c, and d are to be determined. By the definition of an inverse, we have $AA^{-1} = I$, that is,

$$\begin{bmatrix} 1 & 2 \\ -1 & 3 \end{bmatrix} \begin{bmatrix} a & b \\ c & d \end{bmatrix} = \begin{bmatrix} 1 & 0 \\ 0 & 1 \end{bmatrix}$$

which simplifies to

$$\begin{bmatrix} a + 2c & b + 2d \\ -a + 3c & -b + 3d \end{bmatrix} = \begin{bmatrix} 1 & 0 \\ 0 & 1 \end{bmatrix}$$

But this matrix equation is equivalent to the two systems of linear equations

$$\left. \begin{array}{r} a + 2c = 1 \\ -a + 3c = 0 \end{array} \right\} \quad \text{and} \quad \left. \begin{array}{r} b + 2d = 0 \\ -b + 3d = 1 \end{array} \right\}$$

with augmented matrices given by

$$\begin{bmatrix} 1 & 2 & | & 1 \\ -1 & 3 & | & 0 \end{bmatrix} \quad \text{and} \quad \begin{bmatrix} 1 & 2 & | & 0 \\ -1 & 3 & | & 1 \end{bmatrix}$$

Note that the matrices of coefficients of the two systems are identical. This suggests that we solve the two systems of simultaneous linear equations by writing the following augmented matrix, which we obtain by joining the coefficient matrix and the two columns of constants:

$$\begin{bmatrix} 1 & 2 & | & 1 & 0 \\ -1 & 3 & | & 0 & 1 \end{bmatrix}$$

Using the Gauss-Jordan elimination method, we obtain the following sequence of equivalent matrices:

$$\begin{bmatrix} 1 & 2 & | & 1 & 0 \\ -1 & 3 & | & 0 & 1 \end{bmatrix} \xrightarrow{R_2 + R_1} \begin{bmatrix} 1 & 2 & | & 1 & 0 \\ 0 & 5 & | & 1 & 1 \end{bmatrix} \xrightarrow{\frac{1}{5}R_2}$$

$$\begin{bmatrix} 1 & 2 & | & 1 & 0 \\ 0 & 1 & | & \frac{1}{5} & \frac{1}{5} \end{bmatrix} \xrightarrow{R_1 - 2R_2} \begin{bmatrix} 1 & 0 & | & \frac{3}{5} & -\frac{2}{5} \\ 0 & 1 & | & \frac{1}{5} & \frac{1}{5} \end{bmatrix}$$

Thus, $a = 3/5$, $c = 1/5$, $b = -2/5$, and $d = 1/5$, giving

$$A^{-1} = \begin{bmatrix} \frac{3}{5} & -\frac{2}{5} \\ \frac{1}{5} & \frac{1}{5} \end{bmatrix}$$

The following computations verify that A^{-1} is indeed the inverse of A:

$$\begin{bmatrix} 1 & 2 \\ -1 & 3 \end{bmatrix} \begin{bmatrix} \frac{3}{5} & -\frac{2}{5} \\ \frac{1}{5} & \frac{1}{5} \end{bmatrix} = \begin{bmatrix} 1 & 0 \\ 0 & 1 \end{bmatrix} = \begin{bmatrix} \frac{3}{5} & -\frac{2}{5} \\ \frac{1}{5} & \frac{1}{5} \end{bmatrix} \begin{bmatrix} 1 & 2 \\ -1 & 3 \end{bmatrix}$$

The preceding example suggests the general algorithm for computing the inverse of a square matrix of size n when it exists.

FINDING THE INVERSE OF A MATRIX

Given the $n \times n$ matrix A:

1. Adjoin the $n \times n$ identity matrix I to obtain the augmented matrix

$$[A \mid I]$$

2. Use a sequence of row operations to reduce $[A \mid I]$ to the form

$$[I \mid B]$$

if possible.

The matrix B is the inverse of A.

REMARK   Although matrix multiplication is not generally commutative, it is possible to prove that if $AB = I$, then $BA = I$ also. Hence, to verify that B is the inverse of A, it suffices to show that $AB = I$.   ◦ ◦ ◦

EXAMPLE 1   Find the inverse of the matrix

$$A = \begin{bmatrix} 2 & 1 & 1 \\ 3 & 2 & 1 \\ 2 & 1 & 2 \end{bmatrix}$$

Solution   We form the augmented matrix

$$\begin{bmatrix} 2 & 1 & 1 & | & 1 & 0 & 0 \\ 3 & 2 & 1 & | & 0 & 1 & 0 \\ 2 & 1 & 2 & | & 0 & 0 & 1 \end{bmatrix}$$

and use the Gauss-Jordan elimination method to reduce it to the form $[I \mid B]$:

$$\begin{bmatrix} 2 & 1 & 1 & | & 1 & 0 & 0 \\ 3 & 2 & 1 & | & 0 & 1 & 0 \\ 2 & 1 & 2 & | & 0 & 0 & 1 \end{bmatrix} \xrightarrow{R_1 - R_2} \begin{bmatrix} -1 & -1 & 0 & | & 1 & -1 & 0 \\ 3 & 2 & 1 & | & 0 & 1 & 0 \\ 2 & 1 & 2 & | & 0 & 0 & 1 \end{bmatrix}$$

$$\xrightarrow[\substack{R_2 + 3R_1 \\ R_3 + 2R_1}]{-R_1} \begin{bmatrix} 1 & 1 & 0 & | & -1 & 1 & 0 \\ 0 & -1 & 1 & | & 3 & -2 & 0 \\ 0 & -1 & 2 & | & 2 & -2 & 1 \end{bmatrix}$$

$$\xrightarrow[\substack{-R_2 \\ R_3 - R_2}]{R_1 + R_2} \begin{bmatrix} 1 & 0 & 1 & | & 2 & -1 & 0 \\ 0 & 1 & -1 & | & -3 & 2 & 0 \\ 0 & 0 & 1 & | & -1 & 0 & 1 \end{bmatrix}$$

$$\xrightarrow[R_2 + R_3]{R_1 - R_3} \begin{bmatrix} 1 & 0 & 0 & | & 3 & -1 & -1 \\ 0 & 1 & 0 & | & -4 & 2 & 1 \\ 0 & 0 & 1 & | & -1 & 0 & 1 \end{bmatrix}$$

The inverse of A is the matrix

$$A^{-1} = \begin{bmatrix} 3 & -1 & -1 \\ -4 & 2 & 1 \\ -1 & 0 & 1 \end{bmatrix}$$

We leave it to you to verify these results.   ○ ○ ○

Example 2 illustrates what happens to the reduction process when a matrix A does not have an inverse.

EXAMPLE 2   Find the inverse of the matrix

$$A = \begin{bmatrix} 1 & 2 & 3 \\ 2 & 1 & 2 \\ 3 & 3 & 5 \end{bmatrix}$$

Solution   We form the augmented matrix

$$\begin{bmatrix} 1 & 2 & 3 & | & 1 & 0 & 0 \\ 2 & 1 & 2 & | & 0 & 1 & 0 \\ 3 & 3 & 5 & | & 0 & 0 & 1 \end{bmatrix}$$

and use the Gauss-Jordan elimination method:

$$\begin{bmatrix} 1 & 2 & 3 & | & 1 & 0 & 0 \\ 2 & 1 & 2 & | & 0 & 1 & 0 \\ 3 & 3 & 5 & | & 0 & 0 & 1 \end{bmatrix} \xrightarrow[R_3 - 3R_1]{R_2 - 2R_1} \begin{bmatrix} 1 & 2 & 3 & | & 1 & 0 & 0 \\ 0 & -3 & -4 & | & -2 & 1 & 0 \\ 0 & -3 & -4 & | & -3 & 0 & 1 \end{bmatrix}$$

$$\xrightarrow[R_3 - R_2]{-R_2} \begin{bmatrix} 1 & 2 & 3 & | & 1 & 0 & 0 \\ 0 & 3 & 4 & | & 2 & -1 & 0 \\ 0 & 0 & 0 & | & -1 & -1 & 1 \end{bmatrix}$$

Since all entries in the last row of the 3×3 submatrix that comprises the left-hand side of the augmented matrix just obtained are all equal to zero, the latter cannot be reduced to the form $[I \mid B]$. Accordingly, we draw the conclusion that A is singular—that is, does not have an inverse.   ◦ ◦ ◦

More generally, we have the following criterion for determining when the inverse of a matrix does not exist.

MATRICES THAT HAVE NO INVERSES   If there is a row to the left of the vertical line in the augmented matrix containing all zeros, then the matrix does not have an inverse.

A Formula for the Inverse of a 2 × 2 Matrix

Before turning to some applications, we will show an alternative method that employs a formula for finding the inverse of a 2×2 matrix. This method will prove useful in many situations—we will see an application in Example 5. The derivation of this formula is left as an exercise (exercise 40).

FORMULA FOR THE INVERSE OF A 2 × 2 MATRIX   Let

$$A = \begin{bmatrix} a & b \\ c & d \end{bmatrix}$$

Suppose $D = ad - bc$ is not equal to zero. Then A^{-1} exists and is given by

$$A^{-1} = \frac{1}{D}\begin{bmatrix} d & -b \\ -c & a \end{bmatrix} \tag{13}$$

REMARK   As an aid to memorizing the formula, note that D is the product of the elements along the main diagonal minus the product of the elements along the other diagonal:

$$\begin{bmatrix} a & b \\ c & d \end{bmatrix} \qquad D = ad - bc$$

main diagonal

Suppose A is a square matrix with the property that one of its rows is a nonzero constant multiple of another row. What can you say about the existence or nonexistence of A^{-1}? Explain your answer.

Next, the matrix

$$\begin{bmatrix} d & -b \\ -c & a \end{bmatrix}$$

is obtained by interchanging a and d and reversing the signs of b and c. Finally, A^{-1} is obtained by dividing this matrix by D.   ◦ ◦ ◦

EXAMPLE 3   Find the inverse of

$$A = \begin{bmatrix} 1 & 2 \\ 3 & 4 \end{bmatrix}$$

Solution   We first compute $D = (1)(4) - (3)(2) = 4 - 6 = -2$. Next, we write the matrix

$$\begin{bmatrix} 4 & -2 \\ -3 & 1 \end{bmatrix}$$

Finally, dividing this matrix by D, we obtain

$$A^{-1} = \frac{1}{-2} \begin{bmatrix} 4 & -2 \\ -3 & 1 \end{bmatrix} = \begin{bmatrix} -2 & 1 \\ \frac{3}{2} & -\frac{1}{2} \end{bmatrix}$$ ○ ○ ○

Solving Systems of Equations with Inverses

We now show how the inverse of a matrix may be used to solve certain systems of linear equations in which the number of equations in the system is equal to the number of variables. For simplicity, let us illustrate the process for a system of three linear equations in three variables:

$$
\begin{aligned}
a_{11}x_1 + a_{12}x_2 + a_{13}x_3 &= c_1 \\
a_{21}x_1 + a_{22}x_2 + a_{23}x_3 &= c_2 \\
a_{31}x_1 + a_{32}x_2 + a_{33}x_3 &= c_3
\end{aligned}
\qquad (14)
$$

Let us write

$$A = \begin{bmatrix} a_{11} & a_{12} & a_{13} \\ a_{21} & a_{22} & a_{23} \\ a_{31} & a_{32} & a_{33} \end{bmatrix} \qquad X = \begin{bmatrix} x_1 \\ x_2 \\ x_3 \end{bmatrix} \quad \text{and} \quad B = \begin{bmatrix} b_1 \\ b_2 \\ b_3 \end{bmatrix}$$

You should verify that the system (14) of linear equations may be written in the form of the matrix equation

$$AX = B \qquad (15)$$

If A is nonsingular, then the method of this section may be used to compute A^{-1}. Next, multiplying both sides of equation (15) by A^{-1} (on the left), we obtain

$$A^{-1}AX = A^{-1}B \quad \text{or} \quad IX = A^{-1}B \quad \text{or} \quad X = A^{-1}B$$

the desired solution to the problem.

In the case of a system of n equations with n unknowns, we have the following, more general result.

USING INVERSES TO SOLVE SYSTEMS OF EQUATIONS	If $AX = B$ is a linear system of n equations in n unknowns, and if A^{-1} exists, then $$X = A^{-1}B$$ is the unique solution of the system.

The use of inverses to solve systems of equations is particularly advantageous when we are required to solve more than one system of equations, $AX = B$, involving the same coefficient matrix, A, and different matrices of constants, B. As you will see in Examples 4 and 5, we need to compute A^{-1} just once in each case.

Applications

EXAMPLE 4   Solve the following systems of linear equations:

a. $\begin{aligned} 2x + y + z &= 1 \\ 3x + 2y + z &= 2 \\ 2x + y + 2z &= -1 \end{aligned}$ **b.** $\begin{aligned} 2x + y + z &= 2 \\ 3x + 2y + z &= -3 \\ 2x + y + 2z &= 1 \end{aligned}$

Solution   We may write the given systems of equations in the form

$$AX = B \quad \text{and} \quad AX = C$$

respectively, where

$$A = \begin{bmatrix} 2 & 1 & 1 \\ 3 & 2 & 1 \\ 2 & 1 & 2 \end{bmatrix} \quad X = \begin{bmatrix} x \\ y \\ z \end{bmatrix} \quad B = \begin{bmatrix} 1 \\ 2 \\ -1 \end{bmatrix} \quad \text{and} \quad C = \begin{bmatrix} 2 \\ -3 \\ 1 \end{bmatrix}$$

The inverse of the matrix A,

$$A^{-1} = \begin{bmatrix} 3 & -1 & -1 \\ -4 & 2 & 1 \\ -1 & 0 & 1 \end{bmatrix}$$

was found in Example 1. Using this result, we find that the solution of the first system is

$$X = A^{-1}B = \begin{bmatrix} 3 & -1 & -1 \\ -4 & 2 & 1 \\ -1 & 0 & 1 \end{bmatrix} \begin{bmatrix} 1 \\ 2 \\ -1 \end{bmatrix}$$

$$= \begin{bmatrix} (3)(1) + (-1)(2) + (-1)(-1) \\ (-4)(1) + (2)(2) + (1)(-1) \\ (-1)(1) + (0)(2) + (1)(-1) \end{bmatrix} = \begin{bmatrix} 2 \\ -1 \\ -2 \end{bmatrix}$$

or $x = 2$, $y = -1$, and $z = -2$.

The solution of the second system is

$$X = A^{-1}C = \begin{bmatrix} 3 & -1 & -1 \\ -4 & 2 & 1 \\ -1 & 0 & 1 \end{bmatrix} \begin{bmatrix} 2 \\ -3 \\ 1 \end{bmatrix} = \begin{bmatrix} 8 \\ -13 \\ -1 \end{bmatrix}$$

or $x = 8$, $y = -13$, and $z = -1$.   ◐ ◐ ◐

EXAMPLE 5   The management of Checkers Rent-A-Car plans to expand its fleet of rental cars for the next quarter by purchasing compact and full-size cars. The average cost of a compact is $10,000, and the average cost of a full-size car is $24,000.

a. If a total of 800 cars is to be purchased with a budget of $12 million, how many cars of each size will be acquired?

b. If the predicted demand calls for a total purchase of 1000 cars with a budget of $14 million, how many cars of each type will be acquired?

Solution   Let x and y denote the number of compact and full-size cars to be purchased. Furthermore, let n denote the total number of cars to be acquired and b the amount of money budgeted for the purchase of these cars. Then

$$x + y = n$$
$$10,000x + 24,000y = b$$

This system of two equations in two variables may be written in the matrix form

$$AX = B$$

where

$$A = \begin{bmatrix} 1 & 1 \\ 10,000 & 24,000 \end{bmatrix} \qquad X = \begin{bmatrix} x \\ y \end{bmatrix} \qquad \text{and} \qquad B = \begin{bmatrix} n \\ b \end{bmatrix}$$

Therefore,

$$X = A^{-1}B$$

Since A is a 2×2 matrix, its inverse may be found by using formula (13). We find $D = (1)(24,000) - (10,000)(1) = 14,000$, so

$$A^{-1} = \frac{1}{14,000} \begin{bmatrix} 24,000 & -1 \\ -10,000 & 1 \end{bmatrix} = \begin{bmatrix} \frac{24,000}{14,000} & -\frac{1}{14,000} \\ -\frac{10,000}{14,000} & \frac{1}{14,000} \end{bmatrix}$$

Thus

$$X = \begin{bmatrix} \frac{12}{7} & -\frac{1}{14,000} \\ -\frac{5}{7} & \frac{1}{14,000} \end{bmatrix} \begin{bmatrix} n \\ b \end{bmatrix}$$

a. Here $n = 800$ and $b = 12,000,000$, so

$$X = A^{-1}B = \begin{bmatrix} \frac{12}{7} & -\frac{1}{14,000} \\ -\frac{5}{7} & \frac{1}{14,000} \end{bmatrix} \begin{bmatrix} 800 \\ 12,000,000 \end{bmatrix} = \begin{bmatrix} 514.3 \\ 285.7 \end{bmatrix}$$

Therefore, 514 compact cars and 286 full-size cars will be acquired in this case.

b. Here $n = 1000$ and $b = 14,000,000$, so

$$X = A^{-1}B = \begin{bmatrix} \frac{12}{7} & -\frac{1}{14,000} \\ -\frac{5}{7} & \frac{1}{14,000} \end{bmatrix} \begin{bmatrix} 1000 \\ 14,000,000 \end{bmatrix} = \begin{bmatrix} 714.3 \\ 285.7 \end{bmatrix}$$

Therefore, 714 compact cars and 286 full-size cars will be purchased in this case.

○ ○ ○

SELF-CHECK EXERCISES 2.6

1. Find the inverse of the matrix

$$A = \begin{bmatrix} 2 & 1 & -1 \\ 1 & 1 & -1 \\ -1 & -2 & 3 \end{bmatrix}$$

if it exists.

2. Solve the system of linear equations

$$\begin{aligned} 2x + y - z &= b_1 \\ x + y - z &= b_2 \\ -x - 2y + 3z &= b_3 \end{aligned}$$

where (a) $b_1 = 5$, $b_2 = 4$, $b_3 = -8$ and (b) $b_1 = 2$, $b_2 = 0$, $b_3 = 5$, by finding the inverse of the coefficient matrix.

3. Grand Canyon Tours, Inc., offers air and ground scenic tours of the Grand Canyon. Tickets for the $7\frac{1}{2}$-hour tour cost $169 for an adult and $129 for a child, and each tour group is limited to 19 people. On three recent fully booked tours, total receipts were $2931 for the first tour, $3011 for the second tour, and $2771 for the third tour. Determine how many adults and how many children were in each tour.

Solutions to Self-Check Exercises 2.6 can be found on page 154.

2.6 EXERCISES

In exercises 1–4, show that the given matrices are inverses of each other by showing that their product is the identity matrix I.

1. $\begin{bmatrix} 1 & -3 \\ 1 & -2 \end{bmatrix}$ and $\begin{bmatrix} -2 & 3 \\ -1 & 1 \end{bmatrix}$

2. $\begin{bmatrix} 4 & 5 \\ 2 & 3 \end{bmatrix}$ and $\begin{bmatrix} \frac{3}{2} & -\frac{5}{2} \\ -1 & 2 \end{bmatrix}$

3. $\begin{bmatrix} 3 & 2 & 3 \\ 2 & 2 & 1 \\ 2 & 1 & 1 \end{bmatrix}$ and $\begin{bmatrix} -\frac{1}{3} & -\frac{1}{3} & \frac{4}{3} \\ 0 & 1 & -1 \\ \frac{2}{3} & -\frac{1}{3} & -\frac{2}{3} \end{bmatrix}$

4. $\begin{bmatrix} 2 & 4 & -2 \\ -4 & -6 & 1 \\ 3 & 5 & -1 \end{bmatrix}$ and $\begin{bmatrix} \frac{1}{2} & -3 & -4 \\ -\frac{1}{2} & 2 & 3 \\ -1 & 1 & 2 \end{bmatrix}$

In exercises 5–16, find the inverse of the given matrix, if it exists. Verify your answer.

5. $\begin{bmatrix} 2 & 5 \\ 1 & 3 \end{bmatrix}$

6. $\begin{bmatrix} 2 & 3 \\ 3 & 5 \end{bmatrix}$

7. $\begin{bmatrix} 3 & -3 \\ -2 & 2 \end{bmatrix}$

8. $\begin{bmatrix} 4 & 2 \\ 6 & 3 \end{bmatrix}$

9. $\begin{bmatrix} 2 & -3 & -4 \\ 0 & 0 & -1 \\ 1 & -2 & 1 \end{bmatrix}$

10. $\begin{bmatrix} 1 & -1 & 3 \\ 2 & 1 & 2 \\ -2 & -2 & 1 \end{bmatrix}$

11. $\begin{bmatrix} 4 & 2 & 2 \\ -1 & -3 & 4 \\ 3 & -1 & 6 \end{bmatrix}$

12. $\begin{bmatrix} 1 & 2 & 0 \\ -3 & 4 & -2 \\ -5 & 0 & -2 \end{bmatrix}$

13. $\begin{bmatrix} 1 & 4 & -1 \\ 2 & 3 & -2 \\ -1 & 2 & 3 \end{bmatrix}$

14. $\begin{bmatrix} 3 & -2 & 7 \\ -2 & 1 & 4 \\ 6 & -5 & 8 \end{bmatrix}$

15. $\begin{bmatrix} 1 & 1 & -1 & 1 \\ 2 & 1 & 1 & 0 \\ 2 & 1 & 0 & 1 \\ 2 & -1 & -1 & 3 \end{bmatrix}$

16. $\begin{bmatrix} 1 & 1 & 2 & 3 \\ 2 & 3 & 0 & -1 \\ 0 & 2 & -1 & 1 \\ 1 & 2 & 1 & 1 \end{bmatrix}$

In exercises 17–24, (a) write a matrix equation that is equivalent to the given system of linear equations and (b) solve the system using the inverses found in exercises 5–16.

17. $2x + 5y = 3$
$x + 3y = 2$
(See exercise 5.)

18. $2x + 3y = 5$
$3x + 5y = 8$
(See exercise 6.)

19. $2x - 3y - 4z = 4$
$-z = 3$
$x - 2y + z = -8$
(See exercise 9.)

20. $x_1 - x_2 + 3x_3 = 2$
$2x_1 + x_2 + 2x_3 = 2$
$-2x_1 - 2x_2 + x_3 = 3$
(See exercise 10.)

21. $x + 4y - z = 3$
$2x + 3y - 2z = 1$
$-x + 2y + 3z = 7$
(See exercise 13.)

22. $3x_1 - 2x_2 + 7x_3 = 6$
$-2x_1 + x_2 + 4x_3 = 4$
$6x_1 - 5x_2 + 8x_3 = 4$
(See exercise 14.)

23. $x_1 + x_2 - x_3 + x_4 = 6$
$2x_1 + x_2 + x_3 = 4$
$2x_1 + x_2 + x_4 = 7$
$2x_1 - x_2 - x_3 + 3x_4 = 9$
(See exercise 15.)

24. $x_1 + x_2 + 2x_3 + 3x_4 = 4$
$2x_1 + 3x_2 - x_4 = 11$
$2x_2 - x_3 + x_4 = 7$
$x_1 + 2x_2 + x_3 + x_4 = 6$
(See exercise 16.)

In exercises 25–32, (a) write each system of equations as a matrix equation and (b) solve the system of equations by using the inverse of the coefficient matrix.

25. $x + 2y = b_1$
$2x - y = b_2$
where   i. $b_1 = 14, b_2 = 5$
and   ii. $b_1 = 4, b_2 = -1$

26. $3x - 2y = b_1$
$4x + 3y = b_2$
where   i. $b_1 = -6, b_2 = 10$
and   ii. $b_1 = 3, b_2 = -2$

27. $x + 2y + z = b_1$
$x + y + z = b_2$
$3x + y + z = b_3$
where   i. $b_1 = 7, b_2 = 4, b_3 = 2$
and   ii. $b_1 = 5, b_2 = -3, b_3 = -1$

28. $x_1 + x_2 + x_3 = b_1$
$x_1 - x_2 + x_3 = b_2$
$x_1 - 2x_2 - x_3 = b_3$
where   i. $b_1 = 5, b_2 = -3, b_3 = -1$
and   ii. $b_1 = 1, b_2 = 4, b_3 = -2$

29. $3x + 2y - z = b_1$
$2x - 3y + z = b_2$
$x - y - z = b_3$
where   i. $b_1 = 2, b_2 = -2, b_3 = 4$
and   ii. $b_1 = 8, b_2 = -3, b_3 = 6$

30. $2x_1 + x_2 + x_3 = b_1$
$x_1 - 3x_2 + 4x_3 = b_2$
$-x_1 + x_3 = b_3$
where   i. $b_1 = 1, b_2 = 4, b_3 = -3$
and   ii. $b_1 = 2, b_2 = -5, b_3 = 0$

31. $x_1 + x_2 + x_3 + x_4 = b_1$
$x_1 - x_2 - x_3 + x_4 = b_2$
$x_2 + 2x_3 + 2x_4 = b_3$
$x_1 + 2x_2 + x_3 - 2x_4 = b_4$
where   i. $b_1 = 1, b_2 = -1, b_3 = 4, b_4 = 0$
and   ii. $b_1 = 2, b_2 = 8, b_3 = 4, b_4 = -1$

32. $x_1 + x_2 + 2x_3 + x_4 = b_1$
$4x_1 + 5x_2 + 9x_3 + x_4 = b_2$
$3x_1 + 4x_2 + 7x_3 + x_4 = b_3$
$2x_1 + 3x_2 + 4x_3 + 2x_4 = b_4$
where   i. $b_1 = 3, b_2 = 6, b_3 = 5, b_4 = 7$
and   ii. $b_1 = 1, b_2 = -1, b_3 = 0, b_4 = -4$

33. Let

$$A = \begin{bmatrix} 2 & 3 \\ -4 & -5 \end{bmatrix}$$

a. Find A^{-1}.   **b.** Show that $(A^{-1})^{-1} = A$.

34. Let

$$A = \begin{bmatrix} 6 & -4 \\ -4 & 3 \end{bmatrix} \quad \text{and} \quad B = \begin{bmatrix} 3 & -5 \\ 4 & -7 \end{bmatrix}$$

a. Find AB, A^{-1}, B^{-1}.
b. Show that $(AB)^{-1} = B^{-1}A^{-1}$.

35. Let

$$A = \begin{bmatrix} 2 & -5 \\ 1 & -3 \end{bmatrix} \quad B = \begin{bmatrix} 4 & 3 \\ 1 & 1 \end{bmatrix}$$

and

$$C = \begin{bmatrix} 2 & 3 \\ -2 & 1 \end{bmatrix}$$

a. Find ABC, A^{-1}, B^{-1}, C^{-1}.
b. Show that $(ABC)^{-1} = C^{-1}B^{-1}A^{-1}$.

36. Ticket Revenues   Rainbow Harbor Cruises charges $8 per adult and $4 per child for a round-trip ticket. The records show that on a certain weekend, 1000 people took the cruise on Saturday and 800 people took the cruise on Sunday. The total receipts for Saturday were $6400, and the total receipts for Sunday were $4800. Determine how many adults and children took the cruise on Saturday and on Sunday.

37. Pricing   Bel Air Publishing Company publishes a deluxe leather edition and a standard edition of its Daily Organizer. The company's marketing department estimates that x copies of the deluxe edition and y copies of the standard edition will be demanded per month when the unit prices are p dollars and q dollars, respectively, where x, y, p, and q are related by the following system of linear equations:

$$5x + y = 1000(70 - p)$$
$$x + 3y = 1000(40 - q)$$

Find the monthly demand for the deluxe edition and the standard edition when the unit prices are set according to the following schedules:
a. $p = 50$ and $q = 25$ **b.** $p = 45$ and $q = 25$
c. $p = 45$ and $q = 20$

38. Nutrition/Diet Planning   Bob, a nutritionist attached to the University Medical Center, has been asked to prepare special diets for two patients, Susan and Tom. Bob has decided that Susan's meals should contain at least 400 mg of calcium, 20 mg of iron, and 50 mg of vitamin C, whereas Tom's meals should contain at least 350 mg of calcium, 15 mg of iron, and 40 mg of vitamin C. Bob has also decided that the meals are to be prepared from three basic foods: food A, food B, and food C. The special nutritional contents of these foods are summarized in the accompanying table. Find how many ounces of each type of food should be used in a meal so that the minimum requirements of calcium, iron, and vitamin C are met for each patient's meals.

	Contents in mg/oz		
	Calcium	Iron	Vitamin C
Food A	30	1	2
Food B	25	1	5
Food C	20	2	4

39. Research Funding   The Carver Foundation funds three nonprofit organizations engaged in alternate-energy research activities. From past data, the proportion of funds spent by each organization in research on solar energy, energy from harnessing the wind, and energy from the motion of ocean tides is given in the accompanying table.

	Proportion of Money Spent		
	Solar	Wind	Tides
Organization I	0.6	0.3	0.1
Organization II	0.4	0.3	0.3
Organization III	0.2	0.6	0.2

Find the amount awarded to each organization if the total amount spent by all three organizations on solar, wind, and tidal research is
a. $9.2 million, $9.6 million, and $5.2 million, respectively.
b. $8.2 million, $7.2 million, and $3.6 million, respectively.

40. Let

$$A = \begin{bmatrix} a & b \\ c & d \end{bmatrix}$$

a. Find A^{-1}.
b. Find the necessary condition for A to be nonsingular.
c. Verify that $AA^{-1} = A^{-1}A = I$.

USING TECHNOLOGY

FINDING THE INVERSE OF A SQUARE MATRIX

The graphing utility can be used to find the inverse of a square matrix.

EXAMPLE 1 Use a graphing utility to find the inverse of

$$\begin{bmatrix} 1 & 3 & 5 \\ -2 & 2 & 4 \\ 5 & 1 & 3 \end{bmatrix}$$

Solution We first enter the given matrix as

$$A = \begin{bmatrix} 1 & 3 & 5 \\ -2 & 2 & 4 \\ 5 & 1 & 3 \end{bmatrix}$$

Then, recalling the matrix A and using the x^{-1} key, we find

$$A^{-1} = \begin{bmatrix} 0.1 & -0.2 & 0.1 \\ 1.3 & -1.1 & -0.7 \\ -0.6 & 0.7 & 0.4 \end{bmatrix}$$

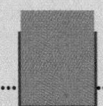

EXERCISES

In exercises 1–6, use a graphing utility to find the inverse of the matrix. Express your answers accurate to two decimal places.

1. $\begin{bmatrix} 1.2 & 3.1 & -2.1 \\ 3.4 & 2.6 & 7.3 \\ -1.2 & 3.4 & -1.3 \end{bmatrix}$

2. $\begin{bmatrix} 4.2 & 3.7 & 4.6 \\ 2.1 & -1.3 & -2.3 \\ 1.8 & 7.6 & -2.3 \end{bmatrix}$

3. $\begin{bmatrix} 1.1 & 2.3 & 3.1 & 4.2 \\ 1.6 & 3.2 & 1.8 & 2.9 \\ 4.2 & 1.6 & 1.4 & 3.2 \\ 1.6 & 2.1 & 2.8 & 7.2 \end{bmatrix}$

4. $\begin{bmatrix} 2.1 & 3.2 & -1.4 & -3.2 \\ 6.2 & 7.3 & 8.4 & 1.6 \\ 2.3 & 7.1 & 2.4 & -1.3 \\ -2.1 & 3.1 & 4.6 & 3.7 \end{bmatrix}$

5. $\begin{bmatrix} 2 & -1 & 3 & 2 & 4 \\ 3 & 2 & -1 & 4 & 1 \\ 3 & 2 & 6 & 4 & -1 \\ 2 & 1 & -1 & 4 & 2 \\ 3 & 4 & 2 & 5 & 6 \end{bmatrix}$

6. $\begin{bmatrix} 1 & 4 & 2 & 3 & 1.4 \\ 6 & 2.4 & 5 & 1.2 & 3 \\ 4 & 1 & 2 & 3 & 1.2 \\ -1 & 2 & -3 & 4 & 2 \\ 1.1 & 2.2 & 3 & 5.1 & 4 \end{bmatrix}$

SOLUTIONS TO SELF–CHECK EXERCISES 2.6

1. We form the augmented matrix

$$\left[\begin{array}{ccc|ccc} 2 & 1 & -1 & 1 & 0 & 0 \\ 1 & 1 & -1 & 0 & 1 & 0 \\ -1 & -2 & 3 & 0 & 0 & 1 \end{array}\right]$$

and row-reduce as follows:

$$\left[\begin{array}{ccc|ccc} 2 & 1 & -1 & 1 & 0 & 0 \\ 1 & 1 & -1 & 0 & 1 & 0 \\ -1 & -2 & 3 & 0 & 0 & 1 \end{array}\right] \xrightarrow{R_1 \leftrightarrow R_2}$$

$$\left[\begin{array}{ccc|ccc} 1 & 1 & -1 & 0 & 1 & 0 \\ 2 & 1 & -1 & 1 & 0 & 0 \\ -1 & -2 & 3 & 0 & 0 & 1 \end{array}\right] \xrightarrow[R_3 + R_1]{R_2 - 2R_1}$$

$$\left[\begin{array}{ccc|ccc} 1 & 1 & -1 & 0 & 1 & 0 \\ 0 & -1 & 1 & 1 & -2 & 0 \\ 0 & -1 & 2 & 0 & 1 & 1 \end{array}\right] \xrightarrow[\substack{-R_2 \\ R_3 - R_2}]{R_1 + R_2}$$

$$\left[\begin{array}{ccc|ccc} 1 & 0 & 0 & 1 & -1 & 0 \\ 0 & 1 & -1 & -1 & 2 & 0 \\ 0 & 0 & 1 & -1 & 3 & 1 \end{array}\right] \xrightarrow{R_2 + R_3}$$

$$\left[\begin{array}{ccc|ccc} 1 & 0 & 0 & 1 & -1 & 0 \\ 0 & 1 & 0 & -2 & 5 & 1 \\ 0 & 0 & 1 & -1 & 3 & 1 \end{array}\right]$$

From the preceding results, we see that

$$A^{-1} = \left[\begin{array}{ccc} 1 & -1 & 0 \\ -2 & 5 & 1 \\ -1 & 3 & 1 \end{array}\right]$$

2. a. We write the systems of linear equations in the matrix form

$$AX = B_1$$

where

$$A = \left[\begin{array}{ccc} 2 & 1 & -1 \\ 1 & 1 & -1 \\ -1 & -2 & 3 \end{array}\right] \qquad X = \left[\begin{array}{c} x \\ y \\ z \end{array}\right]$$

and
$$B_1 = \begin{bmatrix} 5 \\ 4 \\ -8 \end{bmatrix}$$

Now, using the results of exercise 1, we have

$$X = \begin{bmatrix} x \\ y \\ z \end{bmatrix} = A^{-1}B_1 = \begin{bmatrix} 1 & -1 & 0 \\ -2 & 5 & 1 \\ -1 & 3 & 1 \end{bmatrix} \begin{bmatrix} 5 \\ 4 \\ -8 \end{bmatrix} = \begin{bmatrix} 1 \\ 2 \\ -1 \end{bmatrix}$$

Therefore $x = 1$, $y = 2$, and $z = -1$.

b. Here A and X are as in (a), but

$$B_2 = \begin{bmatrix} 2 \\ 0 \\ 5 \end{bmatrix}$$

Therefore,

$$X = \begin{bmatrix} x \\ y \\ z \end{bmatrix} = A^{-1}B_2 = \begin{bmatrix} 1 & -1 & 0 \\ -2 & 5 & 1 \\ -1 & 3 & 1 \end{bmatrix} \begin{bmatrix} 2 \\ 0 \\ 5 \end{bmatrix} = \begin{bmatrix} 2 \\ 1 \\ 3 \end{bmatrix}$$

or $x = 2$, $y = 1$, and $z = 3$.

3. Let x denote the number of adults and y the number of children in a tour. Since the tours are filled to capacity, we have

$$x + y = 19$$

Next, using the fact that the total receipts for the first tour were \$2931 leads to the equation

$$169x + 129y = 2931$$

Therefore, the number of adults and the number of children in the first tour are found by solving the system of linear equations

$$\begin{aligned} x + \quad y &= 19 \\ 169x + 129y &= 2931 \end{aligned} \tag{a}$$

Similarly, we see that the number of adults and the number of children in the second and third tours are found by solving the systems

$$\begin{aligned} x + \quad y &= 19 \\ 169x + 129y &= 3011 \end{aligned} \tag{b}$$

and

$$\begin{aligned} x + \quad y &= 19 \\ 169x + 129y &= 2771 \end{aligned} \tag{c}$$

These systems may be written in the form

$$AX = B_1 \qquad AX = B_2 \qquad \text{and} \qquad AX = B_3$$

where

$$A = \begin{bmatrix} 1 & 1 \\ 169 & 129 \end{bmatrix} \qquad X = \begin{bmatrix} x \\ y \end{bmatrix}$$

$$B_1 = \begin{bmatrix} 19 \\ 2931 \end{bmatrix} \qquad B_2 = \begin{bmatrix} 19 \\ 3011 \end{bmatrix}$$

and

$$B_3 = \begin{bmatrix} 19 \\ 2771 \end{bmatrix}$$

To solve these systems, we first find A^{-1}. Using formula (13), we obtain

$$A^{-1} = \begin{bmatrix} -\frac{129}{40} & \frac{1}{40} \\ \frac{169}{40} & -\frac{1}{40} \end{bmatrix}$$

Then solving each system, we find

$$X = \begin{bmatrix} x \\ y \end{bmatrix} = A^{-1}B_1$$

$$= \begin{bmatrix} -\frac{129}{40} & \frac{1}{40} \\ \frac{169}{40} & -\frac{1}{40} \end{bmatrix} \begin{bmatrix} 19 \\ 2931 \end{bmatrix} = \begin{bmatrix} 12 \\ 7 \end{bmatrix} \tag{a}$$

$$X = \begin{bmatrix} x \\ y \end{bmatrix} = A^{-1}B_2$$

$$= \begin{bmatrix} -\frac{129}{40} & \frac{1}{40} \\ \frac{169}{40} & -\frac{1}{40} \end{bmatrix} \begin{bmatrix} 19 \\ 3011 \end{bmatrix}$$

$$= \begin{bmatrix} 14 \\ 5 \end{bmatrix} \tag{b}$$

$$X = \begin{bmatrix} x \\ y \end{bmatrix} = A^{-1}B_3$$

$$= \begin{bmatrix} -\frac{129}{40} & \frac{1}{40} \\ \frac{169}{40} & -\frac{1}{40} \end{bmatrix} \begin{bmatrix} 19 \\ 2771 \end{bmatrix} = \begin{bmatrix} 8 \\ 11 \end{bmatrix} \tag{c}$$

We conclude that there were
a. 12 adults and 7 children on the first tour.
b. 14 adults and 5 children on the second tour.
c. 8 adults and 11 children on the third tour.

2.7 LEONTIEF INPUT–OUTPUT MODEL (OPTIONAL)

Input–Output Analysis

One of the many important applications of matrix theory to the field of economics is the study of the relationship between industrial production and consumer demand. At the heart of this analysis is the Leontief input–output model, pioneered by Wassily Leontief, who was awarded a Nobel Prize in economics in 1973 for his contributions to the field.

In order to illustrate this concept, let us consider an oversimplified economy consisting of three sectors: agriculture (A), manufacturing (M), and service (S). In general, part of the output of one sector is absorbed by another sector through interindustry purchases, with the excess available to fulfill consumer demands. The relationship governing both intraindustrial and interindustrial sales and purchases is conveniently represented by means of an **input–output matrix:**

$$
\begin{array}{c}
\text{Output (amount produced)}\\
\begin{array}{cccc}
 & A & M & S \\
\begin{array}{c} A \\ \text{Input} \quad M \\ \text{(amount used in production)} \quad S \end{array} &
\left[\begin{array}{ccc}
0.2 & 0.2 & 0.1 \\
0.2 & 0.4 & 0.1 \\
0.1 & 0.2 & 0.3
\end{array}\right]
\end{array}
\end{array}
\tag{16}
$$

The first column (read from top to bottom) tells us that the production of 1 unit of agricultural products requires the consumption of 0.2 unit of agricultural products, 0.2 unit of manufactured goods, and 0.1 unit of services. The second column tells us that the production of 1 unit of manufactured products requires the consumption of 0.2 unit of agricultural products, 0.4 unit of manufactured products, and 0.2 unit of services. Finally, the third column tells us that the production of 1 unit of services requires the consumption of 0.1 unit each of agricultural goods and manufactured products, and 0.3 unit of services.

EXAMPLE 1   Refer to the input–output matrix (16).

a. ·If the units are measured in millions of dollars, determine the amount of agricultural products consumed in the production of $100 million worth of manufactured goods.

b. Determine the dollar amount of manufactured products required to produce $200 million worth of all goods and services in the economy.

Solution

a. The production of 1 unit—that is, $1 million worth of manufactured goods—requires the consumption of 0.2 unit of agricultural products. Thus, the amount of agricultural products consumed in the production of $100 million worth of manufactured goods is given by $(100)(0.2)$, or $20 million.

b. The amount of manufactured goods required to produce 1 unit of all goods and services in the economy is given by adding the numbers of the second row of the input–output matrix—that is, $0.2 + 0.4 + 0.1$, or 0.7 unit. Therefore, the production of $200 million worth of all goods and services in the economy requires $200(0.7)$, or $140 million worth, of manufactured products.   ○ ○ ○

Next, suppose that the total output of goods of the agriculture and manufacturing sectors and the total output from the service sector of the economy are given by x, y, and z units, respectively. What is the value of agricultural products consumed in the internal process of producing this total output of various goods and services?

To answer this question, we first note, by examining the input–output matrix

$$
\begin{array}{c}
\quad\quad\quad\quad\quad\text{Output} \\
\quad\quad\quad A \quad M \quad S \\
\text{Input} \begin{array}{c} A \\ M \\ S \end{array}
\left[
\begin{array}{ccc}
0.2 & 0.2 & 0.1 \\
0.2 & 0.4 & 0.1 \\
0.1 & 0.2 & 0.3
\end{array}
\right]
\end{array}
$$

that 0.2 unit of agricultural products is required to produce 1 unit of agricultural products, so the amount of agricultural goods required to produce x units of agricultural products is given by $0.2x$ unit. Next, again referring to the input–output matrix, we see that 0.2 unit of agricultural products is required to produce 1 unit of manufactured products, so the requirement for producing y units of the latter is $0.2y$ unit of agricultural products. Finally, we see that 0.1 unit of agricultural goods is required to produce 1 unit of services, so the value of agricultural products required to produce z units of services is $0.1z$ unit. Thus, the total amount of agricultural products required to produce the total output of goods and services in the economy is

$$0.2x + 0.2y + 0.1z$$

units. In a similar manner, we see that the total amount of manufactured products and the total value of services to produce the total output of goods and services in the economy are given by

$$0.2x + 0.4y + 0.1z$$

and $$0.1x + 0.2y + 0.3z$$

respectively.

These results could also be obtained using matrix multiplication. To see this, write the total output of goods and services x, y, and z as a 3×1 matrix

$$
X = \begin{bmatrix} x \\ y \\ z \end{bmatrix} \quad \text{(Gross production matrix)}
$$

The matrix X is called the **gross production matrix**. Letting A denote the input–output matrix, we have

$$A = \begin{bmatrix} 0.2 & 0.2 & 0.1 \\ 0.2 & 0.4 & 0.1 \\ 0.1 & 0.2 & 0.3 \end{bmatrix} \quad \text{(Input–output matrix)}$$

Then the product

$$AX = \begin{bmatrix} 0.2 & 0.2 & 0.1 \\ 0.2 & 0.4 & 0.1 \\ 0.1 & 0.2 & 0.3 \end{bmatrix} \begin{bmatrix} x \\ y \\ z \end{bmatrix}$$

$$= \begin{bmatrix} 0.2x + 0.2y + 0.1z \\ 0.2x + 0.4y + 0.1z \\ 0.1x + 0.2y + 0.3z \end{bmatrix} \quad \text{(Internal consumption matrix)}$$

is a 3×1 matrix whose entries represent the respective values of the agricultural products, manufactured products, and services consumed in the internal process of production. The matrix AX is referred to as the **internal consumption matrix**.

Now, since X gives the total production of goods and services in the economy, and AX, as we have just seen, gives the amount of products and services consumed in the production of these goods and services, the 3×1 matrix $X - AX$ gives the net output of goods and services that is exactly enough to satisfy consumer demands. Letting matrix D represent these consumer demands, we are led to the following matrix equation:

$$X - AX = D$$

or

$$(I - A)X = D$$

where I is the 3×3 identity matrix.

Assuming that the inverse of $(I - A)$ exists, multiplying both sides of the last equation by $(I - A)^{-1}$ yields

$$X = (I - A)^{-1}D$$

LEONTIEF INPUT–OUTPUT MODEL

In a Leontief input–output model, the matrix equation giving the net output of goods and services needed to satisfy consumer demand is

$$\begin{array}{ccc} \text{Total} & \text{Internal} & \text{Consumer} \\ \text{output} & \text{consumption} & \text{demand} \\ X & - \quad AX & = \quad D \end{array}$$

where X is the total output matrix, A is the input–output matrix, and D is the matrix representing consumer demand.

The solution to this equation is

$$X = (I - A)^{-1}D \qquad \text{[Assuming that } (I - A)^{-1} \text{ exists]} \tag{17}$$

which gives the amount of goods and services that must be produced to satisfy consumer demand.

Applications

Equation (17) gives us a means of finding the amount of goods and services to be produced in order to satisfy a given level of consumer demand, as illustrated by the following example.

EXAMPLE 2   For the three-sector economy with input–output matrix given by (16), which is reproduced here,

$$\begin{bmatrix} 0.2 & 0.2 & 0.1 \\ 0.2 & 0.4 & 0.1 \\ 0.1 & 0.2 & 0.3 \end{bmatrix} \qquad \text{(Each unit equals \$1 million.)}$$

a. find the gross output of goods and services needed to satisfy a consumer demand of $100 million worth of agricultural products, $80 million worth of manufactured products, and $50 million worth of services.

b. find the value of the goods and services consumed in the internal process of production in order to meet this gross output.

Solution

a. We are required to determine the gross production matrix

$$X = \begin{bmatrix} x \\ y \\ z \end{bmatrix}$$

where x, y, and z denote the value of the agricultural products, the manufactured products, and services. The matrix representing the consumer demand is given by

$$D = \begin{bmatrix} 100 \\ 80 \\ 50 \end{bmatrix}$$

Next, we compute

$$I - A = \begin{bmatrix} 1 & 0 & 0 \\ 0 & 1 & 0 \\ 0 & 0 & 1 \end{bmatrix} - \begin{bmatrix} 0.2 & 0.2 & 0.1 \\ 0.2 & 0.4 & 0.1 \\ 0.1 & 0.2 & 0.3 \end{bmatrix} = \begin{bmatrix} 0.8 & -0.2 & -0.1 \\ -0.2 & 0.6 & -0.1 \\ -0.1 & -0.2 & 0.7 \end{bmatrix}$$

Using the method of Section 2.6, we find (to two decimal places)

$$(I - A)^{-1} = \begin{bmatrix} 1.43 & 0.57 & 0.29 \\ 0.54 & 1.96 & 0.36 \\ 0.36 & 0.64 & 1.57 \end{bmatrix}$$

Finally, using equation (17), we find

$$X = (I - A)^{-1}D = \begin{bmatrix} 1.43 & 0.57 & 0.29 \\ 0.54 & 1.96 & 0.36 \\ 0.36 & 0.64 & 1.57 \end{bmatrix}\begin{bmatrix} 100 \\ 80 \\ 50 \end{bmatrix} = \begin{bmatrix} 203.1 \\ 228.8 \\ 165.7 \end{bmatrix}$$

To fulfill consumer demand, \$203.1 million worth of agricultural products, \$228.8 million worth of manufactured products, and \$165.7 million worth of services should be produced.

b. The amount of goods and services consumed in the internal process of production is given by AX, or equivalently by $X - D$. In this case, it is more convenient to use the latter, which gives the required result of

$$\begin{bmatrix} 203.1 \\ 228.8 \\ 165.7 \end{bmatrix} - \begin{bmatrix} 100 \\ 80 \\ 50 \end{bmatrix} = \begin{bmatrix} 103.1 \\ 148.8 \\ 115.7 \end{bmatrix}$$

or \$103.1 million worth of agricultural products, \$148.8 million worth of manufactured products, and \$115.7 million worth of services.   ○ ○ ○

EXAMPLE 3   The TKK Corporation, a large conglomerate, has three subsidiaries engaged in producing raw rubber, manufacturing tires, and manufacturing other rubber-based goods. The production of 1 unit of raw rubber requires the consumption of 0.08 unit of rubber, 0.04 unit of tires, and 0.02 unit of other rubber-based goods. To produce 1 unit of tires requires 0.6 unit of raw rubber, 0.02 unit of tires, and 0 units of other rubber-based goods. To produce 1 unit of other rubber-based goods requires 0.3 unit of raw rubber, 0.01 unit of tires, and 0.06 unit of other rubber-based goods. Market research indicates that the demand for the following year will be \$200 million for raw rubber, \$800 million for tires, and \$120 million for other rubber-based products. Find the level of production for each subsidiary in order to satisfy this demand.

Solution   View the corporation as an economy having three sectors, with an input–output matrix given by

$$A = \begin{matrix} & & \begin{matrix} \text{Raw} \\ \text{rubber} & \text{Tires} & \text{Goods} \end{matrix} \\ \begin{matrix} \text{Raw rubber} \\ \text{Tires} \\ \text{Goods} \end{matrix} & \begin{bmatrix} 0.08 & 0.60 & 0.30 \\ 0.04 & 0.02 & 0.01 \\ 0.02 & 0 & 0.06 \end{bmatrix} \end{matrix}$$

Using equation (17), we find that the required level of production is given by

$$X = \begin{bmatrix} x \\ y \\ z \end{bmatrix} = (I - A)^{-1}D$$

where x, y, and z denote the outputs of raw rubber, tires, and other rubber-based goods, and

$$D = \begin{bmatrix} 200 \\ 800 \\ 120 \end{bmatrix}$$

Now,

$$I - A = \begin{bmatrix} 0.92 & -0.60 & -0.30 \\ -0.04 & 0.98 & -0.01 \\ -0.02 & 0 & 0.94 \end{bmatrix}$$

You are asked to verify that

$$(I - A)^{-1} = \begin{bmatrix} 1.13 & 0.69 & 0.37 \\ 0.05 & 1.05 & 0.03 \\ 0.02 & 0.02 & 1.07 \end{bmatrix} \qquad \text{(See exercise 7.)}$$

Therefore,

$$X = (I - A)^{-1}D = \begin{bmatrix} 1.13 & 0.69 & 0.37 \\ 0.05 & 1.05 & 0.03 \\ 0.02 & 0.02 & 1.07 \end{bmatrix}\begin{bmatrix} 200 \\ 800 \\ 120 \end{bmatrix} = \begin{bmatrix} 822.4 \\ 853.6 \\ 148.4 \end{bmatrix}$$

To fulfill the predicted demand, $822.4 million worth of raw rubber, $853.6 million worth of tires, and $148.4 million worth of other rubber-based goods should be produced.   ◦ ◦ ◦

SELF-CHECK EXERCISES 2.7

 I. Solve the matrix equation $(I - A)X = D$ for x and y given that

$$A = \begin{bmatrix} 0.4 & 0.1 \\ 0.2 & 0.2 \end{bmatrix}, \quad X = \begin{bmatrix} x \\ y \end{bmatrix}, \quad \text{and} \quad D = \begin{bmatrix} 50 \\ 10 \end{bmatrix}$$

2. A simple economy consists of two sectors: agriculture (A) and transportation (T). The input–output matrix for this economy is given by

$$\begin{array}{cc} & \begin{array}{cc} A & \quad T \end{array} \\ A = \begin{array}{c} A \\ T \end{array} & \begin{bmatrix} 0.4 & 0.1 \\ 0.2 & 0.2 \end{bmatrix} \end{array}$$

a. Find the gross output of agricultural products needed to satisfy a consumer demand for $50 million worth of agricultural products and $10 million worth of transportation.

b. Find the value of agricultural products and transportation consumed in the internal process of production in order to meet the gross output.

Solutions to Self-Check Exercises 2.7 can be found on page 167.

2.7 EXERCISES

1. **An Input–Output Matrix for a Three-Sector Economy** A simple economy consists of three sectors: agriculture (A), manufacturing (M), and transportation (T). The input–output matrix for this economy is given by

$$\begin{array}{cc} & \begin{array}{ccc} A & M & T \end{array} \\ \begin{array}{c} A \\ M \\ T \end{array} & \begin{bmatrix} 0.4 & 0.1 & 0.1 \\ 0.1 & 0.4 & 0.3 \\ 0.2 & 0.2 & 0.2 \end{bmatrix} \end{array}$$

a. If the units are measured in millions of dollars, determine the amount of agricultural products consumed in the production of $100 million worth of manufactured goods.

b. Determine the dollar amount of manufactured products required to produce $200 million worth of all goods in the economy.

c. Which sector consumes the greatest amount of agricultural products in the production of a unit of goods in that sector? The least?

2. **An Input–Output Matrix for a Four-Sector Economy** The relationship governing the intraindustrial and interindustrial sales and purchases of four basic industries— agriculture (A), manufacturing (M), transportation (T), and energy (E)—of a certain economy is given by the following input–output matrix.

$$\begin{array}{cc} & \begin{array}{cccc} A & M & T & E \end{array} \\ \begin{array}{c} A \\ M \\ T \\ E \end{array} & \begin{bmatrix} 0.3 & 0.2 & 0 & 0.1 \\ 0.2 & 0.3 & 0.2 & 0.1 \\ 0.2 & 0.2 & 0.1 & 0.3 \\ 0.1 & 0.2 & 0.3 & 0.2 \end{bmatrix} \end{array}$$

a. How many units of energy are required to produce 1 unit of manufactured goods?

b. How many units of energy are required to produce 3 units of all goods in the economy?

c. Which sector of the economy is least dependent on the cost of energy?

d. Which sector of the economy has the smallest intraindustry purchases (sales)?

A calculator is recommended for the remainder of this exercise set.

In exercises 3–6, solve the matrix equation $(I - A)X = D$ for the given matrices A and D.

3. $A = \begin{bmatrix} 0.4 & 0.2 \\ 0.3 & 0.1 \end{bmatrix}$ $D = \begin{bmatrix} 10 \\ 12 \end{bmatrix}$

4. $A = \begin{bmatrix} 0.2 & 0.3 \\ 0.5 & 0.2 \end{bmatrix}$ $D = \begin{bmatrix} 4 \\ 8 \end{bmatrix}$

5. $A = \begin{bmatrix} 0.5 & 0.2 \\ 0.2 & 0.5 \end{bmatrix}$ $D = \begin{bmatrix} 10 \\ 20 \end{bmatrix}$

6. $A = \begin{bmatrix} 0.6 & 0.2 \\ 0.1 & 0.4 \end{bmatrix}$ $D = \begin{bmatrix} 8 \\ 12 \end{bmatrix}$

7. Let

$$A = \begin{bmatrix} 0.08 & 0.60 & 0.30 \\ 0.04 & 0.02 & 0.01 \\ 0.02 & 0 & 0.06 \end{bmatrix}$$

Show that

$$(I - A)^{-1} = \begin{bmatrix} 1.13 & 0.69 & 0.37 \\ 0.05 & 1.05 & 0.03 \\ 0.02 & 0.02 & 1.07 \end{bmatrix}$$

8. **An Input–Output Model for a Two-Sector Economy** A simple economy consists of two industries: agriculture and manufacturing. The production of 1 unit of agricultural products requires the consumption of 0.2 unit of agricultural products and 0.3 unit of manufactured goods. The production of 1 unit of manufactured products requires the consumption of 0.4 unit of agricultural products and 0.3 unit of manufactured goods.

a. Find the gross output of goods needed to satisfy a consumer demand for $100 million worth of agricultural products and $150 million worth of manufactured products.

b. Find the value of the goods consumed in the internal process of production in order to meet the gross output.

9. Rework exercise 8 if the consumer demand for the output of agricultural goods and the consumer demand for manufactured products are $120 million and $140 million, respectively.

10. Refer to Example 3. Suppose the demand for raw rubber increases by 10%, the demand for tires increases by 20%,

USING TECHNOLOGY

THE LEONTIEF INPUT–OUTPUT MODEL

Since the solution to a problem involving a Leontief input–output model often involves several matrix operations, the graphing utility can be used to facilitate the necessary computations.

EXAMPLE 1 Suppose the input–output matrix associated with an economy is given by A and the matrix D is a demand vector, where

$$A = \begin{bmatrix} 0.2 & 0.4 & 0.15 \\ 0.3 & 0.1 & 0.4 \\ 0.25 & 0.4 & 0.2 \end{bmatrix} \quad \text{and} \quad D = \begin{bmatrix} 20 \\ 15 \\ 40 \end{bmatrix}$$

Find the final outputs of each industry so that the demands of both industry and the open sector are met.

Solution First, we enter the matrices I (the identity matrix), A, and D. We are required to compute the output matrix $X = (I - A)^{-1}D$. Using the matrix operations of the graphing utility, we find

$$X = (I - A)^{-1} * D = \begin{bmatrix} 110.28 \\ 116.95 \\ 142.94 \end{bmatrix}$$

So, the final outputs of the first, second, and third industries are 110.28, 116.95, and 142.94 units, respectively. ○ ○ ○

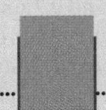 **EXERCISES**

In exercises 1–4, A is an input–output matrix associated with an economy, and D (in units of dollars) is a demand vector. Using a graphing utility, find the final outputs of each industry so that the demands of both industry and the open sector are met.

1.
$$A = \begin{bmatrix} 0.3 & 0.2 & 0.4 & 0.1 \\ 0.2 & 0.1 & 0.2 & 0.3 \\ 0.3 & 0.1 & 0.2 & 0.3 \\ 0.4 & 0.2 & 0.1 & 0.2 \end{bmatrix} \qquad D = \begin{bmatrix} 40 \\ 60 \\ 70 \\ 20 \end{bmatrix}$$

2.
$$A = \begin{bmatrix} 0.12 & 0.31 & 0.40 & 0.05 \\ 0.31 & 0.22 & 0.12 & 0.20 \\ 0.18 & 0.32 & 0.05 & 0.15 \\ 0.32 & 0.14 & 0.22 & 0.05 \end{bmatrix} \qquad D = \begin{bmatrix} 50 \\ 20 \\ 40 \\ 60 \end{bmatrix}$$

3.
$$A = \begin{bmatrix} 0.2 & 0.2 & 0.3 & 0.05 \\ 0.1 & 0.1 & 0.2 & 0.3 \\ 0.3 & 0.2 & 0.1 & 0.4 \\ 0.2 & 0.05 & 0.2 & 0.1 \end{bmatrix} \qquad D = \begin{bmatrix} 25 \\ 30 \\ 50 \\ 40 \end{bmatrix}$$

4.
$$A = \begin{bmatrix} 0.2 & 0.4 & 0.3 & 0.1 \\ 0.1 & 0.2 & 0.1 & 0.3 \\ 0.2 & 0.1 & 0.4 & 0.05 \\ 0.3 & 0.1 & 0.2 & 0.05 \end{bmatrix} \qquad D = \begin{bmatrix} 40 \\ 20 \\ 30 \\ 60 \end{bmatrix}$$

and the demand for rubber-based products decreases by 10%. Find the level of production for each subsidiary in order to meet this demand.

11. **An Input–Output Model for a Three-Sector Economy**   Consider the economy of exercise 1, consisting of three sectors: agriculture (A), manufacturing (M), and transportation (T), with an input–output matrix given by

$$
\begin{array}{c} \\ A \\ M \\ T \end{array}
\begin{array}{ccc} A & M & T \\ \begin{bmatrix} 0.4 & 0.1 & 0.1 \\ 0.1 & 0.4 & 0.3 \\ 0.2 & 0.2 & 0.2 \end{bmatrix} \end{array}
$$

a. Find the gross output of goods needed to satisfy a consumer demand for $200 million worth of agricultural products, $100 million worth of manufactured products, and $60 million worth of transportation.
b. Find the value of goods and transportation consumed in the internal process of production in order to meet this gross output.

12. **An Input–Output Model for a Three-Sector Economy**   Consider a simple economy consisting of three sectors: food, clothing, and shelter. The production of 1 unit of food requires the consumption of 0.4 unit of food, 0.2 unit of clothing, and 0.2 unit of shelter. The production of 1 unit of clothing requires the consumption of 0.1 unit of food, 0.2 unit of clothing, and 0.3 unit of shelter. The production of 1 unit of shelter requires the consumption of 0.3 unit of food, 0.1 unit of clothing, and 0.1 unit of shelter. Find the level of production for each sector in order to satisfy the demand for $100 million worth of food, $30 million worth of clothing, and $250 million worth of shelter.

In exercises 13–16, matrix A is an input–output matrix associated with an economy, and matrix D (units in millions of dollars) is a demand vector. In each problem, find the final outputs of each industry so that the demands of both industry and the open sector are met.

13. $A = \begin{bmatrix} 0.4 & 0.2 \\ 0.3 & 0.5 \end{bmatrix} \qquad D = \begin{bmatrix} 12 \\ 24 \end{bmatrix}$

14. $A = \begin{bmatrix} 0.1 & 0.4 \\ 0.3 & 0.2 \end{bmatrix} \qquad D = \begin{bmatrix} 5 \\ 10 \end{bmatrix}$

15. $A = \begin{bmatrix} \frac{1}{5} & \frac{2}{5} & \frac{1}{5} \\ \frac{1}{2} & 0 & \frac{1}{2} \\ 0 & \frac{1}{5} & 0 \end{bmatrix} \qquad D = \begin{bmatrix} 10 \\ 5 \\ 15 \end{bmatrix}$

16. $A = \begin{bmatrix} 0.2 & 0.4 & 0.1 \\ 0.3 & 0.2 & 0.1 \\ 0.1 & 0.2 & 0.2 \end{bmatrix} \qquad D = \begin{bmatrix} 6 \\ 8 \\ 10 \end{bmatrix}$

SOLUTIONS TO SELF-CHECK EXERCISES 2.7

1. Multiplying both sides of the given equation on the left by $(I - A)^{-1}$, we see that

$$X = (I - A)^{-1}D$$

Now

$$I - A = \begin{bmatrix} 1 & 0 \\ 0 & 1 \end{bmatrix} - \begin{bmatrix} 0.4 & 0.1 \\ 0.2 & 0.2 \end{bmatrix} = \begin{bmatrix} 0.6 & -0.1 \\ -0.2 & 0.8 \end{bmatrix}$$

Next, we use the Gauss-Jordan procedure to compute $(I - A)^{-1}$ (to two decimal places):

$$\begin{bmatrix} 0.6 & -0.1 & | & 1 & 0 \\ -0.2 & 0.8 & | & 0 & 1 \end{bmatrix} \xrightarrow{\frac{1}{0.6}R_1}$$

$$\begin{bmatrix} 1 & -0.17 & | & 1.67 & 0 \\ -0.2 & 0.8 & | & 0 & 1 \end{bmatrix} \xrightarrow{R_2 + 0.2R_1}$$

$$\begin{bmatrix} 1 & -0.17 & | & 1.67 & 0 \\ 0 & 0.77 & | & 0.33 & 1 \end{bmatrix} \xrightarrow{\frac{1}{0.77}R_2}$$

$$\begin{bmatrix} 1 & -0.17 & | & 1.67 & 0 \\ 0 & 1 & | & 0.43 & 1.30 \end{bmatrix} \xrightarrow{R_1 + 0.17R_2}$$

$$\begin{bmatrix} 1 & 0 & | & 1.74 & 0.22 \\ 0 & 1 & | & 0.43 & 1.30 \end{bmatrix}$$

giving

$$(I - A)^{-1} = \begin{bmatrix} 1.74 & 0.22 \\ 0.43 & 1.30 \end{bmatrix}$$

Therefore,

$$X = \begin{bmatrix} x \\ y \end{bmatrix} = (I - A)^{-1}D = \begin{bmatrix} 1.74 & 0.22 \\ 0.43 & 1.30 \end{bmatrix}\begin{bmatrix} 50 \\ 10 \end{bmatrix} = \begin{bmatrix} 89.2 \\ 34.5 \end{bmatrix}$$

or $x = 89.2$ and $y = 34.5$.

2. **a.** Let

$$X = \begin{bmatrix} x \\ y \end{bmatrix}$$

denote the gross production matrix, where x denotes the value of the agricultural products and y the value of transportation. Also, let

$$D = \begin{bmatrix} 50 \\ 10 \end{bmatrix}$$

denote the consumer demand. Then

$$(I - A)X = D$$

or equivalently,

$$X = (I - A)^{-1}D$$

Using the results of exercise 1, we find that $x = 89.2$ and $y = 34.5$. That is, to fulfill consumer demands, $89.2 million worth of agricultural products must be produced and $34.5 million worth of transportation services must be used.

b. The amount of agricultural products consumed and transportation services used is given by

$$X - D = \begin{bmatrix} 89.2 \\ 34.5 \end{bmatrix} - \begin{bmatrix} 50 \\ 10 \end{bmatrix} = \begin{bmatrix} 39.2 \\ 24.5 \end{bmatrix}$$

or $39.2 million worth of agricultural products and $24.5 million worth of transportation services.

 Group projects for each chapter can be found at the Brooks/Cole Web site at
http://www.brookscole.com/math/authors/tans/

CHAPTER 2 SUMMARY OF PRINCIPAL FORMULAS AND TERMS

Formulas

Laws for matrix addition:

Commutative law	$A + B = B + A$
Associative law	$(A + B) + C = A + (B + C)$

Laws for matrix multiplication:

Associative law	$(AB)C = A(BC)$
Distributive law	$A(B + C) = AB + AC$

Inverse of a 2×2 matrix

If $\quad A = \begin{bmatrix} a & b \\ c & d \end{bmatrix}$

and $\quad D = ad - bc \neq 0$

then $\quad A^{-1} = \dfrac{1}{D} \begin{bmatrix} d & -b \\ -c & a \end{bmatrix}$

Solution of system $AX = B$ $\qquad X = A^{-1}B$
(A, nonsingular)

Terms

System of linear equations	Parameter
Solution of a system of linear equations	Dependent system
	Inconsistent system
Gauss-Jordan elimination method	Transpose of a matrix
Equivalent system	Scalar
Matrix	Scalar product
Matrix of coefficients	Matrix product
Augmented matrix	Identity matrix
Row-reduced form of a matrix	Inverse of a matrix
Unit column	Singular matrix
Row operations	Input–output matrix
Size of a matrix	Gross production matrix
Row matrix	Internal consumption matrix
Column matrix	Leontief input–output model
Square matrix	

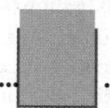

CHAPTER 2 REVIEW EXERCISES

In exercises 1–4, perform the given operations, if possible.

1. $\begin{bmatrix} 1 & 2 \\ -1 & 3 \\ 2 & 1 \end{bmatrix} + \begin{bmatrix} 1 & 0 \\ 0 & 1 \\ 1 & 2 \end{bmatrix}$

2. $\begin{bmatrix} -1 & 2 \\ 3 & 4 \end{bmatrix} - \begin{bmatrix} 1 & 2 \\ 5 & -2 \end{bmatrix}$

3. $[-3 \quad 2 \quad 1] \begin{bmatrix} 2 & 1 \\ -1 & 0 \\ 2 & 1 \end{bmatrix}$

4. $\begin{bmatrix} 1 & 3 & 2 \\ -1 & 2 & 3 \end{bmatrix} \begin{bmatrix} 1 \\ 4 \\ 2 \end{bmatrix}$

In exercises 5–8, find the values of the variables.

5. $\begin{bmatrix} 1 & x \\ y & 3 \end{bmatrix} = \begin{bmatrix} z & 2 \\ 3 & w \end{bmatrix}$

6. $\begin{bmatrix} 3 & x \\ y & 3 \end{bmatrix} \begin{bmatrix} 1 \\ 2 \end{bmatrix} = \begin{bmatrix} 7 \\ 4 \end{bmatrix}$

7. $\begin{bmatrix} 3 & a+3 \\ -1 & b \\ c+1 & d \end{bmatrix} = \begin{bmatrix} 3 & 6 \\ e+2 & 4 \\ -1 & 2 \end{bmatrix}$

8. $\begin{bmatrix} x & 3 & 1 \\ 0 & y & 2 \end{bmatrix} \begin{bmatrix} 1 & 1 \\ 3 & z \\ 4 & 2 \end{bmatrix} = \begin{bmatrix} 12 & 4 \\ 2 & 2 \end{bmatrix}$

In exercises 9–16, compute the given expressions, if possible, given that

$$A = \begin{bmatrix} 1 & 3 & 1 \\ -2 & 1 & 3 \\ 4 & 0 & 2 \end{bmatrix} \qquad B = \begin{bmatrix} 2 & 1 & 3 \\ -2 & -1 & -1 \\ 1 & 4 & 2 \end{bmatrix}$$

and

$$C = \begin{bmatrix} 3 & -1 & 2 \\ 1 & 6 & 4 \\ 2 & 1 & 3 \end{bmatrix}$$

9. $2A + 3B$

10. $3A - 2B$

11. $2(3A)$

12. $2(3A - 4B)$

13. $A(B - C)$ 14. $AB + AC$

15. $A(BC)$ 16. $(\frac{1}{2})(CA - CB)$

In exercises 17–23, solve the system of linear equations using the Gauss-Jordan elimination method.

17. $2x - 3y = 5$
    $3x + 4y = -1$

18. $3x + 2y = 3$
    $2x - 4y = -14$

19. $x - y + 2z = 5$
    $3x + 2y + z = 10$
    $2x - 3y - 2z = -10$

20. $3x - 2y + 4z = 16$
    $2x + y - 2z = -1$
    $x + 4y - 8z = -18$

21. $3x - 2y + 4z = 11$
    $2x - 4y + 5z = 4$
    $x + 2y - z = 10$

22. $x - 2y + 3z + 4w = 17$
    $2x + y - 2z - 3w = -9$
    $3x - y + 2z - 4w = 0$
    $4x + 2y - 3z + w = -2$

23. $3x - 2y + z = 4$
    $x + 3y - 4z = -3$
    $2x - 3y + 5z = 7$
    $x - 8y + 9z = 10$

In exercises 24–31, find the inverse of the given matrix (if it exists).

24. $A = \begin{bmatrix} 2 & 4 \\ 1 & 6 \end{bmatrix}$

25. $A = \begin{bmatrix} 3 & 1 \\ 1 & 2 \end{bmatrix}$

26. $A = \begin{bmatrix} 2 & 4 \\ 1 & -2 \end{bmatrix}$

27. $A = \begin{bmatrix} 3 & 4 \\ 2 & 2 \end{bmatrix}$

28. $A = \begin{bmatrix} 1 & 2 & 4 \\ 2 & 1 & 3 \\ -1 & 0 & 2 \end{bmatrix}$

29. $A = \begin{bmatrix} 2 & 3 & 1 \\ 1 & -1 & 2 \\ 1 & 2 & 1 \end{bmatrix}$

30. $A = \begin{bmatrix} 2 & 1 & -3 \\ 1 & 2 & -4 \\ 3 & 1 & -2 \end{bmatrix}$

31. $A = \begin{bmatrix} 1 & 2 & 4 \\ 3 & 1 & 2 \\ 1 & 0 & -6 \end{bmatrix}$

In exercises 32–35, compute the given expressions, if possible, given that

$$A = \begin{bmatrix} 1 & 2 \\ -1 & 2 \end{bmatrix} \quad B = \begin{bmatrix} 3 & 1 \\ 4 & 2 \end{bmatrix} \quad C = \begin{bmatrix} 1 & 1 \\ -1 & 2 \end{bmatrix}$$

32. $(ABC)^{-1}$

33. $(A^{-1}B)^{-1}$

34. $(A + B)^{-1}$

35. $(2A - C)^{-1}$

In exercises 36–39, write each system of linear equations in the form AX = C. Find A^{-1} and use the result to solve the system.

36. $x - 3y = -1$
    $2x + 4y = 8$

37. $2x + 3y = -8$
    $x - 2y = 3$

38. $2x - 3y + 4z = 17$
    $x + 2y - 4z = -7$
    $3x - y + 2z = 14$

39. $x - 2y + 4z = 13$
    $2x + 3y - 2z = 0$
    $x + 4y - 6z = -15$

40. Mr. Spaulding bought 10,000 shares of stock X, 20,000 shares of stock Y, and 30,000 shares of stock Z at a unit price of $20, $30, and $50 per share, respectively. Six months later, the closing prices of stocks X, Y, and Z were $22, $35, and $51 per share, respectively. Spaulding made no other stock transactions during the period in question. Compare the value of Mr. Spaulding's stock holdings at the time of purchase and six months later.

41. Ms. Newburg operates three self-service gasoline stations in different parts of town. On a certain day, station A sold 600 gallons of premium, 1000 gallons of regular, 800 gallons of super-unleaded, and 1400 gallons of regular-unleaded gasoline; station B sold 700 gallons of premium, 800 gallons of regular, 600 gallons of super-unleaded, and 1200 gallons of regular-unleaded gasoline; station C sold 1200 gallons of premium, 800 gallons of regular, 1000 gallons of super-unleaded, and 900 gallons of regular-unleaded gasoline. Assume that the price of gasoline on that day was $1.60 per gallon for premium, $1.20 per gallon for regular, $1.50 per gallon for super-unleaded, and $1.30 per gallon for regular-unleaded gasoline, and use matrix algebra to find the total revenue at each station.

42. The Wildcat Oil Company has two refineries—one located in Houston and the other in Tulsa. The Houston refinery ships 60% of its petroleum to a Chicago distributor and 40% of its petroleum to a Los Angeles distributor. The Tulsa refinery ships 30% of its petroleum to the Chicago distributor and 70% of its petroleum to the Los Angeles distributor. Assume that, over the year, the Chicago distributor received 240,000 gallons of petroleum and the Los Angeles distributor received 460,000 gallons of petroleum. Find the amount of petroleum produced at each of Wildcat's refineries.

43. Desmond Jewelry, Inc., wishes to produce three types of pendants: type A, type B, and type C. To manufacture a type-A pendant requires 2 minutes on machines I and II and 3 minutes on machine III. A type-B pendant

requires 2 minutes on machine I, 3 minutes on machine II, and 4 minutes on machine III. A type-C pendant requires 3 minutes on machine I, 4 minutes on machine II, and 3 minutes on machine III. There are $3\frac{1}{2}$ hours available on machine I, $4\frac{1}{2}$ hours available on machine II, and 5 hours available on machine III. How many pendants of each type should Desmond make in order to use all the available time?

 Additional study hints and sample chapter tests for each chapter can be found at the Brooks/Cole Web site at http://www.brookscole.com/math/authors/tans/

Many practical problems involve maximizing or minimizing a function subject to certain constraints. For example, we may wish to maximize a profit function subject to certain limitations on the amount of material and labor available. Maximization or minimization problems that can be formulated in terms of a *linear* objective function and constraints in the form of linear inequalities are called *linear programming problems.* In this chapter we look at linear programming problems involving two variables. These problems are amenable to geometric analysis, and the method of solution introduced here will shed much light on the basic nature of a linear programming problem.

How should the aircraft engines be shipped? Curtis-Roe Aviation Industries manufactures jet engines in two different locations. These engines are to be shipped to the company's two main assembly plants. In Example 3, page 185, we will show how many engines should be produced and shipped from each manufacturing plant to each assembly plant in order to minimize shipping costs.

3

LINEAR PROGRAMMING: A GEOMETRIC APPROACH

3.1 GRAPHING SYSTEMS OF LINEAR INEQUALITIES IN TWO VARIABLES

Graphing Linear Inequalities

In Chapter 1, we saw that a linear equation in two variables x and y

$$ax + by + c = 0 \qquad (a, b \text{ not both equal to zero})$$

has a *solution set* that may be exhibited graphically as points on a straight line in the xy-plane. We now show that there is also a simple graphical representation for **linear inequalities** in two variables:

$$ax + by + c < 0 \qquad ax + by + c \leq 0$$
$$ax + by + c > 0 \qquad ax + by + c \geq 0$$

Before turning to a general procedure for graphing such inequalities, let us consider a specific example. Suppose we wish to graph

$$2x + 3y < 6 \tag{1}$$

We first graph the equation $2x + 3y = 6$, which is obtained by replacing the given inequality "<" with an equality "=" (Figure 3.1).

Figure 3.1
A straight line divides the xy-plane into two half-planes.

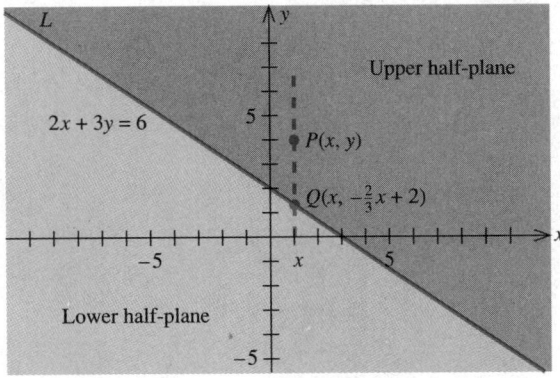

Observe that this line divides the xy-plane into two half-planes: an upper half-plane and a lower half-plane. Let us show that the upper half-plane is the graph of the linear inequality

$$2x + 3y > 6 \tag{2}$$

whereas the lower half-plane is the graph of the linear inequality

$$2x + 3y < 6 \tag{3}$$

To see this, let us write equations (2) and (3) in the equivalent forms

$$y > -\frac{2}{3}x + 2 \tag{4}$$

and

$$y < -\frac{2}{3}x + 2 \qquad\qquad (5)$$

The equation of the line itself is

$$y = -\frac{2}{3}x + 2 \qquad\qquad (6)$$

Now pick any point $P(x, y)$ lying above the line L. Let Q be the point lying on L and directly below P (see Figure 3.1). Since Q lies on L, its coordinates must satisfy equation (6). In other words, Q has representation $Q(x, -\frac{2}{3}x + 2)$. Comparing the y-coordinates of P and Q and recalling that P lies above Q so that its y-coordinate must be larger than that of Q, we have

$$y > -\frac{2}{3}x + 2$$

But this inequality is just equation (4), or equivalently equation (2). Similarly, we can show that any point lying below L must satisfy equation (5) and therefore (3).

This analysis shows that the lower half-plane provides a solution to our problem (Figure 3.2). (The dotted line shows that the points on L do not belong to the solution set.) Observe that the two half-planes in question are mutually exclusive; that is, they do not have any points in common. Because of this, there is an alternative and easier method of determining the solution to the problem.

Figure 3.2

The set of points lying below the dotted line satisfies the given inequality.

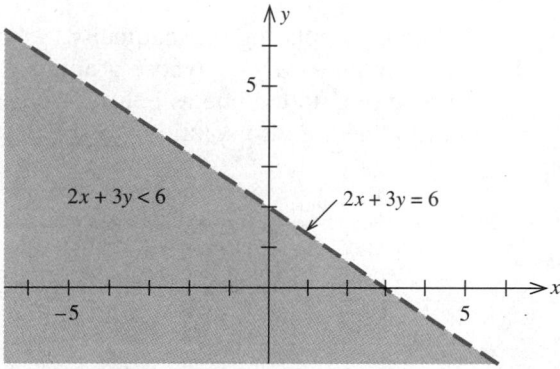

To determine the required half-plane, let us pick *any* point lying in one of the half-planes. For simplicity, let us pick the origin $(0, 0)$, which lies in the lower half-plane. Substituting $x = 0$ and $y = 0$ (the coordinates of this point) into the given inequality (1), we find

$$2(0) + 3(0) < 6$$

or $0 < 6$, which is certainly true. This tells us that the required half-plane is the half-plane containing the test point—namely, the lower half-plane.

Next, let us see what happens if we choose the point (2, 3), which lies in the upper half-plane. Substituting $x = 2$ and $y = 3$ into the given inequality, we find

$$2(2) + 3(3) < 6$$

or $13 < 6$, which is false. This tells us that the upper half-plane is *not* the required half-plane, as expected. Note, too, that no point (x, y) lying on the line constitutes a solution to our problem, because of the *strict* inequality "$<$".

This discussion suggests the following procedure for graphing a linear inequality in two variables.

PROCEDURE FOR GRAPHING LINEAR INEQUALITIES

I. Draw the graph of the equation obtained for the given inequality by replacing the inequality sign with an equals sign. Use a dotted line if the problem involves a strict inequality, "$<$" or "$>$". Otherwise, use a solid line to indicate that the line itself constitutes part of the solution.

2. Pick a test point lying in one of the half-planes determined by the line sketched in part (1) and substitute the values of x and y into the given inequality. Use the origin whenever possible.

3. If the inequality is satisfied, the graph of the inequality includes the half-plane containing the test point. Otherwise, the solution includes the half-plane not containing the test point.

EXAMPLE I   Determine the solution set for the inequality $2x + 3y \geq 6$.

Solution   Replacing the inequality "$\geq$" with an equality "$=$", we obtain the equation $2x + 3y = 6$, whose graph is the straight line shown in Figure 3.3. Instead of a dotted line as before, we use a solid line to show that all points on the line are also solutions to the problem. Picking the origin as our test

Figure 3.3
The set of points lying on the line and in the upper half-plane satisfies the given inequality.

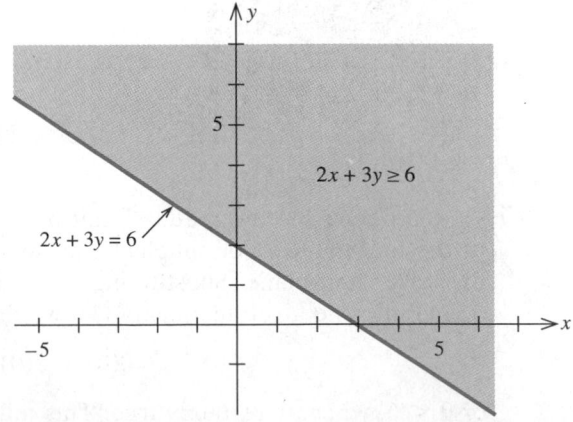

Figure 3.4

The set of points lying on the line $x = -1$ and in the left half-plane satisfies the given inequality.

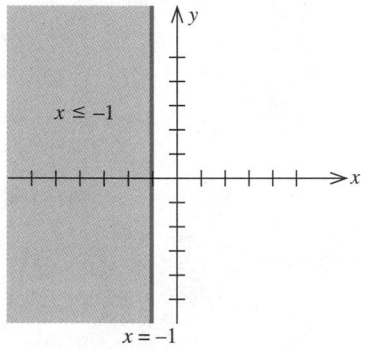

Figure 3.5

The set of points in the lower half-plane satisfies $x - 2y > 0$.

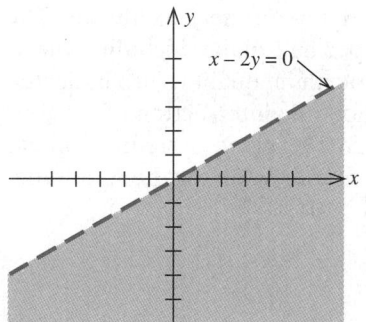

point, we find $2(0) + 3(0) \geq 6$, or $0 \geq 6$, which is impossible. So we conclude that the solution set comprises the half-plane not containing the origin, including, in this case, the line given by $2x + 3y = 6$.   ○ ○ ○

EXAMPLE 2   Graph $x \leq -1$.

Solution   The graph of $x = -1$ is the vertical line shown in Figure 3.4. Picking the origin $(0, 0)$ as a test point, we find $0 \leq -1$, which is false. Therefore, the required solution is the *left* half-plane, which does not contain the origin.

○ ○ ○

EXAMPLE 3   Graph $x - 2y > 0$.

Solution   We first graph the equation $x - 2y = 0$, or $y = (\frac{1}{2})x$ (Figure 3.5). Since the origin lies on the line, we may not use it as a test point. (Why?) Let us pick $(1, 2)$ as a test point. Substituting $x = 1$ and $y = 2$ into the given inequality, we find $1 - 2(2) > 0$, or $-3 > 0$, which is false. Therefore, the required solution is the half-plane that does not contain the test point— namely, the lower half-plane.   ○ ○ ○

Graphing Systems of Linear Inequalities

By the **solution set of a system of linear inequalities** in the two variables x and y, we mean the set of all points (x, y) satisfying each inequality of the system. The graphical solution of such a system may be obtained by graphing the solution set for each inequality independently and then determining the region in common with each solution set.

EXAMPLE 4   Determine the solution set for the system

$$4x + 3y \geq 12$$
$$x - y \leq 0$$

Solution   Proceeding as in the previous examples, you should have no difficulty locating the half-planes determined by each of the linear inequalities that make up the system. These half-planes are shown in Figure 3.6. The intersection of the two half-planes is the shaded region. A point in this region is an element of the solution set for the given system. The point P, the intersection of the two straight lines determined by the equations, is found by solving the simultaneous equations

$$4x + 3y = 12$$
$$x - y = 0$$

Figure 3.6
The set of points in the shaded area satisfies the system
$$\begin{cases} 4x + 3y \geq 12 \\ x - y \leq 0 \end{cases}$$

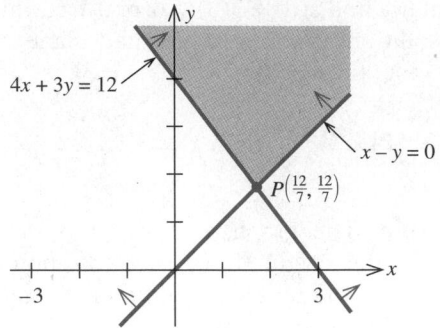

EXAMPLE 5   Sketch the solution set for the system

$$x \geq 0$$
$$y \geq 0$$
$$x + y - 6 \leq 0$$
$$2x + y - 8 \leq 0$$

Solution   The first inequality in the system defines the right half-plane—all points to the right of the y-axis plus all points lying on the y-axis itself. The second inequality in the system defines the upper half-plane, including the x-axis. The half-planes defined by the third and fourth inequalities are indicated by arrows in Figure 3.7. Thus, the required region, the intersection of the four half-planes defined by the four inequalities in the given system of linear inequalities, is the shaded region. The point P is found by solving the simultaneous equations $x + y - 6 = 0$ and $2x + y - 8 = 0$.

Figure 3.7
The set of points in the shaded region, including the x- and y-axes, satisfies the given inequalities.

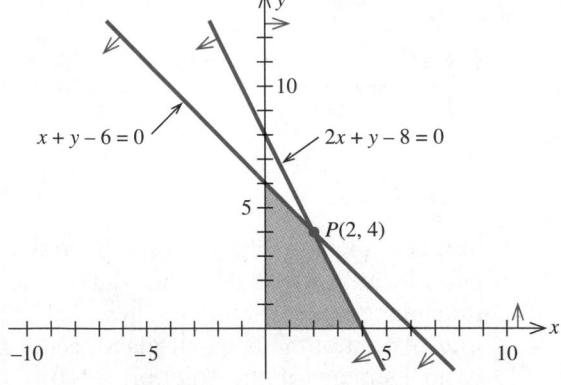

The solution set found in Example 5 is an example of a *bounded set*. Observe that the set can be enclosed by a circle. For example, if you draw a circle of radius 10 with center at the origin, you will see that the set lies entirely inside the circle. On the other hand, the solution set of Example 4 cannot be enclosed by a circle and is said to be *unbounded*.

BOUNDED AND UNBOUNDED SOLUTION SETS

A solution set of a system of linear inequalities is **bounded** if it can be enclosed by a circle. Otherwise, it is **unbounded.**

EXAMPLE 6   Determine the graphical solution set for the following system of linear inequalities:

$$2x + y \geq 50$$
$$x + 2y \geq 40$$
$$x \geq 0$$
$$y \geq 0$$

Solution   The required solution set is the unbounded region shown in Figure 3.8.

Figure 3.8
The solution set is an unbounded region.

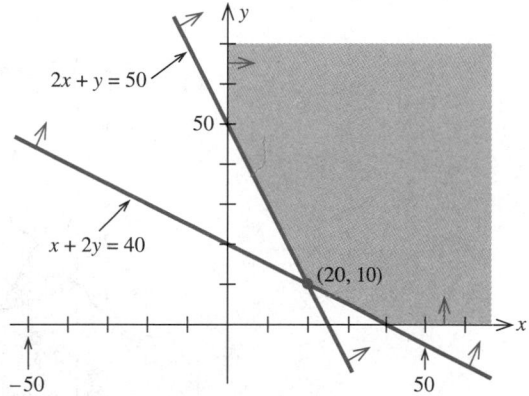

SELF-CHECK EXERCISES 3.1

1. Determine graphically the solution set for the following system of inequalities:

$$x + 2y \leq 10$$
$$5x + 3y \leq 30$$
$$x \geq 0, y \geq 0$$

2. Determine graphically the solution set for the following system of inequalities:

$$5x + 3y \geq 30$$
$$x - 3y \leq 0$$
$$x \geq 2$$

Solutions to Self-Check Exercises 3.1 can be found on page 181.

3.1 EXERCISES

In exercises 1–10, find the graphical solution of each inequality.

1. $4x - 8 < 0$ **2.** $3y + 2 > 0$

3. $x - y \leq 0$ **4.** $3x + 4y \leq -2$

5. $x \leq -3$ **6.** $y \geq -1$

7. $2x + y \leq 4$ **8.** $-3x + 6y \geq 12$

9. $4x - 3y \leq -24$ **10.** $5x - 3y \geq 15$

In exercises 11–18, write a system of linear inequalities that describes the shaded region.

11.

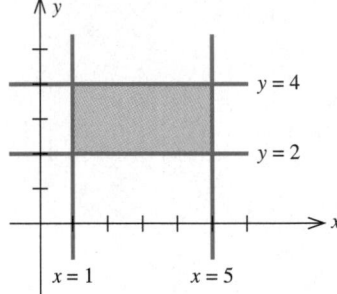

12.

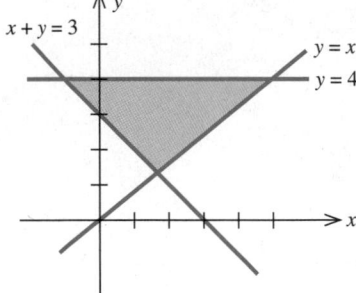

13.

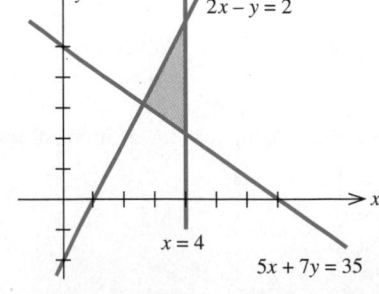

14.

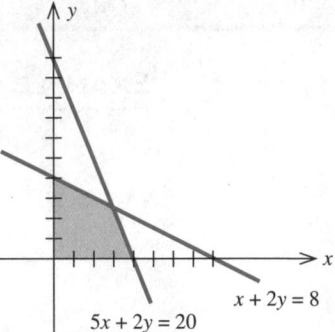

15.

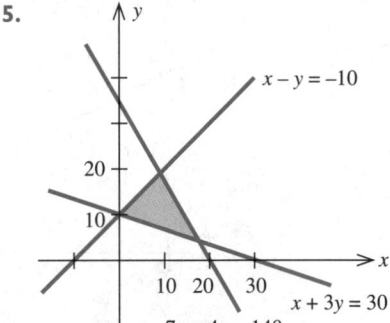

16.

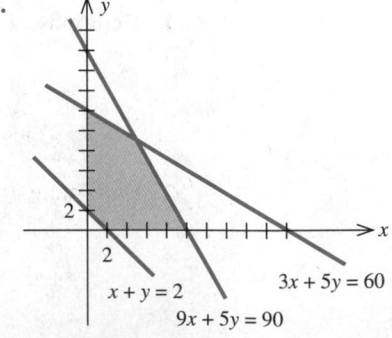

17.

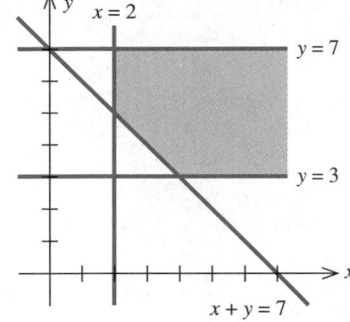

18.

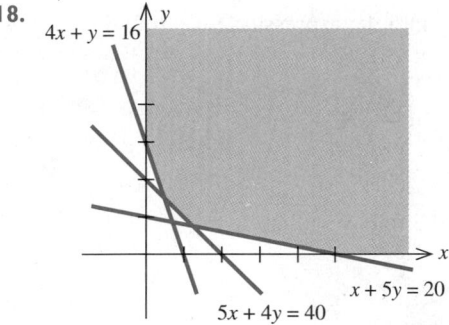

In exercises 19–36, determine graphically the solution set for each system of inequalities and indicate whether the solution set is bounded or unbounded.

19. $2x + 4y > 16$
    $-x + 3y \geq 7$

20. $3x - 2y > -13$
    $-x + 2y > 5$

21. $x - y \leq 0$
    $2x + 3y \geq 10$

22. $x + y \geq -2$
    $3x - y \leq 6$

23. $x + 2y \geq 3$
    $2x + 4y \leq -2$

24. $2x - y \geq 4$
    $4x - 2y < -2$

25. $x + y \leq 6$
    $0 \leq x \leq 3$
    $y \geq 0$

26. $4x - 3y \leq 12$
    $5x + 2y \leq 10$
    $x \geq 0, y \geq 0$

27. $3x - 6y \leq 12$
    $-x + 2y \leq 4$
    $x \geq 0, y \geq 0$

28. $x + y \geq 20$
    $x + 2y \geq 40$
    $x \geq 0, y \geq 0$

29. $3x - 7y \geq -24$
    $x + 3y \geq 8$
    $x \geq 0, y \geq 0$

30. $3x + 4y \geq 12$
    $2x - y \geq -2$
    $0 \leq y \leq 3$
    $x \geq 0$

31. $x + 2y \geq 3$
    $5x - 4y \leq 16$
    $0 \leq y \leq 2$
    $x \geq 0$

32. $x + y \leq 4$
    $2x + y \leq 6$
    $2x - y \geq -1$
    $x \geq 0, y \geq 0$

33. $6x + 5y \leq 30$
    $3x + y \geq 6$
    $x + y \geq 4$
    $x \geq 0, y \geq 0$

34. $6x + 7y \leq 84$
    $12x - 11y \leq 18$
    $6x - 7y \leq 28$
    $x \geq 0, y \geq 0$

35. $x - y \geq -6$
    $x - 2y \leq -2$
    $x + 2y \geq 6$
    $x - 2y \geq -14$
    $x \geq 0, y \geq 0$

36. $x - 3y \geq -18$
    $3x - 2y \geq 2$
    $x - 3y \leq -4$
    $3x - 2y \leq 16$
    $x \geq 0, y \geq 0$

SOLUTIONS TO SELF-CHECK EXERCISES 3.1

1. The required solution set is shown in the following figure.

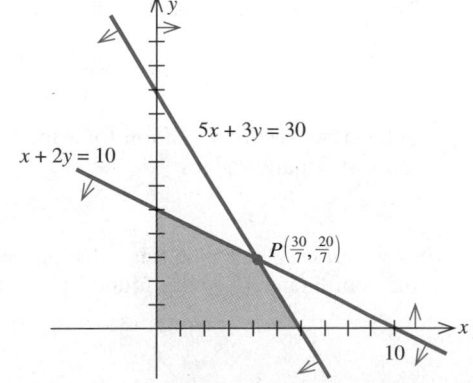

The point P is found by solving the system of equations

$$x + 2y = 10$$
$$5x + 3y = 30$$

Solving the first equation for x in terms of y gives

$$x = 10 - 2y$$

Substituting this value of x into the second equation of the system gives

$$5(10 - 2y) + 3y = \ 30$$
$$50 - 10y + 3y = \ 30$$
$$-7y = -20$$

so $y = 20/7$. Substituting this value of y into the expression for x found earlier, we obtain

$$x = 10 - 2\left(\frac{20}{7}\right) = \frac{30}{7}$$

giving the point of intersection as (30/7, 20/7).

2. The required solution set is shown in the following figure.

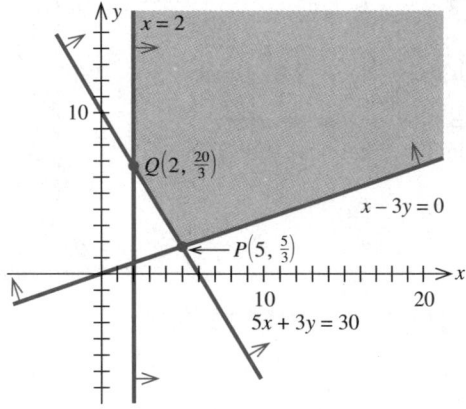

To find the coordinates of P, we solve the system

$$5x + 3y = 30$$
$$x - 3y = \ 0$$

Solving the second equation for x in terms of y and substituting this value of x in the first equation gives

$$5(3y) + 3y = 30$$

or $y = 5/3$. Substituting this value of y into the second equation gives $x = 5$. Next, the coordinates of Q are found by solving the system

$$5x + 3y = 30$$
$$x = 2$$

obtaining $x = 2$ and $y = 20/3$.

3.2 LINEAR PROGRAMMING PROBLEMS

Many business and economic problems are concerned with optimizing (maximizing or minimizing) a function subject to a system of equalities or inequalities. The function to be optimized is called the **objective function.** Profit functions and cost functions are examples of objective functions. The system of equalities or inequalities to which the objective function is subjected reflects the constraints (for example, limitations on resources such as materials and labor) imposed on the solution(s) to the problem. Problems of this nature are called **mathematical programming problems.** In particular, problems in which both the objective function and the constraints are expressed as linear equations or inequalities are called **linear programming problems.**

A LINEAR PROGRAMMING PROBLEM	A linear programming problem consists of a linear objective function to be maximized or minimized subject to certain constraints in the form of linear equalities or inequalities.

Maximization Problems

As an example of a linear programming problem in which the objective function is to be maximized, let us consider the following simplified version of a production problem involving two variables.

EXAMPLE 1   The Ace Novelty Company wishes to produce two types of souvenirs: type A and type B. Each type-A souvenir will result in a profit of $1, and each type-B souvenir will result in a profit of $1.20. To manufacture a type-A souvenir requires 2 minutes on machine I and 1 minute on machine II. A type-B souvenir requires 1 minute on machine I and 3 minutes on machine II. There are 3 hours available on machine I and 5 hours available on machine II for processing the order. How many souvenirs of each type should Ace make in order to maximize profit?

Solution   As a first step toward the mathematical formulation of this problem, we tabulate the given information, as shown in Table 3.1.

Table 3.1

	Type A	**Type B**	**Time Available**
Machine I	2 min	1 min	180 min
Machine II	1 min	3 min	300 min
Profit per Unit	$1	$1.20	

Let x be the number of type-A souvenirs and y be the number of type-B souvenirs to be made. Then the total profit P, in dollars, is given by

$$P = x + 1.2y$$

which is the objective function to be maximized.

The total amount of time that machine I is used is given by $2x + y$ minutes and must not exceed 180 minutes. Thus we have the inequality

$$2x + y \leq 180$$

Similarly, the total amount of time that machine II is used is $x + 3y$ minutes, which cannot exceed 300 minutes, so we are led to the inequality

$$x + 3y \leq 300$$

Finally, neither x nor y can be negative, so

$$x \geq 0$$
$$y \geq 0$$

To summarize, the problem at hand is one of maximizing the objective function $P = x + 1.2y$ subject to the system of inequalities

$$2x + \ y \leq 180$$
$$x + 3y \leq 300$$
$$x \geq 0$$
$$y \geq 0$$

The solution to this problem will be completed in Example 1, Section 3.3.

• • •

Minimization Problems

In the following example of a linear programming problem, the objective function is to be minimized.

EXAMPLE 2   A nutritionist advises an individual who is suffering from
....   iron and vitamin-B deficiency to take at least 2400 mg of iron, 2100 mg of vitamin B-1 (thiamine), and 1500 mg of vitamin B-2 (riboflavin) over a period of time. Two vitamin pills are suitable, brand A and brand B. Each brand A pill contains 40 mg of iron, 10 mg of vitamin B-1, and 5 mg of vitamin B-2, and costs 6 cents. Each brand B pill contains 10 mg of iron and 15 mg each of vitamins B-1 and B-2, and costs 8 cents (see Table 3.2).

Table 3.2

	Brand A	**Brand B**	**Minimum Requirement**
Iron	40 mg	10 mg	2400 mg
Vitamin B-1	10 mg	15 mg	2100 mg
Vitamin B-2	5 mg	15 mg	1500 mg
Cost per Pill	6¢	8¢	

What combination of pills should the individual purchase in order to meet the minimum iron and vitamin requirements at the lowest cost?

Solution   Let x be the number of brand A pills and y be the number of brand B pills to be purchased. The cost C, measured in cents, is given by

$$C = 6x + 8y$$

and is the objective function to be minimized.

The amount of iron contained in x brand A pills and y brand B pills is given by $40x + 10y$ mg, and this must be greater than or equal to 2400 mg. This translates into the inequality

$$40x + 10y \geq 2400$$

Similar considerations involving the minimum requirements of vitamins B-1 and B-2 lead to the inequalities

$$10x + 15y \geq 2100$$

and

$$5x + 15y \geq 1500$$

respectively. Thus, the problem here is to minimize $C = 6x + 8y$ subject to

$$40x + 10y \geq 2400$$
$$10x + 15y \geq 2100$$
$$5x + 15y \geq 1500$$
$$x \geq 0, y \geq 0$$

The solution to this problem will be completed in Example 2, Section 3.3.

　　◦ ◦ ◦

A Transportation Problem

EXAMPLE 3   Curtis-Roe Aviation Industries has two plants, I and II, that produce the Zephyr jet engines used in their light commercial airplanes. The maximum production capacities of these two plants are 100 units and 110 units per month, respectively. The engines are shipped to two of Curtis-Roe's main assembly plants, A and B. The shipping costs (in dollars) per engine from plants I and II to the main assembly plants A and B are as follows:

From	To Assembly Plant	
	A	**B**
Plant I	100	60
Plant II	120	70

In a certain month, assembly plant A needs 80 engines, whereas assembly plant B needs 70 engines. Find how many engines should be shipped from each plant to each main assembly plant if shipping costs are to be kept to a minimum.

Solution   Let x denote the number of engines shipped from plant I to assembly plant A, and let y denote the number of engines shipped from plant I to assembly plant B. Since the requirements of assembly plants A and B are 80 and 70 engines, respectively, the number of engines shipped from plant II to assembly plants A and B are $(80 - x)$ and $(70 - y)$, respectively. These numbers may be displayed in a schematic. With the aid of the following schematic and the shipping cost schedule, we find that the total shipping costs incurred by Curtis-Roe are given by

$$C = 100x + 60y + 120(80 - x) + 70(70 - y)$$
$$= 14{,}500 - 20x - 10y$$

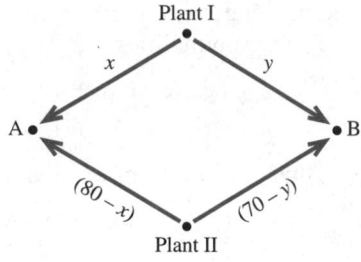

Next, the production constraints on plants I and II lead to the inequalities

$$x + y \leq 100$$

and

$$(80 - x) + (70 - y) \leq 110$$

The last inequality simplifies to

$$x + y \geq 40$$

Also, the requirements of the two main assembly plants lead to the inequalities

$$x \geq 0, \quad y \geq 0, \quad 80 - x \geq 0, \quad \text{and} \quad 70 - y \geq 0$$

The last two may be written as $x \leq 80$ and $y \leq 70$.

Summarizing, we have the following linear programming problem: Minimize the objective (cost) function $C = 14{,}500 - 20x - 10y$ subject to the constraints

$$x + y \geq 40$$
$$x + y \leq 100$$
$$x \leq 80$$
$$y \leq 70$$

where $x \geq 0$ and $y \geq 0$.

You will be asked to complete the solution to this problem in exercise 30, Section 3.3.

○ ○ ○

A Warehouse Problem

EXAMPLE 4 The Acrosonic Company manufactures its model F loudspeaker systems in two separate locations, plant I and plant
II. The output at plant I is at most 400 per month, whereas the output at
plant II is at most 600 per month. These loudspeaker systems are shipped to
three warehouses that serve as distribution centers for the company. In order
for the warehouses to meet their orders, the minimum monthly requirements
of warehouses A, B, and C are 200, 300, and 400, respectively. Shipping costs
from plant I to warehouses A, B, and C are $20, $8, and $10 per loudspeaker
system, respectively, and shipping costs from plant II to each of these warehouses are $12, $22, and $18, respectively. What should the shipping schedule
be if Acrosonic wishes to meet the requirements of the distribution centers
and at the same time keep its shipping costs to a minimum?

Solution The respective shipping costs per loudspeaker system in dollars
may be tabulated as in Table 3.3.

Table 3.3

Plant	Warehouse		
	A	**B**	**C**
I	20	8	10
II	12	22	18

Letting x_1 denote the number of loudspeaker systems shipped from plant I
to warehouse A, x_2 the number shipped from plant I to warehouse B, and so
on, leads to Table 3.4.

Table 3.4

Plant	Warehouse			Max. Prod.
	A	**B**	**C**	
I	x_1	x_2	x_3	400
II	x_4	x_5	x_6	600
Min. req.	200	300	400	

From Tables 3.3 and 3.4 we see that the cost of shipping x_1 loudspeaker
systems from plant I to warehouse A is $\$20x_1$, the cost of shipping x_2 loud-
speaker systems from plant I to warehouse B is $\$8x_2$, and so on. Thus, the
total monthly shipping cost incurred by Acrosonic is given by

$$C = 20x_1 + 8x_2 + 10x_3 + 12x_4 + 22x_5 + 18x_6$$

Next, the production constraints on plants I and II lead to the inequalities

$$x_1 + x_2 + x_3 \le 400$$

and
$$x_4 + x_5 + x_6 \le 600$$

(see Table 3.4). Also, the minimum requirements of each of the three warehouses lead to the three inequalities

$$x_1 + x_4 \geq 200$$
$$x_2 + x_5 \geq 300$$

and
$$x_3 + x_6 \geq 400$$

Summarizing, we have the following linear programming problem: Minimize $C = 20x_1 + 8x_2 + 10x_3 + 12x_4 + 22x_5 + 18x_6$ subject to

$$x_1 + x_2 + x_3 \leq 400$$
$$x_4 + x_5 + x_6 \leq 600$$
$$x_1 + x_4 \geq 200$$
$$x_2 + x_5 \geq 300$$
$$x_3 + x_6 \geq 400$$
$$x_1 \geq 0, x_2 \geq 0, \ldots, x_6 \geq 0$$

The solution to this problem will be completed in Section 4.2, Example 5.

○ ○ ○

SELF-CHECK EXERCISE 3.2

Mr. Balduzzi, proprietor of Luigi's Pizza Palace, allocates $9000 a month for advertising in two newspapers, the *City Tribune* and the *Daily News*. The *City Tribune* charges $300 for a certain advertisement, whereas the *Daily News* charges $100 for the same ad. Mr. Balduzzi has stipulated that the ad is to appear in at least 15 but no more than 30 editions of the *Daily News* per month. The *City Tribune* has a daily circulation of 50,000, and the *Daily News* has a circulation of 20,000. Under these conditions, determine how many ads Balduzzi should place in each newspaper in order to reach the largest number of readers. Formulate but do not solve the problem. (The solution to this problem can be found in exercise 3 of Solutions to Self-Check Exercises 3.3.) *Solution to Self-Check Exercise 3.2 can be found on page 192.*

3.2 EXERCISES

Formulate, but do not solve, each of the following exercises as a linear programming problem.

1. **Manufacturing—Production Scheduling** A company manufactures two products, A and B, on two machines I and II. It has been determined that the company will realize a profit of $3 on each unit of product A and a profit of $4 on each unit of product B. To manufacture a unit of product A requires 6 minutes on machine I and 5 minutes on machine II. To manufacture a unit of product B requires 9 minutes on machine I and 4 minutes on machine II. There are 5 hours of machine time available on machine I and 3 hours of machine time available on machine II in each work shift. How many units of each product should be produced in each shift to maximize the company's profit?

Leanne Jenkins

Title: Associate Media Director/National Broadcast
Institution: Hill, Holliday, Connors, Cosmopulos, Inc.

Jenkins contrasts the freedom she enjoys when making broadcasting media buys with the approach taken in print advertising. Whereas print buyers have to deal with a publication's rate card—which prices ads by size and color—broadcast buyers tell the networks "This is what I'll pay you" for a typical 30-second spot. Like a tourist in a Middle East bazaar, Jenkins enjoys haggling over the price of a prime-time commercial.

Spalding, a sporting goods manufacturer, is a Hill Holliday client. To advertise golf balls, it budgets approximately $10 million for commercial time as well as print ads to reach its primary target audience: men age 35 and older. To get Spalding's message across, Jenkins first determines what percentage of the budget must be allocated to television versus print in order to compete with other golf ball manufacturers. To blanket the Professional Golfers' Association of America (PGA) broadcasts as well as other sports programs, Jenkins figures she will need approximately $6 million for commercials.

As she studies the PGA ratings from previous years, Jenkins estimates that next year's broadcasts will achieve an average of 3 rating points each. On a given weekend afternoon, 3%, or 1,650,000 men (out of 55,000,000 men in the target audience), will watch golf and have the opportunity to see Spalding's commercials.

In a typical scenario, Jenkins contacts CBS to purchase a block of 50 commercials based on her cost-per-point estimate of $14,000 for each rating point. Spread out over an entire year, Spalding's commercials will run three or four times during each broadcast. Jenkins multiplies the cost per point by 50 commercials at 3.5 points per commercial to determine a total cost of $2.625 million. Over 40% of Spalding's television advertising budget will go for commercials on CBS alone. The balance will be divided among golf programs on other networks as well as other sports broadcasts.

Knowing how popular golf is, the network will argue that its coverage will achieve at least 3.5 rating points and is worth at least $15,000 per point. Because the demand for sports events is high, advertisers usually must accept slightly higher costs—although Jenkins works to keep expenses down.

When negotiating for commercials on regularly scheduled programs that are not special events, Jenkins frequently gets lower prices. Popular programs such as "Roseanne," which are shown weekly, mean higher supply and lower demand. With special programming such as the U.S. Open Golf Championship, however, Spalding may have to pay the network's price. The Open happens only once a year, which means very limited supply with much higher demand.

For Jenkins and the networks, ratings, as well as their associated costs, are the key ingredients for reaching millions of viewers each week.

Hill Holliday, New England's largest advertising agency, handles a diverse group of clients from beer makers to insurance companies. As associate media director for national broadcasting, Jenkins purchases blocks of commercial time from ABC, CBS, NBC, FOX, and over ten cable networks. Based on demographic studies, she determines whether a particular program will attract a specific audience interested in a client's goods or services. If a match occurs, Jenkins negotiates the lowest possible price to reach the highest number of viewers.

◦ ◦ ◦

2. **Manufacturing—Production Scheduling** Kane Manufacturing has a division that produces two models of hibachis, model A and model B. To produce each model A hibachi requires 3 pounds of cast iron and 6 minutes of labor. To produce each model B hibachi requires 4 pounds of cast iron and 3 minutes of labor. The profit for each model A hibachi is $2, and the profit for each model B hibachi is $1.50. If 1000 pounds of cast iron and 20 hours of labor are available for the production of hibachis per day, how many hibachis of each model should the division produce per day in order to help maximize Kane's profits?

3. **Manufacturing—Production Scheduling** Refer to exercise 2. Because of a backlog of orders on model A hibachis, the manager of Kane Manufacturing has decided to produce at least 150 of these models a day. Operating under this additional constraint, how many hibachis of each model should Kane produce to maximize profit?

4. **Finance—Allocation of Funds** Madison Finance has a total of $20 million earmarked for homeowner and auto loans. On the average, homeowner loans have a 10% annual rate of return, whereas auto loans yield a 12% annual rate of return. Management has also stipulated that the total amount of homeowner loans should be greater than or equal to four times the total amount of automobile loans. Determine the total amount of loans of each type Madison should extend to each category in order to maximize its returns.

5. **Nutrition—Diet Planning** A nutritionist at the Medical Center has been asked to prepare a special diet for certain patients. She has decided that the meals should contain a minimum of 400 mg of calcium, 10 mg of iron, and 40 mg of vitamin C. She has further decided that the meals are to be prepared from foods A and B. Each ounce of food A contains 30 mg of calcium, 1 mg of iron, 2 mg of vitamin C, and 2 mg of cholesterol. Each ounce of food B contains 25 mg of calcium, 0.5 mg of iron, 5 mg of vitamin C, and 5 mg of cholesterol. Find how many ounces of each type of food should be used in a meal so that the cholesterol content is minimized and the minimum requirements of calcium, iron, and vitamin C are met.

6. **Agriculture—Crop Planning** A farmer has 150 acres of land suitable for cultivating crops A and B. The cost of cultivating crop A is $40 per acre, whereas that of crop B is $60 per acre. The farmer has a maximum of $7400 available for land cultivation. Each acre of crop

A requires 20 hours of labor and each acre of crop B requires 25 hours of labor. The farmer has a maximum of 3300 hours of labor available. If he expects to make a profit of $150 per acre on crop A and $200 per acre on crop B, how many acres of each crop should he plant in order to maximize his profit?

7. **Manufacturing—Shipping Costs** The TMA Company manufactures 19-inch color television picture tubes in two separate locations, location I and location II. The output at location I is at most 6000 tubes per month, whereas the output at location II is at most 5000 per month. TMA is the main supplier of picture tubes to the Pulsar Corporation, its holding company, which has priority in having all its requirements met. In a certain month, Pulsar placed orders for 3000 and 4000 picture tubes to be shipped to two of its factories located in city A and city B, respectively. The shipping costs (in dollars) per picture tube from the two TMA plants to the two Pulsar factories are as follows:

Shipping Costs per Picture Tube		
	To Pulsar Factories	
From	**City A**	**City B**
TMA (Loc. I)	$3	$2
TMA (Loc. II)	$4	$5

Find a shipping schedule that meets the requirements of both companies while keeping costs to a minimum.

8. **Social Programs Planning** AntiFam, a hunger-relief organization, has earmarked between $2 and $2.5 million, inclusive, for aid to two African countries, country A and country B. Country A is to receive between $1 and $1.5 million, inclusive, in aid, and country B is to receive at least $0.75 million in aid. It has been estimated that each dollar spent in country A will yield an effective return of 60 cents, whereas a dollar spent in country B will yield an effective return of 80 cents. How should the aid be allocated if the money is to be utilized most effectively according to these criteria?

[*Hint:* If x and y denote the amount of money to be given to country A and country B, respectively, then the objective function to be maximized is $P = 0.6x + 0.8y$.]

9. **Manufacturing—Production Scheduling** A company manufactures products A, B, and C. Each product is processed in three departments: I, II, and III. The total available labor hours per week for departments I, II, and III are 900, 1080, and 840, respectively. The time

requirements (in hours per unit) and profit per unit for each product are as follows:

	Product A	Product B	Product C
Dept. I	2	1	2
Dept. II	3	1	2
Dept. III	2	2	1
Profit	$18	$12	$15

How many units of each product should the company produce in order to maximize its profit?

10. **Advertising** As part of a campaign to promote its annual clearance sale, the Excelsior Company decided to buy television advertising time on Station KAOS. Excelsior's advertising budget is $102,000. Morning time costs $3000 per minute, afternoon time costs $1000 per minute, and evening (prime) time costs $12,000 per minute. Because of previous commitments, KAOS cannot offer Excelsior more than 6 minutes of prime time or more than a total of 25 minutes of advertising time over the two weeks in which the commercials are to be run. KAOS estimates that morning commercials are seen by 200,000 people, afternoon commercials are seen by 100,000 people, and evening commercials are seen by 600,000 people. How much morning, afternoon, and evening advertising time should Excelsior buy to maximize exposure of its commercials?

11. **Manufacturing—Production Scheduling** The Custom Office Furniture Company is introducing a new line of executive desks made from a specially selected grade of walnut. Initially, three different models—A, B, and C—are to be marketed. Each model A desk requires $1\frac{1}{4}$ hours for fabrication, 1 hour for assembly, and 1 hour for finishing; each model B desk requires $1\frac{1}{2}$ hours for fabrication, 1 hour for assembly, and 1 hour for finishing; each model C desk requires $1\frac{1}{2}$ hours, $\frac{3}{4}$ hour, and $\frac{1}{2}$ hour for fabrication, assembly, and finishing, respectively. The profit on each model A desk is $26, the profit on each model B desk is $28, and the profit on each model C desk is $24. The total time available in the fabrication department, the assembly department, and the finishing department in the first month of production is 310 hours, 205 hours, and 190 hours, respectively. To maximize Custom's profit, how many desks of each model should be made in the month?

12. **Manufacturing—Shipping Costs** The Acrosonic Company of Example 4 also manufactures a model G loudspeaker system in plants I and II. The output at plant I is at most 800 systems per month, whereas the output at plant II is at most 600 per month. These loudspeaker systems are also shipped to the three warehouses—A, B, and C—whose minimum monthly requirements are 500, 400, and 400, respectively. Shipping costs from plant I to warehouse A, warehouse B, and warehouse C are $16, $20, and $22 per loudspeaker system, respectively, and shipping costs from plant II to each of these warehouses are $18, $16, and $14, respectively. What shipping schedule will enable Acrosonic to meet the warehouses' requirements and at the same time keep its shipping costs to a minimum?

13. **Manufacturing—Shipping Costs** Steinwelt Piano manufactures uprights and consoles in two plants, plant I and plant II. The output of plant I is at most 300 per month, whereas the output of plant II is at most 250 per month. These pianos are shipped to three warehouses that serve as distribution centers for the company. In order to fill current and projected future orders, warehouse A requires a minimum of 200 pianos per month, warehouse B requires at least 150 pianos per month, and warehouse C requires at least 200 pianos per month. The shipping cost of each piano from plant I to warehouse A, warehouse B, and warehouse C is $60, $60, and $80, respectively, and the shipping cost of each piano from plant II to warehouse A, warehouse B, and warehouse C is $80, $70, and $50, respectively. What shipping schedule will enable Steinwelt to meet the warehouses' requirements while keeping shipping costs to a minimum?

14. **Manufacturing—Prefabricated Housing Production** Boise lumber has decided to enter the lucrative prefabricated housing business. Initially, it plans to offer three models: a standard model, a deluxe model, and a luxury model. Each house is prefabricated and partially assembled in the factory, and the final assembly is completed on site. The dollar amount of building material required, the amount of labor required in the factory for prefabrication and partial assembly, the amount of on-site labor required, and the profit per unit are as follows:

	Standard Model	Deluxe Model	Luxury Model
Material ($)	6,000	8,000	10,000
Factory Labor (hr)	240	220	200
On-Site Labor (hr)	180	210	300
Profit ($)	3,400	4,000	5,000

For the first year's production, a sum of $8,200,000 is budgeted for the building material; the number of labor hours available for work in the factory (for prefabrication and partial assembly) is not to exceed 218,000 hours; and the amount of labor for on-site work is to be less than or equal to 237,000 labor hours. Determine how many houses of each type Boise should produce (market research has confirmed that there should be no problems with sales) to maximize its profit from this new venture.

15. **Manufacturing—Cold Formula Production**   Bayer Pharmaceutical produces three kinds of cold formulas: formula I, formula II, and formula III. It takes 2.5 hours

to produce 1000 bottles of formula I, 3 hours to produce 1000 bottles of formula II, and 4 hours to produce 1000 bottles of formula III. The profits for each 1000 bottles of formula I, formula II, and formula III are $180, $200, and $300, respectively. Suppose that for a certain production run there are enough ingredients on hand to make at most 9000 bottles of formula I, 12,000 bottles of formula II, and 6000 bottles of formula III. Furthermore, suppose that the time for the production run is limited to a maximum of 70 hours. Find how many bottles of each formula should be produced in this production run so that the profit is maximized.

SOLUTION TO SELF–CHECK EXERCISE 3.2

Let x denote the number of ads to be placed in the *City Tribune* and y the number to be placed in the *Daily News*. The total cost for placing x ads in the *City Tribune* and y ads in the *Daily News* is $300x + 100y$ dollars, and since the monthly budget is $9000, we must have

$$300x + 100y \le 9000$$

Next, the condition that the ad must appear in at least 15 but no more than 30 editions of the *Daily News* translates into the inequalities

$$y \ge 15$$

and $$y \le 30$$

Finally, the objective function to be maximized is

$$P = 50{,}000x + 20{,}000y$$

To summarize, we have the following linear programming problem:

Maximize $P = 50{,}000x + 20{,}000y$
subject to $300x + 100y \le 9000$
$$y \ge 15$$
$$y \le 30$$
$$x \ge 0, y \ge 0$$

3.3 GRAPHICAL SOLUTION OF LINEAR PROGRAMMING PROBLEMS

The Graphical Method

Linear programming problems in two variables have relatively simple geometrical interpretations. For example, the system of linear constraints associated with a two-dimensional linear programming problem, unless it is inconsistent, defines a planar region whose boundary is composed of straight line segments and/or half-lines. Such problems are therefore amenable to graphical analysis.

Consider the following two-dimensional linear programming problem:

Maximize $P = 3x + 2y$

subject to $2x + 3y \leq 12$

$2x + \ y \leq 8$

$x \geq 0, y \geq 0$

(7)

The system of linear inequalities (7) defines the planar region S shown in Figure 3.9. Each point in S is a candidate for the solution of the problem at hand and is referred to as a **feasible solution.** The set S itself is referred to as a **feasible set.** Our goal is to find, from among all the points in the set S, the point(s) that optimizes the objective function P. Such a feasible solution is called an **optimal solution** and constitutes the solution to the linear programming problem under consideration.

Figure 3.9
Each point in the feasible set S is a candidate for the optimal solution.

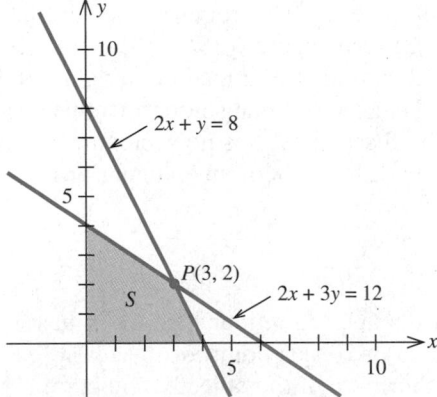

As noted earlier, each point $P(x, y)$ in S is a candidate for the optimal solution to the problem at hand. For example, the point $(1, 3)$ is easily seen to lie in S and is therefore in the running. The value of the objective function P at the point $(1, 3)$ is given by $P = 3(1) + 2(3) = 9$. Now, if we could compute the value of P corresponding to each point in S, then the point(s) in S that gave the largest value to P would constitute the solution set sought. Unfortunately, in most problems, the number of candidates is either too large or, as in this problem, the number is infinite. Thus, this method is at best unwieldy and at worst impractical.

Let us turn the question around. Instead of asking for the value of the objective function P at a feasible point, let us assign a value to the objective function P and ask whether there are feasible points that would correspond to the given value of P. To this end, suppose we assign a value of 6 to P.

Figure 3.10
A family of parallel lines that intersect the feasible set S.

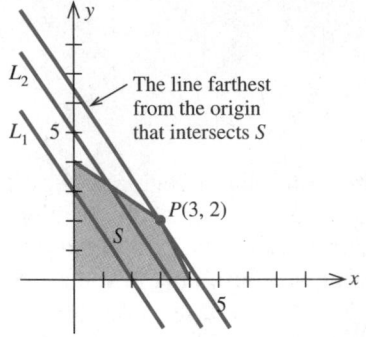

Then the objective function P becomes $3x + 2y = 6$, a linear equation in x and y, and, therefore, has a graph that is a straight line L_1 in the plane. In Figure 3.10 we have drawn the graph of this straight line superimposed on the feasible set S.

It is clear that each point on the straight line segment given by the intersection of the straight line L_1 and the feasible set S corresponds to the given value, 6, of P. For this reason, the line L_1 is called an **isoprofit line.** Let us repeat the process, this time assigning a value of 10 to P. We obtain the equation $3x + 2y = 10$ and the line L_2 (see Figure 3.10), which suggests that there are feasible points that correspond to a larger value of P. Observe that the line L_2 is parallel to the line L_1 since both lines have slope equal to $-3/2$, which is easily seen by casting the corresponding equations in the slope-intercept form.

In general, by assigning different values to the objective function, we obtain a family of parallel lines, each with slope equal to $-3/2$. Furthermore, a line corresponding to a larger value of P lies farther away from the origin than one with a smaller value of P. The implication is clear. To obtain the optimal solution(s) to the problem at hand, find the straight line, from this family of straight lines, that is farthest from the origin and still intersects the feasible set S. The required line is the one that passes through the point $P(3, 2)$ (see Figure 3.10), so the solution to the problem is given by $x = 3$, $y = 2$, resulting in a maximum value of $P = 3(3) + 2(2) = 13$.

That the optimal solution to this problem was found to occur at a vertex of the feasible set S is no accident. In fact, the result is a consequence of the following basic theorem on linear programming, which we state without proof.

THEOREM I

If a linear programming problem has a solution, then it must occur at a vertex, or corner point, of the feasible set S associated with the problem. Furthermore, if the objective function P is optimized at two adjacent vertices of S, then it is optimized at every point on the line segment joining these vertices, in which case there are infinitely many solutions to the problem.

Theorem 1 tells us that our search for the solution(s) to a linear programming problem may be restricted to the examination of the set of vertices of the feasible set S associated with the problem. Since a feasible set S has finitely many vertices, the theorem suggests that the solution(s) to the linear programming problem may be found by inspecting the values of the objective function P at these vertices.

Although Theorem 1 sheds some light on the nature of the solution of a linear programming problem, it does not tell us when a linear programming problem has a solution. The following theorem states some conditions that will guarantee when a linear programming problem has a solution.

THEOREM 2:
EXISTENCE OF A
SOLUTION

Suppose we are given a linear programming problem with a feasible set S and an objective function $P = ax + by$.

1. If S is bounded, then P has both a maximum and a minimum value on S.

2. If S is unbounded and both a and b are nonnegative, then P has a minimum value on S provided that the constraints defining S include the inequalities $x \geq 0$ and $y \geq 0$.

3. If S is the empty set, then the linear programming problem has no solution; that is, P has neither a maximum nor a minimum value.

The **method of corners,** a simple procedure for solving linear programming problems based on Theorem 1, follows.

THE METHOD OF
CORNERS

1. Graph the feasible set.

2. Find the coordinates of all corner points (vertices) of the feasible set.

3. Evaluate the objective function at each corner point.

4. Find the vertex that renders the objective function a maximum (minimum). If there is only one such vertex, then this vertex constitutes a unique solution to the problem. If the objective function is maximized (minimized) at two adjacent corner points of S, there are infinitely many optimal solutions given by the points on the line segment determined by these two vertices.

Applications

EXAMPLE 1   We are now in a position to complete the solution to the production problem posed in Example 1, Section 3.2. Recall that the mathematical formulation led to the following linear programming problem.

$$
\begin{aligned}
\text{Maximize} \quad & P = x + 1.2y \\
\text{subject to} \quad & 2x + y \leq 180 \\
& x + 3y \leq 300 \\
& x \geq 0, y \geq 0
\end{aligned}
$$

Solution   The feasible set S for the problem is shown in Figure 3.11. The vertices of the feasible set are $A(0, 0)$, $B(90, 0)$, $C(48, 84)$, and $D(0, 100)$. The

Figure 3.11
The corner point that yields the max-
imum profit is C(48, 84).

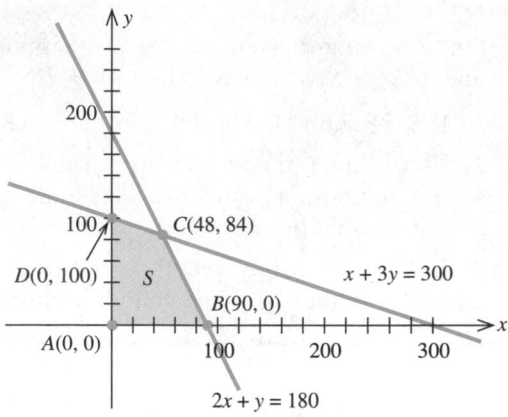

values of P at these vertices may be tabulated as follows:

Vertex	$P = x + 1.2y$
$A(0, 0)$	0
$B(90, 0)$	90
$C(48, 84)$	148.8
$D(0, 100)$	120

From the table, we see that the maximum of $P = x + 1.2y$ occurs at the vertex (48, 84) and has a value of 148.8. Recalling what the symbols x, y, and P represent, we conclude that Ace Novelty would maximize its profit (a figure of $148.80) by producing 48 type-A souvenirs and 84 type-B souvenirs.

○ ○ ○

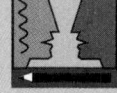

Consider the linear programming problem:

Maximize $P = 4x + 3y$

subject to $2x + \ \ y \le 10$

$2x + 3y \le 18$

$x \ge 0, y \ge 0$

a. Sketch the feasible set S for the linear programming problem.
b. Draw the isoprofit lines superimposed on S corresponding to $P = 12, 16, 20,$ and 24, and show that these lines are parallel to each other.
c. Show that the solution to the linear programming problem is $x = 3$ and $y = 4$. Is this result the same as that found using the method of corners?

EXAMPLE 2   Complete the solution of the nutrition problem posed in Example 2, Section 3.2.

Solution   Recall that the mathematical formulation of the problem led to the following linear programming problem in two variables:

$$\text{Minimize} \quad C = 6x + 8y$$
$$\text{subject to} \quad 40x + 10y \geq 2400$$
$$10x + 15y \geq 2100$$
$$5x + 15y \geq 1500$$
$$x \geq 0, y \geq 0$$

The feasible set S defined by the system of constraints is shown in Figure 3.12. The vertices of the feasible set S are $A(0, 240)$, $B(30, 120)$, $C(120, 60)$,

Figure 3.12

The corner point that yields the minimum cost is $B(30, 120)$.

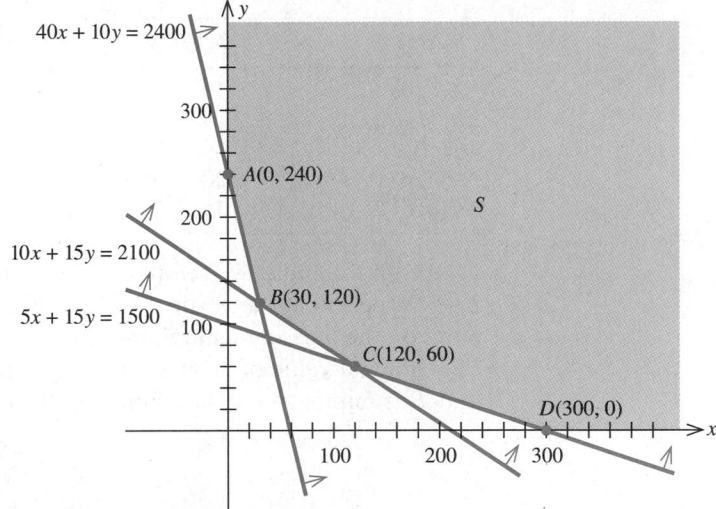

and $D(300, 0)$. The values of the objective function C at these vertices are given in the following table:

Vertex	$C = 6x + 8y$
$A(0, 240)$	1920
$B(30, 120)$	1140
$C(120, 60)$	1200
$D(300, 0)$	1800

From the table, we can see that the minimum for the objective function $C = 6x + 8y$ occurs at the vertex $B(30, 120)$ and has a value of 1140. Thus, the individual should purchase 30 brand A pills and 120 brand B pills at a minimum cost of $11.40.

EXAMPLE 3   Find the maximum and minimum of $P = 2x + 3y$ subject to the following system of linear inequalities:

$$2x + 3y \le 30$$
$$y - x \le 5$$
$$x + y \ge 5$$
$$x \le 10$$
$$x \ge 0, y \ge 0$$

Solution   The feasible set S is shown in Figure 3.13. The vertices of the feasible set S are $A(5, 0)$, $B(10, 0)$, $C(10, 10/3)$, $D(3, 8)$, and $E(0, 5)$. The values of the objective function P at these vertices are given in the following table.

Vertex	$P = 2x + 3y$
$A(5, 0)$	10
$B(10, 0)$	20
$C(10, 10/3)$	30
$D(3, 8)$	30
$E(0, 5)$	15

From the table, we see that the maximum for the objective function $P = 2x + 3y$ occurs at the vertices $C(10, 10/3)$ and $D(3, 8)$. This tells us that every point on the line segment joining the points $C(10, 10/3)$ and $D(3, 8)$ maximizes P, giving it a value of 30 at each of these points. From the table it is also clear that P is minimized at the point $(5, 0)$, where it attains a value of 10.

Figure 3.13
Every point lying on the line segment joining C and D maximizes P.

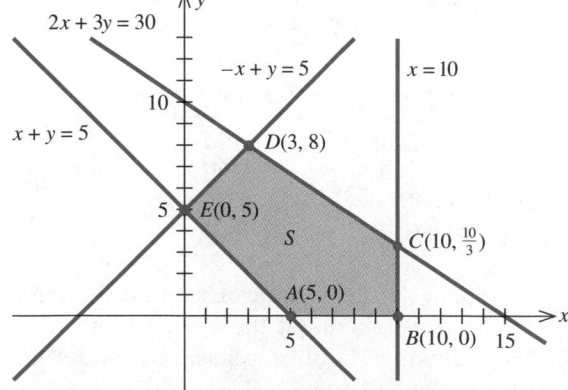

● ● ●

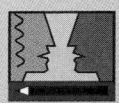

Consider the linear programming problem:

Maximize $P = 2x + 3y$

subject to $2x + \;y \le 10$

$2x + 3y \le 18$

$x \ge 0, y \ge 0$

a. Sketch the feasible set S for the linear programming problem.

b. Draw the isoprofit lines superimposed on S corresponding to $P = 6, 8, 12,$ and 18, and show that these lines are parallel to each other.

c. Show that there are infinitely many solutions to the problem. Is this result as predicted by the method of corners?

Figure 3.14

The maximization problem has no solution, because the feasible set is unbounded.

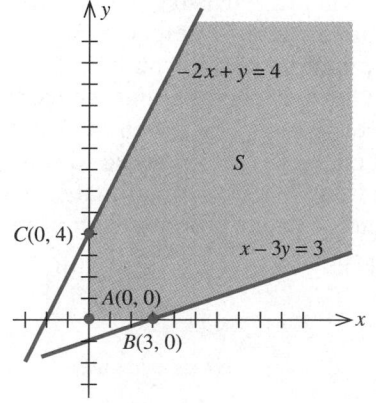

We close this section by examining two situations in which a linear programming problem may have no solution.

EXAMPLE 4   Solve the following linear programming problem:

Maximize $P = x + 2y$

subject to $-2x + \;y \le 4$

$x - 3y \le 3$

$x \ge 0, y \ge 0$

Solution   The feasible set S for this problem is shown in Figure 3.14. Since the set S is unbounded (both x and y can take on arbitrarily large positive values), we see that we can make P as large as we please by making x and y large enough. Therefore, the problem has no solution. In this situation we say that the solution is unbounded.   ○ ○ ○

EXAMPLE 5   Solve the following linear programming problem:

Maximize $P = x + 2y$

subject to $x + 2y \le 4$

$2x + 3y \ge 12$

$x \ge 0, y \ge 0$

Solution   The half-planes described by the constraints (inequalities) have no points in common (Figure 3.15). Therefore, there are no feasible points and the problem has no solution. In this situation, we say that this problem is

Figure 3.15
The problem is inconsistent because there is no point that satisfies all the given inequalities.

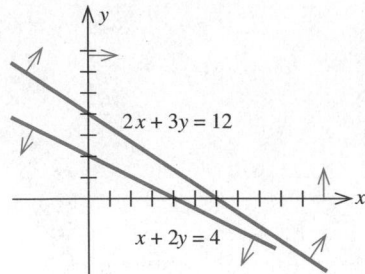

$2x + 3y = 12$

$x + 2y = 4$

infeasible, or **inconsistent.** (These situations are unlikely to occur in well-posed problems arising from practical applications of linear programming.)

○ ○ ○

The method of corners is particularly effective in solving two-variable linear programming problems with a small number of constraints, as the preceding examples have amply demonstrated. Its effectiveness, however, decreases rapidly as the number of variables and/or constraints increases. For example, it may be shown that a linear programming problem in 3 variables and 5 constraints may have up to 10 feasible corner points. The determination of the feasible corner points calls for the solution of ten 3×3 systems of linear equations and then the verification, by the substitution of each of these solutions into the system of constraints, that it is in fact a feasible point. When the number of variables and constraints goes up to 5 and 10, respectively (still a very small system from the standpoint of applications in economics), the number of vertices to be found and checked for feasible corner points increases dramatically to 252, and each of these vertices is found by solving a 5×5 linear system! For this reason, the method of corners is seldom used to solve linear programming problems; its redeeming value lies in the fact that much insight is gained into the nature of the solutions of linear programming problems through its use in solving two-variable problems.

SELF–CHECK EXERCISES 3.3

I. Use the method of corners to solve the following linear programming problem:

$$\text{Maximize} \quad P = 4x + 5y$$
$$\text{subject to} \quad x + 2y \le 10$$
$$5x + 3y \le 30$$
$$x \ge 0, y \ge 0$$

2. Use the method of corners to solve the following linear programming problem:

$$\text{Minimize} \quad C = 5x + 3y$$
$$\text{subject to} \quad 5x + 3y \ge 30$$
$$x - 3y \le 0$$
$$x \ge 2$$

3. Mr. Balduzzi, proprietor of Luigi's Pizza Palace, allocates $9000 a month for advertising in two newspapers, the *City Tribune* and the *Daily News*. The *City Tribune* charges $300 for a certain advertisement, whereas the *Daily News* charges $100 for the same ad. Mr. Balduzzi has stipulated that the ad is to appear in at least 15 but no more than 30 editions of the *Daily News* per month. The *City Tribune* has a daily circulation of 50,000, and the *Daily News* has a circulation of 20,000. Under these conditions, determine how many ads Balduzzi should place in each newspaper in order to reach the largest number of readers.

Solutions to Self-Check Exercises 3.3 can be found on page 205.

3.3 EXERCISES

In exercises 1–6, find the optimal value(s) of the given objective function on the feasible set S.

1. $Z = 2x + 3y$

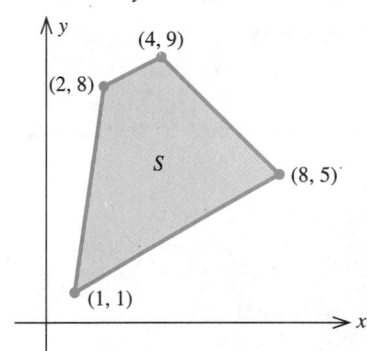

3. $Z = 3x + 4y$

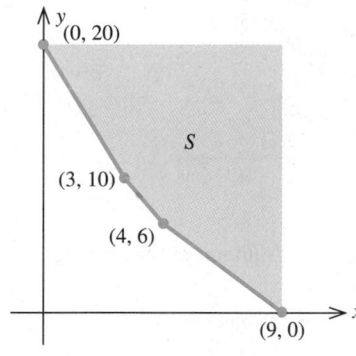

2. $Z = 3x - y$

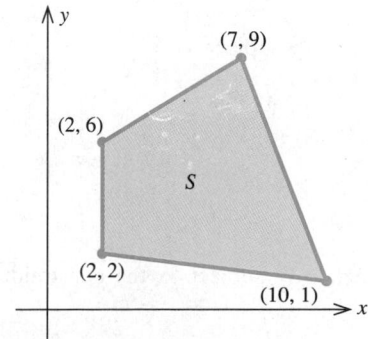

4. $Z = 7x + 9y$

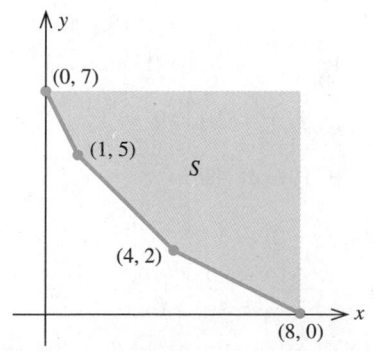

5. $Z = x + 4y$

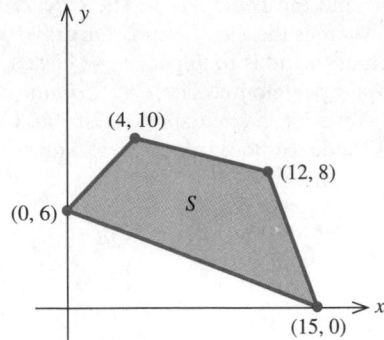

6. $Z = 3x + 2y$

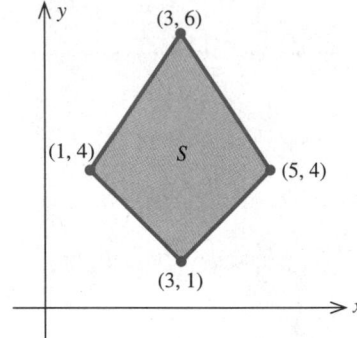

In exercises 7–21, solve each linear programming problem by the method of corners.

7. Maximize $P = 2x + 3y$ subject to

$$x + y \le 6$$
$$x \le 3$$
$$x \ge 0, y \ge 0$$

8. Maximize $P = 3x - 4y$ subject to

$$x + 3y \le 15$$
$$4x + y \le 16$$
$$x \ge 0, y \ge 0$$

9. Minimize $C = 2x + 10y$ subject to

$$5x + 2y \ge 40$$
$$x + 2y \ge 20$$
$$y \ge 3, x \ge 0$$

10. Minimize $C = 2x + 5y$ subject to

$$4x + y \ge 40$$
$$2x + y \ge 30$$
$$x + 3y \ge 30$$
$$x \ge 0, y \ge 0$$

11. Minimize $C = 6x + 3y$ subject to the constraints of exercise 10.

12. Maximize $P = 2x + 5y$ subject to

$$2x + y \le 16$$
$$2x + 3y \le 24$$
$$y \le 6$$
$$x \ge 0, y \ge 0$$

13. Minimize $C = 10x + 15y$ subject to

$$x + y \le 10$$
$$3x + y \ge 12$$
$$-2x + 3y \ge 3$$
$$x \ge 0, y \ge 0$$

14. Maximize $P = 2x + 5y$ subject to the constraints of exercise 13.

15. Maximize $P = 3x + 4y$ subject to

$$x + 2y \le 50$$
$$5x + 4y \le 145$$
$$2x + y \ge 25$$
$$y \ge 5, x \ge 0$$

16. Maximize $P = 4x - 3y$ subject to the constraints of exercise 15.

17. Maximize $P = 2x + 3y$ subject to

$$x + y \le 48$$
$$x + 3y \ge 60$$
$$9x + 5y \le 320$$
$$x \ge 10, y \ge 0$$

18. Minimize $C = 5x + 3y$ subject to the constraints of exercise 17.

19. Find the maximum and minimum of $P = 10x + 12y$ subject to

$$5x + 2y \geq 63$$
$$x + y \geq 18$$
$$3x + 2y \leq 51$$
$$x \geq 0, y \geq 0$$

20. Find the maximum and minimum of $P = 4x + 3y$ subject to

$$3x + 5y \geq 20$$
$$3x + y \leq 16$$
$$-2x + y \leq 1$$
$$x \geq 0, y \geq 0$$

21. Find the maximum and minimum of $P = 2x + 4y$ subject to

$$x + y \leq 20$$
$$-x + y \leq 10$$
$$x \leq 10$$
$$x + y \geq 5$$
$$y \geq 5, x \geq 0$$

22. **Manufacturing—Production Scheduling** A company manufactures two products, A and B, on two machines I and II. It has been determined that the company will realize a profit of $3 on each unit of product A and a profit of $4 on each unit of product B. To manufacture a unit of product A requires 6 minutes on machine I and 5 minutes on machine II. To manufacture a unit of product B requires 9 minutes on machine I and 4 minutes on machine II. There are 5 hours of machine time available on machine I and 3 hours of machine time available on machine II in each work shift. How many units of each product should be produced in each shift to maximize the company's profit?

23. **Manufacturing—Production Scheduling** The National Business Machines Corporation manufactures two models of fax machines: A and B. Each model A costs $200 to make and each model B costs $300. The profits are $25 for each model A and $40 for each model B fax machine. If the total number of fax machines demanded per month does not exceed 2500 and the company has earmarked no more than $600,000 per month for manufacturing costs, find how many units of

each model National should make each month in order to maximize its monthly profits.

24. **Manufacturing—Production Scheduling** The Bata Aerobics Company manufactures two models of steppers used for aerobic exercises. To manufacture each luxury model requires 10 pounds of plastic and 10 minutes of labor. To manufacture each standard model requires 16 pounds of plastic and 8 minutes of labor. The profit for each luxury model is $40 and the profit for each standard model is $30. If 6000 pounds of plastic and 60 hours of labor are available for the production of the steppers per day, how many steppers of each model should Bata produce in order to maximize its profits?

25. **Manufacturing—Production Scheduling** Kane Manufacturing has a division that produces two models of hibachis, model A and model B. To produce each model A hibachi requires 3 pounds of cast iron and 6 minutes of labor. To produce each model B hibachi requires 4 pounds of cast iron and 3 minutes of labor. The profit for each model A hibachi is $2, and the profit for each model B hibachi is $1.50. If 1000 pounds of cast iron and 20 hours of labor are available for the production of hibachis per day, how many hibachis of each model should the division produce in order to help maximize Kane's profits?

26. **Manufacturing—Production Scheduling** Refer to exercise 25. Because of a backlog of orders for model A hibachis, Kane's manager had decided to produce at least 150 of these models a day. Operating under this additional constraint, how many hibachis of each model should Kane produce to maximize profit?

27. **Finance—Allocation of Funds** Madison Finance has a total of $20 million earmarked for homeowner and auto loans. On the average, homeowner loans have a 10% annual rate of return, whereas auto loans yield a 12% annual rate of return. Management has also stipulated that the total amount of homeowner loans should be greater than or equal to four times the total amount of automobile loans. Determine the total amount of loans of each type Madison should extend to each category in order to maximize its returns.

28. **Nutrition—Diet Planning** A nutritionist at the Medical Center has been asked to prepare a special diet for certain patients. She has decided that the meals should contain a minimum of 400 mg of calcium, 10 mg of iron,

and 40 mg of vitamin C. She has further decided that the meals are to be prepared from foods A and B. Each ounce of food A contains 30 mg of calcium, 1 mg of iron, 2 mg of vitamin C, and 2 mg of cholesterol. Each ounce of food B contains 25 mg of calcium, 0.5 mg of iron, 5 mg of vitamin C, and 5 mg of cholesterol. Find how many ounces of each type of food should be used in a meal so that the cholesterol content is minimized and the minimum requirements of calcium, iron, and vitamin C are met.

29. **Agriculture—Crop Planning**   A farmer has 150 acres of land suitable for cultivating crops A and B. The cost of cultivating crop A is $40 per acre, whereas that of crop B is $60 per acre. The farmer has a maximum of $7400 available for land cultivation. Each acre of crop A requires 20 hours of labor and each acre of crop B requires 25 hours of labor. The farmer has a maximum of 3300 hours of labor available. If he expects to make a profit of $150 per acre on crop A and $200 per acre on crop B, how many acres of each crop should he plant in order to maximize his profit?

30. Complete the solution to Example 3, Section 3.2.

31. **Manufacturing—Shipping Costs**   The TMA Company manufactures 19-inch color television picture tubes in two separate locations, locations I and II. The output at location I is at most 6000 tubes per month, whereas the output at location II is at most 5000 per month. TMA is the main supplier of picture tubes to the Pulsar Corporation, its holding company, which has priority in having all its requirements met. In a certain month, Pulsar placed orders for 3000 and 4000 picture tubes to be shipped to two of its factories located in city A and city B, respectively. The shipping costs (in dollars) per picture tube from the two TMA plants to the two Pulsar factories are as follows:

Shipping Costs per Picture Tube		
	To Pulsar Factories	
From	**City A**	**City B**
TMA (Loc. I)	$3	$2
TMA (Loc. II)	$4	$5

Find a shipping schedule that meets the requirements of both companies while keeping costs to a minimum.

32. **Social Programs Planning**   AntiFam, a hunger-relief organization, has earmarked between $2 and $2.5 million, inclusive, for aid to two African countries, country A and country B. Country A is to receive between $1 and $1.5 million, inclusive, in aid, and country B is to receive at least $0.75 million in aid. It has been estimated that each dollar spent in country A will yield an effective return of 60 cents, whereas a dollar spent in country B will yield an effective return of 80 cents. How should the aid be allocated if the money is to be utilized most effectively according to these criteria?

[*Hint:* If x and y denote the amount of money to be given to country A and country B, respectively, then the objective function to be maximized is $P = 0.6x + 0.8y$.]

33. **Investment Planning**   Patricia has at most $30,000 to invest in securities in the form of corporate stocks. She has narrowed her choices to two groups of stocks: growth stocks that she assumes will yield a 15% return (dividends and capital appreciation) within a year, and speculative stocks that she assumes will yield a 25% return (mainly in capital appreciation) within a year. Determine how much she should invest in each group of stocks in order to maximize the return on her investments within a year if she has decided to invest at least three times as much in growth stocks as in speculative stocks.

34. **Veterinary Science**   A veterinarian has been asked to prepare a diet for a group of dogs to be used in a nutrition study at the School of Animal Science. It has been stipulated that each serving should be no larger than 8 ounces and must contain at least 29 units of nutrient I and 20 units of nutrient II. The vet has decided that the diet may be prepared from two brands of dog food: brand A and brand B. Each ounce of brand A contains 3 units of nutrient I and 4 units of nutrient II. Each ounce of brand B contains 5 units of nutrient I and 2 units of nutrient II. Brand A costs 3 cents per ounce and brand B costs 4 cents per ounce. Determine how many ounces of each brand of dog food should be used per serving to meet the given requirements at a minimum cost.

35. **Market Research**   The Trendex Corporation, a telephone survey company, has been hired to conduct a television viewing poll among urban and suburban families in the Los Angeles area. The client has stipulated that a maximum of 1500 families is to be interviewed. At least 500 urban families must be interviewed, and at least half of the total number of families interviewed must be from the suburban area. For this service, Trendex will be paid $1500 plus $2 for each completed interview. From previous experience, Trendex has determined that it will incur an expense of $1.10 for each successful interview with an urban family and $1.25 for each successful interview with a suburban family. Find

how many urban and suburban families Trendex should interview in order to maximize its profit.

36. Suppose you are given the following linear programming problem: Maximize $P = ax + by$, where $a > 0$ and $b > 0$, on the feasible set S shown in the accompanying figure.

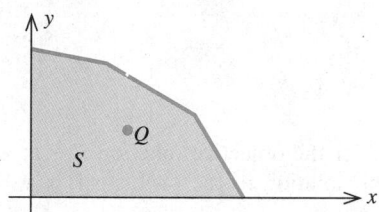

Explain, without using Theorem 1, why the optimal solution of the linear programming problem cannot occur at the point Q.

37. Suppose you are given the following linear programming problem: Maximize $P = ax + by$ on the unbounded feasible set S shown in the accompanying figure.

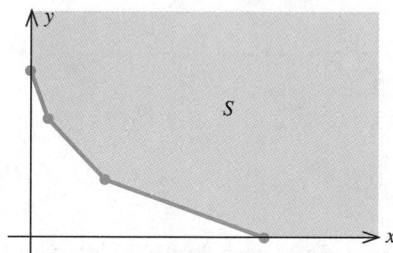

True or false. (Give a reason for your answer.)
a. If $a > 0$ or $b > 0$, then the linear programming problem has no optimal solution.
b. If $a \leq 0$ and $b \leq 0$, then the linear programming problem has at least one optimal solution.

38. Suppose you are given the following linear programming problem: Maximize $P = ax + by$, where $a > 0$ and $b > 0$, on the feasible set S shown in the accompanying figure.

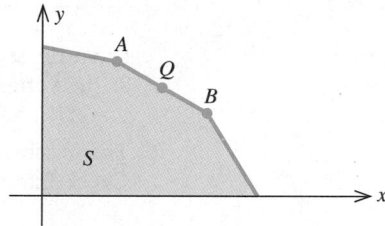

Explain, without using Theorem 1, why the optimal solution of the linear programming problem cannot occur at the point Q unless the problem has infinitely many solutions lying along the line segment joining the vertices A and B.

[*Hint:* Let $A(x_1, y_1)$ and $B(x_2, y_2)$. Then $Q(\bar{x}, \bar{y})$, where $\bar{x} = x_1 + (x_2 - x_1)t$ and $\bar{y} = y_2 + (y_2 - y_1)t$ with $0 < t < 1$. Study the value of P at and near Q.]

SOLUTIONS TO SELF-CHECK EXERCISES 3.3

1. The feasible set S for the problem was graphed in the solution to exercise 1, Self-Check Exercises 3.1. It is reproduced in the accompanying figure.

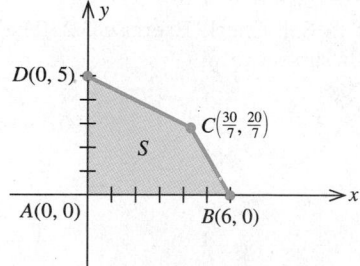

The values of the objective function P at the vertices of S are summarized in the accompanying table.

Vertex	$P = 4x + 5y$
$A(0, 0)$	0
$B(6, 0)$	24
$C(\frac{30}{7}, \frac{20}{7})$	$\frac{220}{7} = 31\frac{3}{7}$
$D(0, 5)$	25

From the table, we see that the maximum for the objective function P is attained at the vertex $C(30/7, 20/7)$. Therefore, the solution to the problem is $x = 30/7$, $y = 20/7$, and $P = 31\frac{3}{7}$.

2. The feasible set S for the problem was graphed in the solution to exercise 2, Self-Check Exercises 3.1. It is reproduced in the accompanying figure.

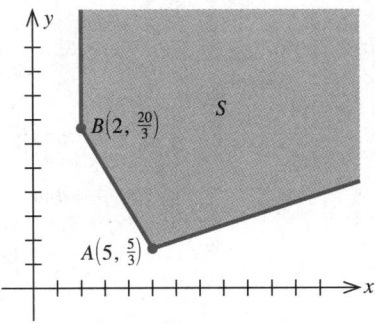

Evaluating the objective function $C = 5x + 3y$ at each corner point, we obtain the table

Vertex	$C = 5x + 3y$
$A(5, \frac{5}{3})$	30
$B(2, \frac{20}{3})$	30

We conclude that the objective function is minimized at every point on the line segment joining the points $(5, 5/3)$ and $(2, 20/3)$ and the minimum value of C is 30.

3. Refer to Self-Check Exercise 3.2. The problem is to maximize $P = 50{,}000x + 20{,}000y$ subject to

$$300x + 100y \leq 9{,}000$$
$$y \geq 15$$
$$y \leq 30$$
$$x \geq 0, y \geq 0$$

The feasible set S for the problem is shown in the accompanying figure.

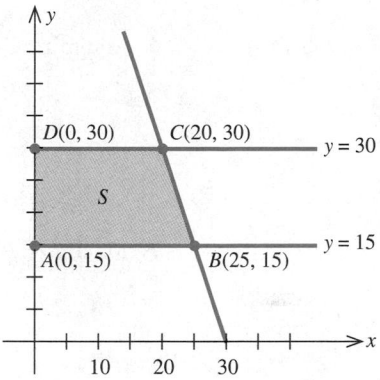

Evaluating the objective function $P = 50{,}000x + 20{,}000y$ at each vertex of S, we obtain

Vertex	$P = 50{,}000x + 20{,}000y$
$A(0, 15)$	300,000
$B(25, 15)$	1,550,000
$C(20, 30)$	1,600,000
$D(0, 30)$	600,000

From the table, we see that P is maximized when $x = 20$ and $y = 30$. Therefore, Balduzzi should place 20 ads in the *City Tribune* and 30 with the *Daily News*.

CHAPTER 3 SUMMARY OF PRINCIPAL TERMS

Terms

Solution set of a system of linear
  inequalities

Bounded solution set

Unbounded solution set

Objective function

Linear programming problem

Feasible solution

Feasible set

Optimal solution

Method of corners

CHAPTER 3 REVIEW EXERCISES

In exercises 1 and 2, find the optimal value(s) of the given objective function on the feasible set S.

1. $Z = 2x + 3y$

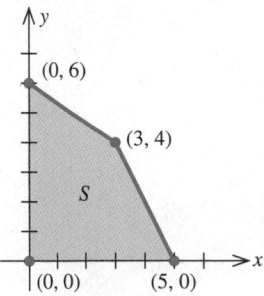

2. $Z = 4x + 3y$

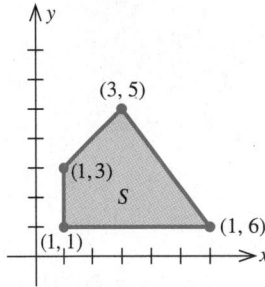

In exercises 3–12, use the method of corners to solve the given linear programming problem.

3. Maximize $P = 3x + 5y$ subject to

$$2x + 3y \leq 12$$
$$x + y \leq 5$$
$$x \geq 0, y \geq 0$$

4. Maximize $P = 2x + 3y$ subject to

$$2x + y \leq 12$$
$$x - 2y \leq 1$$
$$x \geq 0, y \geq 0$$

5. Minimize $C = 2x + 5y$ subject to

$$x + 3y \leq 15$$
$$4x + y \leq 16$$
$$x \geq 0, y \geq 0$$

6. Minimize $C = 3x + 4y$ subject to

$$2x + y \geq 4$$
$$2x + 5y \geq 10$$
$$x \geq 0, y \geq 0$$

7. Maximize $P = 3x + 2y$ subject to

$$2x + y \leq 16$$
$$2x + 3y \leq 36$$
$$4x + 5y \geq 28$$
$$x \geq 0, y \geq 0$$

8. Maximize $P = 6x + 2y$ subject to

$$x + 2y \leq 12$$
$$x + y \leq 8$$
$$2x - 3y \geq 6$$
$$x \geq 0, y \geq 0$$

9. Minimize $C = 2x + 7y$ subject to

$$3x + 5y \geq 45$$
$$3x + 10y \geq 60$$
$$x \geq 0, y \geq 0$$

10. Minimize $C = 4x + y$ subject to

$$6x + y \geq 18$$
$$2x + y \geq 10$$
$$x + 4y \geq 12$$
$$x \geq 0, y \geq 0$$

11. Find the maximum and minimum of $Q = x + y$ subject to

$$5x + 2y \geq 20$$
$$x + 2y \geq 8$$
$$x + 4y \leq 22$$
$$x \geq 0, y \geq 0$$

12. Find the maximum and minimum of $Q = 2x + 5y$ subject to

$$x + \ y \geq 4$$
$$-x + \ y \leq 6$$
$$x + 3y \leq 30$$
$$x \leq 12$$
$$x \geq 0, y \geq 0$$

13. **Financial Analysis**   An investor has decided to commit no more than $80,000 to the purchase of the common stocks of two companies, Company A and Company B. He has also estimated that there is a chance of a 1% capital loss on his investment in Company A and a chance of a 4% loss on his investment in Company B, and he has decided that these losses should not exceed $2000. On the other hand, he expects to make a 14% profit from his investment in Company A and a 20% profit from his investment in Company B. Determine how much he should invest in the stock of each company in order to maximize his investment returns.

14. **Manufacturing—Production Scheduling**   The Soundex Company produces two models of clock radios. Model A requires 15 minutes of work on assembly line I and 10 minutes of work on assembly line II. Model B requires 10 minutes of work on assembly line I and 12 minutes of work on assembly line II. At most, 25 hours of assembly time on line I and 22 hours of assembly time on line II are available per day. It is anticipated that Soundex will realize a profit of $12 on model A and $10 on model B. How many clock radios of each model should be produced per day in order to maximize Soundex's profit?

15. **Manufacturing—Production Scheduling**   Kane Manufacturing has a division that produces two models of hibachis, model A and model B. To produce each model A hibachi requires 3 pounds of cast iron and 6 minutes of labor. To produce each model B hibachi requires 4 pounds of cast iron and 3 minutes of labor. The profit for each model A hibachi is $2 and the profit for each model B hibachi is $1.50. One thousand pounds of cast iron and 20 hours of labor are available for the production of hibachis per day. Because of a backlog of orders for model B hibachis, Kane's manager has decided to produce at least 180 model B hibachis per day. How many hibachis of each model should Kane produce to maximize its profits?

The geometric approach introduced in Chapter 3 may be used to solve linear programming problems involving two or even three variables. But for linear programming problems involving more than two variables, an algebraic approach is preferred. One such technique, the *simplex method*, was developed by George Dantzig in the late 1940s and remains in wide use to this day.

We begin the chapter by developing the simplex method for solving *standard maximization problems*. We then see how, thanks to the principle of duality discovered by the great mathematician John von Neumann, this method can be used to solve a restricted class of *standard minimization problems*. Finally, we see how the simplex method can be adapted to solve *nonstandard problems*—problems that do not belong to the other aforementioned categories.

How much profit? The Ace Novelty Company produces three types of souvenirs. Each type requires a certain amount of time on each of three different machines. Each machine may be operated for a certain amount of time per day. In Example 4, page 223, we will determine how many souvenirs of each type Ace Novelty should make per day in order to maximize its daily profits.

4

LINEAR PROGRAMMING: AN ALGEBRAIC APPROACH

211

4.1 THE SIMPLEX METHOD: STANDARD MAXIMIZATION PROBLEMS

The Simplex Method

As mentioned in Chapter 3, the method of corners is not suitable for solving linear programming problems when the number of variables or constraints is large. Its major shortcoming is that a knowledge of all the corner points of the feasible set S associated with the problem is required. What we need is a method of solution that is based on a judicious selection of the corner points of the feasible set S, thereby reducing the number of points to be inspected. One such technique, called the **simplex method,** was developed in the late 1940s by George Dantzig and is based on the Gauss-Jordan elimination method. The simplex method is readily adaptable to the computer, which makes it ideally suitable for solving linear programming problems involving large numbers of variables and constraints.

Basically, the simplex method is an iterative procedure; that is, it is repeated over and over again. Beginning at some initial feasible solution (a corner point of the feasible set S, usually the origin) each iteration brings us to another corner point of S with an improved (but certainly no worse) value of the objective function. The iteration is terminated when the optimal solution is reached (if it exists).

In this section, we describe the simplex method for solving a large class of problems that are referred to as **standard maximization problems.**

Before stating a formal procedure for solving standard linear programming problems based on the simplex method, let us consider the following analysis of a two-variable problem. The ensuing discussion will clarify the general procedure and at the same time enhance our understanding of the simplex method by examining the motivation that led to the steps of the procedure.

A STANDARD LINEAR PROGRAMMING PROBLEM

A standard maximization problem is one in which

1. the objective function is to be maximized.

2. all the variables involved in the problem are nonnegative.

3. each linear constraint may be written so that the expression involving the variables is less than or equal to a nonnegative constant.

Consider the linear programming problem presented at the beginning of Section 3.3.

$$\text{Maximize} \quad P = 3x + 2y \qquad (1)$$

$$\text{subject to} \quad 2x + 3y \le 12$$

$$2x + \ y \le \ 8 \qquad (2)$$

$$x \ge 0, y \ge 0$$

Figure 4.1

The optimal solution occurs at C(3, 2).

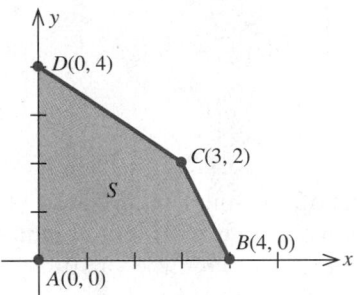

You can easily verify that this is a standard maximization problem. The feasible set S associated with this problem is reproduced in Figure 4.1, where we have labeled the four feasible corner points $A(0, 0)$, $B(4, 0)$, $C(3, 2)$, and $D(0, 4)$. Recall that the optimal solution to the problem occurs at the corner point $C(3, 2)$.

As a first step in the solution using the simplex method, we replace the system of inequality constraints (2) with a system of equality constraints. This may be accomplished by using nonnegative variables called **slack variables.** Let us begin by considering the inequality

$$2x + 3y \leq 12$$

Observe that the left-hand side of this equation is always less than or equal to the right-hand side. Therefore, by adding a nonnegative variable u to the left-hand side to compensate for this difference, we obtain the equality

$$2x + 3y + u = 12$$

For example, if $x = 1$ and $y = 1$ (you can see by referring to Figure 4.1 that the point $(1, 1)$ is a feasible point of S), then $u = 7$. Thus,

$$2(1) + 3(1) + 7 = 12$$

If $x = 2$ and $y = 1$ (the point $(2, 1)$ is also a feasible point of S), then $u = 5$. Thus,

$$2(2) + 3(1) + 5 = 12$$

The variable u is a *slack variable.*

Similarly, the inequality $2x + y \leq 8$ is converted into the equation $2x + y + v = 8$ through the introduction of the slack variable v. The system of linear inequalities (2) may now be viewed as the system of linear equations

$$2x + 3y + u = 12$$
$$2x + y + v = 8$$

where x, y, u, and v are all nonnegative.

Finally, rewriting the objective function (1) in the form $-3x - 2y + P = 0$, where the coefficient of P is $+1$, we are led to the following system of linear equations:

$$
\begin{aligned}
2x + 3y + u &= 12 \\
2x + y + v &= 8 \\
-3x - 2y + P &= 0
\end{aligned}
\qquad (3)
$$

Since the system (3) comprises three linear equations in the five variables x, y, u, v, and P, we may solve for three of the variables in terms of the other two. Thus, there are infinitely many solutions to this system expressible in terms of two parameters. Our linear programming problem is now seen to be equivalent to the following: From among all the solutions of the system (3) for which x, y, u, and v are nonnegative (such solutions are called **feasible solutions**), determine the solution(s) that renders P a maximum.

The augmented matrix associated with system (3) is

nonbasic variables ——————————————— basic variables

$$
\begin{array}{ccccc}
x & y & u & v & P
\end{array}
$$

$$
\left[
\begin{array}{ccc|ccc|c}
2 & 3 & 1 & 0 & 0 & 12 \\
2 & 1 & 0 & 1 & 0 & 8 \\
-3 & -2 & 0 & 0 & 1 & 0
\end{array}
\right]
\tag{4}
$$

Observe that each of the u-, v-, and P-columns of the augmented matrix (4) is a unit column (see page 87). The variables associated with unit columns are called **basic variables;** all other variables are called **nonbasic variables.**

Now, the configuration of the augmented matrix (4) suggests that we solve for the basic variables u, v, and P in terms of the nonbasic variables x and y, obtaining

$$
\begin{aligned}
u &= 12 - 2x - 3y \\
v &= 8 - 2x - y \\
P &= 3x + 2y
\end{aligned}
\tag{5}
$$

Of the infinitely many feasible solutions obtainable by assigning arbitrary nonnegative values to the parameters x and y, a particular solution is obtained by letting $x = 0$ and $y = 0$. In fact, this solution is given by

$$
x = 0, \quad y = 0, \quad u = 12, \quad v = 8, \quad \text{and} \quad P = 0
$$

Such a solution, obtained by setting the nonbasic variables equal to zero, is called a **basic solution** of the system. This particular solution corresponds to the corner point $A(0, 0)$ of the feasible set associated with the linear programming problem (see Figure 4.1). Observe that $P = 0$ at this point.

Now, if the value of P cannot be increased, we have found the optimal solution to the problem at hand. To determine whether the value of P can, in fact, be improved, let us turn our attention to the objective function in (1). Since both the coefficients of x and y are positive, the value of P can be improved by increasing x and/or y—that is, by moving away from the origin. Note that we arrive at the same conclusion by observing that the last row of the augmented matrix (4) contains entries that are *negative*. (Compare the original objective function, $P = 3x + 2y$, with the rewritten objective function, $-3x - 2y + P = 0$.)

Continuing our quest for an optimal solution, our next task is to determine whether it is more profitable to increase the value of x or that of y (increasing x and y simultaneously is more difficult). Since the coefficient of x is greater than that of y, a unit increase in the x-direction will result in a greater increase in the value of the objective function P than a unit increase in the y-direction. Thus, we should increase the value of x while holding y constant. How much can x be increased while holding $y = 0$? Upon setting $y = 0$ in the first two equations of (5), we see that

$$
\begin{aligned}
u &= 12 - 2x \\
v &= 8 - 2x
\end{aligned}
\tag{6}
$$

Since u must be nonnegative, the first equation of (6) implies that x cannot exceed 12/2, or 6. The second equation of (6) and the nonnegativity of v imply that x cannot exceed 8/2, or 4. Thus, we conclude that x can be increased by at most 4.

Now, if we set $y = 0$ and $x = 4$ in system (5), we obtain the solution

$$x = 4, \quad y = 0, \quad u = 4, \quad v = 0, \quad \text{and} \quad P = 12$$

which is a basic solution to system (3), this time with y and v as nonbasic variables. (Recall that the nonbasic variables are precisely the variables that are set equal to zero.)

Let us see how this basic solution may be found by working with the augmented matrix of the system. Since x is to replace v as a basic variable, our aim is to find an augmented matrix that is equivalent to the matrix (4) and has a configuration in which the x-column is in the unit form

$$\begin{bmatrix} 0 \\ 1 \\ 0 \end{bmatrix}$$

replacing what is presently the form of the v-column in (4). Now, this may be accomplished by pivoting about the circled number 2.

$$\begin{array}{ccccc|c} x & y & u & v & P & \\ 2 & 3 & 1 & 0 & 0 & 12 \\ ② & 1 & 0 & 1 & 0 & 8 \\ -3 & -2 & 0 & 0 & 1 & 0 \end{array} \xrightarrow{\frac{1}{2}R_2} \begin{array}{ccccc|c} x & y & u & v & P & \\ 2 & 3 & 1 & 0 & 0 & 12 \\ ① & \frac{1}{2} & 0 & \frac{1}{2} & 0 & 4 \\ -3 & -2 & 0 & 0 & 1 & 0 \end{array} \quad \textbf{(7)}$$

$$\xrightarrow[R_3 + 3R_2]{R_1 - 2R_2} \begin{array}{ccccc|c} x & y & u & v & P & \\ ⓪ & 2 & ① & -1 & ⓪ & 4 \\ ① & \frac{1}{2} & 0 & \frac{1}{2} & 0 & 4 \\ ⓪ & -\frac{1}{2} & ⓪ & \frac{3}{2} & ① & 12 \end{array} \quad \textbf{(8)}$$

Using (8), we now solve for the basic variables x, u, and P in terms of the nonbasic variables y and v, obtaining

$$x = 4 - (1/2)y - (1/2)v$$
$$u = 4 - 2y + v$$
$$P = 12 + (1/2)y - (3/2)v$$

Setting the nonbasic variables y and v equal to zero gives

$$x = 4, \quad y = 0, \quad u = 4, \quad v = 0, \quad \text{and} \quad P = 12$$

as before.

We have now completed one iteration of the simplex procedure, and our search has brought us from the feasible corner point $A(0, 0)$, where $P = 0$, to the feasible corner point $B(4, 0)$, where P attained a value of 12, which is certainly an improvement! (See Figure 4.2.)

Figure 4.2
One iteration has taken us from $A(0, 0)$, where $P = 0$, to $B(4, 0)$, where $P = 12$.

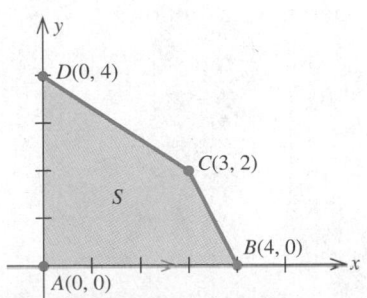

Before going on, let us introduce the following terminology. The circled element 2 in the first augmented matrix of (7), which was to be converted into a 1, is called a *pivot element*. The column containing the pivot element is called the *pivot column*. The pivot column is associated with a nonbasic variable that is to be converted to a basic variable. Note that *the last entry in the pivot column is the negative number with the largest absolute value to the left of the vertical line in the last row*—precisely the criterion for choosing the direction of maximum increase in P.

The row containing the pivot element is called the *pivot row*. The pivot row can also be found by dividing each positive number in the pivot column into the corresponding number in the last column (the column of constants). *The pivot row is the one with the smallest ratio*. In the augmented matrix (7), the pivot row is the second row since the ratio 8/2, or 4, is less than the ratio 12/2, or 6. (Compare this with the earlier analysis pertaining to the determination of the largest permissible increase in the value of x.)

The following is a summary of the procedure for selecting the pivot element:

SELECTING THE PIVOT ELEMENT	**1.** Select the pivot column: Locate the most negative entry to the left of the vertical line in the last row. The column containing this entry is the **pivot column.** (If there is more than one such column, choose any one.)
	2. Select the pivot row: Divide each positive entry in the pivot column into its corresponding entry in the column of constants. The **pivot row** is the row corresponding to the smallest ratio thus obtained. (If there is more than one such entry, choose any one.)
	3. The **pivot element** is the element common to both the pivot column and the pivot row.

Continuing with the solution to our problem, we observe that the last row of the augmented matrix (8) contains a negative number—namely, $-1/2$. This indicates that P is not maximized at the feasible corner point $B(4, 0)$, and another iteration is required. Without once again going into a detailed analysis, we proceed immediately to the selection of a pivot element. In accordance with the rules, we perform the necessary row operations as follows:

$$
\begin{array}{c}
 \\
\text{pivot} \\
\text{row}
\end{array}
\rightarrow
\begin{array}{ccccc}
x & y & u & v & P \\
\end{array}
\left[
\begin{array}{ccccc|c}
0 & \textcircled{2} & 1 & -1 & 0 & 4 \\
1 & \frac{1}{2} & 0 & \frac{1}{2} & 0 & 4 \\
0 & -\frac{1}{2} & 0 & \frac{3}{2} & 1 & 12
\end{array}
\right]
\begin{array}{c}
\text{ratio} \\
\frac{4}{2} = 2 \\
\frac{4}{1/2} = 8 \\

\end{array}
$$

$$
\begin{array}{c}
\uparrow \\
\text{pivot} \\
\text{column}
\end{array}
$$

Figure 4.3

The next iteration has taken us from
$B(4, 0)$, *where* $P = 12$, *to* $C(3, 2)$,
where $P = 13$.

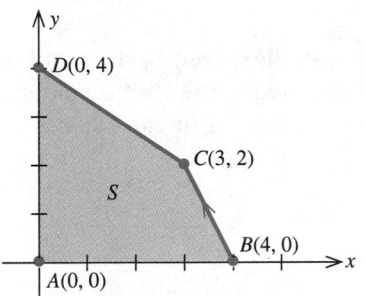

$$\xrightarrow{\frac{1}{2}R_1}
\begin{array}{c}
\begin{array}{cccccc} x & y & u & v & P & \end{array} \\
\begin{bmatrix}
0 & 1 & \frac{1}{2} & -\frac{1}{2} & 0 & 2 \\
1 & \frac{1}{2} & 0 & \frac{1}{2} & 0 & 4 \\
0 & -\frac{1}{2} & 0 & \frac{3}{2} & 1 & 12
\end{bmatrix}
\end{array}$$

$$\xrightarrow[R_3 + \frac{1}{2}R_1]{R_2 - \frac{1}{2}R_1}
\begin{array}{c}
\begin{array}{cccccc} x & y & u & v & P & \end{array} \\
\begin{bmatrix}
0 & 1 & \frac{1}{2} & -\frac{1}{2} & 0 & 2 \\
1 & 0 & -\frac{1}{4} & \frac{3}{4} & 0 & 3 \\
0 & 0 & \frac{1}{4} & \frac{5}{4} & 1 & 13
\end{bmatrix}
\end{array}$$

Interpreting the last augmented matrix in the usual fashion, we find the basic solution $x = 3$, $y = 2$, and $P = 13$. Since there are no negative entries in the last row, the solution is optimal and P cannot be increased further. The optimal solution is the feasible corner point $C(3, 2)$ (Figure 4.3). Observe that this agrees with the solution we found using the method of corners in Section 3.3.

Having seen how the simplex method works, let us list the steps involved in the procedure. The first step is to set up the initial **simplex tableau.**

SETTING UP THE INITIAL SIMPLEX TABLEAU

1. Transform the system of linear inequalities into a system of linear equations by introducing slack variables.

2. Rewrite the objective function

$$P = c_1x_1 + c_2x_2 + \cdots + c_nx_n$$

in the form

$$-c_1x_1 - c_2x_2 - \cdots - c_nx_n + P = 0$$

where all the variables are on the left and the coefficient of P is $+1$. Write this equation below the equations of part (1).

3. Write the augmented matrix associated with this system of linear equations.

EXAMPLE 1 Set up the initial simplex tableau for the linear programming problem posed in Example 1, Section 3.2.

Solution The problem at hand is to maximize

$$P = x + 1.2y$$

or, equivalently,

$$P = x + \frac{6}{5}y$$

subject to

$$2x + y \leq 180$$
$$x + 3y \leq 300 \tag{9}$$
$$x \geq 0, y \geq 0$$

This is a standard maximization problem and may be solved by the simplex method. Since the system (9) comprises two linear inequalities, we introduce the two slack variables u and v to convert it to a system of linear equations:

$$2x + y + u \quad = 180$$
$$x + 3y \quad + v = 300$$

Next, by rewriting the objective function in the form

$$-x - \frac{6}{5}y + P = 0$$

where the coefficient of P is $+1$, and placing it below the system of equations, we obtain the system of linear equations

$$2x + y + u \qquad = 180$$
$$x + 3y \qquad + v \qquad = 300$$
$$-x - \frac{6}{5}y \qquad + P = \quad 0$$

The initial simplex tableau associated with this system is

x	y	u	v	P	Constant
2	1	1	0	0	180
1	3	0	1	0	300
-1	$-\frac{6}{5}$	0	0	1	0

Before completing the solution to the problem posed in Example 1, let us summarize the main steps of the **simplex method.**

THE SIMPLEX METHOD

1. *Set up the initial simplex tableau.*
2. *Determine whether the optimal solution has been reached by examining all of the entries in the last row to the left of the vertical line:*
   a. If all the entries are nonnegative, the optimal solution has been reached. Proceed to step 4.
   b. If there are one or more negative entries, the optimal solution has not been reached. Proceed to step 3.
3. *Perform the pivot operation:*
   Locate the pivot element and convert it to a 1 by dividing all the elements in the pivot row by the pivot element. Using row operations, convert
   (*Continued on next page*)

(*Continued from previous page*)
the pivot column into a unit column by adding suitable multiples of the pivot row to each of the other rows as required. Return to step 2.

4. *Determine the optimal solution(s):*
The value of the variable heading each unit column is given by the entry lying in the column of constants in the row containing the 1. The variables heading columns not in unit form are assigned the value zero.

EXAMPLE 2 Complete the solution to the problem discussed in Example 1.

Solution The first step in our procedure, setting up the initial simplex tableau, was completed in Example 1. We continue with step 2.

Step 2 *Determine whether the optimal solution has been reached.*
First, refer to the initial simplex tableau:

x	y	u	v	P	Constant
2	1	1	0	0	180
1	3	0	1	0	300
-1	$-\frac{6}{5}$	0	0	1	0

(10)

Since there are negative entries in the last row of the initial simplex tableau, the initial solution is not optimal. We proceed to step 3.

Step 3 *Perform the following iterations.*
First, locate the pivot element:

a. Since the entry $-6/5$ is the most negative entry to the left of the vertical line in the last row of the initial simplex tableau, the second column in the tableau is the pivot column.
b. Divide each positive number of the pivot column into the corresponding entry in the column of constants and compare the ratios thus obtained. We see that the ratio 300/3 is less than the ratio 180/1, so the second row in the tableau is the pivot row.
c. The entry 3 lying in the pivot column and the pivot row is the pivot element.

Next, we convert this pivot element into a 1 by multiplying all the entries in the pivot row by 1/3. Then, using elementary row operations, we complete the conversion of the pivot column into a unit column.

The details of the iteration are recorded as follows:

	x	y	u	v	P	Constant
	2	1	1	0	0	180
pivot row →	1	③	0	1	0	300
	-1	$-\frac{6}{5}$	0	0	1	0

$$\begin{bmatrix} \text{ratio} \\ \frac{180}{1} = 180 \\ \frac{300}{3} = 100 \end{bmatrix}$$

↑
pivot
column

	x	y	u	v	P	Constant
$\xrightarrow{\frac{1}{3}R_2}$	2	1	1	0	0	180
	$\frac{1}{3}$	①	0	$\frac{1}{3}$	0	100
	-1	$-\frac{6}{5}$	0	0	1	0

	x	y	u	v	P	Constant
$\xrightarrow{R_1 - R_2}$	$\frac{5}{3}$	0	1	$-\frac{1}{3}$	0	80
$\xrightarrow{R_3 + \frac{6}{5}R_2}$	$\frac{1}{3}$	1	0	$\frac{1}{3}$	0	100
	$-\frac{3}{5}$	0	0	$\frac{2}{5}$	1	120

(11)

This completes one iteration. The last row of the simplex tableau contains a negative number, so an optimal solution has not been reached. Therefore, we repeat the iterative step once again, as follows:

	x	y	u	v	P	Constant
pivot row →	⑤⁄₃	0	1	$-\frac{1}{3}$	0	80
	$\frac{1}{3}$	1	0	$\frac{1}{3}$	0	100
	$-\frac{3}{5}$	0	0	$\frac{2}{5}$	1	120

$$\begin{bmatrix} \text{ratio} \\ \frac{80}{5/3} = 48 \\ \frac{100}{1/3} = 300 \end{bmatrix}$$

↑
pivot
column

	x	y	u	v	P	Constant
$\xrightarrow{\frac{3}{5}R_1}$	①	0	$\frac{3}{5}$	$-\frac{1}{5}$	0	48
	$\frac{1}{3}$	1	0	$\frac{1}{3}$	0	100
	$-\frac{3}{5}$	0	0	$\frac{2}{5}$	1	120

	x	y	u	v	P	Constant
$\xrightarrow{R_2 - \frac{1}{3}R_1}$	1	0	$\frac{3}{5}$	$-\frac{1}{5}$	0	48
$\xrightarrow{R_3 + \frac{3}{5}R_1}$	0	1	$-\frac{1}{5}$	$\frac{2}{5}$	0	84
	0	0	$\frac{9}{25}$	$\frac{7}{25}$	1	$148\frac{4}{5}$

(12)

The last row of the simplex tableau (12) contains no negative numbers, and we therefore conclude that the optimal solution has been reached.

Step 4   *Determine the optimal solution.*

Locate the basic variables in the final tableau. In this case, the basic variables (those heading unit columns) are x, y, and P. The value assigned to the basic variable x is the number 48, which is the entry lying in the column of constants and in the first row (the row that contains the 1).

x	y	u	v	P	Constant
①	0	$\frac{3}{5}$	$-\frac{1}{5}$	0	48
0	①	$-\frac{1}{5}$	$\frac{2}{5}$	0	84
0	0	$\frac{9}{25}$	$\frac{7}{25}$	①	$148\frac{4}{5}$

Similarly, we conclude that $y = 84$ and $P = 148.8$. Next, we note that the variables u and v are nonbasic and are accordingly assigned the values $u = 0$ and $v = 0$. These results agree with those obtained in Example 1, Section 3.3.   ⟐ ⟐ ⟐

EXAMPLE 3   Maximize $P = 2x + 2y + z$ subject to

$$2x + y + 2z \leq 14$$
$$2x + 4y + z \leq 26$$
$$x + 2y + 3z \leq 28$$
$$x \geq 0, y \geq 0, z \geq 0$$

Solution   Introducing the slack variables u, v, and w and rewriting the objective function in the standard form gives the system of linear equations

$$2x + y + 2z + u = 14$$
$$2x + 4y + z + v = 26$$
$$x + 2y + 3z + w = 28$$
$$-2x - 2y - z + P = 0$$

The initial simplex tableau is given by

x	y	z	u	v	w	P	Constant
2	1	2	1	0	0	0	14
2	4	1	0	1	0	0	26
1	2	3	0	0	1	0	28
−2	−2	−1	0	0	0	1	0

Since the most negative entry in the last row (-2) occurs twice, we may choose either the x- or the y-column as the pivot column. Choosing the x-column as the pivot column and proceeding with the first iteration, we obtain the following sequence of tableaus:

	x	y	z	u	v	w	P	Constant	ratio
pivot row →	②	1	2	1	0	0	0	14	$\frac{14}{2} = 7$
	2	4	1	0	1	0	0	26	$\frac{26}{2} = 13$
	1	2	3	0	0	1	0	28	$\frac{28}{1} = 28$
	-2	-2	-1	0	0	0	1	0	

$$\underset{\substack{\uparrow \\ \text{pivot} \\ \text{column}}}{}$$

	x	y	z	u	v	w	P	Constant
$\xrightarrow{\frac{1}{2}R_1}$	①	$\frac{1}{2}$	1	$\frac{1}{2}$	0	0	0	7
	2	4	1	0	1	0	0	26
	1	2	3	0	0	1	0	28
	-2	-2	-1	0	0	0	1	0

	x	y	z	u	v	w	P	Constant
	1	$\frac{1}{2}$	1	$\frac{1}{2}$	0	0	0	7
$\xrightarrow{\substack{R_2 - 2R_1 \\ R_3 - R_1 \\ R_4 + 2R_1}}$	0	3	-1	-1	1	0	0	12
	0	$\frac{3}{2}$	2	$-\frac{1}{2}$	0	1	0	21
	0	-1	1	1	0	0	1	14

Since there is a negative number in the last row of the simplex tableau, we perform another iteration, as follows:

	x	y	z	u	v	w	P	Constant	ratio
	1	$\frac{1}{2}$	1	$\frac{1}{2}$	0	0	0	7	$\frac{7}{1/2} = 14$
pivot row →	0	③	-1	-1	1	0	0	12	$\frac{12}{3} = 4$
	0	$\frac{3}{2}$	2	$-\frac{1}{2}$	0	1	0	21	$\frac{21}{3/2} = 14$
	0	-1	1	1	0	0	1	14	

$$\underset{\substack{\uparrow \\ \text{pivot} \\ \text{column}}}{}$$

	x	y	z	u	v	w	P	Constant
	1	$\frac{1}{2}$	1	$\frac{1}{2}$	0	0	0	7
$\xrightarrow{\frac{1}{3}R_2}$	0	①	$-\frac{1}{3}$	$-\frac{1}{3}$	$\frac{1}{3}$	0	0	4
	0	$\frac{3}{2}$	2	$-\frac{1}{2}$	0	1	0	21
	0	-1	1	1	0	0	1	14

	x	y	z	u	v	w	P	Constant
	1	0	$\frac{7}{6}$	$\frac{2}{3}$	$-\frac{1}{6}$	0	0	5
$\xrightarrow{\begin{array}{c}R_1-\frac{1}{2}R_2\\R_3-\frac{3}{2}R_2\\R_4+R_2\end{array}}$	0	1	$-\frac{1}{3}$	$-\frac{1}{3}$	$\frac{1}{3}$	0	0	4
	0	0	$\frac{5}{2}$	0	$-\frac{1}{2}$	1	0	15
	0	0	$\frac{2}{3}$	$\frac{2}{3}$	$\frac{1}{3}$	0	1	18

All entries in the last row are nonnegative, so we have reached the optimal solution. We conclude that $x = 5$, $y = 4$, $z = 0$, $u = 0$, $v = 0$, $w = 15$, and $P = 18$.

○ ○ ○

Consider the linear programming problem

Maximize $P = x + 2y$
subject to $-2x + y \le 4$
$x - 3y \le 3$
$x \ge 0, y \ge 0$

a. Sketch the feasible set S for the linear programming problem and explain why the problem has an unbounded solution.
b. Use the simplex method to solve the problem as follows:
i. Perform one iteration on the initial simplex tableau and interpret your result. Indicate the point on S corresponding to this (nonoptimal) solution.
ii. Show that the simplex procedure breaks down when you attempt to perform another iteration by demonstrating that there is no pivot element.
iii. Describe what happens if you violate the rule for finding the pivot element by allowing the ratios to be negative and proceeding with the iteration.

Application

EXAMPLE 4   The Ace Novelty Company has determined that the profits are $6, $5, and $4 for each type-A, type-B, and type-C souvenir that it plans to produce. To manufacture a type-A souvenir requires 2 minutes on machine I, 1 minute on machine II, and 2 minutes on machine III. A type-B souvenir requires 1 minute on machine I, 3 minutes on machine II, and 1 minute on machine III. A type-C souvenir requires 1 minute on machine I and 2 minutes on each of machines II and III. Each day there are 3 hours available on machine I, 5 hours available on machine II, and 4 hours available on machine III for manufacturing these souvenirs. How many souvenirs of each type should Ace Novelty make per day in order to maximize its profit? (Compare with Example 1, Section 2.1.)

Solution   The given information may be tabulated as follows:

	Type A	Type B	Type C	Time Available (min)
Machine I	2	1	1	180
Machine II	1	3	2	300
Machine III	2	1	2	240
Profit per Unit ($)	6	5	4	

Let x, y, and z denote the respective numbers of type-A, type-B, and type-C souvenirs to be made. The total amount of time that machine I is used is given by $2x + y + z$ minutes and must not exceed 180 minutes. Thus, we have the inequality

$$2x + y + z \leq 180$$

Similar considerations on the use of machines II and III lead to the inequalities

$$x + 3y + 2z \leq 300$$
$$2x + y + 2z \leq 240$$

The profit resulting from the sale of the souvenirs produced is given by

$$P = 6x + 5y + 4z$$

The mathematical formulation of this problem has led to the following standard linear programming problem: Maximize the objective (profit) function $P = 6x + 5y + 4z$ subject to

$$2x + y + z \leq 180$$
$$x + 3y + 2z \leq 300$$
$$2x + y + 2z \leq 240$$
$$x \geq 0, y \geq 0, z \geq 0$$

Introducing the slack variables u, v, and w gives the system of linear equations

$$2x + y + z + u \qquad\qquad = 180$$
$$x + 3y + 2z \qquad + v \qquad\quad = 300$$
$$2x + y + 2z \qquad\quad + w \qquad = 240$$
$$-6x - 5y - 4z \qquad\qquad + P = 0$$

The tableaus resulting from the use of the simplex algorithm are

	x	y	z	u	v	w	P	Constant	ratio
pivot row →	②	1	1	1	0	0	0	180	$\frac{180}{2} = 90$
	1	3	2	0	1	0	0	300	$\frac{300}{1} = 300$
	2	1	2	0	0	1	0	240	$\frac{240}{2} = 120$
	-6	-5	-4	0	0	0	1	0	

$\uparrow$
pivot
column

	x	y	z	u	v	w	P	Constant
	①	$\frac{1}{2}$	$\frac{1}{2}$	$\frac{1}{2}$	0	0	0	90
	1	3	2	0	1	0	0	300
	2	1	2	0	0	1	0	240
	−6	−5	−4	0	0	0	1	0

$\xrightarrow{\frac{1}{2}R_1}$

	x	y	z	u	v	w	P	Constant	ratio
	1	$\frac{1}{2}$	$\frac{1}{2}$	$\frac{1}{2}$	0	0	0	90	$\frac{90}{1/2} = 180$
	0	⑤⁄₂	$\frac{3}{2}$	$-\frac{1}{2}$	1	0	0	210	$\frac{210}{5/2} = 84$
	0	0	1	−1	0	1	0	60	
	0	−2	−1	3	0	0	1	540	

$\xrightarrow[\substack{R_2 - R_1 \\ R_3 - 2R_1 \\ R_4 + 6R_1}]{}$ pivot row

↑ pivot column

	x	y	z	u	v	w	P	Constant
	1	$\frac{1}{2}$	$\frac{1}{2}$	$\frac{1}{2}$	0	0	0	90
	0	①	$\frac{3}{5}$	$-\frac{1}{5}$	$\frac{2}{5}$	0	0	84
	0	0	1	−1	0	1	0	60
	0	−2	−1	3	0	0	1	540

$\xrightarrow{\frac{2}{5}R_2}$

	x	y	z	u	v	w	P	Constant
	1	0	$\frac{1}{5}$	$\frac{3}{5}$	$-\frac{1}{5}$	0	0	48
	0	1	$\frac{3}{5}$	$-\frac{1}{5}$	$\frac{2}{5}$	0	0	84
	0	0	1	−1	0	1	0	60
	0	0	$\frac{1}{5}$	$\frac{13}{5}$	$\frac{4}{5}$	0	1	708

$\xrightarrow[\substack{R_1 - \frac{1}{2}R_2 \\ R_4 + 2R_2}]{}$

From the final simplex tableau, we read off the solution

$$x = 48, \quad y = 84, \quad z = 0, \quad u = 0, \quad v = 0, \quad w = 60, \quad \text{and} \quad P = 708$$

Thus, in order to maximize its profit, Ace Novelty should produce 48 type-A souvenirs, 84 type-B souvenirs, and no type-C souvenirs. The resulting profit is $708 per day. The value of the slack variable $w = 60$ tells us that 1 hour of the available time on machine III is left unused.   ○ ○ ○

Interpreting Our Results   It is instructive to compare the results obtained here with those obtained in Example 7, Section 2.2. Recall that in order to use all of the machine time available on each of the three machines, Ace Novelty had to produce 36 type-A, 48 type-B, and 60 type-C souvenirs. This would have resulted in a profit of $696. Example 4 shows how, through the optimal use of equipment, a company may boost its profit and at the same time reduce machine wear!

Multiple Solutions and Unbounded Solutions

As we saw in Section 3.3, a linear programming problem may have infinitely many solutions. We also saw that a linear programming problem may have no solution or an unbounded solution. How do we spot each of these phenomena when using the simplex method to solve a problem?

A linear programming problem may have infinitely many solutions if and only if the last row to the left of the vertical line of the final simplex tableau has a zero in a column that is not a unit column. Next, a linear programming problem will have no solution if the simplex method breaks down at some stage. For example, if at some stage there are no nonnegative ratios in our computation, then the linear programming problem has no solution.

Consider the linear programming problem

$$\text{Maximize} \quad P = 4x + 6y$$
$$\text{subject to} \quad 2x + \ y \le 10$$
$$2x + 3y \le 18$$
$$x \ge 0, y \ge 0$$

a. Sketch the feasible set for the linear programming problem.
b. Use the method of corners to show that there are infinitely many optimal solutions. What are they?
c. Use the simplex method to solve the problem as follows:
 i. Perform one iteration on the initial simplex tableau and conclude that you have arrived at an optimal solution. What is the value of P and where is it attained? Compare this result with that obtained in part (b).
 ii. Observe that the tableau obtained in part (i) indicates that there are infinitely many solutions (see the comment on multiple solutions on this page).
Now perform another iteration on the simplex tableau using the x-column as the pivot column. Interpret the final tableau.

SELF-CHECK EXERCISES 4.1

1. Solve the following linear programming problem by the simplex method:

$$\text{Maximize} \quad P = 2x + 3y + 6z$$
$$\text{subject to} \quad 2x + 3y + z \le 10$$
$$x + y + 2z \le 8$$
$$2y + 3z \le 6$$
$$x \ge 0, y \ge 0, z \ge 0$$

2. The LaCrosse Iron Works makes two models of cast iron fireplace grates, model A and model B. Producing one model A grate requires 20 pounds of cast iron and 20 minutes of labor, whereas producing one model B grate requires 30 pounds of cast iron and 15 minutes of labor. The profit for a model A grate is $6 and the profit for a model B grate is $8. There are 7200 pounds of cast iron and 100 hours of labor available per week. Because of a surplus from the previous week, the proprietor has decided that he should make no more than 150 units of model A grates this week. Determine how many of each model he should make in order to maximize his profits.

Solutions to Self-Check Exercises 4.1 can be found on page 234.

4.1 EXERCISES

In exercises 1–9, determine whether the given simplex tableau is in final form. If so, find the solution to the associated regular linear programming problem. If not, find the pivot element to be used in the next iteration of the simplex method.

1.

x	y	u	v	P	Constant
0	1	$\frac{5}{7}$	$-\frac{1}{7}$	0	$\frac{20}{7}$
1	0	$-\frac{3}{7}$	$\frac{2}{7}$	0	$\frac{30}{7}$
0	0	$\frac{13}{7}$	$\frac{3}{7}$	1	$\frac{220}{7}$

2.

x	y	u	v	P	Constant
1	1	1	0	0	6
1	0	-1	1	0	2
3	0	5	0	1	30

3.

x	y	u	v	P	Constant
0	$\frac{1}{2}$	1	$-\frac{1}{2}$	0	2
1	$\frac{1}{2}$	0	$\frac{1}{2}$	0	4
0	$-\frac{1}{2}$	0	$\frac{3}{2}$	1	12

4.

x	y	z	u	v	w	P	Constant
3	0	5	1	1	0	0	28
2	1	3	0	1	0	0	16
2	0	8	0	3	0	1	48

5.

x	y	z	u	v	w	P	Constant
1	$-\frac{1}{3}$	0	$\frac{1}{3}$	0	$-\frac{2}{3}$	0	$\frac{1}{3}$
0	2	0	0	1	1	0	6
0	$\frac{2}{3}$	1	$\frac{1}{3}$	0	$\frac{1}{3}$	0	$\frac{13}{3}$
0	4	0	1	0	2	1	17

6.

x	y	z	u	v	w	P	Constant
$\frac{1}{2}$	0	$\frac{1}{4}$	1	$-\frac{1}{4}$	0	0	$\frac{19}{2}$
$\frac{1}{2}$	1	$\frac{3}{4}$	0	$\frac{1}{4}$	0	0	$\frac{21}{2}$
2	0	3	0	0	1	0	30
-1	0	$-\frac{1}{2}$	6	$\frac{3}{2}$	0	1	63

7.

x	y	z	s	t	u	v	P	Constant
$\frac{5}{2}$	3	0	1	0	0	-4	0	46
1	0	0	0	1	0	0	0	9
0	1	0	0	0	1	0	0	12
0	0	1	0	0	0	1	0	6
-180	-200	0	0	0	0	300	1	1800

8.

x	y	z	s	t	u	v	P	Constant
1	0	0	$\frac{2}{5}$	0	$-\frac{6}{5}$	$-\frac{8}{5}$	0	4
0	0	0	$-\frac{2}{5}$	1	$\frac{6}{5}$	$\frac{8}{5}$	0	5
0	1	0	0	0	1	0	0	12
0	0	1	0	0	0	1	0	6
0	0	0	72	0	-16	12	1	4920

9.

x	y	z	u	v	P	Constant
1	1	$\frac{3}{5}$	0	$\frac{1}{5}$	0	30
0	0	$-\frac{19}{5}$	1	$-\frac{3}{5}$	0	10
0	0	$\frac{26}{5}$	0	$\frac{2}{5}$	1	60

In exercises 10–23, solve each linear programming problem by the simplex method.

10. Maximize $P = 5x + 3y$ subject to
$$x + y \le 80$$
$$3x \le 90$$
$$x \ge 0, y \ge 0$$

11. Maximize $P = 10x + 12y$ subject to
$$x + 2y \le 12$$
$$3x + 2y \le 24$$
$$x \ge 0, y \ge 0$$

12. Maximize $P = 5x + 4y$ subject to
$$3x + 5y \le 78$$
$$4x + y \le 36$$
$$x \ge 0, y \ge 0$$

13. Maximize $P = 4x + 6y$ subject to
$$3x + y \le 24$$
$$2x + y \le 18$$
$$x + 3y \le 24$$
$$x \ge 0, y \ge 0$$

14. Maximize $P = 15x + 12y$ subject to
$$x + y \le 12$$
$$3x + y \le 30$$
$$10x + 7y \le 70$$
$$x \ge 0, y \ge 0$$

15. Maximize $P = 3x + 4y + 5z$ subject to
$$x + y + z \le 8$$
$$3x + 2y + 4z \le 24$$
$$x \ge 0, y \ge 0, z \ge 0$$

16. Maximize $P = 3x + 3y + 4z$ subject to
$$x + y + 3z \le 15$$
$$4x + 4y + 3z \le 65$$
$$x \ge 0, y \ge 0, z \ge 0$$

17. Maximize $P = 3x + 4y + z$ subject to
$$3x + 10y + 5z \le 120$$
$$5x + 2y + 8z \le 6$$
$$8x + 10y + 3z \le 105$$
$$x \ge 0, y \ge 0, z \ge 0$$

18. Maximize $P = x + 2y - z$ subject to
$$2x + y + z \le 14$$
$$4x + 2y + 3z \le 28$$
$$2x + 5y + 5z \le 30$$
$$x \ge 0, y \ge 0, z \ge 0$$

19. Maximize $P = 4x + 6y + 5z$ subject to
$$x + y + z \le 20$$
$$2x + 4y + 3z \le 42$$
$$2x + 3z \le 30$$
$$x \ge 0, y \ge 0, z \ge 0$$

20. Maximize $P = x + 4y - 2z$ subject to
$$3x + y - z \le 80$$
$$2x + y - z \le 40$$
$$-x + y + z \le 80$$
$$x \ge 0, y \ge 0, z \ge 0$$

21. Maximize $P = 12x + 10y + 5z$ subject to
$$2x + y + z \le 10$$
$$3x + 5y + z \le 45$$
$$2x + 5y + z \le 40$$
$$x \ge 0, y \ge 0, z \ge 0$$

22. Maximize $P = 2x + 6y + 6z$ subject to

$$2x + y + 3z \leq 10$$
$$4x + y + 2z \leq 56$$
$$6x + 4y + 3z \leq 126$$
$$2x + y + z \leq 32$$
$$x \geq 0, y \geq 0, z \geq 0$$

23. Maximize $P = 4x + y + 5z$ subject to

$$x + 2y + 5z \leq 30$$
$$2x + y + z \leq 10$$
$$3x + 2y + z \leq 12$$
$$x + y + z \leq 8$$
$$x \geq 0, y \geq 0, z \geq 0$$

24. Rework Example 3 using the y-column as the pivot column in the first iteration of the simplex method.

25. Show that the following linear programming problem: Maximize $P = 2x + 2y - 4z$ subject to

$$3x + 3y - 2z \leq 100$$
$$5x + 5y + 3z \leq 150$$
$$x \geq 0, y \geq 0, z \geq 0$$

has optimal solutions $x = 30, y = 0, z = 0, P = 60$ and $x = 0, y = 30, z = 0$, and $P = 60$.

26. Manufacturing—Production Scheduling A company manufactures two products, A and B, on two machines, I and II. It has been determined that the company will realize a profit of $3 on each unit of product A and a profit of $4 on each unit of product B. To manufacture a unit of product A requires 6 minutes on machine I and 5 minutes on machine II. To manufacture a unit of product B requires 9 minutes on machine I and 4 minutes on machine II. There are 5 hours of machine time available on machine I and 3 hours of machine time available on machine II in each work shift. How many units of each product should be produced in each shift to maximize the company's profit?

27. Manufacturing—Production Scheduling The National Business Machines Corporation manufactures two models of fax machines: A and B. Each model A costs $200 to make and each model B costs $300. The profits are $25 for each model A and $40 for each model B fax machine. If the total number of fax machines demanded per month does not exceed 2500 and the company has earmarked no more than $600,000 per month for manufacturing costs, find how many units of each model National should make each month in order to maximize its monthly profits.

28. Manufacturing—Production Scheduling Kane Manufacturing has a division that produces two models of hibachis, model A and model B. To produce each model A hibachi requires 3 pounds of cast iron and 6 minutes of labor. To produce each model B hibachi requires 4 pounds of cast iron and 3 minutes of labor. The profit for each model A hibachi is $2, and the profit for each model B hibachi is $1.50. If 1000 pounds of cast iron and 20 hours of labor are available for the production of hibachis per day, how many hibachis of each model should the division produce in order to help maximize Kane's profits?

29. Agriculture—Crop Planning A farmer has 150 acres of land suitable for cultivating crops A and B. The cost of cultivating crop A is $40 per acre, whereas that of crop B is $60 per acre. The farmer has a maximum of $7400 available for land cultivation. Each acre of crop A requires 20 hours of labor and each acre of crop B requires 25 hours of labor. The farmer has a maximum of 3300 hours of labor available. If he expects to make a profit of $150 per acre on crop A and $200 per acre on crop B, how many acres of each crop should he plant in order to maximize his profit?

30. Manufacturing—Production Scheduling A company manufactures products A, B, and C. Each product is processed in three departments: I, II, and III. The total available labor hours per week for departments I, II, and III are 900, 1080, and 840, respectively. The time requirements (in hours per unit) and profit per unit for each product are as follows:

	Product A	Product B	Product C
Dept. I	2	1	2
Dept. II	3	1	2
Dept. III	2	2	1
Profit	$18	$12	$15

How many units of each product should the company produce in order to maximize its profit?

31. Advertising—Television Commercials As part of a campaign to promote its annual clearance sale, the Excelsior Company decided to buy television advertising time on Station KAOS. Excelsior's television advertising budget is $102,000. Morning time costs $3000 per minute, afternoon time costs $1000 per minute, and evening (prime) time costs $12,000 per minute. Because of previous commitments, KAOs cannot offer Excelsior more than 6 minutes of prime time or more than a total of 25 minutes of advertising time over the two weeks in which the commercials are to be run. KAOS estimates that morning commercials are seen by 200,000 people, afternoon commercials are seen by 100,000 people, and evening commercials are seen by 600,000 people. How much morning, afternoon, and evening advertising time should Excelsior buy to maximize exposure of its commercials?

32. **Investments—Asset Allocation** Sharon has a total of $200,000 to invest in three types of mutual funds: growth, balanced, and income funds. Growth funds have a rate of return of 12% per year, balanced funds have a rate of return of 10% per year, and income funds have a return of 6% per year. The growth, balanced, and income mutual funds are assigned risk factors of 0.1, 0.06, and 0.02, respectively. Sharon has decided that at least half of her total portfolio is to be in income funds and at least a quarter of it in balanced funds. She has also decided that the average risk factor for her investment should not exceed 0.05. Determine how much Sharon should invest in each type of fund in order to realize a maximum return on her investment.

[*Hint:* The average risk factor for the investment is given by $0.1x + 0.06y + 0.02z \leq 0.05(x + y + z)$.]

33. **Manufacturing—Production Control** The Custom Office Furniture Company is introducing a new line of executive desks made from a specially selected grade of walnut. Initially, three models—A, B, and C—are to be marketed. Each model A desk requires $1\frac{1}{4}$ hours for fabrication, 1 hour for assembly, and 1 hour for finishing; each model B desk requires $1\frac{1}{2}$ hours for fabrication, 1 hour for assembly, and 1 hour for finishing; each model C desk requires $1\frac{1}{2}$ hours, $\frac{3}{4}$ hour, and $\frac{1}{2}$ hour for fabrication, assembly, and finishing, respectively. The profit on each model A desk is $26, the profit on each model B desk is $28, and the profit on each model C desk is $24. The total time available in the fabrication department, the assembly department, and the finishing department in the first month of production is 310 hours, 205 hours, and 190 hours, respectively. To maximize Custom's profit, how many desks of each model should be made in the month?

34. **Manufacturing—Prefabricated Housing Production** Boise Lumber has decided to enter the lucrative prefabricated housing business. Initially, it plans to offer three models: a standard model, a deluxe model, and a luxury model. Each house is prefabricated and partially assembled in the factory, and the final assembly is completed on site. The dollar amount of building material required, the amount of labor required in the factory for prefabrication and partial assembly, the amount of on-site labor required, and the profit per unit are as follows:

	Standard Model	Deluxe Model	Luxury Model
Material ($)	6,000	8,000	10,000
Factory Labor (hr)	240	220	200
On-Site Labor (hr)	180	210	300
Profit ($)	3,400	4,000	5,000

For the first year's production, a sum of $8,200,000 is budgeted for the building material; the number of labor hours available for work in the factory (for prefabrication and partial assembly) is not to exceed 218,000 hours; and the amount of labor for on-site work is to be less than or equal to 237,000 labor hours. Determine how many houses of each type Boise should produce (market research has confirmed that there should be no problems with sales) to maximize its profit from this new venture.

35. **Manufacturing—Cold Formula Production** Bayer Pharmaceutical produces three kinds of cold formulas: formula I, formula II, and formula III. It takes 2.5 hours to produce 1000 bottles of formula I, 3 hours to produce 1000 bottles of formula II, and 4 hours to produce 1000 bottles of formula III. The profits for each 1000 bottles of formula I, formula II, and formula III are $180, $200, and $300, respectively. Suppose that for a certain production run there are enough ingredients on hand to make at most 9000 bottles of formula I, 12,000 bottles of formula II, and 6000 bottles of formula III. Furthermore, suppose that the time for the production run is limited to a maximum of 70 hours. Find how many bottles of each formula should be produced in this production run so that the profit is maximized.

Harley Lance Kaplan

Title: Registered Investment Advisor, Certified Financial Planner
Institution: Beta Industries

Beta Industries' clients are people who "want to invest their money more efficiently," notes Kaplan. He helps them "identify their available assets and define their financial objectives." He then assembles balanced portfolios to meet their needs.

Kaplan must frequently educate clients on the potential risks and rewards investors face. He patiently explains that, based on historical and statistical data and current risk factors, he can *project,* not predict, "possible future consequences." Even then, any investment strategy has to be tempered by an individual client's tolerance for risk.

Age often plays a pivotal role in a client's tolerance level. A 25-year-old investor with only $10,000 and quite a few working years remaining might be more willing to take greater risks, relying on either a growth or an aggressive growth vehicle. (For the client with over $250,000 to invest, Kaplan recommends a personal money manager rather than a vehicle such as a mutual fund, which has many investors contributing relatively small amounts.)

Data show that with a growth investment, a client risks gaining or losing as much as 10% to 25% of the principal. An aggressive growth investment can gain or lose up to 50%. Both types realize a solid rate of return over inflation, from a minimum of 10% to as much as 35% in the best scenarios. The key to success is keeping the funds in place for at least five to seven years to ride out the cycles that typically affect the economy.

Kaplan's strategy for a 62-year-old, who has only a few more years to continue earning, might rely heavily on the safety and liquidity offered by NOW accounts (combined checking and savings) and income vehicles such as bonds with moderate risk and a 3% and 5% rate of return over inflation.

With a $120,000 principal, Kaplan might recommend that 70% be placed into safety and income, whereas the balance be invested as a hedge against inflation (20% into growth and 10% into aggressive growth). Kaplan stresses that inflation robs everyone of purchasing power, especially those on fixed incomes. For example, if inflation remains constant at 3% per year, $10,000 in interest earned by a retiree would buy only $7374 in goods and services after ten years.

Investment vehicles vary widely. When choosing among investments, Kaplan uses a number of comparisons, including the net rate of return, how each investment affects a client's tax situation, and potential risk. For the client in the 40% tax bracket, the net return on a corporate bond paying 10% and a tax-free municipal paying 6% is the same. The only substantial difference may be the level of risk each bond carries. Everything else being equal, the safer bond is the logical choice.

Through the years, Kaplan has learned to weigh not just tangible assets but intangible emotions—to tailor an investment strategy suited to each individual's needs.

Beta Industries is an investment firm, not a broad-based manufacturing conglomerate as its name implies. Its clients range from individuals with significant net worth to Fortune 500 corporations and nonprofit organizations with growing endowments. Kaplan and his partners help clients meet their long-term financial goals by recommending sound investment strategies.

USING TECHNOLOGY

THE SIMPLEX METHOD: SOLVING MAXIMIZATION PROBLEMS

The graphing calculator can be used to solve a linear programming problem by the simplex method as illustrated in Example 1.

EXAMPLE 1 (Refer to Example 4, Section 4.1.) The problem reduces to the following linear programming problem:

$$\text{Maximize} \quad P = 6x + 5y + 4z$$
$$\text{subject to} \quad 2x + \ y + \ z \le 180$$
$$x + 3y + 2z \le 300$$
$$2x + \ y + 2z \le 240$$
$$x \ge 0, y \ge 0, z \ge 0$$

With u, v, and w as slack variables, we are led to the following sequence of simplex tableaus, where the first tableau is entered as the matrix A:

	x	y	z	u	v	w	P	Constant	ratio
pivot row →	②	1	1	1	0	0	0	180	$\frac{180}{2} = 90$
	1	3	2	0	1	0	0	300	$\frac{300}{1} = 300$
	2	1	2	0	0	1	0	240	$\frac{240}{2} = 120$
	−6	−5	−4	0	0	0	1	0	

$\xrightarrow{\ *\textbf{Row}(\frac{1}{2}, A, 1) \blacktriangleright B\ }$

↑ pivot column

x	y	z	u	v	w	P	Constant
①	0.5	0.5	0.5	0	0	0	90
1	3	2	0	1	0	0	300
2	1	2	0	0	1	0	240
−6	−5	−4	0	0	0	1	0

$\xrightarrow{\ *\textbf{Row}+(-1, B, 1, 2) \blacktriangleright C\ }$
$*\textbf{Row}+(-2, C, 1, 3) \blacktriangleright B$
$*\textbf{Row}+(6, B, 1, 4) \ \ \blacktriangleright C$

	x	y	z	u	v	w	P	Constant	ratio
	1	0.5	0.5	0.5	0	0	0	90	$\frac{90}{0.5} = 180$
pivot row →	0	②.5	1.5	−0.5	1	0	0	210	$\frac{210}{2.5} = 84$
	0	0	1	−1	0	1	0	60	
	0	−2	−1	3	0	0	1	540	

$\xrightarrow{\ *\textbf{Row}(\frac{1}{2.5}, C, 2) \blacktriangleright B\ }$

↑ pivot column

232

x	y	z	u	v	w	P	Constant
1	0.5	0.5	0.5	0	0	0	90
0	①	0.6	−0.2	0.4	0	0	84
0	0	1	−1	0	1	0	60
0	−2	−1	3	0	0	1	540

$$\xrightarrow{\begin{array}{l}\textbf{*Row}+(-0.5,B,2,1)\blacktriangleright C\\ \textbf{*Row}+(2,C,2,4)\qquad\blacktriangleright B\end{array}}$$

x	y	z	u	v	w	P	Constant
1	0	0.2	0.6	−0.2	0	0	48
0	1	0.6	−0.2	0.4	0	0	84
0	0	1	−1	0	1	0	60
0	0	0.2	2.6	0.8	0	1	708

The final simplex tableau is the same as the one obtained earlier. We see that $x = 48$, $y = 84$, $z = 0$, and $P = 708$. So Ace Novelty should produce 48 type-A souvenirs, 84 type-B souvenirs, and no type-C souvenirs—resulting in a profit of $708 per day.

EXERCISES

In exercises 1–4, use a graphing calculator to solve the linear programming problem by the simplex method.

1. Maximize $P = 2x + 3y + 4z + 2w$ subject to

$$\begin{aligned} x + 2y + 3z + 2w &\le 6 \\ 2x + 4y + z - w &\le 4 \\ 3x + 2y - 2z + 3w &\le 12 \\ x \ge 0, y \ge 0, z \ge 0, w &\ge 0 \end{aligned}$$

2. Maximize $P = 3x + 2y + 2z + w$ subject to

$$\begin{aligned} 2x + y - z + 2w &\le 8 \\ 2x - y + 2z + 3w &\le 20 \\ x + y + z + 2w &\le 8 \\ 4x - 2y + z + 3w &\le 24 \\ x \ge 0, y \ge 0, z \ge 0, w &\ge 0 \end{aligned}$$

3. Maximize $P = x + y + 2z + 3w$ subject to

$$\begin{aligned} 3x + 6y + 4z + 2w &\le 12 \\ x + 4y + 8z + 4w &\le 16 \\ 2x + y + 4z + w &\le 10 \\ x \ge 0, y \ge 0, z \ge 0, w &\ge 0 \end{aligned}$$

4. Maximize $P = 2x + 4y + 3z + 5w$ subject to

$$\begin{aligned} x - 2y + 3z + 4w &\le 8 \\ 2x + 2y + 4z + 6w &\le 12 \\ 3x + 2y + z + 5w &\le 10 \\ 2x + 8y - 2z + 6w &\le 24 \\ x \ge 0, y \ge 0, z \ge 0, w &\ge 0 \end{aligned}$$

SOLUTIONS TO SELF-CHECK EXERCISES 4.1

1. Introducing the slack variables u, v, and w, we obtain the system of linear equations

$$
\begin{aligned}
2x + 3y + z + u &= 10 \\
x + y + 2z + v &= 8 \\
2y + 3z + w &= 6 \\
-2x - 3y - 6z + P &= 0
\end{aligned}
$$

The initial simplex tableau and the successive tableaus resulting from the use of the simplex procedure follow:

	x	y	z	u	v	w	P	Constant		ratio	
	2	3	1	1	0	0	0	10		$\frac{10}{1} = 10$	$\xrightarrow{\frac{1}{3}R_3}$
	1	1	2	0	1	0	0	8		$\frac{8}{2} = 4$	
pivot → row	0	2	③	0	0	1	0	6		$\frac{6}{3} = 2$	
	−2	−3	−6	0	0	0	1	0			

↑
pivot
column

x	y	z	u	v	w	P	Constant	
2	3	1	1	0	0	0	10	
1	1	2	0	1	0	0	8	$\xrightarrow[\substack{R_2 - 2R_3 \\ R_4 + 6R_3}]{R_1 - R_3}$
0	$\frac{2}{3}$	①	0	0	$\frac{1}{3}$	0	2	
−2	−3	−6	0	0	0	1	0	

	x	y	z	u	v	w	P	Constant		ratio	
	2	$\frac{7}{3}$	0	1	0	$-\frac{1}{3}$	0	8		$\frac{8}{2} = 4$	
pivot → row	①	$-\frac{1}{3}$	0	0	1	$-\frac{2}{3}$	0	4		$\frac{4}{1} = 4$	$\xrightarrow[\substack{R_4 + 2R_2}]{R_1 - 2R_2}$
	0	$\frac{2}{3}$	1	0	0	$\frac{1}{3}$	0	2		—	
	−2	1	0	0	0	2	1	12			

↑
pivot
column

x	y	z	u	v	w	P	Constant
0	3	0	1	−2	1	0	0
1	$-\frac{1}{3}$	0	0	1	$-\frac{2}{3}$	0	4
0	$\frac{2}{3}$	1	0	0	$\frac{1}{3}$	0	2
0	$\frac{1}{3}$	0	0	2	$\frac{2}{3}$	1	20

All entries in the last row are nonnegative and the tableau is final. We conclude that $x = 4$, $y = 0$, $z = 2$, and $P = 20$.

2. Let x denote the number of model A grates and y the number of model B grates to be made this week. Then the profit function to be maximized is given by

$$P = 6x + 8y$$

The limitations on the availability of material and labor may be expressed by the linear inequalities

$$20x + 30y \leq 7200 \quad \text{or} \quad 2x + 3y \leq 720$$

and
$$20x + 15y \leq 6000 \quad \text{or} \quad 4x + 3y \leq 1200$$

Finally, the condition that no more than 150 units of model A grates be made per week may be expressed by the linear inequality

$$x \leq 150$$

Thus, we are led to the following linear programming problem:

Maximize $\quad P = 6x + 8y$

subject to $\quad 2x + 3y \leq 720$

$$4x + 3y \leq 1200$$

$$x \qquad \leq 150$$

$$x \geq 0, y \geq 0$$

To solve this problem, we introduce slack variables u, v, and w and use the simplex method, obtaining the following sequence of simplex tableaus:

x	y	u	v	w	P	Constant		ratio
2	③	1	0	0	0	720		$\frac{720}{3} = 240$
4	3	0	1	0	0	1200		$\frac{1200}{3} = 400$
1	0	0	0	1	0	150		—
-6	-8	0	0	0	1	0		

$\uparrow$
pivot
column

$\xrightarrow{\frac{1}{3}R_1}$

x	y	u	v	w	P	Constant
$\frac{2}{3}$	①	$\frac{1}{3}$	0	0	0	240
4	3	0	1	0	0	1200
1	0	0	0	1	0	150
-6	-8	0	0	0	1	0

$\xrightarrow[R_4 + 8R_1]{R_2 - 3R_1}$

x	y	u	v	w	P	Constant		ratio
$\frac{2}{3}$	1	$\frac{1}{3}$	0	0	0	240		$\frac{240}{2/3} = 360$
2	0	-1	1	0	0	480		$\frac{480}{2} = 240$
①	0	0	0	1	0	150		$\frac{150}{1} = 150$
$-\frac{2}{3}$	0	$\frac{8}{3}$	0	0	1	1920		

$\uparrow$
pivot
column

	x	y	u	v	w	P	Constant
	0	1	$\frac{1}{3}$	0	$-\frac{2}{3}$	0	140
$\dfrac{R_1 - \frac{2}{3}R_3}{\substack{R_2 - 2R_3 \\ R_4 + \frac{2}{3}R_3}}$	0	0	-1	1	-2	0	180
	1	0	0	0	1	0	150
	0	0	$\frac{8}{3}$	0	$\frac{2}{3}$	1	2020

The last tableau is final and we see that $x = 150$, $y = 140$, and $P = 2020$. Therefore, LaCrosse should make 150 model A grates and 140 model B grates this week. The profit will be $2020.

4.2 THE SIMPLEX METHOD: STANDARD MINIMIZATION PROBLEMS

Minimization with ≤ Constraints

In the last section we developed a procedure, called the simplex method, for solving standard linear programming problems. Recall that a standard maximization problem satisfies three conditions:

1. The objective function is to be maximized.
2. All the variables involved are nonnegative.
3. Each linear constraint may be written so that the expression involving the variables is less than or equal to a nonnegative constant.

In this section we see how the simplex method may be used to solve certain classes of problems that are not necessarily standard maximization problems. In particular, we see how the modified procedure may be used to solve problems involving the minimization of objective functions.

We begin by considering the class of linear programming problems that calls for the minimization of objective functions but otherwise satisfies conditions (2) and (3) for standard maximization problems. The method used to solve these problems is illustrated in the following example.

EXAMPLE 1   Minimize $C = -2x - 3y$
                subject to $5x + 4y \leq 32$
                                    $x + 2y \leq 10$
                                    $x \geq 0,\ y \geq 0$

Solution   This problem involves the minimization of the objective function and is accordingly not a standard maximization problem. Note, however, that all other conditions for a standard maximization problem hold true. To solve a problem of this type, we observe that minimizing the objective function C

is equivalent to maximizing the objective function $P = -C$. Thus, the solution to this problem may be found by solving the following associated standard maximization problem: Maximize $P = 2x + 3y$ subject to the given constraints. Using the simplex method with u and v as slack variables, we obtain the following sequence of simplex tableaus:

	x	y	u	v	P	Constant	ratio
	5	4	1	0	0	32	$\frac{32}{4} = 8$
pivot row →	1	②	0	1	0	10	$\frac{10}{2} = 5$
	−2	−3	0	0	1	0	

↑ pivot column

	x	y	u	v	P	Constant
$\frac{1}{2}R_2$ →	5	4	1	0	0	32
	$\frac{1}{2}$	①	0	$\frac{1}{2}$	0	5
	−2	−3	0	0	1	0

	x	y	u	v	P	Constant	ratio
pivot row →	③	0	1	−2	0	12	$\frac{12}{3} = 4$
$R_1 - 4R_2$ →	$\frac{1}{2}$	1	0	$\frac{1}{2}$	0	5	$\frac{1}{\frac{1}{2}} = 10$
$R_3 + 3R_2$ →	$-\frac{1}{2}$	0	0	$\frac{3}{2}$	1	15	

↑ pivot column

Refer to Example 1.

a. Sketch the feasible set S for the linear programming problem.
b. Solve the problem using the method of corners.
c. Indicate on S the corner points corresponding to each iteration of the simplex procedure and trace the path leading to the optimal solution.

	x	y	u	v	P	Constant
$\frac{1}{3}R_1$ →	①	0	$\frac{1}{3}$	$-\frac{2}{3}$	0	4
	$\frac{1}{2}$	1	0	$\frac{1}{2}$	0	5
	$-\frac{1}{2}$	0	0	$\frac{3}{2}$	1	15

	x	y	u	v	P	Constant
$R_2 - \frac{1}{2}R_1$ →	1	0	$\frac{1}{3}$	$-\frac{2}{3}$	0	4
$R_3 + \frac{1}{2}R_1$ →	0	1	$-\frac{1}{6}$	$\frac{5}{6}$	0	3
	0	0	$\frac{1}{6}$	$\frac{7}{6}$	1	17

The last tableau is in final form. The solution to the standard maximization problem associated with the given linear programming problem is $x = 4$, $y = 3$, and $P = 17$, so the required solution is given by $x = 4$, $y = 3$, and $C = -17$. You may verify that the solution is correct by using the method of corners.   ◦ ◦ ◦

The Dual Problem

Another special class of linear programming problems we encounter in practical applications is characterized by the following conditions:

1. The objective function is to be *minimized.*

2. All the variables involved are nonnegative.

3. Each linear constraint may be written so that the expression involving the variables is *greater than* or equal to a constant.

Such problems are called **standard minimization problems.**

A convenient method for solving this type of problem is based on the following observation. Each maximization linear programming problem is associated with a minimization problem, and vice versa. For the purpose of identification, the given problem is called the **primal problem;** the problem related to it is called the **dual problem.** The following example illustrates the technique for constructing the dual of a given linear programming problem.

EXAMPLE 2   Write the dual problem associated with the following problem:

Minimize the objective function $C = 6x + 8y$

subject to $40x + 10y \geq 2400$

$10x + 15y \geq 2100$

$5x + 15y \geq 1500$

$x \geq 0, y \geq 0$

Solution   We first write down the following tableau for the given primal problem:

x	y	Constant
40	10	2400
10	15	2100
5	15	1500
6	8	

Next, we interchange the columns and rows of the foregoing tableau and head the three columns of the resulting array with the three variables u, v, and w, obtaining the tableau

u	v	w	Constant
40	10	5	6
10	15	15	8
2400	2100	1500	

Interpreting the last tableau as if it were part of the initial simplex tableau for a standard maximization problem, with the exception that the signs of the coefficients pertaining to the objective function are not reversed, we construct the required dual problem as follows:

Maximize the objective function $P = 2400u + 2100v + 1500w$

subject to $40u + 10v + 5w \leq 6$

$10u + 15v + 15w \leq 8$

where $u \geq 0$, $v \geq 0$, and $w \geq 0$.   ◦ ◦ ◦

The connection between the solution of the primal problem and that of the dual problem is given by the following theorem. The theorem, attributed to John von Neumann (1903–1957), is stated without proof.

THE FUNDAMENTAL THEOREM OF DUALITY

A primal problem has a solution if and only if the corresponding dual problem has a solution. Furthermore, if a solution exists, then

1. the objective functions of both the primal and the dual problem attain the same optimal value.

2. the optimal solution to the primal problem appears under the slack variables in the last row of the final simplex tableau associated with the dual problem.

Armed with this theorem, we will solve the problem posed in Example 2.

EXAMPLE 3   Complete the solution to the problem posed in Example 2.

Solution   Observe that the dual problem associated with the given (primal) problem is a standard maximization problem. The solution may thus be found using the simplex algorithm. Introducing the slack variables x and y, we obtain the system of linear equations

$$40u + 10v + 5w + x = 6$$
$$10u + 15v + 15w + y = 8$$
$$-2400u - 2100v - 1500w + P = 0$$

Continuing with the simplex algorithm, we obtain the sequence of simplex tableaus

	u	v	w	x	y	P	Constant	ratio
pivot row →	㊵	10	5	1	0	0	6	$\frac{6}{40} = \frac{3}{20}$
	10	15	15	0	1	0	8	$\frac{8}{10} = \frac{4}{5}$
	-2400	-2100	-1500	0	0	1	0	

↑
pivot
column

	u	v	w	x	y	P	Constant
	①	$\frac{1}{4}$	$\frac{1}{8}$	$\frac{1}{40}$	0	0	$\frac{3}{20}$
$\xrightarrow{\frac{1}{40}R_1}$	10	15	15	0	1	0	8
	-2400	-2100	-1500	0	0	1	0

	u	v	w	x	y	P	Constant		ratio
	1	$\frac{1}{4}$	$\frac{1}{8}$	$\frac{1}{40}$	0	0	$\frac{3}{20}$		$\frac{3/20}{1/4} = \frac{3}{5}$
$\xrightarrow{\begin{array}{c}R_2 - 10R_1 \\ R_3 + 2400R_1\end{array}}$	0	$\frac{25}{2}$	$\frac{55}{4}$	$-\frac{1}{4}$	1	0	$\frac{13}{2}$		$\frac{13/2}{25/2} = \frac{13}{25}$
	0	-1500	-1200	60	0	1	360		

	u	v	w	x	y	P	Constant
	1	$\frac{1}{4}$	$\frac{1}{8}$	$\frac{1}{40}$	0	0	$\frac{3}{20}$
$\xrightarrow{\frac{2}{25}R_2}$	0	①	$\frac{11}{10}$	$-\frac{1}{50}$	$\frac{2}{25}$	0	$\frac{13}{25}$
	0	-1500	-1200	60	0	1	360

	u	v	w	x	y	P	Constant
	1	0	$-\frac{3}{20}$	$\frac{3}{100}$	$-\frac{1}{50}$	0	$\frac{1}{50}$
$\xrightarrow{\begin{array}{c}R_1 - \frac{1}{4}R_2 \\ R_3 + 1500R_2\end{array}}$	0	1	$\frac{11}{10}$	$-\frac{1}{50}$	$\frac{2}{25}$	0	$\frac{13}{25}$
	0	0	450	30	120	1	1140

Solution for the
primal problem

    The last tableau is final. The fundamental theorem of duality tells us that the solution to the primal problem is $x = 30$ and $y = 120$ with a minimum value for C of 1140. Observe that the solution to the dual (maximization) problem may be read from the simplex tableau in the usual manner: $u = 1/50$, $v = 13/25$, $w = 0$, and $P = 1140$. Note that the maximum value of P is equal to the minimum value of C as guaranteed by the fundamental theorem of duality. The solution to the primal problem agrees with the solution of the same problem solved in Section 3.3, Example 2, using the method of corners.

○ ○ ○

REMARKS

1. We leave it to you to demonstrate that the dual of a standard minimization problem is always a standard maximization problem provided that the coefficients of the objective function in the primal problem are all nonnegative. Such problems can always be solved by applying the simplex method to solve the dual problem.

2. Standard minimization problems in which the coefficients of the objective function are not all nonnegative do not necessarily have a dual problem that is a standard maximization problem. We will study such problems in Section 4.3.   ◦ ◦ ◦

EXAMPLE 4   Minimize $C = 3x + 2y$

subject to $8x + y \geq 80$

$3x + 2y \geq 100$

$x + 4y \geq 80$

$x \geq 0, y \geq 0$

Solution   We begin by writing the dual problem associated with the given primal problem. First, we write down the following tableau for the primal problem:

x	y	Constant
8	1	80
3	2	100
1	4	80
3	2	

Next, interchanging the columns and rows of this tableau and heading the three columns of the resulting array with the three variables u, v, and w, we obtain the tableau

u	v	w	Constant
8	3	1	3
1	2	4	2
80	100	80	

Interpreting the last tableau as if it were part of the initial simplex tableau for a standard maximization problem, with the exception that the signs of the coefficients pertaining to the objective function are not reversed, we construct the dual problem as follows: Maximize the objective function $P = 80u + 100v + 80w$ subject to the constraints

$$8u + 3v + w \leq 3$$
$$u + 2v + 4w \leq 2$$

where $u \geq 0$, $v \geq 0$, and $w \geq 0$. Having constructed the dual problem, which is a standard maximization problem, we now solve it using the simplex method. Introducing the slack variables x and y, we obtain the system of linear equations

$$8u + 3v + w + x   = 3$$
$$u + 2v + 4w + y = 2$$
$$-80u - 100v - 80w + P = 0$$

Continuing with the simplex algorithm, we obtain the sequence of simplex tableaus

	u	v	w	x	y	P	Constant
	8	3	1	1	0	0	3
pivot row $\rightarrow$	1	②	4	0	1	0	2
	-80	-100	-80	0	0	1	0

ratio
$\frac{3}{3} = 1$
$\frac{2}{2} = 1$

$\uparrow$
pivot column

$\xrightarrow{\frac{1}{2}R_2}$

	u	v	w	x	y	P	Constant
	8	3	1	1	0	0	3
	$\frac{1}{2}$	①	2	0	$\frac{1}{2}$	0	1
	-80	-100	-80	0	0	1	0

pivot row $\hookrightarrow$

$\xrightarrow[R_3 + 100R_2]{R_1 - 3R_2}$

	u	v	w	x	y	P	Constant
	⑬⁄₂	0	-5	1	$-\frac{3}{2}$	0	0
	$\frac{1}{2}$	1	2	0	$\frac{1}{2}$	0	1
	-30	0	120	0	50	1	100

ratio
$\frac{0}{13/2} = 0$
$\frac{1}{1/2} = 2$

$\uparrow$
pivot column

$\xrightarrow{\frac{2}{13}R_1}$

	u	v	w	x	y	P	Constant
	①	0	$-\frac{10}{13}$	$\frac{2}{13}$	$-\frac{3}{13}$	0	0
	$\frac{1}{2}$	1	2	0	$\frac{1}{2}$	0	1
	-30	0	120	0	50	1	100

$\xrightarrow[R_3 + 30R_1]{R_2 - \frac{1}{2}R_1}$

	u	v	w	x	y	P	Constant
	1	0	$-\frac{10}{13}$	$\frac{2}{13}$	$-\frac{3}{13}$	0	0
	0	1	$\frac{31}{13}$	$-\frac{1}{13}$	$\frac{8}{13}$	0	1
	0	0	$\frac{1260}{13}$	$\frac{60}{13}$	$\frac{560}{13}$	1	100

Solution for the primal problem

The last tableau is final. The fundamental theorem of duality tells us that the solution to the primal problem is $x = 60/13$ and $y = 560/13$ with a minimum value for C of 100.

◦ ◦ ◦

Application

Our last example illustrates how the warehouse problem posed in Section 3.2 may be solved by duality.

EXAMPLE 5 Complete the solution to the warehouse problem given in Section 3.2, Example 4 (page 187).

Minimize

$$C = 20x_1 + 8x_2 + 10x_3 + 12x_4 + 22x_5 + 18x_6 \qquad \textbf{(13)}$$

subject to

$$
\begin{aligned}
x_1 + x_2 + x_3 & & \leq 400 \\
x_4 + x_5 + x_6 & \leq 600 \\
x_1 \qquad\quad + x_4 & & \geq 200 \\
x_2 \qquad\quad + x_5 & & \geq 300 \\
x_3 \qquad\quad + x_6 & \geq 400 \\
x_1 \geq 0, x_2 \geq 0, \dots, x_6 \geq 0
\end{aligned}
\qquad \textbf{(14)}
$$

Solution Upon multiplying each of the first two inequalities of (14) by -1, we obtain the following equivalent system of constraints in which each of the expressions involving the variables is greater than or equal to a constant:

$$
\begin{aligned}
-x_1 - x_2 - x_3 & & \geq -400 \\
-x_4 - x_5 - x_6 & \geq -600 \\
x_1 \qquad\quad + x_4 & & \geq \quad 200 \\
x_2 \qquad\quad + x_5 & & \geq \quad 300 \\
x_3 \qquad\quad + x_6 & \geq \quad 400
\end{aligned}
$$

The problem may now be solved by duality. First we write the array of numbers

x_1	x_2	x_3	x_4	x_5	x_6	Constant
-1	-1	-1	0	0	0	-400
0	0	0	-1	-1	-1	-600
1	0	0	1	0	0	200
0	1	0	0	1	0	300
0	0	1	0	0	1	400
20	8	10	12	22	18	

Interchanging the rows and columns of this array of numbers and heading the five columns of the resulting array of numbers by the variables u_1, u_2, u_3,

u_4, and u_5 leads to

u_1	u_2	u_3	u_4	u_5	Constant
-1	0	1	0	0	20
-1	0	0	1	0	8
-1	0	0	0	1	10
0	-1	1	0	0	12
0	-1	0	1	0	22
0	-1	0	0	1	18
-400	-600	200	300	400	

from which we construct the associated dual problem: Maximize $P = -400u_1 - 600u_2 + 200u_3 + 300u_4 + 400u_5$ subject to

$$
\begin{aligned}
-u_1 \quad\quad + u_3 \quad\quad\quad\quad &\le 20 \\
-u_1 \quad\quad\quad\quad + u_4 \quad &\le \ 8 \\
-u_1 \quad\quad\quad\quad\quad + u_5 &\le 10 \\
- u_2 + u_3 \quad\quad\quad\quad &\le 12 \\
- u_2 \quad\quad + u_4 \quad &\le 22 \\
- u_2 \quad\quad\quad\quad + u_5 &\le 18 \\
u_1 \ge 0, u_2 \ge 0, \ldots, u_5 \ge 0
\end{aligned}
$$

Solving the standard maximization problem by the simplex algorithm, we obtain the following sequence of tableaus ($x_1, x_2, \ldots, x_6$ are slack variables):

	u_1	u_2	u_3	u_4	u_5	x_1	x_2	x_3	x_4	x_5	x_6	P	Constant	ratio
	-1	0	1	0	0	1	0	0	0	0	0	0	20	—
	-1	0	0	1	0	0	1	0	0	0	0	0	8	—
pivot row →	-1	0	0	0	①	0	0	1	0	0	0	0	10	10
	0	-1	1	0	0	0	0	0	1	0	0	0	12	—
	0	-1	0	1	0	0	0	0	0	1	0	0	22	—
	0	-1	0	0	1	0	0	0	0	0	1	0	18	18
	400	600	-200	-300	-400	0	0	0	0	0	0	1	0	

↑
pivot
column

	u_1	u_2	u_3	u_4	u_5	x_1	x_2	x_3	x_4	x_5	x_6	P	Constant	ratio
	−1	0	1	0	0	1	0	0	0	0	0	0	20	—
pivot row →	−1	0	0	①1	0	0	1	0	0	0	0	0	8	8
	−1	0	0	0	1	0	0	1	0	0	0	0	10	—
	0	−1	1	0	0	0	0	0	1	0	0	0	12	—
$\dfrac{R_6 - R_3}{R_7 + 400R_3}$	0	−1	0	1	0	0	0	0	0	1	0	0	22	22
	1	−1	0	0	0	0	0	−1	0	0	1	0	8	—
	0	600	−200	−300	0	0	0	400	0	0	0	1	4000	

↑ pivot column

	u_1	u_2	u_3	u_4	u_5	x_1	x_2	x_3	x_4	x_5	x_6	P	Constant	ratio
	−1	0	1	0	0	1	0	0	0	0	0	0	20	—
	−1	0	0	1	0	0	1	0	0	0	0	0	8	—
$\dfrac{R_5 - R_2}{R_7 + 300R_2}$	−1	0	0	0	1	0	0	1	0	0	0	0	10	—
	0	−1	1	0	0	0	0	0	1	0	0	0	12	—
	1	−1	0	0	0	0	−1	0	0	1	0	0	14	14
pivot row →	①1	−1	0	0	0	0	0	−1	0	0	1	0	8	8
	−300	600	−200	0	0	0	300	400	0	0	0	1	6400	

↑ pivot column

	u_1	u_2	u_3	u_4	u_5	x_1	x_2	x_3	x_4	x_5	x_6	P	Constant	ratio
$R_1 + R_6$	0	−1	1	0	0	1	0	−1	0	0	1	0	28	28
	0	−1	0	1	0	0	1	−1	0	0	1	0	16	—
	0	−1	0	0	1	0	0	0	0	0	1	0	18	—
$\dfrac{R_2 + R_6}{R_3 + R_6}$ pivot row	0	−1	①1	0	0	0	0	0	1	0	0	0	12	12
$R_5 - R_6$	0	0	0	0	0	0	−1	1	0	1	−1	0	6	—
$R_7 + 300R_6$	1	−1	0	0	0	0	0	−1	0	0	1	0	8	—
	0	300	−200	0	0	0	300	100	0	0	300	1	8800	

↑ pivot column

u_1	u_2	u_3	u_4	u_5	x_1	x_2	x_3	x_4	x_5	x_6	P	Constant
0	0	0	0	0	1	0	-1	-1	0	1	0	16
0	-1	0	1	0	0	1	-1	0	0	1	0	16
0	-1	0	0	1	0	0	0	0	0	1	0	18
0	-1	1	0	0	0	0	0	1	0	0	0	12
0	0	0	0	0	0	-1	1	0	1	-1	0	6
1	-1	0	0	0	0	0	-1	0	0	1	0	8
0	100	0	0	0	0	300	100	200	0	300	1	11,200

$$\frac{R_1 - R_4}{R_7 + 200R_4}$$

The last tableau is final, and we find that

$$x_1 = 0, \quad x_2 = 300, \quad x_3 = 100, \quad x_4 = 200, \quad x_5 = 0, \quad x_6 = 300,$$
$$\text{and} \quad P = 11,200$$

Thus, to minimize shipping costs, Acrosonic should ship 300 loudspeaker systems from plant I to warehouse B, 100 systems from plant I to warehouse C, 200 systems from plant II to warehouse A, and 300 systems from plant II to warehouse C at a total cost of $11,200.   ◦ ◦ ◦

SELF–CHECK EXERCISES 4.2

1. Write the dual problem associated with the following problem:

$$\begin{aligned} \text{Minimize} \quad & C = 2x + 5y \\ \text{subject to} \quad & 4x + y \geq 40 \\ & 2x + y \geq 30 \\ & x + 3y \geq 30 \\ & x \geq 0, y \geq 0 \end{aligned}$$

2. Solve the primal problem posed in exercise 1.

Solutions to Self-Check Exercises 4.2 can be found on page 249.

4.2 EXERCISES

In exercises 1–6, use the technique developed in this section to solve the given minimization problem.

1. Minimize $C = -2x + y$ subject to

$$\begin{aligned} x + 2y &\leq 6 \\ 3x + 2y &\leq 12 \\ x &\geq 0, y \geq 0 \end{aligned}$$

2. Minimize $C = -2x - 3y$ subject to

$$\begin{aligned} 3x + 4y &\leq 24 \\ 7x - 4y &\leq 16 \\ x &\geq 0, y \geq 0 \end{aligned}$$

3. Minimize $C = -3x - 2y$ subject to the constraints of exercise 2.

4. Minimize $C = x - 2y + z$ subject to

$$x - 2y + 3z \le 10$$
$$2x + y - 2z \le 15$$
$$2x + y + 3z \le 20$$
$$x \ge 0, y \ge 0, z \ge 0$$

5. Minimize $C = 2x - 3y - 4z$ subject to

$$-x + 2y - z \le 8$$
$$x - 2y + 2z \le 10$$
$$2x + 4y - 3z \le 12$$
$$x \ge 0, y \ge 0, z \ge 0$$

6. Minimize $C = -3x - 2y - z$ subject to the constraints of exercise 5.

In exercises 7–10, you are given the final simplex tableau for the dual problem. Give the solution to the primal problem and to the associated dual problem.

7. Problem: Minimize $C = 8x + 12y$ subject to

$$x + 3y \ge 2$$
$$2x + 2y \ge 3$$
$$x \ge 0, y \ge 0$$

Final tableau:

u	v	x	y	P	Constant
0	1	$\frac{3}{4}$	$-\frac{1}{4}$	0	3
1	0	$-\frac{1}{2}$	$\frac{1}{2}$	0	2
0	0	$\frac{5}{4}$	$\frac{1}{4}$	1	13

8. Problem: Minimize $C = 3x + 2y$ subject to

$$5x + y \ge 10$$
$$2x + 2y \ge 12$$
$$x + 4y \ge 12$$
$$x \ge 0, y \ge 0$$

Final tableau:

u	v	w	x	y	P	Constant
1	0	$-\frac{3}{4}$	$\frac{1}{4}$	$-\frac{1}{4}$	0	$\frac{1}{4}$
0	1	$\frac{19}{8}$	$-\frac{1}{8}$	$\frac{5}{8}$	0	$\frac{7}{8}$
0	0	9	1	5	1	13

9. Problem: Minimize $C = 10x + 3y + 10z$ subject to

$$2x + y + 5z \ge 20$$
$$4x + y + z \ge 30$$
$$x \ge 0, y \ge 0, z \ge 0$$

Final tableau:

u	v	x	y	z	P	Constant
0	1	$\frac{1}{2}$	-1	0	0	2
1	0	$-\frac{1}{2}$	2	0	0	1
0	0	2	-9	1	0	3
0	0	5	10	0	1	80

10. Problem: Minimize $C = 2x + 3y$ subject to

$$x + 4y \ge 8$$
$$x + y \ge 5$$
$$2x + y \ge 7$$
$$x \ge 0, y \ge 0$$

Final tableau:

u	v	w	x	y	P	Constant
0	1	$\frac{7}{3}$	$\frac{4}{3}$	$-\frac{1}{3}$	0	$\frac{5}{3}$
1	0	$-\frac{1}{3}$	$-\frac{1}{3}$	$\frac{1}{3}$	0	$\frac{1}{3}$
0	0	2	4	1	1	11

In exercises 11–20, construct the dual problem associated with the given primal problem. Solve the primal problem.

11. Minimize $C = 2x + 5y$ subject to

$$x + 2y \ge 4$$
$$3x + 2y \ge 6$$
$$x \ge 0, y \ge 0$$

12. Minimize $C = 3x + 2y$ subject to

$$2x + 3y \ge 90$$
$$3x + 2y \ge 120$$
$$x \ge 0, y \ge 0$$

13. Minimize $C = 6x + 4y$ subject to

$$6x + y \ge 60$$
$$2x + y \ge 40$$
$$x + y \ge 30$$
$$x \ge 0, y \ge 0$$

14. Minimize $C = 10x + y$ subject to

$$4x + y \geq 16$$
$$x + 2y \geq 12$$
$$x \geq 2$$
$$x \geq 0, y \geq 0$$

15. Minimize $C = 200x + 150y + 120z$ subject to

$$20x + 10y + z \geq 10$$
$$x + y + 2z \geq 20$$
$$x \geq 0, y \geq 0, z \geq 0$$

16. Minimize $C = 40x + 30y + 11z$ subject to

$$2x + y + z \geq 8$$
$$x + y - z \geq 6$$
$$x \geq 0, y \geq 0, z \geq 0$$

17. Minimize $C = 6x + 8y + 4z$ subject to

$$x + 2y + 2z \geq 10$$
$$2x + y + z \geq 24$$
$$x + y + z \geq 16$$
$$x \geq 0, y \geq 0, z \geq 0$$

18. Minimize $C = 12x + 4y + 8z$ subject to

$$2x + 4y + z \geq 6$$
$$3x + 2y + 2z \geq 2$$
$$4x + y + z \geq 2$$
$$x \geq 0, y \geq 0, z \geq 0$$

19. Minimize $C = 30x + 12y + 20z$ subject to

$$2x + 4y + 3z \geq 6$$
$$6x + z \geq 2$$
$$6y + 2z \geq 4$$
$$x \geq 0, y \geq 0, z \geq 0$$

20. Minimize $C = 8x + 6y + 4z$ subject to

$$2x + 3y + z \geq 6$$
$$x + 2y - 2z \geq 4$$
$$x + y + 2z \geq 2$$
$$x \geq 0, y \geq 0, z \geq 0$$

21. Shipping Costs   The Acrosonic Company also manufactures a model G loudspeaker system in plants I and II. The output at plant I is at most 800 per month, and the output at plant II is at most 600 per month. Model G loudspeaker systems are also shipped to the three warehouses—A, B, and C—whose minimum monthly requirements are 500, 400, and 400, respectively. Shipping costs from plant I to warehouse A, warehouse B, and warehouse C are $16, $20, and $22 per loudspeaker system, respectively, and shipping costs from plant II to each of these warehouses are $18, $16, and $14, respectively. What shipping schedule will enable Acrosonic to meet the requirements of the warehouses while keeping its shipping costs to a minimum?

22. Shipping Costs   The Steinwelt Piano Company manufactures uprights and consoles in two plants, plant I and plant II. The output of plant I is at most 300 per month, and the output of plant II is at most 250 per month. These pianos are shipped to three warehouses that serve as distribution centers for Steinwelt. In order to fill current and projected future orders, warehouse A requires a minimum of 200 pianos per month, warehouse B requires at least 150 pianos per month, and warehouse C requires at least 200 pianos per month. The shipping cost of each piano from plant I to warehouse A, warehouse B, and warehouse C is $60, $60, and $80, respectively, and the shipping cost of each piano from plant II to warehouse A, warehouse B, and warehouse C is $80, $70, and $50, respectively. What shipping schedule will enable Steinwelt to meet the requirements of the warehouses while keeping the shipping costs to a minimum?

23. Nutrition—Diet Planning   The owner of the Health Juice-Bar wishes to prepare a low-calorie fruit juice with a high vitamin A and C content by blending orange juice and pink grapefruit juice. Each glass of the blended juice is to contain at least 1200 International Units (I.U.) of vitamin A and 200 I.U. of vitamin C. One ounce of orange juice contains 60 I.U. of vitamin A, 16 I.U. of vitamin C, and 14 calories; each ounce of pink grapefruit juice contains 120 I.U. of vitamin A, 12 I.U. of vitamin C, and 11 calories. How many ounces of each juice should a glass of the blend contain if it is to meet the minimum vitamin requirements and at the same time contain a minimum number of calories?

24. Production Control   An oil company operates two refineries in a certain city. Refinery I has an output of 200, 100, and 100 barrels of low-, medium-, and high-grade oil per day, respectively. Refinery II has an output of 100, 200, and 600 barrels of low-, medium-, and high-grade oil per day, respectively. The company wishes to produce at least 1000, 1400, and 3000 barrels of low-, medium-, and high-grade oil to fill an order. If it costs $200 per day to operate refinery I and $300 per day to operate refinery II, determine how many days each refinery should be operated to meet the requirements of the order at minimum cost to the company.

SOLUTIONS TO SELF-CHECK EXERCISES 4.2

1. We first write down the following tableau for the given (primal) problem:

x	y	Constant
4	1	40
2	1	30
1	3	30
2	5	0

Next, we interchange the columns and rows of the tableau and head the three columns of the resulting array with the three variables u, v, and w, obtaining the tableau

u	v	w	Constant
4	2	1	2
1	1	3	5
40	30	30	0

Interpreting the last tableau as if it were the initial tableau for a standard linear programming problem, with the exception that the signs of the coefficients pertaining to the objective function are not reversed, we construct the required dual problem as follows

$$\text{Maximize} \quad P = 40u + 30v + 30w$$
$$\text{subject to} \quad 4u + 2v + \ w \le 2$$
$$u + \ v + 3w \le 5$$
$$u \ge 0, v \ge 0, w \ge 0$$

2. We introduce slack variables x and y to obtain the system of linear equations

$$4u + 2v + \ w + x \qquad = 2$$
$$u + \ v + 3w \qquad + y \quad = 5$$
$$-40u - 30v - 30w \qquad + P = 0$$

Using the simplex algorithm, we obtain the sequence of simplex tableaus

	u	v	w	x	y	P	Constant	ratio	
pivot row →	④	2	1	1	0	0	2	$\frac{2}{4} = \frac{1}{2}$	$\xrightarrow{\frac{1}{4}R_1}$
	1	1	3	0	1	0	5	$\frac{5}{1} = 5$	
	−40	−30	−30	0	0	1	0		

↑
pivot
column

USING TECHNOLOGY

THE SIMPLEX METHOD: SOLVING MINIMIZATION PROBLEMS

The graphing calculator can be used to solve minimization problems using the simplex method.

EXAMPLE 1

Minimize $C = 2x + 3y$

subject to
$$8x + y \geq 80$$
$$3x + 2y \geq 100$$
$$x + 4y \geq 80$$
$$x \geq 0, y \geq 0$$

Solution We begin by writing the dual problem associated with the given primal problem. From the tableau for the primal problem

x	y	Constant
8	1	80
3	2	100
1	4	80
2	3	

we find, upon changing the columns and rows of this tableau and heading the three columns of the resulting array with the variables u, v, and w, the tableau

u	v	w	Constant
8	3	1	2
1	2	4	3
80	100	80	

This tells us that the dual problem is:

Maximize $P = 80u + 100v + 80w$

subject to
$$8u + 3v + w \leq 2$$
$$u + 2u + 4w \leq 3$$
$$u \geq 0, v \geq 0, w \geq 0$$

To solve this standard maximization problem, we proceed as follows:

	u	v	w	x	y	P	Constant		ratio
pivot row →	8	③	1	1	0	0	2		$\frac{2}{3}$
	1	2	4	0	1	0	3		$\frac{3}{2}$
	−80	−100	−80	0	0	1	0		

$\xrightarrow{\ *\textbf{Row}(\frac{1}{3}, A, 1) \blacktriangleright B\ }$

↑ pivot column

u	v	w	x	y	P	Constant
2.67	①	0.33	0.33	0	0	0.67
1	2	4	0	1	0	3
−80	−100	−80	0	0	1	0

$\xrightarrow{\ *\textbf{Row}+(-2, B, 1, 2) \blacktriangleright C\ }$
$\xrightarrow{\ *\textbf{Row}+(100, C, 1, 3) \blacktriangleright B\ }$

	u	v	w	x	y	P	Constant		ratio
	2.67	1	0.33	0.33	0	0	0.67		2
pivot row →	−4.33	0	③.33	−0.67	1	0	1.67		0.5
	186.67	0	−46.67	33.33	0	1	66.67		

$\xrightarrow{\ *\textbf{Row}(\frac{1}{3.33}, B, 2) \blacktriangleright C\ }$

↑ pivot column

u	v	w	x	y	P	Constant
2.67	1	0.33	0.33	0	0	0.67
−1.30	0	1	−0.2	0.3	0	0.5
186.67	0	−46.67	33.33	0	1	66.67

$\xrightarrow{\ *\textbf{Row}+(-0.33, C, 2, 1) \blacktriangleright B\ }$
$\xrightarrow{\ *\textbf{Row}+(46.67, B, 2, 3) \blacktriangleright C\ }$

u	v	w	x	y	P	Constant
3.1	1	0	0.4	−0.1	0	0.50
−1.3	0	1	−0.2	0.3	0	0.50
125.93	0	0.05	23.99	14.02	1	90.03

Solution for the
primal problem

From the last tableau, we see that $x = 23.99$, $y = 14.02$, and the minimum value of C is 90.03.

◦ ◦ ◦

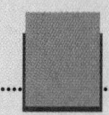 **EXERCISES**

In exercises 1–4, use a graphing calculator to solve the linear programming problem by the simplex method.

1. Minimize $C = x + y + 3z$ subject to

$$2x + y + 3z \geq 6$$
$$x + 2y + 4z \geq 8$$
$$3x + y - 2z \geq 4$$
$$x \geq 0, y \geq 0, z \geq 0$$

2. Minimize $C = 2x + 4y + z$ subject to

$$x + 2y + 4z \geq 7$$
$$3x + y - z \geq 6$$
$$x + 4y + 2z \geq 24$$
$$x \geq 0, y \geq 0, z \geq 0$$

3. Minimize $C = x + 1.2y + 3.5z$ subject to

$$2x + 3y + 5z \geq 12$$
$$3x + 1.2y - 2.2z \geq 8$$
$$1.2x + 3y + 1.8z \geq 14$$
$$x \geq 0, y \geq 0, z \geq 0$$

4. Minimize $C = 2.1x + 1.2y + z$ subject to

$$x + y - z \geq 5.2$$
$$x - 2.1y + 4.2z \geq 8.4$$
$$x \geq 0, y \geq 0, z \geq 0$$

u	v	w	x	y	P	Constant
①	$\frac{1}{2}$	$\frac{1}{4}$	$\frac{1}{4}$	0	0	$\frac{1}{2}$
1	1	3	0	1	0	5
-40	-30	-30	0	0	1	0

$$\xrightarrow[R_3 + 40R_1]{R_2 - R_1}$$

u	v	w	x	y	P	Constant		ratio
1	$\frac{1}{2}$	$\frac{1}{4}$	$\frac{1}{4}$	0	0	$\frac{1}{2}$		$\frac{1/2}{1/4} = 2$
0	$\frac{1}{2}$	⑪⁄₄	$-\frac{1}{4}$	1	0	$\frac{9}{2}$		$\frac{9/2}{11/4} = \frac{18}{11}$
0	-10	-20	10	0	1	20		

pivot row → (row 2)
↑ pivot column (w)
$\xrightarrow{\frac{4}{11}R_2}$

u	v	w	x	y	P	Constant
1	$\frac{1}{2}$	$\frac{1}{4}$	$\frac{1}{4}$	0	0	$\frac{1}{2}$
0	$\frac{2}{11}$	①	$-\frac{1}{11}$	$\frac{4}{11}$	0	$\frac{18}{11}$
0	-10	-20	10	0	1	20

$$\xrightarrow[R_3 + 20R_2]{R_1 - \frac{1}{4}R_2}$$

u	v	w	x	y	P	Constant		ratio
1	⑤⁄₁₁	0	$\frac{3}{11}$	$-\frac{1}{11}$	0	$\frac{1}{11}$		$\frac{1/11}{5/11} = \frac{1}{5}$
0	$\frac{2}{11}$	1	$-\frac{1}{11}$	$\frac{4}{11}$	0	$\frac{18}{11}$		$\frac{18/11}{2/11} = 9$
0	$-\frac{70}{11}$	0	$\frac{90}{11}$	$\frac{80}{11}$	1	$\frac{580}{11}$		

pivot row → (row 1)
↑ pivot column (v)
$\xrightarrow{\frac{11}{5}R_1}$

u	v	w	x	y	P	Constant
$\frac{11}{5}$	①	0	$\frac{3}{5}$	$-\frac{1}{5}$	0	$\frac{1}{5}$
0	$\frac{2}{11}$	1	$-\frac{1}{11}$	$\frac{4}{11}$	0	$\frac{18}{11}$
0	$-\frac{70}{11}$	0	$\frac{90}{11}$	$\frac{80}{11}$	1	$\frac{580}{11}$

$$\xrightarrow[R_3 + \frac{70}{11}R_1]{R_2 - \frac{2}{11}R_1}$$

u	v	w	x	y	P	Constant
$\frac{11}{5}$	1	0	$\frac{3}{5}$	$-\frac{1}{5}$	0	$\frac{1}{5}$
$-\frac{2}{5}$	0	1	$-\frac{1}{5}$	$\frac{2}{5}$	0	$\frac{8}{5}$
14	0	0	12	6	1	54

Solution for the primal problem (under the 12 and 6 entries)

The last tableau is final, and the solution to the primal problem is $x = 12$ and $y = 6$ with a minimum value for C of 54.

4.3 THE SIMPLEX METHOD: NONSTANDARD PROBLEMS (OPTIONAL)

Section 4.1 showed how the simplex method can be used to solve standard maximization problems, and Section 4.2 showed how, thanks to duality, it can be used to solve standard minimization problems provided that the coefficients in the objective function are all nonnegative.

In this section we see how the simplex method can be incorporated into a method for solving **nonstandard problems**—problems that do not fall into either of the two previous categories. We begin by recalling the characteristics of standard problems.

I. Standard Maximization Problem

   1. The objective function is to be maximized.

   2. All the variables involved in the problem are nonnegative.

   3. Each linear constraint may be written so that the expression involving the variables is less than or equal to a nonnegative constant.

II. Standard Minimization Problem (restricted version—see condition 4)

   1. The objective function is to be minimized.

   2. All the variables involved in the problem are nonnegative.

   3. Each linear constraint may be written so that the expression involving the variables is greater than or equal to a constant.

   4. *All the coefficients in the objective function are nonnegative.*

REMARK   Recall that if all the coefficients in the objective function are nonnegative, then a standard minimization problem can be solved by using the simplex method to solve the associated dual problem.   ❍ ❍ ❍

We now give some examples of linear programming problems that do not fit into these two categories of problems.

EXAMPLE 1   Explain why the following linear programming problem is not a standard maximization problem.

Maximize $P = x + 2y$
subject to $4x + 3y \leq 18$
             $-x + 3y \geq 3$
             $x \geq 0, y \geq 0$

Solution   This is not a standard maximization problem because the second constraint inequality,

$$-x + 3y \geq 3$$

violates condition 3. Observe that by multiplying both sides of this inequality by -1, we obtain

$$x - 3y \leq -3$$ (Recall that multiplying both sides of an inequality by a negative number reverses the inequality sign.)

Now the last equation still violates condition 3 because the constant on the right is *negative*.   ◦ ◦ ◦

Observe that the constraints in Example 1 involve both *less than or equal to constraints* ($\leq$) and *greater than or equal to constraints* ($\geq$). Such constraints are called **mixed constraints.** We will solve the problem posed in Example 1 later.

EXAMPLE 2   Explain why the following linear programming problem is not a restricted standard minimization problem.

$$\begin{aligned}
\text{Minimize} \quad & C = 2x - 3y \\
\text{subject to} \quad & x + y \leq 5 \\
& x + 3y \geq 9 \\
& -2x + y \leq 2 \\
& x \geq 0, y \geq 0
\end{aligned}$$

Solution   Observe that the coefficients in the objective function C are not all nonnegative. Therefore, the problem is not a restricted standard minimization problem. By constructing the dual problem, you can convince yourself that the latter is not a standard maximization problem and thus cannot be solved using the methods described in Sections 4.1 and 4.2. Again, we will solve this problem later.   ◦ ◦ ◦

EXAMPLE 3   Explain why the following linear programming problem is not a standard maximization problem. Show that it cannot be rewritten as a restricted standard minimization problem.

$$\begin{aligned}
\text{Maximize} \quad & P = x + 2y \\
\text{subject to} \quad & 2x + 3y \leq 12 \\
& -x + 3y = 3 \\
& x \geq 0, y \geq 0
\end{aligned}$$

Solution   The constraint equation $-x + 3y = 3$ is equivalent to the two inequalities

$$-x + 3y \leq 3 \quad \text{and} \quad -x + 3y \geq 3$$

By multiplying both sides of the second inequality by -1, it can be written in the form

$$x - 3y \leq -3$$

Therefore, the two given constraints are equivalent to the three inequality constraints

$$2x + 3y \leq 12$$
$$-x + 3y \leq 3$$
$$x - 3y \leq -3$$

The third inequality violates condition 3 for a standard maximization problem. Next, we see that the given problem is equivalent to the following:

Minimize $C = -x - 2y$
subject to $-2x - 3y \geq -12$
$$x - 3y \geq -3$$
$$-x + 3y \geq 3$$
$$x \geq 0, y \geq 0$$

Since the coefficients of the objective function are not all nonnegative, we conclude that the given problem cannot be rewritten as a restricted standard minimization problem. You will be asked to solve this problem in exercise 11.

○ ○ ○

The Simplex Method for Solving Nonstandard Problems

In order to describe a technique for solving nonstandard problems, let us consider the problem of Example 1:

Maximize $P = x + 2y$
subject to $4x + 3y \leq 18$
$$-x + 3y \geq 3$$
$$x \geq 0, y \geq 0$$

As a first step, we rewrite the inequality constraints so that the second constraint involves a $\leq$ constraint. As in Example 1, we obtain

$$4x + 3y \leq 18$$
$$x - 3y \leq -3$$
$$x \geq 0, y \geq 0$$

Disregarding the fact that the constant on the right of the second inequality constraint is negative, let us attempt to solve the problem using the simplex method for problems in standard form. Introducing the slack variables u and v gives the system of linear equations

$$4x + 3y + u \qquad = 18$$
$$x - 3y \qquad + v \qquad = -3$$
$$-x - 2y \qquad \qquad + P = 0$$

The initial simplex tableau is

x	y	u	v	P	Constant
4	3	1	0	0	18
1	−3	0	1	0	−3
−1	−2	0	0	1	0

Interpreting the tableau in the usual fashion, we see that

$$x = 0, \quad y = 0, \quad u = 18, \quad \text{and} \quad v = -3$$

Since the value of the slack variable v is negative, we see that this cannot be a feasible solution (remember, all variables must be nonnegative). In fact, you can see from Figure 4.4 that the point $(0, 0)$ does not lie in the feasible set associated with the given problem. Since we must start from a feasible point when using the simplex method for problems in standard form, we see that this method is not applicable at this juncture.

Figure 4.4
S is the feasible set for the problem.

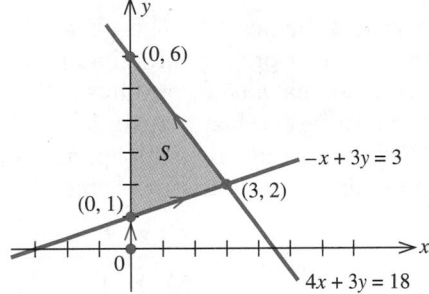

Let us find a way to bring us from the nonfeasible point $(0, 0)$ to *any* feasible point, after which we can switch to the simplex method for problems in standard form. This can be accomplished by pivoting as follows:

Referring to the tableau

x	y	u	v	P	Constant	ratio
4	3	1	0	0	18	$\frac{18}{3} = 6$
1	⊖3	0	1	0	−3	$\frac{-3}{-3} = 1$
−1	−2	0	0	1	0	

↑
pivot
column

notice the negative number −3 lying in the column of constants, above the lower horizontal line. Locate any negative number to the left of this number (there must always be at least one such number if the problem has a solution),

For the problem under consideration there is only one such number, the number -3 in the y-column. This column is designated as the pivot column. To find the pivot element, we form the *positive* ratios of the numbers in the column of constants to the corresponding numbers in the pivot column (above the last row). The pivot row is the row corresponding to the smallest ratio, and the pivot element is the element common to both the pivot row and the pivot column (the number circled in the foregoing tableau). Pivoting about this element, we have

$$\xrightarrow{-\frac{1}{3}R_2}$$

x	y	u	v	P	Constant
4	3	1	0	0	18
$-\frac{1}{3}$	①	0	$-\frac{1}{3}$	0	1
-1	-2	0	0	1	0

$$\xrightarrow[R_3 + 2R_2]{R_1 - 3R_2}$$

x	y	u	v	P	Constant
5	0	1	1	0	15
$-\frac{1}{3}$	1	0	$-\frac{1}{3}$	0	1
$-\frac{5}{3}$	0	0	$-\frac{2}{3}$	1	2

Interpreting the last tableau in the usual fashion, we see that

$$x = 0, \quad y = 1, \quad u = 15, \quad v = 0, \quad \text{and} \quad P = 2$$

Observe that the point $(0, 1)$ is a feasible point (see Figure 4.4). Our iteration has brought us from a nonfeasible point to a feasible point in one iteration. Observe, too, that all the constants in the column of constants are now nonnegative, reflecting the fact that $(0, 1)$ is a feasible point, as we have just noted.

We can now use the simplex method for problems in standard form to complete the solution to our problem.

	x	y	u	v	P	Constant	ratio
pivot row →	⑤	0	1	1	0	15	3
	$-\frac{1}{3}$	1	0	$-\frac{1}{3}$	0	1	—
	$-\frac{5}{3}$	0	0	$-\frac{2}{3}$	1	2	

↑
pivot column

$$\xrightarrow{\frac{1}{5}R_1}$$

x	y	u	v	P	Constant
①	0	$\frac{1}{5}$	$\frac{1}{5}$	0	3
$-\frac{1}{3}$	1	0	$-\frac{1}{3}$	0	1
$-\frac{5}{3}$	0	0	$-\frac{2}{3}$	1	2

$$\xrightarrow[R_3 + \frac{5}{3}R_1]{R_2 + \frac{1}{3}R_1}$$

x	y	u	v	P	Constant	ratio
1	0	$\frac{1}{5}$	⑤ $\frac{1}{5}$	0	3	15
0	1	$\frac{1}{15}$	$-\frac{4}{15}$	0	2	—
0	0	$\frac{1}{3}$	$-\frac{1}{3}$	1	7	

↑
pivot column

x	y	u	v	P	Constant
5	0	1	①	0	15
0	0	$\frac{1}{15}$	$-\frac{4}{15}$	0	2
0	0	$\frac{1}{3}$	$-\frac{1}{3}$	1	7

$\xrightarrow{5R_1}$

$\xrightarrow[R_3 + \frac{1}{3}R_1]{R_2 + \frac{4}{15}R_1}$

x	y	u	v	P	Constant
5	0	1	1	0	15
$\frac{4}{3}$	1	$\frac{1}{3}$	0	0	6
$\frac{5}{3}$	0	$\frac{2}{3}$	0	1	12

All entries in the last row are nonnegative and the tableau is final. We see that the optimal solution is

$$x = 0, \quad y = 6, \quad u = 0, \quad v = 15, \quad \text{and} \quad P = 12$$

Observe that the maximum of P occurs at $(0, 6)$ (see Figure 4.4). The arrows indicate the path that our search for the maximum of P has taken us on.

Before looking at further examples, let us summarize the method for solving nonstandard problems.

THE SIMPLEX METHOD FOR SOLVING NONSTANDARD PROBLEMS

1. If necessary, rewrite the problem as a maximization problem (recall that minimizing C is equivalent to maximizing $-C$).

2. If necessary, rewrite all inequality constraints (except $x \geq 0$, $y \geq 0$) using less than or equal to ($\leq$) inequalities.

3. Introduce slack variables and set up the initial simplex tableau.

4. Scan the upper part of the column of constants of the tableau for negative entries.
   a. If there are no negative entries, complete the solution using the simplex method for problems in standard form.
   b. If there are negative entries, proceed to step 5.

5. **a.** Pick any negative entry in the row in which a negative entry in the column of constants occurs. The column containing this entry is the pivot column.
   b. Compute the positive ratios of the numbers in the column of constants to the corresponding numbers in the pivot column (above the last row). The pivot row corresponds to the smallest ratio. The intersection of the pivot column and the pivot row determines the pivot element.
   c. Pivot the tableau about the pivot element. Then return to step 4.

We now apply the method to solve the nonstandard problem posed in Example 2.

EXAMPLE 4 Solve the problem of Example 2:

Minimize $\quad C = 2x - 3y$

subject to $\quad\quad x + y \leq 5$

$\quad\quad\quad\quad\quad x + 3y \geq 9$

$\quad\quad\quad -2x + y \leq 2$

$\quad\quad\quad\quad\quad x \geq 0, y \geq 0$

Solution   We first rewrite the problem as a maximization problem with inequality constraints using $\leq$, obtaining the following equivalent problem:

Maximize $P = -C = -2x + 3y$
subject to $x + y \leq 5$
$-x - 3y \leq -9$
$-2x + y \leq 2$
$x \geq 0, y \geq 0$

Introducing slack variables u, v, and w, and following the procedure for solving nonstandard problems outlined earlier, we obtain the following sequence of tableaus:

	x	y	u	v	w	P	Constant		ratio	
	1	1	1	0	0	0	5		$\frac{5}{1} = 5$	(Column 1 could have been chosen as the pivot column as well.)
	−1	−3	0	1	0	0	−9		$\frac{-9}{-3} = 3$	
pivot row →	−2	①	0	0	1	0	2		$\frac{2}{1} = 2$	
	2	−3	0	0	0	1	0			

↑ pivot column

	x	y	u	v	w	P	Constant		ratio	
$R_1 - R_3$	3	0	1	0	−1	0	3		$\frac{3}{3} = 1$	
$R_2 + 3R_3$	⑦−7	0	0	1	3	0	−3		$\frac{-3}{-7} = \frac{3}{7}$	
$R_4 + 3R_3$	−2	1	0	0	1	0	2			
	−4	0	0	0	3	1	6			

↑ pivot column

	x	y	u	v	w	P	Constant
	3	0	1	0	−1	0	3
$-\frac{1}{7}R_2$ →	①	0	0	$-\frac{1}{7}$	$-\frac{3}{7}$	0	$\frac{3}{7}$
	−2	1	0	0	1	0	2
	−4	0	0	0	3	1	6

	x	y	u	v	w	P	Constant		ratio	
$R_1 - 3R_2$	0	0	1	⑶⁄₇	$\frac{2}{7}$	0	$\frac{12}{7}$		4	(We now use the simplex method for problems in standard form to complete the problem.)
$R_3 + 2R_2$ →	1	0	0	$-\frac{1}{7}$	$-\frac{3}{7}$	0	$\frac{3}{7}$		—	
$R_4 + 4R_2$	0	1	0	$-\frac{2}{7}$	$\frac{1}{7}$	0	$\frac{20}{7}$		—	
	0	0	0	$-\frac{4}{7}$	$\frac{9}{7}$	1	$\frac{54}{7}$			

↑ pivot column

	x	y	u	v	w	P	Constant
	0	0	$\frac{7}{3}$	①	$\frac{2}{3}$	0	4
$\xrightarrow{\frac{7}{3}R_1}$	1	0	0	$-\frac{1}{7}$	$-\frac{3}{7}$	0	$\frac{3}{7}$
	0	1	0	$-\frac{2}{7}$	$\frac{1}{7}$	0	$\frac{20}{7}$
	0	0	0	$-\frac{4}{7}$	$\frac{9}{7}$	1	$\frac{54}{7}$

	x	y	u	v	w	P	Constant
	0	0	$\frac{7}{3}$	1	$\frac{2}{3}$	0	4
$\xrightarrow[\dfrac{R_3 + \frac{2}{7}R_1}{R_4 + \frac{4}{7}R_1}]{R_2 + \frac{1}{7}R_1}$	1	0	$\frac{1}{3}$	0	$-\frac{1}{3}$	0	1
	0	1	$\frac{2}{3}$	0	$\frac{1}{3}$	0	4
	0	0	$\frac{4}{3}$	0	$\frac{5}{3}$	1	10

All the entries in the last row are nonnegative and the tableau is final. We see that the optimal solution is

$$x = 1, \quad y = 4, \quad u = 0, \quad v = 4, \quad w = 0, \quad \text{and} \quad C = -P = -10$$

The feasible set S for this problem is shown in Figure 4.5. The path leading from the nonfeasible initial point $(0, 0)$ to the optimal point $(1, 4)$ goes through the nonfeasible point $(0, 2)$ and the feasible point $(\frac{3}{7}, \frac{20}{7})$, in that order.

Figure 4.5

The feasible set S and the path leading from the initial nonfeasible point $(0, 0)$ to the optimal point $(1, 4)$.

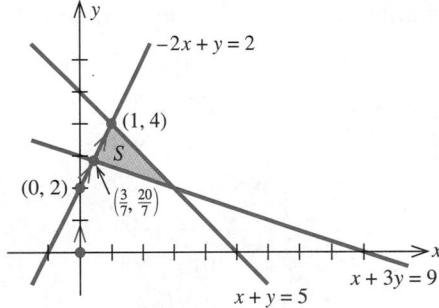

Application

EXAMPLE 5   The Rockford Company manufactures two models of exercise bicycles, a standard model and a deluxe model, in two separate plants—plant I and plant II. The maximum output at plant I is 1200 per month; the maximum output at plant II is 1000 per month. The profit per bike for standard and deluxe models manufactured at plant I is $40 and $60, respectively; the profit per bike for standard and deluxe models manufactured at plant II is $45 and $50, respectively.

For the month of May, Rockford received an order for 1000 standard models and 800 deluxe models. If prior commitments dictate that the number of deluxe models manufactured at plant I may not exceed the number of standard models manufactured there by more than 200, find how many of each model should be produced at each plant so as to satisfy the order and at the same time maximize Rockford's profit.

Solution   Let x and y denote the number of standard and deluxe models to be manufactured at plant I. Since the number of standard and deluxe models required are 1000 and 800, respectively, we see that the number of standard and deluxe models to be manufactured at plant II are $(1000 - x)$ and $(800 - y)$, respectively. Rockford's profit will then be

$$P = 40x + 60y + 45(1000 - x) + 50(800 - y)$$
$$= 85{,}000 - 5x + 10y$$

Since the maximum output of plant I is 1200, we have the constraint

$$x + y \leq 1200$$

Similarly, since the maximum output of plant II is 1000, we have

$$(1000 - x) + (800 - y) \leq 1000$$

or, equivalently,

$$-x - y \leq -800$$

Finally, the additional constraints placed on the production schedule at plants I and II translate into the inequalities

$$y - x \leq 200$$
$$x \leq 1000$$

and
$$y \leq 800$$

To summarize, the problem at hand is the following nonstandard problem:

Maximize $P = 85{,}000 - 5x + 10y$

subject to
$$x + y \leq 1200$$
$$-x - y \leq -800$$
$$-x + y \leq 200$$
$$x \leq 1000$$
$$y \leq 800$$
$$x \geq 0, y \geq 0$$

Let us introduce the slack variables, u, v, w, r, and s. Using the simplex method for nonstandard problems, we obtain the following sequence of tableaus:

x	y	u	v	w	r	s	P	Constant	ratio
1	1	1	0	0	0	0	0	1,200	$\frac{1200}{1} = 1200$
-1	-1	0	1	0	0	0	0	-800	$\frac{-800}{-1} = 800$
-1	1	0	0	1	0	0	0	200	—
1	0	0	0	0	1	0	0	1,000	$\frac{1000}{1} = 1000$
0	1	0	0	0	0	1	0	800	—
5	-10	0	0	0	0	0	1	85,000	

↑
pivot
column

	x	y	u	v	w	r	s	P	Constant
	1	1	1	0	0	0	0	0	1,200
	①	1	0	-1	0	0	0	0	800
$\xrightarrow{-R_2}$	-1	1	0	0	1	0	0	0	200
	1	0	0	0	0	1	0	0	1,000
	0	1	0	0	0	0	1	0	800
	5	-10	0	0	0	0	0	1	85,000

	x	y	u	v	w	r	s	P	Constant	ratio
	0	0	1	1	0	0	0	0	400	—
$R_1 - R_2$	1	1	0	-1	0	0	0	0	800	$\frac{800}{1} = 800$
$R_3 + R_2$	0	②	0	-1	1	0	0	0	1,000	$\frac{1000}{2} = 500$
$\xrightarrow{R_4 - R_2}$										
$R_6 - 5R_2$	0	-1	0	1	0	1	0	0	200	—
	0	1	0	0	0	0	1	0	800	$\frac{800}{1} = 800$
	0	-15	0	5	0	0	0	1	81,000	

(We now use the simplex method for standard problems.)

pivot column

	x	y	u	v	w	r	s	P	Constant
	0	0	1	1	0	0	0	0	400
	1	1	0	-1	0	0	0	0	800
$\xrightarrow{\frac{1}{2}R_3}$	0	①	0	$-\frac{1}{2}$	$\frac{1}{2}$	0	0	0	500
	0	-1	0	1	0	1	0	0	200
	0	1	0	0	0	0	1	0	800
	0	-15	0	5	0	0	0	1	81,000

	x	y	u	v	w	r	s	P	Constant	ratio
	0	0	1	①	0	0	0	0	400	$\frac{400}{1} = 400$
$R_2 - R_3$	1	0	0	$-\frac{1}{2}$	$-\frac{1}{2}$	0	0	0	300	—
$\xrightarrow{R_4 + R_3}$	0	1	0	$-\frac{1}{2}$	$\frac{1}{2}$	0	0	0	500	—
$R_5 - R_3$	0	0	0	$\frac{1}{2}$	$\frac{1}{2}$	1	0	0	700	$\frac{700}{1/2} = 1,400$
$R_6 + 15R_3$	0	0	0	$\frac{1}{2}$	$-\frac{1}{2}$	0	1	0	300	$\frac{300}{1/2} = 600$
	0	0	0	$-\frac{5}{2}$	$\frac{15}{2}$	0	0	1	88,500	

pivot column

	x	y	u	v	w	r	s	P	Constant
	0	0	1	1	0	0	0	0	400
$R_2 + \frac{1}{2}R_1$	1	0	$\frac{1}{2}$	0	$-\frac{1}{2}$	0	0	0	500
$R_3 + \frac{1}{2}R_1$ $\longrightarrow$	0	1	$\frac{1}{2}$	0	$\frac{1}{2}$	0	0	0	700
$R_4 - \frac{1}{2}R_1$	0	0	$-\frac{1}{2}$	0	$\frac{1}{2}$	1	0	0	500
$R_5 - \frac{1}{2}R_1$ $R_6 + \frac{5}{2}R_1$	0	0	$-\frac{1}{2}$	0	$-\frac{1}{2}$	0	1	0	100
	0	0	$\frac{5}{2}$	0	$\frac{15}{2}$	0	0	1	89,500

All the entries in the last row are nonnegative and the tableau is final. We see that $x = 500$, $y = 700$, and $P = 89,500$. This tells us that plant I should manufacture 500 standard and 700 deluxe exercise bicycles and that plant II should manufacture $(1000 - 500)$, or 500, standard and $(800 - 700)$, or 100, deluxe models. Rockford's profit will then be $89,500.   ◦ ◦ ◦

Refer to Example 5.

a. Sketch the feasible set S for the linear programming problem.
b. Solve the problem using the method of corners.
c. Indicate on S the points (both nonfeasible and feasible) corresponding to each iteration of the simplex method and trace the path leading to the optimal solution.

SELF-CHECK EXERCISES 4.3

1. Solve the following nonstandard problem using the method of this section:

Maximize $P = 2x + 3y$
subject to $x + \ y \le \ \ 40$
$-x + 2y \le -10$
$x \ge 0, y \ge 0$

2. A farmer has 150 acres of land suitable for cultivating crops A and B. The cost of cultivating crop A is $40 per acre and that of crop B is $60 per acre. The farmer has a maximum of $7400 available for land cultivation. Each acre of crop A requires 20 hours of labor and each acre of crop B requires 25 hours of labor. The farmer has a maximum of 3300 hours of labor available. He has also decided that he will cultivate at least 70 acres of crop A. If he expects to make a profit of $150 per acre on crop A and $200 per acre on crop B, how many acres of each crop should he plant in order to maximize his profit?

Solutions to Self-Check Exercises 4.3 can be found on page 267.

4.3 EXERCISES

In exercises 1–4, rewrite the given linear programming problem as a maximization problem with constraints involving inequalities of the form ≤ (with the exception of the inequalities $x \geq 0$, $y \geq 0$, and $z \geq 0$).

1. Minimize $P = 2x - 3y$ subject to

$$3x + 5y \geq 20$$
$$3x + y \leq 16$$
$$-2x + y \leq 1$$
$$x \geq 0, y \geq 0$$

2. Minimize $C = 2x + 3y$ subject to

$$x + y \leq 10$$
$$x + 2y \geq 12$$
$$2x + y \geq 12$$
$$x \geq 0, y \geq 0$$

3. Minimize $C = 5x + 10y + z$ subject to

$$2x + y + z \geq 4$$
$$x + 2y + 2z \geq 2$$
$$2x + 4y + 3z \leq 12$$
$$x \geq 0, y \geq 0, z \geq 0$$

4. Maximize $P = 2x + y - 2z$ subject to

$$x + 2y + z \geq 10$$
$$3x + 4y + 2z \geq 5$$
$$2x + 5y + 12z \leq 20$$
$$x \geq 0, y \geq 0, z \geq 0$$

In exercises 5–20, use the method of this section to solve the given linear programming problem.

5. Maximize $P = x + 2y$ subject to

$$2x + 5y \leq 20$$
$$x - 5y \leq -5$$
$$x \geq 0, y \geq 0$$

6. Maximize $P = 2x + 3y$ subject to

$$x + 2y \leq 8$$
$$x - y \leq -2$$
$$x \geq 0, y \geq 0$$

7. Minimize $C = -2x + y$ subject to

$$x + 2y \leq 6$$
$$3x + 2y \leq 12$$
$$x \geq 0, y \geq 0$$

8. Minimize $C = -2x + 3y$ subject to

$$x + 3y \leq 60$$
$$2x + y \geq 45$$
$$x \leq 40$$
$$x \geq 0, y \geq 0$$

9. Maximize $P = x + 4y$ subject to

$$x + 3y \leq 6$$
$$-2x + 3y \leq -6$$
$$x \geq 0, y \geq 0$$

10. Maximize $P = 5x + y$ subject to

$$2x + y \leq 8$$
$$-x + y \geq 2$$
$$x \geq 0, y \geq 0$$

11. Maximize $P = x + 2y$ subject to

$$2x + 3y \leq 12$$
$$-x + 3y = 3$$
$$x \geq 0, y \geq 0$$

12. Minimize $P = x + 2y$ subject to

$$4x + 7y \leq 70$$
$$2x + y = 20$$
$$x \geq 0, y \geq 0$$

13. Maximize $P = 5x + 4y + 2z$ subject to

$$x + 2y + 3z \leq 24$$
$$x - y + z \geq 6$$
$$x \geq 0, y \geq 0, z \geq 0$$

14. Maximize $P = x - 2y + z$ subject to

$$2x + 3y + 2z \leq 12$$
$$x + 2y - 3z \geq 6$$
$$x \geq 0, y \geq 0, z \geq 0$$

15. Minimize $C = x - 2y + z$ subject to

$$x - 2y + 3z \leq 10$$
$$2x + y - 2z \leq 15$$
$$2x + y + 3z \leq 20$$
$$x \geq 0, y \geq 0, z \geq 0$$

16. Minimize $C = 2x - 3y + 4z$ subject to

$$-x + 2y - z \leq 8$$
$$x - 2y + 2z \leq 10$$
$$2x + 4y - 3z \leq 12$$
$$x \geq 0, y \geq 0, z \geq 0$$

17. Maximize $P = 2x + y + z$ subject to

$$x + 2y + 3z \leq 28$$
$$2x + 3y - z \leq 6$$
$$x - 2y + z \geq 4$$
$$x \geq 0, y \geq 0, z \geq 0$$

18. Minimize $C = 2x - y + 3z$ subject to

$$2x + y + z \geq 2$$
$$x + 3y + z \geq 6$$
$$2x + y + 2z \leq 12$$
$$x \geq 0, y \geq 0, z \geq 0$$

19. Maximize $P = x + 2y + 3z$ subject to

$$x + 2y + z \leq 20$$
$$3x + y \leq 30$$
$$2x + y + z = 10$$
$$x \geq 0, y \geq 0, z \geq 0$$

20. Minimize $Q = 3x + 2y + z$ subject to

$$x + 2y + z \leq 20$$
$$3x + y \leq 30$$
$$2x + y + z = 10$$
$$x \geq 0, y \geq 0, z \geq 0$$

21. Agriculture—Crop Planning A farmer has 150 acres of land suitable for cultivating crops A and B. The cost of cultivating crop A is $40 per acre and that of crop B is $60 per acre. The farmer has a maximum of $7400 available for land cultivation. Each acre of crop A requires 20 hours of labor and each acre of crop B requires 25 hours of labor. The farmer has a maximum of 3300 hours of labor available. He has also decided that he will cultivate at least 80 acres of crop A. If he expects

to make a profit of $150 per acre on crop A and $200 per acre on crop B, how many acres of each crop should he plant in order to maximize his profit?

22. Finance—Investments William has at most $50,000 to invest in the common stocks of two companies. He estimates that an investment in company A will yield a return of 10%, whereas an investment in company B, which he feels is a riskier investment, will yield a return of 20%. If he decides that his investment in the stocks of company A is to exceed his investment in the stocks of company B by at least $20,000, determine how much he should invest in the stocks of each company in order to maximize the returns on his investment.

23. Finance—Allocation of Funds The First Street branch of the Capitol Bank has a sum of $60 million earmarked for home and commercial-development loans. The bank expects to realize an 8% annual rate of return on the home loans and a 6% annual rate of return on the commercial-development loans. Management has decided that the total amount of home loans is to be greater than or equal to three times the total amount of commercial-development loans. Owing to prior commitments, at least $10 million of the funds has been designated for commercial-development loans. Determine the amount of each type of loan the bank should extend in order to maximize its returns.

24. Manufacturing—Production Scheduling The Wayland Company manufactures two models of its twin-size futons, standard and deluxe, in two locations, I and II. The maximum output at location I is 600 per week, whereas the maximum output at location II is 400 per week. The profit per futon for standard and deluxe models manufactured at location I is $30 and $20, respectively; the profit per futon for standard and deluxe models manufactured at location II is $34 and $18, respectively. For a certain week, the company has received an order for 600 standard models and 300 deluxe models. If prior commitments dictate that the number of deluxe models manufactured at location II may not exceed the number of standard models manufactured there by more than 50, find how many of each model should be manufactured at each location so as to satisfy the order and at the same time maximize Wayland's profit.

25. Manufacturing—Production Scheduling A company manufactures products A, B, and C. Each product is processed in three departments: I, II, and III. The total available labor hours per week for departments I, II, and III are 900, 1080, and 840, respectively. The time requirements (in hours per unit) and the profit per unit for each product are as follows:

	Product A	Product B	Product C
Dept. I	2	1	2
Dept. II	3	1	2
Dept. III	2	2	1
Profit	$18	$12	$15

If management decides that the number of units of product B manufactured must equal or exceed the number of units of products A and C manufactured, how many units of each product should the company produce in order to maximize its profit?

26. **Manufacturing—Shipping Costs** Steinwelt Piano manufactures uprights and consoles in two plants, plant I and plant II. The output of plant I is at most 300 per month, whereas the output of plant II is at most 250 per month. These pianos are shipped to three warehouses that serve as distribution centers for the company. In order to fill current and projected future orders, warehouse A requires a minimum of 200 pianos per month, warehouse B requires at least 150 pianos per month, and warehouse C requires at least 200 pianos per month. The shipping cost of each piano from plant I to warehouse A, warehouse B, and warehouse C is $60, $60, and $80, respectively, and the shipping cost of each piano from plant II to warehouse A, warehouse B, and warehouse C is $80, $70, and $50, respectively. Use the method of this section to determine the shipping schedule that will enable Steinwelt to meet the warehouses' requirements while keeping the shipping costs to a minimum.

27. **Nutrition—Diet Planning** A nutritionist at the Medical Center has been asked to prepare a special diet for certain patients. She has decided that the meals should contain a minimum of 400 mg of calcium, 10 mg of iron, and 40 mg of vitamin C. She has further decided that the meals are to be prepared from foods A and B. Each ounce of food A contains 30 mg of calcium, 1 mg of iron, 2 mg of vitamin C, and 2 mg of cholesterol. Each ounce of food B contains 25 mg of calcium, 0.5 mg of iron, 5 mg of vitamin C, and 5 mg of cholesterol. Use the method of this section to determine how many ounces of each type of food the nutritionist should use in a meal so the cholesterol content is minimized and the minimum requirements of calcium, iron, and vitamin C are met.

SOLUTIONS TO SELF-CHECK EXERCISES 4.3

1. We are given the problem

$$\text{Maximize} \quad P = 2x + 3y$$
$$\text{subject to} \quad x + y \le 40$$
$$-x + 2y \le -10$$
$$x \ge 0, y \ge 0$$

Using the method of this section and introducing the slack variables u and v, we obtain the following tableaus:

x	y	u	v	P	Constant	ratio
1	1	1	0	0	40	$\frac{40}{1} = 40$
⧀−1⧀	2	0	1	0	−10	$\frac{-10}{-1} = 10$
−2	−3	0	0	1	0	

↑
pivot
column

	x	y	u	v	P	Constant
	1	1	1	0	0	40
$\xrightarrow{-R_2}$	①	−2	0	−1	0	10
	−2	−3	0	0	1	0

	x	y	u	v	P	Constant	ratio
$\xrightarrow[R_3 + 2R_2]{R_1 - R_2}$	0	③	1	1	0	30	$\frac{30}{3} = 10$
	1	−2	0	−1	0	10	—
	0	−7	0	−2	1	20	

$$\underset{\substack{\uparrow \\ \text{pivot} \\ \text{column}}}{}$$

	x	y	u	v	P	Constant
$\xrightarrow{\frac{1}{3}R_1}$	0	①	$\frac{1}{3}$	$\frac{1}{3}$	0	10
	1	−2	0	−1	0	10
	0	−7	0	−2	1	20

	x	y	u	v	P	Constant
$\xrightarrow[R_3 + 7R_1]{R_2 + 2R_1}$	0	1	$\frac{1}{3}$	$\frac{1}{3}$	0	10
	1	0	$\frac{2}{3}$	$-\frac{1}{3}$	0	30
	0	0	$\frac{7}{3}$	$\frac{1}{3}$	1	90

The last tableau is in final form and we obtain the solution

$$x = 30, \quad y = 10, \quad u = 0, \quad v = 0, \quad \text{and} \quad P = 90$$

2. Let x denote the number of acres of crop A to be cultivated and y the number of acres of crop B to be cultivated. Since there is a total of 150 acres of land available for cultivation, we have $x + y \le 150$. Next, the restriction on the amount of money available for land cultivation implies that $40x + 60y \le 7400$. Similarly, the restriction on the amount of time available for labor implies that $20x + 25y \le 3300$. Since the profit on each acre of crop A is \$150 and the profit on each acre of crop B is \$200, we see that the profit realizable by the farmer is $P = 150x + 200y$. Summarizing, we have the following linear programming problem:

$$\text{Maximize} \quad P = 150x + 200y$$

$$\begin{aligned}
\text{subject to} \quad x + \;\;\; y &\le \;\; 150 \\
40x + 60y &\le 7400 \\
20x + 25y &\le 3300 \\
x &\ge \;\;\; 70 \\
x \ge 0, \, y &\ge 0
\end{aligned}$$

In order to solve this nonstandard problem, we rewrite the fourth inequality in the form

$$-x \le -70$$

Using the method of this section with u, v, w, and z as slack variables, we have the following tableaus:

x	y	u	v	w	z	P	Constant		ratio
1	1	1	0	0	0	0	150		$\frac{150}{1} = 150$
40	60	0	1	0	0	0	7,400		$\frac{7400}{40} = 185$
20	25	0	0	1	0	0	3,300		$\frac{3300}{20} = 165$
−1	0	0	0	0	1	0	−70		$\frac{-70}{-1} = 70$
−150	−200	0	0	0	0	1	0		

↑
pivot
column

$\xrightarrow{-R_4}$

x	y	u	v	w	z	P	Constant
1	1	1	0	0	0	0	150
40	60	0	1	0	0	0	7,400
20	25	0	0	1	0	0	3,300
①	0	0	0	0	−1	0	70
−150	−200	0	0	0	0	1	0

$\begin{array}{c} R_1 - R_4 \\ R_2 - 40R_4 \\ \xrightarrow{} \\ R_3 - 20R_4 \\ R_5 + 150R_4 \end{array}$

x	y	u	v	w	z	P	Constant		ratio
0	1	1	0	0	1	0	80		$\frac{80}{1} = 80$
0	60	0	1	0	40	0	4,600		$\frac{4600}{60} = 76\frac{2}{3}$
0	㉕	0	0	1	20	0	1,900		$\frac{1900}{25} = 76$
1	0	0	0	0	−1	0	70		—
0	−200	0	0	0	−150	1	10,500		

↑
pivot
column

$\xrightarrow{\frac{1}{25}R_3}$

x	y	u	v	w	z	P	Constant
0	1	1	0	0	1	0	80
0	60	0	1	0	40	0	4,600
0	①	0	0	$\frac{1}{25}$	$\frac{4}{5}$	0	76
1	0	0	0	0	−1	0	70
0	−200	0	0	0	−150	1	10,500

	x	y	u	v	w	z	P	Constant
	0	0	1	0	$-\frac{1}{25}$	$\frac{1}{5}$	0	4
$\xrightarrow{\begin{array}{c} R_1 - R_3 \\ \hline R_2 - 60R_3 \\ R_5 + 200R_3 \end{array}}$	0	0	0	1	$-\frac{12}{5}$	-8	0	40
	0	1	0	0	$\frac{1}{25}$	$\frac{4}{5}$	0	76
	1	0	0	0	0	-1	0	70
	0	0	0	0	8	10	1	25,700

This last tableau is in final form and the solution is

$$x = 70, \quad y = 76, \quad u = 4, \quad v = 40, \quad w = 0, \quad z = 0, \quad \text{and} \quad P = 25,700$$

Thus, by cultivating 70 acres of crop A and 76 acres of crop B, the farmer will attain a maximum profit of \$25,700.

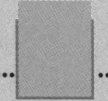

CHAPTER 4 SUMMARY OF PRINCIPAL TERMS

Terms

Standard maximization problem	Simplex tableau
Slack variable	Simplex method
Basic variable	Standard minimization problem
Nonbasic variable	Primal problem
Pivot column	Dual problem
Pivot row	Nonstandard problems
Pivot element	Mixed constraints

CHAPTER 4 REVIEW EXERCISES

In exercises 1–12, use the simplex method to solve the given linear programming problem.

1. Maximize $P = 3x + 4y$ subject to

$$x + 3y \leq 15$$
$$4x + y \leq 16$$
$$x \geq 0, y \geq 0$$

2. Maximize $P = 2x + 5y$ subject to

$$2x + y \leq 16$$
$$2x + 3y \leq 24$$
$$y \leq 6$$
$$x \geq 0, y \geq 0$$

3. Maximize $P = 2x + 3y + 5z$ subject to

$$x + 2y + 3z \leq 12$$
$$x - 3y + 2z \leq 10$$
$$x \geq 0, y \geq 0, z \geq 0$$

4. Maximize $P = x + 2y + 3z$ subject to

$$2x + y + z \leq 14$$
$$3x + 2y + 4z \leq 24$$
$$2x + 5y - 2z \leq 10$$
$$x \geq 0, y \geq 0, z \geq 0$$

5. Minimize $C = 3x + 2y$ subject to

$$2x + 3y \geq 6$$
$$2x + y \geq 4$$
$$x \geq 0, y \geq 0$$

6. Minimize $C = x + 2y$ subject to

$$3x + y \geq 12$$
$$x + 4y \geq 16$$
$$x \geq 0, y \geq 0$$

7. Minimize $C = 24x + 18y + 24z$ subject to

$$3x + 2y + z \geq 4$$
$$x + y + 3z \geq 6$$
$$x \geq 0, y \geq 0, z \geq 0$$

8. Minimize $C = 4x + 2y + 6z$ subject to

$$x + 2y + z \geq 4$$
$$2x + y + 2z \geq 2$$
$$3x + 2y + z \geq 3$$
$$x \geq 0, y \geq 0, z \geq 0$$

9. Maximize $P = 3x - 4y$ subject to

$$x + y \leq 45$$
$$x - 2y \geq 10$$
$$x \geq 0, y \geq 0$$

10. Minimize $C = 2x + 3y$ subject to

$$x + y \leq 10$$
$$x + 2y \geq 12$$
$$2x + y \geq 12$$
$$x \geq 0, y \geq 0$$

11. Maximize $P = 2x + 3y$ subject to

$$2x + 5y \leq 20$$
$$-x + 5y \geq 5$$
$$x \geq 0, y \geq 0$$

12. Minimize $C = -3x - 4y$ subject to

$$x + y \leq 45$$
$$x + y \geq 15$$
$$x \leq 30$$
$$y \leq 25$$
$$x \geq 0, y \geq 0$$

13. A company manufactures three products, A, B, and C, on two machines, I and II. It has been determined that the company will realize a profit of $4 on each unit of product A, $6 on each unit of product B, and $8 on each unit of product C. To manufacture a unit of product A requires 9 minutes on machine I and 6 minutes on machine II; to manufacture a unit of product B requires 12 minutes on machine I and 6 minutes on machine II; and to manufacture a unit of product C requires 18 minutes on machine I and 10 minutes on machine II. There are 6 hours of machine time available on machine I and 4 hours of machine time available on machine II in each work shift. How many units of each product should be produced in each shift in order to maximize the company's profit?

14. Mr. Moody has decided to invest at most $100,000 in securities in the form of corporate stocks. He has classified his options into three groups of stocks: blue-chip stocks that he assumes will yield a 10% return (dividends and capital appreciation) within a year, growth stocks that he assumes will yield a 15% return within a year, and speculative stocks that he assumes will yield a 20% return (mainly due to capital appreciation) within a year. Because of the relative risks involved in his investment, Mr. Moody has further decided that no more than 30% of his investment should be in growth and speculative stocks and at least 50% of his investment should be in blue-chip and speculative stocks. Determine how much Mr. Moody should invest in each group of stocks in the hope of maximizing the return on his investments.

15. Sandra has at most $200,000 to invest in stocks, bonds, and money-market funds. She expects annual yields of 15%, 10%, and 8%, respectively, on these investments. If Sandra wants at least $50,000 to be invested in money-market funds and requires that the amount invested in bonds be greater than or equal to the sum of her investments in stocks and money-market funds, determine how much she should invest in each vehicle in order to maximize the returns on her investments.

Interest that is periodically added to the principal and thereafter itself earns interest is called *compound interest*. We begin this chapter by deriving the *compound interest formula*, which gives the amount of money accumulated when an initial amount of money is invested in an account for a fixed term and earns compound interest.

An *annuity* is a sequence of payments made at regular intervals. We derive formulas giving the *future value of an annuity* (what you end up with) and the *present value of an annuity* (the lump sum that, when invested now, will yield the same future value of the annuity). Then, using these formulas, we answer questions involving the amortization of certain types of installment loans and questions involving *sinking funds* (funds that are set up to be used for a specific purpose at a future date).

How much will the home mortgage payment be? The Blakelys received a bank loan to help finance the purchase of a house. They have agreed to repay the loan in equal monthly installments over a certain period of time. In Example 2, page 300, we will show how to determine the size of the monthly installment so that the loan is fully amortized at the end of the term.

5

MATHEMATICS OF FINANCE

273

5.1 COMPOUND INTEREST

Simple Interest

A natural application of linear functions to the business world is found in the computation of **simple interest**—interest that is computed on the original principal only. Thus, if I denotes the interest on a principal P (in dollars) at an interest rate of r per year for t years, then we have

$$I = Prt$$

The **accumulated amount** A, the sum of the principal and interest after t years, is given by

$$A = P + I = P + Prt$$

or

$$A = P(1 + rt)$$

and is a linear function of t (see exercise 30 at the end of this section). In business applications we are normally interested only in the case where t is positive, so only that part of the line that lies in Quadrant I is of interest to us (Figure 5.1).

Figure 5.1
The accumulated amount is a linear function of t.

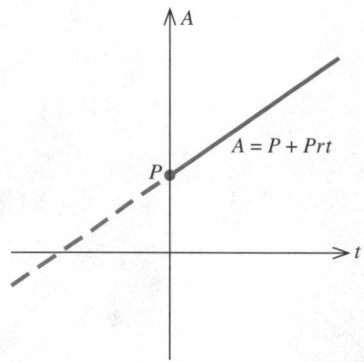

$A = P + Prt$

SIMPLE INTEREST FORMULAS	Interest: $I = Prt$ (1a)
	Accumulated amount: $A = P(1 + rt)$ (1b)

EXAMPLE 1   A bank pays simple interest at the rate of 8% per year for certain deposits. If a customer deposits $1000 and makes no withdrawals for three years, what is the total amount on deposit at the end of three years? What is the interest earned in that period of time?

Solution   Using equation (1b) with $P = 1000$, $r = 0.08$, and $t = 3$, we see that the total amount on deposit at the end of three years is given by

$$A = P(1 + rt)$$
$$= 1000[1 + (0.08)(3)] = 1240$$

or $1240.

The interest earned over the 3-year period is given by

$$I = Prt   \text{(Using 1a)}$$
$$= 1000(0.08)(3) = 240$$

or $240.

○ ○ ○

Refer to Example 1. Use a graphing utility to plot the graph of the function $A = 1000(1 + 0.08t)$ using the viewing rectangle $[0, 10] \times [0, 2000]$.

a. What is the A-intercept of the straight line and what does it represent?
b. What is the slope of the straight line and what does it represent? (See exercise 30.)

EXAMPLE 2 An amount of $2000 is invested in a 10-year trust fund that pays 6% annual simple interest. What is the total amount of the trust fund at the end of ten years?

Solution The total amount of the trust fund at the end of ten years is given by

$$A = P(1 + rt)$$
$$= 2000[1 + (0.06)(10)] = 3200$$

or $3200.

Compound Interest

In contrast to simple interest, earned interest that is periodically added to the principal and thereafter itself earns interest at the same rate is called **compound interest.** In order to find a formula for the accumulated amount, let us consider a numerical example. Suppose that $1000 (the principal) is deposited in a bank for a term of three years, earning interest at the rate of 8% per year (called the **nominal,** or **stated, rate**) compounded annually. Then, using equation (1b) with $P = 1000$, $r = 0.08$, and $t = 1$, we see that the accumulated amount at the end of the first year is

$$A_1 = P(1 + rt)$$
$$= 1000[1 + 0.08(1)] = 1000(1.08) = 1080$$

or $1080.

To find the accumulated amount A_2 at the end of the second year, we use (1b) once again, this time with $P = A_1$. (Remember, the principal *and* interest now earn interest over the second year.) We obtain

$$A_2 = P(1 + rt) = A_1(1 + rt)$$
$$= 1000[1 + 0.08(1)][1 + 0.08(1)]$$
$$= 1000[1 + 0.08]^2 = 1000(1.08)^2 = 1166.40$$

or $1166.40.

Finally, the accumulated amount A_3 at the end of the third year is found using (1b) with $P = A_2$, giving

$$A_3 = P(1 + rt) = A_2(1 + rt)$$
$$= 1000[1 + 0.08(1)]^2[1 + 0.08(1)]$$
$$= 1000[1 + 0.08]^3 = 1000(1.08)^3 \approx 1259.71$$

or approximately \$1259.71.

If you reexamine our calculations, you will see that the accumulated amounts at the end of each year have the following form:

First year: $A_1 = 1000(1 + 0.08)$, or $A_1 = P(1 + r)$

Second year: $A_2 = 1000(1 + 0.08)^2$, or $A_2 = P(1 + r)^2$

Third year: $A_3 = 1000(1 + 0.08)^3$, or $A_3 = P(1 + r)^3$

These observations suggest the following general result: If P dollars is invested over a term of t years, earning interest at the rate of r per year compounded annually, then the accumulated amount is

$$A = P(1 + r)^t \tag{2}$$

Formula (2) was derived under the assumption that interest was compounded *annually*. In practice, however, interest is usually compounded more than once a year. The interval of time between successive interest calculations is called the **conversion period.**

If interest at a nominal rate of r per year is compounded m times a year on a principal of P dollars, then the simple interest rate per conversion period is

$$i = \frac{r}{m} \qquad \frac{\text{(Annual interest rate)}}{\text{(Periods per year)}}$$

For example, if the nominal interest rate is 8% per year ($r = 0.08$) and interest is compounded quarterly ($m = 4$), then

$$i = \frac{r}{m} = \frac{0.08}{4} = 0.02$$

or 2% per period.

In order to find a general formula for the accumulated amount when a principal of P dollars is deposited in a bank for a term of t years and earns interest at the (nominal) rate of r per year compounded m times per year, we proceed as before, using (1b) repeatedly with the interest rate $i = r/m$. We see that the accumulated amount at the end of each period is

First period: $A_1 = P(1 + i)$

Second period: $A_2 = A_1(1 + i)$ $= [P(1 + i)](1 + i) = P(1 + i)^2$

Third period: $A_3 = A_2(1 + i)$ $= [P(1 + i)^2](1 + i) = P(1 + i)^3$

$$\vdots \qquad\qquad \vdots$$

nth period: $A_n = A_{n-1}(1 + i) = [P(1 + i)^{n-1}](1 + i) = P(1 + i)^n$

But there are $n = mt$ periods in t years (number of conversion periods times the term). Therefore, the accumulated amount at the end of t years is given by

$$A = P(1 + i)^n$$

COMPOUND INTEREST FORMULA

$$A = P(1 + i)^n \qquad (3)$$

where $i = r/m$, $n = mt$, and

A = Accumulated amount at the end of n conversion periods
P = Principal
r = Nominal interest rate per year
m = Number of conversion periods per year
t = Term (number of years)

EXPLORING WITH TECHNOLOGY

Let $A_1(t)$ denote the accumulated amount of $100 earning simple interest at the rate of 10% per year over t years, and let $A_2(t)$ denote the accumulated amount of $100 earning interest at the rate of 10% per year compounded monthly over t years.

a. Find expressions for $A_1(t)$ and $A_2(t)$.
b. Use a graphing utility to plot the graphs of A_1 and A_2 on the same set of axes, using the viewing rectangle $[0, 20] \times [0, 800]$.
c. Comment on the growth of $A_1(t)$ and $A_2(t)$ by referring to the graphs of A_1 and A_2.

◐   ◐   ◐

EXAMPLE 3 Find the accumulated amount after three years if $1000 is invested at 8% per year compounded (a) annually, (b) semi-annually, (c) quarterly, and (d) monthly.

Solution
a. Here $P = 1000$, $r = 0.08$, and $m = 1$. Thus, $i = r = 0.08$ and $n = 3$, so equation (3) gives

$$A = 1000(1.08)^3$$
$$= 1259.71$$

or $1259.71.

b. Here $P = 1000$, $r = 0.08$, and $m = 2$. Thus, $i = 0.08/2 = 0.04$ and $n = (3)(2) = 6$, so (3) gives

$$A = 1000(1.04)^6$$
$$= 1265.32$$

or $1265.32.

c. In this case, $P = 1000$, $r = 0.08$, and $m = 4$. Thus, $i = 0.08/4 = 0.02$ and $n = (3)(4) = 12$, so (3) gives

$$A = 1000(1.02)^{12}$$
$$= 1268.24$$

or $1268.24.

d. Here $P = 1000$, $r = 0.08$, and $m = 12$. Thus, $i = 0.08/12$ and $n = (3)(12) = 36$, so (3) gives

$$A = 1000 \left(1 + \frac{0.08}{12} \right)^{36}$$
$$= 1270.24$$

or $1270.24. These results are summarized in Table 5.1.

Table 5.1

Nominal Rate (r)	Conversion Period	Interest Rate/ Conversion Period	Initial Investment	Accumulated Amount
8%	Annual ($m = 1$)	8%	$1000	$1259.71
8%	Semiannual ($m = 2$)	4%	$1000	$1265.32
8%	Quarterly ($m = 4$)	2%	$1000	$1268.24
8%	Monthly ($m = 12$)	2/3%	$1000	$1270.24

◦ ◦ ◦

Effective Rate of Interest

Example 3 showed that the interest actually earned on an investment depends on the frequency with which the interest is compounded. Thus, the stated, or nominal, rate of 8% per year does not reflect the actual rate at which interest is earned. This suggests that we need to find a common basis for comparing interest rates. One such way of comparing interest rates is provided by the use of the *effective rate*. The **effective rate** is the *simple* interest rate that would produce the same accumulated amount in one year as the nominal rate compounded m times a year. The effective rate is also called the **effective annual yield.**

To derive a relationship between the nominal interest rate, r per year compounded m times, and its corresponding effective rate, R per year, let us assume an initial investment of P dollars. Then the accumulated amount after

EXPLORING WITH TECHNOLOGY

Investments allowed to grow over time can increase in value surprisingly fast. Consider the potential growth of $10,000 if earnings are reinvested. More specifically, suppose $A_1(t)$, $A_2(t)$, $A_3(t)$, $A_4(t)$, and $A_5(t)$ denote the accumulated values of an investment of $10,000 over a term of t years and earning interest at the rate of 4%, 6%, 8%, 10%, and 12% per year compounded annually.

a. Find expressions for $A_1(t)$, $A_2(t)$, ..., $A_5(t)$.
b. Use a graphing utility to plot the graphs of A_1, A_2, ..., A_5 on the same set of axes, using the viewing rectangle $[0, 20] \times [0, 100{,}000]$.
c. Use **TRACE** to find $A_1(20)$, $A_2(20)$, ..., $A_5(20)$ and interpret your results.

○   ○   ○

one year at a simple interest rate of R per year is

$$A = P(1 + R)$$

Also, the accumulated amount after one year at an interest rate of r per year compounded m times a year is

$$A = P(1 + i)^n = P\left(1 + \frac{r}{m}\right)^m \qquad \text{(Since } i = r/m\text{)}$$

Equating the two expressions gives

$$P(1 + R) = P\left(1 + \frac{r}{m}\right)^m$$

$$1 + R = \left(1 + \frac{r}{m}\right)^m \qquad \text{(Dividing both sides by } P\text{)}$$

or, upon solving for R, we obtain the following formula for computing the effective rate of interest.

EFFECTIVE RATE OF INTEREST FORMULA

$$r_{\text{eff}} = \left(1 + \frac{r}{m}\right)^m - 1 \qquad (4)$$

where r_{eff} = Effective rate of interest
r = Nominal interest rate per year
m = Number of conversion periods per year

EXAMPLE 4   Find the effective rate of interest corresponding to a nominal
rate of 8% per year compounded (a) annually, (b) semi-
annually, (c) quarterly, and (d) monthly.

Solution

a. The effective rate of interest corresponding to a nominal rate of 8% per
year compounded annually is, of course, given by 8% per year. This result is
also confirmed by using equation (4) with $r = 0.08$ and $m = 1$. Thus,

$$r_{\text{eff}} = (1 + 0.08) - 1 = 0.08$$

b. Let $r = 0.08$ and $m = 2$. Then (4) yields

$$r_{\text{eff}} = \left(1 + \frac{0.08}{2}\right)^2 - 1$$
$$= (1.04)^2 - 1$$
$$= 0.0816$$

so the required effective rate is 8.16% per year.

c. Let $r = 0.08$ and $m = 4$. Then (4) yields

$$r_{\text{eff}} = \left(1 + \frac{0.08}{4}\right)^4 - 1$$
$$= (1.02)^4 - 1$$
$$= 0.08243$$

so the corresponding effective rate in this case is 8.243% per year.

d. Let $r = 0.08$ and $m = 12$. Then (4) yields

$$r_{\text{eff}} = \left(1 + \frac{0.08}{12}\right)^{12} - 1$$
$$= 0.08300$$

so the corresponding effective rate in this case is 8.300% per year.   ◦ ◦ ◦

Now, if the effective rate of interest r_{eff} is known, the accumulated amount
after t years on an investment of P dollars may be more readily computed by
using the formula

$$A = P(1 + r_{\text{eff}})^t$$

The Truth in Lending Act passed by Congress in 1968 requires that the
effective rate of interest be disclosed in all contracts involving interest charges.
The passage of this act has benefited consumers because they now have a
common basis for comparing the various nominal rates quoted by different
financial institutions. Furthermore, knowing the effective rate enables consumers to compute the actual charges involved in a transaction. Thus, if the
effective rates of interest found in Example 4 were known, the accumulated
values of Example 3, shown in Table 5.2, could have been readily found.

Table 5.2

Nominal Rate	Frequency of Interest Payment	Effective Rate	Initial Investment	Accumulated Amount After 3 Years	
8%	Annually	8%	$1000	$1000(1 + 0.08)^3$	= $1259.71
8%	Semiannually	8.16%	$1000	$1000(1 + 0.0816)^3$	= $1265.32
8%	Quarterly	8.243%	$1000	$1000(1 + 0.08243)^3$	= $1268.23
8%	Monthly	8.300%	$1000	$1000(1 + 0.08300)^3$	= $1270.24

Present Value

Let us return to the compound interest formula (3), which expresses the accumulated amount at the end of n periods when interest at the rate of r is compounded m times a year. The principal P in (3) is often referred to as the **present value,** and the accumulated value A is called the **future value,** since it is realized at a future date. In certain instances, an investor may wish to determine how much money he should invest now, at a fixed rate of interest, so that he will realize a certain sum at some future date. This problem may be solved by expressing P in terms of A. Thus, from (3) we find

$$P = A(1 + i)^{-n}$$

Here, as before, $i = r/m$, where m is the number of conversion periods per year.

PRESENT VALUE FORMULA FOR COMPOUND INTEREST

$$P = A(1 + i)^{-n} \tag{5}$$

  EXAMPLE 5   Find how much money should be deposited in a bank paying interest at the rate of 6% per year compounded monthly so that at the end of three years the accumulated amount will be $20,000.

Solution   Here, $r = 0.06$ and $m = 12$, so $i = 0.06/12 = 0.005$ and $n = (3)(12) = 36$. Thus, the problem is to determine P given that $A = 20,000$. Using equation (5), we obtain

$$P = 20,000(1.005)^{-36}$$
$$= 16,713$$

or $16,713.   ⊙ ⊙ ⊙

EXAMPLE 6   Find the present value of $49,158.60 due in five years at an interest rate of 10% per year compounded quarterly.

Solution Using (5) with $r = 0.1$ and $m = 4$, so that $i = 0.1/4 = 0.025$, $n = (4)(5) = 20$, and $A = 49{,}158.6$, we obtain

$$P = (49{,}158.6)(1.025)^{-20} = 30{,}000.07$$

or approximately $30,000.   ○ ○ ○

Applications

The returns on certain investments such as zero coupon certificates of deposit (CDs) and zero coupon bonds are compared by quoting the time it takes for each investment to triple, or even quadruple. These calculations make use of the compound interest formula (3).

EXAMPLE 7   Jane has narrowed her investment options down to two:

a. Purchase a CD that matures in 12 years and pays interest upon maturity at the rate of 10% per year compounded daily (assume 365 days in a year).
b. Purchase a zero coupon CD that will triple her investment in the same period.

Which option will optimize her investment?

Solution   Let us compute the accumulated amount under option (a). Here

$$r = 0.10, \quad m = 365, \quad \text{and} \quad t = 12$$

so $n = 12(365) = 4380$ and $i = 0.10/365 \approx 0.0002740$. The accumulated amount at the end of 12 years (after 4380 conversion periods) is

$$A = P(1.0002740)^{4380} \approx 3.32P$$

or $3.32. If Jane chooses option (b), the accumulated amount of her investment after 12 years will be $3P$. Therefore, she should choose option (a).   ○ ○ ○

EXAMPLE 8   Diane has an Individual Retirement Account (IRA) with a brokerage firm. Her money is invested in a money-market mutual fund that pays interest on a daily basis. Over a 2-year period in which no deposits or withdrawals were made, her account grew from $4500 to $5268.24. Find the effective rate at which Diane's account was earning interest over that period (assume 365 days in a year).

Solution   Let r_{eff} denote the required effective rate of interest. We have

$$5268.24 = 4500(1 + r_{\text{eff}})^2$$
$$(1 + r_{\text{eff}})^2 = 1.17072$$
$$1 + r_{\text{eff}} = 1.081998 \quad \text{(Taking the square root on both sides)}$$

or $r_{\text{eff}} = 0.081998$. Therefore, the required effective rate is 8.20% per year.   ○ ○ ○

SELF-CHECK EXERCISES 5.1

1. Find the present value of $20,000 due in three years at an interest rate of 12% per year compounded monthly.

2. Mr. Baker is a retiree living on social security and the income from his investment. Currently, his $100,000 investment in a 1-year CD is yielding 10.6% interest compounded daily. If he reinvests the principal ($100,000) on the due date of the CD in another 1-year CD paying 9.2% interest compounded daily, find the net decrease in his yearly income from his investment.

Solutions to Self-Check Exercises 5.1 can be found on page 285.

5.1 EXERCISES

 A calculator is recommended for this exercise set.

1. Find the simple interest on a $500 investment made for two years at an interest rate of 8% per year. What is the accumulated amount?

2. Find the simple interest on a $1000 investment made for three years at an interest rate of 5% per year. What is the accumulated amount?

3. Find the accumulated amount at the end of nine months on an $800 deposit in a bank paying simple interest at a rate of 6% per year.

4. Find the accumulated amount at the end of eight months on a $1200 bank deposit paying simple interest at a rate of 7% per year.

5. If the accumulated amount is $1160 at the end of two years and the simple rate of interest is 8% per year, what is the principal?

6. A bank deposit paying simple interest at the rate of 5% per year grew to a sum of $3100 in ten months. Find the principal.

7. How many days will it take for a sum of $1000 to earn $20 interest if it is deposited in a bank paying ordinary simple interest at the rate of 5% per year? (Use a 365-day year.)

8. How many days will it take for a sum of $1500 to earn $25 interest if it is deposited in a bank paying 5% per year? (Use a 365-day year.)

9. A bank deposit paying simple interest grew from an initial sum of $1000 to a sum of $1075 in nine months. Find the interest rate.

10. Determine the simple interest rate at which $1200 will grow to $1250 in eight months.

In exercises 11–20, find the accumulated amount A if the principal P is invested at the interest rate of r per year for t years.

11. $P = \$1000$, $r = 7\%$, $t = 8$, compounded annually

12. $P = \$1000$, $r = 8\frac{1}{2}\%$, $t = 6$, compounded annually

13. $P = \$2500$, $r = 7\%$, $t = 10$, compounded semiannually

14. $P = \$2500$, $r = 9\%$, $t = 10.5$, compounded semiannually

15. $P = \$12,000$, $r = 8\%$, $t = 10.5$, compounded quarterly

16. $P = \$42,000$, $r = 7\frac{3}{4}\%$, $t = 8$, compounded quarterly

17. $P = \$150,000$, $r = 14\%$, $t = 4$, compounded monthly

18. $P = \$180,000$, $r = 9\%$, $t = 6\frac{1}{4}$, compounded monthly

19. $P = \$150,000$, $r = 12\%$, $t = 3$, compounded daily

20. $P = \$200,000$, $r = 8\%$, $t = 4$, compounded daily

In exercises 21–24, find the effective rate corresponding to the given nominal rate.

21. 10% compounded semiannually

22. 9% compounded quarterly

23. 8% compounded monthly

24. 8% compounded daily

In exercises 25–28, find the present value of $40,000 due in four years at the given rate of interest.

25. 8% per year compounded semiannually

26. 8% per year compounded quarterly

27. 7% per year compounded monthly

28. 9% per year compounded daily

29. **Consumer Decisions**   Mr. Mitchell has been given the option of either paying his $300 bill now or settling it for $306 after one month. If he chooses to pay after one month, find the simple interest rate at which he would be charged.

30. Write equation (1b) in the slope-intercept form and interpret the meaning of the slope and the A-intercept in terms of r and P.
    [*Hint:* Refer to Figure 5.1.]

31. **Hospital Costs**   If the cost of a semiprivate room in a hospital was $380 per day five years ago, and hospital costs have risen at the rate of 8% per year since that time, what rate would you expect to pay for a semiprivate room today?

32. **Family Food Expenditure**   Today a typical family of four spends $600 monthly for food. If inflation occurs at the rate of 4% per year over the next six years, how much should the typical family of four expect to spend for food six years from now?

33. **Housing Appreciation**   The Brennans are planning to buy a house four years from now. Housing experts in their area have estimated that the cost of a home will increase at a rate of 5% per year during that period. If this economic prediction holds true, how much can the Brennans expect to pay for a house that currently costs $150,000?

34. **Electricity Consumption**   A utility company in a western city of the United States expects the consumption of electricity to increase by 8% per year during the next decade, due mainly to the expected increase in population. If consumption does increase at this rate, find the amount by which the utility company will have to increase its generating capacity to meet the needs of the area at the end of the decade.

35. **Pension Funds**   The managers of a pension fund have invested $1.5 million in U.S. government certificates of deposit that pay interest at the rate of 9.5% per year compounded semiannually over a period of ten years. At the end of this period, how much will the investment be worth?

36. **Trust Funds**   A young man is the beneficiary of a trust fund established for him 21 years ago at his birth. If the original amount placed in trust was $10,000, how much will he receive if the money has earned interest at the rate of 8% per year (a) compounded annually? (b) compounded quarterly? (c) compounded monthly?

37. **Investment Planning**   Find how much money should be deposited in a bank paying interest at the rate of 8.5% per year compounded quarterly so that at the end of five years the accumulated amount will be $40,000.

38. **Promissory Notes**   An individual purchased a 4-year, $10,000 promissory note with an interest rate of 8.5% per year compounded semiannually. How much did she purchase the note for?

39. **Financing a College Education**   The parents of a child have just come into a large inheritance and wish to establish a trust fund for the child's college education. If they estimate that they will need $100,000 in 13 years, how much should they set aside in the trust now if they can invest the money at $8\frac{1}{2}\%$ per year (a) compounded annually? (b) compounded semiannually? (c) compounded quarterly?

40. **Investments**   Mr. Kaplan invested a sum of money five years ago in a savings account that has since paid interest at the rate of 8% per year compounded quarterly. His investment is now worth $22,289.22. How much did he originally invest?

41. **Loan Consolidation**   The proprietors of The Coachmen Inn secured two loans from the Union Bank: one for $8000 due in three years and one for $15,000 due in six years, both at an interest rate of 10% per year compounded semiannually. The bank has agreed to allow the two loans to be consolidated into one loan payable in five years at the same interest rate. What amount will the proprietors of the inn be required to pay the bank at the end of five years?

42. **Effective Rate of Interest**   Find the effective rate of interest corresponding to a nominal rate of 9% per year compounded (a) annually, (b) semiannually, (c) quarterly, (d) monthly.

43. **Housing Appreciation**   Ms. Collins purchased a house in 1990 for 100,000. In 1996 she sold the house and made a net profit of $28,000. Find the effective annual rate of return on her investment over the 6-year period.

44. **Common Stock Transaction**   Mr. Stevens purchased 1000 shares of a certain stock for $25,250 (including commissions). He sold the shares two years later and received $32,100 after deducting commissions. Find the

effective annual rate of return on his investment over the 2-year period.

45. **Zero Coupon Bonds**   Carol purchased a zero coupon bond for $6595.37. The bond matures in five years and has a face value of $10,000. Find the effective rate of interest for the bond if interest is compounded semi-annually.

[*Hint:* Assume that the purchase price of the bond is the initial investment and that the face value of the bond is the accumulated amount.]

46. **Money-Market Mutual Funds**   John invested $5000 in a money-market mutual fund that pays interest on a daily basis. The balance in his account at the end of eight months (245 days) was $5347.09. Find the effective rate at which John's account earned interest over this period (assume 365 days in a year).

47. The simple interest formula $A = P(1 + rt)$ [formula (1b)] can be written in the form $A = Prt + P$, which is the slope-intercept form of a straight line with slope Pr and A-intercept P.

a. Describe the family of straight lines obtained by keeping the value of r fixed and allowing the value of P to vary. Interpret your results.

b. Describe the family of straight lines obtained by keeping the value of P fixed and allowing the value of r to vary. Interpret your results.

SOLUTIONS TO SELF–CHECK EXERCISES 5.1

1. Using equation (5) with $r = 0.12$ and $m = 12$, so that

$$i = \frac{0.12}{12} = 0.01, \quad n = (12)(3) = 36, \quad \text{and} \quad A = 20,000$$

we find the required present value to be

$$P = 20,000(1.01)^{-36} = 13,978.50$$

or $13,978.50.

2. The accumulated amount of Mr. Baker's current investment is found by using (3) with $P = 100,000$, $r = 0.106$, and $m = 365$. Thus,

$$i = \frac{0.106}{365} \quad \text{and} \quad n = 365$$

so the required accumulated amount is

$$A = 100,000 \left(1 + \frac{0.106}{365}\right)^{365} = 111,180.48$$

or $111,180.48. Next, we compute the accumulated amount of Mr. Baker's reinvestment. Once again, using (3) with $P = 100,000$, $r = 0.092$, and $m = 365$ so that

$$i = \frac{0.092}{365} \quad \text{and} \quad n = 365$$

we find the required accumulated amount in this case to be

$$\overline{A} = 100,000 \left(1 + \frac{0.092}{365}\right)^{365}$$

or $109,635.21. Therefore, Mr. Baker can expect to experience a net decrease in yearly income of

$$111,180.48 - 109,635.21$$

or $1545.27.

Using Technology

The graphing calculator can be programmed to facilitate the calculation of quantities encountered in the mathematics of finance such as the accumulated amount under compound interest and the effective rate of interest corresponding to a nominal rate.

Finding the Accumulated Amount of an Investment

The first program, which we will call **COMPINT,** enables us to compute the accumulated amount A when P dollars is invested for a term of t years and earns interest at the rate of r percent per year compounded m times a year [see formula (3), page 277].

```
: PROGRAM: COMPINT
: Disp "P"
: Input P
: Disp "r"
: Input r
: Disp "t"
: Input t
: Disp "m"
: Input m
: P(1 + r/m)^(m*t)→A
: Disp "AMOUNT IS"
: Disp A
```

EXAMPLE 1 Find the accumulated amount after 10 years if $500 is invested at 10% per year compounded monthly.

Solution First, we call the program **COMPINT.** Next, we enter 500 for the value of P, 0.1 for the value of r, 10 for the value of t, and 12 for the value of m. We find the accumulated amount to be $1353.52. ● ● ●

Finding the Effective Rate of Interest

The second program, which we will call **EFFRATE,** enables us to compute the effective rate of interest r_{eff} corresponding to a nominal interest rate r per year converted m times per year [see formula (4), page 279].

286

```
: PROGRAM: EFFRATE
: Disp "r"
: Input r
: Disp "m"
: Input m
: (1 + r/m)^m - 1 → E
: Disp "EFFECTIVE RATE IS"
: Disp E
```

EXAMPLE 2   Find the effective rate of interest corresponding to a nominal rate of 10% per year compounded monthly.

Solution   First, we call the program **EFFRATE.** Then we enter 0.1 for the nominal rate r and 12 for the number of conversion periods m. We find that the effective rate is 0.1047, or 10.47% per year.   ○ ○ ○

EXERCISES

Use the COMPINT program to solve exercises 1–4.

1. Find the accumulated amount A if $5000 is invested at the interest rate of $5\frac{3}{8}$% per year compounded monthly for three years.

2. Find the accumulated amount A if $2850 is invested at the interest rate of $6\frac{5}{8}$% per year compounded monthly for four years.

3. Find the accumulated amount A if $327.35 is invested at the interest rate of $5\frac{1}{3}$% per year compounded daily for seven years.

4. Find the accumulated amount A if $327.35 is invested at the interest rate of $6\frac{7}{8}$% per year compounded daily for eight years.

Use the EFFRATE program to solve exercises 5–8.

5. Find the effective rate corresponding to $8\frac{2}{3}$% per year compounded quarterly.

6. Find the effective rate corresponding to $10\frac{5}{8}$% per year compounded monthly.

7. Find the effective rate corresponding to $9\frac{3}{4}$% per year compounded monthly.

8. Find the effective rate corresponding to $4\frac{3}{8}$% per year compounded quarterly.

9. Write a program to calculate the present value for compound interest [formula (5), page 281].

Use the program you wrote for exercise 9 to solve exercises 10–13.

10. Find the present value of $38,000 due in three years at $8\frac{1}{4}$% per year compounded quarterly.

11. Find the present value of $150,000 due in five years at $9\frac{3}{8}$% per year compounded monthly.

12. Find the present value of $67,456 due in three years at $7\frac{7}{8}$% per year compounded monthly.

13. Find the present value of $111,000 due in five years at $11\frac{5}{8}$% per year compounded monthly.

5.2 ANNUITIES

Future Value of an Annuity

An **annuity** is a sequence of payments made at regular time intervals. The time period in which these payments are made is called the **term** of the annuity. Depending on whether the term is given by a *fixed time interval,* a time interval that begins at a definite date but extends indefinitely, or one that is not fixed in advance, an annuity is called an **annuity certain,** a *perpetuity,* or a *contingent annuity,* respectively. In general, the payments in an annuity need not be equal, but in many important applications they are equal. In this section, we assume that annuity payments are equal. Examples of annuities are regular deposits to a savings account, monthly home mortgage payments, and monthly insurance payments.

Annuities are also classified by payment dates. An annuity in which the payments are made at the *end* of each payment period is called an **ordinary annuity,** whereas an annuity in which the payments are made at the beginning of each period is called an *annuity due.* Furthermore, an annuity in which the payment period coincides with the interest conversion period is called a **simple annuity,** whereas an annuity in which the payment period differs from the interest conversion period is called a *complex annuity.*

In this section we consider ordinary annuities that are certain and simple, with periodic payments that are equal in size. In other words, we study annuities that are subject to the following conditions:

1. The terms are given by fixed time intervals.

2. The periodic payments are equal in size.

3. The payments are made at the *end* of the payment periods.

4. The payment periods coincide with the interest conversion periods.

In order to find a formula for the accumulated amount S of an annuity, suppose a sum of $100 is paid into an account at the end of each quarter over a period of three years. Furthermore, suppose that the account earns interest on the deposit at the rate of 8% per year, compounded quarterly. Then the first payment of $100 made at the end of the first quarter earns interest at the rate of 8% per year compounded four times a year (or $8/4 = 2\%$ per quarter) over the remaining 11 quarters and therefore, by the compound interest formula, has an accumulated amount of

$$100 \left(1 + \frac{0.08}{4}\right)^{11} \quad \text{or} \quad 100(1 + 0.02)^{11}$$

dollars at the end of the term of the annuity (Figure 5.2).

The second payment of $100 made at the end of the second quarter earns interest at the same rate over the remaining 10 quarters and therefore has an accumulated amount of

$$100(1 + 0.02)^{10}$$

Figure 5.2
The sum of the accumulated amounts is the amount of the annuity.

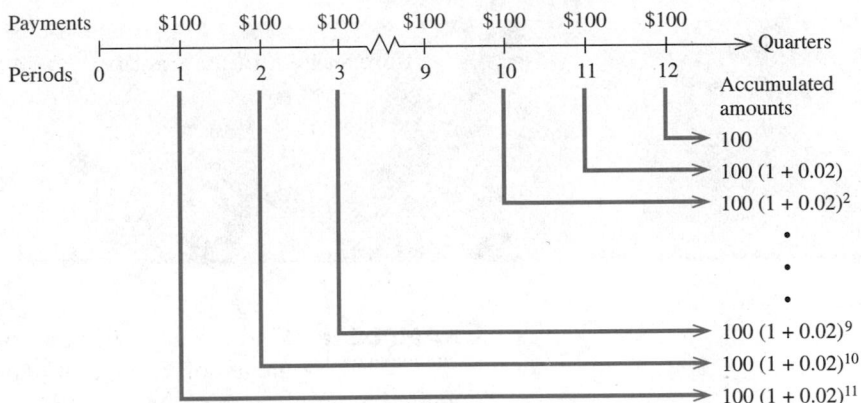

dollars at the end of the term of the annuity, and so on. The last payment earns no interest since it is due at the end of the term. The amount of the annuity is obtained by adding all the terms in Figure 5.2. Thus,

$$S = 100 + 100(1 + 0.02) + 100(1 + 0.02)^2 + \cdots + 100(1 + 0.02)^{11}$$

The sum on the right is the sum of the first n terms of a *geometric progression* with first term R and common ratio $(1 + i)$. We will show in Section 5.4 that the sum S can be written in the more compact form

$$S = 100 \left[\frac{(1 + 0.02)^{12} - 1}{0.02} \right]$$

$$\approx 1341.21$$

or approximately \$1341.21.

In order to find a general formula for the accumulated amount S of an annuity, suppose a sum of \$R is paid into an account at the end of each period for n periods, and suppose the account earns interest at the rate of i per period. Then, proceeding as we did with the numerical example, we see that

$$S = R + R(1 + i) + R(1 + i)^2 + \cdots + R(1 + i)^{n-1}$$

$$= R \left[\frac{(1 + i)^n - 1}{i} \right] \tag{6}$$

The expression inside the brackets is commonly denoted by $s_{\overline{n}|i}$ (read "s angle n at i") and is called the **compound-amount factor**. Extensive tables have been constructed that give values of $s_{\overline{n}|i}$ for different values of i and n (such as Table I, Appendix B). In terms of the compound-amount factor,

$$S = R s_{\overline{n}|i} \tag{7}$$

The quantity S in equations (6) and (7) is realizable at some future date and is accordingly called the *future value of an annuity.*

FUTURE VALUE OF AN ANNUITY

The **future value S of an annuity** of n payments of R dollars each, paid at the end of each investment period into an account that earns interest at the rate of i per period, is

$$S = R\left[\frac{(1+i)^n - 1}{i}\right]$$

EXAMPLE I   Find the amount of an ordinary annuity of 12 monthly payments of $100 that earn interest at 12% per year compounded monthly.

Solution   Since i is the interest rate per *period* and interest is compounded monthly in this case, we have $i = 0.12/12 = 0.01$. Using equation (6) with $R = 100$, $n = 12$, and $i = 0.01$, we have

$$S = \frac{100[(1.01)^{12} - 1]}{0.01}$$

$$= 1268.25 \qquad \text{(Using a calculator)}$$

or $1268.25. The same result is obtained by observing that

$$S = 100 s_{\overline{12}|0.01}$$
$$= 100(12.6825)$$
$$= 1268.25 \qquad \text{(Using Table 1, Appendix C)}$$

○ ○ ○

Present Value of an Annuity

In certain instances, you may want to determine the current value P of a sequence of equal periodic payments that will be made over a certain period of time. After each payment is made, the new balance continues to earn interest at some nominal rate. The amount P is referred to as the *present value of an annuity*.

To derive a formula for determining the present value P of an annuity, we may argue as follows. The amount P invested now and earning interest at the rate of i per period will have an accumulated value of $P(1 + i)^n$ at the end of n periods. But this must be equal to the future value of the annuity S given by (6). Therefore, equating the two expressions, we have

$$P(1+i)^n = R\left[\frac{(1+i)^n - 1}{i}\right]$$

Multiplying both sides of this equation by $(1 + i)^{-n}$ gives

$$P = R(1 + i)^{-n}\left[\frac{(1 + i)^n - 1}{i}\right]$$

$$= R\left[\frac{(1 + i)^n(1 + i)^{-n} - (1 + i)^{-n}}{i}\right] \qquad [(1 + i)^n(1 + i)^{-n} = 1]$$

$$= R\left[\frac{1 - (1 + i)^{-n}}{i}\right]$$

$$= Ra_{\overline{n}|i}$$

where the factor $a_{\overline{n}|i}$ (read "*a* angle *n* at *i*") represents the expression inside the brackets. Extensive tables have also been constructed giving values of $a_{\overline{n}|i}$ for different values of *i* and *n* (see Table I, Appendix C).

PRESENT VALUE OF AN ANNUITY

The **present value** *P* **of an annuity** of *n* payments of *R* dollars each, paid at the end of each investment period into an account that earns interest at the rate of *i* per period, is

$$P = R\left[\frac{1 - (1 + i)^{-n}}{i}\right] \tag{8}$$

EXAMPLE 2 Find the present value of an ordinary annuity of 24 payments of $100 each made monthly and earning interest at 9% per year compounded monthly.

Solution Here $R = 100$, $i = r/m = 0.09/12 = 0.0075$, and $n = 24$, so by formula (8)

$$P = \frac{100[1 - (1.0075)^{-24}]}{0.0075}$$

$$= 2188.92$$

or $2188.92. The same result may be obtained by using Table I, Appendix C. Thus,

$$P = 100a_{\overline{24}|0.0075}$$

$$= 100(21.8892)$$

$$= 2188.92 \qquad \qquad \circ \; \circ \; \circ$$

Applications

EXAMPLE 3 As a savings program toward their child's college education, parents decide to deposit $100 at the end of every month into a bank account paying interest at the rate of 6% per year compounded monthly. If the savings program began when the child was 6 years old, how much money would have accumulated by the time the child turns 18?

Solution   By the time the child turns 18, the parents would have made 144 deposits into the account. Thus, $n = 144$. Furthermore, we have $R = 100$, $r = 0.06$, and $m = 12$, so $i = 0.06/12 = 0.005$. Using equation (6), we find that the amount of money that would have accumulated is given by

$$S = \frac{100[(1.005)^{144} - 1]}{0.005}$$

$$= 21{,}015$$

or $21,015.   ○ ○ ○

EXAMPLE 4   After making a down payment of $2000 for an automobile, Mr. McMurphy paid $200 per month for 36 months with interest charged at 12% per year compounded monthly on the unpaid balance. What was the original cost of the car? What portion of Mr. McMurphy's total car payments went toward interest charges?

Solution   The loan taken up by Mr. McMurphy is given by the present value of the annuity

$$P = \frac{200[1 - (1.01)^{-36}]}{0.01} = 200a_{\overline{36}|0.01}$$

$$= 6021.50$$

or $6021.50. Therefore, the original cost of the automobile is $8021.50 ($6021.50 plus the $2000 down payment). The interest charges paid by Mr. McMurphy are given by $(36)(200) - 6021.50$, or $1178.50.   ○ ○ ○

One important application of annuities is in the area of tax planning. During the 1980s, Congress created many tax-sheltered retirement savings plans, such as Individual Retirement Accounts (IRAs), Keogh plans, and Simplified Employee Pension (SEP) plans. These plans are examples of annuities in which the individual is allowed to make contributions (which are often tax deductible) to an investment account. The amount of the contribution is limited by congressional legislation. The taxes on the contributions and/or the interest accumulated in these accounts are deferred until the money is withdrawn, ideally during retirement, when tax brackets should be lower. In the interim period, the individual has the benefit of tax-free growth on his or her investment.

Suppose, for example, that you are eligible to make a fully deductible contribution to an IRA and that you are in a marginal tax bracket of 28%. Additionally, suppose that you receive a year-end bonus of $2000 from your employer and have the option of depositing the $2000 into either an IRA or a regular savings account, both accounts earning interest at an effective annual rate of 8% per year. If you choose to invest your bonus in a regular savings account, you will first have to pay taxes on the $2000, leaving $1440 to invest.

At the end of one year, you will also have to pay taxes on the interest earned, leaving you with

$$
\begin{array}{ccc}
\text{Accumulated} & \text{Tax on} & \text{Net} \\
\text{amount} & - \quad \text{interest} & = \quad \text{amount}
\end{array}
$$

$$1555.20 - 32.26 = 1522.94$$

or \$1522.94.

On the other hand, if you put the money into the IRA account, the entire sum will earn interest, and at the end of one year you will have (1.08)(\$2000), or \$2160, in your account. Of course, you will still have to pay taxes on this money when you withdraw it, but you will have gained the advantage of tax-free growth of the larger principal over the years. The disadvantage of this option is that if you withdraw the money before you reach the age of $59\frac{1}{2}$, you will be liable for taxes on both your contributions and the interest earned and you will also have to pay a 10% penalty.

REMARK   In practice, the size of the contributions an individual might make to the various retirement plans might vary from year to year. Also, he or she might make the contributions at different payment periods. To simplify our discussion, we will consider examples in which fixed payments are made at regular intervals.   ୦ ୦ ୦

 EXAMPLE 5   Ms. Coughlin is planning to make the maximum contribution of \$2000 on January 31 of each year into an IRA earning interest at an effective rate of 9% per year. After she makes her 25th payment on January 31 of the year following her retirement, how much will she have in her IRA?

Solution   The amount of money Ms. Coughlin will have after her 25th payment into her account is found by using equation (6) with $R = 2000$, $r = 0.09$, and $m = 1$, so $i = r/m = 0.09$ and $n = 25$. The required amount is given by

$$S = \frac{2000[(1.09)^{25} - 1]}{0.09}$$

$$= 169{,}401.79$$

or \$169,401.79.   ୦ ୦ ୦

Tax-deferred annuities are another type of investment vehicle that allows an individual to build assets for retirement, college funds, or other future needs. The advantage gained in this type of investment is that the tax on the accumulated interest is deferred to a later date. Note that in this type of investment the contributions themselves are not tax deductible. At first glance the advantage thus gained may seem to be relatively inconsequential, but its true effect is illustrated by the next example.

EXAMPLE 6 Both Mr. Clark and Mr. Colby are salaried individuals, 45 years of age, who are saving for their retirement 20 years from now. Both Clark and Colby are also in the 28% marginal tax bracket. Mr. Clark makes a $1000 contribution annually on December 31 into a savings account earning an effective rate of 8% per year. At the same time, Mr. Colby makes a $1000 annual payment to an insurance company for a tax-deferred annuity. The annuity also earns interest at an effective rate of 8% per year. (Assume that both men remain in the same tax bracket throughout this period, and disregard state income taxes.)

a. Calculate how much each man will have in his investment account at the end of 20 years.

b. Compute the interest earned on each account.

c. Show that even if the interest on Mr. Colby's investment were subjected to a tax of 28% upon withdrawal of his investment at the end of 20 years, the net accumulated amount of his investment would still be greater than that of Mr. Clark's.

Solution

a. Because Mr. Clark is in the 28% marginal tax bracket, the net yield for his investment is (0.72)(8), or 5.76%, per year.

Using formula (6) with $R = 1000$, $r = 0.0576$, and $m = 1$, so that $i = 0.0576$ and $n = 20$, we see that Clark's investment will be worth

$$S = \frac{1000[(1 + 0.0576)^{20} - 1]}{0.0576}$$

$$= 35,850.49$$

or $35,850.49 at his retirement.

Mr. Colby has a tax-sheltered investment with an effective yield of 8% per year. Using (6) with $R = 1000$, $r = 0.08$, and $m = 1$, so that $i = 0.08$ and $n = 20$, we see that Colby's investment will be worth

$$S = \frac{1000[(1 + 0.08)^{20} - 1]}{0.08}$$

$$= 45,761.96$$

or $45,761.96 at his retirement.

b. Each man will have paid 20(1000), or 20,000 dollars, into his account. Therefore, the total interest earned in Mr. Clark's account will be (35,850.49 − 20,000), or $15,850.49, whereas the total interest earned in Mr. Colby's account will be (45,761.96 − 20,000), or $25,761.96.

c. From (b) we see that the total interest earned in Mr. Colby's account will be $25,761.96. If it were taxed at 28%, he would still end up with (0.72)(25,761.96), or $18,548.61. This is larger than the total interest of $15,850.49 earned by Mr. Clark.   ○ ○ ○

SELF-CHECK EXERCISES 5.2

1. Mr. and Mrs. Fletcher opened a spousal IRA on January 31, 1985, with the maximum permissible contribution of $2250. They plan to make a contribution of $2250 thereafter on January 31 of each year until their retirement in the year 2004 (20 payments). If the account earns interest at the rate of 8% per year compounded yearly, how much will the Fletchers have in their account when they retire?

2. The Denver Wildcatting Company has an immediate need for a loan. In an agreement worked out with its banker, Denver assigns its royalty income of $4800 per month for the next three years from certain oil properties to the bank, with the first payment due at the end of the first month. If the bank charges interest at the rate of 9% per year compounded monthly, what is the amount of the loan negotiated between the parties?

Solutions to Self-Check Exercises 5.2 can be found on page 298.

5.2 EXERCISES

A calculator is recommended for this exercise set.

In exercises 1–6, find the amount (future value) of each ordinary annuity.

1. $1000 a year for ten years at 10% per year compounded annually

2. $1500 per semiannual period for eight years at 9% per year compounded semiannually

3. $1800 per quarter for six years at 8% per year compounded quarterly

4. $500 per semiannual period for 12 years at 11% per year compounded semiannually

5. $600 per quarter for nine years at 12% per year compounded quarterly

6. $150 per month for 15 years at 10% per year compounded monthly

In exercises 7–12, find the present value of each ordinary annuity.

7. $5000 a year for eight years at 8% per year compounded annually

8. $1200 per semiannual period for six years at 10% per year compounded semiannually

9. $4000 a year for five years at 9% per year compounded yearly

10. $3000 per semiannual period for six years at 11% per year compounded semiannually

11. $800 per quarter for seven years at 12% per year compounded quarterly

12. $150 per month for ten years at 8% per year compounded monthly

13. **IRAs** If a merchant deposits $1500 annually at the end of each tax year in an IRA account paying interest at the rate of 8% per year compounded annually, how much will she have in her account at the end of 25 years?

14. **Savings Accounts** If Mr. Pruitt deposits $100 at the end of each month in a savings account earning interest at the rate of 8% per year compounded monthly, how much will he have on deposit in his savings account at the end of six years, assuming that he makes no withdrawals during that period?

15. **Savings Accounts** Mrs. Lynde has joined a "Christmas Fund Club" at her bank. At the end of every month, December through October inclusive, she will make a deposit of $40 in her fund. If the money earns interest at the rate of 7% per year compounded monthly, how much will she have in her account on December 1 of the following year?

USING TECHNOLOGY

FINDING THE AMOUNT OF AN ANNUITY

The following program, which we will call **FVAN,** can be used to calculate the future value S of an annuity of n payments of R dollars each earning interest at the rate of i per period [formula (6), page 289]. When using this formula, recall that $i = r/m$ and $n = mt$, where r is the annual interest rate, m is the number of conversion periods per year, and t is the number of years in the term of the annuity.

```
: PROGRAM: FVAN
: Disp "R"
: Input R
: Disp "i"
: Input i
: Disp "N"
: Input N
: (R/i)((1 + i)^N - 1) → S
: Disp "AMOUNT IS"
: Disp S
```

REMARK   Since the letter n is a reserved-name variable in some graphing calculators, we use N in lieu of n in the program **FVAN.**   ◦ ◦ ◦

EXAMPLE 1   Find the amount of an ordinary annuity of 36 quarterly payments of \$220 each that earns interest at 10% per year compounded quarterly.

Solution   Using the program **FVAN** and entering the values $R = 220$, $i = 0.1/4 = 0.025$, and $n = N = 36$, we find that $S = 12606.3107783$, or \$12,606.31.   ◦ ◦ ◦

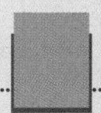

EXERCISES

Use the program **FVAN** *to solve exercises 1–4.*

1. Find the amount of an ordinary annuity of 20 payments of $2500 per quarter at $7\frac{1}{4}$% per year compounded quarterly.

2. Find the amount of an ordinary annuity of 24 payments of $1790 per quarter at $8\frac{3}{4}$% per year compounded quarterly.

3. Find the amount of an ordinary annuity of $120 per month for five years at $6\frac{2}{3}$% per year compounded monthly.

4. Find the amount of an ordinary annuity of $225 per month for six years at $7\frac{5}{8}$% per year compounded monthly.

5. Write a program to calculate the present value P of an annuity of n payments of R dollars each, paid at the end of each investment period into an account that earns interest at the rate of i per period [see formula (8), page 291].

Use the program you wrote in exercise 5 to solve exercises 6–9.

6. Find the present value of an ordinary annuity of $4500 per semiannual period for five years earning interest at 9% per year compounded semiannually.

7. Find the present value of an ordinary annuity of $2100 per quarter for seven years earning interest at $7\frac{1}{8}$% per year compounded quarterly.

8. Find the present value of an ordinary annuity of $245 per month for six years earning interest at $8\frac{3}{8}$% per year compounded monthly.

9. Find the present value of an ordinary annuity of $185 per month for 12 years earning interest at $6\frac{5}{8}$% per year compounded monthly.

16. Keogh Accounts Mr. Robinson, who is self-employed, contributes $5000 a year into a Keogh account. How much will he have in the account after 25 years if the account earns interest at the rate of 8.5% per year compounded yearly?

17. Retirement Planning As a fringe benefit for the past 12 years, Mr. Collins's employer has contributed $100 at the end of each month into an employee retirement account for Collins that pays interest at the rate of 7% per year compounded monthly. Mr. Collins has also contributed $2000 at the end of each of the last eight years into an IRA that pays interest at the rate of 9% per year compounded yearly. How much does Mr. Collins have in his retirement fund at this time?

18. Savings Accounts The Pirerras are planning to go to Europe three years from now and have agreed to set aside $150 each month for their trip. If they deposit this money at the end of each month into a savings account paying interest at the rate of 8% per year compounded monthly, how much money will be in their travel fund at the end of the third year?

19. Auto Leasing Mr. and Mrs. Betz have leased an auto for two years at $450 per month. If money is worth 9% per year compounded monthly, what is the equivalent cash payment (present value) of this annuity?

20. Auto Financing Ms. Newburg made a down payment of $2000 toward the purchase of a new car. In order to pay the balance of the purchase price, she has secured a loan from her bank at the rate of 12% per year compounded monthly. Under the terms of her finance agreement, she is required to make payments of $210 per month for 36 months. What is the cash price of the car?

21. Installment Plans R. G. Pierce Publishing Company sells encyclopedias under two payment plans: cash or installment. Under the installment plan, the customer pays $22 per month over three years with interest charged on the balance at a rate of 18% per year compounded monthly. Find the cash price for a set of encyclopedias if it is equivalent to the price paid by a customer using the installment plan.

22. Lottery Payouts A state lottery commission pays the winner of the "Million Dollar" lottery 20 installments of $50,000 per year. The commission makes the first payment of $50,000 immediately and the other $n = 19$ payments at the end of each of the next 19 years. Determine how much money the commission should have in the bank initially to guarantee the payments, assuming that the balance on deposit with the bank earns interest at the rate of 8% per year compounded yearly.

[*Hint:* Find the present value of an annuity.]

23. Purchasing a Home The Johnsons have accumulated a nest egg of $25,000 that they intend to use as a down payment toward the purchase of a new house. Because their present gross income has placed them in a relatively high tax bracket, they have decided to invest a minimum of $800 per month in monthly payments (to take advantage of their tax deductions) toward the purchase of their house. However, because of other financial obligations, their monthly payments should not exceed $1000. If local mortgage rates are 9.5% per year compounded monthly for a conventional 30-year mortgage, what is the price range of houses that they should consider?

24. Purchasing a Home Refer to exercise 23. If local mortgage rates were increased to 10%, how would this affect the price range of houses the Johnsons should consider?

25. Purchasing a Home Refer to exercise 23. If the Johnsons decide to secure a 15-year mortgage instead of a 30-year mortgage, what is the price range of houses they should consider when the local mortgage rate for this type of loan is 9%?

SOLUTIONS TO SELF–CHECK EXERCISES 5.2

1. The amount the Fletchers will have in their account when they retire may be found by using formula (6) with $R = 2250$, $r = 0.08$, $m = 1$, so that $i = r = 0.08$ and $n = 20$. Thus

$$S = \frac{2250[(1.08)^{20} - 1]}{0.08}$$

$$= 102,964.42$$

or $102,964.42.

2. We want to find the present value of an ordinary annuity of 36 payments of $4800 each made monthly and earning interest at 9% per year compounded monthly. Using formula (8) with $R = 4800$, $m = 12$, so that $i = r/m = 0.09/12 = 0.0075$ and $n = (12)(3) = 36$, we find

$$P = \frac{4800[1 - (1.0075)^{-36}]}{0.0075} = 150{,}944.67$$

or $150,944.67, the amount of the loan negotiated.

5.3 AMORTIZATION AND SINKING FUNDS

Amortization of Loans

The annuity formulas derived in Section 5.2 may be used to answer questions involving the amortization of certain types of installment loans. For example, in a typical housing loan, the mortgagor makes periodic payments toward reducing his indebtedness to the lender, who charges interest at a fixed rate on the unpaid portion of the debt. In practice, the borrower is required to repay the lender in periodic installments, usually of the same size and over a fixed term, so that the loan (principal plus interest charges) is amortized at the end of the term.

By thinking of the monthly loan repayments R as the payments in an annuity, we see that the original amount of the loan is given by P, the present value of the annuity. From equation (8), Section 5.2, we have

$$P = R\left[\frac{1 - (1 + i)^{-n}}{i}\right] = Ra_{\overline{n}|i} \tag{9}$$

A question a financier might ask is: How much should the monthly installment be so that a loan will be amortized at the end of the term of the loan? To answer this question, we simply solve (9) for R in terms of P, obtaining

$$R = \frac{Pi}{1 - (1 + i)^{-n}} = \frac{P}{a_{\overline{n}|i}}$$

AMORTIZATION FORMULA

The periodic payment R on a loan of P dollars to be amortized over n periods with interest charged at the rate of i per period is

$$R = \frac{Pi}{1 - (1 + i)^{-n}} \tag{10}$$

  EXAMPLE 1   A sum of $50,000 is to be repaid over a 5-year period through equal installments made at the end of each year. If an interest rate of 8% per year is charged on the unpaid balance, and interest calculations are made at the end of each year, determine the size of each installment so

that the loan (principal plus interest charges) is amortized at the end of five years. Verify the result by displaying the amortization schedule.

Solution   Substituting $P = 50,000$, $i = r = 0.08$ (here $m = 1$), and $n = 5$ into formula (10), we obtain

$$R = \frac{(50,000)(0.08)}{1 - (1.08)^{-5}} = 12,522.82$$

giving the required yearly installment as $12,522.82.

The amortization schedule is presented in Table 5.3. The outstanding principal at the end of five years is, of course, zero. (The figure of $.01 in Table 5.3 is the result of round-off errors.) Observe that initially the larger portion of the repayment goes toward payment of interest charges, but as time goes by the larger portion of the installment goes toward repayment of the principal.

Table 5.3

End of Period	Interest Charged	Repayment Made	Payment Toward Principal	Outstanding Principal
0				50,000.00
1	4,000.00	12,522.82	8,522.82	41,477.18
2	3,318.17	12,522.82	9,204.65	32,272.53
3	2,581.80	12,522.82	9,941.02	22,331.51
4	1,786.52	12,522.82	10,736.30	11,595.21
5	927.62	12,522.82	11,595.20	0.01

◍ ◍ ◍

Financing a Home

EXAMPLE 2   The Blakelys borrowed $120,000 from a bank to help finance the purchase of a house. The bank charges interest at a rate of 9% per year on the unpaid balance, with interest computations made at the end of each month. The Blakelys have agreed to repay the loan in equal monthly installments over 30 years. How much should each payment be if the loan is to be amortized at the end of the term?

Solution   Here $P = 120,000$, $i = r/m = 0.09/12 = 0.0075$, and $n = (30)(12) = 360$. Using formula (10), we find that the size of each monthly installment required is given by

$$R = \frac{(120,000)(0.0075)}{1 - (1.0075)^{-360}}$$

$$= 965.55$$

or $965.55.

◍ ◍ ◍

EXAMPLE 3   Refer to Example 2 and find the amount of equity the Blakelys will have in their home (disregarding the down payment and appreciation) after the

a. 60th payment (5 years)

b. 240th payment (20 years)

Solution

a. After 60 payments have been made, the outstanding principal is given by the sum of the present values of the remaining installments (that is, $360 - 60 = 300$ installments). But this sum is just the present value of an annuity with $n = 300$, $R = 965.55$, and $i = 0.0075$ (see Example 2). Using formula (8), we find that

$$P = 965.55 \left[\frac{1 - (1 + 0.0075)^{-300}}{0.0075} \right]$$

$$= 115,056.50$$

or $115,056.50. Therefore, we may find how much equity the Blakelys have in their home by subtracting the outstanding principal from the original principal:

$$120,000 - 115,056.50 = 4943.50$$

or $4943.50.

b. As in (a), we compute the present value of the annuity with $n = 360 - 240 = 120$. Thus,

$$P = 965.55 \left[\frac{1 - (1 + 0.0075)^{-120}}{0.0075} \right]$$

$$= 76,222.15$$

or $76,222.15. Therefore, the Blakelys' equity after the 240th payment is

$$120,000 - 76,222.15 = 43,777.85$$

or $43,777.85.   ○ ○ ○

 EXAMPLE 4   The Jacksons have determined that after making a down payment they could afford at most $1000 for a monthly house payment. The bank charges interest at the rate of 9.6% per year on the unpaid balance, with interest computations made at the end of each month. If the loan is to be amortized in equal monthly installments over 30 years, what is the maximum amount that the Jacksons can borrow from the bank?

Solution   Here $i = r/m = 0.096/12 = 0.008$, $n = (30)(12) = 360$, $R = 1000$, and we are required to find P. From equation (8), we have

$$P = \frac{R[1 - (1 + i)^{-n}]}{i}$$

Decker's job is simple: to loan money to people who want to buy a home. The hard part is deciding whether applicants qualify for one of Shawmut's 40 different mortgage plans. Sifting through income figures, current indebtedness, and credit history helps Decker determine which individuals make the best mortgage candidates.

John Decker

Title: Mortgage Counselor
Institution: Shawmut Bank

Decker stresses that he and his colleagues "strive to grant loans. That's our job." But before allowing someone to file a formal application, Decker takes the person over several "pre-qualification hurdles" to gauge his or her ability to handle a mortgage.

To start, Decker relies on a two-tiered, debt-to-income ratio to see whether a person has sufficient gross monthly income to make payments. Under the first tier, the proposed monthly payment (principal and interest, property tax, homeowner's insurance, and, when applicable, a condo fee) cannot exceed 28 percent of an individual's gross monthly income. If Decker's initial calculations are positive, he then must determine the individual's ability to meet monthly mortgage payments while also repaying other debts, such as car and student loans, credit cards, alimony, and so on. These combined payments cannot exceed 36 percent of gross monthly income. A typical person with an $800 mortgage obligation and $550 in other payments would have to earn $3750 per month to clear these first two hurdles.

Contrary to popular belief, bankers *want* to lend money. "The idea is to grant mortgages," says Decker, which contribute substantially to a bank's profitability. But lending money means making sensible decisions about how much to lend as well as a person's ability to repay the loan.

Using a loan-to-value formula, banks might lend 80 percent of a property's value. In such cases, the applicant has to put 20 percent down to make the purchase. Or Shawmut may decide to lend up to 95 percent or as little as 75 percent of the appraised value.

Understandably, banks don't like to see bankruptcies, late payments, or liens on a personal credit history. Decker notes, however, that even this final hurdle doesn't necessarily mean failure in securing a mortgage. He works closely with each individual to overcome any stigma that might prompt a rejection.

Once Decker has put together a successful mortgage application, it is reviewed internally. Then, even though a mortgage is granted, it might come through at a slightly higher interest rate—10.4 percent instead of the expected 10 percent rate. Then he computes the change in monthly payments, for while this might seem like a small change, on a large mortgage it could be enough of a variable to affect the new customer's ability to make monthly payments. On a $100,000, 30-year, fixed-rate mortgage, payments will increase about $35 per month.

Using the debt-to-income ratio, Decker plugs the new variable into his formulas to determine whether a problem exists. Decker notes that such a small increase doesn't usually pose a significant problem. If the customer can't make the new payment, however, Decker explores alternatives until he finds a solution. For Decker, it comes down to this: "If there is any possible way to give a loan, we're going to make it work."

◗　◗　◗

Substituting the numerical values for R, n, and i into this expression for P, we obtain

$$P = \frac{1000[1 - (1.008)^{-360}]}{0.008} \approx 117,902$$

Therefore, the Jacksons can borrow at most $117,902.   ○ ○ ○

Sinking Funds

Sinking funds are another important application of the annuity formulas. Simply stated, a **sinking fund** is a fund that is set up for a specific purpose at some future date. For example, an individual might establish a sinking fund for the purpose of discharging a debt at a future date. A corporation might establish a sinking fund in order to accumulate sufficient capital to replace equipment that is expected to be obsolete at some future date.

By thinking of the amount to be accumulated by a specific date in the future as the future value of an annuity [equation (6), Section 5.2], we can answer questions about a large class of sinking fund problems.

EXAMPLE 5   The proprietor of Carson Hardware has decided to set up a sinking fund for the purpose of purchasing a computer in two years' time. It is expected that the computer will cost $30,000. If the fund earns 10% interest per year compounded quarterly, determine the size of each (equal) quarterly installment the proprietor should pay into the fund. Verify the result by displaying the schedule.

Solution   The problem at hand is to find the size of each quarterly payment R of an annuity given that its future value is $S = 30,000$, the interest earned per conversion period is $i = r/m = 0.1/4 = 0.025$, and the number of payments is $n = (2)(4) = 8$. The formula for the annuity,

$$S = R\left[\frac{(1 + i)^n - 1}{i}\right]$$

when solved for R yields

$$R = \frac{iS}{(1 + i)^n - 1} \tag{11}$$

or, equivalently,

$$R = \frac{S}{s_{\overline{n}|i}}$$

Substituting the appropriate numerical values for i, S, and n into equation (11), we obtain the desired quarterly payment

$$R = \frac{(0.025)(30,000)}{(1.025)^8 - 1} = 3434.02$$

or $3434.02. Table 5.4 shows the required schedule.

Table 5.4

End of Period	Deposit Made	Interest Earned	Addition to Fund	Accumulated Amount in Fund
1	3,434.02	0	3,434.02	3,434.02
2	3,434.02	85.85	3,519.87	6,953.89
3	3,434.02	173.85	3,607.87	10,561.76
4	3,434.02	264.04	3,698.06	14,259.82
5	3,434.02	356.50	3,790.52	18,050.34
6	3,434.02	451.26	3,885.28	21,935.62
7	3,434.02	548.39	3,982.41	25,918.03
8	3,434.02	647.95	4,081.97	30,000.00

The formula derived in this last example is restated below.

SINKING FUND PAYMENT

The periodic payment R required to accumulate a sum of S dollars over n periods with interest charged at the rate of i per period is

$$R = \frac{iS}{(1 + i)^n - 1}$$

SELF-CHECK EXERCISES 5.3

1. The Moores wish to borrow $100,000 from a bank to help finance the purchase of a house. Their banker has offered the following plans for their consideration. In plan I, the Moores have 30 years to repay the loan in monthly installments with interest on the unpaid balance charged at 10.5% per year compounded monthly. In plan II, the loan is to be repaid in monthly installments over 15 years with interest on the unpaid balance charged at 9.75% per year compounded monthly.
   a. Find the monthly repayment for each plan.
   b. What is the difference in total payments made under each plan?

2. Mr. Harris, a self-employed individual who is 46 years old, is setting up a Defined-Benefit Keogh Plan for his retirement. If he wishes to have $250,000 in this Keogh account by age 65, what is the size of each yearly installment he will be required to make into a savings account earning interest at $8\frac{1}{4}$% per year? (Assume that Mr. Harris is eligible to make each of the 20 required contributions.)

Solutions to Self-Check Exercises 5.3 can be found on page 307.

5.3 EXERCISES

A calculator is recommended for this exercise set.

In exercises 1–8, find the periodic payment R required to amortize a loan of P dollars over n periods with interest earned at the rate of i per period.

1. $P = \$100,000$, $i = 0.08$, $n = 10$

2. $P = \$40,000$, $i = 0.015$, $n = 30$

3. $P = \$5,000$, $i = 0.01$, $n = 12$

4. $P = \$16,000$, $i = 0.0075$, $n = 48$

5. $P = \$25,000$, $i = 0.0075$, $n = 48$

6. $P = \$80,000$, $i = 0.00875$, $n = 180$

7. $P = \$80,000$, $i = 0.00875$, $n = 360$

8. $P = \$100,000$, $i = 0.00875$, $n = 300$

In exercises 9–14, find the periodic payment R required to accumulate a sum of S dollars over n periods with interest earned at the rate of i per period.

9. $S = \$20,000$, $i = 0.02$, $n = 12$

10. $S = \$40,000$, $i = 0.01$, $n = 36$

11. $S = \$100,000$, $i = 0.0075$, $n = 120$

12. $S = \$120,000$, $i = 0.0075$, $n = 180$

13. $S = \$250,000$, $i = 0.00875$, $n = 300$

14. $S = \$350,000$, $i = 0.00625$, $n = 120$

15. **Loan Amortization**   A sum of $100,000 is to be repaid over a 10-year period through equal installments made at the end of each year. If an interest rate of 10% per year is charged on the unpaid balance, and interest calculations are made at the end of each year, determine the size of each installment so that the loan (principal plus interest charges) is amortized at the end of 10 years.

16. **Loan Amortization**   What monthly payment is required to amortize a loan of $30,000 over ten years if interest at the rate of 12% per year is charged on the unpaid balance, and interest calculations are made at the end of each month?

17. **Home Mortgages**   Complete the following table, which shows the monthly payments on a $100,000, 30-year mortgage at the interest rates shown. Use this information to answer the following questions.

Amount of Mortgage	Interest Rate	Monthly Payment
$100,000	7%	$665.30
$100,000	8%	...
$100,000	9%	...
$100,000	10%	...
$100,000	11%	...
$100,000	12%	$1,028.61

a. What is the difference in monthly payments between a $100,000, 30-year mortgage secured at 7% per year and one secured at 10% per year?
b. Use the table to calculate the monthly mortgage payments on a $150,000 mortgage at 10% per year over 30 years and a $50,000 mortgage at 10% per year over 30 years.

18. **Financing a Home**   The Flemings secured a bank loan of $96,000 to help finance the purchase of a house. The bank charges interest at a rate of 9% per year on the unpaid balance, and interest computations are made at the end of each month. The Flemings have agreed to repay the loan in equal monthly installments over 25 years. What should the size of each repayment be if the loan is to be amortized at the end of the term?

19. **Financing a Car**   The price of a new car is $16,000. Assume an individual makes a down payment of 25% toward the purchase of the car and secures financing for the balance at the rate of 10% per year compounded monthly.
a. What monthly payment will she be required to make if the car is financed over a period of 36 months? Over a period of 48 months?
b. What will the interest charges be if she elects the 36-month plan? The 48-month plan?

20. **Financial Analysis**   A group of private investors purchased a condominium complex for $2 million. They made an initial down payment of 10% and obtained

financing for the balance. If the loan is to be amortized over 15 years at an interest rate of 12% per year compounded quarterly, find the required quarterly payment.

21. Financing a Home   The Taylors have purchased a $180,000 house. They made an initial down payment of $20,000 and secured a mortgage with interest charged at the rate of 8% per year on the unpaid balance. Interest computations are made at the end of each month. If the loan is to be amortized over 30 years, what monthly payment will the Taylors be required to make? What is their equity (disregarding appreciation) after 5 years? After 10 years? After 20 years?

22. Sinking Funds   A city has $2.5 million worth of school bonds that are due in 20 years and has established a sinking fund to retire this debt. If the fund earns interest at the rate of 7% per year compounded annually, what amount must be deposited annually in this fund?

23. Sinking Funds   The Lowell Corporation wishes to establish a sinking fund to retire a $200,000 debt that is due in ten years. If the investment will earn interest at the rate of 9% per year compounded quarterly, find the amount of the quarterly deposit that must be made in order to accumulate the required sum.

24. Trust Funds   John Carlson is the beneficiary of a $20,000 trust fund set up for him by his grandparents. Under the terms of the trust, he is to receive the money over a 5-year period in equal installments at the end of each year. If the fund earns interest at the rate of 9% per year compounded annually, what amount will he receive each year?

25. Sinking Funds   The management of the Gibraltar Brokerage Services, Inc., anticipates a capital expenditure of $20,000 in three years' time for the purpose of purchasing new fax machines and has decided to set up a sinking fund to finance this purchase. If the fund earns interest at the rate of 10% per year compounded quarterly, determine the size of each (equal) quarterly installment that should be deposited in the fund.

26. Keogh Accounts   Ms. Anderson, a self-employed individual, wishes to accumulate a retirement fund of $250,000. How much should she deposit each month into her Keogh account, which pays interest at the rate of 8.5% per year compounded monthly, to reach her goal upon retirement 25 years from now?

27. IRAs   Mr. Berger has deposited $375 in his IRA at the end of each quarter for the past 20 years. His investment has earned interest at the rate of 8% per year compounded quarterly over this period. Now, at age 60, he is considering retirement. What quarterly payment will he receive over the next 15 years? (Assume that the money is earning interest at the same rate and payments are made at the end of each quarter.) If he continues working and makes quarterly payments of the same amount in his IRA account until age 65, what quarterly payment will he receive from his fund upon retirement over the following 10 years?

28. Financing a Car   Ms. Dwyer purchased a new car during a special sales promotion by the manufacturer. She secured a loan from the manufacturer in the amount of $16,000 at a rate of 7.9% per year compounded monthly. Her bank is now charging 11.5% per year compounded monthly for new car loans. Assuming that each loan would be amortized by 36 equal monthly installments, determine the amount of interest she would have paid at the end of three years for each loan. How much less will she have paid in interest payments over the life of the loan by borrowing from the manufacturer instead of her bank?

29. Financing a Home   The Sandersons are planning to refinance their home. The outstanding principal on their original loan is $100,000 and was to be amortized in 240 equal monthly installments at an interest rate of 10% per year compounded monthly. The new loan they expect to secure is to be amortized over the same period at an interest rate of 7.8% per year compounded monthly. How much less can they expect to pay over the life of the loan in interest payments by refinancing the loan at this time?

30. Financing a Home   After making a down payment of $25,000, the Meyers need to secure a loan of $140,000 to purchase a certain house. Their bank's current rate for 25-year home loans is 11% compounded monthly. The owner has offered to finance the loan at 9.8% compounded monthly. Assuming that both loans would be amortized over a 25-year period by 300 equal monthly installments, determine the difference in the amount of interest the Meyers would pay by choosing the seller's financing rather than their bank's.

SOLUTIONS TO SELF-CHECK EXERCISES 5.3

1. a. We use equation (10) in each instance. Under plan I,

$$P = 100,000, \quad i = \frac{r}{m} = \frac{0.105}{12} = 0.00875, \quad \text{and} \quad n = (30)(12) = 360$$

Therefore, the size of each monthly repayment under plan I is

$$R = \frac{100,000(0.00875)}{1 - (1.00875)^{-360}}$$

$$= 914.74$$

or $914.74.

Under plan II,

$$P = 100,000, \quad i = \frac{r}{m} = \frac{0.0975}{12} = 0.008125, \quad \text{and} \quad n = (15)(12) = 180$$

Therefore, the size of each monthly repayment under plan II is

$$R = \frac{100,000(0.008125)}{1 - (1.008125)^{-180}}$$

$$= 1059.36$$

or $1059.36.

b. Under plan I, the total amount of repayments will be

$$(360)(914.74) \qquad \text{(Number of payments}$$
$$\text{times the size}$$
$$\text{of each installment)}$$

$$= 329,306.40$$

or $329,306.40. Under plan II, the total amount of repayments will be

$$(180)(1059.36)$$

$$= 190,684.80$$

or $190,684.80. Therefore, the difference in payments is

$$329,306.40 - 190,684.80 = 138,621.60$$

or $138,621.60.

2. We use equation (11) with

$$S = 250,000$$

$$i = r = 0.0825 \text{ (since } m = 1)$$

and

$$n = 20$$

giving the required size of each installment as

$$R = \frac{(0.0825)(250,000)}{(1.0825)^{20} - 1}$$

$$= 5313.59$$

or $5313.59.

USING TECHNOLOGY

AMORTIZING A LOAN

The following program, which we will call **AMORT,** enables us to calculate the periodic payment R on a loan of P dollars to be amortized over n periods with interest charged at the rate of i per period [formula (10), page 299].

```
: PROGRAM: AMORT
: Disp "P"
: Input P
: Disp "i"
: Input i
: Disp "N"
: Input N
: P*i/(1 - (1 + i)^-N) → R
: Disp "R IS"
: Disp R
```

EXAMPLE 1   The Wongs are considering obtaining a preapproved 30-year loan of $120,000 to help finance the purchase of a house. The mortgage company charges interest at the rate of 8% per year on the unpaid balance, with interest computations made at the end of each month. What will the monthly installments be if the loan is amortized at the end of the term?

Solution   Using the **AMORT** program and entering the values $P = 120,000$, $i = r/m = 0.08/12$, and $n = N = (30)(12) = 360$, we find that $R = 880.517488654$, and so the monthly installment is $880.52.   ○ ○ ○

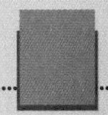 **EXERCISES**

Use the program AMORT to solve exercises 1–4.

1. Find the periodic payment required to amortize a loan of $55,000 over 120 periods with interest earned at the rate of $6\frac{5}{8}\%$ per period.

2. Find the periodic payment required to amortize a loan of $178,000 over 180 periods with interest earned at the rate of $7\frac{1}{8}\%$ per period.

3. Find the periodic payment required to amortize a loan of $227,000 over 360 periods with interest earned at the rate of $8\frac{1}{8}\%$ per period.

4. Find the periodic payment required to amortize a loan of $150,000 over 360 periods with interest earned at the rate of $7\frac{3}{8}\%$ per period.

5. Write a program to calculate the periodic payment R to accumulate a sum of S dollars over n periods with interest charged at the rate of i per period [see formula (11), page 303].

Use the program you wrote in exercise 5 to solve exercises 6–9.

6. Find the periodic payment required to accumulate $25,000 over 12 periods with interest earned at the rate of 2% per period.

7. Find the periodic payment required to accumulate $50,000 over 36 periods with interest earned at the rate of $2\frac{1}{4}\%$ per period.

8. Find the periodic payment required to accumulate $137,000 over 120 periods with interest earned at the rate of $\frac{3}{4}\%$ per period.

9. Find the periodic payment required to accumulate $144,000 over 120 periods with interest earned at the rate of $\frac{5}{8}\%$ per period.

10. A loan of $120,000 is to be repaid over a 10-year period through equal installments made at the end of each year. If an interest rate of 8.5% per year is charged on the unpaid balance and interest calculations are made at the end of each year, determine the size of each installment so that the loan is amortized at the end of ten years. Verify the result by displaying the amortization schedule.

11. A loan of $265,000 is to be repaid over an 8-year period through equal installments made at the end of each year. If an interest rate of 7.4% per year is charged on the unpaid balance and interest calculations are made at the end of each year, determine the size of each installment so that the loan is amortized at the end of eight years. Verify the result by displaying the amortization schedule.

5.4 ARITHMETIC AND GEOMETRIC PROGRESSIONS (OPTIONAL)

Arithmetic Progressions

An **arithmetic progression** is a sequence of numbers in which each term after the first is obtained by adding a constant d to the preceding term. The constant d is called the **common difference.** For example, the sequence

$$3, 6, 9, 12, \ldots$$

is an arithmetic progression with the common difference equal to 3.

Observe that an arithmetic progression is completely determined if the first term and the common difference are known. In fact, if

$$a_1, a_2, a_3, \ldots, a_n, \ldots$$

is an arithmetic progression with the first term given by a and common difference given by d, then by definition,

$$a_1 = a$$
$$a_2 = a_1 + d = a + d$$
$$a_3 = a_2 + d = (a + d) + d = a + 2d$$
$$a_4 = a_3 + d = (a + 2d) + d = a + 3d$$
$$\vdots$$
$$a_n = a_{n-1} + d = a + (n - 2)d + d = a + (n - 1)d$$

Thus we see that the nth term of an arithmetic progression with first term a and common difference d is given by

$$a_n = a + (n - 1)d \qquad \textbf{(12)}$$

nth TERM OF AN ARITHMETIC PROGRESSION	The nth term of an arithmetic progression with first term a and common difference d is given by $$a_n = a + (n - 1)d$$

EXAMPLE I   Find the twelfth term of the arithmetic progression

$$2, 7, 12, 17, 22, \ldots$$

Solution   The first term of the arithmetic progression is $a_1 = a = 2$, and the common difference is $d = 5$, so upon setting $n = 12$ in equation (12), we find

$$a_{12} = 2 + (12 - 1)5 = 57 \qquad \circ \; \circ \; \circ$$

EXAMPLE 2   Write the first five terms of an arithmetic progression whose third and eleventh terms are 21 and 85, respectively.

Solution   Using (12), we obtain

$$a_3 = a + 2d = 21$$
$$a_{11} = a + 10d = 85$$

Subtracting the first equation from the second gives $8d = 64$, or $d = 8$. Substituting this value of d into the first equation yields $a + 16 = 21$, or $a = 5$. Thus, the required arithmetic progression is given by the sequence

$$5, 13, 21, 29, 37, \ldots \qquad \circ \ \circ \ \circ$$

Let S_n denote the sum of the first n terms of an arithmetic progression with first term $a_1 = a$ and common difference d. Then

$$S_n = a + (a + d) + (a + 2d) + \cdots + [a + (n - 1)d] \qquad (13)$$

Rewriting the expression for S_n with the terms in reverse order gives

$$S_n = [a + (n - 1)d] + [a + (n - 2)d] + \cdots + (a + d) + a \qquad (14)$$

Adding equations (13) and (14), we obtain

$$2S_n = [2a + (n - 1)d] + [2a + (n - 1)d]$$
$$+ \cdots + [2a + (n - 1)d]$$
$$= n[2a + (n - 1)d]$$

or
$$S_n = \frac{n}{2}[2a + (n - 1)d]$$

SUM OF TERMS IN AN ARITHMETIC PROGRESSION

The sum of the first n terms of an arithmetic progression with first term a and common difference d is given by

$$S_n = \frac{n}{2}[2a + (n - 1)d] \qquad (15)$$

EXAMPLE 3   Find the sum of the first 20 terms of the arithmetic progression of Example 2.

Solution   Letting $a = 2$, $d = 5$, and $n = 20$ in (15), we obtain

$$S_{20} = \frac{20}{2}[2 \cdot 2 + 19 \cdot 5] = 990 \qquad \circ \ \circ \ \circ$$

EXAMPLE 4   The Madison Electric Company had sales of $200,000 in its first year of operation. If the sales increased by $30,000 per year thereafter, find Madison's sales in the fifth year and its total sales over the first five years of operation.

Solution   Madison's yearly sales follow an arithmetic progression, with the first term given by $a = 200,000$ and the common difference given by $d = 30,000$. The sales in the fifth year are found by using equation (12) with $n = 5$. Thus,

$$a_5 = 200{,}000 + (5 - 1)30{,}000 = 320{,}000$$

or $320,000.

Madison's total sales over the first five years of operation are found by using (15) with $n = 5$. Thus,

$$S_5 = \frac{5}{2}[2(200{,}000) + (5 - 1)30{,}000]$$

$$= 1{,}300{,}000$$

or $1,300,000.

○ ○ ○

Geometric Progressions

A **geometric progression** is a sequence of numbers in which each term after the first is obtained by multiplying the preceding term by a constant r. The constant r is called the **common ratio.**

A geometric progression is completely determined if the first term and the common ratio are known. Thus, if

$$a_1, a_2, a_3, \ldots, a_n, \ldots$$

is a geometric progression with the first term given by a and common ratio given by r, then by definition,

$$a_1 = a$$
$$a_2 = a_1 r = ar$$
$$a_3 = a_2 r = ar^2$$
$$a_4 = a_3 r = ar^3$$
$$\vdots$$
$$a_n = a_{n-1} r = ar^{n-1}$$

Thus, we see that the nth term of a geometric progression with first term a and common ratio r is given by

$$a_n = ar^{n-1} \tag{16}$$

nth TERM OF A GEOMETRIC PROGRESSION

The nth term of a geometric progression with first term a and common ratio r is given by

$$a_n = ar^{n-1}$$

EXAMPLE 5 Find the eighth term of a geometric progression whose first five terms are 162, 54, 18, 6, 2.

Solution The common ratio is found by taking the ratio of any term other than the first to the preceding term. Taking the ratio of the fourth term to the third term, for example, gives $r = 6/18 = 1/3$. To find the eighth term of the geometric progression, use formula (16) with $a = 162$, $r = 1/3$, and $n = 8$, obtaining

$$a_8 = 162 \left(\frac{1}{3}\right)^7$$

$$= \frac{2}{27}$$

○ ○ ○

EXAMPLE 6 Find the tenth term of a geometric progression whose third term is 16 and whose seventh term is 1.

Solution Using equation (16) with $n = 3$ and $n = 7$, respectively, yields

$$a_3 = ar^2 = 16$$

and

$$a_7 = ar^6 = 1$$

Dividing a_7 by a_3 gives

$$\frac{ar^6}{ar^2} = \frac{1}{16}$$

from which we obtain $r^4 = 1/16$, or $r = 1/2$. Substituting this value of r into the expression for a_3, we obtain

$$a\left(\frac{1}{2}\right)^2 = 16 \qquad \text{or} \qquad a = 64$$

Finally, using (16) once again with $a = 64$, $r = 1/2$, and $n = 10$ gives

$$a_{10} = 64 \left(\frac{1}{2}\right)^9 = \frac{1}{8}$$

○ ○ ○

To find the sum of the first n terms of a geometric progression with the first term $a_1 = a$ and common ratio r, denote the required sum by S_n. Then

$$S_n = a + ar + ar^2 + \cdots + ar^{n-2} + ar^{n-1} \qquad \textbf{(17)}$$

Upon multiplying (17) by r, we obtain

$$rS_n = ar + ar^2 + ar^3 + \cdots + ar^{n-1} + ar^n \qquad \textbf{(18)}$$

Subtracting (18) from (17) gives

$$S_n - rS_n = a - ar^n$$

or

$$(1 - r)S_n = a(1 - r^n)$$

If $r \neq 1$, we may divide both sides of the last equation by $(1 - r)$, obtaining

$$S_n = \frac{a(1 - r^n)}{(1 - r)}$$

If $r = 1$, then (17) gives

$$S_n = a + a + a + \cdots + a \qquad (n \text{ terms})$$

$$= na$$

Thus,

$$S_n = \begin{cases} \dfrac{a(1 - r^n)}{1 - r} & \text{if } r \neq 1 \\ na & \text{if } r = 1 \end{cases} \qquad \textbf{(19)}$$

SUM OF TERMS IN A GEOMETRIC PROGRESSION

The sum of the first n terms of a geometric progression with first term a and common ratio r is given by

$$S_n = \begin{cases} \dfrac{a(1 - r^n)}{1 - r} & \text{if } r \neq 1 \\ na & \text{if } r = 1 \end{cases}$$

EXAMPLE 7 Find the sum of the first six terms of the geometric progression

$$3, 6, 12, 24, \ldots$$

Solution Here $a = 3$ and $r = 6/3 = 2$, so formula (19) gives

$$S_6 = \frac{3(1 - 2^6)}{1 - 2} = 189$$

○ ○ ○

EXAMPLE 8 The Michaelson Land Development Company had sales of $1 million in its first year of operation. If sales increased by 10% per year thereafter, find Michaelson's sales in the fifth year and its total sales over the first five years of operation.

Solution Michaelson's yearly sales follow a geometric progression, with the first term given by $a = 1,000,000$ and the common ratio given by $r = 1.1$. The sales in the fifth year are found by using (16) with $n = 5$. Thus,

$$a_5 = 1,000,000(1.1)^4 = 1,464,100$$

or $1,464,100.

Michaelson's total sales over the first five years of operation are found by using equation (19) with $n = 5$. Thus,

$$S_5 = \frac{1,000,000[1 - (1.1)^5]}{1 - 1.1}$$

$$= 6,105,100$$

or $6,105,100.

○ ○ ○

Double Declining-Balance Method of Depreciation

In Section 1.3 we discussed the straight-line, or linear, method of depreciating an asset. Linear depreciation assumes that the asset depreciates at a constant rate. For certain assets, such as machines, whose market values drop rapidly in the early years of usage and thereafter less rapidly, another method of depreciation called the **double declining-balance method** is often used. In practice, a business firm normally employs the double declining-balance method for depreciating such assets for a certain number of years and then switches over to the linear method.

To derive an expression for the book value of an asset being depreciated by the double declining-balance method, let C (in dollars) denote the original cost of the asset and let the asset be depreciated over N years. Using this method, the amount depreciated each year is $2/N$ times the value of the asset at the beginning of that year. Thus, the amount by which the asset is depreciated in its first year of use is given by $2C/N$, so if $V(1)$ denotes the book value of the asset at the end of the first year, then

$$V(1) = C - \frac{2C}{N} = C\left(1 - \frac{2}{N}\right)$$

Next, if $V(2)$ denotes the book value of the asset at the end of the second year, then a similar argument leads to

$$V(2) = C\left(1 - \frac{2}{N}\right) - C\left(1 - \frac{2}{N}\right)\frac{2}{N}$$

$$= C\left(1 - \frac{2}{N}\right)\left(1 - \frac{2}{N}\right)$$

$$= C\left(1 - \frac{2}{N}\right)^2$$

Continuing, we find that if $V(n)$ denotes the book value of the asset at the end of n years, then the terms $C, V(1), V(2), \ldots, V(n)$ form a geometric progression with first term C and common ratio $(1 - 2/N)$. Consequently, the nth term, $V(n)$, is given by

$$V(n) = C\left(1 - \frac{2}{N}\right)^n \qquad (1 \le n \le N) \tag{20}$$

Also, if $D(n)$ denotes the amount by which the asset has been depreciated by the end of the nth year, then

$$D(n) = C - C\left(1 - \frac{2}{N}\right)^n$$

$$= C\left[1 - \left(1 - \frac{2}{N}\right)^n\right] \tag{21}$$

EXAMPLE 9   A tractor purchased at a cost of $60,000 is to be depreciated by the double declining-balance method over ten years. What is the book value of the tractor at the end of five years? By what amount has the tractor been depreciated by the end of the fifth year?

Solution   We have $C = 60,000$ and $N = 10$. Thus, using formula (20) with $n = 5$ gives the book value of the tractor at the end of five years as

$$V(5) = 60,000\left(1 - \frac{2}{10}\right)^5$$

$$= 60,000\left(\frac{4}{5}\right)^5 = 19,660.80$$

or $19,660.80.

The amount by which the tractor has been depreciated by the end of the fifth year is given by

$$60,000 - 19,660.80 = 40,339.20$$

or $40,339.20. You may verify the last result by using equation (21) directly.

○ ○ ○

EXPLORING WITH TECHNOLOGY

A tractor purchased at a cost of $60,000 is to be depreciated over ten years with a residual value of $0. Using the double declining-balance method, its value at the end of n years is $V_1(n) = 60,000(0.8)^n$ dollars. Using straight-line depreciation, its value at the end of n years is $V_2(n) = 60,000 - 6000n$. Use a graphing utility to sketch the graphs of V_1 and V_2 in the viewing rectangle $[0, 10] \times [0, 70,000]$. Comment on the relative merits of each method of depreciation.

◑ ◐ ◑

SELF-CHECK EXERCISES 5.4

1. Find the sum of the first five terms of the geometric progression with first term -24 and common ratio $-1/2$.

2. Office equipment purchased for $75,000 is to be depreciated by the double declining-balance method over five years. Find the book value at the end of three years.

3. Derive the formula for the future value of an annuity [equation (6), Section 5.2].

Solutions to Self-Check Exercises 5.4 can be found on page 319.

5.4 EXERCISES

In exercises 1–4, find the nth term of the arithmetic progression that has the given values of a, d, and n.

1. $a = 6, d = 3, n = 9$

2. $a = -5, d = 3, n = 7$

3. $a = -15, d = 3/2, n = 8$

4. $a = 1.2, d = 0.4, n = 98$

5. Find the first five terms of the arithmetic progression whose fourth and eleventh terms are 30 and 107, respectively.

6. Find the first five terms of the arithmetic progression whose 7th and 23rd terms are -5 and -29, respectively.

7. Find the seventh term of the arithmetic progression: x, $x + y, x + 2y, \ldots$

8. Find the eleventh term of the arithmetic progression:
$a + b, 2a, 3a - b, \ldots$

9. Find the sum of the first 15 terms of the arithmetic progression: $4, 11, 18, \ldots$

10. Find the sum of the first 20 terms of the arithmetic progression: $5, -1, -7, \ldots$

11. Find the sum of the odd integers between 14 and 58.

12. Find the sum of the even integers between 21 and 99.

13. Find $f(1) + f(2) + f(3) + \cdots + f(20)$, given that $f(x) = 3x - 4$.

14. Find $g(1) + g(2) + g(3) + \cdots + g(50)$, given that $g(x) = 12 - 4x$.

15. Show that equation (15) can be written as $S_n = \frac{n}{2}(a + a_n)$, where a_n represents the last term of an arithmetic progression. Use this formula to find:
   a. the sum of the first 11 terms of the arithmetic progression whose 1st and 11th terms are 3 and 47, respectively.
   b. the sum of the first 20 terms of the arithmetic progression whose 1st and 20th terms are 5 and -33, respectively.

16. **Sales Growth**   The Moderne Furniture Company had sales of $1,500,000 during its first year of operation. If the sales increased by $160,000 per year thereafter, find Moderne's sales in the fifth year and its total sales over the first five years of operation.

17. **Exercise Program**   As part of her fitness program, Karen has taken up jogging. If she jogs 1 mile the first day and increases her daily run by 1/4 mile every week, how long will it take her to reach her goal of 10 miles per day?

18. **Cost of Drilling**   A 100-foot oil well is to be drilled. The cost of drilling the first foot is $6 and the cost of drilling each additional foot is $2.50 more than that of the preceding foot. Find the cost of drilling the entire 100 feet.

19. **Consumer Decisions**   A tourist wishes to go from the airport to his hotel, which is 25 miles away. The taxi rate is $1 for the first mile and 60 cents for each additional mile. The airport limousine also goes to his hotel and charges a flat rate of $7.50. How much money will the tourist save by taking the airport limousine?

20. **Salary Comparisons**   A recent college graduate received two job offers. Company A offered her an initial salary of $24,800 with guaranteed annual increases of $1500 per year for the first five years. Company B offered

an initial salary of $26,400 with guaranteed annual increases of $1100 per year for the first five years.
   a. Which company is offering a higher salary for the fifth year of employment?
   b. Which company is offering more money for the first five years of employment?

21. **Sum-of-the-Years'-Digits Method of Depreciation** One of the methods that the Internal Revenue Service allows for computing depreciation of certain business property is the sum-of-the-years'-digits method. If a property valued at C dollars has an estimated useful life of N years and a salvage value of S dollars, then the amount of depreciation D_n allowed during the nth year is given by

$$D_n = (C - S)\frac{N - (n - 1)}{S_N} \qquad (0 \le n \le N)$$

where S_N is the sum of the first N positive integers representing the estimated useful life of the property. Thus,

$$S_N = 1 + 2 + \cdots + N = \frac{N(N + 1)}{2}$$

   a. Verify that the sum of the arithmetic progression $S_N = 1 + 2 + \cdots + N$ is given by

$$\frac{N(N + 1)}{2}$$

   b. If office furniture worth $6000 is to be depreciated by this method over $N = 10$ years, and the salvage value of the furniture is $500, find the depreciation for the third year by computing D_3.

22. **Sum-of-the-Years'-Digits Method of Depreciation** Refer to Example 1, Section 1.3. Suppose the amount of depreciation allowed for a printing machine, which has an estimated useful life of five years and an initial value of $100,000 (with no salvage value), is $20,000 per year using the straight-line method of depreciation. Determine the amount of depreciation that would be allowed for the first year if the printing machine were depreciated using the sum-of-the-years'-digits method described in exercise 21. Which method would result in a larger depreciation of the asset in its first year of use?

In exercises 23–28, determine which of the given sequences are geometric progressions. For each geometric progression, find the seventh term and the sum of the first seven terms.

23. $4, 8, 16, 32, \ldots$

24. $1, -1/2, 1/4, -1/8, \ldots$

25. $1/2, -3/8, 1/4, -9/64, \ldots$

26. 0.004, 0.04, 0.4, 4, . . .

27. 243, 81, 27, 9, . . .

28. −1, 1, 3, 5, . . .

29. Find the 20th term and sum of the first 20 terms of the geometric progression −3, 3, −3, 3, . . .

30. Find the 23rd term in a geometric progression having the first term $a = 0.1$ and ratio $r = 2$.

31. **Population Growth** It has been projected that the population of a certain city in the Southwest will increase by 8% during each of the next five years. If the current population is 200,000, what is the expected population in five years?

32. **Sales Growth** The Metro Cable T.V. Company had sales of $2,500,000 in its first year of operation. If thereafter the sales increased by 12% of the previous year, find the sales of the company in the fifth year and the total sales over the first five years of operation.

33. **COLAs** Suppose that the cost-of-living index had increased by 9% during each of the past six years and that a member of the E.U.W. Union had been guaranteed an annual increase equal to 2% above the cost-of-living index over that period. What would the present salary of a union member be whose salary was $22,000 six years ago?

34. **Savings Plans** The parents of a nine-year-old boy have agreed to deposit $10 in their son's bank account on his tenth birthday and to double the size of their deposit every year thereafter until his eighteenth birthday.
    a. How much will they have to deposit on his eighteenth birthday?
    b. How much will they have deposited by his eighteenth birthday?

35. **Salary Comparisons** Suppose that an employee of the Stenton Printing Company whose current annual salary is $28,000 has the option of taking an annual raise of 8% per year for the next four years or a fixed annual raise of $1500 per year. Which option would be more profitable to him considering his total earnings over the 4-year period?

36. **Bacteria Growth** A culture of a certain bacteria is known to double in number every three hours. If the culture has an initial count of 20, what will the population of the culture be at the end of 24 hours?

37. **Trust Funds** Ms. Simpson is the recipient of a trust fund that she will receive over a period of six years. Under the terms of the trust, she is to receive $10,000 the first year and each succeeding annual payment is to be increased by 15%.
    a. How much will she receive during the sixth year?
    b. What is the total amount of the six payments she will receive?

In exercises 38–40, find the book value of office equipment purchased at a cost C at the end of the nth year if it is to be depreciated by the double declining-balance method over ten years. Assume a salvage value of $0.

38. $C = \$20,000$, $n = 4$

39. $C = \$150,000$, $n = 8$

40. $C = \$80,000$, $n = 7$

41. **Double Declining-Balance Method of Depreciation** Restaurant equipment purchased at a cost of $150,000 is to be depreciated by the double declining-balance method over ten years. What is the book value of the equipment at the end of six years? By what amount has the equipment been depreciated at the end of the sixth year?

42. **Double Declining-Balance Method of Depreciation** Refer to exercise 22. Recall that a printing machine that had an estimated useful life of five years and an initial value of $100,000 (with no salvage value) was to be depreciated. At the end of the first year using the straight-line method of depreciation, the amount of depreciation allowed was $20,000, and when the sum-of-the-years'-digits method was used the depreciation was $33,333. Determine the amount of depreciation that would be allowed for the first year if the printing machine were depreciated by the double declining-balance method. Which of these three methods would result in the largest depreciation of the printing machine at the end of its first year of use?

SOLUTIONS TO SELF–CHECK EXERCISES 5.4

I. Use equation (19) with $a = -24$ and $r = -1/2$, obtaining

$$S_5 = \frac{-24\left[1 - \left(-\frac{1}{2}\right)^5\right]}{1 - \left(-\frac{1}{2}\right)}$$

$$= \frac{-24\left(1 + \frac{1}{32}\right)}{\frac{3}{2}} = -\frac{33}{2}$$

2. Use equation (20) with $C = 75{,}000$, $N = 5$, and $n = 3$, giving the book value of the office equipment at the end of three years as

$$V(3) = 75{,}000\left(1 - \frac{2}{5}\right)^3 = 16{,}200$$

or $16,200.

3. We have

$$S = R + R(1 + i) + R(1 + i)^2 + \cdots + R(1 + i)^{n-1}$$

Now, the sum on the right is easily seen to be the sum of the first n terms of a geometric progression with first term R and common ratio $(1 + i)$, so by virtue of formula (19),

$$S = \frac{R[1 - (1 + i)^n]}{1 - (1 + i)} = R\left[\frac{(1 + i)^n - 1}{i}\right]$$

CHAPTER 5 SUMMARY OF PRINCIPAL FORMULAS AND TERMS

Formulas

1. Simple interest $A = P(1 + rt)$

2. Compound interest:
   a. Accumulated amount $A = P(1 + i)^n$
   b. Present value $P = A(1 + i)^{-n}$
   c. Interest rate per $i = r/m$
      compounding period
   d. Number of conversion $n = mt$
      periods

3. Effective rate of interest $r_{\text{eff}} = \left(1 + \frac{r}{m}\right)^m - 1$

4. Annuities:

a. Future value

$$S = R\left[\frac{(1 + i)^n - 1}{i}\right]$$

b. Present value

$$P = R\left[\frac{1 - (1 + i)^{-n}}{i}\right]$$

5. Amortization payment

$$R = \frac{Pi}{1 - (1 + i)^{-n}}$$

6. Sinking fund payment

$$R = \frac{iS}{(1 + i)^n - 1}$$

Terms

Simple interest	Present value
Accumulated amount (future value)	Annuity
Compound interest	Ordinary annuity
Nominal rate (stated rate)	Future value of an annuity
Effective rate	Present value of an annuity
	Sinking fund

CHAPTER 5 REVIEW EXERCISES

1. Find the accumulated amount after four years if $5000 is invested at 10% per year compounded (a) annually, (b) semiannually, (c) quarterly, and (d) monthly.

2. Find the accumulated amount after eight years if $12,000 is invested at 6.5% per year compounded (a) annually, (b) semiannually, (c) quarterly, and (d) monthly.

3. Find the effective rate of interest corresponding to a nominal rate of 12% per year compounded (a) annually, (b) semiannually, (c) quarterly, and (d) monthly.

4. Find the effective rate of interest corresponding to a nominal rate of 11.5% per year compounded (a) annually, (b) semiannually, (c) quarterly, and (d) monthly.

5. Find the present value of $41,413 due in five years at an interest rate of 6.5% per year compounded quarterly.

6. Find the present value of $64,540 due in six years at an interest rate of 8% per year compounded monthly.

7. Find the amount (future value) of an ordinary annuity of $150 per quarter for seven years at 8% per year compounded quarterly.

8. Find the future value of an ordinary annuity of $120 per month for ten years at 9% per year compounded monthly.

9. Find the present value of an ordinary annuity of 36 payments of $250 each made monthly and earning interest at 9% per year compounded monthly.

10. Find the present value of an ordinary annuity of 60 payments of $5000 each made quarterly and earning interest at 8% per year compounded quarterly.

11. Find the payment R needed to amortize a loan of $22,000 at 8.5% per year compounded monthly with 36 monthly installments over a period of three years.

12. Find the payment R needed to amortize a loan of $10,000 at 9.2% per year compounded monthly with 36 monthly installments over a period of three years.

13. Find the payment R needed to accumulate $18,000 with 48 monthly installments over a period of four years at an interest rate of 6% per year compounded monthly.

14. Find the payment R needed to accumulate $15,000 with

60 monthly installments over a period of five years at an interest rate of 7.2% per year compounded monthly.

15. Find the rate of interest per year compounded on a daily basis that is equivalent to 7.2% per year compounded monthly.

16. Find the rate of interest per year compounded on a daily basis that is equivalent to 9.6% per year compounded monthly.

17. The J.C.N. Media Corporation had sales of $1,750,000 in the first year of operation. If the sales increased by 14% per year thereafter, find the company's sales in the fourth year and the total sales over the first four years of operation.

18. The manager of a money-market fund has invested $4.2 million in certificates of deposit that pay interest at the rate of 5.4% per year compounded quarterly over a period of five years. How much will the investment be worth at the end of five years?

19. Ms. Kim invested a sum of money four years ago in a savings account that has since paid interest at the rate of 6.5% per year compounded monthly. Her investment is now worth $19,440.31. How much did she originally invest?

20. Mr. Sanchez invested $24,000 in a mutual fund five years ago. Today his investment is worth $34,616. Find the effective annual rate of return on his investment over the 5-year period.

21. The Blakes have decided to start a monthly savings program in order to provide for their son's college education. How much should they deposit at the end of each month in a savings account earning interest at the rate of 8% per year compounded monthly so that at the end of the tenth year the accumulated amount will be $40,000?

22. Mrs. Chang has contributed $200 at the end of each month into her company's employee retirement account for the past ten years. Her employer has matched her contribution each month. If the account has earned interest at the rate of 8% per year compounded monthly over the 10-year period, determine how much Mrs. Chang now has in her retirement account.

23. Ms. Lemsky has leased an auto for four years at $300 per month. If money is worth 5% per year compounded monthly, what is the equivalent cash payment (present value) of this annuity? (Assume that the payments are made at the end of each month.)

24. Ms. Cantwell made a down payment of $400 toward the purchase of new furniture. In order to pay the balance of the purchase price she has secured a loan from her bank at 12% per year compounded monthly. Under the terms of her finance agreement she is required to make payments of $75.32 at the end of each month for 24 months. What was the purchase price of the furniture?

25. The Turners have purchased a house for $150,000. They made an initial down payment of $30,000 and secured a mortgage with interest charged at the rate of 9% per year on the unpaid balance. (Interest computations are made at the end of each month.) Assume the loan is amortized over 30 years.
a. What monthly payment will the Turners be required to make?
b. What will their total interest payment be?
c. What will their equity be (disregard depreciation) after ten years?

26. Refer to exercise 25. If the loan is amortized over 15 years,
a. what monthly payment will the Turners be required to make?
b. what will their total interest payment be?
c. what will their equity be (disregard depreciation) after ten years?

27. The management of a corporation anticipates a capital expenditure of $500,000 in five years for the purpose of purchasing replacement machinery. In order to finance this purchase, a sinking fund that earns interest at the rate of 10% per year compounded quarterly will be set up. Determine the amount of each (equal) quarterly installment that should be deposited in the fund. (Assume that the payments are made at the end of each quarter.)

28. The management of a condominium association anticipates a capital expenditure of $120,000 in two years for the purpose of painting the exterior of the condominium. In order to pay for this maintenance, a sinking fund will be set up that will earn interest at the rate of 5.8% per year compounded monthly. Determine the amount of each (equal) monthly installment the association will be required to deposit into the fund at the end of each month for the next two years.

29. The outstanding balance on Mr. Baker's credit card account is $3200. The bank issuing the credit card is charging 18.6% per year compounded monthly. If Mr. Baker decides to pay off this balance in equal monthly installments at the end of each month for the next 18 months, how much will his monthly payment be?

30. Refer to exercise 29. What is the effective rate of interest the bank is charging Mr. Baker?

We often deal with well-defined collections of objects called *sets*. In this chapter we see how sets can be combined algebraically to yield other sets. We also look at some techniques for determining the number of elements in a set and for determining the number of ways the elements of a set can be arranged or combined. These techniques enable us to solve many practical problems, as you will see throughout the chapter.

What are the investment options? An investor has decided to purchase shares of stock from a recommended list of aerospace, energy development, and electronics companies. In Example 5, page 345, we will determine how many ways the investor may select a group of three companies from the list.

6

SETS AND COUNTING

6.1 SETS AND SET OPERATIONS

Set Terminology and Notation

We often deal with collections of different kinds of objects. For example, in conducting a study of the distribution of the weights of newborn infants, we might consider the collection of all infants born in the Massachusetts General Hospital during 1996. In a study of the fuel consumption of compact cars, we might be interested in the collection of compact cars manufactured by General Motors in the 1996 model year. Such collections are examples of *sets*. More specifically, a **set** is a well-defined collection of objects. Thus, a set is not just any collection of objects, but it must be well defined in the sense that if we are given an object, then we should be able to determine whether or not it belongs to the collection.

The objects of a set are called the **elements,** or *members,* of the set and are usually denoted by lowercase letters $a, b, c, \ldots$; the sets themselves are usually denoted by uppercase letters $A, B, C, \ldots$. The elements of a set may be displayed by listing each element between braces. For example, using **roster notation,** the set A consisting of the first three letters of the English alphabet is written

$$A = \{a, b, c\}$$

The set B of all letters of the alphabet may be written

$$B = \{a, b, c, \ldots, z\}$$

Another kind of set notation commonly used is **set-builder notation.** Here, a rule is given that describes the definite property or properties an object x must satisfy to qualify for membership in the set. Using this notation, the set B is written as

$$B = \{x \mid x \text{ is a letter of the English alphabet}\}$$

and is read "B is the set of all elements x such that x is a letter of the English alphabet."

If a is an element of a set A, we write $a \in A$, read "a belongs to A" or "a is an element of A." If, however, the element a does not belong to the set A, then we write $a \notin A$, read "a does not belong to A." For example, if $A = \{1, 2, 3, 4, 5\}$, then $3 \in A$ but $6 \notin A$.

---

a. Let A denote the collection of all the days in August 1996 in which the average daily temperature in San Francisco was approximately 75°F. Is A a set? Explain your answer.
b. Let B denote the collection of all the days in August 1996 in which the average daily temperature in San Francisco was between 73.5°F and 81.2°F, inclusive. Is B a set? Explain your answer.

SET EQUALITY	Two sets A and B are **equal,** written $A = B$, if and only if they have exactly the same elements.

EXAMPLE 1   Let A, B, and C be the sets

$$A = \{a, e, i, o, u\}$$
$$B = \{a, i, o, e, u\}$$
$$C = \{a, e, i, o\}$$

Then $A = B$ since they both contain exactly the same elements. Note that the order in which the elements are displayed is immaterial. Also, $A \neq C$ since $u \in A$ but $u \notin C$. Similarly, we conclude that $B \neq C$.   ০ ০ ০

SUBSET	If every element of a set A is also an element of a set B, then we say that A is a **subset** of B and write $A \subseteq B$.

By this definition, two sets A and B are equal if and only if (1) $A \subseteq B$ and (2) $B \subseteq A$. You may verify this (see exercise 66).

EXAMPLE 2   Referring to Example 1, we find that $C \subseteq B$ since every element of C is also an element of B. Also, if D is the set

$$D = \{a, e, i, o, x\}$$

then D is not a subset of A, written $D \nsubseteq A$, since $x \in D$ but $x \notin A$. Observe that $A \nsubseteq D$ as well since $u \in A$ but $u \notin D$.   ০ ০ ০

If A and B are sets such that $A \subseteq B$ but $A \neq B$, then we say that A is a **proper subset** of B. In other words, a set A is a proper subset of a set B, written $A \subset B$, if (1) $A \subseteq B$ and (2) there exists at least one element in B that is not in A. The latter condition states that the set A is properly "smaller" than the set B.

EXAMPLE 3   Let $A = \{1, 2, 3, 4, 5, 6\}$ and $B = \{2, 4, 6\}$. Then B is a proper subset of A since (1) $B \subseteq A$, which is easily verified, and (2) there exists at least one element in A that is not in B—for example, the element 1.   ০ ০ ০

   Notice that when we are referring to sets and subsets we use the symbols $\subset$, $\subseteq$, $\supset$, and $\supseteq$ to express the idea of "containment." However, when we wish to show that an element is contained in a set, we use the symbol $\in$. Thus, in Example 3, we would write $1 \in A$ and *not* $\{1\} \in A$.

EMPTY SET	The set that contains no elements is called the **empty set** and is denoted by $\varnothing$.

The empty set, $\varnothing$, is a subset of every set. To see this, observe that $\varnothing$ has no elements and, therefore, contains no element that is not also in A.

EXAMPLE 4   List all subsets of the set $A = \{a, b, c\}$.

Solution   There is one subset consisting of no elements—namely, the empty set, $\varnothing$. Next, observe that there are three subsets consisting of one element,

$$\{a\}, \quad \{b\}, \quad \text{and} \quad \{c\}$$

three subsets consisting of two elements,

$$\{a, b\}, \quad \{a, c\}, \quad \text{and} \quad \{b, c\}$$

and one subset consisting of three elements, the set A itself. Therefore, the subsets of A are

$$\varnothing, \quad \{a\}, \quad \{b\}, \quad \{c\}, \quad \{a, b\}, \quad \{a, c\}, \quad \{b, c\}, \quad \{a, b, c\} \qquad \circ \circ \circ$$

In contrast with the empty set, we have, on the other extreme, the notion of a largest, or *universal,* set. A **universal set** is the set of all elements of interest in a particular discussion. It is the largest in the sense that all sets considered in the discussion of the problem are subsets of the universal set. Of course, different universal sets are associated with different problems, as shown in Example 5.

EXAMPLE 5

a. If the problem at hand is to determine the ratio of female to male students in a college, then a logical choice of a universal set is the set comprising the whole student body of the college.

b. If the problem is to determine the ratio of female to male students in the business department of the college in (a), then the set of all students in the business department may be chosen as the universal set. $\circ \circ \circ$

A visual representation of sets is realized through the use of **Venn diagrams,** which are of considerable help in understanding the concepts introduced earlier, as well as in solving problems involving sets. The universal set U is represented by a rectangle, and subsets of U are represented by regions lying inside the rectangle.

Figure 6.1

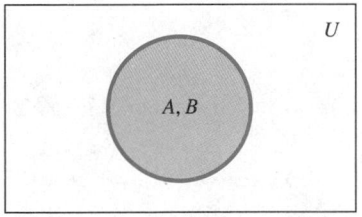

(a) $A = B$.

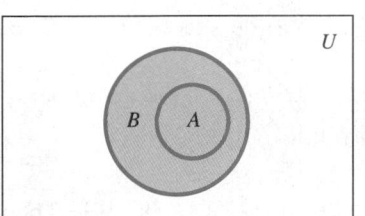

(b) $A \subset B$.

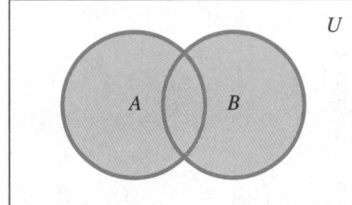 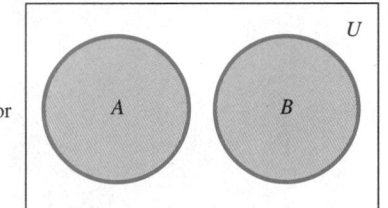

(c) $A \not\subset B$ and $B \not\subset A$.

EXAMPLE 6 Use Venn diagrams to illustrate the following statements:

a. The sets A and B are equal.

b. The set A is a proper subset of the set B.

c. The sets A and B are not subsets of each other.

Solution The respective Venn diagrams are shown in Figures 6.1a, 6.1b, and 6.1c.

○ ○ ○

Set Operations

Having introduced the concept of a set, our next task is to consider operations on sets—that is, to consider ways in which sets may be combined to yield other sets. These operations enable us to combine sets in much the same way the operations of addition and multiplication enable us to combine numbers to obtain other numbers. In what follows, all sets are assumed to be subsets of a given universal set U.

SET UNION

Let A and B be sets. The **union** of A and B, written $A \cup B$, is the set of all elements that belong to either A or B or both.

$$A \cup B = \{x \mid x \in A \text{ or } x \in B \text{ or both}\}$$

The shaded portion of the Venn diagram (Figure 6.2) depicts the set $A \cup B$.

Figure 6.2
$A \cup B$ *set union.*

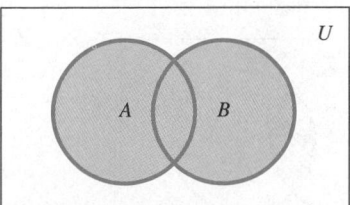

EXAMPLE 7   If $A = \{a, b, c\}$ and $B = \{a, c, d\}$, then $A \cup B = \{a, b, c, d\}$.

◦ ◦ ◦

SET INTERSECTION

Let A and B be sets. The set of elements in common with the sets A and B, written $A \cap B$, is called the **intersection** of A and B.

$$A \cap B = \{x \mid x \in A \text{ and } x \in B\}$$

The shaded portion of the Venn diagram (Figure 6.3) depicts the set $A \cap B$.

EXAMPLE 8   Let $A = \{a, b, c\}$ and $B = \{a, c, d\}$. Then $A \cap B = \{a, c\}$. (Compare this result with Example 7.)

◦ ◦ ◦

EXAMPLE 9   Let $A = \{1, 3, 5, 7, 9\}$ and $B = \{2, 4, 6, 8, 10\}$. Then $A \cap B = \varnothing$.

◦ ◦ ◦

The two sets of Example 9 have null intersection. In general, the sets A and B are said to be **disjoint** if they have no elements in common—that is, if $A \cap B = \varnothing$.

EXAMPLE 10   If U is the set of all students in the classroom and $M = \{x \in U \mid x \text{ is male}\}$ and $F = \{x \in U \mid x \text{ is female}\}$, then $F \cap M = \varnothing$, and F and M are disjoint.

◦ ◦ ◦

COMPLEMENT OF A SET

If U is a universal set and A is a subset of U, then the set of all elements in U that are not in A is called the **complement** of A and is denoted A^c.

$$A^c = \{x \mid x \in U, x \notin A\}$$

The shaded portion of the Venn diagram (Figure 6.4) shows the set A^c.

Figure 6.3
$A \cap B$ set intersection.

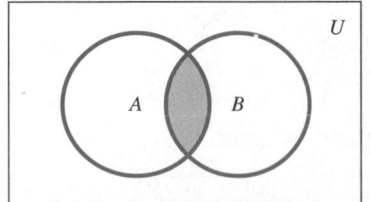

Figure 6.4
Set complementation.

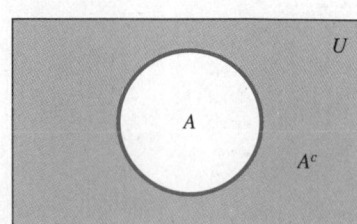

Let A, B, and C be nonempty subsets of a set U.

a. Suppose $A \cap B \neq \varnothing$, $A \cap C \neq \varnothing$, and $B \cap C \neq \varnothing$. Can you conclude that $A \cap B \cap C \neq \varnothing$? Explain your answer with an example.

b. Suppose $A \cap B \cap C \neq \varnothing$. Can you conclude that $A \cap B \neq \varnothing$, $A \cap C \neq \varnothing$, and $B \cap C \neq \varnothing$ simultaneously? Explain your answer.

EXAMPLE 11 Let $U = \{1, 2, 3, 4, 5, 6, 7, 8, 9, 10\}$ and $A = \{2, 4, 6, 8, 10\}$. Then $A^c = \{1, 3, 5, 7, 9\}$.   ∘ ∘ ∘

The following rules hold for the operation of complementation. See whether you can verify them.

SET COMPLEMENTATION

If U is a universal set and A is a subset of U, then

a. $U^c = \varnothing$ **b.** $\varnothing^c = U$ **c.** $(A^c)^c = A$

d. $A \cup A^c = U$ **e.** $A \cap A^c = \varnothing$

The following rules govern the operations on sets.

SET OPERATIONS

Let U be a universal set. If A, B, and C are arbitrary subsets of U, then

$A \cup B = B \cup A$	*Commutative law for union*
$A \cap B = B \cap A$	*Commutative law for intersection*
$A \cup (B \cup C) = (A \cup B) \cup C$	*Associative law for union*
$A \cap (B \cap C) = (A \cap B) \cap C$	*Associative law for intersection*
$A \cup (B \cap C)$ $= (A \cup B) \cap (A \cup C)$	*Distributive law for union*
$A \cap (B \cup C)$ $= (A \cap B) \cup (A \cap C)$	*Distributive law for intersection*

There are two additional rules, referred to as De Morgan's Laws, that govern the operations on sets.

DE MORGAN'S LAWS

Let A and B be sets. Then

$$(A \cup B)^c = A^c \cap B^c \qquad \qquad \textbf{(1)}$$

$$(A \cap B)^c = A^c \cup B^c \qquad \qquad \textbf{(2)}$$

Equation (1) states that the complement of the union of two sets is equal to the intersection of their complements. Equation (2) states that the complement of the intersection of two sets is equal to the union of their complements.

We will not prove De Morgan's Laws here, but the plausibility of (2) is illustrated in the following example.

Figure 6.5
$(A \cap B)^c$.

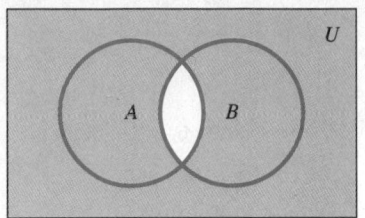

EXAMPLE 12   Using Venn diagrams, show that

$$(A \cap B)^c = A^c \cup B^c$$

Solution $(A \cap B)^c$ is the set of elements in U but not in $A \cap B$ and is thus the shaded region shown in Figure 6.5. Next, A^c and B^c are shown in Figures 6.6a and 6.6b. Their union, $A^c \cup B^c$, is easily seen to be equivalent to $(A \cap B)^c$ by referring once again to Figure 6.5.   ○ ○ ○

Figure 6.6
$A^c \cup B^c$ is the set obtained by joining (a) and (b).

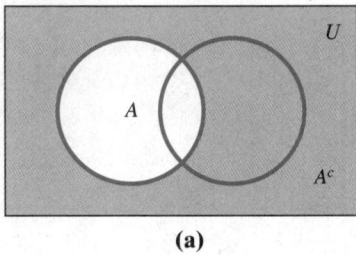

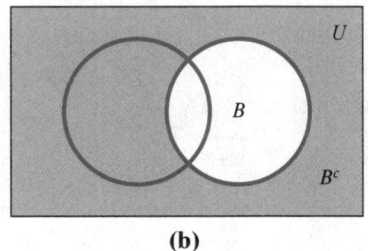

| **(a)** | **(b)** |

EXAMPLE 13   Let $U = \{1, 2, 3, 4, 5, 6, 7, 8, 9, 10\}$, $A = \{1, 2, 4, 8, 9\}$, and $B = \{3, 4, 5, 6, 8\}$. Verify by direct computation that $(A \cup B)^c = A^c \cap B^c$.

Solution $A \cup B = \{1, 2, 3, 4, 5, 6, 8, 9\}$, so $(A \cup B)^c = \{7, 10\}$. However, $A^c = \{3, 5, 6, 7, 10\}$ and $B^c = \{1, 2, 7, 9, 10\}$, so $A^c \cap B^c = \{7, 10\}$. The required result follows.   ○ ○ ○

Application

EXAMPLE 14   Let U denote the set of all cars in a dealer's lot and

$$A = \{x \in U \mid x \text{ is equipped with automatic transmission}\}$$
$$B = \{x \in U \mid x \text{ is equipped with air conditioning}\}$$
$$C = \{x \in U \mid x \text{ is equipped with side air bags}\}$$

Find an expression in terms of A, B, and C for each of the following sets:

a. The set of cars with at least one of the given options

b. The set of cars with exactly one of the given options

c. The set of cars with automatic transmission and side air bags but no air conditioning

Solution

a. The set of cars with at least one of the given options is $A \cup B \cup C$ (Figure 6.7a).

b. The set of cars with automatic transmission only is given by $A \cap B^c \cap C^c$. Similarly, we find that the set of cars with air conditioning only is given by $B \cap C^c \cap A^c$, whereas the set of cars with side air bags only is given by $C \cap A^c \cap B^c$. Thus, the set of cars with exactly one of the given options is $(A \cap B^c \cap C^c) \cup (B \cap C^c \cap A^c) \cup (C \cap A^c \cap B^c)$ (Figure 6.7b).

c. The set of cars with automatic transmission and side air bags but no air conditioning is given by $A \cap C \cap B^c$ (Figure 6.7c).

Figure 6.7

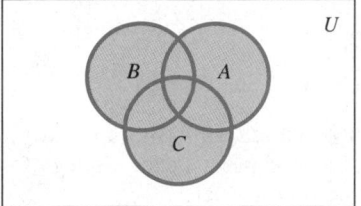

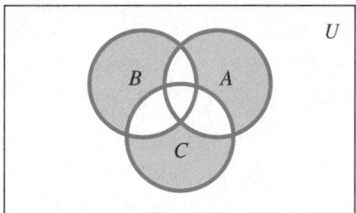

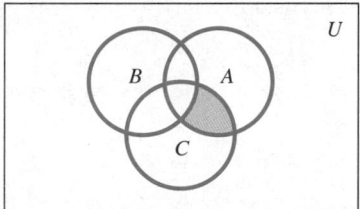

(a) The set of cars with at least one option.

(b) The set of cars with exactly one option.

(c) The set of cars with automatic transmission and side air bags but no air conditioning.

SELF-CHECK EXERCISES 6.1

1. Let $U = \{1, 2, 3, 4, 5, 6, 7\}$, $A = \{1, 2, 3\}$, $B = \{3, 4, 5, 6\}$, and $C = \{2, 3, 4\}$. Find the following sets:

a. A^c **b.** $A \cup B$ **c.** $B \cap C$
d. $(A \cup B) \cap C$ **e.** $(A \cap B) \cup C$ **f.** $A^c \cap (B \cup C)^c$

2. Let U denote the set of all members of the House of Representatives. Let

$$D = \{x \in U \mid x \text{ is a Democrat}\}$$
$$R = \{x \in U \mid x \text{ is a Republican}\}$$
$$F = \{x \in U \mid x \text{ is a female}\}$$
$$L = \{x \in U \mid x \text{ is a lawyer by training}\}$$

Describe each of the following sets in words.

a. $D \cap F$ **b.** $F^c \cap R$ **c.** $D \cap F \cap L^c$

Solutions to Self-Check Exercises 6.1 can be found on page 334.

6.1 EXERCISES

In exercises 1–4, write the given set in set-builder notation.

1. The set of gold medalists in the 1996 Summer Olympic Games

2. The set of football teams in the NFL

3. $\{3, 4, 5, 6, 7\}$

4. $\{1, 3, 5, 7, 9, 11, \ldots, 39\}$

In exercises 5–8, list the elements of the given set in roster notation.

5. $\{x \mid x$ is a digit in the number 352,646$\}$

6. $\{x \mid x$ is a letter in the word *HIPPOPOTAMUS*$\}$

7. $\{x \mid 2 - x = 4; x,$ an integer$\}$

8. $\{x \mid 2 - x = 4; x,$ a fraction$\}$

In exercises 9–14, state whether the given statements are true or false.

9. **a.** $\{a, b, c\} = \{c, a, b\}$ **b.** $A \in A$

10. **a.** $\varnothing \in A$ **b.** $A \subset A$

11. **a.** $0 \in \varnothing$ **b.** $0 = \varnothing$

12. **a.** $\{\varnothing\} = \varnothing$ **b.** $\{a, b\} \in \{a, b, c\}$

13. $\{$Chevrolet, Pontiac, Buick$\} \subset \{$General Motors$\}$

14. $\{x \mid x$ is a silver medalist in the 1996 Summer Olympic Games$\} = \varnothing$

In exercises 15 and 16, let A = {1, 2, 3, 4, 5}. Determine whether the given statements are true or false.

15. **a.** $2 \in A$ **b.** $A \subseteq \{2, 4, 6\}$

16. **a.** $0 \in A$ **b.** $\{1, 3, 5\} \in A$

17. Let $A = \{1, 2, 3\}$. Which of the following sets are equal to A?
    a. $\{2, 1, 3\}$ **b.** $\{3, 2, 1\}$ **c.** $\{0, 1, 2, 3\}$

18. Let $A = \{a, e, l, t, r\}$. Which of the following sets are equal to A?
    a. $\{x \mid x$ is a letter of the word *later*$\}$
    b. $\{x \mid x$ is a letter of the word *latter*$\}$
    c. $\{x \mid x$ is a letter of the word *relate*$\}$

19. List all subsets of the following sets:
    a. $\{1, 2\}$ **b.** $\{1, 2, 3\}$ **c.** $\{1, 2, 3, 4\}$

20. List all subsets of the set $A = \{$IBM, U.S. Steel, Union Carbide, Boeing$\}$. Which of these are proper subsets of A?

In exercises 21–24, find the smallest possible set (that is, the set with the least number of elements) that contains the given sets as subsets.

21. $\{1, 2\}, \{1, 3, 4\}, \{4, 6, 8, 10\}$

22. $\{1, 2, 4\}, \{a, b\}$

23. $\{$Jill, John, Jack$\}, \{$Susan, Sharon$\}$

24. $\{$GM, Ford, Chrysler$\}, \{$Daimler-Benz, Volkswagen$\}, \{$Toyota, Datsun$\}$

25. Use Venn diagrams to represent the following relationships:
    a. $A \subset B$ and $B \subset C$
    b. $A \subset U$ and $B \subset U$, where A and B have no elements in common
    c. The sets A, B, and C are equal.

26. Let U denote the set of all students who applied for admission to the freshman class at Faber College for the upcoming academic year and let

$$A = \{x \in U \mid x \text{ is a successful applicant}\}$$
$$B = \{x \in U \mid x \text{ is a female student who enrolled in the freshman class}\}$$
$$C = \{x \in U \mid x \text{ is a male student who enrolled in the freshman class}\}$$

a. Use Venn diagrams to represent the sets U, A, B, and C.
b. Determine whether the following statements are true or false.
i. $A \subseteq B$ **ii.** $B \subset A$ **iii.** $C \subset B$

In exercises 27 and 28, shade the portion of the accompanying figure that represents each of the given sets.

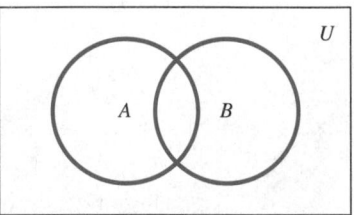

27. **a.** $A \cap B^c$ **b.** $A^c \cap B$

28. **a.** $A^c \cap B^c$ **b.** $(A \cup B)^c$

In exercises 29–32, shade the portion of the accompanying figure that represents each of the given sets.

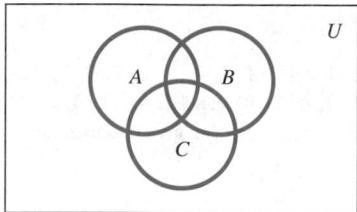

29. **a.** $A \cup B \cup C$ **b.** $A \cap B \cap C$

30. **a.** $A \cap B \cap C^c$ **b.** $A^c \cap B \cap C$

31. **a.** $A^c \cap B^c \cap C^c$ **b.** $(A \cup B)^c \cap C$

32. **a.** $A \cup (B \cap C)^c$ **b.** $(A \cup B \cup C)^c$

In exercises 33–36, let $U = \{1, 2, 3, 4, 5, 6, 7, 8, 9, 10\}$, $A = \{1, 3, 5, 7, 9\}$, $B = \{2, 4, 6, 8, 10\}$, and $C = \{1, 2, 4, 5, 8, 9\}$. Find each of the given sets.

33. **a.** A^c **b.** $B \cup C$ **c.** $C \cup C^c$

34. **a.** $C \cap C^c$ **b.** $(A \cap C)^c$ **c.** $A \cup (B \cap C)$

35. **a.** $(A \cap B) \cup C$ **b.** $(A \cup B \cup C)^c$
    c. $(A \cap B \cap C)^c$

36. **a.** $A^c \cap (B \cap C^c)$ **b.** $(A \cup B^c) \cup (B \cap C^c)$
    c. $(A \cup B)^c \cap C^c$

In exercises 37 and 38, determine whether the given pairs of sets are disjoint.

37. **a.** $\{1, 2, 3, 4\}, \{4, 5, 6, 7\}$
    b. $\{a, c, e, g\}, \{b, d, f\}$

38. **a.** $\varnothing, \{1, 3, 5\}$ **b.** $\{0, 1, 3, 4\}, \{0, 2, 5, 7\}$

In exercises 39–42, let U denote the set of all employees at the Universal Life Insurance Company. Let

$$T = \{x \in U \mid x \text{ drinks tea}\}$$
$$C = \{x \in U \mid x \text{ drinks coffee}\}$$

Describe each of the given sets in words.

39. **a.** T^c **b.** C^c

40. **a.** $T \cup C$ **b.** $T \cap C$

41. **a.** $T \cap C^c$ **b.** $T^c \cap C$

42. **a.** $T^c \cap C^c$ **b.** $(T \cup C)^c$

In exercises 43–46, let U denote the set of all employees in a hospital. Let

$$N = \{x \in U \mid x \text{ is a nurse}\}$$
$$D = \{x \in U \mid x \text{ is a doctor}\}$$
$$A = \{x \in U \mid x \text{ is an administrator}\}$$
$$M = \{x \in U \mid x \text{ is a male}\}$$
$$F = \{x \in U \mid x \text{ is a female}\}$$

Describe each of the given sets in words.

43. **a.** D^c **b.** N^c

44. **a.** $N \cup D$ **b.** $N \cap M$

45. **a.** $D \cap M^c$ **b.** $D \cap A$

46. **a.** $N \cap F$ **b.** $(D \cup N)^c$

In exercises 47 and 48, let U denote the set of all senators in Congress. Let

$$D = \{x \in U \mid x \text{ is a Democrat}\}$$
$$R = \{x \in U \mid x \text{ is a Republican}\}$$
$$F = \{x \in U \mid x \text{ is a female}\}$$
$$L = \{x \in U \mid x \text{ is a lawyer}\}$$

Write the set that represents each of the given statements.

47. **a.** The set of all Democrats who are female
    b. The set of all Republicans who are male and are not lawyers

48. **a.** The set of all Democrats who are female or are lawyers
    b. The set of all senators who are not Democrats or are lawyers

In exercises 49 and 50, let U denote the set of all students in the business college of a certain university. Let

$$A = \{x \in U \mid x \text{ had taken a course in accounting}\}$$
$$B = \{x \in U \mid x \text{ had taken a course in economics}\}$$
$$C = \{x \in U \mid x \text{ had taken a course in marketing}\}$$

Write the set that represents each of the given statements.

49. **a.** The set of students who have not had a course in economics
    b. The set of students who have had courses in accounting and economics
    c. The set of students who have had courses in accounting and economics but not marketing

50. **a.** The set of students who have had courses in economics but not courses in accounting or marketing

b. The set of students who have had at least one of the three courses

c. The set of students who have had all three courses

In exercises 51 and 52, refer to the following diagram where U is the set of all tourists surveyed over a 1-week period in London and

$A = \{x \in U \mid x \text{ has taken the underground (subway)}\}$

$B = \{x \in U \mid x \text{ has taken a cab}\}$

$C = \{x \in U \mid x \text{ has taken a bus}\}$

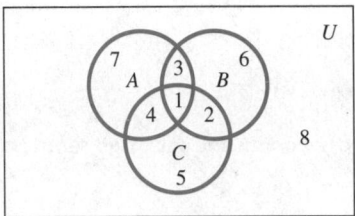

Express the indicated regions in set notation and in words.

51. a. Region 1
   b. Regions 1 and 4 together
   c. Regions 4, 5, 7, and 8 together

52. a. Region 3
   b. Regions 4 and 6 together
   c. Regions 5, 6, and 7 together

In exercises 53–58, use Venn diagrams to illustrate each of the given statements.

53. $A \subset A \cup B; B \subset A \cup B$

54. $A \cap B \subset A; A \cap B \subset B$

55. $A \cup (B \cup C) = (A \cup B) \cup C$

56. $A \cap (B \cap C) = (A \cap B) \cap C$

57. $A \cap (B \cup C) = (A \cap B) \cup (A \cap C)$

58. $(A \cup B)^c = A^c \cap B^c$

In exercises 59 and 60, let $U = \{1, 2, 3, 4, 5, 6, 7, 8, 9, 10\}$, $A = \{1, 3, 5, 7, 9\}$, $B = \{1, 2, 4, 7, 8\}$, and $C = \{2, 4, 6, 8\}$. Verify by direct computation each of the given equations.

59. a. $A \cup (B \cup C) = (A \cup B) \cup C$
   b. $A \cap (B \cap C) = (A \cap B) \cap C$

60. a. $A \cap (B \cup C) = (A \cap B) \cup (A \cap C)$
   b. $(A \cup B)^c = A^c \cap B^c$

In exercises 61–64, refer to the accompanying figure and find the points that belong to each of the given sets.

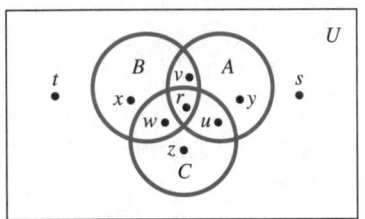

61. a. $A \cup B$ **b.** $A \cap B$

62. a. $A \cap (B \cup C)$ **b.** $(B \cap C)^c$

63. a. $(B \cup C)^c$ **b.** A^c

64. a. $(A \cap B) \cap C^c$ **b.** $(A \cup B \cup C)^c$

65. Suppose that $A \subset B$ and $B \subset C$, where A and B are any two sets. What conclusion can be drawn regarding the sets A and C?

66. Verify the assertion that two sets A and B are equal if and only if (1) $A \subseteq B$ and (2) $B \subseteq A$.

SOLUTIONS TO SELF-CHECK EXERCISES 6.1

1. a. A^c is the set of all elements in U but not in A. Therefore,

$$A^c = \{4, 5, 6, 7\}$$

b. $A \cup B$ consists of all elements in A and/or B. So,

$$A \cup B = \{1, 2, 3, 4, 5, 6\}$$

c. $B \cap C$ is the set of all elements in both B and C. Therefore,

$$B \cap C = \{3, 4\}$$

d. Using the result from (b), we find

$$(A \cup B) \cap C = \{1, 2, 3, 4, 5, 6\} \cap \{2, 3, 4\}$$
$$= \{2, 3, 4\}$$

e. First we compute

$$A \cap B = \{3\}$$

Next, since $(A \cap B) \cup C$ is the set of all elements in $(A \cap B)$ and/or C, we conclude that

$$(A \cap B) \cup C = \{3\} \cup \{2, 3, 4\}$$
$$= \{2, 3, 4\}$$

f. From part (a), we have $A^c = \{4, 5, 6, 7\}$. Next, we compute

$$B \cup C = \{3, 4, 5, 6\} \cup \{2, 3, 4\}$$
$$= \{2, 3, 4, 5, 6\}$$

from which we deduce that

$$(B \cup C)^c = \{1, 7\} \qquad \text{(The set of elements in } U \text{ but not in } B \cup C)$$

Finally, using these results, we obtain

$$A^c \cap (B \cup C)^c = \{4, 5, 6, 7\} \cap \{1, 7\} = \{7\}$$

2. a. $D \cap F$ denotes the set of all elements in both D and F. Since an element in D is a Democrat and an element in F is a female representative, we see that $D \cap F$ is the set of all female Democrats in the House of Representatives.
b. Since F^c is the set of male representatives and R is the set of Republicans, we see that $F^c \cap R$ is the set of male Republicans in the House of Representatives.
c. L^c is the set of representatives who are not lawyers by training. Therefore, $D \cap F \cap L^c$ is the set of female Democratic representatives who are not lawyers by training.

6.2 THE NUMBER OF ELEMENTS IN A FINITE SET

Counting the Elements in a Set

The solution to some problems in mathematics calls for finding the number of elements in a set. Such problems are called **counting problems** and constitute a field of study known as **combinatorics.** Our study of combinatorics is restricted to the results that will be required for our work in probability later on.

The number of elements in a finite set is determined by simply counting the elements in the set. If A is a set, then $n(A)$ denotes the number of elements in A. For example, if

$$A = \{1, 2, 3, \ldots, 20\}, \quad B = \{a, b\}, \quad \text{and} \quad C = \{8\}$$

then $n(A) = 20$, $n(B) = 2$, and $n(C) = 1$.

The empty set has no elements in it, so $n(\varnothing) = 0$. Another result that is easily seen to be true is the following: If A and B are disjoint sets, then

$$n(A \cup B) = n(A) + n(B) \tag{3}$$

EXAMPLE 1 If $A = \{a, c, d\}$ and $B = \{b, e, f, g\}$, then $n(A) = 3$ and $n(B) = 4$, so $n(A) + n(B) = 7$. However, $A \cup B = \{a, b, c, d, e, f, g\}$ and $n(A \cup B) = 7$. Thus, equation (3) holds true in this case. Note that $A \cap B = \varnothing$. ○ ○ ○

In the general case, A and B need not be disjoint, which leads us to the formula

$$n(A \cup B) = n(A) + n(B) - n(A \cap B) \tag{4}$$

Figure 6.8
$n(A \cup B) = x + y + z.$

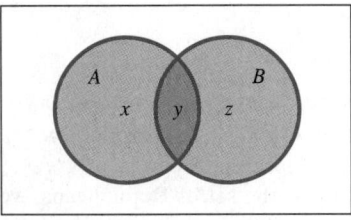

To see this, we observe that the set $A \cup B$ may be viewed as the union of three mutually disjoint sets with x, y, and z elements, respectively (Figure 6.8). This figure shows that

$$n(A \cup B) = x + y + z$$

Also, $\qquad n(A) = x + y$

and $\qquad n(B) = y + z$

so $\qquad n(A) + n(B) = (x + y) + (y + z)$
$$= (x + y + z) + y$$
$$= n(A \cup B) + n(A \cap B) \qquad [n(A \cap B) = y]$$

Thus, solving for $n(A \cup B)$, we obtain

$$n(A \cup B) = n(A) + n(B) - n(A \cap B)$$

the desired result.

EXAMPLE 2 Let $A = \{a, b, c, d, e\}$ and $B = \{b, d, f, h\}$. Verify equation (4) directly.

Solution $\qquad A \cup B = \{a, b, c, d, e, f, h\} \quad$ so $\quad n(A \cup B) = 7$
$\qquad\qquad A \cap B = \{b, d\} \quad$ so $\quad n(A \cap B) = 2$

Furthermore, $\qquad n(A) = 5 \quad$ and $\quad n(B) = 4$

so $\qquad n(A) + n(B) - n(A \cap B) = 5 + 4 - 2 = 7 = n(A \cup B)$ ○ ○ ○

Applications

EXAMPLE 3 In a survey of 100 coffee drinkers, it was found that 70 take sugar, 60 take cream, and 50 take both sugar and cream with their coffee. How many coffee drinkers take sugar or cream with their coffee?

Solution   Let U denote the set of 100 coffee drinkers surveyed, and let

$$A = \{x \in U \mid x \text{ takes sugar}\}$$
$$B = \{x \in U \mid x \text{ takes cream}\}$$

Then $n(A) = 70$, $n(B) = 60$, and $n(A \cap B) = 50$. The set of coffee drinkers who take sugar or cream with their coffee is given by $A \cup B$. Using (4), we find

$$n(A \cup B) = n(A) + n(B) - n(A \cap B)$$
$$= 70 + 60 - 50 = 80$$

so 80 out of the 100 coffee drinkers surveyed take cream or sugar with their coffee.   ◦ ◦ ◦

An equation similar to (4) may be derived for the case that involves any finite number of finite sets. For example, a relationship involving the number of elements in the sets A, B, and C is given by

$$n(A \cup B \cup C) = n(A) + n(B) + n(C) - n(A \cap B)$$
$$- n(A \cap C) - n(B \cap C) + n(A \cap B \cap C) \quad \textbf{(5)}$$

Prove formula (5) using an argument similar to that used to prove formula (4). Another proof is outlined in exercise 32 on page 341.

As useful as equations such as (5) are, in practice it is often easier to attack a problem directly with the aid of Venn diagrams, as shown by the following example.

EXAMPLE 4   A leading cosmetics manufacturer advertises its products in three magazines: *Cosmopolitan, McCalls,* and the *Ladies Home Journal.* A survey of 500 customers by the manufacturer reveals the following information:

180 learned of its products from *Cosmopolitan*

200 learned of its products from *McCalls*

192 learned of its products from the *Ladies Home Journal*

84 learned of its products from *Cosmopolitan* and *McCalls*

52 learned of its products from *Cosmopolitan* and the *Ladies Home Journal*

64 learned of its products from *McCalls* and the *Ladies Home Journal*

38 learned of its products from all three magazines

How many of the customers saw the manufacturer's advertisement in

a. at least one magazine?

b. exactly one magazine?

Solution   Let U denote the set of all customers surveyed, and let

$$C = \{x \in U \mid x \text{ learned of the products from } Cosmopolitan\}$$
$$M = \{x \in U \mid x \text{ learned of the products from } McCalls\}$$
$$L = \{x \in U \mid x \text{ learned of the products from the } Ladies \ Home \ Journal\}$$

Figure 6.9

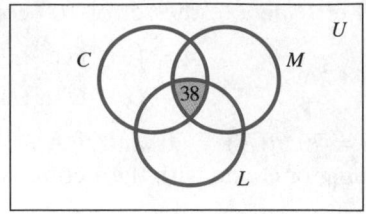

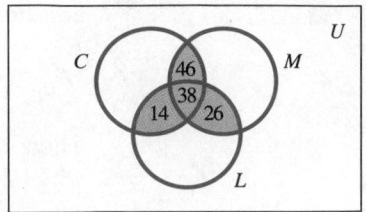

(a) All three magazines.   **(b)** Two or more magazines.

The result that 38 customers learned of the products from all three magazines translates into $n(C \cap M \cap L) = 38$ (Figure 6.9a). Next, the result that 64 learned of the products from *McCalls* and the *Ladies Home Journal* translates into $n(M \cap L) = 64$. This leaves

$$64 - 38 = 26$$

who learned of the products from *only McCalls* and the *Ladies Home Journal* (Figure 6.9b). Similarly, $n(C \cap L) = 52$, so

$$52 - 38 = 14$$

learned of the products from *only Cosmopolitan* and the *Ladies Home Journal,* and $n(C \cap M) = 84$, so

$$84 - 38 = 46$$

learned of the products from *only Cosmopolitan* and *McCalls.* These numbers appear in the appropriate regions in Figure 6.9b.

Continuing, we have $n(L) = 192$, so the number who learned of the products from the *Ladies Home Journal* only is given by

$$192 - 14 - 38 - 26 = 114 \qquad \text{(See Figure 6.10.)}$$

Similarly, $n(M) = 200$, so

$$200 - 46 - 38 - 26 = 90$$

learned of the products from *only McCalls,* and $n(C) = 180$, so

$$180 - 14 - 38 - 46 = 82$$

Figure 6.10
At least one magazine.

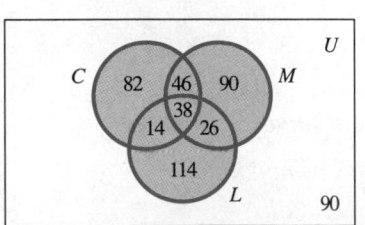

learned of the products from *only Cosmopolitan.* Finally,

$$500 - (90 + 26 + 114 + 14 + 82 + 46 + 38) = 90$$

learned of the products from other sources.

We are now in a position to answer questions (a) and (b).

a. Referring to Figure 6.10, we see that the number of customers who learned of the products from at least one magazine is given by

$$n(C \cup M \cup L) = 90 + 26 + 114 + 14 + 82 + 46 + 38 = 410$$

b. The number of customers who learned of the products from exactly one magazine (Figure 6.11) is given by

$$n(L \cap C^c \cap M^c) + n(M \cap C^c \cap L^c) + n(C \cap L^c \cap M^c)$$
$$= 90 + 114 + 82 = 286$$

Figure 6.11
Exactly one magazine.

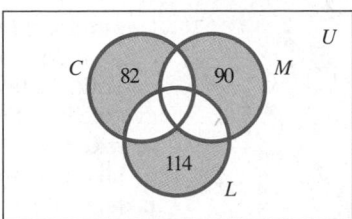

○ ○ ○

SELF–CHECK EXERCISES 6.2

1. Let A and B be subsets of a universal set U and suppose $n(U) = 100$, $n(A) = 60$, $n(B) = 40$, and $n(A \cap B) = 20$. Compute
   a. $n(A \cup B)$ **b.** $n(A \cap B^c)$ **c.** $n(A^c \cap B)$

2. In a recent survey of 1000 readers of *Video Magazine,* it was found that 900 own at least one videocassette recorder (VCR) in the VHS format, 240 own at least one VCR in the Beta format, and 160 own VCRs in both formats. How many of the readers surveyed own VCRs in the VHS format only? How many of the readers surveyed do not own a VCR in either format?

Solutions to Self-Check Exercises 6.2 can be found on page 342.

6.2 EXERCISES

In exercises 1 and 2, verify the equation

$$n(A \cup B) = n(A) + n(B)$$

for the given disjoint sets.

1. $A = \{a, e, i, o, u\}$ and $B = \{g, h, k, l, m\}$

2. $A = \{x \mid x \text{ is a whole number between 0 and 4}\}$
   $B = \{x \mid x \text{ is a negative integer greater than } -4\}$

3. Let $A = \{2, 4, 6, 8\}$ and $B = \{6, 7, 8, 9, 10\}$. Compute:
   a. $n(A)$ **b.** $n(B)$
   c. $n(A \cup B)$ **d.** $n(A \cap B)$

4. Verify directly that $n(A \cup B) = n(A) + n(B) - n(A \cap B)$ for the sets in exercise 3.

5. Let $A = \{a, e, i, o, u\}$ and $B = \{b, d, e, o, u\}$. Verify by direct computation that $n(A \cup B) = n(A) + n(B) - n(A \cap B)$.

6. If $n(A) = 15$, $n(A \cap B) = 5$, and $n(A \cup B) = 30$, what is $n(B)$?

7. If $n(A) = 10$, $n(A \cup B) = 15$, and $n(B) = 8$, what is $n(A \cap B)$?

In exercises 8 and 9, let A and B be subsets of a universal set U and suppose $n(U) = 200$, $n(A) = 100$, $n(B) = 80$, and $n(A \cap B) = 40$. Compute:

8. a. $n(A \cup B)$ **b.** $n(A^c)$
   c. $n(A \cap B^c)$

9. a. $n(A^c \cap B)$ **b.** $n(B^c)$
   c. $n(A^c \cap B^c)$

10. Find $n(A \cup B)$ given that $n(A) = 6$, $n(B) = 10$, and $n(A \cap B) = 3$.

11. If $n(B) = 6$, $n(A \cup B) = 14$, and $n(A \cap B) = 3$, find $n(A)$.

12. If $n(A) = 4$, $n(B) = 5$, and $n(A \cup B) = 9$, find $n(A \cap B)$.

13. If $n(A) = 16$, $n(B) = 16$, $n(C) = 14$, $n(A \cap B) = 6$, $n(A \cap C) = 5$, $n(B \cap C) = 6$, and $n(A \cup B \cup C) = 31$, find $n(A \cap B \cap C)$.

14. If $n(A) = 12$, $n(B) = 12$, $n(A \cap B) = 5$, $n(A \cap C) = 5$, $n(B \cap C) = 4$, $n(A \cap B \cap C) = 2$, and $n(A \cup B \cup C) = 25$, find $n(C)$.

15. A survey of 1000 subscribers to the *Los Angeles Times* revealed that 900 people subscribe to the daily morning edition and 500 subscribe to both the daily morning and the Sunday editions. How many subscribe to the Sunday edition? How many subscribe to the Sunday edition only?

16. Of 100 clock radios sold recently in a department store, 70 had FM circuitry and 90 had AM circuitry. How many radios had both FM and AM circuitry? How many could receive FM transmission only? How many could receive AM transmission only?

17. On a certain day, the Wilton County Jail had 190 prisoners. Of these, 130 were accused of felonies and 121 were accused of misdemeanors. How many prisoners were accused of both a felony and a misdemeanor?

18. Consumer Surveys   In a recent survey of 200 members of a local sports club, 100 members indicated that they plan to attend the next Summer Olympic Games, 60 indicated that they plan to attend the next Winter Olympic Games, and 40 indicated that they plan to attend both games. How many members of the club plan to attend
   a. at least one of the two games?
   b. exactly one of the games?
   c. the Summer Olympic Games only?
   d. none of the games?

19. Consumer Surveys   In a survey of 120 consumers conducted in a shopping mall, 80 consumers indicated that they buy brand A of a certain product, 68 buy brand

B, and 42 buy both brands. Determine the number of consumers participating in the survey who buy
   a. at least one of these brands.
   b. exactly one of these brands.
   c. only brand A.
   d. none of these brands.

20. Commuter Trends   Of 50 employees of a store located in downtown Boston, 18 people take the subway to work, 12 take the bus, and 7 take both the subway and the bus. Determine the number of employees who
   a. take the subway or the bus to work.
   b. take only the bus to work.
   c. take either the bus or the subway to work.
   d. get to work by some other means.

21. Investing   In a poll conducted among 200 active investors, it was found that 120 use discount brokers, 126 use full-service brokers, and 64 use both discount and full-service brokers. Determine the number of investors who
   a. use at least one kind of broker.
   b. use exactly one kind of broker.
   c. use only discount brokers.
   d. don't use a broker.

In exercises 22–25, let A, B, and C be subsets of a universal set U and suppose $n(U) = 100$, $n(A) = 28$, $n(B) = 30$, $n(C) = 34$, $n(A \cap B) = 8$, $n(A \cap C) = 10$, $n(B \cap C) = 15$, and $n(A \cap B \cap C) = 5$. Compute:

22. a. $n(A \cup B \cup C)$ **b.** $n(A^c \cap B \cap C)$

23. a. $n[A \cap (B \cup C)]$ **b.** $n[A \cap (B \cup C)^c]$

24. a. $n(A^c \cap B^c \cap C^c)$ **b.** $n[A^c \cap (B \cup C)]$

25. a. $n[A \cup (B \cap C)]$ **b.** $n(A^c \cap B^c \cap C^c)^c$

26. Student Dropout Rate   Data released by the Department of Education regarding the rate (percentage) of ninth-grade students that don't graduate showed that out of 50 states,

12 states had an increase in the dropout rate during the past two years.

15 states had a dropout rate of at least 30% during the past two years.

21 states had an increase in the dropout rate and/or a dropout rate of at least 30% during the past two years.

How many states
   a. had both a dropout rate of at least 30% and an increase in the dropout rate over the 2-year period?

b. had a dropout rate that was less than 30% but that had increased over the 2-year period?

27. Economic Surveys A survey of the opinions of 10 leading economists in a certain country showed that because oil prices were expected to drop in that country over the next 12 months,

7 had lowered their estimate of the consumer inflation rate.

8 had raised their estimate of the gross national product growth rate.

2 had lowered their estimate of the consumer inflation rate but had not raised their estimate of the gross national product growth rate.

How many economists had both lowered their estimate of the consumer inflation rate and raised their estimate of the gross national product growth rate for that period?

28. SAT Scores Results of a Department of Education survey of SAT test scores in 22 states showed that

10 states had an average composite test score of at least 900.

15 states had an increase of at least 10 points in the average composite score.

8 states had both an average composite SAT score of at least 900 and an increase in the average composite score of at least 10 points.

How many of the 22 states
a. had composite scores less than 900 and showed an increase of at least 10 points over the 3-year period?
b. had composite scores of at least 900 and did not show an increase of at least 10 points over the 3-year period?

29. Student Reading Habits A survey of 100 college students who frequent the reading lounge of a university revealed the following results:

40 read *Time* magazine.

30 read *Newsweek.*

25 read *U.S. News and World Report.*

15 read *Time* magazine and *Newsweek.*

12 read *Time* magazine and *U.S. News and World Report.*

10 read *Newsweek* and *U.S. News and World Report.*

4 read all three magazines.

How many of the students surveyed read
a. at least one magazine?
b. exactly one magazine?
c. exactly two magazines?
d. none of these magazines?

30. Student Surveys To help plan the number of meals to be prepared in a college cafeteria, a survey was conducted and the following data were obtained:

130 students ate breakfast.

180 students ate lunch.

275 students ate dinner.

68 students ate breakfast and lunch.

112 students ate breakfast and dinner.

90 students ate lunch and dinner.

58 students ate all three meals.

How many of the students
a. ate at least one meal in the cafeteria?
b. ate exactly one meal in the cafeteria?
c. ate only dinner in the cafeteria?
d. ate exactly two meals in the cafeteria?

31. Consumer Surveys The 120 consumers of exercise 19 were also asked about their buying preferences concerning another product that is sold in the market under three labels. The results were:

12 buy only those sold under label A.

25 buy only those sold under label B.

26 buy only those sold under label C.

15 buy only those sold under labels A and B.

10 buy only those sold under labels A and C.

12 buy only those sold under labels B and C.

8 buy the product sold under all three labels.

How many of the consumers surveyed buy the product sold under
a. at least one of the three labels?
b. labels A and B but not C?
c. label A?
d. none of these labels?

32. Derive equation (5).

[*Hint:* Equation (4) may be written as $n(D \cup E) = n(D) + n(E) - n(D \cap E)$. Now, put $D = A \cup B$ and $E = C$. Use (4) again if necessary.]

SOLUTIONS TO SELF–CHECK EXERCISES 6.2

1. Refer to the following Venn diagram.

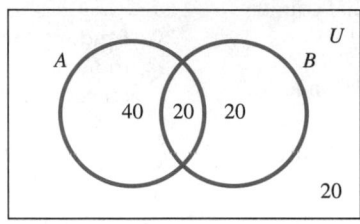

Using this result, we see that
a. $n(A \cup B) = 40 + 20 + 20 = 80$
b. $n(A \cap B^c) = 40$
c. $n(A^c \cap B) = 20$

2. Let U denote the set of all readers surveyed, and let

$$A = \{x \in U \,|\, x \text{ owns at least one VCR in the VHS format}\}$$
$$B = \{x \in U \,|\, x \text{ owns at least one VCR in the Beta format}\}$$

Then the result that 160 of the readers own VCRs in both formats gives $n(A \cap B) = 160$. Also, $n(A) = 900$ and $n(B) = 240$. Using this information, we obtain the following Venn diagram:

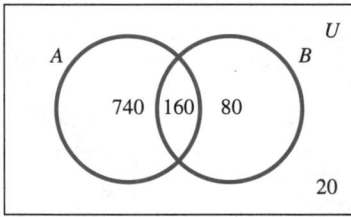

From the Venn diagram we see that the number of readers who own VCRs in the VHS format only is given by

$$n(A \cap B^c) = 740$$

The number of readers who do not own a VCR in either format is given by

$$n(A^c \cap B^c) = 20$$

6.3 THE MULTIPLICATION PRINCIPLE

The Fundamental Principle of Counting

The solution of certain problems requires more sophisticated counting techniques than those developed in the previous section. We look at some such techniques in this and the following section. We begin by stating a fundamental principle of counting called the **multiplication principle.**

THE MULTIPLICATION PRINCIPLE	Suppose there are m ways of performing a task T_1 and n ways of performing a task T_2. Then there are mn ways of performing the task T_1 followed by the task T_2.

EXAMPLE 1 Three trunk roads connect town A and town B, and two trunk roads connect town B and town C.

a. Use the multiplication principle to find the number of ways a journey from town A to town C via town B may be completed.

b. Verify (a) directly by exhibiting all possible routes.

Solution
a. Since there are three ways of performing the first task (going from town A to town B) followed by two ways of performing the second task (going from town B to town C), the multiplication principle says that there are $3 \cdot 2$, or 6, ways to complete a journey from town A to town C via town B.

b. Label the trunk roads connecting town A and town B with the Roman numerals I, II, and III, and the trunk roads connecting town B and town C with the lowercase letters a and b. A schematic of this is shown in Figure

Figure 6.12
Roads from A to C.

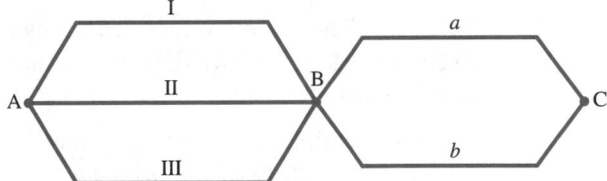

6.12. Then the routes from town A to town C via town B may be exhibited with the aid of a **tree diagram** (Figure 6.13). If we follow all of the branches

Figure 6.13
Tree diagram displaying the possible routes from town A to town C.

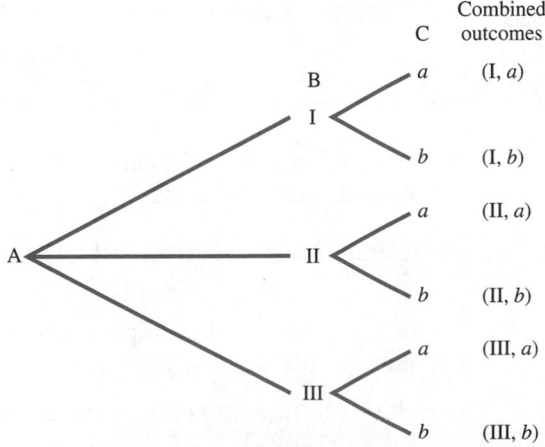

from the initial point A to the right-hand edge of the tree, we obtain the six routes represented by the six ordered pairs

$$(\mathrm{I}, a), \quad (\mathrm{I}, b), \quad (\mathrm{II}, a), \quad (\mathrm{II}, b), \quad (\mathrm{III}, a), \quad \text{and} \quad (\mathrm{III}, b)$$

where (I, a) means that the journey from town A to town B is made on trunk road I with the rest of the journey from town B to town C to be completed on trunk road a, and so forth.   ○ ○ ○

One way of gauging the performance of an airline is to track the arrival times of its flights. Suppose we denote by E, O, and L, a flight that arrives early, on time, or late, respectively.

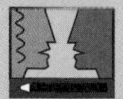

a. Use a tree diagram to exhibit the possible outcomes when you track two successive flights of the airline. How many outcomes are there?
b. How many outcomes are there if you track three successive flights? Justify your answer.

EXAMPLE 2   Diners at Angelo's Spaghetti Bar may select their entree from 6 varieties of pasta and 28 choices of sauce. How many such combinations are there that consist of one variety of pasta and one kind of sauce?

Solution   There are 6 ways of choosing a pasta followed by 28 ways of choosing a sauce, so by the multiplication principle, there are $6 \cdot 28$, or 168, combinations of this pasta dish.   ○ ○ ○

The multiplication principle may be easily extended, which leads to the **generalized multiplication principle:**

GENERALIZED MULTIPLICATION PRINCIPLE

Suppose a task T_1 can be performed in N_1 ways, a task T_2 can be performed in N_2 ways, . . . , and, finally, a task T_n can be performed in N_n ways. Then the number of ways of performing the tasks $T_1, T_2, \ldots, T_n$ in succession is given by the product

$$N_1 N_2 \cdots N_n$$

We now illustrate the application of the generalized multiplication principle to several diverse situations.

EXAMPLE 3   A coin is tossed three times and the sequence of heads and tails is recorded.

a. Use the generalized multiplication principle to determine the number of outcomes of this activity.

b. Exhibit all the sequences by means of a tree diagram.

Solution

a. The coin may land in two ways. Therefore, in three tosses the number of outcomes (sequences) is given by 2 · 2 · 2, or 8.

b. Let H and T denote the outcomes "a head" and "a tail," respectively. Then the required sequences may be obtained as shown in Figure 6.14, giving the sequence as HHH, HHT, HTH, HTT, THH, THT, TTH, and TTT.

Figure 6.14

Tree diagram displaying possible outcomes of three consecutive coin tosses.

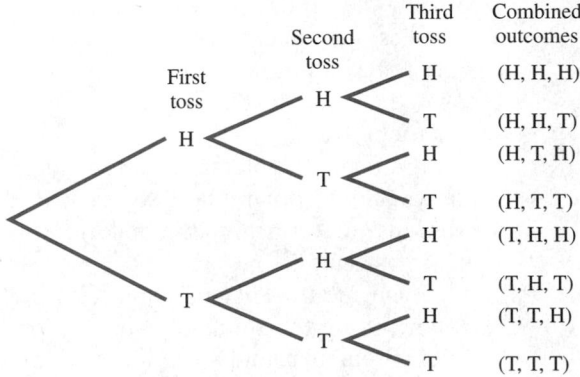

Applications

EXAMPLE 4   A combination lock is unlocked by dialing a sequence of numbers, first to the left, then to the right, and to the left again. If there are ten digits on the dial, determine the number of possible combinations.

Solution   There are ten choices for the first number, followed by ten for the second and ten for the third, so by the generalized multiplication principle, there are 10 · 10 · 10, or 1000, possible combinations.   ◦ ◦ ◦

EXAMPLE 5   An investor has decided to purchase shares in the stock of three companies: one engaged in aerospace activities, one involved in energy development, and one involved in electronics. After some research, the account executive of a brokerage firm has recommended that the investor consider stock from five aerospace companies, three energy development companies, and four electronics companies. In how many ways may the investor select the group of three companies from the executive's list?

Solution   The investor has five choices for selecting an aerospace company, three choices for selecting an energy development company, and four choices for selecting an electronics company. Therefore, by the generalized multiplication principle, there are 5 · 3 · 4, or 60, ways in which she can select a group of three companies, one from each industry group.   ◦ ◦ ◦

As the concierge for a small, 10-story office building in Boston, Higgins supervises daily operations. From his post in the lobby, he resolves tenant problems, oversees building maintenance and security, and coordinates the budget. Part building manager, lobby security guard, and friend, Higgins brings an upbeat quality to his job, joking with tenants as they enter and leave.

John L. Higgins

Title: Concierge
Institution: RVM&G Property Managers

Higgins's responsibility is to provide building services that "assist the tenants in their day-to-day work." Although not the official building manager, he oversees outside contractors, requesting bids and outlining the scope of the work to be done. He ensures that all such work is within budget and performed correctly. Since the building doesn't have a security guard, Higgins also monitors lobby traffic.

If a major problem arises, tenants feel comfortable approaching Higgins, knowing he'll do his best to resolve it. He stresses that "positive tenant relations" are critical to the building's success. In Higgins's book, "the tenant is always right."

Each month Higgins chronicles building operations in a report to the owner, listing the status of capital improvements, ongoing maintenance, significant tenant complaints, and any variances in the building's accounts. To determine a variance, he has to balance each account monthly. If an account's funds are too high or too low, Higgins has to explain the fluctuation. Math at this level requires simple arithmetic skills, but the calculations become more complicated when it comes to computing a tenant's base rent and future increases.

First, the building's operating costs—utility bills, staff salaries, and vendor fees (cleaning, exterminating, window washing, and so on)—have to be calculated. For example, if the costs add up to $420,000 per year, Higgins divides the total by the building's 70,000 square feet to calculate a general operating expense of $6 per square foot. Dividing the annual real estate tax of $280,000 by the square footage yields another $4 per square foot.

If the owner agrees to modify a new tenant's offices, construction costs—which can run into tens of thousands of dollars—have to be factored in. The construction costs are then amortized like a mortgage over the life of a typical 5-year lease. The principal and interest are calculated so that the tenant pays a fixed amount each month until construction costs are repaid.

Miscellaneous costs (broker, legal, and architectural fees) and the owner's expected profit margin round out the tenant's base for the first year. The base is multiplied by the number of square feet a tenant occupies to determine monthly rent. As operating costs and real estate taxes increase in subsequent years, Higgins has to refigure each tenant's monthly rent using the original base as his starting point.

o   o   o

EXAMPLE 6   Tom is planning to leave for New York City from Washington, D.C., on Monday morning and has decided that he will either fly or take the train. There are five flights and two trains departing for New York City from Washington that morning. When he returns on Sunday afternoon, Tom plans to either fly or hitch a ride with a friend. There are two flights departing from New York City to Washington that afternoon. In how many ways can Tom complete this round trip?

Solution   There are seven ways Tom can go from Washington, D.C., to New York City (five by plane and two by train). On the return trip, Tom can travel in three ways (two by plane and one by car). Therefore, by the multiplication principle, Tom can complete the round trip in $7 \cdot 3$, or 21, ways.   ⚬ ⚬ ⚬

SELF-CHECK EXERCISES 6.3

1. Encore Travel, Inc., offers a "Theater Week in London" package originating from New York City. There is a choice of eight flights departing from New York City per week, a choice of five hotel accommodations, and a choice of one complimentary ticket to one of eight shows. How many such travel packages can one choose from?

2. The Cafe Napolean offers a dinner special on Wednesdays consisting of a choice of two entrees (Beef Bourguignon and Chicken Basquaise); one dinner salad; one French roll; a choice of three vegetables; a choice of a carafe of Burgundy, Rosé, or Chablis wine; a choice of coffee or tea; and a choice of six French pastries for dessert. How many combinations of dinner specials are there?

Solutions to Self-Check Exercises 6.3 can be found on page 349.

6.3 EXERCISES

1. **Rental Rates**   Lynbrook West, an apartment complex financed by the State Housing Finance Agency, consists of one-, two-, three-, and four-bedroom units. The rental rate for each type of unit—low, moderate, or market—is determined by the income of the tenant. How many different rates are there?

2. **Commuter Passes**   Five different types of monthly commuter passes are offered by a city's local transit authority for three different groups of passengers: youths, adults, and senior citizens. How many different kinds of passes must be printed each month?

3. **Blackjack**   In the game of Blackjack, a 2-card hand consisting of an ace and either a face card or a ten is called a "blackjack." If a standard 52-card deck is used, determine how many blackjack hands can be dealt.

4. **Coin Tosses**   A coin is tossed four times and the sequence of heads and tails is recorded.
   a. Use the generalized multiplication principle to determine the number of outcomes of this activity.
   b. Exhibit all the sequences by means of a tree diagram.

5. **Wardrobe Selection**   A female executive selecting her wardrobe purchased two blazers, four blouses, and three skirts in coordinating colors. How many ensembles consisting of a blazer, a blouse, and a skirt can she create from this collection?

6. **Commuter Options**   Four commuter trains and three express buses depart from city A to city B in the morning, and three commuter trains and three express buses operate on the return trip in the evening. In how many ways can a commuter from city A to city B complete a daily round trip via bus and/or train?

7. Psychology Experiments A psychologist has constructed the following maze for use in an experiment. The maze is constructed so that a rat must pass through a series of one-way doors. How many different paths are there from start to finish?

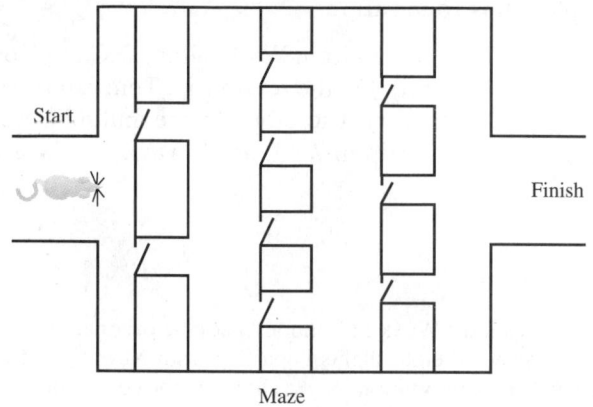

Maze

8. Union Bargaining Issues In a survey conducted by a union, members were asked to rate the importance of the following issues:

Job security

Increased fringe benefits

Improved working conditions

Five different responses were allowed for each issue. Among completed surveys, how many different responses to this survey were possible?

9. Health Plan Options A new state employee is offered a choice of ten basic health plans, three dental plans, and two vision care plans. How many different health care plans are there to choose from if one plan is selected from each category?

10. Code Words How many 3-letter code words can be constructed from the first ten letters of the Greek alphabet if no repetitions are allowed?

11. Computer Dating A computer dating service uses the results of its compatibility survey for arranging dates. The survey consists of 50 questions, each having 5 possible answers. How many different responses are possible if every question is answered?

12. Serial Numbers Computers manufactured by a certain company have a serial number consisting of a letter of the alphabet followed by a 4-digit number. If all the serial numbers of this type have been used, how many sets have already been manufactured?

13. Menu Selections Two soups, five entrees, and three desserts are listed on the "Special" menu at the Neptune Restaurant. How many different selections consisting of one soup, one entree, and one dessert can a customer choose from this menu?

14. Automobile Selection An automobile manufacturer has three different subcompact cars in the line. Customers selecting one of these cars have a choice of three engine sizes, four body styles, and three color schemes. How many different selections can a customer make?

15. ATM Cards To gain access to his account, a customer using an automatic teller machine (ATM) must enter a 4-digit code. If repetition of the same four digits is not allowed (for example, 5555), how many possible combinations are there?

16. Television Viewing Polls An opinion poll is to be conducted among cable TV viewers. Six multiple-choice questions, each with four possible answers, will be asked. In how many different ways can a viewer complete the poll if exactly one response is given to each question?

17. Political Polls An opinion poll was conducted by the Morris Polling Group. Respondents were classified according to their sex (M or F), political affiliation (D, I, R), and the region of the country in which they reside (NW, W, C, S, E, NE).
a. Use the generalized multiplication principle to determine the number of possible classifications.
b. Construct a tree diagram to exhibit all possible classifications of females.

18. License Plate Numbers A certain state uses license plates consisting of three letters of the alphabet followed by three digits. How many different license plate numbers can be formed?

19. Exams An exam consists of ten true-or-false questions. Assuming that every question is answered, in how many different ways can a student complete the exam? In how many ways may the exam be completed if a penalty is imposed for each incorrect answer, so that a student may leave some questions unanswered?

20. Warranty Numbers A warranty identification number for a certain product consists of a letter of the alphabet followed by a 5-digit number. How many possible identification numbers are there if the first digit of the 5-digit number must be nonzero?

21. Lotteries In a state lottery there are 15 finalists eligible for the Big Money Draw. In how many ways can the first, second, and third prizes be awarded if no ticket holder may win more than one prize?

22. Telephone Numbers
a. How many 7-digit telephone numbers are possible if the first digit must be nonzero?
b. How many international direct-dialing numbers are possible if each number consists of a 3-digit area code (the first digit of which must be nonzero) and a number of the type described in (a)?

23. Slot Machines A "lucky dollar" is one of the nine symbols printed on each reel of a slot machine with three reels. A player receives one of various payouts whenever one or more "lucky dollars" appear in the window of the machine. Find the number of winning combinations for which the machine gives a payoff.

[*Hint:* (a) Compute the number of ways in which the nine symbols on the first, second, and third wheels can appear in the window slot and (b) compute the number of ways in which the eight symbols other than the "lucky dollar" can appear in the window slot. The difference $(a - b)$ is the number of ways in which the "lucky dollar" can appear in the window slot. Why?]

SOLUTIONS TO SELF-CHECK EXERCISES 6.3

1. A tourist has a choice of eight flights, five hotel accommodations, and eight tickets. By the generalized multiplication principle, there are $8 \cdot 5 \cdot 8$, or 320, travel packages.

2. There is a choice of two entrees, one dinner salad, one French roll, a choice of three vegetables, a choice of three wines, a choice of two nonalcoholic beverages, and a choice of six pastries. Therefore, by the generalized multiplication principle, there are $2 \cdot 1 \cdot 1 \cdot 3 \cdot 3 \cdot 2 \cdot 6$, or 216, combinations of dinner specials.

6.4 PERMUTATIONS AND COMBINATIONS

Permutations

In this section, we apply the generalized multiplication principle to the solution of two types of counting problems. Both types involve determining the number of ways the elements of a set may be arranged, and both play an important role in the solution of problems in probability.

We begin by considering the permutations of a set. Specifically, given a set of distinct objects, a **permutation** of the set is an arrangement of these objects in a *definite order*. To see why the order in which objects are arranged is important in certain practical situations, suppose that the winning number for the first prize in a raffle is 9237. Then the number 2973, although it contains the same digits as the winning number, cannot be a first prize winner (Figure 6.15). Here the four objects—the numbers 9, 2, 3, and 7—are arranged in a different order; one arrangement is associated with the winning number for the first prize and the other is not.

Figure 6.15
The same digits appear on each ticket, but the order of the digits is different.

EXAMPLE 1   Let $A = \{a, b, c\}$.

a. Find the number of permutations of A.

b. List all the permutations of A with the aid of a tree diagram.

Solution

a. Each permutation of A consists of a sequence of the three letters a, b, c. Therefore, we may think of such a sequence as being constructed by filling in each of the three blanks

$$\underline{\hspace{1.5cm}} \quad \underline{\hspace{1.5cm}} \quad \underline{\hspace{1.5cm}}$$

with one of the three letters. Now, there are three ways in which we may fill the first blank—we may choose a, b, or c. Having selected a letter for the first blank, there are two letters left for the second blank. Finally, there is but one way left to fill the third blank. Schematically, we have

$$\underline{\quad 3 \quad} \quad \underline{\quad 2 \quad} \quad \underline{\quad 1 \quad}$$

Invoking the generalized multiplication principle, we conclude that there are $3 \cdot 2 \cdot 1$, or 6, permutations of the set A.

b. The tree diagram associated with this problem appears in Figure 6.16, and the six permutations of A are *abc*, *acb*, *bac*, *bca*, *cab*, and *cba*.

Figure 6.16
Permutations of three objects.

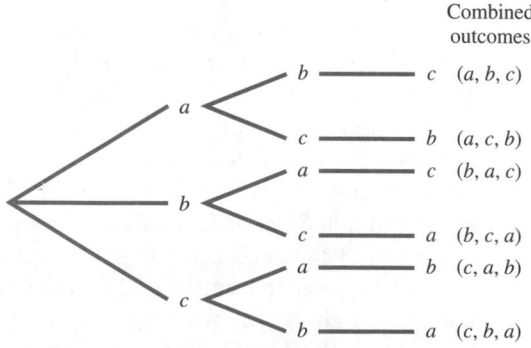

REMARK   Notice that, when the possible outcomes are listed in the tree diagram in Example 1, order is taken into account. Thus, (a, b, c) and (a, c, b) are two different arrangements.   ◌ ◌ ◌

EXAMPLE 2 Find the number of ways a baseball team consisting of nine people can arrange themselves in a line for a group picture.

Solution We want to determine the number of permutations of the nine members of the baseball team. Each permutation in this situation consists of an arrangement of the nine team members in a line. The nine positions can be represented by nine blanks. Thus

| Position | $\underline{1}$ | $\underline{2}$ | $\underline{3}$ | $\underline{4}$ | $\underline{5}$ | $\underline{6}$ | $\underline{7}$ | $\underline{8}$ | $\underline{9}$ |

There are nine ways to choose from among the nine players to fill the first position. When that position is filled, eight players are left, which gives us eight ways to fill the second position. Proceeding in a similar manner, we find that there are seven ways to fill the third position, and so on. Schematically, we have

| Number of ways to fill each position | $\underline{9}$ | $\underline{8}$ | $\underline{7}$ | $\underline{6}$ | $\underline{5}$ | $\underline{4}$ | $\underline{3}$ | $\underline{2}$ | $\underline{1}$ |

Invoking the generalized multiplication principle, we conclude that there are $9 \cdot 8 \cdot 7 \cdot 6 \cdot 5 \cdot 4 \cdot 3 \cdot 2 \cdot 1$, or 362,880, ways the baseball team can be arranged for the picture.   ◦ ◦ ◦

 Whenever we are asked to determine the number of ways the objects of a set can be arranged in a line, order is important. For example, if we take a picture of two baseball players, A and B, then the two players can line up for the picture in two ways, AB or BA, and the two pictures will be different.

Pursuing the same line of argument used in solving the problems in the last two examples, we can derive an expression for the number of ways of permuting a set A of n distinct objects taken n at a time. In fact, each permutation may be viewed as being obtained by filling each of n blanks with one and only one element from the set. There are n ways of filling the first blank, followed by $(n - 1)$ ways of filling the second blank, and so on, so by the generalized multiplication principle there are

$$n(n - 1)(n - 2) \cdot \cdots \cdot 3 \cdot 2 \cdot 1$$

ways of permuting the elements of the set A.

Before stating this result formally, let us introduce a notation that will enable us to write in a compact form many of the expressions that follow. We use the symbol $n!$, read **"n-factorial,"** to denote the product of the first n natural numbers.

n-FACTORIAL	For any natural number n
	$$n! = n(n - 1)(n - 2) \cdot \cdots \cdot 3 \cdot 2 \cdot 1$$
and	$0! = 1$

For example,

$$1! = 1$$
$$2! = 2 \cdot 1 = 2$$
$$3! = 3 \cdot 2 \cdot 1 = 6$$
$$4! = 4 \cdot 3 \cdot 2 \cdot 1 = 24$$
$$5! = 5 \cdot 4 \cdot 3 \cdot 2 \cdot 1 = 120$$
$$\vdots$$

and

$$10! = 10 \cdot 9 \cdot 8 \cdot 7 \cdot 6 \cdot 5 \cdot 4 \cdot 3 \cdot 2 \cdot 1 = 3,628,800$$

Using this notation, we may express *the number of permutations of n distinct objects taken n at a time, P(n, n),* as

$$P(n, n) = n!$$

In many situations we are interested in determining the number of ways of permuting n distinct objects taken r at a time, where $r \le n$. To derive a formula for computing the number of ways of permuting a set consisting of n distinct objects taken r at a time, we observe that each such permutation may be viewed as being obtained by filling each of r blanks with precisely one element from the set. Now there are n ways of filling the first blank, followed by $(n - 1)$ ways of filling the second blank, and so on. Finally, there are $(n - r + 1)$ ways of filling the rth blank. We may represent this argument schematically:

$$\begin{array}{cccccc} \text{Number of ways} & \underline{n} & \underline{n-1} & \underline{n-2} & \cdots & \underline{n-r+1} \\ \text{Position} & \text{1st} & \text{2nd} & \text{3rd} & & r\text{th} \end{array}$$

Using the generalized multiplication principle, we conclude that *the number of ways of permuting n distinct objects taken r at a time, P(n, r), is given by*

$$P(n, r) = \underbrace{n(n - 1)(n - 2) \cdots (n - r + 1)}_{r \text{ factors}}$$

Since

$$n(n - 1)(n - 2) \cdots (n - r + 1)$$

$$= [n(n - 1)(n - 2) \cdots (n - r + 1)] \cdot \frac{[(n - r)(n - r - 1) \cdots \cdots 3 \cdot 2 \cdot 1]}{[(n - r)(n - r - 1) \cdots \cdots 3 \cdot 2 \cdot 1]}$$

$$\uparrow$$

(Here we are multiplying by 1.)

$$= \frac{[n(n - 1)(n - 2) \cdots (n - r + 1)][(n - r)(n - r - 1) \cdots \cdots 3 \cdot 2 \cdot 1]}{(n - r)(n - r - 1) \cdots \cdots 3 \cdot 2 \cdot 1}$$

$$= \frac{n!}{(n - r)!}$$

we have the following formula:

PERMUTATIONS OF *n* DISTINCT OBJECTS	The number of *permutations* of n distinct objects taken r at a time is

$$P(n, r) = \frac{n!}{(n - r)!} \qquad\qquad (6)$$

REMARK   When $r = n$, equation (6) reduces to

$$P(n, n) = \frac{n!}{0!} = \frac{n!}{1} = n! \qquad \text{(Note that } 0! = 1.)$$

In other words, the number of permutations of a set of n distinct objects, taken all together, is $n!$. ⊙ ⊙ ⊙

EXAMPLE 3   Compute (a) $P(4, 4)$ and (b) $P(4, 2)$ and interpret your results.

Solution

a. $P(4, 4) = \dfrac{4!}{(4 - 4)!} = \dfrac{4!}{0!} = \dfrac{4!}{1} = \dfrac{4 \cdot 3 \cdot 2 \cdot 1}{1} = 24$ (Recall that $0! = 1$.)

This gives the number of permutations of 4 objects taken 4 at a time.

b. $P(4, 2) = \dfrac{4!}{(4 - 2)!} = \dfrac{4!}{2!} = \dfrac{4 \cdot 3 \cdot 2 \cdot 1}{2 \cdot 1} = 12$

This is the number of permutations of 4 objects taken 2 at a time. ⊙ ⊙ ⊙

EXAMPLE 4   Let $A = \{a, b, c, d\}$.

a. Use equation (6) to compute the number of permutations of the set A taken two at a time.

b. Display the permutations of (a) with the aid of a tree diagram.

Solution

a. Here $n = 4$ and $r = 2$, so the required number of permutations is given by

$$P(4, 2) = \frac{4!}{(4 - 2)!} = \frac{4!}{2!} = \frac{4 \cdot 3 \cdot 2 \cdot 1}{2 \cdot 1} = 4 \cdot 3$$

or 12.

b. The tree diagram associated with the problem is shown in Figure 6.17, and the permutations of A taken two at a time are

$$ab, \quad ac, \quad ad, \quad ba, \quad bc, \quad bd, \quad ca, \quad cb, \quad cd, \quad da, \quad db, \quad \text{and} \quad dc$$

Figure 6.17
Permutations of four objects taken two at a time.

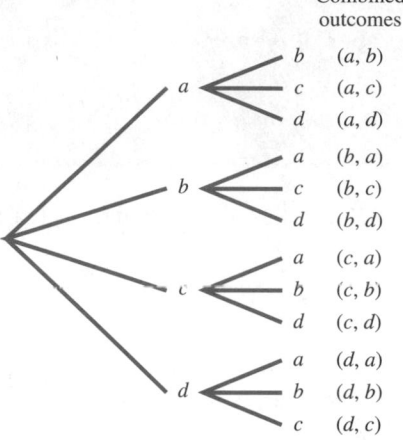

Combined outcomes

b (a, b)
c (a, c)
d (a, d)
a (b, a)
c (b, c)
d (b, d)
a (c, a)
b (c, b)
d (c, d)
a (d, a)
b (d, b)
c (d, c)

EXAMPLE 5   Find the number of ways a chairman, a vice-chairman, a secretary, and a treasurer can be chosen from a committee of eight members.

Solution   The problem is equivalent to finding the number of permutations of eight distinct objects taken four at a time. Therefore, the number of ways of choosing the four officials from the committee of eight members is given by

$$P(8, 4) = \frac{8!}{(8 - 4)!} = \frac{8!}{4!} = 8 \cdot 7 \cdot 6 \cdot 5 = 1680$$

or 1680 ways.

The permutations considered thus far have been those involving sets of *distinct* objects. In many situations we are interested in finding the number of permutations of a set of objects in which not all of the objects are distinct.

PERMUTATIONS OF n OBJECTS, NOT ALL DISTINCT	Given a set of n objects in which n_1 objects are alike and of one kind, n_2 objects are alike and of another kind, . . . , and, finally, n_r objects are alike and of yet another kind so that

$$n_1 + n_2 + \cdots + n_r = n$$

then the number of permutations of these n objects taken n at a time is given by

$$\frac{n!}{n_1!n_2! \cdots n_r!} \qquad \textbf{(7)}$$

To establish equation (7), let us denote the number of such permutations by x. Now, if we *think* of the n_1 objects as being distinct, then they may be permuted in $n_1!$ ways. Similarly, if we *think* of the n_2 objects as being distinct, then they may be permuted in $n_2!$ ways, and so on. Therefore, if we *think* of the n objects as being distinct, then, by the generalized multiplication principle, there are $x \cdot n_1! \cdot n_2! \cdot \cdots \cdot n_r!$ permutations of these objects. But, the number of permutations of a set of n distinct objects taken n at a time is just equal to $n!$. Therefore, we have

$$x(n_1! \cdot n_2! \cdot \cdots \cdot n_r!) = n!$$

from which we deduce that

$$x = \frac{n!}{n_1! n_2! \cdots n_r!}$$

EXAMPLE 6 Find the number of permutations that can be formed from all the letters in the word *ATLANTA*.

Solution There are seven objects (letters) involved, so $n = 7$. However, three of them are alike and of one kind (the three A's), two of them are alike and of another kind (the two T's), so that in this case, $n_1 = 3$, $n_2 = 2$, $n_3 = 1$, and $n_4 = 1$. Therefore, using formula (7), the number of required permutations is given by

$$\frac{7!}{3!2!1!1!} = \frac{7 \cdot 6 \cdot 5 \cdot 4 \cdot 3 \cdot 2 \cdot 1}{3 \cdot 2 \cdot 1 \cdot 2 \cdot 1}$$

or 420.   ◑ ◑ ◑

EXAMPLE 7 Weaver and Kline, a stock brokerage firm, has received nine inquiries regarding new accounts. In how many ways can these inquiries be directed to three of the firm's account executives if each account executive is to handle three inquiries?

Solution If we think of the nine inquiries as being slots arranged in a row with inquiry 1 on the left and inquiry 9 on the right, then the problem can be thought of as one of filling each slot with a business card from an account executive. Then nine business cards would be used, of which three are alike and of one kind, three are alike and of another kind, and three are alike and of yet another kind. Thus, using (7) with $n = 9$, $n_1 = n_2 = n_3 = 3$, we see that the number of ways of assigning the inquiries is given by

$$\frac{9!}{3!3!3!} = \frac{9 \cdot 8 \cdot 7 \cdot 6 \cdot 5 \cdot 4 \cdot 3 \cdot 2 \cdot 1}{3 \cdot 2 \cdot 1 \cdot 3 \cdot 2 \cdot 1 \cdot 3 \cdot 2 \cdot 1}$$

or 1680 ways.   ◑ ◑ ◑

Combinations

Up to now, we have dealt with permutations of a set—that is, with arrangements of the objects of the set in which the *order* of the elements is taken into consideration. In many situations one is interested in determining the number of ways of selecting r objects from a set of n objects without any regard to the order in which the objects are selected. Such a subset is called a **combination.**

For example, if one is interested in knowing the number of 5-card poker hands that can be dealt from a standard deck of 52 cards, then the order in which the poker hand is dealt is unimportant (Figure 6.18). In this situation, we are interested in determining the number of combinations of 5 cards (objects) selected from a deck (set) of 52 cards (objects). (We will solve this problem later.)

Figure 6.18

To derive a formula for determining the number of combinations of *n* objects taken *r* at a time, written

$$C(n, r) \quad \text{or} \quad \binom{n}{r}$$

we observe that each of the $C(n, r)$ combinations of *r* objects can be permuted in *r*! ways (Figure 6.19).

Figure 6.19

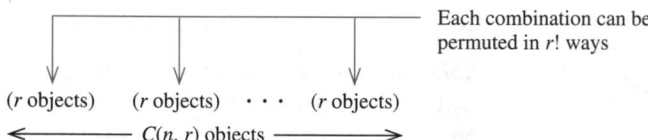

Thus, by the multiplication principle, the product $r!C(n, r)$ gives the number of permutations of *n* objects taken *r* at a time; that is,

$$r!C(n, r) = P(n, r)$$

from which we find

$$C(n, r) = \frac{P(n, r)}{r!}$$

or, using equation (6),

$$C(n, r) = \frac{n!}{r!(n - r)!}$$

COMBINATIONS OF	The number of *combinations* of n distinct objects taken r at a time is given by
n OBJECTS	$$C(n, r) = \frac{n!}{r!(n-r)!} \qquad \text{where } r \leq n \qquad \qquad \textbf{(8)}$$

EXAMPLE 8 Compute (a) $C(4, 4)$ and (b) $C(4, 2)$ and interpret the results.

Solution

a. $C(4, 4) = \dfrac{4!}{4!(4-4)!} = \dfrac{4!}{4!0!} = 1$ (Recall that $0! = 1$.)

This gives 1 as the number of combinations of 4 distinct objects taken 4 at a time.

b. $C(4, 2) = \dfrac{4!}{2!(4-2)!} = \dfrac{4!}{2!2!} = \dfrac{4 \cdot 3}{2} = 6$

This gives 6 as the number of combinations of 4 distinct objects taken 2 at a time.   ◦ ◦ ◦

Applications

EXAMPLE 9 A Senate investigation subcommittee of four members is to be selected from a Senate committee of ten members. Determine the number of ways this can be done.

Solution Since the order in which the members of the subcommittee are selected is unimportant, the number of ways of choosing the subcommittee is given by $C(10, 4)$, the number of combinations of ten objects taken four at a time. But

$$C(10, 4) = \frac{10!}{4!(10-4)!} = \frac{10!}{4!6!} = \frac{10 \cdot 9 \cdot 8 \cdot 7}{4 \cdot 3 \cdot 2 \cdot 1} = 210$$

so there are 210 ways of choosing such a subcommittee.   ◦ ◦ ◦

REMARK   Remember, a combination is a selection of objects *without* regard to order. Thus, in Example 9, we used a combination formula rather than a permutation formula to solve the problem because the order of selection was not important; that is, it did not matter whether a member of the subcommittee was selected first, second, third, or fourth.   ◦ ◦ ◦

EXAMPLE 10 How many poker hands of 5 cards can be dealt from a standard deck of 52 cards?

Solution The order in which the 5 cards are dealt is not important. The number of ways of dealing a poker hand of 5 cards from a standard deck of

52 cards is given by $C(52, 5)$, the number of combinations of 52 objects taken 5 at a time. Now,

$$C(52, 5) = \frac{52!}{5!(52 - 5)!} = \frac{52!}{5!47!}$$
$$= \frac{52 \cdot 51 \cdot 50 \cdot 49 \cdot 48}{5 \cdot 4 \cdot 3 \cdot 2 \cdot 1}$$
$$= 2{,}598{,}960$$

so there are 2,598,960 ways of dealing such a poker hand.   ○ ○ ○

The next several examples show that solving a counting problem often involves the repeated application of equation (6) and/or (8), possibly in conjunction with the multiplication principle.

EXAMPLE 11   The members of a string quartet comprising two violinists, a violist, and a cellist are to be selected from a group of six violinists, three violists, and two cellists, respectively.

a. In how many ways can the string quartet be formed?

b. In how many ways can the string quartet be formed if one of the violinists is to be designated as the first violinist and the other is to be designated as the second violinist?

Solution

a. Since the order in which each musician is selected is not important, we use combinations. The violinists may be selected in $C(6, 2)$, or 15, ways; the violist may be selected in $C(3, 1)$, or 3, ways; and the cellist may be selected in $C(2, 1)$, or 2, ways. By the multiplication principle, there are $15 \cdot 3 \cdot 2$, or 90, ways of forming the string quartet.

b. The order in which the violinists are selected is important here. Consequently, the number of ways of selecting the violinists is given by $P(6, 2)$, or 30, ways. The number of ways of selecting the violist and the cellist are, of course, 3 and 2, respectively. Therefore, the number of ways in which the string quartet may be formed is given by $30 \cdot 3 \cdot 2$, or 180, ways.   ○ ○ ○

REMARK   The solution of Example 11 involves both a permutation and a combination. When we select two violinists from six violinists, order is not important, and we use a combination formula to solve the problem. However, when one of the violinists is designated as a first violinist, order is important, and we use a permutation formula to solve the problem.   ○ ○ ○

EXAMPLE 12   Refer to Example 5, page 345. Suppose that the investor has decided to purchase shares in the stocks of two aerospace companies, two energy development companies, and two electronics companies. In how many ways may the investor select the group of six companies for the investment from the recommended list of five aerospace companies, three energy development companies, and four electronics companies?

Solution There are $C(5, 2)$ ways in which the investor may select the aerospace companies, $C(3, 2)$ ways in which she may select the companies involved in energy development, and $C(4, 2)$ ways in which she may select the electronics companies as investments. By the generalized multiplication principle, there are

$$C(5, 2)C(3, 2)C(4, 2) = \frac{5!}{2!3!} \cdot \frac{3!}{2!1!} \cdot \frac{4!}{2!2!}$$

$$= \frac{5 \cdot 4}{2} \cdot 3 \cdot \frac{4 \cdot 3}{2} = 180$$

ways of selecting the group of six companies for her investment.   ○ ○ ○

EXAMPLE 13 The Futurists, a rock group, are planning a concert tour with performances to be given in five cities: San Francisco, Los Angeles, San Diego, Denver, and Las Vegas. In how many ways can they arrange their itinerary

a. if there are no restrictions?

b. if the three performances in California must be given consecutively?

Solution

a. The order is important here, and we see that there are

$$P(5, 5) = 5! = 120$$

ways of arranging their itinerary.

b. First, note that there are $P(3, 3)$ ways of choosing between performing in California and in the two cities outside that state. Next, there are $P(3, 3)$ ways of arranging their itinerary in the three cities in California. Therefore, by the multiplication principle, there are

$$P(3, 3) \cdot P(3, 3) = \frac{3!}{(3 \cdot 3)!} \cdot \frac{3!}{(3 \cdot 3)!} = (6)(6)$$

or 36, ways of arranging their itinerary.   ○ ○ ○

EXAMPLE 14 The U.N. Security Council consists of 5 permanent members and 10 nonpermanent members. Decisions made by the council require 9 votes for passage. However, any permanent member may veto a measure and thus block its passage. In how many ways can a measure be passed if all 15 members of the Council vote (no abstentions)?

Solution If a measure is to be passed, then all 5 permanent members must vote for passage of that measure. This can be done in $C(5, 5)$, or 1, way.

Next, observe that since 9 votes are required for passage of a measure, *at least* 4 of the 10 nonpermanent members must also vote for its passage. To determine the number of ways this can be done, notice that there are $C(10, 4)$ ways in which exactly 4 of the nonpermanent members can vote for passage of a measure, $C(10, 5)$ ways in which exactly 5 of them can vote for passage of a measure, and so on. Finally, there are $C(10, 10)$ ways in which all 10 nonpermanent members can vote for passage of a measure. Therefore, there are

$$C(10, 4) + C(10, 5) + \cdots + C(10, 10)$$

ways in which at least 4 of the 10 nonpermanent members can vote for a measure. So, by the multiplication principle, the number of ways a measure can be passed is

$$C(5, 5) \cdot [C(10, 4) + C(10, 5) + \cdots + C(10, 10)]$$

$$= (1) \left[\frac{10!}{4!6!} + \frac{10!}{5!5!} + \cdots + \frac{10!}{10!0!} \right]$$

$$= (1)(210 + 252 + 210 + 120 + 45 + 10 + 1) = 848$$

or 848 ways.   ○ ○ ○

SELF-CHECK EXERCISES 6.4

1. Evaluate   **a.** 5!   **b.** $C(7, 4)$ **c.** $P(6, 2)$

2. A space shuttle crew consists of a shuttle commander, a pilot, 3 engineers, a scientist, and a civilian. The shuttle commander and pilot are to be chosen from 8 candidates, the 3 engineers from 12 candidates, the scientist from 5 candidates, and the civilian from 2 candidates. How many such space shuttle crews can be formed?

Solutions to Self-Check Exercises 6.4 can be found on page 366.

6.4 EXERCISES

In exercises 1–22, evaluate the given expression.

1. $3(5!)$

2. $2(7!)$

3. $\dfrac{5!}{2!3!}$

4. $\dfrac{6!}{4!2!}$

5. $P(5, 5)$

6. $P(6, 6)$

7. $P(5, 2)$

8. $P(5, 3)$

9. $P(n, 1)$

10. $P(k, 2)$

11. $C(6, 6)$

12. $C(8, 8)$

13. $C(7, 4)$

14. $C(9, 3)$

15. $C(5, 0)$

16. $C(6, 5)$

17. $C(9, 6)$

18. $C(10, 3)$

19. $C(n, 2)$

20. $C(7, r)$

21. $P(n, n - 2)$

22. $C(n, n - 2)$

In exercises 23–30, classify each problem according to whether it involves a permutation or a combination.

23. In how many ways can the letters of the word *glacier* be arranged?

24. In the eighth-grade dance class, there are 10 girls and 14 boys. In how many ways can these students be paired off to form dance couples consisting of one boy and one girl?

25. As part of a quality-control program, 3 record-o-phones are selected at random for testing from each 100 phones produced by the manufacturer. In how many ways can this test batch be chosen?

26. How many 3-digit numbers can be formed using the numerals in the set $\{3, 2, 7, 9\}$ if repetition is not allowed?

27. In how many ways can nine different books be arranged on a shelf?

28. A member of a book club wishes to purchase two books from a selection of eight books recommended for a certain month. In how many ways can she choose them?

29. How many 5-card poker hands can be dealt consisting of three queens and a pair?

30. Four couples are to be seated at an oval dining table. If each couple must be seated together, how many possible seating arrangements are there?

31. How many four-letter permutations can be formed from the first four letters of the alphabet?

32. How many three-letter permutations can be formed from the first five letters of the alphabet?

33. In how many ways can four students be seated in a row of four seats?

34. In how many ways can five people line up at a checkout counter in a supermarket?

35. How many different batting orders can be formed for a 9-member baseball team?

36. In how many ways can the names of six candidates for political office be listed on a ballot?

37. In how many ways can a member of a hiring committee select 3 of 12 job applicants for further consideration?

38. In how many ways can an investor select four mutual funds for his investment portfolio from a recommended list of eight mutual funds?

39. Find the number of distinguishable permutations that can be formed from the letters of the word *ANTARCTICA*.

40. Find the number of distinguishable permutations that can be formed from the letters of the word *PHILIPPINES*.

41. **Management Decisions**   In how many ways can a supermarket chain select 3 out of 12 possible sites for the construction of new supermarkets?

42. **Book Selections**   A student is given a reading list of ten books from which he must select two for an outside reading requirement. In how many ways can he make his selections?

43. **Quality Control**   In how many ways can a quality-control engineer select a sample of 3 transistors for testing from a batch of 100 transistors?

44. **Study Groups**   A group of five students studying for a bar exam had formed a study group. Each member of the group will be responsible for preparing a study outline for one of five courses. In how many different ways can the five courses be assigned to the members of the group?

45. **Television Programming**   In how many ways can a television programming director schedule six different commercials in the six time slots allocated to commercials during a 1-hour program?

46. **Waiting Lines**   Seven people arrive at the ticket counter of a cinema at the same time. In how many ways can they line up to purchase their tickets?

47. **Management Decisions**   Weaver and Kline, a stock brokerage firm, has received 6 inquiries regarding new accounts. In how many ways can these inquiries be directed to its 12 account executives if each executive handles no more than 1 inquiry?

48. **Car Pools**   A company car that has a seating capacity of six is to be used by six employees who have formed a car pool. If only four of these employees can drive, how many possible seating arrangements are there for the group?

49. **Book Displays**   At a college library exhibition of faculty publications, three mathematics books, four social science books, and three biology books will be displayed on a shelf. (Assume that none of the books is alike.)
    a. In how many ways can the ten books be arranged on the shelf?
    b. In how many ways can the ten books be arranged on the shelf if books on the same subject matter are placed together?

50. **Seating**   In how many ways can four married couples attending a concert be seated in a row of eight seats
    a. if there are no restrictions?
    b. if each married couple is seated together?
    c. if the members of each sex are seated together?

51. **Newspaper Advertisements**   Four items from 5 different departments of the Metro Department Store will be featured in a one-page newspaper advertisement as shown in the following diagram:

Advertisement

1	2	3	4
5	6	7	8
9	10	11	12
13	14	15	16
17	18	19	20

a. In how many different ways can the 20 featured items be arranged on the page?

b. If items from the same department must be in the same row, how many arrangements are possible?

52. Management Decisions The C & J Realty Company has received 12 inquiries from prospective home buyers. In how many ways can the inquiries be directed to 4 of the firm's real estate agents if each agent handles 3 inquiries?

53. Sports In the women's tennis tournament at Wimbledon, two finalists, A and B, are competing for the title, which will be awarded to the first player to win two sets. In how many different ways can the match be completed?

54. Sports In the men's tennis tournament at Wimbledon, two finalists, A and B, are competing for the title, which will be awarded to the first player to win three sets. In how many different ways can the match be completed?

55. U.N. Voting Refer to Example 14. In how many ways can a measure be passed if two particular permanent and two particular nonpermanent members of the Council abstain from voting?

56. Jury Selection In how many different ways can a panel of 12 jurors and 2 alternate jurors be chosen from a group of 30 prospective jurors?

57. Teaching Assistantships Twelve graduate students have applied for three available teaching assistantships. In how many ways can the assistantships be awarded among these applicants
a. if no preference is given to any student?
b. if one particular student must be awarded an assistantship?
c. if the group of applicants includes seven men and

five women and it is stipulated that at least one woman must be awarded an assistantship?

58. Exams A student taking an examination is required to answer 10 out of 15 questions.
a. In how many ways can the 10 questions be selected?
b. In how many ways can the 10 questions be selected if exactly 2 of the first 3 questions must be answered?

59. Contract Bidding The U.B.S. Television Company is considering bids submitted by seven different firms for three different contracts. In how many ways can the contracts be awarded among these firms if no firm is to receive more than two contracts?

60. Senate Committees In how many ways can a subcommittee of four be chosen from a Senate committee of five Democrats and four Republicans
a. if all members are eligible?
b. if the subcommittee must consist of two Republicans and two Democrats?

61. Course Selection A student planning her curriculum for the upcoming year must select one of five business courses, one of three mathematics courses, two of six elective courses, and either one of four history courses or one of three social science courses. How many different curricula are available for her consideration?

62. Personnel Selection The J.C.L. Computer Company has five vacancies in its executive trainee program. In how many ways can the company select five trainees from a group of ten female and ten male applicants
a. if the vacancies may be filled by any combination of men and women?
b. if the vacancies must be filled by two men and three women?

63. Drivers' Tests The State Motor Vehicle Department requires learners to pass a written test on the motor vehicle laws of the state. The exam consists of ten trueor-false questions, of which eight must be answered correctly to qualify for a permit. In how many different ways can a learner who answers all the questions on the exam qualify for a permit?

64. Quality Control The Goodman Tire Company has 32 tires of a particular size and grade in stock, 2 of which are defective. If a set of 4 tires is to be selected,
a. how many different selections can be made?
b. how many different selections can be made that do not include any defective tires?

A list of poker hands ranked in order from the highest to the lowest is shown in the following table, along with a description and example of each hand. Use the table to answer exercises 65–70.

Hand	Description	Example
Straight flush	5 cards in sequence in the same suit	A ♥ 2 ♥ 3 ♥ 4 ♥ 5 ♥
Four of a kind	4 cards of the same rank and any other card	K ♥ K ♦ K ♠ K ♣ 2 ♥
Full house	3 of a kind and a pair	3 ♥ 3 ♦ 3 ♣ 7 ♥ 7 ♦
Flush	5 cards of the same suit that are not all in sequence	5 ♥ 6 ♥ 9 ♥ J ♥ K ♥
Straight	5 cards in sequence but not all of the same suit	10 ♥ J ♦ Q ♣ K ♠ A ♥
Three of a kind	3 cards of the same rank and 2 unmatched cards	K ♥ K ♦ K ♠ 2 ♥ 4 ♦
Two pair	2 cards of the same rank and 2 cards of any other rank with an unmatched card	K ♥ K ♦ 2 ♥ 2 ♠ 4 ♣
One pair	2 cards of the same rank and 3 unmatched cards	K ♥ K ♦ 5 ♥ 2 ♠ 4 ♥

If a 5-card poker hand is dealt from a well-shuffled deck of 52 cards, how many different hands consist of the following:

65. A straight flush? (Note that an ace may be played as either a high or a low card in a straight sequence—that is, A, 2, 3, 4, 5 or 10, J, Q, K, A. Hence, there are 10 possible sequences for a straight in one suit.)

66. A straight (but not a straight flush)?

67. A flush (but not a straight flush)?

68. Four of a kind?

69. A full house?

70. Two pairs?

71. **Bus Routing** The following is a schematic diagram of a city's street system between the points *A* and *B*. The City Transit Authority is in the process of selecting a route from A to B along which to provide bus service. If the company's intention is to keep the route as short as possible, how many routes must be considered?

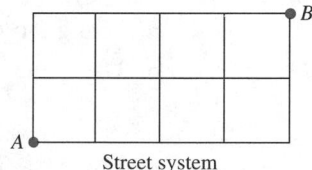

Street system

72. **Sports** In the World Series, one National League team and one American League team compete for the coveted title that is awarded to the first team to win four games. In how many different ways can the series of games be completed?

73. **Voting Quorums** A quorum (minimum) of 6 voting members is required at all meetings of the Curtis Townhomes Owners Association. If there is a total of 12 voting members in the group, find the number of ways this quorum can be formed.

74. **Circular Permutations** Suppose *n* distinct objects are arranged in a circle. Show that the number of (different) circular arrangements of the *n* objects is $(n - 1)!$

[*Hint:* Consider the arrangement of the 5 letters *A, B, C, D,* and *E* in the accompanying figure. The permutations *ABCDE, BCDEA, CDEAB, DEABC,* and *EABCD* are not distinguishable. Generalize this observation to the case of *n* objects.]

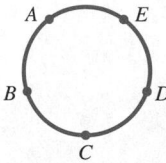

75. Refer to exercise 74. In how many ways can five TV commentators be seated at a round table for a discussion?

USING TECHNOLOGY

EVALUATING $n!$, $P(n, r)$, AND $C(n, r)$

The graphing utility can be used to calculate factorials, permutations, and combinations with relative ease. The graphing utility is, therefore, an indispensable tool in solving counting problems involving large numbers of objects. Here we use the nPr (permutation) and nCr (combination) functions of a graphing utility.

EXAMPLE 1 Use a graphing utility to find (a) 12!, (b) $P(52, 5)$, and (c) $C(38, 10)$.

Solution

a. Using the factorial function, we find that $12! = 479,001,600$.

b. Using the nPr function, we have

$$P(52, 5) = 52 \text{ nPr } 5 = 311,875,200$$

c. Using the nCr function, we obtain

$$C(38, 10) = 38 \text{ nCr } 10 = 472,733,756$$

○ ○ ○

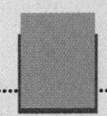

EXERCISES

In exercises 1–10, evaluate the expression.

1. $15!$ 2. $20!$

3. $4(18!)$ 4. $\dfrac{30!}{18!}$

5. $P(52, 7)$ 6. $P(24, 8)$

7. $C(52, 7)$ 8. $C(26, 8)$

9. $P(10, 4) \cdot C(12, 6)$

10. $P(20, 5) \cdot C(9, 3) \cdot C(8, 4)$

11. A mathematics professor uses a computerized test bank to prepare her final exam. If 25 different problems are available for the first 3 exam questions, 40 different problems available for the next 5 questions, and 30 different problems available for the last 2 questions, how many different 10-question exams can she set? (Assume that the order of the questions within each group is not important.)

12. The S & S Brokerage Company has received 100 inquiries from prospective clients. In how many ways can the inquiries be directed to 5 of the firm's brokers if each broker handles 20 inquiries?

76. Refer to exercise 74. In how many ways can four men and four women be seated at a round table at a dinner party if each guest is seated next to members of the opposite sex.

77. At the end of Section 3.3, we mentioned that a linear programming problem in 3 variables and 8 constraints may have up to 56 feasible corner points and that the

determination of these feasible corner points calls for the solution of 56 3×3 systems of linear equations. Prove this assertion.

78. Refer to exercise 77. Show that a linear programming problem in 5 variables and 15 constraints may have up to 3003 feasible corner points. This assertion was also made at the end of Section 3.3.

SOLUTIONS TO SELF-CHECK EXERCISES 6.4

1. **a.** $5! = 5 \cdot 4 \cdot 3 \cdot 2 \cdot 1 = 120$

   b. $C(7, 4) = \dfrac{7!}{4!3!} = \dfrac{7 \cdot 6 \cdot 5}{3 \cdot 2 \cdot 1} = 35$

   c. $P(6, 2) = \dfrac{6!}{4!} = 6 \cdot 5 = 30$

2. There are $P(8, 2)$ ways of picking the shuttle commander and pilot (the order *is* important here), $C(12, 3)$ ways of picking the engineers (the order is not important here), $C(5, 1)$ ways of picking the scientist, and $C(2, 1)$ ways of picking the civilian. By the multiplication principle, there are

$$P(8, 2) \cdot C(12, 3) \cdot C(5, 1) \cdot C(2, 1)$$
$$= \frac{8!}{6!} \cdot \frac{12!}{9!3!} \cdot \frac{5!}{4!1!} \cdot \frac{2!}{1!1!}$$
$$= \frac{(8)(7)(12)(11)(10)(5)(2)}{(3)(2)}$$
$$= 123{,}200$$

or 123,200 ways a crew can be selected.

CHAPTER 6 SUMMARY OF PRINCIPAL FORMULAS AND TERMS

Formulas

1. Commutative laws

$A \cup B = B \cup A$
$A \cap B = B \cap A$

2. Associative laws

$A \cup (B \cup C) = (A \cup B) \cup C$
$A \cap (B \cap C) = (A \cap B) \cap C$

3. Distributive laws

$A \cup (B \cap C)$
$\quad = (A \cup B) \cap (A \cup C)$
$A \cap (B \cup C)$
$\quad = (A \cap B) \cup (A \cap C)$

4. De Morgan's laws

$(A \cup B)^c = A^c \cap B^c$
$(A \cap B)^c = A^c \cup B^c$

5. Number of elements in the union of two finite sets

$$n(A \cup B) = n(A) + n(B) \\ - n(A \cap B)$$

6. Permutation of n distinct objects, taken r at a time

$$P(n, r) = \frac{n!}{(n - r)!}$$

7. Permutation of n objects, not all distinct, taken n at a time

$$\frac{n!}{n_1! n_2! \cdots n_r!}$$

8. Combination of n distinct objects, taken r at a time

$$C(n, r) = \frac{n!}{r!(n - r)!}$$

Terms

Set	Set union
Element of a set	Set intersection
Roster notation	Set complementation
Set-builder notation	Multiplication principle
Set equality	Generalized multiplication principle
Subset	Permutation
Empty set	n-Factorial
Universal set	Combination
Venn diagram	

CHAPTER 6 REVIEW EXERCISES

In exercises 1–4, list the elements of the given set in roster notation.

1. $\{x \mid 3x - 2 = 7; x, \text{ an integer}\}$

2. $\{x \mid x \text{ is a letter of the word } TALLAHASSEE\}$

3. The set whose elements are the even numbers between 3 and 11

4. $\{x \mid (x - 3)(x + 4) = 0; x, \text{ a negative integer}\}$

Let A = {a, c, e, r}. In exercises 5–8, determine whether the given set is equal to A.

5. $\{r, e, c, a\}$

6. $\{x \mid x \text{ is a letter of the word } career\}$

7. $\{x \mid x \text{ is a letter of the word } racer\}$

8. $\{x \mid x \text{ is a letter of the word } cares\}$

In exercises 9–12, shade the portion of the accompanying figure that represents the given set.

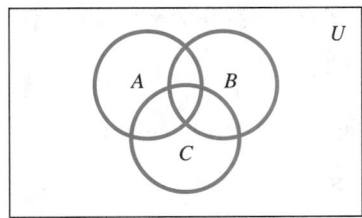

9. $A \cup (B \cap C)$

10. $(A \cap B \cap C)^c$

11. $A^c \cap B^c \cap C^c$

12. $A^c \cap (B^c \cup C^c)$

Let $U = \{a, b, c, d, e\}$, $A = \{a, b\}$, $B = \{b, c, d\}$, and $C = \{a, d, e\}$. In exercises 13–16, verify the given equation by direct computation.

13. $A \cup (B \cup C) = (A \cup B) \cup C$

14. $A \cap (B \cap C) = (A \cap B) \cap C$

15. $A \cap (B \cup C) = (A \cap B) \cup (A \cap C)$

16. $A \cup (B \cap C) = (A \cup B) \cap (A \cup C)$

Let $U = \{$all participants in a consumer behavior survey conducted by a national polling group$\}$

$A = \{$consumers who avoided buying a product because it is not recyclable$\}$

$B = \{$consumers who used cloth rather than disposable diapers$\}$

$C = \{$consumers who boycotted a company's products because of their record on the environment$\}$

and $D = \{$consumers who voluntarily recycled their garbage$\}$

In exercises 17–20, describe each of the given sets in words.

17. $A \cap C$ **18.** $A \cup D$

19. $B^c \cap D$ **20.** $C^c \cup D^c$

Let A and B be subsets of a universal set U and suppose $n(U) = 350$, $n(A) = 120$, $n(B) = 80$, and $n(A \cap B) = 50$. In exercises 21–26, find the number of elements in each of the given sets.

21. $n(A \cup B)$ **22.** $n(A^c)$

23. $n(B^c)$ **24.** $n(A^c \cap B)$

25. $n(A \cap B^c)$ **26.** $n(A^c \cap B^c)$

In exercises 27–30, evaluate each of the given quantities.

27. $C(20, 18)$ **28.** $P(9, 7)$

29. $C(5, 3) \cdot P(4, 2)$ **30.** $4 \cdot P(5, 3) \cdot C(7, 4)$

31. A comparison of 5 major credit cards showed that

3 offered cash advances.

3 offered extended payments for *all* goods and services puchased.

2 required an annual fee of less than $35.

2 offered both cash advances and extended payments.

1 offered extended payments and had an annual fee less than $35.

No card had an annual fee less than $35 and offered both cash advances and extended payments.

How many cards had an annual fee less than $35 and offered cash advances? (Assume that every card had at least 1 of the 3 mentioned features.)

32. The Department of Foreign Languages of a liberal arts college conducted a survey of its recent graduates to determine the foreign language courses they had taken while undergraduates at the college. Of the 480 graduates

200 had at least one year of Spanish.

178 had at least one year of French.

140 had at least one year of German.

 33 had at least one year of Spanish and French.

 24 had at least one year of Spanish and German.

 18 had at least one year of French and German.

  3 had at least one year of all three languages.

How many of the graduates had
a. at least one year of at least one of the three languages?
b. at least one year of exactly one of the three languages?
c. less than a year of any of the three languages?

33. In how many ways can six different compact discs be arranged on a shelf?

34. In how many ways can three pictures be selected from a group of six different pictures?

35. In an election being held by the Associated Students Organization there are six candidates for president, four for vice-president, five for secretary, and six for treasurer. How many different possible outcomes are there for this election?

36. How many three-digit numbers can be formed from the numerals in the set $\{1, 2, 3, 4, 5\}$
a. if repetition of digits is not allowed?
b. if repetition of digits is allowed?

37. Find the number of distinguishable permutations that can be formed from the letters of each word.
a. *CINCINNATI*   **b.** *HONOLULU*

38. From a standard 52-card deck, how many 5-card poker hands can be dealt consisting of
a. 5 clubs?   **b.** 3 kings and a pair?

39. In how many ways can seven students be assigned seats in a row containing seven desks
a. if there are no restrictions?
b. if two of the students must not be seated next to each other?

40. There are eight seniors and six juniors in the Math Club at Jefferson High School. In how many ways can a math team consisting of four seniors and two juniors be selected from the members of the Math Club?

41. A sample of 4 balls is to be selected at random from an urn containing 15 balls numbered 1 to 15. If 6 balls are green, 5 are white, and 4 are black,
a. how many different samples can be selected?
b. how many samples can be selected that contain at least 1 white ball?

42. From a shipment of 60 transistors, 5 of which are defective, a sample of 4 transistors is selected at random.
a. In how many different ways can the sample be selected?
b. How many samples contain 3 defective transistors?
c. How many samples do not contain any defective transistors?

The systematic study of probability began in the seventeenth century, when certain aristocrats wanted to discover superior strategies to use in the gaming rooms of Europe. Some of the best mathematicians of the period were engaged in this pursuit. Since then, probability has evolved in virtually every sphere of human endeavor in which an element of uncertainty is present.

We begin by introducing some of the basic terminology used in the study of the subject. Then, in Section 7.2, we give the technical meaning of the term *probability*. The rest of this chapter is devoted to the development of techniques for computing the probabilities of the occurrence of certain events.

Where did the defective picture tube come from? Picture tubes for the Pulsar 19-inch color television sets are manufactured in three locations and then shipped to the main plant of the Vista Vision Corporation for final assembly. Each location produces a certain number of the picture tubes with different degrees of reliability. In Example 1, page 428, we will determine the likelihood that a defective picture tube is manufactured in a particular location.

7

PROBABILITY

371

7.1 EXPERIMENTS, SAMPLE SPACES, AND EVENTS

Terminology

A number of specialized terms are used in the study of probability. We begin by defining the term *experiment*.

EXPERIMENT   An **experiment** is an activity with observable results.

The results of the experiment are called the **outcomes** of the experiment. Three examples of experiments are:

- Tossing a coin and observing whether it falls "heads" or "tails"
- Casting a die and observing which of the numbers 1, 2, 3, 4, 5, or 6 shows up
- Testing a spark plug from a batch of 100 spark plugs and observing whether or not it is defective

In our discussion of experiments, we use the following terms:

SAMPLE POINT, SAMPLE SPACE, AND EVENT

Sample point: an outcome of an experiment
Sample space: the set consisting of all possible sample points of an experiment
Event: a subset of a sample space of an experiment

The sample space of an experiment is a universal set whose elements are precisely the outcomes, or the sample points, of the experiment; the events of the experiment are the subsets of the universal set. A sample space associated with an experiment that has a finite number of possible outcomes (sample points) is called a **finite sample space.**

Since the events of an experiment are subsets of a universal set (the sample space of the experiment), we may use the results for set theory given in Chapter 6 to help us study probability. We begin by explaining the roles played by the empty set and a universal set when viewed as events associated with an experiment. The empty set, $\varnothing$, is called the *impossible event;* it cannot occur since the $\varnothing$ has no elements (outcomes). Next, the universal set S is referred to as the *certain event;* it must occur since S contains all the outcomes of the experiment.

This terminology is illustrated in the next several examples.

EXAMPLE 1 Describe the sample space associated with the experiment of tossing a coin and observing whether it falls "heads" or "tails." What are the events of this experiment?

Solution The two outcomes are "heads" and "tails," and the required sample space is given by $S = \{H, T\}$, where H denotes the outcome "heads" and T denotes the outcome "tails." The events of the experiment, the subsets of S, are:

$$\varnothing, \quad \{H\}, \quad \{T\}, \quad S$$

Note that we have included the impossible event, $\varnothing$, and the certain event, S.

Since the events of an experiment are subsets of the sample space of the experiment, we may talk about the *union* and *intersection* of any two events; we can also consider the *complement* of an event with respect to the sample space.

THE UNION OF TWO EVENTS	The **union** of the two events E and F is the event $E \cup F$.

Thus, the event $E \cup F$ comprises the set of outcomes of E and/or F.

THE INTERSECTION OF TWO EVENTS	The **intersection** of the two events E and F is the event $E \cap F$.

Thus, the event $E \cap F$ comprises the set of outcomes of E and F.

THE COMPLEMENT OF AN EVENT	The **complement** of an event E is the event E^c.

Thus, the event E^c is the set comprising all the outcomes in the sample space S that are not in E.

Venn diagrams depicting the union, intersection, and complementation of events are shown in Figure 7.1.

These concepts are illustrated in the following example.

Figure 7.1

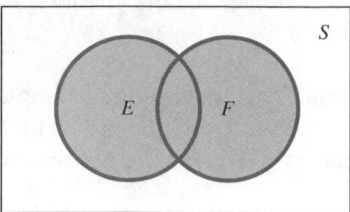

$E \cup F$

(a) The union of two events.

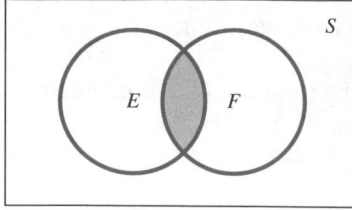

$E \cap F$

(b) The intersection of two events.

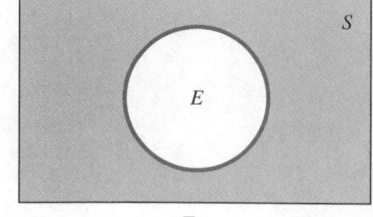

E^c

(c) The complement of the event E.

EXAMPLE 2   Consider the experiment of casting a die and observing the number that falls uppermost. Let $S = \{1, 2, 3, 4, 5, 6\}$ denote the sample space of the experiment and $E = \{2, 4, 6\}$ and $F = \{1, 3\}$ be events of this experiment. Compute (a) $E \cup F$, (b) $E \cap F$, and (c) F^c. Interpret your results.

Solution

a. $E \cup F = \{1, 2, 3, 4, 6\}$ and is the event that the outcome of the experiment is a 1, a 2, a 3, a 4, or a 6.

b. $E \cap F = \varnothing$ is the impossible event; the number appearing uppermost when a die is cast cannot be both even and odd at the same time.

c. $F^c = \{2, 4, 5, 6\}$ is precisely the event that the event F does not occur.

◦ ◦ ◦

If two events cannot occur at the same time, they are said to be mutually exclusive. Using set notation, we have the following definition.

MUTUALLY EXCLUSIVE EVENTS E and F are **mutually exclusive** if $E \cap F = \varnothing$.

a. Suppose E and F are two complementary events. Must E and F be mutually exclusive? Explain your answer.

b. Suppose E and F are mutually exclusive events. Must E and F be complementary? Explain your answer.

As before, we may use Venn diagrams to illustrate these events. In this case, the two mutually exclusive events are depicted as two nonintersecting circles (Figure 7.2).

Figure 7.2
Mutually exclusive events.

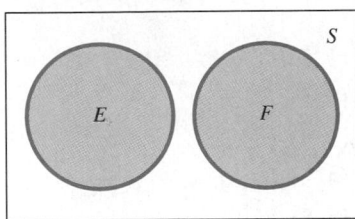

EXAMPLE 3 An experiment consists of tossing a coin three times and observing the resulting sequence of "heads" and "tails."

a. Describe the sample space S of the experiment.

b. Determine the event E that exactly two heads appear.

c. Determine the event F that at least one head appears.

Solution

a. The sample points may be obtained with the aid of a tree diagram (Figure 7.3). The required sample space S is given by

$$S = \{HHH, HHT, HTH, HTT, THH, THT, TTH, TTT\}$$

Figure 7.3

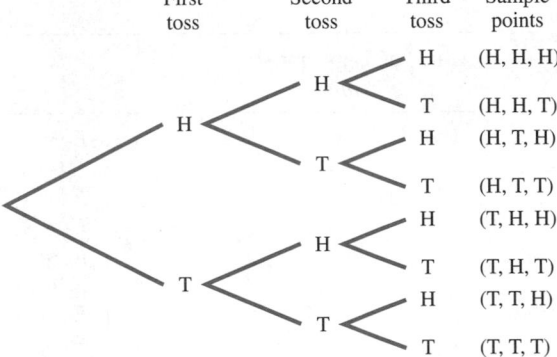

b. By scanning the sample space S obtained in (a), we see that the outcomes in which exactly two heads appear are given by the event

$$E = \{HHT, HTH, THH\}$$

c. Proceeding as in (b), we find

$$F = \{HHH, HHT, HTH, HTT, THH, THT, TTH\} \qquad \circ \; \circ \; \circ$$

EXAMPLE 4 An experiment consists of casting a pair of dice and observing the number that falls uppermost on each die.

a. Describe an appropriate sample space S for this experiment.

b. Determine the events $E_2, E_3, E_4, \ldots, E_{12}$ that the sum of the numbers falling uppermost is $2, 3, 4, \ldots, 12$, respectively.

Solution

a. We may represent each outcome of the experiment by an ordered pair of numbers, the first representing the number that appears uppermost on the first die and the second representing the number that appears uppermost on the second die. In order to distinguish between the two dice, think of the first die as being red and the second as being green. Since there are six possible outcomes for each die, the multiplication principle implies that there are $6 \cdot 6$, or 36, elements in the sample space:

$$S = \{(1, 1), (1, 2), (1, 3), (1, 4), (1, 5), (1, 6),$$
$$(2, 1), (2, 2), (2, 3), (2, 4), (2, 5), (2, 6),$$
$$(3, 1), (3, 2), (3, 3), (3, 4), (3, 5), (3, 6),$$
$$(4, 1), (4, 2), (4, 3), (4, 4), (4, 5), (4, 6),$$
$$(5, 1), (5, 2), (5, 3), (5, 4), (5, 5), (5, 6),$$
$$(6, 1), (6, 2), (6, 3), (6, 4), (6, 5), (6, 6)\}$$

b. With the aid of the results of (a), we obtain the required list of events, shown in Table 7.1.

Table 7.1

Sum of Uppermost Numbers	Event
2	$E_2 = \{(1, 1)\}$
3	$E_3 = \{(1, 2), (2, 1)\}$
4	$E_4 = \{(1, 3), (2, 2), (3, 1)\}$
5	$E_5 = \{(1, 4), (2, 3), (3, 2), (4, 1)\}$
6	$E_6 = \{(1, 5), (2, 4), (3, 3), (4, 2), (5, 1)\}$
7	$E_7 = \{(1, 6), (2, 5), (3, 4), (4, 3), (5, 2), (6, 1)\}$
8	$E_8 = \{(2, 6), (3, 5), (4, 4), (5, 3), (6, 2)\}$
9	$E_9 = \{(3, 6), (4, 5), (5, 4), (6, 3)\}$
10	$E_{10} = \{(4, 6), (5, 5), (6, 4)\}$
11	$E_{11} = \{(5, 6), (6, 5)\}$
12	$E_{12} = \{(6, 6)\}$

○ ○ ○

Applications

EXAMPLE 5 The manager of a local cinema records the number of patrons attending a first-run movie at the 1 P.M. screening. The theater has a seating capacity of 500.

a. What is an appropriate sample space for this experiment?

b. Describe the event E that fewer than 50 people attend the screening.

c. Describe the event F that the theater is more than half full at the screening.

Solution

a. The number of patrons at the screening (the outcome) could run from 0 to 500. Therefore, a sample space for this experiment is

$$S = \{0, 1, 2, 3, \ldots, 500\}$$

b. $E = \{0, 1, 2, 3, \ldots, 49\}$

c. $F = \{251, 252, 253, \ldots, 500\}$

○ ○ ○

EXAMPLE 6 An experiment consists of studying the composition of a three-child family in which the children were born at different times.

a. Describe an appropriate sample space S for this experiment.

b. Describe the event E that there are two girls and a boy in the family.

c. Describe the event F that the oldest child is a girl.

d. Describe the event G that the oldest child is a girl and the youngest child is a boy.

Solution

a. The sample points of the experiment may be obtained with the aid of the tree diagram shown in Figure 7.4, where b denotes a boy and g denotes a girl. We see from the tree diagram that the required sample space is given by

$$S = \{(b, b, b), (b, b, g), (b, g, b), (b, g, g), (g, b, b),$$
$$(g, b, g), (g, g, b), (g, g, g)\}$$

Figure 7.4
Tree diagram for three-child families.

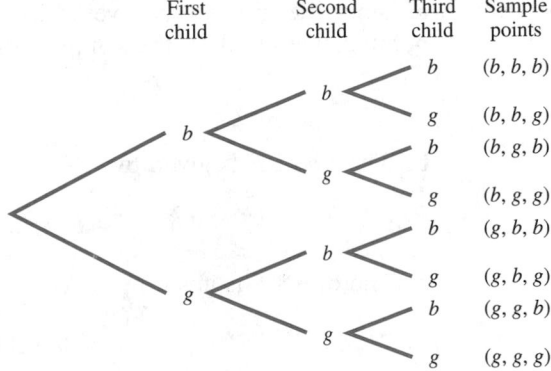

Using the tree diagram, we find that

b. $E = \{(b, g, g), (g, b, g), (g, g, b)\}$

c. $F = \{(g, b, b), (g, b, g), (g, g, b), (g, g, g)\}$

d. $G = \{(g, b, b), (g, g, b)\}$ ○ ○ ○

Think of an experiment.

a. Describe the sample point(s) and sample space of the experiment.

b. Construct two events, E and F, of the experiment.

c. Find the union and intersection of E and F and the complement of E.

d. Are E and F mutually exclusive? Explain your answer.

The next example shows that sample spaces may be infinite.

EXAMPLE 7   The Ever-Brite Battery Company is developing a high-amperage, high-capacity battery as a source for powering electric cars. The battery is tested by installing it in a prototype electric car and running the car with a fully charged battery on a test track at a constant speed of 55 mph until the car runs out of power. The distance covered by the car is then observed.

a. What is the sample space for this experiment?

b. Describe the event E that the driving range under test conditions is less than 150 miles.

c. Describe the event F that the driving range is between 200 and 250 miles, inclusive.

Solution

a. Since the distance d covered by the car in any run may be given by any nonnegative number, the sample space S is given by

$$S = \{d \mid d \geq 0\}$$

b. The event E is given by

$$E = \{d \mid d < 150\}$$

c. The event F is given by

$$F = \{d \mid 200 \leq d \leq 250\}$$ o o o

SELF-CHECK EXERCISES 7.1

1. A sample of three apples taken from Cavallero's Fruit Stand are examined to determine whether they are good or rotten.
   a. What is an appropriate sample space for this experiment?
   b. Describe the event E that exactly one of the apples picked is rotten.
   c. Describe the event F that the first apple picked is rotten.

2. Refer to Self-Check exercise 1.
   a. Find $E \cup F$.
   b. Find $E \cap F$.
   c. Find F^c.
   d. Are the events E and F mutually exclusive?

Solutions to Self-Check Exercises 7.1 can be found on page 381.

7.1 EXERCISES

In exercises 1–6, let S = {a, b, c, d, e, f} be a sample space of an experiment, and let E = {a, b}, F = {a, d, f}, and G = {b, c, e} be events of this experiment.

1. Find the events $E \cup F$ and $E \cap F$.

2. Find the events $F \cup G$ and $F \cap G$.

3. Find the events F^c and $E \cap G^c$.

4. Find the events E^c and $F^c \cap G$.

5. Are the events E and F mutually exclusive?

6. Are the events $E \cup F$ and $E \cap F^c$ mutually exclusive?

In exercises 7–12, let S = {1, 2, 3, 4, 5, 6}, E = {2, 4, 6}, F = {1, 3, 5}, and G = {5, 6}.

7. Find the event $E \cup F \cup G$.

8. Find the event $E \cap F \cap G$.

9. Find the event $(E \cup F \cup G)^c$.

10. Find the event $(E \cap F \cap G)^c$.

11. Are the events E and F mutually exclusive?

12. Are the events F and G mutually exclusive?

In exercises 13–18, let S be any sample space and E, F, and G be any three events associated with the experiment. Describe the given events using the symbols ∪, ∩, and ^{c}.

13. The event that E and/or F occurs.

14. The event that both E and F occur.

15. The event that G does not occur.

16. The event that E but not F occurs.

17. The event that none of the events E, F, and G occurs.

18. The event that E occurs but neither of the events F or G occurs.

19. Let $S = \{a, b, c\}$ be a sample space of an experiment with outcomes $a, b,$ and c. List all the events of this experiment.

20. Let $S = \{1, 2, 3\}$ be a sample space associated with an experiment.
    a. List all events of this experiment.
    b. How many subsets of S contain the number 3?
    c. How many subsets of S contain either the number 2 or the number 3?

21. An experiment consists of selecting a card from a standard deck of playing cards and noting whether it is black (B) or red (R).
    a. Describe an appropriate sample space for this experiment.
    b. What are the events of this experiment?

22. An experiment consists of selecting a letter at random from the letters in the word *MASSACHUSETTS* and observing the outcomes.
    a. What is an appropriate sample space for this experiment?
    b. Describe the event "the letter selected is a vowel."

23. An experiment consists of tossing a coin and casting a die and observing the outcomes.
    a. Describe an appropriate sample space for this experiment.
    b. Describe the event "a head is tossed and an even number is cast."

24. An experiment consists of spinning the hand of the numbered disc shown in the following figure and observing the region in which the pointer stops. (If the needle stops on a line, the result is discounted and the needle is spun again.)

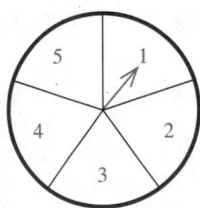

    a. What is an appropriate sample space for this experiment?
    b. Describe the event "the spinner points to the number 2."
    c. Describe the event "the spinner points to an odd number."

25. **Quality Control** A sample of three transistors taken from a local electronics store was examined to determine whether the transistors were defective (d) or nondefective (n). What is an appropriate sample space for this experiment?

26. **Blood Typing** Human blood is classified by the pres-

ence or absence of three main antigens (A, B, and Rh). When a blood specimen is typed, the presence of the A and/or B antigen is indicated by listing the letter A and/or the letter B. If neither the A nor B antigen is present, the letter O is used. The presence or absence of the Rh antigen is indicated by the symbols $+$ or $-$, respectively. Thus, if a blood specimen is classified as AB^+, it contains the A and the B antigens as well as the Rh antigen. Similarly, O^- blood contains none of the three antigens. Using this information, determine the sample space corresponding to the different blood groups.

27. Game Shows In a television game show, the winner is asked to select three prizes from five different prizes, A, B, C, D, and E.
a. Describe a sample space of possible outcomes (order is not important).
b. How many points are there in the sample space corresponding to a selection that includes A?
c. How many points are there in the sample space corresponding to a selection that includes A and B?
d. How many points are there in the sample space corresponding to a selection that includes either A or B?

28. Automatic Tellers The manager of a local bank observes how long it takes a customer to complete his transactions at the automatic bank teller.
a. Describe an appropriate sample space for this experiment.
b. Describe the event that it takes a customer between 2 and 3 minutes to complete his transactions at the automatic bank teller.

29. Common Stocks Mr. Robbins purchased shares of a machine tool company and shares of an airline company. Let E be the event that the shares of the machine tool company increase in value over the next six months, and let F be the event that the shares of the airline company increase in value over the next six months. Using the symbols $\cup$, $\cap$, and ^{c}, describe the following events.
a. The shares in the machine tool company do not increase in value.
b. The shares in both the machine tool company and the airline company do not increase in value.
c. The shares of at least one of the two companies increase in value.
d. The shares of only one of the two companies increase in value.

30. Customer Service Surveys The customer service department of the Universal Instruments Company, manufacturer of the Galaxy home computer, conducted a survey among customers who had returned their purchase registration cards. Purchasers of its deluxe model home computer were asked to report the length of time (t) in days before service was required.
a. Describe a sample space corresponding to this survey.
b. Describe the event E that a home computer required service before a period of 90 days had elapsed.
c. Describe the event F that a home computer did not require service before a period of 1 year had elapsed.

31. Assembly Time Studies A time study was conducted by the production manager of the Vista Vision Corporation to determine the length of time in minutes required by an assembly worker to complete a certain task during the assembly of its Pulsar color television sets.
a. Describe a sample space corresponding to this time study.
b. Describe the event E that an assembly worker took 2 minutes or less to complete the task.
c. Describe the event F that an assembly worker took more than 2 minutes to complete the task.

32. Political Polls An opinion poll is conducted among a state's electorate to determine the relationship between their income levels and their stands on a proposition aimed at reducing state income taxes. Voters are classified as belonging to either the low-, middle-, or upper-income group. They are asked whether they favor, oppose, or are undecided about the proposition. Let the letters L, M, and U represent the low-, middle-, and upper-income groups, respectively, and let the letters f, o, and u represent the responses—favor, oppose, and undecided, respectively.
a. Describe a sample space corresponding to this poll.
b. Describe the event E_1 that a respondent favors the proposition.
c. Describe the event E_2 that a respondent opposes the proposition and does not belong to the low-income group.
d. Describe the event E_3 that a respondent does not favor the proposition and does not belong to the upper-income group.

33. Quality Control As part of a quality-control procedure, an inspector at Bristol Farms randomly selects ten eggs from each consignment of eggs he receives and records the number of broken eggs.
a. What is an appropriate sample space for this experiment?
b. Describe the event E that at most three eggs are broken.
c. Describe the event F that at least five eggs are broken.

34. Political Polls   In the opinion poll of exercise 32, the voters were also asked to indicate their political affiliations—Democrat, Republican, or Independent. As before, let the letters L, M, and U represent the low-, middle-, and upper-income groups, respectively, and let the letters D, R, and I represent Democrat, Republican, and Independent, respectively.
   a. Describe a sample space corresponding to this poll.
   b. Describe the event E_1 that a respondent is a Democrat.
   c. Describe the event E_2 that a respondent belongs to the upper-income group and is a Republican.
   d. Describe the event E_3 that a respondent belongs to the middle-income group and is not a Democrat.

35. Shuttle Bus Usage   A certain airport hotel operates a shuttle bus service between the hotel and the airport. The maximum capacity of a bus is 20 passengers. On alternate trips of the shuttle bus over a period of one week, the hotel manager kept a record of the number of passengers arriving at the hotel in each bus.
   a. What is an appropriate sample space for this experiment?
   b. Describe the event E that a shuttle bus carried fewer than ten passengers.
   c. Describe the event F that a shuttle bus arrived with a full capacity.

36. Sports   Eight players, A, B, C, D, E, F, G, and H, are competing in a series of elimination matches of a tennis

tournament in which the winner of each preliminary match will advance to the semifinals and the winner of the semifinals will advance to the finals. An outline of the scheduled matches follows. Describe a sample space listing the possible participants in the finals.

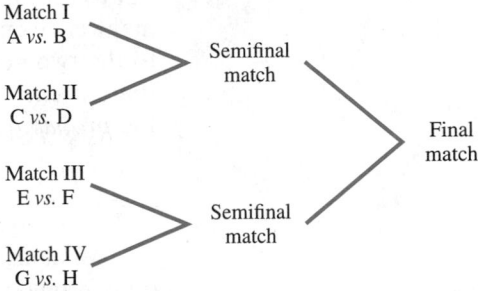

37. Let S be a sample space for an experiment. Show that if E is any event of an experiment, then E and E^c are mutually exclusive.

38. Let S be a sample space for an experiment, and let E and F be events of this experiment. Show that the events $E \cup F$ and $E^c \cap F^c$ are mutually exclusive.
   [*Hint:* Use De Morgan's Law.]

39. Let S be a sample space of an experiment with n outcomes. Determine the number of events of this experiment.

SOLUTIONS TO SELF-CHECK EXERCISES 7.1

1. a. Let g denote a good apple and r a rotten apple. Thus, the required sample points may be obtained with the aid of a tree diagram (compare with Example 3). The required sample space is given by

$$S = \{ggg, ggr, grg, grr, rgg, rgr, rrg, rrr\}$$

b. By scanning the sample space S obtained in (a), we identify the outcomes in which exactly one apple is rotten. We find

$$E = \{ggr, grg, rgg\}$$

c. Proceeding as in (b), we find

$$F = \{rgg, rgr, rrg, rrr\}$$

2. Using the results of Self-Check exercise 1, we find
   a. $E \cup F = \{ggr, grg, rgg, rgr, rrg, rrr\}$
   b. $E \cap F = \{rgg\}$
   c. F^c is the set of outcomes in S but not in F. Thus,

$$F^c = \{ggg, ggr, grg, grr\}$$

   d. Since $E \cap F \neq \varnothing$, we conclude that E and F are not mutually exclusive.

7.2 DEFINITION OF PROBABILITY

Finding the Probability of an Event

Let us return to the coin-tossing experiment. The sample space of this experiment is given by $S = \{H, T\}$, where the sample points H and T correspond to the two possible outcomes, a *head* and a *tail*. If the coin is *unbiased,* then there is *one chance out of two* of obtaining a head (or a tail) and we say that the *probability* of tossing a head (tail) is 1/2, abbreviated

$$P(H) = \frac{1}{2} \qquad P(T) = \frac{1}{2}$$

An alternative method of obtaining the values of $P(H)$ and $P(T)$ is based on continued experimentation and does not depend on the assumption that the two outcomes are equally likely. Table 7.2 summarizes the results of such an exercise.

Table 7.2
As the number of trials increases, the relative frequency approaches .5.

Number of Tosses (*n*)	Number of Heads (*m*)	Relative Frequency of Heads (*m/n*)
10	4	.4000
100	58	.5800
1,000	492	.4920
10,000	5,034	.5034
20,000	10,024	.5012
40,000	20,032	.5008

Observe that the relative frequencies (column 3) differ considerably when the number of trials is small, but as the number of trials becomes very large, the relative frequency approaches the number .5. This result suggests that we assign to $P(H)$ the value 1/2, as before.

More generally, consider an experiment that may be repeated over and over again under independent and similar conditions. Suppose that in n trials, an event E occurs m times. We call the ratio m/n the **relative frequency** of the event E after n repetitions. If this relative frequency approaches some value $P(E)$ as n becomes larger and larger, then $P(E)$ is called the **empirical probability** of E. Thus, the probability $P(E)$ of an event occurring is a measure of the proportion of the time that the event E will occur in the long run. Observe that this method of computing the probability of a head occurring is effective even in the case when a biased coin is used in the experiment.

The **probability of an event** is a number that lies between 0 and 1. In general, the larger the probability of an event, the more likely the event will occur. Thus, an event with a probability of .8 is more likely to occur than an event with a probability of .6. An event with a probability of 1/2, or .5, has a "fifty-fifty," or equal, chance of occurring.

Now suppose we are given an experiment and wish to determine the probabilities associated with certain events of the experiment. This problem could be solved by computing $P(E)$ directly for each event E of interest. However, in practice, the number of events that we may be interested in is usually quite large, so this approach is not satisfactory.

The following approach is particularly suitable when the sample space of an experiment is finite.* Let S be a finite sample space with n outcomes; that is,

$$S = \{s_1, s_2, s_3, \ldots, s_n\}$$

Then the events

$$\{s_1\}, \{s_2\}, \{s_3\}, \ldots, \{s_n\}$$

which consist of exactly one point, are called **simple,** or **elementary,** events of the experiment. They are elementary in the sense that any (nonempty) event of the experiment may be obtained by taking a finite union of suitable elementary events. The simple events of an experiment are also **mutually exclusive;** that is, given any two simple events of the experiment, only one can occur.

By assigning probabilities to each of the simple events, we obtain the results shown in Table 7.3. This table is called a **probability distribution** for the experiment. The function P, which assigns a probability to each of the simple events, is called a **probability function.**

The numbers $P(s_1), P(s_2), \ldots, P(s_n)$ have the following properties:

Table 7.3

A probability distribution.

Simple Event	Probability**
$\{s_1\}$	$P(s_1)$
$\{s_2\}$	$P(s_2)$
$\{s_3\}$	$P(s_3)$
.	.
.	.
.	.
$\{s_n\}$	$P(s_n)$

1. $0 \le P(s_i) \le 1 \quad (i = 1, 2, \ldots, n)$

2. $P(s_1) + P(s_2) + \cdots + P(s_n) = 1$

3. $P(\{s_i\} \cup \{s_j\}) = P(s_i) + P(s_j) \quad i \ne j \quad (i = 1, 2, \ldots, n; j = 1, 2, \ldots, n)$

The first property simply states that the probability of a simple event must be between 0 and 1 inclusive. The second property states that the sum of the probabilities of all simple events of the sample space is 1. This follows from the fact that the event S is certain to occur. The third property states that the probability of the union of two mutually exclusive events is given by the sum of their probabilities.

As we saw earlier, there is no unique method for assigning probabilities to the simple events of an experiment. In practice, the methods used to determine these probabilities may range from theoretical considerations of the problem on the one extreme to the reliance on "educated guesses" on the other.

Sample spaces in which the outcomes are equally likely are called **uniform sample spaces.** Assigning probabilities to the simple events in these spaces is relatively easy.

ο ο ο

* For the remainder of the chapter we assume that all sample spaces are finite.

** For simplicity, we use the notation $P(s_i)$ instead of the technically more correct $P(\{s_i\})$.

PROBABILITY OF AN EVENT IN A UNIFORM SAMPLE SPACE

If

$$S = \{s_1, s_2, \ldots, s_n\}$$

is the sample space for an experiment in which the outcomes are equally likely, then we assign the probabilities

$$P(s_1) = P(s_2) = \cdots = P(s_n) = \frac{1}{n}$$

to each of the simple events $s_1, s_2, \ldots, s_n$.

EXAMPLE 1   A fair die is cast and the number that falls uppermost is observed. Determine the probability distribution for the experiment.

Solution   The sample space for the experiment is $S = \{1, 2, 3, 4, 5, 6\}$ and the simple events are accordingly given by the sets $\{1\}$, $\{2\}$, $\{3\}$, $\{4\}$, $\{5\}$, and $\{6\}$. Since the die is assumed to be fair, the six outcomes are equally likely. We therefore assign a probability of 1/6 to each of the simple events and obtain the probability distribution shown in Table 7.4.

Table 7.4
A probability distribution.

Simple Event	Probability
{1}	$\frac{1}{6}$
{2}	$\frac{1}{6}$
{3}	$\frac{1}{6}$
{4}	$\frac{1}{6}$
{5}	$\frac{1}{6}$
{6}	$\frac{1}{6}$

∘ ∘ ∘

You suspect that a die is biased.

a. Describe a method you might use to prove your assertion.

b. How would you assign the probability to each outcome 1 through 6 of the experiment of casting the die and observing which number lands uppermost?

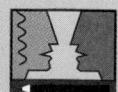

The next example shows how the *relative frequency* interpretation of probability lends itself to the computation of probabilities.

Table 7.5
Data obtained during 200 test runs of an electric car.

Distance Covered in Miles (x)	Frequency of Occurrence
$0 \le x \le 50$	4
$50 < x \le 100$	10
$100 < x \le 150$	30
$150 < x \le 200$	100
$200 < x \le 250$	40
$250 < x$	16

EXAMPLE 2   Refer to Example 7, Section 7.1. The data shown in Table 7.5 were obtained in tests involving 200 test runs. Each run was made with a fully charged battery.

a. Describe an appropriate sample space for this experiment.

b. Find the probability distribution for this experiment.

Solution

a. Let s_1 denote the outcome that the distance covered by the car does not exceed 50 miles; let s_2 denote the outcome that the distance covered by the car is greater than 50 miles but does not exceed 100 miles, and so on. Finally, let s_6 denote the outcome that the distance covered by the car is greater than 250 miles. Then the required sample space is given by

$$S = \{s_1, s_2, s_3, s_4, s_5, s_6\}$$

b. To compute the probability distribution for the experiment, we turn to the relative frequency interpretation of probability. Accepting the inaccuracies inherent in a relatively small number of trials (200 runs), we take the probability of s_1 occurring as

$$P(s_1) = \frac{\text{Number of trials in which } s_1 \text{ occurs}}{\text{Total number of trials}}$$

$$= \frac{4}{200} = .02$$

In a similar manner, we assign probabilities to the other simple events, obtaining the probability distribution shown in Table 7.6.

Table 7.6
A probability distribution.

Simple Event	Probability
$\{s_1\}$	.02
$\{s_2\}$	.05
$\{s_3\}$	.15
$\{s_4\}$	.50
$\{s_5\}$	.20
$\{s_6\}$	.08

○ ○ ○

We are now in a position to give a procedure for computing the probability $P(E)$ of an arbitrary event E of an experiment.

FINDING THE PROBABILITY OF AN EVENT *E*	**1.** Determine a sample space S associated with the experiment. **2.** Assign probabilities to the simple events of S. **3.** If $E = \{s_1, s_2, s_3, \ldots, s_n\}$ where $\{s_1\}, \{s_2\}, \{s_3\}, \ldots, \{s_n\}$ are simple events, then $$P(E) = P(s_1) + P(s_2) + P(s_3) + \cdots + P(s_n)$$ If E is the empty set, $\varnothing$, then $P(E) = 0$.

The principle stated in step 3 is called the **addition principle** and is a consequence of property 3 of the probability function (page 392). This principle allows us to find the probabilities of all the other events once the probabilities of the simple events are known.

⚠️ The addition rule applies *only* to the addition of probabilities of simple events.

Applications

EXAMPLE 3   A pair of fair dice is cast.

a. Calculate the probability that the two dice show the same number.

b. Calculate the probability that the sum of the numbers of the two dice is 6.

Solution   From the results of Example 4, page 375, we see that the sample space S of the experiment consists of 36 outcomes:

$$S = \{(1, 1), (1, 2), \ldots, (6, 5), (6, 6)\}$$

Since both dice are fair, each of the 36 outcomes is equally likely. Accordingly, we assign the probability of 1/36 to each simple event. We are now in a position to answer the questions posed.

a. The event that the two dice show the same number is given by

$$E = \{(1, 1), (2, 2), (3, 3)(4, 4), (5, 5), (6, 6)\} \qquad \text{(See Figure 7.5.)}$$

Therefore, by the addition principle, the probability that the two dice show the same number is given by

$$P(E) = P[(1, 1)] + P[(2, 2)] + \cdots + P[(6, 6)]$$
$$= \frac{1}{36} + \frac{1}{36} + \cdots + \frac{1}{36} \qquad \text{(Six terms)}$$
$$= \frac{1}{6}$$

b. The event that the sum of the numbers of the two dice is 6 is given by

$$E_6 = \{(1, 5), (2, 4), (3, 3), (4, 2), (5, 1)\} \qquad \text{(See Figure 7.6.)}$$

Figure 7.5
The event that the two dice show the same number.

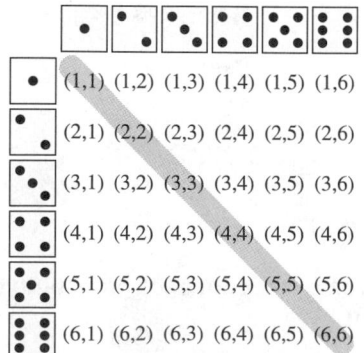

Figure 7.6
The event that the sum of the numbers on the two dice is 6.

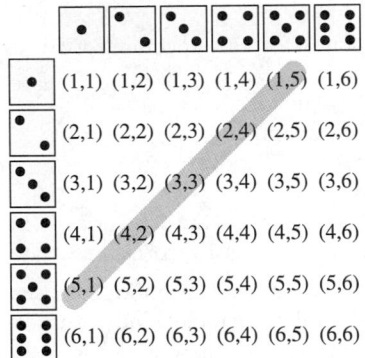

(1,1)	(1,2)	(1,3)	(1,4)	(1,5)	(1,6)
(2,1)	(2,2)	(2,3)	(2,4)	(2,5)	(2,6)
(3,1)	(3,2)	(3,3)	(3,4)	(3,5)	(3,6)
(4,1)	(4,2)	(4,3)	(4,4)	(4,5)	(4,6)
(5,1)	(5,2)	(5,3)	(5,4)	(5,5)	(5,6)
(6,1)	(6,2)	(6,3)	(6,4)	(6,5)	(6,6)

Therefore, the probability that the sum of the numbers on the two dice is 6 is given by

$$P(E_6) = P[(1,5)] + P[(2,4)] + P[(3,3)] + P[(4,2)] + P[(5,1)]$$
$$= \frac{1}{36} + \frac{1}{36} + \cdots + \frac{1}{36} \quad \text{(Five terms)}$$
$$= \frac{5}{36}$$

○ ○ ○

EXAMPLE 4   Consider the experiment by the Ever-Brite Company in Example 2. What is the probability that the prototype car will travel more than 150 miles on a fully charged battery?

Solution   Using the results of Example 2, we see that the event that the car will travel more than 150 miles on a fully charged battery is given by $E = \{s_4, s_5, s_6\}$. Therefore, the probability that the car will travel more than 150 miles on one charge is given by

$$P(E) = P(s_4) + P(s_5) + P(s_6)$$

or, using the probability distribution for the experiment obtained in Example 2,

$$P(E) = .50 + .20 + .08 = .78$$

○ ○ ○

SELF–CHECK EXERCISES 7.2

1. A biased die was cast repeatedly, and the results of the experiment are summarized in the following table:

Outcome	1	2	3	4	5	6
Frequency of Occurrence	142	173	158	175	162	190

Using the relative frequency interpretation of probability, find the probability distribution for this experiment.

2. In an experiment conducted to study the effectiveness of an eye-level third brake light in the prevention of rear-end collisions, 250 of the 500 highway patrol cars of a certain state were equipped with such lights. At the end of the 1-year trial period, the records revealed that of those equipped with a third brake light, there were 14 incidents of rear-end collision. There were 22 such incidents involving the cars not equipped with the accessory. Based on these data, what is the probability that a highway patrol car equipped with a third brake light will be rear-ended within a 1-year period? What is the probability that a car not so equipped will be rear-ended within a 1-year period?

Solutions to Self-Check Exercises 7.2 can be found on page 391.

7.2 EXERCISES

In exercises 1–8, list the simple events associated with each of the given experiments.

1. A nickel and a dime are tossed and the result of heads or tails is recorded for each coin.

2. A card is selected at random from a standard 52-card deck and its suit—hearts (h), diamonds (d), spades (s), or clubs (c)—is recorded.

3. **Opinion Polls** An opinion poll is conducted among a group of registered voters. Their political affiliation, Democrat (D), Republican (R), or Independent (I), and their sex, male (m) or female (f), are recorded.

4. **Quality Control** As part of a quality-control procedure, eight circuit boards are checked and the number of defectives is recorded.

5. **Movie Attendance** In a survey conducted to determine whether movie attendance is increasing (i), decreasing (d), or holding steady (s) among various sectors of the population, participants are classified as follows:

   Group 1: those aged 10–19

   Group 2: those aged 20–29

   Group 3: those aged 30–39

   Group 4: those aged 40–49

   Group 5: those 50 and over

   The response of each participant and his or her age group are recorded.

6. **Durable Goods Orders** Data concerning durable goods orders are obtained each month by an economist. A record is kept for a 1-year period of any increase (i), decrease (d), or unchanged movement (u) in the number of durable goods orders for each month as compared to the number of such orders in the same month in the previous year.

7. **Blood Types** Blood tests are given as a part of the admission procedure at the Monterey Garden Community Hospital. The blood type of each patient (A, B, AB, or O) and the presence or absence of the Rh factor in each patient's blood (Rh^+ or Rh^-) are recorded.

8. **Meteorology** A meteorologist preparing a weather map classifies the expected average temperature in each of five neighboring states for the upcoming week as follows:

   a. More than 10° below average

   b. Normal to 10° below average

   c. Higher than normal to 10° above average

   d. More than 10° above average

   Using each state's abbreviation and the categories—(a), (b), (c), and (d)—the meteorologist records these data.

9. **Grade Distributions** The grade distribution for a certain class is shown in the following table. Find the probability distribution associated with these data.

Grade	A	B	C	D	F
Frequency of Occurrence	4	10	18	6	2

10. **Blood Types** The percentage of the general population that has each blood type is shown in the following table.

Blood Type	A	B	AB	O
Percentage of Population	41%	12%	3%	44%

Determine the probability distribution associated with these data.

11. **Traffic Surveys** The number of cars entering a tunnel leading to an airport in a major city over a period of 200 peak hours was observed and the following data were obtained:

Number of Cars (x)	Frequency of Occurrence
$0 < x \le 200$	15
$200 < x \le 400$	20
$400 < x \le 600$	35
$600 < x \le 800$	70
$800 < x \le 1000$	45
$x > 1000$	15

a. Describe an appropriate sample space for this experiment.

b. Find the probability distribution for this experiment.

12. **Product Surveys** The accompanying data were obtained from a survey of 1500 Americans who were asked: How safe are American-made consumer products?

Rating	A (Very Safe)	B (Somewhat Safe)
Number of Respondents	285	915

Rating	C (Not too Safe)	D (Not Safe at All)
Number of Respondents	225	30

Rating	E (Don't Know)
Number of Respondents	45

Determine the probability distribution associated with these data.

13. **Political Views** In a poll conducted among 2000 college freshmen to ascertain the political views of college students, the accompanying data were obtained.

Political Views	A (Far Left)	B (Liberal)
Number of Respondents	52	398

Political Views	C (Middle-of-the-Road)
Number of Respondents	1140

Political Views	D (Conservative)	E (Far Right)
Number of Respondents	386	24

Determine the probability distribution associated with these data.

14. **Service-Utilization Studies** The Metro Telephone Company of Belmont compiled the accompanying information during a service-utilization study pertaining to the number of customers using their Dial-the-Time service from 7 A.M. to 9 A.M. on a certain weekday morning. Using these data, find the probability distribution associated with the experiment.

Number of Calls Received/Minute	Frequency of Occurrence
10	6
11	15
12	12
13	3
14	12
15	36
16	24
17	0
18	6
19	6

15. **Assembly Time Studies** The results of a time study conducted by the production manager of the Ace Novelty Company are shown in the accompanying table, where the number of space action-figures produced each quarter hour during an 8-hour workday has been tabulated. Find the probability distribution associated with this experiment.

Number of Figures Produced (in dozens)	Frequency of Occurrence
30	4
31	0
32	6
33	8
34	6
35	4
36	4

16. **Electronic Mail Services** The number of subscribers to five leading electronic mail services is shown in the accompanying table.

Company	A	B	C
Subscribers	300,000	200,000	120,000

Company	D	E
Subscribers	80,000	60,000

Find the probability distribution associated with these data.

17. Corrective Lens Use According to Mediamark Research, Inc., 84 million out of 179 million adults in the United States correct their vision by using prescription eyeglasses, bifocals, or contact lenses. (Some respondents use more than one type.) What is the probability that an adult selected at random from the adult population uses corrective lenses?

18. Correctional Supervision A study conducted by the Corrections Department of a certain state revealed that 163,605 people out of a total adult population of 1,778,314 were under correctional supervision (on probation, parole, or in jail). What is the probability that a person selected at random from the adult population in that state is under correctional supervision?

19. Lightning Deaths According to data obtained from the National Weather Service, 376 of the 439 people killed by lightning in the United States between 1985 and 1992 were men. (Job and recreational habits of men make them more vulnerable to lightning.) Assuming that this trend holds in the future, find the probability that
a. a person killed by lightning is a male.
b. a person killed by lightning is a female.

20. Quality Control One light bulb is selected at random from a lot of 120 light bulbs, of which 5% are defective. What is the probability that the light bulb selected is defective?

21. Efforts to Stop Shoplifting According to a survey of 176 retailers, 46% of them use electronic tags as protection against shoplifting and employee theft. If one of these retailers is selected at random, what is the probability that he or she uses electronic tags as antitheft devices?

22. If a ball is selected at random from an urn containing three red balls, two white balls, and five blue balls, what is the probability that it will be a white ball?

23. If one card is drawn at random from a standard 52-card deck, what is the probability that the card drawn is
a. a diamond?   **b.** a black card?
c. an ace?

24. A pair of fair dice is cast. What is the probability that
a. the sum of the numbers shown uppermost is less than 5?
b. at least one 6 is cast?

25. Traffic Lights What is the probability of arriving at a traffic light when it is red if the red signal is flashed for 30 seconds, the yellow signal for 5 seconds, and the green signal for 45 seconds?

26. Roulette What is the probability that a roulette ball will come to rest on an even number other than 0 or 00? (Assume that there are 38 equally likely outcomes consisting of the numbers 1–36, 0, and 00.)

27. Refer to exercise 9. What is the probability that a student selected at random from this class received a passing grade (D or better)?

28. Refer to exercise 11. What is the probability that more than 600 cars will enter the airport tunnel during a peak hour?

29. Disposition of Criminal Cases Of the 98 first-degree murder cases from 1990 through the first half of 1992 in the Suffolk superior court, 9 cases were thrown out of the system, 62 cases were plea-bargained, and 27 cases went to trial. What is the probability that
a. a case selected at random was settled through plea bargaining?
b. a case selected at random went to trial?

30. Sweepstakes One hundred thousand entries have been received in a sweepstakes sponsored by the Gemini Paper Products Company. If 1 grand prize, 5 first prizes, 25 second prizes, and 500 third prizes are to be awarded, what is the probability that
a. a person who has submitted one entry will win the grand prize?
b. a person who has submitted one entry will win a prize?

31. A pair of fair dice is cast and the sum of the two numbers falling uppermost observed. The probability of obtaining a sum of 2 is the same as that of obtaining a 7 since there is only one way of getting a 2—namely, by each die showing a 1; and there is only one way of obtaining a 7—namely, by one die showing a 3 and the other die showing a 4. What is wrong with this argument?

In exercises 32–35, determine whether the given experiment has a sample space with equally likely outcomes.

32. A loaded die is cast and the number appearing uppermost on the die is recorded.

33. Two fair dice are cast and the sum of the numbers appearing uppermost is recorded.

34. A ball is selected at random from an urn containing six black balls and six red balls and the color of the ball is recorded.

35. A weighted coin is thrown and the outcome of heads or tails is recorded.

36. Let $S = \{s_1, s_2, s_3, s_4, s_5\}$ be the sample space associated with an experiment having the following probability distribution.

Outcome	$\{s_1\}$	$\{s_2\}$	$\{s_3\}$	$\{s_4\}$	$\{s_5\}$
Probability	$\frac{1}{14}$	$\frac{3}{14}$	$\frac{6}{14}$	$\frac{2}{14}$	$\frac{2}{14}$

Find the probability of the event
a. $A = \{s_1, s_2, s_4\}$ **b.** $B = \{s_1, s_5\}$
c. $C = S$

37. Let $S = \{s_1, s_2, s_3, s_4, s_5, s_6\}$ be the sample space associated with an experiment having the following probability distribution.

Outcome	$\{s_1\}$	$\{s_2\}$	$\{s_3\}$	$\{s_4\}$	$\{s_5\}$	$\{s_6\}$
Probability	$\frac{1}{12}$	$\frac{1}{4}$	$\frac{1}{12}$	$\frac{1}{6}$	$\frac{1}{3}$	$\frac{1}{12}$

Find the probability of the event
a. $A = \{s_1, s_3\}$ **b.** $B = \{s_2, s_4, s_5, s_6\}$
c. $C = S$

38. Political Polls An opinion poll was conducted among a group of registered voters in a certain state concerning a proposition aimed at limiting state and local taxes.

Results of the poll indicated that 35% of the voters favored the proposition, 32% were against it, and the remaining group were undecided. If the results of the poll are assumed to be representative of the opinions of the state's electorate, what is the probability that
a. a registered voter selected at random from the electorate favors the proposition?
b. a registered voter selected at random from the electorate is undecided about the proposition?

39. Consider the composition of a three-child family in which the children were born at different times. Assume that a girl is as likely as a boy at each birth.
a. What is the probability that there are two girls and a boy in the family?
b. What is the probability that the oldest child is a girl?
c. What is the probability that the oldest child is a girl and the youngest child is a boy?

40. Airfone Usage The number of planes in the fleets of five leading airlines that contain Airfones is shown in the accompanying table.

Airline	Number of Planes with Airfones	Size of Fleet
A	50	295
B	40	325
C	31	167
D	29	50
E	25	248

a. If a plane is selected at random from airline A, what is the probability that it contains an Airfone?
b. If a plane is selected at random from the entire fleet of the five airlines, what is the probability that it contains an Airfone?

SOLUTIONS TO SELF-CHECK EXERCISES 7.2

1. $P(1) = \dfrac{\text{Number of trials in which a 1 appears uppermost}}{\text{Total number of trials}}$

$= \dfrac{142}{1000}$

$= .142$

Similarly, we compute $P(2), \ldots, P(6)$, obtaining the following probability distribution.

Outcome	1	2	3	4	5	6
Probability	.142	.173	.158	.175	.162	.190

2. The probability that a highway patrol car equipped with a third brake light will be rear-ended within a 1-year period is given by

$$\frac{\text{Number of rear-end collisions involving cars equipped with a third brake light}}{\text{Total number of such cars}} = \frac{14}{250} = .056$$

The probability that a highway patrol car not equipped with a third brake light will be rear-ended within a 1-year period is given by

$$\frac{\text{Number of rear-end collisions involving cars not equipped with a third brake light}}{\text{Total number of such cars}} = \frac{22}{250} = .088$$

7.3 RULES OF PROBABILITY

Properties of the Probability Function and Their Applications

In this section we examine some of the properties of the probability function and look at the role they play in solving certain problems. We begin by looking at the generalization of the three properties of the probability function, which were stated for simple events in the last section. Let S be a sample space of an experiment and suppose E and F are events of the experiment. Then we have

PROPERTY 1	$P(E) \geq 0$ for any E
PROPERTY 2	$P(S) = 1$
PROPERTY 3	If E and F are mutually exclusive (that is, only one of them can occur, or equivalently, $E \cap F = \varnothing$), then

$$P(E \cup F) = P(E) + P(F) \qquad \text{(See Figure 7.7.)}$$

Figure 7.7
If E and F are mutually exclusive events, then $P(E \cup F) = P(E) + P(F)$.

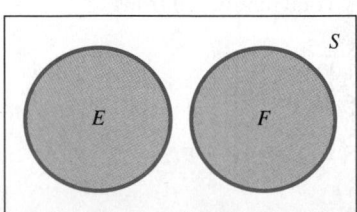

Property 3 may be easily extended to the case involving any finite number of mutually exclusive events. Thus, if $E_1, E_2, \ldots, E_n$ are mutually exclusive events, then

$$P(E_1 \cup E_2 \cup \cdots \cup E_n) = P(E_1) + P(E_2) + \cdots + P(E_n)$$

Table 7.7
Probability distribution.

Score (x)	Probability
$x > 700$	.01
$600 < x \le 700$	.07
$500 < x \le 600$	.19
$400 < x \le 500$	.23
$300 < x \le 400$	.31
$x \le 300$	.19

EXAMPLE 1   The superintendent of a metropolitan school district has estimated the probabilities associated with the SAT verbal scores of students from that district. The results are shown in Table 7.7. If a student is selected at random, what is the probability that his or her SAT verbal score will be

a. more than 400?

b. less than or equal to 500?

c. greater than 400 but less than or equal to 600?

Solution   Let A, B, C, D, E, and F denote, respectively, the event that the score is greater than 700, greater than 600 but less than or equal to 700, greater than 500 but less than or equal to 600, and so forth. Then, these events are mutually exclusive. Therefore

a. the probability that the student's score will be more than 400 is given by

$$P(D \cup C \cup B \cup A) = P(D) + P(C) + P(B) + P(A)$$
$$= .23 + .19 + .07 + .01$$
$$= .5$$

b. the probability that the student's score will be less than or equal to 500 is given by

$$P(D \cup E \cup F) = P(D) + P(E) + P(F)$$
$$= .23 + .31 + .19$$
$$= .73$$

Figure 7.8
$P(E \cup F) = P(E) + P(F) - P(E \cap F)$.

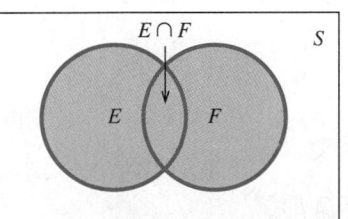

c. the probability that the student's score will be greater than 400 but less than or equal to 600 is given by

$$P(C \cup D) = P(C) + P(D)$$
$$= .19 + .23$$
$$= .42$$

○ ○ ○

Property 3 holds if and only if E and F are mutually exclusive. In the general case, we have the following rule:

**PROPERTY 4
(ADDITION RULE)**

If E and F are any two events of an experiment, then

$$P(E \cup F) = P(E) + P(F) - P(E \cap F) \quad \text{(See Figure 7.8.)}$$

To derive the property for uniform sample spaces, we use equation (4), Section 6.2, to see that

$$n(E \cup F) = n(E) + n(F) - n(E \cap F)$$

where E and F are events of an experiment with sample space S. Dividing both sides of this equation by $n(S)$, we obtain

$$\frac{n(E \cup F)}{n(S)} = \frac{n(E)}{n(S)} + \frac{n(F)}{n(S)} - \frac{n(E \cap F)}{n(S)}$$

Recalling the definition of the probability of an event then leads to

$$P(E \cup F) = P(E) + P(F) - P(E \cap F)$$

as we wish to show.

REMARK   Observe that when E and F are mutually exclusive—that is, when $E \cap F = \emptyset$—then the equation of property 4 reduces to that of property 3. In other words, if E and F are mutually exclusive events, then

$$P(E \cup F) = P(E) + P(F)$$

If E and F are not mutually exclusive events, then

$$P(E \cup F) = P(E) + P(F) - P(E \cap F)$$ ◦ ◦ ◦

EXAMPLE 2   A card is drawn from a well-shuffled deck of 52 playing cards.
....   What is the probability that it is an ace or a spade?

Solution   Let E denote the event that the card drawn is an ace, and let F denote the event that the card drawn is a spade. Then

$$P(E) = \frac{4}{52} \quad \text{and} \quad P(F) = \frac{13}{52}$$

Furthermore, E and F are not mutually exclusive events. In fact $E \cap F$ is the event that the card drawn is an ace of spades. Consequently,

$$P(E \cap F) = \frac{1}{52}$$

The event that a card drawn is an ace or a spade is $E \cup F$, with probability given by

$$P(E \cup F) = P(E) + P(F) - P(E \cap F)$$
$$= \frac{4}{52} + \frac{13}{52} - \frac{1}{52} = \frac{4}{13} \quad \text{(See Figure 7.9.)}$$

This result, of course, can be obtained by arguing that 16 of the 52 cards are either spades or aces of other suits.   ◦ ◦ ◦

Figure 7.9
$P(E \cup F) = P(E) + P(F) - P(E \cap F)$.

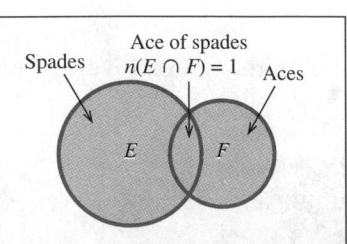

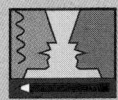

Let E, F, and G be any three events of an experiment. Use formula (5) of Section 6.2 to show that

$$P(E \cup F \cup G) = P(E) + P(F) + P(G) - P(E \cap F) - P(E \cap G) - P(F \cap G)$$
$$+ P(E \cap F \cap G)$$

If E, F, and G are mutually exclusive, what is $P(E \cup F \cup G)$?

EXAMPLE 3 The quality-control department of the Vista Vision Corporation, manufacturer of the Pulsar 19-inch color television set, has determined from records obtained from the company's service centers that 3% of the sets sold experience video problems, 1% experience audio problems, and 0.1% experience both video as well as audio problems before the expiration of the 90-day warranty. Find the probability that a set purchased by a consumer will experience video or audio problems before the warranty expires.

Solution Let E denote the event that a set purchased will experience video problems within 90 days, and let F denote the event that a set purchased will experience audio problems within 90 days. Then

$$P(E) = .03, \quad P(F) = .01, \quad \text{and} \quad P(E \cap F) = .001$$

The event that a set purchased will experience video problems or audio problems before the warranty expires is $E \cup F$, and the probability of this event is given by

$$P(E \cup F) = P(E) + P(F) - P(E \cap F) \qquad \text{(See Figure 7.10.)}$$
$$= .03 + .01 - .001$$
$$= .039 \qquad \qquad \circ \; \circ \; \circ$$

Another property of the probability function that is of considerable aid in computing probability follows.

Figure 7.10
$P(E \cup F) = P(E) + P(F) - P(E \cap F)$.

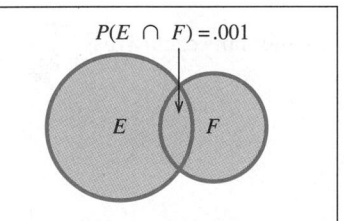

$P(E \cap F) = .001$

PROPERTY 5 (RULE OF COMPLEMENTS)	If E is an event of an experiment, and E^c denotes the complement of E, then $$P(E^c) = 1 - P(E)$$

Property 5 is an immediate consequence of properties 2 and 3. Indeed, we have $E \cup E^c = S$ and $E \cap E^c = \varnothing$, so

$$1 = P(S) = P(E \cup E^c) = P(E) + P(E^c)$$

and, therefore, $\qquad P(E^c) = 1 - P(E)$

EXAMPLE 4   Refer to Example 3. What is the probability that a Pulsar 19-inch color television set bought by a consumer will not experience video or audio difficulties before the warranty expires?

Solution   Let E denote the event that a set bought by a consumer will experience video or audio difficulties before the warranty expires. Then the event that the set will not experience either problem before the warranty expires is given by E^c, with probability

$$P(E^c) = 1 - P(E)$$
$$= 1 - .039$$
$$= .961$$ ○ ○ ○

Computations Involving the Rules of Probability

We close this section by looking at two additional examples that illustrate the rules of probability.

EXAMPLE 5   Let E and F be two mutually exclusive events and suppose $P(E) = .1$ and $P(F) = .6$. Compute

a. $P(E \cap F)$ **b.** $P(E \cup F)$ **c.** $P(E^c)$ **d.** $P(E^c \cap F^c)$

e. $P(E^c \cup F^c)$

Solution

a. Since the events E and F are mutually exclusive—that is, $E \cap F = \emptyset$—we have $P(E \cap F) = 0$.

b. $P(E \cup F) = P(E) + P(F)$ (Since E and F are mutually exclusive)
$$= .1 + .6$$
$$= .7$$

c. $P(E^c) = 1 - P(E)$ (Property 5)
$$= 1 - .1$$
$$= .9$$

d. Observe that, by De Morgan's Law, $E^c \cap F^c = (E \cup F)^c$, so

$$P(E^c \cap F^c) = P[E \cup F)^c]$$ (See Figure 7.11.)
$$= 1 - P(E \cup F)$$ (Property 5)
$$= 1 - .7$$ [Using the result of (b)]
$$= .3$$

e. Again, using De Morgan's Law, we find

$$P(E^c \cup F^c) = P[(E \cap F)^c]$$
$$= 1 - P(E \cap F)$$
$$= 1 - 0$$ [Using the result of (a)]
$$= 1$$ ○ ○ ○

Figure 7.11
$P(E^c \cap F^c) = P[(E \cup F)^c]$.

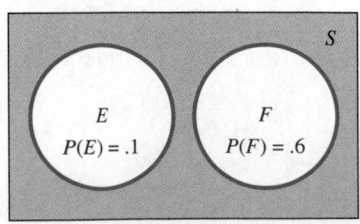

EXAMPLE 6 Let E and F be two events of an experiment with sample space S. Suppose $P(E) = .2$, $P(F) = .1$, and $P(E \cap F) = .05$. Compute

a. $P(E \cup F)$ **b.** $P(E^c \cap F^c)$

c. $P(E^c \cap F)$ [*Hint:* Draw a Venn diagram.]

Solution

a. $\begin{aligned} P(E \cup F) &= P(E) + P(F) - P(E \cap F) \qquad \text{(Property 4)} \\ &= .2 + .1 - .05 \\ &= .25 \end{aligned}$

b. Using De Morgan's Law, we have

$$\begin{aligned} P(E^c \cap F^c) &= P[(E \cup F)^c] \\ &= 1 - P(E \cup F) \qquad \text{(Property 5)} \\ &= 1 - .25 \qquad\qquad \text{[Using the result of (a)]} \\ &= .75 \end{aligned}$$

c. From the Venn diagram describing the relationship between E, F, and S (Figure 7.12), we have

$$P(E^c \cap F) = .05 \qquad \text{(The shaded subset is the event } E^c \cap F.)$$

This result may also be obtained by using the relationship

$$\begin{aligned} P(E^c \cap F) &= P(F) - P(E \cap F) \\ &= .1 - .05 \\ &= .05 \end{aligned}$$

as before.

Figure 7.12
$P(E^c \cap F)$: *the probability that the event F, but not the event E, will occur.*

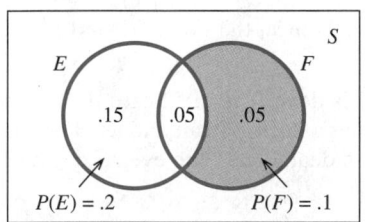

SELF–CHECK EXERCISES 7.3

1. Let E and F be events of an experiment with sample space S. Suppose $P(E) = .4$, $P(F) = .5$, and $P(E \cap F) = .1$. Compute
   a. $P(E \cup F)$ **b.** $P(E \cap F^c)$

2. Ms. Garcia wishes to sell or lease a condominium through a realty company. The realtor estimates that the probability of finding a buyer within a month of the date the property is listed for sale or lease is .3, the probability of finding a lessee is .8, and the probability of finding both a buyer and a lessee is .1. Determine the probability that the property will be sold or leased within one month from the date the property is listed for sale or lease.

Solutions to Self-Check Exercises 7.3 can be found on page 401.

7.3 EXERCISES

A pair of dice is cast and the number that appears uppermost on each die is observed. In exercises 1–6, refer to this experiment and find the probability of the given event.

1. The sum of the numbers is an even number.

2. The sum of the numbers is either 7 or 11.

3. A pair of 1's is thrown.

4. A double is thrown.

5. One die shows a 6 and the other a number less than 3.

6. The sum of the numbers is at least 4.

An experiment consists of selecting a card at random from a 52-card deck. In exercises 7–12, refer to this experiment and find the probability of the given event.

7. A king of diamonds is drawn.

8. A diamond or a king is drawn.

9. A face card is drawn.

10. A red face card is drawn.

11. An ace is not drawn.

12. A black face card is not drawn.

13. Five hundred people have purchased raffle tickets. What is the probability that a person holding one ticket will win the first prize? What is the probability that he or she will not win the first prize?

14. **TV Households**   The results of a recent television survey of American TV households revealed that 87 out of every 100 TV households have at least one remote control. What is the probability that a randomly selected TV household does not have at least one remote control?

In exercises 15–24, explain why the given statement is incorrect.

15. The sample space associated with an experiment is given by $S = \{a, b, c\}$, where $P(a) = .3$, $P(b) = .4$, and $P(c) = .4$.

16. The probability that a bus will arrive late at the Civic Center is .35, and the probability that it will be on time or early is .60.

17. A person participates in a weekly office pool in which he has one chance in ten of winning the purse. If he participates for five weeks in succession, the probability of winning at least one purse is 5/10.

18. The probability that a certain stock will increase in value over a period of one week is .6. Therefore, the probability that the stock will decrease in value is .4.

19. A red die and a green die are tossed. The probability that a 6 will appear uppermost on the red die is 1/6 and the probability that a 1 will appear uppermost on the green die is 1/6. Hence, the probability that the red die will show a 6 or the green die will show a 1 is 1/6 + 1/6.

20. Joanne, a high school senior, has applied for admission to four colleges, A, B, C, and D. She has estimated that the probability that she will be accepted for admission by A, B, C, and D is .5, .3, .1, and .08, respectively. Thus, the probability that she will be accepted for admission by at least one college is $P(A) + P(B) + P(C) + P(D) = .5 + .3 + .1 + .08 = .98$.

21. The sample space associated with an experiment is given by $S = \{a, b, c, d, e\}$. The events $E = \{a, b\}$ and $F = \{c, d\}$ are mutually exclusive. Hence, the events E^c and F^c are mutually exclusive.

22. A 5-card poker hand is dealt from a 52-card deck. Let A denote the event that a flush is dealt and let B be the event that a straight is dealt. Then the events A and B are mutually exclusive.

23. Mr. Owens, an optician, estimates that the probability that a customer coming into his store will purchase one or more pairs of glasses but not contact lenses is .40, and the probability that he will purchase one or more pairs of contact lenses but not glasses is .25. Hence, Owens concludes that the probability that a customer coming into his store will purchase neither a pair of glasses nor a pair of contact lenses is .35.

24. There are eight grades in the Garfield Elementary School. If a student is selected at random from the school, then the probability that the student is in the first grade is 1/8.

25. Let E and F be two events that are mutually exclusive and suppose $P(E) = .2$ and $P(F) = .5$. Compute
   a. $P(E \cap F)$ **b.** $P(E \cup F)$
   c. $P(E^c)$ **d.** $P(E^c \cap F^c)$

26. Let E and F be two events of an experiment with sample space S. Suppose $P(E) = .6$, $P(F) = .4$, and $P(E \cap F) = .2$. Compute
a. $P(E \cup F)$ **b.** $P(E^c)$
c. $P(F^c)$ **d.** $P(E^c \cap F)$

27. Let $S = \{s_1, s_2, s_3, s_4\}$ be the sample space associated with an experiment having the probability distribution shown in the accompanying table. If $A = \{s_1, s_2\}$ and $B = \{s_1, s_3\}$, find
a. $P(A)$, $P(B)$ **b.** $P(A^c)$, $P(B^c)$
c. $P(A \cap B)$ **d.** $P(A \cup B)$

Outcome	Probability
$\{s_1\}$	$\frac{1}{8}$
$\{s_2\}$	$\frac{3}{8}$
$\{s_3\}$	$\frac{1}{4}$
$\{s_4\}$	$\frac{1}{4}$

28. Let $S = \{s_1, s_2, s_3, s_4, s_5, s_6\}$ be the sample space associated with an experiment having the probability distribution shown in the accompanying table. If $A = \{s_1, s_2\}$ and $B = \{s_1, s_5, s_6\}$, find
a. $P(A)$, $P(B)$ **b.** $P(A^c)$, $P(B^c)$
c. $P(A \cap B)$ **d.** $P(A \cup B)$
e. $P(A^c \cap B^c)$ **f.** $P(A^c \cup B^c)$

Outcome	Probability
$\{s_1\}$	$\frac{1}{3}$
$\{s_2\}$	$\frac{1}{8}$
$\{s_3\}$	$\frac{1}{6}$
$\{s_4\}$	$\frac{1}{6}$
$\{s_5\}$	$\frac{1}{12}$
$\{s_6\}$	$\frac{1}{8}$

29. Teacher Attitudes   In a survey of 2140 teachers in a certain metropolitan area conducted by a nonprofit organization regarding teacher attitudes, the following data were obtained:

900 said that lack of parental support is a problem.

890 said that abused or neglected children are problems.

680 said that malnutrition or students in poor health is a problem.

120 said that lack of parental support and abused or neglected children are problems.

110 said that lack of parental support and malnutrition or poor health are problems.

140 said that abused or neglected children and malnutrition or poor health are problems.

40 said that lack of parental support, abuse or neglect, and malnutrition or poor health are problems.

What is the probability that a teacher selected at random from this group said that lack of parental support is the only problem hampering a student's schooling?

[*Hint:* Draw a Venn diagram.]

30. Course Enrollments   Among 500 freshmen pursuing a business degree at a university, 320 are enrolled in an economics course, 225 are enrolled in a mathematics course, and 140 are enrolled in both an economics and a mathematics course. What is the probability that a freshman selected at random from this group is enrolled in
a. an economics and/or a mathematics course?
b. exactly one of these two courses?
c. neither an economics course nor a mathematics course?

31. Consumer Surveys   A leading manufacturer of kitchen appliances advertised its products in two magazines: *Good Housekeeping* and the *Ladies Home Journal.* A survey of 500 customers revealed that 140 learned of its products from *Good Housekeeping,* 130 learned of its products from the *Ladies Home Journal,* and 80 learned of its products from both magazines. What is the probability that a person selected at random from this group saw the manufacturer's advertisement in
a. both magazines?
b. at least one of the two magazines?
c. exactly one magazine?

32. Study Habits   Students at a certain university were asked to state how many hours per week they spent studying in the library. Results of the survey revealed the following information.

Time Spent (in hours) (x)	Percentage of Students
$0 \leq x \leq 1$	32.3
$1 < x \leq 4$	40.7
$4 < x \leq 10$	16.5
$x > 10$	10.5

Find the probability distribution associated with these data. What is the probability that a student selected at random at the university studied in the library
a. more than 4 hours per week?
b. no more than 10 hours per week?

33. **Assembly Time Studies** A time study was conducted by the production manager of the Universal Instruments Company to determine how much time it took an assembly worker to complete a certain task during the assembly of its Galaxy home computers. Results of the study indicated that 20% of the workers were able to complete the task in less than 3 minutes, 60% of the workers were able to complete the task in 4 minutes or less, and 10% of the workers required more than 5 minutes to complete the task. If an assembly-line worker is selected at random from this group, what is the probability that
a. he or she will be able to complete the task in 5 minutes or less?
b. he or she will not be able to complete the task within 4 minutes?
c. the time taken for the worker to complete the task will be between 3 and 4 minutes (inclusive)?

34. **Plans to Keep Cars** In a survey conducted to see how long Americans keep their cars, 2000 automobile owners were asked how long they plan to keep their present cars. The results of the survey follow.

Number of Years Car Is Kept (x)	Number of Respondents
$0 \le x < 1$	60
$1 \le x < 3$	440
$3 \le x < 5$	360
$5 \le x < 7$	340
$7 \le x < 10$	240
$x \ge 10$	560

Find the probability distribution associated with these data. What is the probability that an automobile owner selected at random from those surveyed plans to keep his or her present car
a. less than five years?
b. three or more years?

35. **Gun-Control Laws** A poll was conducted among 250 residents of a certain city regarding tougher gun-control laws. The results of the poll are shown in the accompanying table.

	Own Only a Handgun	Own Only a Rifle	Own a Handgun and a Rifle	Own Neither	Total
Favor Tougher Laws	0	12	0	138	150
Oppose Tougher Laws	58	5	25	0	88
No Opinion	0	0	0	12	12
Total	58	17	25	150	250

If one of the participants in this poll is selected at random, what is the probability that he or she
a. favors tougher gun-control laws?
b. owns a handgun?
c. owns a handgun but not a rifle?
d. favors tougher gun-control laws and does not own a handgun?

36. Suppose the probability that Bill can solve a problem is p_1, and the probability that Mike can solve it is p_2. Show that the probability that Bill and Mike working independently can solve the problem is $p_1 + p_2 - p_1 p_2$.

37. Fifty raffle tickets are numbered 1 through 50 and one of them is drawn at random. What is the probability that the number is a multiple of 5 or 7? Consider the following "solution": Since 10 tickets bear numbers that are multiples of 5 and 7 tickets bear numbers that are multiples of 7, we conclude that the required probability is

$$\frac{10}{50} + \frac{7}{50} = \frac{17}{50}$$

What is wrong with this argument? What is the correct answer?

38. True or False. If A is a subset of B and $P(B) = 0$, then $P(A) = 0$. Explain your answer.

39. True or False. If A is a subset of B, then $P(A) \le P(B)$. Explain your answer.

SOLUTIONS TO SELF-CHECK EXERCISES 7.3

1. a. Using property 4, we find

$$P(E \cup F) = P(E) + P(F) - P(E \cap F)$$
$$= .4 + .5 - .1$$
$$= .8$$

b. From the accompanying Venn diagram, in which the subset $E \cap F^c$ is shaded, we see that

$$P(E \cap F^c) = .3$$

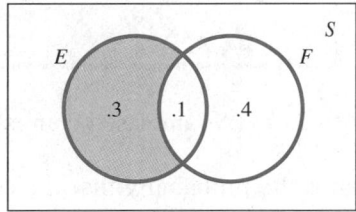

The result may also be obtained by using the relationship

$$P(E \cap F^c) = P(E) - P(E \cap F)$$
$$= .4 - .1 = .3$$

2. Let E denote the event that the property will be sold within one month of the date it is listed for sale or lease, and let F denote the event that the property will be leased within the same time period. Then

$$P(E) = .3, \quad P(F) = .8, \quad \text{and} \quad P(E \cap F) = .1$$

The probability of the event that the property will be sold or leased within one month of the date it is listed for sale or lease is given by

$$P(E \cup F) = P(E) + P(F) - P(E \cap F)$$
$$= .3 + .8 - .1 = 1$$

that is, a certainty.

7.4 USE OF COUNTING TECHNIQUES IN PROBABILITY

Further Applications of Counting Techniques

As we have seen many times before, a problem in which the underlying sample space has a small number of elements may be solved by first determining all such sample points. For problems involving sample spaces with a large number of sample points, however, this approach is neither practical nor desirable.

In this section we see how the counting techniques studied in Chapter 6 may be employed to help us solve problems in which the associated sample spaces contain large numbers of sample points. In particular, we restrict our attention to the study of uniform sample spaces—that is, sample spaces in which the outcomes are equally likely. For such spaces we have the following result:

COMPUTING THE PROBABILITY OF AN EVENT IN A UNIFORM SAMPLE SPACE

Let S be a uniform sample space and let E be any event. Then,

$$P(E) = \frac{\text{Number of favorable outcomes in } E}{\text{Number of possible outcomes in } S} = \frac{n(E)}{n(S)} \qquad (1)$$

EXAMPLE 1   An unbiased coin is tossed six times.

a. What is the probability that the coin will land heads exactly three times?

b. What is the probability that the coin will land heads at most three times?

c. What is the probability that the coin will land heads on the first and the last toss?

Solution

a. Each outcome of the experiment may be represented as a sequence of heads and tails. Using the generalized multiplication principle, we see that the number of outcomes of this experiment is given by 2^6, or 64. Let E denote the event that the coin lands heads exactly three times. Since there are $C(6, 3)$ ways this can occur, we see that the required probability is

$$P(E) = \frac{n(E)}{n(S)} = \frac{C(6, 3)}{64} = \frac{\dfrac{6!}{3!3!}}{64} \qquad (S, \text{sample space of the experiment})$$

$$= \frac{\dfrac{6 \cdot 5 \cdot 4}{3 \cdot 2}}{64} = \frac{20}{64} = \frac{5}{16} = .3125$$

b. Let F denote the event that the coin lands heads at most three times. Then $n(F)$ is given by the sum of the number of ways the coin lands heads zero times (no heads!), the number of ways it lands heads exactly once, the number of ways it lands heads exactly twice, and the number of ways it lands heads exactly three times. That is,

$$n(F) = C(6, 0) + C(6, 1) + C(6, 2) + C(6, 3)$$

$$= \frac{6!}{0!6!} + \frac{6!}{1!5!} + \frac{6!}{2!4!} + \frac{6!}{3!3!}$$

$$= 1 + 6 + \frac{(6)(5)}{2} + \frac{(6)(5)(4)}{(3)(2)} = 42$$

Therefore, the required probability is

$$P(F) = \frac{n(F)}{n(S)} = \frac{42}{64} = \frac{21}{32} \approx .66$$

c. Let F denote the event that the coin lands heads on the first and the last toss. Then $n(F) = 1 \cdot 2 \cdot 2 \cdot 2 \cdot 2 \cdot 1 = 2^4$, so the probability that this event occurs is

$$P(F) = \frac{2^4}{2^6}$$

$$= \frac{1}{2^2}$$

$$= \frac{1}{4}$$

◑ ◑ ◑

EXAMPLE 2 Two cards are selected at random from a well-shuffled pack of 52 playing cards.

a. What is the probability that they are both aces?

b. What is the probability that neither of them is an ace?

Solution

a. The experiment consists of selecting 2 cards from a pack of 52 playing cards. Since the order in which the cards are selected is immaterial, the sample points are combinations of 52 cards taken 2 at a time. Now, there are $C(52, 2)$ ways of selecting 52 cards taken 2 at a time, so the number of elements in the sample space S is given by $C(52, 2)$. Next, we observe that there are $C(4, 2)$ ways of selecting 2 aces from the 4 in the deck. Therefore, if E denotes the event that the cards selected are both aces, then

$$P(E) = \frac{n(E)}{n(S)}$$

$$= \frac{C(4, 2)}{C(52, 2)} = \frac{\frac{4!}{2!2!}}{\frac{52!}{2!50!}}$$

$$= \frac{1}{221}$$

b. Let F denote the event that neither of the two cards selected is an ace. Since there are $C(48, 2)$ ways of selecting two cards, neither of which is an ace, we find that

$$P(F) = \frac{n(F)}{n(S)} = \frac{C(48, 2)}{C(52, 2)} = \frac{\frac{48!}{2!46!}}{\frac{52!}{2!50!}} = \frac{48 \cdot 47}{2} \cdot \frac{2}{52 \cdot 51}$$

$$= \frac{188}{221}$$

◑ ◑ ◑

Figure 7.13
A sample of 6 tapes selected from 90 nondefective tapes and 10 defective tapes.

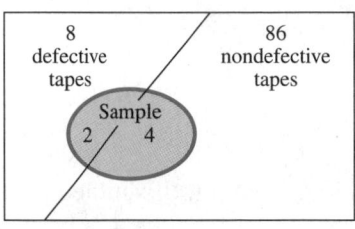

EXAMPLE 3   A bin in the hi-fi department of Building 20, a bargain outlet, contains 100 blank cassette tapes, of which 10 are known to be defective. If a customer selects 6 of these cassette tapes, determine the probability that

a. 2 of them are defective.   **b.** at least 1 of them is defective.

Solution

a. There are $C(100, 6)$ ways of selecting a set of 6 cassette tapes from the 100, and this gives $n(S)$, the number of outcomes in the sample space associated with the experiment. Next, we observe that there are $C(10, 2)$ ways of selecting a set of 2 defective cassette tapes from the 10 defective cassette tapes, and $C(90, 4)$ ways of selecting a set of 4 nondefective cassette tapes from the 90 nondefective cassette tapes (Figure 7.13). Thus, by the multiplication principle, there are $C(10, 2) \cdot C(90, 4)$ ways of selecting 2 defective and 4 nondefective cassette tapes. Therefore, the probability of selecting 6 cassette tapes, of which 2 are defective, is given by

$$\frac{C(10, 2) \cdot C(90, 4)}{C(100, 6)} = \frac{\dfrac{10!}{2!8!} \dfrac{90!}{4!86!}}{\dfrac{100!}{6!94!}}$$

$$= \frac{10 \cdot 9}{2} \cdot \frac{90 \cdot 89 \cdot 88 \cdot 87}{4 \cdot 3 \cdot 2 \cdot 1} \cdot \frac{6 \cdot 5 \cdot 4 \cdot 3 \cdot 2 \cdot 1}{100 \cdot 99 \cdot 98 \cdot 97 \cdot 96 \cdot 95}$$

$$\approx .097$$

b. Let E denote the event that none of the cassette tapes selected is defective. Then E^c gives the event that at least 1 of the cassette tapes is defective. But, by the Rule of Complements,

$$P(E^c) = 1 - P(E)$$

To compute $P(E)$, we observe that there are $C(90, 6)$ ways of selecting a set of 6 cassette tapes that are nondefective. Therefore,

$$P(E) = \frac{C(90, 6)}{C(100, 6)}$$

and

$$P(E^c) = 1 - \frac{C(90, 6)}{C(100, 6)}$$

$$= 1 - \frac{\dfrac{90!}{6!84!}}{\dfrac{100!}{6!94!}}$$

$$= 1 - \frac{90 \cdot 89 \cdot 88 \cdot 87 \cdot 86 \cdot 85}{6 \cdot 5 \cdot 4 \cdot 3 \cdot 2 \cdot 1} \cdot \frac{6 \cdot 5 \cdot 4 \cdot 3 \cdot 2 \cdot 1}{100 \cdot 99 \cdot 98 \cdot 97 \cdot 96 \cdot 95}$$

$$\approx .48$$

The Birthday Problem

 EXAMPLE 4 A group of 5 people is selected at random. What is the probability that at least 2 of them have the same birthday?

Solution For simplicity we assume that none of the 5 people was born on February 29 of a leap year. Since the 5 people were selected at random, we may also assume that each of them is equally likely to have any of the 365 days of a year as his or her birthday. If we let A, B, C, D, and E represent the 5 people, then an outcome of the experiment may be represented by (a, b, c, d, e), where the numbers a, b, c, d, and e give the birthdays of A, B, C, D, and E, respectively.

We first observe that since there are 365 possibilities for each of the dates a, b, c, d, and e, the multiplication principle implies that there are

$$\boxed{365} \cdot \boxed{365} \cdot \boxed{365} \cdot \boxed{365} \cdot \boxed{365}$$
$$\quad a \qquad b \qquad c \qquad d \qquad e$$

or 365^5 outcomes of the experiment. Therefore,

$$n(S) = 365^5$$

where S denotes the sample space of the experiment.

Next, let E denote the event that 2 or more of the 5 people have the same birthday. It is now necessary to compute $P(E)$. However, a direct computation of $P(E)$ is relatively difficult. It is much easier to compute $P(E^c)$, where E^c is the event that no 2 of the 5 people have the same birthday, and then use the relation

$$P(E) = 1 - P(E^c)$$

To compute $P(E^c)$, observe that there are 365 ways (corresponding to the 365 dates) on which A's birthday can occur, followed by 364 ways on which B's birthday could occur if B were not to have the same birthday as A, and so on. Therefore, by the generalized multiplication principle,

$$n(E^c) = \quad 365 \quad \cdot \quad 364 \quad \cdot \quad 363 \quad \cdot \quad 362 \quad \cdot \quad 361$$
$$\quad\quad\quad \text{A's} \qquad \text{B's} \qquad \text{C's} \qquad \text{D's} \qquad \text{E's}$$
$$\quad\quad \text{birthday} \quad \text{birthday} \quad \text{birthday} \quad \text{birthday} \quad \text{birthday}$$

Thus,

$$P(E^c) = \frac{n(E^c)}{n(S)}$$

$$= \frac{365 \cdot 364 \cdot 363 \cdot 362 \cdot 361}{365^5}$$

and

$$P(E) = 1 - P(E^c)$$

$$= 1 - \frac{365 \cdot 364 \cdot 363 \cdot 362 \cdot 361}{365^5}$$

$$\approx .027$$

We can extend the result obtained in Example 4 to the general case involving r people. In fact, if E denotes the event that at least 2 of the r people have the same birthday, an argument similar to that used in Example 4 leads to the result

$$P(E) = 1 - \frac{365 \cdot 364 \cdot 363 \cdot \cdots \cdot (365 - r + 1)}{365^r}$$

By letting r take on the values 5, 10, 15, 20, . . . , 50, in turn, we obtain the probabilities that at least 2 of 5, 10, 15, 20, . . . , 50 people, respectively, have the same birthday. These results are summarized in Table 7.8.

The results show that in a group of 23 randomly selected people the chances are greater than 50% that at least 2 of them will have the same birthday. In a group of 50 people, it is an excellent bet that at least 2 people in the group will have the same birthday.

Table 7.8
Probability that at least two people in a randomly selected group of r people have the same birthday.

r	5	10	15	20	22	23	25	30	40	50
P(E)	.027	.117	.253	.411	.476	.507	.569	.706	.891	.970

During an episode of the "Tonight Show," Johnny Carson related "The Birthday Problem" to the audience—noting that, in a group of 50 or more people, probabilists have calculated that the probability of at least 2 people having the same birthday is very high. To illustrate this point, he proceeded to conduct his own experiment. A person selected at random from the audience was asked to state his birthday. Mr. Carson then asked if anyone in the audience had the same birthday. The response was negative. He repeated the experiment. Once again, the response was negative. These results, observed Mr. Carson, were contrary to expectations. In a later episode of the show Mr. Carson explained why this experiment had been improperly conducted. Explain why Mr. Carson failed to illustrate the point he was trying to make in the earlier episode.

SELF-CHECK EXERCISES 7.4

1. Four balls are selected at random without replacement from an urn containing 10 white balls and 8 red balls. What is the probability that all the chosen balls are white?

2. A box contains 20 microchips, of which 4 are substandard. If 2 of the chips are taken from the box, what is the probability that they are both substandard?

Solutions to Self-Check Exercises 7.4 can be found on page 409.

7.4 EXERCISES

A calculator is recommended for this exercise set. An unbiased coin is tossed five times. In exercises 1–4, find the probability of the given event.

1. The coin lands heads all five times.

2. The coin lands heads exactly once.

3. The coin lands heads at least once.

4. The coin lands heads more than once.

Two cards are selected at random without replacement from a well-shuffled deck of 52 playing cards. In exercises 5–8, find the probability of the given event.

5. A pair is drawn.

6. A pair is not drawn.

7. Two black cards are drawn.

8. Two cards of the same suit are drawn.

Four balls are selected at random without replacement from an urn containing three white balls and five blue balls. In exercises 9–12, find the probability of the given event.

9. Two of the balls are white and two are blue.

10. All of the balls are blue.

11. Exactly three of the balls are blue.

12. Two or three of the balls are white.

Assume that the probability of a boy being born is the same as the probability of a girl being born. In exercises 13–16, find the probability that a family with three children will have the given composition.

13. Two boys and one girl

14. At least one girl

15. No girls

16. The two oldest children are girls.

17. An exam consists of ten true-or-false questions. If a student guesses at every answer, what is the probability that he or she will answer exactly six questions correctly?

18. **Personnel Selection** Jacobs & Johnson, Inc., an accounting firm, employs 14 accountants, of whom 8 are C.P.A.s. If a delegation of 3 accountants is randomly selected from the firm to attend a conference, what is the probability that 3 C.P.A.s will be selected?

19. **Quality Control** Two light bulbs are selected at random from a lot of 24, of which 4 are defective. What is the probability that

a. both of the light bulbs are defective?

b. at least 1 of the light bulbs is defective?

20. A customer at Cavallaro's Fruit Stand picks a sample of 3 oranges at random from a crate containing 60 oranges, of which 4 are rotten. What is the probability that the sample contains 1 or more rotten oranges?

21. **Quality Control** A shelf in the Metro Department Store contains 80 typewriter cartridge ribbons for a popular portable typewriter. Six of the cartridges are defective. If a customer selects 2 of these cartridges at random from the shelf, determine the probability that

a. both are defective.

b. at least 1 is defective.

22. **Quality Control** Electronic baseball games manufactured by Tempco Electronics are shipped in lots of 24. Before shipping, a quality-control inspector randomly selects a sample of 8 from each lot for testing. If the sample contains any defective games, the entire lot is rejected. What is the probability that a lot containing exactly 2 defective games will still be shipped?

23. **Personnel Selection** The City Transit Authority plans to hire 12 new bus drivers. From a group of 100 qualified applicants, of which 60 are men and 40 are women, 12 names are to be selected by lot. If Mary and John Lewis are among the 100 qualified applicants,

a. what is the probability that Mary's name will be selected? That both Mary's and John's names will be selected?

b. and it is stipulated that an equal number of men and women are to be selected (6 men from the group of 60 men and 6 women from the group of 40 women), what is the probability that Mary's name will be selected? That Mary's and John's names will be selected?

24. **Public Housing** The City Housing Authority has received 50 applications from qualified applicants for 8 low-income apartments. Three of the apartments are on the north side of town and 5 are on the south side. If the apartments are to be assigned by means of a lottery, what is the probability that

a. a specific qualified applicant will be selected for one of these apartments?

b. two specific qualified applicants will be selected for apartments on the same side of town?

25. A student studying for a vocabulary test knows the meanings of 12 words from a list of 20 words. If the test contains 10 words from the study list, what is the probability that at least 8 of the words on the test are words that the student knows?

26. **Drivers' Tests**   Four different written driving tests are administered by the City Motor Vehicle Department. One of these four tests is selected at random for each applicant for a driver's license. If a group consisting of two women and three men apply for a license, what is the probability that
   a. exactly two of the five will take the same test?
   b. the two women will take the same test?

27. **Brand Selection**   A druggist wishes to select three brands of aspirin to sell in his store. He has five major brands to choose from: brands A, B, C, D, and E. If he selects the three brands at random, what is the probability that he will select
   a. brand B?   **b.** brands B and C?
   c. at least one of the two brands B and C?

28. **Blackjack**   In the game of blackjack, a 2-card hand consisting of an ace and a face card or a ten is called a blackjack.
   a. If a player is dealt 2 cards from a standard deck of 52 well-shuffled cards, what is the probability that the player will receive a blackjack?
   b. If a player is dealt 2 cards from 2 well-shuffled standard decks, what is the probability that the player will receive a blackjack?

29. **Slot Machines**   Refer to exercise 23, Section 6.3, where the "lucky dollar" slot machine was described. What is the probability that the three "lucky dollar" symbols will appear in the window of the slot machine?

30. **Roulette**   In 1959 a world record was set for the longest run on an ungaffed (fair) roulette wheel at the El San Juan Hotel in Puerto Rico. The number 10 appeared 6 times in a row. What is the probability of the occurrence of this event? (Assume that there are 38 equally likely outcomes consisting of the numbers 1–36, 0, and 00.)

In "The Numbers Game," a state lottery, four numbers are drawn with replacement from an urn containing the digits 0–9, inclusive. In exercises 31–34, find the probability of a ticket holder having the indicated winning ticket.

31. All four digits in exact order (the grand prize)

32. Two specified, consecutive digits in exact order (the first two digits, the middle two digits, or the last two digits)

33. One specified digit in exact order (the first, second, third, or fourth digit)

34. All four digits in any order (including the other winning tickets)

A list of poker hands ranked in order from the highest to the lowest is shown in the accompanying table along with a description and example of each hand. Use the table to answer exercises 35–40.

Hand	Description	Example
Straight flush	5 cards in sequence in the same suit	A ♥ 2 ♥ 3 ♥ 4 ♥ 5 ♥
Four of a kind	4 cards of the same rank and any other card	K ♥ K ♦ K ♠ K ♣ 2 ♥
Full house	3 of a kind and a pair	3 ♥ 3 ♦ 3 ♣ 7 ♥ 7 ♦
Flush	5 cards of the same suit that are not all in sequence	5 ♥ 6 ♥ 9 ♥ J ♥ K ♥
Straight	5 cards in sequence but not all of the same suit	10 ♥ J ♦ Q ♣ K ♠ A ♥
Three of a kind	3 cards of the same rank and 2 unmatched cards	K ♥ K ♦ K ♠ 2 ♥ 4 ♦
Two pair	2 cards of the same rank and 2 cards of any other rank with an unmatched card	K ♥ K ♦ 2 ♥ 2 ♠ 4 ♣
One pair	2 cards of the same rank and 3 unmatched cards	K ♥ K ♦ 5 ♥ 2 ♠ 4 ♥

If a 5-card poker hand is dealt from a well-shuffled deck of 52 cards, what is the probability of being dealt the given hand?

35. A straight flush. (Note that an ace may be played as either a high or low card in a straight sequence—that is, A, 2, 3, 4, 5 or 10, J, Q, K, A. Hence, there are 10 possible sequences for a straight in one suit.)

36. A straight (but not a straight flush)

37. A flush (but not a straight flush)

38. Four of a kind

39. A full house

40. Two pairs

41. Zodiac Signs There are 12 signs of the Zodiac: Aries, Taurus, Gemini, Cancer, Leo, Virgo, Libra, Scorpio, Sagittarius, Capricorn, Aquarius, and Pisces. Each sign corresponds to a different calendar period of approximately one month. Assuming that a person is just as likely to be born under one sign as another, what is the probability that in a group of 5 people
a. at least 2 of them have the same sign?
b. at least 2 of them were born under the sign of Aries?

42. Birthday Problem What is the probability that at least two justices of the U.S. Supreme Court have the same birthday?

43. Birthday Problem Fifty people are selected at random. What is the probability that none of the people in this group have the same birthday?

44. There were 42 different presidents of the United States from 1789 through 1992. What is the probability that at least two of them had the same birthday? Compare your calculation with the facts by checking an almanac or some other source.

SOLUTIONS TO SELF-CHECK EXERCISES 7.4

1. The probability that all 4 balls selected are white is given by

$$\frac{\text{The number of ways of selecting 4 white balls from the 10 in the urn}}{\text{The number of ways of selecting any 4 balls from the 18 balls in the urn}}$$

$$= \frac{C(10, 4)}{C(18, 4)}$$

$$= \frac{\dfrac{10!}{4!6!}}{\dfrac{18!}{4!14!}}$$

$$= \frac{10 \cdot 9 \cdot 8 \cdot 7}{4 \cdot 3 \cdot 2} \cdot \frac{4 \cdot 3 \cdot 2}{18 \cdot 17 \cdot 16 \cdot 15}$$

$$= .069$$

2. The probability that both chips are substandard is given by

$$\frac{\text{The number of ways of choosing any 2 of the 4 substandard chips}}{\text{The number of ways of choosing any 2 of the 20 chips}}$$

$$= \frac{C(4, 2)}{C(20, 2)}$$

$$= \frac{\dfrac{4!}{2!2!}}{\dfrac{20!}{2!18!}}$$

$$= \frac{4 \cdot 2}{2} \cdot \frac{2}{20 \cdot 19}$$

$$= .032$$

7.5 CONDITIONAL PROBABILITY AND INDEPENDENT EVENTS

Conditional Probability

Three cities, A, B, and C, are vying to play host to the Summer Olympic Games in the year 2008. If each city has the same chance of winning the right to host the Games, then the probability of city A hosting the Games is 1/3. Suppose that city B then decides to pull out of contention because of fiscal problems. Then it would seem that city A's chances of playing host will increase. In fact, if each of the two remaining cities have equal chances of winning, then the probability of city A playing host to the Games is 1/2.

In general, the probability of an event is affected by the occurrence of other events and/or by the knowledge of information relevant to the event. Basically, the injection of conditions into a problem modifies the underlying sample space of the original problem. This in turn leads to a change in the probability of the event.

EXAMPLE 1   Two cards are drawn without replacement from a well-shuffled deck of 52 playing cards.

a. What is the probability that the first card drawn is an ace?

b. What is the probability that the second card drawn is an ace given that the first card drawn was not an ace?

c. What is the probability that the second card drawn is an ace given that the first card drawn was an ace?

Solution

a. The sample space here consists of 52 equally likely outcomes, 4 of which are aces. Therefore, the probability that the first card drawn is an ace is 4/52.

b. Having drawn the first card, there are 51 cards left in the deck. In other words, for the second phase of the experiment, we are working in a *reduced* sample space. If the first card drawn was not an ace, then this modified sample space of 51 points contains 4 "favorable" outcomes (the 4 aces), so the probability that the second card drawn is an ace is given by 4/51.

c. If the first card drawn was an ace, then there are 3 aces left in the deck of 51 playing cards, so the probability that the second card drawn is an ace is given by 3/51.   ◦ ◦ ◦

Observe that in Example 1 the occurrence of the first event reduces the size of the original sample space. The information concerning the first card drawn also leads us to the consideration of modified sample spaces: in (b) the deck contained 4 aces, and in (c) the deck contained 3 aces.

The probability found in (b) or (c) of Example 1 is known as a *conditional probability,* since it is the probability of an event occurring given that another event has already occurred. For example, in (b) we computed the probability of the event that the second card drawn is an ace, given the event that the first card drawn was not an ace. In general, given two events A and B of an experiment, one may, under certain circumstances, compute the probability of the event B given that the event A has already occurred. This probability, denoted by $P(B \mid A)$, is called the **conditional probability of B given A.**

A formula for computing the conditional probability of B given A may be discovered with the aid of a Venn diagram. Consider an experiment with a uniform sample space S and suppose A and B are two events of the experiment (Figure 7.14).

Figure 7.14

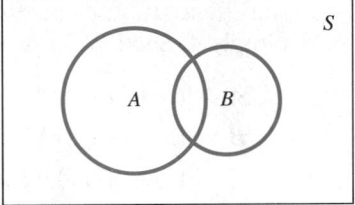 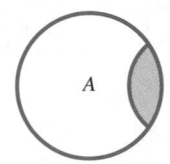

(a) Original sample space.

(b) Reduced sample space A. The shaded area is $A \cap B$.

The condition that the event A has occurred tells us that the possible outcomes of the experiment in the second phase are restricted to those outcomes (elements) in the set A. In other words, we may work with the reduced sample space A instead of the original sample space S in the experiment. Next, we observe that, with respect to the reduced sample space A, the outcomes favorable to the event B are precisely those elements in the set $A \cap B$. Consequently, the conditional probability of B given A is

$$P(B \mid A) = \frac{\text{Number of elements in } A \cap B}{\text{Number of elements in } A}$$

$$= \frac{n(A \cap B)}{n(A)} \qquad (n(A) \neq 0)$$

Dividing the numerator and the denominator by $n(S)$, the number of elements in S, we have

$$P(B \mid A) = \frac{\dfrac{n(A \cap B)}{n(S)}}{\dfrac{n(A)}{n(S)}}$$

which is equivalent to the following formula:

CONDITIONAL PROBABILITY OF AN EVENT

If A and B are events in an experiment and $P(A) \neq 0$, then the **conditional probability** that the event B will occur given that the event A has already occurred is

$$P(B \mid A) = \frac{P(A \cap B)}{P(A)} \qquad \qquad (2)$$

EXAMPLE 2   A pair of fair dice is cast. What is the probability that the sum of the numbers falling uppermost is 7 if it is known that one of the numbers is a 5?

Figure 7.15
$A \cap B = \{(5, 2), (2, 5)\}.$

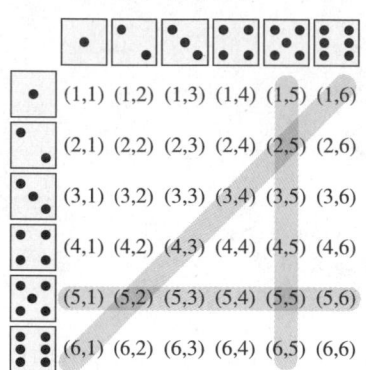

Solution   Let A denote the event that the sum of the numbers falling uppermost is 7, and let B denote the event that one of the numbers is a 5. From the results of Example 4, Section 7.1, we find that

$$A = \{(6, 1), (5, 2), (4, 3), (3, 4), (2, 5), (1, 6)\}$$

and

$$B = \{(5, 1), (5, 2), (5, 3), (5, 4), (5, 5), (5, 6),$$
$$(1, 5), (2, 5), (3, 5), (4, 5), (6, 5)\}$$

so that $A \cap B = \{(5, 2), (2, 5)\}$ (See Figure 7.15.)

Since the dice are fair, each outcome of the experiment is equally likely; therefore

$$P(A \cap B) = \frac{2}{36} \quad \text{and} \quad P(B) = \frac{11}{36} \quad \text{(Recall that } n(S) = 36.\text{)}$$

Thus, the probability that the sum of the numbers falling uppermost is 7 given that one of the numbers is a 5 is, by virtue of equation (2),

$$P(A \mid B) = \frac{\dfrac{2}{36}}{\dfrac{11}{36}} = \frac{2}{11}$$

○ ○ ○

EXAMPLE 3   In a test recently conducted by the U.S. Army, it was found that of 1000 new recruits, 600 men and 400 women, 50 of the men and 4 of the women were red-green color-blind. Given that a recruit selected at random from this group is red-green color-blind, what is the probability that the recruit is a male?

Solution   Let C denote the event that a randomly selected subject is red-green color-blind, and let M denote the event that the subject is a male recruit. Since 54 out of the 1000 subjects are color-blind, we may take

$$P(C) = \frac{54}{1000} = .054$$

Therefore, by equation (2), the probability that a subject is male given that the subject is red-green color-blind is

$$P(M \mid C) = \frac{P(M \cap C)}{P(C)}$$

$$= \frac{.05}{.054} = \frac{25}{27}$$

● ● ●

Let A and B be events in an experiment, and suppose $P(A) \neq 0$. Suppose that in n trials, the event A occurs m times, the event B occurs k times, and the events A and B occur together l times.

a. Explain why it makes good sense to call the ratio l/m the conditional relative frequency of the event B given the event A.

b. Show that the relative frequencies l/m, m/n, and l/n satisfy the equation

$$\frac{l}{m} = \frac{\dfrac{l}{n}}{\dfrac{m}{n}}$$

c. Explain why the result of part (b) suggests that formula (2)

$$P(B \mid A) = \frac{P(A \cap B)}{P(A)} \qquad P(A) \neq 0$$

is plausible.

In certain problems, the probability of an event B occurring given that A has occurred, written $P(B \mid A)$, is known and we wish to find the probability of A *and* B occurring. The solution to such a problem is facilitated by the use of the following formula:

THE PRODUCT RULE $P(A \cap B) = P(A) \cdot P(B \mid A)$ if $P(A) \neq 0$ **(3)**

This formula is obtained from (2) by multiplying both sides of the equation by $P(A)$. We illustrate the use of the Product Rule in the next several examples.

EXAMPLE 4   There are 300 seniors in Jefferson High School, of which 140 are males. It is known that 80% of the males and 60% of the females have their driver's license. If a student is selected at random from this senior class,

a. what is the probability that the student is a male and has a driver's license?

b. what is the probability that the student is a female who does not have a driver's license?

Solution

a. Let M denote the event that the student is a male, and let D denote the event that the student has a driver's license. Then

$$P(M) = \frac{140}{300} \quad \text{and} \quad P(D \,|\, M) = .8$$

Now, the event that the student selected at random is a male and has a driver's license is $M \cap D$, and, by the Product Rule, the probability of this event occurring is given by

$$P(M \cap D) = P(M) \cdot P(D \,|\, M)$$
$$= \left(\frac{140}{300}\right)(.8) = \frac{28}{75}$$

b. Let F denote the event that the student is a female, and let D be as before. Then D^c is the event that the student does not have a driver's license. We have

$$P(F) = \frac{160}{300} \quad \text{and} \quad P(D^c \,|\, F) = 1 - .6 = .4$$

Note that we have used the Rule of Complements in the computation of $P(D^c \,|\, F)$. Now, the event that the student selected at random is a female and does not have a driver's license is $F \cap D^c$, so by the Product Rule, the probability of this event occurring is given by

$$P(F \cap D^c) = P(F) \cdot P(D^c \,|\, F)$$
$$= \left(\frac{160}{300}\right)(.4) = \frac{16}{75}$$

○ ○ ○

EXAMPLE 5   Two cards are drawn without replacement from a well-shuffled deck of 52 playing cards. What is the probability that the first card drawn is an ace and the second card drawn is a face card?

Solution   Let A denote the event that the first card drawn is an ace and let F denote the event that the second card drawn is a face card. Then $P(A) = 4/52$. After drawing the first card, there are 51 cards left in the deck, of which 12 are face cards. Therefore, the probability of drawing a face card given that the first card drawn was an ace is given by

$$P(F \,|\, A) = \frac{12}{51}$$

By the Product Rule, the probability that the first card drawn is an ace and the second card drawn is a face card is given by

$$P(A \cap F) = P(A) \cdot P(F \mid A)$$
$$= \left(\frac{4}{52}\right)\left(\frac{12}{51}\right) = \frac{4}{221}$$

○ ○ ○

The Product Rule can be extended to the case involving three or more events. For example, if A, B, and C are three events in an experiment, then it can be shown that

$$P(A \cap B \cap C) = P(A) \cdot P(B \mid A) \cdot P(C \mid A \cap B)$$

a. Explain the formula in words.
b. Suppose three cards are drawn without replacement from a well-shuffled deck of 52 playing cards. Use the given formula to find the probability that the three cards are aces.

The Product Rule may be generalized to the case involving any finite number of events. For example, in the case involving the three events E, F, and G, it may be shown that

$$P(E \cap F \cap G) = P(E) \cdot P(F \mid E) \cdot P(G \mid E \cap F) \qquad \textbf{(4)}$$

More on Tree Diagrams

Formula (4) and its generalizations may be used to help us solve problems that involve *finite stochastic processes*. More specifically, a **finite stochastic process** is an experiment consisting of a finite number of stages in which the outcomes and associated probabilities of each stage depend on the outcomes and associated probabilities of the preceding stages.

We can use tree diagrams to help us solve problems involving finite stochastic processes. Consider, for example, the experiment consisting of drawing 2 cards without replacement from a well-shuffled deck of 52 playing cards. What is the probability that the second card drawn is a face card?

We may think of this experiment as a stochastic process with two stages. The events associated with the first stage are F, that the card drawn is a face card, and F^c, that the card drawn is not a face card. Since there are 12 face cards, we have

$$P(F) = \frac{12}{52} \quad \text{and} \quad P(F^c) = 1 - \frac{12}{52} = \frac{40}{52}$$

The outcomes of this trial, together with the associated probabilities, may be represented along two "branches" of a tree diagram, as shown in Figure 7.16.

In the second trial, we again have two events: G, that the card drawn is a face card, and G^c, that the card drawn is not a face card. But the outcome

Figure 7.16
F is the probability that a face card is drawn.

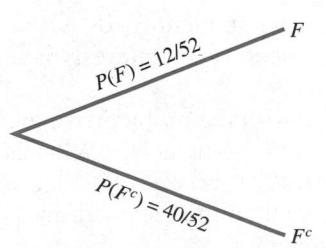

of the second trial depends on the outcome of the first trial. For example, if the first card drawn was a face card, then the event G that the second card drawn is a face card has probability given by the *conditional probability* $P(G \mid F)$. Since the occurrence of a face card in the first draw leaves 11 face cards in a deck of 51 cards for the second draw, we see that

$$P(G \mid F) = \frac{11}{51}$$ (The probability of drawing a face card given that a face card has already been drawn)

Similarly, the occurrence of a face card in the first draw leaves 40 that are other than face cards in a deck of 51 cards for the second draw. Therefore, the probability of drawing other than a face card in the second draw given that the first card drawn is a face card is

$$P(G^c \mid F) = \frac{40}{51}$$

Using these results, we extend the tree diagram of Figure 7.16 by displaying another two branches of the tree growing from its upper branch (Figure 7.17).

Figure 7.17
G is the probability that the second card drawn is a face card.

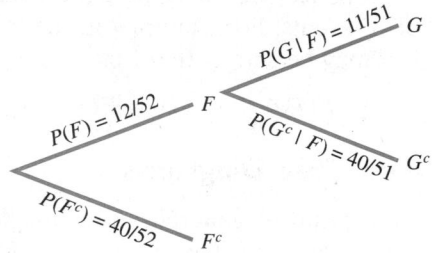

To complete the tree diagram, we compute $P(G \mid F^c)$ and $P(G^c \mid F^c)$, the conditional probabilities that the second card drawn is a face card and other than a face card, respectively, given that the first card drawn is not a face card. We find that

$$P(G \mid F^c) = \frac{12}{51} \quad \text{and} \quad P(G^c \mid F^c) = \frac{39}{51}$$

This leads to the completion of the tree diagram, shown in Figure 7.18, where the branches of the tree that lead to the two outcomes of interest have been highlighted.

Having constructed the tree diagram associated with the problem, we are now in a position to answer the question posed earlier—namely, "What is the probability of the second card being a face card?" Observe that Figure 7.18 shows the two ways in which a face card may result in the second draw—namely, the two Gs on the extreme right of the diagram.

Figure 7.18
Tree diagram showing the two trials of the experiment.

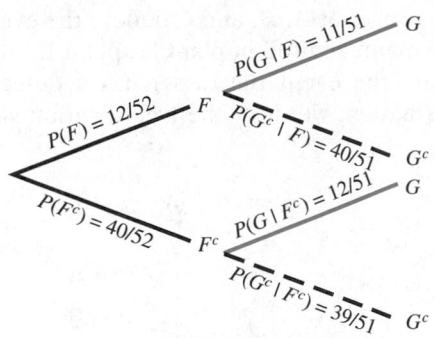

Now, by the Product Rule, the probability that the second card drawn is a face card and the first card drawn is a face card (this is represented by the upper branch) is

$$P(G \cap F) = P(F) \cdot P(G \mid F)$$

Similarly, the probability that the second card drawn is a face card and the first card drawn is other than a face card (this corresponds to the other branch) is

$$P(G \cap F^c) = P(F^c) \cdot P(G \mid F^c)$$

Observe that each of these probabilities is obtained by taking the *product of the probabilities appearing on the respective branch.* Since $G \cap F$ and $G \cap F^c$ are mutually exclusive events (why?), the probability that the second card drawn is a face card is given by

$$P(G \cap F) + P(G \cap F^c) = P(F) \cdot P(G \mid F) + P(F^c) \cdot P(G \mid F^c)$$

or, upon replacing the probabilities on the right of the expression by their numerical values,

$$P(G \cap F) + P(G \cap F^c) = \left(\frac{12}{52}\right)\left(\frac{11}{51}\right) + \left(\frac{40}{52}\right)\left(\frac{12}{51}\right)$$

$$= \frac{3}{13}$$

EXAMPLE 6   The picture tubes for the Pulsar 19-inch color television sets are manufactured in three locations and then shipped to the main plant of the Vista Vision Corporation for final assembly. Plants A, B, and C supply 50%, 30%, and 20%, respectively, of the picture tubes used by the company. The quality-control department of the company has determined that 1% of the picture tubes produced by plant A are defective, whereas 2% of the picture tubes produced by plants B and C are defective. What is the probability that a randomly selected Pulsar 19-inch color television set will have a defective picture tube?

Solution   Let A, B, and C denote the events that the set chosen has a picture tube manufactured in plant A, plant B, and plant C, respectively. Also, let D denote the event that a set has a defective picture tube. Using the given information, we draw the tree diagram shown in Figure 7.19.

Figure 7.19
Tree diagram showing the probabilities of producing defective picture tubes at each plant.

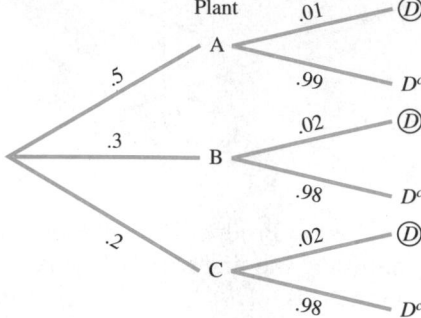

(The events that result in a set with a defective picture tube being selected are circled.) Taking the product of the probabilities along each branch leading to such an event and adding them yields the probability that a set chosen at random has a defective picture tube. Thus, the required probability is given by

$$(.5)(.01) + (.3)(.02) + (.2)(.02) = .005 + .006 + .004$$
$$= .015$$

EXAMPLE 7   A box contains eight 9-volt transistor batteries, of which two are known to be defective. The batteries are selected one at a time without replacement and tested until a nondefective one is found. What is the probability that the number of batteries tested is (a) one, (b) two, (c) three?

Solution   We may view this experiment as a multistage process with up to three stages. In the first stage, a battery is selected with a probability of 6/8 of its being nondefective and a probability of 2/8 of its being defective. If the battery selected is good, the experiment is terminated. Otherwise, a second battery is selected with probabilities of 6/7 and 1/7, respectively, of its being nondefective and defective. If the second battery selected is good, the experiment is terminated. Otherwise, a third battery is selected with probabilities of 1 and 0, respectively, of its being nondefective and defective. The tree diagram associated with this experiment is shown in Figure 7.20, where N denotes the event that the battery selected is nondefective, and D denotes the event that the battery selected is defective.

With the aid of the tree diagram we see that (a) the probability that only one battery is selected is 6/8, (b) the probability that two batteries are selected is (2/8)(6/7), or 3/14, and (c) the probability that three batteries are selected is $(2/8)(1/7)(1) = 1/28$.

Figure 7.20
In this experiment, batteries are selected until a nondefective one is found.

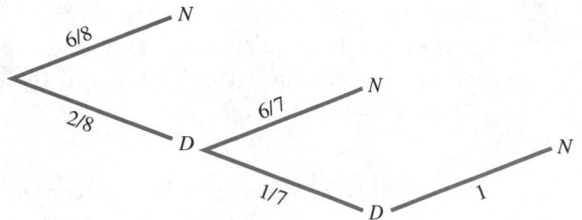

Independent Events

Let us return to the experiment of drawing 2 cards in succession without replacement from a well-shuffled deck of 52 playing cards considered in Example 5. Let E denote the event that the first card drawn is not a face card and let F denote the event that the second card drawn is a face card. It is intuitively clear that the events E and F are *not* independent of each other since whether or not the first card drawn is a face card affects the likelihood that the second card drawn is a face card.

Next, let us consider the experiment of tossing a coin twice and observing the outcomes: If H denotes the event that the first toss produces "heads" and T denotes the event that the second toss produces "tails," then it is intuitively clear that H and T *are* independent of each other since the outcome of the first toss does not affect the outcome of the second.

In general, two events A and B are **independent** if the outcome of one does not affect the outcome of the other. Thus, if A and B are independent events, then

$$P(A \mid B) = P(A) \qquad \text{or} \qquad P(B \mid A) = P(B)$$

Using the Product Rule, we can find a simple test to determine the independence of two events. Suppose that A and B are independent and $P(A) \neq 0$ and $P(B) \neq 0$. Then,

$$P(B \mid A) = P(B)$$

so by the Product Rule, we have

$$P(A \cap B) = P(A) \cdot P(B \mid A) = P(A) \cdot P(B)$$

Conversely, if this equation holds, then it can be seen that $P(B \mid A) = P(B)$; that is, A and B are independent. Accordingly, we have the following *test for the independence of two events.*

TEST FOR THE INDEPENDENCE OF TWO EVENTS

Two events A and B are **independent** if and only if

$$P(A \cap B) = P(A) \cdot P(B) \tag{5}$$

Do not confuse *independent* events with *mutually exclusive* events. The former pertains to how the occurrence of one event affects the occurrence of another event, whereas the latter pertains to the question of whether the events can occur at the same time.

EXAMPLE 8   Consider the experiment consisting of tossing a fair coin twice and observing the outcomes. Show that the event of "heads" in the first toss and "tails" in the second toss are independent events.

Solution   Let A denote the event that the outcome of the first toss is a *head* and let B denote the event that the outcome of the second toss is a *tail*. The sample space of the experiment is

$$S = \{(H, H), (H, T), (T, H), (T, T)\}$$

and $\qquad A = \{(H, H), (H, T)\} \qquad$ and $\qquad B = \{(H, T), (T, T)\}$

so that $\quad A \cap B = \{(H, T)\}$

Next, we compute

$$P(A \cap B) = \frac{1}{4}, \quad P(A) = \frac{1}{2}, \quad \text{and} \quad P(B) = \frac{1}{2}$$

and observe that equation (5) is satisfied in this case so that A and B are independent events, as we set out to show.   ○ ○ ○

 EXAMPLE 9   A survey conducted by an independent agency for the National Lung Society found that of 2000 women, 680 were heavy smokers and 50 had emphysema. Of those who had emphysema, 42 were also heavy smokers. Using the data in this survey, determine whether the events "being a heavy smoker" and "having emphysema" are independent events.

Solution   Let A denote the event that a woman is a heavy smoker and let B denote the event that a woman has emphysema. Then the probability that a woman is a heavy smoker and has emphysema is given by

$$P(A \cap B) = \frac{42}{2000} = .021$$

Next $\qquad P(A) = \frac{680}{2000} = .34 \qquad$ and $\qquad P(B) = \frac{50}{2000} = .025$

so that $\quad P(A) \cdot P(B) = (.34)(.025) = .0085$

Since $P(A \cap B) \neq P(A) \cdot P(B)$, we conclude that A and B are not independent events.   ○ ○ ○

Let E and F be independent events in a sample space S. Are E^c and F^c independent?

The solution of many practical problems involves more than two independent events. In such cases we use the following result.

INDEPENDENCE OF MORE THAN TWO EVENTS

If $E_1, E_2, \ldots, E_n$ are independent events, then

$$P(E_1 \cap E_2 \cap \cdots \cap E_n) = P(E_1) \cdot P(E_2) \cdot \cdots \cdot P(E_n) \qquad \textbf{(6)}$$

Formula (6) states that the probability of the simultaneous occurrence of n independent events is equal to the product of the probabilities of the n events.

 It is important to note that the mere requirement that the n events $E_1, E_2, \ldots, E_n$ satisfy (6) is not sufficient to guarantee that the n events are indeed independent. However, a criterion does exist for determining the independence of n events and may be found in more advanced texts on probability.

EXAMPLE 10   It is known that the three events A, B, and C are independent and $P(A) = .2$, $P(B) = .4$, and $P(C) = .5$.

a. Compute $P(A \cap B)$.   **b.** Compute $P(A \cap B \cap C)$.

Solution   Using formulas (5) and (6), we find

a. $P(A \cap B) = P(A) \cdot P(B)$
$= (.2)(.4) = .08$

b. $P(A \cap B \cap C) = P(A) \cdot P(B) \cdot P(C)$
$= (.2)(.4)(.5) = .04$ ○ ○ ○

EXAMPLE 11   The Acrosonic model F loudspeaker system has four loudspeaker components: a woofer, a midrange, a tweeter, and an electrical crossover. The quality-control manager of Acrosonic has determined that on the average 1% of the woofers, 0.8% of the midranges, and 0.5% of the tweeters are defective, while 1.5% of the electrical crossovers are defective. Determine the probability that an Acrosonic model F loudspeaker system selected at random coming off the assembly line and before final inspection is not defective. Assume that the defects in the manufacturing of the components are unrelated.

Solution   Let A, B, C, and D denote, respectively, the events that the woofer, the midrange, the tweeter, and the electrical crossover are defective. Then,

$$P(A) = .01, \quad P(B) = .008, \quad P(C) = .005, \quad \text{and} \quad P(D) = .015$$

and the probabilities of the corresponding complementary events are

$$P(A^c) = .99, \quad P(B^c) = .992, \quad P(C^c) = .995, \quad \text{and} \quad P(D^c) = .985$$

The event that a loudspeaker system selected at random is not defective is given by $A^c \cap B^c \cap C^c \cap D^c$, and since the events A, B, C, and D (and therefore also A^c, B^c, C^c, and D^c) are assumed to be independent, we find that the required probability is given by

$$P(A^c \cap B^c \cap C^c \cap D^c) = P(A^c) \cdot P(B^c) \cdot P(C^c) \cdot P(D^c)$$
$$= (.99)(.992)(.995)(.985)$$
$$\approx .96$$

○ ○ ○

SELF-CHECK EXERCISES 7.5

1. Let A and B be events in a sample space S such that $P(A) = .4$, $P(B) = .8$, and $P(A \cap B) = .3$. Find
   a. $P(A \mid B)$ **b.** $P(B \mid A)$

2. Three friends—Alice, Betty, and Cathy—are unmarried and are 30, 35, and 40 years old, respectively. While they were having tea one afternoon, Alice recalled reading an article by Yale sociologist Neil Bennett in which he concluded that women who are still unmarried at age 30 have only a 20% chance of marrying, those who are still unmarried at 35 have a 5.4% chance, and those who are still unmarried at 40 have a 1.3% chance. Betty wondered what the probability was that all three of them would eventually "tie the knot." Cathy, a statistician, pulled out her pocket calculator and answered Betty's question. What was her answer?

Solutions to Self-Check Exercises 7.5 can be found on page 426.

7.5 EXERCISES

1. Let A and B be events in a sample space S such that $P(A) = .6$, $P(B) = .5$, and $P(A \cap B) = .2$. Find
   a. $P(A \mid B)$ **b.** $P(B \mid A)$

2. Let A and B be two events in a sample space S such that $P(A) = .4$, $P(B) = .6$, and $P(A \cap B) = .3$. Find
   a. $P(A \mid B)$ **b.** $P(B \mid A)$

3. Let A and B be two events in a sample space S such that $P(A) = .6$ and $P(B \mid A) = .5$. Find $P(A \cap B)$.

4. Let A and B be the events described in exercise 1. Find
   a. $P(A \mid B^c)$ [*Hint:* $(A \cap B^c) \cup (A \cap B) = A$.]
   b. $P(B \mid A^c)$

In exercises 5–8, determine whether the given events A and B are independent.

5. $P(A) = .3$, $P(B) = .6$, $P(A \cap B) = .18$

6. $P(A) = .6$, $P(B) = .8$, $P(A \cap B) = .2$

7. $P(A) = .5$, $P(B) = .7$, $P(A \cup B) = .85$

8. $P(A^c) = .3$, $P(B^c) = .4$, $P(A \cap B) = .42$

9. If A and B are independent events and $P(A) = .4$ and $P(B) = .6$, find
   a. $P(A \cap B)$ **b.** $P(A \cup B)$

10. If A and B are independent events and $P(A) = .35$ and $P(B) = .45$, find
   a. $P(A \cap B)$ **b.** $P(A \cup B)$

11. The accompanying tree diagram represents an experiment consisting of two trials.

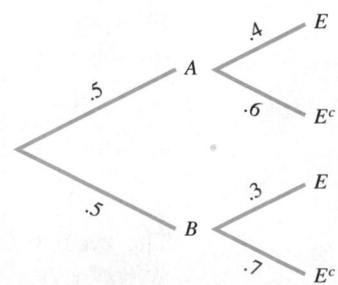

Use the diagram to find
a. $P(A)$ **b.** $P(E \mid A)$
c. $P(A \cap E)$ **d.** $P(E)$
e. Does $P(A \cap E) = P(A) \cdot P(E)$?
f. Are A and E independent events?

12. The accompanying tree diagram represents an experiment consisting of two trials. Use the diagram to find
a. $P(A)$ **b.** $P(E \mid A)$
c. $P(A \cap E)$ **d.** $P(E)$
e. Does $P(A \cap E) = P(A) \cdot P(E)$?
f. Are A and E independent events?

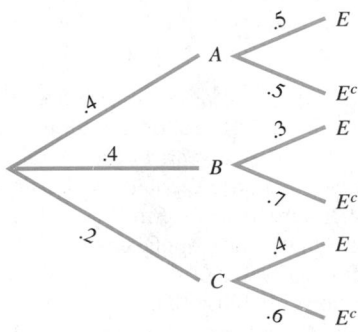

13. An experiment consists of two trials. The outcomes of the first trial are A and B with probabilities of occurring of .4 and .6. There are also two outcomes, C and D, in the second trial with probabilities of .3 and .7. Draw a tree diagram representing this experiment. Use this diagram to find
a. $P(A)$ **b.** $P(C \mid A)$
c. $P(A \cap C)$ **d.** $P(C)$
e. Does $P(A \cap C) = P(A) \cdot P(C)$?
f. Are A and C independent events?

14. An experiment consists of two trials. The outcomes of the first trial are A, B, and C, with probabilities of occurring of .2, .5, and .3, respectively. The outcomes of the second trial are E and F, with probabilities of occurring of .6 and .4. Draw a tree diagram representing this experiment. Use this diagram to find
a. $P(B)$ **b.** $P(F \mid B)$
c. $P(B \cap F)$ **d.** $P(F)$
e. Does $P(B \cap F) = P(B) \cdot P(F)$?
f. Are B and F independent events?

15. A pair of fair dice is cast. What is the probability that the sum of the numbers falling uppermost is less than 9, if it is known that one of the numbers is a 6?

16. **Product Reliability**   The probability that a battery will

last 10 hours or more is .80, and the probability that it will last 15 hours or more is .15. Given that a battery has lasted 10 hours, find the probability that it will last 15 hours or more.

17. Two cards are drawn without replacement from a well-shuffled deck of 52 playing cards.
a. What is the probability that the first card drawn is a heart?
b. What is the probability that the second card drawn is a heart given that the first card drawn was not a heart?
c. What is the probability that the second card drawn is a heart given that the first card drawn was a heart?

18. Five black balls and four white balls are placed in an urn. Two balls are then drawn in succession. What is the probability that the second ball drawn is a white ball
a. if the second ball is drawn without replacing the first?
b. if the first ball is replaced before the second is drawn?

19. **Auditing Tax Returns**   A tax specialist has estimated the probability that a tax return selected at random will be audited is .02. Furthermore, he estimates the probability that an audited return will result in additional assessments being levied on the taxpayer is .60. What is the probability that a tax return selected at random will result in additional assessments being levied on the taxpayer?

20. **Student Enrollment**   At a certain medical school, 1/7 of the students are from a minority group. Of those students who belong to a minority group, 1/3 are black.
a. What is the probability that a student selected at random from this medical school is black?
b. What is the probability that a student selected at random from this medical school is black if it is known that the student is a member of a minority group?

21. **Educational Level of Voters**   In a survey of 1000 eligible voters selected at random, it was found that 80 had a college degree. Additionally, it was found that 80% of those who had a college degree voted in the last presidential election, whereas 55% of the people who did not have a college degree voted in the last presidential election. Assuming that the poll is representative of all eligible voters, find the probability that an eligible voter selected at random
a. had a college degree and voted in the last presidential election.
b. did not have a college degree and did not vote in the last presidential election.
c. voted in the last presidential election.
d. did not vote in the last presidential election.

22. Three cards are drawn without replacement from a well-shuffled deck of 52 playing cards. What is the probability that the third card drawn is a diamond?

23. A coin is tossed three times. What is the probability that the coin will land
a. heads at least twice?
b. heads on the second toss given that heads were thrown on the first toss?
c. heads on the third toss given that tails were thrown on the first toss?

24. In a three-child family, what is the probability that all three children are girls given that one of the children is a girl? (Assume that the probability of a boy being born is the same as the probability of a girl being born.)

25. Quality Control An automobile manufacturer obtains the microprocessors used to regulate fuel consumption in its automobiles from three microelectronic firms: A, B, and C. The quality-control department of the company has determined that 1% of the microprocessors produced by firm A are defective, 2% of those produced by firm B are defective, and 1.5% of those produced by firm C are defective. Firms A, B, and C supply 45%, 25%, and 30%, respectively, of the microprocessors used by the company. What is the probability that a randomly selected automobile manufactured by the company will have a defective microprocessor?

26. Car Theft Figures obtained from a city's police department seem to indicate that, of all motor vehicles reported as stolen, 64% were stolen by professionals whereas 36% were stolen by amateurs (primarily for joy rides). Of those vehicles presumed stolen by professionals, 24% were recovered within 48 hours, 16% were recovered after 48 hours, and 60% were never recovered. Of those vehicles presumed stolen by amateurs, 38% were recovered within 48 hours, 58% were recovered after 48 hours, and 4% were never recovered.
a. Draw a tree diagram representing these data.
b. What is the probability that a vehicle stolen by a professional in this city will be recovered within 48 hours?
c. What is the probability that a vehicle stolen in this city will never be recovered?

27. Housing Loans The chief loan officer of the La Crosse Home Mortgage Company summarized the housing loans extended by the company in 1993 according to type and term of the loan. Her list shows that 70% of the loans were fixed-rate mortgages (*F*), 25% were

adjustable-rate mortgages (*A*), and 5% belong to some other category (*O*) (mostly second trust-deed loans and loans extended under the graduated payment plan). Of the fixed-rate mortgages, 80% were 30-year loans and 20% were 15-year loans; of the adjustable-rate mortgages, 40% were 30-year loans and 60% were 15-year loans; finally, of the other loans extended, 30% were 20-year loans, 60% were 10-year loans, and 10% were for a term of 5 years or less.
a. Draw a tree diagram representing this experiment.
b. What is the probability that a home loan extended by La Crosse has an adjustable rate and is for a term of 15 years?
c. What is the probability that a home loan extended by La Crosse is for a term of 15 years?

28. College Admissions The admissions office of a private university released the following admission data for the preceding academic year: From a pool of 3900 male applicants, 40% were accepted by the university and of these, 40% subsequently enrolled. Additionally, from a pool of 3600 female applicants, 45% were accepted by the university and of these, 40% subsequently enrolled. Find the probability that
a. a male applicant will be accepted by and subsequently will enroll in the university.
b. a student who applies for admissions will be accepted by the university.
c. a student who applies for admission will be accepted by the university and subsequently will enroll.

29. Quality Control Suppose that a box contains two defective Christmas tree lights that have been inadvertently mixed with eight nondefective lights. If the lights are selected one at a time without replacement and tested until both defective lights are found, what is the probability that both defective lights will be found after three trials?

30. Quality Control It is estimated that 0.80% of a large consignment of eggs in a certain supermarket is broken.
a. What is the probability that a customer who randomly selects a dozen of these eggs receives at least one broken egg?
b. What is the probability that a customer who selects these eggs at random will have to check three cartons before finding a carton without any broken eggs? (Each carton contains a dozen eggs.)

31. Student Financial Aid The accompanying data were obtained from the financial aid office of a certain university.

	Receiving Financial Aid	Not Receiving Financial Aid	Total
Undergraduates	4,222	3,898	8,120
Graduates	1,879	731	2,610
Total	6,101	4,629	10,730

Let A be the event that a student selected at random from this university is an undergraduate student and let B be the event that a student selected at random is receiving financial aid.

a. Find each of the following probabilities:

$$P(A), P(B), P(A \cap B), P(B \mid A), \text{ and } P(B \mid A^c)$$

b. Are the events A and B independent events?

32. Employee Education and Income   The personnel department of the Franklin National Life Insurance Company compiled the accompanying data regarding the income and education of its employees.

	Income $40,000 or Below	Income Above $40,000
Noncollege Graduate	2040	840
College Graduate	400	720

Let A be the event that a randomly chosen employee has a college degree and B the event that the chosen employee's income is more than $40,000.

a. Find each of the following probabilities:

$$P(A), P(B), P(A \cap B), P(B \mid A), \text{ and } P(B \mid A^c)$$

b. Are the events A and B independent events?

33. Two cards are drawn without replacement from a well-shuffled deck of 52 cards. Let A be the event that the first card drawn is a heart and let B be the event that the second card drawn is a red card. Show that the events A and B are dependent events.

34. Medical Research   A nationwide survey conducted by the National Cancer Society revealed the following information: Of 10,000 people surveyed, 3200 were "heavy coffee drinkers" and 160 had cancer of the pancreas. Of those who had cancer of the pancreas, 132 were heavy coffee drinkers. Using the data in this survey, determine whether the events "being a heavy coffee drinker" and "having cancer of the pancreas" are independent events.

35. Mail Delivery   Suppose the probability that your mail will be delivered before 2 P.M. on a delivery day is .90. What is the probability that your mail will be delivered before 2 P.M.

a. for two consecutive delivery days?

b. for three consecutive delivery days?

36. Reliability of Security Systems   Before being allowed to enter a maximum-security area at a military installation, a person must pass three identification tests: a voice-pattern test, a fingerprint test, and a handwriting test. If the reliability of the first test is 97%, the reliability of the second test is 98.5%, and that of the third is 98.5%, what is the probability that this security system will allow an improperly identified person to enter the maximum-security area?

37. Quality Control   Copykwik, Inc., has four photocopy machines A, B, C, and D. The probability that a given machine will break down on a particular day is

$$P(A) = \frac{1}{50}, P(B) = \frac{1}{60}, P(C) = \frac{1}{75}, P(D) = \frac{1}{40}$$

Assuming independence, what is the probability on a particular day that

a. all four machines will break down?

b. none of the machines will break down?

c. exactly one machine will break down?

38. Product Reliability   The proprietor of Cunningham's Hardware Store has decided to install floodlights on the premises as a measure against vandalism and theft. If the probability is .01 that a certain brand of floodlight will burn out within a year, find the minimum number of floodlights that must be installed to ensure that the probability that at least one of them will remain functional within the year is at least .99999. (Assume that the floodlights operate independently.)

39. Let E be any event in a sample space S.

a. Are E and S independent? Explain your answer.

b. Are E and $\varnothing$ independent? Explain your answer.

40. Suppose the probability that an event will occur in one trial is p. Show that the probability that the event will occur at least once in n independent trials is $1 - (1 - p)^n$.

41. Suppose A and B are mutually exclusive events and $P(A \cup B) \neq 0$. What is $P(A \mid A \cup B)$?

SOLUTIONS TO SELF-CHECK EXERCISES 7.5

1. a. $P(A \mid B) = \dfrac{P(A \cap B)}{P(B)}$ **b.** $P(B \mid A) = \dfrac{P(A \cap B)}{P(A)}$

$= \dfrac{.3}{.8} = \dfrac{3}{8}$ $= \dfrac{.3}{.4} = \dfrac{3}{4}$

2. Let A, B, and C denote the events that three women who are still unmarried at ages 30, 35, and 40 *will* marry eventually. Then it seems reasonable to assume that these events are independent, with $P(A) = .20$, $P(B) = .054$, and $P(C) = .013$. Based on these figures, the probability that all three women will be married eventually is given by

$$P(A)P(B)P(C) = (.2)(.054)(.013)$$
$$= .00014$$

7.6 BAYES' THEOREM

A Posteriori Probabilities

Suppose three machines, A, B, and C, produce similar engine components. Machine A produces 45% of the total components, machine B produces 30%, and machine C, 25%. For the usual production schedule, 6% of the components produced by machine A do not meet established specifications; for machine B and machine C, the corresponding figures are 4% and 3%. One component is selected at random from the total output and is found to be defective. What is the probability that the component selected was produced by machine A?

The answer to this question is found by calculating the probability *after* the outcomes of the experiment have been observed. Such probabilities are called *a posteriori* **probabilities** as opposed to *a priori* **probabilities**— probabilities that give the likelihood that an event *will* occur, the subject of the last several sections.

Returning to the example under consideration, we need to determine the a posteriori probability for the event that the component selected was produced by machine A. To this end, let A, B, and C denote the event that a component is produced by machine A, machine B, and machine C, respectively. We may represent this experiment with a Venn diagram (Figure 7.21).

The three mutually exclusive events A, B, and C form a **partition** of the sample space S. That is, aside from being mutually exclusive, their union is precisely S. The event D that a component is defective is the shaded area. Again referring to Figure 7.21, we see that

1. the event D may be expressed as

$$D = (A \cap D) \cup (B \cap D) \cup (C \cap D)$$

2. the event that a component is defective and is produced by machine A is given by $A \cap D$.

Figure 7.21
D is the event that a defective component is produced by machine A, machine B, or machine C.

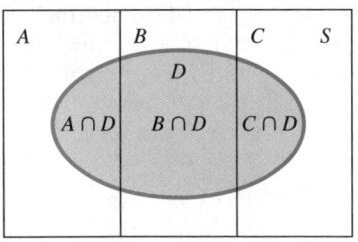

Thus, the a posteriori probability that a defective component selected was produced by machine A is given by

$$P(A \mid D) = \frac{n(A \cap D)}{n(D)}$$

Upon dividing both the numerator and the denominator by $n(S)$ and observing that the events $A \cap D$, $B \cap D$, and $C \cap D$ are mutually exclusive, we obtain

$$P(A \mid D) = \frac{P(A \cap D)}{P(D)}$$

$$= \frac{P(A \cap D)}{P(A \cap D) + P(B \cap D) + P(C \cap D)} \tag{7}$$

Next, using the Product Rule, we may express

$$P(A \cap D) = P(A) \cdot P(D \mid A)$$
$$P(B \cap D) = P(B) \cdot P(D \mid B)$$
and $$P(C \cap D) = P(C) \cdot P(D \mid C)$$

so that equation (7) may be expressed in the form

$$P(A \mid D) = \frac{P(A) \cdot P(D \mid A)}{P(A) \cdot P(D \mid A) + P(B) \cdot P(D \mid B) + P(C) \cdot P(D \mid C)} \tag{8}$$

which is a special case of a result known as **Bayes' Theorem.**

Observe that the expression on the right of (8) involves the probabilities $P(A)$, $P(B)$, $P(C)$ and the conditional probabilities $P(D \mid A)$, $P(D \mid B)$, and $P(D \mid C)$, all of which may be calculated in the usual fashion. In fact, by displaying these quantities on a tree diagram, we obtain Figure 7.22. We may compute the required probability by substituting the relevant quantities into (8), or we may make use of the following device:

$$P(A \mid D) = \frac{\text{Product of probabilities along the limb through } A}{\text{Sum of products of the probabilities along each limb terminating at } D}$$

Figure 7.22

A tree diagram displaying the probabilities that a defective component is produced by machine A, machine B, or machine C.

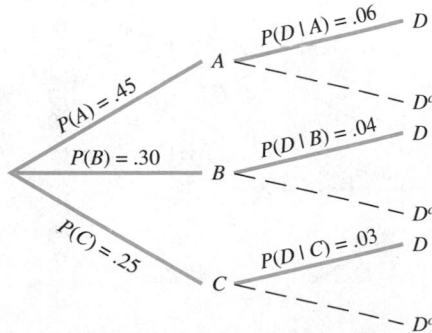

In either case, we obtain

$$P(A \mid D) = \frac{(.45)(.06)}{(.45)(.06) + (.3)(.04) + (.25)(.03)}$$

$$= .58$$

Before looking at any further examples, let us state the general form of Bayes' Theorem.

BAYES' THEOREM

Let $A_1, A_2, \ldots, A_n$ be a partition of a sample space S and let E be an event of the experiment such that $P(E) \neq 0$. Then the a posteriori probability $P(A_i \mid E)$ $(1 \leq i \leq n)$ is given by

$$P(A_i \mid E) = \frac{P(A_i) \cdot P(E \mid A_i)}{P(A_1) \cdot P(E \mid A_1) + P(A_2) \cdot P(E \mid A_2) + \cdots + P(A_n) \cdot P(E \mid A_n)} \quad (9)$$

Applications

EXAMPLE 1   The picture tubes for the Pulsar 19-inch color television sets are manufactured in three locations and then shipped to the main plant of the Vista Vision Corporation for final assembly. Plants A, B, and C supply 50%, 30%, and 20%, respectively, of the picture tubes used by Vista Vision. The quality-control department of the company has determined that 1% of the picture tubes produced by plant A are defective, whereas 2% of the picture tubes produced by plants B and C are defective. If a Pulsar 19-inch color television set is selected at random and the picture tube is found to be defective, what is the probability that the picture tube was manufactured in plant C? Compare with Example 6, page 417.

Solution   Let A, B, and C denote the event that the set chosen has a picture tube manufactured in plant A, plant B, and plant C, respectively. Also, let D denote the event that a set has a defective picture tube. Using the given information, we may draw the tree diagram shown in Figure 7.23. Next, using

Figure 7.23
$P(C \mid D) =$

$$\frac{\text{Product of probabilities of branches to D through C}}{\text{Sum of product of probabilities of branches leading to D}}$$

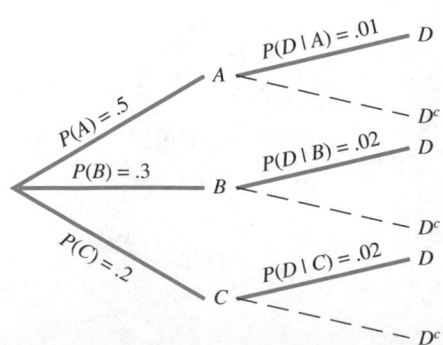

formula (9), we find that the required a posteriori probability is given by

$$P(C \mid D) = \frac{P(C) \cdot P(D \mid C)}{P(A) \cdot P(D \mid A) + P(B) \cdot P(D \mid B) + P(C) \cdot P(D \mid C)}$$

$$= \frac{(.2)(.02)}{(.5)(.01) + (.3)(.02) + (.2)(.02)}$$

$$\approx .27$$

○ ○ ○

EXAMPLE 2   A study was conducted in a large metropolitan area to determine the annual incomes of married couples in which the husbands were the sole providers and of those in which the husbands and wives were both employed. Table 7.9 gives the results of this study.

a. What is the probability that a couple selected at random from this area has two incomes?

b. If a randomly chosen couple has two incomes, what is the probability that the annual income of this couple is over $100,000?

c. If a randomly chosen couple has two incomes, what is the probability that the annual income of this couple is greater than $34,999?

Table 7.9

Annual Family Income ($)	% of Married Couples	% of Income Group with Both Spouses Working
100,000 and over	4	65
75,000–99,999	10	73
50,000–74,999	21	68
35,000–49,999	24	63
25,000–34,999	30	43
Under 25,000	11	28

Solution   Let A denote the event that the annual income of the couple is $100,000 and over; let B denote the event that the annual income is between $75,000 and $99,999; let C denote the event that the annual income is between $50,000 and $74,999, and so on. Finally, let F denote the event that the annual income is less than $25,000, and let T denote the event that both spouses work. The probabilities of the occurrence of these events are displayed in the tree diagram on page 430 (Figure 7.24).

a. The probability that a couple selected at random from this group has two incomes is given by

$$P(T) = P(A) \cdot P(T \mid A) + P(B) \cdot P(T \mid B) + P(C) \cdot P(T \mid C)$$
$$+ P(D) \cdot P(T \mid D) + P(E) \cdot P(T \mid E) + P(F) \cdot P(T \mid F)$$
$$= (.04)(.65) + (.10)(.73) + (.21)(.68) + (.24)(.63)$$
$$+ (.30)(.43) + (.11)(.28)$$
$$= .5528$$

Figure 7.24

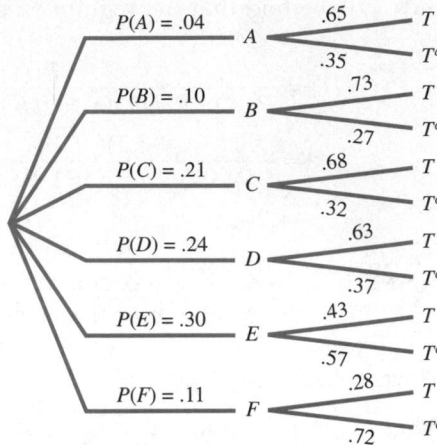

b. Using the results of (a) and Bayes' Theorem, we find that the probability that a randomly chosen couple has an annual income over \$100,000 given that both spouses are working is

$$P(A \mid T) = \frac{P(A) \cdot P(T \mid A)}{P(T)} = \frac{(.04)(.65)}{.5528}$$

$$= .047$$

c. The probability that a random chosen couple has an annual income greater than \$34,999 given that both spouses are working is

$$P(A \mid T) + P(B \mid T) + P(C \mid T) + P(D \mid T)$$

$$= \frac{P(A) \cdot P(T \mid A) + P(B) \cdot P(T \mid B) + P(C) \cdot P(T \mid C) + P(D) \cdot P(T \mid D)}{P(T)}$$

$$= \frac{(.04)(.65) + (.1)(.73) + (.21)(.68) + (.24)(.63)}{.5528}$$

$$= .711$$

◐ ◐ ◐

SELF-CHECK EXERCISES 7.6

1. The accompanying tree diagram represents a two-stage experiment. Use the diagram to find $P(B \mid D)$.

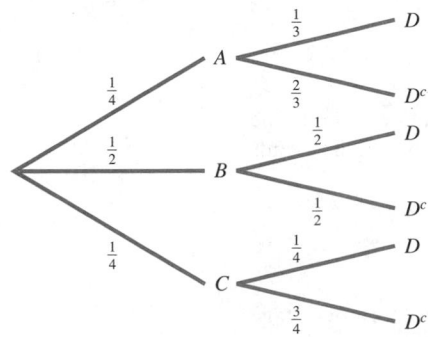

2. In a recent presidential election, it was estimated that the probability that the Republican candidate would be elected was 3/5 and therefore the probability that the Democratic candidate would be elected was 2/5 (the two Independent candidates were given little chance of being elected). It was also estimated that if the Republican candidate were elected, then the probability that research for a new manned bomber would continue was 4/5. But if the Democratic candidate were successful, then the probability that the research would continue was 3/10. Research was terminated shortly after the successful presidential candidate took office. What is the probability that the Republican candidate won that election?

Solutions to Self-Check Exercises 7.6 can be found on page 436.

7.6 EXERCISES

In exercises 1–3, refer to the accompanying Venn diagram. An experiment in which the three mutually exclusive events A, B, and C form a partition of the uniform sample space S is depicted in the diagram.

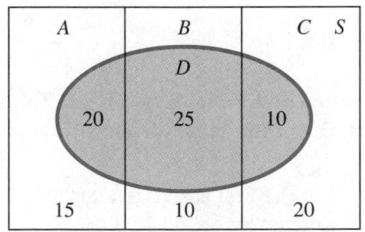

1. Draw a tree diagram using the information given in the Venn diagram illustrating the probabilities of the events A, B, C, and D.

2. Find: **a.** $P(D)$ **b.** $P(A \mid D)$

3. Find: **a.** $P(D^c)$ **b.** $P(B \mid D^c)$

In exercises 4–6, refer to the accompanying Venn diagram. An experiment in which the three mutually exclusive events A, B, and C form a partition of the uniform sample space S is depicted in the diagram.

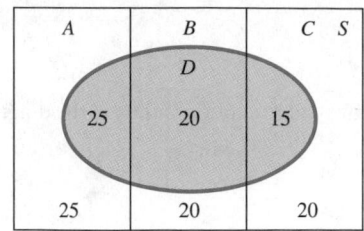

4. Draw a tree diagram using the information given in the Venn diagram illustrating the probabilities of the events A, B, C, and D.

5. Find: **a.** $P(D)$ **b.** $P(B \mid D)$

6. Find: **a.** $P(D^c)$ **b.** $P(B \mid D^c)$

7. The accompanying tree diagram represents a two-stage experiment. Use the diagram to find
a. $P(A) \cdot P(D \mid A)$ **b.** $P(B) \cdot P(D \mid B)$
c. $P(A \mid D)$

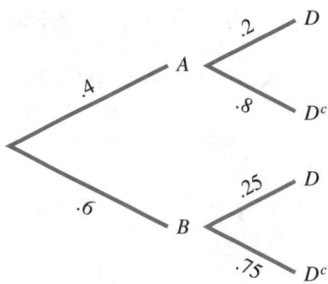

8. The accompanying tree diagram represents a two-stage experiment. Use the diagram to find
a. $P(A) \cdot P(D \mid A)$ **b.** $P(B) \cdot P(D \mid B)$
c. $P(A \mid D)$

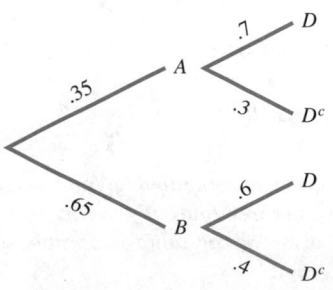

9. The accompanying tree diagram represents a two-stage experiment. Use the diagram to find
a. $P(A) \cdot P(D \mid A)$ **b.** $P(B) \cdot P(D \mid B)$
c. $P(C) \cdot P(D \mid C)$ **d.** $P(A \mid D)$

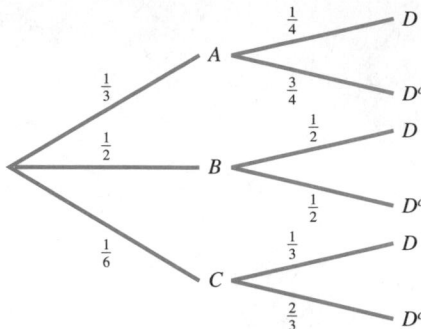

10. The accompanying tree diagram represents a two-stage experiment. Use this diagram to find
a. $P(A \cap D)$ **b.** $P(B \cap D)$
c. $P(C \cap D)$ **d.** $P(D)$
e. Verify:

$P(A \mid D)$

$= \dfrac{P(A \cap D)}{P(D)}$

$= \dfrac{P(A) \cdot P(D \mid A)}{P(A) \cdot P(D \mid A) + P(B) \cdot P(D \mid B) + P(C) \cdot P(D \mid C)}$

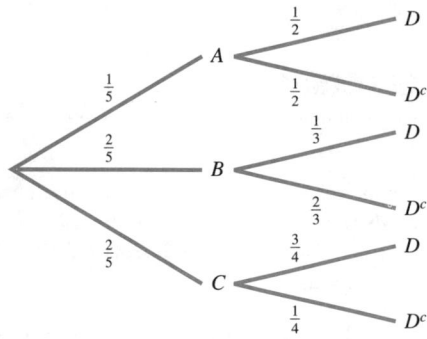

In exercises 11–14, refer to the following experiment: Two cards are drawn in succession without replacement from a standard deck of 52 cards.

11. What is the probability that the first card is a heart given that the second card is a heart?

12. What is the probability that the first card is a heart given that the second card is a diamond?

13. What is the probability that the first card is a jack given that the second card is an ace?

14. What is the probability that the first card is a face card given that the second card is an ace?

In exercises 15–18, refer to the following experiment: Urn A contains four white and six black balls. Urn B contains three white and five black balls. A ball is drawn from urn A and then transferred to urn B. A ball is then drawn from urn B.

15. Represent the probabilities associated with this two-stage experiment in the form of a tree diagram.

16. What is the probability that the transferred ball was white given that the second ball drawn was white?

17. What is the probability that the transferred ball was black given that the second ball drawn was white?

18. What is the probability that the transferred ball was black given that the second ball drawn was black?

19. **Politics** The 1992 U.S. Senate was composed of 57 Democrats and 43 Republicans. Sixty-six percent of the Democrats served in the military, whereas 64 percent of the Republicans had seen military service. If a senator selected at random had served in the military, what is the probability that he was Republican? (*Note:* No congresswoman had served in the military.)

20. **Quality Control** Jansen Electronics has four machines that produce an identical component for use in its videocassette players. The proportion of the components produced by each machine and the probability of that component being defective are shown in the accompanying table. What is the probability that
    a. a component selected at random is defective?
    b. a component selected at random was produced by machine I, given that it is defective?
    c. a component selected at random was produced by machine II, given that it is defective?

Machine	Proportion of Components Produced	Probability of Defective Component
I	.15	.04
II	.30	.02
III	.35	.02
IV	.20	.03

21. An experiment consists of randomly selecting one of three coins, tossing it, and observing the outcome—heads or tails. The first coin is a two-headed coin, the second is a biased coin such that $P(H) = .75$, and the third is a fair coin.
    a. What is the probability that the coin that is tossed will show heads?
    b. If the coin selected shows heads, what is the probability that this coin is the fair coin?

A calculator is recommended for the remainder of this exercise set.

22. **Reliability of Medical Tests** A medical test has been designed to detect the presence of a certain disease. Among those who have the disease, the probability that the disease will be detected by the test is .95. However, the probability that the test will erroneously indicate the presence of the disease in those who do not actually have it is .04. It is estimated that 4% of the population who take this test have the disease.
    a. If the test administered to an individual is positive, what is the probability that the person actually has the disease?
    b. If an individual takes the test twice and both times the test is positive, what is the probability that the person actually has the disease?

23. **Reliability of Medical Tests** Refer to exercise 22. Suppose that 20% of the people who were referred to a clinic for the test did in fact have the disease. If the test administered to an individual from this group is positive, what is the probability that the person actually has the disease?

24. **Quality Control** A desk lamp produced by the Luminar Company was found to be defective. The company has three factories where the lamps are manufactured. The percentage of the total number of desk lamps produced by each factory and the probability that a lamp manufactured by that factory is defective are shown in the accompanying table. What is the probability that the defective lamp was manufactured in factory III?

Factory	Percentage of Total Production	Probability of Defective Component
I	.35	.015
II	.35	.01
III	.30	.02

25. Auto-Accident Rates An insurance company has compiled the accompanying data relating the age of drivers and the accident rate (the probability of being involved in an accident during a one-year period) for drivers within that group.

Age Group	Percentage of Insured Drivers	Accident Rate
Under 25	.16	.055
25–44	.40	.025
45–64	.30	.02
65 and over	.14	.04

a. What is the probability that an insured driver will be involved in an accident during a particular one-year period?
b. What is the probability that an insured driver who is involved in an accident is under 25?

26. Seat-Belt Compliance Data compiled by the Highway Patrol Department regarding the use of seat belts by drivers in a certain area after the passage of a compulsory seat-belt law are shown in the accompanying table.

Drivers	Percentage of Drivers in Group	Percentage of Group Stopped for Moving Violation
Group I (using seat belts)	.64	.002
Group II (not using seat belts)	.36	.005

If a driver in that area is stopped for a moving violation, what is the probability that
a. he or she will have a seat belt on?
b. he or she will not have a seat belt on?

27. Medical Research Based on data obtained from the National Institute of Dental Research, it has been determined that 42% of twelve-year-olds have never had a cavity, 34% of thirteen-year-olds have never had a cavity, and 28% of fourteen-year-olds have never had a cavity. If a child is selected at random from a group of 24 junior high school students comprising 6 twelve-year-olds, 8 thirteen-year-olds, and 10 fourteen-year-olds and this child does not have a cavity, what is the probability that this child is fourteen years old?

28. Voting Patterns In a recent senatorial election, 50% of the voters in a certain district were registered as Democrats, 35% were registered as Republicans, and 15% were registered as Independents. The incumbent Democratic senator was re-elected over her Republican and Independent opponents. Exit polls indicated that she gained 75% of the Democratic vote, 25% of the Republican vote, and 30% of the Independent vote. Assuming that the exit poll is accurate, what is the probability that a vote for the incumbent was cast by a registered Republican?

29. Crime Rates Data compiled by the Department of Justice on the number of people arrested for serious crimes (murder, forcible rape, robbery, and so on) in 1988 revealed that 89% were male and 11% were female. Of the males, 30% were under 18, whereas 27% of the females arrested were under 18.
a. What is the probability that a person arrested for a serious crime in 1988 was under 18?
b. If a person arrested for a serious crime in 1988 was known to be under 18, what is the probability that the person is female?

30. Opinion Polls A poll was conducted among 500 registered voters in a certain area regarding their position on a national lottery to raise revenue for the government. The results of the poll are shown in the accompanying table.

Sex	Percentage of Voters Polled	Percentage Favoring Lottery	Percentage Not Favoring Lottery	Percentage Expressing No Opinion
Male	.51	.62	.32	.06
Female	.49	.68	.28	.04

What is the probability that a registered voter who
a. favored a national lottery was a woman?
b. expressed no opinion regarding the lottery was a woman?

31. Customer Surveys The sales department of the Thompson Drug Company released the accompanying data concerning the sales of a certain pain reliever manufactured by the company.

Pain Reliever	Percentage of Drug Sold	Percentage of Group Sold in Extra-Strength Dosage
Group I (capsule form)	.57	.38
Group II (tablet form)	.43	.31

If a customer purchased the extra-strength dosage of this drug, what is the probability that it was in capsule form?

32. **Selection of Supreme Court Judges** In a past presidential election, it was estimated that the probability that the Republican candidate would be elected was 3/5 and therefore the probability that the Democratic candidate would be elected was 2/5 (the two Independent candidates were given little chance of being elected). It was also estimated that if the Republican candidate were elected, the probabilities that a conservative, moderate, or liberal judge would be appointed to the Supreme Court (one retirement was expected during the presidential term) were 1/2, 1/3, and 1/6, respectively. If the Democratic candidate were elected, the probabilities that a conservative, moderate, or liberal judge would be appointed to the Supreme Court would be 1/8, 3/8, and 1/2, respectively. A conservative judge *was* appointed to the Supreme Court during the presidential term. What is the probability that the Democratic candidate was elected?

33. **Personnel Selection** Applicants for temporary office work at the Carter Temporary Help Agency who have successfully completed a typing test are then placed in suitable positions by Ms. Dwyer and Ms. Newberg. Employers who hire temporary help through the agency return a card indicating satisfaction or dissatisfaction with the work performance of those hired. From past experience it is known that 80% of the employees placed by Ms. Dwyer are rated as satisfactory, whereas 70% of those placed by Ms. Newberg are rated as satisfactory. Ms. Newberg places 55% of the temporary office help at the agency and Ms. Dwyer the remaining 45%. If a Carter office worker is rated unsatisfactory, what is the probability that he or she was placed by Ms. Newberg?

34. **College Majors** The Office of Admissions and Records of a large western university released the accompanying information concerning the contemplated majors of its freshman class.

Major	Percentage of Freshmen Choosing This Major	Percentage of Females	Percentage of Males
Business	.24	.38	.62
Humanities	.08	.60	.40
Education	.08	.66	.34
Social science	.07	.58	.42
Natural sciences	.09	.52	.48
Other	.44	.48	.52

a. What is the probability that a student selected at random from the freshman class is a female?
b. What is the probability that a business student selected at random from the freshman class is a male?
c. What is the probability that a female student selected at random from the freshman class is majoring in business?

35. **Medical Diagnoses** A study was conducted among a certain group of union members whose health insurance policies required second opinions prior to surgery. Of those members whose doctors advised them to have surgery, 20% were informed by a second doctor that no surgery was needed. Of these, 70% took the second doctor's opinion and did not go through with the surgery. Of the members who were advised to have surgery by both doctors, 95% went through with the surgery. What is the probability that a union member who had surgery was advised to do so by a second doctor?

36. **Age Distribution of Renters** A study conducted by the Metro Housing Agency in a midwestern city revealed the accompanying information concerning the age distribution of renters within the city.

Age	Percentage of Adult Population	Percentage of Group Who Are Renters
21–44	.51	.58
45–64	.31	.45
65 and over	.18	.60

a. What is the probability that an adult selected at random from this population is a renter?
b. If a renter is selected at random, what is the probability that he or she is in the 21–44 age bracket?
c. If a renter is selected at random, what is the probability that he or she is 45 years of age or older?

SOLUTIONS TO SELF-CHECK EXERCISES 7.6

1. By Bayes' Theorem, we have, using the probabilities given in the tree diagram,

$$P(B\,|\,D) = \frac{P(B)P(D\,|\,B)}{P(A)P(D\,|\,A) + P(B)P(D\,|\,B) + P(C)P(D\,|\,C)}$$

$$= \frac{\left(\frac{1}{2}\right)\left(\frac{1}{2}\right)}{\left(\frac{1}{4}\right)\left(\frac{1}{3}\right) + \left(\frac{1}{2}\right)\left(\frac{1}{2}\right) + \left(\frac{1}{4}\right)\left(\frac{1}{4}\right)}$$

$$= \frac{12}{19}$$

2. Let R and D, respectively, denote the event that the Republican and the Democratic candidate won the presidential election. Then $P(R) = 3/5$ and $P(D) = 2/5$. Also, let C denote the event that research for the new manned bomber would continue. These data may be exhibited as in the accompanying tree diagram.

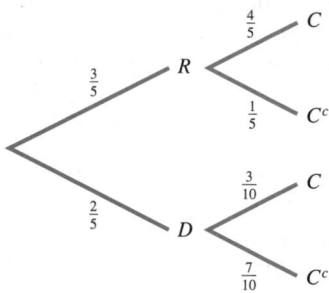

Using Bayes' Theorem, we find that the probability that the Republican candidate had won the election is given by

$$P(R\,|\,C^c) = \frac{P(R)P(C^c\,|\,R)}{P(R)P(C^c\,|\,R) + P(D)P(C^c\,|\,D)}$$

$$= \frac{\left(\frac{3}{5}\right)\left(\frac{1}{5}\right)}{\left(\frac{3}{5}\right)\left(\frac{1}{5}\right) + \left(\frac{2}{5}\right)\left(\frac{7}{10}\right)} = \frac{3}{10}$$

7.7 MARKOV CHAINS (OPTIONAL)

Transitional Probabilities

In this section we look at an important application of mathematics that is based primarily on matrix theory and the theory of probability. Markov chains are a relatively recent development in the field of mathematics and have wide applications in many practical areas.

A finite stochastic process, you may recall, is an experiment consisting of a finite number of stages in which the outcomes and associated probabilities at each stage depend on the outcomes and associated probabilities of the *preceding stages*. In this section we are concerned with a special class of stochastic processes—namely, those in which the probabilities associated with the outcomes at any stage of the experiment depend only on the outcomes of the *preceding stage*. Such a process is called a **Markov process,** or a **Markov chain,** named after the Russian mathematician A. A. Markov (1856–1922).

The outcome at any stage of the experiment in a Markov process is called the **state** of the experiment. In particular, the outcome at the current stage of the experiment is called the **current state** of the process. Here is a typical problem involving a Markov chain:

> *Starting from one state of a process (the current state), determine the probability that the process will be at a particular state at some future time.*

EXAMPLE 1 An analyst at Weaver and Kline, a stock brokerage firm, observes that the closing price of the preferred stock of an airline company over a short span of time depends only on its previous closing price. At the end of each trading day he makes a note of the stock's performance for that day, recording the closing price as "higher," "unchanged," or "lower" according to whether the stock closes higher, unchanged, or lower than the previous day's closing price. This sequence of observations may be viewed as a Markov chain.   ο ο ο

The transition from one state to another in a Markov chain may be studied with the aid of tree diagrams, as in the next example.

EXAMPLE 2 Refer to Example 1. If on a certain day the stock's closing price is higher than that of the previous day, then the probability that it closes higher, unchanged, or lower on the next trading day is .2, .3, and .5, respectively. Next, if the stock's closing price is unchanged from the previous day, then the probability that it closes higher, unchanged, or lower on the next trading day is .5, .2, and .3, respectively. Finally, if the stock's closing price is lower than that of the previous day, then the probability that it closes higher, unchanged, or lower on the next trading day is .4, .4, and

.2, respectively. Describe the transition between states and the probabilities associated with these transitions with the aid of tree diagrams.

Solution   The Markov chain being described has three states: higher, unchanged, and lower. If the current state is higher, then the transition to the other states from this state may be displayed by constructing a tree diagram in which the associated probabilities are shown on the appropriate limbs (Figure 7.25a). Tree diagrams describing the transition from each of the other two possible current states to the other states may be constructed in a similar manner (Figures 7.25b and 7.25c).

Figure 7.25
Tree diagrams showing transition probabilities between states.

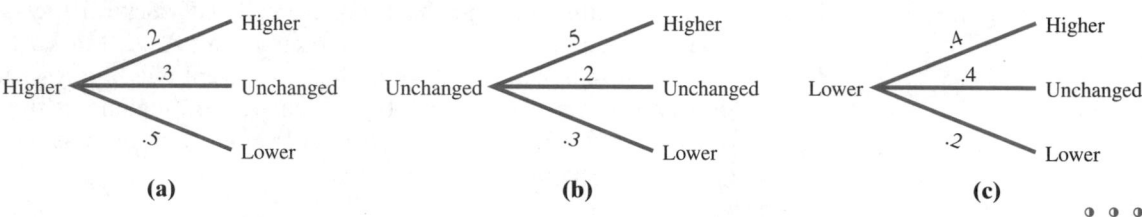

The probabilities encountered in this example are called **transition probabilities** because they are associated with the transition from one state to the next in the Markov process. These transition probabilities may be conveniently represented in the form of a matrix. Suppose for simplicity that we have a Markov chain with three possible outcomes at each stage of the experiment. Let us refer to these outcomes as state 1, state 2, and state 3. Then the transition probabilities associated with the transition from state 1 to each of the states 1, 2, and 3 in the next phase of the experiment are precisely the respective conditional probabilities that the outcome is state 1, state 2, and state 3 *given* that the outcome state 1 has occurred. In short, the desired transition probabilities are $P(\text{state } 1 \mid \text{state } 1)$, $P(\text{state } 2 \mid \text{state } 1)$, and $P(\text{state } 3 \mid \text{state } 1)$, respectively. Thus we write

$$\text{Current state}$$
$$a_{11} = P(\text{state } 1 \mid \text{state } 1)$$
$$a_{12} = P(\text{state } 2 \mid \text{state } 1)$$
$$a_{13} = P(\text{state } 3 \mid \text{state } 1)$$
$$\text{Next state}$$

Note that the first subscript in this notation refers to the current state and the second subscript refers to the state in the next stage of the experiment. Using a tree diagram, we have the following representation:

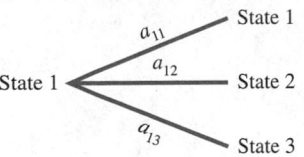

Similarly, the transition probabilities associated with the transition from state 2 and state 3 to each of the states 1, 2, and 3 are

$$
\begin{aligned}
a_{21} &= P(\text{state 1} \mid \text{state 2}) & a_{31} &= P(\text{state 1} \mid \text{state 3}) \\
a_{22} &= P(\text{state 2} \mid \text{state 2}) \quad \text{and} \quad & a_{32} &= P(\text{state 2} \mid \text{state 3}) \\
a_{23} &= P(\text{state 3} \mid \text{state 2}) & a_{33} &= P(\text{state 3} \mid \text{state 3})
\end{aligned}
$$

These observations lead to the following matrix representation of the transition probabilities:

$$
\begin{array}{c}
& \qquad\qquad\quad \text{Next state} \\
& \quad\;\; \text{State 1} \quad \text{State 2} \quad \text{State 3} \\
\begin{array}{c}
\text{State 1} \\
\text{Current state} \quad \text{State 2} \\
\text{State 3}
\end{array}
&
\begin{bmatrix}
a_{11} & a_{12} & a_{13} \\
a_{21} & a_{22} & a_{23} \\
a_{31} & a_{32} & a_{33}
\end{bmatrix}
\end{array}
$$

EXAMPLE 3 Use a matrix to represent the transition probabilities obtained in Example 2.

Solution There are three states at each stage of the Markov chain under consideration. Letting state 1, state 2, and state 3 denote the states "higher," "unchanged," and "lower," respectively, we find that

$$
a_{11} = .2, \quad a_{12} = .3, \quad a_{13} = .5, \quad \text{and so on}
$$

so the required matrix representation is given by

$$
T = \begin{bmatrix}
.2 & .3 & .5 \\
.5 & .2 & .3 \\
.4 & .4 & .2
\end{bmatrix}
$$

○ ○ ○

The matrix obtained in the preceding example is a transition matrix. In the general case, we have the following definition:

TRANSITION MATRIX

A **transition matrix** associated with a Markov chain with n states is an $n \times n$ matrix T with entries a_{ij} $(1 \le i \le n; 1 \le j \le n)$

Next state

$$T = \begin{array}{c} \text{Current} \\ \text{state} \end{array} \begin{array}{c} \\ \text{State 1} \\ \text{State 2} \\ \vdots \\ \text{State } i \\ \vdots \\ \text{State } n \end{array} \begin{array}{ccccccc} \text{State 1} & \text{State 2} & \cdots & \text{State } j & \cdots & \text{State } n \\ \left[\begin{array}{ccccc} a_{11} & a_{12} & \cdots & a_{1j} & \cdots & a_{1n} \\ a_{21} & a_{22} & \cdots & a_{2j} & \cdots & a_{2n} \\ \vdots & \vdots & & \vdots & & \vdots \\ a_{i1} & a_{i2} & \cdots & a_{ij} & \cdots & a_{in} \\ \vdots & \vdots & & \vdots & & \vdots \\ a_{n1} & a_{n2} & \cdots & a_{nj} & \cdots & a_{nn} \end{array}\right] \end{array}$$

having the following properties:

1. $a_{ij} \ge 0$ for all i and j.

2. The sum of the entries in each row of T is 1.

Since $a_{ij} = P(\text{state } j \mid \text{state } i)$ is the probability of the occurrence of an event, it must be nonnegative, and this is precisely what property 1 implies. Property 2 follows from the fact that the transition from any one of the current states must terminate in one of the n states in the next stage of the experiment. Any square matrix satisfying properties 1 and 2 is referred to as a **stochastic matrix.**

One advantage in representing the transition probabilities in the form of a matrix is that we may use the results from matrix theory to help us solve problems involving Markov processes, a fact that will be demonstrated in the next several examples.

Next, for simplicity, let us consider a Markov process where each stage of the experiment has precisely two possible states.

 EXAMPLE 4   Because of the continued successful implementation of an urban renewal program, it is expected that each year 3% of the population currently residing in the city will move to the suburbs, and 6% of the population currently residing in the suburbs will move into the city. At present, 65% of the total population of the metropolitan area lives in the city itself, while the remaining 35% live in the suburbs. Assuming that the total population of the metropolitan area remains constant, what will the distribution of the population be like one year from now?

Solution   This problem may be solved with the aid of a tree diagram and the techniques of this chapter. The required tree diagram describing this process is shown in Figure 7.26. The probability that a person selected at random will be a city dweller one year from now is given by

$$(.65)(.97) + (.35)(.06) = .6515$$

Figure 7.26
Tree diagram showing a Markov
process with two states—living in
the city and living in the suburbs.

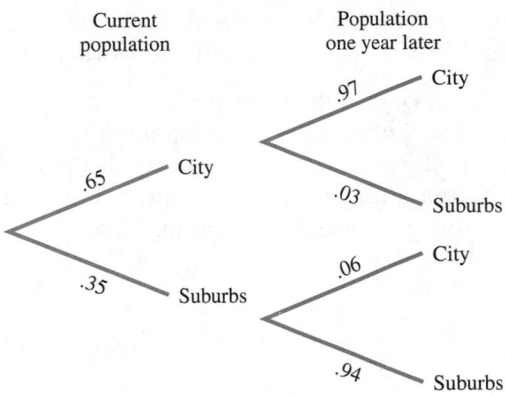

In a similar manner we find that the probability that a person selected at random will reside in the suburbs one year from now is given by

$$(.65)(.03) + (.35)(.94) = .3485$$

Thus, the population of the area one year from now may be expected to be distributed as follows: 65.15% living in the city and 34.85% residing in the suburbs. ◦ ◦ ◦

Let us reexamine the solution to this problem. As noted earlier, the process under consideration may be viewed as a Markov chain with two possible states at each stage of the experiment: "living in the city" (state 1) and "living in the suburbs" (state 2). The transition matrix associated with this Markov chain is

$$T = \begin{array}{c} \text{State 1} \\ \text{State 2} \end{array} \begin{array}{cc} \text{State 1} & \text{State 2} \\ \begin{bmatrix} .97 & .03 \\ .06 & .94 \end{bmatrix} \end{array} \qquad \text{(Transition matrix)}$$

Next, observe that the initial (current) probability distribution of the population may be summarized in the form of a row vector of dimension 2 (that is, a 1×2 matrix). Thus,

$$X_0 = \begin{array}{cc} \text{State 1} & \text{State 2} \\ [.65 & .35] \end{array} \qquad \text{(Initial-state matrix)}$$

Using the results of Example 4, we can write the population distribution one year later as

$$X_1 = \begin{array}{cc} \text{State 1} & \text{State 2} \\ [.6515 & .3485] \end{array} \qquad \text{(Distribution after one year)}$$

You may now verify that

$$X_0 T = [.65 \quad .35] \begin{bmatrix} .97 & .03 \\ .06 & .94 \end{bmatrix}$$

$$= [.6515 \quad .3485] = X_1$$

so this problem may be solved using matrix multiplication.

EXAMPLE 5   Refer to Example 4. What is the population distribution of the city after two years? After three years?

Solution   Let X_2 be the row vector representing the probability population distribution of the metropolitan area after two years. We may view X_1, the vector representing the probability population distribution of the metropolitan area after one year, as representing the "initial" probability distribution in this part of our calculation. Thus,

$$X_2 = X_1 T$$

$$= [.6515 \quad .3485] \begin{bmatrix} .97 & .03 \\ .06 & .94 \end{bmatrix}$$

$$= [.6529 \quad .3471]$$

The vector representing the probability distribution of the metropolitan area after three years is given by

$$X_3 = X_2 T = [.6529 \quad .3471] \begin{bmatrix} .97 & .03 \\ .06 & .94 \end{bmatrix}$$

$$= [.6541 \quad .3459]$$

That is, after three years, the population will be distributed as follows: 65.41% will live in the city and 34.59% will live in the suburbs.   ○ ○ ○

Distribution Vectors

Observe that, in the foregoing computations, we have $X_1 = X_0 T$, $X_2 = X_1 T = X_0 T^2$, and $X_3 = X_2 T = X_0 T^3$. These results are easily generalized. To see this, suppose we have a Markov process in which there are n possible states at each stage of the experiment. Suppose further that the probability of the system being in state 1, state 2, . . . , state n, initially, is given by p_1, p_2, . . . , p_n, respectively. This distribution may be represented as an n-dimensional vector

$$X_0 = [p_1, p_2, \ldots, p_n]$$

called a **distribution vector.** If T represents the $n \times n$ transition matrix associated with the Markov process, then the probability distribution of the system after m observations is given by

$$X_m = X_0 T^m \tag{10}$$

Formula (10) may be used to help us investigate the long-term trends of certain Markov processes, as illustrated in the next example.

EXAMPLE 6   A survey conducted by the National Commission on the Educational Status of Women reveals that 70% of the daughters of women who have completed two or more years of college have also completed two or more years of college, whereas 20% of the daughters of women who have had less than two years of college have completed two or more years of college. If this trend continues, determine, in the long run, the percentage of women in the population who will have completed at least two years of college given that currently only 20% of the women have completed at least two years of college.

Solution   This problem may be viewed as a Markov process with two possible states: "completed two or more years of college" (state 1) and "completed less than two years of college" (state 2). The transition matrix associated with this Markov chain is given by

$$T = \begin{bmatrix} .7 & .3 \\ .2 & .8 \end{bmatrix}$$

The initial distribution vector is given by

$$X_0 = [.2 \quad .8]$$

In order to study the long-term trend pertaining to this particular aspect of the educational status of women, let us compute $X_1, X_2, \ldots,$ the distribution vectors associated with the Markov process under consideration. These vectors give the percentage of women with two or more years of college and that of women with less than two years of college after one generation, after two generations, and so on. With the aid of equation (10), we find (to four decimal places)

After 1 generation
$$X_1 = X_0 T = [.2 \quad .8] \begin{bmatrix} .7 & .3 \\ .2 & .8 \end{bmatrix}$$
$$= [.3 \quad .7]$$

After 2 generations
$$X_2 = X_1 T = [.3 \quad .7] \begin{bmatrix} .7 & .3 \\ .2 & .8 \end{bmatrix}$$
$$= [.35 \quad .65]$$

After 3 generations
$$X_3 = X_2 T = [.35 \quad .65] \begin{bmatrix} .7 & .3 \\ .2 & .8 \end{bmatrix}$$
$$= [.375 \quad .625]$$

Proceeding further, we obtain the following sequence of vectors:

$$X_4 = [.3875 \quad .6125]$$
$$X_5 = [.3938 \quad .6062]$$
$$X_6 = [.3969 \quad .6031]$$
$$X_7 = [.3984 \quad .6016]$$
$$X_8 = [.3992 \quad .6008]$$
$$X_9 = [.3996 \quad .6004]$$
After 10 generations
$$X_{10} = [.3998 \quad .6002]$$

From the results of these computations, we see that as m increases, the probability distribution vector X_m approaches the probability distribution vector

$$[.4 \quad .6] \quad \text{or} \quad [\tfrac{2}{5} \quad \tfrac{3}{5}]$$

Such a vector is called the **limiting,** or **steady-state, distribution vector** for the system. We interpret these results in the following way: Initially, 20% of the women in the population have completed two or more years of college, whereas 80% have completed less than two years of college. After one generation, the former has increased to 30% of the population and the latter has dropped to 70% of the population. The trend continues, and eventually, 40% of all women in future generations will have completed two or more years of college, whereas 60% will have completed less than two years of college.

○ ○ ○

In order to explain the foregoing result, let us analyze equation (10) more closely. Now, the initial distribution vector X_0 is a constant; that is, it remains fixed throughout our computation of X_1, X_2, ... It appears reasonable, therefore, to conjecture that this phenomenon is a result of the behavior of the powers, T^m, of the transition matrix T. Pursuing this line of investigation, we compute

$$T^2 = \begin{bmatrix} .7 & .3 \\ .2 & .8 \end{bmatrix}\begin{bmatrix} .7 & .3 \\ .2 & .8 \end{bmatrix} = \begin{bmatrix} .55 & .45 \\ .3 & .7 \end{bmatrix}$$

$$T^3 = \begin{bmatrix} .7 & .3 \\ .2 & .8 \end{bmatrix}\begin{bmatrix} .55 & .45 \\ .3 & .7 \end{bmatrix} = \begin{bmatrix} .475 & .525 \\ .35 & .65 \end{bmatrix}$$

Proceeding further, we obtain the following sequence of matrices:

$$T^4 = \begin{bmatrix} .4375 & .5625 \\ .375 & .625 \end{bmatrix} \qquad T^5 = \begin{bmatrix} .4188 & .5813 \\ .3875 & .6125 \end{bmatrix}$$

$$T^6 = \begin{bmatrix} .4094 & .5906 \\ .3938 & .6062 \end{bmatrix} \qquad T^7 = \begin{bmatrix} .4047 & .5953 \\ .3969 & .6031 \end{bmatrix}$$

$$T^8 = \begin{bmatrix} .4023 & .5977 \\ .3984 & .6016 \end{bmatrix} \qquad T^9 = \begin{bmatrix} .4012 & .5988 \\ .3992 & .6008 \end{bmatrix}$$

$$T^{10} = \begin{bmatrix} .4006 & .5994 \\ .3996 & .6004 \end{bmatrix} \qquad T^{11} = \begin{bmatrix} .4003 & .5997 \\ .3998 & .6002 \end{bmatrix}$$

These results show that the powers T^m of the transition matrix T tend toward a fixed matrix as m gets larger and larger. In this case, the "limiting matrix" is the matrix

$$L = \begin{bmatrix} .40 & .60 \\ .40 & .60 \end{bmatrix} \quad \text{or} \quad \begin{bmatrix} \frac{2}{5} & \frac{3}{5} \\ \frac{2}{5} & \frac{3}{5} \end{bmatrix}$$

Such a matrix is called the **steady-state matrix** for the system. Thus, as suspected, the long-term behavior of a Markov process such as the one in this example depends on the behavior of the limiting matrix of the powers of the transition matrix—the steady-state matrix for the system. In view of this, the long-term (steady-state) distribution vector for this problem may be found by computing the product

$$X_0 L = \begin{bmatrix} .2 & .8 \end{bmatrix}\begin{bmatrix} .40 & .60 \\ .40 & .60 \end{bmatrix} = \begin{bmatrix} .40 & .60 \end{bmatrix}$$

which agrees with the result obtained earlier.

Next, since the transition matrix T in this situation seems to have a stabilizing effect over the long term, we are led to wonder whether the steady state would be reached regardless of the initial state of the system. To answer this question, suppose that the initial distribution vector is

$$X_0 = [p \quad 1 - p]$$

Then, as before, the steady-state distribution vector is given by

$$X_0 L = [p \quad 1 - p]\begin{bmatrix} .40 & .60 \\ .40 & .60 \end{bmatrix} = [.40 \quad .60]$$

Thus, the steady state is reached regardless of the initial state of the system!

Regular Markov Chains

The transition matrix T of Example 6 has some important properties, which we emphasized in the foregoing discussion. First of all, the sequence T, T^2, T^3, ... approaches a steady-state matrix in which the rows of the limiting matrix are all equal and all entries are positive. A matrix T having this property is given the special name *regular Markov chain.*

REGULAR MARKOV CHAIN

A stochastic matrix T is a **regular Markov chain** if the sequence

$$T, T^2, T^3, \ldots$$

approaches a steady-state matrix in which the rows of the limiting matrix are all equal and all the entries are positive.

It can be shown that *a stochastic matrix T is regular if and only if some power of T has entries that are all positive.* Second, as in the case of Example 6, a Markov chain with a regular transition matrix has a steady-state distribution vector whose elements coincide with those of a row (since they are all the same) of the steady-state matrix; thus, this steady-state distribution vector is always reached regardless of the initial distribution vector.

We will return to computations involving regular Markov chains, but for the moment let us see how one may determine whether a given matrix is indeed regular.

EXAMPLE 7 Determine which of the following matrices are regular:

a. $\begin{bmatrix} .7 & .3 \\ .2 & .8 \end{bmatrix}$ **b.** $\begin{bmatrix} .4 & .6 \\ 1 & 0 \end{bmatrix}$ **c.** $\begin{bmatrix} 0 & 1 \\ 1 & 0 \end{bmatrix}$

Solution

a. Since all the entries of the matrix are positive, the given matrix is regular. Note that this is the transition matrix of Example 6.

b. In this case, one of the entries of the given matrix is equal to zero. Let us compute

$$\begin{bmatrix} .4 & .6 \\ 1 & 0 \end{bmatrix}^2 = \begin{bmatrix} .4 & .6 \\ 1 & 0 \end{bmatrix}\begin{bmatrix} .4 & .6 \\ 1 & 0 \end{bmatrix} = \begin{bmatrix} .76 & .24 \\ .4 & .6 \end{bmatrix}$$

$\uparrow$
(All the entries are positive.)

Since the second power of the matrix has entries that are all positive, we conclude that the given matrix is in fact regular.

c. Denote the given matrix by A. Then

$$A = \begin{bmatrix} 0 & 1 \\ 1 & 0 \end{bmatrix}$$

$$A^2 = \begin{bmatrix} 0 & 1 \\ 1 & 0 \end{bmatrix}\begin{bmatrix} 0 & 1 \\ 1 & 0 \end{bmatrix} = \begin{bmatrix} 1 & 0 \\ 0 & 1 \end{bmatrix}$$

$$A^3 = \begin{bmatrix} 0 & 1 \\ 1 & 0 \end{bmatrix}\begin{bmatrix} 1 & 0 \\ 0 & 1 \end{bmatrix} = \begin{bmatrix} 0 & 1 \\ 1 & 0 \end{bmatrix}$$

(Not all the entries are positive.)

> Find the set of all 2×2 stochastic matrices with elements that are either 0 or 1.

Observe that $A^3 = A$. It follows, therefore, that $A^4 = A^2$, $A^5 = A$, and so on. In other words, any power of A must coincide with either A or A^2. Since not all entries of A and A^2 are positive, the same is true of any power of A. We conclude, accordingly, that the given matrix is not regular.   ● ● ●

We now return to the study of regular Markov chains. In Example 6 we found the steady-state distribution vector associated with a regular Markov chain by studying the limiting behavior of a sequence of distribution vectors. Alternatively, as pointed out in the subsequent discussion, the steady-state distribution vector may also be obtained by first determining the steady-state matrix associated with the regular Markov chain.

Fortunately, there is a relatively simple procedure for finding the steady-state distribution vector associated with a regular Markov process. It does not involve the rather tedious computations required to obtain the sequences in Example 6. The procedure follows.

FINDING THE STEADY-STATE DISTRIBUTION VECTOR

Let T be a regular stochastic matrix. Then the steady-state distribution vector X may be found by solving the vector equation

$$XT = X$$

together with the condition that the sum of the elements of the vector X be equal to 1.

A justification of the foregoing procedure is given in exercise 50, page 453.

EXAMPLE 8   Find the steady-state distribution vector for the regular Markov chain whose transition matrix is

$$T = \begin{bmatrix} .7 & .3 \\ .2 & .8 \end{bmatrix} \qquad \text{(See Example 6.)}$$

Solution   Let $X = [x \quad y]$ be the steady-state distribution vector associated with the Markov process, where the numbers x and y are to be determined. The condition $XT = X$ translates into the matrix equation

$$[x \quad y]\begin{bmatrix} .7 & .3 \\ .2 & .8 \end{bmatrix} = [x \quad y]$$

or, equivalently, the system of linear equations

$$0.7x + 0.2y = x$$
$$0.3x + 0.8y = y$$

But each of the equations that make up this system of equations is equivalent to the single equation

$$0.3x - 0.2y = 0 \qquad \begin{array}{l} (0.7x - x + 0.2y = 0) \\ (0.3x + 0.8y - y = 0) \end{array}$$

Next, the condition that the elements of X add up to 1 gives

$$x + y = 1$$

Thus, the fulfillment of the two conditions simultaneously implies that x and y are the solutions of the system

$$0.3x - 0.2y = 0$$
$$x + y = 1$$

Solving the first equation for x, we obtain

$$x = \frac{2}{3}y$$

which, upon substitution into the second, yields

$$\frac{2}{3}y + y = 1$$

or

$$y = \frac{3}{5}$$

Thus, $x = 2/5$, and the required steady-state distribution vector is given by $X = [2/5 \quad 3/5]$, which agrees with the result obtained earlier.   ○ ○ ○

Application

EXAMPLE 9   In order to keep track of the location of its cabs, the Zephyr
....   Cab Company has divided a town into three zones: zone I, zone II, and zone III. Zephyr's management has determined from company records that of the passengers picked up in zone I, 60% are discharged in the same zone, 30% are discharged in zone II, and 10% are discharged in zone III. Of those picked up in zone II, 40% are discharged in zone I, 30% are discharged in zone II, and 30% are discharged in zone III. Of those picked up in zone III, 30% are discharged in zone I, 30% are discharged in zone II, and 40% are discharged in zone III. Suppose that at the beginning of the day 80% of the cabs are in zone I, 15% are in zone II, and 5% are in zone III, and that a taxi without a passenger will cruise within the zone it is currently in until a pickup is made.

a. Find the transition matrix for the Markov chain that describes the successive locations of a cab.

b. Use the results of (a) to determine the long-term distribution of the taxis in the three zones.

Solution

a. The required transition matrix is given by

$$T = \begin{bmatrix} .6 & .3 & .1 \\ .4 & .3 & .3 \\ .3 & .3 & .4 \end{bmatrix}$$

b. Let $X = [x \ \ y \ \ z]$ be the steady-state distribution vector associated with the Markov process under consideration, where x, y, and z are to be determined. The condition $XT = X$ translates into the matrix equation

$$[x \ \ y \ \ z] \begin{bmatrix} .6 & .3 & .1 \\ .4 & .3 & .3 \\ .3 & .3 & .4 \end{bmatrix} = [x \ \ y \ \ z]$$

or, equivalently, the system of linear equations

$$0.6x + 0.4y + 0.3z = x$$
$$0.3x + 0.3y + 0.3z = y$$
$$0.1x + 0.3y + 0.4z = z$$

This system simplifies to

$$4x - 4y - 3z = 0$$
$$3x - 7y + 3z = 0$$
$$x + 3y - 6z = 0$$

Since $x + y + z = 1$ as well, we are required to solve the system

$$x + \ y + \ \ z = 1$$
$$4x - 4y - 3z = 0$$
$$3x - 7y + 3z = 0$$
$$x + 3y - 6z = 0$$

Using the Gauss-Jordan elimination procedure of Chapter 2, we find that

$$x = \frac{33}{70}, \quad y = \frac{3}{10}, \quad \text{and} \quad z = \frac{8}{35}$$

or $x \approx 0.47$, $y = 0.30$, and $z \approx 0.23$. Thus, in the long run, 47% of the taxis will be in zone I, 30% in zone II, and 23% in zone III.   ๐ ๐ ๐

REMARK   In this simplified model, we do not take into consideration variable demand and variable delivery time.   ๐ ๐ ๐

SELF-CHECK EXERCISES 7.7

I. Three supermarkets serve a certain section of a city. During the upcoming year, supermarket A is expected to retain 80% of its customers, lose 5% of its customers to supermarket B, and lose 15% to supermarket C. Supermarket B is expected to retain 90% of its customers and lose 5% of its customers to each of supermarkets A and C. Supermarket C is expected to retain 75% of its customers, lose 10% to supermarket A, and lose 15% to supermarket B. Construct the transition matrix for the Markov chain that describes the change in the market share of the three supermarkets.

2. Refer to exercise 1. Currently the market shares of supermarket A, supermarket B, and supermarket C are 0.4, 0.3, and 0.3, respectively.
   a. Find the initial distribution vector for this Markov chain.
   b. What share of the market will be held by each supermarket after one year? Assuming that the trend continues, what will the market share be after two years?

3. Find the steady-state distribution vector for the regular Markov chain whose transition matrix is

$$T = \begin{bmatrix} .5 & .5 \\ .8 & .2 \end{bmatrix}$$

Solutions to Self-Check Exercises 7.7 can be found on page 453.

7.7 EXERCISES

In exercises 1–10, determine which of the given matrices are stochastic.

1. $\begin{bmatrix} .4 & .6 \\ .7 & .3 \end{bmatrix}$

2. $\begin{bmatrix} .8 & .3 \\ .2 & .7 \end{bmatrix}$

3. $\begin{bmatrix} \frac{1}{4} & \frac{3}{4} \\ \frac{1}{8} & \frac{7}{8} \end{bmatrix}$

4. $\begin{bmatrix} \frac{1}{3} & \frac{1}{2} & \frac{1}{4} \\ 0 & 1 & 0 \\ \frac{1}{2} & 0 & \frac{1}{2} \end{bmatrix}$

5. $\begin{bmatrix} .3 & .4 & .3 \\ .2 & .7 & .1 \\ .4 & .3 & .2 \end{bmatrix}$

6. $\begin{bmatrix} \frac{1}{3} & \frac{1}{3} & \frac{1}{4} \\ \frac{1}{4} & 0 & \frac{3}{4} \\ \frac{1}{2} & -\frac{1}{2} & \frac{1}{2} \end{bmatrix}$

7. $\begin{bmatrix} .1 & .7 & .2 \\ .4 & .2 & .4 \\ .3 & .1 & .6 \end{bmatrix}$

8. $\begin{bmatrix} 1 & 0 & 0 \\ 0 & 0 & 1 \\ 0 & 1 & 0 \end{bmatrix}$

9. $\begin{bmatrix} .2 & .3 & .5 \\ .3 & .1 & .6 \end{bmatrix}$

10. $\begin{bmatrix} .5 & .2 & .3 & 0 \\ .2 & .3 & .4 & .1 \\ .3 & .2 & .1 & .4 \end{bmatrix}$

11. The transition matrix for a Markov process is given by

$$T = \text{State} \begin{matrix} & \text{State} \\ & \begin{matrix} 1 & 2 \end{matrix} \\ \begin{matrix} 1 \\ 2 \end{matrix} & \begin{bmatrix} .3 & .7 \\ .6 & .4 \end{bmatrix} \end{matrix}$$

a. What does the entry $a_{11} = .3$ represent?
b. Given that the outcome state 1 has occurred, what is the probability that the next outcome of the experiment will be state 2?

c. If the initial-state distribution vector is given by

$$\begin{matrix} \text{State} \\ \begin{matrix} 1 & 2 \end{matrix} \\ X_0 = \begin{bmatrix} .4 & .6 \end{bmatrix} \end{matrix}$$

find $X_0 T$, the probability distribution of the system after one observation.

12. The transition matrix for a Markov process is given by

$$T = \text{State} \begin{matrix} & \text{State} \\ & \begin{matrix} 1 & 2 \end{matrix} \\ \begin{matrix} 1 \\ 2 \end{matrix} & \begin{bmatrix} \frac{1}{6} & \frac{5}{6} \\ \frac{2}{3} & \frac{1}{3} \end{bmatrix} \end{matrix}$$

a. What does the entry $a_{22} = \frac{1}{3}$ represent?
b. Given that the outcome state 1 has occurred, what is the probability that the next outcome of the experiment will be state 2?
c. If the initial-state distribution vector is given by

$$\begin{matrix} \text{State} \\ \begin{matrix} 1 & 2 \end{matrix} \\ X_0 = \begin{bmatrix} \frac{1}{4} & \frac{3}{4} \end{bmatrix} \end{matrix}$$

find $X_0 T$, the probability distribution of the system after one observation.

13. The transition matrix for a Markov process is given by

$$T = \text{State} \begin{matrix} & \text{State} \\ & \begin{matrix} 1 & 2 \end{matrix} \\ \begin{matrix} 1 \\ 2 \end{matrix} & \begin{bmatrix} .6 & .4 \\ .2 & .8 \end{bmatrix} \end{matrix}$$

and the initial-state distribution vector is given by

State
1   2
$$X_0 = [.5 \quad .5]$$

Find $X_0 T$ and interpret your result with the aid of a tree diagram.

14. The transition matrix for a Markov process is given by

State
1   2

$$T = \text{State} \quad \begin{matrix} 1 \\ 2 \end{matrix} \begin{bmatrix} \frac{1}{2} & \frac{1}{2} \\ \frac{3}{4} & \frac{1}{4} \end{bmatrix}$$

and the initial-state distribution vector is given by

State
1   2
$$X_0 = [\tfrac{1}{3} \quad \tfrac{2}{3}]$$

Find $X_0 T$ and interpret your result with the aid of a tree diagram.

In exercises 15–18, find X_2 (the probability distribution of the system after two observations) for the given distribution vector X_0 and the given transition matrix T.

15. $X_0 = [.6 \quad .4]$; $T = \begin{bmatrix} .4 & .6 \\ .8 & .2 \end{bmatrix}$

16. $X_0 = [\tfrac{1}{2} \quad \tfrac{1}{2} \quad 0]$; $T = \begin{bmatrix} \frac{1}{2} & 0 & \frac{1}{2} \\ \frac{1}{3} & \frac{1}{3} & \frac{1}{3} \\ \frac{1}{2} & \frac{1}{4} & \frac{1}{4} \end{bmatrix}$

17. $X_0 = [\tfrac{1}{4} \quad \tfrac{1}{2} \quad \tfrac{1}{4}]$; $T = \begin{bmatrix} \frac{1}{4} & \frac{1}{4} & \frac{1}{2} \\ \frac{1}{4} & \frac{1}{2} & \frac{1}{4} \\ \frac{1}{2} & \frac{1}{2} & 0 \end{bmatrix}$

18. $X_0 = [.25 \quad .40 \quad .35]$; $T = \begin{bmatrix} .1 & .8 & .1 \\ .1 & .7 & .2 \\ .3 & .2 & .5 \end{bmatrix}$

In exercises 19–26, determine which of the given matrices are regular.

19. $\begin{bmatrix} \frac{2}{5} & \frac{3}{5} \\ \frac{3}{4} & \frac{1}{4} \end{bmatrix}$ **20.** $\begin{bmatrix} 0 & 1 \\ .3 & .7 \end{bmatrix}$

21. $\begin{bmatrix} 1 & 0 \\ .8 & .2 \end{bmatrix}$ **22.** $\begin{bmatrix} \frac{1}{3} & \frac{2}{3} \\ 0 & 1 \end{bmatrix}$

23. $\begin{bmatrix} \frac{1}{2} & \frac{1}{2} & 0 \\ \frac{3}{4} & 0 & \frac{1}{4} \\ 0 & \frac{1}{2} & \frac{1}{2} \end{bmatrix}$ **24.** $\begin{bmatrix} 1 & 0 & 0 \\ .3 & .4 & .3 \\ .1 & .8 & .1 \end{bmatrix}$

25. $\begin{bmatrix} .7 & .3 & 0 \\ .2 & .8 & 0 \\ .3 & .3 & .4 \end{bmatrix}$ **26.** $\begin{bmatrix} 0 & 1 & 0 \\ 0 & 0 & 1 \\ \frac{1}{4} & 0 & \frac{3}{4} \end{bmatrix}$

In exercises 27–34, find the steady-state vector for the given transition matrix.

27. $\begin{bmatrix} \frac{1}{3} & \frac{2}{3} \\ \frac{1}{4} & \frac{3}{4} \end{bmatrix}$ **28.** $\begin{bmatrix} \frac{4}{5} & \frac{1}{5} \\ \frac{3}{5} & \frac{2}{5} \end{bmatrix}$

29. $\begin{bmatrix} .5 & .5 \\ .2 & .8 \end{bmatrix}$ **30.** $\begin{bmatrix} .9 & .1 \\ 1 & 0 \end{bmatrix}$

31. $\begin{bmatrix} 0 & 1 & 0 \\ \frac{1}{8} & \frac{5}{8} & \frac{1}{4} \\ 1 & 0 & 0 \end{bmatrix}$ **32.** $\begin{bmatrix} .6 & .4 & 0 \\ .3 & .4 & .3 \\ 0 & .6 & .4 \end{bmatrix}$

33. $\begin{bmatrix} .2 & 0 & .8 \\ 0 & .6 & .4 \\ .3 & .4 & .3 \end{bmatrix}$ **34.** $\begin{bmatrix} .1 & .1 & .8 \\ .2 & .2 & .6 \\ .3 & .3 & .4 \end{bmatrix}$

35. Psychology Experiments   A psychologist conducts an experiment in which a mouse is placed in a T-maze where it has a choice at the T-junction of turning left and receiving a reward (cheese) or turning right and receiving a mild electric shock (see the accompanying figure). At the end of each trial a record is kept of the mouse's response. It is observed that the mouse is as likely to turn left (state 1) as right (state 2) during the first trial. In subsequent trials, however, the observation is made that if the mouse had turned left in the previous trial, then on the next trial the probability that it will turn left is .8, whereas the probability that it will turn right is .2. If the mouse had turned right in the previous trial, then the probability that it will turn right on the next trial is .1, whereas the probability that it will turn left is .9.

a. Using a tree diagram, describe the transitions between states and the probabilities associated with these transitions.

b. Represent the transition probabilities obtained in (a) in terms of a matrix.

c. What is the initial-state probability vector?

d. Use the results of (b) and (c) to find the probability that a mouse will turn left on the second trial.

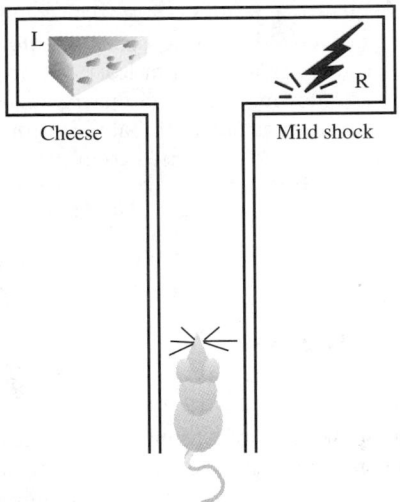

Cheese   Mild shock

36. **Commuter Trends**   Twenty percent of the commuters within a large metropolitan area currently use the public transportation system, whereas the remaining 80% commute via automobile. The city has recently revitalized and expanded its public transportation system. It is expected that six months from now 30% of those who are now commuting to work via automobile will switch to public transportation, and 70% will continue to commute via automobile. At the same time, it is expected that 20% of those now using public transportation will commute via automobile, and 80% will continue to use public transportation.
    a. Construct the transition matrix for the Markov chain that describes the change in the mode of transportation used by these commuters.
    b. Find the initial distribution vector for this Markov chain.
    c. What percentage of the commuters are expected to use public transportation six months from now?

37. **Urbanization of Farmland**   A study conducted by the State Department of Agriculture in a Sunbelt state reveals an increasing trend toward urbanization of the farmland within the state. Ten years ago 50% of the land within the state was used for agricultural purposes (A), 15% was urbanized (U), and the remaining 35% was neither agricultural nor urban (N). Since that time, 10% of the agricultural land has been converted to urban land, 5% has been used for other purposes, while the remaining 85% is still agricultural. Ninety-five percent of the urban land has remained urban, while 5% of it has been used for nonagricultural purposes. Of the land that was neither agricultural nor urban, 10% has been converted to agricultural land, 5% has been urbanized, and the remaining 85% remains unchanged.
    a. Construct the transition matrix for the Markov chain that describes the shift in land use within the state.
    b. Find the probability vector describing the distribution of land within the state ten years ago.
    c. Assuming that this trend continues, find the probability vector describing the distribution of land within the state ten years from now.

38. **Political Polls**   The Morris Polling Group conducted a poll six months before an election in a state in which a Democrat and a Republican were running for governor and found that 60% of the voters intended to vote for the Democrat and 40% intended to vote for the Republican. In a poll conducted three months later it was found that 70% of those who had earlier stated a preference for the Democratic candidate still maintained that preference, whereas 30% of these voters now preferred the Republican candidate. Of those who had earlier stated a preference for the Republican, 80% still maintained that preference, whereas 20% now preferred the Democratic candidate.
    a. If the election were held at this time, who would win?
    b. Assuming that this trend continues, which candidate is expected to win the election?

39. **Market Share of Auto Manufacturers**   In a study of the domestic market share of the three major automobile manufacturers A, B, and C in a certain country, it was found that their current market shares were 60%, 30%, and 10%, respectively. Furthermore, it was found that of the customers who bought a car manufactured by A, 75% would again buy a car manufactured by A, 15% would buy a car manufactured by B, and 10% would buy a car manufactured by C. Of the customers who bought a car manufactured by B, 90% would again buy a car manufactured by B, whereas 5% each would buy cars manufactured by A and C. Finally, of the customers who bought a car manufactured by C, 85% would again buy a car manufactured by C, 5% would buy a car manufactured by A, and 10% would buy a car manufactured by B. Assuming that these sentiments reflect the buying habits of customers in the future, determine the market share that will be held by each manufacturer after the next two model years.

40. **Market Share**   At a certain university, three bookstores—the University Bookstore, the Campus Bookstore, and the Book Mart—currently serve the university community. From a survey conducted at the beginning of the fall quarter it was found that the University Bookstore and the Campus Bookstore each had

40% of the market, whereas the Book Mart had 20% of the market. Each quarter, the University Bookstore retains 80% of its customers but loses 10% to the Campus Bookstore and 10% to the Book Mart. The Campus Bookstore retains 75% of its customers but loses 10% to the University Bookstore and 15% to the Book Mart. The Book Mart retains 90% of its customers but loses 5% to the University Bookstore and 5% to the Campus Bookstore. If these trends continue, what percentage of the market will each store have at the beginning of the second quarter? The third quarter?

41. College Majors Records compiled by the Admissions Office at a state university indicating the percentage of students who change their major each year are shown in the following transition matrix. Of the freshmen now at the university, 30% have chosen their major field in business, 30% in the humanities, 20% in education, and 20% in the natural sciences and other fields. Assuming that this trend continues, find the percentage of these students who will be majoring in each of the given areas in their senior year.
[*Hint:* Find $X_0 T^3$.]

	Bus.	Hum.	Educ.	Nat. sci. and others
Business	.80	.10	.05	.05
Humanities	.10	.70	.10	.10
Education	.20	.10	.60	.10
Nat. sci. and others	.10	.05	.05	.80

42. Homeowners' Choice of Energy A study conducted by the Urban Energy Commission in a large metropolitan area indicates the probabilities that homeowners within the area will use certain heating fuels or solar energy during the next ten years as the major source of heat for their homes. The transition matrix representing the transition probabilities from one state to another is given here.

	Elec.	Gas	Oil	Solar
Electricity	.70	.15	.05	.10
Natural gas	0	.90	.02	.08
Fuel oil	0	.20	.75	.05
Solar energy	0	.05	0	.95

Among homeowners in the area, 20% currently use electricity, 35% use natural gas, 40% use oil, and 5% use solar energy as the major source of heat for their homes. What is the expected distribution of the homeowners that will be using each type of heating fuel or solar energy within the next decade?

43. One- and Two-Income Families From data compiled over a 10-year period by Manpower, Inc., in a statewide study of married couples in which at least one spouse was working, the following transition matrix was constructed. It gives the transitional probabilities for one and two wage earners among married couples.

		Next State	
		1 Wage Earner	2 Wage Earners
Current State	1 Wage Earner	.72	.28
	2 Wage Earners	.12	.88

At the present time, 48% of the married couples (in which at least one spouse is working) have one wage earner and 52% have two wage earners. Assuming that this trend continues, what will the distribution of one- and two-wage-earner families be among married couples in this area (a) ten years from now? (b) over the long run?

44. Professional Women From data compiled over a 5-year period by *Women's Daily* in a study of the number of women in the professions, the following transition matrix was constructed. It gives the transitional probabilities for the number of men and women in the professions.

		Next State	
		Men	Women
Current State	Men	.95	.05
	Women	.04	.96

As of the beginning of 1986, 52.9% of professional jobs were held by men. If this trend continues, what percentage of professional jobs will be held by women in the long run?

45. Buying Trends of Homebuyers From data collected by the Association of Realtors of a certain city, the following transition matrix was obtained. The matrix describes the buying pattern of home buyers who buy single-family homes (S) or condominiums (C).

		Next State	
		S	C
Current State	S	.85	.15
	C	.35	.65

Currently, 80% of the homeowners live in single-family homes, whereas 20% live in condominiums. If this trend continues, determine the percentage of homeowners in this city who will own single-family homes and condominiums (a) two years from now, and (b) in the long run.

46. Network News Viewership A television poll was conducted among regular viewers of the national news in a certain region where the three national networks share the same time slot for the evening news. Results of the poll indicate that 30% of the viewers watched the ABC evening news, 40% watched the CBS evening news, and 30% watched the NBC evening news. Furthermore, it was found that of those viewers who watched the ABC evening news during one week, 80% would again watch the ABC evening news during the next week, 10% would watch the CBS news, and 10% would watch the NBC news. Of those viewers who watched the CBS evening news during one week, 85% would again watch the CBS evening news during the next week, 10% would watch the ABC news, and 5% would watch the NBC news. Of those viewers who watched the NBC evening news during one week, 85% would again watch the NBC news during the next week, 10% would watch ABC, and 5% would watch CBS.

a. What share of the audience consisting of regular viewers of the national news will each network command after two weeks?

b. In the long run, what share of the audience will each network command?

47. Network News Viewership Refer to exercise 46. If the initial distribution vector is

$$\text{ABC} \quad \text{CBS} \quad \text{NBC}$$
$$X_0 = [.40 \quad .40 \quad .20]$$

what share of the audience will each network command in the long run?

48. Genetics In a certain species of roses, a plant with genotype (genetic makeup) AA has red flowers, a plant with genotype Aa has pink flowers, and a plant with genotype aa has white flowers, where A is the dominant gene and a is the recessive gene for color. If a plant

with one genotype is crossed with another plant, then the color of the offspring's flowers is determined by the genotype of the parent plants. If a plant of each genotype is crossed with a pink-flowered plant, then the transition matrix used to determine the color of the offspring's flowers is given by

$$
\begin{array}{c}
 \\
\text{Parent}
\end{array}
\begin{array}{c}
 \\
\text{Red }(AA) \\
\text{Pink }(Aa)\text{ or }(aA) \\
\text{White }(aa)
\end{array}
\overset{\begin{array}{ccc}\text{Offspring} & & \\ \text{Red} & \text{Pink} & \text{White}\end{array}}{\begin{bmatrix} \frac{1}{2} & \frac{1}{2} & 0 \\ \frac{1}{4} & \frac{1}{2} & \frac{1}{4} \\ 0 & \frac{1}{2} & \frac{1}{2} \end{bmatrix}}
$$

If the offspring of each generation are crossed only with pink-flowered plants, in the long run what percentage of the plants will have red flowers? Pink flowers? White flowers?

49. Market Share of Auto Manufacturers In a study of the domestic market share of the three major automobile manufacturers A, B, and C in a certain country, it was found that of the customers who bought a car manufactured by A, 75% would again buy a car manufactured by A, 15% would buy a car manufactured by B, and 10% would buy a car manufactured by C. Of the customers who bought a car manufactured by B, 90% would again buy a car manufactured by B, whereas 5% each would buy cars manufactured by A and C. Finally, of the customers who bought a car manufactured by C, 85% would again buy a car manufactured by C, 5% would buy a car manufactured by A, and 10% would buy a car manufactured by B. Assuming that these sentiments reflect the buying habits of customers in the future model years, determine the market share that will be held by each manufacturer in the long run.

50. Let T be a regular stochastic matrix. Show that the steady-state distribution vector X may be found by solving the vector equation $XT = X$ together with the condition that the sum of the elements of X be equal to 1. [*Hint:* Take the initial distribution vector to be X, the steady-state distribution vector. Then when n is large, $X \approx XT^n$. (Why?) Multiply both sides of the last equation by T (on the right) and consider the resulting equation when n is large.]

SOLUTIONS TO SELF-CHECK EXERCISES 7.7

1. The required transition matrix is

$$T = \begin{bmatrix} .80 & .05 & .15 \\ .05 & .90 & .05 \\ .10 & .15 & .75 \end{bmatrix}$$

USING TECHNOLOGY

FINDING DISTRIBUTION VECTORS

Since the computation of the probability distribution of a system after a certain number of observations involves matrix multiplication, the graphing utility may be used to facilitate the work.

EXAMPLE 1 Consider the problem posed in Example 9, page 447. Suppose that

$$T = \begin{bmatrix} .6 & .3 & .1 \\ .4 & .3 & .3 \\ .3 & .3 & .4 \end{bmatrix}$$

and that the initial distribution vector is given by $X_0 = [.8 \quad .15 \quad .05]$. Verify that $X_2 = [.4965 \quad .3 \quad .2035]$.

Solution First, we enter into the calculator the matrix X_0 as the matrix A and the matrix T as the matrix B. Then, performing the indicated multiplication, we find that

$$A * B^2 = [[.4965 \quad .3 \quad .2035]]$$

That is,

$$X_2 = X_0 T^2 = [.4965 \quad .3 \quad .2035]$$

as was to be shown. ◦ ◦ ◦

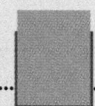

EXERCISES

In exercises 1 and 2, use a graphing utility to find X_5 (the probability distribution of the system after five observations) for the given distribution vector X_0 and the given transition matrix T.

1. $X_0 = [.2 \quad .3 \quad .2 \quad .1 \quad .2]$

$$T = \begin{bmatrix} .2 & .1 & .3 & .2 & .2 \\ .2 & .2 & .4 & .1 & .1 \\ .3 & .1 & .1 & .2 & .3 \\ .2 & .2 & .3 & .2 & .1 \\ .1 & .1 & .3 & .2 & .3 \end{bmatrix}$$

2. $X_0 = [.1 \quad .2 \quad .2 \quad .3 \quad .2]$

$$T = \begin{bmatrix} .3 & .2 & .1 & .1 & .3 \\ .2 & .1 & .2 & .3 & .2 \\ .1 & .2 & .3 & .2 & .2 \\ .3 & .1 & .2 & .3 & .1 \\ .1 & .2 & .2 & .2 & .3 \end{bmatrix}$$

3. Refer to exercise 39, page 451. Using the same data, determine the market share that will be held by each manufacturer five model years after the study began.

2. a. The initial distribution vector is

$$X_0 = [.4 \quad .3 \quad .3]$$

b. The vector representing the market share of each supermarket after one year is

$$X_1 = X_0 T$$

$$= [.4 \quad .3 \quad .3] \begin{bmatrix} .80 & .05 & .15 \\ .05 & .90 & .05 \\ .10 & .15 & .75 \end{bmatrix}$$

$$= [.365 \quad .335 \quad .3]$$

That is, after one year supermarket A will command a 36.5% market share, supermarket B will have a 33.5% market share, and supermarket C will have a 30% market share.

The vector representing the market share of the supermarkets after two years is

$$X_2 = X_1 T$$

$$= [.365 \quad .335 \quad .3] \begin{bmatrix} .80 & .05 & .15 \\ .05 & .90 & .05 \\ .10 & .15 & .75 \end{bmatrix}$$

$$= [.3388 \quad .3648 \quad .2965]$$

That is, two years later the market shares of supermarkets A, B, and C will be 33.88%, 36.48%, and 29.65%, respectively.

3. Let $X = [x \quad y]$ be the steady-state distribution vector associated with the Markov process, where the numbers x and y are to be determined. The condition $XT = X$ translates into the matrix equation

$$[x \quad y] \begin{bmatrix} .5 & .5 \\ .8 & .2 \end{bmatrix} = [x \quad y]$$

which is equivalent to the system of linear equations

$$0.5x + 0.8y = x$$
$$0.5x + 0.2y = y$$

Each equation in the system is equivalent to the equation

$$0.5x - 0.8y = 0$$

Next, the condition that the elements of X add up to 1 gives

$$x + y = 1$$

Thus, the fulfillment of the two conditions simultaneously implies that x and y be the solutions of the system

$$0.5x - 0.8y = 0$$
$$x + \quad y = 1$$

Solving the first equation for x, we obtain

$$x = \frac{8}{5} y$$

which upon substitution into the second yields

$$\frac{8}{5}y + y = 1$$

or

$$y = \frac{5}{13}$$

Therefore, $x = 8/13$, and the required steady-state distribution vector is $[8/13 \quad 5/13]$.

CHAPTER 7 SUMMARY OF PRINCIPAL FORMULAS AND TERMS

Formulas

1. Probability of an event $\qquad$ $P(E) = \dfrac{n(E)}{n(S)}$

2. Probability of the union of two mutually exclusive events $\qquad$ $P(E \cup F) = P(E) + P(F)$

3. Addition Rule $\qquad$ $P(E \cup F) = P(E) + P(F) - P(E \cap F)$

4. Rule of Complements $\qquad$ $P(E^c) = 1 - P(E)$

5. Conditional probability $\qquad$ $P(B \mid A) = \dfrac{P(A \cap B)}{P(A)}$

6. Product Rule $\qquad$ $P(A \cap B) = P(A) \cdot P(B \mid A)$

7. Test for independence $\qquad$ $P(A \cap B) = P(A) \cdot P(B)$

Terms

Experiment	Probability distribution
Sample point	Probability function
Sample space	Uniform sample space
Event of an experiment	Conditional probability
Finite sample space	Finite stochastic process
Union of two events	Independent events
Intersection of two events	Markov chain (process)
Complement of an event	Transition matrix
Mutually exclusive events	Stochastic matrix
Relative frequency	Steady-state distribution vector
Empirical probability	Steady-state matrix
Probability of an event	Regular Markov chain
Elementary (simple) events	

CHAPTER 7 REVIEW EXERCISES

1. Let E and F be two mutually exclusive events and suppose $P(E) = .4$ and $P(F) = .2$. Compute
   a. $P(E \cap F)$ **b.** $P(E \cup F)$
   c. $P(E^c)$ **d.** $P(E^c \cap F^c)$
   e. $P(E^c \cup F^c)$

2. Let E and F be two events of an experiment with sample space S. Suppose $P(E) = .3$, $P(F) = .2$, and $P(E \cap F) = .15$. Compute
   a. $P(E \cup F)$ **b.** $P(E^c \cap F^c)$
   c. $P(E^c \cap F)$

3. Suppose a die is loaded and it was determined that the probability distribution associated with the experiment of casting the die and observing which number falls uppermost is given by

Simple Event	Probability
{1}	.20
{2}	.12
{3}	.16
{4}	.18
{5}	.15
{6}	.19

   a. What is the probability of the number being even?
   b. What is the probability of the number being either a 1 or a 6?
   c. What is the probability of the number being less than 4?

4. An urn contains six red, five black, and four green balls. If two balls are selected at random without replacement from the urn, what is the probability that a red ball and a black ball will be selected?

5. The quality-control department of the Starr Communications Company, the manufacturer of video-game cartridges, has determined from records that 1.5% of the cartridges sold have video defects, 0.8% have audio defects, and 0.4% have both audio and video defects. What is the probability that a cartridge purchased by a customer
   a. will have a video or audio defect?
   b. will not have a video or audio defect?

6. Let E and F be two events and suppose $P(E) = .35$, $P(F) = .55$, and $P(E \cup F) = .70$. Find $P(E \mid F)$.

The accompanying tree diagram represents an experiment consisting of two trials. In exercises 7–11, use the diagram to find the given probability.

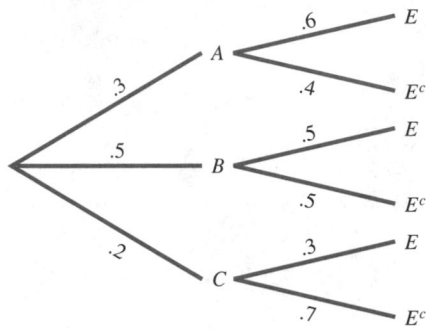

7. $P(A \cap E)$ **8.** $P(B \cap E)$

9. $P(C \cap E)$ **10.** $P(A \mid E)$

11. $P(E)$

12. An experiment consists of tossing a fair coin three times and observing the outcomes. Let A be the event that at least one head is thrown and B the event that at most two tails are thrown.
   a. Find $P(A)$.
   b. Find $P(B)$.
   c. Are A and B independent events?

13. In a group of 20 ballpoint pens on a shelf in the stationery department of the Metro Department Store, 2 are known to be defective. If a customer selects 3 of these pens, what is the probability that
   a. at least 1 is defective?
   b. no more than 1 is defective?

14. Five people are selected at random. What is the probability that none of the people in this group were born on the same day of the week?

15. A pair of fair dice is cast. What is the probability that the sum of the numbers falling uppermost is 8 if it is known that the two numbers are different?

Three cards are drawn at random without replacement from a standard deck of 52 playing cards. In exercises 16–20, find the probability of each of the given events.

16. All three cards are aces.

17. All three cards are face cards.

18. The second and third cards are red.

19. The second card is black, given that the first card was red.

20. The second card is a club, given that the first card was black.

In exercises 21–24, determine which of the following are regular stochastic matrices.

21. $\begin{bmatrix} 1 & 0 \\ -2 & -8 \end{bmatrix}$

22. $\begin{bmatrix} .3 & .7 \\ 1 & 0 \end{bmatrix}$

23. $\begin{bmatrix} \frac{1}{2} & 0 & \frac{1}{2} \\ 0 & 0 & 1 \\ \frac{1}{3} & \frac{1}{3} & \frac{1}{3} \end{bmatrix}$

24. $\begin{bmatrix} .3 & .2 & .1 \\ 0 & 1 & 0 \\ .5 & 0 & .5 \end{bmatrix}$

In exercises 25 and 26, find X_2 (the probability distribution of the system after two observations) for the given distribution vector X_0 and the given transition matrix T.

25. $X_0 = [\frac{1}{2} \ \frac{1}{2} \ 0]$; $T = \begin{bmatrix} 0 & \frac{2}{5} & \frac{3}{5} \\ \frac{1}{4} & \frac{1}{2} & \frac{1}{4} \\ \frac{3}{5} & \frac{1}{5} & \frac{1}{5} \end{bmatrix}$

26. $X_0 = [.35 \quad .25 \quad .40]$; $T = \begin{bmatrix} .2 & .5 & .3 \\ .1 & .4 & .5 \\ .3 & .4 & .3 \end{bmatrix}$

In exercises 27–30, find the steady-state matrix for the given transition matrix.

27. $\begin{bmatrix} .6 & .4 \\ .3 & .7 \end{bmatrix}$

28. $\begin{bmatrix} .5 & .5 \\ .4 & .6 \end{bmatrix}$

29. $\begin{bmatrix} .6 & .2 & .2 \\ .4 & .2 & .4 \\ .3 & .2 & .5 \end{bmatrix}$

30. $\begin{bmatrix} .1 & .3 & .6 \\ .2 & .4 & .4 \\ .6 & .4 & .2 \end{bmatrix}$

31. Of 320 male and 280 female employees at the home office of the Gibraltar Insurance Company, 160 of the men and 190 of the women are on flextime (flexible working hours). Given that an employee selected at random from this group is on flextime, what is the probability that the employee is a male?

32. In a manufacturing plant, three machines, A, B, and C, produce 40%, 35%, and 25%, respectively, of the total production. The company's quality control department has determined that 1% of the items produced by machine A, 1.5% of the items produced by machine B, and 2% of the items produced by machine C are defective. If an item is selected at random and found to be defective, what is the probability that it was produced by machine B?

33. Applicants who wish to be admitted to a certain professional school in a large university are required to take a screening test that was devised by an educational testing service. From past results, the testing service has estimated that 70% of all applicants are eligible for admission and that 92% of those who are eligible for admission pass the exam, whereas 12% of those who are ineligible for admission pass the exam. Using these results, find the probability that
    a. an applicant for admission passed the exam.
    b. an applicant for admission who passed the exam was actually ineligible.

34. *Auto Trend* magazine conducted a survey among automobile owners in a certain area of the country to determine what type of car they now own and what type of car they expect to own four years from now. For purposes of classification, automobiles mentioned in the survey were placed into three categories: large, intermediate, and small. Results of the survey follow:

		Future car		
		Large	Inter.	Small
Present car	Large	.3	.3	.4
	Intermediate	.1	.5	.4
	Small	.1	.2	.7

Assuming that these results indicate the long-term buying trend of car owners within the area, what will the distribution of cars (relative to size) be in this area over the long run?

Statistics is that branch of mathematics concerned with the collection, analysis, and interpretation of data. In Sections 8.1–8.3 of this chapter, we take a look at descriptive statistics; here, our interest lies in the description and presentation of data in the form of tables and graphs. In the rest of the chapter we briefly examine inductive statistics, and we see how mathematical tools such as those developed in Chapter 7 may be used in conjunction with these data to help us draw certain conclusions and make forecasts.

Where do we go from here? Some students in the top 10% of this senior class will further their education at one of the campuses of the State University system. In Example 3, page 524, we will determine the minimum grade point average a senior needs to be eligible for admission to a state university.

PROBABILITY DISTRIBUTIONS AND STATISTICS

DISTRIBUTIONS OF RANDOM VARIABLES

Random Variables

In many situations it is desirable to assign numerical values to the outcomes of an experiment. For example, if an experiment consists of casting a die and observing the face that lands uppermost, then it is natural to assign the numbers 1, 2, 3, 4, 5, and 6, respectively, to the outcomes *one, two, three, four, five,* and *six* of the experiment. If we let X denote the outcome of the experiment, then X assumes one of the numbers. Because the values assumed by X depend on the outcomes of a chance experiment, the outcome X is referred to as a *random variable.*

RANDOM VARIABLE	A **random variable** is a rule that assigns a number to each outcome of a chance experiment.

More precisely, a random variable is a function with domain given by the set of outcomes of a chance experiment and range contained in the set of real numbers.

EXAMPLE 1   A coin is tossed three times. Let the random variable X denote the number of heads that occur in the three tosses.

a. List the outcomes of the experiment; that is, find the domain of the function X.

b. Find the value assigned to each outcome of the experiment by the random variable X.

c. Find the event comprising the outcomes to which a value of 2 has been assigned by X. This event is written $(X = 2)$ and is the event comprising the outcomes in which two heads occur.

Table 8.1
Number of heads in three coin tosses.

Outcome	Value of X
HHH	3
HHT	2
HTH	2
THH	2
HTT	1
THT	1
TTH	1
TTT	0

Solution

a. From the results of Example 3, Section 7.1 (page 375), we see that the set of outcomes of the experiment is given by the sample space

$$S = \{\text{HHH, HHT, HTH, THH, HTT, THT, TTH, TTT}\}$$

b. The outcomes of the experiment are displayed in the first column of Table 8.1. The corresponding value assigned to each such outcome by the random variable X (the number of heads) appears in the second column.

c. With the aid of Table 8.1, we see that the event $(X = 2)$ is given by the set

$$\{\text{HHT, HTH, THH}\}$$ ○ ○ ○

Table 8.2
Number of coin tosses before heads appears.

Outcome	Value of Y
H	1
TH	2
TTH	3
TTTH	4
TTTTH	5
⋮	⋮

EXAMPLE 2   A coin is tossed repeatedly until a head occurs. Let the random variable Y denote the number of coin tosses in the experiment. What are the values of Y?

Solution   The outcomes of the experiment make up the infinite set

$$S = \{H, TH, TTH, TTTH, TTTTH, \ldots\}$$

These outcomes of the experiment are displayed in the first column of Table 8.2. The corresponding values assumed by the random variable Y (the number of tosses) appear in the second column.   ◦ ◦ ◦

EXAMPLE 3   A disposable flashlight is turned on until its battery runs out. Let the random variable Z denote the length (in hours) of the life of the battery. What values may Z assume?

Solution   The values assumed by Z may be any nonnegative real numbers; that is, the possible values of Z comprise the interval $0 \leq Z < \infty$.   ◦ ◦ ◦

One advantage of working with random variables rather than working directly with the outcomes of an experiment is that random variables are functions that may be added, subtracted, and multiplied. Because of this, results developed in the field of algebra and other areas of mathematics may be used freely to help us solve problems in the fields of probability and statistics.

A random variable is classified into three categories depending on the set of values it assumes. A random variable is called **finite discrete** if it assumes only finitely many values. For example, the random variable X of Example 1 is finite discrete since it may assume values only from the finite set $\{0, 1, 2, 3\}$ of numbers. Next, a random variable is said to be **infinite discrete** if it takes on infinitely many values, which may be arranged in a sequence. For example, the random variable Y of Example 2 is infinite discrete since it assumes values from the set $\{1, 2, 3, 4, 5, \ldots\}$, which has been arranged in the form of an infinite sequence. Finally, a random variable is called **continuous** if the values it may assume comprise an interval of real numbers. For example, the random variable Z of Example 3 is continuous since the values it may assume comprise the interval of nonnegative real numbers. For the remainder of this section *all random variables will be assumed to be finite discrete.*

Probability Distributions of Random Variables

In Section 7.2 we learned how to construct the probability distribution for an experiment. There, the probability distribution took the form of a table that gave the probabilities associated with the outcomes of an experiment. Since the random variable associated with an experiment is related to the outcomes of the experiment, it is clear that we should be able to construct a probability distribution associated with the *random variable* rather than one associated with the outcomes of the experiment. Such a distribution is called the

probability distribution of a random variable and may be given in the form of a formula or displayed in a table that gives the distinct (numerical) values of the random variable X and the probabilities associated with these values. Thus, if $x_1, x_2, \ldots, x_n$ are the values assumed by the random variable X with associated probabilities $P(X = x_1), P(X = x_2), \ldots, P(X = x_n)$, respectively, then the required probability distribution of the random variable X, where $p_i = P(X = x_i)$, $i = 1, 2, \ldots, n$, may be expressed in the form of a table:

Probability distribution for the random variable X.

x	$P(X = x)$
x_1	p_1
x_2	p_2
x_3	p_3
$\vdots$	$\vdots$
x_n	p_n

The next several examples illustrate the construction of probability distributions.

EXAMPLE 4   Find the probability distribution of the random variable associated with the experiment of Example 1.

Solution   From the results of Example 1, we see that the values assumed by the random variable X are 0, 1, 2, and 3, corresponding to the events of 0, 1, 2, and 3 heads occurring, respectively. Referring to Table 8.1 once again, we see that the outcome associated with the event $(X = 0)$ is given by the set $\{TTT\}$. Consequently, the probability associated with the random variable X when it assumes the value 0 is given by

$$P(X = 0) = \frac{1}{8} \qquad \text{(Note that } n(S) = 8.)$$

Table 8.3
Probability distribution.

x	$P(X = x)$
0	$\frac{1}{8}$
1	$\frac{3}{8}$
2	$\frac{3}{8}$
3	$\frac{1}{8}$

Next, observe that the event $(X = 1)$ is given by the set $\{HTT, THT, TTH\}$, so

$$P(X = 1) = \frac{3}{8}$$

In a similar manner we may compute $P(X = 2)$ and $P(X = 3)$, which gives the probability distribution shown in Table 8.3.   ● ● ●

EXAMPLE 5   Let X denote the random variable that gives the sum of the faces that fall uppermost when two fair dice are cast. Find the probability distribution of X.

Solution   The values assumed by the random variable X are 2, 3, 4, ..., 12, corresponding to the events $E_2, E_3, E_4, \ldots, E_{12}$ (see Example 4, Section 7.1). Next, the probabilities associated with the random variable X when X assumes the values 2, 3, 4, ..., 12 are precisely the probabilities $P(E_2), P(E_3), \ldots,$

Table 8.4
Probability distribution for the random variable that gives the sum of the faces of two dice.

x	2	3	4	5	6	7	8	9	10	11	12
P(X = x)	1/36	2/36	3/36	4/36	5/36	6/36	5/36	4/36	3/36	2/36	1/36

$P(E_{12})$, respectively, and may be computed in much the same way as the solution to Example 3, Section 7.2. Thus,

$$P(X = 2) = P(E_2) = \frac{1}{36}, \quad P(X = 3) = P(E_3) = \frac{2}{36}, \quad \text{and so on}$$

The required probability distribution of X is given in Table 8.4.

EXAMPLE 6 The following data give the number of cars observed waiting in line at the beginning of 2-minute intervals between 3 and 5 P.M. on a certain Friday at the drive-in teller of the Westwood Savings Bank and the corresponding frequency of occurrence. Find the probability distribution of the random variable X, where X denotes the number of cars observed waiting in line.

Number of Cars	Frequency of Occurrence
0	2
1	9
2	16
3	12
4	8
5	6
6	4
7	2
8	1

Table 8.5
Probability distribution.

x	P(X = x)
0	.03
1	.15
2	.27
3	.20
4	.13
5	.10
6	.07
7	.03
8	.02

Solution Dividing each number on the right of the given table by 60 (the sum of these numbers) gives the respective probabilities (here, we use the relative frequency interpretation of probability) associated with the random variable X when X assumes the values 0, 1, 2, ..., 8. For example,

$$P(X = 0) = \frac{2}{60} \approx .03$$

$$P(X = 1) = \frac{9}{60} = .15 \quad \text{and so on}$$

The resulting probability distribution is shown in Table 8.5.

Histograms

A probability distribution of a random variable may be exhibited graphically by means of a **histogram.** To construct a histogram of a particular probability

Figure 8.1

Histogram showing the probability distribution for the number of heads occurring in three coin tosses.

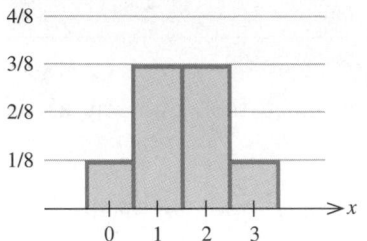

distribution, first locate the values of the random variable on the number line. Then, above each such number, erect a rectangle with width 1 and height equal to the probability associated with that value of the random variable. For example, the histogram of the probability distribution appearing in Table 8.3 is shown in Figure 8.1. The histograms of the probability distributions of Examples 5 and 6 are constructed in a similar manner and are displayed in Figures 8.2 and 8.3, respectively.

Observe that in each histogram, the area of a rectangle associated with a value of a random variable X gives precisely the probability associated with the value of X. This follows because each such rectangle, by construction, has width 1 and height corresponding to the probability associated with the value of the random variable. Another consequence arising from the method of construction of a histogram is that *the probability associated with more than one value of the random variable X is given by the sum of the areas of the*

Figure 8.2

Histogram showing the probability distribution for the sum of the faces of two dice.

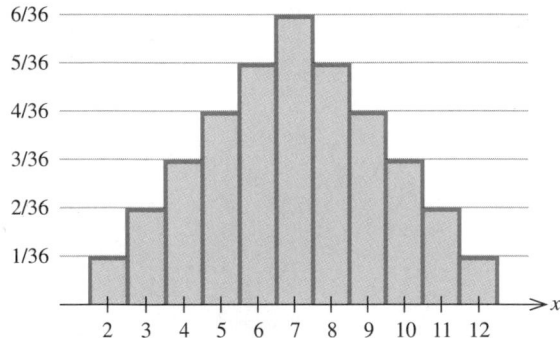

Figure 8.3

Histogram showing the probability distribution for the number of cars waiting in line.

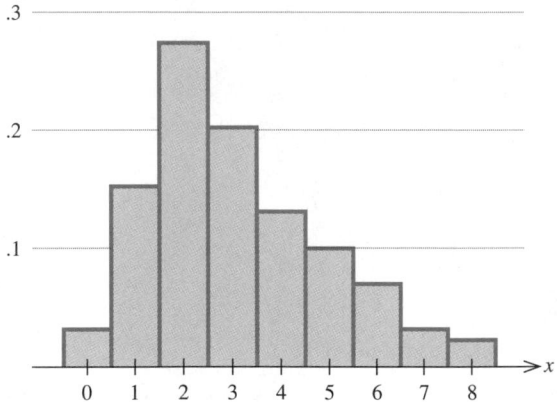

rectangles associated with those values of X. For example, in the coin-tossing experiment of Example 1, the event of obtaining at least two heads, which corresponds to the event $(X = 2)$ or $(X = 3)$, is given by

$$P(X = 2) + P(X = 3)$$

and may be obtained from the histogram depicted in Figure 8.1 by adding the areas associated with the values 2 and 3, respectively, of the random variable X. We obtain

$$P(X = 2) + P(X = 3) = (1)\left(\frac{3}{8}\right) + (1)\left(\frac{1}{8}\right) = \frac{1}{2}$$

This result provides us with a method of computing the probabilities of events directly from the knowledge of a histogram of the probability distribution of the random variable associated with the experiment.

EXAMPLE 7 Suppose that the probability distribution of a random variable X is represented by the histogram shown in Figure 8.4. Identify that part of the histogram whose area gives the probability $P(10 \leq X \leq 20)$. Do not evaluate the result.

Figure 8.4

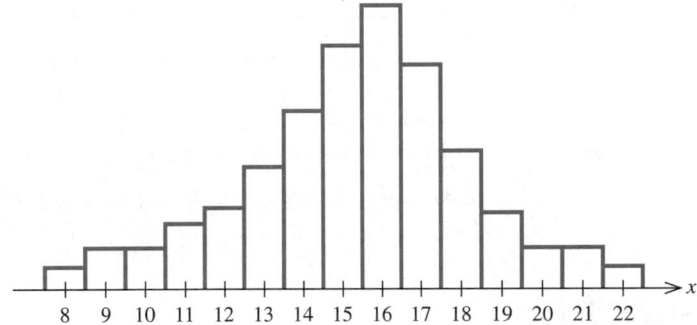

Solution The event $(10 \leq X \leq 20)$ is the event comprising outcomes related to the values 10, 11, 12, ... , 20 of the random variable X. The probability of this event $P(10 \leq X \leq 20)$ is therefore given by the shaded area of the histogram in Figure 8.5.

Figure 8.5
$P(10 \leq X \leq 20)$.

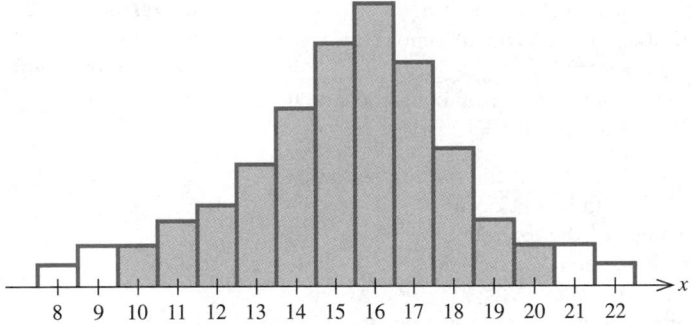

SELF-CHECK EXERCISES 8.1

1. Three balls are selected at random without replacement from an urn containing four black balls and five white balls. Let the random variable X denote the number of black balls drawn.
a. List the outcomes of the experiment.
b. Find the value assigned to each outcome of the experiment by the random variable X.

c. Find the event comprising the outcomes to which a value of 2 has been assigned by X.

2. The following data, extracted from the records of the Dover Public Library, give the number of books borrowed by the library's members over a 1-month period.

Number of Books	0	1	2	3	4	5	6	7	8
Frequency of Occurrence	780	300	412	205	98	54	57	30	6

a. Find the probability distribution of the random variable X, where X denotes the number of books checked out over a 1-month period by a randomly chosen member.

b. Draw the histogram representing this probability distribution.

Solutions to Self-Check Exercises 8.1 can be found on page 472.

8.1 EXERCISES

1. Three balls are selected at random without replacement from an urn containing four green balls and six red balls. Let the random variable X denote the number of green balls drawn.

a. List the outcomes of the experiment.

b. Find the value assigned to each outcome of the experiment by the random variable X.

c. Find the event comprising the outcomes to which a value of 3 has been assigned by X.

2. A coin is tossed four times. Let the random variable X denote the number of tails that occur.

a. List the outcomes of the experiment.

b. Find the value assigned to each outcome of the experiment by the random variable X.

c. Find the event comprising the outcomes to which a value of 2 has been assigned by X.

3. A die is cast repeatedly until a 6 falls uppermost. Let the random variable X denote the number of times the die is cast. What are the values that X may assume?

4. Cards are selected one at a time without replacement from a well-shuffled deck of 52 cards until an ace is drawn. Let X denote the random variable that gives the number of cards drawn. What values may X assume?

5. Let X denote the random variable that gives the sum of the faces that fall uppermost when two fair dice are cast. Find $P(X = 7)$.

6. Two cards are drawn from a well-shuffled deck of 52 playing cards. Let X denote the number of aces drawn. Find $P(X = 2)$.

In exercises 7–12, give the range of values that the random variable X may assume and classify the random variable as finite discrete, infinite discrete, or continuous.

7. X = The number of times a die is thrown until a 2 appears

8. X = The number of defective watches in a sample of eight watches

9. X = The distance a commuter travels to work

10. X = The number of hours a child watches television on a given day

11. X = The number of times an accountant takes the C.P.A. examination before passing

12. X = The number of boys in a four-child family

13. The probability distribution of the random variable X is shown in the accompanying table.

x	−10	−5	0	5
$P(X = x)$	.20	.15	.05	.1

x	10	15	20
$P(X = x)$	.25	.1	.15

Find:
a. $P(X = -10)$ b. $P(X \geq 5)$
c. $P(-5 \leq X \leq 5)$ d. $P(X \leq 20)$

14. The probability distribution of the random variable X is shown in the accompanying table.

x	-5	-3	-2
$P(X = x)$	.17	.13	.33

x	0	2	3
$P(X = x)$	.16	.11	.10

Find:
a. $P(X \leq 0)$ b. $P(X \leq -3)$
c. $P(-2 \leq X \leq 2)$

15. Suppose that a probability distribution of a random variable X is represented by the accompanying histogram. Shade that part of the histogram whose area gives the probability $P(17 \leq X \leq 20)$.

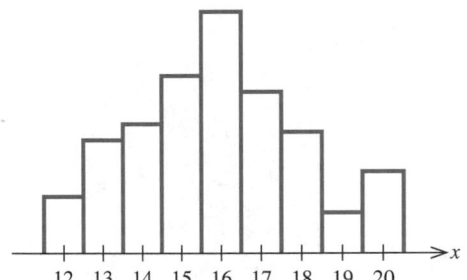

16. **Exams**   An examination consisting of 10 true-or-false questions was taken by a class of 100 students. The

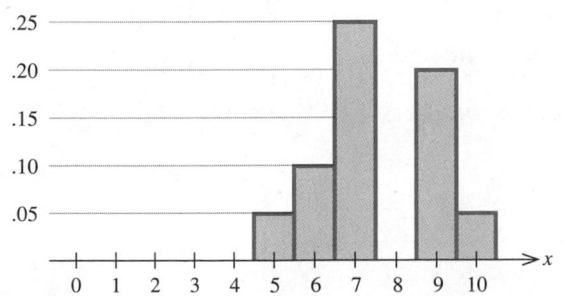

probability distribution of the random variable X, where X denotes the number of questions answered correctly by a randomly chosen student, is represented by the accompanying histogram. The rectangle with base centered on the number 8 is missing. What should the height of this rectangle be?

17. Two dice are cast. Let the random variable X denote the number that falls uppermost on the first die and let Y denote the number that falls uppermost on the second die.
a. Find the probability distributions of X and Y.
b. Find the probability distribution of $X + Y$.

18. **Distribution of Families by Size**   A survey was conducted by the Public Housing Authority in a certain community among 1000 families to determine the distribution of families by size. The results follow.

Family Size	2	3	4	5
Frequency of Occurrence	350	200	245	125

Family Size	6	7	8
Frequency of Occurrence	66	10	4

a. Find the probability distribution of the random variable X, where X denotes the number of persons in a randomly chosen family.
b. Draw the histogram corresponding to the probability distribution found in (a).

19. **Waiting Lines**   The accompanying data were obtained in a study conducted by the manager of the Sav-More Supermarket. In this study the number of customers waiting in line at the express checkout at the beginning of each 3-minute interval between 9 A.M. and 12 noon on Saturday was observed.

Number of Customers	0	1	2	3	4
Frequency of Occurrence	1	4	2	7	14

Number of Customers	5	6	7	8	9	10
Frequency of Occurrence	8	10	6	3	4	1

Using Technology

GRAPHING A HISTOGRAM

The graphing calculator can be used to plot the histogram for a given set of data, as illustrated by the following example.

EXAMPLE 1   A survey of 90,000 households conducted in 1995 revealed the following percentage of women who wear the given shoe size.

Shoe Size	<5	5–5$\frac{1}{2}$	6–6$\frac{1}{2}$	7–7$\frac{1}{2}$	8–8$\frac{1}{2}$	9–9$\frac{1}{2}$	10–10$\frac{1}{2}$	>10$\frac{1}{2}$
Percentage of Women	1	5	15	27	29	14	7	2

Source: Footwear Market Insights survey

a. Plot a histogram for the given data.

b. What is the percentage of women in the survey who wear size 7–7$\frac{1}{2}$ or 8–8$\frac{1}{2}$ shoes?

Solution

a. Let X denote the random variable taking on the values 1 through 8, where 1 corresponds to a shoe size less than 5, 2 corresponds to a shoe size less than 5–5$\frac{1}{2}$, and so on. Entering the values of X as $x_1 = 1$, $x_2 = 2$, ..., $x_8 = 8$, and the corresponding values of Y as $y_1 = 1$, $y_2 = 5$, ..., $y_8 = 2$, and then using the Draw function from the Statistics menu, we draw the histogram shown in Figure T1.

Figure T1
The histogram for the given data using the viewing rectangle [0, 9] × [0, 35].

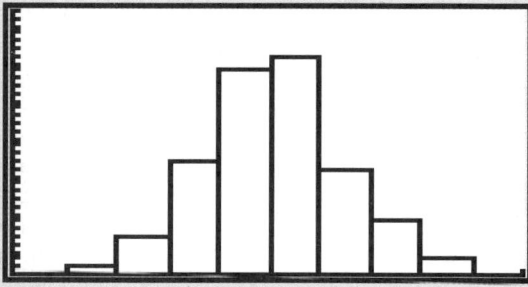

b. The probability that a woman participating in the survey wears size 7–7$\frac{1}{2}$ or 8–8$\frac{1}{2}$ shoes is given by

$$P(X = 4) + P(X = 5) = .27 + .29 = .56$$

which tells us that 56% of the women wear either size 7–7$\frac{1}{2}$ or size 8–8$\frac{1}{2}$ shoes.

○ ○ ○

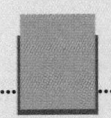

EXERCISES

1. Graph the histogram associated with the data given in Table 8.1, page 462. Compare your graph with that given in Figure 8.1, page 466.

2. Graph the histogram associated with the data given in exercise 18, page 469.

3. Graph the histogram associated with the data given in exercise 19, page 469.

4. Graph the histogram associated with the data given in exercise 21, page 472.

a. Find the probability distribution of the random variable X, where X denotes the number of customers observed waiting in line.

b. Draw the histogram representing this probability distribution.

20. Money-Market Rates   The rates paid by 30 financial institutions on a certain day for money-market deposit accounts are shown in the accompanying table.

Rate (%)	6	6.25	6.55	6.56
Number of Institutions	1	7	7	1

Rate (%)	6.58	6.60	6.65	6.85
Number of Institutions	1	8	3	2

Let the random variable X denote the interest paid by a randomly chosen financial institution on its money-market deposit accounts and find the probability distribution associated with these data.

21. Television Pilots   After the private screening of a new television pilot, audience members were asked to rate the new show on a scale of 1 to 10 (10 being the highest rating). From a group of 140 people, the accompanying responses were obtained.

Rating	1	2	3	4	5
Frequency of Occurrence	1	4	3	11	23

Rating	6	7	8	9	10
Frequency of Occurrence	21	28	29	16	4

Let the random variable X denote the rating given to the show by a randomly chosen audience member. Find the probability distribution associated with these data.

SOLUTIONS TO SELF–CHECK EXERCISES 8.1

1. a. Using the accompanying tree diagram, we see that the outcomes of the experiment are

$$S = \{(B, B, B), (B, B, W), (B, W, B), (B, W, W),$$
$$(W, B, B), (W, B, W), (W, W, B), (W, W, W)\}$$

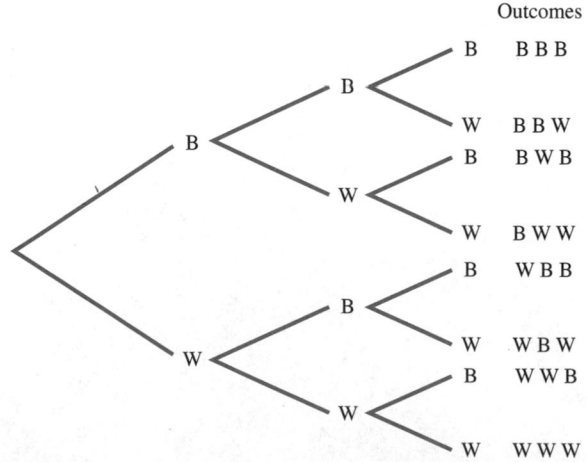

b. Using the results of (a), we obtain the values assigned to the outcomes of the experiment as follows:

Outcome	BBB	BBW	BWB	BWW
Value	3	2	2	1

Outcome	WBB	WBW	WWB	WWW
Value	2	1	1	0

c. The required event is {BBW, BWB, WBB}.

2. a. We divide each number in the bottom row of the given table by 1942 (the sum of these numbers) to obtain the probabilities associated with the random variable X when X takes on the values 0, 1, 2, 3, 4, 5, 6, 7, and 8. For example,

$$P(X = 0) = \frac{780}{1942} \approx .402$$

$$P(X = 1) = \frac{300}{1942} \approx .154$$

The required probability distribution and histogram follow.

x	0	1	2	3	4
$P(X = x)$	.402	.154	.212	.106	.050

x	5	6	7	8
$P(X = x)$	.028	.029	.015	.003

b.

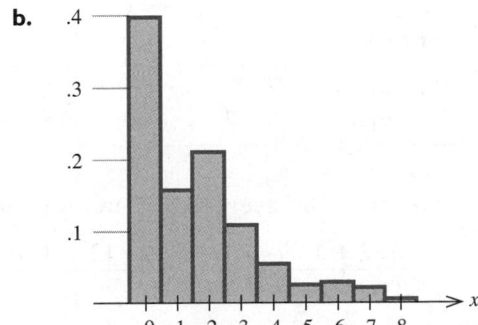

8.2 EXPECTED VALUE

Mean

The average value of a set of numbers is a familiar notion to most people. For example, to compute the average of the four numbers

$$12, \quad 16, \quad 23, \quad 37$$

we simply add these numbers and divide the resulting sum by 4, giving the required average as

$$\frac{12 + 16 + 23 + 37}{4} = \frac{88}{4}$$

or 22. In general, we have the following definition:

AVERAGE OR MEAN

The **average,** or **mean,** of the n numbers

$$x_1, x_2, \ldots, x_n$$

is $\bar{x}$ (read "x bar"), where

$$\bar{x} = \frac{x_1 + x_2 + \cdots + x_n}{n}$$

EXAMPLE 1   Refer to Example 6, Section 8.1. Find the average number of cars waiting in line at the bank's drive-in teller at the beginning of each 2-minute interval during the period in question.

Solution   The number of cars, together with its corresponding frequency of occurrence, are reproduced in Table 8.6. Observe that the number 0 (of cars) occurs twice, the number 1 occurs 9 times, and so on. There are altogether

$$2 + 9 + 16 + 12 + 8 + 6 + 4 + 2 + 1 = 60$$

Table 8.6

Number of Cars	0	1	2	3	4	5	6	7	8
Frequency of Occurrence	2	9	16	12	8	6	4	2	1

numbers to be averaged. Therefore, the required average is given by

$$\frac{0 \cdot 2 + 1 \cdot 9 + 2 \cdot 16 + 3 \cdot 12 + 4 \cdot 8 + 5 \cdot 6 + 6 \cdot 4 + 7 \cdot 2 + 8 \cdot 1}{60} \approx 3.1 \quad \textbf{(1)}$$

or approximately 3.1 cars.

○ ○ ○

Expected Value

Let us reconsider the expression in equation (1) that gives the average of the frequency distribution shown in Table 8.6. Dividing each term by the denominator, the expression may be rewritten in the form

$$0 \cdot \left(\frac{2}{60}\right) + 1 \cdot \left(\frac{9}{60}\right) + 2 \cdot \left(\frac{16}{60}\right) + 3 \cdot \left(\frac{12}{60}\right) + 4 \cdot \left(\frac{8}{60}\right) + 5 \cdot \left(\frac{6}{60}\right)$$

$$+ 6 \cdot \left(\frac{4}{60}\right) + 7 \cdot \left(\frac{2}{60}\right) + 8 \cdot \left(\frac{1}{60}\right)$$

Observe that each term in the sum is a product of two factors; the first factor is the value assumed by the random variable X, where X denotes the number of cars waiting in line, and the second factor is just the probability associated with that value of the random variable. This observation suggests the following general method for calculating the *expected value* (that is, the average, or mean) of a random variable X that assumes a finite number of values from the knowledge of its probability distribution.

EXPECTED VALUE OF A RANDOM VARIABLE X

Let X denote a random variable that assumes the values $x_1, x_2, \ldots, x_n$ with associated probabilities $p_1, p_2, \ldots, p_n$, respectively. Then the **expected value** of X, $E(X)$, is given by

$$E(X) = x_1 p_1 + x_2 p_2 + \cdots + x_n p_n \qquad \text{(2)}$$

Table 8.7
Probability distribution.

x	$P(X = x)$
0	.03
1	.15
2	.27
3	.20
4	.13
5	.10
6	.07
7	.03
8	.02

REMARK   The numbers $x_1, x_2, \ldots, x_n$ may be positive, zero, or negative. For example, such a number will be positive if it represents a profit and negative if it represents a loss.   ◦ ◦ ◦

EXAMPLE 2   Re-solve Example 1 by using the probability distribution associated with the experiment, reproduced in Table 8.7.

Solution   Let X denote the number of cars waiting in line. Then the average number of cars waiting in line is given by the expected value of X—that is, by

$$E(X) = (0)(.03) + (1)(.15) + (2)(.27) + (3)(.20) + (4)(.13)$$
$$+ (5)(.10) + (6)(.07) + (7)(.03) + (8)(.02)$$
$$= 3.1$$

or 3.1 cars, which agrees with the earlier result.   ◦ ◦ ◦

The expected value of a random variable X is a measure of the central tendency of the probability distribution associated with X. In repeated trials of an experiment with random variable X, the average of the observed values of X gets closer and closer to the expected value of X as the number of trials gets larger and larger. Geometrically, the expected value of a random variable

Figure 8.6
• *Expected value of a random variable X.*

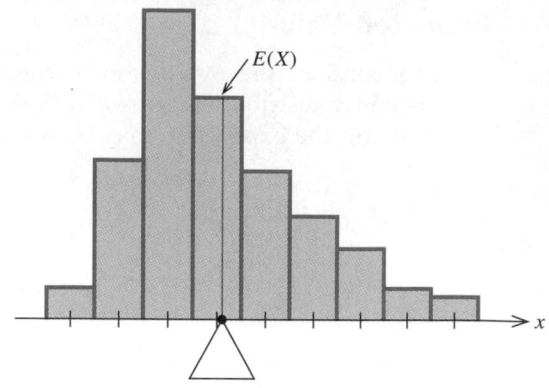

X has the following simple interpretation: If a laminate is made of the histogram of a probability distribution associated with a random variable X, then the expected value of X corresponds to the point on the base of the laminate at which the latter will balance perfectly when the point is directly over a fulcrum (Figure 8.6).

Table 8.8
Probability distribution.

x	$P(X = x)$
2	$\frac{1}{36}$
3	$\frac{2}{36}$
4	$\frac{3}{36}$
5	$\frac{4}{36}$
6	$\frac{5}{36}$
7	$\frac{6}{36}$
8	$\frac{5}{36}$
9	$\frac{4}{36}$
10	$\frac{3}{36}$
11	$\frac{2}{36}$
12	$\frac{1}{36}$

EXAMPLE 3   Let X denote the random variable that gives the sum of the faces that fall uppermost when two fair dice are cast. Find the expected value, $E(X)$, of X.

Solution   The probability distribution of X, reproduced in Table 8.8, was found in Example 5, Section 8.1. Using this result, we find

$$E(X) = 2\left(\frac{1}{36}\right) + 3\left(\frac{2}{36}\right) + 4\left(\frac{3}{36}\right) + 5\left(\frac{4}{36}\right) + 6\left(\frac{5}{36}\right) + 7\left(\frac{6}{36}\right)$$
$$+ 8\left(\frac{5}{36}\right) + 9\left(\frac{4}{36}\right) + 10\left(\frac{3}{36}\right) + 11\left(\frac{2}{36}\right) + 12\left(\frac{1}{36}\right)$$
$$= 7$$

Note that, because of the symmetry of the histogram of the probability distribution with respect to the vertical line $x = 7$, the result could have been obtained by merely inspecting Figure 8.7.

Figure 8.7

Histogram showing the probability distribution for the sum of the faces of two dice.

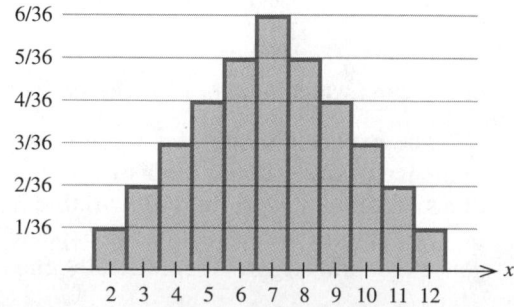

Applications

The next example shows how we can use the concept of expected value to help us make the best investment decision.

EXAMPLE 4   A group of private investors intends to purchase one of two motels currently being offered for sale in a certain city. The terms of sale of the two motels are similar, although the Regina Inn has 52 rooms and is in a slightly better location than the Merlin Motor Lodge, which has 60 rooms. Records obtained for each motel reveal that the occupancy rates, with corresponding probabilities, during the May–September tourist season are as shown in the accompanying tables.

Regina Inn

Occupancy Rate	.80	.85	.90	.95	1.00
Probability	.19	.22	.31	.23	.05

Merlin Motor Lodge

Occupancy Rate	.75	.80	.85	.90	.95	1.00
Probability	.35	.21	.18	.15	.09	.02

The average profit per day for each occupied room at the Regina Inn is $10, whereas the average profit per day for each occupied room at the Merlin Motor Lodge is $9.

a. Find the average number of rooms occupied per day at each motel.

b. If the investors' objective is to purchase the motel that generates the higher daily profit, which motel should they purchase? (Compare the expected daily profit of the two motels.)

Solution

a. Let X denote the occupancy rate at the Regina Inn. Then the average daily occupancy rate at the Regina Inn is given by the expected value of X—that is, by

$$E(X) = (.80)(.19) + (.85)(.22) + (.90)(.31)$$
$$+ (.95)(.23) + (1.00)(.05)$$
$$= .8865$$

The average number of rooms occupied per day at the Regina is

$$(.8865)(52) \approx 46.1$$

or 46.1 rooms. Similarly, letting Y denote the occupancy rate at the Merlin Motor Lodge, we have

$$E(Y) = (.75)(.35) + (.80)(.21) + (.85)(.18) + (.90)(.15)$$
$$+ (.95)(.09) + (1.00)(.02)$$
$$= .8240$$

The average number of rooms occupied per day at the Merlin is

$$(.8240)(60) \approx 49.4$$

or 49.4 rooms.

b. The expected daily profit at the Regina is given by

$$46.1(10) = 461$$

or $461. The expected daily profit at the Merlin is given by

$$(49.4)(9) \approx 445$$

or $445. From these results we conclude that the investors should purchase the Regina Inn, which is expected to yield a higher daily profit.   ◦ ◦ ◦

EXAMPLE 5   The Island Club is holding a fund-raising raffle. Ten thousand tickets have been sold for $2 each. There will be a first prize of $3000, 3 second prizes of $1000 each, 5 third prizes of $500 each, and 20 consolation prizes of $100 each. Letting X denote the net winnings (that is, winnings less the cost of the ticket) associated with the tickets, find $E(X)$. Interpret your results.

Table 8.9
Probability distribution for a raffle.

x	$P(X = x)$
−2	.9971
98	.0020
498	.0005
998	.0003
2998	.0001

Solution   The values assumed by X are $(0 - 2)$, $(100 - 2)$, $(500 - 2)$, $(1000 - 2)$, and $(3000 - 2)$—that is, −2, 98, 498, 998, and 2998—which correspond, respectively, to the value of a losing ticket, a consolation prize, a third prize, and so on. The probability distribution of X may be calculated in the usual manner and appears in Table 8.9. Using the table, we find

$$\begin{aligned} E(X) &= (-2)(.9971) + 98(.0020) + 498(.0005) \\ &\quad + 998(.0003) + 2998(.0001) \\ &= -0.95 \end{aligned}$$

Figure 8.8
Roulette wheel.

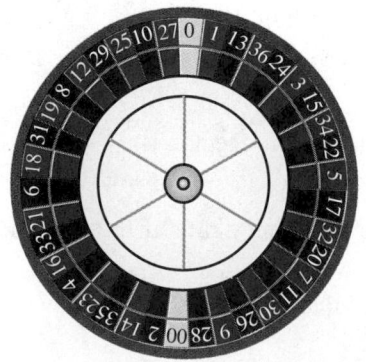

This expected value gives the long-run average loss (negative gain) of a holder of one ticket; that is, if one participated in such a raffle by purchasing one ticket each time, in the long run, one may expect to lose, on the average, 95 cents per raffle.   ◦ ◦ ◦

EXAMPLE 6   In the game of roulette as played in Las Vegas casinos, the wheel is divided into 38 compartments numbered 1 through 36, 0, and 00. One-half of the numbers 1 through 36 are red, the other half black, and 0 and 00 are green (Figure 8.8). Of the many types of bets that may be placed, one type involves betting on the outcome of the color of the winning number. For example, one may place a certain sum of money on *red*. If the winning number is red, one wins an amount equal to the bet placed and loses the bet otherwise. Find the expected value of the winnings on a $1 bet placed on *red*.

Solution Let X be a random variable whose values are 1 and -1, which correspond to a win and a loss. The probabilities associated with the values 1 and -1 are 18/38 and 20/38, respectively. Therefore, the expected value is given by

$$E(X) = 1\left(\frac{18}{38}\right) + (-1)\left(\frac{20}{38}\right) = -\frac{2}{38}$$
$$\approx -0.053$$

Thus, if one places a $1 bet on *red* over and over again, one may expect to lose, on the average, approximately 5 cents per bet in the long run.   ◦ ◦ ◦

Examples 5 and 6 illustrate games that are not "fair." Of course, most participants in such games are aware of this fact and participate in them for other reasons. In a fair game, neither party has an advantage, a condition that translates into the condition that $E(X) = 0$, where X takes on the values of a player's winnings.

EXAMPLE 7 Mike and Bill play a card game with a standard deck of 52 cards. Mike selects a card from a well-shuffled deck and receives A dollars from Bill if the card selected is a diamond; otherwise, Mike pays Bill a dollar. Determine the value of A if the game is to be fair.

Solution Let X denote a random variable whose values are associated with Mike's winnings. Then X takes on the value A with probability $P(X = A) = 1/4$ (since there are 13 diamonds in the deck) if Mike wins and takes on the value -1 with probability $P(X = -1) = 3/4$ if Mike loses. Since the game is to be a fair one, the expected value $E(X)$ of Mike's winnings must be equal to zero, that is,

$$E(X) = A\left(\frac{1}{4}\right) + (-1)\left(\frac{3}{4}\right) = 0$$

Solving this equation for A gives $A = 3$. Thus, the card game will be fair if Bill makes a $3 payoff for a winning bet of $1 placed by Mike.   ◦ ◦ ◦

Odds

In everyday parlance, the probability of the occurrence of an event is often stated in terms of the *odds in favor of* (or *odds against*) the occurrence of the event. For example, one often hears statements such as "the odds that the Dodgers will win the World Series this season are 7 to 5" and "the odds that it will not rain tomorrow are 3 to 2." We will return to these examples later. But first, let us look at a definition that ties these two concepts together.

ODDS IN FAVOR OF AND ODDS AGAINST

If $P(E)$ is the probability of an event E occurring, then

1. the **odds in favor of** E occurring are

$$\frac{P(E)}{1 - P(E)} = \frac{P(E)}{P(E^c)} \qquad P(E) \neq 1 \qquad \text{(3a)}$$

2. the **odds against** E occurring are

$$\frac{1 - P(E)}{P(E)} = \frac{P(E^c)}{P(E)} \qquad P(E) \neq 0 \qquad \text{(3b)}$$

REMARKS

1. Notice that the odds in favor of the occurrence of an event are given by the ratio of the probability of the event occurring to the probability of the event not occurring. The odds against the occurrence of an event are given by the reciprocal of the odds in favor of the occurrence of the event.

2. Whenever possible, odds are expressed as ratios of whole numbers. If the odds in favor of E are a/b, we say the odds in favor of E are a to b. If the odds against E occurring are b/a, we say the odds against E are b to a.

○ ○ ○

EXAMPLE 8   Find the odds in favor of winning a bet on *red* in American roulette. What are the odds against winning a bet on *red*?

Solution   The probability of winning a bet here—the probability that the ball lands in a red compartment—is given by $P = 18/38$. Therefore, using formula (3a), we see that the odds in favor of winning a bet on *red* are

$$\frac{P(E)}{1 - P(E)} = \frac{\dfrac{18}{38}}{1 - \dfrac{18}{38}} \qquad (E, \text{event of winning a bet on } red)$$

$$= \frac{\dfrac{18}{38}}{\dfrac{38 - 18}{38}}$$

$$= \frac{18}{38} \cdot \frac{38}{20}$$

$$= \frac{18}{20} = \frac{9}{10}$$

or 9 to 10. Next, using (3b), we see that the odds against winning a bet on *red* are 10/9, or 10 to 9.

○ ○ ○

Now, suppose the odds in favor of the occurrence of an event are a to b. Then (3a) gives

$$\frac{a}{b} = \frac{P(E)}{1 - P(E)}$$

$$a(1 - P(E)) = bP(E) \qquad \text{(Cross-multiplying)}$$

$$a - aP(E) = bP(E)$$

$$a = (a + b)P(E)$$

and

$$P(E) = \frac{a}{a + b}$$

which leads us to the following result.

PROBABILITY OF AN EVENT (GIVEN THE ODDS)

If the odds in favor of an event E occurring are a to b, then the probability of E occurring is

$$P(E) = \frac{a}{a + b} \qquad \textbf{(4)}$$

Formula (4) is often used to determine subjective probabilities, as the next example shows.

EXAMPLE 9   Consider each of the following statements.

a. "The odds that the Dodgers will win the World Series this season are 7 to 5."

b. "The odds that it will not rain tomorrow are 3 to 2."

Express each of these odds as a probability of the event occurring.

Solution
a. Using formula (4) with $a = 7$ and $b = 5$ gives the required probability as

$$\frac{7}{7 + 5} = \frac{7}{12} \approx .5833$$

b. Here the event is that it will not rain tomorrow. Using (4) with $a = 3$ and $b = 2$, we conclude that the probability that it will not rain tomorrow is

$$\frac{3}{3 + 2} = \frac{3}{5} = .6$$

○ ○ ○

In the movie *Casino,* the executive of the Tangiers Casino, Sam Rothstein (Robert De-Niro), fired the manager of the slot machines in the casino after three gamblers hit three "million dollar" jackpots in a span of 20 minutes. Rothstein claimed that it was a scam and that somebody had gotten into those machines to set the wheels. He was especially annoyed at the slot machine manager's assertion that there was no way to determine this. According to Rothstein the odds of hitting a jackpot in a four-wheel machine is one in a million and a half and the probability of hitting three jackpots in a row is "in the billions." "It cannot happen! It will not happen!" To see why Mr. Rothstein was so indignant, find the odds of hitting the jackpots in three of the machines in quick succession and comment on the likelihood of this happening.

In concluding this section, we should mention that in addition to the mean there are two other measures of central tendency of a set of numerical data: the median and the mode of a set of numbers. The **median** is the middle value in a set of data arranged in increasing or decreasing order (when there is an odd number of entries). If there is an even number of entries, the median is the mean of the two middle numbers. The **mode** is the value that occurs most frequently in a set of data. (See exercises 34 and 35.)

SELF-CHECK EXERCISES 8.2

1. Find the expected value of a random variable X having the following probability distribution.

x	-4	-3	-1	0	1	2
$P(X = x)$	.10	.20	.25	.10	.25	.10

2. The developer of the Shoreline Condominiums has provided the following estimate of the probability that 20, 25, 30, 35, 40, 45, or 50 of the townhouses will be sold within the first month they are offered for sale.

Number of Units	20	25	30	35	40	45	50
Probability	.05	.10	.30	.25	.15	.10	.05

How many townhouses can the developer expect to sell within the first month they are put on the market?

Solutions to Self-Check Exercises 8.2 can be found on page 487.

8.2 EXERCISES

A calculator is recommended for this exercise set.

1. During the first year at a university that uses a four-point grading system, a freshman took ten 3-credit courses and received two A's, three B's, four C's, and one D.
   a. Compute this student's grade point average.
   b. Let the random variable X denote the number of points corresponding to a given letter grade. Find the probability distribution of the random variable X and compute $E(X)$, the expected value of X.

2. Records kept by the chief dietitian at the university cafeteria over a 30-week period show the following weekly consumption of milk (in gallons).

Milk (in Gallons)	200	205	210	215	220
Number of Weeks	3	4	6	5	4

Milk (in Gallons)	225	230	235	240
Number of Weeks	3	2	2	1

   a. Find the average number of gallons of milk consumed per week in the cafeteria.
   b. Let the random variable X denote the number of gallons of milk consumed in a week at the cafeteria. Find the probability distribution of the random variable X and compute $E(X)$, the expected value of X.

3. Find the expected value of a random variable X having the following probability distribution.

x	−5	−1	0
$P(X = x)$	.12	.16	.28

x	1	5	8
$P(X = x)$	.22	.12	.1

4. Find the expected value of a random variable X having the following probability distribution.

x	0	1	2
$P(X = x)$	$\frac{1}{8}$	$\frac{1}{4}$	$\frac{3}{16}$

x	3	4	5
$P(X = x)$	$\frac{1}{4}$	$\frac{1}{16}$	$\frac{1}{8}$

5. The daily earnings X of an employee who works on a commission basis are given by the following probability distribution. Find the employee's expected earnings.

x (in $)	0	25	50	75
$P(X = x)$	.07	.12	.17	.14

x (in $)	100	125	150
$P(X = x)$	.28	.18	.04

6. In a four-child family, what is the expected number of boys? (Assume that the probability of a boy being born is the same as the probability of a girl being born.)

7. Based on past experience, the manager of the Video-Rama Store has compiled the following table, which gives the probabilities that a customer who enters the Video-Rama Store will buy 0, 1, 2, 3, or 4 videocassettes. How many videocassettes can a customer entering this store be expected to buy?

Number of Video-cassettes	0	1	2	3	4
Probability	.42	.36	.14	.05	.03

8. If a sample of three batteries is selected from a lot of ten, of which two are defective, what is the expected number of defective batteries?

9. **Auto Accidents**   The number of accidents that occur at a certain intersection known as "Five Corners" on a Friday afternoon between the hours of 3 P.M. and 6 P.M., along with the corresponding probabilities, are shown in the following table. Find the expected number of accidents during the period in question.

Number of Accidents	0	1	2	3	4
Probability	.935	.03	.02	.01	.005

10. **Expected Demand**   The owner of a newsstand in a college community estimates the weekly demand for a certain magazine as follows:

Quantity Demanded	10	11	12
Probability	.05	.15	.25

Quantity Demanded	13	14	15
Probability	.30	.20	.05

Find the number of issues of the magazine that the newsstand owner can expect to sell per week.

11. **Expected Product Reliability**   A bank has two automatic tellers at its main office and two at each of its three branches. The number of machines that break down on a given day, along with the corresponding probabilities, are shown in the following table.

Number of Machines That Break Down	0	1	2	3	4
Probability	.43	.19	.12	.09	.04

Number of Machines That Break Down	5	6	7	8
Probability	.03	.03	.02	.05

Find the expected number of machines that will break down on a given day.

12. **Lotteries**   In a lottery, 5000 tickets are sold for $1 each. One first prize of $2000, 1 second prize of $500, 3 third prizes of $100, and 10 consolation prizes of $25 are to be awarded. What are the expected net earnings of a person who buys one ticket?

13. **Life Insurance Premiums**   A man wishes to purchase a 5-year term life insurance policy that will pay the beneficiary $20,000 in the event that the man's death occurs during the next five years. Using life insurance tables, he determines that the probability that he will live another five years is .96. What is the minimum amount that he can expect to pay for his premium?

[*Hint:* The minimum premium occurs when the insurance company's expected gain is zero.]

14. **Expected Gain**   A woman purchased a $10,000, 1-year term life insurance policy for $130. Assuming that the probability that she will live another year is .992, find the company's expected gain.

15. **Life Insurance Policies**   As a fringe benefit, Mr. Taylor receives a $25,000 life insurance policy from his employer. The probability that Mr. Taylor will live another year is .9935. If he purchases the same coverage for himself, what is the minimum amount that he can expect to pay for the policy?

16. **Expected Profit**   A buyer for Discount Fashions, an outlet for women's apparel, is considering buying a batch of clothing for $64,000. She estimates that the company will be able to sell it for $80,000, $75,000, or $70,000 with probabilities of .30, .60, and .10, respectively. Based on these estimates, what will the company's expected gross profit be?

17. **Investment Analysis**   The proprietor of Midland Construction Company has to decide between two projects. He estimates that the first project will yield a profit of $180,000 with a probability of .7 or a profit of $150,000 with a probability of .3; the second project will yield a profit of $220,000 with a probability of .6 or a profit of $80,000 with a probability of .4. Which project should the proprietor choose if he wants to maximize his expected profit?

18. **Cable Television**   The management of Multi-Vision, Inc., a cable TV company, intends to submit a bid for the cable television rights in one of two cities, A or B. If the company obtains the rights to city A, the probability of which is .2, the estimated profit over the next ten years is $10 million; if the company obtains the rights to city B, the probability of which is .3, the estimated

profit over the next ten years is $7 million. The cost of submitting a bid for rights in city A is $250,000 and that of city B is $200,000. By comparing the expected profits for each venture, determine whether the company should bid for the rights in city A or city B.

19. **Expected Auto Sales** Mr. Hunt intends to purchase one of two car dealerships currently for sale in a certain city. Records obtained from each of the two dealers reveal that their weekly volume of sales, with corresponding probabilities, are as follows:

Dahl Motors

Number of Cars Sold per Week	5	6	7	8
Probability	.05	.09	.14	.24

Number of Cars Sold per Week	9	10	11	12
Probability	.18	.14	.11	.05

Farthington Auto Sales

Number of Cars Sold per Week	5	6	7
Probability	.08	.21	.31

Number of Cars Sold per Week	8	9	10
Probability	.24	.10	.06

The average profit per car at Dahl Motors is $362, and the average profit per car at Farthington Auto Sales is $436.

a. Find the average number of cars sold per week at each dealership.

b. If Mr. Hunt's objective is to purchase the dealership that generates the higher weekly profit, which dealership should he purchase? (Compare the expected weekly profit for each dealership.)

20. **Expected Home Sales** Ms. Leonard, a real estate broker, is relocating in a large metropolitan area where she has received job offers from realty company A and realty company B. The number of houses she expects to sell in a year at each firm and the associated probabilities are shown in the following tables.

Company A

Number of Houses Sold	12	13	14	15	16
Probability	.02	.03	.05	.07	.07

Number of Houses Sold	17	18	19	20
Probability	.16	.17	.13	.11

Number of Houses Sold	21	22	23	24
Probability	.09	.06	.03	.01

Company B

Number of Houses Sold	6	7	8	9	10
Probability	.01	.04	.07	.06	.11

Number of Houses Sold	11	12	13	14
Probability	.12	.19	.17	.13

Number of Houses Sold	15	16	17	18
Probability	.04	.03	.02	.01

The average price of a house in the locale of company A is $104,000, whereas the average price of a house in the locale of company B is $177,000. If Ms. Leonard will receive a 3% commission on sales at both companies, which job offer should she accept to maximize her expected yearly commission?

21. **Weather Predictions** If the probability that it will rain tomorrow is .3,

a. what are the odds that it will rain tomorrow?

b. what are the odds that it will not rain tomorrow?

Buying women's clothes for a chain of 50 discount clothing stores is a demanding job, requiring Meiselman to be part fashion arbiter and part accountant. She must balance her fashion choices against a bottom line that has to show a profit. To make her job even more difficult, Meiselman's selections are judged by women from northern New England to the midwestern states. What sells in one area may not sell in another. An incorrect choice can be a costly mistake.

Lilli Meiselman

Title: Buyer

Meiselman's job is challenging. She visits New York at least twice a month to buy clothes from a number of different manufacturers. She works with the store's in-house advertising agency, planning advertising as well as approving copy. Meiselman notes that the vast majority of the ads appear in local newspapers, the store's principal advertising medium.

For all the demands on her time, Meiselman enjoys her work. "It's gratifying to see things I bought for the stores being sold," she says. Although Meiselman decides how much to spend on particular items and which styles, sizes, and colors to select, her decisions are the end result of detailed plans that guide her buying decisions.

Based on the previous year's sales, Meiselman will "work with a departmental planner" to develop a seasonal merchandise plan (in retail there are only two seasons, fall and spring, each covering a six-month time span).

As part of that merchandising plan, Meiselman must determine how much inventory is required per month. If she needs $600,000 worth of inventory at the beginning of June, she subtracts projected sales of $400,000 from the May inventory of $800,000, leaving a balance of $400,000. To meet her June inventory goal of $600,000, she has to increase her inventory by an additional $200,000.

The finances have to be coordinated with actual quantities of suits, dresses, blouses, outerwear, and so on. Using a sales-to-stock ratio, Meiselman estimates that perhaps half of a particular item will sell in any given month. To meet her financial goals, she needs sufficient quantity on hand to sell. For example, Meiselman would need 4000 raincoats in April retailing at $99 each to reach a $200,000 sales goal if the stores' sales were expected to reach a 50% sales-to-stock ratio.

Sales volume is the key to success in any discount business, whether clothing or home appliances. Meiselman's merchandise is marked up or down depending on that volume. Her goal is a 45% markup on all merchandise. If a line of suits doesn't sell, she may mark it down, achieving only a 35% markup over the wholesale price. To balance out the loss, she marks up another line of items so that her *average* markup hits 45%.

With its high overhead, the chain has to generate sufficient sales volume and profits to stay in business. Meiselman's fashion choices have helped make that goal a continuing reality.

◗ ◗ ◗

22. **Roulette** In American roulette, as described in Example 6, a player may bet on a split (two adjacent numbers). In this case, if the player bets $1 and either number comes up, the player wins $17 and gets his $1 back. If neither comes up, he loses his $1 bet. Find the expected value of the winnings on a $1 bet placed on a split.

23. **Roulette** If a player placed a $1 bet on red and a $1 bet on black in a single play in American roulette, what would the expected value of his winnings be?

24. **Roulette** In European roulette the wheel is divided into 37 compartments numbered 1 through 36 and 0. (In American roulette there are 38 compartments numbered 1 through 36, 0, and 00.) Find the expected value of the winnings on a $1 bet placed on red in European roulette.

25. The probability of an event E occurring is .8. What are the odds in favor of E occurring? What are the odds against E occurring?

26. The probability of an event E not occurring is .6. What are the odds in favor of E occurring? What are the odds against E occurring?

27. The odds in favor of an event E occurring are 9 to 7. What is the probability of E occurring?

28. The odds against an event E occurring are 2 to 3. What is the probability of E not occurring?

29. **Odds** Joan, a computer sales representative, feels that the odds are 8 to 5 that she will clinch the sale of a minicomputer to a certain company. What is the (subjective) probability that Joan will make the sale?

30. **Sports** Steffi feels that the odds in favor of her winning her tennis match tomorrow are 7 to 5. What is the (subjective) probability that she will win her match tomorrow?

31. **Sports** If a sports forecaster states that the odds of a certain boxer winning a match are 4 to 3, what is the probability that the boxer will win the match?

32. **Odds** Bob, the proprietor of Midland Lumber, feels that the odds in favor of a business deal going through are 9 to 5. What is the (subjective) probability that this deal will not materialize?

33. **Roulette**
   a. Show that for any number c,

$$E(cX) = cE(X)$$

   b. Use this result to find the expected loss if a gambler bets $300 on red in a single play in American roulette.

   [*Hint:* Use the results of Example 6.]

34. **Exam Scores** In an examination given to a class of 20 students, the following test scores were obtained:

$$40, 45, 50, 50, 55, 60, 60, 75, 75, 80,$$
$$80, 85, 85, 85, 85, 90, 90, 95, 95, 100$$

   a. Find the mean, or average, score, the mode, and the median score.
   b. Which of these three measures of central tendency do you think is the least representative of the set of scores?

35. **Wage Rates** The frequency distribution of the hourly wage rates among blue-collar workers in a certain factory is given in the following table. Find the mean (or average) wage rate, the mode, and the median wage rate of these workers.

Wage Rate (in $)	10.70	10.80	10.90	11.00	11.10	11.20
Frequency	60	90	75	120	60	45

SOLUTIONS TO SELF-CHECK EXERCISES 8.2

1. $E(X) = (-4)(.10) + (-3)(.20) + (-1)(.25)$
$$+ (0)(.10) + (1)(.25) + (2)(.10)$$
$$= -0.8$$

2. Let X denote the number of townhouses that will be sold within one month of being put on the market. Then, the number of townhouses the developer expects to sell within one month is given by the expected value of X—that is, by

$$E(X) = 20(.05) + 25(.10) + 30(.30) + 35(.25)$$
$$+ 40(.15) + 45(.10) + 50(.05)$$
$$= 34.25$$

or 34 townhouses.

8.3 VARIANCE AND STANDARD DEVIATION

Figure 8.9
The histograms of two probability distributions.

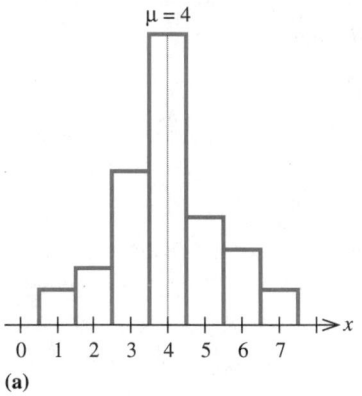

(a)

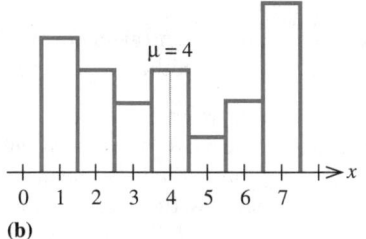

(b)

Variance

The mean, or expected value, of a random variable enables us to express an important property of the probability distribution associated with the random variable in terms of a single number. But the knowledge of the location, or central tendency, of a probability distribution alone is usually not enough to give a reasonably accurate picture of the probability distribution. Consider, for example, the two probability distributions whose histograms appear in Figure 8.9. Both distributions have the same expected value, or mean, $\mu = 4$ (the Greek letter μ is read "mu"). Note that the probability distribution with the histogram shown in Figure 8.9a is closely concentrated about its mean μ, whereas the one with the histogram shown in Figure 8.9b is widely dispersed or spread about its mean.

As another example, suppose Mr. David Horowitz, host of the popular television show "The Consumer Advocate," decides to demonstrate the accuracy of the weights of two popular brands of potato chips. Ten packages of potato chips of each brand are selected at random and weighed carefully. The results are as follows:

Weight in ounces

Brand A	16.1	16	15.8	16	15.9	16.1	15.9	16	16	16.2
Brand B	16.3	15.7	15.8	16.2	15.9	16.1	15.7	16.2	16	16.1

In Example 3 we will verify that the mean weights for each of the two brands is 16 ounces. However, a cursory examination of the data now shows that the weights of the brand B packages exhibit much greater dispersion about the mean than those of brand A.

One measure of the degree of dispersion, or spread, of a probability distribution about its mean is given by the variance of the random variable associated with the probability distribution. A probability distribution with a small spread about its mean will have a small variance, whereas one with a larger spread will have a larger variance. Thus, the variance of the random variable associated with the probability distribution whose histogram appears in Figure 8.9a is smaller than the variance of the random variable associated with the probability distribution whose histogram is shown in Figure 8.9b (see Example 1). Also, as we will see in Example 3, the variance of the random variable associated with the weights of the brand A potato chips is smaller than that of the random variable associated with the weights of the brand B potato chips. (This observation was made earlier.)

We now define the variance of a random variable.

VARIANCE OF A RANDOM VARIABLE X	Suppose a random variable has the probability distribution

x	x_1	x_2	x_3	$\cdots$	x_n
$P(X = x)$	p_1	p_2	p_3	$\cdots$	p_n

and expected value

$$E(X) = \mu$$

Then the **variance** of the random variable X is

$$\mathrm{Var}(X) = p_1(x_1 - \mu)^2 + p_2(x_2 - \mu)^2 + \cdots + p_n(x_n - \mu)^2 \qquad \textbf{(5)}$$

Let's look a little closer at equation (5). First, note that the numbers

$$x_1 - \mu, x_2 - \mu, \ldots, x_n - \mu \qquad \textbf{(6)}$$

measure the **deviations** of $x_1, x_2, \ldots, x_n$ from μ, respectively. Thus the numbers

$$(x_1 - \mu)^2, (x_2 - \mu)^2, \ldots, (x_n - \mu)^2 \qquad \textbf{(7)}$$

measure the squares of the deviations of $x_1, x_2, \ldots, x_n$ from μ, respectively. Next, by multiplying each of the numbers in (7) by the probability associated with each value of the random variable X, the numbers are weighted accordingly so that their sum is a measure of the variance of X about its mean. An attempt to define the variance of a random variable about its mean in a similar manner using the deviations in (6) rather than their squares would not be fruitful since some of the deviations may be positive whereas others may be negative and (because of cancellations) the sum will not give a satisfactory measure of the variance of the random variable.

EXAMPLE 1   Find the variance of the random variable X and of the random variable Y whose probability distributions are shown in the following table. These are the probability distributions associated with the histograms shown in Figures 8.9a and 8.9b.

x	$P(X = x)$	y	$P(Y = y)$
1	.05	1	.2
2	.075	2	.15
3	.2	3	.1
4	.375	4	.15
5	.15	5	.05
6	.1	6	.1
7	.05	7	.25

Solution The mean of the random variable X is given by

$$\mu_X = (1)(.05) + (2)(.075) + (3)(.2) + (4)(.375) + (5)(.15)$$
$$+ (6)(.1) + (7)(.05)$$
$$= 4$$

Therefore, using equation (5) and the data from the probability distribution of X, we find that the variance of X is given by

$$\text{Var}(X) = (.05)(1 - 4)^2 + (.075)(2 - 4)^2 + (.2)(3 - 4)^2$$
$$+ (.375)(4 - 4)^2 + (.15)(5 - 4)^2$$
$$+ (.1)(6 - 4)^2 + (.05)(7 - 4)^2$$
$$= 1.95$$

Next, we find that the mean of the random variable Y is given by

$$\mu_Y = (1)(.2) + (2)(.15) + (3)(.1) + (4)(.15) + (5)(.05)$$
$$+ (6)(.1) + (7)(.25)$$
$$= 4$$

and so the variance of Y is given by

$$\text{Var}(Y) = (.2)(1 - 4)^2 + (.15)(2 - 4)^2 + (.1)(3 - 4)^2$$
$$+ (.15)(4 - 4)^2 + (.05)(5 - 4)^2$$
$$+ (.1)(6 - 4)^2 + (.25)(7 - 4)^2$$
$$= 5.2$$

Note that $\text{Var}(X)$ is smaller than $\text{Var}(Y)$, which confirms the earlier observations about the spread, or dispersion, of the probability distribution of X and Y, respectively.   ◐ ◐ ◐

Standard Deviation

Because equation (5), which gives the variance of the random variable X, involves the squares of the deviations, the unit of measurement of $\text{Var}(X)$ is the square of the unit of measurement of the values of X. For example, if the values assumed by the random variable X are measured in units of a gram, then $\text{Var}(X)$ will be measured in units involving the *square* of a gram. To remedy this situation, one normally works with the square root of $\text{Var}(X)$ rather than $\text{Var}(X)$ itself. The former is called the *standard deviation of X*.

STANDARD DEVIATION OF A RANDOM VARIABLE X

The **standard deviation** of a random variable X, σ (pronounced "sigma"), is defined by

$$\sigma = \sqrt{\text{Var}(X)}$$
$$= \sqrt{p_1(x_1 - \mu)^2 + p_2(x_2 - \mu)^2 + \cdots + p_n(x_n - \mu)^2} \qquad (8)$$

where $x_1, x_2, \ldots, x_n$ denote the values assumed by the random variable X and $p_1 = P(X = x_1)$, $p_2 = P(X = x_2)$, $\ldots$, $p_n = P(X = x_n)$.

EXAMPLE 2 Find the standard deviations of the random variables X and Y of Example 1.

Solution From the results of Example 1, we have $\text{Var}(X) = 1.95$ and $\text{Var}(Y) = 5.2$. Taking their respective square roots, we have

$$\sigma_X = \sqrt{1.95}$$
$$\approx 1.40$$

and

$$\sigma_Y = \sqrt{5.2}$$
$$\approx 2.28$$

 EXAMPLE 3 Let X and Y denote the random variables whose values are the weights of the brand A and brand B potato chips, respectively (see page 488). Compute the means and standard deviations of X and Y and interpret your results.

Solution The probability distributions of X and Y may be computed from the given data as follows:

Brand A

x	Relative Frequency of Occurrence	$P(X = x)$
15.8	1	.1
15.9	2	.2
16.0	4	.4
16.1	2	.2
16.2	1	.1

Brand B

y	Relative Frequency of Occurrence	$P(Y = y)$
15.7	2	.2
15.8	1	.1
15.9	1	.1
16.0	1	.1
16.1	2	.2
16.2	2	.2
16.3	1	.1

The means of X and Y are given by

$$\mu_X = (.1)(15.8) + (.2)(15.9) + (.4)(16.0) + (.2)(16.1)$$
$$+ (.1)(16.2)$$
$$= 16$$
$$\mu_Y = (.2)(15.7) + (.1)(15.8) + (.1)(15.9) + (.1)(16.0)$$
$$+ (.2)(16.1) + (.2)(16.2) + (.1)(16.3)$$
$$= 16$$

Therefore,

$$\text{Var}(X) = (.1)(15.8 - 16)^2 + (.2)(15.9 - 16)^2 + (.4)(16 - 16)^2$$
$$+ (.2)(16.1 - 16)^2 + (.1)(16.2 - 16)^2$$
$$= 0.012$$

and

$$\begin{aligned}
\text{Var}(Y) &= (.2)(15.7 - 16)^2 + (.1)(15.8 - 16)^2 + (.1)(15.9 - 16)^2 \\
&\quad + (.1)(16 - 16)^2 + (.2)(16.1 - 16)^2 + (.2)(16.2 - 16)^2 \\
&\quad + (.1)(16.3 - 16)^2 \\
&= 0.042
\end{aligned}$$

so that the required standard deviations are

$$\begin{aligned}
\sigma_X &= \sqrt{\text{Var}(X)} \\
&= \sqrt{0.012} \\
&\approx 0.11
\end{aligned}$$

and
$$\begin{aligned}
\sigma_Y &= \sqrt{\text{Var}(Y)} \\
&= \sqrt{0.042} \\
&\approx 0.20
\end{aligned}$$

The mean of X and that of Y are both equal to 16. Therefore, the average weight of a package of potato chips of either brand is 16 ounces. However, the standard deviation of Y is greater than that of X. This tells us that the weights of the packages of brand B potato chips are more widely dispersed about the common mean of 16 than are those of brand A.   ○ ○ ○

A useful alternative formula for the variance is

$$\sigma^2 = E(X^2) - \mu^2$$

where $E(X)$ is the expected value of X.

a. Establish the validity of the formula.

b. Use the formula to verify the calculations in Example 3.

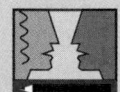

Chebychev's Inequality

The standard deviation of a random variable X may be used in statistical estimations. For example, the following result, derived by the Russian mathematician P. L. Chebychev (1821–1894), gives the proportion of the values of X lying within k standard deviations of the expected value of X.

CHEBYCHEV'S INEQUALITY	Let X be a random variable with expected value μ and standard deviation σ. Then the probability that a randomly chosen outcome of the experiment lies between $\mu - k\sigma$ and $\mu + k\sigma$ is at least $1 - (1/k^2)$, that is,

$$P(\mu - k\sigma \leq X \leq \mu + k\sigma) \geq 1 - \frac{1}{k^2} \tag{9}$$

In order to shed some light on this result, let us take $k = 2$ in inequality (9) and compute

$$P(\mu - 2\sigma \le X \le \mu + 2\sigma) = 1 - \frac{1}{2^2} = 1 - \frac{1}{4} = .75$$

This tells us that at least 75% of the outcomes of the experiment lie within 2 standard deviations of the mean (Figure 8.10). Taking $k = 3$ in formula (9), we have

$$P(\mu - 3\sigma \le X \le \mu + 3\sigma) = 1 - \frac{1}{3^2} = 1 - \frac{1}{9} = \frac{8}{9} \approx .89$$

This tells us that at least 89% of the outcomes of the experiment lie within 3 standard deviations of the mean (Figure 8.11).

Figure 8.10
At least 75% of the outcomes fall within this interval.

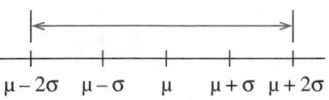

Figure 8.11
At least 89% of the outcomes fall within this interval.

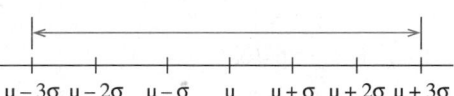

EXAMPLE 4   A probability distribution has a mean of 10 and a standard deviation of 1.5. Use Chebychev's inequality to estimate the probability that an outcome of the experiment lies between 7 and 13.

Solution   Here $\mu = 10$ and $\sigma = 1.5$. Next, to determine the value of k, note that $\mu - k\sigma = 7$ and $\mu + k\sigma = 13$. Substituting the appropriate values for μ and σ, we find $k = 2$. Using Chebychev's inequality (9), we see that the probability that an outcome of the experiment lies between 7 and 13 is given by

$$P(7 \le X \le 13) \ge 1 - \left(\frac{1}{2^2}\right)$$

$$= \frac{3}{4}$$

that is, at least 75%.   ◦ ◦ ◦

REMARK   The results of Example 4 tell us that at least 75% of the outcomes of the experiment lie between $10 - 2\sigma$ and $10 + 2\sigma$ —that is, between 7 and 13.

◦ ◦ ◦

EXAMPLE 5   The Great Northwest Lumber Company employs 400 workers in its mills. It has been estimated that X, the random variable measuring the number of mill workers who have industrial accidents during a 1-year period, is distributed with a mean of 40 and a standard deviation of 6. Using Chebychev's inequality (9), estimate the probability that the number of workers who will have an industrial accident over a 1-year period is between 30 and 50, inclusive.

Solution Here $\mu = 40$ and $\sigma = 6$. We wish to estimate $P(30 \leq X \leq 50)$. In order to use Chebychev's inequality (9), we first determine the value of k from the equation

$$\mu - k\sigma = 30 \qquad \text{or} \qquad \mu + k\sigma = 50$$

Since $\mu = 40$ and $\sigma = 6$ in this case, we see that k satisfies

$$40 - 6k = 30 \qquad \text{and} \qquad 40 + 6k = 50$$

from which we deduce that $k = 5/3$. Thus, the probability that the number of mill workers who will have an industrial accident during a 1-year period is between 30 and 50 is given by

$$P(30 \leq X \leq 50) \geq 1 - \frac{1}{(5/3)^2}$$
$$= \frac{16}{25}$$

that is, at least 64%.

SELF-CHECK EXERCISES 8.3

1. Compute the mean, variance, and standard deviation of the random variable X with probability distribution as follows:

x	−4	−3	−1	0	2	5
$P(X = x)$	.1	.1	.2	.3	.1	.2

2. James recorded the following travel times (the length of time in minutes it took him to drive to work) on ten consecutive days:

55 50 52 48 50 52 46 48 50 51

Calculate the mean and standard deviation of the random variable X associated with these data.

Solutions to Self-Check Exercises 8.3 can be found on page 497.

8.3 EXERCISES

A calculator is recommended for this exercise set.

In exercises 1–6, the probability distribution of a random variable X is given. Compute the mean, variance, and standard deviation of X.

1.

x	1	2	3	4
$P(X = x)$	.4	.3	.2	.1

2.

x	−4	−2	0	2	4
$P(X = x)$	.1	.2	.3	.1	.3

3.

x	−2	−1	0	1	2
$P(X = x)$	1/16	4/16	6/16	4/16	1/16

4.

x	10	11	12	13	14	15
$P(X = x)$	1/8	2/8	1/8	2/8	1/8	1/8

5.

x	430	480	520	565	580
$P(X = x)$	.1	.2	.4	.2	.1

6.

x	-198	-195	-193	-188	-185
$P(X = x)$	.15	.30	.10	.25	.20

7. The following histograms represent the probability distributions of the random variables X and Y. Determine by inspection which probability distribution has the larger variance.

a.

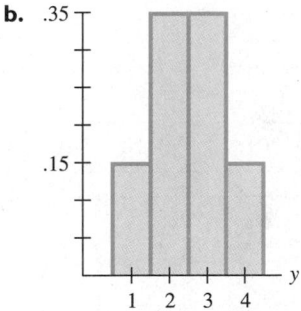

b.

8. The following histograms represent the probability distributions of the random variables X and Y.

a.   **b.**

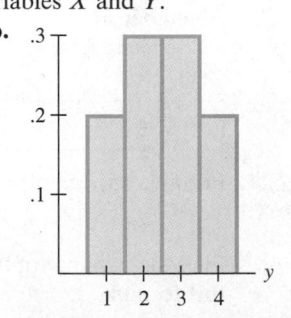

Determine by inspection which probability distribution has the larger variance.

In exercises 9 and 10, find the variance of the probability distribution for the given histogram.

9.

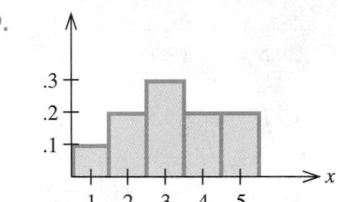

10.

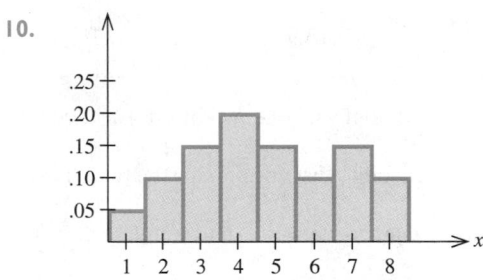

11. An experiment consists of casting an 8-sided die (numbered 1 through 8) and observing the number that appears uppermost. Find the mean and variance of this experiment.

12. Driving Age Requirements   The minimum age requirement for a regular driver's license differs from state to state. The frequency distribution for this age requirement in the fifty states is given in the following table.

Minimum Age	15	16	17	18	19	21
Frequency of Occurrence	1	15	4	28	1	1

a. Describe a random variable X that is associated with these data.
b. Find the probability distribution for the random variable X.
c. Compute the mean, variance, and standard deviation of X.

13. Birth Rates   The birth rates in the United States for the years 1981–1990 are given in the following table. (The birth rate is the number of live births per 1000 population.)

Year	1981	1982	1983	1984
Birth Rate (number/1000)	15.9	15.5	15.5	15.7

Year	1985	1986	1987
Birth Rate (number/1000)	15.7	15.6	15.7

Year	1988	1989	1990
Birth Rate (number/1000)	15.9	16.2	16.7

a. Describe a random variable X that is associated with these data.
b. Find the probability distribution for the random variable X.
c. Compute the mean, variance, and standard deviation of X.

14. **Investment Analysis** Mr. Hunt is considering two business ventures. The anticipated returns (in thousands of dollars) of each venture are described by the following probability distributions.

Venture A

Earnings	Probability
−20	.3
40	.4
50	.3

Venture B

Earnings	Probability
−15	.2
30	.5
40	.3

a. Compute the mean and variance for each venture.
b. Which investment would provide Mr. Hunt with the higher expected return (the greater mean)?
c. In which investment would the element of risk be less (that is, which probability distribution has the smaller variance)?

15. **Investment Analysis** Ms. Walters is considering investing $10,000 in two mutual funds. The anticipated returns from price appreciation and dividends (in hun-

dreds of dollars) are described by the following probability distributions.

Mutual Fund A

Returns	Probability
−4	.2
8	.5
10	.3

Mutual Fund B

Returns	Probability
−2	.2
6	.4
8	.4

a. Compute the mean and variance associated with the returns for each mutual fund.
b. Which investment would provide Ms. Walters with the higher expected return (the greater mean)?
c. In which investment would the element of risk be less (that is, which probability distribution has the smaller variance)?

16. The distribution of the number of chocolate chips in a cookie is shown in the following table. Find the mean and the variance of the number of chocolate chips in a cookie.

Number of Chocolate Chips (x)	0	1	2
$P(X = x)$	.01	.03	.05

Number of Chocolate Chips (x)	3	4	5
$P(X = x)$	.11	.13	.24

Number of Chocolate Chips (x)	6	7	8
$P(X = x)$	.22	.16	.05

17. Formula (5) can also be expressed in the form

$$\text{Var}(X) = (p_1 x_1^2 + p_2 x_2^2 + \cdots + p_n x_n^2) - \mu^2$$

Find the variance of the distribution of exercise 1 using this formula.

18. Find the variance of the distribution of exercise 16 using the formula

$$\text{Var}(X) = (p_1x_1^2 + p_2x_2^2 + \cdots + p_nx_n^2) - \mu^2$$

19. **Housing Prices**   A survey was conducted by the market research department of the National Real Estate Company among 500 prospective buyers in a large metropolitan area to determine the maximum price a prospective buyer would be willing to pay for a house. From the data collected, the distribution that follows was obtained. Compute the mean, variance, and standard deviation of the maximum price that these buyers were willing to pay for a house.

Maximum Price Considered, x (in thousands of dollars)	$P(X = x)$
80	$\frac{10}{500}$
90	$\frac{20}{500}$
100	$\frac{75}{500}$
110	$\frac{85}{500}$
120	$\frac{70}{500}$
150	$\frac{90}{500}$
180	$\frac{90}{500}$
200	$\frac{55}{500}$
250	$\frac{5}{500}$

20. A probability distribution has a mean of 20 and a standard deviation of 3. Use Chebychev's inequality to estimate the probability that an outcome of the experiment lies between
    a. 15 and 25   **b.** 10 and 30

21. A probability distribution has a mean of 42 and a standard deviation of 2. Use Chebychev's inequality to estimate the probability that an outcome of the experiment lies between
    a. 38 and 46   **b.** 32 and 52

22. A probability distribution has a mean of 50 and a standard deviation of 1.4. Use Chebychev's inequality to find the value of c that guarantees that the probability is at least 96% that an outcome of the experiment lies between $50 - c$ and $50 + c$.

23. Suppose X is a random variable with mean μ and standard deviation σ. If a large number of trials is observed, at least what percentage of these values is expected to lie between $\mu - 2\sigma$ and $\mu + 2\sigma$?

24. **Product Reliability**   A Christmas tree light has an expected life of 200 hours and a standard deviation of 2 hours.
    a. Estimate the probability that one of these Christmas tree lights will last between 190 and 210 hours.
    b. Suppose 150,000 of these Christmas tree lights are used by a large city as part of its Christmas decorations. Estimate the number of lights that will require replacement between 180 and 220 hours of use.

25. **Product Reliability**   The expected lifetime of the deluxe model hair dryer produced by the Roland Electric Company has a mean life of 24 months and a standard deviation of 3 months. Find the probability that one of these hair dryers will last between 20 and 28 months.

26. **Starting Salaries**   The mean annual starting salary of a new graduate in a certain profession is $32,000 with a standard deviation of $500. What is the probability that the starting salary of a new graduate in this profession will be between $30,000 and $34,000?

27. **Quality Control**   Sugar packaged by a certain machine has a mean weight of 5 pounds and a standard deviation of 0.02 pound. For what values of c can the manufacturer of the machinery claim that the sugar packaged by this machine has a weight between $5 - c$ and $5 + c$ pounds with probability at least 96%?

SOLUTIONS TO SELF-CHECK EXERCISES 8.3

1. The mean of the random variable X is

$$\mu = (-4)(.1) + (-3)(.1) + (-1)(.2)$$
$$+ (0)(.3) + (2)(.1) + (5)(.2)$$
$$= 0.3$$

USING TECHNOLOGY

FINDING THE MEAN AND STANDARD DEVIATION

The calculation of the mean and standard deviation of a random variable is facilitated by the use of a graphing utility.

EXAMPLE 1 A survey conducted in 1995 of the Fortune 1000 companies revealed the following age distribution of the company directors.

Age	20–25	25–30	30–35	35–40	40–45	45–50	50–55	55–60
Number of Directors	1	6	28	104	277	607	1142	1413

Age	60–65	65–70	70–75	75–80	80–85	85–90
Number of Directors	1424	494	159	62	31	5

Source: Directorship

a. Plot a histogram for the given data.

b. Find the mean age and the standard deviation of the company directors.

Solution

a. Let X denote the random variable taking on the values 1 through 14, where 1 corresponds to the age bracket 20–25, 2 corresponds to the age bracket 25–30, and so on. Entering the values of X as $x_1 = 1$, $x_2 = 2, \ldots, x_{14} = 14$, and the corresponding values of Y as $y_1 = 1$, $y_2 = 6, \ldots, y_{14} = 5$, and then using the Draw function from the Statistics menu of a graphing utility, we obtain the histogram shown in Figure T1.

Figure T1
The histogram for the given data using the viewing rectangle $[0, 16] \times [0, 1500]$.

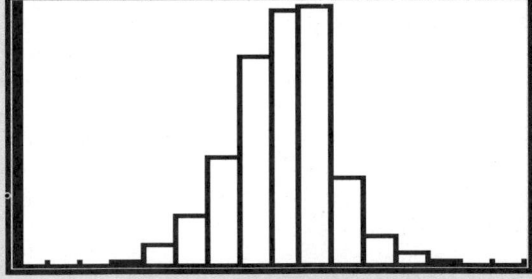

b. Using the appropriate function from the Statistics menu, we find that $\bar{x} = 8.4376$ and $\sigma x = 1.8463$; that is, the mean of X is $\mu \approx 8.4$ and the standard deviation is $\sigma \approx 1.8$. Thus, the average age of the directors is approximately 60 years old.

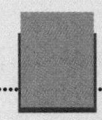

EXERCISES

1. **a.** Graph the histogram associated with the random variable X in Example 1, page 489.
   b. Find the mean and the standard deviation for these data.

2. **a.** Graph the histogram associated with the random variable Y in Example 1, page 489.
   b. Find the mean and the standard deviation for these data.

3. **a.** Graph the histogram associated with the data given in exercise 12, page 495.
   b. Find the mean and the standard deviation for these data.

4. **a.** Graph the histogram associated with the data given in exercise 16, page 496.
   b. Find the mean and the standard deviation for these data.

5. A sugar refiner uses a machine to pack sugar in 5-lb cartons. In order to check the machine's accuracy, cartons are selected at random and weighed. The results follow:

$$
\begin{array}{ccccc}
4.98 & 5.02 & 4.96 & 4.97 & 5.03 \\
4.96 & 4.98 & 5.01 & 5.02 & 5.06 \\
4.97 & 5.04 & 5.04 & 5.01 & 4.99 \\
4.98 & 5.04 & 5.01 & 5.03 & 5.05 \\
4.96 & 4.97 & 5.02 & 5.04 & 4.97 \\
5.03 & 5.01 & 5.00 & 5.01 & 4.98
\end{array}
$$

a. Describe a random variable X that is associated with these data.
b. Find the probability distribution for the random variable X.
c. Compute the mean and standard deviation of X.

6. The scores of 25 students in a mathematics examination follow:

$$
\begin{array}{ccccccc}
90 & 85 & 74 & 92 & 68 & 94 & 66 \\
87 & 85 & 70 & 72 & 68 & 73 & 72 \\
69 & 66 & 58 & 70 & 74 & 88 & 90 \\
98 & 71 & 75 & 68
\end{array}
$$

a. Describe a random variable X that is associated with these data.
b. Find the probability distribution for the random variable X.
c. Compute the mean and standard deviation of X.

The variance of X is

$$\text{Var}(X) = (.1)(-4 - .3)^2 + (.1)(-3 - .3)^2$$
$$+ (.2)(-1 - .3)^2 + (.3)(0 - .3)^2$$
$$+ (.1)(2 - .3)^2 + (.2)(5 - .3)^2$$
$$= 8.01$$

The standard deviation of X is

$$\sigma = \sqrt{\text{Var}(X)} = \sqrt{8.01} \approx 2.83$$

2. We first compute the probability distribution of X from the given data as follows:

x	Relative Frequency of Occurrence	$P(X = x)$
46	1	.1
48	2	.2
50	3	.3
51	1	.1
52	2	.2
55	1	.1

The mean of X is

$$\mu = (.1)(46) + (.2)(48) + (.3)(50)$$
$$+ (.1)(51) + (.2)(52) + (.1)(55)$$
$$= 50.2$$

The variance of X is

$$\text{Var}(X) = (.1)(46 - 50.2)^2 + (.2)(48 - 50.2)^2$$
$$+ (.3)(50 - 50.2)^2 + (.1)(51 - 50.2)^2$$
$$+ (.2)(52 - 50.2)^2 + (.1)(55 - 50.2)^2$$
$$= 5.76$$

from which we deduce the standard deviation

$$\sigma = \sqrt{5.76}$$
$$= 2.4$$

8.4 THE BINOMIAL DISTRIBUTION

Bernoulli Trials

An important class of experiments have (or may be viewed as having) two outcomes. For example, in a coin-tossing experiment, the two outcomes are *heads* and *tails*. In the card game played by Mike and Bill (Example 7, Section 8.2), one may view the selection of a diamond as a *win* (for Mike) and the selection of a card of another suit as a *loss* for Mike. For a third example, consider the experiment in which a person is inoculated with a flu vaccine. Here, the vaccine may be classified as being "effective" or "ineffective" with respect to that particular person.

In general, experiments with two outcomes are called **Bernoulli trials,** or **binomial trials.** It is standard practice to label one of the outcomes of a binomial trial a *success* and the other a *failure*. For example, in a coin-tossing experiment, the outcome *a head* may be called a success, in which case the outcome *a tail* is called a failure. Note that by using the terms *success* and *failure* in this way, we depart from their usual connotations.

A sequence of Bernoulli (binomial) trials is called a *binomial experiment*. More precisely, we have the following definition:

BINOMIAL EXPERIMENT	A **binomial experiment** has the following properties:
	1. The number of trials in the experiment is fixed.
	2. There are two outcomes of the experiment: "success" and "failure."
	3. The probability of success in each trial is the same.
	4. The trials are independent of each other.

In a binomial experiment, it is customary to denote the probability of a success by the letter p and the probability of a failure by the letter q. Because the event of a success and the event of a failure are complementary events, we have the relationship

$$p + q = 1$$

or, equivalently,

$$q = 1 - p$$

The properties of a binomial experiment are illustrated in the following example.

 EXAMPLE 1   A fair die is cast four times. Compute the probability of obtaining exactly one 6 in the four throws.

Solution   There are four trials in this experiment. Each trial consists of casting the die once and observing the face that lands uppermost. We may view each trial as an experiment with two outcomes: a success (S) if the face that lands uppermost is a 6 and a failure (F) if it is any of the other five numbers. Letting p and q denote the probability of success and failure, respectively, of a single trial of the experiment, we find that

$$p = \frac{1}{6} \quad \text{and} \quad q = 1 - \frac{1}{6} = \frac{5}{6}$$

Furthermore, we may assume that the trials of this experiment are independent. Thus, we have a binomial experiment.

Table 8.10

0 Success	1 Success	2 Successes	3 Successes	4 Successes
FFFF	SFFF	SSFF	SSSF	SSSS
	FSFF	SFSF	SSFS	
	FFSF	SFFS	SFSS	
	FFFS	FSSF	FSSS	
		FSFS		
		FFSS		

With the aid of the multiplication principle, we see that the experiment has 2^4, or 16, outcomes. We can obtain these outcomes by constructing the tree diagram associated with the experiment (see Table 8.10, where the outcomes are listed according to the number of successes). From the table, we see that the event of obtaining exactly one success in four trials is given by

$$E = \{\text{SFFF, FSFF, FFSF, FFFS}\}$$

with probability given by

$$P(E) = P(\text{SFFF}) + P(\text{FSFF}) + P(\text{FFSF}) + P(\text{FFFS}) \qquad \textbf{(10)}$$

Since the trials (throws) are independent, the terms on the right-hand side of equation (10) may be computed as follows:

$$P(\text{SFFF}) = P(\text{S})P(\text{F})P(\text{F})P(\text{F}) = p \cdot q \cdot q \cdot q = pq^3$$
$$P(\text{FSFF}) = P(\text{F})P(\text{S})P(\text{F})P(\text{F}) = q \cdot p \cdot q \cdot q = pq^3$$
$$P(\text{FFSF}) = P(\text{F})P(\text{F})P(\text{S})P(\text{F}) = q \cdot q \cdot p \cdot q = pq^3$$
$$P(\text{FFFS}) = P(\text{F})P(\text{F})P(\text{F})P(\text{S}) = q \cdot q \cdot q \cdot p = pq^3$$

Therefore, upon substituting these values in (10), we obtain

$$P(E) = pq^3 + pq^3 + pq^3 + pq^3 = 4pq^3$$
$$= 4\left(\frac{1}{6}\right)\left(\frac{5}{6}\right)^3 = .386$$

● ● ●

Probabilities in Bernoulli Trials

Let us reexamine the computations performed in the last example. There it was found that the probability of obtaining exactly one success in a binomial experiment with four independent trials with probability of success in a single trial p is given by

$$P(E) = 4pq^3 \qquad \text{where } q = 1 - p \qquad \textbf{(11)}$$

Observe that the coefficient 4 of pq^3 appearing in equation (11) is precisely the number of outcomes of the experiment with exactly one success and three failures, the outcomes being

$$\text{SFFF,   FSFF,   FFSF,   and   FFFS}$$

Another way of obtaining this coefficient is to think of the outcomes as arrangements of the letters S and F. Then, the number of ways of selecting one position for S from four possibilities is given by

$$C(4, 1) = \frac{4!}{1!(4 - 1)!}$$
$$= 4$$

Next, observe that, because the trials are independent, each of the four outcomes of the experiment has the same probability, given by

$$pq^3$$

where the exponents 1 and 3 of p and q, respectively, correspond to exactly one success and three failures in the trials that make up each outcome.

As a result of the foregoing discussion, we may write (11) as

$$P(E) = C(4, 1)pq^3 \tag{12}$$

We are also in a position to generalize this result. Suppose that in a binomial experiment the probability of success in any trial is p. What is the probability of obtaining exactly x successes in n independent trials? We start by counting the number of outcomes of the experiment, each of which has exactly x successes. Now, one such outcome involves x successive successes followed by $(n - x)$ failures—that is,

$$\underbrace{SS \cdots S}_{x} \underbrace{FF \cdots F}_{n - x} \tag{13}$$

The other outcomes, each of which has exactly x successes, are obtained by rearranging the S's (x of them) and F's ($n - x$ of them). But there are $C(n, x)$ ways of arranging these letters. Next, arguing as in Example 1, we see that each such outcome has probability given by

$$p^x q^{n-x}$$

For example, for the outcome (13), we find

$$\underbrace{P(SS \cdots S}_{x} \underbrace{FF \cdots F)}_{(n - x)} = \underbrace{P(S)P(S) \cdots P(S)}_{x} \underbrace{P(F)P(F) \cdots P(F)}_{(n - x)}$$

$$= \underbrace{pp \cdots p}_{x} \underbrace{qq \cdots q}_{n - x}$$

$$= p^x q^{n-x}$$

Let us state this important result formally:

COMPUTATION OF PROBABILITIES IN BERNOULLI TRIALS

In a binomial experiment in which the probability of success in any trial is p, the probability of exactly x successes in n independent trials is given by

$$C(n, x)p^x q^{n-x}$$

If we let X be the random variable that gives the number of successes in a binomial experiment, then the probability of exactly x successes in n independent trials may be written

$$P(X = x) = C(n, x)p^x q^{n-x} \qquad (x = 0, 1, 2, \ldots, n) \qquad \textbf{(14)}$$

The random variable X is called a **binomial random variable,** and the probability distribution of X is called a **binomial distribution.**

EXAMPLE 2   A fair die is cast five times. If a 1 or a 6 lands uppermost in a trial, then the throw is considered a success. Otherwise, the throw is considered a failure.

a. Find the probability of obtaining exactly 0, 1, 2, 3, 4, and 5 successes, respectively, in this experiment.

b. Using the results obtained in the solution to (a), construct the binomial distribution for this experiment and draw the histogram associated with it.

Solution

a. This is a binomial experiment with X, the binomial random variable, taking on each of the values 0, 1, 2, 3, 4, and 5 corresponding to exactly 0, 1, 2, 3, 4, and 5 successes, respectively, in five trials. Since the die is fair, the probability of a 1 or a 6 landing uppermost in any trial is given by $p = 2/6 = 1/3$, from which it also follows that $q = 1 - p = 2/3$. Finally, $n = 5$ since there are five trials (throws of the die) in this experiment. Using equation (14), we find that the required probabilities are

$$P(X = 0) = C(5, 0)\left(\frac{1}{3}\right)^0\left(\frac{2}{3}\right)^5 = \frac{5!}{0!5!} \cdot 1 \cdot \frac{32}{243} \approx .132$$

$$P(X = 1) = C(5, 1)\left(\frac{1}{3}\right)^1\left(\frac{2}{3}\right)^4 = \frac{5!}{1!4!} \cdot \frac{16}{243} \approx .329$$

$$P(X = 2) = C(5, 2)\left(\frac{1}{3}\right)^2\left(\frac{2}{3}\right)^3 = \frac{5!}{2!3!} \cdot \frac{8}{243} \approx .329$$

$$P(X = 3) = C(5, 3)\left(\frac{1}{3}\right)^3\left(\frac{2}{3}\right)^2 = \frac{5!}{3!2!} \cdot \frac{4}{243} \approx .165$$

$$P(X = 4) = C(5, 4)\left(\frac{1}{3}\right)^4\left(\frac{2}{3}\right)^1 = \frac{5!}{4!1!} \cdot \frac{2}{243} \approx .041$$

$$P(X = 5) = C(5, 5)\left(\frac{1}{3}\right)^5\left(\frac{2}{3}\right)^0 = \frac{5!}{5!0!} \cdot \frac{1}{243} \approx .004$$

b. Using these results, we find the required binomial distribution associated with this experiment given in Table 8.11. Next, we use this table to construct the histogram associated with the probability distribution (Figure 8.12).

Table 8.11
Probability distribution.

x	$P(X = x)$
0	.132
1	.329
2	.329
3	.165
4	.041
5	.004

Figure 8.12
The probability of the number of successes in five throws.

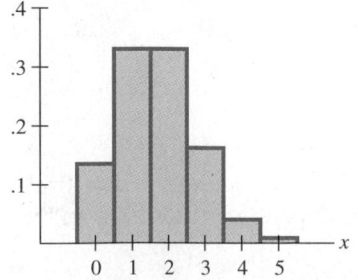

○ ○ ○

EXAMPLE 3   A fair die is cast five times. If a 1 or a 6 lands uppermost in a trial, then the throw is considered a success. Use the results from Example 2 to answer the following questions:

a. What is the probability of obtaining 0 or 1 success in the experiment?

b. What is the probability of obtaining at least 1 success in the experiment?

Solution   Interpreting the probability associated with the random variable X, when X assumes the value $X = a$, as the area of the rectangle centered about $X = a$ (Figure 8.12), or otherwise, we find that

a. the probability of obtaining 0 or 1 success in the experiment is given by

$$P(X = 0) + P(X = 1) = .132 + .329 = .461$$

and

b. the probability of obtaining at least 1 success in the experiment is given by

$$P(X = 1) + P(X = 2) + P(X = 3) + P(X = 4) + P(X = 5)$$
$$= .329 + .329 + .165 + .041 + .004$$
$$= .868$$

○ ○ ○

Consider the equation

$$P(X = x) = C(n, x)p^x q^{n-x}$$

for the binomial distribution.
a. Construct the histogram with $n = 5$ and $p = .2$; the histogram with $n = 5$ and $p = .5$; and the histogram with $n = 5$ and $p = .8$.
b. Comment on the shape of the histograms and give an interpretation.

The following formulas (which we state without proof) will be useful in solving problems involving binomial experiments.

MEAN, VARIANCE, AND STANDARD DEVIATION OF A RANDOM VARIABLE X

If X is a binomial random variable associated with a binomial experiment consisting of n trials with probability of success p and probability of failure q, then the **mean** (expected value), **variance,** and **standard deviation** of X are:

$$\mu = E(X) = np \tag{15a}$$
$$\mathrm{Var}(X) = npq \tag{15b}$$
$$\sigma_X = \sqrt{npq} \tag{15c}$$

EXAMPLE 4   For the experiment in Examples 2 and 3, compute the mean, the variance, and the standard deviation of X, (a) using formulas (15a), (15b), and (15c) and (b) using the definition of each term (Sections 8.2 and 8.3).

Solution

a. We use (15a), (15b), and (15c), with $p = 1/3$, $q = 2/3$, and $n = 5$, obtaining

$$\mu = E(X) = (5)\left(\frac{1}{3}\right) = \frac{5}{3}$$
$$\approx 1.67$$
$$\text{Var}(X) = (5)\left(\frac{1}{3}\right)\left(\frac{2}{3}\right) = \frac{10}{9}$$
$$\approx 1.11$$

and
$$\sigma_X = \sqrt{\text{Var}(X)} = \sqrt{1.11}$$
$$\approx 1.05$$

We leave it to you to interpret the results.

b. Using the definition of expected value (Section 8.2) and the values of the probability distribution shown in Table 8.11, we find

$$\mu = E(X) = (0)(.132) + (1)(.329) + (2)(.329)$$
$$+ (3)(.165) + (4)(.041) + (5)(.004)$$
$$\approx 1.67$$

which agrees with the result obtained in part (a). Next, using the definition of variance and the fact that $\mu = 1.67$, we find

$$\text{Var}(X) = (.132)(-1.67)^2 + (.329)(-0.67)^2 + (.329)(0.33)^2$$
$$+ (.165)(1.33)^2 + (.041)(2.33)^2 + (.004)(3.33)^2$$
$$\approx 1.11$$

and
$$\sigma_X \approx \sqrt{1.11}$$
$$\approx 1.05$$

which again agrees with the earlier results.   ○ ○ ○

Applications

We close this section by looking at several examples involving binomial experiments. In working through these examples, you may use a calculator, or you may consult Table II, Appendix B.

EXAMPLE 5   A division of the Solaron Corporation manufactures photovoltaic cells to use in the company's solar energy converters. It is estimated that 5% of the cells manufactured are defective. If a random

sample of 20 is selected from a large lot of cells manufactured by the company, what is the probability that it will contain at most 2 defective cells?

Solution   We may view this as a binomial experiment. To see this, first note that there are a fixed number of trials ($n = 20$) corresponding to the selection of exactly 20 photovoltaic cells. Second, observe that there are exactly two outcomes in the experiment, defective ("success") and nondefective ("failure"). Third, the probability of success in each trial is .05 ($p = .05$) and the probability of failure in each trial is .95 ($q = .95$). This assumption is justified by virtue of the fact that the lot from which the cells are selected is "large," so the removal of a few cells will not appreciably affect the percentage of defective cells in the lot in each successive trial. Finally, the trials are independent of each other, once again because of the lot size.

Letting X denote the number of defective cells, we find that the probability of finding at most 2 defective cells in the sample of 20 is given by

$$P(X = 0) + P(X = 1) + P(X = 2)$$
$$= C(20, 0)(.05)^0(.95)^{20} + C(20, 1)(.05)^1(.95)^{19}$$
$$+ C(20, 2)(.05)^2(.95)^{18}$$
$$\approx .3585 + .3774 + .1887$$
$$= .9246$$

Thus, for lots of photovoltaic cells manufactured by Solaron, approximately 92% of the samples will have at most 2 defective cells; equivalently, approximately 8% of the samples will contain more than 2 defective cells.   ○ ○ ○

EXAMPLE 6   The probability that a heart transplant performed at the Medical Center is successful (that is, the patient survives a year or more after undergoing such an operation) is .7. Of six patients who have recently undergone such an operation, what is the probability that a year from now

a. none of the heart recipients will be alive?

b. exactly three will be alive?

c. at least three will be alive?

d. all will be alive?

Solution   Here $n = 6$, $p = .7$, and $q = .3$. Let X denote the number of successful operations. Then,

a. the probability that no heart recipients will be alive after one year is given by

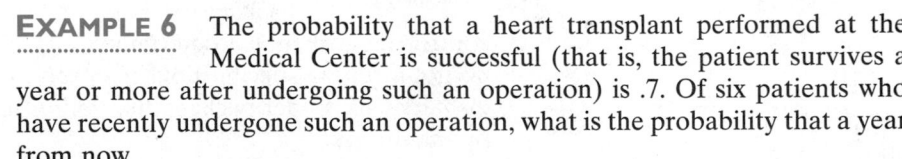

$$P(X = 0) = C(6, 0)(.7)^0(.3)^6$$
$$= \frac{6!}{0!6!} \cdot 1 \cdot (.3)^6$$
$$\approx .0007$$

b. the probability that exactly three will be alive after one year is given by

$$P(X = 3) = C(6, 3)(.7)^3(.3)^3$$

$$= \frac{6!}{3!3!} (.7)^3(.3)^3$$

$$\approx .19$$

c. the probability that at least three will be alive after one year is given by

$$P(X = 3) + P(X = 4) + P(X = 5) + P(X = 6)$$
$$= C(6, 3)(.7)^3(.3)^3 + C(6, 4)(.7)^4(.3)^2$$
$$+ C(6, 5)(.7)^5(.3)^1 + C(6, 6)(.7)^6(.3)^0$$

$$= \frac{6!}{3!3!} (.7)^3(.3)^3 + \frac{6!}{4!2!} (.7)^4(.3)^2 + \frac{6!}{5!1!} (.7)^5(.3)^1$$

$$+ \frac{6!}{6!0!} (.7)^6 \cdot 1$$

$$\approx .93$$

d. the probability that all will be alive after one year is given by

$$P(X = 6) = C(6, 6)(.7)^6(.3)^0 = \frac{6!}{6!0!} (.7)^6$$

$$\approx .12$$

○ ○ ○

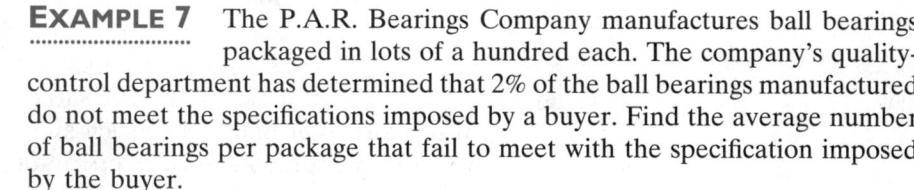

EXAMPLE 7   The P.A.R. Bearings Company manufactures ball bearings packaged in lots of a hundred each. The company's quality-control department has determined that 2% of the ball bearings manufactured do not meet the specifications imposed by a buyer. Find the average number of ball bearings per package that fail to meet with the specification imposed by the buyer.

Solution   The experiment under consideration is binomial. The average number of ball bearings per package that fail to meet with the specifications is therefore given by the expected value of the associated binomial random variable. Using (15a), we find that

$$\mu = E(X) = np = (100)(.02) = 2$$

or 2 substandard ball bearings in a package of 100.

○ ○ ○

SELF–CHECK EXERCISES 8.4

1. A binomial experiment consists of four independent trials. The probability of success in each trial is .2.

a. Find the probability of obtaining exactly 0, 1, 2, 3, and 4 successes, respectively, in this experiment.

b. Construct the binomial distribution and draw the histogram associated with this experiment.

c. Compute the mean and the standard deviation of the random variable associated with this experiment.

2. A recent survey shows that 60% of the households in a large metropolitan area have microwave ovens. If ten households are selected at random, what is the probability that five or fewer of these households have microwave ovens?

Solutions to Self-Check Exercises 8.4 can be found on page 511.

8.4 EXERCISES

A calculator is recommended for this exercise set.

In exercises 1–6, determine whether the given experiment is a binomial experiment. Justify your answer.

1. Casting a fair die three times and observing the number of times a 6 is thrown

2. Casting a fair die and observing the number of times the die is thrown until a 6 appears uppermost

3. Casting a fair die three times and observing the number that appears uppermost

4. A card is selected from a deck of 52 cards and its color is observed. A second card is then drawn (without replacement) and its color is observed.

5. Recording the number of accidents that occur at a given intersection on four clear days and one rainy day

6. Recording the number of hits a baseball player, whose batting average is .325, gets after being up to bat 5 times

In exercises 7–10, find $C(n, x)p^x q^{n-x}$ for the given values of n, x, and p.

7. $n = 4, x = 2, p = 1/3$
8. $n = 6, x = 4, p = 1/4$
9. $n = 5, x = 3, p = .2$
10. $n = 6, x = 5, p = .4$

In exercises 11–16, use the formula $C(n, x)p^x q^{n-x}$ to determine the probability of the given event.

11. The probability of exactly no successes in five trials of a binomial experiment in which $p = 1/3$

12. The probability of exactly three successes in six trials of a binomial experiment in which $p = 1/2$

13. The probability of at least three successes in six trials of a binomial experiment in which $p = 1/2$

14. The probability of no successful outcomes in six trials of a binomial experiment in which $p = 1/3$

15. The probability of no failures in five trials of a binomial experiment in which $p = 1/3$

16. The probability of at least one failure in five trials of a binomial experiment in which $p = 1/3$

17. A fair die is cast four times. Calculate the probability of obtaining exactly two 6's.

18. Let X be the number of successes in five independent trials of a binomial experiment in which the probability of success is $p = 2/5$. Find
a. $P(X = 4)$ **b.** $P(2 \leq X \leq 4)$

19. A binomial experiment consists of five independent trials. The probability of success in each trial is .4.
a. Find the probability of obtaining exactly 0, 1, 2, 3, 4, and 5 successes, respectively, in this experiment.
b. Construct the binomial distribution and draw the histogram associated with this experiment.
c. Compute the mean and the standard deviation of the random variable associated with this experiment.

20. Let the random variable X denote the number of girls in a five-child family. If the probability of a female birth is .5,
a. find the probability of 0, 1, 2, 3, 4, and 5 girls in a five-child family.
b. construct the binomial distribution and draw the histogram associated with this experiment.
c. compute the mean and the standard deviation of the random variable X.

21. The probability that a fuse produced by a certain manufacturing process will be defective is 1/50. Is it correct

to infer from this statement that there is at most 1 defective fuse in each lot of 50 produced by this process? Justify your answer.

22. **Sports**   If the probability that a certain tennis player will serve an ace is 1/4, what is the probability that he will serve exactly two aces out of five serves?

23. **Customer Services**   Mayco, a mail-order department store, has six telephone lines available for customers who wish to place their orders. If the probability that during business hours any one of the six telephone lines is engaged is 1/4, find the probability that when a customer calls to place an order all six lines will be in use.

24. **Sales Predictions**   From experience, the manager of Kramer's Book Mart knows that 40% of the people who are browsing in the store will make a purchase. What is the probability that among ten people who are browsing in the store, at least three will make a purchase?

25. **Advertisements**   An advertisement for brand A chicken noodle soup claims that 60% of all consumers prefer brand A over brand B, the chief competitor's product. To test this claim, Mr. David Horowitz, host of "The Consumer Advocate," selected ten people at random from the audience. After tasting both soups, each person was asked to state his or her preference. Assuming the company's claim is correct, find the probability that
a. the company's claim was supported by the experiment; that is, six or more people stated a preference for brand A.
b. the company's claim was not supported by the experiment; that is, fewer than six people stated a preference for brand A.

26. **Voters**   In a certain congressional district, it is known that 40% of the registered voters classify themselves as conservatives. If ten registered voters are selected at random from this district, what is the probability that four of them will be conservatives?

27. **Blood Types**   It is estimated that one-third of the general population has blood type A^+. If a sample of nine people is selected at random, what is the probability that
a. exactly three of them have blood type A^+?
b. at most three of them have blood type A^+?

28. **Exams**   A biology quiz consists of eight multiple-choice questions. Five must be answered correctly to receive a passing grade. If each question has five possible

answers, of which only one is correct, what is the probability that a student who guesses at random on each question will pass the examination?

29. **Exams**   A psychology quiz consists of ten true-or-false questions. If a student knows the correct answer to six of the questions but determines the answers to the remaining questions by flipping a coin, what is the probability that she will obtain a score of at least 90%?

30. **Quality Control**   The probability that a video-disc player produced by the V.C.A. Television Company is defective is estimated to be .02. If a sample of ten sets is selected at random, what is the probability that the sample contains
a. no defectives?
b. at most two defectives?

31. **Quality Control**   As part of its quality-control program, the video cartridges produced by the Starr Communications Company are subjected to a final inspection before shipment. A sample of six cartridges is selected at random from each lot of cartridges produced and the lot is rejected if the sample contains one or more defective cartridges. If 1.5% of the cartridges produced by Starr is defective, find the probability that a shipment will be accepted.

32. **Robot Reliability**   An automobile manufacturing company uses ten industrial robots as welders on its assembly line. On a given working day, the probability that a robot will be inoperative is .05. What is the probability that on a given working day
a. exactly two robots are inoperative?
b. more than two robots are inoperative?

33. **Engine Failures**   The probability that an airplane engine will fail in a transcontinental flight is .001. Assuming that engine failures are independent of each other, what is the probability that, on a certain transcontinental flight, a four-engine plane will experience
a. exactly one engine failure?
b. exactly two engine failures?
c. more than two engine failures? [*Note:* In this event, the airplane will crash!]

34. **Quality Control**   The manager of Toy World has decided to accept a shipment of electronic games if none of a random sample of 20 is found to be defective.
a. What is the probability that he will accept the shipment if 10% of the electronic games is defective?
b. What is the probability that he will accept the shipment if 5% of the electronic games is defective?

35. **Quality Control** Refer to exercise 34. If the manager's criterion for accepting shipment is that there be no more than 1 defective electronic game in a random sample of 20, what is the probability that he will accept the shipment if 10% of the electronic games is defective?

36. **Quality Control** Refer to exercise 34. If the manager of the store changes his sample size to 10 and decides to accept shipment if none of the games is defective, what is the probability that he will accept the shipment if 10% of the games is defective?

37. How many times must a person toss a coin if the chances of obtaining at least one head are 99% or better?

38. **Drug Testing** A new drug has been found to be effective in treating 75% of the people afflicted by a certain disease. If the drug is administered to 500 people who have this disease, what are the mean and the standard deviation of the number of people for whom the drug can be expected to be effective?

39. **College Graduates** At a certain university the probability that an entering freshman will graduate within four years is .6. From an incoming class of 2000 freshmen, find
    a. the expected number of students who will graduate within four years.
    b. the standard deviation of the number of students who will graduate within four years.

SOLUTIONS TO SELF-CHECK EXERCISES 8.4

1. a. We use formula (14) with $n = 4$, $p = .2$, and $q = 1 - .2 = .8$, obtaining

$$P(X = 0) = C(4, 0)(.2)^0(.8)^4$$
$$= \frac{4!}{0!4!} \cdot 1 \cdot (.8)^4 \approx .410$$
$$P(X = 1) = C(4, 1)(.2)^1(.8)^3$$
$$= \frac{4!}{1!3!}(.2)(.8)^3 \approx .410$$
$$P(X = 2) = C(4, 2)(.2)^2(.8)^2$$
$$= \frac{4!}{2!2!}(.2)^2(.8)^2 \approx .154$$
$$P(X = 3) = C(4, 3)(.2)^3(.8)^1$$
$$= \frac{4!}{3!1!}(.2)^3(.8) \approx .026$$
$$P(X = 4) = C(4, 4)(.2)^4(.8)^0$$
$$= \frac{4!}{4!0!}(.2)^4 \cdot 1 \approx .002$$

b. The required binomial distribution and histogram are as follows:

x	0	1	2	3	4
$P(X = x)$	.410	.410	.154	.026	.002

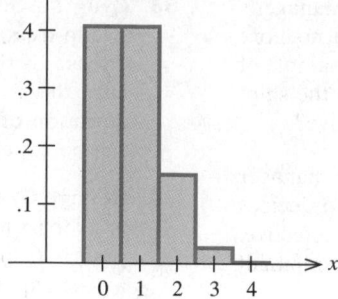

c. The mean is

$$\mu = E(X) = np = (4)(.2)$$
$$= 0.8$$

and the standard deviation is

$$\sigma = \sqrt{npq} = \sqrt{(4)(.2)(.8)}$$
$$= 0.8$$

2. This is a binomial experiment with $n = 10$, $p = .6$, and $q = .4$. Let X denote the number of households that have microwave ovens. Then, the probability that five or fewer households have microwave ovens is given by

$$P(X = 0) + P(X = 1) + P(X = 2) + P(X = 3)$$
$$+ P(X = 4) + P(X = 5)$$
$$= C(10, 0)(.6)^0(.4)^{10} + C(10, 1)(.6)^1(.4)^9$$
$$+ C(10, 2)(.6)^2(.4)^8 + C(10, 3)(.6)^3(.4)^7$$
$$+ C(10, 4)(.6)^4(.4)^6 + C(10, 5)(.6)^5(.4)^5$$
$$\approx 0 + .002 + .011 + .042 + .111 + .201$$
$$\approx .37$$

8.5 THE NORMAL DISTRIBUTION

Probability Density Functions

The probability distributions discussed in the preceding sections were all associated with finite random variables—that is, random variables that take on finitely many values. Such probability distributions are referred to as *finite probability distributions*. In this section, we consider probability distributions associated with a continuous random variable—that is, a random variable that may take on any value lying in an interval of real numbers. Such probability distributions are called **continuous probability distributions.**

Unlike a finite probability distribution, which may be exhibited in the form of a table, a continuous probability distribution is defined by a function f whose domain coincides with the interval of values taken on by the random variable associated with the experiment. Such a function f is called the **probability density function** associated with the probability distribution and has the following properties:

1. $f(x)$ is nonnegative for all values of x.

2. The area of the region between the graph of f and the x-axis is equal to 1. (See Figure 8.13.)

Figure 8.13
A probability density function.

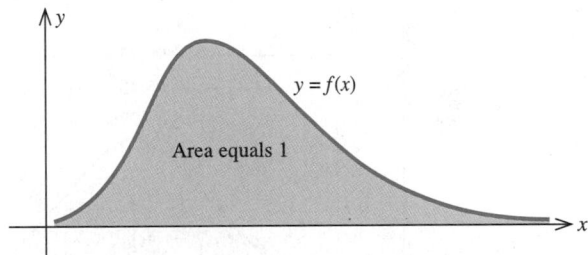

Now suppose we are given a continuous probability distribution defined by a probability density function f. Then the probability that the random variable X assumes a value in an interval $a < x < b$ is given by the area of the region between the graph of f and the x-axis from $x = a$ to $x = b$ (Figure 8.14). We denote the value of this probability by $P(a < X < b)$.* Observe that property 2 of the probability density function states that the probability that a continuous random variable takes on a value lying in its range is 1, a certainty, which is expected. Note the analogy between the areas under the probability density curves and the histograms associated with finite probability distributions (see Section 8.1).

Figure 8.14
$P(a < X < b)$ is given by the area of the shaded region.

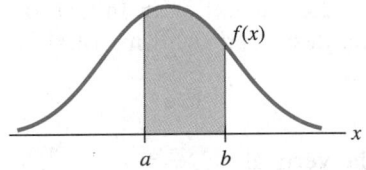

Normal Distributions

The mean μ and the standard deviation σ of a continuous probability distribution have roughly the same meanings as the mean and standard deviation of a finite probability distribution. Thus, the mean of a continuous probability distribution is a measure of the central tendency of the probability distribution, and the standard deviation of the probability distribution measures its spread about its mean. Both of these numbers will play an important role in the following discussion.

For the remainder of this section we will discuss a special class of continuous probability distributions known as **normal distributions.** The normal distribution is without doubt the most important of all the probability distributions. Many phenomena, such as the heights of people in a given population, the weights of newborn infants, the IQs of college students, the actual weights of 16-ounce packages of cereals, and so on, have probability distributions that

* Because the area under one point of the graph of f is equal to zero, we see that $P(a < X < b) = P(a < X \le b) = P(a \le X < b) = P(a \le X \le b)$.

are normal. The normal distribution also provides us with an accurate approximation to the distributions of many random variables associated with random-sampling problems. In fact, in the next section we will see how a normal distribution may be used to approximate a binomial distribution under certain conditions.

The graph of a normal distribution, which is bell shaped, is called a **normal curve** (Figure 8.15).

Figure 8.15
A normal curve.

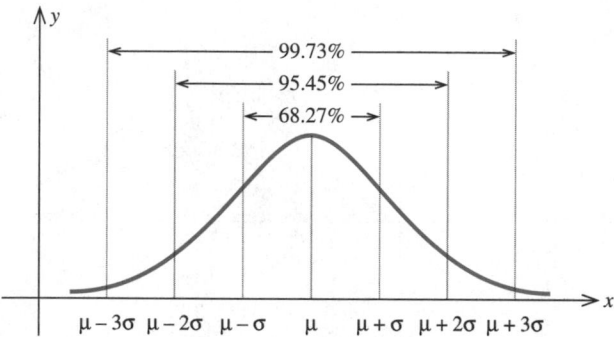

The normal curve (and therefore the corresponding normal distribution) is completely determined by its mean μ and standard deviation σ. In fact, the normal curve has the following characteristics, described in terms of these two parameters:*

1. The curve has a peak at $x = \mu$.

2. The curve is symmetrical with respect to the vertical line $x = \mu$.

3. The curve always lies above the x-axis but approaches the x-axis as x extends indefinitely in either direction.

4. The area under the curve is 1.

5. For any normal curve, 68.27% of the area under the curve lies within 1 standard deviation of the mean (that is, between $\mu - \sigma$ and $\mu + \sigma$), 95.45% of the area lies within 2 standard deviations of the mean, and 99.73% of the area lies within 3 standard deviations of the mean.

Figure 8.16 shows two normal curves with different means μ_1 and μ_2 but the same deviation. Next, Figure 8.17 shows two normal curves with the same mean but different standard deviations σ_1 and σ_2. (Which number is smaller?)

In general, the mean μ of a normal distribution determines where the center of the curve is located, whereas the standard deviation σ of a normal distribution determines the sharpness (or flatness) of the curve.

ο ο ο

* The probability density function associated with this normal curve is given by

$$y = \frac{1}{\sigma\sqrt{2\pi}} e^{-(1/2)[(x-\mu)/\sigma]^2}$$

but the direct use of this formula will not be required in our discussion of the normal distribution.

Figure 8.16
Two normal curves that have the same standard deviation but different means.

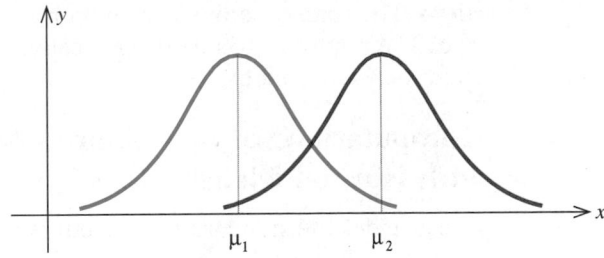

Figure 8.17
Two normal curves that have the same mean but different standard deviations.

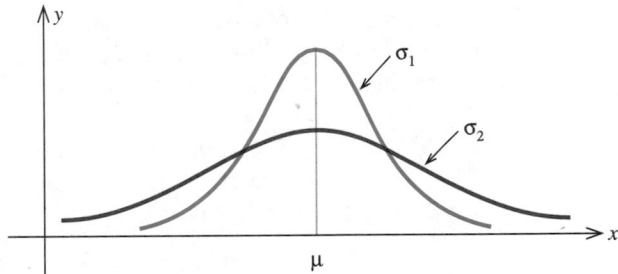

As this discussion reveals, there are infinitely many normal curves corresponding to different choices of the parameters μ and σ, which characterize such curves. Fortunately, any normal curve may be transformed into any other normal curve (as we will see later), so in the study of normal curves it suffices to single out one such particular curve for special attention. The normal curve with mean $\mu = 0$ and standard deviation $\sigma = 1$ is called the **standard normal**

EXPLORING WITH TECHNOLOGY

Consider the probability density function

$$f(x) = \frac{1}{\sqrt{2\pi}} e^{-x^2/2}$$

which is the formula given in the footnote on page 514 with $\mu = 0$ and $\sigma = 1$.
a. Use a graphing utility to plot the graph of f using the viewing rectangle $[-4, 4] \times [0, 0.5]$.
b. Use the numerical integration function of a graphing utility to find the area of the region under the graph of f on the intervals $[-1, 1]$, $[-2, 2]$, and $[-3, 3]$ and thus verify property 5 of normal distributions for the special case where $\mu = 0$ and $\sigma = 1$.

○ ○ ○

curve. The corresponding distribution is called the **standard normal distribu-tion.** The random variable itself is called the **standard normal variable** and is commonly denoted by Z.

Computations of Probabilities Associated with Normal Distributions

Areas under the standard normal curve have been extensively computed and tabulated. Table III, Appendix B, gives the areas of the regions under the standard normal curve to the left of the number z; these areas correspond, of course, to probabilities of the form $P(Z < z)$ or $P(Z \leq z)$. The next several examples illustrate the use of this table in computations involving the probabilities associated with the standard normal variable.

EXAMPLE I   Let Z be the standard normal variable. By first making a sketch of the appropriate region under the standard normal curve, find the values of

a. $P(Z < 1.24)$ **b.** $P(Z > 0.5)$ **c.** $P(0.24 < Z < 1.48)$

d. $P(-1.65 < Z < 2.02)$

Solution

a. The region under the standard normal curve associated with the probability $P(Z < 1.24)$ is shown in Figure 8.18. To find the area of the required region using Table III, Appendix B, we first locate the number 1.2 in the column and the number 0.04 in the row, both headed by z, and read off the number 0.8925 appearing in the body of the table. Thus,

$$P(Z < 1.24) = .8925$$

b. The region under the standard normal curve associated with the probability $P(Z > 0.5)$ is shown in Figure 8.19a. Observe, however, that the required area is, by virtue of the symmetry of the standard normal curve, equal to the shaded area shown in Figure 8.19b. Thus,

$$P(Z > 0.5) = P(Z < -0.5)$$
$$= .3085$$

c. The probability $P(0.24 < Z < 1.48)$ is equal to the shaded area shown in Figure 8.20. But this area is obtained by subtracting the area under the curve

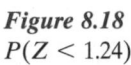

Figure 8.18
$P(Z < 1.24)$.

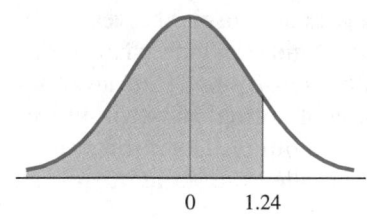

0   1.24

Figure 8.19

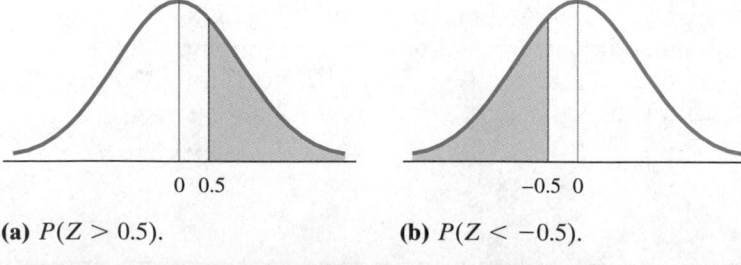

0 0.5   −0.5 0

(a) $P(Z > 0.5)$.   **(b)** $P(Z < -0.5)$.

Figure 8.20
$P(0.24 < Z < 1.48)$.

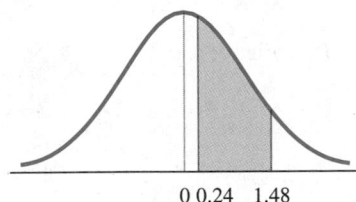

0 0.24 1.48

Figure 8.21
$P(-1.65 < Z < 2.02)$.

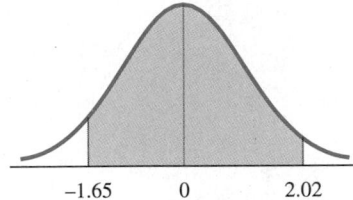

−1.65   0   2.02

Figure 8.22
$P(Z < z) = .9474$.

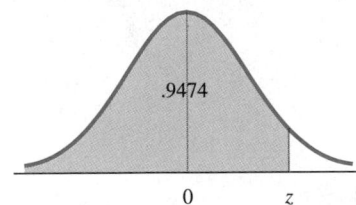

.9474

0   z

to the left of $z = 0.24$ from the area under the curve to the left of $z = 1.48$; that is,

$$P(0.24 < Z < 1.48) = P(Z < 1.48) - P(Z < 0.24)$$
$$= .9306 - .5948$$
$$= .3358$$

d. The probability $P(-1.65 < Z < 2.02)$ is given by the shaded area shown in Figure 8.21. We have

$$P(-1.65 < Z < 2.02) = P(Z < 2.02) - P(Z < -1.65)$$
$$= .9783 - .0495$$
$$= .9288 \qquad \circ \ \circ \ \circ$$

EXAMPLE 2 Let Z be the standard normal variable. Find the value of z if z satisfies

a. $P(Z < z) = .9474$ **b.** $P(Z > z) = .9115$

c. $P(-z < Z < z) = .7888$

Solution

a. Refer to Figure 8.22. We want the value of Z such that the area of the region under the standard normal curve and to the left of $Z = z$ is .9474. Locating the number .9474 in Table III, Appendix B, and reading back, we find that $z = 1.62$.

b. Since $P(Z > z)$, or equivalently, the area of the region to the right of z is greater than 0.5, z must be negative (Figure 8.23). Therefore $-z$ is positive. Furthermore, the area of the region to the right of z is the same as the area of the region to the left of $-z$. Therefore

$$P(Z > z) = P(Z < -z)$$
$$= .9115$$

Looking up the table, we find $-z = 1.35$, so $z = -1.35$.

c. The region associated with $P(-z < Z < z)$ is shown in Figure 8.24. Observe that by symmetry the area of this region is just double that of the area of the

Figure 8.23
$P(Z > z) = .9115$.

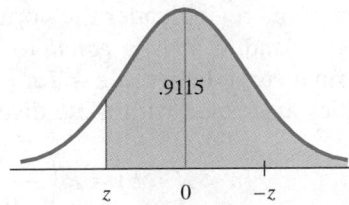

.9115

z   0   −z

Figure 8.24
$P(-z < Z < z) = .7888$.

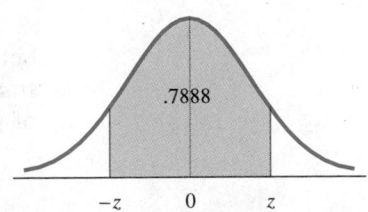

.7888

−z   0   z

region between $Z = 0$ and $Z = z$; that is,

$$P(-z < Z < z) = 2P(0 < Z < z)$$

or $\qquad P(0 < Z < z) = P(Z < z) - \dfrac{1}{2}$ \qquad (See Figure 8.25.)

Figure 8.25

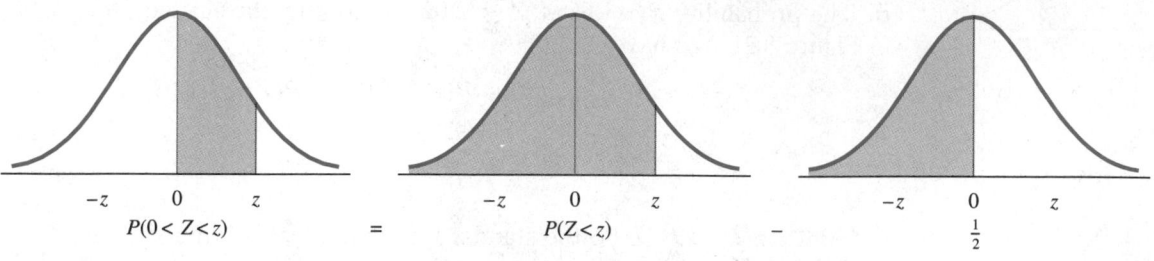

$$P(0 < Z < z) \qquad = \qquad P(Z < z) \qquad - \qquad \dfrac{1}{2}$$

Therefore,

$$\frac{1}{2}P(-z < Z < z) = P(Z < z) - \frac{1}{2}$$

or, solving for $P(Z < z)$,

$$P(Z < z) = \frac{1}{2} + \frac{1}{2}P(-z < Z < z)$$

$$= \frac{1}{2}(1 + .7888)$$

$$= .8944$$

Consulting the table, we find $z = 1.25$.   ◑ ◑ ◑

We now turn our attention to the computation of probabilities associated with normal distributions whose means and standard deviations are not necessarily equal to 0 and 1, respectively. As mentioned earlier, any normal curve may be transformed into the standard normal curve. In particular, it may be shown that if X is a normal random variable with mean μ and standard deviation σ, then it can be transformed into the standard normal random variable Z by means of the substitution

$$Z = \frac{X - \mu}{\sigma}$$

The area of the region under the normal curve (with random variable X) between $x = a$ and $x = b$ is *equal* to the area of the region under the standard normal curve between $z = (a - \mu)/\sigma$ and $z = (b - \mu)/\sigma$. In terms of probabilities associated with these distributions, we have

$$P(a < X < b) = P\left(\frac{a - \mu}{\sigma} < Z < \frac{b - \mu}{\sigma}\right) \qquad \textbf{(16)}$$

(Figure 8.26). Similarly, we have

$$P(X < b) = P\left(Z < \frac{b - \mu}{\sigma}\right) \tag{17}$$

and $$P(X > a) = P\left(Z > \frac{a - \mu}{\sigma}\right) \tag{18}$$

Thus, with the help of equations (16)–(18), computations of probabilities associated with any normal distribution may be reduced to the computations of areas of regions under the standard normal curve.

Figure 8.26

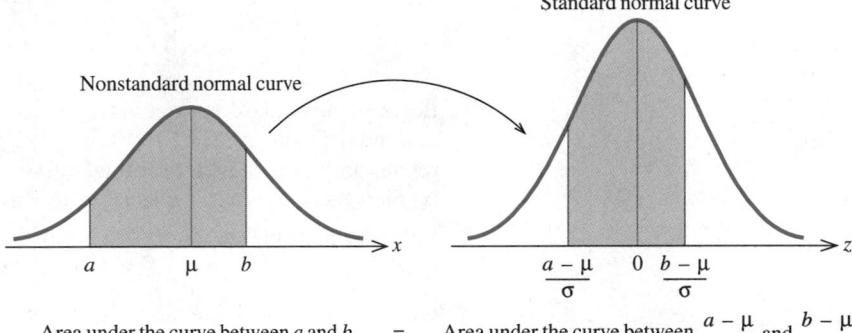

Area under the curve between a and b = Area under the curve between $\dfrac{a - \mu}{\sigma}$ and $\dfrac{b - \mu}{\sigma}$

EXAMPLE 3   Suppose X is a normal random variable with $\mu = 100$ and $\sigma = 20$. Find the values of

a. $P(X < 120)$ **b.** $P(X > 70)$ **c.** $P(75 < X < 110)$

Solution

a. Using formula (17) with $\mu = 100$, $\sigma = 20$, and $b = 120$, we have

$$P(X < 120) = P\left(Z < \frac{120 - 100}{20}\right)$$

$$= P(Z < 1) = .8413 \qquad \text{(Using the table of values of } Z\text{)}$$

b. Using formula (18) with $\mu = 100$, $\sigma = 20$, and $a = 70$, we have

$$P(X > 70)$$

$$= P\left(Z > \frac{70 - 100}{20}\right)$$

$$= P(Z > -1.5) = P(Z < 1.5) = .9332 \qquad \text{(Using the table of values of } Z\text{)}$$

Figure 8.27

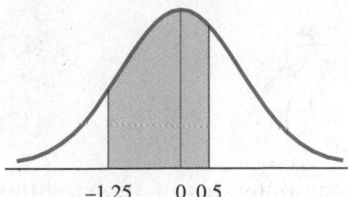

−1.25   0 0.5

c. Using formula (16) with $\mu = 100$, $\sigma = 20$, $a = 75$, and $b = 110$, we have

$$P(75 < X < 110)$$

$$= P\left(\frac{75 - 100}{20} < Z < \frac{110 - 100}{20}\right)$$

$$= P(-1.25 < Z < 0.5)$$

$$= P(Z < 0.5) - P(Z < -1.25) \qquad \text{(See Figure 8.27.)}$$

$$= .6915 - .1056 = .5859 \qquad \text{(Using the table of values of } Z\text{)}$$ ◦ ◦ ◦

SELF-CHECK EXERCISES 8.5

1. Let Z be a standard normal variable.
  a. Find the value of $P(-1.2 < Z < 2.1)$ by first making a sketch of the appropriate region under the standard normal curve.
  b. Find the value of z if z satisfies $P(-z < Z < z) = .8764$.

2. Let X be a normal random variable with $\mu = 80$ and $\sigma = 10$. Find the values of
  a. $P(X < 100)$ **b.** $P(X > 60)$ **c.** $P(70 < X < 90)$

Solutions to Self-Check Exercises 8.5 can be found on page 521.

8.5 EXERCISES

In exercises 1–6, find the value of the probability of the standard normal variable Z corresponding to the shaded area under the standard normal curve.

1. $P(Z < 1.45)$

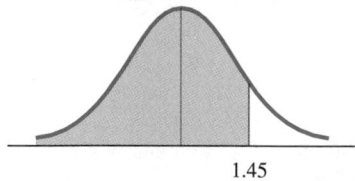

1.45

2. $P(Z > 1.11)$

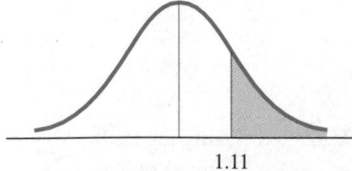

1.11

3. $P(Z < -1.75)$

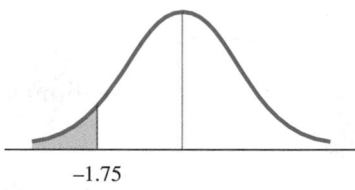

−1.75

4. $P(0.3 < Z < 1.83)$

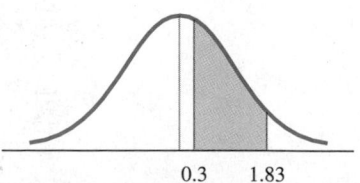

0.3   1.83

5. $P(-1.32 < Z < 1.74)$

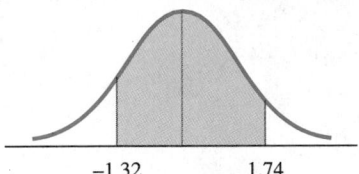

−1.32   1.74

6. $P(-2.35 < Z < -0.51)$

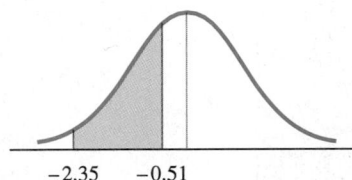

−2.35   −0.51

In exercises 7–14, (a) make a sketch of the area under the standard normal curve corresponding to the given probability and (b) find the value of the probability of the standard normal variable Z corresponding to this area.

7. $P(Z < 1.37)$ **8.** $P(Z > 2.24)$

9. $P(Z < -0.65)$ **10.** $P(0.45 < Z < 1.75)$

11. $P(Z > -1.25)$ **12.** $P(-1.48 < Z < 1.54)$

13. $P(0.68 < Z < 2.02)$

14. $P(-1.41 < Z < -0.24)$

15. Let Z be the standard normal variable. Find the values of z if z satisfies
  a. $P(Z < z) = .8907$
  b. $P(Z < z) = .2090$

16. Let Z be the standard normal variable. Find the values of z if z satisfies
  a. $P(Z > z) = .9678$
  b. $P(-z < Z < z) = .8354$

17. Let Z be the standard normal variable. Find the values of z if z satisfies
  a. $P(Z > -z) = .9713$
  b. $P(Z < -z) = .9713$

18. Suppose X is a normal random variable with $\mu = 380$ and $\sigma = 20$. Find the value of
  a. $P(X < 405)$ **b.** $P(400 < X < 430)$
  c. $P(X > 400)$

19. Suppose X is a normal random variable with $\mu = 50$ and $\sigma = 5$. Find the value of
  a. $P(X < 60)$ **b.** $P(X > 43)$
  c. $P(46 < X < 58)$

20. Suppose X is a normal random variable with $\mu = 500$ and $\sigma = 75$. Find the value of
  a. $P(X < 750)$ **b.** $P(X > 350)$
  c. $P(400 < X < 600)$

SOLUTIONS TO SELF-CHECK EXERCISES 8.5

1. a. The probability $P(-1.2 < Z < 2.1)$ is given by the shaded area in the accompanying figure.

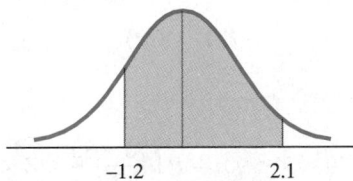

−1.2   2.1

We have

$$P(-1.2 < Z < 2.1) = P(Z < 2.1) - P(Z < -1.2)$$
$$= .9821 - .1151$$
$$= .867$$

b. The region associated with $P(-z < Z < z)$ is shown in the accompanying figure.

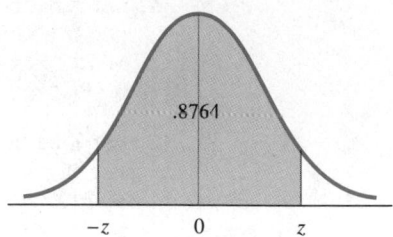

Observe that we have the following relationship:

$$P(Z < z) = \frac{1}{2}[1 + P(-z < Z < z)]$$

(See Example 2c.) With $P(-z < Z < z) = .8764$, we find

$$P(Z < z) = \frac{1}{2}(1 + .8764)$$

$$= .9382$$

Consulting the table, we find $z = 1.54$.

2. Using the transformation (16) and the table of values of Z, we have

a. $P(X < 100) = P\left(Z < \dfrac{100 - 80}{10}\right)$

$\qquad = P(Z < 2)$

$\qquad = .9772$

b. $P(X > 60) = P\left(Z > \dfrac{60 - 80}{10}\right)$

$\qquad = P(Z > -2)$

$\qquad = P(Z < 2)$

$\qquad = .9772$

c. $P(70 < X < 90) = P\left(\dfrac{70 - 80}{10} < Z < \dfrac{90 - 80}{10}\right)$

$\qquad = P(-1 < Z < 1)$

$\qquad = P(Z < 1) - P(Z < -1)$

$\qquad = .8413 - .1587$

$\qquad = .6826$

8.6 APPLICATIONS OF THE NORMAL DISTRIBUTION

Applications Involving Normal Random Variables

In this section we look at some applications involving the normal distribution.

EXAMPLE 1 The medical records of infants delivered at the Kaiser Memorial Hospital show that the infants' birth weights in pounds are normally distributed with a mean of 7.4 and a standard deviation of 1.2. Find the probability that an infant selected at random from among those delivered at the hospital weighed more than 9.2 pounds at birth.

Solution Let X be the normal random variable denoting the birth weights of infants delivered at the hospital. Then the probability that an infant selected at random has a birth weight of more than 9.2 pounds is given by $P(X > 9.2)$. To compute $P(X > 9.2)$ we use formula (18), Section 8.5, with $\mu = 7.4$, $\sigma = 1.2$, and $a = 9.2$. We find

$$P(X > 9.2) = P\left(Z > \frac{9.2 - 7.4}{1.2}\right) \qquad \left[P(X > a) = P\left(Z > \frac{a - \mu}{\sigma}\right)\right]$$
$$= P(Z > 1.5)$$
$$= P(Z < -1.5)$$
$$= .0668$$

Thus, the probability that an infant delivered at the hospital weighs more than 9.2 pounds is .0668.   ◦ ◦ ◦

EXAMPLE 2 The Idaho Natural Produce Corporation ships potatoes to its distributors in bags whose weights are normally distributed with a mean weight of 50 pounds and standard deviation of 0.5 pound. If a bag of potatoes is selected at random from a shipment, what is the probability that it weighs

a. more than 51 pounds?   **b.** less than 49 pounds?

c. between 49 and 51 pounds?

Solution Let X denote the weight of potatoes packed by the company. Then the mean and standard deviation of X are $\mu = 50$ and $\sigma = 0.5$, respectively.

a. The probability that a bag selected at random weighs more than 51 pounds is given by

$$P(X > 51) = P\left(Z > \frac{51 - 50}{0.5}\right) \qquad \left[P(X > a) = P\left(Z > \frac{a - \mu}{\sigma}\right)\right]$$

$$= P(Z > 2)$$
$$= P(Z < -2)$$
$$= .0228$$

b. The probability that a bag selected at random weighs less than 49 pounds is given by

$$P(X < 49) = P\left(Z < \frac{49 - 50}{0.5}\right) \qquad \left[P(X < b) = P\left(Z < \frac{b - \mu}{\sigma}\right)\right]$$

$$= P(Z < -2)$$
$$= .0228$$

c. The probability that a bag selected at random weighs between 49 and 51 pounds is given by

$$P(49 < X < 51) \qquad\qquad \left[P(a < X < b) = \right.$$

$$= P\left(\frac{49 - 50}{0.5} < Z < \frac{51 - 50}{0.5}\right) \qquad \left. P\left(\frac{a - \mu}{\sigma} < Z < \frac{b - \mu}{\sigma}\right)\right]$$

$$= P(-2 < Z < 2)$$
$$= P(Z < 2) - P(Z < -2)$$
$$= .9772 - .0228$$
$$= .9544 \qquad\qquad\qquad\qquad\qquad\qquad\quad \circ\;\circ\;\circ$$

EXAMPLE 3 The grade point average of the senior class of Jefferson High School is normally distributed with a mean of 2.7 and a standard deviation of 0.4 point. If a senior in the top 10% of his or her class is eligible for admission to any of the nine campuses of the State University system, what is the minimum grade point average that a senior should have to ensure eligibility for admission to the State University system?

Solution   Let X denote the grade point average of a randomly selected senior at Jefferson High School and let x denote the minimum grade point average to ensure his or her eligibility for admission to the university. Since only the top 10% is eligible for admission, x must satisfy the equation

$$P(X \geq x) = .1$$

Using formula (18), Section 8.5, with $\mu = 2.7$ and $\sigma = 0.4$, we find

$$P(X \geq x) = P\left(Z \geq \frac{x - 2.7}{0.4}\right) = .1 \qquad \left[P(X > a) = P\left(Z > \frac{a - \mu}{\sigma}\right)\right]$$

But, this is equivalent to the equation

$$P\left(Z \le \frac{x - 2.7}{0.4}\right) = .9 \qquad \text{(Why?)}$$

Consulting Table III, Appendix B, we find

$$\frac{x - 2.7}{0.4} = 1.28$$

Upon solving for x, we obtain

$$x = (1.28)(0.4) + 2.7$$
$$\approx 3.2$$

Thus, to ensure eligibility for admission to one of the nine campuses of the State University system, a senior at Jefferson High School should have a minimum 3.2 grade point average.   o o o

Approximating Binomial Distributions

As mentioned in the last section, one important application of the normal distribution is that it provides us with an accurate approximation of other continuous probability distributions. We now show how a binomial distribution may be approximated by a suitable normal distribution. This technique leads to a convenient and simple solution to certain problems involving binomial probabilities.

Recall that a binomial distribution is a probability distribution of the form

$$P(X = x) = C(n, x)p^x q^{n-x} \qquad (x = 0, 1, 2, \dots, n) \tag{19}$$

(See Section 8.4.) For small values of n, the arithmetic computations of the binomial probabilities may be done with relative ease. However, if n is large, then the work involved becomes prodigious, even when tables of $P(X = x)$ are available. For example, if $n = 50$, $p = .3$, and $q = .7$, then the probability of ten or more successes is given by

$$P(X \ge 10) = P(X = 10) + P(X = 11) + \cdots + P(X = 50)$$

$$= \frac{50!}{10!40!}(.3)^{10}(.7)^{40} + \frac{50!}{11!39!}(.3)^{11}(.7)^{39} + \cdots + \frac{50!}{50!0!}(.3)^{50}(.7)^0$$

To see how the normal distribution helps us in such situations, let us consider a coin-tossing experiment. Suppose a fair coin is tossed 20 times and we wish to compute the probability of obtaining 10 or more heads. The solution to this problem may be obtained, of course, by computing

$$P(X \ge 10) = P(X = 10) + P(X = 11) + \cdots + P(X = 20)$$

The inconvenience of this approach for solving the problem at hand has already been pointed out. As an alternative solution, let us begin by interpreting the solution in terms of finding the area of suitable rectangles of the histogram for the distribution associated with the problem. Using formula (19), we

Table 8.12
Probability distribution.

x	$P(X = x)$
0	.0000
1	.0000
2	.0002
3	.0011
4	.0046
5	.0148
6	.0370
7	.0739
8	.1201
9	.1602
10	.1762
11	.1602
12	.1201
⋮	⋮
20	.0000

compute the probability of obtaining exactly x heads in 20 coin tosses. The results lead to the binomial distribution displayed in Table 8.12.

Using the data from the table, we next construct the histogram for the distribution (Figure 8.28). The probability of obtaining 10 or more heads in 20 coin tosses is equal to the sum of the areas of the shaded rectangles of the histogram of the binomial distribution shown in Figure 8.29.

Next, observe that the shape of the histogram suggests that the binomial distribution under consideration may be approximated by a suitable normal distribution. Since the mean and standard deviation of the binomial distribution are given by

$$\mu = np$$
$$= (20)(.5) = 10$$

and

$$\sigma = \sqrt{npq}$$
$$= \sqrt{(20)(.5)(.5)}$$
$$= 2.24$$

respectively (see Section 8.4), the natural choice of a normal curve for this purpose is one with a mean of 10 and standard deviation of 2.24. Figure 8.30 shows such a normal curve superimposed on the histogram of the binomial distribution.

The good fit suggests that the sum of the areas of the rectangles representing $P(X \geq 10)$, the probability of obtaining 10 or more heads in 20 coin tosses, may be approximated by the area of an appropriate region under the normal curve. To determine this region, let us note that the base of the portion of

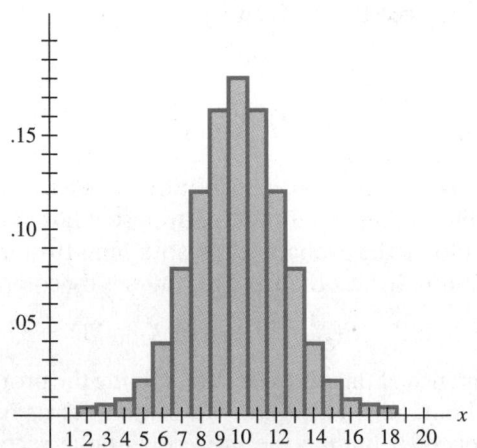

Figure 8.28
Histogram showing the probability of obtaining x heads in 20 coin tosses.

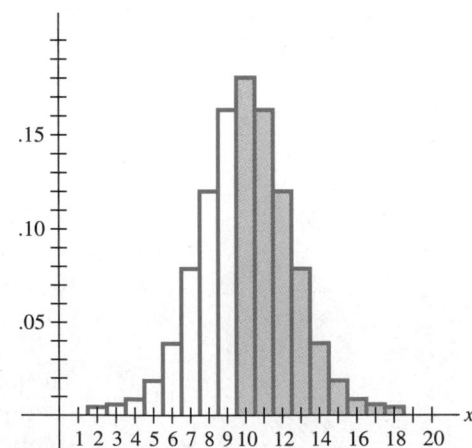

Figure 8.29
The shaded area gives the probability of obtaining 10 or more heads in 20 coin tosses.

Figure 8.30
Normal curve superimposed on the histogram for a binomial distribution.

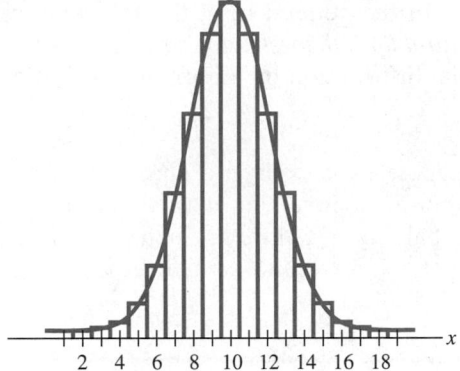

the histogram representing the required probability extends from $x = 9.5$ on, since the base of the leftmost rectangle is centered at $x = 10$ and the base of each rectangle has length 1. (See Figure 8.31.) Therefore, the required region under the normal curve should also have $x \geq 9.5$. Letting Y denote the continuous normal variable, we have

$$
\begin{aligned}
P(X \geq 10) &\approx P(Y \geq 9.5) \\
&= P(Y > 9.5) \\
&= P\left(Z > \frac{9.5 - 10}{2.24}\right) \qquad \left[P(X > a) = P\left(Z > \frac{a - \mu}{\sigma}\right)\right] \\
&= P(Z > -0.22) \\
&= P(Z < 0.22) \\
&= .5871 \qquad\qquad\qquad \text{(Using the table of values of } Z\text{)}
\end{aligned}
$$

The exact value of $P(X \geq 10)$ may be found by computing

$$
P(X = 10) + P(X = 11) + \cdots + P(X = 20)
$$

in the usual fashion and is equal to .5881. Thus, the normal distribution with suitably chosen mean and standard deviation does provide us with a good approximation of the binomial distribution.

Figure 8.31
$P(X \geq 10)$ is approximated by the area under the normal curve.

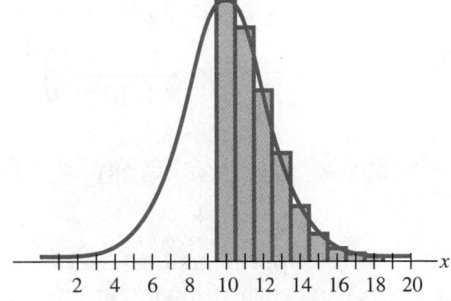

In the general case, the following result, which is a special case of the *central limit theorem,* guarantees the accuracy of the approximation of a binomial distribution by a normal distribution under certain conditions.

THEOREM

Suppose we are given a binomial distribution associated with a binomial experiment involving n trials, each with a probability of success p and probability of failure q. Then if n is large and p is not close to 0 or 1, the binomial distribution may be approximated by a normal distribution with

$$\mu = np \quad \text{and} \quad \sigma = \sqrt{npq}$$

REMARK   It can be shown that if both np and nq are greater than 5, then the error resulting from this approximation is negligible.   ◦ ◦ ◦

Applications Involving Binomial Random Variables

EXAMPLE 4   An automobile manufacturer receives the microprocessors used to regulate fuel consumption in its automobiles in shipments of 1000 each from a certain supplier. It has been estimated that, on the average, 1% of the microprocessors manufactured by the supplier are defective. Determine the probability that more than 20 of the microprocessors in a single shipment are defective.

Solution   Let X denote the number of defective microprocessors in a single shipment. Then X has a binomial distribution with $n = 1000$, $p = .01$, and $q = .99$, so

$$\mu = (1000)(.01) = 10$$

and
$$\sigma = \sqrt{(1000)(.01)(.99)}$$
$$\approx 3.15$$

Approximating the binomial distribution by a normal distribution with a mean of 10 and a standard deviation of 3.15, we find that the probability that more than 20 microprocessors in a shipment are defective is given by

$$P(X > 20) \approx P(Y > 20.5) \qquad \text{(Where } Y \text{ denotes the normal random variable)}$$
$$= P\left(Z > \frac{20.5 - 10}{3.15}\right) \qquad \left[P(X > a) = P\left(Z > \frac{a - \mu}{\sigma}\right)\right]$$
$$= P(Z > 3.33)$$
$$= P(Z < -3.33)$$
$$= .0004$$

In other words, approximately 0.04 percent of the shipments containing 1000 microprocessors each will contain more than 20 defective units.   ◦ ◦ ◦

EXAMPLE 5 The probability that a heart transplant performed at the Medical Center is successful (that is, the patient survives a year or more after undergoing the surgery) is .7. Of 100 patients who have undergone such an operation, what is the probability that

a. fewer than 75 will survive a year or more after the operation?

b. between 80 and 90, inclusive, will survive a year or more after the operation?

Solution Let X denote the number of patients who survive a year or more after undergoing a heart transplant at the Medical Center. Then X is a binomial random variable. Also, $n = 100$, $p = .7$, and $q = .3$, so

$$\mu = (100)(.7) = 70$$

and
$$\sigma = \sqrt{(100)(.7)(.3)}$$
$$\approx 4.58$$

Approximating the binomial distribution by a normal distribution with a mean of 70 and a standard deviation of 4.58, we find, upon letting Y denote the associated normal random variable,

a. the probability that fewer than 75 patients will survive a year or more is given by

$$P(X < 75) \approx P(Y < 74.5) \qquad \text{(Why?)}$$

$$= P\left(Z < \frac{74.5 - 70}{4.58}\right) \qquad \left[P(X < b) = P\left(Z < \frac{b - \mu}{\sigma}\right)\right]$$

$$= P(Z < 0.98)$$

$$= .8365$$

b. the probability that between 80 and 90, inclusive, of the patients will survive a year or more is given by

$$P(80 \le X \le 90)$$
$$\approx P(79.5 < Y < 90.5) \qquad \qquad \left[P(a < X < b)\right.$$
$$= P\left(\frac{79.5 - 70}{4.58} < Z < \frac{90.5 - 70}{4.58}\right) \qquad \left. = P\left(\frac{a - \mu}{\sigma} < Z < \frac{b - \mu}{\sigma}\right)\right]$$
$$= P(2.07 < Z < 4.48)$$
$$= P(Z < 4.48) - P(Z < 2.07)$$
$$= 1 - .9808 \qquad \qquad \text{[\textit{Note:} } P(Z < 4.48) \approx 1.]$$
$$= .0192 \qquad \qquad \qquad \qquad \circ \; \circ \; \circ$$

SELF-CHECK EXERCISES 8.6

1. The serum cholesterol levels in mg/dl in a current Mediterranean population are found to be normally distributed with a mean of 160 and a standard deviation of 50. Scientists at the National Heart, Lung, and Blood Institute consider this pattern ideal for a minimal risk of heart attacks. Find the percentage of the population having blood cholesterol levels between 160 and 180 mg/dl.

2. It has been estimated that 4% of the luggage manufactured by The Luggage Company fails to meet the standards established by the company and is sold as "seconds" to discount and outlet stores. If 500 bags are produced, what is the probability that more than 30 will be classified as "seconds"?

Solutions to Self-Check Exercises 8.6 can be found on page 532.

8.6 EXERCISES

1. Medical Records   The medical records of infants delivered at the Kaiser Memorial Hospital show that the infants' lengths at birth (in inches) are normally distributed with a mean of 20 and a standard deviation of 2.6. Find the probability that an infant selected at random from among those delivered at the hospital measures
a. more than 22 inches.   **b.** less than 18 inches.
c. between 19 and 21 inches.

2. Factory Workers' Wages   According to the data released by the Chamber of Commerce of a certain city, the weekly wages of factory workers are normally distributed with a mean of $400 and a standard deviation of $50. Find the probability that a worker selected at random from the city makes a weekly wage of
a. less than $300.   **b.** more than $460.
c. between $350 and $450.

3. Product Reliability   The TKK Products Corporation manufactures electric light bulbs in the 50-, 60-, 75-, and 100-watt range. Laboratory tests show that the lives of these light bulbs are normally distributed with a mean of 750 hours and a standard deviation of 75 hours. What is the probability that a TKK light bulb selected at random will burn
a. for more than 900 hours?
b. for less than 600 hours?
c. between 750 and 900 hours?
d. between 600 and 800 hours?

4. Education   On the average, a student takes 100 words per minute midway through an advanced court reporting course at the American Institute of Court Reporting. Assuming that the dictation speeds of the students are normally distributed and that the standard deviation is 20 words per minute, find the probability that a student randomly selected from the course can take dictation at a speed of
a. more than 120 words per minute.
b. between 80 and 120 words per minute.
c. less than 80 words per minute.

5. IQs   The IQs of students at the Wilson Elementary School were measured recently and found to be normally distributed with a mean of 100 and a standard deviation of 15. What is the probability that a student selected at random will have an IQ of
a. 140 or higher?   **b.** 120 or higher?
c. between 100 and 120?   **d.** 90 or less?

6. Product Reliability   The tread lives of the Super Titan radial tires under normal driving conditions are normally distributed with a mean of 40,000 miles and a standard deviation of 2000 miles. What is the probability that a tire selected at random will have a tread life of more than 35,000 miles? If four new tires are installed in a car and they experience even wear, determine the probability that all four tires still have useful tread lives after 35,000 miles of driving.

7. **Female Factory Workers' Wages**   According to data released by the Chamber of Commerce of a certain city, the weekly wages (in dollars) of female factory workers are normally distributed with a mean of 375 and a standard deviation of 50. Find the probability that a female factory worker selected at random from the city makes a weekly wage of $350 to $450.

8. **Civil Service Exams**   To be eligible for further consideration, applicants for certain Civil Service positions must first pass a written qualifying examination on which a score of 70 or more must be obtained. In a recent examination, it was found that the scores were normally distributed with a mean of 60 points and a standard deviation of 10 points. Determine the percentage of applicants who passed the written qualifying examination.

9. **Warranties**   The general manager of the Service Department of the MCA Television Company has estimated that the time that elapses between the dates of purchase and the dates on which the 19-inch sets manufactured by the company first require service is normally distributed with a mean of 22 months and a standard deviation of 4 months. If the company gives a 1-year warranty on parts and labor for these sets, determine the percentage of sets manufactured and sold by the company that may require service before the warranty period runs out.

10. **Grade Distributions**   The scores on an economics examination are normally distributed with a mean of 72 and a standard deviation of 16. If the instructor assigns a grade of A to 10% of the class, what is the lowest score a student may have and still obtain an A?

11. **Grade Distributions**   The scores on a sociology examination are normally distributed with a mean of 70 and a standard deviation of 10. If the instructor assigns A's to 15%, B's to 25%, C's to 40%, D's to 15%, and F's to 5% of the class, find the cutoff points for these grades.

In exercises 12–22, use the appropriate normal distributions to approximate the resulting binomial distributions.

12. A fair coin is tossed 20 times. Find the probability of obtaining
   a. fewer than 8 heads.   **b.** more than 6 heads.
   c. between 6 and 10 heads inclusive.

13. A coin is weighted so that the probability of obtaining a head in a single toss is .4. If the coin is tossed 25 times, determine the probability of obtaining

a. fewer than 10 heads.
b. between 10 and 12 heads, inclusive.
c. more than 15 heads.

14. **Sports**   A basketball player has a 75% chance of making a free throw. What is the probability of his making 100 or more free throws in 120 trials?

15. **Sports**   A marksman's chance of hitting a target with each of his shots is 60%. If he fires 30 shots, what is the probability of his hitting the target
   a. at least 20 times?   **b.** fewer than 10 times?
   c. between 15 and 20 times, inclusive?

16. **Quality Control**   The P.A.R. Bearings Company is the principal supplier of ball bearings for the Sperry Gyroscope Company. It has been determined that 6% of the ball bearings shipped are rejected because they fail to meet tolerance requirements. What is the probability that a shipment of 200 ball bearings contains more than 10 rejects?

17. **Quality Control**   The manager of C & R Clothiers, a major manufacturer of men's shirts, has determined that 3% of C & R's shirts do not meet with company standards and are sold as "seconds" to discount and outlet stores. What is the probability that in a day's production of 200 dozen shirts, less than 10 dozen will be classified as "seconds"?

18. **Industrial Accidents**   The Colorado Mining and Mineral Company has 800 employees engaged in its mining operations. It has been estimated that the probability of a worker meeting with an accident during a 1-year period is .1. What is the probability that more than 70 workers will meet with an accident during the 1-year period?

19. **Drug Testing**   An experiment was conducted to test the effectiveness of a new drug in treating a certain disease. The drug was administered to 50 mice that had been previously exposed to the disease. It was found that 35 mice subsequently recovered from the disease. It was determined that the natural recovery rate from the disease is 0.5.
   a. Determine the probability that 35 or more of the mice not treated with the drug would recover from the disease.
   b. Using the results obtained in (a), comment on the effectiveness of the drug in the treatment of the disease.

20. **Loan Delinquencies**   The manager of the Madison Finance Company has estimated that, because of a

recession year, 5% of its 400 loan accounts will be delinquent. If the manager's estimate is correct, what is the probability that 25 or more of the accounts will be delinquent?

21. **Cruise Ship Bookings**   Because of late cancellations, Neptune Lines, an operator of cruise ships, has a policy of accepting more reservations than there are accommodations available. From experience, 8% of the bookings for the 90-day around-the-world cruise on the S.S. *Drion,* which has accommodations for 2000 passengers, are subsequently canceled. If the management of Neptune Lines has decided, for public relations reasons, that a person who has made a reservation should have a

probability of .99 of obtaining accommodation on the ship, determine the largest number of reservations that should be taken for a cruise on the S.S. *Drion.*

22. **Theater Bookings**   The Preview Showcase, a research firm, screens pilots of new TV shows before a randomly selected audience and then solicits their opinions of the shows. Based on past experience, 20% of those who get complimentary tickets are "no-shows." The theater has a seating capacity of 500. Management has decided, for public relations reasons, that a person who has been solicited for a screening should have a probability of .99 of being seated. How many tickets should the company send out to prospective viewers for each screening?

SOLUTIONS TO SELF-CHECK EXERCISES 8.6

1. Let X be the normal random variable denoting the serum cholesterol levels in mg/dl in the current Mediterranean population under consideration. Then the percentage of the population having blood cholesterol levels between 160 and 180 mg/dl is given by $P(160 < X < 180)$. To compute $P(160 < X < 180)$, we use formula (16), Section 8.5, with $\mu = 160$, $\sigma = 50$, $a = 160$, and $b = 180$. We find

$$P(160 < X < 180) = P\left(\frac{160 - 160}{50} < Z < \frac{180 - 160}{50}\right)$$
$$= P(0 < Z < 0.4)$$
$$= P(Z < 0.4) - P(Z < 0)$$
$$= .6554 - .5000$$
$$= .1554$$

So approximately 15.5% of the population has blood cholesterol levels between 160 and 180 mg/dl.

2. Let X denote the number of substandard bags in the production. Then X has a binomial distribution with $n = 500$, $p = .04$, $q = .96$, so

$$\mu = (500)(.04) = 20$$

and
$$\sigma = \sqrt{(500)(.04)(.96)} = 4.38$$

Approximating the binomial distribution by a normal distribution with a mean of 20 and standard deviation of 4.38, we find that the probability that more than 30 bags in the production of 500 will be substandard is given by

$$P(X > 30) \approx P(Y > 30.5)$$

(Where Y denotes the normal random variable)

$$= P\left(Z > \frac{30.5 - 20}{4.38}\right)$$

$$= P(Z > 2.40)$$

$$= P(Z < -2.40)$$

$$= .0082$$

or approximately 0.8%.

CHAPTER 8 SUMMARY OF PRINCIPAL FORMULAS AND TERMS

Formulas

1. Mean of n numbers

$$\bar{x} = \frac{x_1 + x_2 + \cdots + x_n}{n}$$

2. Expected value

$$E(X) = x_1 p_1 + x_2 p_2 + \cdots + x_n p_n$$

3. Odds in favor of E occurring

$$\frac{P(E)}{P(E^c)}$$

4. Odds against E occurring

$$\frac{P(E^c)}{P(E)}$$

5. Probability of an event occurring given the odds

$$\frac{a}{a + b}$$

6. Variance of a random variable

$$Var(X) = p_1(x_1 - \mu)^2 + p_2(x_2 - \mu)^2 + \cdots + p_n(x_n - \mu)^2$$

7. Standard deviation of a random variable

$$\sigma = \sqrt{Var(X)}$$

8. Chebychev's inequality

$$P(\mu - k\sigma \leq X \leq \mu + k\sigma) \geq 1 - \frac{1}{k^2}$$

9. Probability of x successes in n Bernoulli trials

$$C(n, x)p^x q^{n-x}$$

10. Binomial random variable:
    Mean
    Variance
    Standard deviation

$$\mu = E(X) = np$$
$$Var(X) = npq$$
$$\sigma_x = \sqrt{npq}$$

Terms

Random variable	Variance
Finite discrete random variable	Standard deviation
Infinite discrete random variable	Bernoulli (binomial) trial
Continuous random variable	Binomial experiment
Probability distribution of a random variable	Binomial random variable
	Binomial distribution
Histogram	Probability density function
Average (mean)	Normal distribution
Expected value	

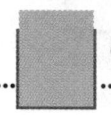

CHAPTER 8 REVIEW EXERCISES

1. Three balls are selected at random without replacement from an urn containing three white balls and four blue balls. Let the random variable X denote the number of blue balls drawn.
   a. List the outcomes of this experiment.
   b. Find the value assigned to each outcome of this experiment by the random variable X.
   c. Find the probability distribution of the random variable associated with this experiment.
   d. Draw the histogram representing this distribution.

2. A man purchased a $25,000, 1-year term life insurance policy for $375. Assuming that the probability that he will live for another year is .989, find the company's expected gain.

3. The probability distribution of a random variable X is shown in the accompanying table.

x	$P(X = x)$
0	.1
1	.1
2	.2
3	.3
4	.2
5	.1

a. Compute $P(1 \le X \le 4)$.
b. Compute the mean and standard deviation of X.

4. A binomial experiment consists of four trials in which the probability of success in any one trial is 2/5.
   a. Construct the probability distribution for the experiment.
   b. Compute the mean and standard deviation of the probability distribution.

In exercises 5–8, let Z be the standard normal variable. Make a rough sketch of the appropriate region under the standard normal curve and find the given probability.

5. $P(Z < 0.5)$ **6.** $P(Z < -0.75)$

7. $P(-0.75 < Z < 0.5)$ **8.** $P(-0.42 < Z < 0.66)$

In exercises 9–12, let Z be the standard normal variable. Find z if z satisfies the given value.

9. $P(Z < z) = .9922$ **10.** $P(Z < z) = .1469$

11. $P(Z > z) = .9788$

12. $P(-z < Z < z) = .8444$

In exercises 13–16, let X be a normal random variable with $\mu = 10$ and $\sigma = 2$. Find the value of the given probability.

13. $P(X < 11)$ **14.** $P(X > 8)$

15. $P(7 < X < 9)$ 16. $P(6.5 < X < 11.5)$

17. If the probability that a bowler will bowl a strike is .7, what is the probability that he will get exactly two strikes in four attempts? At least two strikes in four attempts?

18. The heights of 4000 women who participated in a recent survey were found to be normally distributed with a mean of 64.5 inches and a standard deviation of 2.5 inches. What percentage of these women have heights of 67 inches or greater?

19. Refer to exercise 18. Use Chebychev's inequality to estimate the probability that the height of a woman who participated in the survey will fall within 2 standard deviations of the mean—that is, that her height will be between 59.5 and 69.5 inches.

20. The proprietor of a hardware store will accept a shipment of ceramic wall tiles if no more than 2 of a random sample of 20 are found to be defective. What is the probability that he will accept shipment if exactly 10% of the tiles in a certain shipment is defective?

21. An experimental drug has been found to be effective in treating 15% of the people afflicted by a certain disease. If the drug is administered to 800 people who have this disease, what are the mean and standard deviation of the number of people for whom the drug can be expected to be effective?

22. The Dayton Iron Works Company manufactures steel rods to a specification of 1-inch diameter. These rods are accepted by the buyer if they fall within the tolerance limits of 0.995 and 1.005. Assuming that the diameter of the rods is normally distributed about a mean of 1 inch and a standard deviation of 0.002 inch, estimate the percentage of rods that will be rejected by the buyer.

23. A coin is biased so that the probability of it landing heads is .6. If the coin is tossed 100 times, what is the probability that heads will appear more than 50 times in the 100 tosses?

24. A division of the Solaron Corporation manufactures photovoltaic cells for use in the company's solar energy converters. It is estimated that 5% of the cells manufactured is defective. In a batch of 200 cells manufactured by the company, what is the probability that it will contain at most 20 defective units?

This chapter contains a brief review of the algebra you will need in the study of calculus. To solve many calculus problems, it is often necessary to simplify certain algebraic expressions before moving on to the next step. Thus it is important that you reestablish familiarity with this process. This chapter also contains a short review of inequalities and absolute value; their uses range from describing the domains of functions to formulating practical problems.

How much money is needed to purchase at least 100,000 shares of the Starr Communications Company? Corbyco, a giant conglomerate, wishes to purchase a minimum of 100,000 shares of the company. In Example 4, page 562, you will see how Corbyco's management determines how much money they will need for the acquisition.

9

PRECALCULUS REVIEW

537

9.1 EXPONENTS AND RADICALS

Exponents and Radicals

This section reviews the properties of exponents and radicals. Recall that if b is any real number and n is a positive integer, then the expression b^n (read "b to the power n") is defined as the number

$$b^n = \underbrace{b \cdot b \cdot b \cdot \cdots \cdot b}_{n \text{ factors}}$$

The number b is called the **base,** and the superscript n is called the **power** of the exponential expression b^n. For example,

$$2^5 = 2 \cdot 2 \cdot 2 \cdot 2 \cdot 2 = 32 \quad \text{and} \quad \left(\frac{2}{3}\right)^3 = \left(\frac{2}{3}\right)\left(\frac{2}{3}\right)\left(\frac{2}{3}\right) = \frac{8}{27}$$

If $b \neq 0$, we define

$$b^0 = 1$$

For example, $2^0 = 1$ and $(-\pi)^0 = 1$, but the expression 0^0 is undefined.

Next, recall that if n is a positive integer, then the expression $b^{1/n}$ is defined to be the number that, when raised to the nth power, is equal to b. Thus

$$(b^{1/n})^n = b$$

Such a number, if it exists, is called the **nth root of b,** also written $\sqrt[n]{b}$.

 Observe that the nth root of a negative number is not defined when n is even. For example, the square root of -2 is not defined because there is no real number b such that $b^2 = -2$. Also, given a number b, more than one number might satisfy our definition of the nth root. For example, both 3 and -3 squared equal 9, and each is a square root of 9. So, to avoid ambiguity, we define $b^{1/n}$ to be the positive nth root of b whenever it exists. Thus $\sqrt{9} = 9^{1/2} = 3$.

Recall that if p/q (p and q are integers with $q \neq 0$) is a rational number in lowest terms, then the expression $b^{p/q}$ is defined as the number $(b^{1/q})^p$ or, equivalently, $\sqrt[q]{b^p}$, whenever it exists. Expressions involving negative rational exponents are taken care of by the definition

$$b^{-p/q} = \frac{1}{b^{p/q}}$$

Examples are

$$2^{3/2} = (2^{1/2})^3 \approx (1.4142)^3 \approx 2.8283 \quad \text{and} \quad 4^{-5/2} = \frac{1}{4^{5/2}} = \frac{1}{(4^{1/2})^5} = \frac{1}{2^5} = \frac{1}{32}$$

The rules defining the exponential expression a^n, where $a > 0$ for all rational values of n, are given in Table 9.1.

Table 9.1
Definitions of exponents.

Definition of a^n ($a > 0$)	Example	Definition of a^n ($a > 0$)	Example
Integer exponent: If n is a positive integer, then $$a^n = a \cdot a \cdot a \cdot \cdots \cdot a$$ (n factors of a)	$2^5 = 2 \cdot 2 \cdot 2 \cdot 2 \cdot 2$ (5 factors) $= 32$	**Fractional exponent:** **a.** If n is a positive integer, then $$a^{1/n} \text{ or } \sqrt[n]{a}$$ denotes the **nth root** of a.	$16^{1/2} = \sqrt{16}$ $= 4$
Zero exponent: If n is equal to zero, then $$a^0 = 1$$ (0^0 is not defined.)	$7^0 = 1$	**b.** If m and n are positive integers, then $$a^{m/n} = \sqrt[n]{a^m} = (\sqrt[n]{a})^m$$	$8^{2/3} = (\sqrt[3]{8})^2$ $= 4$
Negative exponent: If n is a positive integer, then $$a^{-n} = \frac{1}{a^n} \quad (a \neq 0)$$	$6^{-2} = \frac{1}{6^2}$ $= \frac{1}{36}$	**c.** If m and n are positive integers, then $$a^{-m/n} = \frac{1}{a^{m/n}} \quad (a \neq 0)$$	$9^{-3/2} = \frac{1}{9^{3/2}}$ $= \frac{1}{27}$

The first three definitions in Table 9.1 are also valid for negative values of a, whereas the fourth definition is valid only for negative values of a when n is odd. Thus,

$$(-8)^{1/3} = \sqrt[3]{-8} = -2 \qquad (n \text{ is odd.})$$
$$(-8)^{1/2} \text{ has no real value} \qquad (n \text{ is even.})$$

Finally, it can be shown that a^n has meaning for *all* real numbers n. For example, using a pocket calculator with a "y^x" key, we see that $2^{\sqrt{2}} \approx 2.665144$.

The Laws of Exponents

The five laws of exponents are listed in Table 9.2. These laws are valid for any real numbers a, b, m, and n whenever the quantities are defined.

Table 9.2
Laws of exponents.

Law	Example
1. $a^m \cdot a^n = a^{m+n}$	$x^2 \cdot x^3 = x^{2+3} = x^5$
2. $\dfrac{a^m}{a^n} = a^{m-n} \quad (a \neq 0)$	$\dfrac{x^7}{x^4} = x^{7-4} = x^3$
3. $(a^m)^n = a^{m \cdot n}$	$(x^4)^3 = x^{4 \cdot 3} = x^{12}$
4. $(ab)^n = a^n \cdot b^n$	$(2x)^4 = 2^4 \cdot x^4 = 16x^4$
5. $\left(\dfrac{a}{b}\right)^n = \dfrac{a^n}{b^n} \quad (b \neq 0)$	$\left(\dfrac{x}{2}\right)^3 = \dfrac{x^3}{2^3} = \dfrac{x^3}{8}$

Remember, $(x^2)^3 \neq x^5$. The correct equation is $(x^2)^3 = x^{2\cdot3} = x^6$.

The next example illustrates the use of the laws of exponents.

EXAMPLE 1   Simplify each expression:

a. $(3x^2)(4x^3)$ **b.** $\dfrac{16^{5/4}}{16^{1/2}}$ **c.** $(6^{2/3})^3$ **d.** $(x^3y^{-2})^{-2}$ **e.** $\left(\dfrac{y^{3/2}}{x^{1/4}}\right)^{-2}$

Solution

a. $(3x^2)(4x^3) = 12x^{2+3} = 12x^5$ (Law 1)

b. $\dfrac{16^{5/4}}{16^{1/2}} = 16^{5/4-1/2} = 16^{3/4} = (\sqrt[4]{16})^3 = 2^3 = 8$ (Law 2)

c. $(6^{2/3})^3 = 6^{2/3\cdot3} = 6^{6/3} = 6^2 = 36$ (Law 3)

d. $(x^3y^{-2})^{-2} = (x^3)^{-2}(y^{-2})^{-2} = x^{(3)(-2)}y^{(-2)(-2)} = x^{-6}y^4 = \dfrac{y^4}{x^6}$ (Law 4)

e. $\left(\dfrac{y^{3/2}}{x^{1/4}}\right)^{-2} = \dfrac{y^{(3/2)(-2)}}{x^{(1/4)(-2)}} = \dfrac{y^{-3}}{x^{-1/2}} = \dfrac{x^{1/2}}{y^3}$ (Law 5)

○ ○ ○

Simplifying Radicals

We can also use the laws of exponents to simplify expressions involving radicals, as illustrated in the next example.

EXAMPLE 2   Simplify the expression. (Assume x, y, and n are positive.)

a. $\sqrt[4]{16x^4y^8}$ **b.** $\sqrt{12m^3n} \cdot \sqrt{3m^5n}$ **c.** $\dfrac{\sqrt[3]{-27x^6}}{\sqrt[3]{8y^3}}$

Solution

a. $\sqrt[4]{16x^4y^8} = (16x^4y^8)^{1/4} = 16^{1/4} \cdot x^{4/4}y^{8/4} = 2xy^2$

b. $\sqrt{12m^3n} \cdot \sqrt{3m^5n} = \sqrt{36m^8n^2} = (36m^8n^2)^{1/2} = 36^{1/2} \cdot m^{8/2}n^{2/2}$
$= 6m^4n$

c. $\dfrac{\sqrt[3]{-27x^6}}{\sqrt[3]{8y^3}} = \dfrac{(-27x^6)^{1/3}}{(8y^3)^{1/3}} = \dfrac{-27^{1/3}x^{6/3}}{8^{1/3}y^{3/3}} = -\dfrac{3x^2}{2y}$

○ ○ ○

When a radical appears in the numerator or denominator of an algebraic expression, we often try to simplify the expression by eliminating the radical from the numerator or denominator. The process, called **rationalization**, is illustrated in the next two examples.

EXAMPLE 3   Rationalize the denominators: **a.** $\dfrac{3x}{2\sqrt{x}}$ **b.** $\sqrt[3]{\dfrac{x}{y}}$

Solution

a. $\dfrac{3x}{2\sqrt{x}} = \dfrac{3x}{2\sqrt{x}} \cdot \dfrac{\sqrt{x}}{\sqrt{x}} = \dfrac{3x\sqrt{x}}{2\sqrt{x^2}} = \dfrac{3x\sqrt{x}}{2x} = \dfrac{3}{2}\sqrt{x}$

b. $\sqrt[3]{\dfrac{x}{y}} = \sqrt[3]{\dfrac{x}{y}} \cdot \sqrt[3]{\dfrac{y^2}{y^2}} = \dfrac{\sqrt[3]{xy^2}}{\sqrt[3]{y^3}} = \dfrac{\sqrt[3]{xy^2}}{y}$ ○ ○ ○

EXAMPLE 4   Rationalize the numerators: **a.** $\dfrac{3\sqrt{x}}{2x}$ **b.** $\sqrt[3]{\dfrac{x^2}{yz^3}}$

Solution

a. $\dfrac{3\sqrt{x}}{2x} = \dfrac{3\sqrt{x}}{2x} \cdot \dfrac{\sqrt{x}}{\sqrt{x}} = \dfrac{3\sqrt{x^2}}{2x\sqrt{x}} = \dfrac{3x}{2x\sqrt{x}} = \dfrac{3}{2\sqrt{x}}$

b. $\sqrt[3]{\dfrac{x^2}{yz^3}} \sqrt[3]{\dfrac{x}{x}} = \dfrac{\sqrt[3]{x^3}}{\sqrt[3]{xyz^3}} = \dfrac{x}{z\sqrt[3]{xy}}$ ○ ○ ○

9.1 | **EXERCISES**

In exercises 1–16, evaluate the expression.

1. $27^{2/3}$

2. $8^{-4/3}$

3. $\left(\dfrac{1}{\sqrt{3}}\right)^{0}$

4. $(7^{1/2})^4$

5. $\left[\left(\dfrac{1}{8}\right)^{1/3}\right]^{-2}$

6. $\left[\left(-\dfrac{1}{3}\right)^2\right]^{-3}$

7. $\left(\dfrac{7^{-5} \cdot 7^2}{7^{-2}}\right)^{-1}$

8. $\left(\dfrac{9}{16}\right)^{-1/2}$

9. $(125^{2/3})^{-1/2}$

10. $\sqrt[3]{2^6}$

11. $\dfrac{\sqrt{32}}{\sqrt{8}}$

12. $\sqrt[3]{\dfrac{-8}{27}}$

13. $\dfrac{16^{5/8}16^{1/2}}{16^{7/8}}$

14. $\left(\dfrac{9^{-3} \cdot 9^5}{9^{-2}}\right)^{-1/2}$

15. $16^{1/4} \cdot (8)^{-1/3}$

16. $\dfrac{6^{2.5} \cdot 6^{-1.9}}{6^{-1.4}}$

In exercises 17–26, determine whether the statement is true or false. Give a reason for your choice.

17. $x^4 + 2x^4 = 3x^4$

18. $3^2 \cdot 2^2 = 6^2$

19. $x^3 \cdot 2x^2 = 2x^6$

20. $3^3 + 3 = 3^4$

21. $\dfrac{2^{4x}}{1^{3x}} = 2^{4x-3x}$

22. $(2^2 \cdot 3^2)^2 = 6^4$

23. $\dfrac{1}{4^{-3}} = \dfrac{1}{64}$

24. $\dfrac{4^{3/2}}{2^4} = \dfrac{1}{2}$

25. $(1.2^{1/2})^{-1/2} = 1$

26. $5^{2/3} \cdot (25)^{2/3} = 25$

In exercises 27–32, rewrite the expression using positive exponents only.

27. $(xy)^{-2}$

28. $3s^{1/3} \cdot s^{-7/3}$

29. $\dfrac{x^{-1/3}}{x^{1/2}}$

30. $\sqrt{x^{-1}} \cdot \sqrt{9x^{-3}}$

31. $12^0(s+t)^{-3}$

32. $(x-y)(x^{-1}+y^{-1})$

In exercises 33–48, simplify the expression.

33. $\dfrac{x^{7/3}}{x^{-2}}$

34. $(49x^{-2})^{-1/2}$

35. $(x^2y^{-3})(x^{-5}y^3)$

36. $\dfrac{5x^6y^3}{2x^2y^7}$

37. $\dfrac{x^{3/4}}{x^{-1/4}}$

38. $\left(\dfrac{x^3y^2}{z^2}\right)^2$

39. $\left(\dfrac{x^3}{-27y^{-6}}\right)^{-2/3}$

40. $\left(\dfrac{e^x}{e^{x-2}}\right)^{-1/2}$

41. $\left(\dfrac{x^{-3}}{y^{-2}}\right)^2\left(\dfrac{y}{x}\right)^4$

42. $\dfrac{(r^n)^4}{r^{5-2n}}$

43. $\sqrt[3]{x^{-2}} \cdot \sqrt{4x^5}$

44. $\sqrt{81x^6y^{-4}}$

45. $-\sqrt[4]{16x^4y^8}$

46. $\sqrt[3]{x^{3a+b}}$

47. $\sqrt[6]{64x^8y^3}$

48. $\sqrt[3]{27r^6} \cdot \sqrt{s^2t^4}$

In exercises 49–52, use the fact that $2^{1/2} \approx 1.414$ and $3^{1/2} \approx 1.732$ to evaluate the expression without using a calculator.

49. $2^{3/2}$

50. $8^{1/2}$

51. $9^{3/4}$

52. $6^{1/2}$

In exercises 53–56, use the fact that $10^{1/2} \approx 3.162$ and $10^{1/3} \approx 2.154$ to evaluate the expression without using a calculator.

53. $10^{3/2}$

54. $1000^{3/2}$

55. $10^{2.5}$

56. $(0.0001)^{-1/3}$

In exercises 57–62, rationalize the denominator of the expression.

57. $\dfrac{3}{2\sqrt{x}}$

58. $\dfrac{3}{\sqrt{xy}}$

59. $\dfrac{2y}{\sqrt{3y}}$

60. $\dfrac{5x^2}{\sqrt{3x}}$

61. $\dfrac{1}{\sqrt[3]{x}}$

62. $\sqrt{\dfrac{2x}{y}}$

In exercises 63–68, rationalize the numerator of the expression.

63. $\dfrac{2\sqrt{x}}{3}$

64. $\dfrac{\sqrt[3]{x}}{24}$

65. $\sqrt{\dfrac{2y}{x}}$

66. $\sqrt[3]{\dfrac{2x}{3y}}$

67. $\dfrac{\sqrt[3]{x^2z}}{y}$

68. $\dfrac{\sqrt[3]{x^2y}}{2x}$

9.2 ALGEBRAIC EXPRESSIONS

In calculus, we often work with algebraic expressions such as

$$2x^{4/3} - x^{1/3} + 1, \quad 2x^2 - x - \frac{2}{\sqrt{x}}, \quad \frac{3xy+2}{x+1}, \quad \text{and} \quad 2x^3 + 2x + 1$$

An algebraic expression of the form ax^n, where the coefficient a is a real number and n is a nonnegative integer, is called a **monomial,** meaning it consists of one term. For example, $7x^2$ is a monomial. A **polynomial** is a monomial or the sum of two or more monomials. For example,

$$x^2 + 4x + 4, \quad x^3 + 5, \quad x^4 + 3x^2 + 3, \quad \text{and} \quad x^2y + xy + y$$

are all polynomials.

Adding and Subtracting Algebraic Expressions

Constant terms and terms that contain the same variable factor are called **like,** or **similar, terms.** Like terms may be combined by adding or subtracting their numerical coefficients. For example,

$$3x + 7x = 10x \quad \text{and} \quad \frac{1}{2}xy + 3xy = \frac{7}{2}xy$$

The distributive property of the real number system,

$$ab + ac = a \cdot (b + c)$$

is used to justify this procedure.

To add or subtract two or more algebraic expressions, first remove the parentheses and then combine like terms. The resulting expression is written in order of decreasing degree from left to right.

$\circledR$* **EXAMPLE I**
....

a. $(3x^4 + 10x^3 + 6x^2 + 10x + 3) + (2x^4 + 10x^3 + 6x^2 + 4x)$

$\quad = 3x^4 + 2x^4 + 10x^3 + 10x^3 + 6x^2 + 6x^2 + 10x$

$\qquad + 4x + 3 \qquad$ (Remove parentheses.)

$\quad = 5x^4 + 20x^3 + 12x^2 + 14x + 3 \qquad$ (Combine like terms.)

b. $(2x^4 + 3x^3 + 4x + 6) - (3x^4 + 9x^3 + 3x^2)$

$\qquad = 2x^4 + 3x^3 + 4x + 6 - 3x^4 - 9x^3 - 3x^2 \qquad$ (Remove parentheses.)

$\qquad = 2x^4 - 3x^4 + 3x^3 - 9x^3 - 3x^2 + 4x + 6$

$\qquad = -x^4 - 6x^3 - 3x^2 + 4x + 6 \qquad$ (Combine like terms.)

c. $2\sqrt{x} + 7y - \dfrac{1}{2}\sqrt{x} + 3y = 2\sqrt{x} - \dfrac{1}{2}\sqrt{x} + 7y + 3y = \dfrac{3}{2}\sqrt{x} + 10y$

d. $\left(2x^{3/2} - x^{1/2} + \dfrac{2}{3}x^{-1/3}\right) - \left(3x^{3/2} + \dfrac{1}{2}x^{1/2} - \dfrac{1}{3}x^{-1/3}\right)$

$\qquad = 2x^{3/2} - x^{1/2} + \dfrac{2}{3}x^{-1/3} - 3x^{3/2} - \dfrac{1}{2}x^{1/2} + \dfrac{1}{3}x^{-1/3} \qquad$ (Remove parentheses.)

$\qquad = 2x^{3/2} - 3x^{3/2} - x^{1/2} - \dfrac{1}{2}x^{1/2} + \dfrac{2}{3}x^{-1/3} + \dfrac{1}{3}x^{-1/3}$

$\qquad = -x^{3/2} - \dfrac{3}{2}x^{1/2} + x^{-1/3} \qquad$ (Combine like terms.)

e. $2t^3 - \{t^2 - [t - (2t - 1)] + 4\}$

$\qquad = 2t^3 - \{t^2 - [t - 2t + 1] + 4\}$

$\qquad = 2t^3 - \{t^2 - [-t + 1] + 4\} \qquad$ (Remove parentheses and combine like terms within brackets.)

$\qquad = 2t^3 - \{t^2 + t - 1 + 4\} \qquad$ (Remove brackets.)

$\qquad = 2t^3 - \{t^2 + t + 3\} \qquad$ (Combine like terms within braces.)

$\qquad = 2t^3 - t^2 - t - 3 \qquad$ (Remove braces.)

○ ○ ○

○ ○ ○

* The symbol $\circledR$ indicates that these examples were selected from the calculus portion of the text in order to help students review the algebraic computations they will *actually* be using in calculus.

An algebraic expression is said to be **simplified** if none of its terms are similar. Observe that when the algebraic expression in (e) of Example 1 was simplified, the innermost grouping symbols were removed first; that is, the parentheses ( ) were removed first, the brackets [ ] second, and the braces { } third.

Multiplying Algebraic Expressions

When we multiply algebraic expressions, each term of one algebraic expression is multiplied by each term of the other. The resulting algebraic expression is then simplified.

EXAMPLE 2 Expand the expression $(x + 3)(x - 2)$.

Solution

$$(x + 3)(x - 2) = x(x - 2) + 3(x - 2)$$
$$= x^2 - 2x + 3x - 6$$
$$= x^2 + x - 6$$

⊙ ⊙ ⊙

Ⓡ **EXAMPLE 3** Perform the indicated operations:

a. $(3x - 4)(3x^2 - 2x + 3)$ **b.** $(x^2 + 1)(3x^2 + 10x + 3)$

c. $(x^2 + x + 1)(x^2 + x + 1)$ **d.** $(e^t + e^{-t})e^t - e^t(e^t - e^{-t})$

Solution

a. $(3x - 4)(3x^2 - 2x + 3)$
$$= 3x(3x^2 - 2x + 3) - 4(3x^2 - 2x + 3)$$
$$= 9x^3 - 6x^2 + 9x - 12x^2 + 8x - 12$$
$$= 9x^3 - 18x^2 + 17x - 12$$

b. $(x^2 + 1)(3x^2 + 10x + 3)$
$$= x^2(3x^2 + 10x + 3) + 1(3x^2 + 10x + 3)$$
$$= 3x^4 + 10x^3 + 3x^2 + 3x^2 + 10x + 3$$
$$= 3x^4 + 10x^3 + 6x^2 + 10x + 3$$

c. $(x^2 + x + 1)(x^2 + x + 1)$
$$= x^2(x^2 + x + 1) + x(x^2 + x + 1) + 1(x^2 + x + 1)$$
$$= x^4 + x^3 + x^2 + x^3 + x^2 + x + x^2 + x + 1$$
$$= x^4 + 2x^3 + 3x^2 + 2x + 1$$

d. $(e^t + e^{-t})e^t - e^t(e^t - e^{-t})$
$$= e^{2t} + e^0 - e^{2t} + e^0$$
$$= e^{2t} - e^{2t} + e^0 + e^0$$
$$= 1 + 1 \qquad \text{(Recall that } e^0 = 1.)$$
$$= 2$$

⊙ ⊙ ⊙

Table 9.3
Product formulas.

Formula	Example
$(a + b)^2 = a^2 + 2ab + b^2$	$(2x + 3y)^2 = (2x)^2 + 2(2x)(3y) + (3y)^2$
	$= 4x^2 + 12xy + 9y^2$
$(a - b)^2 = a^2 - 2ab + b^2$	$(4x - 2y)^2 = (4x)^2 - 2(4x)(2y) + (2y)^2$
	$= 16x^2 - 16xy + 4y^2$
$(a + b)(a - b) = a^2 - b^2$	$(2x + y)(2x - y) = (2x)^2 - (y)^2$
	$= 4x^2 - y^2$

Certain product formulas that are frequently used in algebraic computations are given in Table 9.3.

Factoring

Factoring is the process of expressing an algebraic expression as a product of other algebraic expressions. For example, by applying the distributive property, we may write

$$3x^2 - x = x(3x - 1)$$

The first step in factoring an algebraic expression is to check whether it contains any common terms. If it does, the greatest common term is then factored out. For example, the common factor of the algebraic expression $2a^2x + 4ax + 6a$ is $2a$ because

$$2a^2x + 4ax + 6a = 2a \cdot ax + 2a \cdot 2x + 2a \cdot 3 = 2a(ax + 2x + 3)$$

Ⓡ **EXAMPLE 4**   Factor out the greatest common factor in each of the following expressions:

a. $-0.3t^2 + 3t$ **b.** $6a^4b^4c - 3a^3b^2c - 9a^2b^2$ **c.** $2x^{3/2} - 3x^{1/2}$

d. $2ye^{xy^2} + 2xy^3e^{xy^2}$ **e.** $4x(x + 1)^{1/2} - 2x^2(\frac{1}{2})(x + 1)^{-1/2}$

Solution

a. $-0.3t^2 + 3t = -0.3t(t - 10)$

b. $6a^4b^4c - 3a^3b^2c - 9a^2b^2 = 3a^2b^2(2a^2b^2c - ac - 3)$

c. $2x^{3/2} - 3x^{1/2} = x^{1/2}(2x - 3)$

d. $2ye^{xy^2} + 2xy^3e^{xy^2} = 2ye^{xy^2}(1 + xy^2)$

e. $4x(x + 1)^{1/2} - 2x^2(\frac{1}{2})(x + 1)^{-1/2} = 4x(x + 1)^{1/2} - x^2(x + 1)^{-1/2}$

$$= x(x + 1)^{-1/2}[4(x + 1)^{1/2}(x + 1)^{1/2} - x]$$
$$= x(x + 1)^{-1/2}[4(x + 1) - x]$$
$$= x(x + 1)^{-1/2}(4x + 4 - x) = x(x + 1)^{-1/2}(3x + 4)$$

Here we select $(x + 1)^{-1/2}$ as the common factor because it is "contained" in each algebraic term. In particular, observe that

$$(x + 1)^{-1/2}(x + 1)^{1/2}(x + 1)^{1/2} = (x + 1)^{1/2} \qquad \circ \; \circ \; \circ$$

Sometimes an algebraic expression may be factored by regrouping and rearranging its terms and then factoring out a common term. This technique is illustrated in Example 5.

EXAMPLE 5   Factor:

a. $2ax + 2ay + bx + by$ **b.** $3x\sqrt{y} - 4 - 2\sqrt{y} + 6x$

Solution

a. First, factor the common term $2a$ from the first two terms and the common term b from the last two terms. Thus,

$$2ax + 2ay + bx + by = 2a(x + y) + b(x + y)$$

Since $(x + y)$ is common to both terms of the polynomial, we may factor it out. Hence,

$$2a(x + y) + b(x + y) = (x + y)(2a + b)$$

b. $3x\sqrt{y} - 4 - 2\sqrt{y} + 6x = 3x\sqrt{y} - 2\sqrt{y} + 6x - 4$
$$= \sqrt{y}(3x - 2) + 2(3x - 2)$$
$$= (3x - 2)(\sqrt{y} + 2) \qquad \circ \; \circ \; \circ$$

Factoring Polynomials

A polynomial that is **prime** over a given set of numbers cannot be expressed as a product of two polynomials of positive degree with coefficients belonging to that set. For example, $x^2 + 2x + 2$ is prime over the set of polynomials with integral coefficients because it cannot be expressed as the product of two polynomials of positive degree that have integral coefficients.

The first step in factoring a polynomial is to find the common factors. The next step is to express the polynomial as the product of a constant and/or one or more prime polynomials.

Certain product formulas that are useful in factoring binomials and trinomials are listed in Table 9.4.

Table 9.4

Formula	Example
Difference of two squares $x^2 - y^2 = (x + y)(x - y)$	$x^2 - 36 = (x + 6)(x - 6)$ $8x^2 - 2y^2 = 2(4x^2 - y^2)$ $= 2(2x + y)(2x - y)$ $9 - a^6 = (3 + a^3)(3 - a^3)$

Table 9.4 continued	Formula	Example
	Perfect-square trinomial	
	$x^2 + 2xy + y^2 = (x + y)^2$	$x^2 + 8x + 16 = (x + 4)^2$
	$x^2 - 2xy + y^2 = (x - y)^2$	$4x^2 - 4xy + y^2 = (2x - y)^2$
	Sum of two cubes	
	$x^3 + y^3 = (x + y)(x^2 - xy + y^2)$	$z^3 + 27 = z^3 + (3)^3$
		$= (z + 3)(z^2 - 3z + 9)$
	Difference of two cubes	
	$x^3 - y^3 = (x - y)(x^2 + xy + y^2)$	$8x^3 - y^6 = (2x)^3 - (y^2)^3$
		$= (2x - y^2)(4x^2 + 2xy^2 + y^4)$

The factors of the second-degree polynomial with integral coefficients

$$px^2 + qx + r$$

are $(ax + b)(cx + d)$, where $ac = p$, $ad + bc = q$, and $bd = r$. Since only a limited number of choices are possible, we use a trial-and-error method to factor polynomials having this form.

For example, to factor $x^2 - 2x - 3$, we first observe that the only possible first-degree terms are

$$(x \quad)(x \quad) \qquad \text{(Since the coefficient of } x^2 \text{ is 1)}$$

Next, we observe that the product of the constant terms is (-3). This gives us the following possible factors:

$$(x - 1)(x + 3)$$

$$(x + 1)(x - 3)$$

Looking once again at the polynomial $x^2 - 2x - 3$, we see that the coefficient of x is -2. Checking to see which set of factors yields -2 for the coefficient of x, we find that

Coefficients of inner terms
Coefficients of outer terms

Factors

Outer terms

$$(-1)(1) + (1)(3) = 2$$

$$(x - 1)(x + 3)$$

Inner terms

Coefficients of inner terms
Coefficients of outer terms

Outer terms

$$(1)(1) + (1)(-3) = -2$$

$$(x + 1)(x - 3)$$

Inner terms

and we conclude that the correct factorization is

$$x^2 - 2x - 3 = (x + 1)(x - 3)$$

With practice, you will soon find that you can perform many of these steps mentally, and the need to write out each step will be eliminated.

Ⓡ **EXAMPLE 6**   Factor:

a. $3x^2 + 4x - 4$ **b.** $3x^2 - 6x - 24$

Solution

a. Using trial and error, we find that the correct factorization is

$$3x^2 + 4x - 4 = (3x - 2)(x + 2)$$

b. Since each term has the common factor 3, we have

$$3x^2 - 6x - 24 = 3(x^2 - 2x - 8)$$

Using the trial-and-error method of factorization, we find that

$$x^2 - 2x - 8 = (x - 4)(x + 2)$$

Thus, we have

$$3x^2 - 6x - 24 = 3(x - 4)(x + 2)$$ ◐ ◑ ◐

Roots of Polynomial Equations

A polynomial equation of degree n in the variable x is an equation of the form

$$a_n x^n + a_{n-1} x^{n-1} + \cdots + a_0 = 0$$

where n is a nonnegative integer and $a_0, a_1, \ldots, a_n$ are real numbers with $a_n \neq 0$. For example, the equation

$$-2x^5 + 8x^3 - 6x^2 + 3x + 1 = 0$$

is a polynomial equation of degree 5 in x.

The **roots** of a polynomial equation are precisely the values of x that satisfy the given equation.* One way of finding the roots of a polynomial equation is to first factor the polynomial and then solve the resulting equation. For example, the polynomial equation

$$x^3 - 3x^2 + 2x = 0$$

may be rewritten in the form

$$x(x^2 - 3x + 2) = 0$$

or $$x(x - 1)(x - 2) = 0$$

Since the product of two real numbers can be equal to zero if and only if one (or both) of the factors is equal to zero, we have

$$x = 0, \quad x - 1 = 0, \quad \text{or} \quad x - 2 = 0$$

from which we see that the desired roots are $x = 0, 1$, and 2.

◐ ◑ ◐

* In this book, we are interested only in the *real* roots of an equation.

The Quadratic Formula

In general, the problem of finding the roots of a polynomial equation is a difficult one. But the roots of a quadratic equation (a polynomial equation of degree 2) are easily found either by factoring or by using the following quadratic formula.

THE QUADRATIC FORMULA

The solutions of the equation $ax^2 + bx + c = 0$ ($a \neq 0$) are given by

$$x = \frac{-b \pm \sqrt{b^2 - 4ac}}{2a}$$

EXAMPLE 7 Solve each of the following quadratic equations:

a. $2x^2 + 5x - 12 = 0$ **b.** $x^2 = -3x + 8$

Solution

a. The equation is in standard form, with $a = 2$, $b = 5$, and $c = -12$. Using the quadratic formula, we find

$$x = \frac{-b \pm \sqrt{b^2 - 4ac}}{2a} = \frac{-5 \pm \sqrt{5^2 - 4(2)(-12)}}{2(2)}$$

$$= \frac{-5 \pm \sqrt{121}}{4} = \frac{-5 \pm 11}{4}$$

$$= -4 \text{ or } \frac{3}{2}$$

This equation can also be solved by factoring. Thus

$$2x^2 + 5x - 12 = (2x - 3)(x + 4) = 0$$

from which we see that the desired roots are $x = 3/2$ or $x = -4$, as obtained earlier.

b. We first rewrite the given equation in the standard form $x^2 + 3x - 8 = 0$, from which we see that $a = 1$, $b = 3$, and $c = -8$. Using the quadratic formula, we find

$$x = \frac{-b \pm \sqrt{b^2 - 4ac}}{2a} = \frac{-3 \pm \sqrt{3^2 - 4(1)(-8)}}{2(1)}$$

$$= \frac{-3 \pm \sqrt{41}}{2}$$

That is, the solutions are $\dfrac{-3 + \sqrt{41}}{2} \approx 1.7$ and $\dfrac{-3 - \sqrt{41}}{2} \approx -4.7$. In this case, the quadratic formula proves quite handy!   ๐ ๐ ๐

9.2 EXERCISES

In exercises 1–22, perform the indicated operations and simplify each expression.

1. $(7x^2 - 2x + 5) + (2x^2 + 5x - 4)$

2. $(3x^2 + 5xy + 2y) + (4 - 3xy - 2x^2)$

3. $(5y^2 - 2y + 1) - (y^2 - 3y - 7)$

4. $3(2a - b) - 4(b - 2a)$

5. $x - \{2x - [-x - (1 - x)]\}$

6. $3x^2 - \{x^2 + 1 - x[x - (2x - 1)]\} + 2$

7. $\left(\dfrac{1}{3} - 1 + e\right) - \left(-\dfrac{1}{3} - 1 + e^{-1}\right)$

8. $-\dfrac{3}{4}y - \dfrac{1}{4}x + 100 + \dfrac{1}{2}x + \dfrac{1}{4}y - 120$

9. $3\sqrt{8} + 8 - 2\sqrt{y} + \dfrac{1}{2}\sqrt{x} - \dfrac{3}{4}\sqrt{y}$

10. $\dfrac{8}{9}x^2 + \dfrac{2}{3}x + \dfrac{16}{3}x^2 - \dfrac{16}{3}x - 2x + 2$

11. $(x + 8)(x - 2)$

12. $(5x + 2)(3x - 4)$

13. $(a + 5)^2$

14. $(3a - 4b)^2$

15. $(x + 2y)^2$

16. $(6 - 3x)^2$

17. $(2x + y)(2x - y)$

18. $(3x + 2)(2 - 3x)$

19. $(x^2 - 1)(2x) - x^2(2x)$

20. $(x^{1/2} + 1)\left(\dfrac{1}{2}x^{-1/2}\right) - (x^{1/2} - 1)\left(\dfrac{1}{2}x^{-1/2}\right)$

21. $2(t + \sqrt{t})^2 - 2t^2$

22. $2x^2 + (-x + 1)^2$

In exercises 23–30, factor out the greatest common factor from each expression.

23. $4x^5 - 12x^4 - 6x^3$

24. $4x^2y^2z - 2x^5y^2 + 6x^3y^2z^2$

25. $7a^4 - 42a^2b^2 + 49a^3b$

26. $3x^{2/3} - 2x^{1/3}$

27. $e^{-x} - xe^{-x}$

28. $2ye^{xy^2} + 2xy^3e^{xy^2}$

29. $2x^{-5/2} - \dfrac{3}{2}x^{-3/2}$

30. $\dfrac{1}{2}\left(\dfrac{2}{3}u^{3/2} - 2u^{1/2}\right)$

In exercises 31–44, factor each expression.

31. $6ac + 3bc - 4ad - 2bd$

32. $3x^3 - x^2 + 3x - 1$

33. $4a^2 - b^2$

34. $12x^2 - 3y^2$

35. $10 - 14x - 12x^2$

36. $x^2 - 2x - 15$

37. $3x^2 - 6x - 24$

38. $3x^2 - 4x - 4$

39. $12x^2 - 2x - 30$

40. $(x + y)^2 - 1$

41. $9x^2 - 16y^2$

42. $8a^2 - 2ab - 6b^2$

43. $x^6 + 125$

44. $x^3 - 27$

In exercises 45–52, perform the indicated operations and simplify the algebraic expression.

45. $(x^2 + y^2)x - xy(2y)$

46. $2kr(R - r) - kr^2$

47. $2(x - 1)(2x + 2)^3[4(x - 1) + (2x + 2)]$

48. $5x^2(3x^2 + 1)^4(6x) + (3x^2 + 1)^5(2x)$

49. $4(x - 1)^2(2x + 2)^3(2) + (2x + 2)^4(2)(x - 1)$

50. $(x^2 + 1)(4x^3 - 3x^2 + 2x) - (x^4 - x^3 + x^2)(2x)$

51. $(x^2 + 2)^2[5(x^2 + 2)^2 - 3](2x)$

52. $(x^2 - 4)(x^2 + 4)(2x + 8) - (x^2 + 8x - 4)(4x^3)$

In exercises 53–58, find the real roots of each equation by factoring.

53. $x^2 + x - 12 = 0$

54. $3x^2 - x - 4 = 0$

55. $4t^2 + 2t - 2 = 0$

56. $-6x^2 + x + 12 = 0$

57. $\frac{1}{4}x^2 - x + 1 = 0$

58. $\frac{1}{2}a^2 + a - 12 = 0$

In exercises 59–64, use the quadratic formula to solve the quadratic equation.

59. $4x^2 + 5x - 6 = 0$

60. $3x^2 - 4x + 1 = 0$

61. $8x^2 - 8x - 3 = 0$

62. $x^2 - 6x + 6 = 0$

63. $2x^2 + 4x - 3 = 0$

64. $2x^2 + 7x - 15 = 0$

9.3 ALGEBRAIC FRACTIONS

Rational Expressions

Quotients of polynomials are called **rational expressions.** Examples of rational expressions are

$$\frac{6x - 1}{2x + 3}, \quad \frac{3x^2y^3 - 2xy}{4x}, \quad \text{and} \quad \frac{2}{5ab}$$

Since rational expressions are quotients in which the variables represent real numbers, the properties of the real numbers apply to rational expressions as well, and operations with rational fractions are performed in the same manner as operations with arithmetic fractions. For example, using the properties of the real number system, we may write

$$\frac{ac}{bc} = \frac{a}{b} \cdot \frac{c}{c} = \frac{a}{b} \cdot 1 = \frac{a}{b}$$

where a, b, and c are any real numbers and b and c are not zero.

Similarly, using the same properties of real numbers, we may write

$$\frac{(x + 2)(x - 3)}{(x - 2)(x - 3)} = \frac{x + 2}{x - 2} \quad (x \neq 2, 3)$$

after "canceling" the common factors.

$\frac{\cancel{3} + 4x}{\cancel{3}} \neq 1 + 4x$ is an example of incorrect cancellation. Instead we need to write $\dfrac{3 + 4x}{3} = \dfrac{3}{3} + \dfrac{4x}{3} = 1 + \dfrac{4x}{3}$.

A rational expression is simplified, or in lowest terms, when the numerator and denominator have no common factors other than 1 and -1 and the expression contains no negative exponents.

® **EXAMPLE 1** Simplify the following expressions:

a. $\dfrac{x^2 + 2x - 3}{x^2 + 4x + 3}$ **b.** $\dfrac{3 - 4x - 4x^2}{2x - 1}$

c. $\dfrac{[(t^2 + 4)(2t - 4) - (t^2 - 4t + 4)(2t)]}{(t^2 + 4)^2}$

d. $\dfrac{(1 + x^2)^2(-2x) - (1 - x^2)2(1 + x^2)(2x)}{(1 + x^2)^4}$

e. $\dfrac{\frac{1}{4}(x + h)^2 - \frac{1}{4}x^2}{h}$

Solution

a. $\dfrac{x^2 + 2x - 3}{x^2 + 4x + 3} = \dfrac{(x + 3)(x - 1)}{(x + 3)(x + 1)} = \dfrac{x - 1}{x + 1}$

b. $\dfrac{3 - 4x - 4x^2}{2x - 1} = \dfrac{(1 - 2x)(3 + 2x)}{2x - 1}$

$\qquad\qquad = \dfrac{-(2x - 1)(2x + 3)}{2x - 1} \qquad [1 - 2x = -(2x - 1)]$

$\qquad\qquad = -(2x + 3)$

c. $\dfrac{[(t^2 + 4)(2t - 4) - (t^2 - 4t + 4)(2t)]}{(t^2 + 4)^2}$

$\qquad = \dfrac{2t^3 - 4t^2 + 8t - 16 - 2t^3 + 8t^2 - 8t}{(t^2 + 4)^2}$ (Carry out the indicated multiplication.)

$\qquad = \dfrac{4t^2 - 16}{(t^2 + 4)^2}$ (Combine like terms.)

$\qquad = \dfrac{4(t^2 - 4)}{(t^2 + 4)^2}$ (Factor.)

d. $\dfrac{(1 + x^2)^2(-2x) - (1 - x^2)2(1 + x^2)(2x)}{(1 + x^2)^4}$

$\qquad = \dfrac{(1 + x^2)(2x)[(1 + x^2)(-1) - 2(1 - x^2)]}{(1 + x^2)^4}$ [Factor out $(1 + x^2)(2x)$.]

$\qquad = \dfrac{2x[-1 - x^2 - 2 + 2x^2]}{(1 + x^2)^3}$ [Divide numerator and denominator by $(1 + x^2)$.]

$\qquad = \dfrac{2x(x^2 - 3)}{(1 + x^2)^3}$ (Combine like terms.)

e. $\dfrac{\frac{1}{4}(x + h)^2 - \frac{1}{4}x^2}{h} = \dfrac{\frac{1}{4}x^2 + \frac{1}{2}xh + \frac{1}{4}h^2 - \frac{1}{4}x^2}{h}$

$\qquad = \dfrac{h\left(\frac{1}{2}x + \frac{1}{4}h\right)}{h}$ (Square the binomial and combine like terms.)

$\qquad = \dfrac{1}{2}x + \dfrac{1}{4}h$ (Divide numerator and denominator by h.)   ❍ ❍ ❍

The operations of multiplication and division are performed with algebraic fractions in the same manner as with arithmetic fractions (Table 9.5).

Table 9.5

Operation	Example

If P, Q, R, and S are polynomials, then

Multiplication

$$\frac{P}{Q} \cdot \frac{R}{S} = \frac{PR}{QS} \qquad (Q, S \neq 0)$$

$$\frac{2x}{y} \cdot \frac{(x+1)}{(y-1)} = \frac{2x(x+1)}{y(y-1)} = \frac{2x^2 + 2x}{y^2 - y}$$

Division

$$\frac{P}{Q} \div \frac{R}{S} = \frac{P}{Q} \cdot \frac{S}{R} = \frac{PS}{QR} \qquad (Q, R, S \neq 0)$$

$$\frac{x^2 + 3}{y} \div \frac{y^2 + 1}{x} = \frac{x^2 + 3}{y} \cdot \frac{x}{y^2 + 1}$$

$$= \frac{x^3 + 3x}{y^3 + y}$$

When rational expressions are multiplied and divided, the resulting expressions should be simplified.

EXAMPLE 2   Perform the indicated operations and simplify:

a. $\dfrac{2x - 8}{x + 2} \cdot \dfrac{x^2 + 4x + 4}{x^2 - 16}$ **b.** $\dfrac{x^2 - 6x + 9}{3x + 12} \div \dfrac{x^2 - 9}{6x^2 + 18x}$

Solution

a. $\dfrac{2x - 8}{x + 2} \cdot \dfrac{x^2 + 4x + 4}{x^2 - 16}$

$$= \frac{2(x - 4)}{x + 2} \cdot \frac{(x + 2)^2}{(x + 4)(x - 4)}$$

$$= \frac{2(x - 4)(x + 2)(x + 2)}{(x + 2)(x + 4)(x - 4)} \qquad \text{[Cancel the common factors } (x + 2)(x - 4).]$$

$$= \frac{2(x + 2)}{x + 4}$$

b. $\dfrac{x^2 - 6x + 9}{3x + 12} \div \dfrac{x^2 - 9}{6x^2 + 18x}$

$$= \frac{x^2 - 6x + 9}{3x + 12} \cdot \frac{6x^2 + 18x}{x^2 - 9}$$

$$= \frac{(x - 3)^2}{3(x + 4)} \cdot \frac{6x(x + 3)}{(x + 3)(x - 3)}$$

$$= \frac{(x - 3)(x - 3)(6x)(x + 3)}{3(x + 4)(x + 3)(x - 3)} \qquad \text{[Cancel the common factors } 3(x + 3)(x - 3).]$$

$$= \frac{2x(x - 3)}{x + 4}$$

○ ○ ○

Table 9.6

Operation	Example
If P, Q, and R are polynomials, then	
Addition	
$\dfrac{P}{R} + \dfrac{Q}{R} = \dfrac{P+Q}{R}$ $(R \neq 0)$	$\dfrac{2x}{x+2} + \dfrac{6x}{x+2} = \dfrac{2x+6x}{x+2} = \dfrac{8x}{x+2}$
Subtraction	
$\dfrac{P}{R} - \dfrac{Q}{R} = \dfrac{P-Q}{R}$ $(R \neq 0)$	$\dfrac{3y}{y-x} - \dfrac{y}{y-x} = \dfrac{3y-y}{y-x} = \dfrac{2y}{y-x}$

For rational expressions, the operations of addition and subtraction are performed by finding a common denominator of the fractions and then adding or subtracting the fractions. Table 9.6 shows the rules for fractions with equal denominators.

To add or subtract fractions that have different denominators, first find a common denominator, preferably the least common denominator (LCD). Then carry out the indicated operations following the procedure described in Table 9.6.

To find the least common denominator (LCD) of two or more rational expressions,

1. *find the prime factors of each denominator.*

2. *form the product of the different prime factors that occur in the denominators.* Each prime factor in this product should be raised to the highest power of that factor appearing in the denominators.

$$\frac{x}{2+y} \neq \frac{x}{2} + \frac{x}{y}\,!$$

 EXAMPLE 3   Simplify:

a. $\dfrac{3x+4}{4x} + \dfrac{4y-2}{3y}$ **b.** $\dfrac{2x}{x^2+1} + \dfrac{6(3x^2)}{x^3+2}$ **c.** $\dfrac{1}{x+h} - \dfrac{1}{x}$

Solution

a. $\dfrac{3x+4}{4x} + \dfrac{4y-2}{3y}$

$$= \frac{3x+4}{4x} \cdot \frac{3y}{3y} + \frac{4y-2}{3y} \cdot \frac{4x}{4x} \qquad [\text{LCD} = (4x)(3y)$$
$$= 12xy]$$

$$= \frac{9xy+12y}{12xy} + \frac{16xy-8x}{12xy}$$

$$= \frac{25xy-8x+12y}{12xy}$$

b. $\dfrac{2x}{x^2 + 1} + \dfrac{6(3x^2)}{x^3 + 2}$

$$= \frac{2x(x^3 + 2) + 6(3x^2)(x^2 + 1)}{(x^2 + 1)(x^3 + 2)} \qquad [\text{LCD} = (x^2 + 1)(x^3 + 2)]$$

$$= \frac{2x^4 + 4x + 18x^4 + 18x^2}{(x^2 + 1)(x^3 + 2)}$$

$$= \frac{20x^4 + 18x^2 + 4x}{(x^2 + 1)(x^3 + 2)}$$

$$= \frac{2x(10x^3 + 9x + 2)}{(x^2 + 1)(x^3 + 2)}$$

c. $\dfrac{1}{x + h} - \dfrac{1}{x} = \dfrac{1}{x + h} \cdot \dfrac{x}{x} - \dfrac{1}{x} \cdot \dfrac{x + h}{x + h} \qquad [\text{LCD} = (x)(x + h)]$

$$= \frac{x}{x(x + h)} - \frac{x + h}{x(x + h)}$$

$$= \frac{x - x - h}{x(x + h)}$$

$$= \frac{-h}{x(x + h)}$$

Other Algebraic Fractions

The techniques used to simplify rational expressions may also be used to simplify algebraic fractions in which the numerator and denominator are not polynomials, as illustrated in Example 4.

EXAMPLE 4   Simplify:

a. $\dfrac{1 + \dfrac{1}{x + 1}}{x - \dfrac{4}{x}}$ **b.** $\dfrac{x^{-1} + y^{-1}}{x^{-2} - y^{-2}}$

Solution

a. $\dfrac{1 + \dfrac{1}{x + 1}}{x - \dfrac{4}{x}} = \dfrac{1 \cdot \dfrac{x + 1}{x + 1} + \dfrac{1}{x + 1}}{x \cdot \dfrac{x}{x} - \dfrac{4}{x}} = \dfrac{\dfrac{x + 1 + 1}{x + 1}}{\dfrac{x^2 - 4}{x}}$

$$= \frac{x + 2}{x + 1} \cdot \frac{x}{x^2 - 4} = \frac{x + 2}{x + 1} \cdot \frac{x}{(x + 2)(x - 2)}$$

$$= \frac{x}{(x + 1)(x - 2)}$$

b. $\dfrac{x^{-1} + y^{-1}}{x^{-2} - y^{-2}} = \dfrac{\dfrac{1}{x} + \dfrac{1}{y}}{\dfrac{1}{x^2} - \dfrac{1}{y^2}} = \dfrac{\dfrac{y+x}{xy}}{\dfrac{y^2 - x^2}{x^2 y^2}}$ $\left(x^{-n} = \dfrac{1}{x^n}\right)$

$$= \frac{y+x}{xy} \cdot \frac{x^2 y^2}{y^2 - x^2} = \frac{y+x}{xy} \cdot \frac{(xy)^2}{(y+x)(y-x)}$$

$$= \frac{xy}{y-x} \qquad \qquad \circ \; \circ \; \circ$$

Ⓡ **EXAMPLE 5**   Show that the algebraic expression

$$\frac{3 + \dfrac{8}{x} - \dfrac{4}{x^2}}{2 + \dfrac{4}{x} - \dfrac{5}{x^2}}$$

is equal to $\dfrac{3x^2 + 8x - 4}{2x^2 + 4x - 5}$

(We will work with the first form of this algebraic expression in Chapter 10 when we evaluate limits.)

Solution   The LCD of both the numerator and denominator is x^2. Therefore,

$$\frac{3 + \dfrac{8}{x} - \dfrac{4}{x^2}}{2 + \dfrac{4}{x} - \dfrac{5}{x^2}} = \frac{\dfrac{3(x^2) + 8(x) - 4}{x^2}}{\dfrac{2(x^2) + 4(x) - 5}{x^2}}$$

$$= \frac{3x^2 + 8x - 4}{x^2} \cdot \frac{x^2}{2x^2 + 4x - 5}$$

$$= \frac{3x^2 + 8x - 4}{2x^2 + 4x - 5} \qquad \qquad \circ \; \circ \; \circ$$

EXAMPLE 6   Perform the given operations and simplify:

a. $\dfrac{x^2(2x^2 + 1)^{1/2}}{x - 1} \cdot \dfrac{4x^3 - 6x^2 + x - 2}{x(x - 1)(2x^2 + 1)}$ **b.** $\dfrac{12x^2}{\sqrt{2x^2 + 3}} + 6\sqrt{2x^2 + 3}$

Solution

a. $\dfrac{x^2(2x^2+1)^{1/2}}{x-1} \cdot \dfrac{4x^3-6x^2+x-2}{x(x-1)(2x^2+1)}$

$$= \frac{x(4x^3-6x^2+x-2)}{(x-1)^2(2x^2+1)^{1-1/2}}$$

$$= \frac{x(4x^3-6x^2+x-2)}{(x-1)^2(2x^2+1)^{1/2}}$$

b. $\dfrac{12x^2}{\sqrt{2x^2+3}} + 6\sqrt{2x^2+3} = \dfrac{12x^2}{(2x^2+3)^{1/2}} + 6(2x^2+3)^{1/2}$

$$= \frac{12x^2 + 6(2x^2+3)^{1/2}(2x^2+3)^{1/2}}{(2x^2+3)^{1/2}}$$

$$= \frac{12x^2 + 6(2x^2+3)}{(2x^2+3)^{1/2}}$$

$$= \frac{24x^2+18}{(2x^2+3)^{1/2}} = \frac{6(4x^2+3)}{\sqrt{2x^2+3}}$$

○ ○ ○

Rationalizing Algebraic Fractions

When the denominator of an algebraic fraction contains sums or differences involving radicals, we may **rationalize the denominator**—that is, transform the fraction into an equivalent one with a denominator that does not contain radicals. In doing so, we make use of the fact that

$$(\sqrt{a}+\sqrt{b})(\sqrt{a}-\sqrt{b}) = (\sqrt{a})^2 - (\sqrt{b})^2$$
$$= a - b$$

This procedure is illustrated in Example 7.

EXAMPLE 7   Rationalize the denominator of $\dfrac{1}{1+\sqrt{x}}$.

Solution   Upon multiplying the numerator and the denominator by $(1-\sqrt{x})$, we obtain

$$\frac{1}{1+\sqrt{x}} = \frac{1}{1+\sqrt{x}} \cdot \frac{1-\sqrt{x}}{1-\sqrt{x}}$$

$$= \frac{1-\sqrt{x}}{1-(\sqrt{x})^2}$$

$$= \frac{1-\sqrt{x}}{1-x}$$

○ ○ ○

In other situations, it may be necessary to **rationalize the numerator** of an algebraic expression. In calculus, for example, one encounters the following problem.

EXAMPLE 8   Rationalize the numerator: $\dfrac{\sqrt{1+h}-1}{h}$

Solution

$$\frac{\sqrt{1+h}-1}{h} = \frac{\sqrt{1+h}-1}{h} \cdot \frac{\sqrt{1+h}+1}{\sqrt{1+h}+1}$$

$$= \frac{(\sqrt{1+h})^2 - (1)^2}{h(\sqrt{1+h}+1)}$$

$$= \frac{1+h-1}{h(\sqrt{1+h}+1)} \qquad [(\sqrt{1+h})^2 = \sqrt{1+h} \cdot \sqrt{1+h}$$
$$= 1 + h]$$

$$= \frac{h}{h(\sqrt{1+h}+1)}$$

$$= \frac{1}{\sqrt{1+h}+1}$$

○ ○ ○

9.3   EXERCISES

In exercises 1–10, simplify the expression.

1. $\dfrac{x^2 + x - 2}{x^2 - 4}$

2. $\dfrac{2a^2 - 3ab - 9b^2}{2ab^2 + 3b^3}$

3. $\dfrac{12t^2 + 12t + 3}{4t^2 - 1}$

4. $\dfrac{x^3 + 2x^2 - 3x}{-2x^2 - x + 3}$

5. $\dfrac{(4x - 1)(3) - (3x + 1)(4)}{(4x - 1)^2}$

6. $\dfrac{(1 + x^2)^2(2) - 2x(2)(1 + x^2)(2x)}{(1 + x^2)^4}$

7. $\dfrac{(2x + 3)(1) - (x + 1)(2)}{(2x + 3)^2}$

8. $\dfrac{(x^2 - 1)^4(2x) - x^2(4)(x^2 - 1)^3(2x)}{(x^2 - 1)^8}$

9. $\dfrac{(e^x + 1)e^x - e^x(2)(e^x + 1)(e^x)}{(e^x + 1)^4}$

10. $\dfrac{(u^3 - 1)(3)(u^2 + u - 1)^2(2u + 1) - (u^2 + u - 1)^3(3u^2)}{(u^3 - 1)^2}$

In exercises 11–34, perform the indicated operations and simplify each expression.

11. $\dfrac{2a^2 - 2b^2}{b - a} \cdot \dfrac{4a + 4b}{a^2 + 2ab + b^2}$

12. $\dfrac{x^2 - 6x + 9}{x^2 - x - 6} \cdot \dfrac{3x + 6}{2x^2 - 7x + 3}$

13. $\dfrac{3x^2 + 2x - 1}{2x + 6} \div \dfrac{x^2 - 1}{x^2 + 2x - 3}$

14. $\dfrac{3x^2 - 4xy - 4y^2}{x^2 y} \div \dfrac{(2y - x)^2}{x^3 y}$

15. $\dfrac{58}{3(3t + 2)} + \dfrac{1}{3}$

16. $\dfrac{a + 1}{3a} + \dfrac{b - 2}{5b}$

17. $\dfrac{2x}{2x - 1} - \dfrac{3x}{2x + 5}$

18. $\dfrac{-xe^x}{x + 1} + e^x$

19. $\left(x + \dfrac{1}{x}\right)(x^2 - 1)$

20. $\left(2x - \dfrac{3}{x}\right)(x^3 + 1)$

21. $\dfrac{4}{x^2 - 9} - \dfrac{5}{x^2 - 6x + 9}$

22. $\dfrac{x}{1 - x} + \dfrac{2x + 3}{x^2 - 1}$

23. $2 + \dfrac{1}{a + 2} - \dfrac{2a}{a - 2}$

24. $x - \dfrac{x^2}{x + 2} + \dfrac{2}{x - 2}$

25. $\dfrac{1 + \dfrac{1}{x}}{1 - \dfrac{1}{x}}$

26. $\dfrac{\dfrac{1}{x} + \dfrac{1}{y}}{1 - \dfrac{1}{xy}}$

27. $\dfrac{x^{-2} - y^{-2}}{x + y}$

28. $\dfrac{x^{-3} - y^{-3}}{x^{-1} - y^{-1}}$

29. $\dfrac{4x^2}{2\sqrt{2x^2 + 7}} + \sqrt{2x^2 + 7}$

30. $\dfrac{x^3}{\sqrt{8 - 3x^2}}(-6x) + \sqrt{8 - 3x^2}(6x^2)$

31. $\dfrac{2x(x + 1)^{-1/2} - (x + 1)^{1/2}}{x^2}$

32. $\dfrac{(x^2 + 1)^{1/2} - 2x^2(x^2 + 1)^{-1/2}}{1 - x^2}$

33. $\dfrac{(2x + 1)^{1/2} - (x + 2)(2x + 1)^{-1/2}}{2x + 1}$

34. $\dfrac{2(2x - 3)^{1/3} - (x - 1)(2x - 3)^{-2/3}}{(2x - 3)^{2/3}}$

In exercises 35–38, show that expressions (a) and (b) are equivalent.

35. a. $\dfrac{ax}{x + b}$

b. $\dfrac{a}{1 + \dfrac{b}{x}}$

36. a. $\dfrac{x + 3}{x^2 - x}$

b. $\dfrac{\dfrac{1}{x} + \dfrac{3}{x^2}}{1 - \dfrac{1}{x}}$

37. a. $\dfrac{x^4 + 1}{x^3 - 1}$

b. $\dfrac{x + \dfrac{1}{x^3}}{1 - \dfrac{1}{x^3}}$

38. a. $\sqrt{\dfrac{4x^2 + 2x - 1}{x^2 + 2}}$

b. $\sqrt{\dfrac{4 + \dfrac{2}{x} - \dfrac{1}{x^2}}{1 + \dfrac{2}{x^2}}}$

In exercises 39–44, rationalize the denominator of each expression.

39. $\dfrac{1}{\sqrt{3} - 1}$

40. $\dfrac{1}{\sqrt{x} + 5}$

41. $\dfrac{1}{\sqrt{x} - \sqrt{y}}$

42. $\dfrac{a}{1 - \sqrt{a}}$

43. $\dfrac{\sqrt{a} + \sqrt{b}}{\sqrt{a} - \sqrt{b}}$

44. $\dfrac{2\sqrt{a} + \sqrt{b}}{2\sqrt{a} - \sqrt{b}}$

In exercises 45–50, rationalize the numerator of each expression.

45. $\dfrac{\sqrt{x}}{3}$

46. $\dfrac{\sqrt[3]{y}}{x}$

47. $\dfrac{1 - \sqrt{3}}{3}$

48. $\dfrac{\sqrt{x} - 1}{x}$

49. $\dfrac{1 + \sqrt{x + 2}}{\sqrt{x + 2}}$

50. $\dfrac{\sqrt{x + 3} - \sqrt{x}}{3}$

9.4 INEQUALITIES AND ABSOLUTE VALUE

Intervals

The system of real numbers and some of its properties were mentioned briefly in Section 1.1. In much of our later work we will restrict our attention to certain subsets of the set of real numbers. For example, if x denotes the number of cars rolling off an assembly line each day in an automobile assembly plant, then x must be nonnegative—that is, $x \geq 0$. Taking this example one step further, suppose management decides that the daily production must not exceed 200 cars. Then x must satisfy the inequality $0 \leq x \leq 200$.

More generally, we will be interested in the following subsets of real numbers: open intervals, closed intervals, and half-open intervals. The set of all real numbers that lie *strictly* between two fixed numbers a and b is called an **open interval** (a, b). It consists of all real numbers x that satisfy the inequalities $a < x < b$, and it is called "open" because neither of its endpoints is included in the interval. A **closed interval** contains *both* of its endpoints. Thus, the set of all real numbers x that satisfy the inequalities $a \le x \le b$ is the closed interval $[a, b]$. Notice that square brackets are used to indicate that the endpoints are included in this interval. **Half-open intervals** contain only *one* of their endpoints. Thus, the interval $[a, b)$ is the set of all real numbers x that satisfy $a \le x < b$, whereas the interval $(a, b]$ is described by the inequalities $a < x \le b$. Examples of these **finite intervals** are illustrated in Table 9.7.

Table 9.7
Finite intervals.

Interval		Graph	Example	
Open	(a, b)		$(-2, 1)$	
Closed	$[a, b]$		$[-1, 2]$	
Half-open	$(a, b]$		$(\frac{1}{2}, 3]$	
Half-open	$[a, b)$		$[-\frac{1}{2}, 3)$	

In addition to finite intervals, we will encounter **infinite intervals.** Examples of infinite intervals are the half-lines (a, ∞), $[a, \infty)$, $(-\infty, a)$, and $(-\infty, a]$ defined by the set of all real numbers that satisfy $x > a$, $x \ge a$, $x < a$, and $x \le a$, respectively. The symbol ∞, called *infinity,* is not a real number. It is used here only for notational purposes in conjunction with the definition of infinite intervals. The notation $(-\infty, \infty)$ is used for the set of all real numbers x since, by definition, the inequalities $-\infty < x < \infty$ hold for any real number x. Infinite intervals are illustrated in Table 9.8.

Table 9.8
Infinite intervals.

Interval	Graph	Example	
(a, ∞)		$(2, \infty)$	
$[a, \infty)$		$[-1, \infty)$	
$(-\infty, a)$		$(-\infty, 1)$	
$(-\infty, a]$		$(-\infty, -\frac{1}{2}]$	

Properties of Inequalities

In practical applications, intervals are often found by solving one or more inequalities involving a variable. In such situations, the following properties may be used to advantage.

PROPERTIES OF INEQUALITIES	If a, b, and c are any real numbers, then

		Example
Property 1	If $a < b$ and $b < c$, then $a < c$.	$2 < 3$ and $3 < 8$, so $2 < 8$
Property 2	If $a < b$, then $a + c < b + c$.	$-5 < -3$, so $-5 + 2 < -3 + 2$; that is, $-3 < -1$
Property 3	If $a < b$ and $c > 0$, then $ac < bc$.	$-5 < -3$, and since $2 > 0$, we have $(-5)(2) < (-3)(2)$—that is, $-10 < -6$
Property 4	If $a < b$ and $c < 0$, then $ac > bc$.	$-2 < 4$, and since $-3 < 0$, we have $(-2)(-3) > (4)(-3)$—that is, $6 > -12$

Similar properties hold if each inequality sign, $<$, between a and b is replaced by $\geq$, $>$, or $\leq$.

A real number is a *solution of an inequality* involving a variable if a true statement is obtained when the variable is replaced by that number. The set of all real numbers satisfying the inequality is called the *solution set*.

EXAMPLE 1   Find the set of real numbers that satisfy $-1 \leq 2x - 5 < 7$.

Solution   Add 5 to each member of the given double inequality, obtaining

$$4 \leq 2x < 12$$

Next, multiply each member of the resulting double inequality by 1/2, yielding

$$2 \leq x < 6$$

Thus, the solution is the set of all values of x lying in the interval $[2, 6)$.

$\circ\ \circ\ \circ$

EXAMPLE 2   Solve the inequality $x^2 + 2x - 8 < 0$.

Solution   Observe that $x^2 + 2x - 8 = (x + 4)(x - 2)$, so the given inequality is equivalent to the inequality $(x + 4)(x - 2) < 0$. Since the product of two real numbers is negative if and only if the two numbers have opposite signs, we solve the inequality $(x + 4)(x - 2) < 0$ by studying the signs of the two factors $x + 4$ and $x - 2$. Now, $x + 4 > 0$ when $x > -4$, and $x + 4 < 0$ when $x < -4$. Similarly, $x - 2 > 0$ when $x > 2$, and $x - 2 < 0$ when $x < 2$. These results are summarized graphically in Figure 9.1.

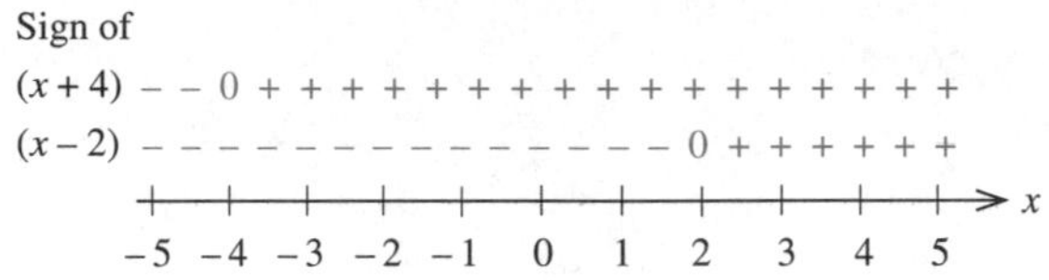

Figure 9.1
Sign diagram for $(x + 4)(x - 2)$.

Sign of
$(x + 4)$ $\quad$ – – 0 + + + + + + + + + + + + + + + + +
$(x - 2)$ $\quad$ – – – – – – – – – – – – – – 0 + + + + + +

$$\xrightarrow{\quad\quad\quad\quad\quad\quad\quad\quad\quad\quad\quad} x$$

$-5\ -4\ -3\ -2\ -1\ \ 0\ \ 1\ \ 2\ \ 3\ \ 4\ \ 5$

From Figure 9.1 we see that the two factors $x + 4$ and $x - 2$ have opposite signs when and only when x lies strictly between -4 and 2. Therefore, the required solution is the interval $(-4, 2)$. $\quad\quad$ ◦ ◦ ◦

EXAMPLE 3 Solve the inequality $\dfrac{x + 1}{x - 1} \geq 0$.

Solution The quotient $(x + 1)/(x - 1)$ is strictly positive if and only if both the numerator and the denominator have the same sign. The signs of $x + 1$ and $x - 1$ are shown in Figure 9.2.

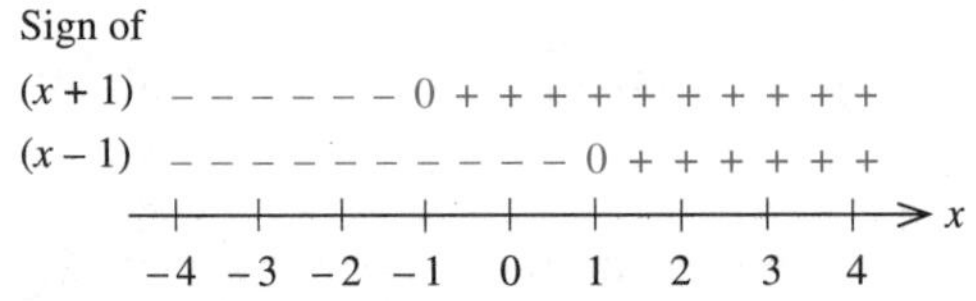

Figure 9.2
Sign diagram for $\dfrac{x + 1}{x - 1}$.

Sign of
$(x + 1)$ $\quad$ – – – – – – 0 + + + + + + + + + +
$(x - 1)$ $\quad$ – – – – – – – – – – 0 + + + + + +

$$\xrightarrow{\quad\quad\quad\quad\quad\quad\quad\quad\quad} x$$

$-4\ -3\ -2\ -1\ \ 0\ \ 1\ \ 2\ \ 3\ \ 4$

From Figure 9.2 we see that $x + 1$ and $x - 1$ have the same sign when and only when $x < -1$ or $x > 1$. The quotient $(x + 1)/(x - 1)$ is equal to zero when and only when $x = -1$. Therefore, the required solution is the set of all x in the intervals $(-\infty, -1]$ and $(1, \infty)$. $\quad\quad$ ◦ ◦ ◦

EXAMPLE 4 The management of Corbyco, a giant conglomerate, has estimated that x thousand dollars is needed to purchase

$$100{,}000(-1 + \sqrt{1 + 0.001x})$$

shares of common stock of the Starr Communications Company. Determine how much money Corbyco needs in order to purchase at least 100,000 shares of Starr's stock.

Solution The amount of cash Corbyco needs to purchase at least 100,000 shares is found by solving the inequality

$$100{,}000(-1 + \sqrt{1 + 0.001x}) \geq 100{,}000$$

Proceeding, we find

$$-1 + \sqrt{1 + 0.001x} \geq 1$$
$$\sqrt{1 + 0.001x} \geq 2$$
$$1 + 0.001x \geq 4 \qquad \text{(Square both sides.)}$$
$$0.001x \geq 3$$
$$x \geq 3000$$

so Corbyco needs at least $3,000,000.

Absolute Value

<table>
<tr><td>ABSOLUTE VALUE</td><td>The absolute value of a number a is denoted by $|a|$ and is defined by

$$|a| = \begin{cases} a & \text{if } a \geq 0 \\ -a & \text{if } a < 0 \end{cases}$$</td></tr>
</table>

Since $-a$ is a positive number when a is negative, it follows that the absolute value of a number is always nonnegative. For example, $|5| = 5$ and $|-5| = -(-5) = 5$. Geometrically, $|a|$ is the distance between the origin and the point on the number line that represents the number a (Figure 9.3).

Figure 9.3
The absolute value of a number.

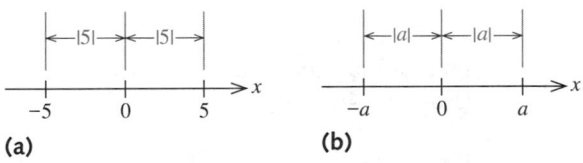

(a)   (b)

<table>
<tr><td>ABSOLUTE VALUE PROPERTIES</td><td colspan="2">If a and b are any real numbers, then</td></tr>
<tr><td></td><td></td><td>Example</td></tr>
<tr><td></td><td>Property 5 $|-a| = |a|$</td><td>$|-3| = -(-3) = 3 = |3|$</td></tr>
<tr><td></td><td>Property 6 $|ab| = |a|\,|b|$</td><td>$|(2)(-3)| = |-6| = 6$
$= |2|\,|-3|$</td></tr>
<tr><td></td><td>Property 7 $\left|\dfrac{a}{b}\right| = \dfrac{|a|}{|b|} \quad (b \neq 0)$</td><td>$\left|\dfrac{(-3)}{(-4)}\right| = \left|\dfrac{3}{4}\right| = \dfrac{3}{4} = \dfrac{|-3|}{|-4|}$</td></tr>
<tr><td></td><td>Property 8 $|a + b| \leq |a| + |b|$</td><td>$|8 + (-5)| = |3| = 3$
$\leq |8| + |-5|$
$= 13$</td></tr>
</table>

Property 8 is called the **triangle inequality**.

EXAMPLE 5   Evaluate each expression:

a. $|\pi - 5| + 3$ **b.** $|\sqrt{3} - 2| + |2 - \sqrt{3}|$

Solution

a. Since $\pi - 5 < 0$, we see that $|\pi - 5| = -(\pi - 5)$. Therefore,

$$|\pi - 5| + 3 = -(\pi - 5) + 3 = 8 - \pi$$

b. Since $\sqrt{3} - 2 < 0$, we see that $|\sqrt{3} - 2| = -(\sqrt{3} - 2)$. Next, observe that $2 - \sqrt{3} > 0$, so $|2 - \sqrt{3}| = 2 - \sqrt{3}$. Therefore,

$$|\sqrt{3} - 2| + |2 - \sqrt{3}| = -(\sqrt{3} - 2) + (2 - \sqrt{3})$$
$$= 4 - 2\sqrt{3} = 2(2 - \sqrt{3})$$

EXAMPLE 6   Solve the inequalities $|x| \le 5$ and $|x| \ge 5$.

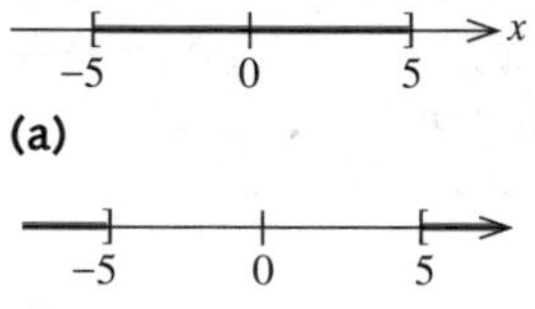

Solution   First we consider the inequality $|x| \le 5$. If $x \ge 0$, then $|x| = x$, so $|x| \le 5$ implies $x \le 5$ in this case. On the other hand, if $x < 0$, then $|x| = -x$, so $|x| \le 5$ implies $-x \le 5$ or $x \ge -5$. Thus, $|x| \le 5$ means $-5 \le x \le 5$ (Figure 9.4a). To obtain an alternative solution, observe that $|x|$ is the distance from the point x to zero, so the inequality $|x| \le 5$ implies immediately that $-5 \le x \le 5$.

Next, the inequality $|x| \ge 5$ states that the distance from x to zero is greater than or equal to 5. This observation yields the result $x \ge 5$ or $x \le -5$ (Figure 9.4b).

Figure 9.4

(a)

(b)

EXAMPLE 7   Solve the inequality $|2x - 3| \le 1$.

Solution   The inequality $|2x - 3| \le 1$ is equivalent to the inequalities $-1 \le 2x - 3 \le 1$ (see Example 6). Thus, $2 \le 2x \le 4$ and $1 \le x \le 2$. The solution is therefore given by the set of all x in the interval $[1, 2]$ (Figure 9.5).

Figure 9.5
$|2x - 3| \le 1$

9.4   EXERCISES

In exercises 1–4, determine whether the statement is true or false.

1. $-3 < -20$ **2.** $-5 \le -5$ **3.** $\dfrac{2}{3} > \dfrac{5}{6}$

4. $-\dfrac{5}{6} < -\dfrac{11}{12}$

In exercises 5–10, show the given interval on a number line.

5. $(3, 6)$ **6.** $(-2, 5]$ **7.** $[-1, 4)$

8. $[-\tfrac{6}{5}, -\tfrac{1}{2}]$ **9.** $(0, \infty)$ **10.** $(-\infty, 5]$

In exercises 11–28, find the values of x that satisfy the inequality (inequalities).

11. $2x + 4 < 8$

12. $-6 > 4 + 5x$

13. $-4x \geq 20$

14. $-12 \leq -3x$

15. $-6 < x - 2 < 4$

16. $0 \leq x + 1 \leq 4$

17. $x + 1 > 4$ or $x + 2 < -1$

18. $x + 1 > 2$ or $x - 1 < -2$

19. $x + 3 > 1$ and $x - 2 < 1$

20. $x - 4 \leq 1$ and $x + 3 > 2$

21. $(x + 3)(x - 5) \leq 0$

22. $(2x - 4)(x + 2) \geq 0$

23. $(2x - 3)(x - 1) \geq 0$

24. $(3x - 4)(2x + 2) \leq 0$

25. $\dfrac{x + 3}{x - 2} \geq 0$

26. $\dfrac{2x - 3}{x + 1} \geq 4$

27. $\dfrac{x - 2}{x - 1} \leq 2$

28. $\dfrac{2x - 1}{x + 2} \leq 4$

In exercises 29–38, evaluate the given expression.

29. $|-6 + 2|$

30. $4 + |-4|$

31. $\dfrac{|-12 + 4|}{|16 - 12|}$

32. $\left|\dfrac{0.2 - 1.4}{1.6 - 2.4}\right|$

33. $\sqrt{3}|-2| + 3|-\sqrt{3}|$

34. $|-1| + |\sqrt{2}| - 2|$

35. $|\pi - 1| + 2$

36. $|\pi - 6| - 3$

37. $|\sqrt{2} - 1| + |3 - \sqrt{2}|$

38. $|2\sqrt{3} - 3| - |\sqrt{3} - 4|$

In exercises 39–44, suppose that a and b are real numbers other than zero and that a > b. State whether the inequality is true or false.

39. $b - a > 0$

40. $\dfrac{a}{b} > 1$

41. $a^2 > b^2$

42. $\dfrac{1}{a} > \dfrac{1}{b}$

43. $a^3 > b^3$

44. $-a < -b$

In exercises 45–50, determine whether the statement is true for all real numbers a and b.

45. $|-a| = a$

46. $|b^2| = b^2$

47. $|a - 4| = |4 - a|$

48. $|a + 1| = |a| + 1$

49. $|a + b| = |a| + |b|$

50. $|a - b| = |a| - |b|$

51. Driving Range of a Car An advertisement for a certain car states that the EPA fuel economy is 20 mpg city and 27 mpg highway and that the car's fuel-tank capacity is 18.1 gallons. Assuming ideal driving conditions, determine the driving range for the car from these data.

52. Find the minimum cost C (in dollars), given that

$$5(C - 25) \geq 1.75 + 2.5C$$

53. Find the maximum profit P (in dollars) given that

$$6(P - 2500) \leq 4(P + 2400)$$

54. Celsius and Fahrenheit Temperatures The relationship between Celsius (°C) and Fahrenheit (°F) temperatures is given by the formula

$$C = \frac{5}{9}(F - 32)$$

a. If the temperature range for Montreal during the month of January is $-15° < °C < -5°$, find the range in degrees Fahrenheit in Montreal for the same period.
b. If the temperature range for New York City during the month of June is $63° < °F < 80°$, find the range in degrees Celsius in New York City for the same period.

55. Meeting Sales Targets A salesman's monthly commission is 15% on all sales over $12,000. If his goal is to make a commission of at least $1000 per month, what minimum monthly sales figures must he attain?

56. Markup on a Car The markup on a used car was at least 30% of its current wholesale price. If the car was sold for $2800, what was the maximum wholesale price?

57. Quality Control The P.A.R. Manufacturing Company manufactures steel rods. Suppose the rods ordered by a customer are manufactured to a specification of 0.5 inch and are acceptable only if they are within the *tolerance limits* of 0.49 inch and 0.51 inch. Letting x denote the diameter of a rod, write an inequality using absolute value to express a criterion involving x that must be satisfied in order for a rod to be acceptable.

58. Quality Control The diameter x (in inches) of a batch of ball bearings manufactured by the P.A.R. Manufacturing Company satisfies the inequality

$$|x - 0.1| \leq 0.01$$

What is the smallest diameter a ball bearing in the batch can have? The largest diameter?

59. Meeting Profit Goals A manufacturer of a certain commodity has estimated that her profit in thousands of dollars is given by the expression

$$-6x^2 + 30x - 10$$

where x (in thousands) is the number of units produced. What production range will enable the manufacturer to realize a profit of at least \$14,000 on the commodity?

60. Distribution of Incomes The distribution of income in a certain city can be described by the exponential model $y = (2.8 \cdot 10^{11})(x)^{-1.5}$, where y is the number of families with an income of x or more dollars.
a. How many families in this city have an income of \$20,000 or more?
b. How many families have an income of \$40,000 or more?
c. How many families have an income of \$100,000 or more?

CHAPTER 9 SUMMARY OF PRINCIPAL FORMULAS AND TERMS

Formulas

1. Product formulas

$$(a + b)^2 = a^2 + 2ab + b^2$$
$$(a - b)^2 = a^2 - 2ab + b^2$$
$$(a + b)(a - b) = a^2 - b^2$$

2. Quadratic formula

$$x = \frac{-b \pm \sqrt{b^2 - 4ac}}{2a}$$

Terms

Roots of polynomial equations

Finite intervals

Infinite intervals

Absolute value

CHAPTER 9 REVIEW EXERCISES

In exercises 1–6, evaluate the expression.

1. $\left(\dfrac{9}{4}\right)^{3/2}$

2. $\dfrac{5^6}{5^4}$

3. $(3 \cdot 4)^{-2}$

4. $(-8)^{5/3}$

5. $\dfrac{(3 \cdot 2^{-3})(4 \cdot 3^5)}{2 \cdot 9^3}$

6. $\dfrac{3\sqrt[3]{54}}{\sqrt[3]{18}}$

In exercises 7–14, simplify the expression.

7. $\dfrac{4(x^2 + y)^3}{x^2 + y}$

8. $\dfrac{a^6 b^{-5}}{(a^3 b^{-2})^{-3}}$

9. $\dfrac{\sqrt[4]{16x^5yz}}{\sqrt[4]{81xyz^5}}$

10. $(2x^3)(-3x^{-2})\left(\dfrac{1}{6}x^{-1/2}\right)$

11. $\left(\dfrac{3xy^2}{4x^3y}\right)^{-2}\left(\dfrac{3xy^3}{2x^2}\right)^3$

12. $(-3a^2b^3)^2(2a^{-1}b^{-2})^{-1}$

13. $\sqrt[3]{81x^5y^{10}} \ \sqrt[3]{9xy^2}$

14. $\left(\dfrac{-x^{1/2}y^{2/3}}{x^{1/3}y^{3/4}} \right)^6$

In exercises 15–20, factor the expression.

15. $-2\pi^2 r^3 + 100\pi r^2$

16. $2v^3 w + 2vw^3 + 2u^2 vw$

17. $16 - x^2$

18. $12t^3 - 6t^2 - 18t$

19. $-2x^2 - 4x + 6$

20. $12x^2 - 92x + 120$

In exercises 21–24, perform the indicated operations and simplify the expression.

21. $\dfrac{(t+6)(60) - (60t + 180)}{(t+6)^2}$

22. $\dfrac{6x}{2(3x^2 + 2)} + \dfrac{1}{4(x+2)}$

23. $\dfrac{2}{3}\left(\dfrac{4x}{2x^2 - 1} \right) + 3\left(\dfrac{3}{3x - 1} \right)$

24. $\dfrac{-2x}{\sqrt{x+1}} + 4\sqrt{x+1}$

In exercises 25–28, solve the equation by factoring.

25. $8x^2 + 2x - 3 = 0$

26. $-6x^2 - 10x + 4 = 0$

27. $-x^3 - 2x^2 + 3x = 0$

28. $2x^4 + x^2 = 1$

In exercises 29–32, find the values of x that satisfy the given inequalities.

29. $-x + 3 \le 2x + 9$

30. $-2 \le 3x + 1 \le 7$

31. $x - 3 > 2$ or $x + 3 < -1$

32. $2x^2 > 50$

In exercises 33–36, evaluate the expression.

33. $|-5 + 7| + |-2|$

34. $\left| \dfrac{5 - 12}{-4 - 3} \right|$

35. $|2\pi - 6| - \pi$

36. $|\sqrt{3} - 4| + |4 - 2\sqrt{3}|$

In exercises 37–40, find the value(s) of x that satisfy the expression.

37. $2x^2 + 3x - 2 \le 0$

38. $\dfrac{1}{x+2} > 2$

39. $|2x - 3| < 5$

40. $\left| \dfrac{x+1}{x-1} \right| = 5$

41. Rationalize the numerator:

$$\dfrac{\sqrt{x} - 1}{x - 1}$$

42. Rationalize the denominator:

$$\dfrac{\sqrt{x} - 1}{2\sqrt{x}}$$

In exercises 43 and 44, use the quadratic formula to solve the quadratic equation.

43. $x^2 - 2x - 5 = 0$

44. $2x^2 + 8x + 7 = 0$

45. Find the minimum cost C (in dollars) given that

$$2(1.5C + 80) \le 2(2.5C - 20)$$

46. Find the maximum revenue R (in dollars) given that

$$12(2R - 320) \le 4(3R + 240)$$

In this chapter we review the concept of a function and begin the study of differential calculus. Historically, differential calculus was developed in response to the problem of finding the tangent line to an arbitrary curve. But it quickly became apparent that solving this problem provided mathematicians with a method for solving many practical problems involving the rate of change of one quantity with respect to another. The basic tool used in differential calculus is the *derivative* of a function. The concept of the derivative is based, in turn, on a more fundamental notion—the *limit* of a function.

How does the change in the demand for a certain make of tires affect the unit price of the tires? The management of the Titan Tire Company has determined the demand function that relates the unit price of its Super Titan tires to the quantity demanded. In Example 7, page 665, you will see how this function can be used to compute the rate of change of the unit price of the Super Titan tires with respect to the quantity demanded.

10

FUNCTIONS, LIMITS, AND THE DERIVATIVE

10.1 FUNCTIONS AND THEIR GRAPHS

Functions

The notion of a *function* was introduced in Section 1.3, where we were concerned primarily with the special class of functions known as linear functions. In the study of calculus, we will be dealing with a more general class of functions called *nonlinear functions*. First, however, let us restate the definition of a function.

FUNCTION	A **function** is a rule that assigns to each element in a set A one and only one element in a set B.

The set A is called the **domain** of the function. It is customary to denote a function by a letter of the alphabet, such as the letter f. If x is an element in the domain of a function f, then the element in B that f associates with x is written $f(x)$ (read "f of x") and is called the value of f at x. The set comprising all the values assumed by $y = f(x)$ as x takes on all possible values in its domain is called the **range** of the function f.

We can think of a function f as a machine. The domain is the set of inputs (raw material) for the machine, the rule describes how the input is to be processed, and the value(s) of the function are the outputs of the machine (Figure 10.1).

We can also think of a function f as a mapping in which an element x in the domain of f is mapped onto a unique element $f(x)$ in B (Figure 10.2).

Figure 10.1
A function machine.

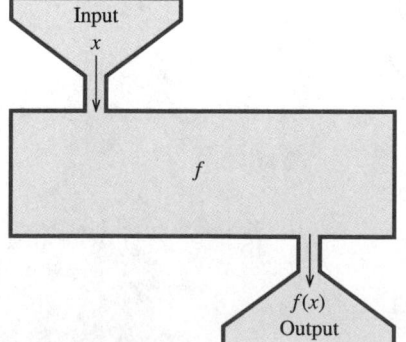

Figure 10.2
The function f viewed as a mapping.

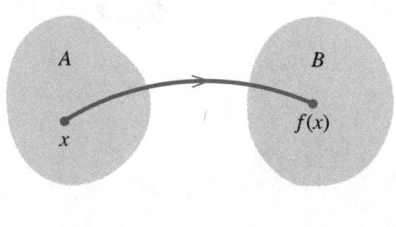

REMARKS

1. It is important to understand that the output $f(x)$ associated with an input x is unique. To appreciate the importance of this uniqueness property, consider a rule that associates with each item x in a department store its selling price y. Then, each x must correspond to *one and only one y*. Notice, however, that different x's may be associated with the same y. In the context of the present example, this says that different items may have the same price.

2. Although the sets A and B that appear in the definition of a function may be quite arbitrary, in this book they will denote sets of real numbers.

◦ ◦ ◦

In general, to evaluate a function at a specific value of x, we replace x with that value, as illustrated in Examples 1 and 2.

EXAMPLE 1 Let the function f be defined by the rule $f(x) = 2x^2 - x + 1$.
Compute:

a. $f(1)$ **b.** $f(-2)$ **c.** $f(a)$ **d.** $f(a + h)$

Solution

a. $f(1) = 2(1)^2 - (1) + 1 = 2 - 1 + 1 = 2$

b. $f(-2) = 2(-2)^2 - (-2) + 1 = 8 + 2 + 1 = 11$

c. $f(a) = 2(a)^2 - (a) + 1 = 2a^2 - a + 1$

d. $f(a + h) = 2(a + h)^2 - (a + h) + 1 = 2a^2 + 4ah + 2h^2 - a - h + 1$

◦ ◦ ◦

EXAMPLE 2 The Thermo-Master Company manufactures an indoor-outdoor thermometer at its Mexican subsidiary. Management estimates that the profit (in dollars) realizable by Thermo-Master in the manufacture and sale of x thermometers per week is

$$P(x) = -0.001x^2 + 8x - 5000$$

Find Thermo-Master's weekly profit if its level of production is (a) 1000 thermometers per week and (b) 2000 thermometers per week.

Solution

a. The weekly profit realizable by Thermo-Master when the level of production is 1000 units per week is found by evaluating the profit function P at $x = 1000$. Thus,

$$P(1000) = -0.001(1000)^2 + 8(1000) - 5000 = 2000$$

or $2000.

b. When the level of production is 2000 units per week, the weekly profit is given by

$$P(2000) = -0.001(2000)^2 + 8(2000) - 5000 = 7000$$

or $7000.

○ ○ ○

Determining the Domain of a Function

Suppose we are given the function $y = f(x)$.* Then the variable x is called the **independent variable.** The variable y, whose value depends on x, is called the **dependent variable.**

In determining the domain of a function, we need to find what restrictions, if any, are to be placed on the independent variable x. In many practical applications, the domain of a function is dictated by the nature of the problem, as illustrated in Example 3.

EXAMPLE 3   An open box is to be made from a rectangular piece of cardboard 16 inches long and 10 inches wide by cutting away identical squares (x by x inches) from each corner and folding up the resulting flaps (Figure 10.3). Find an expression that gives the volume V of the box as a function of x. What is the domain of the function?

Solution   The dimensions of the box are $(16 - 2x)$ inches long, $(10 - 2x)$ inches wide, and x inches high, so its volume (in cubic inches) is given by

$$\begin{aligned}
V = f(x) &= (16 - 2x)(10 - 2x)x \qquad \text{(Length} \cdot \text{Width} \cdot \text{Height)} \\
&= (160 - 52x + 4x^2)x \\
&= 4x^3 - 52x^2 + 160x
\end{aligned}$$

Figure 10.3

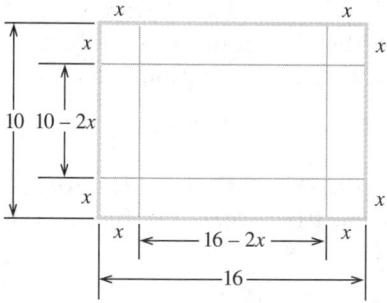

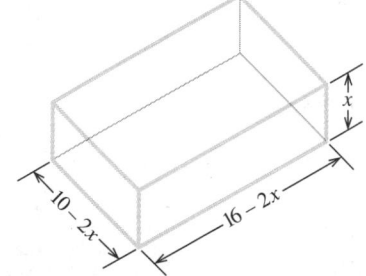

(a) The box is constructed by cutting x-by-x inch squares from each corner.

(b) The dimensions of the resulting box are $(10 - 2x)$ by $(16 - 2x)$ by x inches.

○ ○ ○

* It is customary to refer to a function f as $f(x)$ or by the equation $y = f(x)$ defining it.

Since the length of each side of the box must be greater than or equal to zero, we see that

$$16 - 2x \geq 0, \quad 10 - 2x \geq 0, \quad \text{and} \quad x \geq 0$$

simultaneously; that is,

$$x \leq 8, \quad x \leq 5, \quad \text{and} \quad x \geq 0$$

All three inequalities are satisfied simultaneously provided that $0 \leq x \leq 5$. Thus, the domain of the function f is the interval $[0, 5]$.　　◦ ◦ ◦

In general, if a function is defined by a rule relating x to $f(x)$ without specific mention of its domain, it is understood that the domain will consist of all values of x for which $f(x)$ is a real number. In this connection, you should keep in mind that (1) division by zero is not permitted and (2) the square root of a negative number is not defined.

EXAMPLE 4　Find the domain of each of the functions defined by the following equations:

a. $f(x) = \sqrt{x - 1}$　　**b.** $f(x) = \dfrac{1}{x^2 - 4}$　　**c.** $f(x) = x^2 + 3$

Solution

a. Since the square root of a negative number is undefined, it is necessary that $x - 1 \geq 0$. The inequality is satisfied by the set of real numbers $x \geq 1$. Thus, the domain of f is the interval $[1, \infty)$.

b. The only restriction on x is that $x^2 - 4$ be different from zero since division by zero is not allowed. But $(x^2 - 4) = (x + 2)(x - 2) = 0$ if $x = -2$ or $x = 2$. Thus, the domain of f in this case consists of the intervals $(-\infty, -2)$, $(-2, 2)$, and $(2, \infty)$.

c. Any real number satisfies the equation, so the domain of f is the set of all real numbers.　　◦ ◦ ◦

Graphs of Functions

If f is a function with domain A, then corresponding to each real number x in A there is precisely one real number $f(x)$. We can also express this fact by using **ordered pairs** of real numbers. Write each number x in A as the first member of an ordered pair and each number $f(x)$ corresponding to x as the second member of the ordered pair. This gives exactly one ordered pair $(x, f(x))$ for each x in A. This observation leads to an alternative definition of a function f:

> A function f with domain A is the set of all ordered pairs $(x, f(x))$ where x belongs to A.

Observe that the condition that there be one and only one number $f(x)$ corresponding to each number x in A translates into the requirement that *no two ordered pairs have the same first number.*

Since ordered pairs of real numbers correspond to points in the plane, we have found a way to exhibit a function graphically.

GRAPH OF A FUNCTION OF ONE VARIABLE

The **graph of a function** f is the set of all points (x, y) in the xy-plane such that x is in the domain of f and $y = f(x)$.

Much information about the graph of a function can be gained by plotting a few points on its graph. Later on we will develop more systematic and sophisticated techniques for graphing functions.

Figure 10.4
The graph of $y = x^2 + 1$ is a parabola.

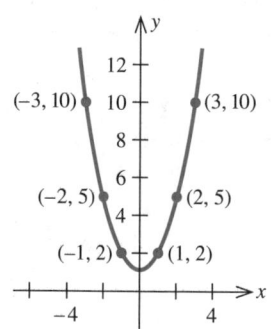

EXAMPLE 5   Sketch the graph of the function defined by the equation
$$y = x^2 + 1.$$

Solution   The domain of the function is the set of all real numbers. By assigning several values to the variable x and computing the corresponding values for y, we obtain the following solutions to the equation $y = x^2 + 1$:

x	-3	-2	-1	0	1	2	3
y	10	5	2	1	2	5	10

By plotting these points and then connecting them with a smooth curve, we obtain the graph of $y = f(x)$, which is a parabola (Figure 10.4).   ◦ ◦ ◦

Let $f(x) = x^2$.

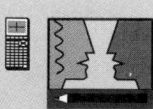

a. Plot the graphs of $F(x) = x^2 + c$ on the same set of axes for $c = -2, -1, -\frac{1}{2}, 0, \frac{1}{2}, 1, 2$.
b. Plot the graphs of $G(x) = (x + c)^2$ on the same set of axes for $c = -2, -1, -\frac{1}{2}, 0, \frac{1}{2}, 1, 2$.
c. Plot the graphs of $H(x) = cx^2$ on the same set of axes for $c = -2, -1, -\frac{1}{2}, -\frac{1}{4}, 0, \frac{1}{4}, \frac{1}{2}, 1, 2$.
d. Study the family of graphs in parts (a), (b), and (c), and describe the relationship between the graph of a function f and the graphs of the functions defined by (i) $y = f(x) + c$, (ii) $y = f(x + c)$, and (iii) $y = cf(x)$, where c is a constant.

Some functions are defined in a piecewise fashion, as Examples 6 and 7 show.

EXAMPLE 6 The Madison Finance Company plans to open two branch offices two years from now in two separate locations: an industrial complex and a newly developed commercial center in the city. As a result of these expansion plans, Madison's total deposits during the next five years are expected to grow in accordance with the rule

$$f(x) = \begin{cases} \sqrt{2x} + 20 & \text{if } 0 \le x \le 2 \\ \frac{1}{2}x^2 + 20 & \text{if } 2 < x \le 5 \end{cases}$$

where $y = f(x)$ gives the total amount of money (in millions of dollars) on deposit with Madison in year x ($x = 0$ corresponds to the present). Sketch the graph of the function f.

Solution The function f is defined in a piecewise fashion on the interval $[0, 5]$. In the subdomain $[0, 2]$, the rule for f is given by $f(x) = \sqrt{2x} + 20$. The values of $f(x)$ corresponding to $x = 0, 1,$ and 2 are tabulated as follows:

x	0	1	2
$f(x)$	20	21.4	22

Next, in the subdomain $(2, 5]$, the rule for f is given by $f(x) = \frac{1}{2}x^2 + 20$. The values of $f(x)$ corresponding to $x = 3, 4,$ and 5 are shown in the following table:

x	3	4	5
$f(x)$	24.5	28	32.5

Using the values of $f(x)$ in the tables, we sketch the graph of the function f as shown in Figure 10.5.

Figure 10.5
We obtain the graph of the function $y = f(x)$ by graphing $y = \sqrt{2x} + 20$ over $[0, 2]$ and $y = \frac{1}{2}x^2 + 20$ over $(2, 5]$.

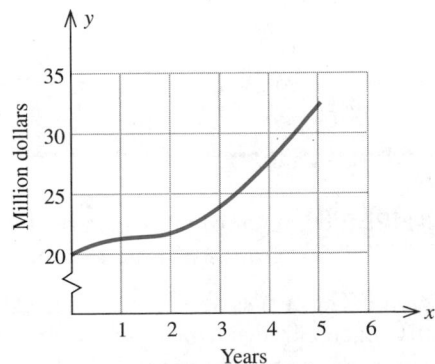

Figure 10.6
The graph of $y = f(x)$ is obtained by graphing $y = -x$ over $(-\infty, 0)$ and $y = \sqrt{x}$ over $[0, \infty)$.

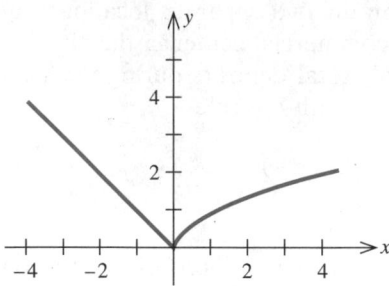

EXAMPLE 7   Sketch the graph of the function f defined by

$$f(x) = \begin{cases} -x & \text{if } x < 0 \\ \sqrt{x} & \text{if } x \geq 0 \end{cases}$$

Solution   The function f is defined in a piecewise fashion on the set of all real numbers. In the subdomain $(-\infty, 0)$, the rule for f is given by $f(x) = -x$. The equation $y = -x$ is a linear equation in the slope-intercept form (with slope -1 and intercept 0). Therefore, the graph of f corresponding to the subdomain $(-\infty, 0)$ is the half-line shown in Figure 10.6. Next, in the subdomain $[0, \infty)$, the rule for f is given by $f(x) = \sqrt{x}$. The values of $f(x)$ corresponding to $x = 0, 1, 2, 3, 4, 9,$ and 16 are shown in the following table:

x	0	1	2	3	4	9	16
$f(x)$	0	1	$\sqrt{2}$	$\sqrt{3}$	2	3	4

Figure 10.7
Since a vertical line passes through the curve at more than one point, we deduce that it is not the graph of a function.

Using these values, we sketch the graph of the function f as shown in Figure 10.6.

● ● ●

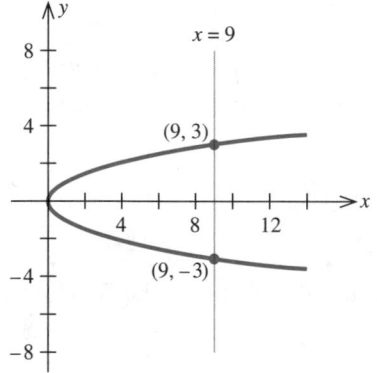

The Vertical-Line Test

Although it is true that every function f of a variable x has a graph in the xy-plane, it is important to realize that not every curve in the xy-plane is the graph of a function. For example, consider the curve depicted in Figure 10.7. This is the graph of the equation $x = y^2$. In general, the **graph of an equation** is the set of all ordered pairs (x, y) that satisfy the given equation. Observe that the points $(9, -3)$ and $(9, 3)$ both lie on the curve. This implies that the number $x = 9$ is associated with *two* numbers: $y = -3$ and $y = 3$. But this clearly violates the uniqueness property of a function. Thus, we conclude that the curve under consideration cannot be the graph of a function.

This example suggests the **vertical-line test** for determining when a curve is the graph of a function.

VERTICAL-LINE TEST   A curve in the xy-plane is the graph of a function $y = f(x)$ if and only if each vertical line intersects it in at most one point.

EXAMPLE 8   Determine which of the curves shown in Figure 10.8 are the graphs of functions of x.

Solution   The curves depicted in Figures 10.8a, c, and d are graphs of functions because each curve satisfies the requirement that each vertical line intersects the curve in at most one point. Note that the vertical line shown in Figure

Figure 10.8
The vertical-line test can be used to determine which of these curves are graphs of functions.

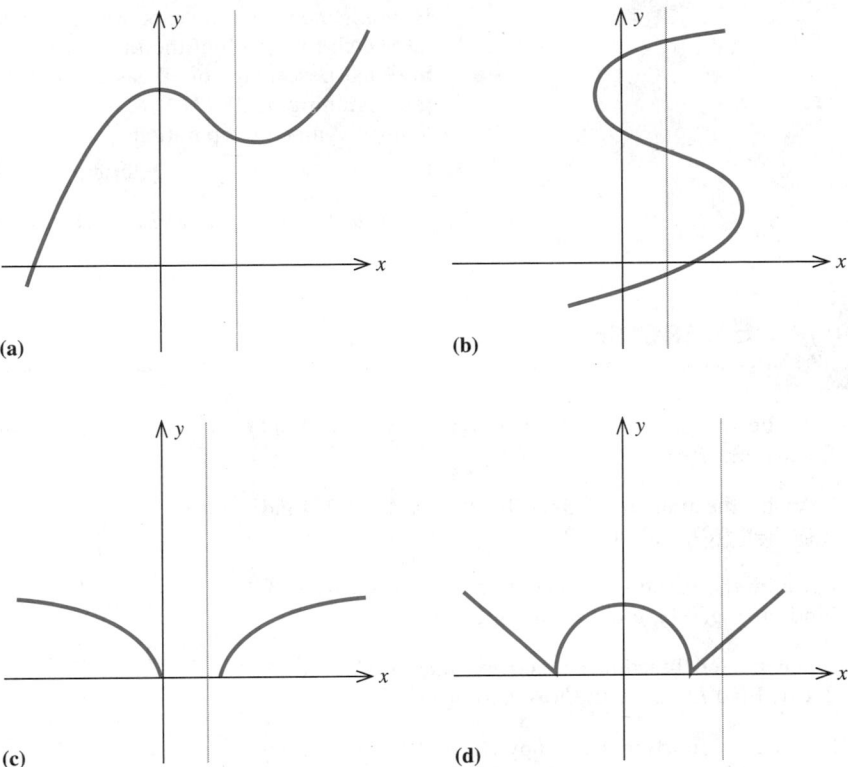

(a)

(b)

(c)

(d)

10.8c does not intersect the graph because the point on the *x*-axis through which this line passes does not lie in the domain of the function. The curve depicted in Figure 10.8b is *not* the graph of a function because the vertical line shown there intersects the graph at three points.   ⦾ ⦾ ⦾

SELF–CHECK EXERCISES 10.1

1. Let *f* be the function defined by

$$f(x) = \frac{\sqrt{x+1}}{x}$$

   a. Find the domain of *f*.
   b. Compute *f*(3).
   c. Compute *f*(*a* + *h*).

 2. Statistics obtained by Amoco Corporation show that more and more motorists are pumping their own gas. The following function gives self-serve sales as a percentage of all U.S. gas sales:

$$f(t) = \begin{cases} 6t + 17 & \text{if } 0 \leq t \leq 6 \\ 15.98(t-6)^{1/4} + 53 & \text{if } 6 < t \leq 20 \end{cases}$$

Here t is measured in years, with $t = 0$ corresponding to the beginning of 1974.

a. Sketch the graph of the function f.

b. What percentage of all gas sales at the beginning of 1978 were self-serve? At the beginning of 1994?

Source: Amoco Corporation

3. Let $f(x) = \sqrt{2x + 1} + 2$. Determine whether the point $(4, 6)$ lies on the graph of f.

Solutions to Self-Check Exercises 10.1 can be found on page 582.

10.1 EXERCISES

1. Let f be the function defined by $f(x) = 5x + 6$. Find $f(3)$, $f(-3)$, $f(a)$, $f(-a)$, and $f(a + 3)$.

2. Let f be the function defined by $f(x) = 4x - 3$. Find $f(4)$, $f(\frac{1}{4})$, $f(0)$, $f(a)$, and $f(a + 1)$.

3. Let g be the function defined by $g(x) = 3x^2 - 6x - 3$. Find $g(0)$, $g(-1)$, $g(a)$, $g(-a)$, and $g(x + 1)$.

4. Let h be the function defined by $h(x) = x^3 - x^2 + x + 1$. Find $h(-5)$, $h(0)$, $h(a)$, and $h(-a)$.

5. Let s be the function defined by $s(t) = 2t/(t^2 - 1)$. Find $s(4)$, $s(0)$, $s(a)$, $s(2 + a)$, and $s(t + 1)$.

6. Let g be the function defined by $g(u) = (3u - 2)^{3/2}$. Find $g(1)$, $g(6)$, $g(\frac{11}{3})$, and $g(u + 1)$.

7. Let f be the function defined by $f(t) = 2t^2/\sqrt{t - 1}$. Find $f(2)$, $f(a)$, $f(x + 1)$, and $f(x - 1)$.

8. Let f be the function defined by $f(x) = 2 + 2\sqrt{5 - x}$. Find $f(-4)$, $f(1)$, $f(\frac{11}{4})$, and $f(x + 5)$.

9. Let f be the function defined by

$$f(x) = \begin{cases} x^2 + 1 & \text{if } x \le 0 \\ \sqrt{x} & \text{if } x > 0 \end{cases}$$

Find $f(-2)$, $f(0)$, and $f(1)$.

10. Let g be the function defined by

$$g(x) = \begin{cases} -\dfrac{1}{2}x + 1 & \text{if } x < 2 \\ \sqrt{x - 2} & \text{if } x \ge 2 \end{cases}$$

Find $g(-2)$, $g(0)$, $g(2)$, and $g(4)$.

11. Let f be the function defined by

$$f(x) = \begin{cases} -\dfrac{1}{2}x^2 + 3 & \text{if } x < 1 \\ 2x^2 + 1 & \text{if } x \ge 1 \end{cases}$$

Find $f(-1)$, $f(0)$, $f(1)$, and $f(2)$.

12. Let f be the function defined by

$$f(x) = \begin{cases} 2 + \sqrt{1 - x} & \text{if } x \le 1 \\ \dfrac{1}{1 - x} & \text{if } x > 1 \end{cases}$$

Find $f(0)$, $f(1)$, and $f(2)$.

In exercises 13–16, determine whether the point lies on the graph of the function.

13. $(2, \sqrt{3})$; $g(x) = \sqrt{x^2 - 1}$

14. $(3, 3)$; $f(x) = \dfrac{x + 1}{\sqrt{x^2 + 7}} + 2$

15. $(-2, -3)$; $f(t) = \dfrac{|t - 1|}{t + 1}$

16. $\left(-3, -\dfrac{1}{13}\right)$; $h(t) = \dfrac{|t + 1|}{t^3 + 1}$

In exercises 17–30, find the domain of the function.

17. $f(x) = x^2 + 3$ 18. $f(x) = 7 - x^2$

19. $f(x) = \dfrac{3x + 1}{x^2}$ 20. $g(x) = \dfrac{2x + 1}{x - 1}$

21. $f(x) = \sqrt{x^2 + 1}$ 22. $f(x) = \sqrt{x - 5}$

23. $f(x) = \sqrt{5 - x}$ 24. $g(x) = \sqrt{2x^2 + 3}$

25. $f(x) = \dfrac{x}{x^2 - 1}$ 26. $f(x) = \dfrac{1}{x^2 + x - 2}$

27. $f(x) = (x + 3)^{3/2}$ 28. $g(x) = 2(x - 1)^{5/2}$

29. $f(x) = \dfrac{\sqrt{1 - x}}{x^2 - 4}$ 30. $f(x) = \dfrac{\sqrt{x - 1}}{(x + 2)(x - 3)}$

31. Let f be a function defined by the rule $f(x) = x^2 - x - 6$.
    a. Find the domain of f.
    b. Compute $f(x)$ for $x = -3, -2, -1, 0, \frac{1}{2}, 1, 2, 3$.
    c. Use the results obtained in (a) and (b) to sketch the graph of f.

32. Let f be a function defined by the rule $f(x) = 2x^2 + x - 3$.
    a. Find the domain of f.
    b. Compute $f(x)$ for $x = -3, -2, -1, -\frac{1}{2}, 0, 1, 2, 3$.
    c. Use the results obtained in (a) and (b) to sketch the graph of f.

In exercises 33–44, sketch the graph of the function with the given rule.

33. $f(x) = 2x^2 + 1$ 34. $f(x) = 9 - x^2$

35. $f(x) = 2 + \sqrt{x}$ 36. $g(x) = 4 - \sqrt{x}$

37. $f(x) = \sqrt{1 - x}$ 38. $f(x) = \sqrt{x - 1}$

39. $f(x) = |x| - 1$ 40. $f(x) = |x| + 1$

41. $f(x) = \begin{cases} x & \text{if } x < 0 \\ 2x + 1 & \text{if } x \geq 0 \end{cases}$

42. $f(x) = \begin{cases} 4 - x & \text{if } x < 2 \\ 2x - 2 & \text{if } x \geq 2 \end{cases}$

43. $f(x) = \begin{cases} -x + 1 & \text{if } x \leq 1 \\ x^2 - 1 & \text{if } x > 1 \end{cases}$

44. $f(x) = \begin{cases} -x - 1 & \text{if } x < -1 \\ 0 & \text{if } -1 \leq x \leq 1 \\ x + 1 & \text{if } x > 1 \end{cases}$

In exercises 45–52, use the vertical-line test to determine whether the graph represents y as a function of x.

45. 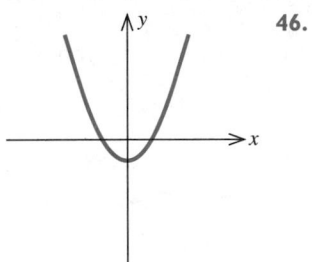   46.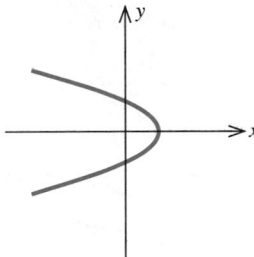

47. 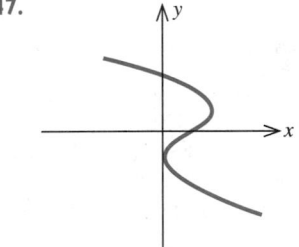   48.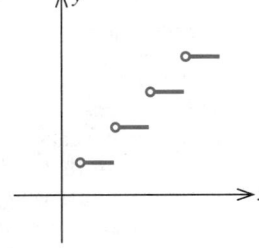

49. 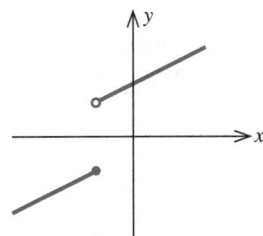   50.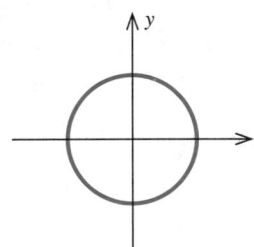

51. 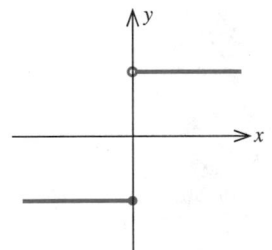   52.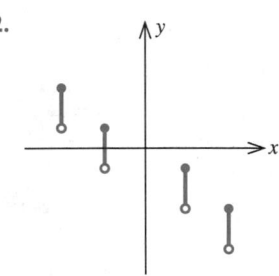

53. The circumference of a circle is given by $C(r) = 2\pi r$, where r is the radius of the circle. What is the circumference of a circle with a 5-inch radius?

54. The volume of a sphere of radius r is given by $V(r) = \frac{4}{3}\pi r^3$. Compute $V(2.1)$ and $V(2)$. What does the quantity $V(2.1) - V(2)$ measure?

55. **Sales of Prerecorded Music** The following graphs show the sales y of prerecorded music (in billions of

dollars) by format as a function of time t (in years) with $t = 0$ corresponding to 1985.

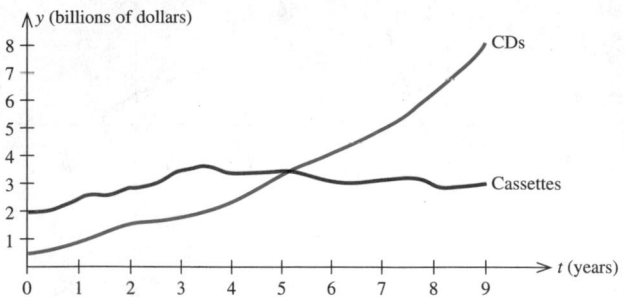

Source: Recording Industry Association of America

a. In what years were the sales of prerecorded cassettes greater than those of prerecorded CDs?
b. In what years were the sales of prerecorded CDs greater than those of prerecorded cassettes?
c. In what year were the sales of prerecorded cassettes the same as those of prerecorded CDs? Estimate the level of sales in each format at that time.

56. The Gender Gap The following graph shows the ratio of women's earnings to men's from 1960 through 1990.

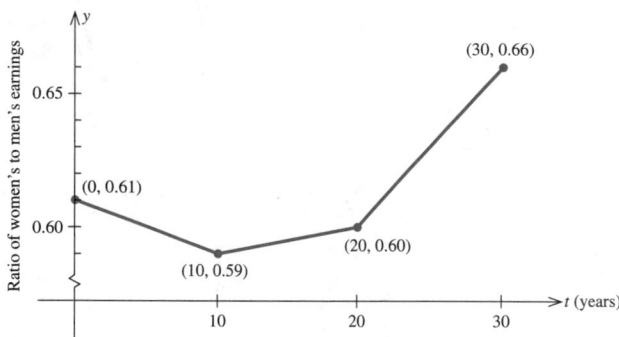

Source: U.S. Bureau of Labor Statistics

a. Write the rule for the function f giving the ratio of women's earnings to men's in year t, with $t = 0$ corresponding to 1960.

[*Hint:* The function f is defined piecewise and is linear over each of three subintervals.]

b. In what decade(s) was the gender gap expanding? Shrinking?
c. Refer to (b). How fast was the gender gap (the ratio per year) expanding or shrinking in each of these decades?

57. Closing the Gender Gap in Education The following graph shows the ratio of bachelor's degrees earned by women to men from 1960 through 1990.

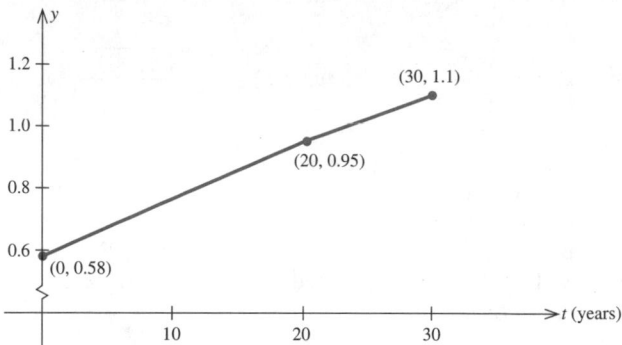

Source: U.S. Department of Education

a. Write the rule for the function f giving the ratio of bachelor's degrees earned by women to men in year t, with $t = 0$ corresponding to 1960.

[*Hint:* The function f is defined piecewise and is linear over each of two subintervals.]

b. How fast was the ratio changing in the period from 1960 through 1980? In the decade from 1980 to 1990?
c. In what year (approximately) was the number of bachelor's degrees earned by women equal to that earned by men for the first time?

58. Consumption Function The consumption function in a certain economy is given by the equation

$$C(y) = 0.75y + 6$$

where $C(y)$ is the personal consumption expenditure, y is the disposable personal income, and both $C(y)$ and y are measured in billions of dollars. Find $C(0)$, $C(50)$, and $C(100)$.

59. Sales Taxes In a certain state, the sales tax T on the amount of taxable goods is 6% of the value of the goods purchased (x), where both T and x are measured in dollars.

a. Express T as a function of x.
b. Find $T(200)$ and $T(5.65)$.

60. Surface Area of a Single-Celled Organism The surface area S of a single-celled organism may be found by multiplying 4π times the square of the radius r of the cell. Express S as a function of r.

61. Friend's Rule   Friend's Rule, a method for calculating pediatric drug dosages, is based on a child's age. If a denotes the adult dosage (in mg) and if t is the age of the child (in years), then the child's dosage is given by

$$D(t) = \frac{2}{25} ta$$

If the adult dose of a substance is 500 mg, how much should a four-year-old child receive?

62. COLAs   Social security recipients receive an automatic cost-of-living adjustment (COLA) once each year. Their monthly benefit is increased by the amount that consumer prices increased during the preceding year. Suppose that consumer prices increased by 5.3% during the preceding year.
a. Express the adjusted monthly benefit of a social security recipient as a function of his or her current monthly benefit.
b. If Mr. Harrington's monthly social security benefit is now $620, what will his adjusted monthly benefit be?

63. Cost of Renting a Truck   The Ace Truck Leasing Company leases a certain size truck at $30 a day and 15 cents a mile, whereas the Acme Truck Leasing Company leases the same size truck at $25 a day and 20 cents a mile.
a. Find the daily cost of leasing from each company as a function of the number of miles driven.
b. Sketch the graphs of the two functions on the same set of axes.
c. Which company should a customer rent a truck from for one day if she plans to drive at most 70 miles and wishes to minimize her cost?

64. Linear Depreciation   A new machine was purchased by the National Textile Company for $120,000. For income tax purposes, the machine is depreciated linearly over ten years; that is, the book value of the machine decreases at a constant rate, so that at the end of ten years the book value is zero.
a. Express the book value of the machine (V) as a function of the age, in years, of the machine (n).
b. Sketch the graph of the function in (a).
c. Find the book value of the machine at the end of the sixth year.
d. Find the rate at which the machine is being depreciated each year.

65. Linear Depreciation   Refer to exercise 64. An office building worth $1,000,000 when completed in 1985 was depreciated linearly over 50 years. What will the book value of the building be in 2004? In 2008? (Assume that the book value of the building will be zero at the end of the 50th year.)

66. Boyle's Law   As a consequence of Boyle's Law, the pressure P of a fixed sample of gas held at a constant temperature is related to the volume V of the gas by the rule

$$P = f(V) = \frac{k}{V}$$

where k is a constant. What is the domain of the function f? Sketch the graph of the function f.

67. Poiseuille's Law   According to a law discovered by the nineteenth-century physician Poiseuille, the velocity (in centimeters per second) of blood r centimeters from the central axis of an artery is given by

$$v(r) = k(R^2 - r^2)$$

where k is a constant and R is the radius of the artery. Suppose that for a certain artery, $k = 1000$ and $R = 0.2$ so that $v(r) = 1000(0.04 - r^2)$.
a. What is the domain of the function $v(r)$?
b. Compute $v(0)$, $v(0.1)$, and $v(0.2)$, and interpret your results.

68. Population Growth   A study prepared for a certain Sunbelt town's Chamber of Commerce projected that the population of the town in the next three years will grow according to the rule

$$P(x) = 50,000 + 30x^{3/2} + 20x$$

where $P(x)$ denotes the population x months from now. By how much will the population increase during the next 9 months? During the next 16 months?

69. Worker Efficiency   An efficiency study conducted for the Elektra Electronics Company showed that the number of "Space Commander" walkie-talkies assembled by the average worker t hours after starting work at 8 A.M. is given by

$$N(t) = -t^3 + 6t^2 + 15t \qquad (0 \le t \le 4)$$

How many walkie-talkies can an average worker be expected to assemble between 8 and 9 A.M.? Between 9 and 10 A.M.?

70. Learning Curves The Emory Secretarial School finds from experience that the average student taking advanced typing will progress according to the rule

$$N(t) = \frac{60t + 180}{t + 6}$$

where $N(t)$ measures the number of words per minute the student can type after t weeks in the course. How fast can the average student be expected to type after two weeks in the course? After four weeks in the course?

71. Politics Political scientists have discovered the following empirical rule, known as the "cube rule," which gives the relationship between the proportion of seats in the House of Representatives won by Democratic candidates $s(x)$ and the proportion of popular votes x received by the Democratic presidential candidate:

$$s(x) = \frac{x^3}{x^3 + (1 - x)^3} \qquad (0 \le x \le 1)$$

Compute $s(0.6)$ and interpret your result.

72. Home Shopping Industry According to industry sources, revenue from the home shopping industry for the years since its inception may be approximated by the function

$$R(t) = \begin{cases} -0.03t^3 + 0.25t^2 - 0.12t & \text{if } 0 \le t \le 3 \\ 0.57t - 0.63 & \text{if } 3 < t \le 11 \end{cases}$$

where $R(t)$ measures the revenue in billions of dollars and t is measured in years, with $t = 0$ corresponding to the beginning of 1984. What was the revenue at the beginning of 1985? At the beginning of 1993?
Source: Paul Kagan Associates

73. Postal Regulations The postage for first-class mail is 32 cents for the first ounce or fraction thereof, and 23 cents for each additional ounce or fraction thereof. Any parcel not exceeding 12 ounces may be sent by first-class mail. Letting x denote the weight of a parcel in ounces and $f(x)$ the postage in cents, complete the following description of the "postage function" f:

$$f(x) = \begin{cases} 32 & \text{if } 0 < x \le 1 \\ 55 & \text{if } 1 < x \le 2 \\ \cdots \\ ? & \text{if } 11 < x \le 12 \end{cases}$$

a. What is the domain of f?
b. Sketch the graph of f.

74. Harbor Cleanup The amount of solids discharged from the MWRA (Massachusetts Water Resources Authority) sewage treatment plant on Deer Island (near Boston Harbor) is given by the function

$$f(t) = \begin{cases} 130 & \text{if } 0 \le t \le 1 \\ -30t + 160 & \text{if } 1 < t \le 2 \\ 100 & \text{if } 2 < t \le 4 \\ -5t^2 + 25t + 80 & \text{if } 4 < t \le 6 \\ 1.25t^2 - 26.25t + 162.5 & \text{if } 6 < t \le 10 \end{cases}$$

where $f(t)$ is measured in tons per day and t is measured in years, with $t = 0$ corresponding to 1989.
Source: Metropolitan District Commission

a. What is the amount of solids discharged per day in 1989? In 1992? In 1996?
b. Sketch the graph of f.

SOLUTIONS TO SELF-CHECK EXERCISES 10.1

1. a. The expression under the radical sign must be nonnegative, so $x + 1 \ge 0$ or $x \ge -1$. Also, $x \ne 0$ because division by zero is not permitted. Therefore, the domain of f is $[-1, 0) \cup (0, \infty)$.

b. $f(3) = \dfrac{\sqrt{3 + 1}}{3} = \dfrac{\sqrt{4}}{3} = \dfrac{2}{3}$

c. $f(a + h) = \dfrac{\sqrt{(a + h) + 1}}{a + h} = \dfrac{\sqrt{a + h + 1}}{a + h}$

2. a. For t in the subdomain $[0, 6]$, the rule for f is given by $f(t) = 6t + 17$. The equation $y = 6t + 17$ is a linear equation, so that portion of the graph of f is the line segment joining the points $(0, 17)$ and $(6, 53)$. Next, in the subdomain $(6, 20]$, the rule for f is given by $f(t) = 15.98(t - 6)^{1/4} + 53$. Using a calculator, we construct the following table of values of $f(t)$ for selected values of t:

t	6	8	10	12	14	16	18	20
$f(t)$	53	72	75.6	78	79.9	81.4	82.7	83.9

We have included $t = 6$ in the table, although it does not lie in the subdomain of the function under consideration, in order to help us obtain a better sketch of that portion of the graph of f in the subdomain $(6, 20]$. The graph of f is shown here.

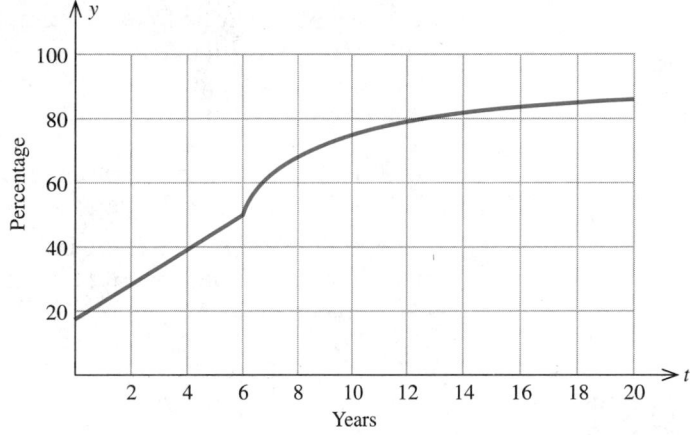

b. The percentage of all self-serve gas sales at the beginning of 1978 is found by evaluating f at $t = 4$. Since this point lies in the interval $[0, 6]$, we use the rule $f(t) = 6t + 17$ and find

$$f(4) = 6(4) + 17$$

giving 41% as the required figure. The percentage of all self-serve gas sales at the beginning of 1994 is given by

$$f(20) = 15.98(20 - 6)^{1/4} + 53$$

or approximately 83.9%.

3. A point (x, y) lies on the graph of the function f if and only if the coordinates satisfy the equation $y = f(x)$. Now

$$f(4) = \sqrt{2(4) + 1} + 2 = \sqrt{9} + 2 = 5 \neq 6$$

and we conclude that the given point does not lie on the graph of f.

USING TECHNOLOGY

Graphing a Function

Most of the graphs of functions in this book can be plotted with the help of a graphing utility. Furthermore, a graphing utility can be used to analyze the nature of a function. However, the amount and accuracy of the information obtained using a graphing utility depend on the experience and sophistication of the user. As you progress through this book, you will see that the more knowledge of calculus you gain, the more effective the graphing utility will prove as a tool in problem solving.

Finding a Suitable Viewing Rectangle

The first step in plotting the graph of a function with a graphing utility is to select a suitable viewing rectangle. We usually do this by experimenting. For example, you might first plot the graph using the *standard viewing rectangle* $[-10, 10]$ by $[-10, 10]$. If necessary, you then might adjust the viewing rectangle by enlarging it or reducing it to obtain a sufficiently complete view of the graph or at least the portion of the graph that is of interest.

EXAMPLE 1   Plot the graph of $f(x) = 2x^2 - 4x - 5$ in the standard viewing rectangle.

Solution   The graph of f, shown in Figure T1, is a parabola. From our previous work (Example 5, Section 10.1), we know that the figure does give a good view of the graph.

Figure T1
The graph of $y = 2x^2 - 4x - 5$ in the viewing rectangle $[-10, 10] \times [-10, 10]$.

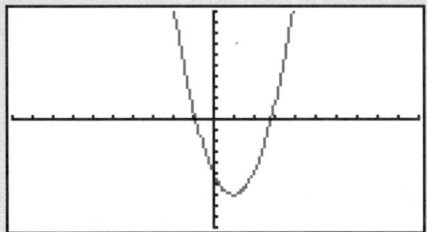

$\bullet\ \bullet\ \bullet$

EXAMPLE 2   Let $f(x) = x^3(x - 3)^4$.

a. Plot the graph of f in the standard viewing rectangle.

b. Plot the graph of f in the rectangle $[-1, 5] \times [-40, 40]$.

Solution

a. The graph of f in the standard viewing rectangle is shown in Figure T2a. Since the graph does not appear to be complete, we need to adjust the viewing rectangle.

584

Figure T2
An incomplete sketch of $f(x) =$
$x^3(x - 3)^4$ *is shown in (a) and a*
complete sketch is shown in (b).

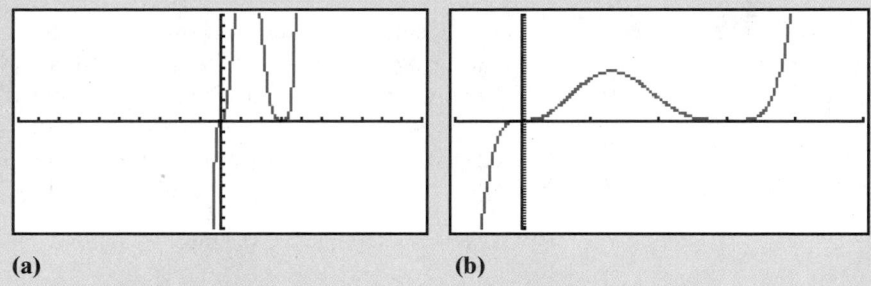

(a)　　　　　　　　　　　　　**(b)**

b. The graph of f in the rectangle $[-1, 5] \times [-40, 40]$, shown in Figure T2b, is an improvement over the previous graph. (Later we will be able to show that the figure does, in fact, give a rather complete view of the graph of f.)

⊙ ⊙ ⊙

Evaluating a Function

A graphing utility can be used to find the value of a function with minimal effort, as the next example shows.

EXAMPLE 3　Let $f(x) = x^3 - 4x^2 + 4x + 2$.

a. Plot the graph of f in the standard viewing rectangle.

b. Find $f(3)$ and verify your result by direct computation.

c. Find $f(4.215)$.

Solution

a. The graph of f is shown in Figure T3.

Figure T3
The graph of $f(x) = x^3 - 4x^2 +$
$4x + 2$ *in the standard viewing*
rectangle.

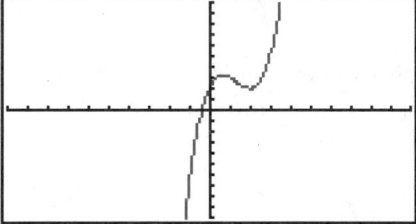

b. Using the evaluation function of the graphing utility and the value 3 for x, we find $y = 5$. This result is verified by computing

$$f(3) = 3^3 - 4(3^2) + 4(3) + 2 = 27 - 36 + 12 + 2 = 5$$

585

c. Using the evaluation function of the graphing utility and the value 4.215 for x, we find $y = 22.679738375$. Thus, $f(4.215) = 22.679738375$. The efficacy of the graphing utility is clearly demonstrated here!　　 o o o

EXAMPLE 4 The anticipated rise in the number of people who have Alzheimer's disease in the United States is given by

$$f(t) = -0.0277t^4 + 0.3346t^3 - 1.1261t^2 + 1.7575t + 3.7745 \qquad (0 \le t \le 6)$$

where $f(t)$ is measured in millions and t is measured in decades, with $t = 0$ corresponding to the beginning of 1990.

a. Use a graphing utility to plot the graph of f in the viewing rectangle $[0, 7] \times [0, 12]$.

b. What is the anticipated number of Alzheimer's patients in the United States at the beginning of the year 2000 ($t = 1$)? At the beginning of 2030 ($t = 4$)?
Source: Alzheimer's Association

Solution

a. The graph of f in the viewing rectangle $[0, 7] \times [0, 12]$ is shown in Figure T4.

Figure T4
The graph of f in the viewing rectangle $[0, 7] \times [0, 12]$.

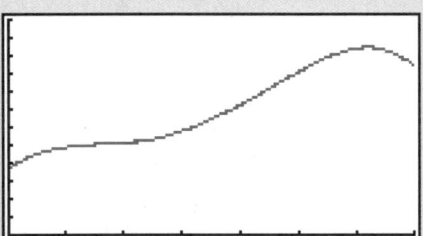

b. Using the evaluation function of the graphing utility and the value 1 for x, we see that the anticipated number of Alzheimer's patients at the beginning of the year 2000 is given by

$$f(1) = 4.7128$$

or approximately 4.7 million. The anticipated number of Alzheimer's patients at the beginning of 2030 is given by

$$f(4) = 7.1101$$

or approximately 7.1 million.　　o o o

EXERCISES

In exercises 1–8, plot the graph of the function f in the standard viewing window.

1. $f(x) = 2x^2 - 16x + 29$ 2. $f(x) = -x^2 - 10x - 20$

3. $f(x) = x^3 - 2x^2 + x - 2$

4. $f(x) = -2.01x^3 + 1.21x^2 - 0.78x + 1$

5. $f(x) = 0.2x^4 - 2.1x^2 + 1$

6. $f(x) = -0.4x^4 + 1.2x - 1.2$

7. $f(x) = 2x\sqrt{x^2 + 1}$ 8. $f(x) = \dfrac{\sqrt{x} + 1}{\sqrt{x} - 1}$

In exercises 9–20, plot the graph of the function f in (a) the standard viewing window and (b) the indicated window.

9. $f(x) = 2x^2 - 32x + 125$; $[5, 15] \times [-5, 10]$

10. $f(x) = x^2 + 20x + 95$; $[-20, 10] \times [-15, -5]$

11. $f(x) = x^3 - 20x^2 + 8x - 10$; $[-20, 20] \times [-1200, 100]$

12. $f(x) = -2x^3 + 10x^2 - 15x - 5$; $[-10, 10] \times [-100, 100]$

13. $f(x) = x^4 - 2x^2 + 8$; $[-2, 2] \times [6, 10]$

14. $f(x) = x^4 - 2x^3$; $[-1, 3] \times [-2, 2]$

15. $f(x) = x + \dfrac{1}{x}$; $[-1, 3] \times [-5, 5]$

16. $f(x) = \dfrac{4}{x^2 - 8}$; $[-5, 5] \times [-5, 5]$

17. $f(x) = 2 - \dfrac{1}{x^2 + 1}$; $[-3, 3] \times [0, 3]$

18. $f(x) = x - 2\sqrt{x}$; $[0, 20] \times [-2, 10]$

19. $f(x) = x\sqrt{4 - x^2}$; $[-3, 3] \times [-2, 2]$

20. $f(x) = \dfrac{\sqrt{x} - 1}{x}$; $[0, 50] \times [-0.25, 0.25]$

In exercises 21–30, plot the graph of the function f in an appropriate viewing window. (Note: The answer is not unique.)

21. $f(x) = x^2 - 4x + 16$ 22. $f(x) = -x^2 + 2x - 11$

23. $f(x) = 2x^3 - 10x^2 + 5x - 10$

24. $f(x) = -x^3 + 5x^2 - 14x + 20$

25. $f(x) = 2x^4 - 3x^3 + 5x^2 - 20x + 40$

26. $f(x) = -2x^4 + 5x^2 - 4$

27. $f(x) = \dfrac{x^3}{x^3 + 1}$ 28. $f(x) = \dfrac{2x^4 - 3x}{x^2 - 1}$

29. $f(x) = 0.2\sqrt{x} - 0.3x^3$ 30. $f(x) = \sqrt{x}(2x - 1)^3$

In exercises 31–34, use the evaluation function of your graphing utility to find the value of f at the given value of x and verify your result by direct computation.

31. $f(x) = -3x^3 + 5x^2 - 2x + 8$; $x = -1$

32. $f(x) = 2x^4 - 3x^3 + 2x^2 + x - 5$; $x = 2$

33. $f(x) = \dfrac{x^4 - 3x^2}{x - 2}$; $x = 1$ 34. $f(x) = \dfrac{\sqrt{x^2 - 1}}{3x + 4}$; $x = 2$

In exercises 35–42, use the evaluation function of your graphing utility to find the value of f at the indicated value of x. Express your answer accurate to four decimal places.

35. $f(x) = 3x^3 - 2x^2 + x - 4$; $x = 2.145$

36. $f(x) = 2x^3 + 5x^2 + 3x + 1$; $x = -0.27$

37. $f(x) = 5x^4 - 2x^2 + 8x - 3$; $x = 1.28$

38. $f(x) = 4x^4 - 3x^3 + 1$; $x = -2.42$

39. $f(x) = \dfrac{2x^3 - 3x + 1}{3x - 2}$; $x = 2.41$

40. $f(x) = \dfrac{2x + 5}{3x^2 - 4x + 1}$; $x = -1.72$

41. $f(x) = \sqrt{2x^2 + 1} + \sqrt{3x^2 - 1}$; $x = 0.62$

42. $f(x) = 2x(3x^3 + 5)^{1/3}$; $x = -6.24$

43. **Manufacturing Capacity** Data obtained from the Federal Reserve show that the annual increase in manufacturing capacity between 1988 and 1994 is given by

$$f(t) = 0.0388889t^3 - 0.283333t^2 + 0.477778t + 2.04286$$
$$(0 \leq t \leq 6)$$

where $f(t)$ is a percentage and t is measured in years, with $t = 0$ corresponding to the beginning of 1988.

a. Use a graphing utility to plot the graph of f in the viewing rectangle $[0, 8] \times [0, 4]$.

b. What is the annual increase in manufacturing capacity at the beginning of 1990 ($t = 2$)? At the beginning of 1992 ($t = 4$)?

Source: Federal Reserve

44. Decline of Union Membership The total union membership as a percentage of the private work force is given by

$$f(t) = 0.00017t^4 - 0.00921t^3 + 0.15437t^2 - 1.360723t$$
$$+ 16.8028 \quad (0 \leq t \leq 10)$$

where t is measured in years, with $t = 0$ corresponding to the beginning of 1983.

a. Plot the graph of f in the viewing rectangle $[0, 11] \times [8, 20]$.

b. What is the total union membership as a percentage of the private force at the beginning of 1986? At the beginning of 1993?

Source: American Federation of Labor and Congress of Industrial Organizations

45. Keeping with the Traffic Flow By driving at a speed to match the prevailing traffic speed, you decrease the chances of an accident. According to a University of Virginia School of Engineering and Applied Science study, the number of accidents per 100 million vehicle miles, y, is related to the deviation from the mean speed, x, in mph by the equation

$$y = 1.05x^3 - 21.95x^2 + 155.9x - 327.3 \quad (6 \leq x \leq 11)$$

a. Plot the graph of y in the viewing rectangle $[6, 11] \times [20, 150]$.

b. What is the number of accidents per 100 million vehicle miles if the deviation from the mean speed is 6 mph? 8 mph? 11 mph?

Source: University of Virginia School of Engineering and Applied Science

10.2 THE ALGEBRA OF FUNCTIONS

The Sum, Difference, Product, and Quotient of Functions

Let $S(t)$ and $R(t)$ denote, respectively, the federal government's spending and revenue at any time t, measured in billions of dollars. The graphs of these functions for the period between 1981 and 1991 are shown in Figure 10.9.

Figure 10.9
$S(t) - R(t)$ gives the federal budget deficit at any time t.

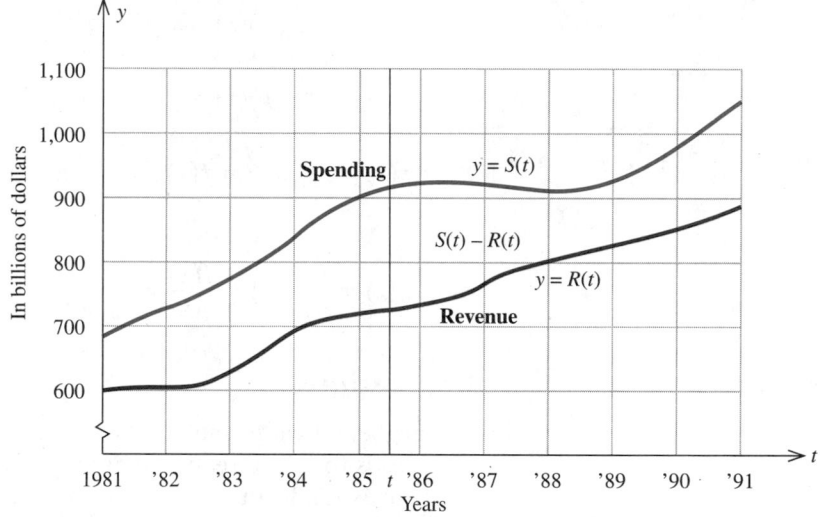

Source: Office of Management and Budget

The budget deficit at any time t is given by $S(t) - R(t)$ billion dollars. This observation suggests that we can define a function D whose value at any time t is given by $D(t) = S(t) - R(t)$. The function D, the *difference* of the two functions S and R, is written $D = S - R$ and may be called the "deficit function" since it gives the budget deficit at any time t. It has the same domain as the functions S and R.

Most functions are built up from other, generally simpler, functions. For example, we may view the function $f(x) = 2x + 4$ as the sum of the two functions $g(x) = 2x$ and $h(x) = 4$. The function $g(x) = 2x$ may, in turn, be viewed as the product of the functions $p(x) = 2$ and $q(x) = x$.

In general, given the functions f and g, we define the **sum, difference, product,** and **quotient** functions as $f + g$, $f - g$, fg, and f/g, respectively, with values defined by

$$(f + g)(x) = f(x) + g(x) \qquad \text{(Sum)}$$
$$(f - g)(x) = f(x) - g(x) \qquad \text{(Difference)}$$
$$(fg)(x) = f(x)g(x) \qquad \text{(Product)}$$
$$\left(\frac{f}{g}\right)(x) = \frac{f(x)}{g(x)} \qquad [g(x) \neq 0] \qquad \text{(Quotient)}$$

Thus, the rules for the sum, difference, product, and quotient of two functions f and g are obtained by simply adding, subtracting, multiplying, and dividing the respective expressions for $f(x)$ and $g(x)$.

EXAMPLE 1   Let f and g be functions defined by the rules $f(x) = \frac{1}{2}x^2$ and $g(x) = 2x + 1$, respectively. Find the rules for the sum s, the difference d, the product p, and the quotient q of the functions f and g.

Solution   The required rules are given by

$$s(x) = (f + g)(x) = f(x) + g(x) = \frac{1}{2}x^2 + 2x + 1$$

$$d(x) = (f - g)(x) = f(x) - g(x) = \frac{1}{2}x^2 - (2x + 1) = \frac{1}{2}x^2 - 2x - 1$$

$$p(x) = (fg)(x) = f(x)g(x) = \frac{1}{2}x^2(2x + 1)$$

and $q(x) = \left(\frac{f}{g}\right)(x) = \frac{f(x)}{g(x)} = \frac{\frac{1}{2}x^2}{2x + 1} = \frac{x^2}{2(2x + 1)} \qquad \left(x \neq -\frac{1}{2}\right)$ ○ ○ ○

Applications

The mathematical formulation of a problem arising from a practical situation often leads to an expression that involves the combination of functions. Consider, for example, the costs incurred in operating a business. Costs that remain more or less constant regardless of the firm's level of activity are called **fixed costs**. Examples of fixed costs are rental fees and executive salaries. On the other hand, costs that vary with production or sales are called **variable costs**. Examples of variable costs are wages and costs of raw materials. The **total cost** of operating a business is thus given by the *sum* of the variable costs and the fixed costs, as illustrated in the next example.

EXAMPLE 2   Suppose that Puritron, a manufacturer of water filters, has a monthly fixed cost of $10,000 and a variable cost of

$$-0.0001x^2 + 10x \qquad (0 \leq x \leq 40,000)$$

dollars, where x denotes the number of filters manufactured per month. Find a function C that gives the total cost incurred by Puritron in the manufacture of x filters.

Solution   Puritron's monthly fixed cost is always $10,000, regardless of the level of production, and it is described by the constant function $F(x) = 10,000$. Next, the variable cost is described by the function $V(x) = -0.0001x^2 + 10x$. Since the total cost incurred by Puritron at any level of production is the sum of the variable cost and the fixed cost, we see that the required total cost function is given by

$$C(x) = V(x) + F(x)$$
$$= -0.0001x^2 + 10x + 10,000 \qquad (0 \le x \le 40,000) \qquad \circ \; \circ \; \circ$$

Next, the **total profit** realized by a firm in operating a business is the *difference* between the total revenue realized and the total cost incurred; that is, $P(x) = R(x) - C(x)$.

EXAMPLE 3   Refer to Example 2. Suppose that the total revenue in dollars realized by Puritron from the sale of x water filters is given by the total revenue function

$$R(x) = -0.0005x^2 + 20x \qquad (0 \le x \le 40,000)$$

a. Find the total profit function—that is, the function that describes the total profit Puritron realizes in manufacturing and selling x water filters per month.

b. What is the profit when the level of production is 10,000 filters per month?

Solution

a. The total profit realized by Puritron in manufacturing and selling x water filters per month is the difference between the total revenue realized and the total cost incurred. Thus, the required total profit function is given by

$$P(x) = R(x) - C(x)$$
$$= (-0.0005x^2 + 20x) - (-0.0001x^2 + 10x + 10,000)$$
$$= -0.0004x^2 + 10x - 10,000$$

b. The profit realized by Puritron when the level of production is 10,000 filters per month is

$$P(10,000) = -0.0004(10,000)^2 + 10(10,000) - 10,000 = 50,000$$

or $50,000 per month. $\circ \; \circ \; \circ$

Composition of Functions

Another way to build up a function from other functions is through a process known as the *composition of functions.* Consider, for example, the function h, whose rule is given by $h(x) = \sqrt{x^2 - 1}$. Let f and g be functions defined by the rules $f(x) = x^2 - 1$ and $g(x) = \sqrt{x}$. Evaluating the function g at the point $f(x)$ [remember that for each real number x in the domain of f, $f(x)$ is simply a real number], we find that

$$g(f(x)) = \sqrt{f(x)} = \sqrt{x^2 - 1}$$

which is just the rule defining the function h!

In general, the **composition of a function g with a function f** is defined as the function h whose value is found by evaluating the function g at the point $f(x)$; that is,

$$h(x) = g(f(x))$$

Figure 10.10
The composite function $h = g \circ f$ is found by evaluating f at x and then g at $f(x)$, in that order.

Figure 10.11
The function $h = g \circ f$ viewed as a mapping.

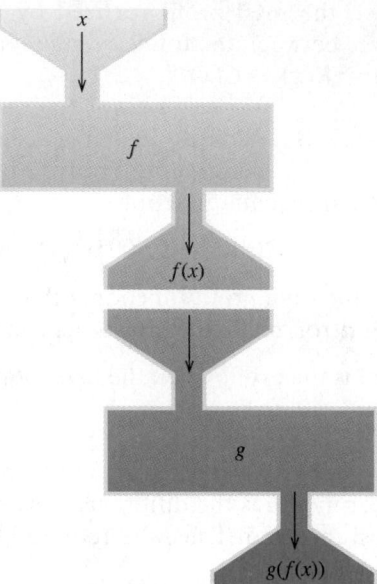

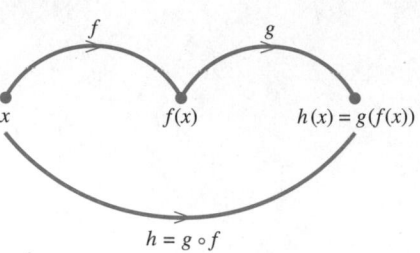

The function h is called a **composite function** and is commonly denoted by $h = g \circ f$ (read "g circle f" or "g composed with f"). The interpretation of the function h as a machine is illustrated in Figure 10.10, and its interpretation as a mapping is shown in Figure 10.11.

EXAMPLE 4   Let $f(x) = x^2 - 1$ and $g(x) = \sqrt{x} + 1$.

a. Find the rule for the composite function $g \circ f$.

b. Find the rule for the composite function $f \circ g$.

Solution

a. To find the rule for the composite function $g \circ f$, evaluate the function g at $f(x)$. Therefore,

$$(g \circ f)(x) = g(f(x)) = \sqrt{f(x)} + 1 = \sqrt{x^2 - 1} + 1$$

b. To find the rule for the composite function $f \circ g$, evaluate the function f at $g(x)$. Thus,

$$(f \circ g)(x) = f(g(x)) = (g(x))^2 - 1 = (\sqrt{x} + 1)^2 - 1$$
$$= x + 2\sqrt{x} + 1 - 1 = x + 2\sqrt{x}$$

○ ○ ○

Let $f(x) = \sqrt{x} + 1$ for $x \geq 0$ and let $g(x) = (x - 1)^2$ for $x \geq 1$.
a. Show that $(g \circ f)(x)$ and $(f \circ g)(x) = x$. [*Remark.* The function g is said to be the *inverse* of f and vice versa.]
b. Plot the graphs of f and g together with the straight line $y = x$. Describe the relationship between the graphs of f and g.

 Example 4 reminds us that, in general, $g \circ f$ is different from $f \circ g$, and care must be taken regarding the order when computing a composite function.

 EXAMPLE 5 An environmental impact study conducted for Oxnard's City Environmental Management Department indicates that, under existing environmental protection laws, the level of carbon monoxide present in the air due to pollution from automobile exhaust will be $0.01x^{2/3}$ parts per million when the number of motor vehicles is x thousand. A separate study conducted by a state government agency estimates that t years from now the number of motor vehicles in Oxnard will be $0.2t^2 + 4t + 64$ thousand.

a. Find an expression for the concentration of carbon monoxide in the air due to automobile exhaust t years from now.

b. What will the level of concentration be five years from now?

Solution

a. The level of carbon monoxide present in the air due to pollution from automobile exhaust is described by the function $g(x) = 0.01x^{2/3}$, where x is the number (in thousands) of motor vehicles. But the number of motor vehicles x (in thousands) t years from now may be estimated by the rule $f(t) = 0.2t^2 + 4t + 64$. Therefore, the concentration of carbon monoxide due to automobile exhaust t years from now is given by

$$C(t) = g(f(t)) = 0.01(0.2t^2 + 4t + 64)^{2/3}$$

parts per million.

b. The level of concentration five years from now will be

$$C(5) = 0.01[0.2(5)^2 + 4(5) + 64]^{2/3}$$
$$= (0.01)89^{2/3} \approx 0.20$$

or 0.20 part per million. ○ ○ ○

Ebasco is a diversified engineering and construction company. In addition to designing and building electric generating facilities, the company provides clients with safety and risk assessment studies. Their studies include analyses of potential hazards, recommendations for safe work practices, and evaluations of responses to emergencies in chemical plants, fuel-storage depots, nuclear facilities, and hazardous waste sites. As a project manager, Marchlik deals directly with clients to help them understand and implement Ebasco's recommendations to ensure a safer environment.

Michael Marchlik

Title: Project Manager
Institution: Ebasco Services Incorporated

Calculus wasn't Mike Marchlik's favorite college subject. In fact, it was not until he started his first job that the "lights went on" and he realized how using calculus allowed him to solve real problems in his everyday work.

Marchlik emphasizes that he doesn't do "number crunching" himself. "I don't work out integrals, but in my work I use computer models that do that." The important issue is not computation but how the answers relate to client problems.

Marchlik's clients typically process highly toxic or explosive materials. Ebasco evaluates a client site, such as a chemical plant, to determine how safety systems might fail and what the probable consequences might be.

To avoid a major disaster like the one that occurred in Bhopal, India, in 1984, Marchlik and his team might be asked to determine how quickly a poisonous chemical would spread if a leak occurred. Or they might help avert a disaster like the one that rocked the Houston area in 1989. In that incident, hydrocarbon vapor exploded at a Phillips 66 chemical plant, shaking office buildings in Houston, 12 miles away. Several people were killed and hundreds were injured. Property damage totaled $1.39 billion.

In assessing risks for a fuel-storage depot, Ebasco considers variables such as weather conditions, including probable wind speed, the flow properties of a gas, and possible ignition sources. With today's more powerful computers, the models can involve a system of very complex equations to project likely scenarios.

Mathematical models vary, however. One model might forecast how much gas will flow out of a hole and how quickly it will disperse. Another model might project where the gas will go depending on local factors such as temperature and wind speed. Choosing the right model is essential. A model based on flat terrain when the client's storage depot is set among hills is going to produce the wrong answer.

Marchlik and his team run several models together to come up with their projections. Each model uses "equations that have to be integrated to come up with solutions." The bottom line? Marchlik stresses that "calculus is at the very heart" of Ebasco's risk-assessment work.

◐   ◐   ◐

SELF-CHECK EXERCISES 10.2

1. Let f and g be functions defined by the rules $f(x) = \sqrt{x} + 1$ and $g(x) = \dfrac{x}{1 + x}$, respectively. Find the rules for
   a. the sum s, the difference d, the product p, and the quotient q of f and g
   b. the composite functions $f \circ g$ and $g \circ f$

2. Health-care spending per person by the private sector comprises payments by individuals, corporations, and their insurance companies, and is approximated by the function

$$f(t) = 2.5t^2 + 31.3t + 406 \qquad (0 \le t \le 20)$$

where $f(t)$ is measured in dollars and t is measured in years, with $t = 0$ corresponding to the beginning of 1975. The corresponding government spending, comprising expenditures for medicaid, medicare, and other federal, state, and local government public health care, is

$$g(t) = 1.4t^2 + 29.6t + 251 \qquad (0 \le t \le 20)$$

where t has the same meaning as before.
   a. Find a function that gives the difference between private and government health-care spending per person at any time t.
   b. What was the difference between private and government expenditures per person at the beginning of 1985? At the beginning of 1993?
   Source: Health Care Financing Administration

Solutions to Self-Check Exercises 10.2 can be found on page 597.

10.2 EXERCISES

In exercises 1–8, let $f(x) = x^3 + 5$, $g(x) = x^2 - 2$, and $h(x) = 2x + 4$. Find the rule for each function.

1. $f + g$ 2. $f - g$ 3. fg 4. gf

5. $\dfrac{f}{g}$ 6. $\dfrac{f - g}{h}$ 7. $\dfrac{fg}{h}$ 8. fgh

In exercises 9–18, let $f(x) = x - 1$, $g(x) = \sqrt{x + 1}$, and $h(x) = 2x^3 - 1$. Find the rule for each function.

9. $f + g$ 10. $g - f$ 11. fg 12. gf

13. $\dfrac{g}{h}$ 14. $\dfrac{h}{g}$ 15. $\dfrac{fg}{h}$ 16. $\dfrac{fh}{g}$

17. $\dfrac{f - h}{g}$ 18. $\dfrac{gh}{g - f}$

In exercises 19–24, find the rules for $f + g$, $f - g$, fg, and $\dfrac{f}{g}$.

19. $f(x) = x^2 + 5$; $g(x) = \sqrt{x} - 2$

20. $f(x) = \sqrt{x - 1}$; $g(x) = x^3 + 1$

21. $f(x) = \sqrt{x + 3}$; $g(x) = \dfrac{1}{x - 1}$

22. $f(x) = \dfrac{1}{x^2 + 1}$; $g(x) = \dfrac{1}{x^2 - 1}$

23. $f(x) = \dfrac{x + 1}{x - 1}$; $g(x) = \dfrac{x + 2}{x - 2}$

24. $f(x) = x^2 + 1$; $g(x) = \sqrt{x + 1}$

In exercises 25–30, find the rules for the composite functions f ∘ g *and* g ∘ f.

25. $f(x) = x^2 + x + 1$; $g(x) = x^2$

26. $f(x) = 3x^2 + 2x + 1$; $g(x) = x + 3$

27. $f(x) = \sqrt{x} + 1$; $g(x) = x^2 - 1$

28. $f(x) = 2\sqrt{x} + 3$; $g(x) = x^2 + 1$

29. $f(x) = \dfrac{x}{x^2 + 1}$; $g(x) = \dfrac{1}{x}$

30. $f(x) = \sqrt{x + 1}$; $g(x) = \dfrac{1}{x - 1}$

In exercises 31–34, evaluate h(2), *where* h = g ∘ f.

31. $f(x) = x^2 + x + 1$; $g(x) = x^2$

32. $f(x) = \sqrt[3]{x^2 - 1}$; $g(x) = 3x^3 + 1$

33. $f(x) = \dfrac{1}{2x + 1}$; $g(x) = \sqrt{x}$

34. $f(x) = \dfrac{1}{x - 1}$; $g(x) = x^2 + 1$

In exercises 35–42, find functions f *and* g *such that* h = g ∘ f. *(Note: The answer is not unique.)*

35. $h(x) = (2x^3 + x^2 + 1)^5$ 36. $h(x) = (3x^2 - 4)^{-3}$

37. $h(x) = \sqrt{x^2 - 1}$ 38. $h(x) = (2x - 3)^{3/2}$

39. $h(x) = \dfrac{1}{x^2 - 1}$ 40. $h(x) = \dfrac{1}{\sqrt{x^2 - 4}}$

41. $h(x) = \dfrac{1}{(3x^2 + 2)^{3/2}}$

42. $h(x) = \dfrac{1}{\sqrt{2x + 1}} + \sqrt{2x + 1}$

In exercises 43–46, find f(a + h) − f(a) *for each function. Simplify your answer.*

43. $f(x) = 3x + 4$ 44. $f(x) = -\frac{1}{2}x + 3$

45. $f(x) = 4 - x^2$ 46. $f(x) = x^2 - 2x + 1$

47. If $f(x) = x^2 + 1$, find and simplify

$$\frac{f(a + h) - f(a)}{h} \qquad (h \neq 0)$$

48. If $f(x) = 1/x$, find and simplify

$$\frac{f(a + h) - f(a)}{h} \qquad (h \neq 0)$$

49. **Manufacturing Costs** TMI, Inc., a manufacturer of blank audiocassette tapes, has a monthly fixed cost of

$12,100 and a variable cost of 60 cents per tape. Find a function C that gives the total cost incurred by TMI in the manufacture of x tapes per month.

50. **Cost of Producing Word Processors** Apollo, Inc., manufactures its word processors at a variable cost of

$$V(x) - 0.000003x^3 \quad 0.03x^2 + 200x$$

dollars, where x denotes the number of units manufactured per month. The monthly fixed cost attributable to the division that produces these word processors is $100,000. Find a function C that gives the total cost incurred by the manufacture of x word processors. What is the total cost incurred in producing 2000 units per month?

51. **Cost of Producing Word Processors** Refer to exercise 50. Suppose the total revenue realized by Apollo, Inc., from the sale of x word processors is given by the total revenue function

$$R(x) = -0.1x^2 + 500x \qquad (0 \leq x \leq 5000)$$

where x is measured in dollars.
a. Find the total profit function.
b. What is the profit when 1500 units are produced and sold each month?

52. **Overcrowding of Prisons** The 1980s saw a trend toward old-fashioned punitive deterrence as opposed to the more liberal penal policies and community-based corrections popular in the 1960s and early 1970s. As a result, prisons became more crowded and the gap between the number of people in prison and the prison capacity widened. Based on figures from the U.S. Department of Justice, the number of prisoners (in thousands) in federal and state prisons is approximated by the function

$$N(t) = 3.5t^2 + 26.7t + 436.2 \qquad (0 \leq t \leq 10)$$

where t is measured in years and $t = 0$ corresponds to 1983. The number of inmates for which prisons were designed is given by

$$C(t) = 24.3t + 365 \qquad (0 \leq t \leq 10)$$

where $C(t)$ is measured in thousands and t has the same meaning as before.
a. Find an expression that shows the gap between the number of prisoners and the number of inmates for which the prisons were designed at any time t.
b. Find the gap at the beginning of 1983. At the beginning of 1986.
Source: U.S. Department of Justice

53. **Effect of Mortgage Rates on Housing Starts** A study prepared for the National Association of Realtors estimated that the number of housing starts per year over the next five years will be

$$N(r) = \frac{7}{1 + 0.02r^2}$$

million units, where r (percent) is the mortgage rate. Suppose that the mortgage rate over the next t months is

$$r(t) = \frac{10t + 150}{t + 10} \qquad (0 \le t \le 24)$$

percent per year.
a. Find an expression for the number of housing starts per year as a function of t, t months from now.
b. Using the result from (a), determine the number of housing starts at present, 12 months from now, and 18 months from now.

54. **Hotel Occupancy Rate** The occupancy rate of the all-suite Wonderland Hotel, located near an amusement park, is given by the function

$$r(t) = \frac{10}{81}t^3 - \frac{10}{3}t^2 + \frac{200}{9}t + 55 \qquad (0 \le t \le 11)$$

where t is measured in months and $t = 0$ corresponds to the beginning of January. Management has estimated

that the monthly revenue (in thousands of dollars) is approximated by the function

$$R(r) = -\frac{3}{5000}r^3 + \frac{9}{50}r^2 \qquad (0 \le r \le 100)$$

where r is the occupancy rate.
a. What is the hotel's occupancy rate at the beginning of January? At the beginning of June?
b. What is the hotel's monthly revenue at the beginning of January? At the beginning of June?
[*Hint:* Compute $R(r(0))$ and $R(r(5))$.]

55. **Housing Starts and Construction Jobs** The president of a major housing construction firm reports that the number of construction jobs (in millions) created is given by

$$N(x) = 1.42x$$

where x denotes the number of housing starts. Suppose that the number of housing starts in the next t months is expected to be

$$x(t) = \frac{7(t + 10)^2}{(t + 10)^2 + 2(t + 15)^2}$$

million units per year. Find an expression for the number of jobs created per month in the next t months. How many jobs will have been created 6 months from now? 12 months from now?

SOLUTIONS TO SELF-CHECK EXERCISES 10.2

1. **a.** $s(x) = f(x) + g(x) = \sqrt{x} + 1 + \dfrac{x}{1 + x}$

$d(x) = f(x) - g(x) = \sqrt{x} + 1 - \dfrac{x}{1 + x}$

$p(x) = f(x)g(x) = (\sqrt{x} + 1) \cdot \dfrac{x}{1 + x} = \dfrac{x(\sqrt{x} + 1)}{1 + x}$

$q(x) = \dfrac{f(x)}{g(x)} = \dfrac{\sqrt{x} + 1}{\dfrac{x}{1 + x}} = \dfrac{(\sqrt{x} + 1)(1 + x)}{x}$

b. $(f \circ g)(x) = f(g(x)) = \sqrt{\dfrac{x}{1 + x} + 1}$

$(g \circ f)(x) = g(f(x)) = \dfrac{\sqrt{x} + 1}{1 + (\sqrt{x} + 1)} = \dfrac{\sqrt{x} + 1}{\sqrt{x} + 2}$

2. a. The difference between private and government health-care spending per person at any time t is given by the function d with the rule

$$d(t) = f(t) - g(t) = (2.5t^2 + 31.3t + 406) - (1.4t^2 + 29.6t + 251)$$
$$= 1.1t^2 + 1.7t + 155$$

b. The difference between private and government expenditures per person at the beginning of 1985 is given by

$$d(10) = 1.1(10)^2 + 1.7(10) + 155$$

or \$282 per person.

The difference between private and government expenditures per person at the beginning of 1993 is given by

$$d(18) = 1.1(18)^2 + 1.7(18) + 155$$

or \$542 per person.

10.3 FUNCTIONS AND MATHEMATICAL MODELS IN CALCULUS

More on Mathematical Models

In calculus, we are concerned primarily with how one (dependent) variable depends on one or more (independent) variables. Consequently, most of our mathematical models involve functions of one or more variables.* Once a function has been constructed to describe a specific real-world problem, a host of questions pertaining to the problem may be answered by analyzing the function (mathematical model). For example, if we have a function that gives the population of a certain culture of bacteria at any time t, then we can determine how fast the population is increasing or decreasing at any time t, and so on. Conversely, if we have a model that gives the rate of change of the cost of producing a certain item as a function of the level of production, and if we know the fixed cost incurred in producing this item, then we can find the total cost incurred in producing a certain number of those items.

Before going on, let us look at two mathematical models. The first is used to estimate spending by business on computer security and the second is used to project the growth of the number of people enrolled in health maintenance organizations (HMOs). These models are derived from data using the *least-squares technique*. (We will not, however, pursue the actual construction of these models.)

o o o

* Functions of more than one variable will be studied later.

EXAMPLE 1 The estimated spending by businesses on computer security equipment and services from 1987 to 1993 is given in the following table. The figures include spending for protection against computer criminals who steal, erase, or alter data, along with protection against fires, electrical failures, and natural disasters.

Year	1987	1988	1989	1990	1991	1992	1993
Spending (in billions of dollars)	0.49	0.59	0.66	0.73	0.81	0.93	1.02

A mathematical model approximating the amount of spending over the period in question is given by

$$S(t) = 0.0864t + 0.4879$$

where t is measured in years, with $t = 0$ corresponding to 1987.

a. Sketch the graph of the function S and the given data on the same set of axes.

b. Assuming this trend continued, what was the estimated spending by business on computer security equipment and services in 1995 ($t = 8$)?

c. What is the rate of increase of the annual expenditure over the period in question?
Source: Frost & Sullivan Inc.

Solution

a. The graph of S is shown in Figure 10.12.

b. The estimated spending in 1995 is

$$S(8) = 0.0864(8) + 0.4879$$
$$\approx 1.1791$$

or approximately $1.18 billion.

Figure 10.12
Estimated spending by businesses on computer security equipment and services.

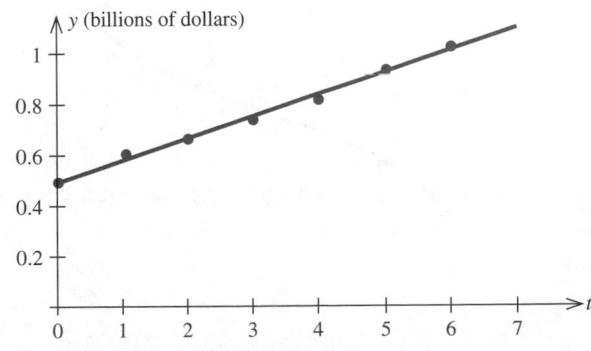

c. The function S is linear, and so we conclude that the annual increase in the expenditure is given by the slope of the straight line represented by S, which is approximately $0.09 billion per year.   ○ ○ ○

 EXAMPLE 2   The number of people enrolled in health maintenance organizations (HMOs) from 1984 to 1994 is given in the following table.

Year	1984	1986	1988	1990	1992	1994
Number of People (in millions)	15.8	25.8	33	37	41.6	50

A mathematical model approximating the number of people, $N(t)$, enrolled in HMOs during this period is

$$N(t) = 0.0514t^3 - 0.853t^2 + 6.8147t + 15.6524 \qquad (0 \le t \le 10)$$

where t is measured in years and $t = 0$ corresponds to 1984.

a. Sketch the graph of the function N to see how the model compares with the actual data.

b. Assuming that this trend continued, how many people were enrolled in HMOs at the beginning of 1998?
Source: Group Health Association of America

Solution

a. The graph of the function N is shown in Figure 10.13.

b. The number of people enrolled in HMOs at the beginning of 1998 is given by

$$N(14) = 0.0514(14)^3 - 0.853(14)^2 + 6.8147(14) + 15.6524$$
$$= 84.9118$$

or approximately 84.91 million people.   ○ ○ ○

Figure 10.13
The graph of $y = N(t)$ approximates the number of people enrolled in HMOs from 1984 to 1994.

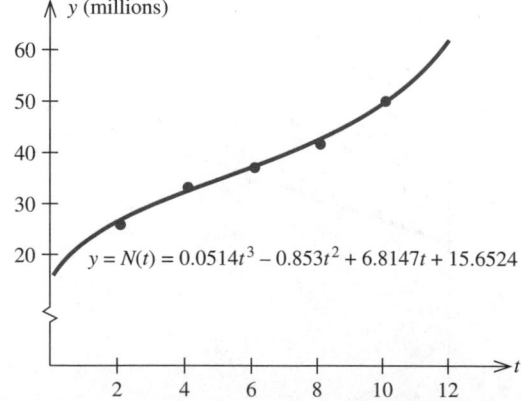

y (millions)

$y = N(t) = 0.0514t^3 - 0.853t^2 + 6.8147t + 15.6524$

We will discuss several mathematical models from the field of economics later in this section, but first we review some important functions that are the bases for many mathematical models.

Polynomial Functions

We begin by recalling a special class of functions, *polynomial functions.*

A POLYNOMIAL FUNCTION

A **polynomial function** of degree n is a function of the form

$$f(x) = a_0x^n + a_1x^{n-1} + \cdots + a_{n-1}x + a_n \qquad (a_0 \neq 0)$$

where $a_0, a_1, \ldots, a_n$ are constants and n is a nonnegative integer.

For example, the functions

$$f(x) = 4x^5 - 3x^4 + x^2 - x + 8$$

and
$$g(x) = 0.001x^3 - 2x^2 + 20x + 400$$

are polynomial functions of degrees 5 and 3, respectively. Observe that a polynomial function is defined everywhere so that it has domain $(-\infty, \infty)$.

A polynomial function of degree 1 $(n = 1)$

$$f(x) = a_0x + a_1 \qquad (a_0 \neq 0)$$

is the equation of a straight line in the slope-intercept form with slope $m = a_0$ and y-intercept $b = a_1$ (see Section 1.2). For this reason, a polynomial function of degree 1 is called a **linear function.** For example, the linear function $f(x) = 2x + 3$ may be written as a linear equation in x and y—namely, $y = 2x + 3$ or $2x - y + 3 = 0$. Conversely, the linear equation $2x - 3y + 4 = 0$ can be solved for y in terms of x to yield the linear function $y = f(x) = \frac{2}{3}x + \frac{4}{3}$.

A polynomial function of degree 2 is referred to as a **quadratic function.** A polynomial function of degree 3 is called a **cubic function,** and so on. The mathematical model in Example 2 involves a cubic function.

Rational and Power Functions

Another important class of functions is rational functions. A **rational function** is simply the quotient of two polynomials. Examples of rational functions are

$$F(x) = \frac{3x^3 + x^2 - x + 1}{x - 2}$$

and
$$G(x) = \frac{x^2 + 1}{x^2 - 1}$$

In general, a rational function has the form

$$R(x) = \frac{f(x)}{g(x)}$$

where $f(x)$ and $g(x)$ are polynomial functions. Since division by zero is not allowed, we conclude that the domain of a rational function is the set of all real numbers except the zeros of g—that is, the roots of the equation $g(x) = 0$. Thus, the domain of the function F is the set of all numbers except $x = 2$, whereas the domain of the function G is the set of all numbers except those that satisfy $x^2 - 1 = 0$ or $x = \pm 1$.

Functions of the form

$$f(x) = x^r$$

where r is any real number, are called **power functions.** We encountered examples of power functions earlier. For example, the functions

$$f(x) = \sqrt{x} = x^{1/2} \quad \text{and} \quad g(x) = \frac{1}{x^2} = x^{-2}$$

are power functions.

Many of the functions we will encounter later involve combinations of the functions introduced here. For example, the following functions may be viewed as suitable combinations of such functions:

$$f(x) = \sqrt{\frac{1 - x^2}{1 + x^2}}$$

$$g(x) = \sqrt{x^2 - 3x + 4}$$

$$h(x) = (1 + 2x)^{1/2} + \frac{1}{(x^2 + 2)^{3/2}}$$

As with polynomials of degree 3 or greater, analyzing the properties of these functions is facilitated by using the tools of calculus, to be developed later.

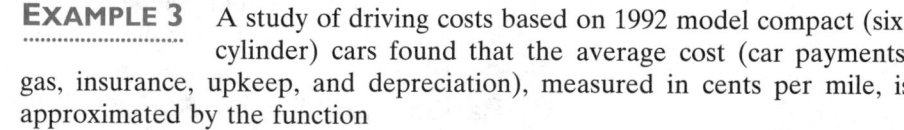

EXAMPLE 3   A study of driving costs based on 1992 model compact (six-cylinder) cars found that the average cost (car payments, gas, insurance, upkeep, and depreciation), measured in cents per mile, is approximated by the function

$$C(x) = \frac{2095}{x^{2.2}} + 20.08$$

where x (in thousands) denotes the number of miles the car is driven in one year. Using this model, estimate the average cost of driving a compact car 10,000 miles a year; 20,000 miles a year.

Source: Runzheimer International Study

Solution   The average cost of driving a compact car 10,000 miles a year is given by

$$C(10) = \frac{2095}{10^{2.2}} + 20.08 \approx 33.3$$

or approximately 33 cents per mile. The average cost of driving 20,000 miles a year is given by

$$C(20) = \frac{2095}{20^{2.2}} + 20.08 \approx 23.0$$

or approximately 23 cents per mile.   ◦ ◦ ◦

Some Economic Models

In the remainder of this section, we look at some examples of *nonlinear* supply and demand functions. Before proceeding, however, you might want to review the concepts of supply and demand equations and of market equilibrium (see Sections 1.3 and 1.4).

EXAMPLE 4   The *demand function* for a certain brand of videocassette is given by

$$p = d(x) = -0.01x^2 - 0.2x + 8$$

where p is the wholesale unit price in dollars and x is the quantity demanded each week, measured in units of a thousand. Sketch the corresponding demand curve. Above what price will there be no demand? What is the maximum quantity demanded per week?

Figure 10.14
A demand curve.

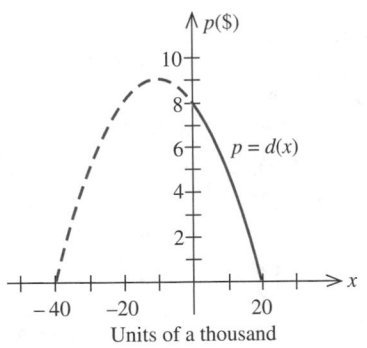

Units of a thousand

Solution   The given function is quadratic, and its graph, which appears in Figure 10.14, may be sketched using the methods just developed. The p-intercept, 8, gives the wholesale unit price above which there will be no demand. To obtain the maximum quantity demanded, we set $p = 0$, which gives

$$-0.01x^2 - 0.2x + 8 = 0$$
$$x^2 + 20x - 800 = 0 \qquad \text{(Multiplying both sides}$$
$$\text{of the equation by } -100)$$

or

$$(x + 40)(x - 20) = 0$$

that is, $x = -40$ or $x = 20$. Since x must be nonnegative, we reject the root $x = -40$. Thus, the maximum number of videocassettes demanded per week is 20,000.   ◦ ◦ ◦

EXAMPLE 5   The *supply function* for a certain brand of videocassette is given by

$$p = s(x) = 0.01x^2 + 0.1x + 3$$

Figure 10.15
A supply curve.

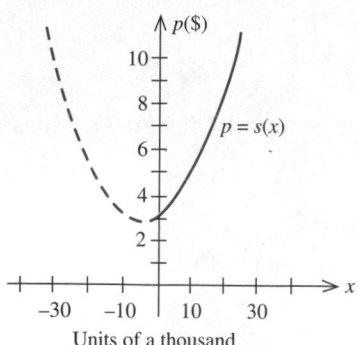

Units of a thousand

where p is the unit wholesale price in dollars and x stands for the quantity that will be made available in the market by the supplier, measured in units of a thousand. Sketch the corresponding supply curve. What is the lowest price at which the supplier will make the videocassettes available in the market?

Solution Figure 10.15 is a sketch of the supply curve. The p-intercept, 3, gives the lowest price at which the supplier will make the videocassettes available in the market.   ◦ ◦ ◦

EXAMPLE 6 In Examples 4 and 5, the demand function for a certain brand of videocassette is given by

$$p = d(x) = -0.01x^2 - 0.2x + 8$$

and the corresponding supply function is given by

$$p = s(x) = 0.01x^2 + 0.1x + 3$$

where p is expressed in dollars and x is measured in units of a thousand. Find the equilibrium quantity and price.

Solution We solve the following system of equations:

$$p = -0.01x^2 - 0.2x + 8$$

and

$$p = 0.01x^2 + 0.1x + 3$$

Substituting the first equation into the second yields

$$-0.01x^2 - 0.2x + 8 = 0.01x^2 + 0.1x + 3$$

which is equivalent to

$$0.02x^2 + 0.3x - 5 = 0$$
$$2x^2 + 30x - 500 = 0$$
$$x^2 + 15x - 250 = 0$$

or

$$(x + 25)(x - 10) = 0$$

Thus, $x = -25$ or $x = 10$. Since x must be nonnegative, the root $x = -25$ is rejected. Therefore, the equilibrium quantity is 10,000 videocassettes. The equilibrium price is given by

$$p = 0.01(10)^2 + 0.1(10) + 3 = 5$$

or $5 per videocassette (Figure 10.16).   ◦ ◦ ◦

Figure 10.16
The supply curve and the demand curve intersect at the point (10, 5).

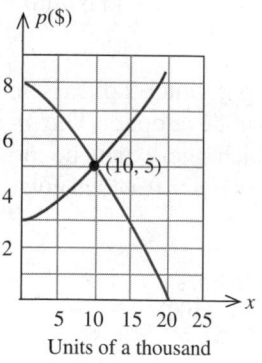

Units of a thousand

SELF-CHECK EXERCISES 10.3

1. Thomas Young has suggested the following rule for calculating the dosage of medicine for children from ages 1 to 12 years. If a denotes the adult dosage (in mg) and t is the age of the child (in years), then the child's dosage is given by

$$D(t) = \frac{at}{t + 12}$$

If the adult dose of a substance is 500 mg, how much should a four-year-old child receive?

2. The demand function for Mrs. Baker's cookies is given by

$$d(x) = -\frac{2}{15}x + 4$$

where $d(x)$ is the wholesale price in dollars per pound and x is the quantity demanded each week, measured in thousands of pounds. The supply function for the cookies is given by

$$s(x) = \frac{1}{75}x^2 + \frac{1}{10}x + \frac{3}{2}$$

where $s(x)$ is the wholesale price in dollars per pound and x is the quantity, in thousands of pounds, that will be made available in the market per week by the supplier.
a. Sketch the graphs of the functions d and s.
b. Find the equilibrium quantity and price.

Solutions to Self-Check Exercises 10.3 can be found on page 608.

10.3 EXERCISES

In exercises 1–6, determine whether the given function is a polynomial function, a rational function, or some other function. State the degree of each polynomial function.

1. $f(x) = 3x^6 - 2x^2 + 1$ 2. $f(x) = \frac{x^2 - 9}{x - 3}$

3. $G(x) = 2(x^2 - 3)^3$ 4. $H(x) = 2x^{-3} + 5x^{-2} + 6$

5. $f(t) = 2t^2 + 3\sqrt{t}$ 6. $f(r) = \frac{6r}{(r^3 - 8)}$

7. **Disposable Income** Economists define the *disposable annual income* for an individual by the equation $D = (1 - r)T$, where T is the individual's total income and r is the net rate at which he or she is taxed. What is the disposable income for an individual whose income is \$40,000 and whose net tax rate is 28%?

8. **Drug Dosages** A method sometimes used by pediatricians to calculate the dosage of medicine for children is based on the child's surface area. If a denotes the adult dosage (in mg) and S is the surface area of the child (in square meters), then the child's dosage is given by

$$D(S) = \frac{Sa}{1.7}$$

If the adult dose of a substance is 500 mg, how much should a child whose surface area is 0.4 square meter receive?

9. **Cowling's Rule** Cowling's Rule is a method for calculating pediatric drug dosages. If a denotes the adult dosage (in mg) and t is the age of the child (in years), then the child's dosage is given by

$$D(t) = \left(\frac{t + 1}{24}\right)a$$

If the adult dose of a substance is 500 mg, how much should a four-year-old child receive?

10. **Worker Efficiency** An efficiency study showed that the average worker at Delphi Electronics assembled cordless telephones at the rate of

$$f(t) = -\frac{3}{2}t^2 + 6t + 10 \qquad (0 \le t \le 4)$$

phones per hour, t hours after starting work during the morning shift. At what rate does the average worker assemble telephones 2 hours after starting work?

11. **Revenue Functions** The revenue (in dollars) realized by Apollo, Inc., from the sale of its electronic correcting

typewriters is given by

$$R(x) = -0.1x^2 + 500x$$

where x denotes the number of units manufactured per month. What is Apollo's revenue when 1000 units are produced?

12. Effect of Advertising on Sales The quarterly profit of Cunningham Realty depends on the amount of money x spent on advertising per quarter according to the rule

$$P(x) = -\frac{1}{8}x^2 + 7x + 30 \qquad (0 \le x \le 50)$$

where $P(x)$ and x are measured in thousands of dollars. What is Cunningham's profit when its quarterly advertising budget is $28,000?

13. Reaction of a Frog to a Drug Experiments conducted by A. J. Clark suggest that the response $R(x)$ of a frog's heart muscle to the injection of x units of acetylcholine (as a percentage of the maximum possible effect of the drug) may be approximated by the rational function

$$R(x) = \frac{100x}{b + x} \qquad (x \ge 0)$$

where b is a positive constant that depends on the particular frog.
a. If a concentration of 40 units of acetylcholine produces a response of 50% for a certain frog, find the "response function" for this frog.
b. Using the model found in (a), find the response of the frog's heart muscle when 60 units of acetylcholine are administered.

14. Linear Depreciation In computing income tax, business firms are allowed by law to depreciate certain assets such as buildings, machines, furniture, automobiles, and so on, over a period of time. The linear depreciation, or straight-line method, is often used for this purpose. Suppose that an asset has an initial value of $\$C$ and is to be depreciated linearly over n years with a scrap value of $\$S$. Show that the book value of the asset at any time t $(0 \le t \le n)$ is given by the linear function

$$V(t) = C - \frac{(C - S)}{n}t$$

[*Hint:* Find an equation of the straight line that passes through the points $(0, C)$ and (n, S). Then rewrite the equation in the slope-intercept form.]

15. Linear Depreciation Using the linear depreciation model of exercise 14, find the book value of a printing machine at the end of the second year if its initial value is $100,000 and it is depreciated linearly over five years with a scrap value of $30,000.

16. Price of Ivory According to the World Wildlife Fund, a group in the forefront of the fight against illegal ivory trade, the price of ivory (in dollars per kilo) compiled from a variety of legal and black market sources is approximated by the function

$$f(t) = \begin{cases} 8.37t + 7.44 & \text{if } 0 \le t \le 8 \\ 2.84t + 51.68 & \text{if } 8 < t \le 30 \end{cases}$$

where t is measured in years and $t = 0$ corresponds to the beginning of 1970.
a. Sketch the graph of the function f.
b. What was the price of ivory at the beginning of 1970? At the beginning of 1990?

17. Price of Automobile Parts For years, automobile manufacturers had a monopoly on the replacement-parts market, particularly for sheet metal parts such as fenders, doors, and hoods, the parts most often damaged in a crash. Beginning in the late 1970s, however, competition appeared on the scene. In a report conducted by an insurance company to study the effects of the competition, the price of an OEM (original equipment manufacturer) fender for a particular 1983 model car was found to be

$$f(t) = \frac{110}{\frac{1}{2}t + 1} \qquad (0 \le t \le 2)$$

where $f(t)$ is measured in dollars and t is in years. Over the same period of time, the price of a non-OEM fender for the car was found to be

$$g(t) = 26 \left(\frac{1}{4}t^2 - 1 \right)^2 + 52 \qquad (0 \le t \le 2)$$

where $g(t)$ is also measured in dollars. Find a function $h(t)$ that gives the difference in price between an OEM fender and a non-OEM fender. Compute $h(0)$, $h(1)$, and $h(2)$. What does the result of your computation seem to say about the price gap between OEM and non-OEM fenders over the two years?

For the demand equations in exercises 18–21, where x represents the quantity demanded in units of a thousand and p is the unit price in dollars, (a) sketch the demand curve and (b) determine the quantity demanded when the unit price is set at $p.

18. $p = -x^2 + 36; p = 11$

19. $p = -x^2 + 16; p = 7$

20. $p = \sqrt{9 - x^2}; p = 2$

21. $p = \sqrt{18 - x^2}; p = 3$

22. **Demand for Smoke Alarms** The demand function for the Sentinel smoke alarm is given by

$$p = \frac{30}{0.02x^2 + 1} \qquad (0 \le x \le 10)$$

where x (measured in units of a thousand) is the quantity demanded per week and p is the unit price in dollars. Sketch the graph of the demand function. What is the unit price that corresponds to a quantity demanded of 10,000 units?

23. **Demand for Commodities** Assume that the demand function for a certain commodity has the form

$$p = \sqrt{-ax^2 + b} \qquad (a \ge 0, b \ge 0)$$

where x is the quantity demanded, measured in units of a thousand, and p is the unit price in dollars. Suppose that the quantity demanded is 6000 ($x = 6$) when the unit price is $8 and 8000 ($x = 8$) when the unit price is $6. Determine the demand equation. What is the quantity demanded when the unit price is set at $7.50?

For the supply equations in exercises 24–27, where x is the quantity supplied in units of a thousand and p is the unit price in dollars, (a) sketch the supply curve and (b) determine the price at which the supplier will make 2000 units of the commodity available in the market.

24. $p = 2x^2 + 18$

25. $p = x^2 + 16x + 40$

26. $p = x^3 + x + 10$

27. $p = x^3 + 2x + 3$

28. **Supply Functions** The supply function for the Luminar desk lamp is given by

$$p = 0.1x^2 + 0.5x + 15$$

where x is the quantity supplied (in thousands) and p is the unit price in dollars. Sketch the graph of the supply function. What unit price will induce the supplier to make 5000 lamps available in the marketplace?

29. **Supply Functions** Suppliers of transistor radios will market 10,000 units when the unit price is $20 and 62,500 units when the unit price is $35. Determine the supply function if it is known to have the form

$$p = a\sqrt{x} + b \qquad (a \ge 0, b \ge 0)$$

where x is the quantity supplied and p is the unit price in dollars. Sketch the graph of the supply function. What unit price will induce the supplier to make 40,000 transistor radios available in the marketplace?

For each pair of supply and demand equations in exercises 30–33, where x represents the quantity demanded in units of a thousand and p the unit price in dollars, find the equilibrium quantity and the equilibrium price.

30. $p = -2x^2 + 80$ and $p = 15x + 30$

31. $p = -x^2 - 2x + 100$ and $p = 8x + 25$

32. $11p + 3x - 66 = 0$ and $2p^2 + p - x = 10$

33. $p = 60 - 2x^2$ and $p = x^2 + 9x + 30$

34. **Market Equilibrium** The weekly demand and supply functions for Sportsman 5×7 tents are given by

$$p = -0.1x^2 - x + 40$$

and

$$p = 0.1x^2 + 2x + 20$$

respectively, where p is measured in dollars and x is measured in units of a hundred. Find the equilibrium quantity and price.

35. **Market Equilibrium** The management of the Titan Tire Company has determined that the weekly demand and supply functions for their Super Titan tires are given by

$$p = 144 - x^2$$

and

$$p = 48 + \frac{1}{2}x^2$$

respectively, where p is measured in dollars and x is measured in units of a thousand. Find the equilibrium quantity and price.

SOLUTIONS TO SELF-CHECK EXERCISES 10.3

1. Since the adult dose of the substance is 500 mg, $a = 500$; thus the rule in this case is

$$D(t) = \frac{500t}{t + 12}$$

A four-year-old should receive

$$D(4) = \frac{500(4)}{4 + 12}$$

or 125 mg of the substance.

2. a. The graphs of the functions d and s are shown in the figure:

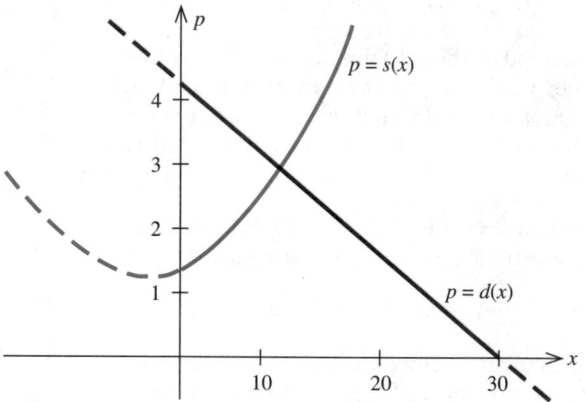

b. Solve the following system of equations:

$$p = -\frac{2}{15}x + 4$$

$$p = \frac{1}{75}x^2 + \frac{1}{10}x + \frac{3}{2}$$

Substituting the first equation into the second yields

$$\frac{1}{75}x^2 + \frac{1}{10}x + \frac{3}{2} = -\frac{2}{15}x + 4$$

$$\frac{1}{75}x^2 + \left(\frac{1}{10} + \frac{2}{15}\right)x - \frac{5}{2} = 0$$

$$\frac{1}{75}x^2 + \frac{7}{30}x - \frac{5}{2} = 0$$

Multiplying both sides of the last equation by 150, we have

$$2x^2 + 35x - 375 = 0$$

or $(2x - 15)(x + 25) = 0$

Thus, $x = -25$ or $x = 15/2 = 7.5$. Since x must be nonnegative, we take $x = 7.5$, and the equilibrium quantity is 7500 pounds. The equilibrium price is given by

$$p = -\frac{2}{15}\left(\frac{15}{2}\right) + 4$$

or $3 per pound.

10.4 LIMITS

Introduction to Calculus

Historically, the development of calculus by Isaac Newton (1642–1727) and Gottfried Wilhelm Leibniz (1646–1716) resulted from the investigation of these problems:

1. Finding the tangent line to a curve at a given point on the curve (Figure 10.17a)

2. Finding the area of a planar region bounded by an arbitrary curve (Figure 10.17b)

Figure 10.17

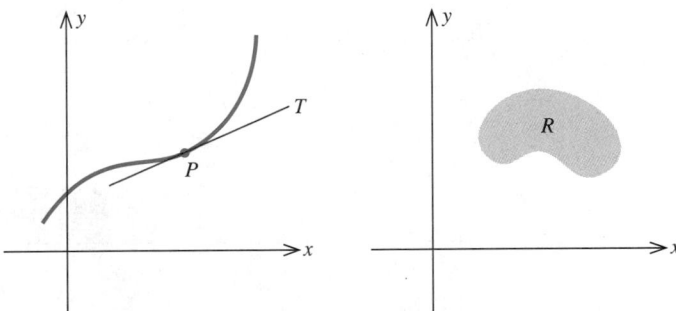

(a) What is the slope of the tangent line T at point P?

(b) What is the area of the region R?

The tangent-line problem might appear to be unrelated to any practical applications of mathematics, but as you will see later, the problem of finding the *rate of change* of one quantity with respect to another is mathematically equivalent to the geometric problem of finding the slope of the *tangent line* to a curve at a given point on the curve. It is precisely the discovery of the relationship between these two problems that spurred the development of calculus in the seventeenth century and made it such an indispensable tool for solving practical problems. Here are a few examples of such problems:

- Finding the velocity of an object
- Finding the rate of change of a bacteria population with respect to time
- Finding the rate of change of a company's profit with respect to time
- Finding the rate of change of a travel agency's revenue with respect to the agency's expenditure for advertising

The study of the tangent-line problem led to the creation of *differential calculus,* which relies on the concept of the *derivative* of a function. The study of the area problem led to the creation of *integral calculus,* which relies on the concept of the *antiderivative,* or *integral,* of a function. (The derivative of a function and the integral of a function are intimately related, as you will see in Section 14.4.) Both the derivative of a function and the integral of a function are defined in terms of a more fundamental concept—the limit, our next topic.

A Real-Life Example

From data obtained in a test run conducted on a prototype of a maglev (magnetic levitation train), which moves along a straight monorail track, engineers have determined that the position of the maglev from the origin at time t is given by

$$s = f(t) = 4t^2 \qquad (0 \le t \le 30) \tag{1}$$

where f is called the **position function** of the maglev. The position of the maglev at time $t = 0, 1, 2, 3, \ldots, 10$, measured from its initial position, is

$$f(0) = 0, \quad f(1) = 4, \quad f(2) = 16, \quad f(3) = 36, \ldots, f(10) = 400$$

feet (see Figure 10.18).

Figure 10.18
A maglev moving along an elevated monorail track.

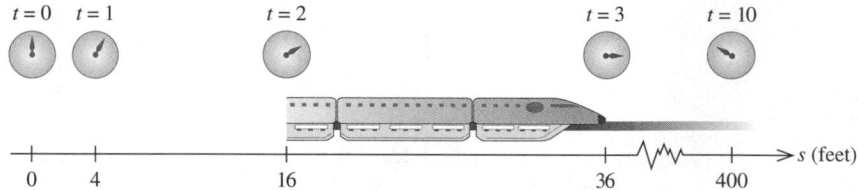

Suppose we want to find the velocity of the maglev at $t = 2$. This is just the velocity of the maglev as shown on its speedometer at that precise instant of time. Offhand, calculating this quantity using only equation (1) appears to be an impossible task, but consider what quantities we *can* compute using this relationship. Obviously, we can compute the position of the maglev at any time t as we did earlier for some selected values of t. Using these values, we can then compute the *average velocity* of the maglev over an interval of time. For example, the average velocity of the train over the time interval [2, 4] is given by

$$\frac{\text{Distance covered}}{\text{Time elapsed}} = \frac{f(4) - f(2)}{4 - 2}$$

$$= \frac{4\,(4^2) - 4(2^2)}{2}$$

$$= \frac{64 - 16}{2} = 24$$

or 24 ft/sec. Although this is not quite the velocity of the maglev at $t = 2$, it does provide us with an approximation of its velocity at that time.

Can we do better? Intuitively, the smaller the time interval we pick (with $t = 2$ as the left endpoint), the better the average velocity over that time interval will approximate the actual velocity of the maglev at $t = 2$.*

Now, let us describe this process in general terms. Let $t > 2$. Then the average velocity of the maglev over the time interval $[2, t]$ is given by

$$\frac{f(t) - f(2)}{t - 2} = \frac{4t^2 - 4(2^2)}{t - 2} = \frac{4(t^2 - 4)}{t - 2} \tag{2}$$

By choosing the values of t closer and closer to 2, we obtain a sequence of numbers that gives the average velocities of the maglev over smaller and smaller time intervals. As we observed earlier, this sequence of numbers should approach the *instantaneous velocity* of the train at $t = 2$.

Let's try some sample calculations. Using equation (2) and taking the sequence $t = 2.5, 2.1, 2.01, 2.001,$ and $2.0001,$ which approaches 2, we find

The average velocity over $[2, 2.5]$ is $\dfrac{4(2.5^2 - 4)}{2.5 - 2} = 18$ ft/sec

The average velocity over $[2, 2.1]$ is $\dfrac{4(2.1^2 - 4)}{2.1 - 2} = 16.4$ ft/sec

and so forth. These results are summarized in Table 10.1.

Table 10.1

			t approaches 2 from the right.		
t	2.5	2.1	2.01	2.001	2.0001
Average Velocity over $[2, t]$	18	16.4	16.04	16.004	16.0004

Average velocity approaches 16 from the right.

From Table 10.1, we see that the average velocity of the maglev seems to approach the number 16 as it is computed over smaller and smaller time intervals. These computations suggest that the instantaneous velocity of the train at $t = 2$ is 16 ft/sec.

REMARK   Notice that we cannot obtain the instantaneous velocity for the maglev at $t = 2$ by substituting $t = 2$ into equation (2) because this value of t is not in the domain of the average velocity function.   ☉ ☉ ☉

☉ ☉ ☉
* Actually, any interval containing $t = 2$ will do.

Intuitive Definition of a Limit

Consider the function g defined by

$$g(t) = \frac{4(t^2 - 4)}{t - 2}$$

which gives the average velocity of the maglev [see equation (2)]. Suppose that we are required to determine the value that $g(t)$ approaches as t approaches the (fixed) number 2. If we take the sequence of values of t approaching 2 from the right-hand side, as we did earlier, we see that $g(t)$ approaches the number 16. Similarly, if we take a sequence of values of t approaching 2 from the left, such as $t = 1.5, 1.9, 1.99, 1.999,$ and 1.9999, we obtain the results shown in Table 10.2.

Table 10.2

			t approaches 2 from the left.		
t	1.5	1.9	1.99	1.999	1.9999
$g(t)$	14	15.6	15.96	15.996	15.9996

$g(t)$ approaches 16 from the left.

Observe that $g(t)$ approaches the number 16 as t approaches 2—this time from the left-hand side. In other words, as t approaches 2 from *either* side of 2, $g(t)$ approaches 16. In this situation, we say that the **limit** of $g(t)$ as t approaches 2 is 16, written

$$\lim_{t \to 2} g(t) = \lim_{t \to 2} \frac{4(t^2 - 4)}{t - 2} = 16$$

The graph of the function g, shown in Figure 10.19, confirms this observation.

Figure 10.19

As t approaches t = 2 from either direction, g(t) approaches y = 16.

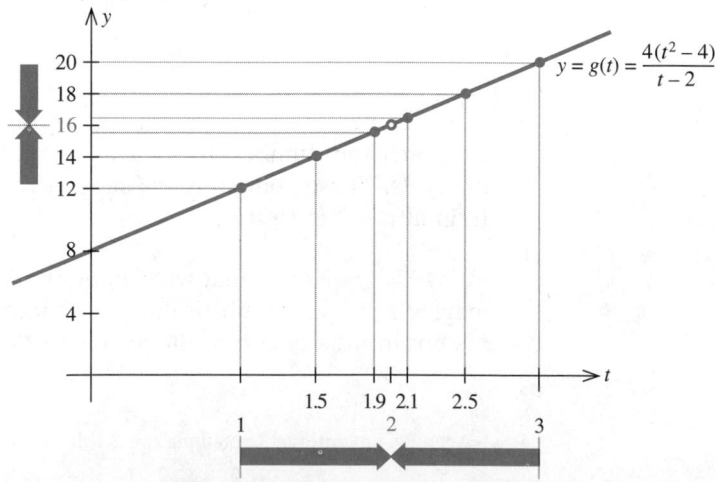

Observe that the point $t = 2$ is not in the domain of the function g [for this reason, the point $(2, 16)$ is missing from the graph of g]. This, however, is inconsequential because the value, if any, of $g(t)$ at $t = 2$ plays no role in computing the limit.

This example leads to the following informal definition.

LIMIT OF A FUNCTION	The function f has the limit L as x approaches a, written $$\lim_{x \to a} f(x) = L$$ if the value $f(x)$ can be made as close to the number L as we please by taking x sufficiently close to (but not equal to) a.

EXPLORING WITH TECHNOLOGY

1. Use a graphing utility to plot the graph of $g(x) = \dfrac{4(x^2 - 4)}{x - 2}$ in the viewing rectangle $[0, 3] \times [0, 20]$.

2. Use **ZOOM** and **TRACE** to describe what happens to the values of $f(x)$ as x approaches 2, first from the right and then from the left.

3. What happens to the y-value when x takes on the value 2? Explain.

4. Reconcile your results with those of the preceding example.

● ● ●

Figure 10.20
$f(x)$ is close to 8 whenever x is close to 2.

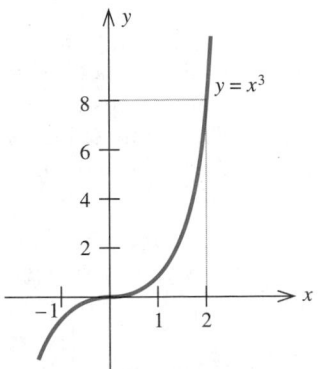

Evaluating the Limit of a Function

Let us now consider some examples involving the computation of limits.

EXAMPLE 1   Let $f(x) = x^3$ and evaluate $\lim_{x \to 2} f(x)$.

Solution   The graph of f is shown in Figure 10.20. You can see that $f(x)$ can be made as close to the number 8 as we please by taking x sufficiently close to 2. Therefore,

$$\lim_{x \to 2} x^3 = 8$$

● ● ●

Figure 10.21
$\lim_{x \to 1} g(x) = 3.$

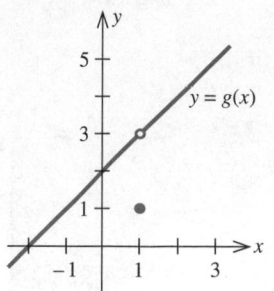

EXAMPLE 2   Let

$$g(x) = \begin{cases} x + 2 & \text{if } x \neq 1 \\ 1 & \text{if } x = 1 \end{cases}$$

Evaluate $\lim_{x \to 1} g(x)$.

Solution   The domain of g is the set of all real numbers. From the graph of g shown in Figure 10.21, we see that $g(x)$ can be made as close to 3 as we please by taking x sufficiently close to 1. Therefore

$$\lim_{x \to 1} g(x) = 3$$

Observe that $g(1) = 1$, which is not equal to the limit of the function g as x approaches 1. [Once again, the value of $g(x)$ at $x = 1$ has no bearing on the existence or value of the limit of g as x approaches 1.]   ◦ ◦ ◦

EXAMPLE 3   Evaluate the limit of the following functions as x approaches the indicated point.

a. $f(x) = \begin{cases} -1 & \text{if } x < 0 \\ 1 & \text{if } x \geq 0 \end{cases};$ $x = 0$ **b.** $g(x) = \dfrac{1}{x^2};$ $x = 0$

Solution   The graphs of the functions f and g are shown in Figure 10.22.

Figure 10.22

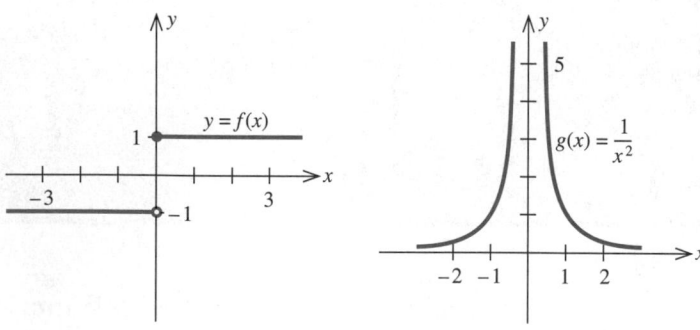

(a) $\lim_{x \to 0} f(x)$ does not exist.   **(b)** $\lim_{x \to 0} g(x)$ does not exist.

a. Referring to Figure 10.22a, we see that no matter how close x is to $x = 0$, $f(x)$ takes on the values 1 or -1 depending on whether x is positive or negative. Thus, there is no *single* real number L that $f(x)$ approaches as x approaches zero. We conclude that the limit of $f(x)$ does *not* exist as x approaches zero.

b. Referring to Figure 10.22b, we see that as x approaches $x = 0$ (from either side), $g(x)$ increases without bound and thus does not approach any specific real number. We conclude, accordingly, that the limit of $g(x)$ does not exist as x approaches zero.   ◦ ◦ ◦

Consider the graph of the function h whose graph is depicted in the following figure:

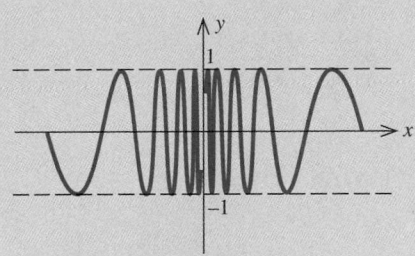

It has the property that as x approaches zero from either the right or the left, the curve oscillates more and more frequently between the lines $y = -1$ and $y = 1$.

a. Explain why $\lim\limits_{x \to 0} h(x)$ does not exist.

b. Compare this function with those in Example 3. More specifically, discuss the different ways the functions fail to have a limit at $x = 0$.

Until now, we have relied on knowing the actual values of a function or the graph of a function near $x = a$ to help us evaluate the limit of the function $f(x)$ as x approaches a. The following properties of limits, which we list without proof, enable us to evaluate limits of functions algebraically.

THEOREM 1

Properties of Limits

Suppose that

$$\lim_{x \to a} f(x) = L \quad \text{and} \quad \lim_{x \to a} g(x) = M$$

Then

i. $\lim\limits_{x \to a}[f(x)]^r = [\lim\limits_{x \to a} f(x)]^r = L^r$ (r, a real number)

ii. $\lim\limits_{x \to a} cf(x) = c \lim\limits_{x \to a} f(x) = cL$ (c, a real number)

iii. $\lim\limits_{x \to a}[f(x) \pm g(x)] = \lim\limits_{x \to a} f(x) \pm \lim\limits_{x \to a} g(x) = L \pm M$

iv. $\lim\limits_{x \to a}[f(x)g(x)] = [\lim\limits_{x \to a} f(x)][\lim\limits_{x \to a} g(x)] = LM$

v. $\lim\limits_{x \to a} \dfrac{f(x)}{g(x)} = \dfrac{\lim\limits_{x \to a} f(x)}{\lim\limits_{x \to a} g(x)} = \dfrac{L}{M}$ (Provided $M \neq 0$)

EXAMPLE 4   Use Theorem 1 to evaluate the following limits:

a. $\lim\limits_{x \to 2} x^3$

b. $\lim\limits_{x \to 4} 5x^{3/2}$

c. $\lim\limits_{x \to 1} (5x^4 - 2)$

d. $\lim\limits_{x \to 3} 2x^3 \sqrt{x^2 + 7}$

e. $\lim\limits_{x \to 2} \dfrac{2x^2 + 1}{x + 1}$

Solution

a. $\lim\limits_{x \to 2} x^3 = [\lim\limits_{x \to 2} x]^3$ (Property i)

$\qquad = 2^3 = 8$ ($\lim\limits_{x \to 2} x = 2$)

b. $\lim\limits_{x \to 4} 5x^{3/2} = 5[\lim\limits_{x \to 4} x^{3/2}]$ (Property ii)

$\qquad = 5(4)^{3/2} = 40$

c. $\lim\limits_{x \to 1}(5x^4 - 2) = \lim\limits_{x \to 1} 5x^4 - \lim\limits_{x \to 1} 2$ (Property iii)

To evaluate $\lim\limits_{x \to 1} 2$, observe that the constant function $g(x) = 2$ has value 2 for all values of x. Therefore, $g(x)$ must approach the limit 2 as x approaches $x = 1$ (or any other point for that matter!). Therefore,

$$\lim\limits_{x \to 1}(5x^4 - 2) = 5(1)^4 - 2 = 3$$

d. $\lim\limits_{x \to 3} 2x^3 \sqrt{x^2 + 7} = 2 \lim\limits_{x \to 3} x^3 \sqrt{x^2 + 7}$ (Property ii)

$\qquad = 2 \lim\limits_{x \to 3} x^3 \lim\limits_{x \to 3} \sqrt{x^2 + 7}$ (Property iv)

$\qquad = 2(3)^3 \sqrt{3^2 + 7}$ (Property i)

$\qquad = 2(27) \sqrt{16} = 216$

e. $\lim\limits_{x \to 2} \dfrac{2x^2 + 1}{x + 1} = \dfrac{\lim\limits_{x \to 2}(2x^2 + 1)}{\lim\limits_{x \to 2}(x + 1)}$ (Property v)

$\qquad = \dfrac{2(2)^2 + 1}{2 + 1} = \dfrac{9}{3} = 3$ ○ ○ ○

Indeterminate Forms

Let us emphasize once again that property v of limits is valid only when the limit of the function that appears in the denominator is not equal to zero at the point in question.

If the numerator has a limit different from zero and the denominator has a limit equal to zero, then the limit of the quotient does not exist at the point in question. This is the case with the function $g(x) = 1/x^2$ in Example 3b. Here, as x approaches zero, the numerator approaches 1 but the denominator approaches zero, so that the quotient becomes arbitrarily large. Thus, as observed earlier, the limit does not exist.

Next, consider

$$\lim_{x \to 2} \frac{4(x^2 - 4)}{x - 2}$$

which we evaluated earlier by looking at the values of the function for x near $x = 2$. If we attempt to evaluate this expression by applying property v of limits, we see that both the numerator and denominator of the function

$$\frac{4(x^2 - 4)}{x - 2}$$

approach zero as x approaches 2; that is, we obtain an expression of the form 0/0. In this event, we say that the limit of the quotient $f(x)/g(x)$ as x approaches 2 has the **indeterminate form 0/0.**

We will need to evaluate limits of this type when we discuss the derivative of a function, a fundamental concept in the study of calculus. As the name suggests, the meaningless expression 0/0 does not provide us with a solution to our problem. One strategy that can be used to solve this type of problem follows.

STRATEGY FOR EVALUATING INDETERMINATE FORMS

1. Replace the given function with an appropriate one that takes on the same values as the original function everywhere except at $x = a$.

2. Evaluate the limit of this function as x approaches a.

Examples 5 and 6 illustrate this strategy.

EXAMPLE 5   Evaluate

$$\lim_{x \to 2} \frac{4(x^2 - 4)}{x - 2}$$

Solution   Since both the numerator and the denominator of this expression approach zero as x approaches 2, we have the indeterminate form 0/0. We rewrite

$$\frac{4(x^2 - 4)}{x - 2} = \frac{4(x - 2)(x + 2)}{(x - 2)}$$

which, upon canceling the common factors, is equivalent to $4(x + 2)$. Next, we replace $4(x^2 - 4)/(x - 2)$ with $4(x + 2)$ and take the limit as x approaches 2, obtaining

$$\lim_{x \to 2} \frac{4(x^2 - 4)}{x - 2} = \lim_{x \to 2} 4(x + 2) = 16$$

The graphs of the functions

$$f(x) = \frac{4(x^2 - 4)}{x - 2} \quad \text{and} \quad g(x) = 4(x + 2)$$

are shown in Figure 10.23a and b. Observe that the graphs are identical except when $x = 2$. The function g is defined for all values of x and, in particular, its value at $x = 2$ is $g(2) = 4(2 + 2) = 16$. Thus, the point $(2, 16)$ is on the graph of g. However, the function f is not defined at $x = 2$. Since $f(x) = g(x)$ for all values of x except $x = 2$, it follows that the graph of f must look exactly like the graph of g, with the exception that the point $(2, 16)$ is missing from the graph of f. This illustrates graphically why we can evaluate the limit of f by evaluating the limit of the "equivalent" function g.

Figure 10.23
The graphs of $f(x)$ and $g(x)$ are identical except at the point $(2, 16)$.

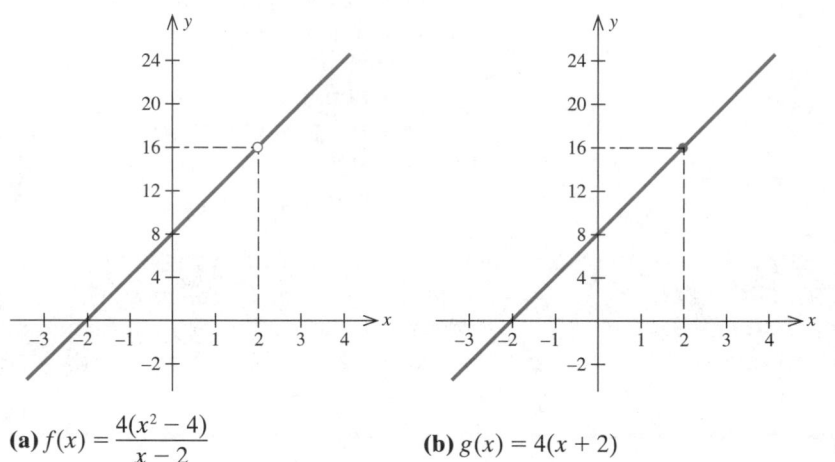

(a) $f(x) = \dfrac{4(x^2 - 4)}{x - 2}$ **(b)** $g(x) = 4(x + 2)$

EXPLORING WITH TECHNOLOGY

1. Use a graphing utility to plot the graph of $f(x) = \dfrac{4(x^2 - 4)}{x - 2}$ in the viewing rectangle $[0, 3] \times [0, 20]$.

   Then use **ZOOM** and **TRACE** to find $\displaystyle\lim_{x \to 2} \frac{4(x^2 - 4)}{x - 2}$.

2. Use a graphing utility to plot the graph of $g(x) = 4(x + 2)$ in the viewing rectangle $[0, 3] \times [0, 20]$. Then use **ZOOM** and **TRACE** to find $\displaystyle\lim_{x \to 2} 4(x + 2)$. What happens to the y-value when x takes on the value 2? Explain.

3. Can you distinguish between the graphs of f and g?

4. Reconcile your results with those of Example 5.

EXPLORING WITH TECHNOLOGY

1. Use a graphing utility to plot the graph of $g(x) = \dfrac{\sqrt{1+x}-1}{x}$ in the viewing rectangle $[-1, 2] \times$ $[0, 1]$. Then use **ZOOM** and **TRACE** to find $\displaystyle\lim_{x \to 0} \dfrac{\sqrt{1+x}-1}{x}$ by observing the values of $g(x)$ as x approaches zero from the left and from the right.

2. Use a graphing utility to plot the graph of $f(x) = \dfrac{1}{\sqrt{1+x}+1}$ in the viewing rectangle $[-1, 2] \times$ $[0, 1]$. Then use **ZOOM** and **TRACE** to find $\displaystyle\lim_{x \to 0} \dfrac{1}{\sqrt{1+x}+1}$. What happens to the y-value when x takes on the value zero? Explain.

3. Can you distinguish between the graphs of f and g?

4. Reconcile your results with those of Example 6.

REMARK   Notice that the limit in Example 5 is the same limit that we evaluated earlier when we discussed the instantaneous velocity of a maglev at a specified time.

EXAMPLE 6   Evaluate

$$\lim_{h \to 0} \frac{\sqrt{1+h}-1}{h}$$

Solution   Letting h approach zero, we obtain the indeterminate form 0/0. Next, we rationalize the numerator of the quotient (see page 558) by multiplying both the numerator and the denominator by the expression $(\sqrt{1+h}+1)$, obtaining

$$\frac{\sqrt{1+h}-1}{h} = \frac{(\sqrt{1+h}-1)(\sqrt{1+h}+1)}{h(\sqrt{1+h}+1)}$$

$$= \frac{1+h-1}{h(\sqrt{1+h}+1)} \qquad [(\sqrt{a}-\sqrt{b})(\sqrt{a}+\sqrt{b}) = a - b]$$

$$= \frac{h}{h(\sqrt{1+h}+1)}$$

$$= \frac{1}{\sqrt{1+h}+1}$$

Therefore,

$$\lim_{h \to 0} \frac{\sqrt{1+h}-1}{h} = \lim_{h \to 0} \frac{1}{\sqrt{1+h}+1}$$

$$= \frac{1}{\sqrt{1}+1} = \frac{1}{2} \qquad \circ \; \circ \; \circ$$

Limits at Infinity

Up to now we have studied the limit of a function as x approaches a (finite) number a. There are occasions, however, when we want to know whether $f(x)$ approaches a unique number as x increases without bound. Consider, for example, the function P, giving the number of fruit flies (*Drosophila*) in a container under controlled laboratory conditions, as a function of a time t. The graph of P is shown in Figure 10.24. You can see from the graph of P that, as t increases without bound (gets larger and larger), $P(t)$ approaches the number 400. This number, called the *carrying capacity* of the environment, is determined by the amount of living space and food available, as well as other environmental factors.

Figure 10.24
The graph of P(t) gives the population of fruit flies in a laboratory experiment.

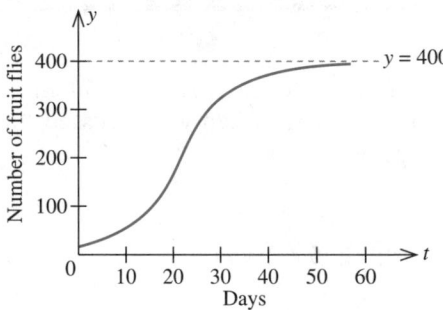

As another example, suppose that we are given the function

$$f(x) = \frac{2x^2}{1+x^2}$$

and we want to determine what happens to $f(x)$ as x gets larger and larger. Picking the sequence of numbers 1, 2, 5, 10, 100, and 1000 and computing the corresponding values of $f(x)$, we obtain the following table of values:

x	1	2	5	10	100	1000
$f(x)$	1	1.6	1.92	1.98	1.9998	1.999998

Figure 10.25

The graph of $y = \dfrac{2x^2}{1 + x^2}$ has a horizontal asymptote at $y = 2$.

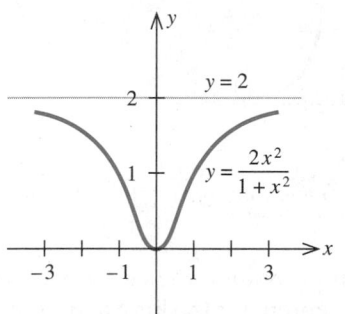

From the table, we see that as x gets larger and larger, $f(x)$ gets closer and closer to 2. The graph of the function f shown in Figure 10.25 confirms this observation. We call the line $y = 2$ a **horizontal asymptote.***

In this situation we say that the limit of the function

$$f(x) = \frac{2x^2}{1 + x^2}$$

as x increases without bound is 2, written

$$\lim_{x \to \infty} \frac{2x^2}{1 + x^2} = 2$$

In the general case, the following definition is applicable:

LIMIT OF A FUNCTION AT INFINITY

The function f has the limit L as x increases without bound (or, as x approaches infinity), written

$$\lim_{x \to \infty} f(x) = L$$

if $f(x)$ can be made arbitrarily close to L by taking x large enough.

Similarly, the function f has the limit M as x decreases without bound (or as x approaches negative infinity), written

$$\lim_{x \to -\infty} f(x) = M$$

if $f(x)$ can be made arbitrarily close to M by taking x to be negative and sufficiently large in absolute value.

EXAMPLE 7 Let f and g be the functions

$$f(x) = \begin{cases} -1 & \text{if } x < 0 \\ 1 & \text{if } x \geq 0 \end{cases} \quad \text{and} \quad g(x) = \frac{1}{x^2}$$

Evaluate

a. $\lim\limits_{x \to \infty} f(x)$ and $\lim\limits_{x \to -\infty} f(x)$ **b.** $\lim\limits_{x \to \infty} g(x)$ and $\lim\limits_{x \to -\infty} g(x)$

Solution The graphs of $f(x)$ and $g(x)$ are shown in Figure 10.26. Referring to the graphs of the respective functions, we see that

a. $\lim\limits_{x \to \infty} f(x) = 1$ and $\lim\limits_{x \to -\infty} f(x) = -1$

b. $\lim\limits_{x \to \infty} \dfrac{1}{x^2} = 0$ and $\lim\limits_{x \to -\infty} \dfrac{1}{x^2} = 0$

○ ○ ○

* We will discuss asymptotes in greater detail in Section 12.3.

Figure 10.26

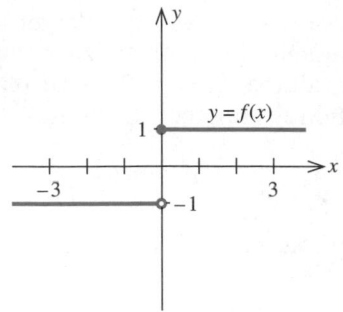

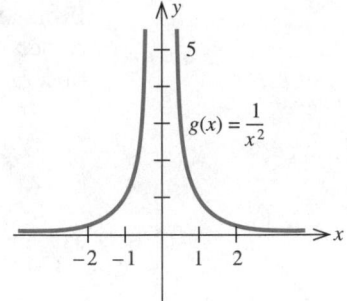

(a) $\lim\limits_{x \to \infty} f(x) = 1$ and $\lim\limits_{x \to -\infty} f(x) = -1$ **(b)** $\lim\limits_{x \to \infty} g(x) = 0$ and $\lim\limits_{x \to -\infty} g(x) = 0$

All the properties of limits listed in Theorem 1 are valid when a is replaced by ∞ or $-\infty$. In addition, we have the following theorem for the limit at infinity.

THEOREM 2

For all $n > 0$

$$\lim_{x \to \infty} \frac{1}{x^n} = 0 \quad \text{and} \quad \lim_{x \to -\infty} \frac{1}{x^n} = 0$$

provided that $\dfrac{1}{x^n}$ is defined.

EXPLORING WITH TECHNOLOGY

1. Use a graphing utility to plot the graphs of

$$y_1 = \frac{1}{x^{0.5}}, \quad y_2 = \frac{1}{x}, \quad \text{and} \quad y_3 = \frac{1}{x^{1.5}}$$

in the viewing rectangle $[0, 200] \times [0, 0.5]$. What can you say about $\lim\limits_{x \to \infty} \dfrac{1}{x^n}$ if $n = 0.5$, $n = 1$, and $n = 1.5$? Are these results predicted by Theorem 2?

2. Use a graphing utility to plot the graphs of

$$y_1 = \frac{1}{x} \quad \text{and} \quad y_2 = \frac{1}{x^{5/3}}$$

in the viewing rectangle $[-50, 0] \times [-0.5, 0]$. What can you say about $\lim\limits_{x \to -\infty} \dfrac{1}{x^n}$ if $n = 1$ and $n = \dfrac{5}{3}$? Are these results predicted by Theorem 2?

[*Hint:* To graph y_2, write it in the form $y2 = 1/(x^{\wedge}(1/3))^{\wedge}5$.]

To evaluate the limit at infinity of a rational function, the following technique is often used: *Divide the numerator and denominator of the expression by x^n, where n is the highest power present in the denominator of the expression.*

EXAMPLE 8   Evaluate

$$\lim_{x \to \infty} \frac{x^2 - x + 3}{2x^3 + 1}$$

Solution   Since the limits of both the numerator and the denominator do not exist as x approaches infinity, the property pertaining to the limit of a quotient (property v) is not applicable. Let us divide the numerator and denominator of the rational expression by x^3, obtaining

$$\lim_{x \to \infty} \frac{x^2 - x + 3}{2x^3 + 1} = \lim_{x \to \infty} \frac{\dfrac{1}{x} - \dfrac{1}{x^2} + \dfrac{3}{x^3}}{2 + \dfrac{1}{x^3}}$$

$$= \frac{0}{2} \qquad \text{(Using Theorem 2)}$$

$$= 0 \qquad\qquad\qquad\qquad \circ \; \circ \; \circ$$

EXAMPLE 9   Let

$$f(x) = \frac{3x^2 + 8x - 4}{2x^2 + 4x - 5}$$

Compute $\lim_{x \to \infty} f(x)$ if it exists.

Solution   Again, we see that property v is not applicable. Dividing the numerator and the denominator by x^2, we obtain

$$\lim_{x \to \infty} \frac{3x^2 + 8x - 4}{2x^2 + 4x - 5} = \lim_{x \to \infty} \frac{3 + \dfrac{8}{x} - \dfrac{4}{x^2}}{2 + \dfrac{4}{x} - \dfrac{5}{x^2}}$$

$$= \frac{\lim\limits_{x \to \infty} 3 + 8 \lim\limits_{x \to \infty} \dfrac{1}{x} - 4 \lim\limits_{x \to \infty} \dfrac{1}{x^2}}{\lim\limits_{x \to \infty} 2 + 4 \lim\limits_{x \to \infty} \dfrac{1}{x} - 5 \lim\limits_{x \to \infty} \dfrac{1}{x^2}}$$

$$= \frac{3 + 0 - 0}{2 + 0 - 0}$$

$$= \frac{3}{2} \qquad\qquad\qquad \text{(Using Theorem 2)} \quad \circ \; \circ \; \circ$$

EXAMPLE 10   Let $f(x) = \dfrac{2x^3 - 3x^2 + 1}{x^2 + 2x + 4}$ and evaluate

a. $\lim\limits_{x \to \infty} f(x)$ **b.** $\lim\limits_{x \to -\infty} f(x)$

Solution

a. Dividing the numerator and the denominator of the rational expression by x^2, we obtain

$$\lim_{x \to \infty} \frac{2x^3 - 3x^2 + 1}{x^2 + 2x + 4} = \lim_{x \to \infty} \frac{2x - 3 + \dfrac{1}{x^2}}{1 + \dfrac{2}{x} + \dfrac{4}{x^2}}$$

Since the numerator becomes arbitrarily large whereas the denominator approaches 1 as x approaches infinity, we see that the quotient $f(x)$ gets larger and larger as x approaches infinity. In other words, the limit does not exist. We indicate this by writing

$$\lim_{x \to \infty} \frac{2x^3 - 3x^2 + 1}{x^2 + 2x + 4} = \infty$$

b. Once again, dividing both the numerator and the denominator by x^2, we obtain

$$\lim_{x \to -\infty} \frac{2x^3 - 3x^2 + 1}{x^2 + 2x + 4} = \lim_{x \to -\infty} \frac{2x - 3 + \dfrac{1}{x^2}}{1 + \dfrac{2}{x} + \dfrac{4}{x^2}}$$

In this case, the numerator becomes arbitrarily large in magnitude but negative in sign, whereas the denominator approaches 1 as x approaches negative infinity. Therefore, the quotient $f(x)$ decreases without bound and the limit does not exist. We indicate this by writing

$$\lim_{x \to -\infty} \frac{2x^3 - 3x^2 + 1}{x^2 + 2x + 4} = -\infty$$

Example 11 gives an application of the limit of a function at infinity.

EXAMPLE 11   The Custom Office Company makes a line of executive desks. It is estimated that the total cost of making x Senior Executive Model desks is $C(x) = 100x + 200{,}000$ dollars per year, so that the average cost of making x desks is given by

$$\overline{C}(x) = \frac{C(x)}{x} = \frac{100x + 200{,}000}{x} = 100 + \frac{200{,}000}{x}$$

dollars per desk. Evaluate $\lim\limits_{x \to \infty} \overline{C}(x)$ and interpret your results.

Figure 10.27
As the level of production increases, the average cost approaches $100 per desk.

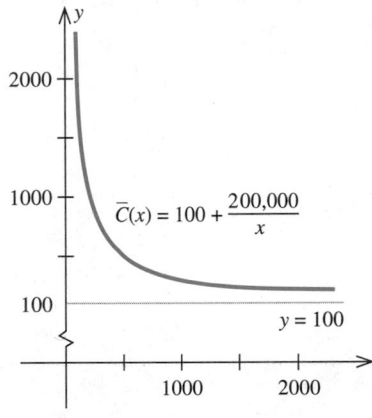

Solution

$$\lim_{x\to\infty} \overline{C}(x) = \lim_{x\to\infty}\left(100 + \frac{200{,}000}{x}\right)$$

$$= \lim_{x\to\infty} 100 + \lim_{x\to\infty}\frac{200{,}000}{x} = 100$$

A sketch of the graph of the function $\overline{C}(x)$ appears in Figure 10.27. The result we obtained is fully expected if we consider its economic implications. Note that as the level of production increases, the fixed cost per desk produced, represented by the term $(200{,}000/x)$, drops steadily. The average cost should approach a constant unit cost of production—$100 in this case.   ⊙ ⊙ ⊙

Consider the graph of the function f depicted in the figure:

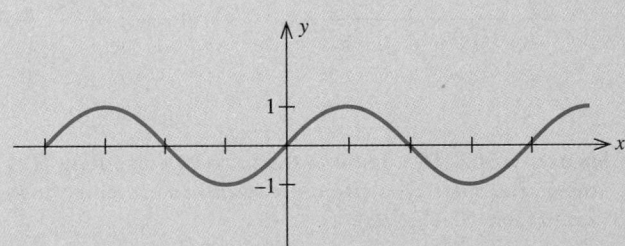

The curve oscillates between $y = -1$ and $y = 1$ indefinitely in either direction.

a. Explain why $\lim_{x\to-\infty} f(x)$ and $\lim_{x\to\infty} f(x)$ do not exist.

b. Compare this function with those of Example 10. More specifically, discuss the different ways each function fails to have a limit at infinity or minus infinity.

SELF-CHECK EXERCISES 10.4

1. Find the indicated limit if it exists.

a. $\displaystyle\lim_{x\to3}\frac{\sqrt{x^2+7}+\sqrt{3x-5}}{x+2}$

b. $\displaystyle\lim_{x\to-1}\frac{x^2-x-2}{2x^2-x-3}$

2. The average cost per disc (in dollars) incurred by the Herald Record Company in pressing x compact audio discs is given by the average cost function

$$\overline{C}(x)=1.8+\frac{3000}{x}$$

Evaluate $\displaystyle\lim_{x\to\infty}\overline{C}(x)$ and interpret your result.

Solutions to Self-Check Exercises 10.4 can be found on page 633.

10.4 EXERCISES

In exercises 1–8, use the graph of the given function f to determine $\displaystyle\lim_{x\to a}f(x)$ *at the indicated value of a, if it exists.*

1.
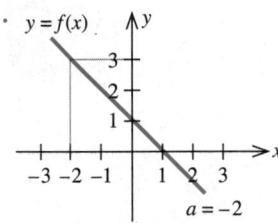
$a=-2$

2.
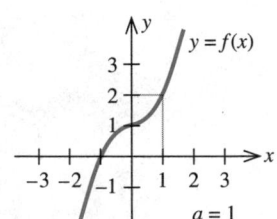
$a=1$

3.
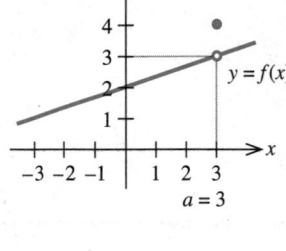
$a=3$

4.
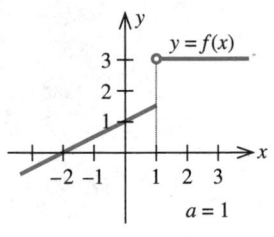
$a=1$

5.
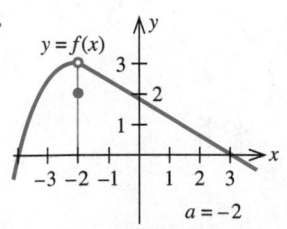
$a=-2$

6.
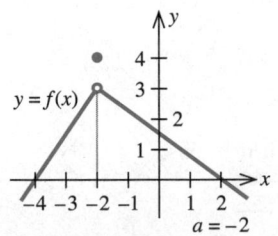
$a=-2$

7.
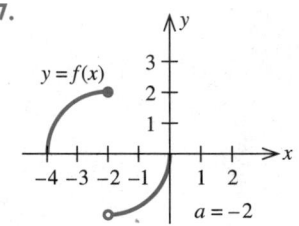
$a=-2$

8.
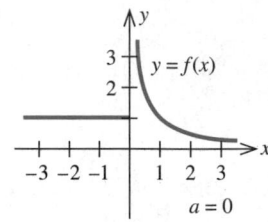
$a=0$

In exercises 9–16, complete the table by computing f(x) at the given values of x. Use these results to estimate the indicated limit (if it exists).

9. $f(x)=x^2+1$; $\displaystyle\lim_{x\to2}f(x)$

x	1.9	1.99	1.999	2.001	2.01	2.1
$f(x)$						

10. $f(x)=2x^2-1$; $\displaystyle\lim_{x\to1}f(x)$

x	0.9	0.99	0.999	1.001	1.01	1.1
$f(x)$						

11. $f(x)=\dfrac{|x|}{x}$; $\displaystyle\lim_{x\to0}f(x)$

x	−0.1	−0.01	−0.001	0.001	0.01	0.1
$f(x)$						

12. $f(x) = \dfrac{|x - 1|}{x - 1}$; $\displaystyle\lim_{x \to 1} f(x)$

x	0.9	0.99	0.999	1.001	1.01	1.1
$f(x)$						

13. $f(x) = \dfrac{1}{(x - 1)^2}$; $\displaystyle\lim_{x \to 1} f(x)$

x	0.9	0.99	0.999	1.001	1.01	1.1
$f(x)$						

14. $f(x) = \dfrac{1}{x - 2}$; $\displaystyle\lim_{x \to 2} f(x)$

x	1.9	1.99	1.999	2.001	2.01	2.1
$f(x)$						

15. $f(x) = \dfrac{x^2 + x - 2}{x - 1}$; $\displaystyle\lim_{x \to 1} f(x)$

x	0.9	0.99	0.999	1.001	1.01	1.1
$f(x)$						

16. $f(x) = \dfrac{x - 1}{x - 1}$; $\displaystyle\lim_{x \to 1} f(x)$

x	0.9	0.99	0.999	1.001	1.01	1.1
$f(x)$						

In exercises 17–22, sketch the graph of the function f and evaluate $\displaystyle\lim_{x \to a} f(x)$, if it exists, for the given values of a.

17. $f(x) = \begin{cases} x - 1 & \text{if } x \le 0 \\ -1 & \text{if } x > 0 \end{cases}$ $a = 0$

18. $f(x) = \begin{cases} x - 1 & \text{if } x \le 3 \\ -2x + 8 & \text{if } x > 3 \end{cases}$ $a = 3$

19. $f(x) = \begin{cases} x & \text{if } x < 1 \\ 0 & \text{if } x = 1 \\ -x + 2 & \text{if } x > 1 \end{cases}$ $a = 1$

20. $f(x) = \begin{cases} -2x + 4 & \text{if } x < 1 \\ 4 & \text{if } x = 1 \\ x^2 + 1 & \text{if } x > 1 \end{cases}$ $a = 1$

21. $f(x) = \begin{cases} |x| & \text{if } x \ne 0 \\ 1 & \text{if } x = 0 \end{cases}$ $a = 0$

22. $f(x) = \begin{cases} |x - 1| & \text{if } x \ne 1 \\ 0 & \text{if } x = 1 \end{cases}$ $a = 1$

In exercises 23–40, find the indicated limit.

23. $\displaystyle\lim_{x \to 2} 3$

24. $\displaystyle\lim_{x \to -2} -3$

25. $\displaystyle\lim_{x \to 3} x$

26. $\displaystyle\lim_{x \to -2} -3x$

27. $\displaystyle\lim_{x \to 1} (1 - 2x^2)$

28. $\displaystyle\lim_{t \to 3} (4t^2 - 2t + 1)$

29. $\displaystyle\lim_{x \to 1} (2x^3 - 3x^2 + x + 2)$

30. $\displaystyle\lim_{x \to 0} (4x^5 - 20x^2 + 2x + 1)$

31. $\displaystyle\lim_{s \to 0} (2s^2 - 1)(2s + 4)$

32. $\displaystyle\lim_{x \to 2} (x^2 + 1)(x^2 - 4)$

33. $\displaystyle\lim_{x \to 2} \dfrac{2x + 1}{x + 2}$

34. $\displaystyle\lim_{x \to 1} \dfrac{x^3 + 1}{2x^3 + 2}$

35. $\displaystyle\lim_{x \to 2} \sqrt{x + 2}$

36. $\displaystyle\lim_{x \to -2} \sqrt[3]{5x + 2}$

37. $\displaystyle\lim_{x \to -3} \sqrt{2x^4 + x^2}$

38. $\displaystyle\lim_{x \to 2} \sqrt{\dfrac{2x^3 + 4}{x^2 + 1}}$

39. $\displaystyle\lim_{x \to -1} \dfrac{\sqrt{x^2 + 8}}{2x + 4}$

40. $\displaystyle\lim_{x \to 3} \dfrac{x\sqrt{x^2 + 7}}{2x - \sqrt{2x + 3}}$

In exercises 41–48, find the indicated limit given that $\displaystyle\lim_{x \to a} f(x) = 3$ and $\displaystyle\lim_{x \to a} g(x) = 4$.

41. $\displaystyle\lim_{x \to a} [f(x) - g(x)]$

42. $\displaystyle\lim_{x \to a} 2f(x)$

43. $\displaystyle\lim_{x \to a} [2f(x) - 3g(x)]$

44. $\displaystyle\lim_{x \to a} [f(x)g(x)]$

45. $\displaystyle\lim_{x \to a} \sqrt{g(x)}$

46. $\displaystyle\lim_{x \to a} \sqrt[3]{5f(x) + 3g(x)}$

47. $\displaystyle\lim_{x \to a} \dfrac{2f(x) - g(x)}{f(x)g(x)}$

48. $\displaystyle\lim_{x \to a} \dfrac{g(x) - f(x)}{f(x) + \sqrt{g(x)}}$

In exercises 49–62, find the indicated limit, if it exists.

49. $\displaystyle\lim_{x \to 1} \dfrac{x^2 - 1}{x - 1}$

50. $\displaystyle\lim_{x \to -2} \dfrac{x^2 - 4}{x + 2}$

51. $\lim\limits_{x\to 0}\dfrac{x^2-x}{x}$

52. $\lim\limits_{x\to 0}\dfrac{2x^2-3x}{x}$

53. $\lim\limits_{x\to -5}\dfrac{x^2-25}{x+5}$

54. $\lim\limits_{b\to -3}\dfrac{b+1}{b+3}$

55. $\lim\limits_{x\to 1}\dfrac{x}{x-1}$

56. $\lim\limits_{x\to 2}\dfrac{x+2}{x-2}$

57. $\lim\limits_{x\to -2}\dfrac{x^2-x-6}{x^2+x-2}$

58. $\lim\limits_{z\to 2}\dfrac{z^3-8}{z-2}$

59. $\lim\limits_{x\to 1}\dfrac{\sqrt{x}-1}{x-1}$

$\left[\textit{Hint: Multiply by }\dfrac{\sqrt{x}+1}{\sqrt{x}+1}.\right]$

60. $\lim\limits_{x\to 4}\dfrac{x-4}{\sqrt{x}-2}$

[*Hint:* See exercise 59.]

61. $\lim\limits_{x\to 1}\dfrac{x-1}{x^3+x^2-2x}$

62. $\lim\limits_{x\to -2}\dfrac{4-x^2}{2x^2+x^3}$

In exercises 63–68, use the graph of the function f to determine $\lim\limits_{x\to\infty} f(x)$ *and* $\lim\limits_{x\to -\infty} f(x)$, *if they exist.*

63.
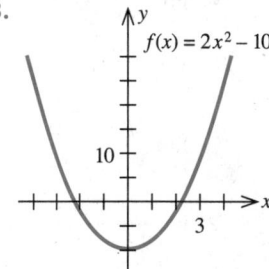

64.
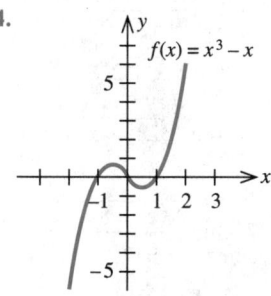

65.
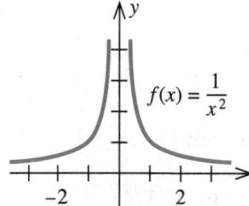

66.
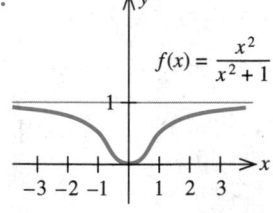

67.
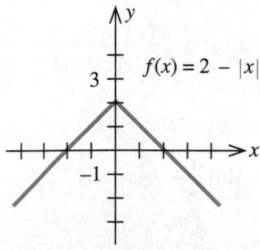

68.
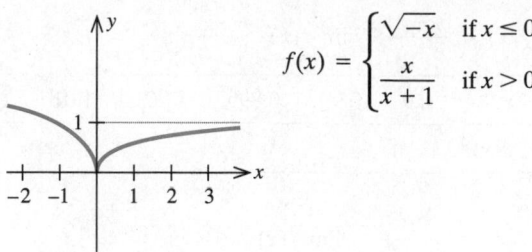

$$f(x) = \begin{cases} \sqrt{-x} & \text{if } x \le 0 \\ \dfrac{x}{x+1} & \text{if } x > 0 \end{cases}$$

In exercises 69–72, complete the table by computing f(x) at the given values of x. Use the results to guess at the indicated limits, if they exist.

69. $f(x) = \dfrac{1}{x^2+1}$; $\lim\limits_{x\to\infty} f(x)$ and $\lim\limits_{x\to -\infty} f(x)$

x	1	10	100	1000
$f(x)$				

x	-1	-10	-100	-1000
$f(x)$				

70. $f(x) = \dfrac{2x}{x+1}$; $\lim\limits_{x\to\infty} f(x)$ and $\lim\limits_{x\to -\infty} f(x)$

x	1	10	100	1000
$f(x)$				

x	-5	-10	-100	-1000
$f(x)$				

71. $f(x) = 3x^3 - x^2 + 10$; $\lim\limits_{x\to\infty} f(x)$ and $\lim\limits_{x\to -\infty} f(x)$

x	1	5	10	100	1000
$f(x)$					

x	-1	-5	-10	-100	-1000
$f(x)$					

72. $f(x) = \dfrac{|x|}{x}$; $\lim\limits_{x\to\infty} f(x)$ and $\lim\limits_{x\to -\infty} f(x)$

x	1	10	100	-1	-10	-100
$f(x)$						

In exercises 73–80, find the indicated limits, if they exist.

73. $\lim\limits_{x \to \infty} \dfrac{3x + 2}{x - 5}$

74. $\lim\limits_{x \to -\infty} \dfrac{4x^2 - 1}{x + 2}$

75. $\lim\limits_{x \to -\infty} \dfrac{3x^3 + x^2 + 1}{x^3 + 1}$

76. $\lim\limits_{x \to \infty} \dfrac{2x^2 + 3x + 1}{x^4 - x^2}$

77. $\lim\limits_{x \to -\infty} \dfrac{x^4 + 1}{x^3 - 1}$

78. $\lim\limits_{x \to \infty} \dfrac{4x^4 - 3x^2 + 1}{2x^4 + x^3 + x^2 + x + 1}$

79. $\lim\limits_{x \to \infty} \dfrac{x^5 - x^3 + x - 1}{x^6 + 2x^2 + 1}$

80. $\lim\limits_{x \to \infty} \dfrac{2x^2 - 1}{x^3 + x^2 + 1}$

81. Toxic Waste A city's main well was recently found to be contaminated with trichloroethylene, a cancer-causing chemical, as a result of an abandoned chemical dump leaching chemicals into the water. A proposal submitted to the selectmen of the city indicates that the cost, measured in millions of dollars, of removing x percent of the toxic pollutant is given by

$$C(x) = \frac{0.5x}{100 - x} \qquad (0 < x < 100)$$

a. Find the cost of removing 50% of the pollutant; 60%; 70%; 80%; 90%; 95%.
b. Evaluate

$$\lim_{x \to 100} \frac{0.5x}{100 - x}$$

and interpret your result.

82. A Doomsday Situation The population of a certain breed of rabbits introduced into an isolated island is given by

$$P(t) = \frac{72}{9 - t} \qquad (0 < t < 9)$$

where t is measured in months.
a. Find the number of rabbits present on the island initially.
b. Show that the population of rabbits is increasing without bound.
c. Sketch the graph of the function P.

REMARK This phenomenon is referred to as a *doomsday* situation. ● ● ●

83. Average Cost The average cost per disc in dollars incurred by the Herald Record Company in pressing x video discs is given by the average cost function

$$\overline{C}(x) = 2.2 + \frac{2500}{x}$$

Evaluate $\lim\limits_{x \to \infty} \overline{C}(x)$ and interpret your result.

84. Concentration of a Drug in the Bloodstream The concentration of a certain drug in a patient's bloodstream t hours after injection is given by

$$C(t) = \frac{0.2t}{t^2 + 1}$$

milligrams per cubic centimeter. Evaluate $\lim\limits_{t \to \infty} C(t)$ and interpret your result.

85. Box Office Receipts The total worldwide box office receipts for a long-running blockbuster movie are approximated by the function

$$T(x) = \frac{120x^2}{x^2 + 4}$$

where $T(x)$ is measured in millions of dollars and x is the number of months since the movie's release.
a. What are the total box office receipts after the first month? The second month? The third month?
b. What will the movie gross in the long run?

86. Population Growth A major corporation is building a 4325-acre complex of homes, offices, stores, schools, and churches in the rural community of Glen Cove. As a result of this development, the planners have estimated that Glen Cove's population t years from now (in thousands) will be given by

$$P(t) = \frac{25t^2 + 125t + 200}{t^2 + 5t + 40}$$

a. What is the current population of Glen Cove?
b. What will the population be in the long run?

87. Driving Costs A study of driving costs of 1992 model subcompact (four-cylinder) cars found that the average cost (car payments, gas, insurance, upkeep, and depreciation), measured in cents per mile, is approximated by the function

$$C(x) = \frac{2010}{x^{2.2}} + 17.80$$

where x denotes the number of miles (in thousands) the car is driven in a year.

USING TECHNOLOGY

FINDING THE LIMIT OF A FUNCTION

A graphing utility can be used to help us find the limit of a function, if it exists, as illustrated in the following examples.

EXAMPLE 1 Let $f(x) = \dfrac{x^3 - 1}{x - 1}$.

a. Plot the graph of f in the viewing rectangle $[-2, 2] \times [0, 4]$.

b. Use **ZOOM-IN** to find $\displaystyle\lim_{x \to 1} \frac{x^3 - 1}{x - 1}$.

c. Verify your result by evaluating the limit algebraically.

Solution

a. The graph of f in the viewing rectangle $[-2, 2] \times [0, 4]$ is shown in Figure T1.

Figure T1

The graph of $f(x) = \dfrac{x^3 - 1}{x - 1}$ in the viewing rectangle $[-2, 2] \times [0, 4]$.

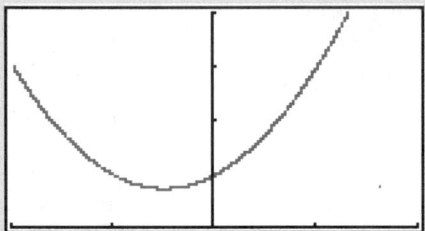

b. Using **ZOOM-IN** repeatedly, we see that the y-value approaches 3 as the x-value approaches 1. We conclude, accordingly, that

$$\lim_{x \to 1} \frac{x^3 - 1}{x - 1} = 3$$

c. We compute

$$\lim_{x \to 1} \frac{x^3 - 1}{x - 1} = \lim_{x \to 1} \frac{(x - 1)(x^2 + x + 1)}{x - 1}$$
$$= \lim_{x \to 1} (x^2 + x + 1) = 3$$

REMARK If you attempt to find the limit in Example 1 by using the evaluation function of your graphing utility to find the value of $f(x)$ when $x = 1$, you will see that the graphing utility does not display the y-value. This happens because the point $x = 1$ is not in the domain of f.

630

EXAMPLE 2 Use **ZOOM-IN** to find $\lim\limits_{x \to 0} (1 + x)^{1/x}$.

Solution We first plot the graph of $f(x) = (1 + x)^{1/x}$ in a suitable viewing rectangle. Figure T2 shows a plot of f in the rectangle $[-1, 1] \times [0, 4]$. Using **ZOOM-IN** repeatedly, we see that $\lim\limits_{x \to 0} (1 + x)^{1/x} \approx 2.71828$.

Figure T2
The graph of $f(x) = (1 + x)^{1/x}$ in the viewing rectangle $[-1, 1] \times [0, 4]$.

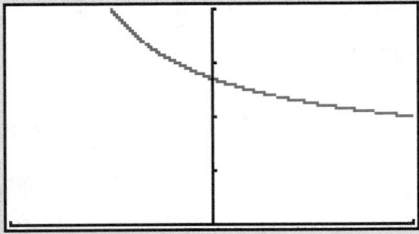

The limit of $f(x) = (1 + x)^{1/x}$ as x approaches zero, denoted by the letter e, plays a very important role in the study of mathematics and its applications (see Section 13.5). Thus

$$\lim_{x \to 0} (1 + x)^{1/x} = e$$

where, as we have just seen, $e \approx 2.71828$. ◦ ◦ ◦

EXAMPLE 3 When organic waste is dumped into a pond, the oxidation process that takes place reduces the pond's oxygen content. However, given time, nature will restore the oxygen content to its natural level. Suppose that the oxygen content t days after the organic waste has been dumped into the pond is given by

$$f(t) = 100 \left(\frac{t^2 + 10t + 100}{t^2 + 20t + 100} \right)$$

percent of its normal level.

a. Plot the graph of f in the viewing rectangle $[0, 200] \times [70, 100]$.

b. What can you say about $f(t)$ when t is very large?

c. Verify your observation in (b) by evaluating $\lim\limits_{t \to \infty} f(t)$.

Solution

a. The graph of f is shown in Figure T3.

b. From the graph of f it appears that $f(t)$ approaches 100 steadily as t gets

Figure T3
The graph of f in the viewing rectangle [0, 200] × [70, 100].

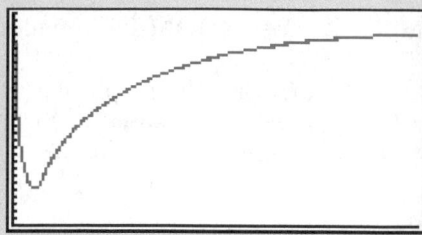

larger and larger. This observation tells us that eventually the oxygen content of the pond will be restored to its natural level.

c. To verify the observation made in (b), we compute

$$\lim_{t \to \infty} f(t) = \lim_{t \to \infty} 100 \left(\frac{t^2 + 10t + 100}{t^2 + 20t + 100} \right)$$

$$= 100 \lim_{t \to \infty} \left(\frac{1 + \dfrac{10}{t} + \dfrac{100}{t^2}}{1 + \dfrac{20}{t} + \dfrac{100}{t^2}} \right) = 100$$

⊙ ⊙ ⊙

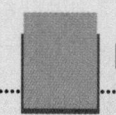 **EXERCISES**

In exercises 1–10, use a graphing utility to find the indicated limit by first plotting the graph of the function in a suitable viewing rectangle and then using the ZOOM-IN feature of the calculator.

1. $\displaystyle\lim_{x \to 1} \frac{2x^3 - 2x^2 + 3x - 3}{x - 1}$

2. $\displaystyle\lim_{x \to -2} \frac{2x^3 + 3x^2 - x + 2}{x + 2}$

3. $\displaystyle\lim_{x \to -1} \frac{x^3 + 1}{x + 1}$

4. $\displaystyle\lim_{x \to -1} \frac{x^4 - 1}{x - 1}$

5. $\displaystyle\lim_{x \to 1} \frac{x^3 - x^2 - x + 1}{x^3 - 3x + 2}$

6. $\displaystyle\lim_{x \to 2} \frac{x^3 + 2x^2 - 16}{2x^3 - x^2 + 2x - 16}$

7. $\displaystyle\lim_{x \to 0} \frac{\sqrt{x + 1} - 1}{x}$

8. $\displaystyle\lim_{x \to 0} \frac{(x + 4)^{3/2} - 8}{x}$

9. $\displaystyle\lim_{x \to 0} (1 + 2x)^{1/x}$

10. $\displaystyle\lim_{x \to 0} \frac{2^x - 1}{x}$

11. Use a graphing utility to show that $\displaystyle\lim_{x \to 3} \frac{2}{x - 3}$ does not exist.

12. Use a graphing utility to show that $\displaystyle\lim_{x \to 2} \frac{x^3 - 2x + 1}{x - 2}$ does not exist.

13. **City Planning** A major developer is building a 5000-acre complex of homes, offices, stores, schools, and churches in the rural community of Marlboro. As a result of this development, the planners have estimated that Marlboro's population (in thousands) t years from now will be given by

$$P(t) = \frac{25t^2 + 125t + 200}{t^2 + 5t + 40}$$

a. Plot the graph of P in the viewing rectangle [0, 50] × [0, 30].

b. What will the population of Marlboro be in the long run?

[*Hint:* Find $\displaystyle\lim_{t \to \infty} P(t)$.]

a. What is the average cost of driving a subcompact car 5000 miles a year? 10,000 miles a year? 15,000 miles a year? 20,000 miles a year? 25,000 miles a year?
b. Use (a) to help sketch the graph of the function C.
c. What happens to the average cost as the number of miles driven increases without bound?

88. **Speed of a Chemical Reaction** Certain proteins, known as enzymes, serve as catalysts for chemical reactions in living things. In 1913, Leonor Michaelis and L. M. Menten discovered the following formula giving the initial speed V (in moles per liter per second) at which the reaction begins in terms of the amount of substrate x (the substance being acted upon, measured in moles per liter):

$$V = \frac{ax}{x+b}$$

where a and b are positive constants. Evaluate

$$\lim_{x \to \infty} \frac{ax}{x+b}$$

and interpret your result.

89. Show by means of an example that $\lim_{x \to a} [f(x) + g(x)]$ may exist even though neither $\lim_{x \to a} f(x)$ nor $\lim_{x \to a} g(x)$ exists. Does this example contradict Theorem 1?

90. Show by means of an example that $\lim_{x \to a} [f(x)g(x)]$ may exist even though neither $\lim_{x \to a} f(x)$ nor $\lim_{x \to a} g(x)$ exists. Does this example contradict Theorem 1?

SOLUTIONS TO SELF-CHECK EXERCISES 10.4

1. a. $\lim_{x \to 3} \dfrac{\sqrt{x^2+7} + \sqrt{3x-5}}{x+2} = \dfrac{\sqrt{9+7}+\sqrt{3(3)-5}}{3+2}$

$= \dfrac{\sqrt{16}+\sqrt{4}}{5}$

$= \dfrac{6}{5}$

b. Letting x approach -1 leads to the indeterminate form $0/0$. Thus we proceed as follows:

$\lim_{x \to -1} \dfrac{x^2-x-2}{2x^2-x-3} = \lim_{x \to -1} \dfrac{(x+1)(x-2)}{(x+1)(2x-3)}$

$= \lim_{x \to -1} \dfrac{x-2}{2x-3}$ (Canceling the common factors)

$= \dfrac{-1-2}{2(-1)-3}$

$= \dfrac{3}{5}$

2. $\lim_{x \to \infty} \overline{C}(x) = \lim_{x \to \infty} \left(1.8 + \dfrac{3000}{x}\right)$

$= \lim_{x \to \infty} 1.8 + \lim_{x \to \infty} \dfrac{3000}{x}$

$= 1.8$

Our computation reveals that, as the production of audio discs increases "without bound," the average cost drops and approaches a unit cost of $1.80 per disc.

10.5 ONE-SIDED LIMITS AND CONTINUITY

One-Sided Limits

Figure 10.28
The function f does not have a limit as x approaches zero.

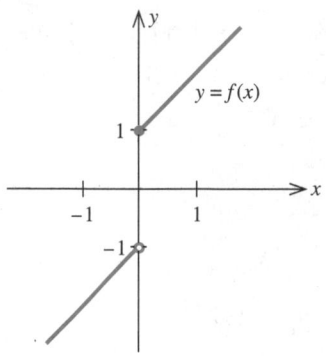

Consider the function f defined by

$$f(x) = \begin{cases} x - 1 & \text{if } x < 0 \\ x + 1 & \text{if } x \geq 0 \end{cases}$$

From the graph of f shown in Figure 10.28, we see that the function f does not have a limit as x approaches zero because no matter how close x is to zero, $f(x)$ takes on values that are close to 1 if x is positive and values that are close to -1 if x is negative. Therefore, $f(x)$ cannot be close to a single number L—no matter how close x is to zero. Now, if we restrict x to be greater than zero (to the right of zero), then we see that $f(x)$ can be made as close to the number 1 as we please by taking x sufficiently close to zero. In this situation, we say that the right-hand limit of f as x approaches zero (from the right) is 1, written

$$\lim_{x \to 0^+} f(x) = 1$$

Similarly, we see that $f(x)$ can be made as close to the number -1 as we please by taking x sufficiently close to, but to the left of, zero. In this situation, we say that the left-hand limit of f as x approaches zero (from the left) is -1, written

$$\lim_{x \to 0^-} f(x) = -1$$

These limits are called **one-sided limits.** More generally, we have the following definitions.

ONE-SIDED LIMITS

The function f has the **right-hand limit** L as x approaches a from the right, written

$$\lim_{x \to a^+} f(x) = L$$

if the values $f(x)$ can be made as close to L as we please by taking x sufficiently close to (but not equal to) a and to the right of a.

Similarly, the function f has the **left-hand limit** M as x approaches a from the left, written

$$\lim_{x \to a^-} f(x) = M$$

if the values $f(x)$ can be made as close to M as we please by taking x sufficiently close to (but not equal to) a and to the left of a.

The connection between one-sided limits and the two-sided limit defined earlier is given by the following theorem.

THEOREM 3

Let f be a function that is defined for all values of x close to $x = a$ with the possible exception of a itself. Then

$$\lim_{x \to a} f(x) = L \quad \text{if and only if} \quad \lim_{x \to a^+} f(x) = \lim_{x \to a^-} f(x) = L$$

Thus, the two-sided limit exists if and only if the one-sided limits exist and are equal.

EXAMPLE I   Let

$$f(x) = \begin{cases} \sqrt{x} & \text{if } x > 0 \\ -x & \text{if } x \le 0 \end{cases} \quad \text{and} \quad g(x) = \begin{cases} -1 & \text{if } x < 0 \\ 1 & \text{if } x \ge 0 \end{cases}$$

a. Show that $\lim_{x \to 0} f(x)$ exists by studying the one-sided limits of f as x approaches $x = 0$.

b. Show that $\lim_{x \to 0} g(x)$ does not exist.

Solution

a. We find

$$\lim_{x \to 0^+} f(x) = \lim_{x \to 0^+} \sqrt{x} = 0$$

and

$$\lim_{x \to 0^-} f(x) = \lim_{x \to 0^-} (-x) = 0$$

so

$$\lim_{x \to 0} f(x) = 0$$

See Figure 10.29a.

Figure 10.29

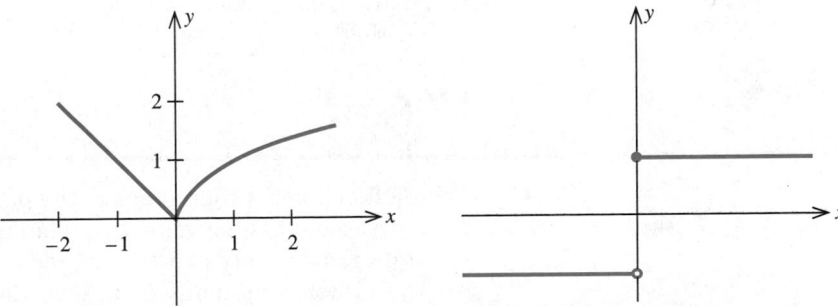

(a) $\lim_{x \to 0} f(x)$ exists.   **(b)** $\lim_{x \to 0} g(x)$ does not exist.

b. We have

$$\lim_{x \to 0^-} g(x) = -1 \quad \text{and} \quad \lim_{x \to 0^+} g(x) = 1$$

and since these one-sided limits are not equal, we conclude that $\lim_{x \to 0} g(x)$ does not exist (Figure 10.29b).   ○ ○ ○

Continuous Functions

Continuous functions will play an important role throughout most of our study of calculus. Loosely speaking, a function is continuous at a point if the graph of the function at that point is devoid of holes, gaps, jumps, or breaks. Consider, for example, the graph of the function f depicted in Figure 10.30.

Figure 10.30
The graph of this function is not continuous at x = a, x = b, x = c, and x = d.

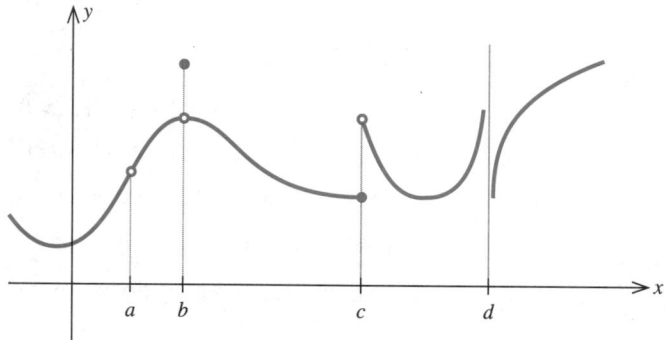

Let us take a closer look at the behavior of f at or near each of the points $x = a$, $x = b$, $x = c$, and $x = d$. First, note that f is not defined at $x = a$; that is, the point $x = a$ is not in the domain of f, thereby resulting in a "hole" in the graph of f. Next, observe that the value of f at b, $f(b)$, is not equal to the limit of $f(x)$ as x approaches b, resulting in a "jump" in the graph of f at that point. The function f does not have a limit at $x = c$ since the left-hand and right-hand limits of $f(x)$ are not equal, also resulting in a jump in the graph of f at that point. Finally, the limit of f does not exist at $x = d$, resulting in a break in the graph of f. The function f is *discontinuous* at each of these points. It is *continuous* at all other points.

CONTINUITY AT A POINT	A function f is **continuous at the point** $x = a$ if the following conditions are satisfied:

1. $f(a)$ is defined.   **2.** $\lim_{x \to a} f(x)$ exists.   **3.** $\lim_{x \to a} f(x) = f(a)$.

Thus, a function f is continuous at the point $x = a$ if the limit of f at the point $x = a$ exists and has the value $f(a)$. Geometrically, f is continuous at the point $x = a$ if the proximity of x to a implies the proximity of $f(x)$ to $f(a)$.

If f is not continuous at $x = a$, then f is said to be **discontinuous** at $x = a$. Also, f is **continuous on an interval** if f is continuous at every point in the interval.

Figure 10.31 depicts the graph of a continuous function on the interval (a, b). Notice that the graph of the function over the stated interval can be sketched without lifting one's pencil from the paper.

Figure 10.31
The graph of f is continuous on the interval (a, b).

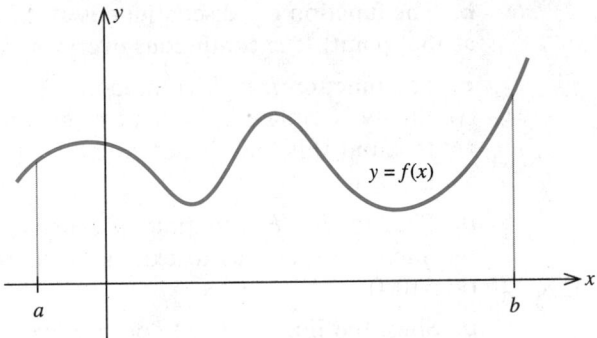

EXAMPLE 2   Find the values of x for which each of the following functions is continuous.

a. $f(x) = x + 2$ **b.** $g(x) = \dfrac{x^2 - 4}{x - 2}$ **c.** $h(x) = \begin{cases} x + 2 & \text{if } x \neq 2 \\ 1 & \text{if } x = 2 \end{cases}$

d. $F(x) = \begin{cases} -1 & \text{if } x < 0 \\ 1 & \text{if } x \geq 0 \end{cases}$ **e.** $G(x) = \begin{cases} \dfrac{1}{x} & \text{if } x > 0 \\ -1 & \text{if } x \leq 0 \end{cases}$

The graph of each function is shown in Figure 10.32.

Figure 10.32

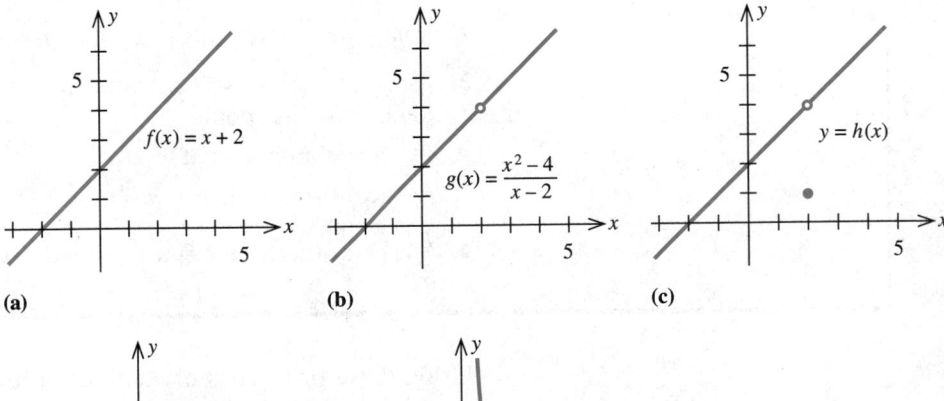

(a)   (b)   (c)

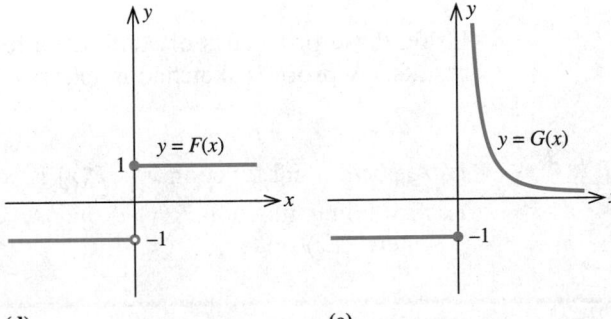

(d)   (e)

Solution

a. The function f is continuous everywhere because the three conditions for continuity are satisfied for all values of x.

b. The function g is discontinuous at the point $x = 2$ because g is not defined at that point. It is continuous everywhere else.

c. The function h is discontinuous at $x = 2$ because the third condition for continuity is violated; the limit of $h(x)$ as x approaches 2 exists and has the value 4, but this limit is not equal to $h(2) = 1$. It is continuous for all other values of x.

d. The function F is continuous everywhere except at the point $x = 0$, where the limit of $F(x)$ fails to exist as x approaches zero (see Example 3a, Section 10.4).

e. Since the limit of $G(x)$ does not exist as x approaches zero, we conclude that G fails to be continuous at $x = 0$. The function G is continuous at all other points.   ⊙ ⊙ ⊙

Properties of Continuous Functions

The following properties of continuous functions follow directly from the definition of continuity and the corresponding properties of limits. They are stated without proof.

PROPERTIES OF CONTINUOUS FUNCTIONS	**1.** The constant function $f(x) = c$ is continuous everywhere.

2. The identity function $f(x) = x$ is continuous everywhere.

If f and g are continuous at x = a, then

3. $[f(x)]^n$, where n is a real number, is continuous at $x = a$ whenever it is defined at that point.

4. $f \pm g$ is continuous at $x = a$.

5. fg is continuous at $x = a$.

6. $\dfrac{f}{g}$ is continuous at $x = a$ provided $g(a) \neq 0$.

Using these properties of continuous functions, we can prove the following results. (A proof is sketched in exercise 98, page 649.)

CONTINUITY OF POLYNOMIAL AND RATIONAL FUNCTIONS	**1.** A polynomial function $y = P(x)$ is continuous at every point x.

2. A rational function $R(x) = p(x)/q(x)$ is continuous at every point x where $q(x) \neq 0$.

EXAMPLE 3   Find the values of x for which each of the following functions is continuous.

a. $f(x) = 3x^3 + 2x^2 - x + 10$ **b.** $g(x) = \dfrac{8x^{10} - 4x + 1}{x^2 + 1}$

c. $h(x) = \dfrac{4x^3 - 3x^2 + 1}{x^2 - 3x + 2}$

Solution

a. The function f is a polynomial function of degree 3, so $f(x)$ is continuous for all values of x.

b. The function g is a rational function. Observe that the denominator of g—namely, $x^2 + 1$—is never equal to zero. Therefore, we conclude that g is continuous for all values of x.

c. The function h is a rational function. In this case, however, the denominator of h is equal to zero at $x = 1$ and $x = 2$, which can be seen by factoring it. Thus,

$$x^2 - 3x + 2 = (x - 2)(x - 1)$$

We conclude, therefore, that h is continuous everywhere except at $x = 1$ and $x = 2$, where it is discontinuous.   ◐ ◐ ◐

Application

Up to this point, most of the applications we have discussed involved functions that are continuous everywhere. In Example 4 we consider an application from the field of educational psychology that involves a discontinuous function.

EXAMPLE 4   Figure 10.33 depicts the learning curve associated with a certain individual. Beginning with no knowledge of the subject being taught, the individual makes steady progress toward understanding

Figure 10.33
A learning curve that is discontinuous at $t = t_1$.

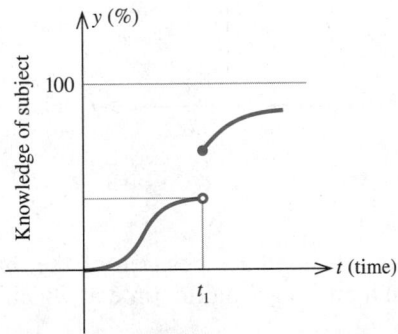

it over the time interval $0 \le t < t_1$. In this instance, the individual's progress slows down as we approach time t_1 because he fails to grasp a particularly difficult concept. All of a sudden, a breakthrough occurs at time t_1, propelling his knowledge of the subject to a higher level. The curve is discontinuous at t_1.

○ ○ ○

Intermediate Value Theorem

Let's look again at our model of the motion of the maglev on a straight stretch of track. We know that the train cannot vanish at any instant of time and it cannot skip portions of the track and reappear someplace else. To put it another way, the train cannot occupy the positions s_1 and s_2 without at least, at some time, occupying an intermediate position (Figure 10.34).

Figure 10.34
The position of the maglev.

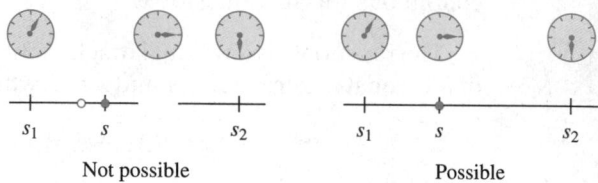

In order to state this fact mathematically, recall that the position of the maglev as a function of time is described by

$$f(t) = 4t^2 \qquad (0 \le t \le 10)$$

Suppose the position of the maglev is s_1 at some time t_1 and its position is s_2 at some time t_2 (Figure 10.35). Then, if s_3 is any number between s_1 and s_2

Figure 10.35
If $s_1 \le s_3 \le s_2$, then there must be at least one t_3 $(t_1 \le t_3 \le t_2)$ such that $f(t_3) = s_3$.

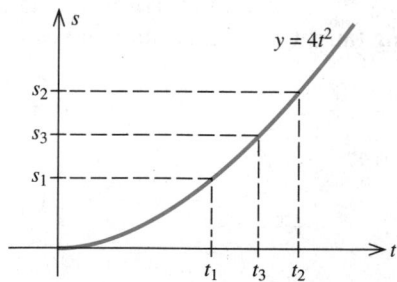

giving an intermediate position of the maglev, there must be at least one t_3 between t_1 and t_2 giving the time at which the train is at s_3—that is, $f(t_3) = s_3$.

This discussion carries the gist of the Intermediate Value Theorem. The proof of this theorem can be found in most advanced calculus texts.

THEOREM 4

The Intermediate Value Theorem

If f is a continuous function on a closed interval $[a, b]$ and M is any number between $f(a)$ and $f(b)$, then there is at least one number c in $[a, b]$ such that $f(c) = M$ (see Figure 10.36).

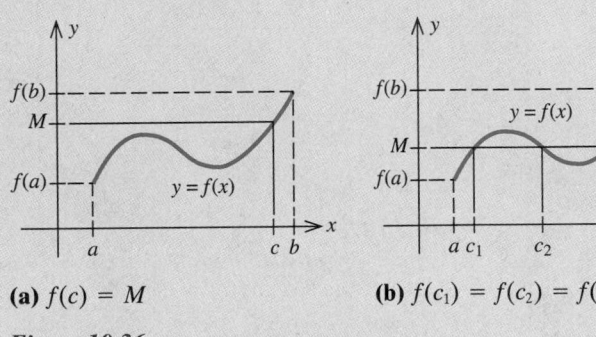

(a) $f(c) = M$

(b) $f(c_1) = f(c_2) = f(c_3) = M$

Figure 10.36

To illustrate the Intermediate Value Theorem, let's look at the example involving the motion of the maglev again (see Figure 10.18, page 610). Notice that the initial position of the train is $f(0) = 0$, and the position at the end of its test run is $f(10) = 400$. Furthermore, the function f is continuous on $[0, 10]$. So, the Intermediate Value Theorem guarantees that if we arbitrarily pick a number between 0 and 400, say 100, giving the position of the maglev, then there must be a $\bar{t}$ (read "t bar") between 0 and 10 at which time the train is at the position $s = 100$. To find the value of $\bar{t}$, we solve the equation $f(\bar{t}) = s$, or

$$4\bar{t}^2 = 100$$

giving $\bar{t} = 5$. (t must lie between 0 and 10.)

 It is important to remember when we use Theorem 4 that the function f must be continuous. The conclusion of the Intermediate Value Theorem may not hold if f is not continuous (see exercise 99, page 649).

The next theorem is an immediate consequence of the Intermediate Value Theorem. It not only tells us when a zero of a function f [root of the equation $f(x) = 0$] exists but also provides the basis for a method of approximating it.

THEOREM 5

Existence of Zeros of a Continuous Function

If f is a continuous function on a closed interval $[a, b]$, and if $f(a)$ and $f(b)$ have opposite signs, then there is at least one solution of the equation $f(x) = 0$ in the interval (a, b). (See Figure 10.37.)

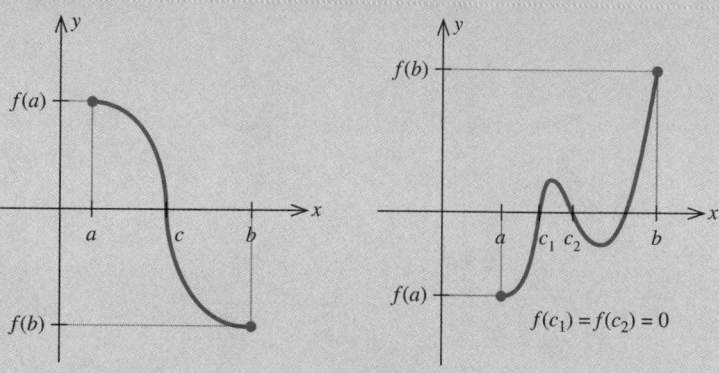

Figure 10.37
If $f(a)$ and $f(b)$ have opposite signs, there must be at least one number c ($a < c < b$) such that $f(c) = 0$.

Figure 10.38
$f(a) < 0$ and $f(b) > 0$, but the graph of f does not cross the x-axis between a and b because f is discontinuous.

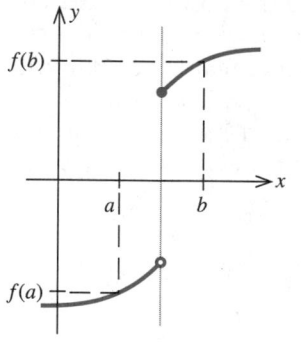

Geometrically, this property states that if the graph of a continuous function goes from above the x-axis to below the x-axis, or vice versa, it must *cross* the x-axis. This is not necessarily true if the function is discontinuous (Figure 10.38).

EXAMPLE 5   Let $f(x) = x^3 + x + 1$.

a. Show that f is continuous for all values of x.

b. Compute $f(-1)$ and $f(1)$ and use the results to deduce that there must be at least one point $x = c$, where c lies in the interval $(-1, 1)$ and $f(c) = 0$.

Solution

a. The function f is a polynomial function of degree 3 and is therefore continuous everywhere.

b. $f(-1) = (-1)^3 + (-1) + 1 = -1$
    $f(1) = 1^3 + 1 + 1 = 3$

Since $f(-1)$ and $f(1)$ have opposite signs, Theorem 5 tells us that there must be at least one point $x = c$ with $-1 < c < 1$ such that $f(c) = 0$.   ○ ○ ○

The next example shows how the Intermediate Value Theorem can be used to help us find the zero of a function.

EXAMPLE 6 Let $f(x) = x^3 + x - 1$. Since f is a polynomial function it is continuous everywhere. Observe that $f(0) = -1$ and $f(1) = 1$ so that Theorem 5 guarantees the existence of at least one root of the equation $f(x) = 0$ in $(0, 1)$.*

We can locate the root more precisely by using Theorem 5 once again as follows: Evaluate $f(x)$ at the midpoint of $[0, 1]$, obtaining

$$f(0.5) = -0.375$$

Because $f(0.5) < 0$ and $f(1) > 0$, Theorem 5 now tells us that the root must lie in $(0.5, 1)$.

Repeat the process: Evaluate $f(x)$ at the midpoint of $[0.5, 1]$, which is

$$\frac{0.5 + 1}{2} = 0.75$$

Thus,

$$f(0.75) = 0.1719$$

Because $f(0.5) < 0$ and $f(0.75) > 0$, Theorem 5 tells us that the root is in $(0.5, 0.75)$. This process can be continued. Table 10.3 summarizes the results of our computations through nine steps.

From Table 10.3 we see that the root is approximately 0.68, accurate to two decimal places. By continuing the process through a sufficient number of steps, we can obtain as accurate an approximation to the root as we please.

 o o o

Table 10.3

Step	Root of $f(x) = 0$ Lies In
1	(0, 1)
2	(0.5, 1)
3	(0.5, 0.75)
4	(0.625, 0.75)
5	(0.625, 0.6875)
6	(0.65625, 0.6875)
7	(0.671875, 0.6875)
8	(0.6796875, 0.6875)
9	(0.6796875, 0.6835937)

REMARK   The process of finding the root of $f(x) = 0$ used in Example 6 is called the **method of bisection.** It is crude but effective.   o o o

SELF-CHECK EXERCISES 10.5

1. Evaluate $\lim\limits_{x \to -1^-} f(x)$ and $\lim\limits_{x \to -1^+} f(x)$, where

$$f(x) = \begin{cases} 1 & \text{if } x < -1 \\ 1 + \sqrt{x + 1} & \text{if } x \geq -1 \end{cases}$$

Does $\lim\limits_{x \to -1} f(x)$ exist?

o o o

* It can be shown that f has precisely one zero in $(0, 1)$ (see exercise 90, Section 12.1).

2. Determine the values of x for which the given function is discontinuous. At each point of discontinuity, indicate which condition(s) for continuity are violated. Sketch the graph of the function.

a. $f(x) = \begin{cases} -x^2 + 1 & \text{if } x \leq 1 \\ x - 1 & \text{if } x > 1 \end{cases}$

b. $g(x) = \begin{cases} -x + 1 & \text{if } x < -1 \\ 2 & \text{if } -1 < x \leq 1 \\ -x + 3 & \text{if } x > 1 \end{cases}$

Solutions to Self-Check Exercises 10.5 can be found on page 649.

10.5 EXERCISES

In exercises 1–8, use the graph of the function f to find $\lim_{x \to a^-} f(x), \lim_{x \to a^+} f(x),$ *and* $\lim_{x \to a} f(x)$ *at the indicated value of a, if the limit exists.*

1.

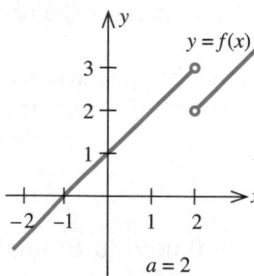

$a = 2$

2.

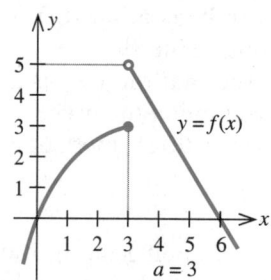

$a = 3$

3.

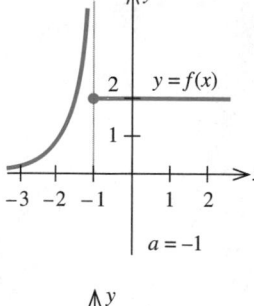

$a = -1$

4.

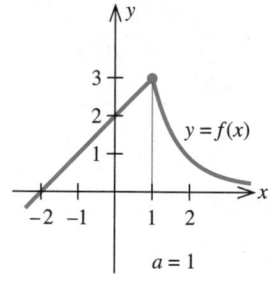

$a = 1$

5.

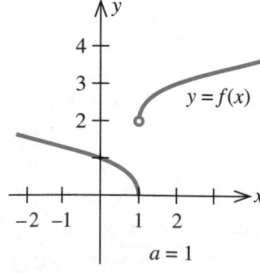

$a = 1$

6.

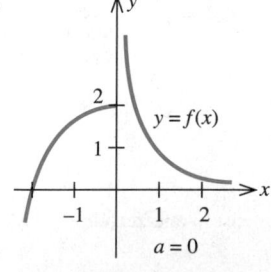

$a = 0$

7.

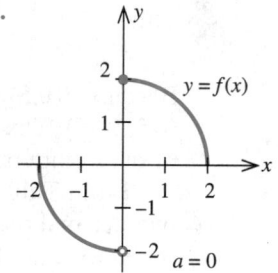

$a = 0$

8.

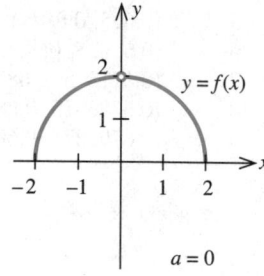

$a = 0$

In exercises 9–14, refer to the graph of the function f and determine whether each statement is true or false.

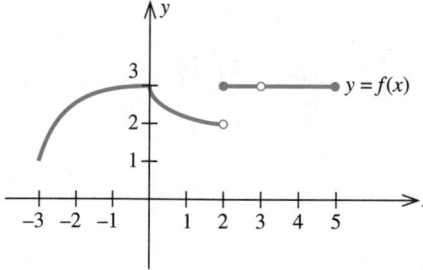

9. $\lim_{x \to -3^+} f(x) = 1$

10. $\lim_{x \to 0} f(x) = f(0)$

11. $\lim_{x \to 2^-} f(x) = 2$

12. $\lim_{x \to 2^+} f(x) = 3$

13. $\lim_{x \to 3} f(x)$ does not exist.

14. $\lim_{x \to 5^-} f(x) = 3$

In exercises 15–20, refer to the graph of the function f and determine whether each statement is true or false.

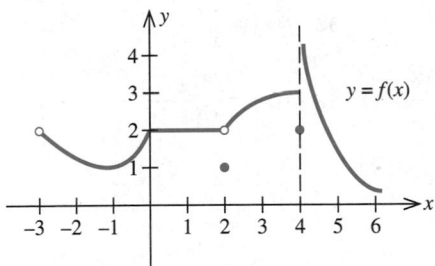

15. $\lim\limits_{x \to -3^+} f(x) = 2$

16. $\lim\limits_{x \to 0} f(x) = 2$

17. $\lim\limits_{x \to 2} f(x) = 1$

18. $\lim\limits_{x \to 4^-} f(x) = 3$

19. $\lim\limits_{x \to 4^+} f(x)$ does not exist.

20. $\lim\limits_{x \to 4} f(x) = 2$

In exercises 21–42 find the indicated one-sided limit, if it exists.

21. $\lim\limits_{x \to 1^+} (2x + 4)$

22. $\lim\limits_{x \to 1^-} (3x - 4)$

23. $\lim\limits_{x \to 2^-} \dfrac{x - 3}{x + 2}$

24. $\lim\limits_{x \to 1^+} \dfrac{x + 2}{x + 1}$

25. $\lim\limits_{x \to 0^+} \dfrac{1}{x}$

26. $\lim\limits_{x \to 0^-} \dfrac{1}{x}$

27. $\lim\limits_{x \to 0^+} \dfrac{x - 1}{x^2 + 1}$

28. $\lim\limits_{x \to 2^+} \dfrac{x + 1}{x^2 - 2x + 3}$

29. $\lim\limits_{x \to 0^+} \sqrt{x}$

30. $\lim\limits_{x \to 2^+} 2\sqrt{x - 2}$

31. $\lim\limits_{x \to -2^+} (2x + \sqrt{2 + x})$

32. $\lim\limits_{x \to -5^+} x(1 + \sqrt{5 + x})$

33. $\lim\limits_{x \to 1^-} \dfrac{1 + x}{1 - x}$

34. $\lim\limits_{x \to 1^+} \dfrac{1 + x}{1 - x}$

35. $\lim\limits_{x \to 2^-} \dfrac{x^2 - 4}{x - 2}$

36. $\lim\limits_{x \to -3^+} \dfrac{\sqrt{x + 3}}{x^2 + 1}$

37. $\lim\limits_{x \to 3^+} \dfrac{x^2 - 9}{x + 3}$

38. $\lim\limits_{x \to -2^-} \dfrac{\sqrt[3]{x + 10}}{2x^2 + 1}$

39. $\lim\limits_{x \to 0^+} f(x)$ and $\lim\limits_{x \to 0^-} f(x)$, where

$$f(x) = \begin{cases} 2x & \text{if } x < 0 \\ x^2 & \text{if } x \geq 0 \end{cases}$$

40. $\lim\limits_{x \to 0^+} f(x)$ and $\lim\limits_{x \to 0^-} f(x)$, where

$$f(x) = \begin{cases} -x + 1 & \text{if } x \leq 0 \\ 2x + 3 & \text{if } x > 0 \end{cases}$$

41. $\lim\limits_{x \to 1^+} f(x)$ and $\lim\limits_{x \to 1^-} f(x)$, where

$$f(x) = \begin{cases} \sqrt{x + 3} & \text{if } x \geq 1 \\ 2 + \sqrt{x} & \text{if } x < 1 \end{cases}$$

42. $\lim\limits_{x \to 1^+} f(x)$ and $\lim\limits_{x \to 1^-} f(x)$, where

$$f(x) = \begin{cases} x + 2\sqrt{x - 1} & \text{if } x \geq 1 \\ 1 - \sqrt{1 - x} & \text{if } x < 1 \end{cases}$$

In exercises 43–50, determine the values of x, if any, at which each function is discontinuous. At each point of discontinuity, state the condition(s) for continuity that are violated.

43.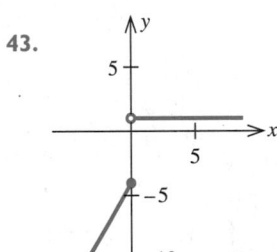

$$f(x) = \begin{cases} 2x - 4 & \text{if } x \leq 0 \\ 1 & \text{if } x > 0 \end{cases}$$

44.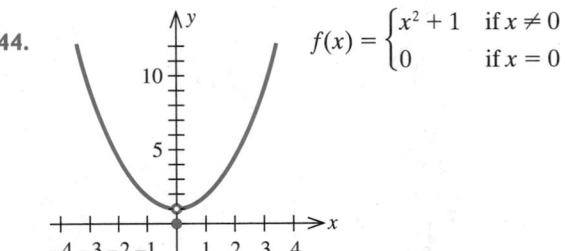

$$f(x) = \begin{cases} x^2 + 1 & \text{if } x \neq 0 \\ 0 & \text{if } x = 0 \end{cases}$$

45.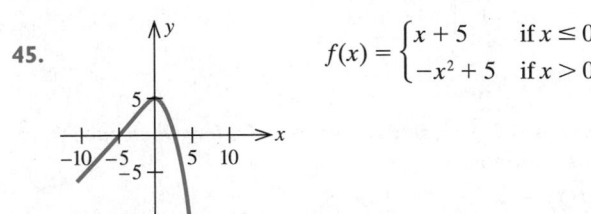

$$f(x) = \begin{cases} x + 5 & \text{if } x \leq 0 \\ -x^2 + 5 & \text{if } x > 0 \end{cases}$$

46.

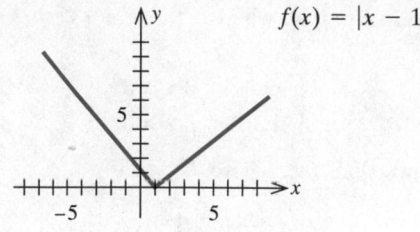

$$f(x) = |x - 1|$$

47.

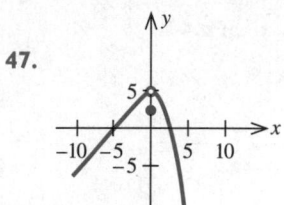

$$f(x) = \begin{cases} x + 5 & \text{if } x < 0 \\ 2 & \text{if } x = 0 \\ -x^2 + 5 & \text{if } x > 0 \end{cases}$$

48.

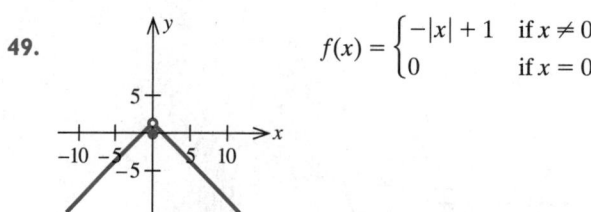

$$f(x) = \begin{cases} \dfrac{x^2 - 1}{x + 1} & \text{if } x \neq -1 \\ 1 & \text{if } x = -1 \end{cases}$$

49.

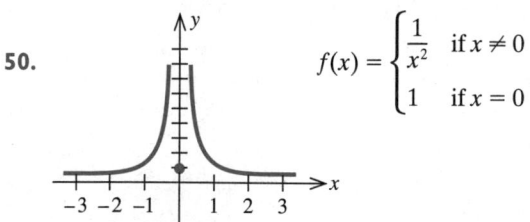

$$f(x) = \begin{cases} -|x| + 1 & \text{if } x \neq 0 \\ 0 & \text{if } x = 0 \end{cases}$$

50.

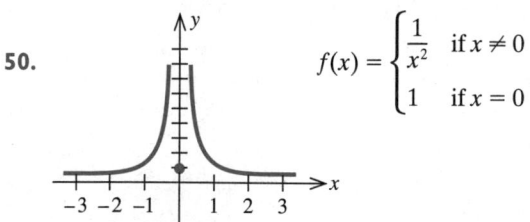

$$f(x) = \begin{cases} \dfrac{1}{x^2} & \text{if } x \neq 0 \\ 1 & \text{if } x = 0 \end{cases}$$

In exercises 51–66, find the values of x for which each function is continuous.

51. $f(x) = 2x^2 + x - 1$

52. $f(x) = x^3 - 2x^2 + x - 1$

53. $f(x) = \dfrac{2}{x^2 + 1}$ **54.** $f(x) = \dfrac{x}{2x^2 + 1}$

55. $f(x) = \dfrac{2}{2x - 1}$ **56.** $f(x) = \dfrac{x + 1}{x - 1}$

57. $f(x) = \dfrac{2x + 1}{x^2 + x - 2}$ **58.** $f(x) = \dfrac{x - 1}{x^2 + 2x - 3}$

59. $f(x) = \begin{cases} x & \text{if } x \leq 1 \\ 2x - 1 & \text{if } x > 1 \end{cases}$

60. $f(x) = \begin{cases} -x + 1 & \text{if } x \leq -1 \\ x + 1 & \text{if } x > -1 \end{cases}$

61. $f(x) = \begin{cases} -2x + 1 & \text{if } x < 0 \\ x^2 + 1 & \text{if } x \geq 0 \end{cases}$

62. $f(x) = \begin{cases} x + 1 & \text{if } x \leq 1 \\ -x^2 + 1 & \text{if } x > 1 \end{cases}$

63. $f(x) = \begin{cases} \dfrac{x^2 - 1}{x - 1} & \text{if } x \neq 1 \\ 2 & \text{if } x = 1 \end{cases}$

64. $f(x) = \begin{cases} \dfrac{x^2 - 4}{x + 2} & \text{if } x \neq -2 \\ 1 & \text{if } x = -2 \end{cases}$

65. $f(x) = |x + 1|$ **66.** $f(x) = \dfrac{|x - 1|}{x - 1}$

In exercises 67–70 determine all values of x at which the function is discontinuous.

67. $f(x) = \dfrac{2x}{x^2 - 1}$ **68.** $f(x) = \dfrac{1}{(x - 1)(x - 2)}$

69. $f(x) = \dfrac{x^2 - 2x}{x^2 - 3x + 2}$ **70.** $f(x) = \dfrac{x^2 - 3x + 2}{x^2 - 2x}$

71. The Postage Function The graph of the "postage function"

$$f(x) = \begin{cases} 32 & \text{if } 0 < x \leq 1 \\ 55 & \text{if } 1 < x \leq 2 \\ \vdots \\ 285 & \text{if } 11 < x \leq 12 \end{cases}$$

where x denotes the weight of a parcel in ounces and $f(x)$ the postage in cents, is shown in the accompanying figure. Determine the values of x for which f is discontinuous.

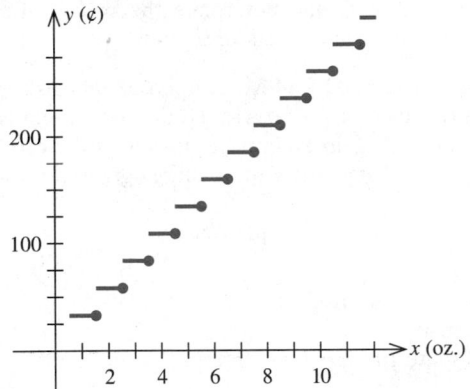

72. Inventory Control As part of an optimal inventory policy, the manager of an office supply company orders 500 reams of xerographic copy paper every 20 days. The accompanying graph shows the *actual* inventory level of copy paper in an office supply store during the first 60 business days of 1999. Determine the values of *t* for which the "inventory function" is discontinuous and give an interpretation of the graph.

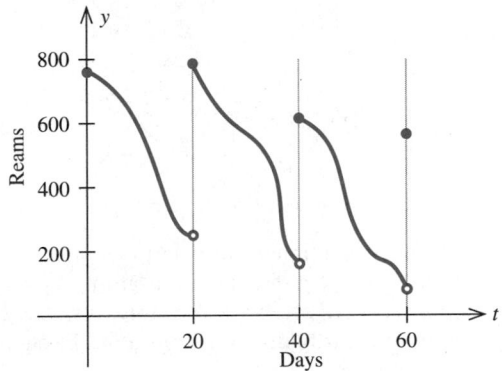

73. Learning Curves The following graph describes the progress Michael made in solving a problem correctly during a mathematics quiz. Here *y* denotes the percentage of work completed and *x* is measured in minutes. Give an interpretation of the graph.

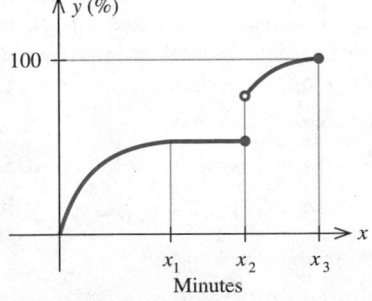

74. Ailing Financial Institutions The Franklin Savings and Loan Company acquired two ailing financial institutions in 1992. One of them was acquired at time $t = T_1$, and the other was acquired at time $t = T_2$ ($t = 0$ corresponds to the beginning of 1992). The graph shows the total amount of money on deposit with Franklin. Explain the significance of the discontinuities of the function at T_1 and T_2.

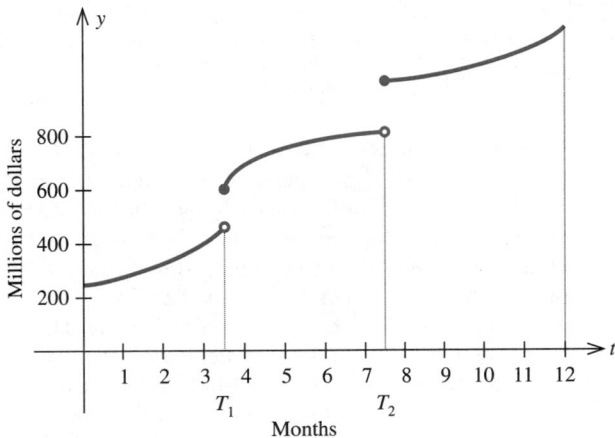

75. Energy Consumption The following graph shows the amount of home heating oil remaining in a 200-gallon tank over a 120-day period ($t = 0$ corresponds to October 1). Explain why the function is discontinuous at $t = 40$, $t = 70$, $t = 95$, and $t = 110$.

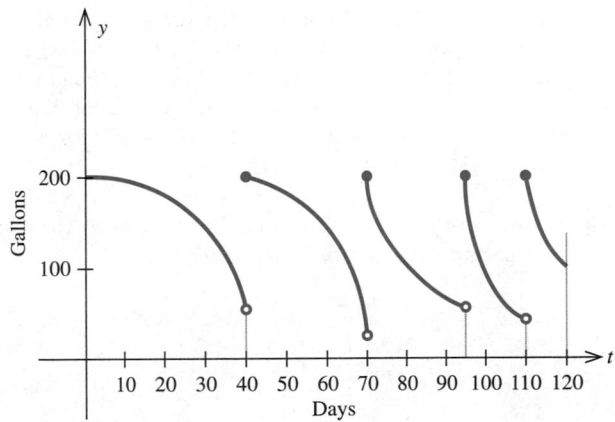

76. Prime Interest Rate The function P shown in the graph gives the prime rate (the interest rate banks charge their best corporate customers) as a function of time for the first 32 weeks in 1989. Determine the values of t for which P is discontinuous and interpret your results.

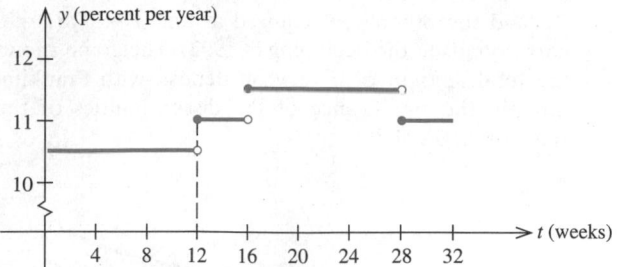

77. **Administration of an Intravenous Solution** A dextrose solution is being administered to a patient intravenously. The 1-liter bottle holding the solution is removed and replaced by another as soon as the contents drop to approximately 5% of the initial (1-liter) amount. The rate of discharge is constant, and it takes 6 hours to discharge 95% of the contents of a full bottle. Draw a graph showing the amount of dextrose solution in a bottle in the IV system over a 24-hour period, assuming that we started with a full bottle.

78. **Commissions** The base salary of a salesman working on commission is $12,000. For each $50,000 of sales beyond $100,000 he is paid a $1000 commission. Sketch a graph showing his earnings as a function of the level of his sales x. Determine the values of x for which the function f is discontinuous.

79. **Parking Fees** The fee charged per car in a downtown parking lot is $1 for the first half hour and 50 cents for each additional half hour or part thereof, subject to a maximum of $5. Derive a function f relating the parking fee to the length of time a car is left in the lot. Sketch the graph of f and determine the values of x for which the function f is discontinuous.

80. **Commodity Prices** The function that gives the cost of a certain commodity is defined by

$$C(x) = \begin{cases} 5x & \text{if } 0 < x < 10 \\ 4x & \text{if } 10 \le x < 30 \\ 3.5x & \text{if } 30 \le x < 60 \\ 3.25x & \text{if } x \ge 60 \end{cases}$$

where x is the number of pounds of a certain commodity sold and $C(x)$ is measured in dollars. Sketch the graph

of the function C and determine the values of x for which the function C is discontinuous.

81. **Energy Expended by a Fish** Suppose that a fish swimming a distance of L ft at a speed of v ft/sec relative to the water and against a current flowing at the rate of u ft/sec ($u < v$) expends a total energy given by

$$E(v) = \frac{aLv^3}{v - u}$$

where E is measured in foot-pounds (ft-lb) and a is a constant.
a. Evaluate $\lim_{v \to u^+} E(v)$ and interpret your result.
b. Evaluate $\lim_{v \to \infty} E(v)$ and interpret your result.

82. Let

$$f(x) = \begin{cases} x + 2 & \text{if } x \le 1 \\ kx^2 & \text{if } x > 1 \end{cases}$$

Find the value of k that will make f continuous on $(-\infty, \infty)$.

83. Let

$$f(x) = \begin{cases} \dfrac{x^2 - 4}{x + 2} & \text{if } x \ne -2 \\ k & \text{if } x = -2 \end{cases}$$

For what value of k will f be continuous on $(-\infty, \infty)$?

84. a. Suppose f is continuous at a and g is discontinuous at a. Is the sum $f + g$ discontinuous at a? Explain.
b. Suppose f and g are both discontinuous at a. Is the sum $f + g$ necessarily discontinuous at a? Explain.

85. a. Suppose f is continuous at a and g is discontinuous at a. Is the product fg necessarily discontinuous at a? Explain.
b. Suppose f and g are both discontinuous at a. Is the product fg necessarily discontinuous at a? Explain.

In exercises 86–89, (a) show that the function f is continuous for all values of x in the interval [a, b] and (b) prove that f must have at least one zero in the interval (a, b) by showing that f(a) and f(b) have opposite signs.

86. $f(x) = x^2 - 6x + 8$; $a = 1$, $b = 3$

87. $f(x) = x^3 - 2x^2 + 3x + 2$; $a = -1$, $b = 1$

88. $f(x) = 2x^3 - 3x^2 - 36x + 14$; $a = 0$, $b = 1$

89. $f(x) = 2x^{5/3} - 5x^{4/3}$; $a = 14$, $b = 16$

In exercises 90 and 91, use the Intermediate Value Theorem to find the value of c such that f(c) = M.

90. $f(x) = x^2 - x + 1$ on $[-1, 4]$; $M = 7$

91. $f(x) = x^2 - 4x + 6$ on $[0, 3]$; $M = 2$

92. Use the method of bisection (see Example 6) to find the root of the equation $x^3 - x + 1 = 0$ accurate to two decimal places.

93. Use the method of bisection to find the root of the equation $x^5 + 2x - 7 = 0$ accurate to two decimal places.

94. Joan is looking straight out a window of an apartment building at a height of 32 ft from the ground. A boy throws a tennis ball straight up by the side of the building where the window is located. Suppose the height of the ball (measured in feet) from the ground at time t is $h(t) = 4 + 64t - 16t^2$.
a. Show that $h(0) = 4$ and $h(2) = 68$.
b. Use the Intermediate Value Theorem to conclude that the ball must cross Joan's line of sight at least once.
c. At what time(s) does the ball cross Joan's line of sight? Interpret your results.

95. True or false? Suppose the function f is defined on the interval $[a, b]$. If $f(a)$ and $f(b)$ have the same sign, then f has no zero in $[a, b]$. Explain your answer.

96. Let $f(x) = x - \sqrt{1 - x^2}$.
a. Show that f is continuous for all values of x in the interval $[-1, 1]$.
b. Show that f has at least one zero in $[-1, 1]$.
c. Find the zeros of f in $[-1, 1]$ by solving the equation $f(x) = 0$.

97. Let $f(x) = \dfrac{x^2}{x^2 + 1}$.
a. Show that f is continuous for all values of x.
b. Show that $f(x)$ is nonnegative for all values of x.
c. Show that f has a zero at $x = 0$. Does this contradict Theorem 5?

98. **a.** Prove that a polynomial function $y = P(x)$ is continuous at every point x. Follow these steps:
1. Use properties 2 and 3 of continuous functions (page 638) to establish that the function $g(x) = x^n$, where n is a positive integer, is continuous everywhere.
2. Use properties 1 and 5 to show that $f(x) = cx^n$, where c is a constant and n is a positive integer, is continuous everywhere.
3. Use property 4 to complete the proof of the result.
b. Prove that a rational function $R(x) = p(x)/q(x)$ is continuous at every point x, where $q(x) \neq 0$. [*Hint:* Use the result of (a) and property 6.]

99. Show that the conclusion of the Intermediate Value Theorem does not hold if f is discontinuous on $[a, b]$.

SOLUTIONS TO SELF-CHECK EXERCISES 10.5

1. For $x < -1$, $f(x) = 1$, and so

$$\lim_{x \to -1^-} f(x) = \lim_{x \to -1^-} 1 = 1$$

For $x \geq -1$, $f(x) = 1 + \sqrt{x + 1}$, and so

$$\lim_{x \to -1^+} f(x) = \lim_{x \to -1^+} (1 + \sqrt{x + 1}) = 1$$

Since the left-hand and right-hand limits of f exist as x approaches $x = -1$ and both are equal to 1, we conclude that

$$\lim_{x \to -1} f(x) = 1$$

USING TECHNOLOGY

Finding the Points of Discontinuity of a Function

You can very often recognize the points of discontinuity of a function f by examining its graph. For example, Figure T1 shows the graph of $f(x) = x/(x^2 - 1)$ obtained using a graphing utility. It is evident that f is discontinuous at $x = -1$ and $x = 1$. This observation is also borne out by the fact that both these points are not in the domain of f.

Figure T1

The graph of $f(x) = \dfrac{x}{x^2 - 1}$ in the viewing rectangle $[-4, 4] \times [-10, 10]$.

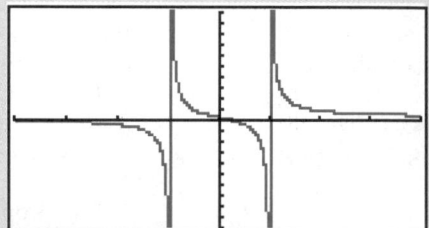

Consider the function

$$g(x) = \frac{2x^3 + x^2 - 7x - 6}{x^2 - x - 2}$$

Using a graphing utility, we obtain the graph of g shown in Figure T2. An examination of this graph does not reveal any points of discontinuity. However, if we factor both the numerator and the denominator of the rational expression, we see that

$$g(x) = \frac{(x + 1)(x - 2)(2x + 3)}{(x + 1)(x - 2)}$$

$$= 2x + 3$$

provided $x \neq -1$ and $x \neq 2$, so that its graph, in fact, looks like that shown in Figure T3.

Figure T2
The graph of $g(x) = \dfrac{2x^3 + x^2 - 7x - 6}{x^2 - x - 2}$ in the standard viewing rectangle.

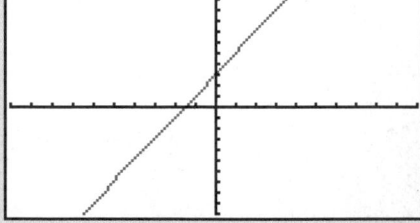

Figure T3
The graph of g has holes at $(-1, 1)$
and $(2, 7)$.

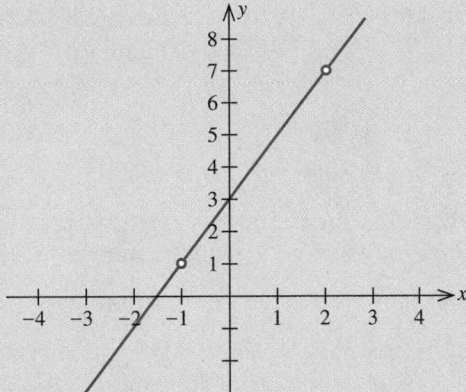

This example shows the limitation of the graphing utility and reminds us of the importance of studying functions analytically!

Graphing Functions Defined Piecewise

The next examples illustrate how to plot the graphs of functions defined in a piecewise manner on a graphing utility.

EXAMPLE I   Plot the graph of

$$f(x) = \begin{cases} x + 1 & \text{if } x \leq 1 \\ \dfrac{2}{x} & \text{if } x > 1 \end{cases}$$

Solution   We enter the function

$$y1 = (x + 1)(x \leq 1) + (2/x)(x > 1)$$

The graph of the function in the viewing rectangle $[-5, 5] \times [-2, 4]$ is shown in Figure T4.

Figure T4
The graph of f in the viewing rec-
tangle $[-5, 5] \times [-2, 4]$.

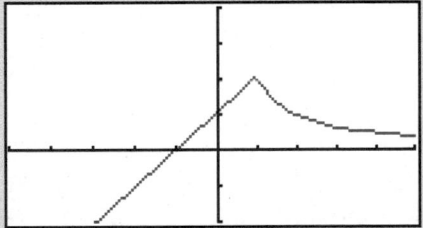

o o o

EXAMPLE 2 The percentage of U.S. households, $P(t)$, watching television during weekdays between the hours of 4 P.M. and 4 A.M. is given by

$$P(t) = \begin{cases} 0.01354t^4 - 0.49375t^3 + 2.58333t^2 + 3.8t + 31.60704 & \text{if } 0 \leq t \leq 8 \\ 1.35t^2 - 33.05t + 208 & \text{if } 8 < t \leq 12 \end{cases}$$

where t is measured in hours, with $t = 0$ corresponding to 4 P.M. Plot the graph of P in the viewing rectangle $[0, 12] \times [0, 80]$.

Solution We enter the function

$$y2 = (0.01354x^4 - 0.49375x^3 + 2.58333x^2 + 3.8x + 31.60704)(x \geq 0)(x \leq 8)$$
$$+ (1.35x^2 - 33.05x + 208)(x > 8)(x \leq 12)$$

The graph of P is shown in Figure T5.

Figure T5
The graph of P in the viewing rectangle $[0, 12] \times [0, 80]$.

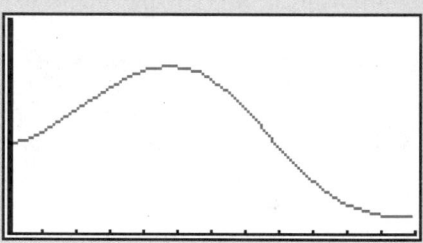

Source: A. C. Nielsen Co.

○ ○ ○

EXERCISES

In exercises 1–10, use a graphing utility to plot the graph of f and to spot the points of discontinuity of f. Then use analytical means to verify your observation and find all points of discontinuity.

1. $f(x) = \dfrac{2}{x^2 - x}$

2. $f(x) = \dfrac{2x + 1}{x^2 + x - 2}$

3. $f(x) = \dfrac{\sqrt{x}}{x^2 - x - 2}$

4. $f(x) = \dfrac{3}{\sqrt{x}(x + 1)}$

5. $f(x) = \dfrac{6x^3 + x^2 - 2x}{2x^2 - x}$

6. $f(x) = \dfrac{2x^3 - x^2 - 13x - 6}{2x^2 - 5x - 3}$

7. $f(x) = \dfrac{2x^4 - 3x^3 - 2x^2}{2x^2 - 3x - 2}$

8. $f(x) = \dfrac{6x^4 - x^3 + 5x^2 - 1}{6x^2 - x - 1}$

9. $f(x) = \dfrac{x^3 + x^2 - 2x}{x^4 + 2x^3 - x - 2}$

[*Hint:* $x^4 + 2x^3 - x - 2 = (x^3 - 1)(x + 2)$]

10. $f(x) = \dfrac{x^3 - x}{x^{4/3} - x + x^{1/3} - 1}$

[*Hint:* $x^{4/3} - x + x^{1/3} - 1 = (x^{1/3} - 1)(x + 1)$.
Can you explain why part of the graph is missing?]

In exercises 11–14, use a graphing utility to plot the graph of f in the indicated viewing rectangle.

11. $f(x) = \begin{cases} -1 & \text{if } x \le 1 \\ x+1 & \text{if } x > 1; \end{cases}$ $[-5,5] \times [-2,8]$

12. $f(x) = \begin{cases} \frac{1}{3}x^2 - 2x & \text{if } x \le 3 \\ -x+6 & \text{if } x > 3; \end{cases}$ $[0,7] \times [-5,5]$

13. $f(x) = \begin{cases} 2 & \text{if } x \le 0 \\ \sqrt{4-x^2} & \text{if } x > 0; \end{cases}$ $[-2,2] \times [-4,4]$

14. $f(x) = \begin{cases} -x^2 + x + 2 & \text{if } x \le 1 \\ 2x^3 - x^2 - 4 & \text{if } x > 1; \end{cases}$ $[-4,4] \times [-5,5]$

15. **Flight Path of a Plane**   The function

$$f(x) = \begin{cases} 0 & \text{if } 0 \le x < 1 \\ -0.00411523x^3 + 0.0679012x^2 & \\ \quad -0.123457x + 0.0596708 & \text{if } 1 \le x < 10 \\ 1.5 & \text{if } 10 \le x \le 100 \end{cases}$$

where both x and $f(x)$ are measured in units of 1000 ft, describes the flight path of a plane taking off from the origin and climbing to an altitude of 15,000 ft. Plot the graph of f to visualize the trajectory of the plane.

16. **Home Shopping Industry**   According to industry sources, revenue from the home shopping industry for the years since its inception may be approximated by the function

$$R(t) = \begin{cases} -0.03t^3 + 0.25t^2 - 0.12t & \text{if } 0 \le t \le 3 \\ 0.57t - 0.63 & \text{if } 3 < t \le 11 \end{cases}$$

where $R(t)$ measures the revenue in billions of dollars and t is measured in years, with $t = 0$ corresponding to the beginning of 1984. Plot the graph of R.
Source: Paul Kagan Associates

653

2. a. The graph of *f* is as follows:

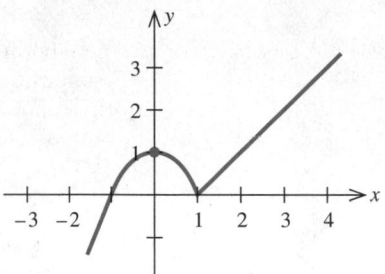

We see that *f* is continuous everywhere.
b. The graph of *g* is as follows:

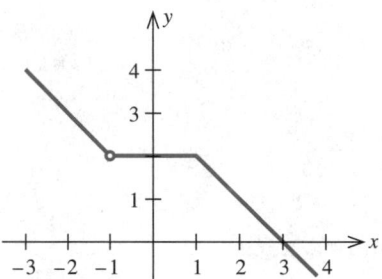

Since *g* is not defined at $x = -1$, it is discontinuous there. It is continuous everywhere else.

10.6 THE DERIVATIVE

An Intuitive Example

We mentioned in Section 10.4 that the problem of finding the *rate of change*
of one quantity with respect to another is mathematically equivalent to the
problem of finding the *slope of the tangent line* to a curve at a given point on
the curve. Before going on to establish this relationship, let us show its plausibility by looking at it from an intuitive point of view.

Consider the motion of the maglev discussed in Section 10.4. Recall that
the position of the maglev at any time *t* is given by

$$s = f(t) = 4t^2 \qquad (0 \le t \le 30)$$

where *s* is measured in feet and *t* in seconds. The graph of the function *f* is
sketched in Figure 10.39.

Observe that the graph of *f* rises slowly at first but more rapidly as *t*
increases, reflecting the fact that the speed of the maglev is increasing with
time. This observation suggests a relationship between the speed of the maglev
at any time *t* and the *steepness* of the curve at the point corresponding to this
value of *t*. Thus, it would appear that we can solve the problem of finding the
speed of the maglev at any time if we can find a way to measure the steepness
of the curve at any point on the curve.

Figure 10.39
Graph showing the position s of a
maglev at time t.

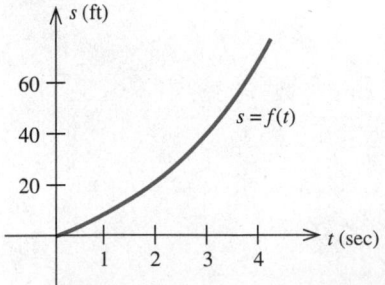

To discover a yardstick that will measure the steepness of a curve, consider the graph of a function f such as the one shown in Figure 10.40a. Think of the curve as representing a stretch of roller coaster track (Figure 10.40b). When the car is at the point P on the curve, a passenger sitting erect in the car and looking straight ahead will have a line of sight that is parallel to the line T, the tangent to the curve at P.

As Figure 10.40a suggests, the steepness of the curve—that is, the rate at which y is increasing or decreasing with respect to x—is given by the slope of the tangent line to the graph of f at the point $P(x, f(x))$. For now we will show how this relationship can be used to estimate the rate of change of a function from its graph.

Figure 10.40

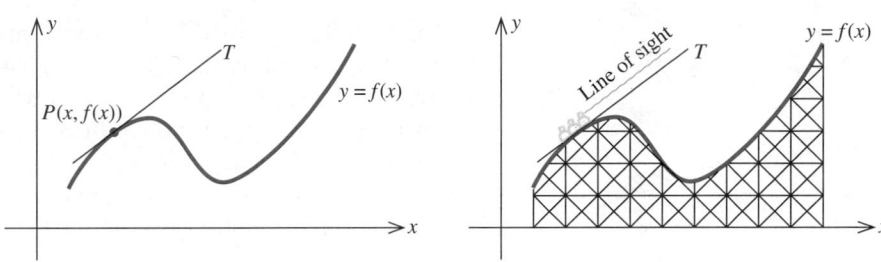

(a) T is the tangent line to the curve at P.

(b) T is parallel to the line of sight.

EXAMPLE 1   The graph of the function $y = N(t)$, shown in Figure 10.41, gives the number of social security beneficiaries from the beginning of 1990 ($t = 0$) through the year 2045 ($t = 55$). Use the graph of

Figure 10.41

The number of social security beneficiaries from 1990 through 2045. We can use the slope of the tangent line at the indicated points to estimate the rate at which the number of social security beneficiaries will be changing.

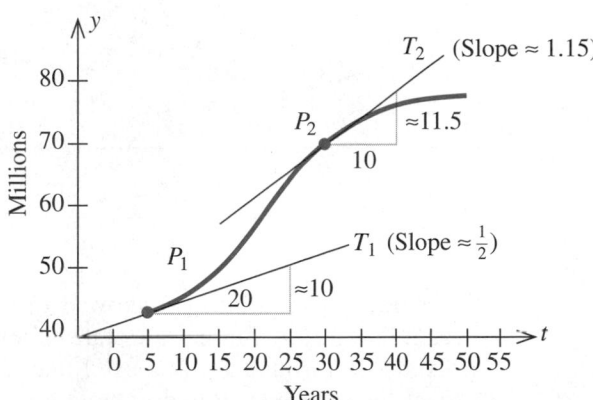

$y = N(t)$ to estimate the rate at which the number of social security beneficiaries will be growing at the beginning of the year 2000 ($t = 10$) and at the beginning of 2025 ($t = 35$). [Assume that the rate of change of the function N at any value of t is given by the slope of the tangent line at the point $P(t, N(t))$.]

Source: Social Security Administration

Solution From the figure, we see that the slope of the tangent line T_1 to the graph of $y = N(t)$ at $P_1(10, 44.7)$ is approximately 0.5. This tells us that the quantity y is increasing at the rate of 1/2 unit per unit increase in t, when $t = 10$. In other words, at the beginning of the year 2000, the number of social security beneficiaries will be increasing at the rate of approximately 0.5 million, or 500,000, per year.

The slope of the tangent line T_2 at $P_2(35, 71.9)$ is approximately 1.15. This tells us that at the beginning of 2025, the number of social security beneficiaries will be growing at the rate of approximately 1.15 million, or 1,150,000, per year.

○ ○ ○

Slope of a Tangent Line

In Example 1, we answered the questions raised by drawing the graph of the function N and estimating the position of the tangent lines. Ideally, however, we would like to solve a problem analytically whenever possible. In order to do this, we need a precise definition of the slope of a tangent line to a curve.

To define the tangent line to a curve C at a point P on the curve, fix P and let Q be any point on C distinct from P (Figure 10.42). The straight line passing through P and Q is called a **secant line.**

Figure 10.42
As Q approaches P along the curve C, the secant lines approach the tangent line T.

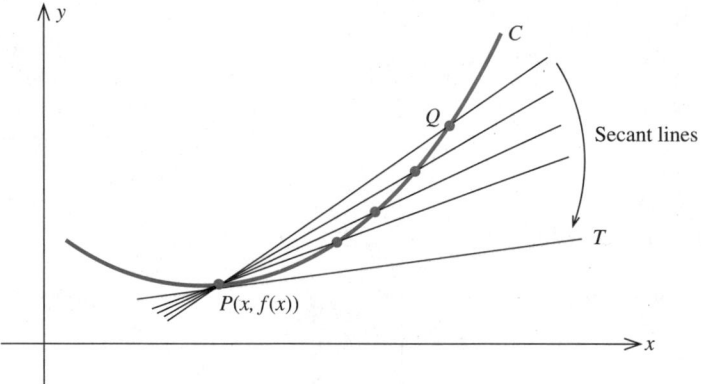

Now, as the point Q is allowed to move toward P along the curve, the secant line through P and Q rotates about the fixed point P and approaches a fixed line through P. This fixed line, which is the limiting position of the secant lines through P and Q as Q approaches P, is the **tangent line to the graph of f** at the point P.

We can describe the process more precisely as follows. Suppose the curve C is the graph of a function f defined by $y = f(x)$. Then the point P is described by $P(x, f(x))$ and the point Q by $Q(x + h, f(x + h))$, where h is some appropriate nonzero number (Figure 10.43a). Observe that we can make Q approach P along the curve C by letting h approach zero (Figure 10.43b).

Figure 10.43

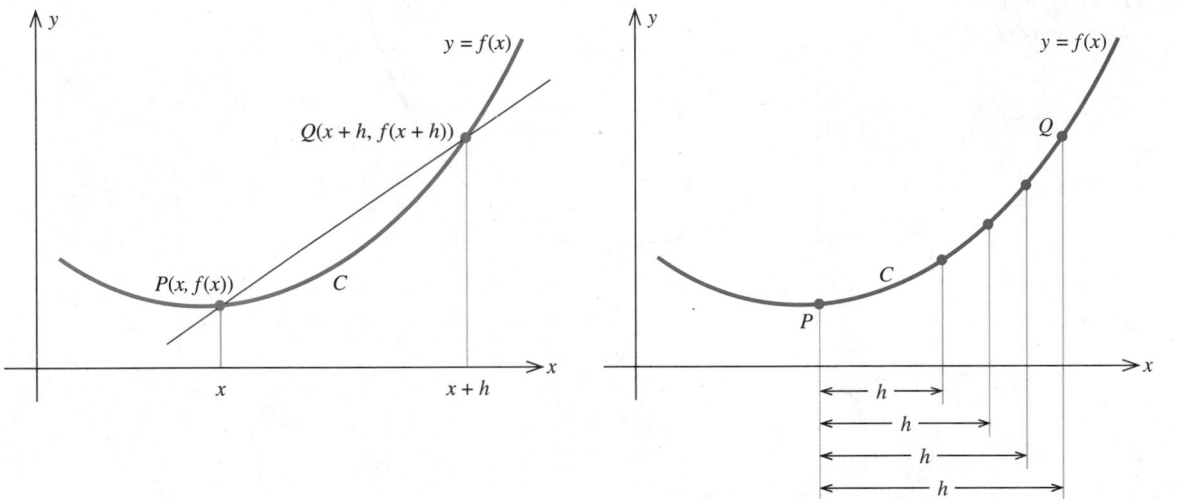

(a) The points $P(x, f(x))$ and $Q(x + h, f(x + h))$.   **(b)** As h approaches zero, Q approaches P.

Next, using the formula for the slope of a line, we can write the slope of the secant line passing through $P(x, f(x))$ and $Q(x + h, f(x + h))$ as

$$\frac{f(x + h) - f(x)}{(x + h) - x} = \frac{f(x + h) - f(x)}{h} \tag{3}$$

As observed earlier, Q approaches P and, therefore, the secant line through P and Q approaches the tangent line T as h approaches zero. Consequently, we might expect that the slope of the secant line would approach the slope of the tangent line T as h approaches zero. This leads to the following definition.

**SLOPE OF A
TANGENT LINE**

The **slope of the tangent line** to the graph f at the point $P(x, f(x))$ is given by

$$\lim_{h \to 0} \frac{f(x + h) - f(x)}{h} \tag{4}$$

if it exists.

Rates of Change

We now show that the problem of finding the slope of the tangent line to the graph of a function f at the point $P(x, f(x))$ is mathematically equivalent to the problem of finding the rate of change of f at x. To see this, suppose that we are given a function f that describes the relationship between the two quantities x and y:

$$y = f(x)$$

Figure 10.44
$f(x + h) - f(x)$ *is the change in y*
that corresponds to a change h in x.

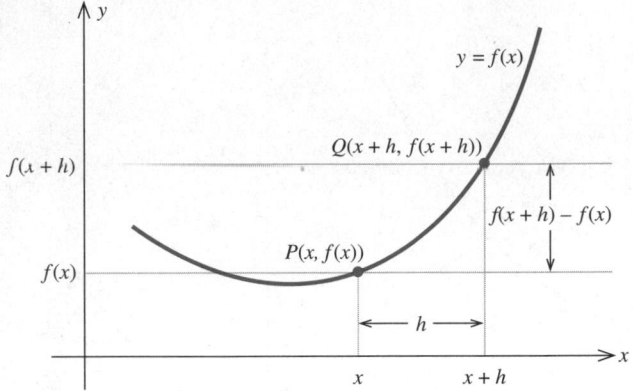

The number $f(x + h) - f(x)$ measures the change in y that corresponds to a change h in x (Figure 10.44).

Then, the **difference quotient**

$$\frac{f(x + h) - f(x)}{h} \tag{5}$$

measures the **average rate of change of y with respect to x** over the interval $[x, x + h]$. For example, if y measures the position of a car at time x, then quotient (5) gives the average velocity of the car over the time interval $[x, x + h]$.

Observe that the difference quotient (5) is the same as (3). We conclude that the difference quotient (5) also measures the slope of the secant line that passes through the two points $P(x, f(x))$ and $Q(x + h, f(x + h))$ lying on the graph of $y = f(x)$. Next, by taking the limit of the difference quotient (5) as h goes to zero—that is, by evaluating

$$\lim_{h \to 0} \frac{f(x + h) - f(x)}{h} \tag{6}$$

we obtain the **rate of change of f at x.** For example, if y measures the position of a car at time x, then the limit (6) gives the velocity of the car at time x. For emphasis, the rate of change of a function f at x is often called the **instantaneous rate of change of f at x.** This distinguishes it from the average rate of change of f, which is computed over an *interval* $[x, x + h]$ rather than at a *point* x.

Observe that the limit (6) is the same as (4). Therefore, the limit of the difference quotient also measures the slope of the tangent line to the graph of $y = f(x)$ at the point $(x, f(x))$.

The following summarizes this discussion.

AVERAGE AND INSTANTANEOUS RATES OF CHANGE

Average rate of change of f over the interval $[x, x + h]$ or **slope of the secant line** to the graph of f through the points $(x, f(x))$ and $(x + h, f(x + h))$ is

$$\frac{f(x + h) - f(x)}{h} \tag{7}$$

Instantaneous rate of change of f at x or **slope of the tangent line** to the graph of f at $(x, f(x))$ is

$$\lim_{h \to 0} \frac{f(x + h) - f(x)}{h} \tag{8}$$

Explain the difference between the average rate of change of a function and the instantaneous rate of change of a function.

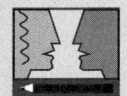

The Derivative

The limit (4), or (8), which measures both the slope of the tangent line to the graph of $y = f(x)$ at the point $P(x, f(x))$ and the (instantaneous) rate of change of f at x is given a special name: the **derivative of f at x.**

THE DERIVATIVE OF A FUNCTION

The **derivative** of a function f with respect to x is the function f' (read "f prime"), defined by

$$f'(x) = \lim_{h \to 0} \frac{f(x + h) - f(x)}{h} \tag{9}$$

The domain of f' is the set of all x where the limit exists.

Thus, the derivative of a function f is a function f' that gives the slope of the tangent line to the graph of f at *any* point $(x, f(x))$ and also the rate of change of f at x (Figure 10.45).

Figure 10.45
The slope of the tangent line at
$P(x, f(x))$ is $f'(x)$; f changes at the
rate of $f'(x)$ units per unit change
in x at x.

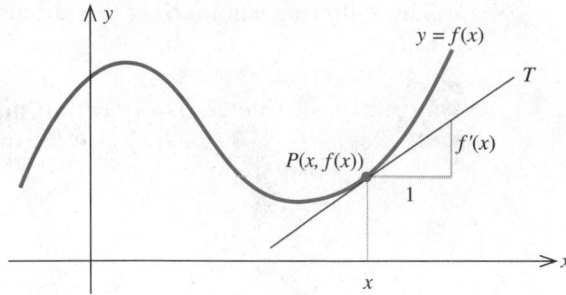

Other notations for the derivative of f include:

$$D_x f(x) \qquad \text{(Read "dee sub } x \text{ of } f \text{ of } x")$$

$$\frac{dy}{dx} \qquad \text{(Read "dee } y \text{ dee } x")$$

and $\qquad\qquad y' \qquad$ (Read "y prime")

The last two are used when the rule for f is written in the form $y = f(x)$.

The calculation of the derivative of f is facilitated using the following four-step process.

THE FOUR-STEP PROCESS FOR FINDING $f'(x)$

1. Compute $f(x + h)$.

2. Form the difference $f(x + h) - f(x)$.

3. Form the quotient $\dfrac{f(x + h) - f(x)}{h}$.

4. Compute $f'(x) = \lim\limits_{h \to 0} \dfrac{f(x + h) - f(x)}{h}$.

EXAMPLE 2   Find the slope of the tangent line to the graph of $f(x) = 3x + 5$ at any point $(x, f(x))$.

Solution   The slope of the tangent line at any point on the graph of f is given by the derivative of f at x. To find the derivative, we use the four-step process:

Step 1 $f(x + h) = 3(x + h) + 5 = 3x + 3h + 5$

Step 2 $f(x + h) - f(x) = (3x + 3h + 5) - (3x + 5) = 3h$

Step 3 $\dfrac{f(x + h) - f(x)}{h} = \dfrac{3h}{h} = 3$

Step 4 $f'(x) = \lim\limits_{h \to 0} \dfrac{f(x + h) - f(x)}{h} = \lim\limits_{h \to 0} 3 = 3$

We expect this result since the tangent line to any point on a straight line must coincide with the line itself and, therefore, must have the same slope as the line. In this case, the graph of f is a straight line with slope 3.   ◦ ◦ ◦

EXAMPLE 3   Let $f(x) = x^2$.

a. Compute $f'(x)$.

b. Compute $f'(2)$ and interpret your result.

Solution

a. To find $f'(x)$, we use the four-step process:

Step I $f(x + h) = (x + h)^2 = x^2 + 2xh + h^2$

Step 2 $f(x + h) - f(x) = x^2 + 2xh + h^2 - x^2 = 2xh + h^2 = h(2x + h)$

Step 3 $\dfrac{f(x + h) - f(x)}{h} = \dfrac{h(2x + h)}{h} = 2x + h$

Step 4 $f'(x) = \lim\limits_{h \to 0} \dfrac{f(x + h) - f(x)}{h} = \lim\limits_{h \to 0} (2x + h) = 2x$

b. $f'(2) = 2(2) = 4$. This result tells us that the slope of the tangent line to the graph of f at the point $(2, 4)$ is 4. It also tells us that the function f is changing at the rate of 4 units per unit change in x at $x = 2$. The graph of f and the tangent line at $(2, 4)$ are shown in Figure 10.46.   ○ ○ ○

Figure 10.46
The tangent line to the graph of
$f(x) = x^2$ *at* $(2, 4)$.

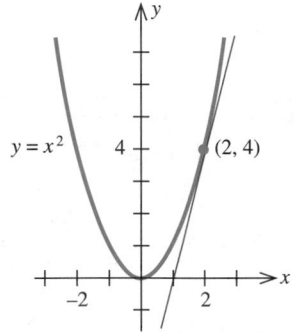

EXPLORING WITH TECHNOLOGY

I. Consider the function $f(x) = x^2$ of Example 3. Suppose we want to compute $f'(2)$ using equation (9). Thus,

$$f'(2) = \lim\limits_{h \to 0} \frac{f(2 + h) - f(2)}{h} = \lim\limits_{h \to 0} \frac{(2 + h)^2 - 2^2}{h}$$

Use a graphing utility to plot the graph of

$$g(x) = \frac{(2 + x)^2 - 4}{x}$$

in the viewing rectangle $[-3, 3] \times [-2, 6]$.

2. Use **ZOOM** and **TRACE** to find $\lim\limits_{x \to 0} g(x)$.

3. Explain why the limit found in part (2) is $f'(2)$.

○ ○ ○

EXAMPLE 4 Let $f(x) = x^2 - 4x$.

a. Compute $f'(x)$.

b. Find the point on the graph of f where the tangent line to the curve is horizontal.

c. Sketch the graph of f and the tangent line to the curve at the point found in (b).

d. What is the rate of change of f at this point?

Solution

a. To find $f'(x)$, we use the four-step process:

Step 1 $f(x + h) = (x + h)^2 - 4(x + h) = x^2 + 2xh + h^2 - 4x - 4h$

Step 2 $f(x + h) - f(x) = x^2 + 2xh + h^2 - 4x - 4h - (x^2 - 4x)$
$$= 2xh + h^2 - 4h = h(2x + h - 4)$$

Step 3 $\dfrac{f(x + h) - f(x)}{h} = \dfrac{h(2x + h - 4)}{h} = 2x + h - 4$

Step 4 $f'(x) = \lim\limits_{h \to 0} \dfrac{f(x + h) - f(x)}{h} = \lim\limits_{h \to 0}(2x + h - 4) = 2x - 4$

b. At a point on the graph of f where the tangent line to the curve is horizontal and hence has slope zero, the derivative f' of f is zero. Accordingly, to find such point(s) we set $f'(x) = 0$, which gives $2x - 4 = 0$, or $x = 2$. The corresponding value of y is given by $y = f(2) = -4$, and the required point is $(2, -4)$.

c. The graph of f and the tangent line are shown in Figure 10.47.

d. The rate of change of f at $x = 2$ is zero.

Figure 10.47
The tangent line to the graph of $y = x^2 - 4x$ at $(2, -4)$ is $y = -4$.

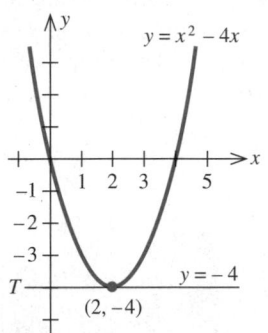

EXAMPLE 5 Let $f(x) = 1/x$.

a. Compute $f'(x)$.

b. Find the slope of the tangent line T to the graph of f at the point where $x = 1$.

c. Find an equation of the tangent line T in (b).

Solution

a. To find $f'(x)$, we use the four-step process:

Step 1 $f(x + h) = \dfrac{1}{x + h}$

Step 2 $f(x + h) - f(x) = \dfrac{1}{x + h} - \dfrac{1}{x} = \dfrac{x - (x + h)}{x(x + h)} = -\dfrac{h}{x(x + h)}$

Can the tangent line to the graph of a function intersect the graph at more than one point? Explain your answer using illustrations.

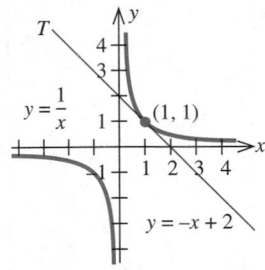

Figure 10.48
The tangent line to the graph of
$f(x) = 1/x$ *at* (1, 1).

Step 3 $\dfrac{f(x + h) - f(x)}{h} = -\dfrac{h}{x(x + h)} \cdot \dfrac{1}{h} = -\dfrac{1}{x(x + h)}$

Step 4 $f'(x) = \lim_{h \to 0} \dfrac{f(x + h) - f(x)}{h} = \lim_{h \to 0} -\dfrac{1}{x(x + h)} = -\dfrac{1}{x^2}$

b. The slope of the tangent line T to the graph of f where $x = 1$ is given by $f'(1) = -1$.

c. When $x = 1$, $y = f(1) = 1$ and T is tangent to the graph of f at the point (1, 1). From (b), we know that the slope of T is -1. Thus, an equation of T is

$$y - 1 = -1(x - 1)$$

or

$$y = -x + 2$$

See Figure 10.48.   ○ ○ ○

Exploring with Technology

1. Use the results of Example 5 to draw the graph of $f(x) = 1/x$ and its tangent line at the point (1, 1) by plotting the graphs of $y_1 = 1/x$ and $y_2 = -x + 2$ in the viewing rectangle $[-4, 4] \times [-4, 4]$.
2. Some graphing utilities draw the tangent line to the graph of a function at a given point automatically—you need only specify the function and give the x-coordinate of the point of tangency. If your graphing utility has this feature, verify the result of part (1) without finding an equation of the tangent line.

                                        ○ ○ ○

Consider the following alternative approach to the definition of the derivative of a function: Let h be a positive number and suppose $P(x - h, f(x - h))$ and $Q(x + h, f(x + h))$ are any two points on the graph of f.

a. Give a geometric and a physical interpretation of the quotient

$$\frac{f(x + h) - f(x - h)}{2h}$$

Make a sketch to illustrate your answer.

(Continued on next page)

(*Continued from previous page*)

b. Give a geometric and a physical interpretation of the limit

$$\lim_{h \to 0} \frac{f(x+h) - f(x-h)}{2h}$$

Make a sketch to illustrate your answer.

c. Explain why it makes sense to define

$$f'(x) = \lim_{h \to 0} \frac{f(x+h) - f(x-h)}{2h}$$

d. Using the definition given in part (c), formulate a four-step process for finding $f'(x)$ similar to that given on page 660 and use it to find the derivative of $f(x) = x^2$. Compare your answer with that obtained in Example 3 on page 661.

Applications

EXAMPLE 6   Suppose that the distance (in feet) covered by a car moving along a straight road t seconds after starting from rest is given by the function $f(t) = 2t^2 \ (0 \le t \le 30)$.

a. Calculate the average velocity of the car over the time intervals [22, 23], [22, 22.1], and [22, 22.01].

b. Calculate the (instantaneous) velocity of the car when $t = 22$.

c. Compare the results obtained in (a) with that obtained in (b).

Solution

a. We first compute the average velocity (average rate of change of f) over the interval $[t, t + h]$ using (7). We find

$$\frac{f(t+h) - f(t)}{h} = \frac{2(t+h)^2 - 2t^2}{h}$$

$$= \frac{2t^2 + 4th + 2h^2 - 2t^2}{h}$$

$$= 4t + 2h$$

Next, using $t = 22$ and $h = 1$, we find that the average velocity of the car over the time interval [22, 23] is

$$4(22) + 2(1) = 90$$

or 90 feet per second. Similarly, using $t = 22$, $h = 0.1$, and $h = 0.01$, we find that its average velocities over the time intervals [22, 22.1] and [22, 22.01] are 88.2 and 88.02 feet per second, respectively.

b. Using the limit (8), we see that the instantaneous velocity of the car at any time t is given by

$$\lim_{h \to 0} \frac{f(t + h) - f(t)}{h} = \lim_{h \to 0} (4t + 2h) \qquad \text{[Using the results from (a)]}$$

$$= 4t$$

In particular, the velocity of the car 22 seconds from rest ($t = 22$) is given by

$$v = 4(22)$$

Figure 10.49
The graph of the demand function $p = 144 - x^2$.

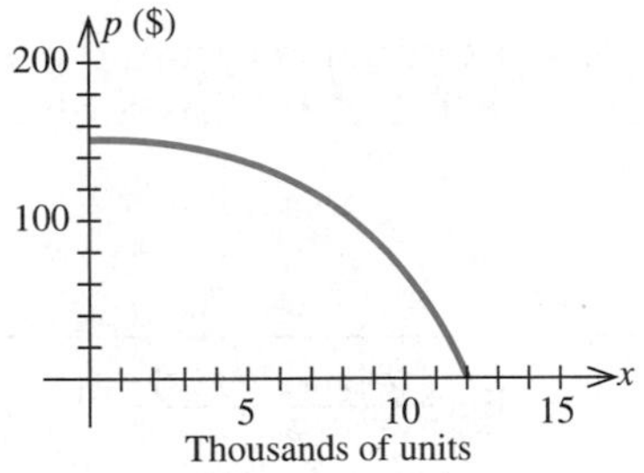

or 88 feet per second.

c. The computations in (a) show that as the time intervals over which the average velocity of the car are computed become smaller and smaller, the average velocities over these intervals do approach 88 ft/sec, the instantaneous velocity of the car at $t = 22$.   ○ ○ ○

EXAMPLE 7   The management of the Titan Tire Company has determined that the weekly demand function for their Super Titan tires is given by

$$p = f(x) = 144 - x^2$$

where p is measured in dollars and x is measured in units of a thousand (Figure 10.49).

a. Find the average rate of change in the unit price of a tire if the quantity demanded is between 5000 and 6000 tires. Between 5000 and 5100 tires. Betweeen 5000 and 5010 tires.

b. What is the instantaneous rate of change of the unit price when the quantity demanded is 5000 units?

Solution

a. The average rate of change of the unit price of a tire if the quantity demanded is between x and $x + h$ is

$$\frac{f(x + h) - f(x)}{h} = \frac{[144 - (x + h)^2] - (144 - x^2)}{h}$$

$$= \frac{144 - x^2 - 2x(h) - h^2 - 144 + x^2}{h}$$

$$= -2x - h$$

To find the average rate of change of the unit price of a tire when the quantity demanded is between 5000 and 6000 tires (that is, over the interval [5, 6]), we take $x = 5$ and $h = 1$, obtaining

$$-2(5) - 1 = -11$$

or $-\$11$ per 1000 tires. (Remember, x is measured in units of a thousand.) Similarly, taking $h = 0.1$ and $h = 0.01$ with $x = 5$, we find that the average rates of change of the unit price when the quantities demanded are between 5000 and 5100 and between 5000 and 5010 are $-\$10.10$ and $-\$10.01$ per 1000 tires, respectively.

b. The instantaneous rate of change of the unit price of a tire when the quantity demanded is x units is given by

$$\lim_{h \to 0} \frac{f(x + h) - f(x)}{h} = \lim_{h \to 0} (-2x - h) \qquad \text{[Using the results from (a)]}$$
$$= -2x$$

In particular, the instantaneous rate of change of the unit price per tire when the quantity demanded is 5000 is given by $-2(5)$, or $-\$10$ per 1000 tires.

The derivative of a function provides us with a tool for measuring the rate of change of one quantity with respect to another. Table 10.4 lists several other applications involving this limit.

Table 10.4

x Stands for	y Stands for	$\dfrac{f(a + h) - f(a)}{h}$ Measures the	$\lim\limits_{h \to 0} \dfrac{f(a + h) - f(a)}{h}$ Measures the
time	**concentration of a drug** in the bloodstream at time x	average rate of change in the concentration of the drug over the time interval $[a, a + h]$	instantaneous rate of change in the concentration of the drug in the bloodstream at time $x = a$
number of items sold	**revenue** at a sales level of x units	average rate of change in the revenue when the sales level is between $x = a$ and $x = a + h$	instantaneous rate of change in the revenue when the sales level is a units
time	**volume of sales** at time x	average rate of change in the volume of sales over the time interval $[a, a + h]$	instantaneous rate of change in the volume of sales at time $x = a$
time	**population** of Drosophila (fruit flies) at time x	average rate of growth of the fruit fly population over the time interval $[a, a + h]$	instantaneous rate of change of the fruit fly population at time $x = a$
temperature in a chemical reaction	**amount of product formed in the chemical reaction** when the temperature is x degrees	average rate of formation of chemical product over the temperature range $[a, a + h]$	instantaneous rate of formation of chemical product when the temperature is a degrees

Differentiability and Continuity

In practical applications, one encounters functions that fail to be **differentiable**—that is, do not have a derivative at certain values in the domain of the function f. It can be shown that a continuous function f fails to be differentiable at a point $x = a$ when the graph of f makes an abrupt change of direction at that point. We call such a point a "corner." A function also fails to be differentiable at a point where the tangent line is vertical since the slope of a vertical line is undefined. These cases are illustrated in Figure 10.50.

Figure 10.50

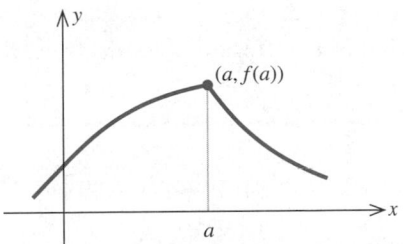

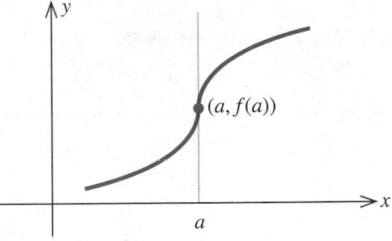

(a) The graph makes an abrupt change of direction at $x = a$.

(b) The slope at $x = a$ is undefined.

Figure 10.51
The function f is not differentiable at (8, 48).

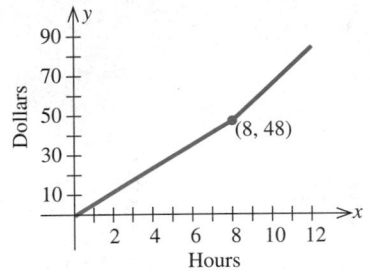

The next example illustrates a function that is not differentiable at a point.

EXAMPLE 8   Mary works at the B&O department store, where, on a weekday, she is paid $6 per hour for the first 8 hours and $9 per hour for overtime. The function

$$f(x) = \begin{cases} 6x & \text{if } 0 \le x \le 8 \\ 9x - 24 & \text{if } 8 < x \end{cases}$$

gives Mary's earnings on a weekday in which she worked x hours. Sketch the graph of the function f and explain why it is not differentiable at $x = 8$.

Solution   The graph of f is shown in Figure 10.51. Observe that the graph of f has a corner at $x = 8$ and consequently is not differentiable at $x = 8$.

We close this section by mentioning the connection between the continuity and the differentiability of a function at a given value $x = a$ in the domain of f. If we reexamine the function of Example 8, it becomes clear that f is continuous everywhere and, in particular, when $x = 8$. This shows that, in general, the continuity of a function at a point $x = a$ does not necessarily imply the differentiability of the function at that point. The converse, however, is true: If a function f is differentiable at a point $x = a$, then it is continuous there.

Exploring with Technology

1. Use a graphing utility to plot the graph of $f(x) = x^{1/3}$ in the viewing rectangle $[-2, 2] \times [-2, 2]$.
2. Use the graphing utility to draw the tangent line to the graph of f at the point $(0, 0)$. Can you explain why the process breaks down?

| **DIFFERENTIABILITY AND CONTINUITY** | If a function is differentiable at $x = a$, then it is continuous at $x = a$. |

For a proof of this result, see exercise 57, page 673.

EXAMPLE 9   Figure 10.52 depicts a portion of the graph of a function. Explain why the function fails to be differentiable at each of the points $x = a, b, c, d, e, f$, and g.

Figure 10.52
The graph of this function is not differentiable at the points a–g.

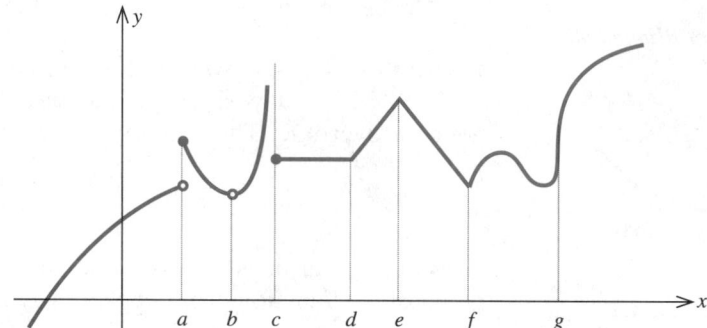

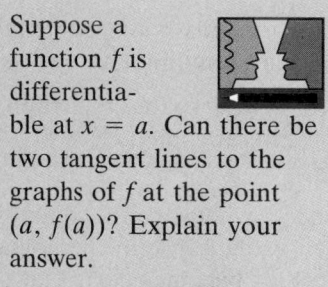

Suppose a function f is differentiable at $x = a$. Can there be two tangent lines to the graphs of f at the point $(a, f(a))$? Explain your answer.

Solution   The function fails to be differentiable at the points $x = a, b$, and c because it is discontinuous at each of these points. The derivative of the function does not exist at $x = d, e$, and f because it has a kink at each of these points. Finally, the function is not differentiable at $x = g$ because the tangent line is vertical at that point.   ◦ ◦ ◦

SELF-CHECK EXERCISES 10.6

1. Let $f(x) = -x^2 - 2x + 3$.
   a. Find the derivative f' of f using the definition of the derivative.
   b. Find the slope of the tangent line to the graph of f at the point $(0, 3)$.
   c. Find the rate of change of f when $x = 0$.
   d. Find an equation of the tangent line to the graph of f at the point $(0, 3)$.
   e. Sketch the graph of f and the tangent line to the curve at the point $(0, 3)$.

2. The losses (in millions of dollars) due to bad loans extended chiefly in agriculture, real estate, shipping, and energy by the Franklin Bank are estimated to be

$$A = f(t) = -t^2 + 10t + 30 \qquad (0 \le t \le 10)$$

where t is the time in years ($t = 0$ corresponds to the beginning of 1994). How fast were the losses mounting at the beginning of 1997? At the beginning of 1999? How fast will the losses be mounting at the beginning of 2001? Interpret your results.

Solutions to Self-Check Exercises 10.6 can be found on page 673.

10.6 EXERCISES

1. **Average Weight of an Infant** The following graph shows the weight measurements of the average infant from the time of birth ($t = 0$) through age 2 ($t = 24$). By computing the slopes of the respective tangent lines, estimate the rate of change of the average infant's weight when $t = 3$ and when $t = 18$. What is the average rate of change in the average infant's weight over its first year of life?

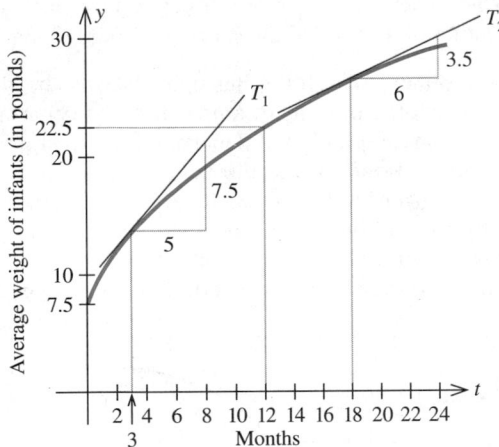

2. **Forestry** The following graph shows the volume of wood produced in a single-species forest. Here $f(t)$ is measured in cubic meters per hectare and t is measured in years. By computing the slopes of the respective tangent lines, estimate the rate at which the wood grown is changing at the beginning of the 10th year and at the beginning of the 30th year.

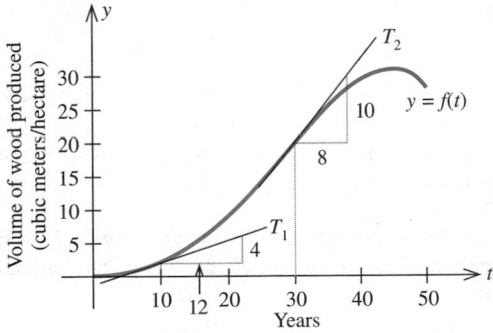

Source: The Random House Encyclopedia

3. **TV Viewing Patterns** The following graph, based on data supplied by the A. C. Nielsen Company, shows the percentage of U.S. households watching television during a 24-hr period on a weekday ($t = 0$ corresponds to 6 A.M.). By computing the slopes of the respective tangent lines, estimate the rate of change of the percentage of households watching television at 4 P.M. and 11 P.M.

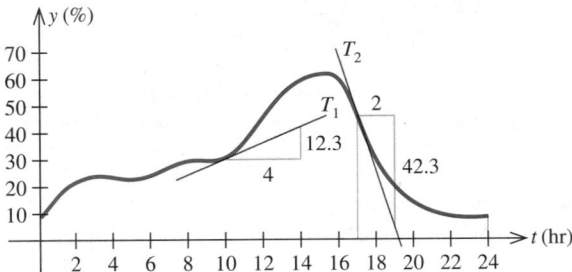

Source: A. C. Nielsen Company

4. **Crop Yield** Productivity and yield of cultivated crops are often reduced by insect pests. The following graph shows the relationship between the yield of a certain crop, $f(x)$, as a function of the density of aphids, x. (Aphids are small insects that suck plant juices.) Here $f(x)$ is measured in kg/4000 sq meters, and x is measured in hundreds of aphids per bean stem. By computing the slopes of the respective tangent lines, estimate the rate of change of the crop yield with respect to the density of aphids when that density is 200 aphids per bean stem. When it is 800 aphids per bean stem.

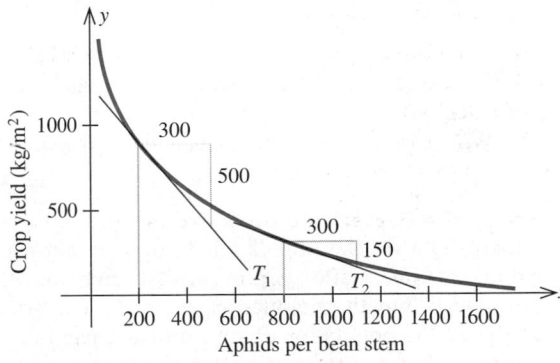

Source: The Random House Encyclopedia

5. The position of car A and car B, starting out side by side and traveling along a straight road, is given by $s = f(t)$ and $s = g(t)$, respectively, where s is measured in feet and t is measured in seconds (see the accompanying figure).

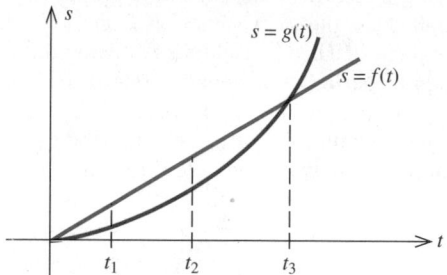

a. Which car is traveling faster at t_1?

b. What can you say about the speed of the cars at t_2?

[*Hint:* Compare tangent lines.]

c. Which car is traveling faster at t_3?

d. What can you say about the positions of the cars at t_3?

6. The velocity of car A and car B, starting out side by side and traveling along a straight road, is given by $v = f(t)$ and $v = g(t)$, respectively, where v is measured in ft/sec and t is measured in seconds (see the accompanying figure).

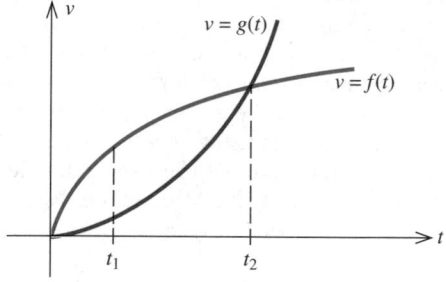

a. What can you say about the velocity and acceleration of the two cars at t_1? (Acceleration is the rate of change of velocity.)

b. What can you say about the velocity and acceleration of the two cars at t_2?

7. Effect of a Bactericide on Bacteria   In the following figure, $f(t)$ gives the population P_1 of a certain bacteria culture at time t after a portion of bactericide A was introduced into the population at $t = 0$. The graph of $g(t)$ gives the population P_2 of a similar bacteria culture at time t after a portion of bactericide B was introduced into the population at $t = 0$.

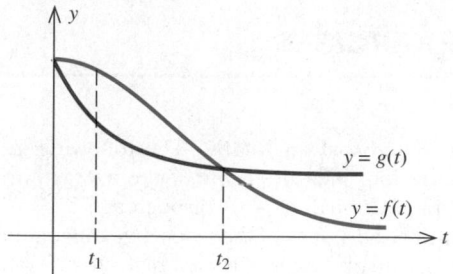

a. Which population is decreasing faster at t_1?

b. Which population is decreasing faster at t_2?

c. Which bactericide is more effective in reducing the population of bacteria in the short run? In the long run?

8. Market Share   The following figure shows the devastating effect the opening of a new discount department store had on an established department store in a small town. The revenue of the discount store at time t (in months) is given by $f(t)$ million dollars, whereas the revenue of the established department store at time t is given by $g(t)$ million dollars. Answer the following questions by giving the value of t at which the specified event took place.

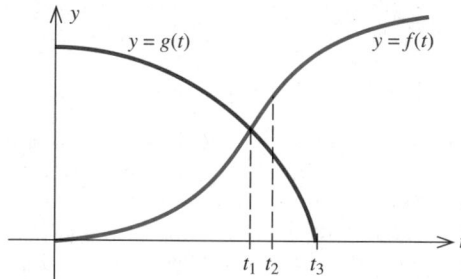

a. The revenue of the established department store is decreasing at the slowest rate.

b. The revenue of the established department store is decreasing at the fastest rate.

c. The revenue of the discount store first overtakes that of the established store.

d. The revenue of the discount store is increasing at the fastest rate.

In exercises 9–16, use the four-step process to find the slope of the tangent line to the graph of the given function at any point.

9. $f(x) = 13$

10. $f(x) = -6$

11. $f(x) = 2x + 7$

12. $f(x) = 8 - 4x$

13. $f(x) = 3x^2$

14. $f(x) = -\frac{1}{2}x^2$

15. $f(x) = -x^2 + 3x$

16. $f(x) = 2x^2 + 5x$

In exercises 17–22, find the slope of the tangent line to the graph of each function at the given point and determine an equation of the tangent line.

17. $f(x) = 2x + 7$ at $(2, 11)$

18. $f(x) = -3x + 4$ at $(-1, 7)$

19. $f(x) = 3x^2$ at $(1, 3)$

20. $f(x) = 3x - x^2$ at $(-2, -10)$

21. $f(x) = -\dfrac{1}{x}$ at $\left(3, -\dfrac{1}{3}\right)$

22. $f(x) = \dfrac{3}{2x}$ at $\left(1, \dfrac{3}{2}\right)$

23. Let $f(x) = 2x^2 + 1$.
   a. Find the derivative f' of f.
   b. Find an equation of the tangent line to the curve at the point $(1, 3)$.
   c. Sketch the graph of f.

24. Let $f(x) = x^2 + 6x$.
   a. Find the derivative f' of f.
   b. Find the point on the graph of f where the tangent line to the curve is horizontal.
   [*Hint:* Find the value of x for which $f'(x) = 0$.]
   c. Sketch the graph of f and the tangent line to the curve at the point found in (b).

25. Let $f(x) = x^2 - 2x + 1$.
   a. Find the derivative f' of f.
   b. Find the point on the graph of f where the tangent line to the curve is horizontal.
   c. Sketch the graph of f and the tangent line to the curve at the point found in (b).
   d. What is the rate of change of f at this point?

26. Let $f(x) = \dfrac{1}{x - 1}$.
   a. Find the derivative f' of f.
   b. Find an equation of the tangent line to the curve at the point $(-1, -\frac{1}{2})$.
   c. Sketch the graph of f.

27. Let $y = f(x) = x^2 + x$.
   a. Find the average rate of change of y with respect to x in the interval from $x = 2$ to $x = 3$. In the interval from $x = 2$ to $x = 2.5$. In the interval from $x = 2$ to $x = 2.1$.

   b. Find the (instantaneous) rate of change of y at $x = 2$.
   c. Compare the results obtained in (a) with that of (b).

28. Let $y = f(x) = x^2 - 4x$.
   a. Find the average rate of change of y with respect to x in the interval from $x = 3$ to $x = 4$. In the interval from $x = 3$ to $x = 3.5$. In the interval from $x = 3$ to $x = 3.1$.
   b. Find the (instantaneous) rate of change of y at $x = 3$.
   c. Compare the results obtained in (a) with that of (b).

29. **Velocity of a Car** Suppose that the distance s (in feet) covered by a car moving along a straight road t seconds after starting from rest is given by the function $f(t) = 2t^2 + 48t$.
   a. Calculate the average velocity of the car over the time intervals $[20, 21]$, $[20, 20.1]$, and $[20, 20.01]$.
   b. Calculate the (instantaneous) velocity of the car when $t = 20$.
   c. Compare the results of (a) with that of (b).

30. **Velocity of a Ball Thrown into the Air** A ball is thrown straight up with an initial velocity of 128 ft/sec, so that its height (in feet) after t seconds is given by $s(t) = 128t - 16t^2$.
   a. What is the average velocity of the ball over the time intervals $[2, 3]$, $[2, 2.5]$, and $[2, 2.1]$?
   b. What is the instantaneous velocity at time $t = 2$?
   c. What is the instantaneous velocity at time $t = 5$? Is the ball rising or falling at this time?
   d. When will the ball hit the ground?

31. During the construction of a high-rise building, a worker accidentally dropped his portable electric screwdriver from a height of 400 ft. After t seconds, the screwdriver had fallen a distance of $s = 16t^2$ ft.
   a. How long did it take the screwdriver to reach the ground?
   b. What was the average velocity of the screwdriver between the time it was dropped and the time it hit the ground?
   c. What was the velocity of the screwdriver at the time it hit the ground?

32. A hot air balloon rises vertically from the ground so that its height after t seconds is $h = \frac{1}{2}t^2 + \frac{1}{2}t$ ft ($0 \le t \le 60$).
   a. What is the height of the balloon at the end of 40 sec?
   b. What is the average velocity of the balloon between $t = 0$ and $t = 40$?
   c. What is the velocity of the balloon at the end of 40 sec?

33. At a temperature of 20°C, the volume V (in liters) of 1.33 g of O_2 is related to its pressure p (in atmospheres) by the formula $V = 1/p$.
    a. What is the average rate of change of V with respect to p as p increases from $p = 2$ to $p = 3$?
    b. What is the rate of change of V with respect to p when $p - 2$?

34. **Cost of Producing Surfboards**   The total cost $C(x)$ (in dollars) incurred by the Aloha Company in manufacturing x surfboards a day is given by

    $$C(x) = -10x^2 + 300x + 130 \qquad (0 \le x \le 15)$$

    a. Find $C'(x)$.
    b. What is the rate of change of the total cost when the level of production is ten surfboards a day?
    c. What is the average cost Aloha incurs in manufacturing ten surfboards a day?

35. **Effect of Advertising on Profit**   The quarterly profit of Cunningham Realty (in thousands of dollars) is given by

    $$P(x) = -\frac{1}{3}x^2 + 7x + 30 \qquad (0 \le x \le 50)$$

    where x (in thousands of dollars) is the amount of money Cunningham spends on advertising per quarter.
    a. Find $P'(x)$.
    b. What is the rate of change of Cunningham's quarterly profit if the amount it spends on advertising is $10,000 per quarter ($x = 10$)? $30,000 per quarter ($x = 30$)?

36. **Demand for Tents**   The demand function for the Sportsman 5 × 7 tents is given by

    $$p = f(x) = -0.1x^2 - x + 40$$

    where p is measured in dollars and x is measured in units of a thousand.
    a. Find the average rate of change in the unit price of a tent if the quantity demanded is between 5000 and 5050 tents. Between 5000 and 5010 tents.
    b. What is the rate of change in the unit price if the quantity demanded is 5000?

37. **A Country's GDP**   The gross domestic product (GDP) of a certain country is projected to be

    $$N(t) = t^2 + 2t + 50 \qquad (0 \le t \le 5)$$

    billion dollars t years from now. What will the rate of change of the country's GDP be two years from now? Four years from now?

38. **Growth of Bacteria**   Under a set of controlled laboratory conditions, the size of the population of a certain bacteria culture at time t (in minutes) is described by the function

    $$P = f(t) = 3t^2 + 2t + 1$$

    Find the rate of population growth at $t - 10$ minutes.

In exercises 39–43, let x and f(x) represent the given quantities. Fix x = a and let h be a small positive number. Give an interpretation of the quantities

$$\frac{f(a + h) - f(a)}{h} \quad \text{and} \quad \lim_{h \to 0} \frac{f(a + h) - f(a)}{h}$$

39. x denotes time and $f(x)$ denotes the population of seals at time x

40. x denotes time and $f(x)$ denotes the prime interest rate at time x

41. x denotes time and $f(x)$ denotes a country's industrial production

42. x denotes the level of production of a certain commodity and $f(x)$ denotes the total cost incurred in producing x units of the commodity

43. x denotes altitude and $f(x)$ denotes atmospheric pressure

In each of exercises 44–49, the graph of a function is shown. For each function, state whether or not (a) f(x) has a limit at x = a, (b) f(x) is continuous at x = a, and (c) f(x) is differentiable at x = a. Justify your answers.

44.

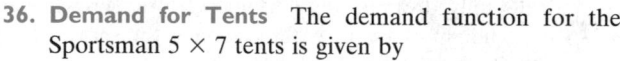

45.

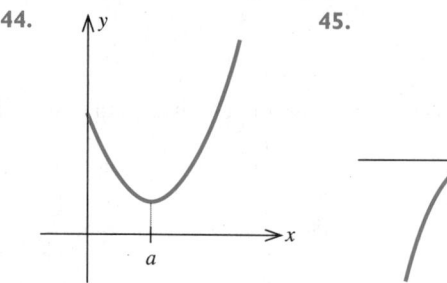

46.

47.

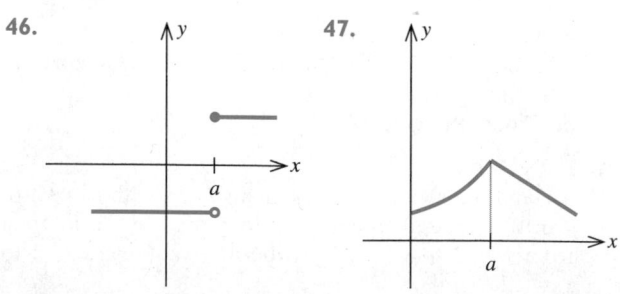

48.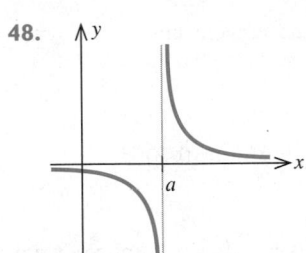

49.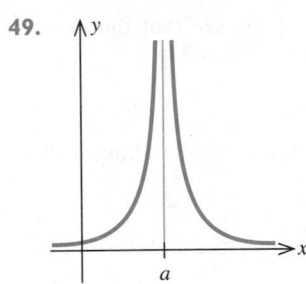

50. The distance s (in feet) covered by a motorcycle traveling in a straight line and starting from rest in t seconds is given by the function

$$s(t) = -0.1t^3 + 2t^2 + 24t$$

Calculate the motorcycle's average velocity over the time interval $[2, 2 + h]$ for $h = 1, 0.1, 0.01, 0.001, 0.0001,$ and $0.00001,$ and use your results to guess at the motorcycle's instantaneous velocity at $t = 2$.

51. The daily total cost $C(x)$ incurred by Trappee and Sons, Inc., for producing x cases of Texa-Pep hot sauce is given by

$$C(x) = 0.000002x^3 + 5x + 400$$

Calculate

$$\frac{C(100 + h) - C(100)}{h}$$

for $h = 1, 0.1, 0.01, 0.001,$ and $0.0001,$ and use your results to estimate the rate of change of the total cost function when the level of production is 100 cases a day.

52. Sketch the graph of the function $f(x) = |x + 1|$ and show that the function does not have a derivative at $x = -1$.

53. Sketch the graph of the function $f(x) = 1/(x - 1)$ and show that the function does not have a derivative at $x = 1$.

54. Let

$$f(x) = \begin{cases} x^2 & \text{if } x \le 1 \\ ax + b & \text{if } x > 1 \end{cases}$$

Find the values of a and b so that f is continuous and has a derivative at $x = 1$. Sketch the graph of f.

55. Sketch the graph of the function $f(x) = x^{2/3}$. Is the function continuous at $x = 0$? Does $f'(0)$ exist? Why or why not?

56. Prove that the derivative of the function $f(x) = |x|$ for $x \ne 0$ is given by

$$f'(x) = \begin{cases} 1 & \text{if } x > 0 \\ -1 & \text{if } x < 0 \end{cases}$$

[*Hint:* Recall the definition of the absolute value of a number.]

57. Show that if a function f is differentiable at a point $x = a$, then f must be continuous at that point.

[*Hint:* Write

$$f(x) - f(a) = \left[\frac{f(x) - f(a)}{x - a} \right] (x - a)$$

Use the product rule for limits and the definition of the derivative to show that

$$\lim_{x \to a} [f(x) - f(a)] = 0]$$

SOLUTIONS TO SELF-CHECK EXERCISES 10.6

1. a. $f'(x) = \lim_{h \to 0} \dfrac{f(x + h) - f(x)}{h}$

$\qquad = \lim_{h \to 0} \dfrac{[-(x + h)^2 - 2(x + h) + 3] - (-x^2 - 2x + 3)}{h}$

$\qquad = \lim_{h \to 0} \dfrac{-x^2 - 2xh - h^2 - 2x - 2h + 3 + x^2 + 2x - 3}{h}$

$\qquad = \lim_{h \to 0} \dfrac{h(-2x - h - 2)}{h}$

$\qquad = \lim_{h \to 0} (-2x - h - 2) = -2x - 2$

b. From the result of (a), we see that the slope of the tangent line to the graph of f at any point $(x, f(x))$ is given by

$$f'(x) = -2x - 2$$

In particular, the slope of the tangent line to the graph of f at $(0, 3)$ is

$$f'(0) = \quad 2$$

c. The rate of change of f when $x = 0$ is given by $f'(0) = -2$, or -2 units per unit change in x.

d. Using the result from (b), we see that an equation of the required tangent line is

$$y - 3 = -2(x - 0)$$

or $$y = -2x + 3$$

e.

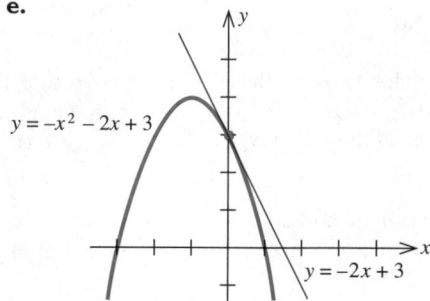

2. The rate of change of the losses at any time t is given by

$$
\begin{aligned}
f'(t) &= \lim_{h \to 0} \frac{f(t + h) - f(t)}{h} \\
&= \lim_{h \to 0} \frac{[-(t + h)^2 + 10(t + h) + 30] - (-t^2 + 10t + 30)}{h} \\
&= \lim_{h \to 0} \frac{-t^2 - 2th - h^2 + 10t + 10h + 30 + t^2 - 10t - 30}{h} \\
&= \lim_{h \to 0} \frac{h(-2t - h + 10)}{h} \\
&= \lim_{h \to 0} (-2t - h + 10) \\
&= -2t + 10
\end{aligned}
$$

Therefore, the rate of change of the losses suffered by the bank at the beginning of 1997 ($t = 3$) was

$$f'(3) = -2(3) + 10 = 4$$

That is, the losses were increasing at the rate of $4 million per year. At the beginning of 1999 ($t = 5$),

$$f'(5) = -2(5) + 10 = 0$$

USING TECHNOLOGY

Graphing a Function and Its Tangent Lines

We can use a graphing utility to plot the graph of a function f and the tangent line at any point on the graph.

EXAMPLE I   Let $f(x) = x^2 - 4x$.

a. Find an equation of the tangent line to the graph of f at the point $(3, -3)$.

b. Plot both the graph of f and the tangent line found in (a) on the same set of axes.

Solution

a. The slope of the tangent line at any point on the graph of f is given by $f'(x)$. But from Example 4 (page 662) we find $f'(x) = 2x - 4$. Using this result, we see that the slope of the required tangent line is

$$f'(3) = 2(3) - 4 = 2$$

Finally, using the point-slope form of the equation of a line, we find that an equation of the tangent line is

$$y - (-3) = 2(x - 3)$$
$$y + 3 = 2x - 6$$

or $y = 2x - 9$.

b. The graph of f in the standard viewing rectangle and the tangent line of interest are shown in Figure T1.

Figure T1
The graph of $f(x) = x^2 - 4x$ and the tangent line $y = 2x - 9$ in the standard viewing rectangle.

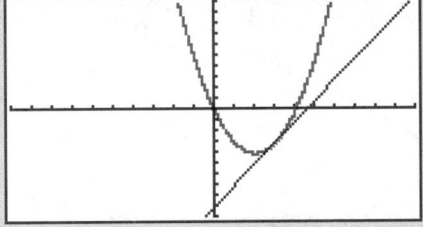

REMARK   Some graphing utilities will draw both the graph of a function f and the tangent line to the graph of f at a specified point when the function and the specified value of x are entered.   ◦ ◦ ◦

675

Finding the Derivative of a Function at a Given Point

The numerical derivative operation of a graphing utility can be used to give an approximate value of the derivative of a function for a given value of x.

EXAMPLE 2   Let $f(x) = \sqrt{x}$.

a. Use the numerical derivative operation of a graphing utility to find the derivative of f at $(4, 2)$.

b. Find an equation of the tangent line to the graph of f at $(4, 2)$.

c. Plot the graph of f and the tangent line on the same set of axes.

Solution

a. Using the numerical derivative operation of a graphing utility, we find that

$$f'(4) = \frac{1}{4}$$

b. An equation of the required tangent line is

$$y - 2 = \frac{1}{4}(x - 4)$$

or $y = \frac{1}{4}x + 1$.

c. The graph of f and the tangent line in the viewing rectangle $[0, 15] \times [0, 4]$ is shown in Figure T2.

Figure T2
The graph of $f(x) = \sqrt{x}$ and the tangent line $y = \frac{1}{4}x + 1$ in the viewing rectangle $[0, 15] \times [0, 4]$.

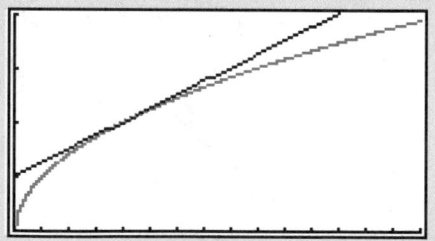

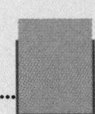

EXERCISES

In exercises 1–10, (a) find an equation of the tangent line to the graph of f at the indicated point, and (b) use a graphing utility to plot the graph of f and the tangent line on the same set of axes. Use a suitable viewing rectangle.

1. $f(x) = 4x - 3$; $(2, 5)$

2. $f(x) = -2x + 5$; $(1, 3)$

3. $f(x) = 2x^2 + x$; $(-2, 6)$

4. $f(x) = -x^2 + 2x$; $(1, 1)$

5. $f(x) = 2x^2 + x - 3$; $(2, 7)$

6. $f(x) = -3x^2 + 2x - 1$; $(1, -2)$

7. $f(x) = x + \dfrac{1}{x}$; $(1, 2)$

8. $f(x) = x - \dfrac{1}{x}$; $(1, 0)$

9. $f(x) = \sqrt{x}$; $(4, 2)$

10. $f(x) = \dfrac{1}{\sqrt{x}}$; $\left(4, \dfrac{1}{2}\right)$

In exercises 11–20, (a) use the numerical derivative operation of a graphing utility to find the derivative of f for the given value of x (to two desired places of accuracy), (b) find an equation of the tangent line to the graph of f at the indicated point, and (c) plot the graph of f and the tangent line on the same set of axes. Use a suitable viewing rectangle.

11. $f(x) = x^3 + x + 1$; $x = 1$; $(1, 3)$

12. $f(x) = -2x^3 + 3x^2 + 2$; $x = -1$; $(-1, 7)$

13. $f(x) = x^4 - 3x^2 + 1$; $x = 2$; $(2, 5)$

14. $f(x) = -x^4 + 3x + 1$; $x = 1$; $(1, 3)$

15. $f(x) = x - \sqrt{x}$; $x = 4$; $(4, 2)$

16. $f(x) = x^{3/2} - x$; $x = 4$; $(4, 4)$

17. $f(x) = \dfrac{1}{x + 1}$; $x = 1$; $\left(1, \dfrac{1}{2}\right)$

18. $f(x) = \dfrac{x}{x + 1}$; $x = 3$; $\left(3, \dfrac{3}{4}\right)$

19. $f(x) = x\sqrt{x^2 + 1}$; $x = 2$; $(2, 2\sqrt{5})$

20. $f(x) = \dfrac{x}{\sqrt{x^2 + 1}}$; $x = 1$; $\left(1, \dfrac{\sqrt{2}}{2}\right)$

and we see that the growth in losses due to bad loans was zero at this point. At the beginning of 2001 ($t = 7$),

$$f'(7) = -2(7) + 10 = -4$$

and we conclude that the losses will be decreasing at the rate of $4 million per year.

 Group projects for this chapter can be found at the Brooks/Cole Web site: http://www.brookscole.com/math/authors/tans/

CHAPTER 10 SUMMARY OF PRINCIPAL FORMULAS AND TERMS

Formulas

1. Average rate of change of f over $[x, x + h]$
   or
   Slope of the secant line to the graph of f through $(x, f(x))$ and $(x + h, f(x + h))$
   or
   Difference quotient

   $$\frac{f(x + h) - f(x)}{h}$$

2. Instantaneous rate of change of f at $(x, f(x))$
   or
   Slope of the tangent line to the graph of f at $(x, f(x))$ at x
   or
   Derivative of f

   $$\lim_{h \to 0} \frac{f(x + h) - f(x)}{h}$$

Terms

Function	Supply function
Domain	Market equilibrium
Independent variable	Equilibrium quantity
Dependent variable	Equilibrium price
Graph of a function	Limit of a function
Graph of an equation	Indeterminate form
Vertical-line test	Limit of a function at infinity
Composite function	Right-hand limit of a function
Polynomial function	Left-hand limit of a function
Linear function	Continuity of a function at a point
Quadratic function	Intermediate Value Theorem
Cubic function	Secant line
Rational function	Tangent line to the graph of f
Power function	Differentiable function
Demand function	

CHAPTER 10 REVIEW EXERCISES

1. Find the domain of each function:

   a. $f(x) = \sqrt{9 - x}$ **b.** $f(x) = \dfrac{x + 3}{2x^2 - x - 3}$

2. Let $f(x) = 3x^2 + 5x - 2$. Find
   a. $f(-2)$ **b.** $f(a + 2)$
   c. $f(2a)$ **d.** $f(a + h)$

3. Let $y^2 = 2x + 1$.
   a. Sketch the graph of this equation.
   b. Is y a function of x? Why?
   c. Is x a function of y? Why?

4. Sketch the graph of the function defined by

   $$f(x) = \begin{cases} x + 1 & \text{if } x < 1 \\ -x^2 + 4x - 1 & \text{if } x \geq 1 \end{cases}$$

5. Let $f(x) = 1/x$ and $g(x) = 2x + 3$. Find
   a. $f(x)g(x)$ **b.** $f(x)/g(x)$
   c. $f(g(x))$ **d.** $g(f(x))$

In exercises 6–19, find the indicated limits, if they exist.

6. $\lim\limits_{x \to 0} (5x - 3)$ 7. $\lim\limits_{x \to 1} (x^2 + 1)$

8. $\lim\limits_{x \to -1} (3x^2 + 4)(2x - 1)$

9. $\lim\limits_{x \to 3} \dfrac{x - 3}{x + 4}$ 10. $\lim\limits_{x \to 2} \dfrac{x + 3}{x^2 - 9}$

11. $\lim\limits_{x \to -2} \dfrac{x^2 - 2x - 3}{x^2 + 5x + 6}$ 12. $\lim\limits_{x \to 3} \sqrt{2x^3 - 5}$

13. $\lim\limits_{x \to 3} \dfrac{4x - 3}{\sqrt{x + 1}}$ 14. $\lim\limits_{x \to 1^+} \dfrac{x - 1}{x(x - 1)}$

15. $\lim\limits_{x \to 1^-} \dfrac{\sqrt{x} - 1}{x - 1}$ 16. $\lim\limits_{x \to \infty} \dfrac{x^2}{x^2 - 1}$

17. $\lim\limits_{x \to -\infty} \dfrac{x + 1}{x}$ 18. $\lim\limits_{x \to \infty} \dfrac{3x^2 + 2x + 4}{2x^2 - 3x + 1}$

19. $\lim\limits_{x \to -\infty} \dfrac{x^2}{x + 1}$

20. Sketch the graph of the function

    $$f(x) = \begin{cases} 2x - 3 & \text{if } x \leq 2 \\ -x + 3 & \text{if } x > 2 \end{cases}$$

    and evaluate $\lim\limits_{x \to a^+} f(x)$, $\lim\limits_{x \to a^-} f(x)$, and $\lim\limits_{x \to a} f(x)$ at the point $a = 2$, if the limits exist.

21. Sketch the graph of the function

    $$f(x) = \begin{cases} 4 - x & \text{if } x \leq 2 \\ x + 2 & \text{if } x > 2 \end{cases}$$

    and evaluate $\lim\limits_{x \to a^+} f(x)$, $\lim\limits_{x \to a^-} f(x)$, and $\lim\limits_{x \to a} f(x)$ at the point $a = 2$, if the limits exist.

In exercises 22–25, determine all values of x for which each function is discontinuous.

22. $g(x) = \begin{cases} x + 3 & \text{if } x \neq 2 \\ 0 & \text{if } x = 2 \end{cases}$

23. $f(x) = \dfrac{3x + 4}{4x^2 - 2x - 2}$

24. $f(x) = \begin{cases} \dfrac{1}{(x + 1)^2} & \text{if } x \neq -1 \\ 2 & \text{if } x = -1 \end{cases}$

25. $f(x) = \dfrac{|2x|}{x}$

26. Let $y = x^2 + 2$.
    a. Find the average rate of change of y with respect to x in the intervals $[1, 2]$, $[1, 1.5]$, and $[1, 1.1]$.
    b. Find the (instantaneous) rate of change of y at $x = 1$.

27. Use the definition of the derivative to find the slope of the tangent line to the graph of the function $f(x) = 3x + 5$ at any point $P(x, f(x))$ on the graph.

28. Use the definition of the derivative to find the slope of the tangent line to the graph of the function $f(x) = -1/x$ at any point $P(x, f(x))$ on the graph.

29. Use the definition of the derivative to find the slope of the tangent line to the graph of the function $f(x) = \frac{3}{2}x + 5$ at the point $(-2, 2)$ and determine an equation of the tangent line.

30. Use the definition of the derivative to find the slope of the tangent line to the graph of the function $f(x) = -x^2$ at the point $(2, -4)$ and determine an equation of the tangent line.

31. The graph of the function f is shown in the accompanying figure.

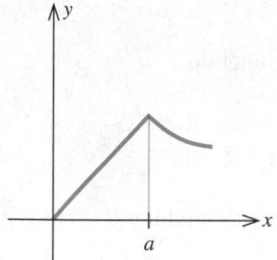

a. Is f continuous at $x = a$? Why?
b. Is f differentiable at $x = a$? Justify your answers.

32. Sales of a certain clock radio are approximated by the relationship $S(x) = 6000x + 30,000$ $(0 \le x \le 5)$, where $S(x)$ denotes the number of clock radios sold in year x ($x = 0$ corresponds to the year 1998). Find the number of clock radios expected to be sold in the year 2002.

33. A company's total sales (in millions of dollars) are approximately linear as a function of time (in years). Sales in 1994 were \$2.4 million, whereas sales in 1999 amounted to \$7.4 million.
a. Find an equation that gives the company's sales as a function of time.
b. What were the sales in 1997?

34. A company has a fixed cost of \$30,000 and a production cost of \$6 for each unit it manufactures. A unit sells for \$10.
a. What is the cost function?
b. What is the revenue function?
c. What is the profit function?
d. Compute the profit (loss) corresponding to production levels of 6000, 8000, and 12,000 units, respectively.

35. Find the point of intersection of the two straight lines having the equations $y = \frac{3}{4}x + 6$ and $3x - 2y + 3 = 0$.

36. The cost and revenue functions for a certain firm are given by $C(x) = 12x + 20,000$ and $R(x) = 20x$, respectively. Find the company's break-even point.

37. Given the demand equation $3x + p - 40 = 0$ and the supply equation $2x - p + 10 = 0$, where p is the unit price in dollars and x represents the quantity in units of a thousand, determine the equilibrium quantity and the equilibrium price.

38. Clark's Rule is a method for calculating pediatric drug dosages based on a child's weight. If a denotes the adult dosage (in mg) and if w is the weight of the child (in pounds), then the child's dosage is given by

$$D(w) = \frac{aw}{150}$$

If the adult dose of a substance is 500 mg, how much should a child who weighs 35 pounds receive?

39. The monthly revenue R (in hundreds of dollars) realized in the sale of Royal electric shavers is related to the unit price p (in dollars) by the equation

$$R(p) = -\frac{1}{2}p^2 + 30p$$

Find the revenue when an electric shaver is priced at \$30.

40. The membership of the newly opened Venus Health Club is approximated by the function

$$N(x) = 200(4 + x)^{1/2} \qquad (1 \le x \le 24)$$

where $N(x)$ denotes the number of members x months after the club's grand opening. Find $N(0)$ and $N(12)$ and interpret your results.

41. Psychologist L. L. Thurstone discovered the following model for the relationship between the learning time T and the length of a list n:

$$T = f(n) = An\sqrt{n - b}$$

where A and b are constants that depend on the person and the task. Suppose that for a certain person and a certain task, $A = 4$ and $b = 4$. Compute $f(4), f(5), \ldots,$ $f(12)$ and use this information to sketch the graph of the function f. Interpret your results.

42. The monthly demand and supply functions for the Luminar desk lamp are given by

$$p = d(x) = -1.1x^2 + 1.5x + 40$$
and
$$p = s(x) = 0.1x^2 + 0.5x + 15$$

respectively, where p is measured in dollars and x in units of a thousand. Find the equilibrium quantity and price.

43. The Photo-Mart transfers movie films to videocassettes. The fees charged for this service are shown in the following table. Find a function C relating the cost $C(x)$ to the number of feet x of film transferred. Sketch the graph of the function C and discuss its continuity.

Length of Film in Feet (x)	Price ($) for Conversion
$1 \leq x \leq 100$	5.00
$100 < x \leq 200$	9.00
$200 < x \leq 300$	12.50
$300 < x \leq 400$	15.00
$x > 400$	$7 + 0.02x$

44. The average cost (in dollars) of producing x units of a certain commodity is given by

$$\overline{C}(x) = 20 + \frac{400}{x}$$

Evaluate $\lim_{x \to \infty} \overline{C}(x)$ and interpret your results.

Additional study hints and sample chapter tests can be found at the Brooks/Cole Web site: http://www.brookscole.com/math/authors/tans/

This chapter gives several rules that will greatly simplify the task of finding the derivative of a function, thus enabling us to study how fast one quantity is changing with respect to another in many real-world situations. For example, we will be able to find how fast the population of an endangered species of whales grows after certain conservation measures have been implemented, how fast an economy's consumer price index (CPI) is changing at any time, and how fast a person's learning time changes with respect to the length of a list. We also see how these rules of differentiation facilitate the study of marginal analysis, the study of the rate of change of economic quantities. Finally, we introduce the notion of the differential of a function. Using differentials is a relatively easy way of approximating the change in one quantity in response to a small change in a related quantity.

How is a pond's oxygen content affected by organic waste? In Example 7, page 703, you will see how to find the rate at which oxygen is being restored to the pond after organic waste has been dumped into it.

DIFFERENTIATION

11.1 BASIC RULES OF DIFFERENTIATION

Four Basic Rules

The method used in Chapter 10 for computing the derivative of a function is based on a faithful interpretation of the definition of the derivative as the limit of a quotient. Thus, to find the rule for the derivative f' of a function f, we first computed the difference quotient

$$\frac{f(x + h) - f(x)}{h}$$

and then evaluated its limit as h approached zero. As you have probably observed, this method is tedious even for relatively simple functions.

The main purpose of this chapter is to derive certain rules that will simplify the process of finding the derivative of a function. Throughout this book we will use the notation

$$\frac{d}{dx}[f(x)]$$

[read "dee, dee x of f of x"] to mean "the derivative of f with respect to x at x." In stating the rules of differentiation, we assume that the functions f and g are differentiable.

RULE 1: DERIVATIVE OF A CONSTANT	$\dfrac{d}{dx}(c) = 0$ (c, a constant)

The derivative of a constant function is equal to zero.

Figure 11.1
The slope of the tangent line to the graph of $f(x) = c$, where c is a constant, is zero.

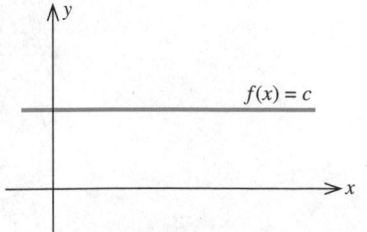

We can see this from a geometric viewpoint by recalling that the graph of a constant function is a straight line parallel to the x-axis (Figure 11.1). Since the tangent line to a straight line at any point on the line coincides with the straight line itself, its slope [as given by the derivative of $f(x) = c$] must be zero. We can also use the definition of the derivative to prove this result by computing

$$f'(x) = \lim_{h \to 0} \frac{f(x + h) - f(x)}{h}$$

$$= \lim_{h \to 0} \frac{c - c}{h}$$

$$= \lim_{h \to 0} 0 = 0$$

EXAMPLE 1

a. If $f(x) = 28$, then

$$f'(x) = \frac{d}{dx}(28) = 0$$

b. If $f(x) = -2$, then

$$f'(x) = \frac{d}{dx}(-2) = 0$$

 o o o

RULE 2: THE POWER RULE	If n is any real number, then $\dfrac{d}{dx}(x^n) = nx^{n-1}$.

Let us verify the Power Rule for the special case $n = 2$. If $f(x) = x^2$, then

$$f'(x) = \frac{d}{dx}(x^2) = \lim_{h \to 0} \frac{f(x+h) - f(x)}{h}$$

$$= \lim_{h \to 0} \frac{(x+h)^2 - x^2}{h}$$

$$= \lim_{h \to 0} \frac{x^2 + 2xh + h^2 - x^2}{h}$$

$$= \lim_{h \to 0} \frac{2xh + h^2}{h} = \lim_{h \to 0}(2x + h) = 2x$$

as we set out to show.

The proof of the Power Rule for the general case is not easy and will be omitted. However, you will be asked to prove the rule for the special case $n = 3$ in exercise 67, page 696.

EXAMPLE 2

a. If $f(x) = x$, then

$$f'(x) = \frac{d}{dx}(x) = 1 \cdot x^{1-1} = x^0 = 1$$

b. If $f(x) = x^8$, then

$$f'(x) = \frac{d}{dx}(x^8) = 8x^7$$

c. If $f(x) = x^{5/2}$, then

$$f'(x) = \frac{d}{dx}(x^{5/2}) = \frac{5}{2}x^{3/2}$$

o o o

In order to differentiate a function whose rule involves a radical, we first rewrite the rule using fractional powers. The resulting expression can then be differentiated using the Power Rule.

EXAMPLE 3   Find the derivative of the given function.

a. $f(x) = \sqrt{x}$ **b.** $g(x) = \dfrac{1}{\sqrt[3]{x}}$

Solution

a. Rewriting $\sqrt{x}$ in the form $x^{1/2}$, we obtain

$$f'(x) = \frac{d}{dx}(x^{1/2})$$

$$= \frac{1}{2}x^{-1/2} = \frac{1}{2x^{1/2}} = \frac{1}{2\sqrt{x}}$$

b. Rewriting $\dfrac{1}{\sqrt[3]{x}}$ in the form $x^{-1/3}$, we obtain

$$g'(x) = \frac{d}{dx}(x^{-1/3})$$

$$= -\frac{1}{3}x^{-4/3} = -\frac{1}{3x^{4/3}}$$ ● ● ●

RULE 3: DERIVATIVE OF A CONSTANT MULTIPLE OF A FUNCTION	$\dfrac{d}{dx}[cf(x)] = c\dfrac{d}{dx}[f(x)]$ (c, a constant)

The derivative of a constant times a differentiable function is equal to the constant times the derivative of the function.

This result follows from these computations: If $g(x) = cf(x)$, then

$$g'(x) = \lim_{h \to 0} \frac{g(x+h) - g(x)}{h} = \lim_{h \to 0} \frac{cf(x+h) - cf(x)}{h}$$

$$= c \lim_{h \to 0} \frac{f(x+h) - f(x)}{h}$$

$$= cf'(x)$$

EXAMPLE 4

a. If $f(x) = 5x^3$, then

$$f'(x) = \frac{d}{dx}(5x^3) = 5\frac{d}{dx}(x^3)$$

$$= 5(3x^2) = 15x^2$$

b. If $f(x) = \dfrac{3}{\sqrt{x}}$, then

$$f'(x) = \frac{d}{dx}(3x^{-1/2})$$

$$= 3\left(-\frac{1}{2}x^{-3/2}\right) = -\frac{3}{2x^{3/2}}$$

⊙ ⊙ ⊙

RULE 4: THE SUM RULE

$$\frac{d}{dx}[f(x) \pm g(x)] = \frac{d}{dx}[f(x)] \pm \frac{d}{dx}[g(x)]$$

The derivative of the sum or difference of two differentiable functions is equal to the sum or difference of their derivatives.

This result may be extended to the sum and difference of any finite number of differentiable functions. Let us verify the rule for a sum of two functions. If $s(x) = f(x) + g(x)$, then

$$s'(x) = \lim_{h \to 0} \frac{s(x + h) - s(x)}{h}$$

$$= \lim_{h \to 0} \frac{[f(x + h) + g(x + h)] - [f(x) + g(x)]}{h}$$

$$= \lim_{h \to 0} \frac{[f(x + h) - f(x)] + [g(x + h) - g(x)]}{h}$$

$$= \lim_{h \to 0} \frac{f(x + h) - f(x)}{h} + \lim_{h \to 0} \frac{g(x + h) - g(x)}{h}$$

$$= f'(x) + g'(x)$$

EXAMPLE 5   Find the derivatives of the following functions:

a. $f(x) = 4x^5 + 3x^4 - 8x^2 + x + 3$ **b.** $g(t) = \dfrac{t^2}{5} + \dfrac{5}{t^3}$

Solution

a. $f'(x) = \dfrac{d}{dx}(4x^5 + 3x^4 - 8x^2 + x + 3)$

$$= \dfrac{d}{dx}(4x^5) + \dfrac{d}{dx}(3x^4) - \dfrac{d}{dx}(8x^2) + \dfrac{d}{dx}(x) + \dfrac{d}{dx}(3)$$

$$= 20x^4 + 12x^3 - 16x + 1$$

b. Here the independent variable is t instead of x, so we differentiate with respect to t. Thus,

$$g'(t) = \dfrac{d}{dt}\left(\dfrac{1}{5}t^2 + 5t^{-3}\right) \quad \left(\text{Rewriting } \dfrac{1}{t^3} \text{ as } t^{-3}\right)$$

$$= \dfrac{2}{5}t - 15t^{-4}$$

$$= \dfrac{2t^5 - 75}{5t^4} \quad \left(\text{Rewriting } t^{-4} \text{ as } \dfrac{1}{t^4} \text{ and simplifying}\right) \quad \circ \ \circ \ \circ$$

EXAMPLE 6   Find the slope and an equation of the tangent line to the graph of $f(x) = 2x + \dfrac{1}{\sqrt{x}}$ at the point $(1, 3)$.

Solution   The slope of the tangent line at any point on the graph of f is given by

$$f'(x) = \dfrac{d}{dx}\left(2x + \dfrac{1}{\sqrt{x}}\right)$$

$$= \dfrac{d}{dx}(2x + x^{-1/2}) \quad \left(\text{Rewriting } \dfrac{1}{\sqrt{x}} = \dfrac{1}{x^{1/2}} = x^{-1/2}\right)$$

$$= 2 - \dfrac{1}{2}x^{-3/2} \quad \text{(Using the Sum Rule)}$$

$$= 2 - \dfrac{1}{2x^{3/2}}$$

In particular, the slope of the tangent line to the graph of f at $(1, 3)$ (where $x = 1$) is

$$f'(1) = 2 - \dfrac{1}{2(1^{3/2})} = 2 - \dfrac{1}{2} = \dfrac{3}{2}$$

Using the point-slope form of the equation of a line with slope 3/2 and the point $(1, 3)$, we see that an equation of the tangent line is

$$y - 3 = \dfrac{3}{2}(x - 1) \quad [(y - y_1) = m(x - x_1)]$$

or, upon simplification,

$$y = \dfrac{3}{2}x + \dfrac{3}{2} \qquad\qquad \circ \ \circ \ \circ$$

Applications

EXAMPLE 7 A group of marine biologists at the Neptune Institute of Oceanography recommended that a series of conservation measures be carried out over the next decade to save a certain species of whale from extinction. After the conservation measures are implemented, the population of this species is expected to be

$$N(t) = 3t^3 + 2t^2 - 10t + 600 \qquad (0 \le t \le 10)$$

where $N(t)$ denotes the population at the end of year t. Find the rate of growth of the whale population when $t = 2$ and $t = 6$. How large will the whale population be eight years after implementing the conservation measures?

Solution The rate of growth of the whale population at any time t is given by

$$N'(t) = 9t^2 + 4t - 10$$

In particular, when $t = 2$ and $t = 6$, we have

$$N'(2) = 9(2)^2 + 4(2) - 10$$
$$= 34$$
$$N'(6) = 9(6)^2 + 4(6) - 10$$
$$= 338$$

so the whale population's rate of growth will be 34 whales per year after two years and 338 per year after six years.

The whale population at the end of the eighth year will be

$$N(8) = 3(8)^3 + 2(8)^2 - 10(8) + 600$$
$$= 2184$$

or 2184 whales. The graph of the function N appears in Figure 11.2. Note the rapid growth of the population in the later years, as the conservation measures begin to pay off, compared to the growth in the early years. ○ ○ ○

Figure 11.2
The whale population after year t is given by N(t).

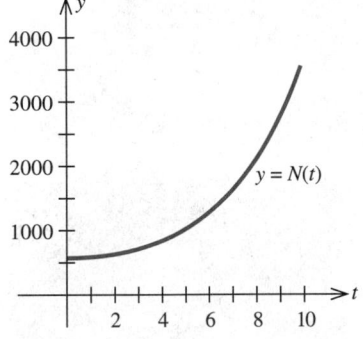

EXAMPLE 8   The altitude of a rocket (in feet) t seconds into flight is given by

$$s = f(t) = -t^3 + 96t^2 + 195t + 5 \qquad (t \ge 0)$$

a. Find an expression v for the rocket's velocity at any time t.

b. Compute the rocket's velocity when $t = 0, 30, 50, 65,$ and 70. Interpret your results.

c. Using the results from the solution to (b) and the observation that at the highest point in its trajectory the rocket's velocity is zero, find the maximum altitude attained by the rocket.

Solution

a. The rocket's velocity at any time t is given by

$$v = f'(t) = -3t^2 + 192t + 195$$

b. The rocket's velocity when $t = 0, 30, 50, 65,$ and 70 is given by

$$f'(0) = -3(0)^2 + 192(0) + 195 = 195$$

$$f'(30) = -3(30)^2 + 192(30) + 195 = 3255$$

$$f'(50) = -3(50)^2 + 192(50) + 195 = 2295$$

$$f'(65) = -3(65)^2 + 192(65) + 195 = 0$$

$$f'(70) = -3(70)^2 + 192(70) + 195 = -1065$$

or 195, 3255, 2295, 0, and −1065 feet per second.

Thus, the rocket has an initial velocity of 195 ft/sec at $t = 0$ and accelerates to a velocity of 3255 ft/sec at $t = 30$. Fifty seconds into the flight, the rocket's velocity is 2295 ft/sec, which is less than the velocity at $t = 30$. This means that the rocket begins to decelerate after an initial period of acceleration. (Later on we will learn how to determine the rocket's maximum velocity.)

The deceleration continues: The velocity is 0 ft/sec at $t = 65$ and −1065 ft/sec when $t = 70$. This number tells us that 70 seconds into flight the rocket is heading back to Earth with a speed of 1065 ft/sec.

c. The results of (b) show that the rocket's velocity is zero when $t = 65$. At this instant, the rocket's maximum altitude is

$$s = f(65) = -(65)^3 + 96(65)^2 + 195(65) + 5$$

$$= 143,655$$

or 143,655 feet. A sketch of the graph of f appears in Figure 11.3.   ○ ○ ○

Figure 11.3

The rocket's altitude t seconds into flight is given by f(t).

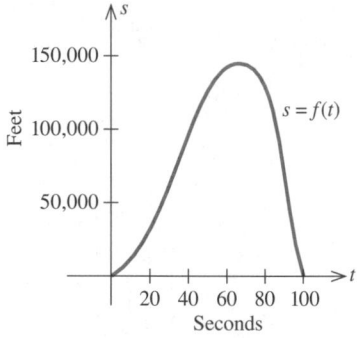

EXPLORING WITH TECHNOLOGY

Refer to Example 8.

1. Use a graphing utility to plot the graph of the velocity function

$$v = f'(t) = -3t^2 + 192t + 195$$

using the viewing rectangle $[0, 120] \times [-5000, 5000]$. Then using **ZOOM** and **TRACE** or the root-finding capability of your graphing utility, verify that $f'(65) = 0$.

2. Plot the graph of the position function of the rocket

$$s = f(t) = -t^3 + 96t^2 + 195t + 5$$

using the viewing rectangle $[0, 120] \times [0, 150,000]$. Then, using **ZOOM** and **TRACE** repeatedly, verify that the maximum altitude of the rocket is 143,655 ft.

(Continued on next page)

(*Continued from previous page*)
3. Use **ZOOM** and **TRACE** or the root-finding capability of your graphing utility to find when the rocket returns to Earth.

○   ○   ○

SELF–CHECK EXERCISES 11.1

1. Find the derivative of each of the following functions using the rules of differentiation:
   a. $f(x) = 1.5x^2 + 2x^{1.5}$
   b. $g(x) = 2\sqrt{x} + \dfrac{3}{\sqrt{x}}$

2. Let $f(x) = 2x^3 - 3x^2 + 2x - 1$.
   a. Compute $f'(x)$.
   b. What is the slope of the tangent line to the graph of f when $x = 2$?
   c. What is the rate of change of the function f at $x = 2$?

3. A certain country's gross domestic product (GDP) (in millions of dollars) is described by the function

$$G(t) = -2t^3 + 45t^2 + 20t + 6000 \qquad (0 \le t \le 11)$$

where $t = 0$ corresponds to the beginning of 1988.
   a. At what rate was the GDP changing at the beginning of 1993? At the beginning of 1995? At the beginning of 1998?
   b. What was the average rate of growth of the GDP over the period 1993 to 1998?

Solutions to Self-Check Exercises 11.1 can be found on page 696.

11.1 EXERCISES

In exercises 1–34, find the derivative of the function f by using the rules of differentiation.

1. $f(x) = -3$

2. $f(x) = 365$

3. $f(x) = x^5$

4. $f(x) = x^7$

5. $f(x) = x^{2.1}$

6. $f(x) = x^{0.8}$

7. $f(x) = 3x^2$

8. $f(x) = -2x^3$

9. $f(r) = \pi r^2$

10. $f(r) = \frac{4}{3}\pi r^3$

11. $f(x) = 9x^{1/3}$

12. $f(x) = \frac{5}{4}x^{4/5}$

13. $f(x) = 3\sqrt{x}$

14. $f(u) = \dfrac{2}{\sqrt{u}}$

15. $f(x) = 7x^{-12}$

16. $f(x) = 0.3x^{-1.2}$

17. $f(x) = 5x^2 - 3x + 7$

18. $f(x) = x^3 - 3x^2 + 1$

19. $f(x) = -x^3 + 2x^2 - 6$

20. $f(x) = x^4 - 2x^2 + 5$

21. $f(x) = 0.03x^2 - 0.4x + 10$

22. $f(x) = 0.002x^3 - 0.05x^2 + 0.1x - 20$

23. $f(x) = \dfrac{x^3 - 4x^2 + 3}{x}$

USING TECHNOLOGY

FINDING THE RATE OF CHANGE OF A FUNCTION

We can use the numerical derivative operation of a graphing utility to obtain the value of the derivative at a given value of x. Since the derivative of a function $f(x)$ measures the rate of change of the function with respect to x, the numerical derivative operation can be used to answer questions pertaining to the rate of change of one quantity y with respect to another quantity x, where $y = f(x)$, for a specific value of x.

EXAMPLE 1 Let $y = 3t^3 + 2\sqrt{t}$.

a. Use the numerical derivative operation of a graphing utility to find how fast y is changing with respect to t when $t = 1$.

b. Verify the result of (a) using the rules of differentiation of this section.

Solution

a. Write $f(t) = 3t^3 + 2\sqrt{t}$. Using the numerical derivative operation of a graphing utility, we find that the rate of change of y with respect to t when $t = 1$ is given by $f'(1) = 10$.

b. Here $f(t) = 3t^3 + 2t^{1/2}$ and

$$f'(t) = 9t^2 + 2\left(\frac{1}{2}t^{-1/2}\right) = 9t^2 + \frac{1}{\sqrt{t}}$$

Using this result, we see that when $t = 1$, y is changing at the rate of

$$f'(1) = 9(1^2) + \frac{1}{\sqrt{1}} = 10$$

or 10 units per unit change in t, as obtained earlier.　　　　○ ○ ○

EXAMPLE 2 According to the U.S. Department of Energy and the Shell Development Company, a typical car's fuel economy depends on the speed it is driven and is approximated by the function

$$f(x) = 0.00000310315x^4 - 0.000455174x^3$$
$$+ 0.00287869x^2 + 1.25986x \qquad (0 \leq x \leq 75)$$

where x is measured in mph and $f(x)$ is measured in miles per gallon (mpg).

a. Use a graphing utility to graph the function f on the interval $[0, 75]$.

b. Find the rate of change of f when $x = 20$. When $x = 50$.

c. Interpret your results.

Source: U.S. Department of Energy and the Shell Development Company

Solution

a. The result is shown in Figure T1.

b. Using the numerical derivative operation of a graphing utility, we see that $f'(20) = 0.9280996$. The rate of change of f when $x = 50$ is given by $f'(50) = -0.314501$.

Figure T1

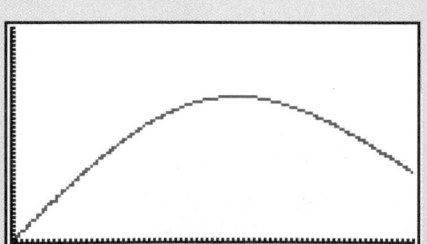

c. The results of part (b) tell us that when a typical car is being driven at 20 mph, its fuel economy increases at the rate of approximately 0.9 mpg per 1 mph increase in its speed. At a speed of 50 mph, its fuel economy decreases at the rate of approximately 0.3 mpg per 1 mph increase in its speed. ○ ○ ○

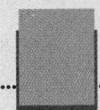

EXERCISES

In exercises 1–6, use the numerical derivative operation of a graphing utility to find the rate of change of f(x) at the given value of x. Give your answer accurate to four decimal places.

1. $f(x) = 4x^5 - 3x^3 + 2x^2 + 1$; $x = 0.5$

2. $f(x) = -x^5 + 4x^2 + 3$; $x = 0.4$

3. $f(x) = x - 2\sqrt{x}$; $x = 3$

4. $f(x) = \dfrac{\sqrt{x} - 1}{x}$; $x = 2$

5. $f(x) = x^{1/2} - x^{1/3}$; $x = 1.2$

6. $f(x) = 2x^{5/4} + x$; $x = 2$

7. **Carbon Monoxide in the Atmosphere** The projected average global atmospheric concentration of carbon monoxide is approximated by the function

$$f(t) = 0.881443t^4 - 1.45533t^3 + 0.695876t^2$$
$$+ 2.87801t + 293 \qquad (0 \le t \le 4)$$

where t is measured in 40-year intervals, with t = 0 corresponding to the beginning of 1860 and f(t) is measured in parts per million by volume.
a. Use a graphing utility to plot the graph of f in the viewing rectangle [0, 4] × [280, 400].
b. Use a graphing utility to estimate how fast the projected average global atmospheric concentration of carbon monoxide was changing at the beginning of the year 1900 (t = 1). At the beginning of 1990 (t = 3.5).
Source: "Beyond the Limits," Meadows et al.

8. **Growth of HMOs** Based on data compiled by the Group Health Association of America, the number of people receiving their care in an HMO (health maintenance organization) from the beginning of 1984 through 1994 is approximated by the function

$$f(t) = 0.0514t^3 - 0.853t^2 + 6.8147t$$
$$+ 15.6524 \qquad (0 \le t \le 11)$$

where f(t) gives the number of people in millions and t is measured in years, with t = 0 corresponding to the beginning of 1984.
a. Use a graphing utility to plot the graph of f in the viewing window [0, 12] × [0, 80].
b. How fast was the number of people receiving their care in an HMO growing at the beginning of 1992?
Source: Group Health Association of America

9. **Home Sales** According to the Greater Boston Real Estate Board—Multiple Listing Service, the average number of days a single-family home remains for sale from listing to accepted offer is approximated by the function

$$f(t) = 0.0171911t^4 - 0.662121t^3 + 6.18083t^2$$
$$- 8.97086t + 53.3357 \qquad (0 \le t \le 10)$$

where t is measured in years, with t = 0 corresponding to the beginning of 1984.
a. Use a graphing utility to plot the graph of f in the viewing rectangle [0, 12] × [0, 120].
b. How fast is the average number of days a single-family home remains for sale from listing to accepted offer changing at the beginning of 1984 (t = 0)? At the beginning of 1988 (t = 4)?
Source: Greater Boston Real Estate Board—Multiple Listing Service

10. **Manufacturing Capacity** Data obtained from the Federal Reserve show that the annual change in manufacturing capacity between 1988 and 1994 is given by

$$f(t) = 0.0388889t^3 - 0.283333t^2$$
$$+ 0.477778t + 2.04286 \qquad (0 \le t \le 6)$$

where f(t) is a percentage and t is measured in years, with t = 0 corresponding to the beginning of 1988.
a. Use a graphing utility to plot the graph of f in the viewing rectangle [0, 8] × [0, 4].
b. How fast was f(t) changing at the beginning of 1990 (t = 2)? At the beginning of 1992 (t = 4)?
Source: Federal Reserve

24. $f(x) = \dfrac{x^3 + 2x^2 + x - 1}{x}$

25. $f(x) = 4x^4 - 3x^{5/2} + 2$

26. $f(x) = 5x^{4/3} - \frac{2}{3}x^{3/2} + x^2 - 3x + 1$

27. $f(x) = 3x^{-1} + 4x^{-2}$ **28.** $f(x) = -\frac{1}{3}(x^{-3} - x^6)$

29. $f(t) = \dfrac{4}{t^4} - \dfrac{3}{t^3} + \dfrac{2}{t}$

30. $f(x) = \dfrac{5}{x^3} - \dfrac{2}{x^2} - \dfrac{1}{x} + 200$

31. $f(x) = 2x - 5\sqrt{x}$ **32.** $f(t) = 2t^2 + \sqrt{t^3}$

33. $f(x) = \dfrac{2}{x^2} - \dfrac{3}{x^{1/3}}$ **34.** $f(x) = \dfrac{3}{x^3} + \dfrac{4}{\sqrt{x}} + 1$

35. Let $f(x) = 2x^3 - 4x$. Find:
  a. $f'(-2)$ **b.** $f'(0)$ **c.** $f'(2)$

36. Let $f(x) = 4x^{5/4} + 2x^{3/2} + x$. Find:
  a. $f'(0)$ **b.** $f'(16)$

In exercises 37–40, find the given limit by evaluating the derivative of a suitable function at an appropriate point. [Hint: Look at the definition of the derivative.]

37. $\displaystyle\lim_{h\to 0} \dfrac{(1+h)^3 - 1}{h}$ **38.** $\displaystyle\lim_{x\to 1} \dfrac{x^5 - 1}{x - 1}$

[*Hint:* Let $h = x - 1$.]

39. $\displaystyle\lim_{h\to 0} \dfrac{3(2+h)^2 - (2+h) - 10}{h}$

40. $\displaystyle\lim_{t\to 0} \dfrac{1 - (1+t)^2}{t(1+t)^2}$

In exercises 41–44, find the slope and an equation of the tangent line to the graph of the function f at the specified point.

41. $f(x) = 2x^2 - 3x + 4$; $(2, 6)$

42. $f(x) = -\frac{5}{3}x^2 + 2x + 2$; $(-1, -\frac{5}{3})$

43. $f(x) = x^4 - 3x^3 + 2x^2 - x + 1$; $(1, 0)$

44. $f(x) = \sqrt{x} + \dfrac{1}{\sqrt{x}}$; $\left(4, \dfrac{5}{2}\right)$

45. Let $f(x) = x^3$.
  a. Find the point on the graph of f where the tangent line is horizontal.
  b. Sketch the graph of f and draw the horizontal tangent line.

46. Let $f(x) = x^3 - 4x^2$. Find the point(s) on the graph of f where the tangent line is horizontal.

47. Let $f(x) = x^3 + 1$.
  a. Find the point(s) on the graph of f where the slope of the tangent line is equal to 12.
  b. Find the equation(s) of the tangent line(s) of (a).
  c. Sketch the graph of f showing the tangent line(s).

48. Let $f(x) = \frac{2}{3}x^3 + x^2 - 12x + 6$. Find the values of x for which:
  a. $f'(x) = -12$ **b.** $f'(x) = 0$
  c. $f'(x) = 12$

49. Let $f(x) = \frac{1}{4}x^4 - \frac{1}{3}x^3 - x^2$. Find the point(s) on the graph of f where the slope of the tangent line is equal to:
  a. $-2x$ **b.** 0 **c.** $10x$

50. A straight line perpendicular to and passing through the point of tangency of the tangent line is called the *normal* to the curve. Find an equation of the tangent line and the normal to the curve $y = x^3 - 3x + 1$ at the point $(2, 3)$.

51. Growth of a Cancerous Tumor The volume of a spherical cancer tumor is given by the function

$$V(r) = \frac{4}{3}\pi r^3$$

where r is the radius of the tumor in centimeters. Find the rate of change in the volume of the tumor when:
  a. $r = \frac{2}{3}$ cm   **b.** $r = \frac{5}{4}$ cm

52. Velocity of Blood in an Artery The velocity (in centimeters per second) of blood r centimeters from the central axis of an artery is given by

$$v(r) = k(R^2 - r^2)$$

where k is a constant and R is the radius of the artery (see the accompanying figure). Suppose that $k = 1000$ and $R = 0.2$ cm. Find $v(0.1)$ and $v'(0.1)$ and interpret your results.

R

r

Blood vessel

53. Effect of Stopping on Average Speed According to data from a study by General Motors, the average speed of your trip A (in mph) is related to the number of stops per mile you make on the trip x by the equation

$$A = \frac{26.5}{x^{0.45}}$$

Compute dA/dx for $x = 0.25$ and $x = 2$, and interpret your results.
Source: General Motors

54. **Worker Efficiency**   An efficiency study conducted for the Elektra Electronics Company showed that the number of "Space Commander" walkie-talkies assembled by the average worker t hours after starting work at 8 A.M. is given by

$$N(t) = -t^3 + 6t^2 + 15t$$

   a. Find the rate at which the average worker will be assembling walkie-talkies t hours after starting work.
   b. At what rate will the average worker be assembling walkie-talkies at 10 A.M.? At 11 A.M.?
   c. How many walkie-talkies will the average worker assemble between 10 A.M. and 11 A.M.?

55. **Consumer Price Index**   An economy's consumer price index (CPI) is described by the function

$$I(t) = -0.2t^3 + 3t^2 + 100 \qquad (0 \le t \le 10)$$

   where $t = 0$ corresponds to 1988.
   a. At what rate was the CPI changing in 1993? In 1995? In 1998?
   b. What was the average rate of increase in the CPI over the period from 1993 to 1998?

56. **Effect of Advertising on Sales**   The relationship between the amount of money x that the Cannon Precision Instruments Corporation spends on advertising and the company's total sales $S(x)$ is given by the function

$$S(x) = -0.002x^3 + 0.6x^2 + x + 500 \qquad (0 \le x \le 200)$$

   where x is measured in thousands of dollars. Find the rate of change of the sales with respect to the amount of money spent on advertising. Are Cannon's total sales increasing at a faster rate when the amount of money spent on advertising is $100,000 or $150,000?

57. **Population Growth**   A study prepared for a Sunbelt town's Chamber of Commerce projected that the town's population in the next three years will grow according to the rule

$$P(t) = 50,000 + 30t^{3/2} + 20t$$

where $P(t)$ denotes the population t months from now. How fast will the population be increasing nine months from now? Sixteen months from now?

58. **Curbing Population Growth**   Five years ago, the government of a Pacific island state launched an extensive propaganda campaign toward curbing the country's population growth. According to the Census Department, the population (measured in thousands of people) for the following four years was

$$P(t) = -\frac{1}{3}t^3 + 64t + 3000$$

   where t is measured in years and $t = 0$ at the start of the campaign. Find the rate of change of the population at the end of years 1, 2, 3, and 4. Was the plan working?

59. **Conservation of Species**   A certain species of turtle faces extinction because dealers collect truckloads of turtle eggs to be sold as aphrodisiacs. After severe conservation measures are implemented, it is hoped that the turtle population will grow according to the rule

$$N(t) = 2t^3 + 3t^2 - 4t + 1000 \qquad (0 \le t \le 10)$$

   where $N(t)$ denotes the population at the end of year t. Find the rate of growth of the turtle population when $t = 2$ and $t = 8$. What will the population be ten years after the conservation measures are implemented?

60. **Flight of a Rocket**   The altitude (in feet) of a rocket t seconds into flight is given by

$$s = f(t) = -2t^3 + 114t^2 + 480t + 1 \qquad (t \ge 0)$$

   a. Find an expression v for the rocket's velocity at any time t.
   b. Compute the rocket's velocity when $t = 0, 20, 40$, and 60. Interpret your results.
   c. Using the results from the solution to (b), find the maximum altitude attained by the rocket.
   [*Hint:* At its highest point, the velocity of the rocket is zero.]

61. **Stopping Distance of a Racing Car**   During a test by the editors of an auto magazine, the stopping distance s (in feet) of the MacPherson X-2 racing car conformed to the rule

$$s = f(t) = 120t - 15t^2 \qquad (t \ge 0)$$

where t was the time (in seconds) after the brakes were applied.
   a. Find an expression for the car's velocity v at any time t.

b. What was the car's velocity when the brakes were first applied?

c. What was the car's stopping distance for that particular test?

[*Hint:* The stopping time is found by setting $v = 0$.]

62. Demand Functions   The demand function for the Luminar desk lamp is given by

$$p = f(x) = -0.1x^2 - 0.4x + 35$$

where x is the quantity demanded (measured in thousands) and p is the unit price in dollars.

a. Find $f'(x)$.

b. What is the rate of change of the unit price when the quantity demanded is 10,000 units ($x = 10$)? What is the unit price at that level of demand?

63. Increase in Temporary Workers   According to the Labor Department, the number of temporary workers (in millions) is estimated to be

$$N(t) = 0.025t^2 + 0.255t + 1.505 \qquad (0 \le t \le 5)$$

where t is measured in years, with $t = 0$ corresponding to 1991.

a. How many temporary workers were there at the beginning of 1994?

b. How fast was the number of temporary workers growing at the beginning of 1994?

Source: U.S. Department of Labor

64. Supply Functions   The supply function for a certain make of transistor radio is given by

$$p = f(x) = 0.1\sqrt{x} + 10$$

where x is the quantity supplied and p is the unit price in dollars.

a. Find $f'(x)$.

b. What is the rate of change of the unit price if the quantity supplied is 40,000 transistor radios?

65. Average Speed of a Vehicle on a Highway   The average speed of a vehicle on a stretch of Route 134 between

6 A.M. and 10 A.M. on a typical weekday is approximated by the function

$$f(t) = 20t - 40\sqrt{t} + 50 \qquad (0 \le t \le 4)$$

where $f(t)$ is measured in miles per hour and t is measured in hours, with $t - 0$ corresponding to 6 A.M.

a. Compute $f'(t)$.

b. Compute $f(0)$, $f(1)$, and $f(2)$, and interpret your results.

c. Compute $f'(\frac{1}{2})$, $f'(1)$, and $f'(2)$, and interpret your results.

66. Health-Care Spending   Despite efforts at cost containment, the cost of the Medicare program is increasing at a high rate. Two major reasons for this increase are an aging population and the constant development and extensive use by physicians of new technologies. Based on data from the Health Care Financing Administration and the U.S. Census Bureau, health-care spending through the year 2000 may be approximated by the function

$$S(t) = 0.02836t^3 - 0.05167t^2 + 9.60881t$$
$$+ 41.9 \qquad (0 \le t \le 35)$$

where $S(t)$ is the spending in billions of dollars and t is measured in years, with $t = 0$ corresponding to the beginning of 1965.

a. Find an expression for the rate of change of health-care spending at any time t.

b. How fast was health-care spending changing at the beginning of 1980? How fast will health-care spending be changing at the beginning of 2000?

c. What was the amount of health-care spending at the beginning of 1980? What will the amount of health-care spending be at the beginning of 2000?

Source: Health Care Financing Administration

67. Prove the Power Rule (Rule 2) for the special case $n = 3$.

$$\left[\textit{Hint: Compute } \lim_{h \to 0} \left[\frac{(x + h)^3 - x^3}{h} \right]. \right]$$

SOLUTIONS TO SELF-CHECK EXERCISES 11.1

1. a. $f'(x) = \dfrac{d}{dx}(1.5x^2) + \dfrac{d}{dx}(2x^{1.5})$

$$= (1.5)(2x) + (2)(1.5x^{0.5})$$

$$= 3x + 3\sqrt{x} = 3(x + \sqrt{x})$$

b. $g'(x) = \dfrac{d}{dx}(2x^{1/2}) + \dfrac{d}{dx}(3x^{-1/2})$

$\qquad = (2)\left(\dfrac{1}{2}x^{-1/2}\right) + (3)\left(-\dfrac{1}{2}x^{-3/2}\right)$

$\qquad = x^{-1/2} - \dfrac{3}{2}x^{-3/2}$

$\qquad = \dfrac{1}{2}x^{-3/2}(2x - 3) = \dfrac{2x - 3}{2x^{3/2}}$

2. a. $f'(x) = \dfrac{d}{dx}(2x^3) - \dfrac{d}{dx}(3x^2) + \dfrac{d}{dx}(2x) - \dfrac{d}{dx}(1)$

$\qquad = (2)(3x^2) - (3)(2x) + 2$

$\qquad = 6x^2 - 6x + 2$

b. The slope of the tangent line to the graph of f when $x = 2$ is given by

$$f'(2) = 6(2)^2 - 6(2) + 2 = 14$$

c. The rate of change of f at $x = 2$ is given by $f'(2)$. Using the results of (b), we see that $f'(2)$ is 14 units per unit change in x.

3. a. The rate at which the GDP was changing at any time t $(0 < t < 11)$ is given by

$$G'(t) = -6t^2 + 90t + 20$$

In particular, the rates of change of the GDP at the beginning of the years 1993 $(t = 5)$, 1995 $(t = 7)$, and 1998 $(t = 10)$ are given by

$$G'(5) = 320, \quad G'(7) = 356, \quad \text{and} \quad G'(10) = 320$$

respectively; that is, by \$320 million per year, \$356 million per year, and \$320 million per year.

b. The average rate of growth of the GDP over the period from the beginning of 1993 $(t = 5)$ to the beginning of 1998 $(t = 10)$ is given by

$$\frac{G(10) - G(5)}{10 - 5} = \frac{[-2(10)^3 + 45(10)^2 + 20(10) + 6000]}{5}$$

$$- \frac{[-2(5)^3 + 45(5)^2 + 20(5) + 6000]}{5}$$

$$= \frac{8700 - 6975}{5}$$

or \$345 million per year.

11.2 THE PRODUCT AND QUOTIENT RULES

In this section, we study two more rules of differentiation: the **Product Rule** and the **Quotient Rule.**

The Product Rule

The derivative of the product of two differentiable functions is given by the following rule:

RULE 5: THE PRODUCT RULE	$$\frac{d}{dx}[f(x)g(x)] = f(x)g'(x) + g(x)f'(x)$$

The derivative of the product of two functions is the first function times the derivative of the second plus the second function times the derivative of the first.

The Product Rule may be extended to the case involving the product of any finite number of functions (see exercise 59, page 707). We prove the Product Rule at the end of this section.

 The derivative of the product of two functions is not given by the product of the derivatives of the functions; that is, in general

$$\frac{d}{dx}[f(x)g(x)] \neq f'(x)g'(x)$$

EXAMPLE 1   Find the derivative of the function

$$f(x) = (2x^2 - 1)(x^3 + 3)$$

Solution   By the Product Rule,

$$f'(x) = (2x^2 - 1)\frac{d}{dx}(x^3 + 3) + (x^3 + 3)\frac{d}{dx}(2x^2 - 1)$$
$$= (2x^2 - 1)(3x^2) + (x^3 + 3)(4x)$$
$$= 6x^4 - 3x^2 + 4x^4 + 12x$$
$$= 10x^4 - 3x^2 + 12x$$
$$= x(10x^3 - 3x + 12)$$

⬤ ⬤ ⬤

EXAMPLE 2   Differentiate (that is, find the derivative of) the function

$$f(x) = x^3(\sqrt{x} + 1)$$

Solution   First, we express the function in exponential form, obtaining

$$f(x) = x^3(x^{1/2} + 1)$$

By the Product Rule,

$$f'(x) = x^3 \frac{d}{dx}(x^{1/2} + 1) + (x^{1/2} + 1)\frac{d}{dx}x^3$$

$$= x^3\left(\frac{1}{2}x^{-1/2}\right) + (x^{1/2} + 1)(3x^2)$$

$$= \frac{1}{2}x^{5/2} + 3x^{5/2} + 3x^2$$

$$= \frac{7}{2}x^{5/2} + 3x^2$$

⊙ ⊙ ⊙

REMARK   We can also solve the problem by first expanding the product before differentiating *f*. Examples for which this is not possible will be considered in Section 11.3, where the true value of the Product Rule will be appreciated.

⊙ ⊙ ⊙

The Quotient Rule

The derivative of the quotient of two differentiable functions is given by the following rule:

RULE 6: THE QUOTIENT RULE	$\dfrac{d}{dx}\left[\dfrac{f(x)}{g(x)}\right] = \dfrac{g(x)f'(x) - f(x)g'(x)}{[g(x)]^2} \qquad [g(x) \neq 0]$

As an aid to remembering this expression, observe that it has the following form:

$$\frac{d}{dx}\left[\frac{f(x)}{g(x)}\right] = \frac{(\text{Denominator})\left(\begin{array}{c}\text{Derivative of}\\\text{numerator}\end{array}\right) - (\text{Numerator})\left(\begin{array}{c}\text{Derivative of}\\\text{denominator}\end{array}\right)}{(\text{Square of denominator})}$$

For a proof of the Quotient Rule, see exercise 60, page 707.

⚠ The derivative of a quotient is not equal to the quotient of the derivatives; that is,

$$\frac{d}{dx}\left[\frac{f(x)}{g(x)}\right] \neq \frac{f'(x)}{g'(x)}$$

For example, if $f(x) = x^3$ and $g(x) = x^2$, then

$$\frac{d}{dx}\left[\frac{f(x)}{g(x)}\right] = \frac{d}{dx}\left(\frac{x^3}{x^2}\right) = \frac{d}{dx}(x) = 1$$

which is not equal to

$$\frac{f'(x)}{g'(x)} = \frac{\dfrac{d}{dx}(x^3)}{\dfrac{d}{dx}(x^2)} = \frac{3x^2}{2x} = \frac{3}{2}x$$

EXAMPLE 3   Find $f'(x)$ if $f(x) = \dfrac{x}{2x-4}$.

Solution   Using the Quotient Rule, we obtain

$$f'(x) = \frac{(2x-4)\dfrac{d}{dx}(x) - x\dfrac{d}{dx}(2x-4)}{(2x-4)^2}$$

$$= \frac{(2x-4)(1) - x(2)}{(2x-4)^2}$$

$$= \frac{2x-4-2x}{(2x-4)^2} = -\frac{4}{(2x-4)^2}$$

EXAMPLE 4   Find $f'(x)$ if $f(x) = \dfrac{x^2+1}{x^2-1}$.

Solution   By the Quotient Rule,

$$f'(x) = \frac{(x^2-1)\dfrac{d}{dx}(x^2+1) - (x^2+1)\dfrac{d}{dx}(x^2-1)}{(x^2-1)^2}$$

$$= \frac{(x^2-1)(2x) - (x^2+1)(2x)}{(x^2-1)^2}$$

$$= \frac{2x^3 - 2x - 2x^3 - 2x}{(x^2-1)^2}$$

$$= -\frac{4x}{(x^2-1)^2}$$

EXAMPLE 5   Find $h'(x)$ if

$$h(x) = \frac{\sqrt{x}}{x^2+1}$$

Solution Rewrite $h(x)$ in the form $h(x) = \dfrac{x^{1/2}}{x^2 + 1}$. By the Quotient Rule, we find

$$h'(x) = \frac{(x^2 + 1)\dfrac{d}{dx}(x^{1/2}) - x^{1/2}\dfrac{d}{dx}(x^2 + 1)}{(x^2 + 1)^2}$$

$$= \frac{(x^2 + 1)\left(\dfrac{1}{2}x^{-1/2}\right) - x^{1/2}(2x)}{(x^2 + 1)^2}$$

$$= \frac{\dfrac{1}{2}x^{-1/2}(x^2 + 1 - 4x^2)}{(x^2 + 1)^2} \qquad \text{(Factoring out } \tfrac{1}{2}x^{-1/2} \text{ from the numerator)}$$

$$= \frac{1 - 3x^2}{2\sqrt{x}(x^2 + 1)^2} \qquad\qquad\qquad\qquad \text{◗ ◗ ◗}$$

Application

EXAMPLE 6 The sales (in millions of dollars) of a laser disc recording of a hit movie t years from the date of release are given by

$$S(t) = \frac{5t}{t^2 + 1}$$

a. Find the rate at which the sales are changing at time t.

b. How fast are the sales changing at the time the laser discs are released ($t = 0$)? Two years from the date of release?

Solution

a. The rate at which the sales are changing at time t is given by $S'(t)$. Using the Quotient Rule, we obtain

$$S'(t) = \frac{d}{dt}\left[\frac{5t}{t^2 + 1}\right] = 5\frac{d}{dt}\left[\frac{t}{t^2 + 1}\right]$$

$$= 5\left[\frac{(t^2 + 1)(1) - t(2t)}{(t^2 + 1)^2}\right]$$

$$= 5\left[\frac{t^2 + 1 - 2t^2}{(t^2 + 1)^2}\right] = \frac{5(1 - t^2)}{(t^2 + 1)^2}$$

b. The rate at which the sales are changing at the time the laser discs are released is given by

$$S'(0) = \frac{5(1 - 0)}{(0 + 1)^2} = 5$$

That is, they are increasing at the rate of $5 million per year.

Two years from the date of release, the sales are changing at the rate of

$$S'(2) = \frac{5(1-4)}{(4+1)^2} = -\frac{3}{5} = -0.6$$

That is, they are decreasing at the rate of $600,000 per year.
The graph of the function S is shown in Figure 11.4.

Figure 11.4
After a spectacular rise, the sales begin to taper off.

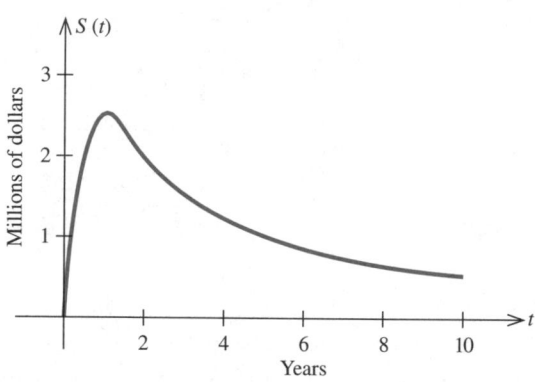

○ ○ ○

Suppose the revenue of a company is given by $R(x) = xp(x)$, where x is the number of units of the product sold at a unit price of $p(x)$ dollars.

a. Compute $R'(x)$ and explain, in words, the relationship between $R'(x)$ and $p(x)$ and/or its derivative.

b. What can you say about $R'(x)$ if $p(x)$ is constant? Is this expected?

EXAMPLE 7   When organic waste is dumped into a pond, the oxidation
process that takes place reduces the pond's oxygen content.
However, given time, nature will restore the oxygen content to its natural
level. Suppose the oxygen content t days after organic waste has been dumped
into the pond is given by

$$f(t) = 100 \left[\frac{t^2 - 4t + 4}{t^2 + 4} \right] \qquad (0 \le t < \infty)$$

percent of its normal level.

a. Derive a general expression that gives the rate of change of the pond's
oxygen level at any time t.

b. How fast is the pond's oxygen content changing one day after the organic
waste has been dumped? Two days after? Three days after?

Solution

a. The rate of change of the pond's oxygen level at any time t is given by the
derivative of the function f. Thus, the required expression is

$$f'(t) = 100 \frac{d}{dt} \left[\frac{t^2 - 4t + 4}{t^2 + 4} \right]$$

$$= 100 \left[\frac{(t^2 + 4)(2t - 4) - (t^2 - 4t + 4)(2t)}{(t^2 + 4)^2} \right]$$

$$= 100 \left[\frac{2t^3 - 4t^2 + 8t - 16 - 2t^3 + 8t^2 - 8t}{(t^2 + 4)^2} \right]$$

$$= \frac{100(4t^2 - 16)}{(t^2 + 4)^2} = \frac{400(t^2 - 4)}{(t^2 + 4)^2}$$

b. The rate at which the pond's oxygen content is changing one day after the
organic waste has been dumped is given by

$$f'(1) = \frac{400(1 - 4)}{(1 + 4)^2} = -48$$

That is, it is dropping at the rate of 48% per day. Two days after, the rate is

$$f'(2) = \frac{400(4 - 4)}{(4 + 4)^2} = 0$$

That is, it is neither increasing nor decreasing. Three days after, the rate is

$$f'(3) = \frac{400(3^2 - 4)}{(3^2 + 4)^2} = 11.83$$

That is, the oxygen content is increasing at the rate of 11.83% per day, and
the restoration process has indeed begun.   ๐ ๐ ๐

Consider a particle moving along a straight line. Newton's second law of motion states that the external force F acting on the particle is equal to the rate of change of its momentum. Thus,

$$F = \frac{d}{dt}(mv)$$

where m, the mass of the particle, and v, its velocity, are both functions of time t.

a. Use the Product Rule to show that

$$F = m\frac{dv}{dt} + v\frac{dm}{dt}$$

and explain the expression on the right-hand side in words.

b. Use the results of part (a) to show that if the mass of a particle is constant, then $F = ma$, where a is the acceleration of the particle.

Verification of the Product Rule

We will now verify the Product Rule. If $p(x) = f(x)g(x)$, then

$$p'(x) = \lim_{h \to 0} \frac{p(x+h) - p(x)}{h}$$

$$= \lim_{h \to 0} \frac{f(x+h)g(x+h) - f(x)g(x)}{h}$$

By adding $-f(x+h)g(x) + f(x+h)g(x)$ (which is zero!) to the numerator and factoring, we have

$$p'(x) = \lim_{h \to 0} \frac{f(x+h)[g(x+h) - g(x)] + g(x)[f(x+h) - f(x)]}{h}$$

$$= \lim_{h \to 0} \left\{ f(x+h) \left[\frac{g(x+h) - g(x)}{h} \right] + g(x) \left[\frac{f(x+h) - f(x)}{h} \right] \right\}$$

$$= \lim_{h \to 0} f(x+h) \left[\frac{g(x+h) - g(x)}{h} \right]$$

$$+ \lim_{h \to 0} g(x) \left[\frac{f(x+h) - f(x)}{h} \right] \qquad \text{(By property iii of limits)}$$

$$= \lim_{h \to 0} f(x+h) \cdot \lim_{h \to 0} \frac{g(x+h) - g(x)}{h}$$

$$+ \lim_{h \to 0} g(x) \cdot \lim_{h \to 0} \frac{f(x+h) - f(x)}{h} \qquad \text{(By property iv of limits)}$$

$$= f(x)g'(x) + g(x)f'(x)$$

Observe that in the second from the last link in the chain of equalities, we have used the fact that $\lim_{h \to 0} f(x + h) = f(x)$ because f is continuous at x.

SELF–CHECK EXERCISES 11.2

1. Find the derivative of $f(x) = \dfrac{2x + 1}{x^2 - 1}$.

2. What is the slope of the tangent line to the graph of

$$f(x) = (x^2 + 1)(2x^3 - 3x^2 + 1)$$

at the point $(2, 25)$? How fast is the function f changing when $x = 2$?

3. The total sales of the Security Products Corporation in its first two years of operation are given by

$$S = f(t) = \frac{0.3t^3}{1 + 0.4t^2} \qquad (0 \le t \le 2)$$

where S is measured in millions of dollars and $t = 0$ corresponds to the date Security Products began operations. How fast were the sales increasing at the beginning of the company's second year of operation?

Solutions to Self-Check Exercises 11.2 can be found on page 710.

11.2 EXERCISES

In exercises 1–30, find the derivative of the given function.

1. $f(x) = 2x(x^2 + 1)$ **2.** $f(x) = 3x^2(x - 1)$

3. $f(t) = (t - 1)(2t + 1)$

4. $f(x) = (2x + 3)(3x - 4)$

5. $f(x) = (3x + 1)(x^2 - 2)$

6. $f(x) = (x + 1)(2x^2 - 3x + 1)$

7. $f(x) = (x^3 - 1)(x + 1)$

8. $f(x) = (x^3 - 12x)(3x^2 + 2x)$

9. $f(w) = (w^3 - w^2 + w - 1)(w^2 + 2)$

10. $f(x) = \dfrac{1}{5}x^5 + (x^2 + 1)(x^2 - x - 1) + 28$

11. $f(x) = (5x^2 + 1)(2\sqrt{x} - 1)$

12. $f(t) = (1 + \sqrt{t})(2t^2 - 3)$

13. $f(x) = (x^2 - 5x + 2)\left(x - \dfrac{2}{x}\right)$

14. $f(x) = (x^3 + 2x + 1)\left(2 + \dfrac{1}{x^2}\right)$

15. $f(x) = \dfrac{1}{x - 2}$

16. $g(x) = \dfrac{3}{2x + 4}$

17. $f(x) = \dfrac{x - 1}{2x + 1}$

18. $f(t) = \dfrac{1 - 2t}{1 + 3t}$

19. $f(x) = \dfrac{1}{x^2 + 1}$

20. $f(u) = \dfrac{u}{u^2 + 1}$

21. $f(s) = \dfrac{s^2 - 4}{s + 1}$

22. $f(x) = \dfrac{x^3 - 2}{x^2 + 1}$

23. $f(x) = \dfrac{\sqrt{x}}{x^2 + 1}$

24. $f(x) = \dfrac{x^2 + 1}{\sqrt{x}}$

25. $f(x) = \dfrac{x^2 + 2}{x^2 + x + 1}$

26. $f(x) = \dfrac{x + 1}{2x^2 + 2x + 3}$

27. $f(x) = \dfrac{(x + 1)(x^2 + 1)}{x - 2}$

28. $f(x) = (3x^2 - 1)\left(x^2 - \dfrac{1}{x}\right)$

29. $f(x) = \dfrac{x}{x^2 - 4} - \dfrac{x - 1}{x^2 + 4}$

30. $f(x) = \dfrac{x + \sqrt{3x}}{3x - 1}$

In exercises 31–34, suppose f and g are functions that are differentiable at $x = 1$ and that $f(1) = 2$, $f'(1) = -1$, $g(1) = -2$, $g'(1) = 3$. Find the value of $h'(1)$.

31. $h(x) = f(x)g(x)$ **32.** $h(x) = (x^2 + 1)g(x)$

33. $h(x) = \dfrac{xf(x)}{x + g(x)}$ **34.** $h(x) = \dfrac{f(x)g(x)}{f(x) - g(x)}$

In exercises 35–38, find the derivative of each of the given functions and evaluate $f'(x)$ at the given value of x.

35. $f(x) = (2x - 1)(x^2 + 3)$; $x = 1$

36. $f(x) = \dfrac{2x + 1}{2x - 1}$; $x = 2$

37. $f(x) = \dfrac{x}{x^4 - 2x^2 - 1}$; $x = -1$

38. $f(x) = (\sqrt{x} + 2x)(x^{3/2} - x)$; $x = 4$

In exercises 39–42, find the slope and an equation of the tangent line to the graph of the function f at the specified point.

39. $f(x) = (x^3 + 1)(x^2 - 2)$; $(2, 18)$

40. $f(x) = \dfrac{x^2}{x + 1}$; $\left(2, \dfrac{4}{3}\right)$

41. $f(x) = \dfrac{x + 1}{x^2 + 1}$; $(1, 1)$

42. $f(x) = \dfrac{1 + 2x^{1/2}}{1 + x^{3/2}}$; $\left(4, \dfrac{5}{9}\right)$

43. Find an equation of the tangent line to the graph of the function $f(x) = (x^3 + 1)(3x^2 - 4x + 2)$ at the point $(1, 2)$.

44. Find an equation of the tangent line to the graph of the function $f(x) = \dfrac{3x}{x^2 - 2}$ at the point $(2, 3)$.

45. Let $f(x) = (x^2 + 1)(2 - x)$. Find the point(s) on the graph of f where the tangent line is horizontal.

46. Let $f(x) = \dfrac{x}{x^2 + 1}$. Find the point(s) on the graph of f where the tangent line is horizontal.

47. Find the point(s) on the graph of the function $f(x) =$ $(x^2 + 6)(x - 5)$ where the slope of the tangent line is equal to -2.

48. Find the point(s) on the graph of the function $f(x) = \dfrac{x + 1}{x - 1}$ where the slope of the tangent line is equal to $-1/2$.

49. A straight line perpendicular to and passing through the point of tangency of the tangent line is called the *normal* to the curve. Find the equation of the tangent line and the normal to the curve $y = \dfrac{1}{1 + x^2}$ at the point $(1, \frac{1}{2})$.

50. Concentration of a Drug in the Bloodstream   The concentration of a certain drug in a patient's bloodstream t hours after injection is given by

$$C(t) = \frac{0.2t}{t^2 + 1}$$

a. Find the rate at which the concentration of the drug is changing with respect to time.
b. How fast is the concentration changing $\frac{1}{2}$ hour after the injection? One hour after the injection? Two hours after the injection?

51. Cost of Removing Toxic Waste   A city's main well was recently found to be contaminated with trichloro-ethylene, a cancer-causing chemical, as a result of an abandoned chemical dump leaching chemicals into the water. A proposal submitted to the city's selectmen indicates that the cost, measured in millions of dollars, of removing x percent of the toxic pollutant is given by

$$C(x) = \frac{0.5x}{100 - x}$$

Find $C'(80)$, $C'(90)$, $C'(95)$, and $C'(99)$, and interpret your results.

52. Drug Dosages   Thomas Young has suggested the following rule for calculating the dosage of medicine for children 1 to 12 years old. If a denotes the adult dosage (in mg), and if t is the child's age (in years), then the child's dosage is given by

$$D(t) = \frac{at}{t + 12}$$

Suppose that the adult dosage of a substance is 500 mg. Find an expression that gives the rate of change of a child's dosage with respect to the child's age. What is the rate of change of a child's dosage with respect to his or her age for a six-year-old child? For a ten-year-old child?

53. **Effect of Bactericide** The number of bacteria $N(t)$ in a certain culture t minutes after an experimental bactericide is introduced obeys the rule

$$N(t) = \frac{10{,}000}{1 + t^2} + 2000$$

Find the rate of change of the number of bacteria in the culture 1 minute after and 2 minutes after the bactericide is introduced. What is the population of the bacteria in the culture 1 minute after the bactericide is introduced? 2 minutes after?

54. **Demand Functions** The demand function for the Sicard wristwatch is given by

$$d(x) = \frac{50}{0.01x^2 + 1} \qquad (0 \le x \le 20)$$

where x (measured in units of a thousand) is the quantity demanded per week and $d(x)$ is the unit price in dollars.
a. Find $d'(x)$.
b. Find $d'(5)$, $d'(10)$, and $d'(15)$, and interpret your results.

55. **Learning Curves** From experience, the Emory Secretarial School knows that the average student taking advanced typing will progress according to the rule

$$N(t) = \frac{60t + 180}{t + 6} \qquad (t \ge 0)$$

where $N(t)$ measures the number of words per minute the student can type after t weeks in the course.
a. Find an expression for $N'(t)$.
b. Compute $N'(t)$ for $t = 1, 3, 4$, and 7, and interpret your results.
c. Sketch the graph of the function N. Does it confirm the results obtained in (b)?
d. What will the average student's typing speed be at the end of the 12-week course?

56. **Box Office Receipts** The total worldwide box office receipts for a long-running movie are approximated by the function

$$T(x) = \frac{120x^2}{x^2 + 4}$$

where $T(x)$ is measured in millions of dollars and x is the number of years since the movie's release. How fast are the total receipts changing one year after the movie's release? Three years after its release? Five years after its release?

57. **Formaldehyde Levels** A study on formaldehyde levels in 900 homes indicates that emissions of various chemicals can decrease over time. The formaldehyde level (parts per million) in an average home in the study is given by

$$f(t) = \frac{0.055t + 0.26}{t + 2} \qquad (0 \le t \le 12)$$

where t is the age of the house in years. How fast is the formaldehyde level of the average house dropping when it is new? At the beginning of its fourth year?
Source: Bonneville Power Administration

58. **Population Growth** A major corporation is building a 4325-acre complex of homes, offices, stores, schools, and churches in the rural community of Glen Cove. As a result of this development, the planners have estimated that Glen Cove's population (in thousands) t years from now will be given by

$$P(t) = \frac{25t^2 + 125t + 200}{t^2 + 5t + 40}$$

a. Find the rate at which Glen Cove's population is changing with respect to time.
b. What will the population be after ten years? At what rate will the population be increasing when $t = 10$?

59. Extend the Product Rule for differentiation to the following case involving the product of three differentiable functions: Let $h(x) = u(x)v(x)w(x)$ and show that $h'(x) = u(x)v(x)w'(x) + u(x)v'(x)w(x) + u'(x)v(x)w(x)$.
[*Hint:* Let $f(x) = u(x)v(x)$, $g(x) = w(x)$, and $h(x) = f(x)g(x)$, and apply the Product Rule to the function h.]

60. Prove the Quotient Rule for differentiation (Rule 6). Verify the following steps:

a. $\dfrac{k(x + h) - k(x)}{h} = \dfrac{f(x + h)g(x) - f(x)g(x + h)}{hg(x + h)g(x)}$

b. By adding $[-f(x)g(x) + f(x)g(x)]$ to the numerator and simplifying,

$$\frac{k(x + h) - k(x)}{h} = \frac{1}{g(x + h)g(x)}$$
$$\times \left\{ \left[\frac{f(x + h) - f(x)}{h} \right] \cdot g(x) \right.$$
$$\left. - \left[\frac{g(x + h) - g(x)}{h} \right] \cdot f(x) \right\}$$

c. $k'(x) = \lim\limits_{h \to 0} \dfrac{k(x + h) - k(x)}{h}$

$$= \frac{g(x)f'(x) - f(x)g'(x)}{[g(x)]^2}$$

USING TECHNOLOGY

THE PRODUCT AND QUOTIENT RULES

EXAMPLE 1 Let $f(x) = (2\sqrt{x} + 0.5x)(0.3x^3 + 2x - 0.3/x)$. Find $f'(0.2)$.

Solution Using the numerical derivative operation of a graphing utility, we find

$$f'(0.2) = 6.47974948127$$

° ° °

EXAMPLE 2 **Importance of Time in Treating Heart Attacks**

According to the American Heart Association, the treatment benefit for heart attacks depends on the time to treatment and is described by the function

$$f(t) = \frac{-16.94t + 203.28}{t + 2.0328} \qquad (0 \le t \le 12)$$

where t is measured in hours and $f(t)$ is a percentage.

a. Use a graphing utility to graph the function f using the viewing rectangle $[0, 13] \times [0, 120]$.

b. Use a graphing utility to find the derivative of f when $t = 0$. When $t = 2$.

c. Interpret the results obtained in (b).
Source: American Heart Association

Solution

a. The graph of f is shown in Figure T1.

Figure T1

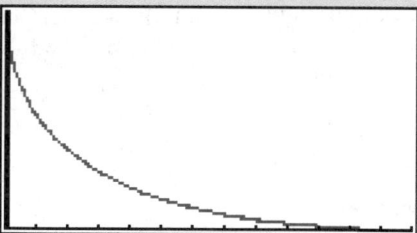

708

b. Using the numerical derivative operation of a graphing utility, we find

$$f'(0) \approx -57.5266$$

$$f'(2) \approx -14.6165$$

c. The results of (b) show that the treatment benefit drops off at the rate of 58% per hour at the time when the heart attack first occurs and falls off at the rate of 15% per hour when the time to treatment is 2 hours. Thus, it is extremely urgent that a patient suffering a heart attack receive medical attention as soon as possible. ○ ○ ○

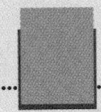

EXERCISES

In exercises 1–6, use the numerical derivative operation of a graphing utility to find the rate of change of f(x) at the given value of x. Give your answer accurate to four decimal places.

1. $f(x) = (2x^2 + 1)(x^3 + 3x + 4); x = -0.5$

2. $f(x) = (\sqrt{x} + 1)(2x^2 + x - 3); x = 1.5$

3. $f(x) = \dfrac{\sqrt{x} - 1}{\sqrt{x} + 1}; x = 3$

4. $f(x) = \dfrac{\sqrt{x}(x^2 + 4)}{x^3 + 1}; x = 4$

5. $f(x) = \dfrac{\sqrt{x}(1 + x^{-1})}{x + 1}; x = 1$

6. $f(x) = \dfrac{x^2(2 + \sqrt{x})}{1 + \sqrt{x}}; x = 1$

7. New Construction Jobs The president of a major housing construction company claims that the number of construction jobs created in the next t months is given by

$$f(t) = 1.42 \left(\frac{7t^2 + 140t + 700}{3t^2 + 80t + 550} \right)$$

where $f(t)$ is measured in millions of jobs per year. At what rate will construction jobs be created one year from now, assuming her projection is correct?

8. Population Growth A major corporation is building a 4325-acre complex of homes, offices, stores, schools, and churches in the rural community of Glen Cove. As a result of this development, the planners have estimated that Glen Cove's population (in thousands) t years from now will be given by

$$P(t) = \frac{25t^2 + 125t + 200}{t^2 + 5t + 40}$$

a. What will the population be ten years from now?
b. At what rate will the population be increasing ten years from now?

SOLUTIONS TO SELF-CHECK EXERCISES 11.2

1. We use the Quotient Rule to obtain

$$f'(x) = \frac{(x^2 - 1)\frac{d}{dx}(2x + 1) - (2x + 1)\frac{d}{dx}(x^2 - 1)}{(x^2 - 1)^2}$$

$$= \frac{(x^2 - 1)(2) - (2x + 1)(2x)}{(x^2 - 1)^2}$$

$$= \frac{2x^2 - 2 - 4x^2 - 2x}{(x^2 - 1)^2}$$

$$= \frac{-2x^2 - 2x - 2}{(x^2 - 1)^2}$$

$$= -\frac{2(x^2 + x + 1)}{(x^2 - 1)^2}$$

2. The slope of the tangent line to the graph of f at any point is given by

$$f'(x) = (x^2 + 1)\frac{d}{dx}(2x^3 - 3x^2 + 1)$$

$$+ (2x^3 - 3x^2 + 1)\frac{d}{dx}(x^2 + 1)$$

$$= (x^2 + 1)(6x^2 - 6x) + (2x^3 - 3x^2 + 1)(2x)$$

In particular, the slope of the tangent line to the graph of f when $x = 2$ is

$$f'(2) = (2^2 + 1)[6(2^2) - 6(2)] + [2(2^3) - 3(2^2) + 1][2(2)]$$
$$= 60 + 20 = 80$$

Note that it is not necessary to simplify the expression for $f'(x)$ since we are required only to evaluate the expression at $x = 2$. We also conclude, from this result, that the function f is changing at the rate of 80 units per unit change in x when $x = 2$.

3. The rate at which the company's total sales are changing at any time t is given by

$$S'(t) = \frac{(1 + 0.4t^2)\frac{d}{dt}(0.3t^3) - (0.3t^3)\frac{d}{dt}(1 + 0.4t^2)}{(1 + 0.4t^2)^2}$$

$$= \frac{(1 + 0.4t^2)(0.9t^2) - (0.3t^3)(0.8t)}{(1 + 0.4t^2)^2}$$

Therefore, at the beginning of the second year of operation, Security Products' sales were increasing at the rate of

$$S'(1) = \frac{(1 + 0.4)(0.9) - (0.3)(0.8)}{(1 + 0.4)^2} = 0.520408$$

or $520,408 per year.

11.3 THE CHAIN RULE

This section introduces another rule of differentiation called the **Chain Rule.** When used in conjunction with the rules of differentiation developed in the last two sections, the Chain Rule greatly enlarges the class of functions we are able to differentiate.

The Chain Rule

Consider the function $h(x) = (x^2 + x + 1)^2$. If we were to compute $h'(x)$ using only the rules of differentiation from the previous sections, then our approach might be to expand $h(x)$. Thus,

$$h(x) = (x^2 + x + 1)^2 = (x^2 + x + 1)(x^2 + x + 1)$$
$$= x^4 + 2x^3 + 3x^2 + 2x + 1$$

from which we find

$$h'(x) = 4x^3 + 6x^2 + 6x + 2$$

But what about the function $H(x) = (x^2 + x + 1)^{100}$? The same technique may be used to find the derivative of the function H, but the amount of work involved would be prodigious! Consider, also, the function $G(x) = \sqrt{x^2 + 1}$. For each of the two functions H and G, the rules of differentiation of the previous sections cannot be applied directly to compute the derivatives H' and G'.

Observe that both H and G are **composite** functions; that is, each is composed of, or built up from, simpler functions. For example, the function H is composed of the two simpler functions $f(x) = x^2 + x + 1$ and $g(x) = x^{100}$ as follows:

$$H(x) = g[f(x)] = [f(x)]^{100}$$
$$= (x^2 + x + 1)^{100}$$

In a similar manner, we see that the function G is composed of the two simpler functions $f(x) = x^2 + 1$ and $g(x) = \sqrt{x}$. Thus,

$$G(x) = g[f(x)] = \sqrt{f(x)}$$
$$= \sqrt{x^2 + 1}$$

As a first step toward finding the derivative h' of a composite function $h = g \circ f$ defined by $h(x) = g[f(x)]$, we write

$$u = f(x) \quad \text{and} \quad y = g[f(x)] = g(u)$$

The dependency of h on g and f is illustrated in Figure 11.5. Since u is a function of x, we may compute the derivative of u with respect to x, if f is a differentiable function, obtaining $du/dx = f'(x)$. Next, if g is a differentiable function of u, we may compute the derivative of g with respect to u, obtaining $dy/du = g'(u)$. Now, since the function h is composed of the function g and

Figure 11.5
The composite function $h(x) = g[f(x)]$.

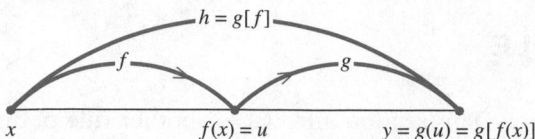

the function f, we might suspect that the rule $h'(x)$ for the derivative h' of h will be given by an expression that involves the rules for the derivatives of f and g. But how do we combine these derivatives to yield h'?

This question can be answered by interpreting the derivative of each function as giving the rate of change of that function. For example, suppose that $u = f(x)$ changes three times as fast as x—that is,

$$f'(x) = \frac{du}{dx} = 3$$

And suppose that $y = g(u)$ changes twice as fast as u—that is,

$$g'(u) = \frac{dy}{du} = 2$$

Then we would expect $y = h(x)$ to change six times as fast as x—that is,
$$h'(x) = g'(u)f'(x) = (2)(3) = 6$$

or equivalently,

$$\frac{dy}{dx} = \frac{dy}{du} \cdot \frac{du}{dx} = (2)(3) = 6$$

This observation suggests the following result, which we state without proof.

RULE 7: THE CHAIN RULE

If $h(x) = g[f(x)]$, then

$$h'(x) = \frac{d}{dx}g(f(x)) = g'(f(x))f'(x) \qquad \textbf{(1)}$$

Equivalently, if we write $y = h(x) = g(u)$, where $u = f(x)$, then

$$\frac{dy}{dx} = \frac{dy}{du} \cdot \frac{du}{dx} \qquad \textbf{(2)}$$

REMARKS

I. If we label the composite function h in the following manner

Inside function
↓
$$h(x) = g[f(x)]$$
↑
Outside function

then $h'(x)$ is just the *derivative* of the "outside function" *evaluated at* the "inside function" times the *derivative* of the "inside function."

2. Equation (2) can be remembered by observing that if we "cancel" the du's, then

$$\frac{dy}{dx} = \frac{dy}{du} \cdot \frac{du}{dx} = \frac{dy}{dx}$$

○ ○ ○

The Chain Rule for Powers of Functions

Many composite functions have the special form $h(x) = g(f(x))$, where g is defined by the rule $g(x) = x^n$ (n, a real number)—that is,

$$h(x) = [f(x)]^n$$

In other words, the function h is given by the power of a function f. The functions

$$h(x) = (x^2 + x + 1)^2, \quad H = (x^2 + x + 1)^{100}, \quad \text{and} \quad G = \sqrt{x^2 + 1}$$

discussed earlier are examples of this type of composite function. By using the following corollary of the Chain Rule, the General Power Rule, we are able to find the derivative of this type of function much more easily than by using the Chain Rule directly.

THE GENERAL POWER RULE	If the function f is differentiable and $h(x) = [f(x)]^n$ (n, a real number), then $$h'(x) = \frac{d}{dx}[f(x)]^n = n[f(x)]^{n-1}f'(x) \qquad (3)$$

To see this, we observe that $h(x) = g(f(x))$, where $g(x) = x^n$, so that, by virtue of the Chain Rule, we have

$$h'(x) = g'(f(x))f'(x)$$
$$= n[f(x)]^{n-1}f'(x)$$

since $g'(x) = nx^{n-1}$.

EXAMPLE 1 Find $H'(x)$ if

$$H(x) = (x^2 + x + 1)^{100}$$

Solution Using the General Power Rule, we obtain

$$H'(x) = 100(x^2 + x + 1)^{99}\frac{d}{dx}(x^2 + x + 1)$$

$$= 100(x^2 + x + 1)^{99}(2x + 1)$$

Observe the bonus that comes from using the Chain Rule: The answer is completely factored.

○ ○ ○

EXAMPLE 2 Differentiate the function $G(x) = \sqrt{x^2 + 1}$.

Solution We rewrite the function $G(x)$ as

$$G(x) = (x^2 + 1)^{1/2}$$

and apply the General Power Rule, obtaining

$$G'(x) = \frac{1}{2}(x^2 + 1)^{-1/2}\frac{d}{dx}(x^2 + 1)$$

$$= \frac{1}{2}(x^2 + 1)^{-1/2} \cdot 2x = \frac{x}{\sqrt{x^2 + 1}}$$ o o o

EXAMPLE 3 Differentiate the function $f(x) = x^2(2x + 3)^5$.

Solution Applying the Product Rule followed by the General Power Rule, we obtain

$$f'(x) = x^2\frac{d}{dx}(2x + 3)^5 + (2x + 3)^5\frac{d}{dx}(x^2)$$

$$= (x^2)5(2x + 3)^4 \cdot \frac{d}{dx}(2x + 3) + (2x + 3)^5(2x)$$

$$= 5x^2(2x + 3)^4(2) + 2x(2x + 3)^5$$

$$= 2x(2x + 3)^4(5x + 2x + 3) = 2x(7x + 3)(2x + 3)^4$$ o o o

EXAMPLE 4 Find $f'(x)$ if $f(x) = \dfrac{1}{(4x^2 - 7)^2}$.

Solution Rewriting $f(x)$ and then applying the General Power Rule, we obtain

$$f'(x) = \frac{d}{dx}\left[\frac{1}{(4x^2 - 7)^2}\right] = \frac{d}{dx}(4x^2 - 7)^{-2}$$

$$= -2(4x^2 - 7)^{-3}\frac{d}{dx}(4x^2 - 7)$$

$$= -2(4x^2 - 7)^{-3}(8x) = -\frac{16x}{(4x^2 - 7)^3}$$ o o o

EXAMPLE 5 Find the slope of the tangent line to the graph of the function

$$f(x) = \left(\frac{2x + 1}{3x + 2}\right)^3$$

at the point $(0, \frac{1}{8})$.

Solution The slope of the tangent line to the graph of f at any point is given by $f'(x)$. To compute $f'(x)$, we use the General Power Rule followed by the Quotient Rule, obtaining

$$f'(x) = 3\left(\frac{2x+1}{3x+2}\right)^2 \frac{d}{dx}\left(\frac{2x+1}{3x+2}\right)$$

$$= 3\left(\frac{2x+1}{3x+2}\right)^2\left[\frac{(3x+2)(2)-(2x+1)(3)}{(3x+2)^2}\right]$$

$$= 3\left(\frac{2x+1}{3x+2}\right)^2\left[\frac{6x+4-6x-3}{(3x+2)^2}\right]$$

$$= \frac{3(2x+1)^2}{(3x+2)^4}$$

In particular, the slope of the tangent line to the graph of f at $(0, \frac{1}{8})$ is given by

$$f'(0) = \frac{3(0+1)^2}{(0+2)^4} = \frac{3}{16}$$

ο ο ο

EXPLORING WITH TECHNOLOGY

Refer to Example 5.

1. Use a graphing utility to plot the graph of the function f using the viewing rectangle $[-2, 1] \times [-1, 2]$. Then draw the tangent line to the graph of f at the point $(0, \frac{1}{8})$.

2. For a better picture, repeat part (1) using the viewing rectangle $[-1, 1] \times [-0.1, 0.3]$.

3. Use the numerical differentiation capability of the graphing utility to verify that the slope of the tangent line at $(0, \frac{1}{8})$ is $\frac{3}{16}$.

ο ο ο

Applications

EXAMPLE 6 The membership of The Fitness Center, which opened a few years ago, is approximated by the function

$$N(t) = 100(64 + 4t)^{2/3} \qquad (0 \le t \le 52)$$

where $N(t)$ gives the number of members at the beginning of week t.

a. Find $N'(t)$.

b. How fast was the center's membership increasing initially ($t = 0$)?

c. How fast was the membership increasing at the beginning of the 40th week?

d. What was the membership when the center opened? At the beginning of the 40th week?

Solution

a. Using the General Power Rule, we obtain

$$N'(t) = \frac{d}{dt}[100(64 + 4t)^{2/3}]$$

$$= 100\frac{d}{dt}(64 + 4t)^{2/3}$$

$$= 100\left(\frac{2}{3}\right)(64 + 4t)^{-1/3}\frac{d}{dt}(64 + 4t)$$

$$= \frac{200}{3}(64 + 4t)^{-1/3}(4)$$

$$= \frac{800}{3(64 + 4t)^{1/3}}$$

b. The rate at which the membership was increasing when the center opened is given by

$$N'(0) = \frac{800}{3(64)^{1/3}} \approx 66.7$$

or approximately 67 people per week.

c. The rate at which the membership was increasing at the beginning of the 40th week is given by

$$N'(40) = \frac{800}{3(64 + 160)^{1/3}} \approx 43.9$$

or approximately 44 people per week.

d. The membership when the center opened is given by

$$N(0) = 100(64)^{2/3} = 100(16)$$

or 1600 people. The membership at the beginning of the 40th week is given by

$$N(40) = 100(64 + 160)^{2/3} \approx 3688.3$$

or approximately 3688 people.   ○ ○ ○

The profit P of a one-product software manufacturer depends on the number of units of its products sold. The manufacturer estimates that it will sell x units of its product per week. Suppose $P = g(x)$ and $x = f(t)$, where g and f are differentiable functions.

a. Write an expression giving the rate of change of the profit with respect to the number of units sold.
b. Write an expression giving the rate of change of the number of units sold per week.
c. Write an expression giving the rate of change of the profit per week.

EXAMPLE 7 Arteriosclerosis begins during childhood when plaque (soft masses of fatty material) forms in the arterial walls, blocking the flow of blood through the arteries and leading to heart attacks, strokes, and gangrene. Suppose the idealized cross section of the aorta is circular with radius a cm, and by year t the thickness of the plaque (assume that it is uniform) is $h = f(t)$ cm (Figure 11.6). Then the area of the opening is given by $A = \pi(a - h)^2$ square centimeters.

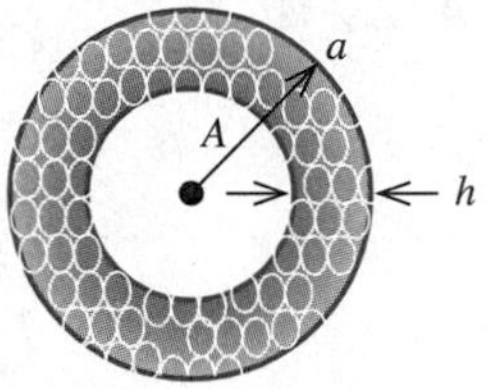

Suppose that the radius of an individual's artery is 1 cm ($a = 1$) and the thickness of the plaque in year t is given by

$$h = g(t) = 1 - 0.01(10{,}000 - t^2)^{1/2} \text{ cm}$$

Since the area of the arterial opening is given by

$$A = f(h) = \pi(1 - h)^2$$

the rate at which A is changing with respect to time is given by

$$\frac{dA}{dt} = \frac{dA}{dh} \cdot \frac{dh}{dt} = f'(h) \cdot g'(t) \qquad \text{(By the Chain Rule)}$$

$$= 2\pi(1 - h)(-1)[-0.01(\tfrac{1}{2})(10{,}000 - t^2)^{-1/2}(-2t)] \qquad \begin{array}{l}\text{(Using the Chain}\\\text{Rule twice)}\end{array}$$

$$= -2\pi(1 - h)\left[\frac{0.01t}{(10{,}000 - t^2)^{1/2}}\right]$$

$$= -\frac{0.02\pi(1 - h)t}{\sqrt{10{,}000 - t^2}}$$

For example, when $t = 50$,

$$h = g(50) = 1 - 0.01(10{,}000 - 2{,}500)^{1/2} \approx 0.134$$

so that

$$\frac{dA}{dt} = -\frac{0.02\pi(1 - 0.134)50}{\sqrt{10{,}000 - 2{,}500}} \approx -0.03$$

that is, the area of the arterial opening is decreasing at the rate of 0.03 cm²/year.

○ ○ ○

Figure 11.6
Cross section of the aorta.

Suppose the population P of a certain bacteria culture is given by $P = f(T)$, where T is the temperature of the medium. Furthermore, suppose the temperature T is a function of time t in seconds—that is, $T = g(t)$. Give an interpretation of each of these quantities:

a. $\dfrac{dP}{dT}$ **b.** $\dfrac{dT}{dt}$ **c.** $\dfrac{dP}{dt}$ **d.** $(f \circ g)(t)$ **e.** $f'(g(t))g'(t)$

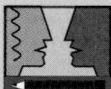

Using Technology

FINDING THE DERIVATIVE OF A COMPOSITE FUNCTION

EXAMPLE 1 Find the rate of change of $f(x) = \sqrt{x}(1 + 0.02x^2)^{3/2}$ when $x = 2.1$.

Solution Using the numerical derivative operation of a graphing utility, we find

$$f'(2.1) = 0.582146320119$$

or approximately 0.58 unit per unit change in x.   ⊙ ⊙ ⊙

EXAMPLE 2 The management of Astro World ("The Amusement Park of the Future") estimates that the total number of visitors (in thousands) to the amusement park t hours after opening time at 9 A.M. is given by

$$N(t) = \frac{30t}{\sqrt{2 + t^2}}$$

What is the rate at which visitors are admitted to the amusement park at 10:30 A.M.?

Solution Using the numerical derivative operation of a graphing utility, we find

$$N'(1.5) \approx 6.8481$$

or approximately 6848 visitors per hour.   ⊙ ⊙ ⊙

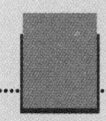

EXERCISES

In exercises 1–6, use the numerical derivative operation of a graphing utility to find the rate of change of f(x) at the given value of x. Give your answer accurate to four decimal places.

1. $f(x) = \sqrt{x^2 - x^4}; x = 0.5$

2. $f(x) = x - \sqrt{1 - x^2}; x = 0.4$

3. $f(x) = x\sqrt{1 - x^2}; x = 0.2$

4. $f(x) = (x + \sqrt{x^2 + 4})^{3/2}; x = 1$

5. $f(x) = \dfrac{\sqrt{1 + x^2}}{x^3 + 2}; x = -1$

6. $f(x) = \dfrac{x^3}{1 + (1 + x^2)^{3/2}}; x = 3$

7. Worldwide Production of Vehicles According to *Automotive News,* the worldwide production of vehicles between 1960 and 1990 is given by the function

$$f(t) = 16.5\sqrt{1 + 2.2t} \qquad (0 \le t \le 3)$$

where $f(t)$ is measured in units of a million and t is measured in decades, with $t = 0$ corresponding to the beginning of 1960. What was the rate of change of the worldwide production of vehicles at the beginning of 1970? At the beginning of 1980?
Source: Automotive News

SELF-CHECK EXERCISES 11.3

1. Find the derivative of

$$f(x) = -\frac{1}{\sqrt{2x^2 - 1}}$$

2. Suppose that the life expectancy at birth (in years) of a female in a certain country is described by the function

$$g(t) = 50.02(1 + 1.09t)^{0.1} \qquad (0 \le t \le 150)$$

where t is measured in years and $t = 0$ corresponds to the beginning of 1900.
a. What is the life expectancy at birth of a female born at the beginning of 1980? At the beginning of the year 2000?
b. How fast is the life expectancy at birth of a female born at any time t changing?

Solutions to Self-Check Exercises 11.3 can be found on page 723.

11.3 EXERCISES

In exercises 1–46, find the derivative of the given function.

1. $f(x) = (2x - 1)^4$

2. $f(x) = (1 - x)^3$

3. $f(x) = (x^2 + 2)^5$

4. $f(t) = 2(t^3 - 1)^5$

5. $f(x) = (2x - x^2)^3$

6. $f(x) = 3(x^3 - x)^4$

7. $f(x) = (2x + 1)^{-2}$

8. $f(t) = \frac{1}{2}(2t^2 + t)^{-3}$

9. $f(x) = (x^2 - 4)^{3/2}$

10. $f(t) = (3t^2 - 2t + 1)^{3/2}$

11. $f(x) = \sqrt{3x - 2}$

12. $f(t) = \sqrt{3t^2 - t}$

13. $f(x) = \sqrt[3]{1 - x^2}$

14. $f(x) = \sqrt{2x^2 - 2x + 3}$

15. $f(x) = \frac{1}{(2x + 3)^3}$

16. $f(x) = \frac{2}{(x^2 - 1)^4}$

17. $f(t) = \frac{1}{\sqrt{2t - 3}}$

18. $f(x) = \frac{1}{\sqrt{2x^2 - 1}}$

19. $y = \frac{1}{(4x^4 + x)^{3/2}}$

20. $f(t) = \frac{4}{\sqrt[3]{2t^2 + t}}$

21. $f(x) = (3x^2 + 2x + 1)^{-2}$

22. $f(t) = (5t^3 + 2t^2 - t + 4)^{-3}$

23. $f(x) = (x^2 + 1)^3 - (x^3 + 1)^2$

24. $f(t) = (2t - 1)^4 + (2t + 1)^4$

25. $f(t) = (t^{-1} - t^{-2})^3$

26. $f(v) = (v^{-3} + 4v^{-2})^3$

27. $f(x) = \sqrt{x + 1} + \sqrt{x - 1}$

28. $f(u) = (2u + 1)^{3/2} + (u^2 - 1)^{-3/2}$

29. $f(x) = 2x^2(3 - 4x)^4$

30. $h(t) = t^2(3t + 4)^3$

31. $f(x) = (x - 1)^2(2x + 1)^4$

32. $g(u) = (1 + u^2)^5(1 - 2u^2)^8$

33. $f(x) = \left(\frac{x + 3}{x - 2}\right)^3$

34. $f(x) = \left(\frac{x + 1}{x - 1}\right)^5$

35. $s(t) = \left(\frac{t}{2t + 1}\right)^{3/2}$

36. $g(s) = \left(s^2 + \frac{1}{s}\right)^{3/2}$

37. $g(u) = \sqrt{\frac{u + 1}{3u + 2}}$

38. $g(x) = \sqrt{\frac{2x + 1}{2x - 1}}$

39. $f(x) = \frac{x^2}{(x^2 - 1)^4}$

40. $g(u) = \frac{2u^2}{(u^2 + u)^3}$

41. $h(x) = \frac{(3x^2 + 1)^3}{(x^2 - 1)^4}$

42. $g(t) = \frac{(2t - 1)^2}{(3t + 2)^4}$

43. $f(x) = \frac{\sqrt{2x + 1}}{x^2 - 1}$

44. $f(t) = \frac{4t^2}{\sqrt{2t^2 + 2t - 1}}$

45. $g(t) = \dfrac{\sqrt{t+1}}{\sqrt{t^2+1}}$ 46. $f(x) = \dfrac{\sqrt{x^2+1}}{\sqrt{x^2-1}}$

In exercises 47–52, find dy/du, du/dx, and dy/dx.

47. $y = u^{4/3}$ and $u = 3x^2 - 1$

48. $y = \sqrt{u}$ and $u = 7x - 2x^2$

49. $y = u^{-2/3}$ and $u = 2x^3 - x + 1$

50. $y = 2u^2 + 1$ and $u = x^2 + 1$

51. $y = \sqrt{u} + \dfrac{1}{\sqrt{u}}$ and $u = x^3 - x$

52. $y = \dfrac{1}{u}$ and $u = \sqrt{x} + 1$

In exercises 53–56, find an equation of the tangent line to the graph of the function at the given point.

53. $f(x) = (1 - x)(x^2 - 1)^2$; $(2, -9)$

54. $f(x) = \left(\dfrac{x+1}{x-1}\right)^2$; $(3, 4)$

55. $f(x) = x\sqrt{2x^2 + 7}$; $(3, 15)$

56. $f(x) = \dfrac{8}{\sqrt{x^2 + 6x}}$; $(2, 2)$

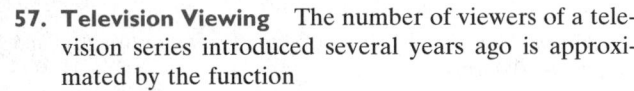

57. **Television Viewing** The number of viewers of a television series introduced several years ago is approximated by the function

$$N(x) = (60 + 2x)^{2/3} \qquad (1 \le x \le 26)$$

where $N(x)$ (measured in millions) denotes the number of weekly viewers of the series in the xth week. Find the rate of increase of the weekly audience at the end of the second week and at the end of the twelfth week. How many viewers were there in the second week? In the 24th week?

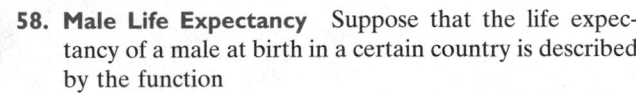

58. **Male Life Expectancy** Suppose that the life expectancy of a male at birth in a certain country is described by the function

$$f(t) = 46.9(1 + 1.09t)^{0.1} \qquad (0 \le t \le 150)$$

where t is measured in years and $t = 0$ corresponds to the beginning of 1900. How long can a male born at the beginning of the year 2000 in that country expect to live? What is the rate of change of the life expectancy of a male born in that country at the beginning of the year 2000?

59. **Concentration of Carbon Monoxide in the Air** According to a joint study conducted by Oxnard's Environmental Management Department and a state government agency, the concentration of carbon monoxide in the air due to automobile exhaust t years from now is given by

$$C(t) = 0.01(0.2t^2 + 4t + 64)^{2/3}$$

parts per million.
a. Find the rate at which the level of carbon monoxide is changing with respect to time.
b. Find the rate at which the level of carbon monoxide will be changing five years from now.

60. **Continuing Education Enrollment** The registrar of Kellogg University estimates that the total student enrollment in the Continuing Education division will be given by

$$N(t) = -\dfrac{20,000}{\sqrt{1 + 0.2t}} + 21,000$$

where $N(t)$ denotes the number of students enrolled in the division t years from now. Find an expression for $N'(t)$. How fast is the student enrollment increasing currently? How fast will it be increasing five years from now?

61. **Air Pollution** According to the South Coast Air Quality Management District, the level of nitrogen dioxide, a brown gas that impairs breathing, present in the atmosphere on a certain May day in downtown Los Angeles is approximated by

$$A(t) = 0.03t^3(t - 7)^4 + 60.2 \qquad (0 \le t \le 7)$$

where $A(t)$ is measured in pollutant standard index and t is measured in hours, with $t = 0$ corresponding to 7 A.M.
a. Find $A'(t)$.
b. Find $A'(1)$, $A'(3)$, and $A'(4)$, and interpret your results.

62. **Effect of Luxury Tax on Consumption** Government economists of a developing country determined that the purchase of imported perfume is related to a proposed "luxury tax" by the formula

$$N(x) = \sqrt{10,000 - 40x - 0.02x^2} \qquad (0 \le x \le 200)$$

where $N(x)$ measures the percentage of normal consumption of perfume when a "luxury tax" of x percent is imposed on it. Find the rate of change of $N(x)$ for taxes of 10%, 100%, and 150%.

63. Pulse Rate of an Athlete   The pulse rate (the number of heartbeats per minute) of a long-distance runner t seconds after leaving the starting line is given by

$$P(t) = \frac{300\sqrt{\frac{1}{2}t^2 + 2t + 25}}{t + 25} \qquad (t \geq 0)$$

Compute $P'(t)$. How fast is the athlete's pulse rate increasing 10 seconds into the run? 60 seconds into the run? 2 minutes into the run? What is her pulse rate 2 minutes into the run?

64. Thurstone Learning Model   Psychologist L. L. Thurstone suggested the following relationship between learning time T and the length of a list n:

$$T = f(n) = An\sqrt{n - b}$$

where A and b are constants that depend on the person and the task.
a. Compute dT/dn and interpret your result.
b. Suppose that for a certain person and a certain task, $A = 4$ and $b = 4$. Compute $f'(13)$ and $f'(29)$ and interpret your results.

65. Oil Spills   In calm waters, the oil spilling from the ruptured hull of a grounded tanker spreads in all directions. Assuming that the area polluted is a circle and that its radius is increasing at a rate of 2 ft/sec, determine how fast the area is increasing when the radius of the circle is 40 feet.

66. Arteriosclerosis   Refer to Example 7, page 717. Suppose that the radius of an individual's artery is 1 cm and the thickness of the plaque (in cm) t years from now is given by

$$h = g(t) = \frac{0.5t^2}{t^2 + 10} \qquad (0 \leq t \leq 10)$$

How fast will the arterial opening be decreasing five years from now?

67. Traffic Flow   Opened in the late 1950s, the Central Artery in downtown Boston was designed to move 75,000 vehicles a day. The number of vehicles moved per day is approximated by the function

$$x = f(t) = 6.25t^2 + 19.75t + 74.75 \qquad (0 \leq t \leq 5)$$

where x is measured in thousands and t in decades, with $t = 0$ corresponding to the beginning of 1959. Suppose the average speed of traffic flow in mph is given by

$$S = g(x) = -0.00075x^2 + 67.5 \qquad (75 \leq x \leq 350)$$

where x has the same meaning as before. What was the rate of change of the average speed of traffic flow at the beginning of 1999? What was the average speed of traffic flow at that time?

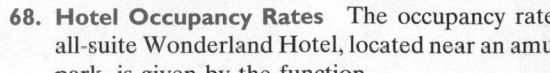

68. Hotel Occupancy Rates   The occupancy rate of the all-suite Wonderland Hotel, located near an amusement park, is given by the function

$$r(t) = \frac{10}{81}t^3 - \frac{10}{3}t^2 + \frac{200}{9}t + 60 \qquad (0 \leq t \leq 12)$$

where t is measured in months and $t = 0$ corresponds to the beginning of January. Management has estimated that the monthly revenue (in thousands of dollars per month) is approximated by the function

$$R(r) = -\frac{3}{5000}r^3 + \frac{9}{50}r^2 \qquad (0 \leq r \leq 100)$$

where r is the occupancy rate.
a. Find an expression that gives the rate of change of Wonderland's occupancy rate with respect to time.
b. Find an expression that gives the rate of change of Wonderland's monthly revenue with respect to the occupancy rate.
c. What is the rate of change of Wonderland's monthly revenue with respect to time at the beginning of January? At the beginning of June?

[*Hint:* Use the Chain Rule to find $R'(r(0))r'(0)$ and $R'(r(6))r'(6)$.]

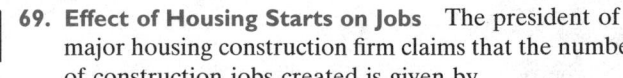

69. Effect of Housing Starts on Jobs   The president of a major housing construction firm claims that the number of construction jobs created is given by

$$N(x) = 1.42x$$

where x denotes the number of housing starts. Suppose that the number of housing starts in the next t months is expected to be

$$x(t) = \frac{7t^2 + 140t + 700}{3t^2 + 80t + 550}$$

million units per year. Find an expression that gives the rate at which the number of construction jobs will be created t months from now. At what rate will construction jobs be created one year from now?

70. Demand for PCs   The quantity demanded per month, x, of a certain make of personal computer (PC) is related to the average unit price, p (in dollars), of personal computers by the equation

$$x = f(p) = \frac{100}{9}\sqrt{810,000 - p^2}$$

It is estimated that t months from now, the average price of a PC will be given by

$$p(t) = \frac{400}{1 + \frac{1}{8}\sqrt{t}} + 200 \qquad (0 \le t \le 60)$$

dollars. Find the rate at which the quantity demanded per month of the PCs will be changing 16 months from now.

71. **Cruise Ship Bookings** The management of Cruise World, operators of Caribbean luxury cruises, expects that the percentage of young adults booking passage on their cruises in the years ahead will rise dramatically. They have constructed the following model, which gives the percentage of young adult passengers in year t:

$$p = f(t) = 50\left(\frac{t^2 + 2t + 4}{t^2 + 4t + 8}\right) \qquad (0 \le t \le 5)$$

Young adults normally pick shorter cruises and generally spend less on their passage. The following model gives an approximation of the average amount of money R (in dollars) spent per passenger on a cruise when the

percentage of young adults is p:

$$R(p) = 1000\left(\frac{p + 4}{p + 2}\right)$$

Find the rate at which the price of the average passage will be changing two years from now.

72. In Section 11.1, we proved that

$$\frac{d}{dx}(x^n) = nx^{n-1}$$

for the special case when $n = 2$. Use the Chain Rule to show that

$$\frac{d}{dx}(x^{1/n}) = \frac{1}{n}x^{1/n - 1}$$

for any nonzero integer n, assuming that $f(x) = x^{1/n}$ is differentiable.

[*Hint:* Let $f(x) = x^{1/n}$ so that $[f(x)]^n = x$. Differentiate both sides with respect to x.]

73. With the aid of exercise 72, prove that

$$\frac{d}{dx}(x^r) = rx^{r-1}$$

for any rational number r.

[*Hint:* Let $r = m/n$, where m and n are integers with $n \ne 0$, and write $x^r = (x^m)^{1/n}$.]

SOLUTIONS TO SELF-CHECK EXERCISES 11.3

1. Rewriting, we have

$$f(x) = -(2x^2 - 1)^{-1/2}$$

Using the General Power Rule, we find

$$f'(x) = -\frac{d}{dx}(2x^2 - 1)^{-1/2}$$

$$= -\left(-\frac{1}{2}\right)(2x^2 - 1)^{-3/2}\frac{d}{dx}(2x^2 - 1)$$

$$= \frac{1}{2}(2x^2 - 1)^{-3/2}(4x)$$

$$= \frac{2x}{(2x^2 - 1)^{3/2}}$$

2. a. The life expectancy at birth of a female born at the beginning of 1980 is given by

$$g(80) = 50.02[1 + 1.09(80)]^{0.1} \approx 78.29$$

or approximately 78 years. Similarly, the life expectancy at birth of a female born at the beginning of the year 2000 is given by

$$g(100) = 50.02[1 + 1.09(100)]^{0.1} \approx 80.04$$

or approximately 80 years.

b. The rate of change of the life expectancy at birth of a female born at any time t is given by $g'(t)$. Using the General Power Rule, we have

$$g'(t) = 50.02 \frac{d}{dt}(1 + 1.09t)^{0.1}$$

$$= (50.02)(0.1)(1 + 1.09t)^{-0.9} \frac{d}{dt}(1 + 1.09t)$$

$$= (50.02)(0.1)(1.09)(1 + 1.09t)^{-0.9}$$

$$= 5.45218(1 + 1.09t)^{-0.9}$$

$$= \frac{5.45218}{(1 + 1.09t)^{0.9}}$$

11.4 MARGINAL FUNCTIONS IN ECONOMICS

Marginal analysis is the study of the rate of change of economic quantities. For example, an economist is concerned not merely with the value of an economy's gross domestic product (GDP) at a given time, but also with the rate at which it is growing or declining. In the same vein, a manufacturer is interested not only in the total cost corresponding to a certain level of production of a commodity, but also in the rate of change of the total cost with respect to the level of production, and so on. Let us begin with an example to explain the meaning of the adjective *marginal*, as used by economists.

Cost Functions

 EXAMPLE 1 Suppose that the total cost in dollars incurred per week by the Polaraire Company for manufacturing x refrigerators is given by the total cost function

$$C(x) = 8000 + 200x - 0.2x^2 \qquad (0 \le x \le 400)$$

a. What is the actual cost incurred for manufacturing the 251st refrigerator?

b. Find the rate of change of the total cost function with respect to x when $x = 250$.

c. Compare the results obtained in (a) and (b).

Solution

a. The actual cost incurred in producing the 251st refrigerator is the difference between the total cost incurred in producing the first 251 refrigerators and the total cost of producing the first 250 refrigerators:

$$C(251) - C(250) = [8000 + 200(251) - 0.2(251)^2]$$
$$- [8000 + 200(250) - 0.2(250)^2]$$
$$= 45{,}599.8 - 45{,}500$$
$$= 99.80$$

or $99.80.

b. The rate of change of the total cost function C with respect to x is given by the derivative of C—that is, $C'(x) = 200 - 0.4x$. Thus, when the level of production is 250 refrigerators, the rate of change of the total cost with respect to x is given by

$$C'(250) = 200 - 0.4(250)$$
$$= 100$$

or $100.

c. From the solution to (a), we know that the actual cost for producing the 251st refrigerator is $99.80. This answer is very closely approximated by the answer to (b), $100. To see why this is so, observe that the difference $C(251) - C(250)$ may be written in the form

$$\frac{C(251) - C(250)}{1} = \frac{C(250 + 1) - C(250)}{1} = \frac{C(250 + h) - C(250)}{h}$$

where $h = 1$. In other words, the difference $C(251) - C(250)$ is precisely the average rate of change of the total cost function C over the interval $[250, 251]$ or, equivalently, the slope of the secant line through the points $(250, 45{,}500)$ and $(251, 45{,}599.8)$. However, the number $C'(250) = 100$ is the instantaneous rate of change of the total cost function C at $x = 250$ or, equivalently, the slope of the tangent line to the graph of C at $x = 250$.

Now when h is small, the average rate of change of the function C is a good approximation to the instantaneous rate of change of the function C, or, equivalently, the slope of the secant line through the points in question is a good approximation to the slope of the tangent line through the point in question. Thus, we may expect

$$C(251) - C(250) = \frac{C(251) - C(250)}{1} = \frac{C(250 + h) - C(250)}{h}$$
$$\approx \lim_{h \to 0} \frac{C(250 + h) - C(250)}{h} = C'(250)$$

which is precisely the case in this example.   ◦ ◦ ◦

The actual cost incurred in producing an additional unit of a certain commodity given that a plant is already at a certain level of operation is called the **marginal cost.** Knowing this cost is very important to the management decision-making process. As we saw in Example 1, the marginal cost is approximated by the rate of change of the total cost function evaluated at the appropriate point. For this reason, economists have defined the **marginal cost function** to be the derivative of the corresponding total cost function. In other words, if C is a total cost function, then the marginal cost function is defined to be its derivative C'. Thus, the adjective *marginal* is synonymous with *derivative of.*

 EXAMPLE 2 A subsidiary of the Elektra Electronics Company manufactures a programmable pocket calculator. Management determined that the daily total cost of producing these calculators (in dollars) is given by

$$C(x) = 0.0001x^3 - 0.08x^2 + 40x + 5000$$

where x stands for the number of calculators produced.

a. Find the marginal cost function.

b. What is the marginal cost when $x = 200, 300, 400$, and 600?

c. Interpret your results.

Solution

a. The marginal cost function C' is given by the derivative of the total cost function C. Thus,

$$C'(x) = 0.0003x^2 - 0.16x + 40$$

b. The marginal cost when $x = 200, 300, 400$, and 600 is given by

$$C'(200) = 0.0003(200)^2 - 0.16(200) + 40 = 20$$
$$C'(300) = 0.0003(300)^2 - 0.16(300) + 40 = 19$$
$$C'(400) = 0.0003(400)^2 - 0.16(400) + 40 = 24$$
$$C'(600) = 0.0003(600)^2 - 0.16(600) + 40 = 52$$

or $20, $19, $24, and $52, respectively.

c. From the results of (b), we see that Elektra's actual cost for producing the 201st calculator is approximately $20. The actual cost incurred for producing one additional calculator when the level of production is already 300 calculators is approximately $19, and so on. Observe that when the level of production is already 600 units, the actual cost of producing one additional unit is approximately $52. The higher cost for producing this additional unit when the level of production is 600 units may be the result of several factors, among them excessive costs incurred because of overtime or higher maintenance, production breakdown caused by greater stress and strain on the equipment, and so on. The graph of the total cost function appears in Figure 11.7.   ○ ○ ○

Figure 11.7
The cost of producing x calculators is given by C(x).

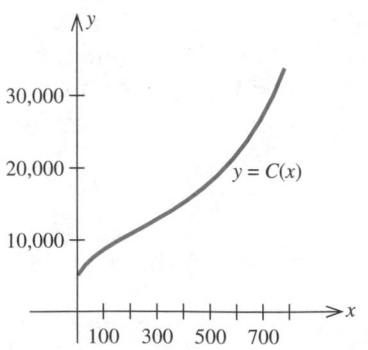

Average Cost Functions

Let us now introduce another marginal concept closely related to the marginal cost. Let $C(x)$ denote the total cost incurred in producing x units of a certain commodity. Then the **average cost** of producing x units of the commodity is obtained by dividing the total production cost by the number of units produced. This leads to the following definition.

AVERAGE COST FUNCTION	Suppose $C(x)$ is a total cost function. Then the **average cost function,** denoted by $\overline{C}(x)$ and read "C bar of x," is $$\frac{C(x)}{x} \qquad (4)$$

The derivative $\overline{C}'(x)$ of the average cost function, called the **marginal average cost function,** measures the rate of change of the average cost function with respect to the number of units produced.

EXAMPLE 3   The total cost of producing x units of a certain commodity is given by

$$C(x) = 400 + 20x$$

dollars.

a. Find the average cost function $\overline{C}$.

b. Find the marginal average cost function $\overline{C}'$.

c. Interpret the results obtained in (a) and (b).

Solution

a. The average cost function is given by

$$\overline{C}(x) = \frac{C(x)}{x} = \frac{400 + 20x}{x}$$

$$= 20 + \frac{400}{x}$$

b. The marginal average cost function is

$$\overline{C}'(x) = -\frac{400}{x^2}$$

c. Since the marginal average cost function is negative for all admissible values of x, the rate of change of the average cost function is negative for all $x > 0$; that is, $\overline{C}(x)$ decreases as x increases. However, the graph of $\overline{C}$ always

Figure 11.8
As the level of production increases, the average cost approaches $20.

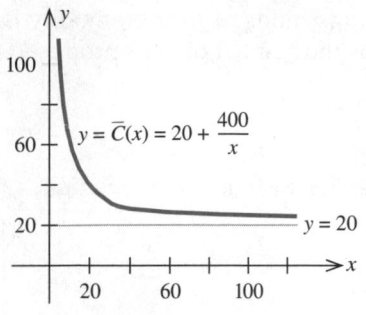

lies above the horizontal line $y = 20$, but it approaches the line since

$$\lim_{x \to \infty} \overline{C}(x) = \lim_{x \to \infty} \left(20 + \frac{400}{x} \right) = 20$$

A sketch of the graph of the function $\overline{C}(x)$ appears in Figure 11.8. This result is fully expected if we consider the economic implications. Note that as the level of production increases, the fixed cost per unit of production, represented by the term $(400/x)$, drops steadily. The average cost approaches the constant unit of production, which is $20 in this case.   ○ ○ ○

EXAMPLE 4 Once again consider the subsidiary of the Elektra Electronics Company. The daily total cost for producing its programmable calculators is given by

$$C(x) = 0.0001x^3 - 0.08x^2 + 40x + 5000$$

dollars, where x stands for the number of calculators produced (see Example 2).

a. Find the average cost function $\overline{C}$.

b. Find the marginal average cost function $\overline{C}'$. Compute $\overline{C}'(500)$.

c. Sketch the graph of the function $\overline{C}$, and interpret the results obtained in (a) and (b).

Solution

a. The average cost function is given by

$$\overline{C}(x) = \frac{C(x)}{x} = 0.0001x^2 - 0.08x + 40 + \frac{5000}{x}$$

b. The marginal average cost function is given by

$$\overline{C}'(x) = 0.0002x - 0.08 - \frac{5000}{x^2}$$

Figure 11.9
The average cost reaches a minimum of $35 when 500 calculators are produced.

Also $$\overline{C}'(500) = 0.0002(500) - 0.08 - \frac{5000}{(500)^2} = 0$$

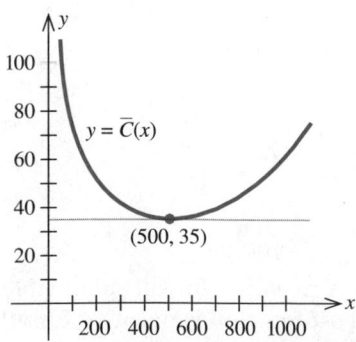

c. To sketch the graph of the function $\overline{C}$, observe that if x is a small positive number, then $\overline{C}(x) > 0$. Furthermore, $\overline{C}(x)$ becomes arbitrarily large as x approaches zero from the right, since the term $(5000/x)$ becomes arbitrarily large as x approaches zero. Next, the result $\overline{C}'(500) = 0$ obtained in (b) tells us that the tangent line to the graph of the function $\overline{C}$ is horizontal at the point $(500, 35)$ on the graph. Finally, plotting the points on the graph corresponding to, say, $x = 100, 200, 300, \ldots, 900$, we obtain the sketch in Figure 11.9. As expected, the average cost drops as the level of production increases. But in this case, as opposed to the case in Example 3, the average cost reaches a minimum value of $35, corresponding to a production level of 500, and *increases* thereafter.

This phenomenon is typical in situations where the marginal cost increases from some point on as production increases, as in Example 2. This situation is in contrast to that of Example 3, in which the marginal cost remains constant at any level of production. ◐ ◐ ◑

EXPLORING WITH TECHNOLOGY

Refer to Example 4.

1. Use a graphing utility to plot the graph of the average cost function

$$\overline{C}(x) = 0.0001x^2 - 0.08x + 40 + \frac{5000}{x}$$

in the viewing rectangle $[0, 1000] \times [0, 100]$. Then, using **ZOOM** and **TRACE,** show that the lowest point on the graph of $\overline{C}$ is (500, 35).

2. Draw the tangent line to the graph of $\overline{C}$ at (500, 35). What is its slope? Is this expected?

3. Plot the graph of the marginal average cost function

$$\overline{C}'(x) = 0.0002x - 0.08 - \frac{5000}{x^2}$$

in the viewing rectangle $[0, 2000] \times [-1, 1]$. Then use **ZOOM** and **TRACE** to show that the zero of the function $\overline{C}'$ occurs at $x = 500$. Verify this result using the root-finding capability of your graphing utility. Is this result compatible with that obtained in part (2)? Explain your answer.

◐ ◐ ◑

Revenue Functions

Another marginal concept, the *marginal revenue function,* is associated with the revenue function R, given by

$$R(x) = px \tag{5}$$

where x is the number of units sold of a certain commodity and p is the unit selling price. In general, however, the unit selling price p of a commodity is related to the quantity x of the commodity demanded. This relationship, $p = f(x)$, is called a *demand equation* (see Section 1.3). Solving the demand equation for p in terms of x, we obtain the unit price function f, given by

$$p = f(x)$$

Thus, the revenue function R is given by

$$R(x) = px = xf(x)$$

where f is the unit price function. The derivative R' of the function R, called the **marginal revenue function,** measures the rate of change of the revenue function.

EXAMPLE 5 Suppose that the relationship between the unit price p in dollars and the quantity demanded x of the Acrosonic model F loudspeaker system is given by the equation

$$p = -0.02x + 400 \qquad (0 \le x \le 20{,}000)$$

a. Find the revenue function R.

b. Find the marginal revenue function R'.

c. Compute $R'(2000)$ and interpret your result.

Solution

a. The revenue function R is given by

$$\begin{aligned} R(x) &= px \\ &= x(-0.02x + 400) \\ &= -0.02x^2 + 400x \qquad (0 \le x \le 20{,}000) \end{aligned}$$

b. The marginal revenue function R' is given by

$$R'(x) = -0.04x + 400$$

c. $\qquad\qquad R'(2000) = -0.04(2000) + 400 = 320$

Thus, the actual revenue to be realized from the sale of the 2001st loudspeaker system is approximately $320.   ◐ ◐ ◐

Profit Functions

Our final example of a marginal function involves the profit function. The profit function P is given by

$$P(x) = R(x) - C(x) \tag{6}$$

where R and C are the revenue and cost functions and x is the number of units of a commodity produced and sold. The **marginal profit function** $P'(x)$ measures the rate of change of the profit function P and provides us with a good approximation of the actual profit or loss realized from the sale of the $(x + 1)$st unit of the commodity (assuming the xth unit has been sold).

EXAMPLE 6 Refer to Example 5. Suppose that the cost of producing x units of the Acrosonic model F loudspeaker is

$$C(x) = 100x + 200{,}000$$

dollars.

a. Find the profit function P.

b. Find the marginal profit function P'.

c. Compute $P'(2000)$ and interpret your result.

d. Sketch the graph of the profit function P.

Solution

a. From the solution to Example 5a, we have

$$R(x) = -0.02x^2 + 400x$$

Thus, the required profit function P is given by

$$P(x) = R(x) - C(x)$$
$$= (-0.02x^2 + 400x) - (100x + 200{,}000)$$
$$= -0.02x^2 + 300x - 200{,}000$$

b. The marginal profit function P' is given by

$$P'(x) = -0.04x + 300$$

c.
$$P'(2000) = -0.04(2000) + 300 = 220$$

Thus, the actual profit realized from the sale of the 2001st loudspeaker system is approximately $220.

d. The graph of the profit function P appears in Figure 11.10.   ◦ ◦ ◦

Figure 11.10

The total profit made when x loud-speakers are produced is given by $P(x)$.

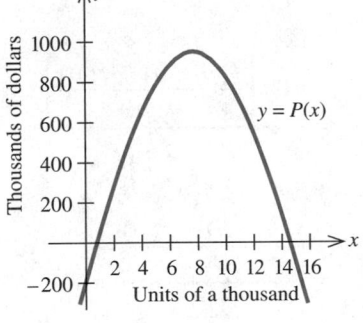

Elasticity of Demand

Finally, let us use the marginal concepts introduced in this section to derive an important criterion used by economists to analyze the demand function: *elasticity of demand*.

In what follows, it will be convenient to write the demand function f in the form $x = f(p)$; that is, we will think of the quantity demanded of a certain commodity as a function of its unit price. Since the quantity demanded of a commodity usually decreases as its unit price increases, the function f is typically a decreasing function of p (see Figure 11.11a).

Figure 11.11

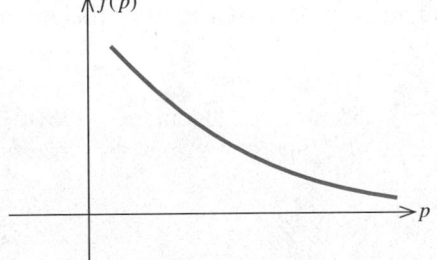

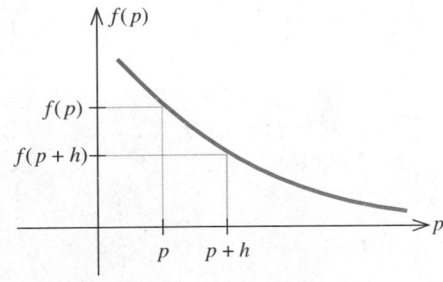

(a) A demand function.

(b) $f(p + h)$ is the quantity demanded when the unit price increases from p to $p + h$ dollars.

Suppose that the unit price of a commodity is increased by h dollars from p dollars to $(p + h)$ dollars (Figure 11.11b). Then the quantity demanded drops from $f(p)$ units to $f(p + h)$ units, a change of $[f(p + h) - f(p)]$ units. The percentage change in the unit price is

$$\frac{h}{p}(100) \qquad \left(\frac{\text{Change in unit price}}{\text{Price } p}\right)(100)$$

and the corresponding percentage change in the quantity demanded is

$$100\left[\frac{f(p + h) - f(p)}{f(p)}\right] \qquad \left(\frac{\text{Change in quantity demanded}}{\text{Quantity demanded at price } p}\right)(100)$$

Now, one good way to measure the effect that a percentage change in price has on the percentage change in the quantity demanded is to look at the ratio of the latter to the former. We find

$$\frac{\text{Percentage change in the quantity demanded}}{\text{Percentage change in the unit price}} = \frac{100\left[\dfrac{f(p + h) - f(p)}{f(p)}\right]}{100\left(\dfrac{h}{p}\right)}$$

$$= \frac{\dfrac{f(p + h) - f(p)}{h}}{\dfrac{f(p)}{p}}$$

If f is differentiable at p, then

$$\frac{f(p + h) - f(p)}{h} \approx f'(p)$$

when h is small. Therefore, if h is small, then the ratio is approximately equal to

$$\frac{f'(p)}{\dfrac{f(p)}{p}} = \frac{pf'(p)}{f(p)}$$

Economists call the negative of this quantity the *elasticity of demand*.

ELASTICITY OF DEMAND

If f is a differentiable demand function defined by $x = f(p)$, then the **elasticity of demand** at price p is given by

$$E(p) = -\frac{pf'(p)}{f(p)} \qquad\qquad (7)$$

REMARK   It will be shown later (Section 12.1) that if f is decreasing on an interval, then $f'(p) < 0$ for p in that interval. In light of this, we see that since both p and $f(p)$ are positive, the quantity $pf'(p)/f(p)$ is negative. Because economists would rather work with a positive value, the elasticity of demand $E(p)$ is defined to be the negative of this quantity.   ◦ ◦ ◦

EXAMPLE 7   Consider the demand equation

$$p = -0.02x + 400 \qquad (0 \le x \le 20{,}000)$$

which describes the relationship between the unit price in dollars and the quantity demanded x of the Acrosonic model F loudspeaker systems.

a. Find the elasticity of demand $E(p)$.

b. Compute $E(100)$ and interpret your result.

c. Compute $E(300)$ and interpret your result.

Solution

a. Solving the given demand equation for x in terms of p, we find

$$x = f(p) = -50p + 20{,}000$$

from which we see that

$$f'(p) = -50$$

Therefore,

$$E(p) = -\frac{pf'(p)}{f(p)} = -\frac{p(-50)}{-50p + 20{,}000}$$

$$= \frac{p}{400 - p}$$

b. $E(100) = 100/(400 - 100) = 1/3$, which is the elasticity of demand when $p = 100$. To interpret this result, recall that $E(100)$ is the negative of the ratio of the percentage change in the quantity demanded to the percentage change in the unit price when $p = 100$. Therefore, our result tells us that when the unit price p is set at \$100 per speaker, an increase of 1% in the unit price will cause a decrease of approximately 0.33% in the quantity demanded.

c. $E(300) = 300/(400 - 300) = 3$, which is the elasticity of demand when $p = 300$. It tells us that when the unit price is set at \$300 per speaker, an increase of 1% in the unit price will cause a decrease of approximately 3% in the quantity demanded.   ◦ ◦ ◦

Economists often use the following terminology to describe demand in terms of elasticity.

ELASTICITY OF DEMAND

The demand is said to be **elastic** if $E(p) > 1$.

The demand is said to be **unitary** if $E(p) = 1$.

The demand is said to be **inelastic** if $E(p) < 1$.

As an illustration, our computations in Example 7 revealed that demand for Acrosonic loudspeakers is elastic when $p = 300$ but inelastic when $p = 100$. These computations confirm that when demand is elastic, a small percentage change in the unit price will result in a greater percentage change in the quantity demanded; and when demand is inelastic, a small percentage change in the unit price will cause a smaller percentage change in the quantity demanded. Finally, when demand is unitary, a small percentage change in the unit price will result in the same percentage change in the quantity demanded.

We can describe the way revenue responds to changes in the unit price using the notion of elasticity. If the quantity demanded of a certain commodity is related to its unit price by the equation $x = f(p)$, then the revenue realized through the sale of x units of the commodity at a price of p dollars each is

$$R(p) = px = pf(p)$$

The rate of change of the revenue with respect to the unit price p is given by

$$R'(p) = f(p) + pf'(p)$$

$$= f(p)\left[1 + \frac{pf'(p)}{f(p)}\right]$$

$$= f(p)[1 - E(p)]$$

Now, suppose the demand is elastic when the unit price is set at a dollars. Then $E(a) > 1$, and so $1 - E(a) < 0$. Since $f(p)$ is positive for all values of p, we see that

$$R'(a) = f(a)[1 - E(a)] < 0$$

and so $R(p)$ is decreasing at $p = a$. This implies that a small increase in the unit price when $p = a$ results in a decrease in the revenue, whereas a small decrease in the unit price will result in an increase in the revenue. Similarly, you can show that if the demand is inelastic when the unit price is set at a dollars, then a small increase in the unit price will cause the revenue to increase, and a small decrease in the unit price will cause the revenue to decrease. Finally, if the demand is unitary when the unit price is set at a dollars, then $E(a) = 1$ and $R'(a) = 0$. This implies that a small increase or decrease in the unit price will not result in a change in the revenue. The following statements summarize this discussion.

1. If the demand is elastic at p [$E(p) > 1$], then an increase in the unit price will cause the revenue to decrease, whereas a decrease in the unit price will cause the revenue to increase.

2. If the demand is inelastic at p [$E(p) < 1$], then an increase in the unit price will cause the revenue to increase, and a decrease in the unit price will cause the revenue to decrease.

3. If the demand is unitary at p [$E(p) = 1$], then an increase in the unit price will cause the revenue to stay about the same.

These results are illustrated in Figure 11.12.

Figure 11.12
The revenue is increasing on an interval where the demand is inelastic, decreasing on an interval where the demand is elastic, and stationary at the point where the demand is unitary.

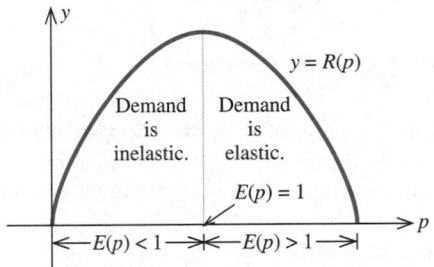

REMARK   As an aid to remembering this, note that

1. if demand is elastic, then the change in revenue and the change in the unit price move in opposite directions.

2. if demand is inelastic, then they move in the same direction.   ○ ○ ○

EXAMPLE 8   Refer to Example 7.

a. Is demand elastic, unitary, or inelastic when $p = 100$? When $p = 300$?

b. If the price is $100, will raising the unit price slightly cause the revenue to increase or decrease?

Solution

a. From the results of Example 7, we see that $E(100) = \frac{1}{3} < 1$ and $E(300) = 3 > 1$. We conclude accordingly that demand is inelastic when $p = 100$ and elastic when $p = 300$.

b. Since demand is inelastic when $p = 100$, raising the unit price slightly will cause the revenue to increase.   ○ ○ ○

SELF-CHECK EXERCISES 11.4

 1. The weekly demand for Pulsar VCRs (videocassette recorders) is given by the demand equation

$$p = -0.02x + 300 \qquad (0 \le x \le 15{,}000)$$

where p denotes the wholesale unit price in dollars and x denotes the quantity demanded. The weekly total cost function associated with manufacturing these VCRs is

$$C(x) = 0.000003x^3 - 0.04x^2 + 200x + 70{,}000$$

dollars.
a. Find the revenue function R and the profit function P.
b. Find the marginal cost function C', the marginal revenue function R', and the marginal profit function P'.
c. Find the marginal average cost function $\overline{C}'$.
d. Compute $C'(3000)$, $R'(3000)$, and $P'(3000)$, and interpret your results.

2. Refer to the preceding exercise. Determine whether the demand is elastic, unitary, or inelastic when $p = 100$ and when $p = 200$.

Solutions to Self-Check Exercises 11.4 can be found on page 739.

11.4 EXERCISES

A calculator is recommended for exercises 2–33.

1. Production Costs The graph of a typical total cost function $C(x)$ associated with the manufacture of x units of a certain commodity is shown in the following figure.
a. Explain why the function C is always increasing.
b. As the level of production x increases, the cost per unit drops, so that $C(x)$ increases but at a slower pace. However, a level of production is soon reached at which the cost per unit begins to increase dramatically (due to a shortage of raw material, overtime, breakdown of machinery due to excessive stress and strain) so that $C(x)$ continues to increase at a faster pace. Use the graph of C to find the approximate level of production x_0 where this occurs.

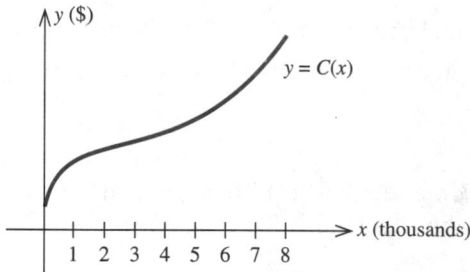

2. Marginal Cost The total weekly cost in dollars incurred by the Lincoln Record Company in pressing x long-playing records is

$$C(x) = 2000 + 2x - 0.0001x^2 \quad (0 \le x \le 6000)$$

a. What is the actual cost incurred in producing the 1001st record? The 2001st record?
b. What is the marginal cost when $x = 1000$? When $x = 2000$?

3. Marginal Cost A division of Ditton Industries manufactures the Futura model microwave oven. The daily cost (in dollars) of producing these microwave ovens is

$$C(x) = 0.0002x^3 - 0.06x^2 + 120x + 5000$$

where x stands for the number of units produced.
a. What is the actual cost incurred in manufacturing the 101st oven? The 201st oven? The 301st oven?
b. What is the marginal cost when $x = 100$? When $x = 200$? When $x = 300$?

4. Marginal Average Cost The Custom Office Company makes a line of executive desks. It is estimated that the total cost for making x units of their Senior Executive Model is

$$C(x) = 100x + 200,000$$

dollars per year.
a. Find the average cost function $\overline{C}$.
b. Find the marginal average cost function $\overline{C}'$.
c. What happens to $\overline{C}(x)$ when x is very large? Interpret your results.

5. Marginal Average Cost The management of the Thermo-Master Company, whose Mexican subsidiary manufactures an indoor-outdoor thermometer, has estimated that the total weekly cost in dollars for producing x thermometers is

$$C(x) = 5000 + 2x$$

a. Find the average cost function $\overline{C}$.
b. Find the marginal average cost function $\overline{C}'$.
c. Interpret your results.

6. Find the average cost function $\overline{C}$ and the marginal average cost function $\overline{C}'$ associated with the total cost function C of exercise 2.

7. Find the average cost function $\overline{C}$ and the marginal average cost function $\overline{C}'$ associated with the total cost function C of exercise 3.

8. Marginal Revenue The Williams Commuter Air Service realizes a monthly revenue of

$$R(x) = 8000x - 100x^2$$

dollars when the price charged per passenger is x dollars.
a. Find the marginal revenue R'.
b. Compute $R'(39)$, $R'(40)$, and $R'(41)$. What do your results imply?

9. Marginal Revenue The management of the Acrosonic Company plans to market the Electro-Stat, an electro-static speaker system. The marketing department has determined that the demand for these speakers is

$$p = -0.04x + 800 \quad (0 \le x \le 20,000)$$

where p denotes the speaker's unit price (in dollars) and x denotes the quantity demanded.
a. Find the revenue function R.
b. Find the marginal revenue function R'.
c. Compute $R'(5000)$ and interpret your result.

10. Marginal Profit Lynbrook West, an apartment complex, has 100 two-bedroom units. The monthly profit (in dollars) realized from renting x apartments is

$$P(x) = -10x^2 + 1760x - 50{,}000$$

a. What is the actual profit realized from renting the 51st unit, assuming that 50 units have already been rented?
b. Compute the marginal profit when $x = 50$ and compare your results with that obtained in (a).

11. Marginal Profit Refer to exercise 9. Acrosonic's production department estimates that the total cost (in dollars) incurred in manufacturing x Electro-Stat speaker systems in the first year of production will be

$$C(x) = 200x + 300{,}000$$

a. Find the profit function P.
b. Find the marginal profit function P'.
c. Compute $P'(5000)$ and $P'(8000)$.
d. Sketch the graph of the profit function and interpret your results.

12. Marginal Cost, Revenue, and Profit The weekly demand for the Pulsar 25 color console television is

$$p = 600 - 0.05x \qquad (0 \le x \le 12{,}000)$$

where p denotes the wholesale unit price in dollars and x denotes the quantity demanded. The weekly total cost function associated with manufacturing the Pulsar 25 is given by

$$C(x) = 0.000002x^3 - 0.03x^2 + 400x + 80{,}000$$

where $C(x)$ denotes the total cost incurred in producing x sets.
a. Find the revenue function R and the profit function P.
b. Find the marginal cost function C', the marginal revenue function R', and the marginal profit function P'.
c. Compute $C'(2000)$, $R'(2000)$, and $P'(2000)$, and interpret your results.
d. Sketch the graphs of the functions C, R, and P, and interpret (b) and (c) using the graphs obtained.

13. Marginal Cost, Revenue, and Profit The Pulsar Corporation also manufactures a series of 19-inch color television sets. The quantity x of these sets demanded each week is related to the wholesale unit price p by the equation

$$p = -0.006x + 180$$

The weekly total cost incurred by Pulsar for producing x sets is

$$C(x) = 0.000002x^3 - 0.02x^2 + 120x + 60{,}000$$

dollars. Answer the questions in exercise 12 for these data.

14. Marginal Average Cost Find the average cost function $\bar{C}$ associated with the total cost function C of exercise 12.
a. What is the marginal average cost function $\bar{C}'$?
b. Compute $\bar{C}'(5{,}000)$ and $\bar{C}'(10{,}000)$ and interpret your results.
c. Sketch the graph of $\bar{C}$.

15. Marginal Average Cost Find the average cost function $\bar{C}$ associated with the total cost function C of exercise 13.
a. What is the marginal average cost function $\bar{C}'$?
b. Compute $\bar{C}'(5{,}000)$ and $\bar{C}'(10{,}000)$ and interpret your results.

16. Marginal Revenue The quantity of Sicard wristwatches demanded per month is related to the unit price by the equation

$$p = \frac{50}{0.01x^2 + 1} \qquad (0 \le x \le 20)$$

where p is measured in dollars and x in units of a thousand.
a. Find the revenue function R.
b. Find the marginal revenue function R'.
c. Compute $R'(2)$ and interpret your result.

17. Marginal Propensity to Consume The consumption function of the U.S. economy for 1929 to 1941 is

$$C(x) = 0.712x + 95.05$$

where $C(x)$ is the personal consumption expenditure and x is the personal income, both measured in billions of dollars. Find the rate of change of consumption with respect to income, dC/dx. This quantity is called the *marginal propensity to consume*.

18. Marginal Propensity to Consume   Refer to exercise 17. Suppose that a certain economy's consumption function is

$$C(x) = 0.873x^{1.1} + 20.34$$

where $C(x)$ and x are measured in billions of dollars. Find the marginal propensity to consume when $x = 10$.

19. Marginal Propensity to Save   Suppose that $C(x)$ measures an economy's personal consumption expenditure and x the personal income, both in billions of dollars. Then

$$S(x) = x - C(x) \qquad \text{(Income minus consumption)}$$

measures the economy's savings corresponding to an income of x billion dollars. Show that

$$\frac{dS}{dx} = 1 - \frac{dC}{dx}$$

The quantity dS/dx is called the *marginal propensity to save*.

20. Refer to exercise 19. For the consumption function of exercise 17, find the marginal propensity to save.

21. Refer to exercise 19. For the consumption function of exercise 18, find the marginal propensity to save when $x = 10$.

For each demand equation in exercises 22–27, compute the elasticity of demand and determine whether the demand is elastic, unitary, or inelastic at the indicated price.

22. $x = -\frac{3}{2}p + 9;\quad p = 2$

23. $x = -\frac{5}{4}p + 20;\quad p = 10$

24. $x + \frac{1}{3}p - 20 = 0;\quad p = 30$

25. $0.4x + p - 20 = 0;\quad p = 10$

26. $p = 144 - x^2;\quad p = 96$

27. $p = 169 - x^2;\quad p = 29$

28. Elasticity of Demand   The management of the Titan Tire Company has determined that the quantity demanded x of their Super Titan tires per week is related to the unit price p by the equation

$$x = \sqrt{144 - p}$$

where p is measured in dollars and x in units of a thousand.

a. Compute the elasticity of demand when $p = 63$, when $p = 96$, and when $p = 108$.
b. Interpret the results obtained in (a).
c. Is the demand elastic, unitary, or inelastic when $p = 63$? When $p = 96$? When $p = 108$?

29. Elasticity of Demand   The demand equation for the Roland portable hair dryer is given by

$$x = \frac{1}{5}(225 - p^2) \qquad (0 \le p \le 15)$$

where x (measured in units of a hundred) is the quantity demanded per week and p is the unit price in dollars.
a. Is the demand elastic or inelastic when $p = 8$ and when $p = 10$?
b. When is the demand unitary?

[*Hint:* Solve $E(p) = 1$ for p.]

c. If the unit price is lowered slightly from \$10, will the revenue increase or decrease?
d. If the unit price is increased slightly from \$8, will the revenue increase or decrease?

30. Elasticity of Demand   The quantity demanded per week x (in units of a hundred) of the Mikado miniature camera is related to the unit price p (in dollars) by the demand equation

$$x = \sqrt{400 - 5p} \qquad (0 \le p \le 80)$$

a. Is the demand elastic or inelastic when $p = 40$? When $p = 60$?
b. When is the demand unitary?
c. If the unit price is lowered slightly from \$60, will the revenue increase or decrease?
d. If the unit price is increased slightly from \$40, will the revenue increase or decrease?

31. Elasticity of Demand   The proprietor of the Showplace, a video club, has estimated that the rental price p (in dollars) of prerecorded videocassette tapes is related to the quantity x rented per week by the demand equation

$$x = \frac{2}{3}\sqrt{36 - p^2} \qquad (0 \le p \le 6)$$

Currently, the rental price is \$2 per tape.
a. Is the demand elastic or inelastic at this rental price?
b. If the rental price is increased, will the revenue increase or decrease?

32. **Elasticity of Demand**   The demand function for a certain make of exercise bicycle sold exclusively through cable television is

$$p = \sqrt{9 - 0.02x} \qquad (0 \le x \le 450)$$

where p is the unit price in hundreds of dollars and x is the quantity demanded per week. Compute the elasticity of demand and determine the range of prices corresponding to inelastic, unitary, and elastic demand.
[*Hint:* Solve the equation $E(p) = 1$.]

33. **Elasticity of Demand**   The demand equation for the Sicard wristwatch is given by

$$x = 10 \sqrt{\frac{50 - p}{p}} \qquad (0 < p \le 50)$$

where x (measured in units of a thousand) is the quantity demanded per week and p is the unit price in dollars. Compute the elasticity of demand and determine the range of prices corresponding to inelastic, unitary, and elastic demand.

SOLUTIONS TO SELF-CHECK EXERCISES 11.4

1. a.
$$\begin{aligned}
R(x) &= px \\
&= x(-0.02x + 300) \\
&= -0.02x^2 + 300x \qquad (0 \le x \le 15,000)
\end{aligned}$$

$$\begin{aligned}
P(x) &= R(x) - C(x) \\
&= -0.02x^2 + 300x \\
&\quad -(0.000003x^3 - 0.04x^2 + 200x + 70,000) \\
&= -0.000003x^3 + 0.02x^2 + 100x - 70,000
\end{aligned}$$

b.
$$\begin{aligned}
C'(x) &= 0.000009x^2 - 0.08x + 200 \\
R'(x) &= -0.04x + 300
\end{aligned}$$

and

$$P'(x) = -0.000009x^2 + 0.04x + 100$$

c. The average cost function is

$$\begin{aligned}
\overline{C}(x) &= \frac{C(x)}{x} \\
&= \frac{0.000003x^3 - 0.04x^2 + 200x + 70,000}{x} \\
&= 0.000003x^2 - 0.04x + 200 + \frac{70,000}{x}
\end{aligned}$$

Therefore, the marginal average cost function is

$$\overline{C}'(x) = 0.000006x - 0.04 - \frac{70,000}{x^2}$$

d. Using the results from (b), we find

$$\begin{aligned}
C'(3000) &= 0.000009(3000)^2 - 0.08(3000) + 200 \\
&= 41
\end{aligned}$$

That is, when the level of production is already 3000 VCRs, the actual cost of producing one additional VCR is approximately \$41. Next,

$$R'(3000) = -0.04(3000) + 300 = 180$$

That is, the actual revenue to be realized from selling the 3001st VCR is approximately \$180. Finally,

$$P'(3000) = -0.000009(3000)^2 + 0.04(3000) + 100$$
$$= 139$$

That is, the actual profit realized from selling the 3001st VCR is approximately \$139.

2. We first solve the given demand equation for x in terms of p, obtaining

$$x = f(p) = -50p + 15{,}000$$

and $$f'(p) = -50$$

Therefore,

$$E(p) = -\frac{pf'(p)}{f(p)} = -\frac{p}{-50p + 15{,}000}(-50)$$

$$= \frac{p}{300 - p} \qquad (0 \le p < 300)$$

Next, we compute

$$E(100) = \frac{100}{300 - 100} = \frac{1}{2} < 1$$

and we conclude that demand is inelastic when $p = 100$. Also,

$$E(200) = \frac{200}{300 - 200} = 2 > 1$$

and we see that demand is elastic when $p = 200$.

11.5 HIGHER-ORDER DERIVATIVES

The derivative f' of a function f is also a function. As such, the differentiability of f' may be considered. Thus, the function f' has a derivative f'' at a point x in the domain of f' if the limit of the quotient

$$\frac{f'(x + h) - f'(x)}{h}$$

exists as h approaches zero. In other words, it is the derivative of the first derivative.

The function f'' obtained in this manner is called the **second derivative** of the function f, just as the derivative f' of f is often called the first derivative of f. Continuing in this fashion, we are led to considering the third, fourth, and higher-order derivatives of f whenever they exist. Notations for the first, second, third, and, in general, nth derivatives of a function f at a point x are

$$f'(x), \, f''(x), \, f'''(x), \, \ldots, \, f^{(n)}(x)$$

or

$$D^1 f(x), \, D^2 f(x), \, D^3 f(x), \, \ldots, \, D^n f(x)$$

If f is written in the form $y = f(x)$, then the notations for its derivatives are

$$y', y'', y''', \ldots, y^{(n)}$$

$$\frac{dy}{dx}, \frac{d^2 y}{dx^2}, \frac{d^3 y}{dx^3}, \ldots, \frac{d^n y}{dx^n}$$

or

$$D^1 y, D^2 y, D^3 y, \ldots, D^n y$$

respectively.

EXAMPLE 1   Find the derivatives of all orders of the polynomial function $f(x) = x^5 - 3x^4 + 4x^3 - 2x^2 + x - 8$.

Solution   We have

$$f'(x) = 5x^4 - 12x^3 + 12x^2 - 4x + 1$$

$$f''(x) = \frac{d}{dx} f'(x) = 20x^3 - 36x^2 + 24x - 4$$

$$f'''(x) = \frac{d}{dx} f''(x) = 60x^2 - 72x + 24$$

$$f^{(4)}(x) = \frac{d}{dx} f'''(x) = 120x - 72$$

$$f^{(5)}(x) = \frac{d}{dx} f^{(4)}(x) = 120$$

and, in general,

$$f^{(n)}(x) = 0 \quad \text{for } n > 5$$

EXAMPLE 2   Find the third derivative of the function f defined by $y = x^{2/3}$. What is its domain?

Solution   We have

$$y' = \frac{2}{3} x^{-1/3}$$

$$y'' = \left(\frac{2}{3} \right) \left(-\frac{1}{3} \right) x^{-4/3} = -\frac{2}{9} x^{-4/3}$$

Figure 11.13
The graph of the function $y = x^{2/3}$.

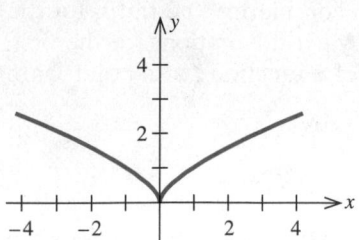

so the required derivative is

$$y''' = \left(-\frac{2}{9}\right)\left(-\frac{4}{3}\right)x^{-7/3} = \frac{8}{27}x^{-7/3} = \frac{8}{27x^{7/3}}$$

The common domain of the functions f', f'', and f''' is the set of all real numbers except $x = 0$. The domain of $y = x^{2/3}$ is the set of all real numbers. The graph of the function $y = x^{2/3}$ appears in Figure 11.13.   ○ ○ ○

REMARK   Always simplify an expression before differentiating it to obtain the next order derivative.   ○ ○ ○

EXAMPLE 3   Find the second derivative of the function $y = (2x^2 + 3)^{3/2}$.

Solution   We have, using the General Power Rule,

$$y' = \frac{3}{2}(2x^2 + 3)^{1/2}(4x) = 6x(2x^2 + 3)^{1/2}$$

Next, using the Product Rule and then the Chain Rule, we find

$$y'' = (6x) \cdot \frac{d}{dx}(2x^2 + 3)^{1/2} + \left[\frac{d}{dx}(6x)\right](2x^2 + 3)^{1/2}$$

$$= (6x)\left(\frac{1}{2}\right)(2x^2 + 3)^{-1/2}(4x) + 6(2x^2 + 3)^{1/2}$$

$$= 12x^2(2x^2 + 3)^{-1/2} + 6(2x^2 + 3)^{1/2}$$

$$= 6(2x^2 + 3)^{-1/2}[2x^2 + (2x^2 + 3)]$$

$$= \frac{6(4x^2 + 3)}{\sqrt{2x^2 + 3}}$$ ○ ○ ○

Applications

Just as the derivative of a function f at a point x measures the rate of change of the function f at that point, the second derivative of f (the derivative of f') measures the rate of change of the derivative f' of the function f. The third derivative of the function f, f''', measures the rate of change of f'', and so on.

In Chapter 12 we will discuss applications involving the geometric interpretation of the second derivative of a function. The following example gives an interpretation of the second derivative in a familiar role.

EXAMPLE 4   Refer to the example on page 654. The distance s (in feet) covered by a maglev moving along a straight track t seconds after starting from rest is given by the function $s = 4t^2$ ($0 \le t \le 10$). What is the maglev's acceleration at the end of 30 seconds?

Solution   The velocity of the maglev t seconds from rest is given by

$$v = \frac{ds}{dt} = \frac{d}{dt}(4t^2) = 8t$$

The acceleration of the maglev t seconds from rest is given by the rate of change of the velocity of t—that is,

$$a = \frac{d}{dt}v = \frac{d}{dt}\left(\frac{ds}{dt}\right) = \frac{d^2s}{dt^2} = \frac{d}{dt}(8t) = 8$$

or 8 feet per second per second, normally abbreviated 8 ft/sec². ◦ ◦ ◦

EXAMPLE 5   A ball is thrown straight up into the air from the roof of a building. The height of the ball as measured from the ground is given by

$$s = -16t^2 + 24t + 120$$

where s is measured in feet and t in seconds. Find the velocity and acceleration of the ball 3 seconds after it is thrown into the air.

Solution   The velocity v and acceleration a of the ball at any time t are given by

$$v = \frac{ds}{dt} = \frac{d}{dt}(-16t^2 + 24t + 120) = -32t + 24$$

and

$$a = \frac{d^2t}{dt^2} = \frac{d}{dt}\left(\frac{ds}{dt}\right) = \frac{d}{dt}(-32t + 24) = -32$$

Therefore, the velocity of the ball 3 seconds after it is thrown into the air is

$$v = -32(3) + 24 = -72$$

That is, the ball is falling downward at a speed of 72 ft/sec. The acceleration of the ball is 32 ft/sec² downward at any time during the motion. ◦ ◦ ◦

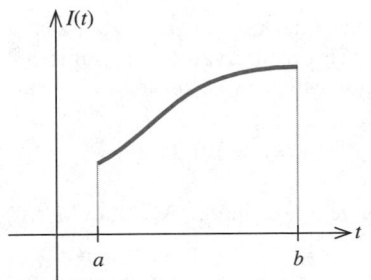

Figure 11.14
The CPI of a certain economy from year a to year b is given by I(t).

Another interpretation of the second derivative of a function—this time from the field of economics—follows. Suppose that the consumer price index (CPI) of an economy between the years a and b is described by the function $I(t)$ ($a \leq t \leq b$) (Figure 11.14). Then the first derivative of I, $I'(t)$, gives the rate of inflation of the economy at any time t. The second derivative of I, $I''(t)$, gives the *rate of change of the inflation rate* at any time t. Thus, when the economist or politician claims that "inflation is slowing," what he or she is saying is that the rate of inflation is decreasing. Mathematically, this is equivalent to noting that the second derivative $I''(t)$ is negative at the time t under consideration. Observe that $I'(t)$ could be positive at a time when $I''(t)$ is negative (see Example 6). Thus, one may not draw the conclusion from the aforementioned quote that prices of goods and services are about to drop!

EXAMPLE 6 An economy's consumer price index (CPI) is described by the function

$$I(t) = -0.2t^3 + 3t^2 + 100 \qquad (0 \le t \le 9)$$

where $t = 0$ corresponds to the year 1990. Compute $I'(6)$ and $I''(6)$ and use these results to show that even though the CPI was rising at the beginning of 1996, "inflation was moderating" at that time.

Figure 11.15
The CPI of an economy is given by $I(t)$.

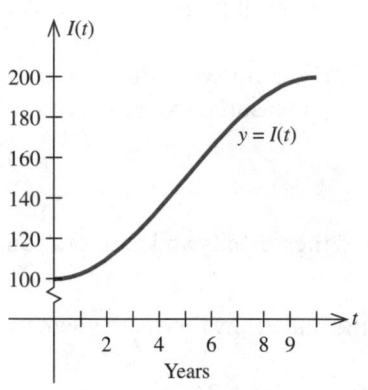

Solution We find

$$I'(t) = -0.6t^2 + 6t$$

and

$$I''(t) = -1.2t + 6$$

so

$$I'(6) = -0.6(6)^2 + 6(6) = 14.4$$

and

$$I''(6) = -1.2(6) + 6 = -1.2$$

Our computations reveal that at the beginning of 1996 ($t = 6$), the CPI was increasing at the rate of 14.4 points per year, whereas the rate of the inflation rate was decreasing by 1.2 points per year. Thus, inflation was moderating at that time (Figure 11.15). In Section 12.2, we will see that relief actually began in early 1995. ○ ○ ○

SELF–CHECK EXERCISES 11.5

1. Find the third derivative of

$$f(x) = 2x^5 - 3x^3 + x^2 - 6x + 10$$

2. Let

$$f(x) = \frac{1}{1 + x}$$

Find $f'(x)$, $f''(x)$, and $f'''(x)$.

3. A certain species of turtle faces extinction because dealers collect truckloads of turtle eggs to be sold as aphrodisiacs. After severe conservation measures are implemented, it is hoped that the turtle population will grow according to the rule

$$N(t) = 2t^3 + 3t^2 - 4t + 1000 \qquad (0 \le t \le 10)$$

where $N(t)$ denotes the population at the end of year t. Compute $N''(2)$ and $N''(8)$ and interpret your results.

Solutions to Self-Check Exercises 11.5 can be found on page 748.

11.5 EXERCISES

In exercises 1–20, find the first and second derivatives of the given function.

1. $f(x) = 4x^2 - 2x + 1$

2. $f(x) = -0.2x^2 + 0.3x + 4$

3. $f(x) = 2x^3 - 3x^2 + 1$

4. $g(x) = -3x^3 + 24x^2 + 6x - 64$

5. $h(t) = t^4 - 2t^3 + 6t^2 - 3t + 10$

6. $f(x) = x^5 - x^4 + x^3 - x^2 + x - 1$

7. $f(x) = (x^2 + 2)^5$ **8.** $g(t) = t^2(3t + 1)^4$

9. $g(t) = (2t^2 - 1)^2(3t^2)$

10. $h(x) = (x^2 + 1)^2(x - 1)$

11. $f(x) = (2x^2 + 2)^{7/2}$

12. $h(w) = (w^2 + 2w + 4)^{5/2}$

13. $f(x) = x(x^2 + 1)^2$ **14.** $g(u) = u(2u - 1)^3$

15. $f(x) = \dfrac{x}{2x + 1}$ **16.** $g(t) = \dfrac{t^2}{t - 1}$

17. $f(s) = \dfrac{s - 1}{s + 1}$ **18.** $f(u) = \dfrac{u}{u^2 + 1}$

19. $f(u) = \sqrt{4 - 3u}$ **20.** $f(x) = \sqrt{2x - 1}$

In exercises 21–28, find the third derivative of the given function.

21. $f(x) = 3x^4 - 4x^3$

22. $f(x) = 3x^5 - 6x^4 + 2x^2 - 8x + 12$

23. $f(x) = \dfrac{1}{x}$ **24.** $f(x) = \dfrac{2}{x^2}$

25. $g(s) = \sqrt{3s - 2}$ **26.** $g(t) = \sqrt{2t + 3}$

27. $f(x) = (2x - 3)^4$ **28.** $g(t) = (\tfrac{1}{2}t^2 - 1)^5$

29. Acceleration of a Falling Object   During the construction of an office building, a hammer is accidentally dropped from a height of 256 ft. The distance the hammer falls in t seconds is $s = 16t^2$. What is the hammer's velocity when it strikes the ground? What is its acceleration?

30. Acceleration of a Car   The distance s (in feet) covered by a car t seconds after starting from rest is given by

$$s = 20t + 8t^2 - t^3 \qquad (0 \le t \le 6)$$

Find a general expression for the car's acceleration at any time t ($0 \le t \le 6$). Show that the car is decelerating $2\tfrac{2}{3}$ seconds after starting from rest.

31. Crime Rates   The number of major crimes committed in Bronxville between 1988 and 1995 is approximated by the function

$$N(t) = -0.1t^3 + 1.5t^2 + 100 \qquad (0 \le t \le 7)$$

where $N(t)$ denotes the number of crimes committed in year t and $t = 0$ corresponds to the year 1988. Enraged by the dramatic increase in the crime rate, Bronxville's citizens, with the help of the local police, organized "Neighborhood Crime Watch" groups in early 1992 to combat this menace.

a. Verify that the crime rate was increasing from 1988 through 1995.

[*Hint:* Compute $N'(0)$, $N'(1)$, ..., $N'(7)$.]

b. Show that the Neighborhood Crime Watch program was working by computing $N''(4)$, $N''(5)$, $N''(6)$, and $N''(7)$.

32. GDP of a Developing Country   A developing country's gross domestic product (GDP) from 1988 to 1996 is approximated by the function

$$G(t) = -0.2t^3 + 2.4t^2 + 60 \qquad (0 \le t \le 8)$$

where $G(t)$ is measured in billions of dollars and $t = 0$ corresponds to the year 1988.

a. Compute $G'(0)$, $G'(1)$, ..., $G'(8)$.

b. Compute $G''(0)$, $G''(1)$, ..., $G''(8)$.

c. Using the results obtained in (a) and (b), show that after a spectacular growth rate in the early years, the growth of the GDP cooled off.

33. Test Flight of a VTOL   In a test flight of the McCord Terrier, McCord Aviation's experimental VTOL (vertical take-off and landing) aircraft, it was determined that t seconds after lift-off, when the craft was operated in

USING TECHNOLOGY

FINDING THE SECOND DERIVATIVE OF A FUNCTION AT A GIVEN POINT

Some graphing utilities have the capability of numerically computing the second derivative of a function at a point. If your graphing utility has this capability, use it to work through the examples and exercises of this section.

EXAMPLE 1 Use the (second) numerical derivative operation of a graphing utility to find the second derivative of $f(x) = \sqrt{x}$ when $x = 4$.

Solution Using the (second) numerical derivative operation of a graphing utility, we find

$$f''(4) = \text{der2}\ (x\verb|^|.5,\ x,\ 4) = -0.03125 \qquad \circ\ \circ\ \circ$$

EXAMPLE 2 The anticipated rise in the number of persons with Alzheimer's disease in the United States is given by
$$f(t) = -0.02765t^4 + 0.3346t^3 - 1.1261t^2$$
$$+ 1.7575t + 3.7745 \qquad (0 \le t \le 6)$$

where $f(t)$ is measured in millions and t is measured in decades, with $t = 0$ corresponding to the beginning of 1990.

a. How fast is the number of Alzheimer's patients in the United States anticipated to be changing at the beginning of 2030?

b. How fast is the rate of change of the number of Alzheimer's patients in the United States anticipated to be changing at the beginning of 2030?

c. Plot the graph of f in the viewing rectangle $[0, 7] \times [0, 12]$.
Source: Alzheimer's Association

Solution

a. Using the numerical derivative operation of a graphing utility, we find that the number of Alzheimer's patients at the beginning of 2030 can be anticipated to be changing at the rate of

$$f'(4) = 1.7311$$

That is, the number is increasing at the rate of approximately 1.7 million patients per decade.

b. Using the (second) numerical derivative operation of a graphing utility, we find that

$$f''(4) = 0.4694$$

That is, the rate of change of the number of Alzheimer's patients is increasing at the rate of approximately 0.5 million patients per decade per decade at the beginning of 2030.

746

c. The graph is shown in Figure T1.

Figure T1
The graph of f in the viewing window [0, 7] × [0, 12].

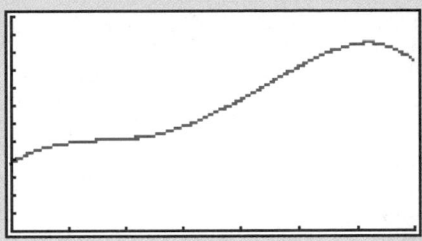

EXERCISES

In exercises 1–8, find the value of the second derivative of f at the given value of x. Express your answer correct to four decimal places.

1. $f(x) = 2x^3 - 3x^2 + 1; x = -1$

2. $f(x) = 2.5x^5 - 3x^3 + 1.5x + 4; x = 2.1$

3. $f(x) = 2.1x^{3.1} - 4.2x^{1.7} + 4.2; x = 1.4$

4. $f(x) = 1.7x^{4.2} - 3.2x^{1.3} + 4.2x - 3.2; x = 2.2$

5. $f(x) = \dfrac{x^2 + 2x - 5}{x^3 + 1}; x = 2.1$

6. $f(x) = \dfrac{x^3 + x + 2}{2x^2 - 5x + 4}; x = 1.2$

7. $f(x) = \dfrac{x^{1/2} + 2x^{3/2} + 1}{2x^{1/2} + 3}; x = 0.5$

8. $f(x) = \dfrac{\sqrt{x} - 1}{2x + \sqrt{x} + 4}; x = 2.3$

9. **Rate of Bank Failures**   The Federal Deposit Insurance Corporation (FDIC) estimates that the rate at which banks were failing between 1982 and 1994 is given by

$$f(t) = -0.063447t^4 - 1.953283t^3 + 14.632576t^2$$
$$- 6.684704t + 47.458874 \qquad (0 \le t \le 12)$$

where $f(t)$ is measured in the number of banks per year and t is measured in years, with $t = 0$ corresponding to the beginning of 1982. Compute $f''(6)$ and interpret your results.
Source: Federal Deposit Insurance Corporation

10. **Multimedia Sales**   According to the Electronics Industries Association, sales in the multimedia market (hardware and software) are expected to be

$$S(t) = -0.0094t^4 + 0.1204t^3 - 0.0868t^2$$
$$+ 0.0195t + 3.3325 \qquad (0 \le t \le 10)$$

where $S(t)$ is measured in billions of dollars and t is measured in years, with $t = 0$ corresponding to 1990. Compute $S''(7)$ and interpret your results.
Source: Electronics Industries Association

the vertical take-off mode, its altitude (in feet) was

$$h(t) = \frac{1}{16}t^4 - t^3 + 4t^2 \qquad (0 \le t \le 8)$$

a. Find an expression for the craft's velocity at time t.
b. Find the craft's velocity when $t = 0$ (the initial velocity), $t = 4$, and $t = 8$.
c. Find an expression for the craft's acceleration at time t.
d. Find the craft's acceleration when $t = 0$, $t = 4$, and $t = 8$.
e. Find the craft's height when $t = 0$, $t = 4$, and $t = 8$.

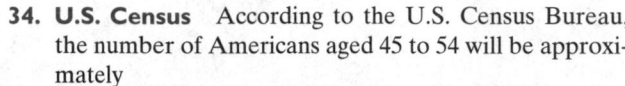 **34. U.S. Census** According to the U.S. Census Bureau, the number of Americans aged 45 to 54 will be approximately

$$N(t) = -0.00233t^4 + 0.00633t^3 - 0.05417t^2$$
$$+ 1.3467t + 25$$

million people in year t, where $t = 0$ corresponds to the beginning of 1990. Compute $N'(10)$ and $N''(10)$ and interpret your results.
Source: U.S. Census Bureau

 35. Air Purification During testing of a certain brand of air purifier, it was determined that the amount of smoke remaining t minutes after the start of the test was

$$A(t) = 100 - 17.63t + 1.915t^2 - 0.1316t^3$$
$$+ 0.00468t^4 - 0.00006t^5$$

percent of the original amount. Compute $A'(10)$ and $A''(10)$ and interpret your results.
Source: Consumer Reports

36. Let f be the function defined by the rule $f(x) = x^{7/3}$. Show that f has first- and second-order derivatives at all points x, in particular at $x = 0$. Show also that the third derivative of f does not exist at $x = 0$.

37. Construct a function f that has derivatives of order up through and including n at a point a but fails to have the $(n + 1)$st derivative there.
[*Hint:* See exercise 36.]

38. Show that a polynomial function has derivatives of all orders.
[*Hint:* Let $P(x) = a_0x^n + a_1x^{n-1} + a_2x^{n-2} + \cdots + a_n$ be a polynomial of degree n, where n is a positive integer and a_0, $a_1, \ldots, a_n$ are constants with $a_0 \ne 0$. Compute $P'(x), P''(x), \ldots$.]

SOLUTIONS TO SELF-CHECK EXERCISES 11.5

1. $f'(x) = 10x^4 - 9x^2 + 2x - 6$
$f''(x) = 40x^3 - 18x + 2$
$f'''(x) = 120x^2 - 18$

2. We write $f(x) = (1 + x)^{-1}$ and use the General Power Rule, obtaining

$$f'(x) = (-1)(1 + x)^{-2}\frac{d}{dx}(1 + x) = -(1 + x)^{-2}(1)$$

$$= -(1 + x)^{-2} = -\frac{1}{(1 + x)^2}$$

Continuing, we find

$$f''(x) = -(-2)(1 + x)^{-3}$$

$$= 2(1 + x)^{-3} = \frac{2}{(1 + x)^3}$$

and

$$f'''(x) = 2(-3)(1 + x)^{-4}$$
$$= -6(1 + x)^{-4}$$

$$= -\frac{6}{(1 + x)^4}$$

3. $N'(t) = 6t^2 + 6t - 4$

$N''(t) = 12t + 6 = 6(2t + 1)$

Therefore, $N''(2) = 30$ and $N''(8) = 102$. The results of our computations reveal that at the end of year 2, the *rate* of growth of the turtle population is increasing at the rate of 30 turtles per year per year. At the end of year 8, the rate is increasing at the rate of 102 turtles per year per year. Clearly, the conservation measures are paying off handsomely.

11.6 IMPLICIT DIFFERENTIATION AND RELATED RATES

Differentiating Implicitly

Up to now we have dealt with functions expressed in the form $y = f(x)$; that is, the dependent variable y is expressed *explicitly* in terms of the independent variable x. However, not all functions are expressed in this form. Consider, for example, the equation

$$x^2y + y - x^2 + 1 = 0 \qquad (8)$$

This equation expresses y *implicitly* as a function of x. In fact, solving equation (8) for y in terms of x, we obtain

$$(x^2 + 1)y = x^2 - 1 \qquad \text{(Implicit equation)}$$

or $\qquad y = f(x) = \dfrac{x^2 - 1}{x^2 + 1} \qquad \text{(Explicit equation)}$

which gives an explicit representation of f.

Next, consider the equation

$$y^4 - y^3 - y + 2x^3 - x = 8$$

When certain restrictions are placed on x and y, this equation defines y as a function of x. But in this instance, we would be hard pressed to find y explicitly in terms of x. The following question arises naturally: How does one go about computing dy/dx in this case?

As it turns out, thanks to the Chain Rule, a method *does* exist for computing the derivative of a function directly from the implicit equation defining the function. This method is called **implicit differentiation** and is demonstrated in the next several examples.

EXAMPLE 1 Find dy/dx given the equation $y^2 = x$.

Solution Differentiating both sides of the equation with respect to x, we obtain

$$\frac{d}{dx}(y^2) = \frac{d}{dx}(x)$$

In order to carry out the differentiation of the term $\dfrac{d}{dx}(y^2)$, we note that y is a function of x. Writing $y = f(x)$ to remind us of this fact, we find that

$$\frac{d}{dx}(y^2) = \frac{d}{dx}[f(x)]^2 \qquad [\text{Writing } y = f(x)]$$

$$= 2f(x)f'(x) \qquad [\text{Using the Chain Rule}]$$

$$= 2y\frac{dy}{dx} \qquad [\text{Returning to using } y \text{ instead of } f(x)]$$

Therefore, the equation

$$\frac{d}{dx}(y^2) = \frac{d}{dx}(x)$$

is equivalent to

$$2y\frac{dy}{dx} = 1$$

Solving for dy/dx yields

$$\frac{dy}{dx} = \frac{1}{2y}$$

○ ○ ○

Before considering other examples, let us summarize the important steps involved in implicit differentiation. (Here we assume that dy/dx exists.)

FINDING *dy/dx*
BY IMPLICIT
DIFFERENTIATION

1. Differentiate both sides of the equation *with respect to x.* (Make sure that the derivative of any term involving y includes the factor dy/dx.)
2. Solve the resulting equation for dy/dx in terms of x and y.

EXAMPLE 2   Find dy/dx given the equation

$$y^3 - y + 2x^3 - x = 8$$

Solution   Differentiating both sides of the given equation with respect to x, we obtain

$$\frac{d}{dx}(y^3 - y + 2x^3 - x) = \frac{d}{dx}(8)$$

or $$\frac{d}{dx}(y^3) - \frac{d}{dx}(y) + \frac{d}{dx}(2x^3) - \frac{d}{dx}(x) = 0$$

Now, recalling that y is a function of x, we apply the Chain Rule to the first two terms on the left. Thus,

$$3y^2 \frac{dy}{dx} - \frac{dy}{dx} + 6x^2 - 1 = 0$$

$$(3y^2 - 1)\frac{dy}{dx} = 1 - 6x^2$$

$$\frac{dy}{dx} = \frac{1 - 6x^2}{3y^2 - 1}$$ ○ ○ ○

Refer to Example 2. Suppose we think of the equation $y^3 - y + 2x^3 - x = 8$ as defining x implicitly as a function of y. Find dx/dy and justify your method of solution.

EXAMPLE 3   Consider the equation $x^2 + y^2 = 4$.

a. Find dy/dx by implicit differentiation.

b. Find the slope of the tangent line to the graph of the function $y = f(x)$ at the point $(1, \sqrt{3})$.

c. Find an equation of the tangent line in (b).

Solution

a. Differentiating both sides of the equation with respect to x, we obtain

$$\frac{d}{dx}(x^2 + y^2) = \frac{d}{dx}(4)$$

$$\frac{d}{dx}(x^2) + \frac{d}{dx}(y^2) = 0$$

$$2x + 2y\frac{dy}{dx} = 0$$

$$\frac{dy}{dx} = -\frac{x}{y} \qquad (y \neq 0)$$

b. The slope of the tangent line to the graph of the function at the point $(1, \sqrt{3})$ is given by

$$\left.\frac{dy}{dx}\right|_{(1,\sqrt{3})} = -\left.\frac{x}{y}\right|_{(1,\sqrt{3})} = -\frac{1}{\sqrt{3}}$$

[*Note:* This notation is read "dy/dx evaluated at the point $(1, \sqrt{3})$."]

c. An equation of the tangent line in question is found by using the point-slope form of the equation of a line with the slope $m = -1/\sqrt{3}$ and the point $(1, \sqrt{3})$. Thus,

$$y - \sqrt{3} = -\frac{1}{\sqrt{3}}(x - 1)$$

$$\sqrt{3}y - 3 = -x + 1$$

$$x + \sqrt{3}y - 4 = 0$$

A sketch of this tangent line is shown in Figure 11.16.

Figure 11.16
The line $x + \sqrt{3}y - 4 = 0$ is tangent to the graph of the function $y = f(x)$.

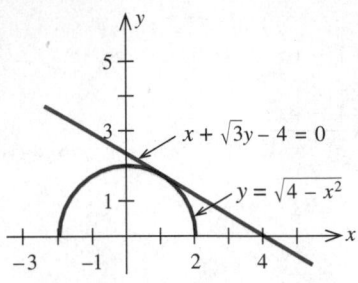

We can also solve the equation $x^2 + y^2 = 4$ explicitly for y in terms of x. If we do this, we obtain

$$y = \pm\sqrt{4 - x^2}$$

From this, we see that the equation $x^2 + y^2 = 4$ defines the two functions

$$y = f(x) = \sqrt{4 - x^2}$$

and

$$y = g(x) = -\sqrt{4 - x^2}$$

Since the point $(1, \sqrt{3})$ does not lie on the graph of $y = g(x)$, we conclude that

$$y = f(x) = \sqrt{4 - x^2}$$

is the required function. The graph of f is the upper semicircle shown in Figure 11.16.   ◦ ◦ ◦

To find dy/dx at a *specific* point (a, b), differentiate the given equation implicitly with respect to x and then replace x and y by a and b, respectively, *before* solving the equation for dy/dx. This often simplifies the amount of algebra involved.

EXAMPLE 4 Find dy/dx given that x and y are related by the equation

$$x^2y^3 + 6x^2 = y + 12$$

and that $y = 2$ when $x = 1$.

Solution Differentiating both sides of the given equation with respect to x, we obtain

$$\frac{d}{dx}(x^2y^3) + \frac{d}{dx}(6x^2) = \frac{d}{dx}(y) + \frac{d}{dx}(12)$$

$$x^2 \cdot \frac{d}{dx}(y^3) + y^3 \cdot \frac{d}{dx}(x^2) + 12x = \frac{dy}{dx} \qquad \text{[Using the Product Rule on } \frac{d}{dx}(x^2y^3)\text{]}$$

$$3x^2y^2\frac{dy}{dx} + 2xy^3 + 12x = \frac{dy}{dx}$$

Substituting $x = 1$ and $y = 2$ into this equation gives

$$3(1)^2(2)^2\frac{dy}{dx} + 2(1)(2)^3 + 12(1) = \frac{dy}{dx}$$

$$12\frac{dy}{dx} + 16 + 12 = \frac{dy}{dx}$$

and, solving for dy/dx,

$$\frac{dy}{dx} = -\frac{28}{11}$$

Note that it is not necessary to find an explicit expression for dy/dx.   ◦ ◦ ◦

REMARK   In Examples 3 and 4, you can verify that the points at which we evaluated dy/dx actually lie on the curve in question by showing that the coordinates of the points satisfy the given equations.   ◦ ◦ ◦

EXAMPLE 5   Find dy/dx given that x and y are related by the equation

$$\sqrt{x^2 + y^2} - x^2 = 5$$

Solution   Differentiating both sides of the given equation with respect to x, we obtain

$$\frac{d}{dx}(x^2 + y^2)^{1/2} - \frac{d}{dx}(x^2) = \frac{d}{dx}(5)$$

[Writing $\sqrt{x^2 + y^2} = (x^2 + y^2)^{1/2}$]

$$\frac{1}{2}(x^2 + y^2)^{-1/2} \frac{d}{dx}(x^2 + y^2) - 2x = 0$$

[Using the General Power Rule on the first term]

$$\frac{1}{2}(x^2 + y^2)^{-1/2}\left(2x + 2y\frac{dy}{dx}\right) - 2x = 0$$

$$2x + 2y\frac{dy}{dx} = 4x(x^2 + y^2)^{1/2}$$

[Transposing $2x$ and multiplying both sides by $2(x^2 + y^2)^{1/2}$]

$$2y\frac{dy}{dx} = 4x(x^2 + y^2)^{1/2} - 2x$$

$$\frac{dy}{dx} = \frac{2x\sqrt{x^2 + y^2} - x}{y}$$

◦ ◦ ◦

Related Rates

Implicit differentiation is a useful technique for solving a class of problems known as **related rates** problems. For example, suppose that x and y are each functions of a third variable t. Here, x might denote the mortgage rate and y the number of single-family homes sold at any time t. Furthermore, suppose that we have an equation that gives the relationship between x and y (the number of houses sold y is related to the mortgage rate x). Differentiating both sides of this equation implicitly with respect to t, we obtain an equation that gives a relationship between dx/dt and dy/dt. In the context of our example, this equation gives us a relationship between the rate of change of the mortgage rate and the rate of change of the number of houses sold, as a function of time. Thus, knowing

$$\frac{dx}{dt} \qquad \text{(How fast the mortgage rate is changing at time } t\text{)}$$

we can determine

$$\frac{dy}{dt} \qquad \text{(How fast the sale of houses is changing at that instant of time)}$$

EXAMPLE 6   A study prepared for the National Association of Realtors estimates that the number of housing starts in the Southwest, $N(t)$ (in units of a million), over the next five years is related to the mortgage rate $r(t)$ (percent per year) by the equation

$$9N^2 + r = 36$$

What is the rate of change in the number of housing starts with respect to time when the mortgage rate is 11% per year and is increasing at the rate of 1.5% per year?

Solution   We are given that

$$r = 11 \quad \text{and} \quad \frac{dr}{dt} = 1.5$$

at a certain instant of time, and we are required to find dN/dt. First, by substituting $r = 11$ into the given equation, we find

$$9N^2 + 11 = 36$$

$$N^2 = \frac{25}{9}$$

or $N = 5/3$ (we reject the negative root). Next, differentiating the given equation implicitly on both sides with respect to t, we obtain

$$\frac{d}{dt}(9N^2) + \frac{d}{dt}(r) = \frac{d}{dt}(36)$$

$$18N\frac{dN}{dt} + \frac{dr}{dt} = 0 \qquad \text{(Using the Chain Rule on the first term)}$$

Then, substituting $N = 5/3$ and $dr/dt = 1.5$ into this equation gives

$$18\left(\frac{5}{3}\right)\frac{dN}{dt} + 1.5 = 0$$

Solving this equation for dN/dt then gives

$$\frac{dN}{dt} = -\frac{1.5}{30} \approx -0.05$$

Thus, at the instant of time under consideration, the number of housing starts is decreasing at the rate of 50,000 units per year.   ○ ○ ○

EXAMPLE 7   A major audio-tape manufacturer is willing to make x thousand 10-packs of metal alloy audiocassette tapes per week available in the marketplace when the wholesale price is $\$p$ per 10-pack. It is known that the relationship between x and p is governed by the supply equation

$$x^2 - 3xp + p^2 = 5$$

How fast is the supply of tapes changing when the price per 10-pack is \$11, the quantity supplied is 4000 10-packs, and the wholesale price per 10-pack is increasing at the rate of 10 cents per 10-pack per week?

Solution   We are given that

$$p = 11, \quad x = 4, \quad \text{and} \quad \frac{dp}{dt} = 0.1$$

at a certain instant of time, and we are required to find dx/dt. Differentiating the given equation on both sides with respect to t, we obtain

$$\frac{d}{dt}(x^2) - \frac{d}{dt}(3xp) + \frac{d}{dt}(p^2) = \frac{d}{dt}(5)$$

$$2x\frac{dx}{dt} - 3\left(p\frac{dx}{dt} + x\frac{dp}{dt}\right) + 2p\frac{dp}{dt} = 0 \qquad \text{(Using the Product Rule on the second term)}$$

Substituting the given values of $p, x,$ and dp/dt into the last equation, we have

$$2(4)\frac{dx}{dt} - 3\left[(11)\frac{dx}{dt} + 4(0.1)\right] + 2(11)(0.1) = 0$$

$$8\frac{dx}{dt} - 33\frac{dx}{dt} - 1.2 + 2.2 = 0$$

$$25\frac{dx}{dt} = 1$$

$$\frac{dx}{dt} = 0.04$$

Thus, at the instant of time under consideration the supply of 10-pack audiocassettes is increasing at the rate of $(0.04)(1000)$, or 40, 10-packs per week.

○ ○ ○

In certain related problems, we need to formulate the problem mathematically before analyzing it. The following guidelines can be used to help solve problems of this type.

SOLVING RELATED RATES PROBLEMS

1. Assign a variable to each quantity. Draw a diagram if needed.
2. Write the *given* values of the variables and their rates of change with respect to t.
3. Find an equation giving the relationship between the variables.
4. Differentiate both sides of this equation implicitly with respect to t.
5. Replace the variables and their derivatives with the numerical data found in step 2, and solve the equation for the required rate of change.

EXAMPLE 8   At a distance of 4000 feet from the launch site, a spectator
is observing a rocket being launched. If the rocket lifts off
vertically and is rising at a speed of 600 ft/sec when it is at an altitude of 3000
ft, how fast is the distance between the rocket and the spectator changing at
that instant?

Solution

Step 1   Let y = the altitude of the rocket and x = the distance between the
rocket and the spectator at any time t (see Figure 11.17).

Figure 11.17
*The rate at which x is changing with
respect to time is related to the rate
of change of y with respect to time.*

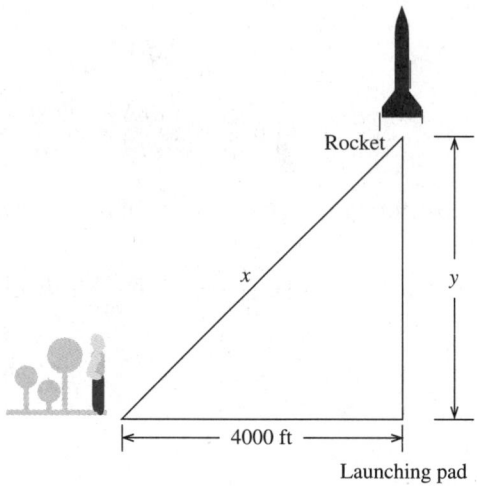

Rocket

x

y

4000 ft

Launching pad

Step 2   We are given that at a certain instant of time

$$y = 3000 \quad \text{and} \quad \frac{dy}{dt} = 600$$

and we are asked to find dx/dt at that instant.

Step 3   Applying the Pythagorean Theorem to the right triangle in Figure
11.17, we find that

$$x^2 = y^2 + 4000^2$$

Therefore, when $y = 3000$,

$$x = \sqrt{3000^2 + 4000^2} = 5000$$

Step 4   Next, we differentiate the equation $x^2 = y^2 + 4000^2$ with respect to
t, obtaining

$$2x\frac{dx}{dt} = 2y\frac{dy}{dt}$$

(Remember that both x and y are functions of t.)

Step 5 Substituting $x = 5000$, $y = 3000$, and $dy/dt = 600$, we find

$$2(5000)\frac{dx}{dt} = 2(3000)(600)$$

$$\frac{dx}{dt} = 360$$

Therefore, the distance between the rocket and the spectator is changing at a rate of 360 ft/sec.   ◦ ◦ ◦

 Be sure that you do *not* replace the variables in the equation found in step 3 by their numerical values before differentiating the equation.

SELF–CHECK EXERCISES 11.6

1. Given the equation $x^3 + 3xy + y^3 = 4$, find dy/dx by implicit differentiation.

2. Find an equation of the tangent line to the graph of $16x^2 + 9y^2 = 144$ at the point $\left(2, -\frac{4\sqrt{5}}{3}\right)$.

3. A passenger ship and an oil tanker left port sometime in the morning, the former heading north and the latter heading east. At noon, the passenger ship was 40 miles from port and sailing at 30 mph, while the oil tanker was 30 miles from port and sailing at 20 mph. How fast was the distance between the two ships changing at that instant of time?

Solutions to Self-Check Exercises 11.6 can be found on page 759.

11.6 EXERCISES

In exercises 1–8, find the derivative dy/dx (a) by solving each of the given implicit equations for y explicitly in terms of x and (b) by differentiating each of the given equations implicitly. Show that, in each case, the results are equivalent.

1. $x + 2y = 5$

2. $3x + 4y = 6$

3. $xy = 1$

4. $xy - y - 1 = 0$

5. $x^3 - x^2 - xy = 4$

6. $x^2y - x^2 + y - 1 = 0$

7. $\dfrac{x}{y} - x^2 = 1$

8. $\dfrac{y}{x} - 2x^3 = 4$

In exercises 9–30, find dy/dx by implicit differentiation.

9. $x^2 + y^2 = 16$

10. $2x^2 + y^2 = 16$

11. $x^2 - 2y^2 = 16$

12. $x^3 + y^3 + y - 4 = 0$

13. $x^2 - 2xy = 6$

14. $x^2 + 5xy + y^2 = 10$

15. $x^2y^2 - xy = 8$

16. $x^2y^3 - 2xy^2 = 5$

17. $x^{1/2} + y^{1/2} = 1$

18. $x^{1/3} + y^{1/3} = 1$

19. $\sqrt{x + y} = x$

20. $(2x + 3y)^{1/3} = x^2$

21. $\dfrac{1}{x^2} + \dfrac{1}{y^2} = 1$

22. $\dfrac{1}{x^3} + \dfrac{1}{y^3} = 5$

23. $\sqrt{xy} = x + y$

24. $\sqrt{xy} = 2x + y^2$

25. $\dfrac{x + y}{x - y} = 3x$

26. $\dfrac{x - y}{2x + 3y} = 2x$

27. $xy^{3/2} = x^2 + y^2$

28. $x^2y^{1/2} = x + 2y^3$

29. $(x + y)^3 + x^3 + y^3 = 0$

30. $(x + y^2)^{10} = x^2 + 25$

In exercises 31–34, find an equation of the tangent line to the graph of the function f defined by the given equation at the indicated point.

31. $4x^2 + 9y^2 = 36$; $(0, 2)$

32. $y^2 - x^2 = 16$; $(2, 2\sqrt{5})$

33. $x^2y^3 - y^2 + xy - 1 = 0$; $(1, 1)$

34. $(x - y - 1)^3 = x$; $(1, -1)$

In exercises 35–38, find the second derivative d^2y/dx^2 of each of the functions defined implicitly by the given equation.

35. $xy = 1$ **36.** $x^3 + y^3 = 28$

37. $y^2 - xy = 8$ **38.** $x^{1/3} + y^{1/3} = 1$

39. The volume of a right-circular cylinder of radius r and height h is $V = \pi r^2 h$. Suppose that the radius and height of the cylinder are changing with respect to time t.
a. Find a relationship between dV/dt, dr/dt, and dh/dt.
b. At a certain instant of time, the radius and height of the cylinder are 2 and 6 inches and are increasing at the rate of 0.1 and 0.3 in./sec, respectively. How fast is the volume of the cylinder increasing?

40. A car leaves an intersection traveling west. Its position 4 seconds later is 20 ft from the intersection. At the same time, another car leaves the same intersection heading north so that its position 4 seconds later is 28 ft from the intersection. If the speed of the cars at that instant of time is 9 ft/sec and 11 ft/sec, respectively, find the rate at which the distance between the two cars is changing.

41. Price-Demand Suppose that the quantity demanded weekly of the Super Titan radial tires is related to the unit price by the equation

$$p + x^2 = 144$$

where p is measured in dollars and x is measured in units of a thousand. How fast is the quantity demanded changing when $x = 9$, $p = 63$, and the price per tire is increasing at the rate of $2 per week?

42. Price-Supply Suppose the quantity x of Super Titan radial tires made available per week in the marketplace by the Titan Tire Company is related to the unit selling price by the equation

$$p - \frac{1}{2}x^2 = 48$$

where x is measured in units of a thousand and p is in dollars. How fast is the weekly supply of Super Titan radial tires being introduced into the marketplace when $x = 6$, $p = 66$, and the price per tire is decreasing at the rate of $3 per week?

43. Price-Demand The demand equation for a certain brand of metal alloy audiocassette tape is

$$100x^2 + 9p^2 = 3600$$

where x represents the number (in thousands) of 10-packs demanded per week when the unit price is p. How fast is the quantity demanded increasing when the unit price per 10-pack is $14 and the selling price is dropping at the rate of 15 cents per 10-pack per week?
[*Hint:* To find the value of x when $p = 14$, solve the equation $100x^2 + 9p^2 = 3600$ for x when $p = 14$.]

44. Effect of Price on Supply Suppose that the wholesale price of a certain brand of medium-size eggs p (in dollars per carton) is related to the weekly supply x (in thousands of cartons) by the equation

$$625p^2 - x^2 = 100$$

If 25,000 cartons of eggs are available at the beginning of a certain week and the price is falling at the rate of 2 cents per carton per week, at what rate is the supply falling?
[*Hint:* To find the value of p when $x = 25$, solve the supply equation for p when $x = 25$.]

45. Supply-Demand Refer to exercise 44. If 25,000 cartons of eggs are available at the beginning of a certain week and the supply is falling at the rate of 1000 cartons per week, at what rate is the wholesale price changing?

46. Elasticity of Demand The demand function for a certain make of cartridge typewriter ribbon is

$$p = -0.01x^2 - 0.1x + 6$$

where p is the unit price in dollars and x is the quantity demanded each week, measured in units of a thousand. Compute the elasticity of demand and determine whether the demand is inelastic, unitary, or elastic when $x = 10$.

47. Elasticity of Demand The demand function for a certain brand of compact disc is

$$p = -0.01x^2 - 0.2x + 8$$

where p is the wholesale unit price in dollars and x is the quantity demanded each week, measured in units of a thousand. Compute the elasticity of demand, and determine whether the demand is inelastic, unitary, or elastic when $x = 15$.

48. The volume V of a cube with sides of length x inches is changing with respect to time. At a certain instant of time, the sides of the cube are 5 inches long and increasing at the rate of 0.1 in./sec. How fast is the volume of the cube changing at that instant of time?

49. Oil Spills In calm waters, oil spilling from the ruptured hull of a grounded tanker spreads in all directions. If the area polluted is a circle and its radius is increasing at a rate of 2 ft/sec, determine how fast the area is increasing when the radius of the circle is 40 feet.

50. Two ships leave the same port at noon. Ship A sails north at 15 mph and ship B sails east at 12 mph. How fast is the distance between them changing at 1 P.M.?

51. A car leaves an intersection traveling east. Its position t seconds later is given by $x = t^2 + t$ feet. At the same time, another car leaves the same intersection heading north, traveling $y = t^2 + 3t$ feet in t seconds. Find the rate at which the distance between the two cars will be changing 5 seconds later.

52. At a distance of 50 feet from the pad, a man observes a helicopter taking off from a heliport. If the helicopter lifts off vertically and is rising at a speed of 44 ft/sec when it is at an altitude of 120 feet, how fast is the distance between the helicopter and the man changing at that instant?

53. A spectator watches a rowing race from the edge of a river bank. The lead boat is moving in a straight line

that is 120 feet from the river bank. If the boat is moving at a constant speed of 20 ft/sec, how fast is the boat moving away from the spectator when it is 50 feet past her?

54. A man 6 feet tall is walking away from a street light 18 feet high at a speed of 6 ft/sec. How fast is the tip of his shadow moving along the ground?

55. A 20-foot ladder is leaning against a wall. If the bottom of the ladder is pulled away from the wall at a rate of 2 ft/sec, how fast is the top of the ladder sliding down the wall when the bottom of the ladder is 12 feet from the wall?

[*Hint:* Refer to the figure. By the Pythagorean Theorem, $x^2 + y^2 = 400$. Find dy/dt when $x = 12$ and $dx/dt = 2$.]

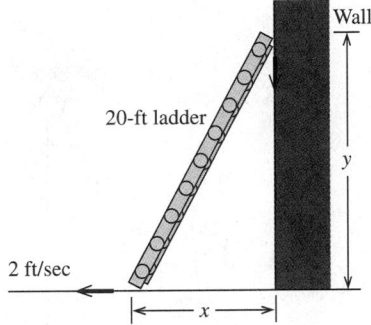

56. A 13-foot ladder is leaning against a wall. If the bottom of the ladder is pulled away from the wall at a rate of 2.5 ft/sec, how fast is the top of the ladder sliding down the wall when the bottom of the ladder is 12 feet from the wall?

SOLUTIONS TO SELF-CHECK EXERCISES 11.6

1. Differentiating both sides of the equation with respect to x, we have

$$3x^2 + 3y + 3xy' + 3y^2y' = 0$$
$$(x^2 + y) + (x + y^2)y' = 0$$
$$y' = -\frac{x^2 + y}{x + y^2}$$

2. To find the slope of the tangent line to the graph of the function at any point, we differentiate the equation implicitly with respect to x, obtaining

$$32x + 18yy' = 0$$
$$y' = -\frac{16x}{9y}$$

In particular, the slope of the tangent line at $\left(2, -\dfrac{4\sqrt{5}}{3}\right)$ is

$$m = -\frac{16(2)}{9\left(-\dfrac{4\sqrt{5}}{3}\right)} = \frac{8}{3\sqrt{5}}$$

Using the point-slope form of the equation of a line, we find

$$y - \left(-\frac{4\sqrt{5}}{3}\right) = \frac{8}{3\sqrt{5}}(x - 2)$$

$$y = \frac{8\sqrt{5}}{15}x - \frac{36\sqrt{5}}{15}$$

3. **Step 1** Let x = the distance of the oil tanker from port
y = the distance of the passenger ship from port
and z = the distance between the two ships (see Figure).

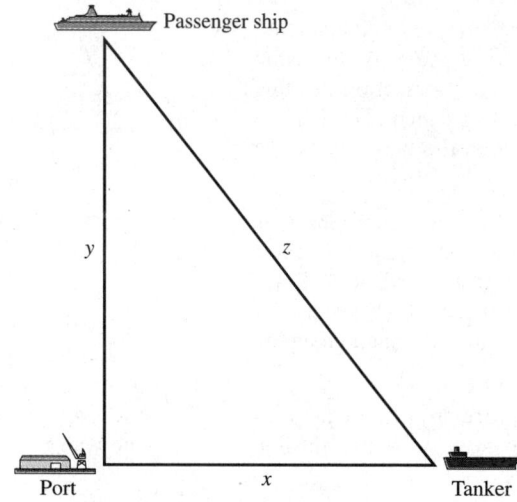

We want to find dz/dt, the rate at which the distance between the two ships is changing at a certain instant of time.

Step 2 We are given that at a certain instant of time (noon),

$$x = 30, \quad y = 40, \quad \frac{dx}{dt} = 20, \quad \text{and} \quad \frac{dy}{dt} = 30$$

and we are required to find dz/dt at that instant of time.

Step 3 Applying the Pythagorean Theorem to the right triangle in the figure, we find that

$$z^2 = x^2 + y^2 \tag{1}$$

In particular, when $x = 30$ and $y = 40$, we have

$$z^2 = 30^2 + 40^2 = 2500$$

or $z = 50$.

Step 4   Differentiating (1) implicitly with respect to t, we obtain

$$2z\frac{dz}{dt} = 2x\frac{dx}{dt} + 2y\frac{dy}{dt}$$

or

$$z\frac{dz}{dt} = x\frac{dx}{dt} + y\frac{dy}{dt}$$

Step 5   Finally, substituting $x = 30$, $y = 40$, $z = 50$, $dx/dt = 20$, and $dy/dt = 30$ into the last equation, we find

$$50\frac{dz}{dt} = (30)(20) + (40)(30)$$

and

$$\frac{dz}{dt} = 36$$

Therefore, at noon on the day in question, the ships are moving apart at the rate of 36 mph.

11.7 DIFFERENTIALS

The Millers are planning to buy a house in the near future and estimate that they will need a 30-year fixed-rate mortgage for $120,000. If the interest rate increases from the present rate of 9% per year to 9.4% per year between now and the time the Millers decide to secure the loan, approximately how much more per month will their mortgage be? (You will be asked to answer this question in exercise 43, page 772.)

Questions such as this, in which one wishes to *estimate* the change in the dependent variable (monthly mortgage payment) corresponding to a small change in the independent variable (interest rate per year), occur in many real-life applications. For example:

- An economist would like to know how a small increase in a country's capital expenditure will affect the country's gross domestic output.

- A sociologist would like to know how a small increase in the amount of capital investment in a housing project will affect the crime rate.

- A businesswoman would like to know how raising a product's unit price by a small amount will affect her profit.

- A bacteriologist would like to know how a small increase in the amount of a bactericide will affect a population of bacteria.

In order to calculate these changes and estimate their effects, we use the *differential* of a function, a concept that will be introduced shortly.

Increments

Let x denote a variable quantity and suppose that x changes from x_1 to x_2. This change in x is called the **increment in x** and is denoted by the symbol Δx (read "delta x"). Thus,

$$\Delta x = x_2 - x_1 \qquad \text{(Final value minus initial value)} \qquad \textbf{(9)}$$

EXAMPLE 1   Find the increment in x:

a. as x changes from 3 to 3.2   **b.** as x changes from 3 to 2.7

Solution

a. Here $x_1 = 3$ and $x_2 = 3.2$, so

$$\Delta x = x_2 - x_1 = 3.2 - 3 = 0.2$$

b. Here $x_1 = 3$ and $x_2 = 2.7$. Therefore,

$$\Delta x = x_2 - x_1 = 2.7 - 3 = -0.3 \qquad \text{o o o}$$

Observe that Δx plays the same role that h played in Section 10.4.

Now, suppose two quantities, x and y, are related by an equation $y = f(x)$, where f is a function. If x changes from x to $x + \Delta x$, then the corresponding change in y is called the *increment in y*. It is denoted by Δy and is defined in Figure 11.18 by

$$\Delta y = f(x + \Delta x) - f(x) \tag{10}$$

Figure 11.18
An increment of Δx in x induces an increment of $\Delta y = f(x + \Delta x) - f(x)$ in y.

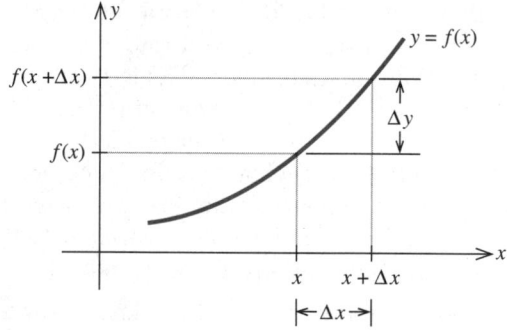

EXAMPLE 2   Let $y = x^3$. Find Δx and Δy when

a. x changes from 2 to 2.01   **b.** x changes from 2 to 1.98

Solution   Let $f(x) = x^3$.

a. Here $\Delta x = 2.01 - 2 = 0.01$. Next,

$$\Delta y = f(x + \Delta x) - f(x) = f(2.01) - f(2)$$
$$= (2.01)^3 - 2^3 = 8.120601 - 8 = 0.120601$$

b. Here $\Delta x = 1.98 - 2 = -0.02$. Next,

$$\Delta y = f(x + \Delta x) - f(x) = f(1.98) - f(2)$$
$$= (1.98)^3 - 2^3 = 7.762392 - 8 = -0.237608 \qquad \text{o o o}$$

Differentials

A relatively quick and simple way of approximating Δy, the change in y due to a small change Δx, may be found by examining the graph of the function f shown in Figure 11.19. Observe that near the point of tangency P, the tangent

Figure 11.19
If Δx is small, dy is a good approximation of Δy.

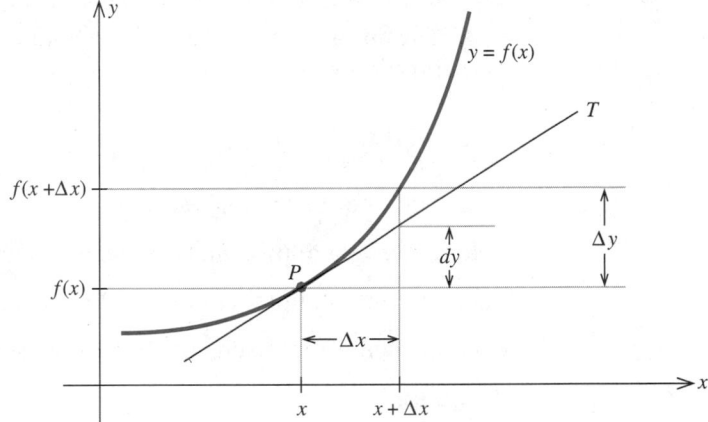

line T is close to the graph of f. Therefore if Δx is small, then dy is a good approximation of Δy. We can find an expression for dy as follows: Notice that the slope of T is given by

$$\frac{dy}{\Delta x} \qquad \text{(Rise divided by run)}$$

However, the slope of T is given by $f'(x)$. Therefore, we have

$$\frac{dy}{\Delta x} = f'(x)$$

or $dy = f'(x)\,\Delta x$. Thus, we have the approximation

$$\Delta y \approx dy = f'(x)\,\Delta x$$

in terms of the derivative of f at x. The quantity dy is called the *differential of y.*

THE DIFFERENTIAL

Let $y = f(x)$ define a differentiable function of x. Then:

1. The **differential** dx of the independent variable x is $dx = \Delta x$.

2. The **differential** dy of the dependent variable y is

$$dy = f'(x)\,\Delta x = f'(x)\,dx \qquad \textbf{(11)}$$

REMARKS

1. For the independent variable x: There is no difference between Δx and dx; both measure the change in x from x to $x + \Delta x$.

2. For the dependent variable y: Δy measures the *actual* change in y as x changes from x to $x + \Delta x$, whereas dy measures the *approximate* change in y corresponding to the same change in x.

3. The differential dy depends on both x and dx, but for fixed x, dy is a linear function of dx.   ◦ ◦ ◦

EXAMPLE 3   Let $y = x^3$.

a. Find the differential dy of y.

b. Use dy to approximate Δy when x changes from 2 to 2.01.

c. Use dy to approximate Δy when x changes from 2 to 1.98.

d. Compare the results of (b) with those of Example 2.

Solution

a. Let $f(x) = x^3$. Then

$$dy = f'(x)\, dx = 3x^2\, dx$$

b. Here $x = 2$ and $dx = 2.01 - 2 = 0.01$. Therefore,

$$dy = 3x^2\, dx = 3(2)^2(0.01) = 0.12$$

c. Here $x = 2$ and $dx = 1.98 - 2 = -0.02$. Therefore,

$$dy = 3x^2\, dx = 3(2)^2(-0.02) = -0.24$$

d. As you can see, both approximations 0.12 and -0.24 are quite close to the actual changes of Δy obtained in Example 2: 0.120601 and -0.237608.
                                                                        ◦ ◦ ◦

Observe how much easier it is to find an approximation to the exact change in a function with the help of the differential, rather than calculating the exact change in the function itself. In the following example, we take advantage of this fact.

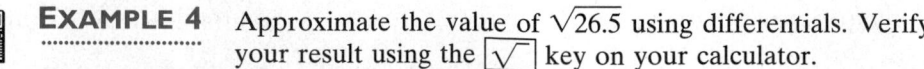

EXAMPLE 4   Approximate the value of $\sqrt{26.5}$ using differentials. Verify your result using the $\boxed{\sqrt{}}$ key on your calculator.

Solution   Since we want to compute the square root of a number, let us consider the function $y = f(x) = \sqrt{x}$. Since 25 is the number nearest 26.5 whose square root is readily recognized, let us take $x = 25$. We want to know the change in y, Δy, as x changes from $x = 25$ to $x = 26.5$, an increase of

$\Delta x = 1.5$ units. Using equation (11), we find

$$\Delta y \approx dy = f'(x)\,\Delta x$$

$$= \left[\left.\frac{1}{2\sqrt{x}}\right|_{x=25}\right] \cdot (1.5) = \left(\frac{1}{10}\right)(1.5) = 0.15$$

Therefore,

$$\sqrt{26.5} - \sqrt{25} = \Delta y \approx 0.15$$

and

$$\sqrt{26.5} \approx \sqrt{25} + 0.15 = 5.15$$

The exact value of $\sqrt{26.5}$, rounded off to five decimal places, is 5.14782. Thus, the error incurred in the approximation is 0.00218.   ○ ○ ○

Applications

EXAMPLE 5 The total cost incurred in operating a certain type of truck on a 500-mile trip, traveling at an average speed of v miles per hour, is estimated to be

$$C(v) = 125 + v + \frac{4500}{v}$$

dollars. Find the approximate change in the total operating cost when the average speed is increased from 55 mph to 58 mph.

Solution With $v = 55$ and $\Delta v = dv = 3$, we find

$$\Delta C \approx dC = C'(v)\,dv = \left.\left(1 - \frac{4500}{v^2}\right)\right|_{v=55} \cdot 3$$

$$= \left(1 - \frac{4500}{3025}\right)(3) \approx -1.46$$

so the total operating cost is found to decrease by $1.46. This might explain why so many independent truckers often exceed the 55-mph speed limit.   ○ ○ ○

EXAMPLE 6 The relationship between the amount of money x spent by Cannon Precision Instruments on advertising and Cannon's total sales $S(x)$ is given by the function

$$S(x) = -0.002x^3 + 0.6x^2 + x + 500 \qquad (0 \le x \le 200)$$

where x is measured in thousands of dollars. Use differentials to estimate the change in Cannon's total sales if advertising expenditures are increased from $100,000 ($x = 100$) to $105,000 ($x = 105$).

Solution The required change in sales is given by

$$\Delta S \approx dS = S'(100)\, dx$$
$$= -0.006x^2 + 1.2x + 1|_{x=100} \cdot (5) \qquad (dx = 105 - 100 = 5)$$
$$= (-60 + 120 + 1)(5) = 305$$

that is, an increase of $305,000.

○ ○ ○

EXAMPLE 7   The Rings of Neptune
....

a. A ring has an inner radius of r units and an outer radius of R units, where $(R - r)$ is small in comparison to r (see Figure 11.20a). Use differentials to estimate the area of the ring.

Figure 11.20

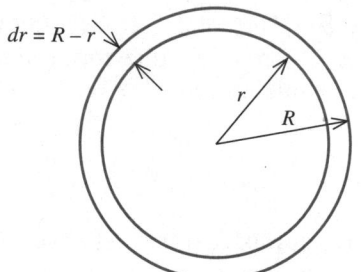

$dr = R - r$

r

R

(a) The area of the ring is the circumference of the inner circle times the thickness.

(b) Neptune and its rings.

b. Recent observations, including those of Voyager I and II, showed that Neptune's ring system is considerably more complex than had been believed. For one thing, it is made up of a large number of distinguishable rings rather than one continuous great ring as previously thought (see Figure 11.20b). The outermost ring, 1989N1R, has an inner radius of approximately 62,900 km (measured from the center of the planet) and a radial width of approximately 50 km. Using these data, estimate the area of the ring.

Solution

a. Using the fact that the area of a circle of radius x is $A = f(x) = \pi x^2$, we find

$$\pi R^2 - \pi r^2 = f(R) - f(r)$$
$$= \Delta A \qquad \text{(Remember } \Delta A \text{ is the change in } f \text{ when}$$
$$\approx dA \qquad\qquad x \text{ changes from } x = r \text{ to } x = R.)$$
$$= f'(r)\, dr$$

where $dr = R - r$. So, we see that the area of the ring is approximately $2\pi r(R - r)$ square units. In words, the area of the ring is approximately equal to

Circumference of the inner circle × Thickness of the ring

b. Applying the results of (a) with $r = 62,900$ and $dr = 50$, we find that the area of the ring is approximately $2\pi(62,900)(50)$, or $19,760,618$ km², which is roughly 4% of Earth's surface.   ⚬ ⚬ ⚬

Before looking at the next example, we need to familiarize ourselves with some terminology. If a quantity with exact value q is measured or calculated with an error of Δq, then the quantity $\Delta q/q$ is called the *relative error* in the measurement or calculation of q. If the quantity $\Delta q/q$ is expressed as a percentage, it is then called the *percentage error*. Because Δq is approximated by dq, we normally approximate the relative error $\Delta q/q$ by dq/q.

EXAMPLE 8   Suppose the radius of a ball-bearing is measured to be 0.5
in., with a maximum error of ±0.0002 in. Then the relative error in r is

$$\frac{dr}{r} = \frac{\pm0.0002}{0.5} = \pm0.0004$$

and the percentage error is $\pm0.04\%$.   ⚬ ⚬ ⚬

EXAMPLE 9   Suppose the side of a cube is measured with a maximum
percentage error of 2%. Use differentials to estimate the maximum percentage error in the calculated volume of the cube.

Solution   Suppose the side of the cube is x, so that its volume is

$$V = x^3$$

We are given that $\left|\dfrac{dx}{x}\right| \leq 0.02$. Now,

$$dV = 3x^2\, dx$$

and so

$$\frac{dV}{V} = \frac{3x^2\, dx}{x^3} = 3\frac{dx}{x}$$

Therefore,

$$\left|\frac{dV}{V}\right| = 3\left|\frac{dx}{x}\right| \leq 3(0.02) = 0.06$$

and we see that the maximum percentage error in the measurement of the volume of the cube is 6%.   ⚬ ⚬ ⚬

Finally, we want to point out that if at some point in reading this section you have a sense of déjà vu, do not be surprised because the notion of the differential was first used in Section 11.4 (see Example 1). There we took $\Delta x = 1$ since we were interested in finding the marginal cost when the level of production was increased from $x = 250$ to $x = 251$. If we had used differentials, we would have found

$$C(251) - C(250) \approx C'(250) \, dx$$

so that taking $dx = \Delta x = 1$, we have $C(251) - C(250) \approx C'(250)$, which agrees with the result obtained in Example 1. Thus, in Section 11.4, we touched upon the notion of the differential, albeit in the special case in which $dx = 1$.

SELF-CHECK EXERCISES 11.7

1. Find the differential of $f(x) = \sqrt{x} + 1$.

2. A certain country's government economists have determined that the demand equation for corn in that country is given by

$$p = f(x) = \frac{125}{x^2 + 1}$$

where p is expressed in dollars per bushel and x, the quantity demanded per year, is measured in billions of bushels. The economists are forecasting a harvest of 6 billion bushels for the year. If the actual production of corn were 6.2 billion bushels for the year instead, what would be the approximate drop in the predicted price of corn per bushel?

Solutions to Self-Check Exercises 11.7 can be found on page 773.

11.7 EXERCISES

In exercises 1–14, find the differential of the given function.

1. $f(x) = 2x^2$

2. $f(x) = 3x^2 + 1$

3. $f(x) = x^3 - x$

4. $f(x) = 2x^3 + x$

5. $f(x) = \sqrt{x + 1}$

6. $f(x) = \dfrac{3}{\sqrt{x}}$

7. $f(x) = 2x^{3/2} + x^{1/2}$

8. $f(x) = 3x^{5/6} + 7x^{2/3}$

9. $f(x) = x + \dfrac{2}{x}$

10. $f(x) = \dfrac{3}{x - 1}$

11. $f(x) = \dfrac{x - 1}{x^2 + 1}$

12. $f(x) = \dfrac{2x^2 + 1}{x + 1}$

13. $f(x) = \sqrt{3x^2 - x}$

14. $f(x) = (2x^2 + 3)^{1/3}$

15. Let f be a function defined by

$$y = f(x) = x^2 - 1$$

a. Find the differential of f.
b. Use your result from (a) to find the approximate change in y if x changes from 1 to 1.02.
c. Find the actual change in y if x changes from 1 to 1.02, and compare your result with that obtained in (b).

16. Let f be a function defined by

$$y = f(x) = 3x^2 - 2x + 6$$

a. Find the differential of f.

b. Use your result from (a) to find the approximate change in y if x changes from 2 to 1.97.

c. Find the actual change in y if x changes from 2 to 1.97, and compare your result with that obtained in (b).

17. Let f be a function defined by

$$y = f(x) = \frac{1}{x}$$

a. Find the differential of f.

b. Use your result from (a) to find the approximate change in y if x changes from -1 to -0.95.

c. Find the actual change in y if x changes from -1 to -0.95, and compare your result with that obtained in (b).

18. Let f be a function defined by

$$y = f(x) = \sqrt{2x + 1}$$

a. Find the differential of f.

b. Use your result from (a) to find the approximate change in y if x changes from 4 to 4.1.

c. Find the actual change in y if x changes from 4 to 4.1, and compare your result with that obtained in (b).

In exercises 19–26, use differentials to approximate the given quantity.

19. $\sqrt{10}$ **20.** $\sqrt{17}$ **21.** $\sqrt{49.5}$

22. $\sqrt{99.7}$ **23.** $\sqrt[3]{7.8}$ **24.** $\sqrt[4]{81.6}$

25. $\sqrt{0.089}$ **26.** $\sqrt[3]{0.00096}$

27. Use a differential to approximate $\sqrt{4.02} + \dfrac{1}{\sqrt{4.02}}$.

[*Hint:* Let $f(x) = \sqrt{x} + \dfrac{1}{\sqrt{x}}$ and compute dy with $x = 4$ and $dx = 0.02$.]

28. Use a differential to approximate $\dfrac{2(4.98)}{(4.98)^2 + 1}$.

[*Hint:* Study the hint for exercise 27.]

 A calculator is recommended for the remainder of this exercise set.

29. Error Estimation The length of each edge of a cube is 12 cm, with a possible error in measurement of 0.02 cm. Use differentials to estimate the error that might occur when the volume of the cube is calculated.

30. Error Estimation A hemisphere-shaped dome of radius 60 ft is to be coated with a layer of rust-proofer before painting. Use differentials to estimate the amount of rust-proofer needed if the coat is to be 0.01 inch thick. [*Hint:* The volume of a hemisphere of radius r is $V = \frac{2}{3}\pi r^3$.]

31. Growth of a Cancerous Tumor The volume of a spherical cancerous tumor is given by

$$V(r) = \frac{4}{3}\pi r^3$$

If the radius of a tumor is estimated at 1.1 cm, with a maximum error in measurement of 0.005 cm, determine the error that might occur when the volume of the tumor is calculated.

32. Unclogging Arteries Research done in the 1930s by the French physiologist Jean Poiseuille showed that the resistance R of a blood vessel of length l and radius r is $R = kl/r^4$, where k is a constant. Suppose a dose of the drug TPA increases r by 10%. How will this affect the resistance R? Assume that l is constant.

33. Gross Domestic Product An economist has determined that a certain country's gross domestic output (GDP) is approximated by the function $f(x) = 640x^{1/5}$, where $f(x)$ is measured in billions of dollars and x is the capital outlay in billions of dollars. Use differentials to estimate the change in the country's GDP if the country's capital expenditure changes from \$243 billion to \$248 billion.

34. Learning Curves The length of time (in seconds) a certain individual takes to learn a list of n items is approximated by

$$f(n) = 4n\sqrt{n - 4}$$

Use differentials to approximate the additional time it takes the individual to learn the items on a list when n is increased from 85 to 90 items.

35. Effect of Advertising on Profits The relationship between Cunningham Realty's quarterly profits, $P(x)$, and the amount of money x spent on advertising per quarter is described by the function

$$P(x) = -\frac{1}{8}x^2 + 7x + 30 \qquad (0 \le x \le 50)$$

where both $P(x)$ and x are measured in thousands of dollars. Use differentials to estimate the increase in profits when advertising expenditure each quarter is increased from \$24,000 to \$26,000.

Using Technology

FINDING THE DIFFERENTIAL OF A FUNCTION

The calculation of the differential of f at a given value of x involves the evaluation of the derivative of f at that point and can be facilitated through the use of the numerical derivative function.

EXAMPLE 1 Use dy to approximate Δy if $y = x^2(2x^2 + x + 1)^{2/3}$ and x changes from 2 to 1.98.

Solution Let $f(x) = x^2(2x^2 + x + 1)^{2/3}$. Since $dx = 1.98 - 2 = -0.02$, we find the required approximation to be

$$dy = f'(2) \cdot (-0.02)$$

But using the numerical derivative operation, we find

$$f'(2) = 30.5758132855$$

and so
$$dy = (-0.02)(30.5758132855) = -0.611516266 \qquad \circ\ \circ\ \circ$$

EXAMPLE 2 The Meyers are considering the purchase of a house in the near future and estimate that they will need a loan of $120,000. Based on a 30-year conventional mortgage with an interest rate of r per year, their monthly repayment will be

$$P = \frac{10{,}000r}{1 - \left(1 + \dfrac{r}{12}\right)^{-360}}$$

dollars. If the interest rate increases from the present rate of 10% per year to 10.2% per year between now and the time the Meyers decide to secure the loan, approximately how much more per month will their mortgage payment be?

Solution Let us write

$$P = f(r) = \frac{10{,}000r}{1 - \left(1 + \dfrac{r}{12}\right)^{-360}}$$

Then the increase in the mortgage payment will be approximately

$$dP = f'(0.1)\, dr = f'(0.1)(0.002) \qquad \text{(Since } dr = 0.102 - 0.1\text{)}$$
$$= (8867.59947979)(0.002) \approx 17.7352 \qquad \text{(Using the numerical derivative operation)}$$

or approximately $17.74 per month. $\qquad \circ\ \circ\ \circ$

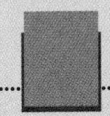

EXERCISES

In exercises 1–6, use dy to approximate Δy for the function $y = f(x)$ when x changes from $x = a$ to $x = b$.

1. $f(x) = 0.21x^7 - 3.22x^4 + 5.43x^2 + 1.42x + 12.42$; $a = 3$, $b = 3.01$

2. $f(x) = \dfrac{0.2x^2 + 3.1}{1.2x + 1.3}$; $a = 2$, $b = 1.96$

3. $f(x) = \sqrt{2.2x^2 + 1.3x + 4}$; $a = 1$, $b = 1.03$

4. $f(x) = x\sqrt{2x^3 - x + 4}$; $a = 2$, $b = 1.98$

5. $f(x) = \dfrac{\sqrt{x^2 + 4}}{x - 1}$; $a = 4$, $b = 4.1$

6. $f(x) = 2.1x^2 + \dfrac{3}{\sqrt{x}} + 5$; $a = 3$, $b = 2.95$

7. **Calculating Mortgage Payments** Refer to Example 2. How much more per month will the Meyers' mortgage payment be if the interest rate increases from 10% to 10.3% per year? To 10.4% per year? To 10.5% per year?

8. **Estimating the Area of a Ring of Neptune** The ring 1989N2R of the planet Neptune has an inner radius of approximately 53,200 km (measured from the center of the planet) and a radial width of 15 km. Use differentials to estimate the area of the ring.

9. **Effect of Price Increase on Quantity Demanded** The quantity demanded per week of the Alpha Sports Watch, x (in thousands), is related to its unit price of p dollars by the equation

$$x = f(p) = 10\sqrt{\frac{50 - p}{p}} \qquad (0 \le p \le 50)$$

Use differentials to find the decrease in the quantity of the watches demanded per week if the unit price is increased from \$40 to \$42.

771

36. Effect of Mortgage Rates on Housing Starts A study prepared for the National Association of Realtors estimates that the number of housing starts per year over the next five years will be

$$N(r) = \frac{7}{1 + 0.02r^2}$$

million units, where r (percent) is the mortgage rate. Use differentials to estimate the decrease in the number of housing starts when the mortgage rate is increased from 12% to 12.5%.

37. Supply-Price The supply equation for a certain brand of transistor radio is given by

$$p = s(x) = 0.3\sqrt{x} + 10$$

where x is the quantity supplied and p is the unit price in dollars. Use differentials to approximate the change in price when the quantity supplied is increased from 10,000 units to 10,500 units.

38. Demand-Price The demand function for the Sentinel smoke alarm is given by

$$p = d(x) = \frac{30}{0.02x^2 + 1}$$

where x is the quantity demanded (in units of a thousand) and p is the unit price in dollars. Use differentials to estimate the change in the price p when the quantity demanded changes from 5000 to 5500 units per week.

39. Surface Area of an Animal Animal physiologists use the formula

$$S = kW^{2/3}$$

to calculate an animal's surface area (in square meters) from its weight W (in kilograms), where k is a constant that depends on the animal under consideration. Suppose a physiologist calculates the surface area of a horse ($k = 0.1$). If the horse's weight is estimated at 300 kg, with a maximum error in measurement of 0.6 kg, determine the percentage error in the calculation of the horse's surface area.

40. Forecasting Profits The management of Trappee and Sons, Inc., forecast that they will sell 200,000 cases of their Texa-Pep hot sauce next year. Their annual profit is described by

$$P(x) = -0.000032x^3 + 6x - 100$$

thousand dollars, where x is measured in thousands of cases. If the maximum error in the forecast is 15%, determine the corresponding error in Trappee's profits.

41. Forecasting Commodity Prices A certain country's government economists have determined that the demand equation for soybeans in that country is given by

$$p = f(x) = \frac{55}{2x^2 + 1}$$

where p is expressed in dollars per bushel and x, the quantity demanded per year, is measured in billions of bushels. The economists are forecasting a harvest of 1.8 billion bushels for the year, with a maximum error of 15% in their forecast. Determine the corresponding maximum error in the predicted price per bushel of soybeans.

42. Crime Studies A sociologist has found that the number of serious crimes in a certain city per year is described by the function

$$N(x) = \frac{500(400 + 20x)^{1/2}}{(5 + 0.2x)^2}$$

where x (in cents per dollar deposited) is the level of reinvestment in the area in conventional mortgages by the city's ten largest banks. Use differentials to estimate the change in the number of crimes if the level of reinvestment changes from 20 cents per dollar deposited to 22 cents per dollar deposited.

43. Financing a Home The Millers are planning to buy a home in the near future and estimate that they will need a 30-year fixed-rate mortgage for $120,000. Their monthly payment P (in dollars) can be computed using the formula

$$P = \frac{10{,}000r}{1 - \left(1 + \dfrac{r}{12}\right)^{-360}}$$

where r is the interest rate per year.
a. Find the differential of P.
b. If the interest rate increases from the present rate of 9% per year to 9.2% per year between now and the time the Millers decide to secure the loan, approximately how much more will their monthly mortgage payment be? How much more will it be if the interest rate increases to 9.3% per year? To 9.4% per year? To 9.5% per year?

SOLUTIONS TO SELF-CHECK EXERCISES 11.7

1. We find

$$f'(x) = \frac{1}{2}x^{-1/2} = \frac{1}{2\sqrt{x}}$$

Therefore, the required differential of f is

$$dy = \frac{1}{2\sqrt{x}}\,dx$$

2. We first compute the differential

$$dp = -\frac{250x}{(x^2 + 1)^2}\,dx$$

Next, using equation (11) with $x = 6$ and $dx = 0.2$, we find

$$\Delta p \approx dp = -\frac{250(6)}{(36 + 1)^2}(0.2) = -0.22$$

or a drop in price of 22 cents per bushel.

CHAPTER 11 SUMMARY OF PRINCIPAL FORMULAS AND TERMS

Formulas

1. Derivative of a constant $\dfrac{d}{dx}(c) = 0 \qquad (c, \text{a constant})$

2. Power Rule $\dfrac{d}{dx}(x^n) = nx^{n-1}$

3. Constant Multiple Rule $\dfrac{d}{dx}(cu) = c\dfrac{du}{dx} \qquad (c, \text{a constant})$

4. Sum Rule $\dfrac{d}{dx}(u \pm v) = \dfrac{du}{dx} \pm \dfrac{dv}{dx}$

5. Product Rule $\dfrac{d}{dx}(uv) = u\dfrac{dv}{dx} + v\dfrac{du}{dx}$

6. Quotient Rule $\dfrac{d}{dx}\left(\dfrac{u}{v}\right) = \dfrac{v\dfrac{du}{dx} - u\dfrac{dv}{dx}}{v^2}$

7. Chain Rule $\dfrac{dy}{dx} = \dfrac{dy}{du} \cdot \dfrac{du}{dx}$

8. General Power Rule $\dfrac{d}{dx}(u^n) = nu^{n-1}\dfrac{du}{dx}$

9. Average cost function $\overline{C}(x) = \dfrac{C(x)}{x}$

10. Revenue function $\qquad$ $R(x) = px$

11. Profit function $\qquad$ $P(x) = R(x) - C(x)$

12. Elasticity of demand $\qquad$ $E(p) = -\dfrac{pf'(p)}{f(p)}$

13. Differential of y $\qquad$ $dy = f'(x)\,dx$

Terms

Marginal cost function	Unitary demand
Marginal average cost function	Inelastic demand
Marginal revenue function	Second derivative of f
Marginal profit function	Implicit differentiation
Elastic demand	Related rates

CHAPTER 11 REVIEW EXERCISES

In exercises 1–30, find the derivative of the given function.

1. $f(x) = 3x^5 - 2x^4 + 3x^2 - 2x + 1$

2. $f(x) = 4x^6 + 2x^4 + 3x^2 - 2$

3. $g(x) = -2x^{-3} + 3x^{-1} + 2$

4. $f(t) = 2t^2 - 3t^3 - t^{-1/2}$

5. $g(t) = 2t^{-1/2} + 4t^{-3/2} + 2$

6. $h(x) = x^2 + \dfrac{2}{x}$

7. $f(t) = t + \dfrac{2}{t} + \dfrac{3}{t^2}$

8. $g(s) = 2s^2 - \dfrac{4}{s} + \dfrac{2}{\sqrt{s}}$

9. $h(x) = x^2 - \dfrac{2}{x^{3/2}}$

10. $f(x) = \dfrac{x+1}{2x-1}$

11. $g(t) = \dfrac{t^2}{2t^2 + 1}$

12. $h(t) = \dfrac{\sqrt{t}}{\sqrt{t} + 1}$

13. $f(x) = \dfrac{\sqrt{x} - 1}{\sqrt{x} + 1}$

14. $f(t) = \dfrac{t}{2t^2 + 1}$

15. $f(x) = \dfrac{x^2(x^2 + 1)}{x^2 - 1}$

16. $f(x) = (2x^2 + x)^3$

17. $f(x) = (3x^3 - 2)^8$

18. $h(x) = (\sqrt{x} + 2)^5$

19. $f(t) = \sqrt{2t^2 + 1}$

20. $g(t) = \sqrt[3]{1 - 2t^3}$

21. $s(t) = (3t^2 - 2t + 5)^{-2}$

22. $f(x) = (2x^3 - 3x^2 + 1)^{-3/2}$

23. $h(x) = \left(x + \dfrac{1}{x}\right)^2$

24. $h(x) = \dfrac{1+x}{(2x^2 + 1)^2}$

25. $h(t) = (t^2 + t)^4(2t^2)$

26. $f(x) = (2x + 1)^3(x^2 + x)^2$

27. $g(x) = \sqrt{x}(x^2 - 1)^3$

28. $f(x) = \dfrac{x}{\sqrt{x^3 + 2}}$

29. $h(x) = \dfrac{\sqrt{3x + 2}}{4x - 3}$

30. $f(t) = \dfrac{\sqrt{2t + 1}}{(t + 1)^3}$

In exercises 31–36, find the second derivative of the given function.

31. $f(x) = 2x^4 - 3x^3 + 2x^2 + x + 4$

32. $g(x) = \sqrt{x} + \dfrac{1}{\sqrt{x}}$

33. $h(t) = \dfrac{t}{t^2 + 4}$

34. $f(x) = (x^3 + x + 1)^2$

35. $f(x) = \sqrt{2x^2 + 1}$

36. $f(t) = t(t^2 + 1)^3$

In exercises 37–42, find dy/dx by implicit differentiation.

37. $6x^2 - 3y^2 = 9$

38. $2x^3 - 3xy = 4$

39. $y^3 + 3x^2 = 3y$ **40.** $x^2 + 2x^2y^2 + y^2 = 10$

41. $x^2 - 4xy - y^2 = 12$

42. $3x^2y - 4xy + x - 2y = 6$

43. Let $f(x) = 2x^3 - 3x^2 - 16x + 3$.
   a. Find the points on the graph of f at which the slope of the tangent line is equal to -4.
   b. Find the equation(s) of the tangent line(s) of (a).

44. Let $f(x) = \frac{1}{3}x^3 + \frac{1}{2}x^2 - 4x + 1$.
   a. Find the points on the graph of f at which the slope of the tangent line is equal to -2.
   b. Find the equation(s) of the tangent line(s) of (a).

45. Find an equation of the tangent line to the graph of $y = \sqrt{4 - x^2}$ at the point $(1, \sqrt{3})$.

46. Find an equation of the tangent line to the graph of $y = x(x + 1)^5$ at the point $(1, 32)$.

47. Find the third derivative of the function

$$f(x) = \frac{1}{2x - 1}$$

What is its domain?

48. The number of subscribers to CNC Cable Television in the town of Randolph is approximated by the function

$$N(x) = 1000(1 + 2x)^{1/2} \qquad (1 \le x \le 30)$$

where $N(x)$ denotes the number of subscribers to the service in the xth week. Find the rate of increase in the number of subscribers at the end of the twelfth week.

49. The total weekly cost in dollars incurred by the Herald Record Company in pressing x video discs is given by the total cost function

$$C(x) = 2500 + 2.2x \qquad (0 \le x \le 8000)$$

   a. What is the marginal cost when $x = 1000$? When $x = 2000$?
   b. Find the average cost function $\overline{C}$ and the marginal average cost function $\overline{C}'$.
   c. Using the results from (b), show that the average cost incurred by Herald in pressing a video disc approaches $2.20 per disc when the level of production is high enough.

50. The marketing department of Telecon Corporation has determined that the demand for their cordless phones obeys the relationship

$$p = -0.02x + 600 \qquad (0 \le x \le 30{,}000)$$

where p denotes the phone's unit price (in dollars) and x denotes the quantity demanded.
   a. Find the revenue function R.
   b. Find the marginal revenue function R'.
   c. Compute $R'(10{,}000)$ and interpret your result.

51. The weekly demand for the Lectro-Copy photocopying machine is given by the demand equation

$$p = 2000 - 0.04x \qquad (0 \le x \le 50{,}000)$$

where p denotes the wholesale unit price in dollars and x denotes the quantity demanded. The weekly total cost function for manufacturing these copiers is given by

$$C(x) = 0.000002x^3 - 0.02x^2 + 1000x + 120{,}000$$

where $C(x)$ denotes the total cost incurred in producing x units.
   a. Find the revenue function R, the profit function P, and the average cost function $\overline{C}$.
   b. Find the marginal cost function C', the marginal revenue function R', the marginal profit function P', and the marginal average cost function $\overline{C}'$.
   c. Compute $C'(3000)$, $R'(3000)$, and $P'(3000)$.
   d. Compute $\overline{C}'(5000)$ and $\overline{C}'(8000)$ and interpret your results.

This chapter further explores the power of the derivative, which we use to help analyze the properties of functions. The information obtained can then be used to sketch accurate graphs of functions. We also see how the derivative is used to solve a large class of optimization problems, including finding what level of production will yield a maximum profit for a company, finding what level of production will result in minimal cost to a company, finding the maximum height attained by a rocket, finding the maximum velocity at which air is expelled when a person coughs, and a host of other problems.

What is the maximum altitude and what is the maximum velocity attained by the rocket? In Example 7, page 846, you will see how the techniques of calculus can be used to help answer these questions.

12

APPLICATIONS
OF THE DERIVATIVE

○ ○ ○

12.1 APPLICATIONS OF THE FIRST DERIVATIVE

Determining the Intervals Where a Function Is Increasing or Decreasing

According to a study by the U.S. Department of Energy and the Shell Development Company, a typical car's fuel economy as a function of its speed is described by the graph shown in Figure 12.1. Observe that the fuel economy $f(x)$ in miles per gallon (mpg) improves as x, the vehicle's speed in miles per hour (mph), increases from 0 to 42, and then drops as the speed increases beyond 42 mph. We use the terms *increasing* and *decreasing* to describe the behavior of a function as we move from left to right along its graph.

Figure 12.1

A typical car's fuel economy improves as the speed at which it is driven increases from 0 mph to 42 mph and drops at speeds greater than 42 mph.

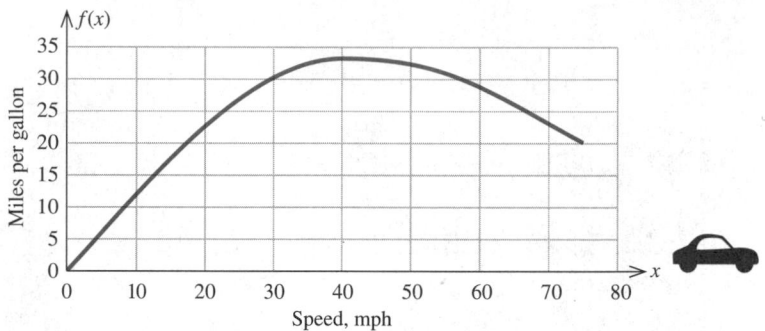

Source: U.S. Department of Energy and Shell Development Company

More precisely, we have the following definitions.

INCREASING AND DECREASING FUNCTIONS

A function f is **increasing** on an interval (a, b) if, for any two numbers x_1 and x_2 in (a, b), $f(x_1) < f(x_2)$ whenever $x_1 < x_2$ (Figure 12.2a).
A function f is **decreasing** on an interval (a, b) if, for any two numbers x_1 and x_2 in (a, b), $f(x_1) > f(x_2)$ whenever $x_1 < x_2$ (Figure 12.2b).

Figure 12.2

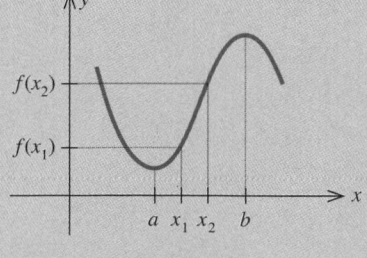

(a) f is increasing on (a, b).

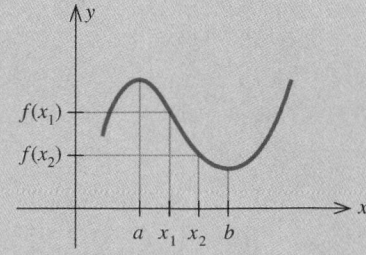

(b) f is decreasing on (a, b).

We say that f is *increasing at a point c* if there exists an interval (a, b) containing c such that f is increasing on (a, b). Similarly, we say that f is *decreasing at a point c* if there exists an interval (a, b) containing c such that f is decreasing on (a, b).

Since the rate of change of a function at a point $x = c$ is given by the derivative of the function at that point, the derivative lends itself naturally to being a tool for determining the intervals where a differentiable function is increasing or decreasing. Indeed, as we saw in Chapter 10, the derivative of a function at a point measures both the slope of the tangent line to the graph of the function at that point and the rate of change of the function at the same point. In fact, at a point where the derivative is positive, the slope of the tangent line to the graph is positive and the function is increasing. At a point where the derivative is negative, the slope of the tangent line to the graph is negative and the function is decreasing (Figure 12.3).

Figure 12.3

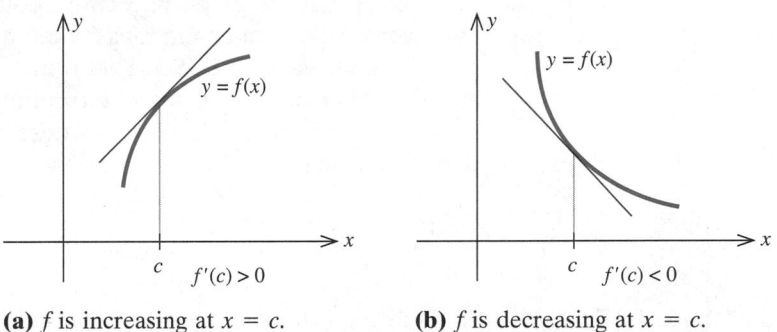

(a) f is increasing at $x = c$.   **(b)** f is decreasing at $x = c$.

These observations lead to the following important theorem, which we state without proof.

THEOREM I

a. If $f'(x) > 0$ for each value of x in an interval (a, b), then f is increasing on (a, b).
b. If $f'(x) < 0$ for each value of x in an interval (a, b), then f is decreasing on (a, b).
c. If $f'(x) = 0$ for each value of x in an interval (a, b), then f is constant on (a, b).

EXAMPLE I   Find the interval where the function $f(x) = x^2$ is increasing and the interval where it is decreasing.

Solution   The derivative of $f(x) = x^2$ is $f'(x) = 2x$. Since

$$f'(x) = 2x > 0 \quad \text{if } x > 0$$

and $$f'(x) = 2x < 0 \quad \text{if } x < 0$$

Figure 12.4
The graph of f falls on $(-\infty, 0)$
where $f'(x) < 0$ and rises on $(0, \infty)$
where $f'(x) > 0$.

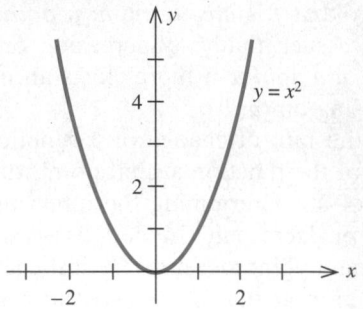

f is increasing on the interval $(0, \infty)$ and decreasing on the interval $(-\infty, 0)$ (Figure 12.4).   ◦ ◦ ◦

Recall that the graph of a continuous function cannot have any breaks. As a consequence, a continuous function cannot change sign unless it equals zero for some value of x. (See Theorem 5, page 642.) This observation suggests the following procedure for determining the sign of the derivative f' of a function f, and hence the intervals where the function f is increasing and where it is decreasing.

DETERMINING THE INTERVALS WHERE A FUNCTION IS INCREASING OR DECREASING

1. Find all values of x for which $f'(x) = 0$ or f' is discontinuous, and identify the open intervals determined by these points.

2. Select a test point c in each interval found in step 1 and determine the sign of $f'(c)$ in that interval.
   a. If $f'(c) > 0$, f is increasing on that interval.
   b. If $f'(c) < 0$, f is decreasing on that interval.

EXAMPLE 2   Determine the intervals where the function $f(x) = x^3 - 3x^2 - 24x + 32$ is increasing and where it is decreasing.

Solution

1. The derivative of f is

$$f'(x) = 3x^2 - 6x - 24 = 3(x + 2)(x - 4)$$

and it is continuous everywhere. The zeros of $f'(x)$ are $x = -2$ and $x = 4$, and these points divide the real line into the intervals $(-\infty, -2)$, $(-2, 4)$, and $(4, \infty)$.

2. To determine the sign of $f'(x)$ in the intervals $(-\infty, -2)$, $(-2, 4)$, and $(4, \infty)$, compute $f'(x)$ at a convenient test point in each interval. The results

Figure 12.5
Sign diagram for f'.

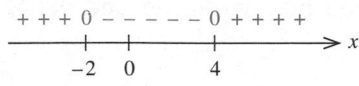

are shown in the following table:

Interval	Test Point c	$f'(c)$	Sign of $f'(x)$
$(-\infty, -2)$	-3	21	$+$
$(-2, 4)$	0	-24	$-$
$(4, \infty)$	5	21	$+$

Using these results, we obtain the sign diagram shown in Figure 12.5. We conclude that f is increasing on the intervals $(-\infty, -2)$ and $(4, \infty)$ and is decreasing on the interval $(-2, 4)$. Figure 12.6 shows the graph of f.

Figure 12.6
The graph of f rises on $(-\infty, -2)$, falls on $(-2, 4)$, and rises again on $(4, \infty)$.

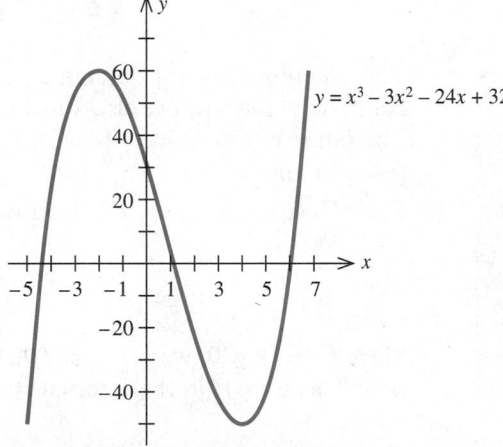

$y = x^3 - 3x^2 - 24x + 32$

EXPLORING WITH TECHNOLOGY

Refer to Example 2.
1. Use a graphing utility to plot the graphs of

$$f(x) = x^3 - 3x^2 - 24x + 32$$

and its derivative function

$$f'(x) = 3x^2 - 6x - 24$$

in the viewing rectangle $[-10, 10] \times [-50, 70]$.
2. By looking at the graph of f', determine the intervals where $f'(x) > 0$ and the intervals where $f'(x) < 0$. Next, look at the graph of f and determine the intervals where it is increasing and the intervals where it is decreasing. Describe the relationship. Is it what you expected?

REMARK   Do not be concerned with how the graphs in this section are obtained. We will learn how to sketch these graphs later. However, if you are familiar with the use of a graphing utility, you may go ahead and verify each graph.   ○ ○ ○

EXAMPLE 3   Find the interval where the function $f(x) = x^{2/3}$ is increasing and the interval where it is decreasing.

Solution

1. The derivative of f is

$$f'(x) = \frac{2}{3}x^{-1/3} = \frac{2}{3x^{1/3}}$$

The function f' is not defined at $x = 0$, so f' is discontinuous there. It is continuous everywhere else. Furthermore, f' is not equal to zero anywhere. The point $x = 0$ divides the real line (the domain of f) into the intervals $(-\infty, 0)$ and $(0, \infty)$.

2. Pick a test point, say $x = -1$, in the interval $(-\infty, 0)$ and compute

$$f'(-1) = -\frac{2}{3}$$

Since $f'(-1) < 0$, we see that $f'(x) < 0$ on $(-\infty, 0)$. Next, we pick a test point, say $x = 1$, in the interval $(0, \infty)$ and compute

$$f'(1) = \frac{2}{3}$$

Since $f'(1) > 0$, we see that $f'(x) > 0$ on $(0, \infty)$. Figure 12.7 shows these results in the form of a sign diagram. We conclude that f is decreasing on the interval $(-\infty, 0)$ and increasing on the interval $(0, \infty)$. The graph of f, shown in Figure 12.8, confirms these results.

> **True or false?**
> If f is continuous at c and f is increasing at c, then $f'(c) \neq 0$. Explain your answer. [*Hint:* Consider $f(x) = x^3$ and $c = 0$.]

Figure 12.7
Sign diagram for f'.

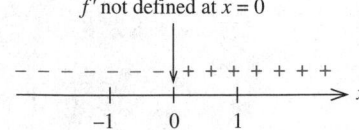

Figure 12.8
f decreases on $(-\infty, 0)$ and increases on $(0, \infty)$.

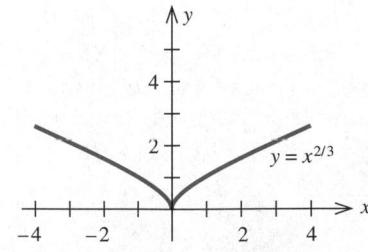

○ ○ ○

EXAMPLE 4 Find the intervals where the function $f(x) = x + 1/x$ is increasing and where it is decreasing.

Solution

1. The derivative of f is

$$f'(x) = 1 - \frac{1}{x^2} = \frac{x^2 - 1}{x^2}$$

Since f' is not defined at $x = 0$, it is discontinuous there. Furthermore, $f'(x)$ is equal to zero when $x^2 - 1 = 0$ or $x = \pm 1$. These values of x partition the domain of f' into the open intervals $(-\infty, -1)$, $(-1, 0)$, $(0, 1)$, and $(1, \infty)$, where the sign of f' is different from zero.

2. To determine the sign of f' in each of these intervals, we compute $f'(x)$ at the test points $x = -2, -\frac{1}{2}, \frac{1}{2}$, and 2, obtaining $f'(-2) = \frac{3}{4}, f'(-\frac{1}{2}) = -3$, $f'(\frac{1}{2}) = -3$, and $f'(2) = \frac{3}{4}$. From the sign diagram for f' (Figure 12.9), we conclude that f is increasing on $(-\infty, -1)$ and $(1, \infty)$ and decreasing on $(-1, 0)$ and $(0, 1)$. The graph of f appears in Figure 12.10. Note that f' does not change sign as we move across the point of discontinuity, $x = 0$. (Compare this with Example 3.)

Figure 12.9
f' does not change sign as we move across $x = 0$.

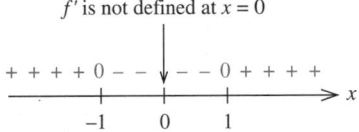

Figure 12.10
The graph of f rises on $(-\infty, -1)$, falls on $(-1, 0)$ and $(0, 1)$, and rises again on $(1, \infty)$.

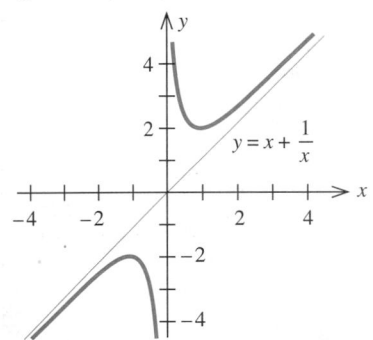

 Example 4 reminds us that we must not automatically conclude that the derivative f' must change sign when we move across a point of discontinuity or a zero of f'.

Consider the profit function P associated with a certain commodity defined by

$$P(x) = R(x) - C(x) \qquad (x \geq 0)$$

where R is the revenue function, C is the total cost function, and x is the number of units of the product produced and sold.

a. Find an expression for $P'(x)$.

b. Find relationships in terms of the derivatives of R and C so that
   i. P is increasing at $x = a$
   ii. P is decreasing at $x = a$
   iii. P is neither increasing nor decreasing at $x = a$
   [*Hint:* Recall that the derivative of a function at $x = a$ measures the rate of change of the function at that point.]

c. Explain the results of part (b) in economic terms.

Relative Extrema

In addition to helping us determine where the graph of a function is increasing and decreasing, the first derivative may be used to help us locate certain "high points" and "low points" on the graph of f. Knowing these points is invaluable in sketching the graphs of functions and solving optimization problems. These "high points" and "low points" correspond to the *relative (local) maxima* and *relative minima* of a function. They are so called because they are the highest or the lowest points when compared to points nearby.

The graph in Figure 12.11 gives the U.S. budget deficit from 1980 ($t = 0$) through 1991. The relative maxima and the relative minima of the function f are indicated on the graph.

Figure 12.11
U.S. budget deficit from 1980 to 1991.

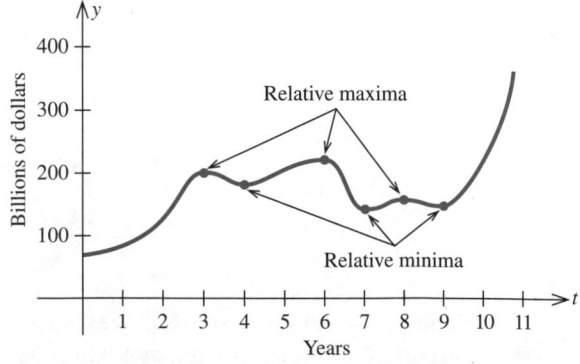

Source: Office of Management and Budget

More generally, we have the following definition:

RELATIVE MAXIMUM	A function f has a **relative maximum** at $x = c$ if there exists an open interval (a, b) containing c such that $f(x) \leq f(c)$ for all x in (a, b).

Geometrically, this means that there is *some* interval containing $x = c$ such that no point on the graph of f with its x-coordinate in that interval can lie above the point $(c, f(c))$; that is, $f(c)$ is the largest value of $f(x)$ in some interval around $x = c$. Figure 12.12 depicts the graph of a function f that has a relative maximum at $x = x_1$ and another at $x = x_3$.

Observe that all the points on the graph of f with x-coordinates in the interval I_1 containing x_1 (shown in blue) lie on or below the point $(x_1, f(x_1))$. This is also true for the point $(x_3, f(x_3))$ and the interval I_3. Thus, even though there are points on the graph of f that are "higher" than the points $(x_1, f(x_1))$ and $(x_3, f(x_3))$, the latter points are "highest" relative to points in their respective neighborhoods (intervals). Points on the graph of a function f that are "highest" and "lowest" with respect to *all* points in the domain of f will be studied in Section 12.4.

The definition of the relative minimum of a function parallels that of the relative maximum of a function.

RELATIVE MINIMUM	A function f has a **relative minimum** at $x = c$ if there exists an open interval (a, b) containing c such that $f(x) \geq f(c)$ for all x in (a, b).

The graph of the function f depicted in Figure 12.12 has a relative minimum at $x = x_2$ and another at $x = x_4$.

Figure 12.12
f has a relative maximum at $x = x_1$ and at $x = x_3$. It has a relative minimum at $x = x_2$ and at $x = x_4$.

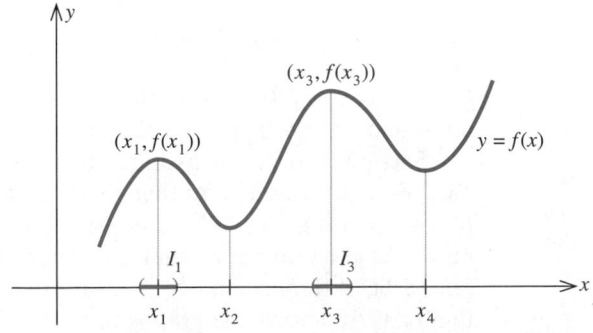

Finding the Relative Extrema

We refer to the relative maximum and relative minimum of a function as the **relative extrema** of that function. As a first step in our quest to find the relative extrema of a function, we consider functions that have derivatives at such points. Suppose that f is a function that is differentiable in some interval (a, b) that contains a point $x = c$, and that f has a relative maximum at $x = c$ (Figure 12.13a).

Figure 12.13

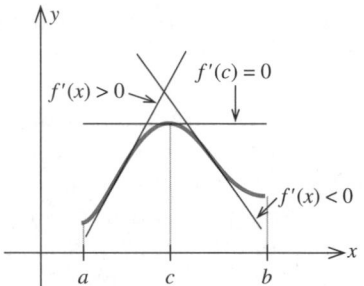

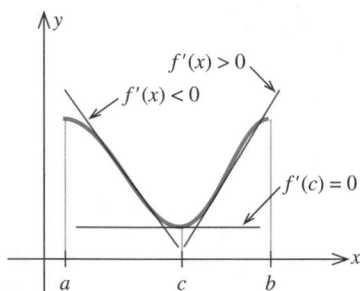

(a) f has a relative maximum at $x = c$.

(b) f has a relative minimum at $x = c$.

Observe that the slope of the tangent line to the graph of f must change from positive to negative as we move across the point $x = c$ from left to right. Therefore, the tangent line to the graph of f at the point $(c, f(c))$ must be horizontal; that is, $f'(c) = 0$ (Figure 12.13a).

Using a similar argument, we can show that the derivative f' of a differentiable function f must also be equal to zero at a point $x = c$, where f has a relative minimum (Figure 12.13b).

This analysis reveals an important characteristic of the relative extrema of a differentiable function f: *At any point c where f has a relative extremum, $f'(c) = 0$.*

Figure 12.14
$f'(0) = 0$, *but f does not have a relative extremum at (0, 0).*

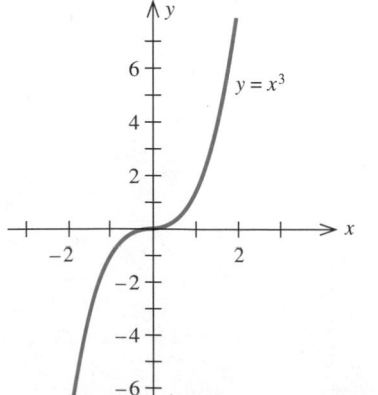

Before we develop a procedure for finding such points, a few words of caution are in order.

First, this result tells us that if a differentiable function f has a relative extremum at a point $x = c$, then $f'(c) = 0$. The converse of this statement—if $f'(c) = 0$ at some point $x = c$, then f must have a relative extremum at that point—is not true. Consider, for example, the function $f(x) = x^3$. Here $f'(x) = 3x^2$, so $f'(0) = 0$. Yet f has neither a relative maximum nor a relative minimum at $x = 0$ (Figure 12.14).

Second, our result assumes that the function is differentiable and thus has a derivative at a point that gives rise to a relative extremum. The functions $f(x) = |x|$ and $g(x) = x^{2/3}$ demonstrate that a relative extremum of a function may exist at a point at which the derivative does not exist. Both these functions fail to be differentiable at $x = 0$, but each has a relative minimum there. Figure 12.15 shows the graphs of these functions. Note that the slopes of the tangent lines change from negative to positive as we move across $x = 0$, just

Figure 12.15
Each of these functions has a rela-
tive extremum at (0, 0), but the deriv-
ative does not exist there.

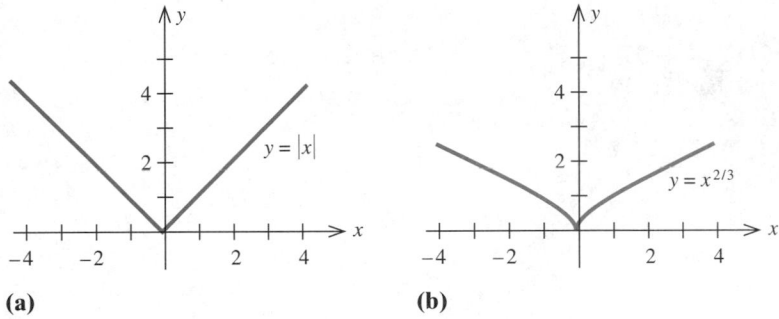

(a)   **(b)**

as in the case of a function that is differentiable at a value of x that gives rise to a relative minimum.

We refer to a point in the domain of f that *may* give rise to a relative extremum as a *critical point*.

CRITICAL POINT OF f	A **critical point** of a function f is any point x in the domain of f such that $f'(x) = 0$ or $f'(x)$ does not exist.

Figure 12.16 depicts the graph of a function that has critical points at $x = a$, b, c, d, and e. Observe that $f'(x) = 0$ at $x = a$, b, and c. Next, since there is a corner at $x = d$, $f'(x)$ does not exist there. Finally, $f'(x)$ does not exist at $x = e$ because the tangent line there is vertical. Also, observe that the critical points $x = a$, b, and d give rise to relative extrema of f, whereas the critical points $x = c$ and $x = e$ do not.

Figure 12.16
Critical points of f.

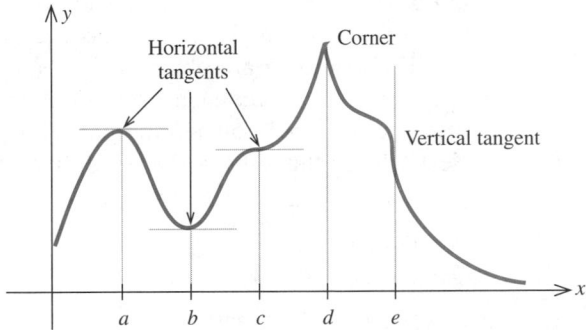

Having defined what a critical point is, we can now state a formal procedure for finding the relative extrema of a continuous function that is differentiable everywhere except at isolated values of x. Incorporated into the procedure is the so-called **First Derivative Test,** which enables us to determine whether or not a point gives rise to a relative maximum or a relative minimum of the function f.

EXAMPLE 5 Find the relative maxima and relative minima of the function $f(x) = x^2$.

Solution The derivative of $f(x) = x^2$ is given by $f'(x) = 2x$. Setting $f'(x) = 0$ yields $x = 0$ as the only critical point of f. Since

$$f'(x) < 0 \quad \text{if } x < 0$$
and
$$f'(x) > 0 \quad \text{if } x > 0$$

we see that $f'(x)$ changes sign from negative to positive as we move across the critical point $x = 0$. Thus we conclude that $f(0) = 0$ is a relative minimum of f (Figure 12.17).   ● ● ●

Figure 12.17
f has a relative minimum at $x = 0$.

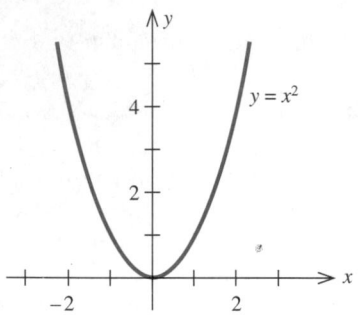

EXAMPLE 6 Find the relative maxima and relative minima of the function $f(x) = x^{2/3}$ (see Example 3).

Solution The derivative of f is $f'(x) = (2/3)x^{-1/3}$. As noted in Example 3, f' is not defined at $x = 0$, is continuous everywhere else, and is not equal to zero in its domain. Thus, $x = 0$ is the only critical point of the function f.

The sign diagram obtained in Example 3 is reproduced in Figure 12.18. We can see that the sign of $f'(x)$ changes from negative to positive as we move across $x = 0$ from left to right. Thus, an application of the First Derivative Test tells us that $f(0) = 0$ is a relative minimum of f (Figure 12.19).

Figure 12.18
Sign diagram for f'.

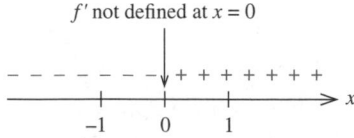

Figure 12.19
f has a relative minimum at $x = 0$.

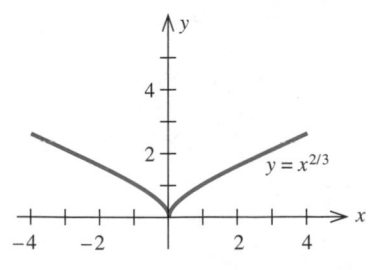

● ● ●

Recall that the average cost function $\overline{C}$ is defined by $\overline{C} = \dfrac{C(x)}{x}$, where $C(x)$ is the total cost function and x is the number of units of a commodity manufactured (see Section 11.4).

a. Show that $\overline{C}'(x) = \dfrac{C'(x) - \overline{C}(x)}{x}$, where $x > 0$.

b. Use the result of part (a) to conclude that $\overline{C}$ is decreasing for values of x at which $C'(x) < \overline{C}(x)$. Find similar conditions for which $\overline{C}$ is increasing and for which $\overline{C}$ is constant.

c. Explain the results of part (b) in economic terms.

EXAMPLE 7 Find the relative maxima and the relative minima of the function

$$f(x) = x^3 - 3x^2 - 24x + 32$$

Solution The derivative of f is

$$f'(x) = 3x^2 - 6x - 24 = 3(x + 2)(x - 4)$$

Figure 12.20
Sign diagram for f'.

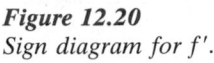
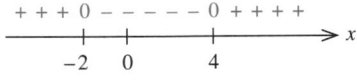

and it is continuous everywhere. The zeros of $f'(x)$, $x = -2$ and $x = 4$, are the only critical points of the function f. The sign diagram for f' is shown in Figure 12.20. We examine the two critical points $x = -2$ and $x = 4$ for a relative extremum using the First Derivative Test and the sign diagram for f':

1. *The critical point $x = -2$:* Since the function $f'(x)$ changes sign from positive to negative as we move across $x = -2$ from left to right, we conclude that a relative maximum of f occurs at $x = -2$. The value of $f(x)$ when $x = -2$ is

$$f(-2) = (-2)^3 - 3(-2)^2 - 24(-2) + 32 = 60$$

2. *The critical point $x = 4$:* $f'(x)$ changes sign from negative to positive as we move across $x = 4$ from left to right, so $f(4) = -48$ is a relative minimum of f. The graph of f appears in Figure 12.21.

Figure 12.21
f has a relative maximum at $x = -2$ and a relative minimum at $x = 4$.

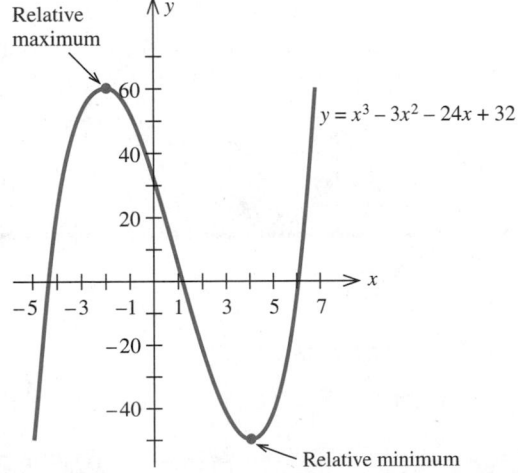

Figure 12.22
$x = 0$ *is not a critical point because f is not defined at $x = 0$.*

f' is not defined at $x = 0$

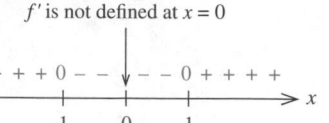

Figure 12.23
$f(x) = x + \dfrac{1}{x}$

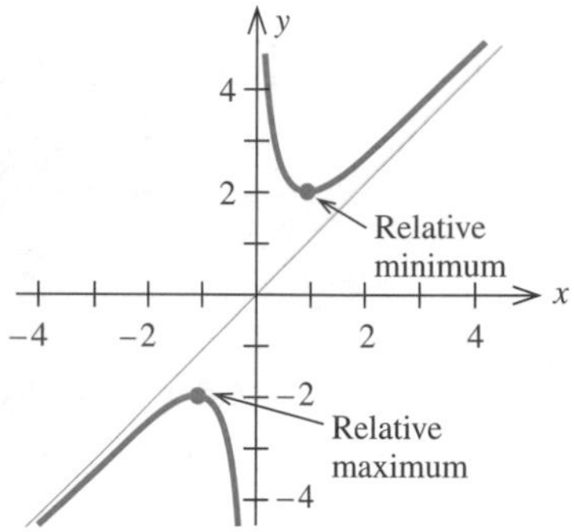

Relative minimum

Relative maximum

EXAMPLE 8 Find the relative maxima and the relative minima of the function

$$f(x) = x + \frac{1}{x}$$

Solution The derivative of f is

$$f'(x) = 1 - \frac{1}{x^2} = \frac{x^2 - 1}{x^2} = \frac{(x + 1)(x - 1)}{x^2}$$

Since f' is equal to zero at $x = -1$ and $x = 1$, these are critical points for the function f. Next, observe that f' is discontinuous at $x = 0$. However, because *f is not defined at that point,* the point $x = 0$ does not qualify as a critical point of f. Figure 12.22 shows the sign diagram for f'.

Since $f'(x)$ changes sign from positive to negative as we move across $x = -1$ from left to right, the First Derivative Test implies that $f(-1) = -2$ is a relative maximum of the function f. Next, $f'(x)$ changes sign from negative to positive as we move across $x = 1$ from left to right, so $f(1) = 2$ is a relative minimum of the function f. The graph of f appears in Figure 12.23. Note that this function has a relative maximum that lies below its relative minimum.

○ ○ ○

EXPLORING WITH TECHNOLOGY

Refer to Example 8.
1. Use a graphing utility to plot the graphs of $f(x) = x + 1/x$ and its derivative function $f'(x) = 1 - 1/x^2$ in the viewing rectangle $[-4, 4] \times [-8, 8]$.
2. By studying the graph of f', determine the critical points of f. Next, note the sign of $f'(x)$ immediately to the left and to the right of each critical point. What can you conclude about each critical point? Are your conclusions borne out by the graph of f?

◐ ◐ ◐

Applications

EXAMPLE 9 The profit function of the Acrosonic Company is given by

$$P(x) = -0.02x^2 + 300x - 200{,}000$$

Figure 12.24
The profit function is increasing on (0, 7500) and decreasing on (7500, ∞).

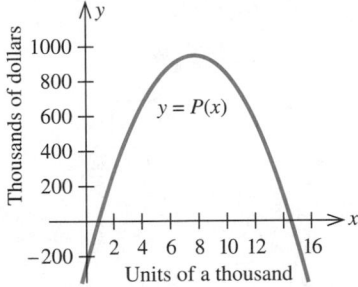

dollars, where x is the number of Acrosonic model F loudspeaker systems produced. Find where the function P is increasing and where it is decreasing.

Solution The derivative P' of the function P is

$$P'(x) = -0.04x + 300 = -0.04(x - 7500)$$

Thus, $P'(x) = 0$ when $x = 7500$. Furthermore, $P'(x) > 0$ for x in the interval (0, 7500), and $P'(x) < 0$ for x in the interval (7500, ∞). This means that the profit function P is increasing on (0, 7500) and decreasing on (7500, ∞) (Figure 12.24).   ◦ ◦ ◦

EXAMPLE 10 The number of major crimes committed in the city of Bronxville from 1991 to 1998 is approximated by the function

$$N(t) = -0.1t^3 + 1.5t^2 + 100 \qquad (0 \le t \le 7)$$

where $N(t)$ denotes the number of crimes committed in year t and $t = 0$ corresponds to the beginning of 1991. Find where the function N is increasing and where it is decreasing.

Solution The derivative N' of the function N is

$$N'(t) = -0.3t^2 + 3t = -0.3t(t - 10)$$

Since $N'(t) > 0$ for t in the interval (0, 7), the function N is increasing throughout that interval (Figure 12.25).   ◦ ◦ ◦

Figure 12.25
The number of crimes, $N(t)$, is increasing over the 7-year interval.

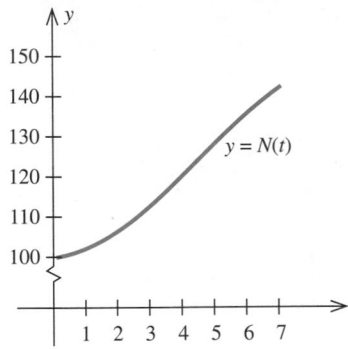

SELF–CHECK EXERCISES 12.1

1. Find the intervals where the function $f(x) = \frac{2}{3}x^3 - x^2 - 12x + 3$ is increasing and the intervals where it is decreasing.

2. Find the relative extrema of $f(x) = \dfrac{x^2}{1 - x^2}$.

Solutions to Self-Check Exercises 12.1 can be found on page 800.

12.1 EXERCISES

In exercises 1–8, you are given the graph of a function f. Determine the intervals where f is increasing, constant, or decreasing.

1.

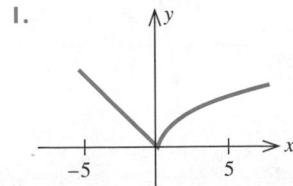

2.

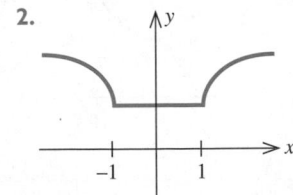

3.

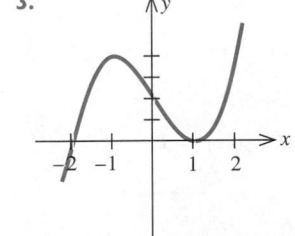

4.

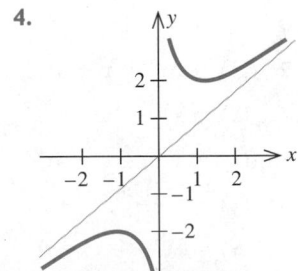

5.

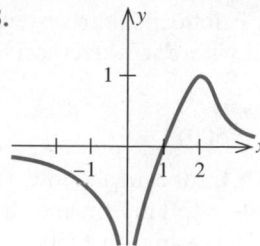

6.

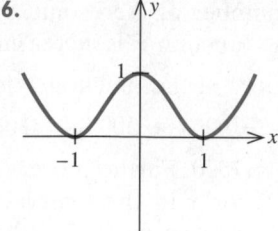

7.

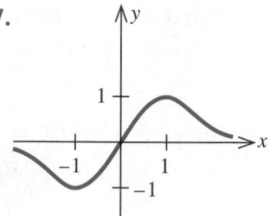

8.

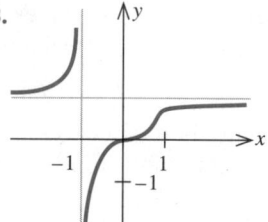

9. The Boston Marathon   The graph of the function f shown in the accompanying figure gives the elevation of that part of the Boston Marathon course that includes the notorious Heartbreak Hill. Determine the intervals (stretches of the course) where the function f is increasing (the runner is laboring), where it is constant (the runner is taking a breather), and where it is decreasing (the runner is coasting).

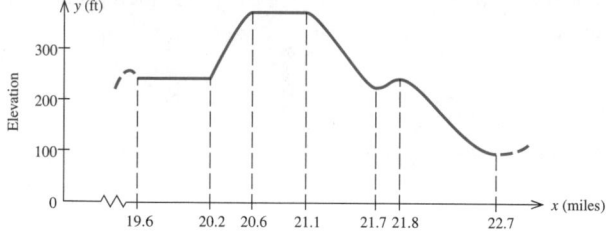

In exercises 10–35, find the interval(s) where each function is increasing and the interval(s) where it is decreasing.

10. $f(x) = 4 - 5x$

11. $f(x) = 3x + 5$

12. $f(x) = 2x^2 + x + 1$

13. $f(x) = x^2 - 3x$

14. $f(x) = x^3 - 3x^2$

15. $g(x) = x - x^3$

16. $f(x) = x^3 - 3x + 4$

17. $g(x) = x^3 + 3x^2 + 1$

18. $f(x) = \frac{2}{3}x^3 - 2x^2 - 6x - 2$

19. $f(x) = \frac{1}{3}x^3 - 3x^2 + 9x + 20$

20. $g(x) = x^4 - 2x^2 + 4$

21. $h(x) = x^4 - 4x^3 + 10$

22. $h(x) = \dfrac{1}{2x + 3}$

23. $f(x) = \dfrac{1}{x - 2}$

24. $g(t) = \dfrac{2t}{t^2 + 1}$

25. $h(t) = \dfrac{t}{t - 1}$

26. $f(x) = x^{2/3} + 5$

27. $f(x) = x^{3/5}$

28. $f(x) = (x - 5)^{2/3}$

29. $f(x) = \sqrt{x + 1}$

30. $g(x) = x\sqrt{x + 1}$

31. $f(x) = \sqrt{16 - x^2}$

32. $h(x) = \dfrac{x^2}{x - 1}$

33. $f(x) = \dfrac{x^2 - 1}{x}$

34. $g(x) = \dfrac{x}{(x + 1)^2}$

35. $f(x) = \dfrac{1}{(x - 1)^2}$

In exercises 36–43, you are given the graph of a function f. Determine the relative maxima and relative minima, if any.

36.

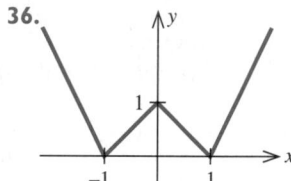

37.

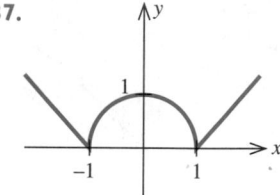

38.

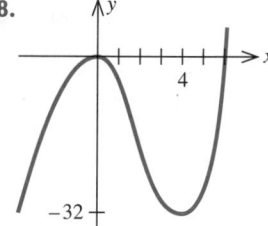

39.

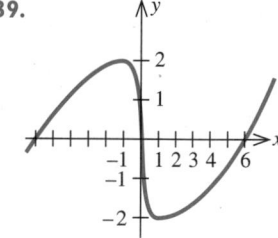

40.

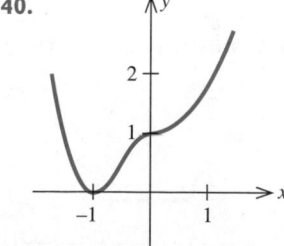

41.

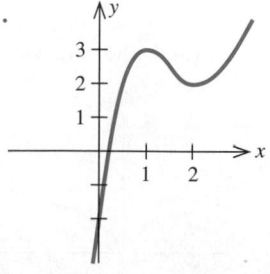

42.

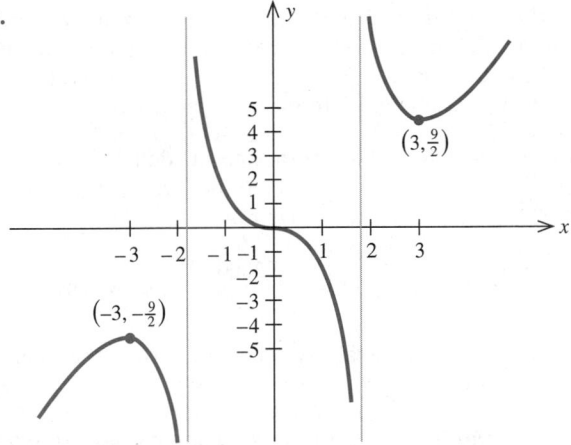

(a)

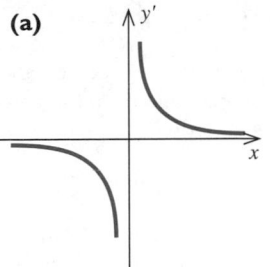

(b)

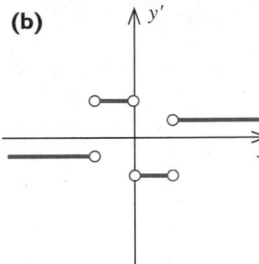

(c)

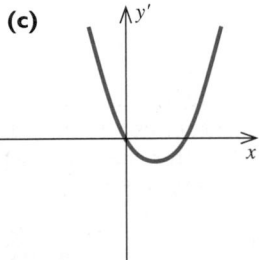

(d)

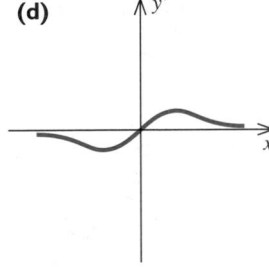

43.

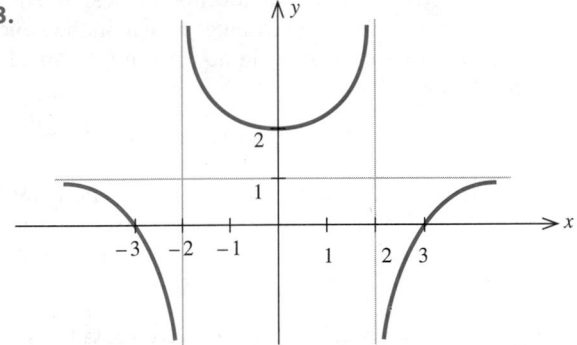

In exercises 44–47, match the graph of the function with the graph of its derivative in (a)–(d).

44. 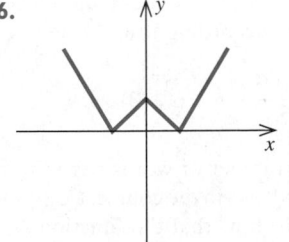   **45.**

46.   **47.**

In exercises 48–71, find the relative maxima and relative minima, if any, of each function.

48. $g(x) = x^2 + 3x + 8$

49. $f(x) = x^2 - 4x$

50. $h(t) = -t^2 + 6t + 6$

51. $f(x) = \frac{1}{2}x^2 - 2x + 4$

52. $f(x) = x^{5/3}$

53. $f(x) = x^{2/3} + 2$

54. $f(x) = x^3 - 3x + 6$

55. $g(x) = x^3 - 3x^2 + 4$

56. $f(x) = \frac{1}{2}x^4 - x^2$

57. $F(x) = \frac{1}{3}x^3 - x^2 - 3x + 4$

58. $h(x) = \frac{1}{2}x^4 - 3x^2 + 4x - 8$

59. $g(x) = x^4 - 4x^3 + 8$

60. $F(t) = 3t^5 - 20t^3 + 20$

61. $f(x) = 3x^4 - 2x^3 + 4$

62. $h(x) = \dfrac{x}{x+1}$

63. $g(x) = \dfrac{x+1}{x}$

64. $g(x) = 2x^2 + \dfrac{4000}{x} + 10$

65. $f(x) = x + \dfrac{9}{x} + 2$

66. $g(x) = \dfrac{x}{x^2 - 1}$

67. $f(x) = \dfrac{x}{1 + x^2}$

68. $g(t) = \dfrac{t^2}{1 + t^2}$

69. $f(x) = \dfrac{x^2}{x^2 - 4}$

70. $g(x) = x\sqrt{x-4}$ **71.** $f(x) = (x-1)^{2/3}$

72. A stone is thrown straight up from the roof of an 80-foot building. The distance of the stone from the ground at any time t (in seconds) is given by

$$h(t) = -16t^2 + 64t + 80$$

feet. When is the stone rising and when is it falling? If the stone were to miss the building, when would it hit the ground? Sketch the graph of h.
[*Hint:* The stone is on the ground when $h(t) = 0$.]

73. Profit Functions The Mexican subsidiary of the Thermo-Master Company manufactures an indoor-outdoor thermometer. Management estimates that the profit realizable by the company for the manufacture and sale of x units of thermometers per week is

$$P(x) = -0.001x^2 + 8x - 5000$$

dollars. Find the intervals where the profit function P is increasing and the intervals where P is decreasing.

74. Flight of a Rocket The height (in feet) attained by a rocket t seconds into flight is given by the function

$$h(t) = -\frac{1}{3}t^3 + 16t^2 + 33t + 10$$

When is the rocket rising and when is it descending?

75. Environment of Forests Following the lead of the National Wildlife Federation, the Department of the Interior of a South American country began to record an index of environmental quality that measured progress and decline in the environmental quality of its forests. The index for the years 1984 through 1994 is approximated by the function

$$I(t) = \frac{1}{3}t^3 - \frac{5}{2}t^2 + 80 \qquad (0 \le t \le 10)$$

where $t = 0$ corresponds to the year 1984. Find the intervals where the function I is increasing and the intervals where it is decreasing. Interpret your results.
Source: World Almanac

76. Average Speed of a Highway Vehicle The average speed of a vehicle on a stretch of Route 134 between 6 A.M. and 10 A.M. on a typical weekday is approximated by the function

$$f(t) = 20t - 40\sqrt{t} + 50 \qquad (0 \le t \le 4)$$

where $f(t)$ is measured in miles per hour and t is measured in hours, with $t = 0$ corresponding to 6 A.M. Find the interval where f is increasing and the interval where f is decreasing and interpret your results.

77. Average Cost The average cost in dollars incurred by the Lincoln Record Company per week in pressing x long-playing records is given by

$$\overline{C}(x) = -0.0001x + 2 + \frac{2000}{x} \qquad (0 < x \le 6000)$$

Show that $\overline{C}(x)$ is always decreasing over the interval $(0, 6000)$.

78. Air Pollution According to the South Coast Air Quality Management District, the level of nitrogen dioxide, a brown gas that impairs breathing, present in the atmosphere on a certain May day in downtown Los Angeles is approximated by

$$A(t) = 0.03t^3(t-7)^4 + 60.2 \qquad (0 \le t \le 7)$$

where $A(t)$ is measured in pollutant standard index (PSI) and t is measured in hours, with $t = 0$ corresponding to 7 A.M. At what time of day is the air pollution increasing and at what time is it decreasing?

79. Projected Retirement Funds Based on data from the Central Provident Fund of a certain country (a government agency similar to the Social Security Administration), the estimated cash in the fund in 1995 is given by

$$A(t) = -96.6t^4 + 403.6t^3 + 660.9t^2 + 250 \qquad (0 \le t \le 5)$$

where $A(t)$ is measured in billions of dollars and t is measured in decades, with $t = 0$ corresponding to the year 1995. Find the interval where A is increasing and the interval where A is decreasing and interpret your results.
[*Hint:* Use the quadratic formula.]

80. Learning Curves The Emory Secretarial School finds from experience that the average student taking advanced typing will progress according to the rule

$$N(t) = \frac{60t + 180}{t + 6} \qquad (t \ge 0)$$

where $N(t)$ measures the number of words per minute the student can type after t weeks in the course. Compute $N'(t)$ and use this result to show that the function N is increasing on the interval $(0, \infty)$.

81. **Drug Concentration in the Bloodstream** The concentration of a certain drug in a patient's body t hours after injection is given by

$$C(t) = \frac{t^2}{2t^3 + 1} \qquad (0 \le t \le 4)$$

milligrams per cubic centimeter. When is the concentration of the drug increasing and when is it decreasing?

82. **Air Pollution** The amount of nitrogen dioxide, a brown gas that impairs breathing, present in the atmosphere on a certain May day in the city of Long Beach is approximated by

$$A(t) = \frac{136}{1 + 0.25(t - 4.5)^2} + 28 \qquad (0 \le t \le 11)$$

where $A(t)$ is measured in pollutant standard index (PSI) and t is measured in hours, with $t = 0$ corresponding to 7 A.M. Find the intervals where A is increasing and where A is decreasing and interpret your results.

83. **Prison Overcrowding** The 1980s saw a trend toward old-fashioned punitive deterrence as opposed to the more liberal penal policies and community-based corrections popular in the 1960s and early 1970s. As a result, prisons became more crowded and the gap between the number of people in prison and prison capacity widened. Based on figures from the U.S. Department of Justice, the number of prisoners (in thousands) in federal and state prisons is approximated by the function

$$N(t) = 3.5t^2 + 26.7t + 436.2 \qquad (0 \le t \le 10)$$

where t is measured in years and $t = 0$ corresponds to 1984. The number of inmates for which prisons were designed is given by

$$C(t) = 24.3t + 365 \qquad (0 \le t \le 10)$$

where $C(t)$ is measured in thousands and t has the same meaning as before. Show that the gap between the number of prisoners and the number for which the prisons were designed has been widening at any time t.

[*Hint:* First write a function G that gives the gap between the number of prisoners and the number for which the prisons were designed at any time t. Then show that $G'(t) > 0$ for all values of t in the interval $(0, 10)$.]
Source: U.S. Department of Justice

84. Using Theorem 1, verify that the linear function $f(x) = mx + b$ is (a) increasing everywhere if $m > 0$, (b) decreasing everywhere if $m < 0$, and (c) constant if $m = 0$.

85. In what interval is the quadratic function

$$f(x) = ax^2 + bx + c \qquad (a \ne 0)$$

increasing? In what interval is f decreasing?

86. Let

$$f(x) = \begin{cases} -3x & \text{if } x < 0 \\ 2x + 4 & \text{if } x \ge 0 \end{cases}$$

a. Compute $f'(x)$ and show that it changes sign from negative to positive as we move across $x = 0$.
b. Show that f does not have a relative minimum at $x = 0$. Does this contradict the First Derivative Test? Explain your answer.

87. Let

$$f(x) = \begin{cases} -x^2 + 3 & \text{if } x \ne 0 \\ 2 & \text{if } x = 0 \end{cases}$$

a. Compute $f'(x)$ and show that it changes sign from positive to negative as we move across $x = 0$.
b. Show that f does not have a relative maximum at $x = 0$. Does this contradict the First Derivative Test? Explain your answer.

88. Show that the quadratic function

$$f(x) = ax^2 + bx + c \qquad (a \ne 0)$$

has a relative extremum when $x = -b/2a$. Also, show that the relative extremum is a relative maximum if $a < 0$ and a relative minimum if $a > 0$.

89. Show that the cubic function

$$f(x) = ax^3 + bx^2 + cx + d \qquad (a \ne 0)$$

has no relative extremum if and only if $b^2 - 3ac \le 0$.

90. Refer to Example 6, page 643.
a. Show that f is increasing on the interval $(0, 1)$.
b. Show that $f(0) = -1$ and $f(1) = 1$, and use the result of (a) together with the Intermediate Value Theorem to conclude that there is exactly one root of $f(x) = 0$ in $(0, 1)$.

91. Show that the function

$$f(x) = \frac{ax + b}{cx + d}$$

does not have a relative extremum if $ad - bc \ne 0$. What can you say about f if $ad - bc = 0$?

Using Technology

Using the First Derivative to Analyze a Function

A graphing utility is an effective tool for analyzing the properties of functions. This is especially true when we also bring into play the power of calculus, as the following examples show.

EXAMPLE 1   Let $f(x) = 2.4x^4 - 8.2x^3 + 2.7x^2 + 4x + 1$.

a. Use a graphing utility to plot the graph of f.

b. Find the intervals where f is increasing and the intervals where f is decreasing.

c. Find the relative extrema of f.

Solution

a. The graph of f in the viewing rectangle $[-2, 4] \times [-10, 10]$ is shown in Figure T1.

Figure T1
The graph of f in the viewing rectangle $[-2, 4] \times [-10, 10]$.

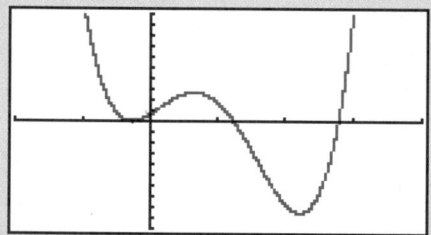

b. We compute

$$f'(x) = 9.6x^3 - 24.6x^2 + 5.4x + 4$$

and observe that f' is continuous everywhere. So the critical points of f occur at values of x where $f'(x) = 0$. In order to solve this last equation, observe that $f'(x)$ is a *polynomial function* of degree 3. The easiest way to solve the polynomial equation

$$9.6x^3 - 24.6x^2 + 5.4x + 4 = 0$$

is to use the function on a graphing utility for solving polynomial equations. (Not all graphing utilities have this function.) You can also use **TRACE** and **ZOOM,** but this will not give the same accuracy without a much greater effort. We find

$$x_1 \approx 2.22564943249, \quad x_2 \approx 0.63272944121, \quad x_3 \approx -0.295878873696$$

Referring to Figure T1, we conclude that f is decreasing on $(-\infty, -0.2959)$ and $(0.6327, 2.2256)$ (correct to four decimal places), and f is increasing on $(-0.2959, 0.6327)$ and $(2.2256, \infty)$.

c. Using the evaluation function of a graphing utility, we find the value of f at each of the critical points found in (b). Upon referring to Figure T1 once again, we see that $f(x_3) \approx 0.2836$ and $f(x_1) \approx -8.2366$ are relative minimum values of f, and $f(x_2) \approx 2.9194$ is a relative maximum value of f. ○ ○ ○

REMARK   The equation $f'(x) = 0$ in Example 1 is a polynomial equation, and so it is easily solved using the function for solving polynomial equations. We could also solve the equation using the function for finding the roots of equations, but that would require much more work. For equations that are *not* polynomial equations, however, our only choice is to use the function for finding the roots of equations. ○ ○ ○

If the derivative of a function is difficult to compute or simplify and we do not require great precision in the solution, we can find the relative extrema of the function using a combination of **ZOOM** and **TRACE.** This technique, which does not require the use of the derivative of f, is illustrated in the following example.

EXAMPLE 2   Let $f(x) = x^{1/3}(x^2 + 1)^{-3/2}3^{-x}$.

a. Use a graphing utility to plot the graph of f.

b. Find the relative extrema of f.

Solution

a. The graph of f in the viewing rectangle $[-4, 2] \times [-2, 1]$ is shown in Figure T2.

Figure T2
The graph of f in the viewing rectangle $[-4, 2] \times [-2, 1]$.

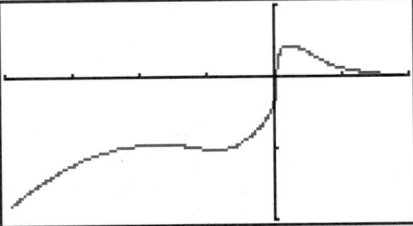

b. From the graph of f in Figure T2, we see that f has relative maxima when $x \approx -2$ and $x \approx 0.25$, and a relative minimum when $x \approx -0.75$. To obtain a better approximation of the first relative maximum, we zoom in with the cursor at approximately the point on the graph corresponding to $x \approx -2$. Then, using **TRACE,** we see that a relative maximum occurs when $x \approx -1.76$ with value $y \approx -1.01$. Similarly, we find the other relative maximum where $x \approx 0.20$ with value $y \approx 0.44$. Repeating the procedure, we find the relative minimum at the point where $x \approx -0.86$ and $y \approx -1.07$.   ◦ ◦ ◦

If you have access to a computer and software such as Derive, Maple, or Mathematica, then symbolic differentiation will yield the derivative $f'(x)$ of any differentiable function. This software will also solve the equation $f'(x) = 0$ with ease. Thus, the use of a computer will simplify even more greatly the analysis of functions.

EXERCISES

In exercises 1–4, use a graphing utility to find (a) the intervals where f is increasing and the intervals where f is decreasing, and (b) the relative extrema of f. Express your answers accurate to four decimal places.

1. $f(x) = 3.4x^4 - 6.2x^3 + 1.8x^2 + 3x - 2$

2. $f(x) = 1.8x^4 - 9.1x^3 + 5x - 4$

3. $f(x) = 2x^5 - 5x^3 + 8x^2 - 3x + 2$

4. $f(x) = 3x^5 - 4x^2 + 3x - 1$

In exercises 5–8, use the ZOOM and TRACE features of a graphing utility to find (a) the intervals where f is increasing and the intervals where f is decreasing, and (b) the relative extrema of f. Express your answers accurate to two decimal places.

5. $f(x) = (2x + 1)^{1/3}(x^2 + 1)^{-2/3}$

6. $f(x) = [x^2(x^3 - 1)]^{1/3} + \dfrac{1}{x}$

7. $f(x) = x - \sqrt{1 - x^2}$

8. $f(x) = \dfrac{\sqrt{x}(x^2 - 1)^2}{x - 2}$

9. **Rate of Bank Failures**   The Federal Deposit Insurance Corporation (FDIC) estimates that the rate at which banks were failing between 1982 and 1994 is given by

$$f(t) = 0.063447t^4 - 1.953283t^3 + 14.632576t^2$$
$$- 6.684704t + 47.458874 \qquad (0 \le t \le 12)$$

where $f(t)$ is measured in the number of banks per year and t is measured in years, with $t = 0$ corresponding to the beginning of 1982.

a. Use a graphing utility to plot the graph of f in the viewing rectangle $[0, 13] \times [0, 220]$.

b. Determine the intervals where f is increasing and where f is decreasing and interpret your result.

c. Find the relative maximum of f and interpret your result.

Source: Federal Deposit Insurance Corporation

10. **Manufacturing Capacity**   Data obtained from the Federal Reserve show that the annual increase in manufacturing capacity between 1988 and 1994 is given by

$$f(t) = 0.0388889t^3 - 0.283333t^2 + 0.477778t$$
$$+ 2.04286 \qquad (0 \le t \le 6)$$

where $f(t)$ is a percentage and t is measured in years, with $t = 0$ corresponding to the beginning of 1988.

a. Use a graphing utility to plot the graph of f in the viewing rectangle $[0, 8] \times [0, 4]$.

b. Determine the intervals where f is increasing and where f is decreasing and interpret your result.

Source: Federal Reserve

11. **Home Sales** According to the Greater Boston Real Estate Board—Multiple Listing Service, the average number of days a single-family home remains for sale from listing to accepted offer is approximated by the function

$$f(t) = 0.0171911t^4 - 0.662121t^3 + 6.18083t^2$$
$$- 8.97086t + 53.3357 \qquad (0 \le t \le 10)$$

where t is measured in years, with $t = 0$ corresponding to the beginning of 1984.

a. Use a graphing utility to plot the graph of f in the viewing rectangle $[0, 12] \times [0, 120]$.

b. Use the graph of f to find, approximately, the intervals where f is increasing and the intervals where f is decreasing. What does this result tell us about the sales of single-family homes in the greater Boston area from 1984 through 1994?

Source: Greater Boston Real Estate Board—Multiple Listing Service

12. **Morning Traffic Rush** The speed of traffic flow on a certain stretch of Route 123 between 6 A.M. and 10 A.M. on a typical weekday is approximated by the function

$$f(t) = 20t - 40\sqrt{t} + 52 \qquad (0 \le t \le 4)$$

where $f(t)$ is measured in mph and t is measured in hours, with $t = 0$ corresponding to 6 A.M. Find the interval where f is increasing, the interval where f is decreasing, and the relative extrema of f. Interpret your results.

13. **Air Pollution** The amount of nitrogen dioxide, a brown gas that impairs breathing, present in the atmosphere on a certain May day in the city of Long Beach is approximated by

$$A(t) = \frac{136}{1 + 0.25(t - 4.5)^2} + 28 \qquad (0 \le t \le 11)$$

where $A(t)$ is measured in pollutant standard index (PSI) and t is measured in hours, with $t = 0$ corresponding to 7 A.M. When is the PSI increasing and when is it decreasing? At what time is the PSI highest, and what is its value at that time?

SOLUTIONS TO SELF-CHECK EXERCISES 12.1

1. The derivative of f is

$$f'(x) = 2x^2 - 2x - 12 = 2(x + 2)(x - 3)$$

and it is continuous everywhere. The zeros of $f'(x)$ are $x = -2$ and $x = 3$. The sign diagram of f' is shown in the accompanying figure. We conclude that f is increasing on the intervals $(-\infty, -2)$ and $(3, \infty)$ and decreasing on the interval $(-2, 3)$.

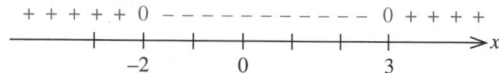

2. The derivative of f is

$$f'(x) = \frac{(1 - x^2)\dfrac{d}{dx}(x^2) - x^2\dfrac{d}{dx}(1 - x^2)}{(1 - x^2)^2}$$

$$= \frac{(1 - x^2)(2x) - x^2(-2x)}{(1 - x^2)^2} = \frac{2x}{(1 - x^2)^2}$$

and it is continuous everywhere except at $x = \pm 1$. Since $f'(x)$ is equal to zero at $x = 0$, $x = 0$ is a critical point of f. Next, observe that $f'(x)$ is discontinuous at $x = \pm 1$, but since these points are not in the domain of f, they do not qualify as critical points of f. Finally, from the sign diagram of f' shown in the accompanying figure, we conclude that $f(0) = 0$ is a relative minimum of f.

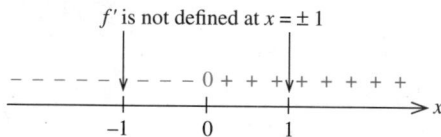

12.2 APPLICATIONS OF THE SECOND DERIVATIVE

Determining the Intervals of Concavity

Consider the graphs shown in Figure 12.26, which give the estimated population of the world and of the United States through the year 2000. Both graphs are rising, indicating that both the U.S. population and the world population will continue to increase through the year 2000. But observe that the graph in Figure 12.26a opens upward, whereas the graph in Figure 12.26b opens downward. What is the significance of this? To answer this question, let

Figure 12.26

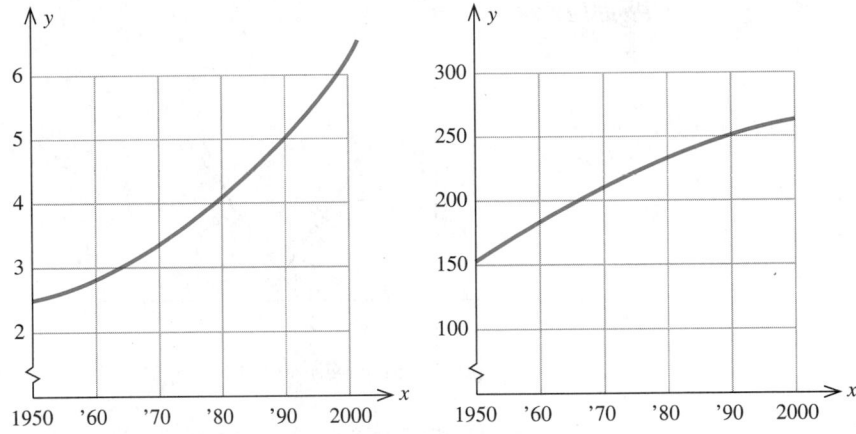

(a) World population in billions.   **(b)** U.S. population in millions.

Source: U.S. Dept. of Commerce and Worldwatch Institute

us look at the slopes of the tangent lines to various points on each graph (Figure 12.27).

In Figure 12.27a, we see that the slopes of the tangent lines to the graph are increasing as we move from left to right. Since the slope of the tangent line to the graph at a point on the graph measures the rate of change of the function at that point, we conclude that the world population not only is increasing through the year 2000 but is increasing at an *increasing* pace. A similar analysis of Figure 12.27b reveals that the U.S. population is increasing, but at a *decreasing* pace.

Figure 12.27

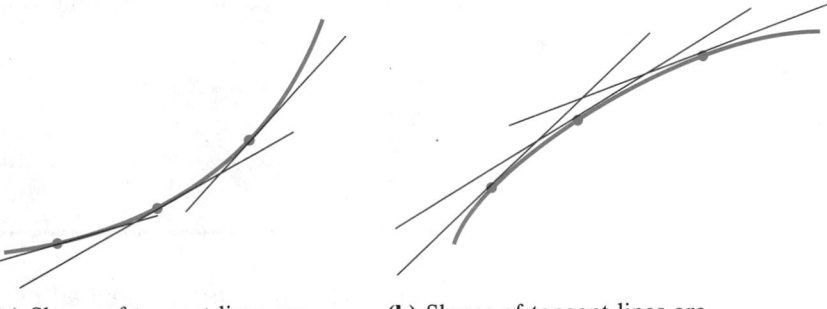

(a) Slopes of tangent lines are increasing.   **(b)** Slopes of tangent lines are decreasing.

The shape of a curve can be described using the notion of *concavity:*

CONCAVITY OF A FUNCTION *f*	Let the function *f* be differentiable on an interval (a, b). Then
	1. *f* is **concave upward** on (a, b) if f' is increasing on (a, b).
	2. *f* is **concave downward** on (a, b) if f' is decreasing on (a, b).

Figure 12.28

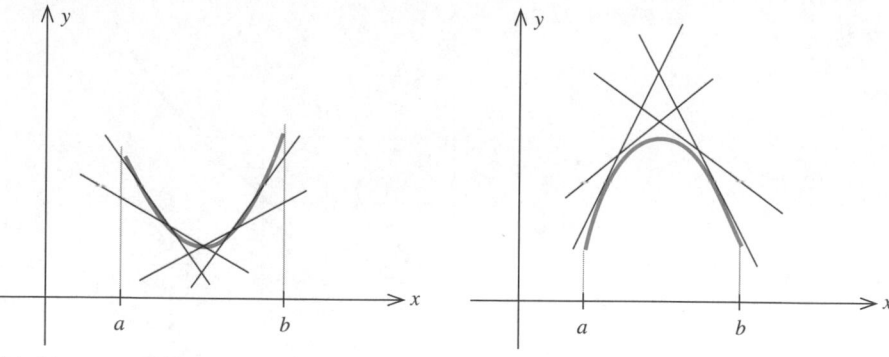

(a) f is concave upward on (a, b).   **(b)** f is concave downward on (a, b).

Geometrically, a curve is concave upward if it lies above its tangent lines (Figure 12.28a). Similarly, a curve is concave downward if it lies below its tangent lines (Figure 12.28b).

We also say that f is *concave upward at a point* $x = c$ if there exists an interval (a, b) containing c in which f is concave upward. Similarly, we say that f is *concave downward at a point* $x = c$ if there exists an interval (a, b) containing c in which f is concave downward.

If a function f has a second derivative f'', we can use f'' to determine the intervals of concavity of the function. Recall that $f''(x)$ measures the rate of change of the slope $f'(x)$ of the tangent line to the graph of f at the point $(x, f(x))$. Thus, if $f''(x) > 0$ on an interval (a, b), then the slopes of the tangent lines to the graph of f are increasing on (a, b) and so f is concave upward on (a, b). Similarly, if $f''(x) < 0$ on an interval (a, b), then f is concave downward on (a, b). These observations suggest the following theorem.

THEOREM 2

a. If $f''(x) > 0$ for each value of x in (a, b), then f is concave upward on (a, b).

b. If $f''(x) < 0$ for each value of x in (a, b), then f is concave downward on (a, b).

The following procedure, based on the conclusions of Theorem 2, may be used to determine the intervals of concavity of a function.

DETERMINING THE INTERVALS OF CONCAVITY OF f

1. Determine the values of x for which f'' is zero or where f'' is not defined, and identify the open intervals determined by these points.

2. Determine the sign of f'' in each interval found in step 1. To do this, compute $f''(c)$, where c is any conveniently chosen test point in the interval.

   a. If $f''(c) > 0$, f is concave upward on that interval.

   b. If $f''(c) < 0$, f is concave downward on that interval.

Figure 12.29
Sign diagram for f″.

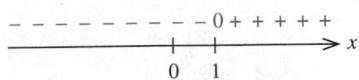

Figure 12.30
f is concave downward on $(-\infty, 1)$ and concave upward on $(1, \infty)$.

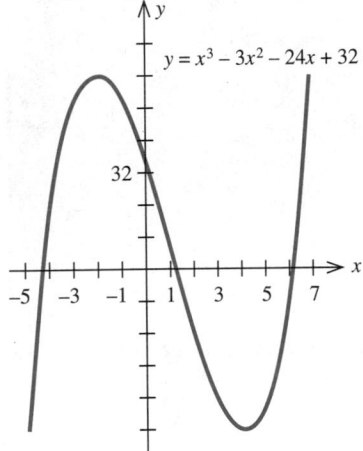

EXAMPLE 1 Determine where the function $f(x) = x^3 - 3x^2 - 24x + 32$ is concave upward and where it is concave downward.

Solution Here,

$$f'(x) = 3x^2 - 6x - 24$$

so

$$f''(x) = 6x - 6 = 6(x - 1)$$

and f'' is defined everywhere. Setting $f''(x) = 0$ gives $x = 1$. The sign diagram of f'' appears in Figure 12.29. We conclude that f is concave downward on the interval $(-\infty, 1)$ and concave upward on the interval $(1, \infty)$. Figure 12.30 shows the graph of f.   ⊙ ⊙ ⊙

EXAMPLE 2 Determine the intervals where the function $f(x) = x + 1/x$ is concave upward and where it is concave downward.

Solution We have

$$f'(x) = 1 - \frac{1}{x^2}$$

and

$$f''(x) = \frac{2}{x^3}$$

We deduce from the sign diagram for f'' (Figure 12.31) that the function f is concave downward on the interval $(-\infty, 0)$ and concave upward on the interval $(0, \infty)$. The graph of f is sketched in Figure 12.32.

Figure 12.31
The sign diagram for f″.

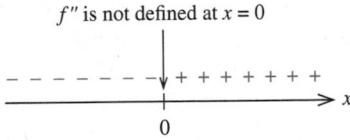

Figure 12.32
f is concave downward on $(-\infty, 0)$ and concave upward on $(0, \infty)$.

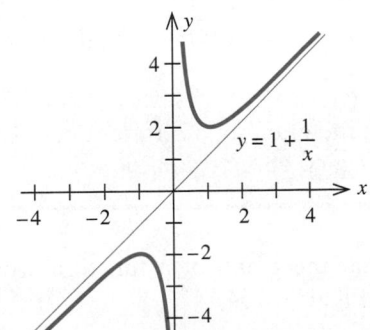

⊙ ⊙ ⊙

EXPLORING WITH TECHNOLOGY

Refer to Example 1.

1. Use a graphing utility to plot the graphs of

$$f(x) = x^3 - 3x^2 - 24x + 32$$

and its second derivative

$$f''(x) = 6x - 6$$

in the viewing rectangle $[-10, 10] \times [-80, 90]$.

2. By studying the graph of f'', determine the intervals where $f''(x) > 0$ and the intervals where $f''(x) < 0$. Next, look at the graph of f and determine the intervals where the graph of f is concave upward and the intervals where the graph of f is concave downward. Are these observations what you might have expected?

○   ○   ○

Inflection Points

Figure 12.33

The graph of S has a point of inflection at (50, 2700).

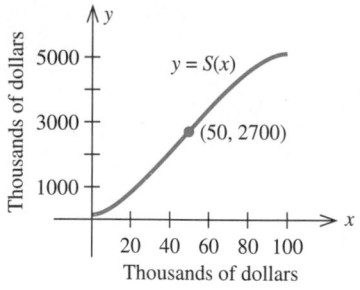

Figure 12.33 shows the total sales S of a manufacturer of automobile air conditioners versus the amount of money x that the company spends on advertising its product. Notice that the graph of the continuous function $y = S(x)$ changes concavity—from upward to downward—at the point $(50, 2700)$. This point is called an *inflection point* of S. To understand the significance of this inflection point, observe that the total sales increase rather slowly at first, but as more money is spent on advertising, the total sales increase rapidly. This rapid increase reflects the effectiveness of the company's ads. However, a point is soon reached after which any additional advertising expenditure results in increased sales but at a slower rate of increase. This point, commonly known as the *point of diminishing returns,* is the point of inflection of the function S. We will return to this example later.

Let us now state formally the definition of an inflection point.

INFLECTION POINT   A point on the graph of a differentiable function f at which the concavity changes is called an **inflection point.**

Observe that the graph of a function crosses its tangent line at a point of inflection (Figure 12.34).

Figure 12.34
At each point of inflection, the graph of a function crosses its tangent line.

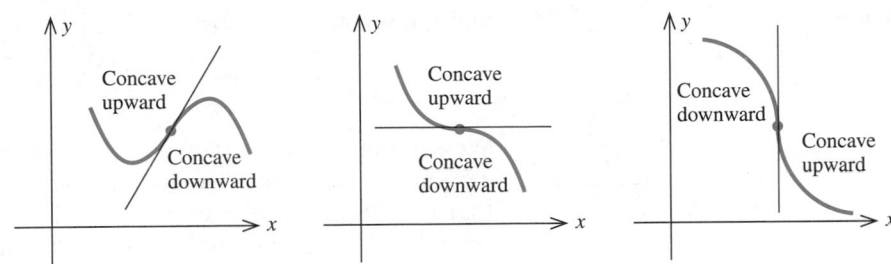

The following procedure may be used to find inflection points.

FINDING INFLECTION POINTS

1. Compute $f''(x)$.
2. Determine the points in the domain of f for which $f''(x) = 0$ or $f''(x)$ does not exist.
3. Determine the sign of $f''(x)$ to the left and right of each point $x = c$ found in step 2. If there is a change in the sign of $f''(x)$ as we move across the point $x = c$, then $(c, f(c))$ is an inflection point of f.

Figure 12.35
Sign diagram for f''.

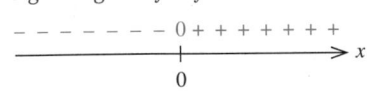

The points determined in step 2 are only *candidates* for the inflection points of f. For example, you can easily verify that $f''(0) = 0$ if $f(x) = x^4$, but a sketch of the graph of f will show that $(0, 0)$ is not an inflection point of f.

EXAMPLE 3   Find the points of inflection of the function $f(x) = x^3$.

Solution

$$f'(x) = 3x^2$$

so

$$f''(x) = 6x$$

Observe that f'' is continuous everywhere and is zero if $x = 0$. The sign diagram of f'' is shown in Figure 12.35. From this diagram, we see that $f''(x)$ changes sign as we move across $x = 0$. Thus the point $(0, 0)$ is an inflection point of the function f (Figure 12.36).   ● ● ●

Figure 12.36
f has an inflection point at $(0, 0)$.

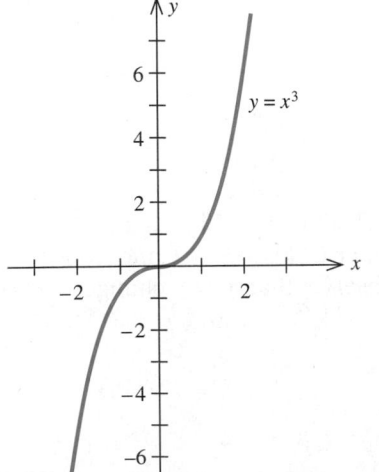

EXAMPLE 4   Determine the intervals where the function $f(x) = (x - 1)^{5/3}$ is concave upward and where it is concave downward, and find the inflection points of f.

Solution   The first derivative of f is

$$f'(x) = \frac{5}{3}(x - 1)^{2/3}$$

Figure 12.37
Sign diagram for f".

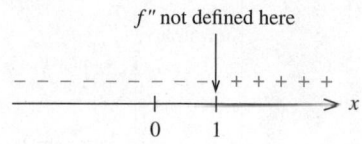

and the second derivative of f is

$$f''(x) = \frac{10}{9}(x-1)^{-1/3} = \frac{10}{9(x-1)^{1/3}}$$

We see that f'' is not defined at $x = 1$. Furthermore, $f''(x)$ is not equal to zero anywhere. From the sign diagram of f'' shown in Figure 12.37, we see that f is concave downward on $(-\infty, 1)$ and concave upward on $(1, \infty)$. Next, since $x = 1$ does lie in the domain of f, our computations also reveal that the point $(1, 0)$ is an inflection point of f (Figure 12.38).

Figure 12.38
f has an inflection point at (1, 0).

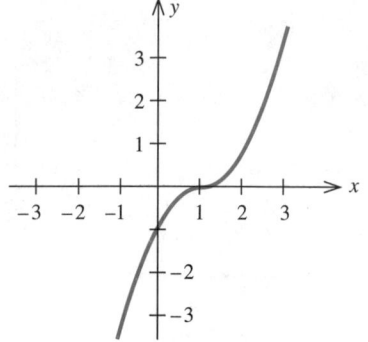

EXAMPLE 5   Determine the intervals where the function

$$f(x) = \frac{1}{x^2 + 1}$$

is concave upward and where it is concave downward, and find the inflection points of f.

Solution   The first derivative of f is

$$f'(x) = \frac{d}{dx}(x^2 + 1)^{-1} = -2x(x^2 + 1)^{-2} \qquad \text{(Using the General Power Rule)}$$

$$= -\frac{2x}{(x^2 + 1)^2}$$

Next, using the Quotient Rule, we find

$$f''(x) = \frac{(x^2 + 1)^2(-2) + (2x)2(x^2 + 1)(2x)}{(x^2 + 1)^4}$$

$$= \frac{(x^2 + 1)[-2(x^2 + 1) + 8x^2]}{(x^2 + 1)^4} = \frac{(x^2 + 1)(6x^2 - 2)}{(x^2 + 1)^4}$$

$$= \frac{2(3x^2 - 1)}{(x^2 + 1)^3} \qquad \text{(Canceling the common factors)}$$

Observe that f'' is continuous everywhere and is zero if

$$3x^2 - 1 = 0$$

$$x^2 = \frac{1}{3}$$

Figure 12.39
Sign diagram for f".

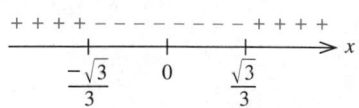

or $x = \pm\sqrt{3}/3$. From the sign diagram for f'' shown in Figure 12.39, we see that f is concave upward on $(-\infty, -\sqrt{3}/3) \cup (\sqrt{3}/3, \infty)$ and concave downward on $(-\sqrt{3}/3, \sqrt{3}/3)$. Also, we observe that $f''(x)$ changes sign as we move across the points $x = -\sqrt{3}/3$ and $x = \sqrt{3}/3$. Since

$$f\left(-\frac{\sqrt{3}}{3}\right) = \frac{1}{\frac{1}{3} + 1} = \frac{3}{4} \quad \text{and} \quad f\left(\frac{\sqrt{3}}{3}\right) = \frac{3}{4}$$

Figure 12.40

The graph of $f(x) = \dfrac{1}{x^2 + 1}$ is concave upward on $\left(-\infty, -\sqrt{3}/3\right) \cup \left(\sqrt{3}/3, \infty\right)$ and concave downward on $\left(-\sqrt{3}/3, \sqrt{3}/3\right)$.

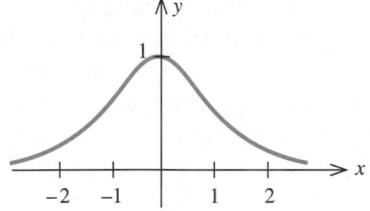

we see that the points $(-\sqrt{3}/3, 3/4)$ and $(\sqrt{3}/3, 3/4)$ are inflection points of f. The graph of f is shown in Figure 12.40.   ◦ ◦ ◦

Applications

Examples 6 and 7 illustrate familiar interpretations of the significance of the inflection point of a function.

EXAMPLE 6 The total sales S (in thousands of dollars) of the Arctic Air Corporation, a manufacturer of automobile air conditioners, is related to the amount of money x the company spends on advertising its products by the formula

$$S = -0.01x^3 + 1.5x^2 + 200 \qquad (0 \le x \le 100)$$

where x is measured in thousands of dollars. Find the inflection point of the function S.

Solution The first two derivatives of S are given by

$$S' = -0.03x^2 + 3x$$
$$S'' = -0.06x + 3$$

Setting $S'' = 0$ gives $x = 50$ as the only candidate for an inflection point of S. Moreover, since

$$S'' > 0 \quad \text{for} \quad x < 50$$
$$S'' < 0 \quad \text{for} \quad x > 50$$

the point $(50, 2700)$ is an inflection point of the function S. The graph of S appears in Figure 12.41. Notice that this is the graph of the function we discussed earlier.   ◦ ◦ ◦

Figure 12.41
The graph of $S(x)$ has a point of inflection at $(50, 2700)$.

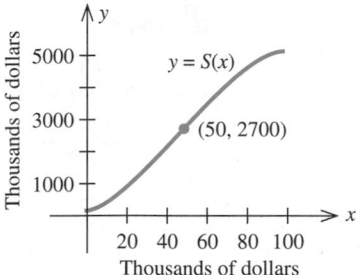

EXAMPLE 7 An economy's consumer price index (CPI) is described by the function

$$I(t) = -0.2t^3 + 3t^2 + 100 \qquad (0 \le t \le 9)$$

where $t = 0$ corresponds to the year 1990. Find the point of inflection of the function I and discuss its significance.

Solution The first two derivatives of I are given by

$$I'(t) = -0.6t^2 + 6t$$
$$I''(t) = -1.2t + 6 = -1.2(t - 5)$$

Setting $I''(t) = 0$ gives $t = 5$ as the only candidate for an inflection point of I. Next, we observe that

$$I'' > 0 \quad \text{for} \quad t < 5$$
$$I'' < 0 \quad \text{for} \quad t > 5$$

so the point $(5, 150)$ is an inflection point of I. The graph of I is sketched in Figure 12.42.

Figure 12.42
The graph of $I(t)$ has a point of inflection at $(5, 150)$.

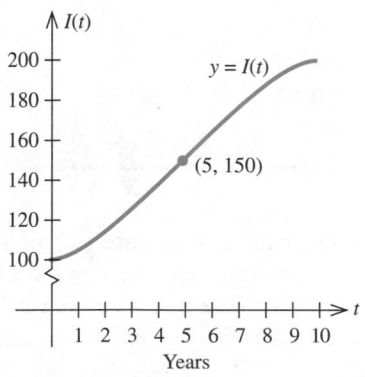

Since the second derivative of I measures the rate of change of the inflation rate, our computations reveal that the rate of inflation had, in fact, peaked at $t = 5$. Thus, relief actually began at the beginning of 1995.   ◐ ◐ ◐

The Second Derivative Test

We now show how the second derivative f'' of a function f can be used to help us determine whether a critical point of f is a relative extremum of f. Figure 12.43a shows the graph of a function that has a relative maximum at $x = c$. Observe that f is concave downward at that point. Similarly, Figure 12.43b shows that at a relative minimum of f the graph is concave upward. But from our previous work we know that f is concave downward at $x = c$ if $f''(c) < 0$ and f is concave upward at $x = c$ if $f''(c) > 0$. These observations suggest the following alternative procedure for determining whether a critical point of f gives rise to a relative extremum of f. This result is called the **Second Derivative Test** and is applicable when f'' exists.

Figure 12.43

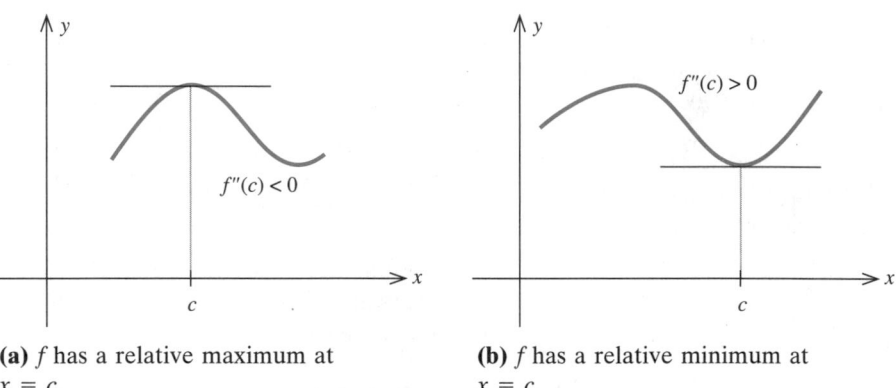

(a) f has a relative maximum at $x = c$.

(b) f has a relative minimum at $x = c$.

THE SECOND DERIVATIVE TEST	
	1. Compute $f'(x)$ and $f''(x)$.
	2. Find all the critical points of f at which $f'(x) = 0$.
	3. Compute $f''(c)$ for each such critical point c.
	a. If $f''(c) < 0$, then f has a relative maximum at c.
	b. If $f''(c) > 0$, then f has a relative minimum at c.
	c. If $f''(c) = 0$, the test fails; that is, it is inconclusive.

REMARK   As stated in step 3(c), the Second Derivative Test does not yield a conclusion if $f''(c) = 0$ or if $f''(c)$ does not exist. In other words, $x = c$ may give rise to a relative extremum or an inflection point (see exercise 86, page 818). In such cases, you should revert to the First Derivative Test.   ◐ ◐ ◐

EXAMPLE 8   Determine the relative extrema of the function

$$f(x) = x^3 - 3x^2 - 24x + 32$$

using the Second Derivative Test. (See Example 7, page 789.)

Solution   We have

$$f'(x) = 3x^2 - 6x - 24 = 3(x + 2)(x - 4)$$

so $f'(x) = 0$ gives $x = -2$ and $x = 4$, the critical points of f, as in Example 7. Next, we compute

$$f''(x) = 6x - 6 = 6(x - 1)$$

Since

$$f''(-2) = 6(-2 - 1) = -18 < 0$$

the Second Derivative Test implies that $f(-2) = 60$ is a relative maximum of f. Also,

$$f''(4) = 6(4 - 1) = 18 > 0$$

and the Second Derivative Test implies that $f(4) = -48$ is a relative minimum of f, which confirms the results obtained earlier.   ◦ ◦ ◦

Suppose a function f has the following properties:

1. $f''(x) > 0$ for all x in an interval (a, b).

2. There is a point c between a and b such that $f'(c) = 0$.

What special property can you ascribe to the point $(c, f(c))$? Answer the question if property 1 is replaced by the property that $f''(x) < 0$ for all x in (a, b).

Comparing the First and Second Derivative Tests

Notice that both the First Derivative Test and the Second Derivative Test are used to classify the critical points of f. What are the pros and cons of the two tests? Since the Second Derivative Test is applicable only when f'' exists, it is less versatile than the First Derivative Test. For example, it cannot be used to locate the relative minimum $f(0) = 0$ of the function $f(x) = x^{2/3}$.

Furthermore, the Second Derivative Test is inconclusive when f'' is equal to zero at a critical point of f, whereas the First Derivative Test always yields positive conclusions. The Second Derivative Test is also inconvenient to use when f'' is difficult to compute. On the plus side, if f'' is computed easily, then we use the Second Derivative Test since it involves just the evaluation of f'' at the critical point(s) of f. Also, the conclusions of the Second Derivative Test are important in theoretical work.

We close this section by summarizing the different roles played by the first derivative f' and the second derivative f'' of a function f in determining the properties of the graph of f. The first derivative f' tells us where f is increasing and where f is decreasing, whereas the second derivative f'' tells us where f is concave upward and where f is concave downward. These different properties of f are reflected by the signs of f' and f'' in the interval of interest. The following table shows the general characteristics of the function f for various possible combinations of the signs of f' and f'' in the interval (a, b).

Signs of f' and f''	Properties of the Graph of f	General Shape of the Graph of f
$f'(x) > 0$ $f''(x) > 0$	f increasing f concave upward	
$f'(x) > 0$ $f''(x) < 0$	f increasing f concave downward	
$f'(x) < 0$ $f''(x) > 0$	f decreasing f concave upward	
$f'(x) < 0$ $f''(x) < 0$	f decreasing f concave downward	

SELF-CHECK EXERCISES 12.2

1. Determine where the function $f(x) = 4x^3 - 3x^2 + 6$ is concave upward and where it is concave downward.

2. Using the Second Derivative Test, if applicable, find the relative extrema of the function $f(x) = 2x^3 - \frac{1}{2}x^2 - 12x - 10$.

3. A certain country's gross domestic product (GDP) (in millions of dollars) in year t is described by the function

$$G(t) = -2t^3 + 45t^2 + 20t + 6000 \qquad (0 \le t \le 11)$$

where $t = 0$ corresponds to the beginning of the year 1988. Find the inflection point of the function G and discuss its significance.

Solutions to Self-Check Exercises 12.2 can be found on page 818.

12.2 EXERCISES

In exercises 1–8, you are given the graph of a function f. Determine the intervals where f is concave upward and where it is concave downward. Also, find all inflection points of f, if any.

1.

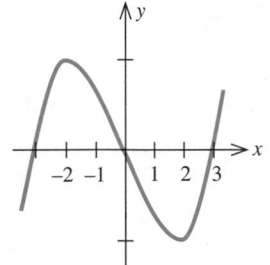

2.

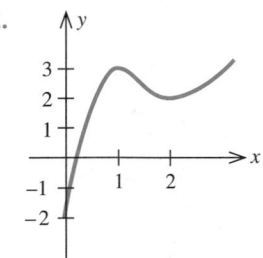

3.

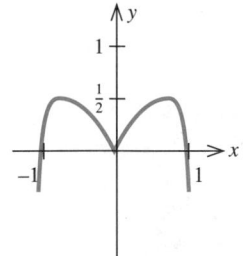

4.

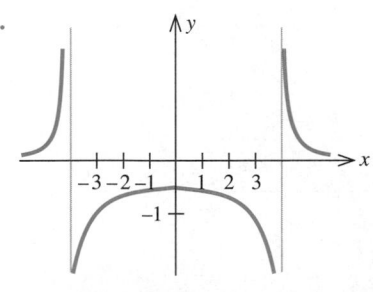

5.

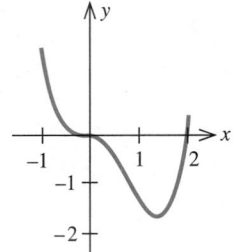

6.

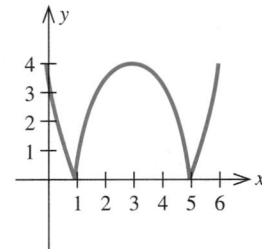

7.

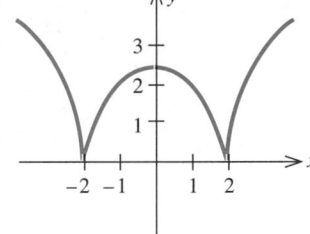

8.

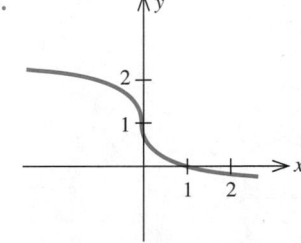

In exercises 9–12, determine which graph—a, b, or c—is the graph of the function f with the specified properties.

9. $f(2) = 1$, $f'(2) > 0$, and $f''(2) < 0$

a.

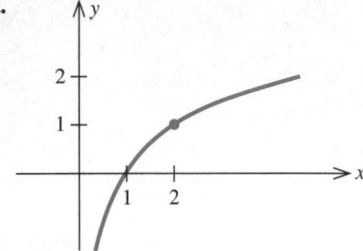

b.

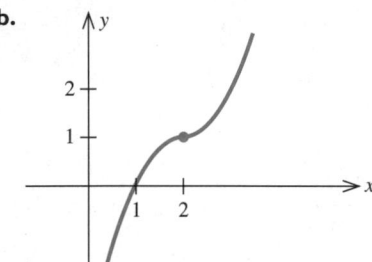

c.

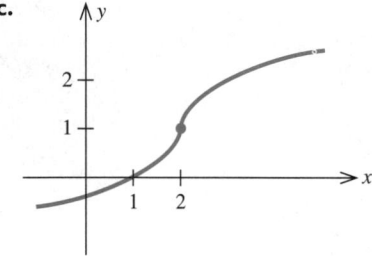

10. $f(1) = 2$, $f'(x) > 0$ on $(-\infty, 1) \cup (1, \infty)$, and $f''(1) = 0$

a.

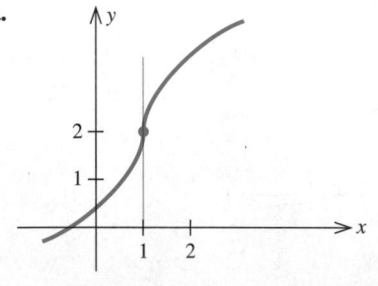

b.

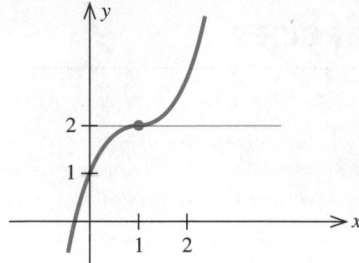

c.

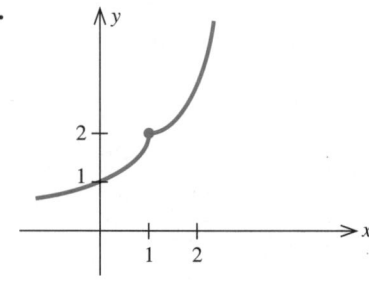

11. $f'(0)$ is undefined, f is decreasing on $(-\infty, 0)$, f is concave downward on $(0, 3)$, and f has an inflection point at $x = 3$.

a.

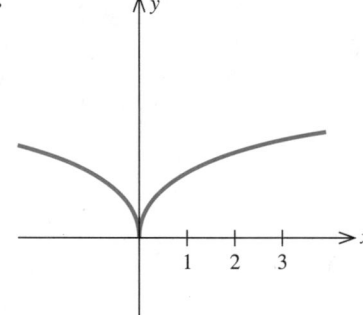

b.

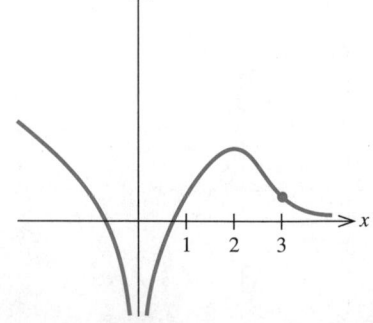

c.

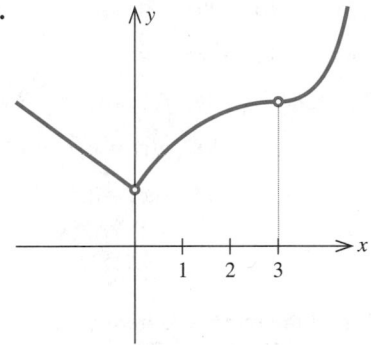

12. f is decreasing on $(-\infty, 2)$ and increasing on $(2, \infty)$, f is concave upward on $(1, \infty)$ and has inflection points at $x = 0$ and $x = 1$.

a.

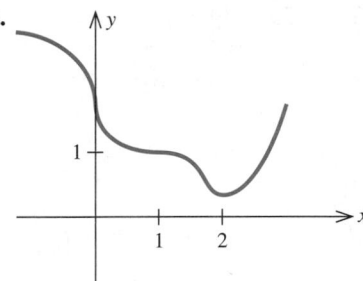

b.

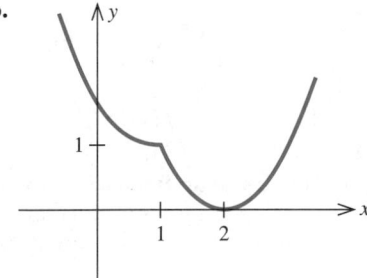

c.

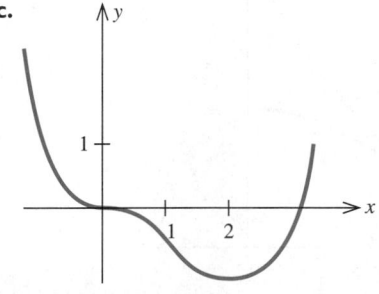

13. Effect of Advertising on Bank Deposits The following graphs were used by the CEO of the Madison Savings Bank to illustrate what effect a projected promotional campaign would have on its deposits over the next year. The functions D_1 and D_2 give the projected amount of money on deposit with the bank over the next 12 months with and without the proposed promotional campaign, respectively.
a. Determine the signs of $D_1'(t)$, $D_2'(t)$, $D_1''(t)$, and $D_2''(t)$ on the interval $(0, 12)$.
b. Explain the significance of your results in the context of the problem.

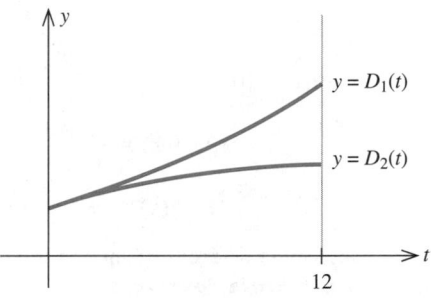

14. Assembly Time of a Worker In the following graph, $N(t)$ gives the number of transistor radios assembled by the average worker by the tth hour, where $t = 0$ corresponds to 8 A.M. and $0 \le t \le 4$. Explain the significance of the inflection point P shown on the graph.

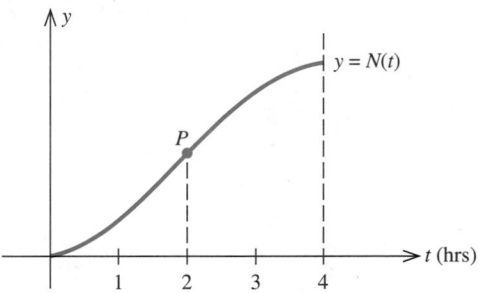

15. Water Pollution When organic waste is dumped into a pond, the oxidation process that takes place reduces the pond's oxygen content. However, given time, nature will restore the oxygen content to its natural level. In the accompanying graph on page 814, $P(t)$ gives the oxygen content (as a percentage of its normal level) t days after organic waste has been dumped into the pond. Explain the significance of the inflection point Q.

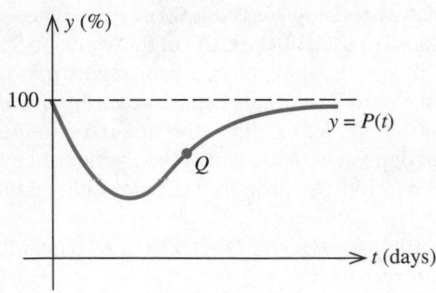

In exercises 16–21 show that the function is concave upward wherever it is defined.

16. $f(x) = 4x^2 - 12x + 7$

17. $g(x) = x^4 + \frac{1}{2}x^2 + 6x + 10$

18. $h(x) = \dfrac{1}{x^2}$ 19. $f(x) = \dfrac{1}{x^4}$

20. $h(x) = \sqrt{x^2 + 4}$ 21. $g(x) = -\sqrt{4 - x^2}$

In exercises 22–43, determine where the function is concave upward and where it is concave downward.

22. $g(x) = -x^2 + 3x + 4$ 23. $f(x) = 2x^2 - 3x + 4$

24. $g(x) = x^3 - x$ 25. $f(x) = x^3 - 1$

26. $f(x) = 3x^4 - 6x^3 + x - 8$

27. $f(x) = x^4 - 6x^3 + 2x + 8$

28. $f(x) = \sqrt[3]{x}$ 29. $f(x) = x^{4/7}$

30. $g(x) = \sqrt{x - 2}$ 31. $f(x) = \sqrt{4 - x}$

32. $g(x) = \dfrac{x}{x + 1}$ 33. $f(x) = \dfrac{1}{x - 2}$

34. $g(x) = \dfrac{x}{1 + x^2}$ 35. $f(x) = \dfrac{1}{2 + x^2}$

36. $f(x) = \dfrac{x + 1}{x - 1}$ 37. $h(t) = \dfrac{t^2}{t - 1}$

38. $h(r) = -\dfrac{1}{(r - 2)^2}$ 39. $g(x) = x + \dfrac{1}{x^2}$

40. $f(x) = (x - 2)^{2/3}$ 41. $g(t) = (2t - 4)^{1/3}$

42. $f(x) = \dfrac{x^2 - 2}{x^3}$ 43. $f(x) = \dfrac{x^2}{x^2 - 1}$

In exercises 44–55, find the inflection points, if any, of each function.

44. $g(x) = x^3 - 6x$ 45. $f(x) = x^3 - 2$

46. $g(x) = 2x^3 - 3x^2 + 18x - 8$

47. $f(x) = 6x^3 - 18x^2 + 12x - 15$

48. $f(x) = x^4 - 2x^3 + 6$ 49. $f(x) = 3x^4 - 4x^3 + 1$

50. $f(x) = \sqrt[5]{x}$ 51. $g(t) = \sqrt[3]{t}$

52. $f(x) = (x - 2)^{4/3}$ 53. $f(x) = (x - 1)^3 + 2$

54. $f(x) = 2 + \dfrac{3}{x}$ 55. $f(x) = \dfrac{2}{1 + x^2}$

In exercises 56–73, find the relative extrema, if any, of each function. Use the Second Derivative Test, if applicable.

56. $g(x) = 2x^2 + 3x + 7$ 57. $f(x) = -x^2 + 2x + 4$

58. $g(x) = x^3 - 6x$ 59. $f(x) = 2x^3 + 1$

60. $f(x) = 2x^3 + 3x^2 - 12x - 4$

61. $f(x) = \frac{1}{3}x^3 - 2x^2 - 5x - 10$

62. $f(t) = 2t + \dfrac{3}{t}$ 63. $g(t) = t + \dfrac{9}{t}$

64. $f(x) = \dfrac{2x}{x^2 + 1}$ 65. $f(x) = \dfrac{x}{1 - x}$

66. $g(x) = x^2 + \dfrac{2}{x}$ 67. $f(t) = t^2 - \dfrac{16}{t}$

68. $g(x) = \dfrac{1}{1 + x^2}$ 69. $g(s) = \dfrac{s}{1 + s^2}$

70. $f(x) = \dfrac{x^2}{x^2 + 1}$ 71. $f(x) = \dfrac{x^4}{x - 1}$

72. $f(x) = \dfrac{x^2 + 4}{x^2 - 1}$ 73. $g(x) = \dfrac{2 - x}{(x + 2)^3}$

74. **Effect of Budget Cuts on Drug-Related Crimes** The following graphs were used by a police commissioner to illustrate what effect a budget cut would have on

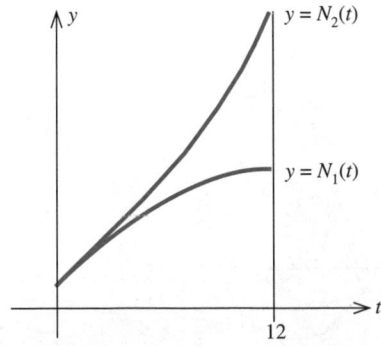

crime in the city. The number $N_1(t)$ gives the projected number of drug-related crimes in the next 12 months. The number $N_2(t)$ gives the projected number of drug-related crimes in the same time frame if next year's budget is cut.

a. Explain why $N_1'(t)$ and $N_2'(t)$ are both positive on the interval (0, 12).

b. What are the signs of $N_1''(t)$ and $N_2''(t)$ on the interval (0, 12)?

c. Interpret the results of (b).

75. **Demand for RNs** The following graph gives the total number of "help wanted" ads for RNs (registered nurses) in 22 cities over the last 12 months as a function of time t (t measured in months).

a. Explain why $N'(t)$ is positive on the interval (0, 12).

b. Determine the signs of $N''(t)$ on the interval (0, 6) and the interval (6, 12).

c. Interpret the results of (b).

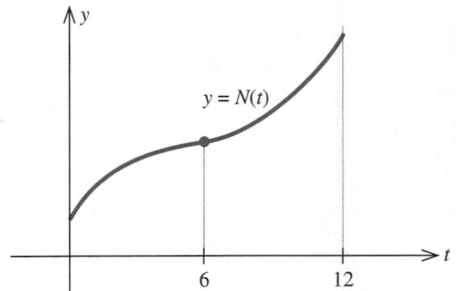

76. **Effect of Advertising on Hotel Revenue** The total annual revenue R of the Miramar Resorts Hotel is related to the amount of money x the hotel spends on advertising its services by the function

$$R(x) = -0.003x^3 + 1.35x^2 + 2x + 8000 \qquad (0 \le x \le 400)$$

where both R and x are measured in thousands of dollars. Find the inflection point of R and discuss its significance.

77. **Effect of Advertising on Sales** The total sales S (in thousands of dollars) of the Cannon Precision Instruments Corporation is related to the amount of money x that Cannon spends on advertising its products by the function

$$S(x) = -0.002x^3 + 0.6x^2 + x + 500 \qquad (0 \le x \le 200)$$

where x is measured in thousands of dollars. Find the inflection point of the function S and discuss its significance.

78. **Forecasting Profits** As a result of increasing energy costs, the growth rate of the profit of the four-year-old Venice Glassblowing Company has begun to decline. Venice's management, after consulting with energy experts, decides to implement certain energy-conservation measures aimed at cutting energy bills. The general manager reports that, according to his calculations, the growth rate of Venice's profit should be on the increase again within four years. If Venice's profit (in hundreds of dollars) x years from now is given by the function

$$P(x) = x^3 - 9x^2 + 40x + 50 \qquad (0 \le x \le 8)$$

determine whether the general manager's forecast will be accurate.

[*Hint:* Find the inflection point of the function P and study the concavity of P.]

79. **Worker Efficiency** An efficiency study conducted for the Elektra Electronics Company showed that the number of Space Commander walkie-talkies assembled by the average worker t hours after starting work at 8 A.M. is given by

$$N(t) = -t^3 + 6t^2 + 15t \qquad (0 \le t \le 4)$$

At what time during the morning shift is the average worker performing at peak efficiency?

80. **Cost of Producing Calculators** A subsidiary of Elektra Electronics manufactures programmable calculators. Management determines that the daily cost $C(x)$ of producing these calculators (in dollars) is

$$C(x) = 0.0001x^3 - 0.08x^2 + 40x + 5000$$

where x is the number of calculators produced. Find the inflection point of the function C and interpret your result.

81. **Flight of a Rocket** The altitude (in feet) of a rocket t seconds into flight is given by

$$s = f(t) = -t^3 + 54t^2 + 480t + 6$$

Find the point of inflection of the function f and interpret your result. What is the maximum velocity attained by the rocket?

USING TECHNOLOGY

FINDING THE INFLECTION POINTS OF A FUNCTION

A graphing utility can be used to find the inflection points of a function and hence the intervals where the graph of the function is concave upward and the intervals where it is concave downward. Some graphing utilities have an operation for finding inflection points directly. If your graphing utility has this capability, use it to work through the example and exercises in this section.

EXAMPLE 1   Let $f(x) = 2.5x^5 - 12.4x^3 + 4.2x^2 - 5.2x + 4$.

a. Use a graphing utility to plot the graph of f.

b. Find the inflection points of f.

c. Find the intervals where f is concave upward and where it is concave downward.

Solution

a. The graph of f in the viewing rectangle $[-3, 3] \times [-25, 60]$ is shown in Figure T1.

Figure T1
The graph of f in the viewing rectangle $[-3, 3] \times [-25, 60]$.

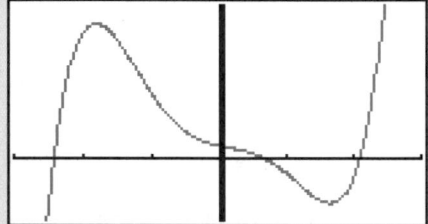

b. From Figure T1, we see that f has three inflection points—one occurring at the point where the x-coordinate is approximately -1, another at the point where $x \approx 0$, and the third at the point where $x \approx 1$. To find the first inflection point, we use the inflection operation, moving the cursor to the point on the graph of f where $x \approx -1$. We obtain the point $(-1.2728, 34.6395)$ (accurate to four decimal places). Next, setting the cursor near the point $x = 0$ yields the inflection point $(0.1139, 3.4440)$. Finally, with the cursor set at $x = 1$, we obtain the third inflection point $(1.1589, -10.4594)$.

c. From the results of (b), we see that f is concave upward on the intervals $(-1.2728, 0.1139)$ and $(1.1589, \infty)$ and concave downward on $(-\infty, -1.2728)$ and $(0.1139, 1.1589)$.   ⊙ ⊙ ⊙

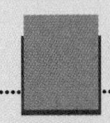

EXERCISES

In exercises 1–8, use a graphing utility to find (a) the intervals where f is concave upward and the intervals where f is concave downward, and (b) the inflection points of f. Express your answers accurate to four decimal places.

1. $f(x) = 1.8x^4 - 4.2x^3 + 2.1x + 2$

2. $f(x) = -2.1x^4 + 3.1x^3 + 2x^2 - x + 1.2$

3. $f(x) = 1.2x^5 - 2x^4 + 3.2x^3 - 4x + 2$

4. $f(x) = -2.1x^5 + 3.2x^3 - 2.2x^2 + 4.2x - 4$

5. $f(x) = x^3(x^2 + 1)^{-1/3}$

6. $f(x) = x^2(x^3 - 1)^3$

7. $f(x) = \dfrac{x^2 - 1}{x^3}$

8. $f(x) = \dfrac{x + 1}{\sqrt{x}}$

9. Growth of HMOs Based on data compiled by the Group Health Association of America, the number of people receiving their care in an HMO (health maintenance organization) from the beginning of 1984 through 1994 is approximated by the function

$$f(t) = 0.0514t^3 - 0.853t^2 + 6.8147t + 15.6524$$
$$(0 < t \le 11)$$

where $f(t)$ gives the number of people in millions and t is measured in years, with $t = 0$ corresponding to the beginning of 1984.
a. Use a graphing utility to plot the graph of f in the viewing rectangle $[0, 12] \times [0, 120]$.
b. Find the points of inflection of f.
c. At what time in the given time interval was the number of people receiving their care at an HMO increasing fastest?
Source: Group Health Association of America

10. Manufacturing Capacity Data obtained from the Federal Reserve show that the annual increase in manu-

facturing capacity between 1988 and 1994 is given by

$$f(t) = 0.0388889t^3 - 0.283333t^2 + 0.477778t$$
$$+ 2.04286 \qquad (0 \le t \le 6)$$

where $f(t)$ is a percentage and t is measured in years, with $t = 0$ corresponding to the beginning of 1988.
a. Use a graphing utility to plot the graph of f in the viewing rectangle $[0, 8] \times [0, 4]$.
b. Find the point of inflection and interpret your result.
Source: Federal Reserve

11. Time on the Market According to the Greater Boston Real Estate Board—Multiple Listing Service, the average number of days a single-family home remains for sale from listing to accepted offer is approximated by the function

$$f(t) = 0.0171911t^4 - 0.662121t^3 + 6.18083t^2$$
$$- 8.97086t + 53.3357 \qquad (0 \le t \le 10)$$

where t is measured in years, with $t = 0$ corresponding to the beginning of 1984.
a. Use a graphing utility to plot the graph of f in the viewing rectangle $[0, 12] \times [0, 120]$.
b. Find the points of inflection and interpret your result.
Source: Greater Boston Real Estate Board—Multiple Listing Service

12. Multimedia Sales According to the Electronic Industries Association, sales in the multimedia market (hardware and software) are expected to be

$$S(t) = -0.0094t^4 + 0.1204t^3 - 0.0868t^2$$
$$+ 0.0195t + 3.3325 \qquad (0 \le t \le 10)$$

where $S(t)$ is measured in billions of dollars and t is measured in years, with $t = 0$ corresponding to 1990.
a. Plot the graph of S in the viewing rectangle $[0, 12] \times [0, 25]$.
b. Find the inflection point of S and interpret your result.
Source: Electronic Industries Association

82. Air Pollution The level of ozone, an invisible gas that irritates and impairs breathing, present in the atmosphere on a certain May day in the city of Riverside was approximated by

$$A(t) = 1.0974t^3 - 0.0915t^4 \qquad (0 \le t \le 11)$$

where $A(t)$ is measured in pollutant standard index (PSI) and t is measured in hours, with $t = 0$ corresponding to 7 A.M. Use the Second Derivative Test to show that the function A has a relative maximum at approximately $t = 9$. Interpret your results.

83. Show that the quadratic function

$$f(x) = ax^2 + bx + c \qquad (a \ne 0)$$

is concave upward if $a > 0$ and concave downward if $a < 0$. Thus, by examining the sign of the coefficient of x^2, one can tell immediately whether the parabola opens upward or downward.

84. Suppose f has an inflection point at $(a, f(a))$. Must the function f' have a relative extremum at $x = a$? Explain your answer.

85. Show that the cubic function

$$f(x) = ax^3 + bx^2 + cx + d \qquad (a \ne 0)$$

has one and only one inflection point. Find the coordinates of this point.

86. Consider the functions $f(x) = x^3$, $g(x) = x^4$, and $h(x) = -x^4$.
  a. Show that $x = 0$ is a critical point of each of the functions f, g, and h.
  b. Show that the second derivative of each of the functions f, g, and h equals zero at $x = 0$.
  c. Show that f has neither a relative maximum nor a relative minimum at $x = 0$, that g has a relative minimum at $x = 0$, and that h has a relative maximum at $x = 0$.

SOLUTIONS TO SELF-CHECK EXERCISES 12.2

1. We first compute

$$f'(x) = 12x^2 - 6x$$
$$f''(x) = 24x - 6 = 6(4x - 1)$$

Observe that f'' is continuous everywhere and has a zero at $x = \frac{1}{4}$. The sign diagram of f'' is shown in the accompanying figure.

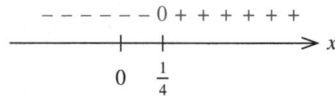

From the sign diagram, we see that f is concave upward on $(\frac{1}{4}, \infty)$ and concave downward on $(-\infty, \frac{1}{4})$.

2. First, we find the critical points of f by solving the equation

$$f'(x) = 6x^2 - x - 12 = 0$$
$$(3x + 4)(2x - 3) = 0$$

giving $x = -\frac{4}{3}$ and $x = \frac{3}{2}$. Next, we compute

$$f''(x) = 12x - 1$$

Since

$$f''\left(-\frac{4}{3}\right) = 12\left(-\frac{4}{3}\right) - 1 = -17 < 0$$

the Second Derivative Test implies that $f(-\frac{4}{3}) = \frac{10}{27}$ is a relative maximum of f. Also

$$f''\left(\frac{3}{2}\right) = 12\left(\frac{3}{2}\right) - 1 = 17 > 0$$

and we see that $f(\frac{3}{2}) = -\frac{179}{8}$ is a relative minimum.

3. We compute the second derivative of G. Thus,

$$G'(t) = -6t^2 + 90t + 20$$
$$G''(t) = -12t + 90$$

Now G'' is continuous everywhere and $G''(t) = 0$, where $t = \frac{15}{2}$, giving $t = \frac{15}{2}$ as the only candidate for an inflection point of G. Since $G''(t) > 0$ for $t < \frac{15}{2}$ and $G''(t) < 0$ for $t > \frac{15}{2}$, we see that the point $(\frac{15}{2}, \frac{15675}{2})$ is an inflection point of G. The results of our computations tell us that the country's GDP was increasing most rapidly at the beginning of July 1995.

12.3 CURVE SKETCHING

A Real-Life Example

As we have see on numerous occasions, the graph of a function is a useful aid for visualizing the function's properties. From a practical point of view, the graph of a function also gives, at one glance, a complete summary of all the information captured by the function.

Consider, for example, the graph of the function giving the Dow Jones Industrial Average (DJIA) on Black Monday, October 19, 1987 (Figure 12.44). Here $t = 0$ corresponds to 8:30 A.M., when the market was open for business, and $t = 7.5$ corresponds to 4 P.M., the closing time. The following information may be gleaned from studying the graph.

The graph is *decreasing* rapidly from $t = 0$ to $t = 1$, reflecting the sharp drop in the index in the first hour of trading. The point (1, 2047) is a *relative minimum* point of the function, and this turning point coincides with the start of an aborted recovery. The short-lived rally, represented by the portion of the graph that is *increasing* on the interval (1, 2), quickly fizzled out at $t = 2$ (10:30 A.M.). The *relative maximum* point (2, 2150) marks the highest point of the recovery. The function is decreasing in the rest of the interval. The point (4, 2006) is an *inflection point* of the function; it shows that there was a temporary respite at $t = 4$ (12:30 P.M.). However, selling pressure continued unabated, and the Dow Jones Industrial Average continued to fall until the closing bell. Finally, the graph also shows that the index opened at the high

Figure 12.44

The Dow Jones Industrial Average on Black Monday.

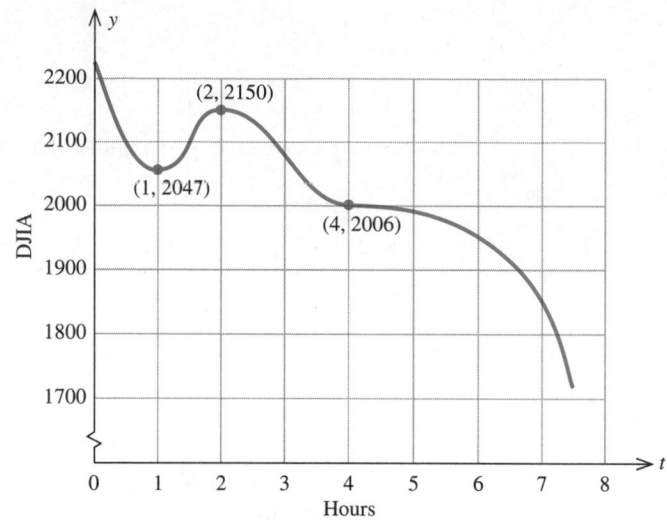

Source: *The Wall Street Journal*

of the day [$f(0) = 2247$ is the *absolute maximum* of the function] and closed at the low of the day [$f(15/2) = 1739$ is the *absolute minimum* of the function], a drop of 508 points!*

Before we turn our attention to the actual task of sketching the graph of a function, let us look at some properties of graphs that will be helpful in this connection.

Vertical Asymptotes

Before going on, you might want to review the material on one-sided limits and the limit at infinity of a function (Sections 10.4 and 10.5).

Consider the graph of the function

$$f(x) = \frac{x + 1}{x - 1}$$

Figure 12.45

The graph of f has a vertical asymptote at x = 1.

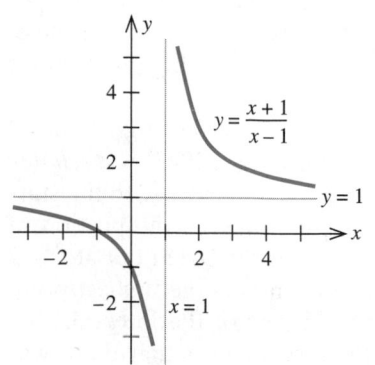

shown in Figure 12.45. Observe that $f(x)$ increases without bound (tends to infinity) as x approaches $x = 1$ from the right; that is,

$$\lim_{x \to 1^+} \frac{x + 1}{x - 1} = \infty$$

You can verify this by taking a sequence of values of x approaching $x = 1$ from the right and looking at the corresponding values of $f(x)$.

Here is another way of looking at the situation: Observe that if x is a number that is a little larger than 1, then both $(x + 1)$ and $(x - 1)$ are positive, so that $(x + 1)/(x - 1)$ is also positive. As x approaches $x = 1$, the numerator

○ ○ ○

* Absolute maxima and absolute minima of functions are covered in Section 12.4.

$(x + 1)$ approaches the number 2, but the denominator $(x - 1)$ approaches zero, so the quotient $(x + 1)/(x - 1)$ approaches infinity, as observed earlier. The line $x = 1$ is called a *vertical asymptote* of the graph of f.

For the function $f(x) = (x + 1)/(x - 1)$, you can show that

$$\lim_{x \to 1^-} \frac{x + 1}{x - 1} = -\infty$$

and this tells us how $f(x)$ approaches the asymptote $x = 1$ from the left. More generally, we have the following definition:

VERTICAL ASYMPTOTES	The line $x = a$ is a **vertical asymptote** of the graph of a function f if either $$\lim_{x \to a^+} f(x) = \infty \quad \text{or} \quad -\infty$$ or $$\lim_{x \to a^-} f(x) = \infty \quad \text{or} \quad -\infty$$

REMARK   Although a vertical asymptote of a graph is not part of the graph, it serves as a useful aid for sketching the graph.   ο ο ο

For rational functions

$$f(x) = \frac{P(x)}{Q(x)}$$

there is a simple criterion for determining whether the graph of f has any vertical asymptotes.

FINDING VERTICAL ASYMPTOTES OF RATIONAL FUNCTIONS	Suppose that f is a rational function $$f(x) = \frac{P(x)}{Q(x)}$$ where P and Q are polynomial functions. Then the line $x = a$ is a vertical asymptote of the graph of f if $Q(a) = 0$ but $P(a) \neq 0$.

For the function

$$f(x) = \frac{x + 1}{x - 1}$$

considered earlier, $P(x) = x + 1$ and $Q(x) = x - 1$. Observe that $Q(1) = 0$ but $P(1) = 2 \neq 0$, so $x = 1$ is a vertical asymptote of the graph of f.

EXAMPLE I Find the vertical asymptotes of the graph of the function

$$f(x) = \frac{x^2}{4 - x^2}$$

Solution The function f is a rational function with $P(x) = x^2$ and $Q(x) = 4 - x^2$. The zeros of Q are found by solving

$$4 - x^2 = 0$$

that is,

$$(2 - x)(2 + x) = 0$$

giving $x = -2$ and $x = 2$. These are candidates for the vertical asymptotes of the graph of f. Examining $x = -2$, we compute $P(-2) = (-2)^2 = 4 \neq 0$, and we see that $x = -2$ is indeed a vertical asymptote of the graph of f. Similarly, we find $P(2) = 2^2 = 4 \neq 0$, and so $x = 2$ is also a vertical asymptote of the graph of f. The graph of f sketched in Figure 12.46 confirms these results.

Figure 12.46
$x = -2$ and $x = 2$ are vertical asymptotes of the graph of f.

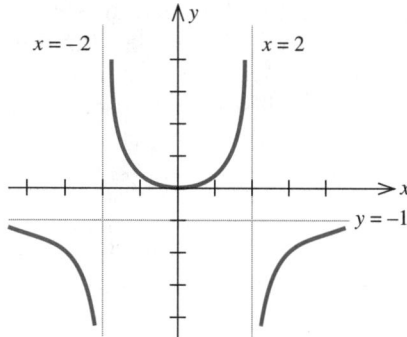

o o o

 Recall that in order for the line $x = a$ to be a vertical asymptote of the graph of a rational function f, *only* the denominator of $f(x)$ must be equal to zero at $x = 0$. If *both* $P(a)$ and $Q(a)$ are equal to zero, then $x = a$ need *not* be a vertical asymptote. For example, look at the function $f(x) = 4(x^2 - 4)/(x - 2)$, whose graph appears in Figure 10.23a, page 618.

Horizontal Asymptotes

Let us return to the function f defined by

$$f(x) = \frac{x + 1}{x - 1}$$

(Figure 12.47). Observe that $f(x)$ approaches the horizontal line $y = 1$ as x approaches infinity, and, in this case, $f(x)$ approaches $y = 1$ as x approaches

Figure 12.47
The graph of f has a horizontal asymptote at $y = 1$.

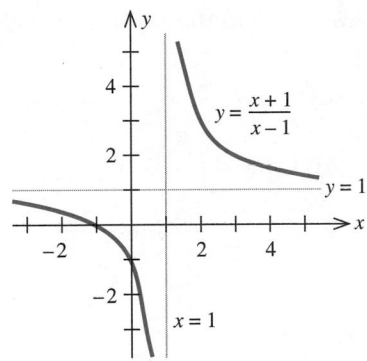

minus infinity as well. The line $y = 1$ is called a *horizontal asymptote* of the graph of f. More generally, we have the following definition.

HORIZONTAL ASYMPTOTES

The line $y = b$ is a **horizontal asymptote** of the graph of a function f if either

$$\lim_{x \to \infty} f(x) = b \quad \text{or} \quad \lim_{x \to -\infty} f(x) = b$$

For the function

$$f(x) = \frac{x + 1}{x - 1}$$

we see that

$$\lim_{x \to \infty} \frac{x + 1}{x - 1} = \lim_{x \to \infty} \frac{1 + \dfrac{1}{x}}{1 - \dfrac{1}{x}} \qquad \text{(Dividing numerator and denominator by } x\text{)}$$

$$= 1$$

Also,

$$\lim_{x \to -\infty} \frac{x + 1}{x - 1} = \lim_{x \to -\infty} \frac{1 + \dfrac{1}{x}}{1 - \dfrac{1}{x}}$$

$$= 1$$

In either case, we conclude that $y = 1$ is a horizontal asymptote of the graph of f, as observed earlier.

EXAMPLE 2   Find the horizontal asymptotes of the graph of the function

$$f(x) = \frac{x^2}{4 - x^2}$$

Solution   We compute

$$\lim_{x \to \infty} \frac{x^2}{4 - x^2} = \lim_{x \to \infty} \frac{1}{\dfrac{4}{x^2} - 1} \qquad \text{(Dividing numerator and denominator by } x^2)$$

$$= -1$$

and so $y = -1$ is a horizontal asymptote, as before. The graph of f sketched in Figure 12.48 confirms this result.

Figure 12.48
The graph of f has a horizontal asymptote at y = −1.

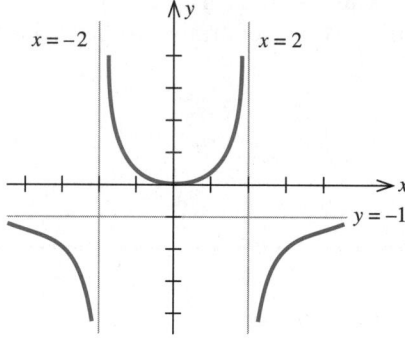

We next state an important property of polynomial functions.

A polynomial function has no vertical or horizontal asymptotes.

To see this, note that a polynomial function $P(x)$ can be written as a rational function with denominator equal to 1. Thus,

$$P(x) = \frac{P(x)}{1}$$

Since the denominator is never equal to zero, P has no vertical asymptotes. Next, if P is a polynomial of degree greater than or equal to 1, then

$$\lim_{x \to \infty} P(x) \quad \text{and} \quad \lim_{x \to -\infty} P(x)$$

are either infinity or minus infinity; that is, they do not exist. Therefore, P has no horizontal asymptotes.

In the last two sections we saw how the first and second derivatives of a function are used to reveal various properties of the graph of a function f. We now show how this information can be used to help us sketch the graph of f. We begin by giving a general procedure for curve sketching.

A GUIDE TO CURVE SKETCHING	

A GUIDE TO CURVE SKETCHING

1. Determine the domain of f.
2. Find the x- and y-intercepts of f.*
3. Determine the behavior of f for large absolute values of x.
4. Find all horizontal and vertical asymptotes of f.
5. Determine the intervals where f is increasing and where f is decreasing.
6. Find the relative extrema of f.
7. Determine the concavity of f.
8. Find the inflection points of f.
9. Plot a few additional points to help further identify the shape of the graph of f, and sketch the graph.

We now illustrate the techniques of curve sketching with examples.

Two Step-by-Step Examples

EXAMPLE 3   Sketch the graph of the function

$$y = f(x) = x^3 - 6x^2 + 9x + 2$$

Solution   Obtain the following information on the graph of f.

1. The domain of f is the interval $(-\infty, \infty)$.
2. By setting $x = 0$, we find that the y-intercept is 2. The x-intercept is found by setting $y = 0$, which, in this case, leads to a cubic equation. Since the solution is not readily found, we will not use this information.
3. Since

$$\lim_{x \to -\infty} f(x) = \lim_{x \to -\infty} (x^3 - 6x^2 + 9x + 2) = -\infty$$

$$\lim_{x \to \infty} f(x) = \lim_{x \to \infty} (x^3 - 6x^2 + 9x + 2) = \infty$$

we see that f decreases without bound as x decreases without bound, and f increases without bound as x increases without bound.

○ ○ ○

* The equation $f(x) = 0$ may be difficult to solve, in which case one may either decide against finding the x-intercepts or use technology, if available, for assistance.

Figure 12.49
Sign diagram for f'.

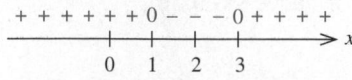

Figure 12.50
Sign diagram for f".

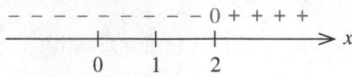

Figure 12.51
We first plot the intercept, the relative extrema, and the inflection point.

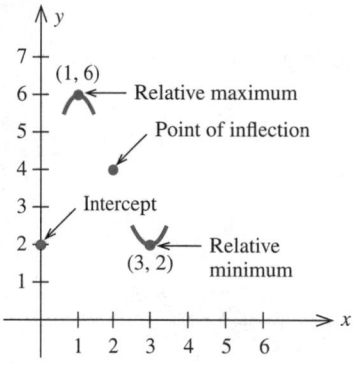

Figure 12.52
The graph of
$y = x^3 - 6x^2 + 9x + 2.$

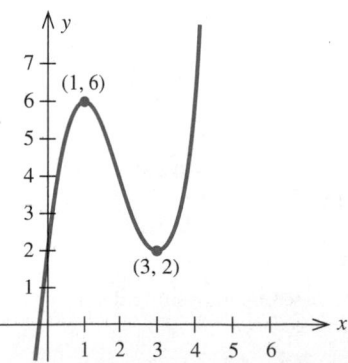

4. Since f is a polynomial function, there are no asymptotes.

5.
$$f'(x) = 3x^2 - 12x + 9 = 3(x^2 - 4x + 3)$$
$$= 3(x - 3)(x - 1)$$

Setting $f'(x) = 0$ gives $x = 1$ or $x = 3$. The sign diagram for f' (Figure 12.49) shows that f is increasing on the intervals $(-\infty, 1)$ and $(3, \infty)$ and decreasing on the interval $(1, 3)$.

6. From the results of step 5, we see that $x = 1$ and $x = 3$ are critical points of f. Furthermore, f' changes sign from positive to negative as we move across $x = 1$, so a relative maximum of f occurs at $x = 1$. Similarly, we see that a relative minimum of f occurs at $x = 3$. Now,

$$f(1) = 1 - 6 + 9 + 2 = 6$$

and
$$f(3) = 3^3 - 6(3)^2 + 9(3) + 2 = 2$$

so $f(1) = 6$ is a relative maximum of f and $f(3) = 2$ is a relative minimum of f.

7.
$$f''(x) = 6x - 12 = 6(x - 2)$$

which is equal to zero when $x = 2$. The sign diagram of f'' (Figure 12.50) shows that f is concave downward on the interval $(-\infty, 2)$ and concave upward on the interval $(2, \infty)$.

8. From the results of step 7, we see that f'' changes sign as we move across the point $x = 2$. Next,

$$f(2) = 2^3 - 6(2)^2 + 9(2) + 2 = 4$$

and so the required inflection point of f is $(2, 4)$.

9. Summarizing, we have

Domain	$(-\infty, \infty)$
Intercept	$(0, 2)$
$\lim\limits_{x \to -\infty} f(x); \lim\limits_{x \to \infty} f(x)$	$-\infty; \infty$
Asymptotes	None
Intervals where f is ↗ or ↘	↗ on $(-\infty, 1) \cup (3, \infty)$; ↘ on $(1, 3)$
Relative extrema	Rel. max. at $(1, 6)$; rel. min. at $(3, 2)$
Concavity	Downward on $(-\infty, 2)$; upward on $(2, \infty)$
Point of inflection	$(2, 4)$

It is a good idea to start graphing by plotting the intercept, relative extrema, and inflection point (Figure 12.51). Then, using the rest of the information, we complete the graph of f, as sketched in Figure 12.52.   ◦ ◦ ◦

The average price of gasoline at the pump over a period of 3 months during which there was a temporary shortage of oil is described by the function *f* defined on the interval [0, 3]. During the first month, the price was increasing at an increasing rate. Starting with the second month, the good news was that the rate of increase was slowing down, although the price of gas was still increasing. This pattern continued until the end of the second month. The price of gas peaked at the end of *t* = 2 and began to fall at an increasing rate until *t* = 3.

a. Describe the signs of $f'(t)$ and $f''(t)$ over each of the intervals (0, 1), (1, 2), and (2, 3).

b. Make a sketch showing a plausible graph of *f* over [0, 3].

EXAMPLE 4 Sketch the graph of the function

$$y = f(x) = \frac{x + 1}{x - 1}$$

Solution Obtain the following information:

1. *f* is undefined when *x* = 1, so the domain of *f* is the set of all real numbers other than *x* = 1.

2. Setting *y* = 0 gives −1, the *x*-intercept of *f*. Next, setting *x* = 0 gives −1 as the *y*-intercept of *f*.

3. Earlier we found that

$$\lim_{x \to \infty} \frac{x + 1}{x - 1} = 1 \quad \text{and} \quad \lim_{x \to -\infty} \frac{x + 1}{x - 1} = 1$$

(see page 823). Consequently, we see that $f(x)$ approaches the line *y* = 1 as |*x*| becomes arbitrarily large. For *x* > 1, $f(x) > 1$ and $f(x)$ approaches the line *y* = 1 from above. For *x* < 1, $f(x) < 1$, so $f(x)$ approaches the line *y* = 1 from below.

4. The straight line *x* = 1 is a vertical asymptote of the graph of *f*. Also, from the results of step 3, we conclude that *y* = 1 is a horizontal asymptote of the graph of *f*.

5.
$$f'(x) = \frac{(x - 1)(1) - (x + 1)(1)}{(x - 1)^2} = -\frac{2}{(x - 1)^2}$$

and is discontinuous at *x* = 1. The sign diagram of f' shows that $f'(x) < 0$ whenever it is defined. Thus, *f* is decreasing on the intervals $(-\infty, 1)$ and $(1, \infty)$ (Figure 12.53).

6. From the results of step 5, we see that there are no critical points of *f* since $f'(x)$ is never equal to zero for any value of *x* in the domain of *f*.

7.
$$f''(x) = \frac{d}{dx}[-2(x - 1)^{-2}] = 4(x - 1)^{-3} = \frac{4}{(x - 1)^3}$$

The sign diagram of f'' (Figure 12.54) shows immediately that *f* is concave downward on the interval $(-\infty, 1)$ and concave upward on the interval $(1, \infty)$.

Figure 12.53
The sign diagram for f'.

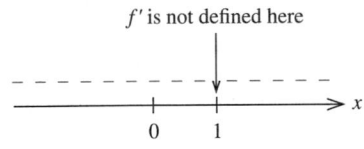

f' is not defined here

Figure 12.54
The sign diagram for f''.

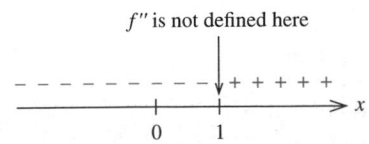

f'' is not defined here

8. From the results of step 7, we see that there are no candidates for inflection points of f since $f''(x)$ is never equal to zero for any value of x in the domain of f. Hence, f has no inflection points.

9. Summarizing, we have

Domain	$(-\infty, 1) \cup (1, \infty)$
Intercepts	$(-1, 0); (0, -1)$
$\lim\limits_{x \to -\infty} f(x); \lim\limits_{x \to \infty} f(x)$	$1; 1$
Asymptotes	$x = 1$ is a vertical asymptote; $y = 1$ is a horizontal asymptote
Intervals where f is ↗ or ↘	↘ on $(-\infty, 1) \cup (1, \infty)$
Relative extrema	None
Concavity	Downward on $(-\infty, 1)$; upward on $(1, \infty)$
Points of inflection	None

The graph of f is sketched in Figure 12.55.

Figure 12.55
The graph of f has a horizontal asymptote at $y = 1$ and a vertical asymptote at $x = 1$.

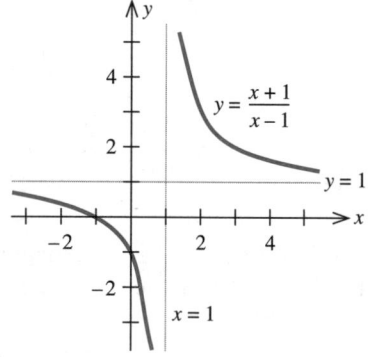

SELF–CHECK EXERCISES 12.3

1. Find the horizontal and vertical asymptotes of the graph of the function

$$f(x) = \frac{2x^2}{x^2 - 1}$$

2. Sketch the graph of the function

$$f(x) = \tfrac{2}{3}x^3 - 2x^2 - 6x + 4$$

Solutions to Self-Check Exercises 12.3 can be found on page 837.

12.3 EXERCISES

In exercises 1–10, find the horizontal and vertical asymptotes of the graph.

1.

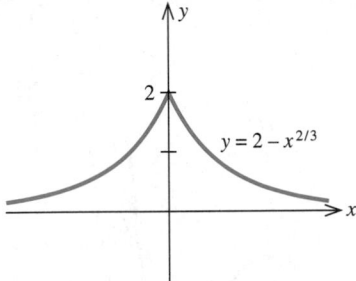

$y = 2 - x^{2/3}$

2.

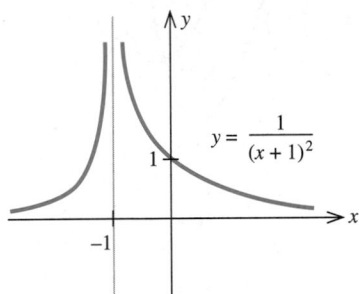

$y = \dfrac{1}{(x+1)^2}$

3.

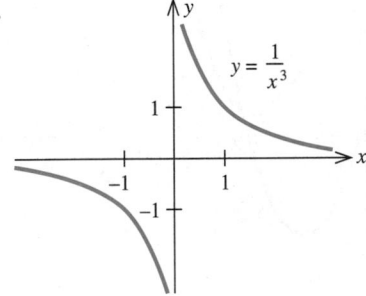

$y = \dfrac{1}{x^3}$

4.

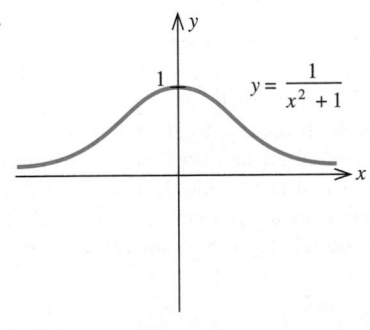

$y = \dfrac{1}{x^2 + 1}$

5.

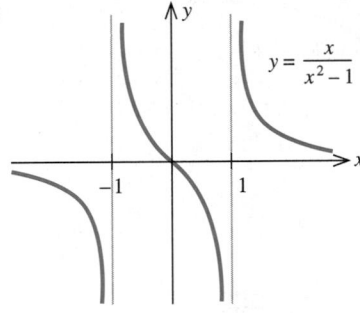

$y = \dfrac{x}{x^2 - 1}$

6.

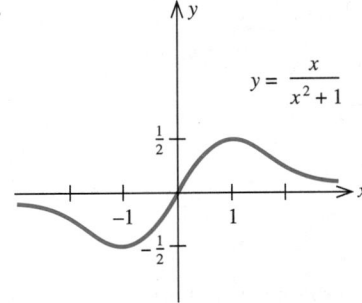

$y = \dfrac{x}{x^2 + 1}$

7.

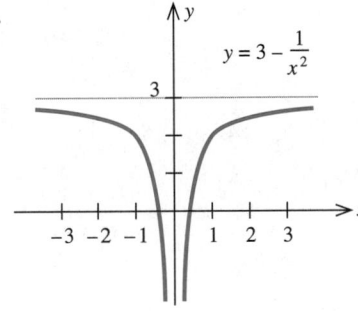

$y = 3 - \dfrac{1}{x^2}$

8.

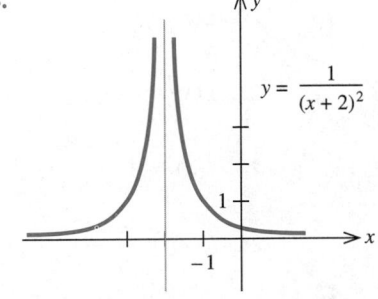

$y = \dfrac{1}{(x+2)^2}$

9.

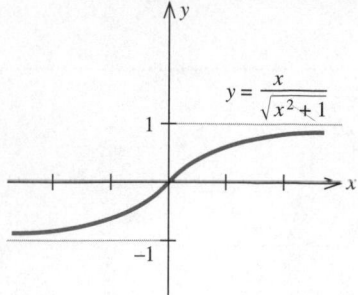

10.

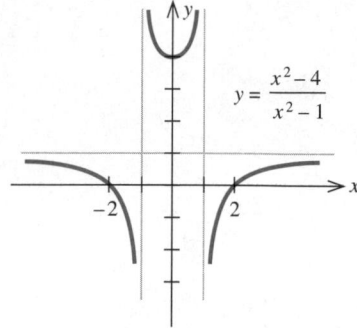

In exercises 11–28, find the horizontal and vertical asymptotes of the graph of each function. (You need not sketch the graph.)

11. $f(x) = \dfrac{1}{x}$

12. $f(x) = \dfrac{1}{x+2}$

13. $f(x) = -\dfrac{2}{x^2}$

14. $g(x) = \dfrac{1}{1+2x^2}$

15. $f(x) = \dfrac{x-1}{x+1}$

16. $g(t) = \dfrac{t+1}{2t-1}$

17. $h(x) = x^3 - 3x^2 + x + 1$

18. $g(x) = 2x^3 + x^2 + 1$

19. $f(t) = \dfrac{t^2}{t^2-9}$

20. $g(x) = \dfrac{x^3}{x^2-4}$

21. $f(x) = \dfrac{3x}{x^2-x-6}$

22. $g(x) = \dfrac{2x}{x^2+x-2}$

23. $g(t) = 2 + \dfrac{5}{(t-2)^2}$

24. $f(x) = 1 + \dfrac{2}{x-3}$

25. $f(x) = \dfrac{x^2-2}{x^2-4}$

26. $h(x) = \dfrac{2-x^2}{x^2+x}$

27. $g(x) = \dfrac{x^3-x}{x(x+1)}$

28. $f(x) = \dfrac{x^4-x^2}{x(x-1)(x+2)}$

In exercises 29 and 30, you are given the graphs of two functions f and g. One function is the derivative function of the other. Identify each of them.

29.

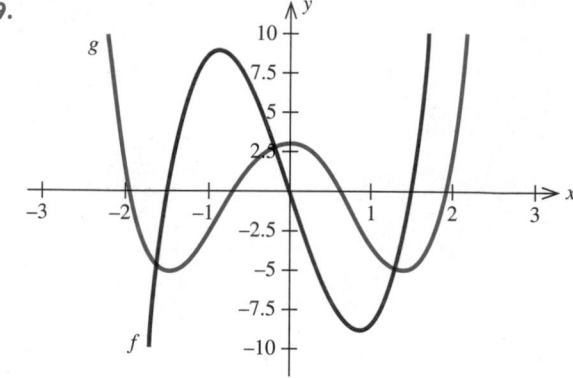

30.

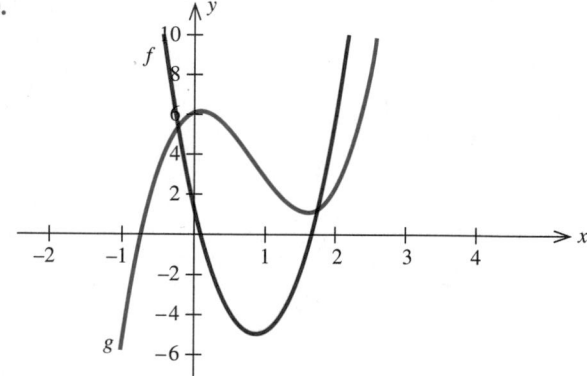

31. Terminal Velocity   A skydiver leaps from the gondola of a hot-air balloon. As she free-falls, air resistance, which is proportional to her velocity, builds up to a point where it balances the force due to gravity. The resulting motion may be described in terms of her velocity as follows: Starting at rest (zero velocity), her velocity increases and approaches a constant velocity, called the *terminal velocity*. Sketch a graph of her velocity v versus time t.

In exercises 32–35, use the information summarized in the table to sketch the graph of f.

32. $f(x) = x^3 - 3x^2 + 1$

Domain	$(-\infty, \infty)$
Intercept	y-intercept: 1
Asymptotes	None
Intervals where f is ↗ and ↘	↗ on $(-\infty, 0) \cup (2, \infty)$; ↘ on $(0, 2)$
Relative extrema	Rel. max. at $(0, 1)$; rel. min. at $(2, -3)$
Concavity	Downward on $(-\infty, 1)$; upward on $(1, \infty)$
Point of inflection	$(1, -1)$

33. $f(x) = \frac{1}{9}(x^4 - 4x^3)$

Domain	$(-\infty, \infty)$
Intercepts	x-intercepts: 0, 4; y-intercept: 0
Asymptotes	None
Intervals where f is ↗ and ↘	↗ on $(3, \infty)$; ↘ on $(-\infty, 3)$
Relative extrema	Rel. min. at $(3, -3)$
Concavity	Downward on $(0, 2)$; upward on $(-\infty, 0) \cup (2, \infty)$
Points of inflection	$(0, 0)$ and $(2, -16/9)$

34. $f(x) = \frac{4x - 4}{x^2}$

Domain	$(-\infty, 0) \cup (0, \infty)$
Intercept	x-intercept: 1
Asymptotes	x-axis and y-axis
Intervals where f is ↗ and ↘	↗ on $(0, 2)$; ↘ on $(-\infty, 0) \cup (2, \infty)$
Relative extrema	Rel. max. at $(2, 1)$
Concavity	Downward on $(-\infty, 0) \cup (0, 3)$; upward on $(3, \infty)$
Points of inflection	$(3, 8/9)$

35. $f(x) = x - 3x^{1/3}$

Domain	$(-\infty, \infty)$
Intercepts	x-intercepts: $\pm 3\sqrt{3}$
Asymptotes	None
Intervals where f is ↗ and ↘	↗ on $(-\infty, -1) \cup (1, \infty)$; ↘ on $(-1, 1)$
Relative extrema	Rel. max. at $(-1, 2)$; rel. min. at $(1, -2)$
Concavity	Downward on $(-\infty, 0)$; upward on $(0, \infty)$
Points of inflection	$(0, 0)$

In exercises 36–59, sketch the graph of the function using the curve-sketching guide of this section.

36. $f(x) = x^2 - 2x + 3$

37. $g(x) = 4 - 3x - 2x^3$

38. $f(x) = 2x^3 + 1$

39. $h(x) = x^3 - 3x + 1$

40. $f(t) = 2t^3 - 15t^2 + 36t - 20$

41. $f(x) = -2x^3 + 3x^2 + 12x + 2$

42. $f(t) = 3t^4 + 4t^3$

43. $h(x) = \frac{3}{2}x^4 - 2x^3 - 6x^2 + 8$

44. $f(x) = \sqrt{x^2 + 5}$

45. $f(t) = \sqrt{t^2 - 4}$

46. $f(x) = \sqrt[3]{x^2}$

47. $g(x) = \frac{1}{2}x - \sqrt{x}$

48. $f(x) = \frac{1}{x + 1}$

49. $g(x) = \frac{2}{x - 1}$

50. $g(x) = \frac{x}{x - 1}$

51. $h(x) = \frac{x + 2}{x - 2}$

52. $g(x) = \frac{x}{x^2 - 4}$

53. $f(t) = \frac{t^2}{1 + t^2}$

54. $f(x) = \frac{x^2 - 9}{x^2 - 4}$

55. $g(t) = -\frac{t^2 - 2}{t - 1}$

56. $h(x) = \frac{1}{x^2 - x - 2}$

57. $g(t) = \frac{t + 1}{t^2 - 2t - 1}$

58. $g(x) = (x + 2)^{3/2} + 1$

59. $h(x) = (x - 1)^{2/3} + 1$

60. Cost of Removing Toxic Pollutants A city's main well was recently found to be contaminated with trichloroethylene (a cancer-causing chemical) as a result of an abandoned chemical dump leaching chemicals into the water. A proposal submitted to the city's selectmen indicated that the cost, measured in millions of dollars, of removing x% of the toxic pollutants is given by

$$C(x) = \frac{0.5x}{100 - x}$$

a. Find the vertical asymptote of $C(x)$.
b. Is it possible to remove 100% of the toxic pollutant from the water?

61. Average Cost of Producing Video Discs The average cost per disc (in dollars) incurred by the Herald Record Company in pressing x video discs is given by the average cost function

$$\overline{C}(x) = 2.2 + \frac{2500}{x}$$

a. Find the horizontal asymptote of $\overline{C}(x)$.
b. What is the limiting value of the average cost?

USING TECHNOLOGY

ANALYZING THE PROPERTIES OF A FUNCTION

One of the main purposes of studying Section 12.3 is to see how the many concepts of calculus come together to paint a picture of a function. The techniques of graphing also play a very practical role. For example, using the techniques of graphing developed in Section 12.3, you can tell whether the graph of a function generated by a graphing utility is reasonably complete. Furthermore, these techniques can often reveal details that are missing from a graph.

EXAMPLE 1 Consider the function $f(x) = 2x^3 - 3.5x^2 + x - 10$. A plot of the graph of f in the standard viewing rectangle is shown in Figure T1. Since the domain of f is the interval $(-\infty, \infty)$, we see that Figure T1 does not reveal the part of the graph to the left of the y-axis. This suggests that we enlarge the viewing rectangle accordingly. Figure T2 shows the graph of f in the viewing rectangle $[-10, 10] \times [-20, 10]$.

Figure T1
The graph of f in the standard viewing rectangle.

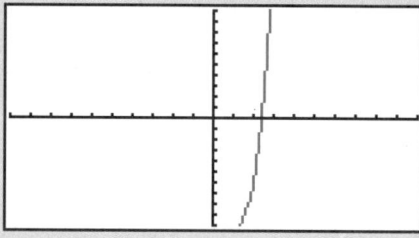

Figure T2
The graph of f in the viewing rectangle $[-10, 10] \times [-20, 10]$.

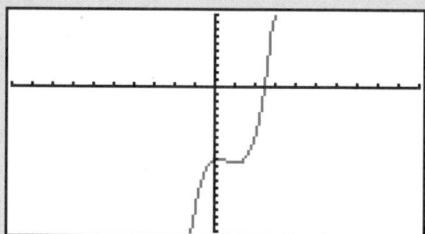

The behavior of f for large values of f [$\lim_{x \to -\infty} f(x) = -\infty$ and $\lim_{x \to \infty} f(x) = \infty$] suggests that this viewing rectangle has captured a sufficiently complete picture of f.

Next, an analysis of the first derivative of f,

$$f'(x) = 6x^2 - 7x + 1 = (6x - 1)(x - 1)$$

reveals that f has critical values at $x = 1/6$ and $x = 1$. In fact, a sign diagram of f' shows that f has a relative maximum at $x = 1/6$ and a relative minimum at $x = 1$, details that are not revealed in the graph of f shown in Figure T2. In order to examine this portion of the graph of f, we use, say, the viewing rectangle $[-1, 2] \times [-11, -9]$. The resulting graph of f is shown in Figure T3, which certainly reveals the hitherto missing details! Thus, through an interaction of calculus and a graphing utility, we are able to obtain a good picture of the properties of f.

Figure T3
The graph of f in the viewing rectangle $[-1, 2] \times [-11, -9]$.

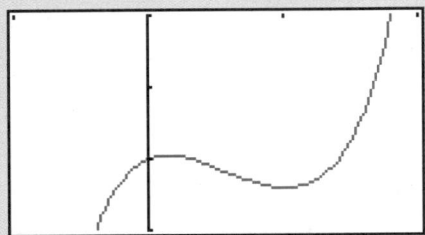

Finding x-Intercepts

As noted in Section 12.3, it is not always easy to find the x-intercepts of the graph of a function. But this information is very important in applications. By using the function for solving polynomial equations or the function for finding the roots of an equation, we can solve the equation $f(x) = 0$ quite easily and hence yield the x-intercepts of the graph of a function.

EXAMPLE 2   Let $f(x) = x^3 - 3x^2 + x + 1.5$.

a. Use the function for solving polynomial equations on a graphing utility to find the x-intercepts of the graph of f.

b. Use the function for finding the roots of an equation on a graphing utility to find the x-intercepts of the graph of f.

Solution

a. Observe that f is a polynomial function of degree 3, and so we may use the function for solving polynomial equations to solve the equation $x^3 - 3x^2 + x + 1.5 = 0$ $[f(x) = 0]$. We find that the solutions (x-intercepts) are

$$x_1 \approx -0.525687120865, \quad x_2 \approx 1.2586520225, \quad x_3 \approx 2.26703509836$$

Figure T4
The graph of
$f(x) = x^3 - 3x^2 + x + 1.5$.

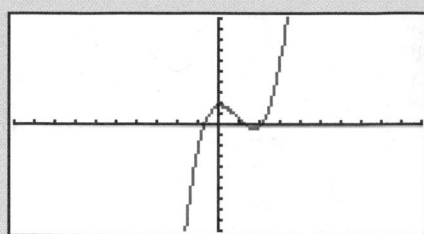

b. Using the graph of f (Figure T4), we see that $x_1 \approx -0.5$, $x_2 \approx 1$, and $x_3 \approx 2$. Using the function for finding the roots of an equation on a graphing utility and these values of x as initial guesses, we find

$$x_1 \approx -0.5256871209, \quad x_2 \approx 1.2586520225, \quad x_3 \approx 2.2670350984 \quad \circ \circ \circ$$

REMARK   The function for solving polynomial equations on a graphing utility will solve a polynomial equation $f(x) = 0$, where f is a polynomial function. The function for finding the roots of a polynomial, however, will solve equations $f(x) = 0$ even if f is not a polynomial. $\circ \circ \circ$

EXAMPLE 3   Unless payroll taxes are increased significantly and/or bene-
fits are scaled back drastically, it is only a matter of time before the current social security system goes broke. Based on data from the Board of Trustees of the Social Security Administration, the assets of the system—the social security "trust fund"—may be approximated by

$$f(t) = -0.0129t^4 + 0.3087t^3 + 2.1760t^2 + 62.8466t + 506.2955 \qquad (0 \le t \le 35)$$

where $f(t)$ is measured in millions of dollars and t is measured in years, with $t = 0$ corresponding to 1995.

a. Use a graphing calculator to sketch the graph of f.

b. Based on this model, when can the social security system be expected to go broke?
Source: Social Security Administration

Solution

a. The graph of f in the window $[0, 35] \times [-1000, 3500]$ is shown in Figure T5.

b. Using the function for finding the roots on a graphing utility, we find that $y = 0$ when $t \approx 34.1$, and this tells us that the system is expected to go broke around 2029.

834

Figure T5

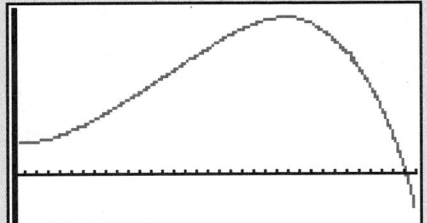

EXERCISES

In exercises 1–4, use the method of Example 1 to analyze the function. (Note: Your answers will not be unique.)

1. $f(x) = 4x^3 - 4x^2 + x + 10$

2. $f(x) = x^3 + 2x^2 + x - 12$

3. $f(x) = \frac{1}{2}x^4 + x^3 + \frac{1}{2}x^2 - 10$

4. $f(x) = 2.25x^4 - 4x^3 + 2x^2 + 2$

In exercises 5–10, find the x-intercepts of the graph of f. Give your answer accurate to four decimal places.

5. $f(x) = 0.2x^3 - 1.2x^2 + 0.8x + 2.1$

6. $f(x) = -0.5x^3 + 1.7x^2 - 1.2$

7. $f(x) = 0.3x^4 - 1.2x^3 + 0.8x^2 + 1.1x - 2$

8. $f(x) = -0.2x^4 + 0.8x^3 - 2.1x + 1.2$

9. $f(x) = 2x^2 - \sqrt{x+1} - 3$

10. $f(x) = x - \sqrt{1 - x^2}$

62. Concentration of a Drug in the Bloodstream   The concentration of a certain drug in a patient's bloodstream t hours after injection is given by

$$C(t) = \frac{0.2t}{t^2 + 1}$$

milligrams per cubic centimeter.
a. Find the horizontal asymptote of $C(t)$.
b. Interpret your result.

63. Effect of Enzymes on Chemical Reactions   Certain proteins, known as enzymes, serve as catalysts for chemical reactions in living things. In 1913, Leonor Michaelis and L. M. Menten discovered the following formula giving the initial speed V (in moles per liter per second) at which the reaction begins in terms of the amount of substrate x (the substance that is being acted upon, measured in moles per liter):

$$V = \frac{ax}{x + b}$$

where a and b are positive constants.
a. Find the horizontal asymptote of V.
b. Interpret your result.

64. Worker Efficiency   An efficiency study showed that the total number of cordless telephones assembled by an average worker at Delphi Electronics t hours after starting work at 8 A.M. is given by

$$N(t) = -\frac{1}{2}t^3 + 3t^2 + 10t \qquad (0 \le t \le 4)$$

Sketch the graph of the function N and interpret your results.

65. GDP of a Developing Country   A developing country's gross domestic product (GDP) from 1991 to 1999 is approximated by the function

$$G(t) = -0.2t^3 + 2.4t^2 + 60 \qquad (0 \le t \le 8)$$

where $G(t)$ is measured in billions of dollars and $t = 0$ corresponds to the year 1991. Sketch the graph of the function G and interpret your results.

66. Concentration of a Drug in the Bloodstream   The concentration of a certain drug in a patient's bloodstream t hours after injection is given by

$$C(t) = \frac{0.2t}{t^2 + 1}$$

milligrams per cubic centimeter. Sketch the graph of the function C and interpret your results.

67. Oxygen Content of a Pond   When organic waste is dumped into a pond, the oxidation process that takes place reduces the pond's oxygen content. However, given time, nature will restore the oxygen content to its natural level. Suppose that the oxygen content t days after organic waste has been dumped into the pond is given by

$$f(t) = 100 \left(\frac{t^2 - 4t + 4}{t^2 + 4} \right) \qquad (0 \le t < \infty)$$

percent of its normal level. Sketch the graph of the function f and interpret your results.

68. Box Office Receipts   The total worldwide box office receipts for a long-running movie are approximated by the function

$$T(x) = \frac{120x^2}{x^2 + 4}$$

where $T(x)$ is measured in millions of dollars and x is the number of years since the movie's release. Sketch the graph of the function T and interpret your results.

69. Cost of Removing Toxic Pollutants   A city's main well was recently found to be contaminated with trichloroethylene, a cancer-causing chemical, as a result of an abandoned chemical dump leaching chemicals into the water. A proposal submitted to the city's selectmen indicates that the cost, measured in millions of dollars, of removing $x\%$ of the toxic pollutant is given by

$$C(x) = \frac{0.5x}{100 - x}$$

Sketch the graph of the function C and interpret your results.

SOLUTIONS TO SELF-CHECK EXERCISES 12.3

1. Since

$$\lim_{x \to \infty} \frac{2x^2}{x^2 - 1} = \lim_{x \to \infty} \frac{2}{1 - \dfrac{1}{x^2}} \qquad \text{(Dividing the numerator and denominator by } x^2)$$

$$= 2$$

we see that $y = 2$ is a horizontal asymptote. Next, since

$$x^2 - 1 = (x + 1)(x - 1) = 0$$

implies $x = -1$ or $x = 1$, these are candidates for the vertical asymptotes of f. Since the numerator of f is not equal to zero for $x = -1$ or $x = 1$, we conclude that $x = -1$ and $x = 1$ are vertical asymptotes of the graph of f.

2. We obtain the following information on the graph of f.
  (1) The domain of f is the interval $(-\infty, \infty)$.
  (2) By setting $x = 0$, we find the y-intercept is 4.
  (3) Since

$$\lim_{x \to -\infty} f(x) = \lim_{x \to -\infty} \left(\frac{2}{3}x^3 - 2x^2 - 6x + 4 \right) = -\infty$$

$$\lim_{x \to \infty} f(x) = \lim_{x \to \infty} \left(\frac{2}{3}x^3 - 2x^2 - 6x + 4 \right) = \infty$$

we see that $f(x)$ decreases without bound as x decreases without bound, and $f(x)$ increases without bound as x increases without bound.
  (4) Since f is a polynomial function, there are no asymptotes.

  (5) $$f'(x) = 2x^2 - 4x - 6 = 2(x^2 - 2x - 3)$$
  $$= 2(x + 1)(x - 3)$$

Setting $f'(x) = 0$ gives $x = -1$ or $x = 3$. The accompanying sign diagram for f' shows that f is increasing on the intervals $(-\infty, -1)$ and $(3, \infty)$ and decreasing on $(-1, 3)$.

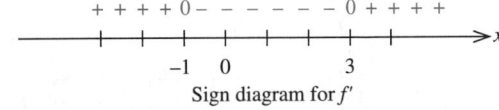

Sign diagram for f'

  (6) From the results of step 5, we see that $x = -1$ and $x = 3$ are critical points of f. Furthermore, the sign diagram of f' tells us that $x = -1$ gives rise to a relative maximum of f and $x = 3$ gives rise to a relative minimum of f. Now,

$$f(-1) = \frac{2}{3}(-1)^3 - 2(-1)^2 - 6(-1) + 4 = \frac{22}{3}$$

and

$$f(3) = \frac{2}{3}(3)^3 - 2(3)^2 - 6(3) + 4 = -14$$

so $f(-1) = 22/3$ is a relative maximum of f and $f(3) = -14$ is a relative minimum of f.

(7) $$f''(x) = 4x - 4 = 4(x - 1)$$

which is equal to zero when $x = 1$. The accompanying sign diagram of f'' shows that f is concave downward on the interval $(-\infty, 1)$ and concave upward on the interval $(1, \infty)$.

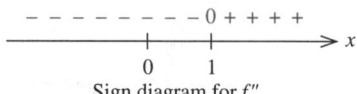

Sign diagram for f''

(8) From the results of step 7, we see that $x = 1$ is the only candidate for an inflection point of f. Since $f''(x)$ changes sign as we move across the point $x = 1$ and

$$f(1) = \frac{2}{3}(1)^3 - 2(1)^2 - 6(1) + 4 = -\frac{10}{3}$$

we see that the required inflection point is $(1, -10/3)$.
(9) Summarizing this information, we have

Domain	$(-\infty, \infty)$
Intercept	$(0, 4)$
Intervals where f is ↗ or ↘	↗ on $(-\infty, -1) \cup (3, \infty)$; ↘ on $(-1, 3)$
Relative extrema	Rel. max. at $(-1, 22/3)$; rel. min. at $(3, -14)$
Concavity	Downward on $(-\infty, 1)$; upward on $(1, \infty)$
Point of inflection	$(1, -10/3)$
$\lim_{x \to -\infty} f(x); \lim_{x \to \infty} f(x)$	$-\infty; \infty$
Asymptotes	None

The graph of f is sketched in the accompanying figure.

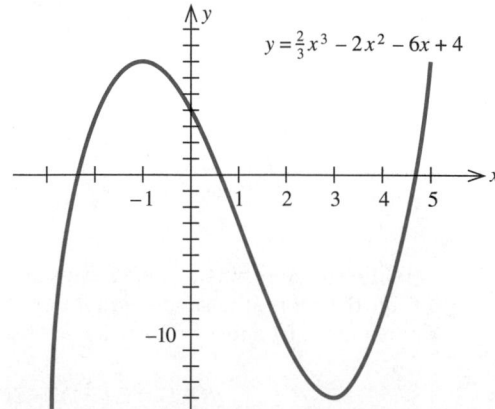

12.4 OPTIMIZATION I

Absolute Extrema

The graph of the function f in Figure 12.56 shows the average age of cars in use in the United States from the beginning of 1946 ($t = 0$) to the beginning of 1990 ($t = 44$). Observe that the highest average age of cars in use during this period is 9 years, whereas the lowest average age of cars in use during the same period is $5\frac{1}{2}$ years. The number 9, the largest value of $f(t)$ for all values of t in the interval [0, 44] (the domain of f), is called the *absolute maximum value of f* on that interval. The number $5\frac{1}{2}$, the smallest value of $f(t)$ for all values of t in [0, 44], is called the *absolute minimum value of f* on that interval. Notice, too, that the absolute maximum value of f is attained at the endpoint $t = 0$ of the interval, whereas the absolute minimum value of f is attained at the two interior points $t = 12$ (corresponding to 1958) and $t = 23$ (corresponding to 1969).

Figure 12.56
$f(t)$ gives the average age of cars in use in year t, t in [0, 44].

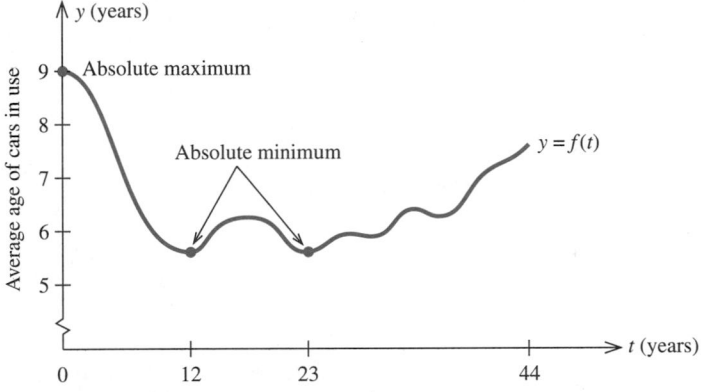

Source: American Automobile Association

A precise definition of the **absolute extrema** (absolute maximum or absolute minimum) of a function follows:

| **THE ABSOLUTE EXTREMA OF A FUNCTION f** | If $f(x) \leq f(c)$ for all x in the domain of f, then $f(c)$ is called the **absolute maximum value** of f. |
| | If $f(x) \geq f(c)$ for all x in the domain of f, then $f(c)$ is called the **absolute minimum value** of f. |

Figure 12.57 shows the graphs of several functions and gives the absolute maximum and absolute minimum of each function, if they exist.

Figure 12.57

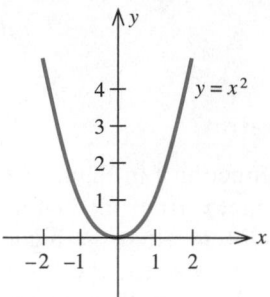

(a) $f(0) = 0$ is the absolute minimum of f; f has no absolute maximum.

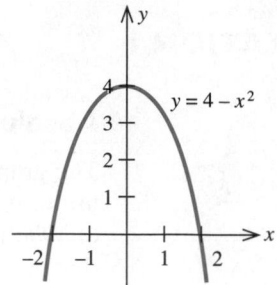

(b) $f(0) = 4$ is the absolute maximum of f; f has no absolute minimum.

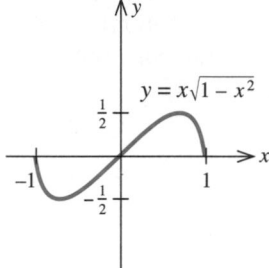

(c) $f\left(\dfrac{\sqrt{2}}{2}\right) = \dfrac{1}{2}$ is the absolute

maximum of f. $f\left(-\dfrac{\sqrt{2}}{2}\right) = -\dfrac{1}{2}$

is the absolute minimum of f.

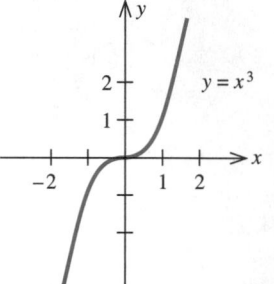

(d) f has no absolute extrema.

Absolute Extrema on a Closed Interval

As the examples in Figure 12.57 show, a continuous function defined on an arbitrary interval does not always have an absolute maximum or an absolute minimum. But an important case arises often in practical applications in which both the absolute maximum and the absolute minimum of a function are guaranteed to exist. This occurs when a continuous function is defined on a *closed* interval. Let us state this important result in the form of a theorem, whose proof we will omit.

THEOREM 3

If a function f is continuous on a closed interval $[a, b]$, then f has both an absolute maximum value and an absolute minimum value on $[a, b]$.

Observe that if an absolute extremum of a continuous function f occurs at a point in an open interval (a, b), then it must be a relative extremum of f and hence its x-coordinate must be a critical point of f. Otherwise, the

Figure 12.58
The relative minimum of f at x_3 is the absolute minimum of f. The right endpoint b gives rise to the absolute maximum value $f(b)$ of f.

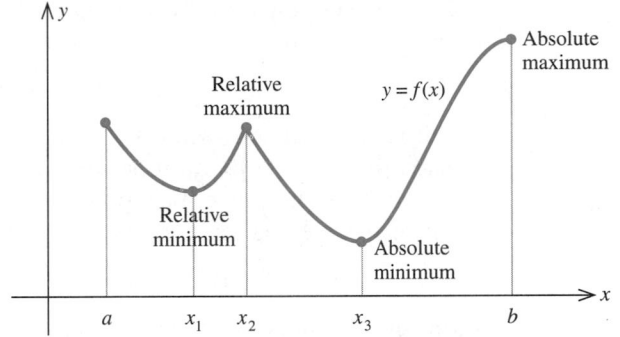

absolute extremum of f must occur at one or both of the endpoints of the interval $[a, b]$. A typical situation is illustated in Figure 12.58.

Here x_1, x_2, and x_3 are critical points of f. The absolute minimum of f occurs at x_3, which lies in the open interval (a, b) and is a critical point of f. The absolute maximum of f occurs at b, an endpoint. This observation suggests the following procedure for finding the absolute extrema of a continuous function on a closed interval.

FINDING THE ABSOLUTE EXTREMA OF f ON A CLOSED INTERVAL

1. Find the critical points of f that lie in (a, b).

2. Compute the value of f at each of these critical points of f and compute $f(a)$ and $f(b)$.

3. The absolute maximum value and absolute minimum value of f will correspond to the largest and smallest numbers, respectively, found in step 2.

Figure 12.59
F has an absolute minimum value of 0 and an absolute maximum value of 4.

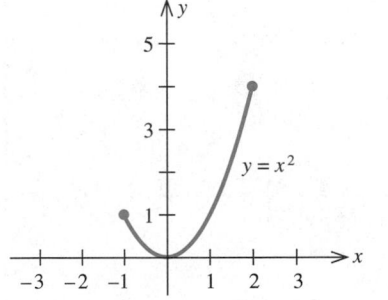

EXAMPLE 1 Find the absolute extrema of the function $F(x) = x^2$ defined on the interval $[-1, 2]$.

Solution The function F is continuous on the closed interval $[-1, 2]$ and differentiable on the open interval $(-1, 2)$. The derivative of F is

$$F'(x) = 2x$$

so $x = 0$ is the only critical point of F. Next, evaluate $F(x)$ at $x = -1$, $x = 0$, and $x = 2$. Thus,

$$F(-1) = 1, \quad F(0) = 0, \quad F(2) = 4$$

It follows that 0 is the absolute minimum value of F and 4 is the absolute maximum value of F. The graph of F, which appears in Figure 12.59, confirms our results.   ○ ○ ○

EXAMPLE 2 Find the absolute extrema of the function

$$f(x) = x^3 - 2x^2 - 4x + 4$$

defined on the interval $[0, 3]$.

Solution   The function f is continuous on the closed interval $[0, 3]$ and differentiable on the open interval $(0, 3)$. The derivative of f is

$$f'(x) = 3x^2 - 4x - 4 = (3x + 2)(x - 2)$$

and it is equal to zero when $x = -\frac{2}{3}$ and $x = 2$. Since the point $x = -\frac{2}{3}$ lies outside the interval $[0, 3]$, it is dropped from further consideration and $x = 2$ is seen to be the sole critical point of f. Next, we evaluate $f(x)$ at the critical point of f as well as the endpoints of f, obtaining

$$f(0) = 4, \quad f(2) = -4, \quad f(3) = 1$$

From these results, we conclude that -4 is the absolute minimum value of f and 4 is the absolute maximum value of f. The graph of f, which appears in Figure 12.60, confirms our results. Observe that the absolute maximum of f occurs at the endpoint $x = 0$ of the interval $[0, 3]$, while the absolute minimum of f occurs at $x = 2$, which is a point in the interval $(0, 3)$.

Figure 12.60
f has an absolute maximum value of 4 and an absolute minimum value of −4.

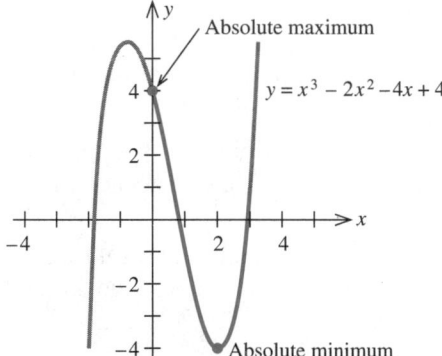

$y = x^3 - 2x^2 - 4x + 4$

○ ○ ○

EXPLORING WITH TECHNOLOGY

Let $f(x) = x^3 - 2x^2 - 4x + 4$. (This is the function of Example 2.)
1. Use a graphing utility to plot the graph of f in the viewing rectangle $[0, 3] \times [-5, 5]$. Use **TRACE** to find the absolute extrema of f on the interval $[0, 3]$ and thus verify the results obtained analytically in Example 2.
2. Plot the graph of f in the viewing rectangle $[-2, 1] \times [-5, 6]$. Use **ZOOM** and **TRACE** to find the absolute extrema of f on the interval $[-2, 1]$. Verify your results analytically.

○ ○ ○

EXAMPLE 3 Find the absolute maximum and absolute minimum values of the function $f(x) = x^{2/3}$ on the interval $[-1, 8]$.

Solution The derivative of f is

$$f'(x) = \frac{2}{3}x^{-1/3} = \frac{2}{3x^{1/3}}$$

Note that f' is not defined at $x = 0$, is continuous everywhere else, and does not equal zero for all x. Therefore, $x = 0$ is the only critical point of f. Evaluating $f(x)$ at $x = -1$, 0, and 8, we obtain

$$f(-1) = 1, \quad f(0) = 0, \quad f(8) = 4$$

We conclude that the absolute minimum value of f is 0, attained at $x = 0$, and the absolute maximum value of f is 4, attained at $x = 8$ (Figure 12.61).

○ ○ ○

Figure 12.61
f has an absolute minimum value of $f(0) = 0$ and an absolute maximum value of $f(8) = 4$.

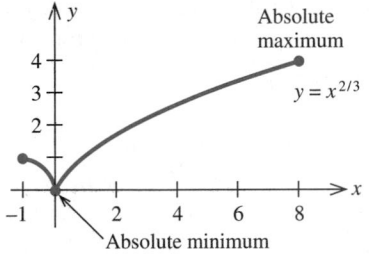

Applications

Many real-world applications call for finding the absolute maximum value or the absolute minimum value of a given function. For example, management is interested in finding what level of production will yield the maximum profit for a company; a farmer is interested in finding the right amount of fertilizer to maximize crop yield; a doctor is interested in finding the maximum concentration of a drug in a patient's body and the time at which it occurs; and an engineer is interested in finding the dimension of a container with a specified shape and volume that can be constructed at a minimum cost.

EXAMPLE 4 The Acrosonic Company's total profit from manufacturing and selling x units of their model F loudspeaker systems is given by

$$P(x) = -0.02x^2 + 300x - 200{,}000 \qquad (0 \le x \le 20{,}000)$$

dollars. How many units of the loudspeaker system must Acrosonic produce to maximize its profits?

Solution To find the absolute maximum of P on $[0, 20{,}000]$, first find the critical points of P on the interval $(0, 20{,}000)$. To do this, compute

$$P'(x) = -0.04x + 300$$

Solving the equation $P'(x) = 0$ gives $x = 7500$. Next, evaluate $P(x)$ at $x = 7500$ as well as the endpoints $x = 0$ and $x = 20{,}000$ of the interval $[0, 20{,}000]$, obtaining

$$P(0) = -200{,}000$$
$$P(7500) = 925{,}000$$
$$P(20{,}000) = -2{,}200{,}000$$

Figure 12.62
P has an absolute maximum at (7500, 925,000).

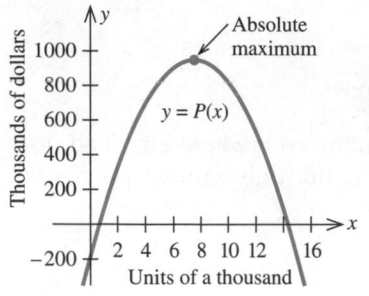

From these computations we see that the absolute maximum value of the function P is 925,000. Thus, by producing 7500 units, Acrosonic will realize a maximum profit of \$925,000. The graph of P is sketched in Figure 12.62.

○ ○ ○

EXAMPLE 5 When a person coughs, the trachea, or windpipe, contracts, allowing air to be expelled at a maximum velocity. It can be shown that, during a cough, the velocity v of airflow is given by the function

$$v = f(r) = kr^2(R - r)$$

where r is the trachea's radius (in centimeters) during a cough, R is the trachea's normal radius (in centimeters), and k is a positive constant that depends on the length of the trachea. Find the radius r for which the velocity of airflow is greatest.

Figure 12.63
The velocity of airflow is greatest when the radius of the contracted trachea is $\frac{2}{3}R$.

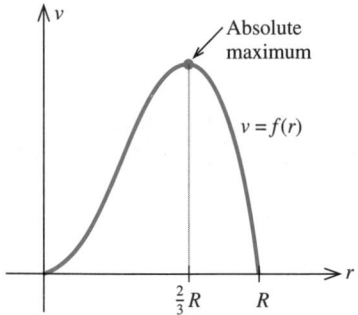

Solution To find the absolute maximum of f on $[0, R]$, first find the critical points of f on the interval $(0, R)$. We compute

$$f'(r) = 2kr(R - r) - kr^2 \quad \text{(Using the Product Rule)}$$
$$= -3kr^2 + 2kRr = kr(-3r + 2R)$$

Setting $f'(r) = 0$ gives $r = 0$ or $r = \frac{2}{3}R$, and so $r = \frac{2}{3}R$ is the sole critical point of f ($r = 0$ is an endpoint). Evaluating $f(r)$ at $r = \frac{2}{3}R$ as well as at the endpoints $r = 0$ and $r = R$, we obtain

$$f(0) = 0$$
$$f\left(\frac{2}{3}R\right) = \frac{4k}{27}R^3$$
$$f(R) = 0$$

from which we deduce that the velocity of airflow is greatest when the radius of the contracted trachea is $\frac{2}{3}R$—that is, when the radius is contracted by approximately 33%. The graph of the function f is shown in Figure 12.63.

○ ○ ○

Recall that the total profit function P is defined as $P(x) = R(x) - C(x)$, where R is the total revenue function, C is the total cost function, and x is the number of units of a product produced and sold. (Assume all derivatives exist.)

a. Show that at the level of production x_0 that yields the maximum profit for the company, the following two conditions are satisfied:

$$R'(x_0) = C'(x_0) \quad \text{and} \quad R''(x_0) < C''(x_0)$$

b. Interpret the two conditions in part (a) in economic terms and explain why they make sense.

Prove that if a cost function $C(x)$ is concave upward $[C''(x) > 0]$, then the level of production that will result in the smallest average production cost occurs when

$$\overline{C}(x) = C'(x)$$

that is, when the average cost $\overline{C}(x)$ is equal to the marginal cost $C'(x)$.

[*Hints:*

1. Show that

$$\overline{C}'(x) = \frac{xC'(x) - C(x)}{x^2}$$

so that the critical point of the function $\overline{C}$ occurs when

$$xC'(x) - C(x) = 0$$

2. Show that at a critical point of $\overline{C}$

$$\overline{C}''(x) = \frac{C''(x)}{x}$$

Use the Second Derivative Test to reach the desired conclusion.]

EXAMPLE 6   The daily average cost function (in dollars per unit) of the Elektra Electronics Company is given by

$$\overline{C}(x) = 0.0001x^2 - 0.08x + 40 + \frac{5000}{x} \qquad (x > 0)$$

Figure 12.64
The minimum average cost is $35 per unit.

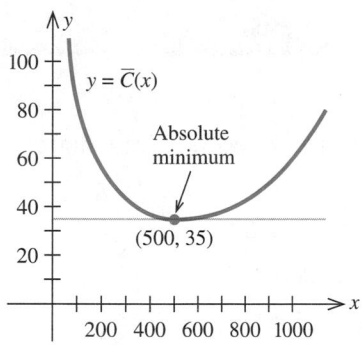

where x stands for the number of programmable calculators Elektra produces. Show that a production level of 500 units per day results in a minimum average cost for the company.

Solution   The domain of the function $\overline{C}$ is the interval $(0, \infty)$, which is not closed. In order to solve the problem, we resort to the graphic method. Using the techniques of graphing from the last section, we sketch the graph of $\overline{C}$ (Figure 12.64). Now,

$$\overline{C}'(x) = 0.0002x - 0.08 - \frac{5000}{x^2}$$

Substituting the given value of x, 500, into $\overline{C}'(x)$ gives $\overline{C}'(500) = 0$, so $x = 500$ is a critical point of $\overline{C}$. Next,

$$\overline{C}''(x) = 0.0002 + \frac{10,000}{x^3}$$

Thus, $$\overline{C}''(500) = 0.0002 + \frac{10,000}{(500)^3} > 0$$

and by the Second Derivative Test, a relative minimum of the function $\overline{C}$ occurs at the point $x = 500$. Furthermore, $\overline{C}''(x) > 0$ for $x > 0$, which implies that the graph of $\overline{C}$ is concave upward everywhere, so the relative minimum of $\overline{C}$ must be the absolute minimum of $\overline{C}$. The minimum average cost is given by

$$\overline{C}(500) = 0.0001(500)^2 - 0.08(500) + 40 + \frac{5000}{500}$$

$$= 35$$

or $35 per unit.

○ ○ ○

EXPLORING WITH TECHNOLOGY

Refer to the preceding Group Discussion question and Example 6.
1. Using a graphing utility, plot the graphs of

$$\overline{C}(x) = 0.0001x^2 - 0.08x + 40 + \frac{5000}{x}$$

and $$C'(x) = 0.0003x^2 - 0.16x + 40$$

in the viewing rectangle $[0, 1000] \times [0, 150]$.
[*Note:* $C(x) = 0.0001x^3 - 0.08x^2 + 40x + 5000$ (Why?)]
2. Find the point of intersection of the graphs of $\overline{C}$ and C' and thus verify the assertion in the Group Discussion question for the special case studied in Example 6.

○ ○ ○

EXAMPLE 7   The altitude (in feet) of a rocket t seconds into flight is given by

$$s = f(t) = -t^3 + 96t^2 + 195t + 5 \qquad (t \geq 0)$$

a. Find the maximum altitude attained by the rocket.

b. Find the maximum velocity attained by the rocket.

Solution

a. The maximum altitude attained by the rocket is given by the largest value of the function f in the closed interval $[0, T]$, where T denotes the time the rocket hits Earth. We know that such a number exists because the dominant term in the expression for the continuous function f is $-t^3$. So for t large enough, the value of $f(t)$ must change from positive to negative and, in particular, it must attain the value 0 for some T.

Figure 12.65
The maximum altitude of the rocket is 143,655 feet.

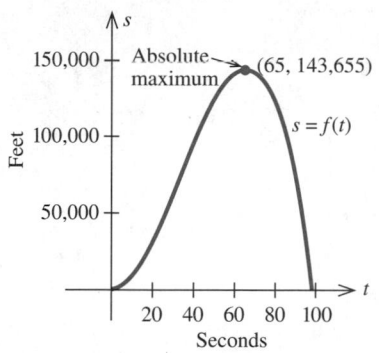

To find the absolute maximum of f, compute

$$f'(t) = -3t^2 + 192t + 195$$
$$= -3(t - 65)(t + 1)$$

and solve the equation $f'(t) = 0$, obtaining $t = -1$ and $t = 65$. Ignore $t = -1$ since it lies outside the interval $[0, T]$. This leaves the critical point $t = 65$ of f. Continuing, we compute

$$f(0) = 5, \quad f(65) = 143,655, \quad f(T) = 0$$

and conclude, accordingly, that the absolute maximum value of f is 143,655. Thus, the maximum altitude of the rocket is 143,655 feet, attained 65 seconds into flight. The graph of f is sketched in Figure 12.65.

b. To find the maximum velocity attained by the rocket, find the largest value of the function that describes the rocket's velocity at any time t—namely,

$$v = f'(t) = -3t^2 + 192t + 195 \qquad (t \geq 0)$$

We find the critical point of v by setting $v' = 0$. But

$$v' = -6t + 192$$

and the critical point of v is $t = 32$. Since

$$v'' = -6 < 0$$

Figure 12.66
The maximum velocity of the rocket is 3267 ft/sec.

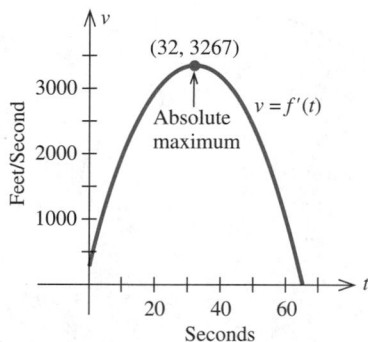

the Second Derivative Test implies that a relative maximum of v occurs at $t = 32$. Our computation has, in fact, clarified the property of the "velocity curve." Since $v'' < 0$ everywhere, the velocity curve is concave downward everywhere. With this observation, we assert that the relative maximum must, in fact, be the absolute maximum of v. The maximum velocity of the rocket is given by evaluating v at $t = 32$,

$$f'(32) = -3(32)^2 + 192(32) + 195$$

or 3267 ft/sec. The graph of the velocity function v is sketched in Figure 12.66.

○ ○ ○

SELF-CHECK EXERCISES 12.4

1. Let $f(x) = x - 2\sqrt{x}$.
  a. Find the absolute extrema of f on the interval $[0, 9]$.
  b. Find the absolute extrema of f.

2. Find the absolute extrema of $f(x) = 3x^4 + 4x^3 + 1$ on $[-2, 1]$.

3. The operating rate (expressed as a percentage) of factories, mines, and utilities in a certain region of the country on the tth day of the year 1999 is given by the function

$$f(t) = 80 + \frac{1200t}{t^2 + 40,000} \qquad (0 \leq t \leq 250)$$

On which day of the first 250 days of 1999 was the manufacturing capacity operating rate highest?

Solutions to Self-Check Exercises 12.4 can be found on page 854.

12.4 EXERCISES

In exercises 1–8, you are given the graph of some function f defined on the indicated interval. Find the absolute maximum and the absolute minimum of f, if they exist.

1.

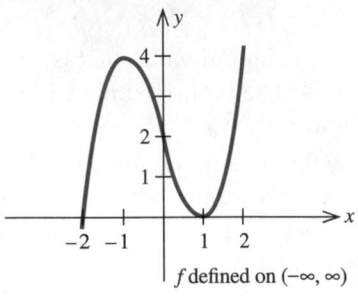

f defined on $(-\infty, \infty)$

2.

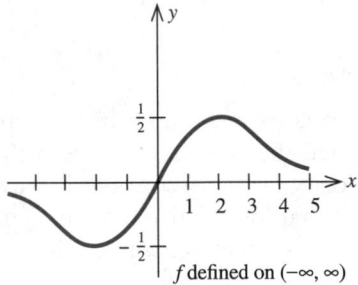

f defined on $(-\infty, \infty)$

3.

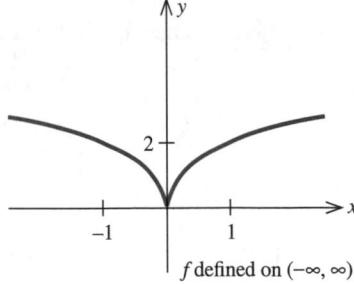

f defined on $(-\infty, \infty)$

4.

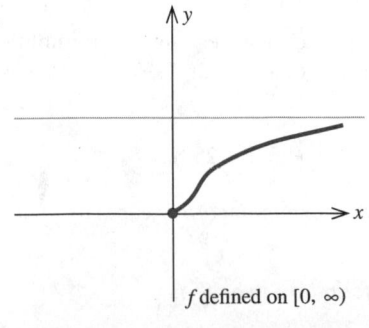

f defined on $[0, \infty)$

5.

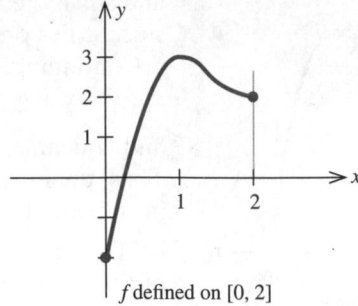

f defined on $[0, 2]$

6.

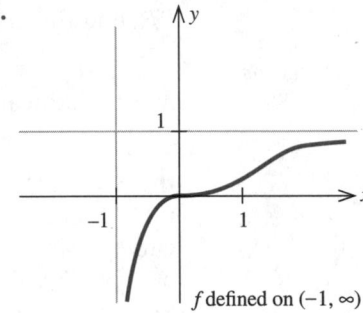

f defined on $(-1, \infty)$

7.

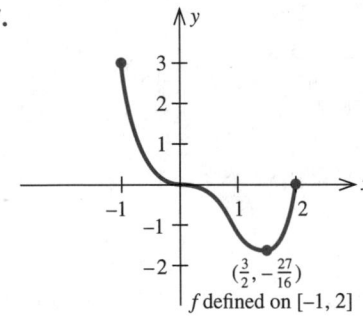

$(\frac{3}{2}, -\frac{27}{16})$

f defined on $[-1, 2]$

8.

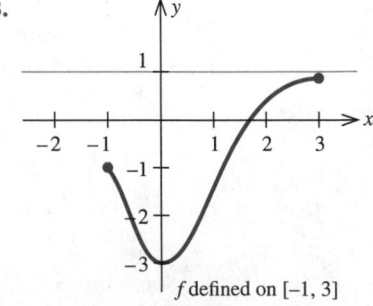

f defined on $[-1, 3]$

In exercises 9–38, find the absolute maximum value and the absolute minimum value, if any, of the given function.

9. $f(x) = 2x^2 + 3x - 4$

10. $g(x) = -x^2 + 4x + 3$

11. $h(x) = x^{1/3}$

12. $f(x) = x^{2/3}$

13. $f(x) = \dfrac{1}{1 + x^2}$

14. $f(x) = \dfrac{x}{1 + x^2}$

15. $f(x) = x^2 - 2x - 3$ on $[-2, 3]$

16. $g(x) = x^2 - 2x - 3$ on $[0, 4]$

17. $f(x) = -x^2 + 4x + 6$ on $[0, 5]$

18. $f(x) = -x^2 + 4x + 6$ on $[3, 6]$

19. $f(x) = x^3 + 3x^2 - 1$ on $[-3, 2]$

20. $g(x) = x^3 + 3x^2 - 1$ on $[-3, 1]$

21. $g(x) = 3x^4 + 4x^3$ on $[-2, 1]$

22. $f(x) = \frac{1}{2}x^4 - \frac{2}{3}x^3 - 2x^2 + 3$ on $[-2, 3]$

23. $f(x) = \dfrac{x + 1}{x - 1}$ on $[2, 4]$ **24.** $g(t) = \dfrac{t}{t - 1}$ on $[2, 4]$

25. $f(x) = 4x + \dfrac{1}{x}$ on $[1, 3]$ **26.** $f(x) = 9x - \dfrac{1}{x}$ on $[1, 3]$

27. $f(x) = \frac{1}{2}x^2 - 2\sqrt{x}$ on $[0, 3]$

28. $g(x) = \frac{1}{8}x^2 - 4\sqrt{x}$ on $[0, 9]$

29. $f(x) = \dfrac{1}{x}$ on $(0, \infty)$ **30.** $g(x) = \dfrac{1}{x + 1}$ on $(0, \infty)$

31. $f(x) = 3x^{2/3} - 2x$ on $[0, 3]$

32. $g(x) = x^2 + 2x^{2/3}$ on $[-2, 2]$

33. $f(x) = x^{2/3}(x^2 - 4)$ on $[-1, 2]$

34. $f(x) = x^{2/3}(x^2 - 4)$ on $[-1, 3]$

35. $f(x) = \dfrac{x}{x^2 + 2}$ on $[-1, 2]$

36. $f(x) = \dfrac{1}{x^2 + 2x + 5}$ on $[-2, 1]$

37. $f(x) = \dfrac{x}{\sqrt{x^2 + 1}}$ on $[-1, 1]$

38. $g(x) = x\sqrt{4 - x^2}$ on $[0, 2]$

39. A stone is thrown straight up from the roof of an 80-foot building. The height of the stone at any time t (in seconds), measured from the ground, is given by

$$h(t) = -16t^2 + 64t + 80$$

feet. What is the maximum height the stone reaches?

40. Maximizing Profits   Lynbrook West, an apartment complex, has 100 two-bedroom units. The monthly profit realized from renting out x apartments is given by

$$P(x) = -10x^2 + 1760x - 50,000$$

dollars. Find how many units should be rented out in order to maximize the monthly rental profit. What is the maximum monthly profit realizable?

41. Maximizing Profits   The estimated monthly profit realizable by the Cannon Precision Instruments Corporation for manufacturing and selling x units of its model M1 camera is

$$P(x) = -0.04x^2 + 240x - 10,000$$

dollars. Determine how many cameras Cannon should produce per month in order to maximize its profits.

42. Flight of a Rocket   The altitude (in feet) attained by a model rocket t seconds into flight is given by the function

$$h(t) = -\frac{1}{3}t^3 + 4t^2 + 20t + 2$$

Find the maximum altitude attained by the rocket.

43. Female Self-Employed Work Force   Based on data obtained from the U.S. Department of Labor, the number of nonfarm, full-time, self-employed women can be approximated by

$$N(t) = 0.81t - 1.14\sqrt{t} + 1.53 \qquad (0 \le t \le 6)$$

where $N(t)$ is measured in millions and t is measured in 5-year intervals, with $t = 0$ corresponding to the beginning of 1963. Determine the absolute extrema of the function N on the interval $[0, 6]$. Interpret your results.

44. Maximizing Profits   The management of Trappee and Sons, Inc., producers of the famous Texa-Pep hot sauce, estimate that their profit from the daily production and sale of x cases (each case consisting of 24 bottles) of the hot sauce is given by

$$P(x) = -0.000002x^3 + 6x - 400$$

dollars. What is the largest possible profit Trappee can make in one day?

45. Maximizing Profits   The quantity demanded per month of the Walter Serkin recording of Beethoven's *Moonlight Sonata,* manufactured by Phonola Record Industries, is related to the price per record. The equation

$$p = -0.00042x + 6 \qquad (0 \le x \le 12,000)$$

where p denotes the unit price in dollars and x is the number of records demanded, relates the demand to the price. The total monthly cost for pressing and packaging x copies of this classical recording is given by

$$C(x) = 600 + 2x - 0.00002x^2 \qquad (0 \le x \le 20,000)$$

dollars. Determine how many copies Phonola should produce per month in order to maximize its profits. [*Hint:* The revenue is $R(x) = px$ and the profit is $P(x) = R(x) - C(x)$.]

46. Maximizing Profit   A manufacturer of tennis rackets finds that the total cost $C(x)$ (in dollars) of manufacturing x rackets per day is given by $C(x) = 400 + 4x + 0.0001x^2$. Each racket can be sold at a price of p dollars, where p is related to x by the demand equation $p = 10 - 0.0004x$. If all the rackets that are manufactured can be sold, find the daily level of production that will yield a maximum profit for the manufacturer.

47. Maximizing Profit   The weekly demand for the Pulsar 25-inch color console television is given by the demand equation

$$p = -0.05x + 600 \qquad (0 \le x \le 12,000)$$

where p denotes the wholesale unit price in dollars and x denotes the quantity demanded. The weekly total cost function associated with manufacturing these sets is given by

$$C(x) = 0.000002x^3 - 0.03x^2 + 400x + 80,000$$

where $C(x)$ denotes the total cost incurred in producing x sets. Find the level of production that will yield a maximum profit for the manufacturer. [*Hint:* Use the quadratic formula.]

48. Minimizing Average Costs   Suppose that the total cost function for manufacturing a certain product is $C(x) = 0.2(0.01x^2 + 120)$ dollars, where x represents the number of units produced. Find the level of production that will minimize the average cost.

49. Minimizing Production Costs   The total monthly cost in dollars incurred by Cannon Precision Instruments Corporation for manufacturing x units of the model M1

camera is given by the function

$$C(x) = 0.0025x^2 + 80x + 10,000$$

a. Find the average cost function $\overline{C}$.
b. Find the level of production that results in the smallest average production cost.
c. Find the level of production for which the average cost is equal to the marginal cost.
d. Compare the result of (c) with that of (b).

50. Minimizing Production Costs   The daily total cost in dollars incurred by Trappee and Sons, Inc., for producing x cases of Texa-Pep hot sauce is given by the function

$$C(x) = 0.000002x^3 + 5x + 400$$

Using this function, answer the questions posed in exercise 49.

51. Maximizing Revenue   Suppose that the quantity demanded per week of a certain dress is related to the unit price p by the demand equation $p = \sqrt{800 - x}$, where p is in dollars and x is the number of dresses made. How many dresses should be made and sold per week in order to maximize the revenue? [*Hint:* $R(x) = px$.]

52. Maximizing Revenue   The quantity demanded per month of the Sicard wristwatch is related to the unit price by the equation

$$p = \frac{50}{0.01x^2 + 1} \qquad (0 \le x \le 20)$$

where p is measured in dollars and x is measured in units of a thousand. How many watches must be sold to yield a maximum revenue?

53. Oxygen Content of a Pond   When organic waste is dumped into a pond, the oxidation process that takes place reduces the pond's oxygen content. However, given time, nature will restore the oxygen content to its natural level. Suppose that the oxygen content t days after organic waste has been dumped into the pond is given by

$$f(t) = 100 \left[\frac{t^2 - 4t + 4}{t^2 + 4} \right] \qquad (0 \le t < \infty)$$

percent of its normal level.
a. When is the level of oxygen content lowest?
b. When is the rate of oxygen regeneration greatest?

54. Air Pollution   The amount of nitrogen dioxide, a brown gas that impairs breathing, present in the atmosphere on a certain May day in the city of Long Beach is approximated by

$$A(t) = \frac{136}{1 + 0.25(t - 4.5)^2} + 28 \qquad (0 \le t \le 11)$$

where $A(t)$ is measured in pollutant standard index (PSI) and t is measured in hours, with $t = 0$ corresponding to 7 A.M. Determine the time of day when the pollution is at its highest level.

55. Maximizing Revenue The average revenue is defined as the function

$$\overline{R}(x) = \frac{R(x)}{x} \qquad (x > 0)$$

Prove that if a revenue function $R(x)$ is concave downward [$R''(x) < 0$], then the level of sales that will result in the largest average revenue occurs when $\overline{R}(x) = R'(x)$.

56. Velocity of Blood According to a law discovered by the nineteenth-century physician Jean Louis Marie Poiseuille, the velocity (in centimeters per second) of blood r centimeters from the central axis of an artery is given by

$$v(r) = k(R^2 - r^2)$$

where k is a constant and R is the radius of the artery. Show that the velocity of blood is greatest along the central axis.

57. GDP of a Developing Country A developing country's gross domestic product (GDP) from 1988 to 1996 is approximated by the function

$$G(t) = -0.2t^3 + 2.4t^2 + 60 \qquad (0 \le t \le 8)$$

where $G(t)$ is measured in billions of dollars and $t = 0$ corresponds to the year 1988. Show that the growth rate of the country's GDP was maximal in 1992.

58. Crime Rates The number of major crimes committed in the city of Bronxville between 1987 and 1994 is approximated by the function

$$N(t) = -0.1t^3 + 1.5t^2 + 100 \qquad (0 \le t \le 7)$$

where $N(t)$ denotes the number of crimes committed in year t ($t = 0$ corresponds to the year 1987). Enraged by the dramatic increase in the crime rate, the citizens of Bronxville, with the help of the local police, organized "Neighborhood Crime Watch" groups in early 1991 to combat this menace. Show that the growth in the crime rate was maximal in 1992, giving credence to the claim

that the Neighborhood Crime Watch program was working.

59. Social Security Surplus Based on data from the Social Security Administration, the estimated cash in the social security retirement and disability trust funds may be approximated by

$$f(t) = -0.0129t^4 + 0.3087t^3 + 2.1760t^2 + 62.8466t$$
$$+ 506.2955 \qquad (0 \le t \le 35)$$

where $f(t)$ is measured in billions of dollars and t is measured in years, with $t = 0$ corresponding to the year 1995. Show that the social security surplus will be at its highest level at approximately the middle of the year 2018.

[*Hint:* Show that $t = 23.6811$ is an approximate critical point of $f'(t)$.]
Source: Social Security Administration

60. Energy Expended by a Fish It has been conjectured that a fish swimming a distance of L ft at a speed of v ft/sec relative to the water and against a current flowing at the rate of u ft/sec ($u < v$) expends a total energy given by

$$E(v) = \frac{aLv^3}{v - u}$$

where E is measured in foot-pounds (ft-lbs) and a is a constant. Find the speed v at which the fish must swim in order to minimize the total energy expended. (*Note:* This result has been verified by biologists.)

61. Let f be a constant function; that is, let $f(x) = c$, where c is some real number. Show that every point $x = a$ is an absolute maximum and, at the same time, an absolute minimum of f.

62. Show that a polynomial function defined on the interval $(-\infty, \infty)$ cannot have both an absolute maximum and an absolute minimum unless it is a constant function.

63. One condition that must be satisfied before Theorem 3 (page 840) is applicable is that the function f must be continuous on the closed interval $[a, b]$. Define a function f on the closed interval $[-1, 1]$ by

$$f(x) = \begin{cases} \dfrac{1}{x} & \text{if } x \in [-1, 1] \qquad (x \ne 0) \\ 0 & \text{if } x = 0 \end{cases}$$

a. Show that f is not continuous at $x = 0$.
b. Show that $f(x)$ does not attain an absolute maximum or an absolute minimum on the interval $[-1, 1]$.
c. Confirm your results by sketching the function f.

USING TECHNOLOGY

FINDING THE ABSOLUTE EXTREMA OF A FUNCTION

Some graphing utilities have a function for finding the absolute maximum and the absolute minimum values of a continuous function on a closed interval. If your graphing utility has this capability, use it to work through the example and exercises of this section.

EXAMPLE 1 Let $f(x) = \dfrac{2x + 4}{(x^2 + 1)^{3/2}}$.

a. Use a graphing utility to plot the graph of f in the viewing rectangle $[-3, 3] \times [-1, 5]$.

b. Find the absolute maximum and absolute minimum values of f on the interval $[-3, 3]$. Express your answers accurate to four decimal places.

Solution

a. The graph of f is shown in Figure T1.

Figure T1
The graph of f in the viewing rectangle $[-3, 3] \times [-1, 5]$.

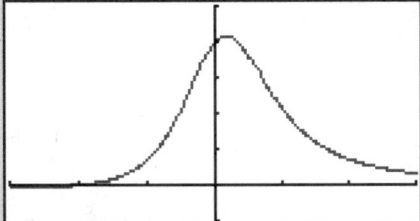

b. Using the function on a graphing utility for finding the absolute minimum value of a continuous function on a closed interval, we find the absolute minimum value of f to be -0.0632. Similarly, using the function for finding the absolute maximum value, we find the absolute maximum value to be 4.1593.

○ ○ ○

REMARK   Some graphing utilities will enable you to find the absolute minimum and absolute maximum values of a continuous function on a closed interval without having to graph the function.

○ ○ ○

852

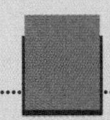

EXERCISES

In exercises 1–6, use a graphing utility to find the absolute maximum and the absolute minimum values of f in the given interval using the method of Example 1. Express your answers accurate to four decimal places.

1. $f(x) = 3x^4 - 4.2x^3 + 6.1x - 2; [-2, 3]$

2. $f(x) = 2.1x^4 - 3.2x^3 + 4.1x^2 + 3x - 4; [-1, 2]$

3. $f(x) = \dfrac{2x^3 - 3x^2 + 1}{x^2 + 2x - 8}; [-3, 1]$

4. $f(x) = \sqrt{x}(x^3 - 4)^2; [0.5, 1]$

5. $f(x) = \dfrac{x^3 - 1}{x^2}$ on $[1, 3]$

6. $f(x) = \dfrac{x^3 - x^2 + 1}{x - 2}$ on $[1, 3]$

7. Rate of Bank Failures The Federal Deposit Insurance Corporation (FDIC) estimates that the rate at which banks were failing between 1982 and 1994 is given by

$$f(t) = 0.063447t^4 - 1.953283t^3 + 14.632576t^2$$
$$- 6.684704t + 47.458874 \qquad (0 \le t \le 12)$$

where $f(t)$ is the number of banks per year and t is measured in years, with $t = 0$ corresponding to the beginning of 1982.
a. Use a graphing utility to plot the graph of f in the viewing rectangle $[0, 12] \times [0, 220]$.
b. What is the highest rate of bank failures during the period in question?
Source: Federal Deposit Insurance Corporation

8. Boston's Daily Temperature for 1993 The average daily temperature in Boston for 1993 is given by

$$f(t) = 0.0434841t^4 - 1.18523t^3 + 9.38548t^2$$
$$- 17.7553t + 38.9272 \qquad (0 \le t \le 12)$$

where $f(t)$ is measured in degrees Fahrenheit and t is in months, with $t = 0$ corresponding to the beginning

of 1993. Find the absolute extrema for f and interpret your results.
Source: Robert Lautzenheiser, climatologist

9. Time on the Market According to the Greater Boston Real Estate Board—Multiple Listing Service, the average number of days a single-family home remains for sale from listing to accepted offer is approximated by the function

$$f(t) = 0.0171911t^4 - 0.662121t^3 + 6.18083t^2$$
$$- 8.97086t + 53.3357 \qquad (0 \le t \le 10)$$

where t is measured in years, with $t = 0$ corresponding to the beginning of 1984.
a. Use a graphing utility to plot the graph of f in the viewing rectangle $[0, 12] \times [0, 120]$.
b. Find the absolute maximum value and the absolute minimum value of f in the interval $[0, 12]$. Interpret your results.
Source: Greater Boston Real Estate Board—Multiple Listing Service

10. Why Social Security Benefits May Exceed Payroll Taxes Unless payroll taxes are increased significantly and/or benefits are scaled back drastically, it is only a matter of time before the current social security system goes broke. Based on data from the Board of Trustees of the Social Security Administration, the assets of the system—the social security "trust fund"—may be approximated by

$$f(t) = -0.0129t^4 + 0.3087t^3 + 2.1760t^2$$
$$+ 62.8466t + 506.2955 \qquad (0 \le t \le 35)$$

where $f(t)$ is measured in millions of dollars and t is measured in years, with $t = 0$ corresponding to 1995.
Source: Social Security Administration
a. Use a graphing calculator to sketch the graph of f.
b. Based on this model, when will the social security system start to pay out more benefits than it gets in payroll taxes?

853

SOLUTIONS TO SELF-CHECK EXERCISES 12.4

1. a. The function f is continuous in its domain and differentiable in the interval $(0, 9)$. The derivative of f is

$$f'(x) = 1 - x^{-1/2} = \frac{x^{1/2} - 1}{x^{1/2}}$$

and it is equal to zero when $x = 1$. Evaluating $f(x)$ at the endpoints $x = 0$ and $x = 9$ and at the critical point $x = 1$ of f, we have

$$f(0) = 0, \quad f(1) = -1, \quad f(9) = 3$$

From these results, we see that -1 is the absolute minimum value of f and 3 is the absolute maximum value of f.

b. In this case, the domain of f is the interval $[0, \infty)$, which is not closed. Therefore, we resort to the graphic method. Using the techniques of graphing, we sketch in the accompanying figure the graph of f. The graph shows that -1 is the absolute

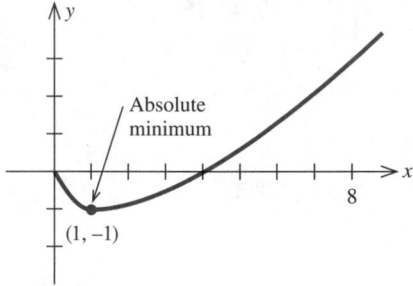

minimum value of f, but f has no absolute maximum since $f(x)$ increases without bound as x increases without bound.

2. The function f is continuous on the interval $[-2, 1]$. It is also differentiable on the open interval $(-2, 1)$. The derivative of f is

$$f'(x) = 12x^3 + 12x^2 = 12x^2(x + 1)$$

and it is continuous on $(-2, 1)$. Setting $f'(x) = 0$ gives $x = -1$ and $x = 0$ as critical points of f. Evaluating $f(x)$ at these critical points of f as well as at the endpoints of the interval $[-2, 1]$, we obtain

$$f(-2) = 17, \quad f(-1) = 0, \quad f(0) = 1, \quad f(1) = 8$$

From these results, we see that 0 is the absolute minimum value of f and 17 is the absolute maximum value of f.

3. The problem is solved by finding the absolute maximum of the function f on $[0, 250]$. Differentiating $f(t)$, we obtain

$$f'(t) = \frac{(t^2 + 40{,}000)(1200) - 1200t(2t)}{(t^2 + 40{,}000)^2}$$

$$= \frac{-1200\,(t^2 - 40{,}000)}{(t^2 + 40{,}000)^2}$$

Upon setting $f'(t) = 0$ and solving the resulting equation, we obtain $t = -200$ or 200. Since -200 lies outside the interval $[0, 250]$, we are interested only in the critical point $t = 200$ of f. Evaluating $f(t)$ at $t = 0$, $t = 200$, and $t = 250$, we find

$$f(0) = 80, \quad f(200) = 83, \quad f(250) = 82.93$$

We conclude that the manufacturing capacity operating rate was the highest on the 200th day of 1999—that is, a little past the middle of July 1999.

12.5 OPTIMIZATION II

Section 12.4 outlined how to find the solution to certain optimization problems in which the objective function is given. In this section, we consider problems in which we are required to first find the appropriate function to be optimized. The following guidelines will be useful for solving these problems.

GUIDELINES FOR SOLVING OPTIMIZATION PROBLEMS	**1.** Assign a letter to each variable mentioned in the problem. If appropriate, draw and label a figure.
	2. Find an expression for the quantity to be optimized.
	3. Use the conditions given in the problem to write the quantity to be optimized as a function f of *one* variable. Note any restrictions to be placed on the domain of f from physical considerations of the problem.
	4. Optimize the function f over its domain using the methods of Section 12.4.

REMARK   In carrying out step 4, remember that if the function f to be optimized is continuous on a closed interval, then the absolute maximum and absolute minimum of f are, respectively, the largest and smallest values of $f(x)$ on the set comprising the critical points of f and the endpoints of the interval. If the domain of f is not a closed interval, then we resort to the graphical method.   ○ ○ ○

Maximization Problems

EXAMPLE 1   A man wishes to have a rectangular garden in his backyard. He has 50 feet of fencing material with which to enclose his garden. Find the dimensions for the largest garden he can have if he uses all of the fencing material.

Solution

Step 1   Let x and y denote the dimensions (in feet) of two adjacent sides of the garden (Figure 12.67) and let A denote its area.

Figure 12.67
What is the maximum rectangular area that can be enclosed with 50 feet of fencing?

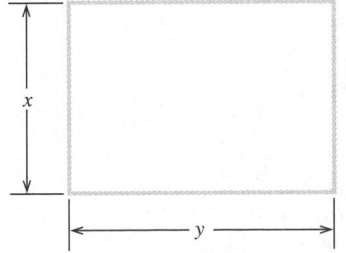

Step 2 The area of the garden,

$$A = xy \qquad (1)$$

is the quantity to be maximized.

Step 3 The perimeter of the rectangle, $(2x + 2y)$ feet, must equal 50 feet. Therefore, we have the equation

$$2x + 2y = 50$$

Next, solving this equation for y in terms of x yields

$$y = 25 - x \qquad (2)$$

which, when substituted into equation (1), gives

$$A = x(25 - x)$$
$$= -x^2 + 25x$$

(Remember, the function to be optimized must involve just one variable.) Since the sides of the rectangle must be nonnegative, we must have $x \geq 0$ and $y = 25 - x \geq 0$; that is, we must have $0 \leq x \leq 25$. Thus, the problem is reduced to finding the absolute maximum of $A = f(x) = -x^2 + 25x$ on the closed interval $[0, 25]$.

Step 4 Observe that f is continuous on $[0, 25]$. So the absolute maximum value of f must occur at the endpoint(s) or at the critical point(s) of f. The derivative of the function A is given by

$$A' = f'(x) = -2x + 25$$

Setting $A' = 0$ gives

$$-2x + 25 = 0$$

or $x = 12.5$, as the critical point of A. Next, we evaluate the function $A = f(x)$ at $x = 12.5$ and at the endpoints $x = 0$ and $x = 25$ of the interval $[0, 25]$, obtaining

$$f(0) = 0, \quad f(12.5) = 156.25, \quad f(25) = 0$$

We see that the absolute maximum value of the function f is 156.25. From equation (2) we see that when $x = 12.5$, the value of y is given by $y = 12.5$. Thus, the garden would be of maximum area (156.25 square feet) if it were in the form of a square with sides 12.5 feet long.

◑ ◑ ◑

EXAMPLE 2 By cutting away identical squares from each corner of a rectangular piece of cardboard and folding up the resulting flaps, we can turn the cardboard into an open box. If the cardboard is 16 inches long and 10 inches wide, find the dimensions of the box that will yield the maximum volume.

Solution

Step 1 Let x denote the length (in inches) of one side of each of the identical squares to be cut out of the cardboard (Figure 12.68), and let V denote the volume of the resulting box.

Figure 12.68

The dimensions of the open box are $(16 - 2x)$ by $(10 - 2x)$ by x inches.

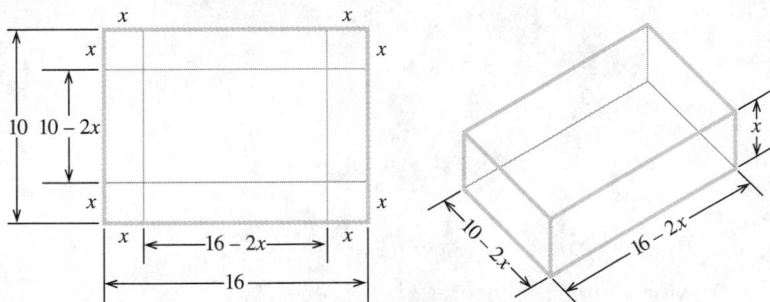

Step 2 The dimensions of the box are $(16 - 2x)$ inches long, $(10 - 2x)$ inches wide, and x inches high. Therefore, its volume,

$$V = (16 - 2x)(10 - 2x)x$$
$$= 4(x^3 - 13x^2 + 40x) \qquad \text{(Expanding the expression)}$$

cubic inches, is the quantity to be maximized.

Step 3 Since each dimension of the box must be nonnegative, x must satisfy the inequalities $x \geq 0$, $16 - 2x \geq 0$, and $10 - 2x \geq 0$. This set of inequalities is satisfied if $0 \leq x \leq 5$. Thus, the problem at hand is equivalent to finding the absolute maximum of

$$V = f(x) = 4(x^3 - 13x^2 + 40x)$$

on the closed interval $[0, 5]$.

Step 4 Observe that f is continuous on $[0, 5]$, so the absolute maximum value of f must be attained at the endpoint(s) or at the critical point(s) of f. Differentiating $f(x)$, we obtain

$$f'(x) = 4(3x^2 - 26x + 40)$$
$$= 4(3x - 20)(x - 2)$$

Upon setting $f'(x) = 0$ and solving the resulting equation for x, we obtain $x = 20/3$ or $x = 2$. Since $20/3$ lies outside the interval $[0, 5]$, it is no longer considered and we are interested only in the critical point $x = 2$ of f. Next, evaluating $f(x)$ at $x = 0$, $x = 5$ (the endpoints of the interval $[0, 5]$), and $x = 2$, we obtain

$$f(0) = 0, \quad f(2) = 144, \quad f(5) = 0$$

Thus, the volume of the box is maximized by taking $x = 2$. The dimensions of the box are $12'' \times 6'' \times 2''$, and the volume is 144 cubic inches.   ○ ○ ○

EXPLORING WITH TECHNOLOGY

Refer to Example 2.

1. Use a graphing utility to plot the graph of

$$f(x) = 4(x^3 - 13x^2 + 40x)$$

in the viewing rectangle $[0, 5] \times [0, 150]$. Explain what happens to $f(x)$ as x increases from $x = 0$ to $x = 5$ and give a physical interpretation.

2. Using **ZOOM** and **TRACE,** find the absolute maximum of f on the interval $[0, 5]$, and thus verify the solution for Example 2 obtained analytically.

○ ○ ○

EXAMPLE 3   A city's Metropolitan Transit Authority (MTA) operates a subway line for commuters from a certain suburb to the downtown metropolitan area. Currently, an average of 6000 passengers a day take the trains, paying a fare of $1.50 per ride. The board of the MTA, contemplating raising the fare to $1.75 per ride in order to generate a larger revenue, engages the services of a consulting firm. The firm's study reveals that for each 25-cent increase in fare, the ridership will be reduced by an average of 1000 passengers a day. Thus the consulting firm recommends that MTA stick to the current fare of $1.50 per ride, which already yields a maximum revenue. Show that the consultants are correct.

Solution

Step 1   Let x denote the number of passengers per day, p denote the fare per ride, and R be MTA's revenue.

Step 2   In order to find a relationship between x and p, observe that the given data imply that when $x = 6000$, $p = 1.5$, and when $x = 5000$, $p = 1.75$. Therefore, the points $(6000, 1.5)$ and $(5000, 1.75)$ lie on a straight line. (Why?) To find the linear relationship between p and x, use the point-slope form of the equation of a straight line. Now, the slope of the line is

$$m = \frac{1.75 - 1.50}{5000 - 6000} = -0.00025$$

Therefore, the required equation is

$$p - 1.5 = -0.00025(x - 6000)$$
$$= -0.00025x + 1.5$$
$$p = -0.00025x + 3$$

The revenue,

$$R = f(x) = xp = -0.00025x^2 + 3x \qquad \text{(Number of riders}$$
$$\text{times unit fare)}$$

is the quantity to be maximized.

Step 3 Since both p and x must be nonnegative, we see that $0 \leq x \leq 12{,}000$, and the problem is to find the absolute maximum of the function f on the closed interval $[0, 12{,}000]$.

Step 4 Observe that f is continuous on $[0, 12{,}000]$. To find the critical point of R, we compute

$$f'(x) = -0.0005x + 3$$

and set it equal to zero, giving $x = 6000$. Evaluating the function f at $x = 6000$, as well as at the endpoints $x = 0$ and $x = 12{,}000$, yields

$$f(0) = 0$$
$$f(6000) = 9000$$
$$f(12{,}000) = 0$$

We conclude that a maximum revenue of $9000 per day is realized when the ridership is 6000 per day. The optimum price of the fare per ride is therefore $1.50, as recommended by the consultants. The graph of the revenue function R is shown in Figure 12.69.   ○ ○ ○

Figure 12.69
f has an absolute maximum of 9000 when x = 6000.

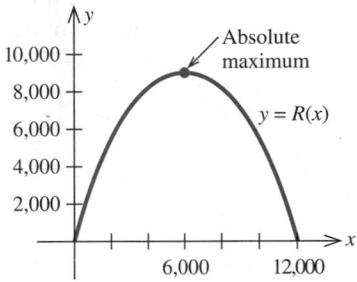

Minimization Problems

EXAMPLE 4 The Betty Moore Company requires that its corned beef hash containers have a capacity of 54 cubic inches, have the shape of right-circular cylinders, and be made of tin. Determine the radius and height of the container that requires the least amount of metal.

Solution

Figure 12.70
We want to minimize the amount of material used to construct the container.

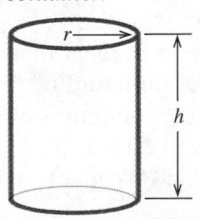

Step 1 Let the radius and height of the container be r and h inches, respectively, and let S denote the surface area of the container (Figure 12.70).

Step 2 The amount of tin used to construct the container is given by the total surface area of the cylinder. Now, the areas of the base and the top of the cylinder are each πr^2 square inches and the area of the side is $2\pi rh$ square inches. Therefore,

$$S = 2\pi r^2 + 2\pi rh \qquad \qquad \textbf{(3)}$$

the quantity to be minimized.

Step 3 The requirement that the volume of a container be 54 cubic inches implies that

$$\pi r^2 h = 54 \qquad (4)$$

Solving equation (4) for h, we obtain

$$h = \frac{54}{\pi r^2} \qquad (5)$$

which, when substituted into (3), yields

$$S = 2\pi r^2 + 2\pi r \left(\frac{54}{\pi r^2}\right)$$

$$= 2\pi r^2 + \frac{108}{r}$$

Figure 12.71
The total surface area of the right cylindrical container is graphed as a function of r.

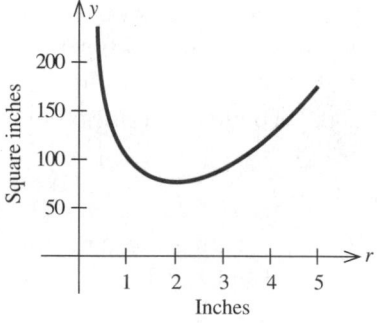

Clearly, the radius r of the container must satisfy the inequality $r > 0$. The problem now is reduced to finding the absolute minimum of the function $S = f(r)$ on the interval $(0, \infty)$.

Step 4 Using the curve-sketching techniques of Section 12.3, we obtain the graph of f in Figure 12.71. To find the critical point of f, we compute

$$S' = 4\pi r - \frac{108}{r^2}$$

and solve the equation $S' = 0$ for r:

$$4\pi r - \frac{108}{r^2} = 0$$

$$4\pi r^3 - 108 = 0$$

$$r^3 = \frac{27}{\pi}$$

$$r = \frac{3}{\sqrt[3]{\pi}} \approx 2 \qquad (6)$$

Next, let us show that this value of r gives rise to the absolute minimum of f. To show this, we first compute

$$S'' = 4\pi + \frac{216}{r^3}$$

Since $S'' > 0$ if $r = 3/\sqrt[3]{\pi}$, the Second Derivative Test implies that the value of r in equation (6) gives rise to a relative minimum of f. Finally, this relative minimum of f is also the absolute minimum of f since f is always concave upward ($S'' > 0$ for all $r > 0$). To find the height of the given container, we substitute the value of r given in

(6) into (5). Thus

$$h = \frac{54}{\pi r^2} = \frac{54}{\pi \left(\dfrac{3}{\pi^{1/3}}\right)^2}$$

$$= \frac{54\pi^{2/3}}{(\pi)9}$$

$$= \frac{6}{\pi^{1/3}} = \frac{6}{\sqrt[3]{\pi}}$$

$$= 2r$$

We conclude that the required container has a radius of approximately 2 inches and a height of approximately 4 inches, or twice the size of the radius. ◦ ◦ ◦

An Inventory Problem

One problem faced by many companies is controlling the inventory of goods carried. Ideally, the manager must ensure that the company has sufficient stock to meet customer demand at all times. At the same time, she must make sure that this is accomplished without overstocking (incurring unnecessary storage costs) and also without having to place orders too frequently (incurring reordering costs).

EXAMPLE 5 The Dixie Import-Export Company is the sole agent for the Excalibur 250-cc motorcycle. Management estimates that the demand for these motorcycles is 10,000 per year and that they will sell at a uniform rate throughout the year. The cost incurred in ordering each shipment of motorcycles is $10,000, and the cost per year of storing each motorcycle is $200.

Dixie's management faces the following problem: Ordering too many motorcycles at one time ties up valuable storage space and increases the storage cost. On the other hand, placing orders too frequently increases the ordering costs. How large should each order be, and how often should orders be placed, to minimize ordering and storage costs?

Solution Let x denote the number of motorcycles in each order (the lot size). Then, assuming that each shipment arrives just as the previous shipment has been sold, the average number of motorcycles in storage during the year is $x/2$. You can see that this is the case by examining Figure 12.72. Thus, Dixie's storage cost for the year is given by $200(x/2)$, or $100x$ dollars.

Next, since the company requires 10,000 motorcycles for the year and since each order is for x motorcycles, the number of orders required is

$$\frac{10,000}{x}$$

Figure 12.72
As each lot is depleted, the new lot arrives. The average inventory level is x/2 if x is the lot size.

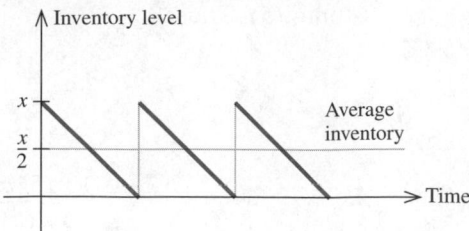

This gives an ordering cost of

$$10,000 \left(\frac{10,000}{x} \right) = \frac{100,000,000}{x}$$

dollars for the year. Thus, the total yearly cost incurred by Dixie, which includes the ordering and storage costs attributed to the sale of these motorcycles, is given by

$$C(x) = 100x + \frac{100,000,000}{x}$$

The problem is reduced to finding the absolute minimum of the function C in the interval $(0, 10,000]$. To accomplish this, we compute

$$C'(x) = 100 - \frac{100,000,000}{x^2}$$

Setting $C'(x) = 0$ and solving the resulting equation, we obtain $x = \pm 1000$. Since the number -1000 is outside the domain of the function C, it is rejected, leaving $x = 1000$ as the only critical point of C. Next, we find

$$C''(x) = \frac{200,000,000}{x^3}$$

Since $C''(1000) > 0$, the Second Derivative Test implies that the critical point $x = 1000$ is a relative minimum of the function C (Figure 12.73). Also, since $C''(x) > 0$ for all x in $(0, 10,000]$, the function C is concave upward everywhere so that the point $x = 1000$ also gives the absolute minimum of C. Thus, in order to minimize the ordering and storage costs, Dixie should place $10,000/1000$, or 10, orders a year, each for a shipment of 1000 motorcycles.

Figure 12.73
C has an absolute minimum at (1000, 200,000).

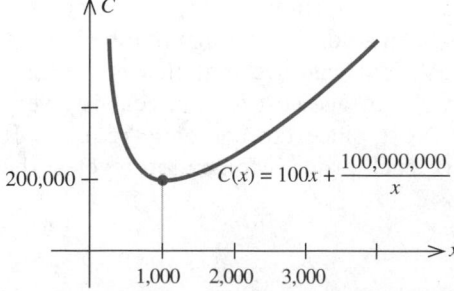

○ ○ ○

SELF-CHECK EXERCISES 12.5

 1. A man wishes to have an enclosed vegetable garden in his backyard. If the garden is to be a rectangular area of 300 square feet, find the dimensions of the garden that will minimize the amount of fencing material needed.

 2. The demand for the Super Titan tires is 1,000,000 per year. The set-up cost for each production run is $4000, and the manufacturing cost is $20 per tire. The cost of storing each tire over the year is $2. Assuming uniformity of demand throughout the year and instantaneous production, determine how many tires should be manufactured per production run in order to keep the production cost to a minimum.

Solutions to Self-Check Exercises 12.5 can be found on page 866.

12.5 EXERCISES

1. Enclosing the Largest Area The owner of the Rancho Los Feliz has 3000 yards of fencing material with which to enclose a rectangular piece of grazing land along the straight portion of a river. If fencing is not required along the river, what are the dimensions of the largest area he can enclose? What is this area?

2. Enclosing the Largest Area Refer to exercise 1. As an alternative plan, the owner of the Rancho Los Feliz might use the 3000 yards of fencing material to enclose the rectangular piece of grazing land along the straight portion of the river and then subdivide it by means of a fence running parallel to the sides. Again, no fencing is required along the river. What are the dimensions of the largest area that can be enclosed? What is this area? (See the accompanying figure.)

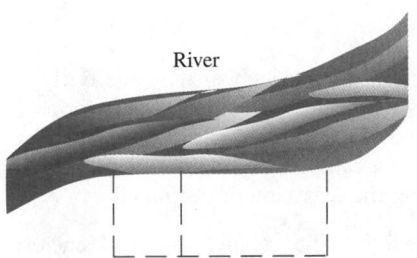

3. Minimizing Construction Costs The management of the UNICO department store has decided to enclose an 800-square-foot area outside the building for displaying potted plants and flowers. One side will be formed by the external wall of the store, two sides will be constructed of pine boards, and the fourth side will be made of galvanized steel fencing material. If the pine board fencing costs $6 per running foot and the steel fencing costs $3 per running foot, determine the dimensions of the enclosure that can be erected at minimum cost.

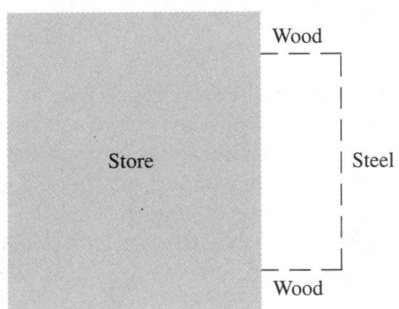

4. Packaging By cutting away identical squares from each corner of a rectangular piece of cardboard and folding up the resulting flaps, we can make an open box. If the cardboard is 15 inches long and 8 inches wide, find the dimensions of the box that will yield the maximum volume.

5. Metal Fabrication If an open box is made from a tin sheet 8 inches square by cutting out identical squares from each corner and bending up the resulting flaps, determine the dimensions of the largest box that can be made.

6. **Minimizing Construction Costs**   If an open box has a square base and a volume of 108 cubic inches, and is constructed from a tin sheet, find the dimensions of the box, assuming a minimum amount of material is used in its construction.

7. **Minimizing Construction Costs**   What are the dimensions of a closed rectangular box that has a square cross section and a capacity of 128 cubic inches, and is constructed using the least amount of material?

8. **Minimizing Construction Costs**   A rectangular box is to have a square base and a volume of 20 cubic feet. If the material for the base costs 30 cents per square foot, the material for the sides costs 10 cents per square foot, and the material for the top costs 20 cents per square foot, determine the dimensions of the box that can be constructed at minimum cost.

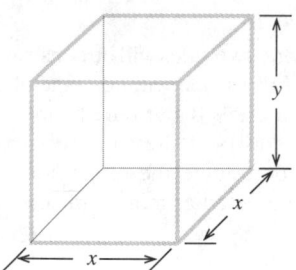

9. **Parcel Post Regulations**   Postal regulations specify that a parcel sent by parcel post may have a combined length and girth of no more than 108 inches. Find the dimensions of a rectangular package that has a square cross section and the largest volume that may be sent through the mail. What is the volume of such a package?
[*Hint:* The length plus the girth is $4x + h$ (see the accompanying figure).]

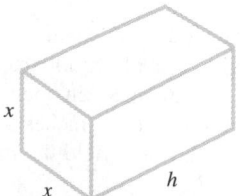

10. **Book Design**   A production editor at Saunders-Roe Publishing decided that the pages of a book should have

1-inch margins at the top and bottom and $\frac{1}{2}$-inch margins on the sides. She further stipulated that each page should have an area of 50 square inches (see the accompanying figure). Determine the page dimensions that will result in the maximum printed area on the page.

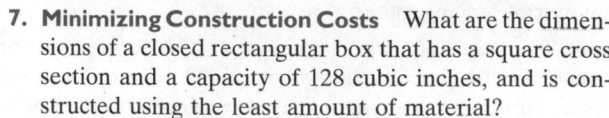

11. **Parcel Post Regulations**   Postal regulations specify that a parcel sent by parcel post may have a combined length and girth of no more than 108 inches. Find the dimensions of the cylindrical package of greatest volume that may be sent through the mail. What is the volume of such a package? Compare with exercise 9.
[*Hint:* The length plus the girth is $2\pi r + l$.]

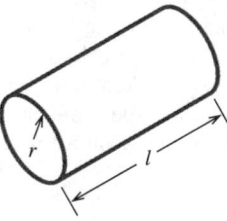

12. **Minimizing Costs**   For its beef stew, the Betty Moore Company uses tin containers that have the form of right circular cylinders. Find the radius and height of a container if it has a capacity of 36 cubic inches and is constructed using the least amount of metal.

13. **Product Design**   The cabinet that will enclose the Acrosonic model D loudspeaker system will be rectangular and will have an internal volume of 2.4 cubic feet. For aesthetic reasons, it has been decided that the height of the cabinet is to be 1.5 times its width. If the top, bottom, and sides of the cabinet are constructed of ve-

neer costing 40 cents per square foot, and the front (ignore the cutouts in the baffle) and rear are constructed of particle board costing 20 cents per square foot, what are the dimensions of the enclosure that can be constructed at a minimum cost?

14. **Designing a Norman Window** A Norman window has the shape of a rectangle surmounted by a semicircle (see the accompanying figure). If a Norman window is to have a perimeter of 28 ft, what should its dimensions be in order to allow the maximum amount of light through the window?

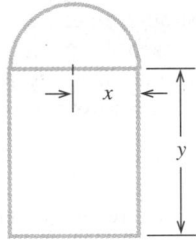

15. **Optimal Charter Flight Fare** If exactly 200 people sign up for a charter flight, the Leisure World Travel Agency charges $300 per person. However, if more than 200 people sign up for the flight (assume this is the case), then each fare is reduced by $1 for each additional person. Determine how many passengers will result in a maximum revenue for the travel agency. What is the maximum revenue? What would the fare per passenger be in this case?

[*Hint:* Let x denote the number of passengers beyond 200. Show that the revenue function R is given by $R(x) = (200 + x) \cdot (300 - x)$.]

16. **Maximizing Yield** An apple orchard has an average yield of 36 bushels of apples per tree if tree density is 22 trees per acre. For each unit increase in tree density, the yield decreases by 2 bushels. Find how many trees should be planted in order to maximize the yield.

17. **Strength of a Beam** A wooden beam has a rectangular cross section of height h inches and width w inches (see the accompanying figure). The strength S of the beam is directly proportional to its width and the square of its height. What are the dimensions of the cross section of the strongest beam that can be cut from a round log of diameter 24 inches?

[*Hint:* $S = kh^2w$, where k is a constant of proportionality.]

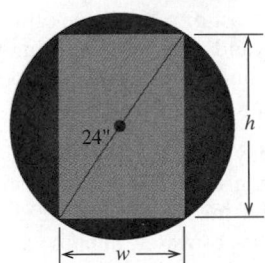

18. **Designing a Grain Silo** A grain silo has the shape of a right circular cylinder surmounted by a hemisphere (see the accompanying figure). If the silo is to have a capacity of 504π cubic feet, find the radius and height of the silo that requires the least amount of material to construct.

[*Hint:* The volume of the silo is $\pi r^2 h + \frac{2}{3}\pi r^3$, and the surface area (including the floor) is $\pi(3r^2 + 2rh)$.]

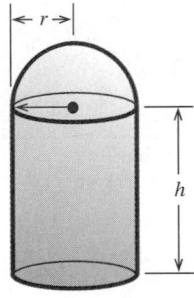

19. **Minimizing Construction Costs** In the following diagram, S represents the position of a power relay station located on a straight coast and E shows the location of a marine biology experimental station on an island. A cable is to be laid connecting the relay station with the

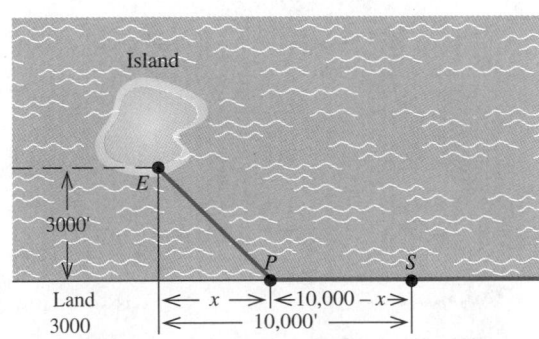

experimental station. If the cost of running the cable on land is $1 per running foot and the cost of running the cable under water is $3 per running foot, locate the point P that will result in a minimum cost (solve for x).

20. **Flights of Birds** During daylight hours, some birds fly more slowly over water than over land because some of their energy is expended in overcoming the down drafts of air over open bodies of water. Suppose a bird that flies at a constant speed of 4 mph over water and 6 mph over land starts its journey at the point E on an island and ends at its nest N on the shore of the mainland, as shown in the accompanying figure. Find the location of the point P that allows the bird to complete its journey in the minimum time (solve for x).

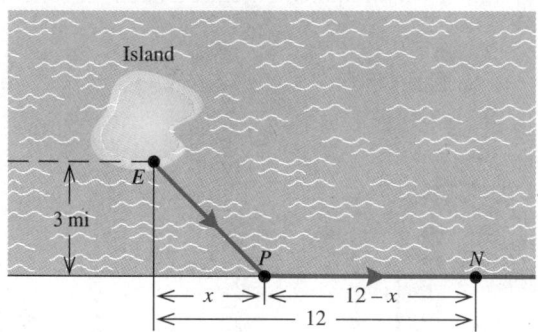
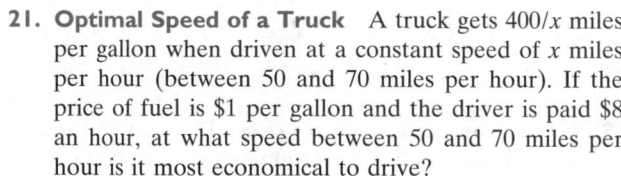
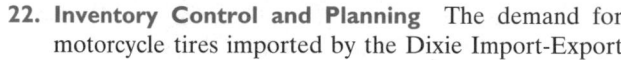

21. **Optimal Speed of a Truck** A truck gets $400/x$ miles per gallon when driven at a constant speed of x miles per hour (between 50 and 70 miles per hour). If the price of fuel is $1 per gallon and the driver is paid $8 an hour, at what speed between 50 and 70 miles per hour is it most economical to drive?

22. **Inventory Control and Planning** The demand for motorcycle tires imported by the Dixie Import-Export

Company is 40,000 per year and may be assumed to be uniform throughout the year. The cost of ordering a shipment of tires is $400, and the cost of storing each tire for a year is $2. Determine how many tires should be in each shipment if the ordering and storage costs are to be minimized. (Assume that each shipment arrives just as the previous one has been sold.)

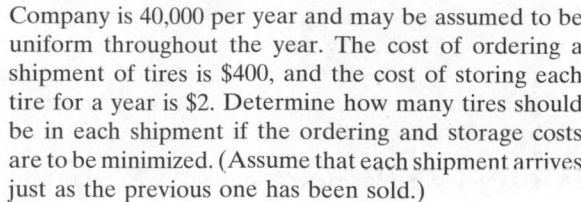

23. **Inventory Control and Planning** The McDuff Preserves Company expects to bottle and sell 2,000,000 32-ounce jars of jam. The company orders its containers from the Consolidated Bottle Company. The cost of ordering a shipment of bottles is $200, and the cost of storing each empty bottle for a year is 40 cents. Determine how many orders McDuff should place per year and how many bottles should be in each shipment if the ordering and storage costs are to be minimized. (Assume that each shipment of bottles is used up before the next shipment arrives.)

24. **Inventory Control and Planning** The Neilsen Cookie Company sells its assorted butter cookies in containers that have a net content of 1 pound. The estimated demand for the cookies is 1,000,000 units. The set-up cost for each production run is $500, and the manufacturing cost is 50 cents for each container of cookies. The cost of storing each container of cookies over the year is 40 cents. Assuming uniformity of demand throughout the year and instantaneous production, determine how many containers of cookies Neilsen should produce per production run in order to minimize the production cost. [*Hint:* Following the method of Example 5, show that the total production cost is given by the function

$$C(x) = \frac{500,000,000}{x} + 0.2x + 500,000$$

Then minimize the function C on the interval (0, 1,000,000).]

SOLUTIONS TO SELF-CHECK EXERCISES 12.5

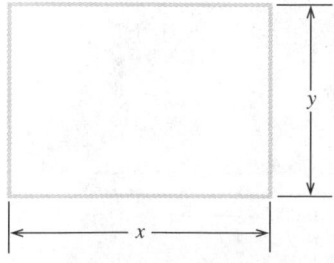

1. Let x and y (measured in feet) denote the length and width of the rectangular garden. Since the area is to be 300 square feet, we have

$$xy = 300$$

Next, the amount of fencing to be used is given by the perimeter, and this quantity is to be minimized. Thus, we want to minimize

$$2x + 2y$$

or, since $y = 300/x$ (obtained by solving for y in the first equation), we see that the expression to be minimized is

$$f(x) = 2x + 2\left(\frac{300}{x}\right)$$

$$= 2x + \frac{600}{x}$$

for positive values of x. Now

$$f'(x) = 2 - \frac{600}{x^2}$$

Setting $f'(x) = 0$ yields $x = -\sqrt{300}$ or $x = \sqrt{300}$. We consider only the critical point $x = \sqrt{300}$ since $-\sqrt{300}$ lies outside the interval $(0, \infty)$. We then compute

$$f''(x) = \frac{1200}{x^3}$$

Since

$$f''(\sqrt{300}) > 0$$

the Second Derivative Test implies that a relative minimum of f occurs at $x = \sqrt{300}$. In fact, since $f''(x) > 0$ for all x in $(0, \infty)$, we conclude that $x = \sqrt{300}$ gives rise to the absolute minimum of f. The corresponding value of y, obtained by substituting this value of x into the equation $xy = 300$, is $y = \sqrt{300}$. Therefore, the required dimensions of the vegetable garden are approximately 17.3 ft × 17.3 ft.

2. Let x denote the number of tires in each production run. Then the average number of tires in storage is $x/2$, so the storage cost incurred by the company is $2(x/2)$, or x dollars. Next, since the company needs to manufacture 1,000,000 tires for the year in order to meet the demand, the number of production runs is $1,000,000/x$. This gives set-up costs amounting to

$$4000\left(\frac{1,000,000}{x}\right) = \frac{4,000,000,000}{x}$$

dollars for the year. The total manufacturing cost is \$20,000,000. Thus, the total yearly cost incurred by the company is given by

$$C(x) = x + \frac{4,000,000,000}{x} + 20,000,000$$

Differentiating $C(x)$, we find

$$C'(x) = 1 - \frac{4,000,000,000}{x^2}$$

Setting $C'(x) = 0$ gives $x = 63,246$ as the critical point in the interval $(0, 1,000,000)$. Next, we find

$$C''(x) = \frac{8,000,000,000}{x^3}$$

Since $C''(x) > 0$ for all $x > 0$, we see that C is concave upward for all $x > 0$. Furthermore, $C''(63,246) > 0$ implies that $x = 63,246$ gives rise to a relative minimum of C (by the Second Derivative Test). Since C is always concave upward for $x > 0$, $x = 63,246$ gives the absolute minimum of C. Therefore, the company should manufacture 63,246 tires in each production run.

CHAPTER 12 SUMMARY OF PRINCIPAL TERMS

Terms

Increasing function	Concave downward
Decreasing function	Inflection point
Relative maximum	Second Derivative Test
Relative minimum	Vertical asymptote
Relative extrema	Horizontal asymptote
Critical point	Absolute extrema
First Derivative Test	Absolute maximum value
Concave upward	Absolute minimum value

CHAPTER 12 REVIEW EXERCISES

In exercises 1–10, (a) find the intervals where the given function f is increasing and where it is decreasing, (b) find the relative extrema of f, (c) find the intervals where f is concave upward and where it is concave downward, and (d) find the inflection points, if any, of f.

1. $f(x) = \frac{1}{3}x^3 - x^2 + x - 6$

2. $f(x) = (x - 2)^3$ **3.** $f(x) = x^4 - 2x^2$

4. $f(x) = x + \dfrac{4}{x}$ **5.** $f(x) = \dfrac{x^2}{x - 1}$

6. $f(x) = \sqrt{x - 1}$ **7.** $f(x) = (1 - x)^{1/3}$

8. $f(x) = x\sqrt{x - 1}$ **9.** $f(x) = \dfrac{2x}{x + 1}$

10. $f(x) = \dfrac{-1}{1 + x^2}$

In exercises 11–18, obtain as much information as possible on each of the given functions. Then use this information to sketch the graph of the function.

11. $f(x) = x^2 - 5x + 5$ **12.** $f(x) = -2x^2 - x + 1$

13. $g(x) = 2x^3 - 6x^2 + 6x + 1$

14. $g(x) = \frac{1}{3}x^3 - x^2 + x - 3$

15. $h(x) = x\sqrt{x - 2}$ **16.** $h(x) = \dfrac{2x}{1 + x^2}$

17. $f(x) = \dfrac{x - 2}{x + 2}$ **18.** $f(x) = x - \dfrac{1}{x}$

In exercises 19–22, find the horizontal and vertical asymptotes of the graphs of the given functions. Do not sketch the graphs.

19. $f(x) = \dfrac{1}{2x + 3}$ **20.** $f(x) = \dfrac{2x}{x + 1}$

21. $f(x) = \dfrac{5x}{x^2 - 2x - 8}$ **22.** $f(x) = \dfrac{x^2 + x}{x(x - 1)}$

In exercises 23–32, find the absolute maximum value and the absolute minimum value, if any, of the given function.

23. $f(x) = 2x^2 + 3x - 2$ **24.** $g(x) = x^{2/3}$

25. $g(t) = \sqrt{25 - t^2}$

26. $f(x) = \frac{1}{3}x^3 - x^2 + x + 1$ on $[0, 2]$

27. $h(t) = t^3 - 6t^2$ on $[2, 5]$

28. $g(x) = \dfrac{x}{x^2 + 1}$ on $[0, 5]$

29. $f(x) = x - \dfrac{1}{x}$ on $[1, 3]$

30. $h(t) = 8t - \dfrac{1}{t^2}$ on $[1, 3]$

31. $f(s) = s\sqrt{1 - s^2}$ on $[-1, 1]$

32. $f(x) = \dfrac{x^2}{x - 1}$ on $[-1, 3]$

33. Odyssey Travel Agency's monthly profit (in thousands of dollars) depends on the amount of money x spent on advertising per month according to the rule

$$P(x) = -x^2 + 8x + 20$$

where x is also measured in thousands of dollars. What should Odyssey's monthly advertising budget be in order to maximize its monthly profits?

34. The Department of the Interior of an African country began to record an index of environmental quality to measure progress or decline in the environmental quality of its wildlife. The index for the years 1984 through 1994 is approximated by the function

$$I(t) = \dfrac{50t^2 + 600}{t^2 + 10} \qquad (0 \le t \le 10)$$

a. Compute $I'(t)$ and show that $I(t)$ is decreasing on the interval $(0, 10)$.
b. Compute $I''(t)$. Study the concavity of the graph of I.
c. Sketch the graph of I.
d. Interpret your results.

35. The weekly demand for video discs manufactured by the Herald Record Company is given by

$$p = -0.0005x^2 + 60$$

where p denotes the unit price in dollars and x denotes the quantity demanded. The weekly total cost function associated with producing these discs is given by

$$C(x) = -0.001x^2 + 18x + 4000$$

where $C(x)$ denotes the total cost incurred in pressing x discs. Find the production level that will yield a maximum profit for the manufacturer.
[*Hint:* Use the quadratic formula.]

36. The total monthly cost (in dollars) incurred by the Carlota Music Company in manufacturing x units of its Professional Series guitars is given by the function

$$C(x) = 0.001x^2 + 100x + 4000$$

a. Find the average cost function $\overline{C}$.
b. Determine the production level that will result in the smallest average production cost.

37. The average worker at Wakefield Avionics, Inc., can assemble

$$N(t) = -2t^3 + 12t^2 + 2t \qquad (0 \le t \le 4)$$

ready-to-fly radio-controlled model airplanes t hours into the 8 A.M. to 12 noon morning shift. At what time during this shift is the average worker performing at peak efficiency?

38. You wish to construct a closed rectangular box that has a volume of 4 cubic feet. The length of the base of the box will be twice as long as its width. The material for the top and bottom of the box costs 30 cents per square foot. The material for the sides of the box costs 20 cents per square foot. Find the dimensions of the least expensive box that can be constructed.

39. The Lehen Vinters Company imports a certain brand of beer. The demand, which may be assumed to be uniform, is 800,000 cases per year. The cost of ordering a shipment of beer is $500, and the cost of storing each case of beer for a year is $2. Determine how many cases of beer should be in each shipment if the ordering and storage costs are to be kept to a minimum. (Assume that each shipment of beer arrives just as the previous one has been sold.)

40. Let

$$f(x) = \begin{cases} x^3 + 1 & \text{if } x \ne 0 \\ 2 & \text{if } x = 0 \end{cases}$$

a. Compute $f'(x)$ and show that it does not change sign as we move across $x = 0$.
b. Show that f has a relative maximum at $x = 0$. Does this contradict the First Derivative Test? Explain your answer.

The exponential function is, without doubt, the most important function in mathematics and its applications. After a brief introduction to the exponential function and its *inverse*, the logarithmic function, we discuss how to differentiate such functions. This lays the foundation for exploring the many applications involving exponential functions. For example, we look at the role played by exponential functions in computing earned interest on a bank account, in studying the growth of a bacteria population in the laboratory, in studying the way radioactive matter decays, in studying the rate at which a factory worker learns a certain process, and in studying the rate at which a communicable disease is spread over time.

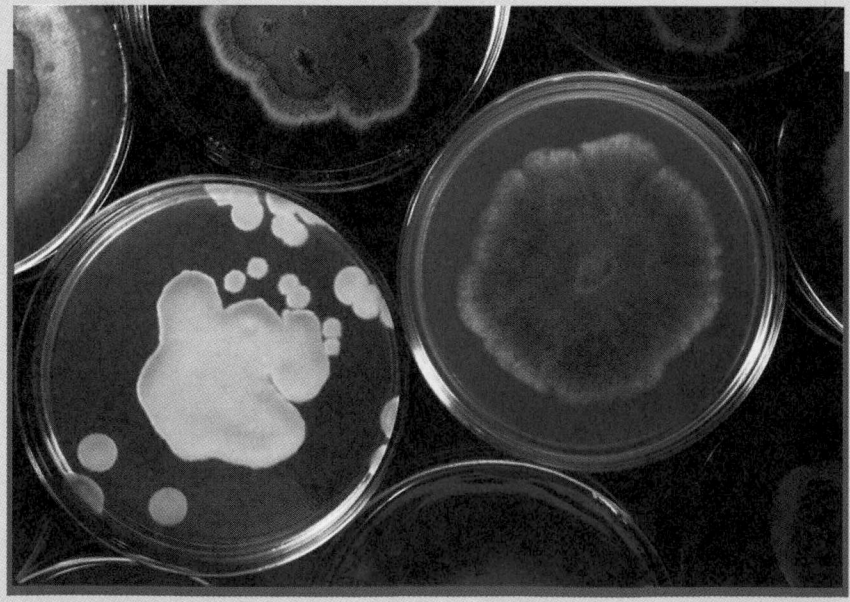

How many bacteria will there be in a culture at the end of a certain period of time? How fast will the bacteria population be growing at the end of that time? Example 1, page 915, answers these questions.

EXPONENTIAL AND LOGARITHMIC FUNCTIONS

13.1 EXPONENTIAL FUNCTIONS

Exponential Functions and Their Graphs

Suppose that you deposit a sum of $1000 in an account earning interest at the rate of 10% per year *compounded continuously* (the way most financial institutions compute interest). The accumulated amount at the end of t years $(0 \leq t \leq 20)$ is described by the function f, whose graph appears in Figure 13.1.* Such a function is called an *exponential function*. Observe that the graph of f rises rather slowly at first but very rapidly as time goes by. For purposes of comparison, we also show the graph of the function $y = g(t) = 1000(1 + 0.10t)$, which gives the accumulated amount for the same principal ($1000) earning *simple* interest at the rate of 10% per year. The moral of the story: It is never too early to start saving.

Figure 13.1

Under continuous compounding, a sum of money grows exponentially.

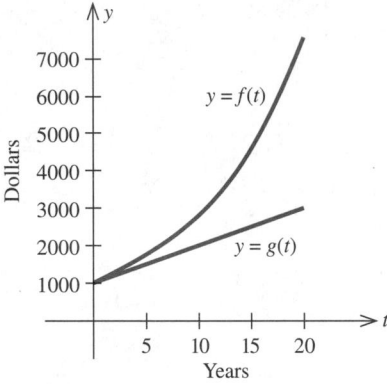

Exponential functions play an important role in many real-world applications, as you will see throughout this chapter. Observe that whenever b is a positive number and n is any real number, the expression b^n is a real number. This enables us to define an *exponential function* as follows:

EXPONENTIAL FUNCTION	The function defined by $$f(x) = b^x \qquad (b > 0, b \neq 1)$$ is called an **exponential function with base b and exponent x.** The domain of f is the set of all real numbers.

* We derive the rule for f later in this section.

For example, the exponential function with base 2 is the function

$$f(x) = 2^x$$

with domain $(-\infty, \infty)$. The values of $f(x)$ for selected values of x follow:

$$f(3) = 2^3 = 8, \quad f\left(\frac{3}{2}\right) = 2^{3/2} = 2 \cdot 2^{1/2} = 2\sqrt{2}, \quad f(0) = 2^0 = 1,$$

$$f(-1) = 2^{-1} = \frac{1}{2}, \quad f\left(-\frac{2}{3}\right) = 2^{-2/3} = \frac{1}{2^{2/3}} = \frac{1}{\sqrt[3]{4}}$$

Computations involving exponentials are facilitated by the laws of exponents. These laws were stated in Section 9.1, and you might want to review the material there. For convenience, however, we will restate these laws.

THE LAWS OF EXPONENTS

Let a and b be positive numbers and let x and y be real numbers. Then,

1. $b^x \cdot b^y = b^{x+y}$

2. $\dfrac{b^x}{b^y} = b^{x-y}$

3. $(b^x)^y = b^{xy}$

4. $(ab)^x = a^x b^x$

5. $\left(\dfrac{a}{b}\right)^x = \dfrac{a^x}{b^x}$

The use of the laws of exponents is illustrated in the next example.

EXAMPLE 1

a. $16^{7/4} \cdot 16^{-1/2} = 16^{7/4-1/2} = 16^{5/4} = 2^5 = 32$ (Law 1)

b. $\dfrac{8^{5/3}}{8^{-1/3}} = 8^{5/3-(-1/3)} = 8^2 = 64$ (Law 2)

c. $(64^{4/3})^{-1/2} = 64^{(4/3)(-1/2)} = 64^{-2/3}$

$$= \frac{1}{64^{2/3}} = \frac{1}{(64^{1/3})^2} = \frac{1}{4^2} = \frac{1}{16} \quad \text{(Law 3)}$$

d. $(16 \cdot 81)^{-1/4} = 16^{-1/4} \cdot 81^{-1/4} = \dfrac{1}{16^{1/4}} \cdot \dfrac{1}{81^{1/4}} = \dfrac{1}{2} \cdot \dfrac{1}{3} = \dfrac{1}{6}$ (Law 4)

e. $\left(\dfrac{3^{1/2}}{2^{1/3}}\right)^4 = \dfrac{3^{4/2}}{2^{4/3}} = \dfrac{9}{2^{4/3}}$ (Law 5)

       ○ ○ ○

EXAMPLE 2   Let $f(x) = 2^{2x-1}$. Find the value of x for which $f(x) = 16$.

Solution   We want to solve the equation

$$2^{2x-1} = 16 = 2^4$$

This equation holds if and only if

$$2x - 1 = 4 \quad\quad (b^m = b^n \Rightarrow m = n)$$

giving $x = \frac{5}{2}$.

       ○ ○ ○

Exponential functions play an important role in mathematical analysis. Because of their special characteristics, they are some of the most useful functions and are found in virtually every field where mathematics is applied. To mention a few examples: Under ideal conditions the number of bacteria present at any time t in a culture may be described by an exponential function of t; radioactive substances decay over time in accordance with an "exponential" law of decay; money left on fixed deposit and earning compound interest grows exponentially; and some of the most important distribution functions encountered in statistics are exponential.

Let us begin our investigation into the properties of exponential functions by studying their graphs.

EXAMPLE 3   Sketch the graph of the exponential function $y = 2^x$.

Figure 13.2
The graph of $y = 2^x$.

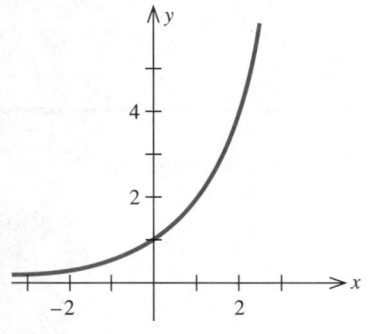

Solution   First, as discussed earlier, the domain of the exponential function $y = f(x) = 2^x$ is the set of real numbers. Next, putting $x = 0$ gives $y = 2^0 = 1$, the y-intercept of f. There is no x-intercept since there is no value of x for which $y = 0$. To find the range of f, consider the following table of values:

x	-5	-4	-3	-2	-1	0	1	2	3	4	5
y	1/32	1/16	1/8	1/4	1/2	1	2	4	8	16	32

We see from these computations that 2^x decreases and approaches zero as x decreases without bound and that 2^x increases without bound as x increases without bound. Thus, the range of f is the interval $(0, \infty)$—that is, the set of positive real numbers. Finally, we sketch the graph of $y = f(x) = 2^x$ in Figure 13.2.   ◐ ◐ ◐

EXAMPLE 4   Sketch the graph of the exponential function $y = (1/2)^x$.

Figure 13.3
The graph of $y = (1/2)^x$.

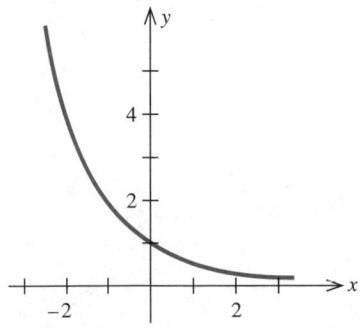

Solution   The domain of the exponential function $y = (1/2)^x$ is the set of all real numbers. The y-intercept is $(1/2)^0 = 1$; there is no x-intercept since there is no value of x for which $y = 0$. From the following table of values

x	-5	-4	-3	-2	-1	0	1	2	3	4	5
y	32	16	8	4	2	1	1/2	1/4	1/8	1/16	1/32

we deduce that $(1/2)^x = 1/2^x$ increases without bound as x decreases without bound and that $(1/2)^x$ decreases and approaches zero as x increases without bound. Thus, the range of f is the interval $(0, \infty)$. The graph of $y = f(x) = (1/2)^x$ is sketched in Figure 13.3.   ◐ ◐ ◐

The functions $y = 2^x$ and $y = (1/2)^x$, whose graphs you studied in Examples 3 and 4, are special cases of the exponential function $y = f(x) = b^x$, obtained by setting $b = 2$ and $b = 1/2$, respectively. In general, the exponential function $y = b^x$ with $b > 1$ has a graph similar to $y = 2^x$, whereas the graph of $y = b^x$ for $0 < b < 1$ is similar to that of $y = (1/2)^x$ (exercises 21 and 22). When $b = 1$, the function $y = b^x$ reduces to the constant function $y = 1$. For comparison, the graphs of all three functions are sketched in Figure 13.4.

Figure 13.4
$y = b^x$ is an increasing function of x if $b > 1$, a constant function if $b = 1$, and a decreasing function if $0 < b < 1$.

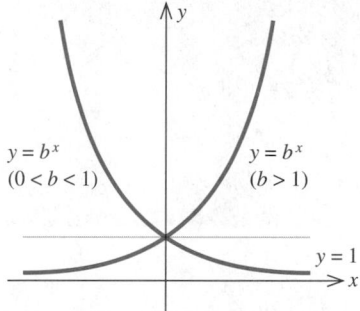

PROPERTIES OF THE EXPONENTIAL FUNCTION

The exponential function $y = b^x$ $(b > 0, b \neq 1)$ has the following properties:

1. Its domain is $(-\infty, \infty)$.
2. Its range is $(0, \infty)$.
3. Its graph passes through the point $(0, 1)$.
4. It is continuous on $(-\infty, \infty)$.
5. It is increasing on $(-\infty, \infty)$ if $b > 1$ and decreasing on $(-\infty, \infty)$ if $b < 1$.

The Base e

Exponential functions to the base e, where e is an irrational number whose value is $2.7182818 \ldots$, play an important role in both theoretical and applied problems. It can be shown, although we will not do so here, that

$$e = \lim_{m \to \infty} \left(1 + \frac{1}{m}\right)^m \tag{1}$$

However, you may convince yourself of the plausibility of this definition of the number e by examining Table 13.1, which may be constructed with the help of a calculator.

Table 13.1

m	10	100	1000	10,000	100,000	1,000,000
$\left(1 + \dfrac{1}{m}\right)^m$	2.59374	2.70481	2.71692	2.71815	2.71827	2.71828

EXPLORING WITH TECHNOLOGY

In order to obtain a visual confirmation of the fact that the expression $(1 + 1/m)^m$ approaches the number $e = 2.71828\ldots$ as m increases without bound, plot the graph of $f(x) = (1 + 1/x)^x$ in a suitable viewing rectangle and observe that $f(x)$ approaches $2.71828\ldots$ as x increases without bound. Use **ZOOM** and **TRACE** to find the value of $f(x)$ for large values of x.

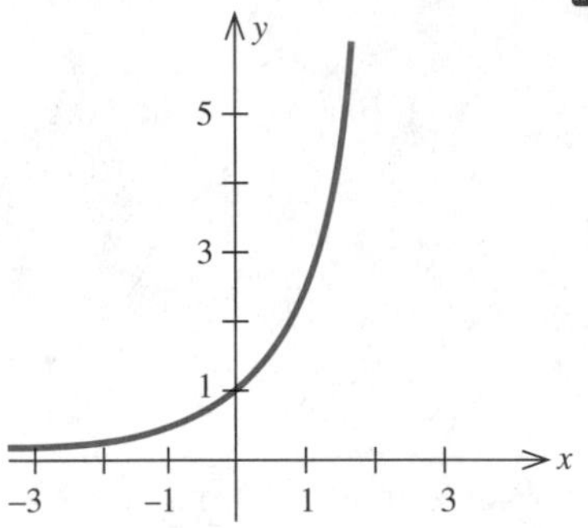

Figure 13.5
The graph of $y = e^x$.

EXAMPLE 5   Sketch the graph of the function $y = e^x$.

Solution   Since $e > 1$, it follows from our previous discussion that the graph of $y = e^x$ is similar to the graph of $y = 2^x$ (see Figure 13.2). With the aid of a calculator we obtain the following table.

x	-3	-2	-1	0	1	2	3
y	0.05	0.14	0.37	1	2.72	7.39	20.09

The graph of $y = e^x$ is sketched in Figure 13.5.

Next, we consider another exponential function to the base e that is closely related to the previous function and is particularly useful in constructing models that describe "exponential decay."

Figure 13.6
The graph of $y = e^{-x}$.

EXAMPLE 6   Using a calculator, sketch the graph of the function $y = e^{-x}$.

Solution   Since $e > 1$, it follows that $0 < 1/e < 1$, so $f(x) = e^{-x} = 1/e^x = (1/e)^x$ is an exponential function with base less than 1. Therefore it has a graph similar to that of the exponential function $y = (1/2)^x$. As before, we construct the following table of values of $y = e^{-x}$ for selected values of x.

x	-3	-2	-1	0	1	2	3
y	20.09	7.39	2.72	1	0.37	0.14	0.05

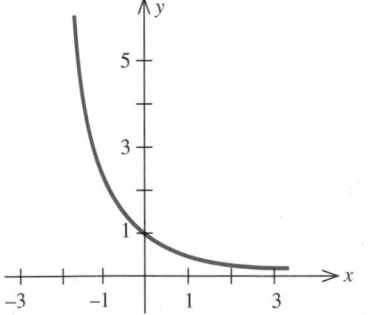

Using this table, we sketch the graph of $y = e^{-x}$ in Figure 13.6.

Continuous Compounding of Interest

One question that arises naturally in the study of compound interest is: What happens to the accumulated amount over a fixed period of time if the interest is computed more and more frequently? Intuition suggests that the more often interest is compounded, the larger the accumulated amount will be. This is confirmed by the results of Example 3, Section 5.1, page 277, where we found that the accumulated amounts did, in fact, increase when we increased the number of conversion periods per year.

This leads us to another question: Does the accumulated amount approach a limit when the interest is computed more and more frequently over a fixed period of time? To answer this question, let us look again at the compound interest formula:

$$A = P\left(1 + \frac{r}{m}\right)^{mt} \tag{2}$$

Recall that m is the number of conversion periods per year. So to find an answer to our problem, we should let m approach infinity (get larger and larger) in (2). But first we rewrite this equation in the form

$$A = P\left[\left(1 + \frac{r}{m}\right)^m\right]^t \qquad [\text{Since } b^{xy} = (b^x)^y]$$

Now, letting $m \to \infty$, we find that

$$\lim_{m\to\infty}\left[P\left(1 + \frac{r}{m}\right)^m\right]^t = P\left[\lim_{m\to\infty}\left(1 + \frac{r}{m}\right)^m\right]^t \qquad (\text{Why?})$$

Next, upon making the substitution $u = m/r$ and observing that $u \to \infty$ as $m \to \infty$, we reduce the foregoing expression to

$$P\left[\lim_{u\to\infty}\left(1 + \frac{1}{u}\right)^{ur}\right]^t = P\left[\lim_{u\to\infty}\left(1 + \frac{1}{u}\right)^u\right]^{rt}$$

But

$$\lim_{u\to\infty}\left(1 + \frac{1}{u}\right)^u = e \qquad [\text{Using (1)}]$$

so

$$\lim_{m\to\infty} P\left[\left(1 + \frac{r}{m}\right)^m\right]^t = Pe^{rt}$$

Our computations tell us that as the frequency with which interest is compounded increases without bound, the accumulated amount approaches Pe^{rt}. In this situation, we say that interest is *compounded continuously*. Let us summarize this important result.

CONTINUOUS COMPOUND INTEREST FORMULA

$$A = Pe^{rt} \tag{3}$$

where

P = Principal
r = Annual interest rate compounded continuously
t = Time in years
A = Accumulated amount at the end of t years

EXPLORING WITH TECHNOLOGY

In the opening paragraph of Section 13.1, we pointed out that the accumulated amount of an account earning interest *compounded continuously* will eventually outgrow by far the accumulated amount of an account earning interest at the same nominal rate but earning simple interest. Illustrate this fact using the following example.

Suppose you deposit $1000 in account I, earning interest at the rate of 10% per year compounded continuously so that the accumulated amount at the end of t years is $A_1(t) = 1000e^{0.1t}$. Suppose you also deposit $1000 in account II, earning simple interest at the rate of 10% per year so that the accumulated amount at the end of t years is $A_2(t) = 1000(1 + 0.1t)$. Use a graphing utility to sketch the graphs of the functions A_1 and A_2 in the viewing rectangle $[0, 20] \times [0, 10{,}000]$ to see the accumulated amounts $A_1(t)$ and $A_2(t)$ over a 20-year period.

◐   ◐   ◐

 EXAMPLE 7   Find the accumulated amount after three years if $1000 is invested at 8% per year compounded (a) daily (take the number of days in a year to be 365) and (b) continuously.

Solution

a. Using the compound interest formula with $P = 1000$, $r = 0.08$, $m = 365$, $t = 3$, and $n = (365)(3) = 1095$, we find

$$A = 1000 \left(1 + \frac{0.08}{365} \right)^{1095} \approx 1271.22 \qquad \left[A = P\left(1 + \frac{r}{m} \right)^n \right]$$

or $1271.22.

b. Here we use formula (3) with $P = 1000$, $r = 0.08$, and $t = 3$, obtaining

$$A = 1000e^{(0.08)(3)}$$
$$\approx 1271.25 \qquad \text{(Using the "}e^x\text{" key)}$$

or $1271.25.   ◐ ◐ ◐

Observe that the accumulated amounts corresponding to interest compounded daily and interest compounded continuously differ by very little. The continuous compound interest formula is a very important tool in theoretical work in financial analysis.

If we solve formula (3) for P, we obtain

$$P = Ae^{-rt} \tag{4}$$

which gives the present value in terms of the future (accumulated) value for the case of continuous compounding.

EXAMPLE 8   The Blakely Investment Company owns an office building in the commercial district of a city. As a result of the continued success of an urban renewal program, local business is booming. The market value of Blakely's property is

$$V(t) = 300,000e^{\sqrt{t}/2}$$

where $V(t)$ is measured in dollars and t is the time in years from the present. If the expected rate of inflation is 9% compounded continuously for the next ten years, find an expression for the present value $P(t)$ of the market price of the property valid for the next ten years. Compute $P(7)$, $P(8)$, and $P(9)$, and interpret your results.

Solution   Using formula (4) with $A = V(t)$ and $r = 0.09$, we find that the present value of the market price of the property t years from now is

$$
\begin{aligned}
P(t) &= V(t)e^{-0.09t} \\
&= 300,000e^{-0.09t + \sqrt{t}/2} \qquad (0 \le t \le 10)
\end{aligned}
$$

Letting $t = 7$, 8, and 9, respectively, we find

$$
\begin{aligned}
P(7) &= 300,000e^{-0.09(7) + \sqrt{7}/2} = 599,837 \quad \text{or} \quad \$599,837 \\
P(8) &= 300,000e^{-0.09(8) + \sqrt{8}/2} = 600,640 \quad \text{or} \quad \$600,640 \\
P(9) &= 300,000e^{-0.09(9) + \sqrt{9}/2} = 598,115 \quad \text{or} \quad \$598,115
\end{aligned}
$$

From the results of these computations, we see that the present value of the property's market price seems to decrease after a certain period of growth. This suggests that there is an optimal time for the owners to sell. Later we will show that the highest present value of the property's market price is $600,779, which occurs at time $t = 7.72$ years.   ○ ○ ○

SELF-CHECK EXERCISES 13.1

1. Solve the equation $2^{2x+1} \cdot 2^{-3} = 2^{x-1}$.

2. Sketch the graph of $y = e^{0.4x}$.

3. **a.** What is the accumulated amount after five years if $10,000 is invested at 10% per year compounded continuously?
   b. Find the present value of $10,000 due in five years at an interest rate of 10% per year compounded continuously.

Solutions to Self-Check Exercises 13.1 can be found on page 885.

EXPLORING WITH TECHNOLOGY

The effective rate of interest is given by

$$r_{\text{eff}} = \left(1 + \frac{r}{m}\right)^m - 1$$

where the number of conversion periods per year is m. In exercise 37 on page 884 you will be asked to show that the effective rate of interest r_{eff} corresponding to a nominal interest rate r per year compounded continuously is given by

$$\hat{r}_{\text{eff}} = e^r - 1$$

To obtain a visual confirmation of this result, consider the special case where $r = 0.1$ (10% per year).

1. Use a graphing utility to plot the graph of both

$$y_1 = \left(1 + \frac{0.1}{x}\right)^x - 1 \quad \text{and} \quad y_2 = e^{0.1} - 1$$

in the viewing rectangle $[0, 3] \times [0, 0.12]$.

2. Does your result seem to imply that

$$\left(1 + \frac{r}{m}\right)^m - 1$$

approaches

$$\hat{r}_{\text{eff}} = e^r - 1$$

as m increases without bound for the special case $r = 0.1$?

◐   ◐   ◐

13.1 EXERCISES

In exercises 1–6, evaluate each expression.

1. a. $4^{-3} \cdot 4^5$ **b.** $3^{-3} \cdot 3^6$

2. a. $(2^{-1})^3$ **b.** $(3^{-2})^3$

3. a. $9(9)^{-1/2}$ **b.** $5(5)^{-1/2}$

4. a. $[(-\frac{1}{2})^3]^{-2}$ **b.** $[(-\frac{1}{3})^2]^{-3}$

5. a. $\dfrac{(-3)^4(-3)^5}{(-3)^8}$ **b.** $\dfrac{(2^{-4})(2^6)}{2^{-1}}$

6. a. $3^{1/4} \cdot (9)^{-5/8}$ **b.** $2^{3/4} \cdot (4)^{-3/2}$

In exercises 7–12, simplify each expression.

7. a. $(64x^9)^{1/3}$ **b.** $(25x^3y^4)^{1/2}$

8. a. $(2x^3)(-4x^{-2})$ **b.** $(4x^{-2})(-3x^5)$

PORTFOLIO

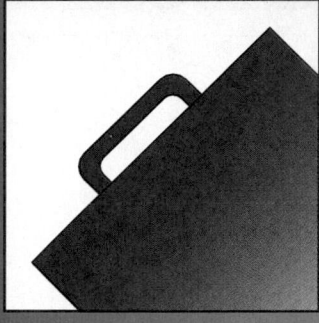

Misato Nakazaki

Title: Assistant Vice President
Institution: A large investment corporation

In the securities industry, buying and selling stocks and bonds has always required a mastery of concepts and formulas that outsiders find confusing. As a bond seller, Nakazaki routinely uses terms such as *issue, maturity, current yield,* and *callable* and *convertible bonds.*

These terms, however, are easily defined. When corporations issue bonds, they are borrowing money at a fixed rate of interest. The bonds are scheduled to mature—to be paid back—on a specific date as far as 30 years into the future. Callable bonds allow the issuer to pay off the loans prior to their expected maturity, reducing overall interest payments. In its simplest terms, current yield is the price of a bond multiplied by the interest rate at which the bond is issued. For example, a bond with a face value of $1000 and an interest rate of 10% yields $100 per year in interest payments. When that same bond is resold at a premium on the secondary market for $1200, its current yield nets only an 8.3% rate of return based on the higher purchase price.

Bonds attract investors for many reasons. A key variable is the sensitivity of the bond's price to future changes in interest rates. If investors get locked into a low-paying bond when future bonds pay higher yields, they lose money. Nakazaki stresses that "no one knows for sure what rates will be over time." Using differentials allows her to calculate interest-rate sensitivity for clients as they ponder purchase decisions.

Computerized formulas, "whose basis is calculus," says Nakazaki, help her factor the endless stream of numbers flowing across her desk.

On a typical day, Nakazaki might be given a bid on "10 million, GMAC, 8.5%, January 2003." Translation: Her customer wants her to buy General Motors Acceptance Corporation bonds with a face value of $10 million and an interest rate of 8.5%, maturing in January 2003.

After she calls her firm's trader to find out the yield on the bond in question, Nakazaki enters the price and other variables, such as the interest rate and date of maturity, and the computer prints out the answers. Nakazaki can then relay to her client the bond's current yield, accrued interest, and so on. In Nakazaki's rapid-fire work environment, such speed is essential. Nakazaki cautions that "computer users have to understand what's behind the formulas." The software "relies on the basics of calculus. If people don't understand the formula, it's useless for them to use the calculations."

◗ ◗ ◗

With an MBA from New York University, Nakazaki typifies the younger generation of Japanese women who have chosen to succeed in the business world. Since earning her degree, she has sold bonds for a global securities firm in New York City.

Nakazaki's client list reads like a who's who of the leading Japanese banks, insurance companies, mutual funds, and corporations. As institutional buyers, her clients purchase large blocks of American corporate bonds and mortgage-backed securities such as Ginnie Maes.

USING TECHNOLOGY

Although the proof is outside the scope of this book, it can be proved that an exponential function of the form $f(x) = b^x$, where $b > 1$, will ultimately grow faster than the power function $g(x) = x^n$ for *any* positive real number n. To give a visual demonstration of this result for the special case of the exponential function $f(x) = e^x$, we can use a graphing calculator to plot the graphs of both f and g (for selected values of n) on the same set of axes in an appropriate viewing rectangle and observe that the graph of f ultimately lies above that of g.

EXAMPLE 1 Use a graphing utility to plot the graphs of (a) $f(x) = e^x$ and $g(x) = x^3$ on the same set of axes in the viewing rectangle $[0, 6] \times [0, 250]$ and (b) $f(x) = e^x$ and $g(x) = x^5$ in the viewing rectangle $[0, 20] \times [0, 1,000,000]$.

Solution

a. The graphs of $f(x) = e^x$ and $g(x) = x^3$ in the viewing rectangle $[0, 6] \times [0, 250]$ are shown in Figure T1a.

b. The graphs of $f(x) = e^x$ and $g(x) = x^5$ in the viewing rectangle $[0, 20] \times [0, 1,000,000]$ are shown in Figure T1b.

Figure T1

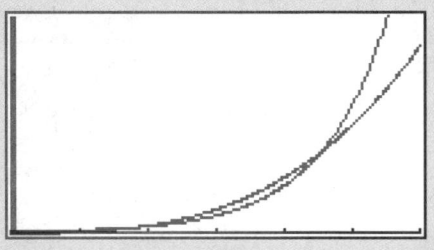
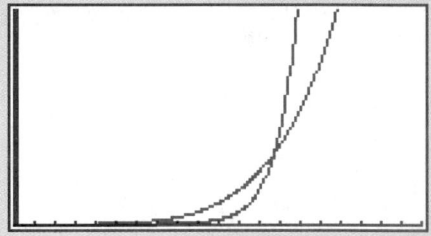

(a) The graphs of $f(x) = e^x$ and $g(x) = x^3$ in the viewing rectangle $[0, 6] \times [0, 250]$.

(b) The graphs of $f(x) = e^x$ and $g(x) = x^5$ in the viewing rectangle $[0, 20] \times [0, 1,000,000]$.

In the exercises that follow, you are asked to use a graphing utility to reveal the properties of exponential functions.

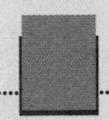

EXERCISES

In exercises 1 and 2, use a graphing utility to plot the graphs of the functions f and g on the same set of axes in the specified viewing rectangle.

1. $f(x) = e^x$ and $g(x) = x^2$; $[0, 4] \times [0, 30]$

2. $f(x) = e^x$ and $g(x) = x^4$; $[0, 15] \times [0, 20{,}000]$

In exercises 3 and 4, use a graphing utility to plot the graphs of the functions f and g on the same set of axes in an appropriate viewing rectangle to demonstrate that f ultimately grows faster than g. [Note: Your answer will not be unique.]

3. $f(x) = 2^x$ and $g(x) = x^{2.5}$

4. $f(x) = 3^x$ and $g(x) = x^3$

5. Use a graphing utility to plot the graphs of $f(x) = 2^x$, $g(x) = 3^x$, and $h(x) = 4^x$ on the same set of axes in the viewing rectangle $[0, 5] \times [0, 100]$. Comment on the relationship between the base b and the growth of the function $f(x) = b^x$.

6. Use a graphing utility to plot the graphs of $f(x) = (1/2)^x$, $g(x) = (1/3)^x$, and $h(x) = (1/4)^x$ on the same set of axes in the viewing rectangle $[0, 4] \times [0, 1]$. Comment on the relationship between the base b and the growth of the function $f(x) = b^x$.

7. Use a graphing utility to plot the graphs of $f(x) = e^x$, $g(x) = 2e^x$, and $h(x) = 3e^x$ on the same set of axes in the viewing rectangle $[-3, 3] \times [0, 10]$. Comment on the role played by the constant k in the graph of $f(x) = ke^x$.

8. Use a graphing utility to plot the graphs of $f(x) = -e^x$, $g(x) = -2e^x$, and $h(x) = -3e^x$ on the same set of axes in the viewing rectangle $[-3, 3] \times [-10, 0]$. Comment on the role played by the constant k in the graph of $f(x) = ke^x$.

9. Use a graphing utility to plot the graphs of $f(x) = e^{0.5x}$, $g(x) = e^x$, and $h(x) = e^{1.5x}$ on the same set of axes in the viewing rectangle $[-2, 2] \times [0, 4]$. Comment on the role played by the constant k in the graph of $f(x) = e^{kx}$.

10. Use a graphing utility to plot the graphs of $f(x) = e^{-0.5x}$, $g(x) = e^{-x}$, and $h(x) = e^{-1.5x}$ on the same set of axes in the viewing rectangle $[-2, 2] \times [0, 4]$. Comment on the role played by the constant k in the graph of $f(x) = e^{kx}$.

9. a. $\dfrac{6a^{-5}}{3a^{-3}}$ **b.** $\dfrac{4b^{-4}}{12b^{-6}}$

10. a. $y^{-3/2}y^{5/3}$ **b.** $x^{-3/5}x^{8/3}$

11. a. $(2x^3y^2)^3$ **b.** $(4x^2y^2z^3)^2$

12. a. $\dfrac{5^0}{(2^{-3}x^{-3}y^2)^2}$ **b.** $\dfrac{(x+y)(x-y)}{(x-y)^0}$

In exercises 13–20, solve the equation for x.

13. $6^{2x} = 6^4$ **14.** $5^{-x} = 5^3$

15. $3^{3x-4} = 3^5$ **16.** $10^{2x-1} = 10^{x+3}$

17. $(2.1)^{x+2} = (2.1)^5$ **18.** $(-1.3)^{x-2} = (-1.3)^{2x+1}$

19. $8^x = \left(\dfrac{1}{32}\right)^{x-2}$ **20.** $3^{x-x^2} = \dfrac{1}{9^x}$

In exercises 21–29, sketch the graphs of the given functions on the same axes. A calculator is recommended for these exercises.

21. $y = 2^x$, $y = 3^x$, and $y = 4^x$

22. $y = (\tfrac{1}{2})^x$, $y = (\tfrac{1}{3})^x$, and $y = (\tfrac{1}{4})^x$

23. $y = 2^{-x}$, $y = 3^{-x}$, and $y = 4^{-x}$

24. $y = 4^{0.5x}$ and $y = 4^{-0.5x}$

25. $y = 4^{0.5x}$, $y = 4^x$, and $y = 4^{2x}$

26. $y = e^x$, $y = 2e^x$, and $y = 3e^x$

27. $y = e^{0.5x}$, $y = e^x$, and $y = e^{1.5x}$

28. $y = e^{-0.5x}$, $y = e^{-x}$, and $y = e^{-1.5x}$

29. $y = 0.5e^{-x}$, $y = e^{-x}$, and $y = 2e^{-x}$

30. Find the accumulated amount after four years if $5000 is invested at 8% per year compounded continuously.

31. Investment Options Investment A offers a 10% return compounded semiannually, and investment B offers a 9.75% return compounded continuously. Which investment has a higher rate of return over a four-year period?

32. Present Value Find the present value of $59,673 due in five years at an interest rate of 8% per year compounded continuously.

33. Saving for College Having received a large inheritance, a child's parents wish to establish a trust for the child's college education. If they need an estimated $70,000 seven years from now, how much should they set aside in trust now, if they invest the money at 10.5% compounded (a) quarterly? (b) continuously?

34. Effect of Inflation on Salaries Mr. Lyons's current annual salary is $25,000. Ten years from now, how much will he need to earn in order to retain his present purchasing power if the rate of inflation over that period is 6% per year? Assume that inflation is continuously compounded.

35. Pensions Ms. Lindstrom, who is now 50 years old, is employed by a firm that guarantees her a pension of $40,000 per year at age 65. What is the present value of her first year's pension if inflation over the next 15 years is (a) 6%? (b) 8%? (c) 12%? Assume that inflation is continuously compounded.

36. Real Estate Investments An investor purchased a piece of waterfront property. Because of the development of a marina in the vicinity, the market value of the property is expected to increase according to the rule

$$V(t) = 80{,}000e^{\sqrt{t}/2}$$

where $V(t)$ is measured in dollars and t is the time in years from the present. If the rate of inflation is expected to be 9% compounded continuously for the next eight years, find an expression for the present value $P(t)$ of the property's market price valid for the next eight years. What is $P(t)$ expected to be in four years?

37. Show that the effective rate of interest $\hat{r}_{\text{eff}}$ that corresponds to a nominal interest rate r per year compounded continuously is given by

$$\hat{r}_{\text{eff}} = e^r - 1$$

[*Hint:* From formula (4) in Section 5.1, page 279, we see that the effective rate $\hat{r}_{\text{eff}}$ corresponding to a nominal interest rate r per year compounded m times a year is given by

$$\hat{r}_{\text{eff}} = \left(1 + \frac{r}{m}\right)^m - 1$$

Let m tend to infinity in this expression.]

38. Refer to exercise 37. Find the effective rate of interest that corresponds to a nominal rate of 10% per year compounded (a) quarterly, (b) monthly, and (c) continuously.

39. Investment Analysis Refer to exercise 37. Bank A pays interest on deposits at a 7% annual rate compounded quarterly, and bank B pays interest on deposits at a $7\frac{1}{8}$% annual rate compounded continuously. Which bank has the higher effective rate of interest?

SOLUTIONS TO SELF-CHECK EXERCISES 13.1

1. $2^{2x+1} \cdot 2^{-3} = 2^{x-1}$

    $\dfrac{2^{2x+1}}{2^{x-1}} \cdot 2^{-3} = 1$ (Dividing both sides by 2^{x-1})

    $2^{(2x+1)-(x-1)-3} = 1$

    $\qquad\quad 2^{x-1} = 1$

    This is true if and only if $x - 1 = 0$ or $x = 1$.

2. We first construct the following table of values.

x	-3	-2	-1	0	1	2	3	4
$y = e^{0.4x}$	0.3	0.5	0.7	1	1.5	2.2	3.3	5

Next, we plot these points and join them by a smooth curve to obtain the graph of f shown in the accompanying figure.

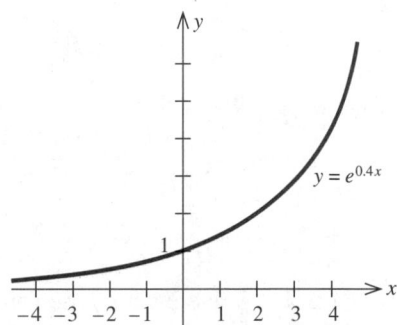

3. **a.** Using formula (3) with $P = 10,000$, $r = 0.1$, and $t = 5$, we find that the required accumulated amount is given by

    $$A = 10,000e^{(0.1)(5)}$$
    $$= 16,487.21$$

    or $16,487.21.

    b. Using formula (4) with $A = 10,000$, $r = 0.1$, and $t = 5$, we see that the required present value is given by

    $$P = 10,000e^{-(0.1)(5)}$$
    $$= 6065.31$$

    or $6065.31.

13.2 LOGARITHMIC FUNCTIONS

Logarithms

You are already familiar with exponential equations of the form

$$b^y = x \qquad (b > 0, b \neq 1)$$

where the variable x is expressed in terms of a real number b and a variable y. But what about solving this same equation for y? You may recall from your study of algebra that the number y is called the **logarithm of x to the base b** and is denoted by $\log_b x$. It is the exponent to which the base b must be raised in order to obtain the number x.

THE LOGARITHM OF x TO THE BASE b	$y = \log_b x$ if and only if $x = b^y$ $(x > 0)$

 Observe that the logarithm $\log_b x$ is defined for only positive values of x.

EXAMPLE 1

a. $\log_{10} 100 = 2$ since $100 = 10^2$

b. $\log_5 125 = 3$ since $125 = 5^3$

c. $\log_3 \dfrac{1}{27} = -3$ since $\dfrac{1}{27} = \dfrac{1}{3^3} = 3^{-3}$

d. $\log_{20} 20 = 1$ since $20 = 20^1$ o o o

EXAMPLE 2   Solve each of the following equations for x.

a. $\log_3 x = 4$ **b.** $\log_{16} 4 = x$ **c.** $\log_x 8 = 3$

Solution

a. By definition, $\log_3 x = 4$ implies $x = 3^4 = 81$.

b. $\log_{16} 4 = x$ is equivalent to $4 = 16^x = (4^2)^x = 4^{2x}$, from which we deduce that $x = \frac{1}{2}$.

c. Referring once again to the definition, we see that the equation $\log_x 8 = 3$ is equivalent to the equation $8 = x^3$, so $x = 2$.   o o o

The two widely used systems of logarithms are the system of **common logarithms,** which uses the number 10 as the base, and the system of **natural logarithms,** which uses the irrational number $e = 2.71828 \ldots$ as the base. It is standard practice to write **log** for $\log_{10}$ and **ln** for $\log_e$.

LOGARITHMIC NOTATION	$\log x = \log_{10} x$	(Common logarithm)
	$\ln x = \log_e x$	(Natural logarithm)

The system of natural logarithms is widely used in theoretical work. Using natural logarithms rather than logarithms to other bases often leads to simpler expressions.

Laws of Logarithms

Computations involving logarithms are facilitated by the following **laws of logarithms.**

LAWS OF LOGARITHMS	If m and n are positive numbers, then
	1. $\log_b mn = \log_b m + \log_b n$
	2. $\log_b \dfrac{m}{n} = \log_b m - \log_b n$
	3. $\log_b m^n = n \log_b m$
	4. $\log_b 1 = 0$
	5. $\log_b b = 1$

You will be asked to prove these laws in exercises 49–51. Their derivations are based on the definition of a logarithm and the corresponding laws of exponents. The following examples illustrate the properties of logarithms.

EXAMPLE 3

a. $\log(2 \cdot 3) = \log 2 + \log 3$

b. $\ln \dfrac{5}{3} = \ln 5 - \ln 3$ **c.** $\log \sqrt{7} = \log 7^{1/2} = \dfrac{1}{2} \log 7$

d. $\log_5 1 = 0$ **e.** $\log_{45} 45 = 1$ o o o

EXAMPLE 4
Given that $\log 2 \approx 0.3010$, $\log 3 \approx 0.4771$, and $\log 5 \approx 0.6990$, use the laws of logarithms to find

a. $\log 15$ **b.** $\log 7.5$ **c.** $\log 81$ **d.** $\log 50$

Solution

a. Note that $15 = 3 \cdot 5$, so by Law 1 for logarithms,

$$\begin{aligned}
\log 15 &= \log 3 \cdot 5 \\
&= \log 3 + \log 5 \\
&\approx 0.4771 + 0.6990 \\
&= 1.1761
\end{aligned}$$

b. Observing that $7.5 = 15/2 = (3 \cdot 5)/2$, we apply Laws 1 and 2, obtaining

$$\begin{aligned}
\log 7.5 &= \log \frac{(3)(5)}{2} \\
&= \log 3 + \log 5 - \log 2 \\
&\approx 0.4771 + 0.6990 - 0.3010 \\
&= 0.8751
\end{aligned}$$

c. Since $81 = 3^4$, we apply Law 3 to obtain

$$\begin{aligned}
\log 81 &= \log 3^4 \\
&= 4 \log 3 \\
&\approx 4(0.4771) \\
&= 1.9084
\end{aligned}$$

d. We write $50 = 5 \cdot 10$ and find

$$\begin{aligned}
\log 50 &= \log(5)(10) \\
&= \log 5 + \log 10 \\
&\approx 0.6990 + 1 \qquad \text{(Using Law 5)} \\
&= 1.6990
\end{aligned}$$

EXAMPLE 5 Expand and simplify the following expressions:

a. $\log_3 x^2 y^3$ **b.** $\log_2 \dfrac{x^2 + 1}{2^x}$ **c.** $\ln \dfrac{x^2 \sqrt{x^2 - 1}}{e^x}$

Solution

a.
$$\begin{aligned}
\log_3 x^2 y^3 &= \log_3 x^2 + \log_3 y^3 \qquad &\text{(Law 1)} \\
&= 2 \log_3 x + 3 \log_3 y \qquad &\text{(Law 3)}
\end{aligned}$$

b.
$$\begin{aligned}
\log_2 \frac{x^2 + 1}{2^x} &= \log_2(x^2 + 1) - \log_2 2^x \qquad &\text{(Law 2)} \\
&= \log_2(x^2 + 1) - x \log_2 2 \qquad &\text{(Law 3)} \\
&= \log_2(x^2 + 1) - x \qquad &\text{(Law 5)}
\end{aligned}$$

c.
$$\begin{aligned}
\ln \frac{x^2 \sqrt{x^2 - 1}}{e^x} &= \ln \frac{x^2(x^2 - 1)^{1/2}}{e^x} \qquad &\text{(Rewriting)} \\
&= \ln x^2 + \ln(x^2 - 1)^{1/2} - \ln e^x \qquad &\text{(Laws 1 and 2)} \\
&= 2 \ln x + \tfrac{1}{2} \ln(x^2 - 1) - x \ln e \qquad &\text{(Law 3)} \\
&= 2 \ln x + \tfrac{1}{2} \ln(x^2 - 1) - x \qquad &\text{(Law 5)}
\end{aligned}$$

Logarithmic Functions and Their Graphs

The definition of the logarithm implies that if b and n are positive numbers and b is different from 1, then the expression $\log_b n$ is a real number. This enables us to define a *logarithmic function* as follows:

LOGARITHMIC FUNCTION	The function defined by $$f(x) = \log_b x \qquad (b > 0, b \neq 1)$$ is called the **logarithmic function with base b.** The domain of f is the set of all positive numbers.

Figure 13.7
The points (u, v) and (v, u) are mirror reflections of each other.

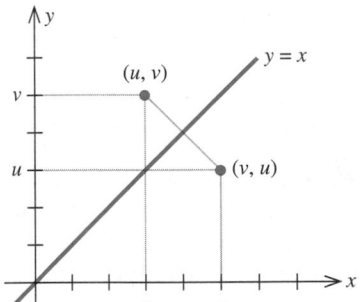

One easy way to obtain the graph of the logarithmic function $y = \log_b x$ is to construct a table of values of the logarithm (base b). However, another method—and a more instructive one—is based on exploiting the intimate relationship between logarithmic and exponential functions.

If a point (u, v) lies on the graph of $y = \log_b x$, then

$$v = \log_b u$$

But we can also write this equation in exponential form as

$$u = b^v$$

So the point (v, u) also lies on the graph of the function $y = b^x$. Let us look at the relationship between the points (u, v) and (v, u) and the line $y = x$ (Figure 13.7). If we think of the line $y = x$ as a mirror, then the point (v, u) is the mirror reflection of the point (u, v). Similarly, the point (u, v) is a mirror reflection of the point (v, u). We can take advantage of this relationship to help us draw the graph of logarithmic functions. For example, if we wish to draw the graph of $y = \log_b x$, where $b > 1$, then we need only draw the mirror reflection of the graph of $y = b^x$ with respect to the line $y = x$ (Figure 13.8).

Figure 13.8
The graphs of $y = b^x$ and $y = \log_b x$ are mirror reflections of each other.

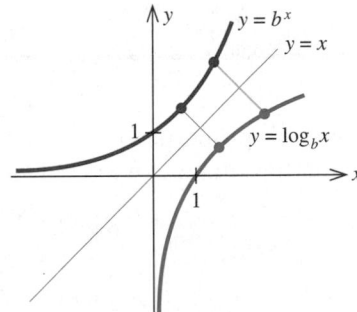

You may discover the following properties of the logarithmic function by taking the reflection of the graph of an appropriate exponential function (exercises 31 and 32).

**PROPERTIES OF THE
LOGARITHMIC
FUNCTION**

The logarithmic function $y = \log_b x$ $(b > 0, b \neq 1)$ has the following properties:

1. Its domain is $(0, \infty)$.
2. Its range is $(-\infty, \infty)$.
3. Its graph passes through the point $(1, 0)$.
4. It is continuous on $(0, \infty)$.
5. It is increasing on $(0, \infty)$ if $b > 1$ and decreasing on $(0, \infty)$ if $b < 1$.

Figure 13.9
*The graph of $y = \ln x$ is the mirror
reflection of the graph of $y = e^x$.*

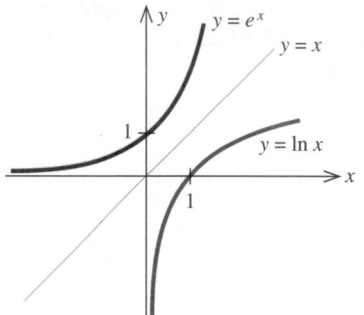

EXAMPLE 6   Sketch the graph of the function $y = \ln x$.

Solution   We first sketch the graph of $y = e^x$. Then the required graph is obtained by tracing the mirror reflection of the graph of $y = e^x$ with respect to the line $y = x$ (Figure 13.9).

Properties Relating the Exponential and Logarithmic Functions

We made use of the relationship that exists between the exponential function $f(x) = e^x$ and the logarithmic function $g(x) = \ln x$ when we sketched the graph of g in Example 6. This relationship is further described by the following properties, which are an immediate consequence of the definition of the logarithm of a number.

**PROPERTIES
RELATING e^x
AND ln x**

$$e^{\ln x} = x \quad \text{for } x > 0 \tag{5}$$
$$\ln e^x = x \quad \text{for any real number } x \tag{6}$$

(Try to verify these properties.)
    From properties 5 and 6, we conclude that the composite function

$$(f \circ g)(x) = f[g(x)]$$
$$= e^{\ln x} = x$$

and

$$(g \circ f)(x) = g[f(x)]$$
$$= \ln e^x = x$$

Thus,

$$f[g(x)] = g[f(x)]$$
$$= x$$

Any two functions f and g that satisfy this relationship are said to be **inverses** of each other. Note that the function f undoes what the function g does, and vice versa, so that the composition of the two functions in any order results in the identity function $F(x) = x$.

The relationships expressed in equations (5) and (6) are useful in solving equations that involve exponentials and logarithms.

 EXAMPLE 7   Solve the equation $2e^{x+2} = 5$.

Solution   We first divide both sides of the equation by 2 to obtain

$$e^{x+2} = \frac{5}{2} = 2.5$$

Next, taking the natural logarithm of each side of the equation and using equation (6), we have

$$\ln e^{x+2} = \ln 2.5$$
$$x + 2 = \ln 2.5$$
$$x = -2 + \ln 2.5$$
$$\approx -1.08$$

○ ○ ○

EXAMPLE 8   Solve the equation $5 \ln x + 3 = 0$.

Solution   Adding -3 to both sides of the equation leads to

$$5 \ln x = -3$$
$$\ln x = -\frac{3}{5} = -0.6$$

and so
$$e^{\ln x} = e^{-0.6}$$

Using equation (5), we conclude that

$$x = e^{-0.6}$$
$$\approx 0.55$$

○ ○ ○

Consider the equation $y = y_0 b^{kx}$, where y_0 and k are positive constants and $b > 0$, $b \neq 1$. Suppose we want to express y in the form $y = y_0 e^{px}$. Use the laws of logarithms to show that $p = k \ln b$ and hence that $y = y_0 e^{(k \ln b)x}$ is an alternative form of $y = y_0 b^{kx}$ using the base e.

SELF-CHECK EXERCISES 13.2

1. Sketch the graph of $y = 3^x$ and $y = \log_3 x$ on the same set of axes.
2. Solve the equation $3e^{x+1} - 2 = 4$.

Solutions to Self-Check Exercises 13.2 can be found on page 893.

13.2 EXERCISES

In exercises 1–10, express the given equation in logarithmic form.

1. $2^6 = 64$

2. $3^5 = 243$

3. $3^{-2} = \frac{1}{9}$

4. $5^{-3} = \frac{1}{125}$

5. $\left(\frac{1}{3}\right)^1 = \frac{1}{3}$

6. $\left(\frac{1}{2}\right)^{-4} = 16$

7. $32^{3/5} = 8$

8. $81^{3/4} = 27$

9. $10^{-3} = 0.001$

10. $16^{-1/4} = 0.5$

In exercises 11–16, use the facts that log 3 = 0.4771 and log 4 = 0.6021 to find the value of the given logarithm.

11. $\log 12$

12. $\log \frac{3}{4}$

13. $\log 16$

14. $\log \sqrt{3}$

15. $\log 48$

16. $\log \frac{1}{300}$

In exercises 17–26, use the laws of logarithms to simplify the given expression.

17. $\log x(x + 1)^4$

18. $\log x(x^2 + 1)^{-1/2}$

19. $\log \dfrac{\sqrt{x + 1}}{x^2 + 1}$

20. $\ln \dfrac{e^x}{1 + e^x}$

21. $\ln xe^{-x^2}$

22. $\ln x(x + 1)(x + 2)$

23. $\ln \dfrac{x^{1/2}}{x^2\sqrt{1 + x^2}}$

24. $\ln \dfrac{x^2}{\sqrt{x}(1 + x)^2}$

25. $\ln x^x$

26. $\ln x^{x^2+1}$

In exercises 27–30, sketch the graph of the given equation.

27. $y = \log_3 x$

28. $y = \log_{1/3} x$

29. $y = \ln 2x$

30. $y = \ln \frac{1}{2}x$

In exercises 31 and 32, sketch the graphs of the given equations on the same coordinate axes.

31. $y = 2^x$ and $y = \log_2 x$

32. $y = e^{3x}$ and $y = \ln 3x$

In exercises 33–42, use logarithms to solve the given equation for t.

33. $e^{0.4t} = 8$

34. $\frac{1}{3}e^{-3t} = 0.9$

35. $5e^{-2t} = 6$

36. $4e^{t-1} = 4$

37. $2e^{-0.2t} - 4 = 6$

38. $12 - e^{0.4t} = 3$

39. $\dfrac{50}{1 + 4e^{0.2t}} = 20$

40. $\dfrac{200}{1 + 3e^{-0.3t}} = 100$

41. $A = Be^{-t/2}$

42. $\dfrac{A}{1 + Be^{t/2}} = C$

43. **Blood Pressure**   A normal child's systolic blood pressure may be approximated by the function

$$p(x) = m(\ln x) + b$$

where $p(x)$ is measured in millimeters of mercury, x is measured in pounds, and m and b are constants. Given that $m = 19.4$ and $b = 18$, determine the systolic blood pressure of a child who weighs 92 pounds.

44. **Magnitude of Earthquakes**   On the Richter scale, the magnitude R of an earthquake is given by the formula

$$R = \log \dfrac{I}{I_0}$$

where I is the intensity of the earthquake being measured and I_0 is the standard reference intensity.
a. Express the intensity I of an earthquake of magnitude $R = 5$ in terms of the standard intensity I_0.
b. Express the intensity I of an earthquake of magnitude $R = 8$ in terms of the standard intensity I_0. How many times greater is the intensity of an earthquake of magnitude 8 than one of magnitude 5?
c. In modern times, the greatest loss of life attributable to an earthquake occurred in eastern China in 1976. Known as the Tangshan earthquake, it registered 8.2 on the Richter scale. How does the intensity of this

earthquake compare with the intensity of an earthquake of magnitude $R = 5$?

45. Sound Intensity The relative loudness of a sound D of intensity I is measured in decibels, where

$$D = 10 \log \frac{I}{I_0}$$

and I_0 is the standard threshold of audibility.
a. Express the intensity I of a 30-decibel sound (the sound level of normal conversation) in terms of I_0.
b. Determine how many times greater the intensity of an 80-decibel sound (rock music) is than that of a 30-decibel sound.
c. Prolonged noise above 150 decibels causes immediate and permanent deafness. How does the intensity of a 150-decibel sound compare with the intensity of an 80-decibel sound?

46. Barometric Pressure Halley's Law states that the barometric pressure (in inches of mercury) at an altitude of x miles above sea level is approximately given by the equation

$$p(x) = 29.92 e^{-0.2x} \qquad (x \geq 0)$$

If the barometric pressure as measured by a hot-air balloonist is 20 inches of mercury, what is the balloonist's altitude?

47. Forensic Science Forensic scientists use the following law to determine the time of death of accident or murder victims. If T denotes the temperature of a body t hours after death, then

$$T = T_0 + (T_1 - T_0)(0.97)^t$$

where T_0 is the air temperature and T_1 is the body temperature at the time of death. John Doe was found murdered at midnight in his house, when the room temperature was 70°F and his body temperature was 80°F. When was he killed? Assume that the normal body temperature is 98.6°F.

48. a. Given that $2^x = e^{kx}$, find k.
b. Show that, in general, if b is a nonnegative real number, then any equation of the form $y = b^x$ may be written in the form $y = e^{kx}$, for some real number k.

49. Use the definition of a logarithm to prove
a. $\log_b mn = \log_b m + \log_b n$
b. $\log_b \dfrac{m}{n} = \log_b m - \log_b n$

[*Hint:* Let $\log_b m = p$ and $\log_b n = q$. Then $b^p = m$ and $b^q = n$.]

50. Use the definition of a logarithm to prove

$$\log_b m^n = n \log_b m$$

51. Use the definition of a logarithm to prove
a. $\log_b 1 = 0$ **b.** $\log_b b = 1$

SOLUTIONS TO SELF-CHECK EXERCISES 13.2

1. First, sketch the graph of $y = 3^x$ with the help of the following table of values:

x	-3	-2	-1	0	1	2	3
$y = 3^x$	1/27	1/9	1/3	0	3	9	27

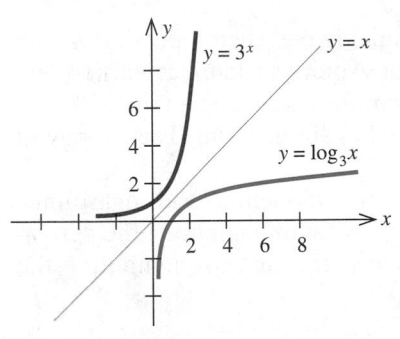

Next, take the mirror reflection of this graph with respect to the line $y = x$ to obtain the graph of $y = \log_3 x$.

2.
$$3e^{x+1} - 2 = 4$$
$$3e^{x+1} = 6$$
$$e^{x+1} = 2$$
$$\ln e^{x+1} = \ln 2$$
$$(x+1)\ln e = \ln 2 \qquad \text{(Law 3)}$$
$$x + 1 = \ln 2 \qquad \text{(Law 5)}$$
$$x = \ln 2 - 1$$
$$\approx -0.3069$$

13.3 DIFFERENTIATION OF EXPONENTIAL FUNCTIONS

The Derivative of the Exponential Function

In order to study the effects of budget deficit-reduction plans at different income levels, it is important to know the income distribution of American families. Based on data from the House Budget Committee, the House Ways and Means Committee, and the U.S. Census Bureau, the graph of f shown in Figure 13.10 gives the number of American families y (in millions) as a function of their annual income x (in thousands of dollars) in 1990.

Figure 13.10
The graph of f shows the number of families versus their annual income.

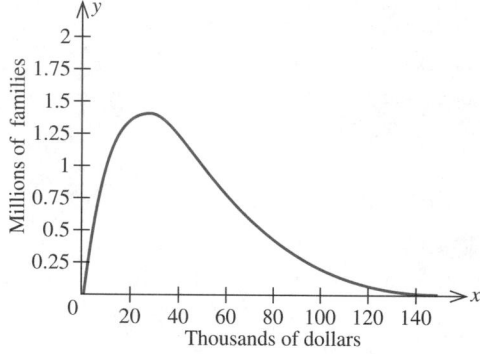

Source: House Budget Committee, House Ways and Means Committee, and U.S. Census Bureau

Observe that the graph of f rises very quickly and then tapers off. From the graph of f, you can see that the bulk of American families earned less than $100,000 per year. In fact, 95% of U.S. families earned less than $102,358 per year in 1990. (We will refer to this model again in Using Technology at the end of this section.)

To analyze mathematical models involving exponential and logarithmic functions in greater detail, we need to develop rules for computing the derivative of these functions. We begin by looking at the rule for computing the derivative of the exponential function.

RULE 1: DERIVATIVE OF THE EXPONENTIAL FUNCTION

$$\frac{d}{dx} e^x = e^x$$

Thus, the derivative of the exponential function with base e is equal to the function itself. To demonstrate the validity of this rule, we compute

$$f'(x) = \lim_{h \to 0} \frac{f(x+h) - f(x)}{h}$$

$$= \lim_{h \to 0} \frac{e^{x+h} - e^x}{h}$$

$$= \lim_{h \to 0} \frac{e^x(e^h - 1)}{h} \qquad \text{(Writing } e^{x+h} = e^x e^h \text{ and factoring)}$$

$$= e^x \lim_{h \to 0} \frac{e^h - 1}{h} \qquad \text{(Why?)}$$

To evaluate

$$\lim_{h \to 0} \frac{e^h - 1}{h}$$

let us refer to Table 13.2, which is constructed with the aid of a calculator. From the table, we see that

$$\lim_{h \to 0} \frac{e^h - 1}{h} = 1$$

(Although a rigorous proof of this fact is possible, it is beyond the scope of this book. Also see Example 1, Using Technology, page 904.) Using this result, we conclude that

$$f'(x) = e^x \cdot 1 = e^x$$

as we set out to show.

Table 13.2

h	0.1	0.01	0.001	-0.1	-0.01	-0.001
$\dfrac{e^h - 1}{h}$	1.0517	1.0050	1.0005	0.9516	0.9950	0.9995

EXAMPLE 1   Compute the derivative of each of these functions:

a. $f(x) = x^2 e^x$ **b.** $g(t) = (e^t + 2)^{3/2}$

Solution

a. The Product Rule gives

$$f'(x) = \frac{d}{dx}(x^2 e^x)$$

$$= x^2 \frac{d}{dx}(e^x) + e^x \frac{d}{dx}(x^2)$$

$$= x^2 e^x + e^x (2x)$$

$$= x e^x (x + 2)$$

b. Using the General Power Rule, we find

$$g'(t) = \frac{3}{2}(e^t + 2)^{1/2} \frac{d}{dt}(e^t + 2)$$

$$= \frac{3}{2}(e^t + 2)^{1/2} e^t$$

$$= \frac{3}{2} e^t(e^t + 2)^{1/2}$$

○ ○ ○

EXPLORING WITH TECHNOLOGY

Consider the exponential function $f(x) = b^x$ $(b > 0, b \neq 1)$.

1. Use the definition of the derivative of a function to show that

$$f'(x) = b^x \cdot \lim_{h \to 0} \frac{b^h - 1}{h}$$

2. Use the result of part (1) to show that

$$\frac{d}{dx}(2^x) = 2^x \cdot \lim_{h \to 0} \frac{2^h - 1}{h}$$

and

$$\frac{d}{dx}(3^x) = 3^x \cdot \lim_{h \to 0} \frac{3^h - 1}{h}$$

3. Use the technique in Using Technology, page 904, to show that (to two decimal places)

$$\lim_{h \to 0} \frac{2^h - 1}{h} = 0.69 \quad \text{and} \quad \lim_{h \to 0} \frac{3^h - 1}{h} = 1.10$$

4. Conclude from the results of parts (2) and (3) that

$$\frac{d}{dx}(2^x) \approx (0.69)2^x \quad \text{and} \quad \frac{d}{dx}(3^x) \approx (1.10)3^x$$

Thus,

$$\frac{d}{dx}(b^x) = k \cdot b^x$$

where k is an appropriate constant.

5. The results of part (4) suggest that, for convenience, we pick the base b, where $2 < b < 3$, so that $k = 1$. This value of b is $e \approx 2.718281828. \ldots$ Thus

$$\frac{d}{dx}(e^x) = e^x$$

This is why we prefer to work with the exponential function $f(x) = e^x$.

○ ○ ○

Applying the Chain Rule to Exponential Functions

To enlarge the class of exponential functions to be differentiated, we appeal to the Chain Rule to obtain the following rule for differentiating composite functions of the form $h(x) = e^{f(x)}$. An example of such a function is $h(x) = e^{x^2-2x}$. Here $f(x) = x^2 - 2x$.

RULE 2: THE CHAIN RULE FOR EXPONENTIAL FUNCTIONS

If $f(x)$ is a differentiable function, then

$$\frac{d}{dx}(e^{f(x)}) = e^{f(x)}f'(x)$$

To see this, observe that if $h(x) = g[f(x)]$, where $g(x) = e^x$, then by virtue of the Chain Rule,

$$h'(x) = g'(f(x))f'(x) = e^{f(x)}f'(x)$$

since $g'(x) = e^x$.

As an aid to remembering the Chain Rule for exponential functions, observe that it has the following form:

$$\frac{d}{dx}(e^{f(x)}) = e^{f(x)} \cdot \text{Derivative of exponent}$$
$$\underset{\text{same}}{\llcorner \qquad \lrcorner}$$

EXAMPLE 2   Find the derivative of each of the following functions.

a. $f(x) = e^{2x}$ **b.** $y = e^{-3x}$ **c.** $g(t) = e^{2t^2+t}$

Solution

a. $f'(x) = e^{2x} \dfrac{d}{dx}(2x) = e^{2x} \cdot 2 = 2e^{2x}$

b. $\dfrac{dy}{dx} = e^{-3x} \dfrac{d}{dx}(-3x) = -3e^{-3x}$

c. $g'(t) = e^{2t^2+t} \cdot \dfrac{d}{dt}(2t^2 + t) = (4t + 1)e^{2t^2+t}$

EXAMPLE 3   Differentiate the function $y = xe^{-2x}$.

Solution   Using the Product Rule followed by the Chain Rule, we find

$$\frac{dy}{dx} = x \frac{d}{dx} e^{-2x} + e^{-2x} \frac{d}{dx}(x)$$

$$= xe^{-2x} \frac{d}{dx}(-2x) + e^{-2x} \qquad \text{(Using the Chain Rule on the first term)}$$

$$= -2xe^{-2x} + e^{-2x}$$

$$= e^{-2x}(1 - 2x) \qquad\qquad\qquad\qquad\qquad \circ\ \circ\ \circ$$

EXAMPLE 4   Differentiate the function $g(t) = \dfrac{e^t}{e^t + e^{-t}}$.

Solution   Using the Quotient Rule followed by the Chain Rule, we find

$$g'(t) = \frac{(e^t + e^{-t}) \dfrac{d}{dt}(e^t) - e^t \dfrac{d}{dt}(e^t + e^{-t})}{(e^t + e^{-t})^2}$$

$$= \frac{(e^t + e^{-t})e^t - e^t(e^t - e^{-t})}{(e^t + e^{-t})^2}$$

$$= \frac{e^{2t} + 1 - e^{2t} + 1}{(e^t + e^{-t})^2} \qquad (e^\circ = 1)$$

$$= \frac{2}{(e^t + e^{-t})^2} \qquad\qquad\qquad\qquad \circ\ \circ\ \circ$$

EXAMPLE 5   In Section 13.5 we will discuss some practical applications of the exponential function

$$Q(t) = Q_0 e^{kt}$$

where Q_0 and k are positive constants and $t \in [0, \infty)$. A quantity $Q(t)$ growing according to this law experiences exponential growth. Show that for a quantity $Q(t)$ experiencing exponential growth, the rate of growth of the quantity $Q'(t)$ at any time t is directly proportional to the amount of the quantity present.

Solution   Using the Chain Rule for exponential functions, we compute the derivative Q' of the function Q. Thus,

$$Q'(t) = Q_0 e^{kt} \frac{d}{dt}(kt)$$

$$= Q_0 e^{kt}(k)$$

$$= kQ_0 e^{kt}$$

$$= kQ(t) \qquad [Q(t) = Q_0 e^{kt}]$$

which is the desired conclusion. $\qquad\qquad\qquad\qquad\qquad\qquad \circ\ \circ\ \circ$

EXAMPLE 6   Find the points of inflection of the function $f(x) = e^{-x^2}$.

Solution   The first derivative of f is

$$f'(x) = -2xe^{-x^2}$$

Differentiating $f'(x)$ with respect to x yields

$$f''(x) = (-2x)(-2xe^{-x^2}) - 2e^{-x^2}$$
$$= 2e^{-x^2}(2x^2 - 1)$$

Setting $f''(x) = 0$ gives

$$2e^{-x^2}(2x^2 - 1) = 0$$

Figure 13.11
Sign diagram for f''.

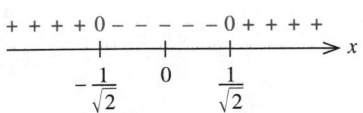

Since e^{-x^2} never equals zero for any real value of x, we see that $x = \pm 1/\sqrt{2}$ are the only candidates for inflection points of f. The sign diagram of f'', shown in Figure 13.11, tells us that both $x = -1/\sqrt{2}$ and $x = 1/\sqrt{2}$ give rise to inflection points of f. Next,

$$f\left(-\frac{1}{\sqrt{2}}\right) = f\left(\frac{1}{\sqrt{2}}\right) = e^{-1/2}$$

and the inflection points of f are $(-1/\sqrt{2}, e^{-1/2})$ and $(1/\sqrt{2}, e^{-1/2})$. The graph of f appears in Figure 13.12.

● ● ●

Figure 13.12
The graph of $y = e^{-x^2}$ has two inflection points.

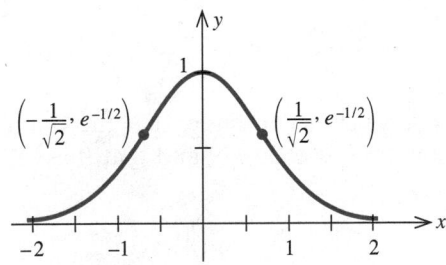

Application

Our final example involves finding the absolute maximum of an exponential function.

EXAMPLE 7   Refer to Example 8, Section 13.1. The present value of the market price of the Blakely Office Building is given by

$$P(t) = 300{,}000e^{-0.09t+\sqrt{t}/2} \qquad (0 \le t \le 10)$$

Find the optimal present value of the building's market price.

Solution   To find the maximum value of P over $[0, 10]$, we compute

$$P'(t) = 300,000e^{-0.09t+\sqrt{t}/2}\frac{d}{dt}\left(-0.09t + \frac{1}{2}t^{1/2}\right)$$

$$= 300,000e^{-0.09t+\sqrt{t}/2}\left(-0.09 + \frac{1}{4}t^{-1/2}\right)$$

Setting $P'(t) = 0$ gives

$$-0.09 + \frac{1}{4t^{1/2}} = 0$$

since $e^{-0.09t+\sqrt{t}/2}$ is never zero for any value of t. Solving this equation, we find

$$\frac{1}{4t^{1/2}} = 0.09$$

$$t^{1/2} = \frac{1}{4(0.09)}$$

$$= \frac{1}{0.36}$$

or $$t \approx 7.72$$

the sole critical point of the function P. Finally, evaluating $P(t)$ at the critical point as well as at the endpoints of $[0, 10]$, we have

t	$P(t)$
0	300,000
7.72	600,779
10	592,838

We conclude, accordingly, that the optimal present value of the property's market price is $600,779 and that this will occur 7.72 years from now.

⊙ ⊙ ⊙

SELF-CHECK EXERCISES 13.3

1. Let $f(x) = xe^{-x}$.
   a. Find the first and second derivatives of f.
   b. Find the relative extrema of f.
   c. Find the inflection points of f.

2. An industrial asset is being depreciated at a rate so that its book value t years from now will be

$$V(t) = 50,000e^{-0.4t}$$

dollars. How fast will the book value of the asset be changing three years from now?

Solutions to Self-Check Exercises 13.3 can be found on page 903.

13.3 EXERCISES

In exercises 1–28, find the derivative of the function.

1. $f(x) = e^{3x}$

2. $f(x) = 3e^x$

3. $g(t) = e^{-t}$

4. $f(x) = e^{-2x}$

5. $f(x) = e^x + x$

6. $f(x) = 2e^x - x^2$

7. $f(x) = x^3 e^x$

8. $f(u) = u^2 e^{-u}$

9. $f(x) = \dfrac{2e^x}{x}$

10. $f(x) = \dfrac{x}{e^x}$

11. $f(x) = 3(e^x + e^{-x})$

12. $f(x) = \dfrac{e^x + e^{-x}}{2}$

13. $f(w) = \dfrac{e^w + 1}{e^w}$

14. $f(x) = \dfrac{e^x}{e^x + 1}$

15. $f(x) = 2e^{3x-1}$

16. $f(t) = 4e^{3t+2}$

17. $h(x) = e^{-x^2}$

18. $f(x) = e^{x^2 - 1}$

19. $f(x) = 3e^{-1/x}$

20. $f(x) = e^{1/(2x)}$

21. $f(x) = (e^x + 1)^{25}$

22. $f(x) = (4 - e^{-3x})^3$

23. $f(x) = e^{\sqrt{x}}$

24. $f(t) = -e^{-\sqrt{2t}}$

25. $f(x) = (x - 1)e^{3x+2}$

26. $f(s) = (s^2 + 1)e^{-s^2}$

27. $f(x) = \dfrac{e^x - 1}{e^x + 1}$

28. $g(t) = \dfrac{e^{-t}}{1 + t^2}$

In exercises 29–32, find the second derivative of the function.

29. $f(x) = e^{-4x} + 2e^{3x}$

30. $f(t) = 3e^{-2t} - 5e^{-t}$

31. $f(x) = 2xe^{3x}$

32. $f(t) = t^2 e^{-2t}$

33. Find an equation of the tangent line to the graph of $y = e^{2x-3}$ at the point $(\frac{3}{2}, 1)$.

34. Find an equation of the tangent line to the graph of $y = e^{-x^2}$ at the point $(1, 1/e)$.

35. Determine the intervals where the function $f(x) = e^{-x^2/2}$ is increasing and where it is decreasing.

36. Determine the intervals where the function $f(x) = x^2 e^{-x}$ is increasing and where it is decreasing.

37. Determine the intervals of concavity for the function $f(x) = (e^x - e^{-x})/2$.

38. Determine the intervals of concavity for the function $f(x) = xe^x$.

39. Find the inflection point of the function $f(x) = xe^{-2x}$.

40. Find the inflection point(s) of the function $f(x) = 2e^{-x^2}$.

In exercises 41–44, find the absolute extrema of the function.

41. $f(x) = e^{-x^2}$ on $[-1, 1]$

42. $h(x) = e^{x^2 - 4}$ on $[-2, 2]$

43. $g(x) = (2x - 1)e^{-x}$ on $[0, \infty)$

44. $f(x) = xe^{-x^2}$ on $[0, 2]$

In exercises 45–48, use the curve-sketching guidelines of Chapter 12, page 825, to sketch the graph of the function.

45. $f(t) = e^t - t$

46. $h(x) = \dfrac{e^x + e^{-x}}{2}$

47. $f(x) = 2 - e^{-x}$

48. $f(x) = \dfrac{3}{1 + e^{-x}}$

A calculator is recommended for the remainder of these exercises.

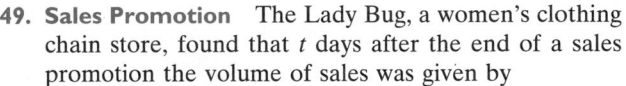

49. Sales Promotion The Lady Bug, a women's clothing chain store, found that t days after the end of a sales promotion the volume of sales was given by

$$S(t) = 20{,}000(1 + e^{-0.5t}) \qquad (0 \le t \le 5)$$

dollars. Find the rate of change of The Lady Bug's sales volume when $t = 1$, $t = 2$, $t = 3$, and $t = 4$.

50. Energy Consumption of Appliances The average energy consumption of the typical refrigerator/freezer manufactured by York Industries is approximately

$$C(t) = 1486e^{-0.073t} + 500 \qquad (0 \le t \le 20)$$

kilowatt hours per year, where t is measured in years, with $t = 0$ corresponding to 1972.

a. What was the average energy consumption of the York refrigerator/freezer at the beginning of 1972?

b. Prove that the average energy consumption of the York refrigerator/freezer is decreasing over the years in question.

c. All refrigerator/freezers manufactured as of January 1, 1990, must meet the 950-kilowatt-hours-per-year maximum energy-consumption standard set by the National Appliance Conservation Act. Show that the York refrigerator/freezer satisfies this requirement.

51. Polio Immunization Polio, a once-feared killer, declined markedly in the United States in the 1950s after Jonas Salk developed the inactivated polio vaccine and mass immunization of children took place. The number of polio cases in the United States from the beginning of 1959 to the beginning of 1963 is approximated by the function

$$N(t) = 5.3e^{0.095t^2 - 0.85t} \qquad (0 \le t \le 4)$$

where $N(t)$ gives the number of polio cases (in thousands) and t is measured in years, with $t = 0$ corresponding to the beginning of 1959.
a. Show that the function N is decreasing over the time interval under consideration.
b. How fast was the number of polio cases decreasing at the beginning of 1959? At the beginning of 1962? [*Comment:* Following the introduction of the oral vaccine developed by Dr. Albert B. Sabin in 1963, polio in the United States has, for all practical purposes, been eliminated.]

52. Price of Perfume The monthly demand for a certain brand of perfume is given by the demand equation

$$p = 100e^{-0.0002x} + 150$$

where p denotes the retail unit price in dollars and x denotes the quantity (in 1-ounce bottles) demanded.
a. Find the rate of change of the price per bottle when $x = 1000$. When $x = 2000$.
b. What is the price per bottle when $x = 1000$? When $x = 2000$?

53. Price of Wine The monthly demand for a certain brand of table wine is given by the demand equation

$$p = 240 \left(1 - \frac{3}{3 + e^{-0.0005x}} \right)$$

where p denotes the wholesale price per case (in dollars) and x denotes the number of cases demanded.
a. Find the rate of change of the price per case when $x = 1000$.
b. What is the price per case when $x = 1000$?

54. Spread of an Epidemic During a flu epidemic, the total number of students on a state university campus who had contracted influenza by the xth day was given

by

$$N(x) = \frac{3000}{1 + 99e^{-x}} \qquad (x \ge 0)$$

a. How many students had influenza initially?
b. Derive an expression for the rate at which the disease was being spread, and prove that the function N is increasing on the interval $(0, \infty)$.
c. Sketch the graph of N. What was the total number of students who contracted influenza during that particular epidemic?

55. Maximum Oil Production It has been estimated that the total production of oil from a certain oil well is given by

$$T(t) = -1000(t + 10)e^{-0.1t} + 10{,}000$$

thousand barrels t years after production has begun. Determine the year when the oil well will be producing at maximum capacity.

56. Optimal Selling Time Refer to exercise 36, page 884. The present value of a piece of waterfront property purchased by an investor is given by the function

$$P(t) = 80{,}000e^{\sqrt{t/2} - 0.09t} \qquad (0 \le t \le 8)$$

Determine the optimal time (based on present value) for the investor to sell the property. What is the property's optimal present value?

57. Oil Used to Fuel Productivity A study on worldwide oil use was prepared for a major oil company. The study predicted that the amount of oil used to fuel productivity in a certain country is given by

$$f(t) = 1.5 + 1.8te^{-1.2t} \qquad (0 \le t \le 4)$$

where $f(t)$ denotes the number of barrels per \$1000 of economic output and t is measured in decades ($t = 0$ corresponds to 1965). Compute $f'(0)$, $f'(1)$, $f'(2)$, and $f'(3)$, and interpret your results.

58. Percentage of Population Relocating Based on data obtained from the Census Bureau, the manager of Plymouth Van Lines estimates that the percentage of the total population relocating in year t ($t = 0$ corresponds to the year 1960) may be approximated by the formula

$$P(t) = 20.6e^{-0.009t} \qquad (0 \le t \le 35)$$

Compute $P'(10)$, $P'(20)$, and $P'(30)$ and interpret your results.

SOLUTIONS TO SELF-CHECK EXERCISES 13.3

1. a. Using the Product Rule, we obtain

$$f'(x) = x\frac{d}{dx}e^{-x} + e^{-x}\frac{d}{dx}x$$

$$= -xe^{-x} + e^{-x} = (1 - x)e^{-x}$$

Using the Product Rule once again, we obtain

$$f''(x) = (1 - x)\frac{d}{dx}e^{-x} + e^{-x}\frac{d}{dx}(1 - x)$$

$$= (1 - x)(-e^{-x}) + e^{-x}(-1)$$

$$= -e^{-x} + xe^{-x} - e^{-x} = (x - 2)e^{-x}$$

b. Setting $f'(x) = 0$ gives

$$(1 - x)e^{-x} = 0$$

Since $e^{-x} \neq 0$, we see that $1 - x = 0$, and this gives $x = 1$ as the only critical point of f. The sign diagram of f' shown in the accompanying figure tells us that the point $(1, e^{-1})$ is a relative maximum of f.

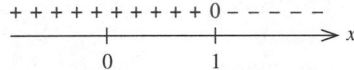

c. Setting $f''(x) = 0$ gives $x - 2 = 0$, so $x = 2$ is a candidate for an inflection point of f. The sign diagram of f'' (accompanying figure) shows that $(2, 2e^{-2})$ is an inflection point of f.

2. The rate of change of the book value of the asset t years from now is

$$V'(t) = 50{,}000\frac{d}{dt}e^{-0.4t}$$

$$= 50{,}000(-0.4)e^{-0.4t} = -20{,}000e^{-0.4t}$$

Therefore, three years from now the book value of the asset will be changing at the rate of

$$V'(3) = -20{,}000e^{-0.4(3)} = -20{,}000e^{-1.2} \approx -6023.88$$

that is, decreasing at the rate of approximately \$6024 per year.

USING TECHNOLOGY

EXAMPLE 1 At the beginning of Section 13.3, we demonstrated via a table of values of $(e^h - 1)/h$ for selected values of h the plausibility of the result

$$\lim_{h \to 0} \frac{e^h - 1}{h} = 1$$

In order to obtain a visual confirmation of this result, we plot the graph of

$$f(x) = \frac{e^x - 1}{x}$$

in the viewing rectangle $[-1, 1] \times [0, 2]$ (see Figure T1). From the graph of f, we see that $f(x)$ appears to approach 1 as x approaches 0.

Figure T1
The graph of f in the viewing rectangle $[-1, 1] \times [0, 2]$.

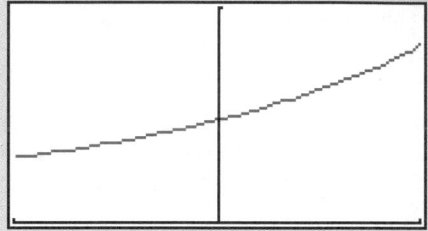

● ● ●

The numerical derivative function of a graphing utility will yield the derivative of an exponential or logarithmic function for any value of x, just as it did for algebraic functions.*

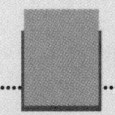

EXERCISES

In exercises 1–6, use the numerical derivative operation of a graphing utility to find the rate of change of f(x) at the given value of x. Give your answer accurate to four decimal places.

1. $f(x) = x^3 e^{-1/x}$; $x = -1$

2. $f(x) = (\sqrt{x} + 1)^{3/2} e^{-x}$; $x = 0.5$

3. $f(x) = x^3 \sqrt{\ln x}$; $x = 2$

4. $f(x) = \dfrac{\sqrt{x} \ln x}{x + 1}$; $x = 3.2$

5. $f(x) = e^{-x} \ln(2x + 1)$; $x = 0.5$

6. $f(x) = \dfrac{e^{-\sqrt{x}}}{\ln(x^2 + 1)}$; $x = 1$

7. **An Extinction Situation** The number of saltwater crocodiles in a certain area of northern Australia is given by

$$P(t) = \frac{300 e^{-0.024t}}{5 e^{-0.024t} + 1}$$

a. How many crocodiles were in the population initially?

b. Show that $\lim_{t \to \infty} P(t) = 0$.

● ● ●

* The rules for differentiating logarithmic functions will be covered in Section 13.4. However, the exercises given here can be done without using these rules.

904

c. Use a graphing calculator to plot the graph of P in the viewing rectangle $[0, 200] \times [0, 70]$.

REMARK   This phenomenon is referred to as an *extinction* situation.   ◦ ◦ ◦

8. **Income of American Families**   Based on data compiled by the House Budget Committee, the House Ways and Means Committee, and the U.S. Census Bureau, it is estimated that the number of American families y (in millions) who earned x thousand dollars in 1990 is related by the equation

$$y = 0.1584xe^{-0.0000016x^3 + 0.00011x^2 - 0.04491x} \qquad (x > 0)$$

a. Use a graphing utility to plot the graph of the equation in the viewing rectangle $[0, 150] \times [0, 2]$.
b. How fast is y changing with respect to x when $x = 10$? When $x = 50$? Interpret your results.
Source: House Budget Committee, House Ways and Means Committee, and U.S. Census Bureau

9. **World Population Growth**   According to a study conducted by the United Nations Population Division, the world population in billions is approximated by the function

$$f(t) = \frac{12}{1 + 3.74914e^{-1.42804t}} \qquad (0 \le t \le 4)$$

where t is measured in half-centuries, with $t = 0$ corresponding to the beginning of 1950.
a. Use a graphing utility to plot the graph of f in the viewing rectangle $[0, 5] \times [0, 14]$.
b. How fast is the world population expected to increase at the beginning of the year 2000?
Source: United Nations Population Division

10. **Increase in Juvenile Offenders**   The number of youths aged 15 to 19 will increase by 21% between 1994 and 2005, pushing the crime rate up. According to the National Council on Crime and Delinquency, the number of violent crime arrests of juveniles under age 18 in year t is given by

$$f(t) = -0.438t^2 + 9.002t + 107 \qquad (0 \le t \le 13)$$

where $f(t)$ is measured in thousands and t in years, with $t = 0$ corresponding to 1989. According to the same source, if trends like inner-city drug use and wider availability of guns continue, then the number of violent crime arrests of juveniles under age 18 in year t will be

given by

$$g(t) = \begin{cases} -0.438t^2 + 9.002t + 107 & \text{if } 0 \le t < 4 \\ 99.456e^{0.07824t} & \text{if } 4 \le t \le 13 \end{cases}$$

where $g(t)$ is measured in thousands and $t = 0$ corresponds to 1989.
a. Compute $f(11)$ and $g(11)$ and interpret your results.
b. Compute $f'(11)$ and $g'(11)$ and interpret your results.
Source: National Council on Crime and Delinquency

11. **Increasing Crop Yields**   If left untreated on bean stems, aphids (small insects that suck plant juices) will multiply at an increasing rate during the summer months and reduce the productivity and crop yield of cultivated crops. But if the aphids are treated in mid-June, the numbers decrease sharply to less than 100 per bean stem, allowing for steep rises in crop yield. The function

$$F(t) = \begin{cases} 62e^{1.152t} & \text{if } 0 \le t < 1.5 \\ 349e^{-1.324(t-1.5)} & \text{if } 1.5 \le t \le 3 \end{cases}$$

gives the number of aphids in a typical bean stem at time t, where t is measured in months, with $t = 0$ corresponding to the beginning of May.
Source: The Random House Encyclopedia
a. How many aphids are there on a typical bean stem at the beginning of June ($t = 1$)? At the beginning of July ($t = 2$)?
b. How fast is the population of aphids changing at the beginning of June? At the beginning of July?

12. **Percentage of Females in the Labor Force**   Based on data from the U.S. Census Bureau, the chief economist of Manpower, Inc., constructed the following formula giving the percentage of the total female population in the civilian labor force, $P(t)$, at the beginning of the tth decade ($t = 0$ corresponds to the year 1900):

$$P(t) = \frac{74}{1 + 2.6e^{-0.166t + 0.04536t^2 - 0.0066t^3}} \qquad (0 \le t \le 11)$$

Assume this trend continues for the rest of the 20th century.
a. What will the percentage of the total female population in the civilian labor force be at the beginning of the year 2000?
b. What will the growth rate of the percentage of the total female population in the civilian labor force be at the beginning of the year 2000?
Source: U.S. Census Bureau

13.4 DIFFERENTIATION OF LOGARITHMIC FUNCTIONS

The Derivative of ln x

Let us now turn our attention to the differentiation of logarithmic functions.

RULE 3: DERIVATIVE OF ln x	$\dfrac{d}{dx}\ln|x| = \dfrac{1}{x}$ $(x \neq 0)$

To derive Rule 3, suppose that $x > 0$ and write $f(x) = \ln x$ in the equivalent form

$$x = e^{f(x)}$$

Differentiating both sides of the equation with respect to x, we find, using the Chain Rule,

$$1 = e^{f(x)} \cdot f'(x)$$

from which we see that $f'(x) = \dfrac{1}{e^{f(x)}}$

or, since $e^{f(x)} = x$, $f'(x) = \dfrac{1}{x}$

as we set out to show. You are asked to prove the rule for the case $x < 0$ in exercise 59, page 912.

EXAMPLE 1 Compute the derivative of each of these functions:

a. $f(x) = x \ln x$ **b.** $g(x) = \dfrac{\ln x}{x}$

Solution

a. Using the Product Rule, we obtain

$$f'(x) = \frac{d}{dx}(x \ln x) = x \frac{d}{dx}(\ln x) + (\ln x)\frac{d}{dx}(x)$$

$$= x\left(\frac{1}{x}\right) + \ln x = 1 + \ln x$$

b. Using the Quotient Rule, we obtain

$$g'(x) = \frac{x\dfrac{d}{dx}(\ln x) - (\ln x)\dfrac{d}{dx}(x)}{x^2} = \frac{x\left(\dfrac{1}{x}\right) - \ln x}{x^2} = \frac{1 - \ln x}{x^2} \quad \circ \ \circ \ \circ$$

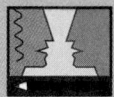

You can derive the formula for the derivative of $f(x) = \ln x$ directly from the definition of the derivative as follows.

a. Show that

$$f'(x) = \lim_{h \to 0} \frac{f(x+h) - f(x)}{h} = \lim_{h \to 0} \ln\left(1 + \frac{h}{x}\right)^{1/h}$$

b. Put $m = x/h$ and note that $m \to \infty$ as $h \to 0$. Furthermore, $f'(x)$ can be written in the form

$$f'(x) = \lim_{m \to \infty} \ln\left(1 + \frac{1}{m}\right)^{m/x}$$

c. Finally, use both the fact that the natural logarithmic function is continuous and the definition of the number e to show that

$$f'(x) = \frac{1}{x} \ln\left[\lim_{m \to \infty} \left(1 + \frac{1}{m}\right)^{m}\right] = \frac{1}{x}$$

The Chain Rule for Logarithmic Functions

To enlarge the class of logarithmic functions to be differentiated, we appeal once more to the Chain Rule to obtain the following rule for differentiating composite functions of the form $h(x) = \ln f(x)$, where $f(x)$ is assumed to be a positive differentiable function.

RULE 4: THE CHAIN RULE FOR LOGARITHMIC FUNCTIONS	If $f(x)$ is a differentiable function, then $$\frac{d}{dx}[\ln f(x)] = \frac{f'(x)}{f(x)} \qquad [f(x) > 0]$$

To see this, observe that $h(x) = g[f(x)]$, where $g(x) = \ln x$ $(x > 0)$. Since $g'(x) = 1/x$, we have, using the Chain Rule,

$$h'(x) = g'(f(x))f'(x)$$
$$= \frac{1}{f(x)} f'(x) = \frac{f'(x)}{f(x)}$$

Observe that in the special case $f(x) = x$, $h(x) = \ln x$, so the derivative of h is, by Rule 3, given by $h'(x) = 1/x$.

EXAMPLE 2   Find the derivative of the function $f(x) = \ln(x^2 + 1)$.

Solution   Using Rule 4, we see immediately that

$$f'(x) = \frac{\dfrac{d}{dx}(x^2 + 1)}{x^2 + 1} = \frac{2x}{x^2 + 1}$$

○ ○ ○

When differentiating functions involving logarithms, we may use the rules of logarithms to advantage, as shown in Examples 3 and 4.

EXAMPLE 3   Differentiate the function $y = \ln[(x^2 + 1)(x^3 + 2)^6]$.

Solution   We first rewrite the given function using the properties of logarithms:

$$
\begin{aligned}
y &= \ln[(x^2 + 1)(x^3 + 2)^6] \\
&= \ln(x^2 + 1) + \ln(x^3 + 2)^6 \qquad (\ln mn = \ln m + \ln n) \\
&= \ln(x^2 + 1) + 6\ln(x^3 + 2) \qquad (\ln m^n = n\ln m)
\end{aligned}
$$

Differentiating and using Rule 4, we obtain

$$
\begin{aligned}
y' &= \frac{\dfrac{d}{dx}(x^2 + 1)}{x^2 + 1} + \frac{6\dfrac{d}{dx}(x^3 + 2)}{x^3 + 2} \\[2mm]
&= \frac{2x}{x^2 + 1} + \frac{6(3x^2)}{x^3 + 2} = \frac{2x}{x^2 + 1} + \frac{18x^2}{x^3 + 2}
\end{aligned}
$$

○ ○ ○

EXPLORING WITH TECHNOLOGY

Use a graphing utility to plot the graphs of $f(x) = \ln x$; its first derivative function, $f'(x) = 1/x$; and its second derivative function, $f''(x) = -1/x^2$, in the same viewing rectangle $[0, 4] \times [-3, 3]$.

1. Describe the properties of the graph of f revealed by studying the graph of $f'(x)$. What can you say about the rate of increase of f for large values of x?

2. Describe the properties of the graph of f revealed by studying the graph of $f''(x)$. What can you say about the concavity of f for large values of x?

○ ○ ○

EXAMPLE 4   Find the derivative of the function $g(t) = \ln(t^2 e^{-t^2})$.

Solution   Here again, to save a lot of work, we first simplify the given expression using the properties of logarithms:

$$g(t) = \ln(t^2 e^{-t^2})$$
$$= \ln t^2 + \ln e^{-t^2} \qquad (\ln mn = \ln m + \ln n)$$
$$= 2 \ln t - t^2 \qquad (\ln m^n = n \ln m \text{ and } \ln e = 1)$$

Therefore, $$g'(t) = \frac{2}{t} - 2t = \frac{2(1 - t^2)}{t}$$

● ● ●

Logarithmic Differentiation

As we saw in the last two examples, the task of finding the derivative of a given function can be made easier by first applying the laws of logarithms to simplify the function. We now illustrate a process called **logarithmic differentiation,** which not only simplifies the calculation of the derivatives of certain functions but also enables us to compute the derivatives of functions we could not otherwise differentiate using the techniques developed thus far.

EXAMPLE 5   Differentiate $y = x(x + 1)(x^2 + 1)$ using logarithmic differentiation.

Solution   First we take the natural logarithm on both sides of the given equation, obtaining

$$\ln y = \ln x(x + 1)(x^2 + 1)$$

Next, we use the properties of logarithms to rewrite the right-hand side of this equation, obtaining

$$\ln y = \ln x + \ln(x + 1) + \ln(x^2 + 1)$$

If we differentiate both sides of this equation, we have

$$\frac{d}{dx} \ln y = \frac{d}{dx} [\ln x + \ln(x + 1) + \ln(x^2 + 1)]$$

$$= \frac{1}{x} + \frac{1}{x + 1} + \frac{2x}{x^2 + 1} \qquad \text{(Using Rule 4)}$$

In order to evaluate the expression on the left-hand side, note that y is a function of x. Therefore, writing $y = f(x)$ to remind us of this fact, we have

$$\frac{d}{dx} \ln y = \frac{d}{dx} \ln[f(x)] \qquad [\text{Writing } y = f(x)]$$

$$= \frac{f'(x)}{f(x)} \qquad [\text{Using Rule 4}]$$

$$= \frac{y'}{y} \qquad [\text{Returning to using } y \text{ instead of } f(x)]$$

Then we have

$$\frac{y'}{y} = \frac{1}{x} + \frac{1}{x+1} + \frac{2x}{x^2+1}$$

Finally, solving for y', we have

$$y' = y\left(\frac{1}{x} + \frac{1}{x+1} + \frac{2x}{x^2+1}\right)$$

$$= x(x+1)(x^2+1)\left(\frac{1}{x} + \frac{1}{x+1} + \frac{2x}{x^2+1}\right) \quad \circ \circ \circ$$

Before considering other examples, let us summarize the important steps involved in logarithmic differentiation.

FINDING dy/dx BY LOGARITHMIC DIFFERENTIATION

1. Take the natural logarithm on both sides of the equation and use the properties of logarithms to write any "complicated expression" as a sum of simpler terms.
2. Differentiate both sides of the equation with respect to x.
3. Solve the resulting equation for dy/dx.

EXAMPLE 6   Differentiate $y = x^2(x-1)(x^2+4)^3$.

Solution   Taking the natural logarithm on both sides of the given equation and using the laws of logarithms, we obtain

$$\ln y = \ln x^2(x-1)(x^2+4)^3$$
$$= \ln x^2 + \ln(x-1) + \ln(x^2+4)^3$$
$$= 2\ln x + \ln(x-1) + 3\ln(x^2+4)$$

Differentiating both sides of the equation with respect to x, we have

$$\frac{d}{dx}\ln y = \frac{y'}{y} = \frac{2}{x} + \frac{1}{x-1} + 3\cdot\frac{2x}{x^2+4}$$

Finally, solving for y', we have

$$y' = y\left(\frac{2}{x} + \frac{1}{x-1} + \frac{6x}{x^2+4}\right)$$

$$= x^2(x-1)(x^2+4)^3\left(\frac{2}{x} + \frac{1}{x-1} + \frac{6x}{x^2+4}\right) \quad \circ \circ \circ$$

EXAMPLE 7   Find the derivative of $f(x) = x^x \ (x > 0)$.

Solution   A word of caution! This function is neither a power function nor an exponential function. Taking the natural logarithm on both sides of the equation gives

$$\ln f(x) = \ln x^x = x \ln x$$

Differentiating both sides of the equation with respect to x, we obtain

$$\frac{f'(x)}{f(x)} = x \frac{d}{dx} \ln x + (\ln x) \frac{d}{dx} x$$

$$= x \left(\frac{1}{x}\right) + \ln x$$

$$= 1 + \ln x$$

Therefore,

$$f'(x) = f(x)(1 + \ln x) = x^x(1 + \ln x) \qquad \circ \ \circ \ \circ$$

EXPLORING WITH TECHNOLOGY

Refer to Example 7.

1. Use a graphing utility to plot the graph of $f(x) = x^x$ in the viewing rectangle $[0, 2] \times [0, 2]$. Then use **ZOOM** and **TRACE** to show that

$$\lim_{x \to 0^+} f(x) = 1$$

2. Use the results of part (1) and Example 7 to show that $\lim_{x \to 0^+} f'(x) = -\infty$. Justify your answer.

$$\circ \quad \circ \quad \circ$$

SELF–CHECK EXERCISES 13.4

1. Find an equation of the tangent line to the graph of $f(x) = x \ln(2x + 3)$ at the point $(-1, 0)$.

2. Use logarithmic differentiation to compute y' given $y = (2x + 1)^3(3x + 4)^5$.

Solutions to Self-Check Exercises 13.4 can be found on page 913.

13.4 EXERCISES

In exercises 1–32, find the derivative of the function.

1. $f(x) = 5 \ln x$

2. $f(x) = \ln 5x$

3. $f(x) = \ln(x + 1)$

4. $g(x) = \ln(2x + 1)$

5. $f(x) = \ln x^8$

6. $h(t) = 2 \ln t^5$

7. $f(x) = \ln \sqrt{x}$

8. $f(x) = \ln(\sqrt{x} + 1)$

9. $f(x) = \ln \dfrac{1}{x^2}$

10. $f(x) = \ln \dfrac{1}{2x^3}$

11. $f(x) = \ln(4x^2 - 6x + 3)$

12. $f(x) = \ln(3x^2 - 2x + 1)$

13. $f(x) = \ln \dfrac{2x}{x + 1}$

14. $f(x) = \ln \dfrac{x + 1}{x - 1}$

15. $f(x) = x^2 \ln x$

16. $f(x) = 3x^2 \ln 2x$

17. $f(x) = \dfrac{2 \ln x}{x}$

18. $f(x) = \dfrac{3 \ln x}{x^2}$

19. $f(u) = \ln(u - 2)^3$

20. $f(x) = \ln(x^3 - 3)^4$

21. $f(x) = \sqrt{\ln x}$

22. $f(x) = \sqrt{\ln x} + x$

23. $f(x) = (\ln x)^3$

24. $f(x) = 2(\ln x)^{3/2}$

25. $f(x) = \ln(x^3 + 1)$

26. $f(x) = \ln\sqrt{x^2 - 4}$

27. $f(x) = e^x \ln x$

28. $f(x) = e^x \ln\sqrt{x + 3}$

29. $f(t) = e^{2t} \ln(t + 1)$

30. $g(t) = t^2 \ln(e^{2t} + 1)$

31. $f(x) = \dfrac{\ln x}{x}$

32. $g(t) = \dfrac{t}{\ln t}$

In exercises 33–36, find the second derivative of the function.

33. $f(x) = \ln 2x$

34. $f(x) = \ln(x + 5)$

35. $f(x) = \ln(x^2 + 2)$

36. $f(x) = (\ln x)^2$

In exercises 37–46, use logarithmic differentiation to find the derivative of the function.

37. $y = (x + 1)^2(x + 2)^3$

38. $y = (3x + 2)^4(5x - 1)^2$

39. $y = (x - 1)^2(x + 1)^3(x + 3)^4$

40. $y = \sqrt{3x + 5}(2x - 3)^4$

41. $y = \dfrac{(2x^2 - 1)^5}{\sqrt{x + 1}}$

42. $y = \dfrac{\sqrt{4 + 3x^2}}{\sqrt[3]{x^2 + 1}}$

43. $y = 3^x$

44. $y = x^{x+2}$

45. $y = (x^2 + 1)^x$

46. $y = x^{\ln x}$

47. Find an equation of the tangent line to the graph of $y = x \ln x$ at the point $(1, 0)$.

48. Find an equation of the tangent line to the graph of $y = \ln x^2$ at the point $(2, \ln 4)$.

49. Determine the intervals where the function $f(x) = \ln x^2$ is increasing and where it is decreasing.

50. Determine the intervals where the function $f(x) = (\ln x)/x$ is increasing and where it is decreasing.

51. Determine the intervals of concavity for the function $f(x) = x^2 + \ln x^2$.

52. Determine the intervals of concavity for the function $f(x) = (\ln x)/x$.

53. Find the inflection points of the function $f(x) = \ln(x^2 + 1)$.

54. Find the inflection points of the function $f(x) = x^2 \ln x$.

55. Find the absolute extrema of the function $f(x) = x - \ln x$ on $[\frac{1}{2}, 3]$.

56. Find the absolute extrema of the function $g(x) = x/(\ln x)$ on $[2, \infty)$.

In exercises 57 and 58, use the guidelines on page 825 to sketch the graph of the given function.

57. $f(x) = \ln(x - 1)$

58. $f(x) = 2x - \ln x$

59. Prove that $\dfrac{d}{dx} \ln |x| = \dfrac{1}{x}$ $(x \neq 0)$ for the case $x < 0$.

SOLUTIONS TO SELF-CHECK EXERCISES 13.4

1. The slope of the tangent line to the graph of f at any point $(x, f(x))$ lying on the graph of f is given by $f'(x)$. Using the Product Rule, we find

$$f'(x) = \frac{d}{dx}[x \ln(2x + 3)]$$

$$= x\frac{d}{dx}\ln(2x + 3) + \ln(2x + 3) \cdot \frac{d}{dx}(x)$$

$$= x\left(\frac{2}{2x + 3}\right) + \ln(2x + 3) \cdot 1$$

$$= \frac{2x}{2x + 3} + \ln(2x + 3)$$

In particular, the slope of the tangent line to the graph of f at the point $(-1, 0)$ is

$$f'(-1) = \frac{-2}{-2 + 3} + \ln 1 = -2$$

Therefore, using the point-slope form of the equation of a line, we see that a required equation is

$$y - 0 = -2(x + 1)$$
$$y = -2x - 2$$

2. Taking the logarithm on both sides of the equation gives

$$\ln y = \ln(2x + 1)^3(3x + 4)^5$$
$$= \ln(2x + 1)^3 + \ln(3x + 4)^5$$
$$= 3\ln(2x + 1) + 5\ln(3x + 4)$$

Differentiating both sides of the equation with respect to x, keeping in mind that y is a function of x, we obtain

$$\frac{d}{dx}(\ln y) = \frac{y'}{y} = 3 \cdot \frac{2}{2x + 1} + 5 \cdot \frac{3}{3x + 4}$$

$$= 3\left[\frac{2}{2x + 1} + \frac{5}{3x + 4}\right]$$

$$= \left(\frac{6}{2x + 1} + \frac{15}{3x + 4}\right)$$

and

$$y' = (2x + 1)^3(3x + 4)^5 \cdot \left(\frac{6}{2x + 1} + \frac{15}{3x + 4}\right)$$

13.5 EXPONENTIAL FUNCTIONS AS MATHEMATICAL MODELS

Figure 13.13
Exponential growth.

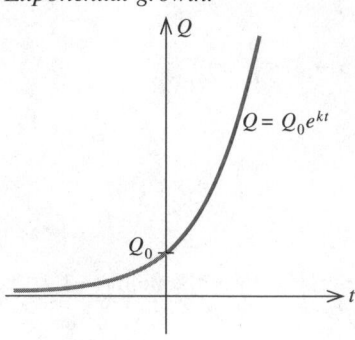

Exponential Growth

Many problems arising from practical situations can be described mathematically in terms of exponential functions or functions closely related to the exponential function. In this section, we look at some applications involving exponential functions from the fields of the life and social sciences.

In Section 13.1, we saw that the exponential function $f(x) = b^x$ is an increasing function when $b > 1$. In particular, the function $f(x) = e^x$ shares this property. From this result one may deduce that the function $Q(t) = Q_0 e^{kt}$, where Q_0 and k are positive constants, has the following properties:

1. $Q(0) = Q_0$

2. $Q(t)$ increases "rapidly" without bound as t increases without bound (Figure 13.13).

Property 1 follows from the computation

$$Q(0) = Q_0 e^0 = Q_0$$

Next, to study the rate of change of the function $Q(t)$, we differentiate it with respect to t, obtaining

$$Q'(t) = \frac{d}{dt}(Q_0 e^{kt})$$

$$= Q_0 \frac{d}{dt}(e^{kt})$$

$$= kQ_0 e^{kt}$$

$$= kQ(t) \qquad \textbf{(7)}$$

Since $Q(t) > 0$ (because Q_0 is assumed to be positive) and $k > 0$, we see that $Q'(t) > 0$ and so $Q(t)$ is an increasing function of t. Our computation has, in fact, shed more light on an important property of the function $Q(t)$. Equation (7) says that the rate of increase of the function $Q(t)$ is proportional to the amount $Q(t)$ of the quantity present at time t. The implication is that as $Q(t)$ increases, so does the *rate of increase* of $Q(t)$, resulting in a very rapid increase in $Q(t)$ as t increases without bound.

Thus, the exponential function

$$Q(t) = Q_0 e^{kt} \qquad (0 \le t < \infty) \qquad \textbf{(8)}$$

provides us with a mathematical model of a quantity $Q(t)$ that is initially present in the amount of $Q(0) = Q_0$ and whose rate of growth at any time t is directly proportional to the amount of the quantity present at time t. Such a quantity is said to exhibit **exponential growth,** and the constant k is called the **growth constant.** Interest earned on a fixed deposit when compounded continuously exhibits exponential growth. Another example of exponential growth follows.

EXAMPLE 1 Under ideal laboratory conditions, the number of bacteria in a culture increases in accordance with the law $Q(t) = Q_0 e^{kt}$, where Q_0 denotes the number of bacteria initially present in the culture, k is some constant determined by the strain of bacteria under consideration, and t is the elapsed time measured in hours. Suppose that 10,000 bacteria are present initially in the culture and 60,000 are present 2 hours later.

a. How many bacteria will there be in the culture at the end of 4 hours?

b. What is the rate of growth of the population after 4 hours?

Solution

a. We are given that $Q(0) = Q_0 = 10{,}000$, so $Q(t) = 10{,}000 e^{kt}$. Next, the fact that 60,000 bacteria are present 2 hours later translates into $Q(2) = 60{,}000$. Thus

$$60{,}000 = 10{,}000 e^{2k}$$

$$e^{2k} = 6$$

Taking the natural logarithm on both sides of the equation, we obtain

$$\ln e^{2k} = \ln 6$$

$$2k = \ln 6 \qquad \text{(Since } \ln e = 1\text{)}$$

$$k \approx 0.8959$$

Thus, the number of bacteria present at any time t is given by

$$Q(t) = 10{,}000 e^{0.8959t}$$

In particular, the number of bacteria present in the culture at the end of 4 hours is given by

$$Q(4) = 10{,}000 e^{0.8959(4)}$$
$$= 360{,}029$$

b. The rate of growth of the bacteria population at any time t is given by

$$Q'(t) = kQ(t)$$

Thus, using the result from (a), we find that the rate at which the population is growing at the end of 4 hours is

$$Q'(4) = kQ(4)$$
$$\approx (0.8959)(360{,}029)$$
$$\approx 322{,}550$$

or approximately 322,550 bacteria per hour.   ○ ○ ○

Figure 13.14
Exponential decay.

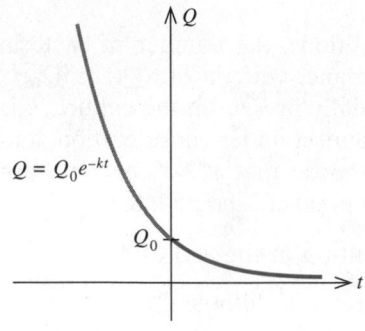

Exponential Decay

In contrast to exponential growth, a quantity exhibits **exponential decay** if it decreases at a rate that is directly proportional to its size. Such a quantity may be described by the exponential function

$$Q(t) = Q_0 e^{-kt} \qquad (t \in [0, \infty)) \qquad \textbf{(9)}$$

where the positive constant Q_0 measures the amount present initially ($t = 0$) and k is some suitable positive number, called the **decay constant.** The choice of this number is determined by the nature of the substance under consideration. The graph of this function is sketched in Figure 13.14.

To verify the properties ascribed to the function $Q(t)$, we simply compute

$$Q(0) = Q_0 e^0 = Q_0$$

$$Q'(t) = \frac{d}{dt}(Q_0 e^{-kt})$$

$$= Q_0 \frac{d}{dt}(e^{-kt})$$

$$= -k Q_0 e^{-kt} = -kQ(t)$$

 EXAMPLE 2   Radioactive substances decay exponentially. For example, the amount of radium present at any time t obeys the law $Q(t) = Q_0 e^{-kt}$, where Q_0 is the initial amount present and k is a suitable positive constant. The **half-life of a radioactive substance** is the time required for a given amount to be reduced by one-half. Now, it is known that the half-life of radium is approximately 1600 years. Suppose that initially there are 200 milligrams of pure radium. Find the amount left after t years. What is the amount left after 800 years?

Solution   The initial amount of radium present is 200 milligrams, so $Q(0) = Q_0 = 200$. Thus $Q(t) = 200 e^{-kt}$. Next, the datum concerning the half-life of radium implies that $Q(1600) = 100$, and this gives

$$100 = 200 e^{-1600k}$$

$$e^{-1600k} = \frac{1}{2}$$

Taking the natural logarithm on both sides of this equation yields

$$-1600k \ln e = \ln \frac{1}{2}$$

$$-1600k = \ln \frac{1}{2} \qquad (\ln e = 1)$$

$$k = -\frac{1}{1600} \ln\left(\frac{1}{2}\right) = 0.0004332$$

Therefore, the amount of radium left after t years is

$$Q(t) = 200e^{-0.0004332t}$$

In particular, the amount of radium left after 800 years is

$$Q(800) = 200e^{-0.0004332(800)} \approx 141.42$$

or approximately 141 milligrams.   ○ ○ ○

EXAMPLE 3 Carbon 14, a radioactive isotope of carbon, has a half-life of 5770 years. What is its decay constant?

Solution We have $Q(t) = Q_0e^{-kt}$. Since the half-life of the element is 5770 years, half of the substance is left at the end of that period; that is,

$$Q(5770) = Q_0e^{-5770k} = \frac{1}{2}Q_0$$

$$e^{-5770k} = \frac{1}{2}$$

Taking the natural logarithm on both sides of this equation, we have

$$\ln e^{-5770k} = \ln\frac{1}{2}$$

$$-5770k = -0.693147$$

$$\approx 0.00012$$ ○ ○ ○

Carbon-14 dating is a well-known method used by anthropologists to establish the age of animal and plant fossils. This method assumes that the proportion of carbon 14 (C-14) present in the atmosphere has remained constant over the past 50,000 years. Professor Willard Libby, recipient of the Nobel Prize in chemistry in 1960, proposed this theory.

The amount of C-14 in the tissues of a living plant or animal is constant. However, when an organism dies, it stops absorbing new quantities of C-14 and the amount of C-14 in the remains diminishes because of the natural decay of the radioactive substance. Thus, the approximate age of a plant or animal fossil can be determined by measuring the amount of C-14 present in the remains.

EXAMPLE 4 A skull from an archeological site has one-tenth the amount of C-14 that it originally contained. Determine the approximate age of the skull.

Solution Here

$$Q(t) = Q_0e^{-kt}$$

$$= Q_0e^{-0.00012t}$$

where Q_0 is the amount of C-14 present originally and k, the decay constant, is equal to 0.00012 (see Example 3). Since $Q(t) = (1/10)Q_0$, we have

$$\frac{1}{10}Q_0 = Q_0 e^{-0.00012t}$$

$$\ln\frac{1}{10} = -0.00012t \qquad \text{(Taking the natural logarithm on both sides)}$$

$$t = \frac{\ln\dfrac{1}{10}}{-0.00012}$$

$$\approx 19{,}200$$

or approximately 19,200 years.   ○ ○ ○

Learning Curves

The next example shows how the exponential function may be applied to describe certain types of learning processes. Consider the function

$$Q(t) = C - Ae^{-kt}$$

where C, A, and k are positive constants. To sketch the graph of the function Q, observe that its y-intercept is given by $Q(0) = C - A$. Next, we compute

$$Q'(t) = kAe^{-kt}$$

Since both k and A are positive, we see that $Q'(t) > 0$ for all values of t. Thus $Q(t)$ is an increasing function of t. Also,

$$\lim_{t\to\infty} Q(t) = \lim_{t\to\infty}(C - Ae^{-kt})$$
$$= \lim_{t\to\infty} C - \lim_{t\to\infty} Ae^{-kt}$$
$$= C$$

Figure 13.15
A learning curve.

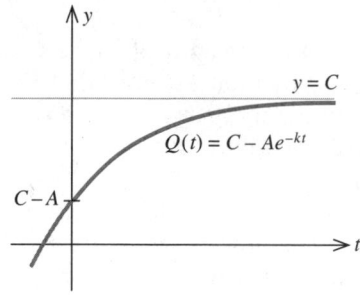

so $y = C$ is a horizontal asymptote of Q. Thus, $Q(t)$ increases and approaches the number C as t increases without bound. The graph of the function Q is shown in Figure 13.15, where that part of the graph corresponding to the negative values of t is drawn with a gray line since, in practice, one normally restricts the domain of the function to the interval $[0, \infty)$.

Observe that $Q(t)$ ($t > 0$) increases rather rapidly initially but the rate of increase slows down considerably after a while. To see this, we compute

$$\lim_{t\to\infty} Q'(t) = \lim_{t\to\infty} kAe^{-kt} = 0$$

This behavior of the graph of the function Q closely resembles the learning pattern experienced by workers engaged in highly repetitive work. For example, the productivity of an assembly-line worker increases very rapidly in the early stages of the training period. This productivity increase is a direct result of the worker's training and accumulated experience. But the rate of increase of productivity slows down as time goes by and the worker's productivity level approaches some fixed level due to the limitations of the worker and the machine. Because of this characteristic, the graph of the function $Q(t) = C - Ae^{-kt}$ is often called a **learning curve.**

 EXAMPLE 5 The Camera Division of the Eastman Optical Company produces a 35-mm single-lens reflex camera. Eastman's training department determines that after completing the basic training program, a new, previously inexperienced employee will be able to assemble

$$Q(t) = 50 - 30e^{-0.5t}$$

model F cameras per day, t months after the employee starts work on the assembly line.

a. How many model F cameras can a new employee assemble per day after basic training?

b. How many model F cameras can an employee with 1 month of experience assemble per day? An employee with 2 months of experience? An employee with 6 months of experience?

c. How many model F cameras can the average experienced employee assemble per day?

Solution

a. The number of model F cameras a new employee can assemble is given by

$$Q(0) = 50 - 30 = 20$$

b. The number of model F cameras that an employee with 1 month of experience, 2 months of experience, and 6 months of experience can assemble per day is given by

$$Q(1) = 50 - 30e^{-0.5} \approx 31.80$$
$$Q(2) = 50 - 30e^{-1} \approx 38.96$$
$$Q(6) = 50 - 30e^{-3} \approx 48.51$$

or approximately 32, 39, and 49, respectively.

c. As t increases without bound, $Q(t)$ approaches 50. Hence, the average experienced employee can ultimately be expected to assemble 50 model F cameras per day. ◦ ◦ ◦

Other applications of the learning curve are found in models that describe the dissemination of information about a product or the velocity of an object dropped into a viscous medium.

Logistic Growth Functions

Our last example of an application of exponential functions to the description of natural phenomena involves the **logistic** (also called the **S-shaped,** or

Figure 13.16
A logistic curve.

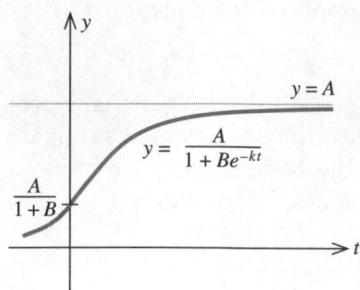

sigmoidal) **curve,** which is the graph of the function

$$Q(t) = \frac{A}{1 + Be^{-kt}}$$

where A, B, and k are positive constants. The function Q is called a **logistic growth function,** and the graph of the function Q is sketched in Figure 13.16.

Observe that $Q(t)$ increases rather rapidly for small values of t. In fact, for small values of t, the logistic curve resembles an exponential growth curve. However, the *rate of growth* of $Q(t)$ decreases quite rapidly as t increases and $Q(t)$ approaches the number A as t increases without bound.

Thus, the logistic curve exhibits both the property of rapid growth of the exponential growth curve and the "saturation" property of the learning curve. Because of these characteristics, the logistic curve serves as a suitable mathematical model for describing many natural phenomena. For example, if a small number of rabbits were introduced to a tiny island in the South Pacific, the rabbit population might be expected to grow very rapidly at first, but the growth rate would decrease quickly as overcrowding, scarcity of food, and other environmental factors affected it. The population would eventually stabilize at a level compatible with the life-support capacity of the environment. Models describing the spread of rumors and epidemics are other examples of the application of the logistic curve.

EXAMPLE 6   The number of soldiers at Fort MacArthur who contracted influenza after t days during a flu epidemic is approximated by the exponential model

$$Q(t) = \frac{5000}{1 + 1249e^{-kt}}$$

If 40 soldiers contracted the flu by the seventh day, find how many soldiers contracted the flu by the fifteenth day.

Solution   The given information implies that

$$Q(7) = 40$$

and

$$Q(7) = \frac{5000}{1 + 1249e^{-7k}}$$

$$= 40$$

Thus,

$$40(1 + 1249e^{-7k}) = 5000$$

$$1 + 1249e^{-7k} = \frac{5000}{40} = 125$$

$$e^{-7k} = \frac{124}{1249}$$

$$-7k = \ln \frac{124}{1249}$$

$$k = -\frac{\ln \dfrac{124}{1249}}{7} \approx 0.33$$

Therefore, the number of soldiers who contracted the flu after t days is given by

$$Q(t) = \frac{5000}{1 + 1249e^{-0.33t}}$$

In particular, the number of soldiers who contracted the flu by the fifteenth day is given by

$$Q(15) = \frac{5000}{1 + 1249e^{-15(0.33)}}$$

$$\approx 508$$

or approximately 508 soldiers.   ◦ ◦ ◦

EXPLORING WITH TECHNOLOGY

Refer to Example 6.
1. Use a graphing utility to plot the graph of the function Q in the viewing rectangle $[0, 40] \times [0, 5000]$.
2. Find how long it takes for the first 1000 soldiers to contract the flu. [*Suggestion:* Plot the graphs of $y_1 = Q(t)$ and $y_2 = 1000$ and find the point of intersection of the two graphs.]

◦   ◦   ◦

SELF-CHECK EXERCISE 13.5

Suppose that the population of a country (in millions) at any time t grows in accordance with the rule

$$P = \left(P_0 + \frac{I}{k}\right)e^{kt} - \frac{I}{k}$$

where P denotes the population at any time t, k is a constant reflecting the natural growth rate of the population, I is a constant giving the (constant) rate of immigration into the country, and P_0 is the total population of the country at time $t = 0$. The population of the United States in the year 1980 ($t = 0$) was 226.5 million. If the natural growth rate is 0.8% annually ($k = 0.008$) and net immigration is allowed at the rate of half a million people per year ($I = 0.5$) until the end of the century, what will the population of the United States be in the year 2000?

Solution to Self-Check Exercise 13.5 can be found on page 925.

13.5 EXERCISES

 A calculator is recommended for this exercise set.

1. Exponential Growth   Given that a quantity $Q(t)$ is described by the exponential growth function

$$Q(t) = 400e^{0.05t}$$

where t is measured in minutes, answer the following questions.
a. What is the growth constant?
b. What quantity is present initially?
c. Use a calculator to complete the following table of values:

t	0	10	20	100	1000
Q					

2. Exponential Decay   Given that a quantity $Q(t)$ exhibiting exponential decay is described by the function

$$Q(t) = 2000e^{-0.06t}$$

where t is measured in years, answer the following questions.
a. What is the decay constant?
b. What quantity is present initially?
c. Use a calculator to complete the following table of values:

t	0	5	10	20	100
Q					

3. Growth of Bacteria   The growth rate of the bacterium *Escherichia coli*, a common bacterium found in the human intestine, is proportional to its size. Under ideal laboratory conditions, when this bacterium is grown in a nutrient broth medium, the number of cells in a culture doubles approximately every 20 minutes.
a. If the initial cell population is 100, determine the function $Q(t)$ that expresses the exponential growth of the number of cells of this bacterium as a function of time t (in minutes).
b. How long will it take for a colony of 100 cells to increase to a population of one million?

c. If the initial cell population were 1000, how would this alter our model?

4. World Population   The world population at the beginning of 1990 was 5.3 billion. Assume that the population continues to grow at its present rate of approximately 2% per year, and find the function $Q(t)$ that expresses the world population (in billions) as a function of time t (in years) where $t = 0$ corresponds to the beginning of 1990.
a. Using this function, complete the following table of values and sketch the graph of the function Q.

Year	1990	1995	2000	2005
World Population				

Year	2010	2015	2020	2025
World Population				

b. Find the estimated rate of growth in the year 2000.

5. World Population   Refer to exercise 4.
a. If the world population continues to grow at its present rate of approximately 2% per year, find the length of time t_0 required for the world population to triple in size.
b. Using the time t_0 found in (a), what would the world population be if the growth rate were reduced to 1.8%?

6. Resale Value   A certain piece of machinery was purchased three years ago by the Garland Mills Company for $500,000. Its present resale value is $320,000. Assuming that the machine's resale value decreases exponentially, what will it be four years from now?

7. Atmospheric Pressure   If the temperature is constant, then the atmospheric pressure P (in pounds per square inch) varies with the altitude above sea level h in accordance with the law

$$P = p_0 e^{-kh}$$

where p_0 is the atmospheric pressure at sea level and k is a constant. If the atmospheric pressure is 15 pounds per square inch at sea level and 12.5 pounds per square inch at 4000 feet, find the atmospheric pressure at an altitude of 12,000 feet.

8. Radioactive Decay The radioactive element polonium decays according to the law

$$Q(t) = Q_0 \cdot 2^{-(t/140)}$$

where Q_0 is the initial amount and the time t is measured in days. If the amount of polonium left after 280 days is 20 milligrams, what was the initial amount present?

9. Radioactive Decay Phosphorus-32 has a half-life of 14.2 days. If 100 grams of this substance are present initially, find the amount present after t days. What amount will be left after 7.1 days?

10. Nuclear Fallout Strontium-90, a radioactive isotope of strontium, is present in the fallout resulting from nuclear explosions. It is especially hazardous to animal life, including humans, because, upon ingestion of contaminated food, it is absorbed into the bone structure. Its half-life is 27 years. If the amount of strontium-90 in a certain area is found to be four times the "safe" level, find how much time must elapse before an "acceptable level" is reached.

11. Carbon-14 Dating Wood deposits recovered from an archeological site contain 20% of the carbon-14 they originally contained. How long ago did the tree from which the wood was obtained die?

12. Carbon-14 Dating Skeletal remains of the so-called "Pittsburgh Man," unearthed in Pennsylvania, had lost 82% of the carbon-14 they originally contained. Determine the approximate age of the bones.

13. Learning Curves The American Stenographic Institute finds that the average student taking advanced shorthand, an intensive 20-week course, progresses according to the function

$$Q(t) = 120(1 - e^{-0.05t}) + 60 \qquad (0 \le t \le 20)$$

where $Q(t)$ measures the number of words (per minute) of dictation that the student can take in shorthand after t weeks in the course. Sketch the graph of the function Q and answer the following questions.
a. What is the beginning shorthand speed for the average student in this course?

b. What shorthand speed does the average student attain halfway through the course?
c. How many words per minute can the average student take after completing this course?

14. Effect of Advertising on Sales The Metro Department Store found that t weeks after the end of a sales promotion the volume of sales was given by a function of the form

$$S(t) = B + Ae^{-kt} \qquad (0 \le t \le 4)$$

where $B = 50,000$ and is equal to the average weekly volume of sales before the promotion. The sales volumes at the end of the first and third weeks were $83,515 and $65,055, respectively. Assume that the sales volume is decreasing exponentially.
a. Find the decay constant k.
b. What is the sales volume at the end of the fourth week?

15. Demand for Computers The Universal Instruments Company found that the monthly demand for its new line of Galaxy Home Computers t months after placing the line on the market was given by

$$D(t) = 2000 - 1500e^{-0.05t} \qquad (t > 0)$$

Graph this function and answer the following questions.
a. What is the demand after 1 month? One year? Two years? Five years?
b. At what level is the demand expected to stabilize?
c. Find the rate of growth of the demand after the tenth month.

16. Newton's Law of Cooling Newton's law of cooling states that the rate at which the temperature of an object changes is proportional to the difference in temperature between the object and that of the surrounding medium. Thus, the temperature $F(t)$ of an object that is greater than the temperature of its surrounding medium is given by

$$F(t) = T + Ae^{-kt}$$

where t is the time expressed in minutes, T is the temperature of the surrounding medium, and A and k are constants. Suppose that a cup of instant coffee is prepared with boiling water (212°F) and left to cool on the counter in a room where the temperature is 72°F. If $k = 0.1865$, determine when the coffee will be cool enough to drink (say, 110°F).

17. Spread of an Epidemic   During a flu epidemic, the number of children in the Woodbridge Community School System who contracted influenza after t days was given by

$$Q(t) = \frac{1000}{1 + 199e^{-0.8t}}$$

a. How many children were stricken by the flu after the first day?
b. How many children had the flu after 10 days?
c. How many children eventually contracted the disease?

18. Growth of a Fruit-Fly Population   On the basis of data collected during an experiment, a biologist found that the population growth of the fruit fly (*Drosophila*) with a limited food supply could be approximated by the exponential model

$$N(t) = \frac{400}{1 + 39e^{-0.16t}}$$

where t denotes the number of days since the beginning of the experiment.
a. What was the initial fruit-fly population in the experiment?
b. What was the maximum fruit-fly population that could be expected under this laboratory condition?
c. What was the population of the fruit-fly colony on the twentieth day?

19. Percentage of Households with VCRs   According to estimates by Paul Kroger Associates, the percentage of households that own videocassette recorders (VCRs) is given by

$$P(t) = \frac{68}{1 + 21.67e^{-0.62t}} \qquad (0 \le t \le 12)$$

where t is measured in years, with $t = 0$ corresponding to the beginning of 1985. What percentage of households owned VCRs at the beginning of 1985? At the beginning of 1995?

20. Population Growth in the 21st Century   The population of the United States is approximated by the function

$$P(t) = \frac{616.5}{1 + 4.02e^{-0.5t}}$$

where $P(t)$ is measured in millions of people and t is measured in 30-year intervals, with $t = 0$ corresponding

to 1930. What is the expected population of the United States in 2020 ($t = 3$)?

21. Spread of a Rumor   Three hundred students attended the dedication ceremony of a new building on a college campus. The president of the traditionally female college announced a new expansion program, which included plans to make the college coeducational. The number of students who learned of the new program t hours later is given by the function

$$f(t) = \frac{3000}{1 + Be^{-kt}}$$

If 600 students on campus had heard about the new program 2 hours after the ceremony, how many students had heard about the policy after 4 hours?

22. Concentration of Glucose in the Bloodstream   A glucose solution is administered intravenously into the bloodstream at a constant rate of r mg/hr. As the glucose is being administered, it is converted into other substances and removed from the bloodstream. Suppose the concentration of the glucose solution at time t is given by

$$C(t) = \frac{r}{k} - \left[\left(\frac{r}{k}\right) - C_0\right]e^{-kt}$$

where C_0 is the concentration at time $t = 0$ and k is a constant.
a. Assuming that $C_0 < r/k$, evaluate

$$\lim_{t \to \infty} C(t)$$

and interpret your result.
b. Sketch the graph of the function C.

23. Gompertz Growth Curve   Consider the function

$$Q(t) = Ce^{-Ae^{-kt}}$$

where $Q(t)$ is the size of a quantity at time t and A, C, and k are positive constants. The graph of this function, called the **Gompertz growth curve**, is used by biologists to describe restricted population growth.
a. Show that the function Q is always increasing.
b. Find the time t at which the growth rate $Q'(t)$ is increasing most rapidly.
[*Hint:* Find the inflection point of Q.]
c. Show that $\lim_{t \to \infty} Q(t) = C$ and interpret your result.

SOLUTION TO SELF-CHECK EXERCISE 13.5

We are given that $P_0 = 226.5$, $k = 0.008$, and $I = 0.5$. So

$$P = \left(226.5 + \frac{0.5}{0.008}\right) e^{0.008t} - \frac{0.5}{0.008}$$

$$= 289e^{0.008t} - 62.5$$

Therefore, the population in the year 2000 will be given by

$$P(20) = 289e^{0.16} - 62.5$$

$$\approx 276.6$$

or approximately 276.6 million.

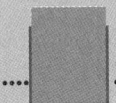

CHAPTER 13 SUMMARY OF PRINCIPAL FORMULAS AND TERMS

Formulas

1. Exponential function with base b

$y = b^x$

2. The number e

$e = \lim\limits_{m \to \infty} \left(1 + \dfrac{1}{m}\right)^m = 2.71828\ldots$

3. Exponential function with base e

$y = e^x$

4. Logarithmic function with base b

$y = \log_b x$

5. Logarithmic function with base e

$y = \ln x$

6. Inverse properties of $\ln x$ and e

$\ln e^x = x$ and $e^{\ln x} = x$

7. Continuous compound interest

$A = Pe^{rt}$

8. Derivative of the exponential function

$\dfrac{d}{dx}(e^x) = e^x$

9. Chain rule for exponential functions

$\dfrac{d}{dx}(e^u) = e^u \dfrac{du}{dx}$

10. Derivative of the logarithmic function

$\dfrac{d}{dx} \ln |x| = \dfrac{1}{x}$

11. Chain rule for logarithmic functions

$\dfrac{d}{dx}(\ln u) = \dfrac{1}{u} \dfrac{du}{dx}$

Terms

Exponential function	Growth constant
Common logarithm	Exponential decay
Natural logarithm	Decay constant
Logarithmic differentiation	Half-life of a radioactive element
Exponential growth	Logistic growth function

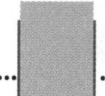 **CHAPTER 13 REVIEW EXERCISES**

1. Sketch on the same set of coordinate axes the graphs of the exponential functions defined by the equations.
   a. $y = 2^{-x}$ **b.** $y = (\frac{1}{2})^x$

In exercises 2 and 3, express each in logarithmic form.

2. $(\frac{2}{3})^{-3} = \frac{27}{8}$ 3. $16^{-3/4} = 0.125$

In exercises 4 and 5, solve each equation for x.

4. $\log_4(2x + 1) = 2$

5. $\ln(x - 1) + \ln 4 = \ln(2x + 4) - \ln 2$

In exercises 6–8, given that $\ln 2 = x$, $\ln 3 = y$, and $\ln 5 = z$, express each of the given logarithmic values in terms of x, y, and z.

6. $\ln 30$ 7. $\ln 3.6$ 8. $\ln 75$

9. Sketch the graph of the function $y = \log_2(x + 3)$.

10. Sketch the graph of the function $y = \log_3(x + 1)$.

In exercises 11–28, find the derivative of the function.

11. $f(x) = xe^{2x}$

12. $f(t) = \sqrt{t}e^t + t$

13. $g(t) = \sqrt{t}e^{-2t}$

14. $g(x) = e^x\sqrt{1 + x^2}$

15. $y = \dfrac{e^{2x}}{1 + e^{-2x}}$

16. $f(x) = e^{2x^2 - 1}$

17. $f(x) = xe^{-x^2}$

18. $g(x) = (1 + e^{2x})^{3/2}$

19. $f(x) = x^2e^x + e^x$

20. $g(t) = t \ln t$

21. $f(x) = \ln(e^{x^2} + 1)$

22. $f(x) = \dfrac{x}{\ln x}$

23. $f(x) = \dfrac{\ln x}{x + 1}$

24. $y = (x + 1)e^x$

25. $y = \ln(e^{4x} + 3)$

26. $f(r) = \dfrac{re^r}{1 + r^2}$

27. $f(x) = \dfrac{\ln x}{1 + e^x}$

28. $g(x) = \dfrac{e^{x^2}}{1 + \ln x}$

29. Find the second derivative of the function $y = \ln(3x + 1)$.

30. Find the second derivative of the function $y = x \ln x$.

31. Find $h'(0)$ if $h(x) = g(f(x))$, $g(x) = x + \dfrac{1}{x}$, and $f(x) = e^x$.

32. Find $h'(1)$ if $h(x) = g(f(x))$, $g(x) = \dfrac{x + 1}{x - 1}$, and $f(x) = \ln x$.

33. Use logarithmic differentiation to find the derivative of $f(x) = (2x^3 + 1)(x^2 + 2)^3$.

34. Use logarithmic differentiation to find the derivative of $f(x) = \dfrac{x(x^2 - 2)^2}{(x - 1)}$.

35. Find an equation of the tangent line to the graph of $y = e^{-2x}$ at the point $(1, e^{-2})$.

36. Find an equation of the tangent line to the graph of $y = xe^{-x}$ at the point $(1, e^{-1})$.

37. Sketch the graph of the function $f(x) = xe^{-2x}$.

38. Sketch the graph of the function $f(x) = x^2 - \ln x$.

39. Find the absolute extrema of the function $f(t) = te^{-t}$.

40. Find the absolute extrema of the function $g(t) = (\ln t)/t$ on $[1, 2]$.

41. A hotel was purchased by a conglomerate for $4.5 million and sold five years later for $8.2 million. Find the annual rate of return (compounded continuously).

42. Find the present value of $119,346 due in four years at an interest rate of 10% per year compounded continuously.

43. A culture of bacteria that initially contained 2000 bacteria has a count of 18,000 bacteria after 2 hours.
a. Determine the function $Q(t)$ that expresses the exponential growth of the number of cells of this bacterium as a function of time t (in minutes).
b. Find the number of bacteria present after 4 hours.

44. The radioactive element radium has a half-life of 1600 years. What is its decay constant?

45. The V.C.A. Television Company found that the monthly demand for its new line of video disc players t months after placing the players on the market is given by

$$D(t) = 4000 - 3000e^{-0.06t} \quad (t \geq 0)$$

Graph this function and answer the following questions.
a. What was the demand after 1 month? After one year? After two years?
b. At what level is the demand expected to stabilize?

46. During a flu epidemic, the number of students at a certain university who contracted influenza after t days could be approximated by the exponential model

$$Q(t) = \frac{3000}{1 + 499e^{-kt}}$$

If 90 students contracted the flu by the tenth day, how many students contracted the flu by the twentieth day?

Differential calculus is concerned with the problem of finding the rate of change of one quantity with respect to another. In this chapter we begin the study of the other branch of calculus, known as integral calculus. Here we are interested in precisely the opposite problem: If we know the rate of change of one quantity with respect to another, can we find the relationship between the two quantities? The principal tool used in the study of integral calculus is the *antiderivative* of a function, and we develop rules for antidifferentiation, or *integration*, as the process of finding the antiderivative is called. We also show that a link is established between differential and integral calculus—via the Fundamental Theorem of Calculus.

How much will the solar cell panels cost? The head of Soloron Corporation's research and development department has projected that the cost of producing solar cell panels will drop at a certain rate in the next several years. In Example 7, page 950, you will see how this information can be used to predict the cost of solar cell panels in the coming years.

14

INTEGRATION

14.1 ANTIDERIVATIVES AND THE RULES OF INTEGRATION

Antiderivatives

Let us return, once again, to the motion of the maglev (Figure 14.1). In Chapter 10, we discussed this problem:

If we know the position of the maglev at any time t, can we find its velocity at time t?

Figure 14.1
A maglev moving along an elevated monorail track.

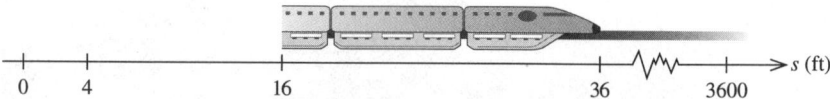

As it turns out, if the position of the maglev is described by the position function f, then its velocity at any time t is given by $f'(t)$. Here f', the velocity function of the maglev, is just the derivative of f.

Now, in Chapters 14 and 15, we will consider precisely the opposite problem:

If we know the velocity of the maglev at any time t, can we find its position at time t?

Stated another way, if we know the velocity function f' of the maglev, can we find its position function f? In order to solve this problem, we need the concept of an *antiderivative* of a function.

ANTIDERIVATIVE

A function F is an **antiderivative** of f on an interval I if $F'(x) = f(x)$ for all x in I.

Thus, an antiderivative of a function f is a function F whose derivative is f. For example, $F(x) = x^2$ is an antiderivative of $f(x) = 2x$ because

$$F'(x) = \frac{d}{dx}(x^2) = 2x = f(x)$$

and $F(x) = x^3 + 2x + 1$ is an antiderivative of $f(x) = 3x^2 + 2$ because

$$F'(x) = \frac{d}{dx}(x^3 + 2x + 1) = 3x^2 + 2 = f(x)$$

EXAMPLE I Let $F(x) = \frac{1}{3}x^3 - 2x^2 + x - 1$. Show that F is an antiderivative of $f(x) = x^2 - 4x + 1$.

Solution Differentiating the function F, we obtain

$$F'(x) = x^2 - 4x + 1 = f(x)$$

and the desired result follows.   ○ ○ ○

EXAMPLE 2 Let $F(x) = x$, $G(x) = x + 2$, and $H(x) = x + C$, where C is a constant. Show that F, G, and H are all antiderivatives of the function f defined by $f(x) = 1$.

Solution Since

$$F'(x) = \frac{d}{dx}(x) = 1 = f(x)$$

$$G'(x) = \frac{d}{dx}(x + 2) = 1 = f(x)$$

$$H'(x) = \frac{d}{dx}(x + C) = 1 = f(x)$$

we see that F, G, and H are indeed antiderivatives of f.   ○ ○ ○

Example 2 shows that once an antiderivative G of a function f is known, then another antiderivative of f may be found by adding an arbitrary constant to the function G. The following theorem states that no function other than one obtained in this manner can be an antiderivative of f. (We omit the proof.)

THEOREM I

Let G be an antiderivative of a function f. Then every antiderivative F of f must be of the form $F(x) = G(x) + C$, where C is a constant.

Returning to Example 2, we see that there are infinitely many antiderivatives of the function $f(x) = 1$. We obtain each one by specifying the constant C in the function $F(x) = x + C$. Figure 14.2 shows the graphs of some of

Figure 14.2
The graphs of some antiderivatives
of $f(x) = 1$.

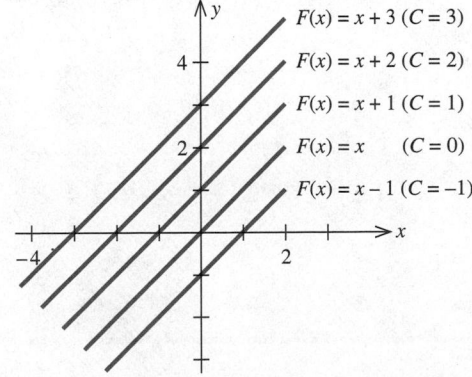

$F(x) = x + 3 \ (C = 3)$

$F(x) = x + 2 \ (C = 2)$

$F(x) = x + 1 \ (C = 1)$

$F(x) = x \quad (C = 0)$

$F(x) = x - 1 \ (C = -1)$

these antiderivatives for selected values of C. These graphs constitute part of a family of infinitely many parallel straight lines, each having a slope equal to 1. This result is expected since there are infinitely many curves (straight lines) with a given slope equal to 1. The antiderivatives $F(x) = x + C$ (C, a constant) are precisely the functions representing this family of straight lines.

EXAMPLE 3   Prove that the function $G(x) = x^2$ is an antiderivative of the function $f(x) = 2x$. Write a general expression for the antiderivatives of f.

Solution   Since $G'(x) = 2x = f(x)$, we have shown that $G(x) = x^2$ is an antiderivative of $f(x) = 2x$. By Theorem 1, every antiderivative of the function $f(x) = 2x$ has the form $F(x) = x^2 + C$, where C is some constant. The graphs of a few of the antiderivatives of f are shown in Figure 14.3.

Figure 14.3
The graphs of some antiderivatives of $f(x) = 2x$.

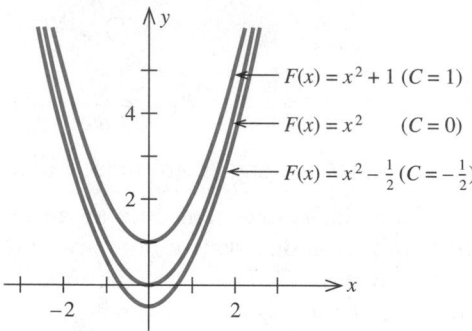

$F(x) = x^2 + 1$ $(C = 1)$

$F(x) = x^2$ $(C = 0)$

$F(x) = x^2 - \frac{1}{2}$ $(C = -\frac{1}{2})$

○ ○ ○

EXPLORING WITH TECHNOLOGY

Let $f(x) = x^2 - 1$.

1. Show that $F(x) = \frac{1}{3}x^3 - x + C$, where C is an arbitrary constant, is an antiderivative of f.

2. Use a graphing utility to plot the graphs of the antiderivatives of f corresponding to $C = -2$, $C = -1$, $C = 0$, $C = 1$, and $C = 2$ on the same set of axes in the viewing rectangle $[-4, 4] \times [-4, 4]$.

3. If your graphing utility has the capability, draw the tangent line to each of the graphs in part (2) at the point whose x-coordinate is 2. What can you say about this family of tangent lines?

4. What is the slope of a tangent line in this family? Explain how you obtained your answer.

○ ○ ○

The Indefinite Integral

The process of finding all the antiderivatives of a function is called **antidifferentiation,** or **integration.** We use the symbol $\int$, called an **integral sign,** to indicate that the operation of integration is to be performed on some function f. Thus,

$$\int f(x)\,dx = F(x) + C$$

[read "the indefinite integral of $f(x)$ with respect to x equals $F(x)$ plus C"] tells us that the **indefinite integral** of f is the family of functions given by $F(x) + C$, where $F'(x) = f(x)$. The function f to be integrated is called the **integrand,** and the constant C is called a **constant of integration.** The expression dx following the integrand $f(x)$ reminds us that the operation is performed with respect to x. If the independent variable is t, we write $\int f(t)\,dt$ instead. In this sense both t and x are "dummy variables."

Using this notation, we can write the results of Examples 2 and 3 as

$$\int 1\,dx = x + C \quad \text{and} \quad \int 2x\,dx = x^2 + K$$

where C and K are arbitrary constants.

Basic Integration Rules

Our next task is to develop some rules for finding the indefinite integral of a given function f. Because integration and differentiation are reverse operations, we discover many of the rules of integration by first making an "educated guess" at the antiderivative F of the function f to be integrated. Then this result is verified by demonstrating that $F' = f$.

RULE 1: INDEFINITE INTEGRAL OF A CONSTANT	$\int k\,dx = kx + C \qquad (k,\text{ a constant})$

To prove this result, observe that

$$F'(x) = \frac{d}{dx}(kx + C) = k$$

EXAMPLE 4   Evaluate each of these indefinite integrals.

a. $\int 2\,dx$ **b.** $\int \pi^2\,dx$

Solution   Each of the integrands has the form $f(x) = k$, where k is a constant. Applying Rule 1 in each case yields

a. $\int 2\,dx = 2x + C$ **b.** $\int \pi^2\,dx = \pi^2 x + C$

○ ○ ○

Next, from the rule of differentiation,

$$\frac{d}{dx}x^n = nx^{n-1}$$

we obtain the following rule of integration.

RULE 2: THE POWER RULE	$\int x^n \, dx = \frac{1}{n+1}x^{n+1} + C \qquad (n \neq -1)$

An antiderivative of a power function is another power function obtained from the integrand by increasing its power by 1 and dividing the resulting expression by the new power.

To prove this result, observe that

$$F'(x) = \frac{d}{dx}\left[\frac{1}{n+1}x^{n+1} + C\right]$$

$$= \frac{n+1}{n+1}x^n$$

$$= x^n$$

$$= f(x)$$

EXAMPLE 5   Evaluate each of the following indefinite integrals:

a. $\int x^3 \, dx$ **b.** $\int x^{3/2} \, dx$ **c.** $\int \frac{1}{x^{3/2}} \, dx$

Solution   Each integrand is a power function with exponent $n \neq -1$. We apply Rule 2 in each case.

a. $\int x^3 \, dx = \frac{1}{4}x^4 + C$

b. $\int x^{3/2} \, dx = \frac{1}{5/2}x^{5/2} + C = \frac{2}{5}x^{5/2} + C$

c. $\int \frac{1}{x^{3/2}} \, dx = \int x^{-3/2} \, dx = \frac{1}{-1/2}x^{-1/2} + C = -2x^{-1/2} + C$

These results may be verified by differentiating each of the antiderivatives and showing that the result is equal to the corresponding integrand.

The next rule tells us that a constant factor may be moved through an integral sign.

RULE 3: INDEFINITE INTEGRAL OF A CONSTANT MULTIPLE OF A FUNCTION

$$\int cf(x)\, dx = c \int f(x)\, dx \qquad (c, \text{a constant})$$

The indefinite integral of a constant multiple of a function is equal to the constant multiple of the indefinite integral of the function.

This result follows from the corresponding rule of differentiation (see Rule 3, Section 11.1).

 Only a constant can be "moved out" of an integral sign. For example, it is incorrect to write

$$\int x^2\, dx = x^2 \int 1\, dx$$

In fact, $\int x^2\, dx = \frac{1}{3}x^3 + C$, whereas $x^2 \int 1\, dx = x^2(x + C) = x^3 + Cx^2$.

EXAMPLE 6   Evaluate each of the following indefinite integrals.

a. $\int 2t^3\, dt$ **b.** $\int -3x^{-2}\, dx$

Solution   Each integrand has the form $cf(x)$, where c is a constant. Applying Rule 3, we obtain

a. $\int 2t^3\, dt = 2 \int t^3\, dt = 2\left[\frac{1}{4}t^4 + K\right] = \frac{1}{2}t^4 + 2K = \frac{1}{2}t^4 + C$

where $C = 2K$. From now on, we will write the constant of integration as C since any nonzero multiple of an arbitrary constant is an arbitrary constant.

b. $\int -3x^{-2}\, dx = -3 \int x^{-2}\, dx = (-3)(-1)x^{-1} + C = \frac{3}{x} + C$

RULE 4: THE SUM RULE

$$\int [f(x) + g(x)]\, dx = \int f(x)\, dx + \int g(x)\, dx$$

$$\int [f(x) - g(x)]\, dx = \int f(x)\, dx - \int g(x)\, dx$$

The indefinite integral of a sum (difference) of two integrable functions is equal to the sum (difference) of their indefinite integrals.

This result is easily extended to the case involving the sum and difference of any finite number of functions. As in Rule 3, the proof of Rule 4 follows from the corresponding rule of differentiation (see Rule 4, Section 11.1).

EXAMPLE 7   Evaluate the indefinite integral

$$\int (3x^5 + 4x^{3/2} - 2x^{-1/2})\, dx$$

Solution   Applying the extended version of Rule 4, we find that

$$\int (3x^5 + 4x^{3/2} - 2x^{-1/2})\, dx$$

$$= \int 3x^5\, dx + \int 4x^{3/2}\, dx - \int 2x^{-1/2}\, dx$$

$$= 3\int x^5\, dx + 4\int x^{3/2}\, dx - 2\int x^{-1/2}\, dx \qquad \text{(Rule 3)}$$

$$= (3)\left(\frac{1}{6}\right)x^6 + (4)\left(\frac{2}{5}\right)x^{5/2} - (2)(2)x^{1/2} + C \qquad \text{(Rule 2)}$$

$$= \frac{1}{2}x^6 + \frac{8}{5}x^{5/2} - 4x^{1/2} + C$$

Observe in Example 7 that we combined the three constants of integration, which arise from evaluating the three indefinite integrals, to obtain one constant C. After all, the sum of three arbitrary constants is also an arbitrary constant.

RULE 5: THE INDEFINITE INTEGRAL OF THE EXPONENTIAL FUNCTION	$\int e^x\, dx = e^x + C$

The indefinite integral of the exponential function with base e is equal to the function itself (except, of course, for the constant of integration).

EXAMPLE 8   Evaluate the indefinite integral

$$\int (2e^x - x^3)\, dx$$

Solution   We have

$$\int (2e^x - x^3)\, dx = \int 2e^x\, dx - \int x^3\, dx$$

$$= 2\int e^x\, dx - \int x^3\, dx$$

$$= 2e^x - \frac{1}{4}x^4 + C$$

The last rule of integration in this section covers the integration of the function $f(x) = x^{-1}$. Remember that this function was the only exceptional case in the integration of the power function $f(x) = x^n$ (see Rule 2).

RULE 6: THE INDEFINITE INTEGRAL OF THE FUNCTION $f(x) = x^{-1}$

$$\int x^{-1}\, dx = \int \frac{1}{x}\, dx = \ln |x| + C \qquad (x \neq 0)$$

To prove Rule 6, observe that

$$\frac{d}{dx} \ln |x| = \frac{1}{x} \qquad \text{(See Rule 3, Section 13.4.)}$$

EXAMPLE 9   Evaluate the indefinite integral

$$\int \left(2x + \frac{3}{x} + \frac{4}{x^2} \right) dx$$

Solution

$$\int \left(2x + \frac{3}{x} + \frac{4}{x^2} \right) dx = \int 2x\, dx + \int \frac{3}{x}\, dx + \int \frac{4}{x^2}\, dx$$

$$= 2 \int x\, dx + 3 \int \frac{1}{x}\, dx + 4 \int x^{-2}\, dx$$

$$= 2 \left(\frac{1}{2} \right) x^2 + 3 \ln |x| + 4(-1)x^{-1} + C$$

$$= x^2 + 3 \ln |x| - \frac{4}{x} + C \qquad \qquad \circ\ \circ\ \circ$$

Differential Equations

Let us return to the problem posed at the beginning of the section: *Given the derivative of a function f', can we find the function f?* As an example, suppose we are given the function

$$f'(x) = 2x - 1 \tag{1}$$

and we wish to find $f(x)$. From what we now know, we can find f by integrating equation (1). Thus,

$$f(x) = \int f'(x)\, dx = \int (2x - 1)\, dx = x^2 - x + C \tag{2}$$

where C is an arbitrary constant. So there are infinitely many functions that have the derivative f', each differing from the other by a constant.

Equation (1) is called a *differential equation*. In general, a **differential equation** is an equation that involves the derivative or differential of an unknown function. [In the case of equation (1), the unknown function is f.] A **solution** of a differential equation is any function that satisfies the differential equation. Thus, equation (2) gives *all* the solutions of the differential equation (1) and, accordingly, it is called the **general solution** of the differential equation $f'(x) = 2x - 1$.

Figure 14.4
The graphs of some of the functions having the derivative $f'(x) = 2x - 1$. Observe that the slopes of the tangent lines to the graphs are the same for a fixed value of x.

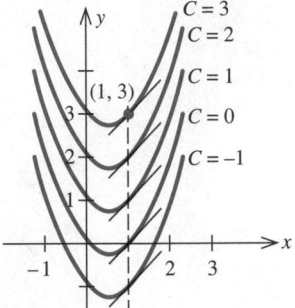

The graphs of $f(x) = x^2 - x + C$ for selected values of C are shown in Figure 14.4. These graphs have one property in common: For any fixed value of x, the tangent lines to these graphs have the same slope. This follows because any member of the family $f(x) = x^2 - x + C$ must have the same slope at x—namely, $2x - 1$!

Although there are infinitely many solutions to the differential equation $f'(x) = 2x - 1$, we can obtain a **particular solution** by specifying the value the function must assume at a certain value of x. For example, suppose we stipulate that the function f under consideration must satisfy the condition $f(1) = 3$ or, equivalently, the graph of f must pass through the point $(1, 3)$. Then, using the condition on the general solution $f(x) = x^2 - x + C$, we find that

$$f(1) = 1 - 1 + C = 3$$

and $C = 3$. Thus, the particular solution is $f(x) = x^2 - x + 3$ (see Figure 14.4).

The condition $f(1) = 3$ is an example of an *initial condition*. More generally, an **initial condition** is a condition imposed on the value of f at a point $x = a$.

Initial Value Problems

An **initial value problem** is one in which we are required to find a function satisfying (1) a differential equation and (2) one or more initial conditions. Example 10 is an initial value problem.

EXAMPLE 10   Find the function f if it is known that

$$f'(x) = 3x^2 - 4x + 8 \quad \text{and} \quad f(1) = 9$$

Solution   We are required to solve the initial value problem

$$\left. \begin{array}{l} f'(x) = 3x^2 - 4x + 8 \\ f(1) = 9 \end{array} \right\}$$

Integrating the function f', we find

$$f(x) = \int f'(x)\, dx$$
$$= \int (3x^2 - 4x + 8)\, dx$$
$$= x^3 - 2x^2 + 8x + C$$

Using the condition $f(1) = 9$, we have

$$9 = f(1) = 1^3 - 2(1)^2 + 8(1) + C = 7 + C$$

or $$C = 2$$

Therefore, the required function f is given by $f(x) = x^3 - 2x^2 + 8x + 2$.

○ ○ ○

Applications

EXAMPLE 11 In a test run of a maglev along a straight elevated monorail track, data obtained from its speedometer indicate that the velocity of the maglev at time t can be described by the velocity function

$$v(t) = 8t \qquad (0 \leq t \leq 30)$$

Find the position function of the maglev. Assume that initially the maglev is located at the origin of a coordinate line.

Solution Let $s(t)$ denote the position of the maglev at any time t $(0 \leq t \leq 30)$. Then $s'(t) = v(t)$. So, we have the initial value problem

$$\left. \begin{array}{l} s'(t) = 8t \\ s(0) = 0 \end{array} \right\}$$

Integrating both sides of the differential equation $s'(t) = 8t$, we obtain

$$s(t) = \int s'(t)\,dt = \int 8t\,dt = 4t^2 + C$$

where C is an arbitrary constant. To evaluate C, we use the initial condition $s(0) = 0$ to write

$$s(0) = 4(0) + C = 0 \qquad \text{or} \qquad C = 0$$

Therefore, the required position function is $s(t) = 4t^2$ $(0 \leq t \leq 30)$.

○ ○ ○

EXAMPLE 12 The current circulation of *Investor's Digest* is 3000 copies per week. The managing editor of the weekly projects a growth rate of

$$4 + 5t^{2/3}$$

copies per week, t weeks from now, for the next three years. Based on her projection, what will the circulation of the digest be 125 weeks from now?

Solution Let $S(t)$ denote the circulation of the digest t weeks from now. Then $S'(t)$ is the rate of change in the circulation in the tth week and is given by

$$S'(t) = 4 + 5t^{2/3}$$

Furthermore, the current circulation of 3000 copies per week translates into the initial condition $S(0) = 3000$. Integrating the differential equation with respect to t gives

$$S(t) = \int S'(t)\,dt = \int (4 + 5t^{2/3})\,dt$$

$$= 4t + 5\left(\frac{t^{5/3}}{\frac{5}{3}}\right) + C = 4t + 3t^{5/3} + C$$

To determine the value of C, we use the condition $S(0) = 3000$ to write

$$S(0) = 4(0) + 3(0) + C = 3000$$

which gives $C = 3000$. Therefore, the circulation of the digest t weeks from now will be

$$S(t) = 4t + 3t^{5/3} + 3000$$

In particular, the circulation 125 weeks from now will be

$$S(125) = 4(125) + 3(125)^{5/3} + 3000 = 12{,}875$$

or 12,875 copies per week.   ○ ○ ○

SELF-CHECK EXERCISES 14.1

1. Evaluate $\int \left(\dfrac{1}{\sqrt{x}} - \dfrac{2}{x} + 3e^x \right) dx$.

2. Find the rule for the function f given that (1) the slope of the tangent line to the graph of f at any point $P(x, f(x))$ is given by the expression $3x^2 - 6x + 3$ and (2) the graph of f passes through the point $(2, 9)$.

3. Suppose that United Motors' share of the new cars sold in a certain country is changing at the rate of

$$f(t) = -0.01875t^2 + 0.15t - 1.2 \qquad (0 \le t \le 12)$$

percent at year t ($t = 0$ corresponds to the beginning of 1987). The company's market share at the beginning of 1987 was 48.4%. What was United Motors' market share at the beginning of 1999?

Solutions to Self-Check Exercises 14.1 can be found on page 944.

14.1 EXERCISES

In exercises 1–4, verify directly that F is an antiderivative of f.

1. $F(x) = \frac{1}{3}x^3 + 2x^2 - x + 2; f(x) = x^2 + 4x - 1$

2. $F(x) = xe^x + \pi; f(x) = e^x(1 + x)$

3. $F(x) = \sqrt{2x^2 - 1}; f(x) = \dfrac{2x}{\sqrt{2x^2 - 1}}$

4. $F(x) = x \ln x - x; f(x) = \ln x$

In exercises 5–8, (a) verify that G is an antiderivative of f, (b) find all antiderivatives of f, and (c) sketch the graphs of a few of the family of antiderivatives found in (b).

5. $G(x) = 2x; f(x) = 2$

6. $G(x) = 2x^2; f(x) = 4x$

7. $G(x) = \frac{1}{3}x^3; f(x) = x^2$

8. $G(x) = e^x; f(x) = e^x$

In exercises 9–50, find the indefinite integral.

9. $\displaystyle\int 6\,dx$

10. $\displaystyle\int \sqrt{2}\,dx$

11. $\displaystyle\int x^3\,dx$

12. $\displaystyle\int 2x^5\,dx$

13. $\displaystyle\int x^{-4}\,dx$

14. $\displaystyle\int 3t^{-7}\,dt$

15. $\displaystyle\int x^{2/3}\,dx$

16. $\displaystyle\int 2u^{3/4}\,du$

17. $\displaystyle\int x^{-5/4}\,dx$

18. $\displaystyle\int 3x^{-2/3}\,dx$

19. $\displaystyle\int \dfrac{2}{x^2}\,dx$

20. $\displaystyle\int \dfrac{1}{3x^5}\,dx$

21. $\displaystyle\int \pi\sqrt{t}\, dt$

22. $\displaystyle\int \frac{3}{\sqrt{t}}\, dt$

23. $\displaystyle\int (3 - 2x)\, dx$

24. $\displaystyle\int (1 + u + u^2)\, du$

25. $\displaystyle\int (x^2 + x + x^{-3})\, dx$

26. $\displaystyle\int (0.3t^2 + 0.02t + 2)\, dt$

27. $\displaystyle\int 4e^x\, dx$

28. $\displaystyle\int (1 + e^x)\, dx$

29. $\displaystyle\int (1 + x + e^x)\, dx$

30. $\displaystyle\int (2 + x + 2x^2 + e^x)\, dx$

31. $\displaystyle\int \left(4x^3 - \frac{2}{x^2} - 1\right) dx$

32. $\displaystyle\int \left(6x^3 + \frac{3}{x^2} - x\right) dx$

33. $\displaystyle\int (x^{5/2} + 2x^{3/2} - x)\, dx$

34. $\displaystyle\int (t^{3/2} + 2t^{1/2} - 4t^{-1/2})\, dt$

35. $\displaystyle\int \left(\sqrt{x} + \frac{3}{\sqrt{x}}\right) dx$

36. $\displaystyle\int \left(\sqrt[3]{x^2} - \frac{1}{x^2}\right) dx$

37. $\displaystyle\int \left(\frac{u^3 + 2u^2 - u}{3u}\right) du$

$\left[\text{Hint: } \dfrac{u^3 + 2u^2 - u}{3u} = \dfrac{1}{3}u^2 + \dfrac{2}{3}u - \dfrac{1}{3}\right]$

38. $\displaystyle\int \frac{x^4 - 1}{x^2}\, dx$

$\left[\text{Hint: } \dfrac{x^4 - 1}{x^2} = x^2 - x^{-2}\right]$

39. $\displaystyle\int (2t + 1)(t - 2)\, dt$

40. $\displaystyle\int u^{-2}(1 - u^2 + u^4)\, du$

41. $\displaystyle\int \frac{1}{x^2}(x^4 - 2x^2 + 1)\, dx$

42. $\displaystyle\int \sqrt{t}(t^2 + t - 1)\, dt$

43. $\displaystyle\int \frac{ds}{(s + 1)^{-2}}$

44. $\displaystyle\int \left(\sqrt{x} + \frac{3}{x} - 2e^x\right) dx$

45. $\displaystyle\int (e^t + t^e)\, dt$

46. $\displaystyle\int \left(\frac{1}{x^2} - \frac{1}{\sqrt[3]{x^2}} + \frac{1}{\sqrt{x}}\right) dx$

47. $\displaystyle\int \left(\frac{x^3 + x^2 - x + 1}{x^2}\right) dx$

[*Hint:* Simplify the integrand first.]

48. $\displaystyle\int \frac{t^3 + \sqrt[3]{t}}{t^2}\, dt$

[*Hint:* Simplify the integrand first.]

49. $\displaystyle\int \frac{(\sqrt{x} - 1)^2}{x^2}\, dx$

[*Hint:* Simplify the integrand first.]

50. $\displaystyle\int (x + 1)^2 \left(1 - \frac{1}{x}\right) dx$

[*Hint:* Simplify the integrand first.]

In exercises 51–58, find f(x) by solving the initial value problem.

51. $f'(x) = 2x + 1;\ f(1) = 3$

52. $f'(x) = 3x^2 - 6x;\ f(2) = 4$

53. $f'(x) = 3x^2 + 4x - 1;\ f(2) = 9$

54. $f'(x) = \dfrac{1}{\sqrt{x}};\ f(4) = 2$

55. $f'(x) = 1 + \dfrac{1}{x^2};\ f(1) = 2$

56. $f'(x) = e^x - 2x;\ f(0) = 2$

57. $f'(x) = \dfrac{x + 1}{x};\ f(1) = 1$

58. $f'(x) = 1 + e^x + \dfrac{1}{x};\ f(1) = 3 + e$

In exercises 59–62, find the function f given that the slope of the tangent line to the graph of f at any point (x, f(x)) is f'(x) and that the graph of f passes through the given point.

59. $f'(x) = \frac{1}{2}x^{-1/2};\ (2, \sqrt{2})$

60. $f'(t) = t^2 - 2t + 3;\ (1, 2)$

61. $f'(x) = e^x + x;\ (0, 3)$ **62.** $f'(x) = \dfrac{2}{x} + 1;\ (1, 2)$

63. Velocity of a Car   The velocity of a car (in ft/sec) t seconds after starting from rest is given by the function

$$f(t) = 2\sqrt{t} \qquad (0 \le t \le 30)$$

Find the car's position at any time t.

64. Velocity of a Maglev   The velocity (in ft/sec) of a maglev is

$$v(t) = 0.2t + 3 \qquad (0 \le t \le 120)$$

At $t = 0$, it is at the station. Find the function giving the position of the maglev at time t, assuming that the motion takes place along a straight stretch of track.

65. Cost of Producing Clocks   The Lorimar Watch Company manufactures travel clocks. The daily marginal cost function associated with producing these clocks is

$$C'(x) = 0.000009x^2 - 0.009x + 8$$

where $C'(x)$ is measured in dollars per unit and x denotes the number of units produced. Management has determined that the daily fixed cost incurred in producing these clocks is $120. Find the total cost incurred by Lorimar in producing the first 500 travel clocks per day.

66. **Revenue Functions**   The management of the Lorimar Watch Company has determined that the daily marginal revenue function associated with producing and selling their travel clocks is given by

$$R'(x) = -0.009x + 12$$

where x denotes the number of units produced and sold and $R'(x)$ is measured in dollars per unit.
a. Determine the revenue function $R(x)$ associated with producing and selling these clocks.
b. What is the demand equation that relates the whole-sale unit price with the quantity of travel clocks demanded?

67. **Profit Functions**   The Cannon Precision Instruments Corporation makes an automatic electronic flash with Thyrister circuitry. The estimated marginal profit associated with producing and selling these electronic flashes is

$$(-0.004x + 20)$$

dollars per unit per month when the production level is x units per month. Cannon's fixed cost for producing and selling these electronic flashes is $16,000 per month. At what level of production does Cannon realize a maximum profit? What is the maximum monthly profit?

68. **Cost of Producing Guitars**   The Carlota Music Company estimates that the marginal cost of manufacturing its Professional Series guitars is

$$C'(x) = 0.002x + 100$$

dollars per month when the level of production is x guitars per month. The fixed costs incurred by Carlota are $4000 per month. Find the total monthly cost incurred by Carlota in manufacturing x guitars per month.

69. **Quality Control**   As part of a quality-control program, the chess sets manufactured by the Jones Brothers Company are subjected to a final inspection before packing. The rate of increase in the number of sets checked per hour by an inspector t hours into the 8 A.M. to 12 noon morning shift is approximately

$$N'(t) = -3t^2 + 12t + 45 \qquad (0 \le t \le 4)$$

a. Find an expression $N(t)$ that approximates the number of sets inspected at the end of t hours.
[*Hint: N(0) = 0.*]

b. How many sets does the average inspector check during a morning shift?

70. **Ballast Dropped from a Balloon**   A ballast is dropped from a stationary hot-air balloon that is hovering at an altitude of 400 ft. Its velocity after t seconds is $-32t$ ft/sec.
a. Find the height $h(t)$ of the ballast from the ground at time t.
[*Hint: h'(t) = -32t and h(0) = 400.*]

b. When will the ballast strike the ground?
c. Find the velocity of the ballast when it hits the ground.

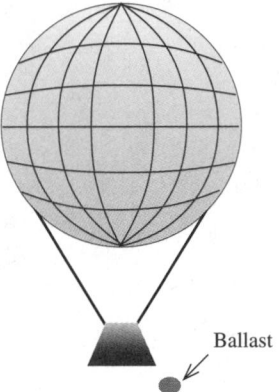

Ballast

71. **Cable TV Subscribers**   A study conducted by Tele-Cable, Inc., estimates that the number of cable television subscribers will grow at the rate of

$$100 + 210t^{3/4}$$

new subscribers per month t months from the start date of the service. If 5000 subscribers signed up for the service before the starting date, how many subscribers will there be 16 months from that date?

72. **Air Pollution**   On an average summer day, the level of carbon monoxide in a city's air is 2 parts per million. An environmental protection agency's study predicts that, unless more stringent measures are taken to protect the city's atmosphere, the concentration of carbon monoxide present in the air will increase at the rate of

$$0.003t^2 + 0.06t + 0.1$$

parts per million per year t years from now. If no further pollution-control efforts are made, what will the concentration of carbon monoxide be on an average summer day five years from now?

73. **Population Growth** The development of Astro World ("The Amusement Park of the Future") on the outskirts of a city will increase the city's population at the rate of

$$4500\sqrt{t} + 1000$$

people per year t years from the start of construction. The population before construction is 30,000. Determine the projected population nine years after construction of the park has begun.

74. **Ozone Pollution** The rate of change of the level of ozone, an invisible gas that is an irritant and impairs breathing, present in the atmosphere on a certain May day in the city of Riverside is given by

$$R(t) = 3.2922t^2 - 0.366t^3 \qquad (0 < t < 11)$$

(measured in pollutant standard index per hour). Here t is measured in hours, with $t = 0$ corresponding to 7 A.M. Find the ozone level $A(t)$ at any time t assuming that at 7 A.M. it is zero.

[Hint: $A'(t) = R(t)$ and $A(0) = 0$.]

75. **Surface Area of a Human** Empirical data suggest that the surface area of a 180-cm-tall human body changes at the rate of

$$S'(W) = 0.131773W^{-0.575}$$

square meters per kilogram, where W is the weight of the body in kg. If the surface area of a 180-cm-tall human body weighing 70 kg is 1.886277 square meters, what is the surface area of a human body of the same height weighing 75 kg?

76. **Flight of a Rocket** The velocity, in feet per second, of a rocket t seconds into vertical flight is given by

$$v(t) = -3t^2 + 192t + 120$$

Find an expression $h(t)$ that gives the rocket's altitude, in feet, t seconds after lift-off. What is the altitude of the rocket 30 seconds after lift-off?

[Hint: $h'(t) = v(t)$; $h(0) = 0$.]

77. **Flow of Blood in an Artery** Nineteenth-century physician Jean Louis Marie Poiseuille discovered that the rate of change of the velocity of blood r cm from the central axis of an artery (in cm/sec/cm) is given by

$$a(r) = -kr$$

where k is a constant. If the radius of an artery is R cm, find an expression for the velocity of blood as a function of r (see the accompanying figure).

[Hint: $v'(r) = a(r)$ and $v(R) = 0$. (Why?)]

Blood vessel

78. **Acceleration of a Car** A car traveling along a straight road at 66 ft/sec accelerated to a speed of 88 ft/sec over a distance of 440 ft. What was the acceleration of the car, assuming it was constant?

79. **Deceleration of a Car** What constant deceleration would a car moving along a straight road have to be subjected to if it were brought to rest from a speed of 88 ft/sec in 9 seconds? What would the stopping distance be?

80. A tank has a constant cross-sectional area of 50 sq ft and an orifice of constant cross-sectional area of $\frac{1}{2}$ sq ft located at the bottom of the tank (see the accompanying figure). If the tank is filled with water to a height of h

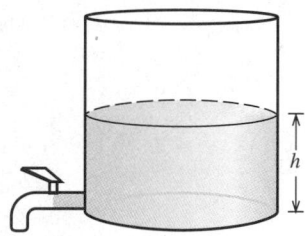

ft and allowed to drain, then the height of the water decreases at a rate that is described by the equation

$$\frac{dh}{dt} = -\frac{1}{25}\left(\sqrt{20} - \frac{t}{50}\right) \qquad (0 \le t \le 50\sqrt{20})$$

Find an expression for the height of the water at any time t if its height initially is 20 ft.

81. **Launching a Fighter Aircraft** A fighter aircraft is launched from the deck of a Nimitz class aircraft carrier with the help of a steam catapult. If the aircraft is to attain a take-off speed of at least 240 ft/sec after traveling 800 ft along the flight deck, find the minimum acceleration it must be subjected to, assuming it is constant.

SOLUTIONS TO SELF-CHECK EXERCISES 14.1

1. $\int \left(\dfrac{1}{\sqrt{x}} - \dfrac{2}{x} + 3e^x \right) dx = \int \left(x^{-1/2} - \dfrac{2}{x} + 3e^x \right) dx$

$\qquad\qquad = \int x^{-1/2}\, dx - 2\int \dfrac{1}{x}\, dx + 3\int e^x\, dx$

$\qquad\qquad = 2x^{1/2} - 2 \ln|x| + 3e^x + C$

$\qquad\qquad = 2\sqrt{x} - 2 \ln|x| + 3e^x + C$

2. The slope of the tangent line to the graph of the function f at any point $P(x, f(x))$ is given by the derivative f' of f. Thus, the first condition implies that

$$f'(x) = 3x^2 - 6x + 3$$

which, upon integration, yields

$$f(x) = \int (3x^2 - 6x + 3)\, dx$$
$$= x^3 - 3x^2 + 3x + k$$

where k is the constant of integration. To evaluate k, we use the initial condition (2), which implies that $f(2) = 9$, or

$$9 = f(2) = 2^3 - 3(2)^2 + 3(2) + k$$

or $k = 7$. Hence, the required rule of definition of the function f is

$$f(x) = x^3 - 3x^2 + 3x + 7$$

3. Let $M(t)$ denote United Motors' market share at year t. Then

$$M(t) = \int f(t)\, dt$$
$$= \int (-0.01875t^2 + 0.15t - 1.2)\, dt$$
$$= -0.00625t^3 + 0.075t^2 - 1.2t + C$$

To determine the value of C, we use the initial condition $M(0) = 48.4$, obtaining $C = 48.4$. Therefore,

$$M(t) = -0.00625t^3 + 0.075t^2 - 1.2t + 48.4$$

In particular, United Motors' market share of new cars at the beginning of 1999 is given by

$$M(12) = -0.00625(12)^3 + 0.075(12)^2 - 1.2(12) + 48.4 = 34$$

or 34%.

14.2 INTEGRATION BY SUBSTITUTION

In Section 14.1 we developed certain rules of integration that are closely related to the corresponding rules of differentiation in Chapters 11 and 13. In this section we introduce a method of integration called the **method of substitution,** which is related to the Chain Rule for differentiating functions. When used in conjunction with the rules of integration developed earlier, the method of substitution is a powerful tool for integrating a large class of functions.

How the Method of Substitution Works

Consider the indefinite integral

$$\int 2(2x + 4)^5 \, dx \tag{3}$$

One way of evaluating this integral is to expand the expression $(2x + 4)^5$ and then integrate the resulting integrand term by term. As an alternative approach, let us see if we can simplify the integral by making a change of variable. Write

$$u = 2x + 4$$

with differential

$$du = 2 \, dx$$

If we substitute these quantities into (3), we obtain

$$\int 2(2x + 4)^5 \, dx = \int (2x + 4)^5(2 \, dx) = \int u^5 \, du$$

$$\underset{\text{Rewriting}}{\uparrow} \qquad\qquad \underset{\begin{cases} u = 2x + 4 \\ du = 2 \, dx \end{cases}}{\uparrow}$$

Now, the last integral involves a power function and is easily evaluated using Rule 2 of Section 14.1. Thus,

$$\int u^5 \, du = \frac{1}{6} u^6 + C$$

Therefore, using this result and replacing u by $u = 2x + 4$, we obtain

$$\int 2(2x + 4)^5 \, dx = \frac{1}{6}(2x + 4)^6 + C$$

We can verify that the foregoing result is indeed correct by computing

$$\frac{d}{dx}\left[\frac{1}{6}(2x + 4)^6\right] = \frac{1}{6} \cdot 6(2x + 4)^5(2) \qquad \text{(Using the Chain Rule)}$$

$$= 2(2x + 4)^5$$

and observing that the last expression is just the integrand of (3).

The Method of Integration by Substitution

To show why the approach used in evaluating the integral in (3) is successful, we write

$$f(x) = x^5 \quad \text{and} \quad g(x) = 2x + 4$$

Then $g'(x) = 2\,dx$. Furthermore, the integrand of (3) is just the composition of f and g. Thus,

$$(f \circ g)(x) = f(g(x))$$
$$= [g(x)]^5 = (2x + 4)^5$$

Therefore, (3) can be written as

$$\int f(g(x))g'(x)\,dx \qquad \qquad \text{(4)}$$

Next, let us show that an integral having the form (4) can always be written as

$$\int f(u)\,du \qquad \qquad \text{(5)}$$

Suppose F is an antiderivative of f. By the Chain Rule, we have

$$\frac{d}{dx}[F(g(x))] = F'(g(x))g'(x)$$

and therefore,

$$\int F'(g(x))g'(x)\,dx = F(g(x)) + C$$

Letting $F' = f$ and making the substitution $u = g(x)$, we have

$$\int f(g(x))g'(x)\,dx = F(u) + C = \int F'(u)\,du = \int f(u)\,du$$

as we wished to show. Thus, if the transformed integral is readily evaluated, as is the case with the integral (3), then the method of substitution will prove successful.

Before we look at more examples, let us summarize the steps involved in integration by substitution.

INTEGRATION BY SUBSTITUTION	**Step 1** Let $u = g(x)$, where $g(x)$ is part of the integrand, usually the "inside function" of the composite function $f(g(x))$.
	Step 2 Compute $du = g'(x)\,dx$.
	Step 3 Use the substitution $u = g(x)$ and $du = g'(x)\,dx$ to convert the *entire* integral into one involving *only u*.
	Step 4 Evaluate the resulting integral.
	Step 5 Replace u by $g(x)$ to obtain the final solution as a function of x.

REMARK   Sometimes we need to consider different choices of g for the substitution $u = g(x)$ in order to carry out step 3 and/or step 4.   o o o

EXAMPLE 1   Evaluate $\int 2x(x^2 + 3)^4 \, dx$.

Solution

Step 1   Observe that the integrand involves the composite function $(x^2 + 3)^4$ with "inside function" $g(x) = x^2 + 3$. So, we choose $u = x^2 + 3$.

Step 2   Compute $du = 2x \, dx$.

Step 3   Making the substitution $u = x^2 + 3$ and $du = 2x \, dx$, we obtain

$$\int 2x(x^2 + 3)^4 \, dx = \int (x^2 + 3)^4 (2x \, dx) = \int u^4 \, du$$

$$\underset{\text{Rewriting}}{\uparrow}$$

an integral involving only the variable u.

Step 4   Evaluate

$$\int u^4 \, du = \frac{1}{5} u^5 + C$$

Step 5   Replacing u by $x^2 + 3$, we obtain

$$\int 2x(x^2 + 3)^4 \, dx = \frac{1}{5}(x^2 + 3)^5 + C$$ o o o

EXAMPLE 2   Evaluate $\int 3\sqrt{3x + 1} \, dx$.

Solution

Step 1   The integrand involves the composite function $\sqrt{3x + 1}$ with "inside function" $g(x) = 3x + 1$. So, let $u = 3x + 1$.

Step 2   Compute $du = 3 \, dx$.

Step 3   Making the substitution $u = 3x + 1$ and $du = 3 \, dx$, we obtain

$$\int 3\sqrt{3x + 1} \, dx = \int \sqrt{3x + 1}(3 \, dx) = \int \sqrt{u} \, du$$

an integral involving only the variable u.

Step 4   Evaluate

$$\int \sqrt{u} \, du = \int u^{1/2} \, du = \frac{2}{3} u^{3/2} + C$$

Step 5   Replacing u by $3x + 1$, we obtain

$$\int 3\sqrt{3x + 1} \, dx = \frac{2}{3}(3x + 1)^{3/2} + C$$ o o o

EXAMPLE 3   Evaluate $\int x^2(x^3 + 1)^{3/2} \, dx$.

Solution

Step 1   The integrand contains the composite function $(x^3 + 1)^{3/2}$ with "inside function" $g(x) = x^3 + 1$. So, let $u = x^3 + 1$.

Step 2   Compute $du = 3x^2 \, dx$.

Step 3   Making the substitution $u = x^3 + 1$ and $du = 3x^2 \, dx$, or $x^2 \, dx = \frac{1}{3} \, du$, we obtain

$$\int x^2(x^3 + 1)^{3/2} \, dx = \int (x^3 + 1)^{3/2}(x^2 \, dx)$$

$$= \int u^{3/2} \left(\frac{1}{3} \, du \right) = \frac{1}{3} \int u^{3/2} \, du$$

an integral involving only the variable u.

Step 4   We evaluate

$$\frac{1}{3} \int u^{3/2} \, du = \frac{1}{3} \cdot \frac{2}{5} u^{5/2} + C = \frac{2}{15} u^{5/2} + C$$

Step 5   Replacing u by $x^3 + 1$, we obtain

$$\int x^2(x^3 + 1)^{3/2} \, dx = \frac{2}{15}(x^3 + 1)^{5/2} + C$$

Let $f(x) = x^2(x^3 + 1)^{3/2}$. Using the result of Example 3, we see that an antiderivative of f is $F(x) = \frac{2}{15}(x^3 + 1)^{5/2}$. However, in terms of u (where $u = x^3 + 1$), an antiderivative of f is $G(u) = \frac{2}{15}u^{5/2}$. Compute $F(2)$. Next, suppose we want to compute $F(2)$ using the function G instead. At what value of u should you evaluate $G(u)$ in order to obtain the desired result? Explain your answer.

In the remaining examples, we drop the practice of labeling the steps involved in evaluating each integral.

EXAMPLE 4   Evaluate $\int e^{-3x} \, dx$.

Solution   Let $u = -3x$ so that $du = -3 \, dx$, or $dx = -\frac{1}{3} \, du$. Then

$$\int e^{-3x} \, dx = \int e^u \left(-\frac{1}{3} \, du \right) = -\frac{1}{3} \int e^u \, du$$

$$= -\frac{1}{3} e^u + C = -\frac{1}{3} e^{-3x} + C$$

EXAMPLE 5 Evaluate $\int \dfrac{x}{3x^2 + 1}\, dx$.

Solution Let $u = 3x^2 + 1$. Then $du = 6x\, dx$, or $x\, dx = \frac{1}{6}\, du$. Making the appropriate substitutions, we have

$$\int \frac{x}{3x^2 + 1}\, dx = \int \frac{1/6}{u}\, du$$

$$= \frac{1}{6}\int \frac{1}{u}\, du$$

$$= \frac{1}{6}\ln |u| + C$$

$$= \frac{1}{6}\ln(3x^2 + 1) + C \qquad (\text{Since } 3x^2 + 1 > 0)$$

○ ○ ○

EXAMPLE 6 Evaluate $\int \dfrac{(\ln x)^2}{2x}\, dx$.

Solution Let $u = \ln x$. Then

$$du = \frac{d}{dx}(\ln x)\, dx = \frac{1}{x}\, dx$$

Thus,

$$\int \frac{(\ln x)^2}{2x}\, dx = \frac{1}{2}\int \frac{(\ln x)^2}{x}\, dx$$

$$= \frac{1}{2}\int u^2\, du$$

$$= \frac{1}{6}u^3 + C$$

$$= \frac{1}{6}(\ln x)^3 + C$$

○ ○ ○

Suppose $\int f(u)\, du = F(u) + C$.

a. Show that $\int f(ax + b)\, dx = \dfrac{1}{a} F(ax + b) + C$.

b. How can you use this result to facilitate the evaluation of integrals such as $\int (2x + 3)^5\, dx$ and $\int e^{3x-2}\, dx$? Explain your answer.

Applications

Examples 7 and 8 show how the method of substitution can be used in practical situations.

EXAMPLE 7   In 1990, the head of the research and development department of the Soloron Corporation claimed that the cost of
producing solar cell panels would drop at the rate of

$$\frac{58}{(3t + 2)^2} \qquad (0 \le t \le 10)$$

dollars per peak watt for the next t years, with $t = 0$ corresponding to the
beginning of the year 1990. (A peak watt is the power produced at noon on
a sunny day.) In 1990, the panels, which are used for photovoltaic power
systems, cost $10 per peak watt. Find an expression giving the cost per peak
watt of producing solar cell panels at the beginning of year t. What will the
cost be at the beginning of 2000?

Solution   Let $C(t)$ denote the cost per peak watt for producing solar cell
panels at the beginning of year t. Then

$$C'(t) = -\frac{58}{(3t + 2)^2}$$

Integrating, we find that

$$C(t) = \int \frac{-58}{(3t + 2)^2}\, dt$$

$$= -58 \int (3t + 2)^{-2}\, dt$$

Let $u = 3t + 2$ so that

$$du = 3\, dt \quad \text{or} \quad dt = \frac{1}{3}\, du$$

Then

$$C(t) = -58 \left(\frac{1}{3}\right) \int u^{-2}\, du$$

$$= -\frac{58}{3}(-1)u^{-1} + k$$

$$= \frac{58}{3(3t + 2)} + k$$

where k is an arbitrary constant. To determine the value of k, note that the
cost per peak watt of producing solar cell panels at the beginning of 1990
($t = 0$) was 10, or $C(0) = 10$. This gives

$$C(0) = \frac{58}{3(2)} + k = 10$$

or $k = \frac{1}{3}$. Therefore, the required expression is given by

$$C(t) = \frac{58}{3(3t + 2)} + \frac{1}{3}$$

$$= \frac{58 + (3t + 2)}{3(3t + 2)}$$

$$= \frac{t + 20}{3t + 2}$$

The cost per peak watt for producing solar cell panels at the beginning of 2000 is given by

$$C(10) = \frac{10 + 20}{3(10) + 2} \approx 0.94$$

or approximately $.94 per peak watt.   ○ ○ ○

EXPLORING WITH TECHNOLOGY

Refer to Example 7.

1. Use a graphing utility to plot the graph of

$$C(t) = \frac{t + 20}{3t + 2}$$

in the viewing rectangle $[0, 10] \times [0, 5]$. Then use the numerical differentiation capability of the graphing utility to compute $C'(10)$.

2. Plot the graph of

$$C'(t) = -\frac{58}{(3t + 2)^2}$$

in the viewing rectangle $[0, 10] \times [-10, 0]$. Then use the evaluation capability of the graphing utility to find $C'(10)$. Is this value of $C'(10)$ the same as that obtained in part (1)? Explain your answer.

○   ○   ○

EXAMPLE 8   A study prepared by the marketing department of Universal Instruments Company forecasts that, after its new line of Galaxy Home Computers is introduced into the market, sales will grow at the rate of

$$2000 - 1500e^{-0.05t} \qquad (0 \le t \le 60)$$

units per month. Find an expression that gives the total number of computers that will sell t months after they become available on the market. How many computers will Universal sell in the first year they are on the market?

Solution   Let $N(t)$ denote the total number of computers that may be expected to be sold t months after their introduction in the market. Then the rate of growth of sales is given by $N'(t)$ units per month. Thus,

$$N'(t) = 2000 - 1500e^{-0.05t}$$

so that
$$N(t) = \int (2000 - 1500e^{-0.05t})\, dt$$
$$= \int 2000\, dt - 1500 \int e^{-0.05t}\, dt$$

Upon integrating the second integral by the method of substitution, we obtain

$$N(t) = 2000t + \frac{1500}{0.05} e^{-0.05t} + C \qquad \text{(Let } u = -0.05t,$$
$$\text{then } du = -0.05\, dt.)$$
$$= 2000t + 30{,}000e^{-0.05t} + C$$

To determine the value of C, note that the number of computers sold at the end of month 0 is nil, so $N(0) = 0$. This gives

$$N(0) = 30{,}000 + C = 0 \qquad \text{(Since } e^0 = 1\text{)}$$

or $C = -30{,}000$. Therefore, the required expression is given by

$$N(t) = 2000t + 30{,}000e^{-0.05t} - 30{,}000$$
$$= 2000t + 30{,}000(e^{-0.05t} - 1)$$

The number of computers that Universal can expect to sell in the first year is given by

$$N(12) = 2000(12) + 30{,}000(e^{-0.05(12)} - 1)$$
$$= 10{,}464$$

or 10,464 units.

⊙ ⊙ ⊙

SELF-CHECK EXERCISES 14.2

1. Evaluate $\int \sqrt{2x + 5}\, dx$.

2. Evaluate $\int \dfrac{x^2}{(2x^3 + 1)^{3/2}}\, dx$.

3. Evaluate $\int x e^{2x^2 - 1}\, dx$.

4. According to a joint study conducted by Oxnard's Environmental Management Department and a state government agency, the concentration of carbon monoxide in the air due to automobile exhaust is increasing at the rate given by

$$f(t) = \frac{8(0.1t + 1)}{300(0.2t^2 + 4t + 64)^{1/3}}$$

parts per million per year t. Currently, the concentration of carbon monoxide due to automobile exhaust is 0.16 part per million. Find an expression giving the concentration of carbon monoxide t years from now.

Solutions to Self-Check Exercises 14.2 can be found on page 955.

14.2 EXERCISES

In exercises 1–50, evaluate the indefinite integral.

1. $\displaystyle\int 4(4x + 3)^4 \, dx$

2. $\displaystyle\int 4x(2x^2 + 1)^7 \, dx$

3. $\displaystyle\int (x^3 - 2x)^2(3x^2 - 2) \, dx$

4. $\displaystyle\int (3x^2 - 2x + 1)(x^3 - x^2 + x)^4 \, dx$

5. $\displaystyle\int \frac{4x}{(2x^2 + 3)^3} \, dx$

6. $\displaystyle\int \frac{3x^2 + 2}{(x^3 + 2x)^2} \, dx$

7. $\displaystyle\int 3t^2 \sqrt{t^3 + 2} \, dt$

8. $\displaystyle\int 3t^2(t^3 + 2)^{3/2} \, dt$

9. $\displaystyle\int (x^2 - 1)^9 x \, dx$

10. $\displaystyle\int x^2(2x^3 + 3)^4 \, dx$

11. $\displaystyle\int \frac{x^4}{1 - x^5} \, dx$

12. $\displaystyle\int \frac{x^2}{\sqrt{x^3 - 1}} \, dx$

13. $\displaystyle\int \frac{2}{x - 2} \, dx$

14. $\displaystyle\int \frac{x^2}{x^3 - 3} \, dx$

15. $\displaystyle\int \frac{0.3x - 0.2}{0.3x^2 - 0.4x + 2} \, dx$

16. $\displaystyle\int \frac{2x^2 + 1}{0.2x^3 + 0.3x} \, dx$

17. $\displaystyle\int \frac{x}{3x^2 - 1} \, dx$

18. $\displaystyle\int \frac{x^2 - 1}{x^3 - 3x + 1} \, dx$

19. $\displaystyle\int e^{-2x} \, dx$

20. $\displaystyle\int e^{-0.02x} \, dx$

21. $\displaystyle\int e^{2-x} \, dx$

22. $\displaystyle\int e^{2t+3} \, dt$

23. $\displaystyle\int xe^{-x^2} \, dx$

24. $\displaystyle\int x^2 e^{x^3-1} \, dx$

25. $\displaystyle\int (e^x - e^{-x}) \, dx$

26. $\displaystyle\int (e^{2x} + e^{-3x}) \, dx$

27. $\displaystyle\int \frac{e^x}{1 + e^x} \, dx$

28. $\displaystyle\int \frac{e^{2x}}{1 + e^{2x}} \, dx$

29. $\displaystyle\int \frac{e^{\sqrt{x}}}{\sqrt{x}} \, dx$

30. $\displaystyle\int \frac{e^{-1/x}}{x^2} \, dx$

31. $\displaystyle\int \frac{e^{3x} + x^2}{(e^{3x} + x^3)^3} \, dx$

32. $\displaystyle\int \frac{e^x - e^{-x}}{(e^x + e^{-x})^{3/2}} \, dx$

33. $\displaystyle\int e^{2x}(e^{2x} + 1)^3 \, dx$

34. $\displaystyle\int e^{-x}(1 + e^{-x}) \, dx$

35. $\displaystyle\int \frac{\ln 5x}{x} \, dx$

36. $\displaystyle\int \frac{(\ln u)^3}{u} \, du$

37. $\displaystyle\int \frac{1}{x \ln x} \, dx$

38. $\displaystyle\int \frac{1}{x(\ln x)^2} \, dx$

39. $\displaystyle\int \frac{\sqrt{\ln x}}{x} \, dx$

40. $\displaystyle\int \frac{(\ln x)^{7/2}}{x} \, dx$

41. $\displaystyle\int \left(xe^{x^2} - \frac{x}{x^2 + 2} \right) dx$

42. $\displaystyle\int \left(xe^{-x^2} + \frac{e^x}{e^x + 3} \right) dx$

43. $\displaystyle\int \frac{x + 1}{\sqrt{x} - 1} \, dx$

[*Hint:* Let $u = \sqrt{x} - 1$.]

44. $\displaystyle\int \frac{e^{-u} - 1}{e^{-u} + u} \, du$

[*Hint:* Let $v = e^{-u} + u$.]

45. $\displaystyle\int x(x - 1)^5 \, dx$

[*Hint:* $u = x - 1$ implies $x = u + 1$.]

46. $\displaystyle\int \frac{t}{t + 1} \, dt$

$\left[\text{Hint: } \dfrac{t}{t + 1} = 1 - \dfrac{1}{t + 1}. \right]$

47. $\displaystyle\int \frac{1 - \sqrt{x}}{1 + \sqrt{x}} \, dx$

[*Hint:* Let $u = 1 + \sqrt{x}$.]

48. $\displaystyle\int \frac{1 + \sqrt{x}}{1 - \sqrt{x}} \, dx$

[*Hint:* Let $u = 1 - \sqrt{x}$.]

49. $\displaystyle\int v^2(1 - v)^6 \, dv$

[*Hint:* Let $u = 1 - v$.]

50. $\displaystyle\int x^3(x^2 + 1)^{3/2} \, dx$

[*Hint:* Let $u = x^2 + 1$.]

In exercises 51–54, find the function f given that the slope of the tangent line to the graph of f at any point (x, f(x)) is f'(x) and that the graph of f passes through the given point.

51. $f'(x) = 5(2x - 1)^4$; $(1, 3)$

52. $f'(x) = \dfrac{3x^2}{2\sqrt{x^3 - 1}}$; $(1, 1)$

53. $f'(x) = -2xe^{-x^2+1}$; $(1, 0)$

54. $f'(x) = 1 - \dfrac{2x}{x^2 + 1}$; $(0, 2)$

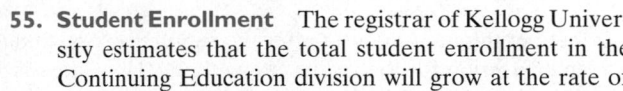

55. **Student Enrollment** The registrar of Kellogg University estimates that the total student enrollment in the Continuing Education division will grow at the rate of

$$N'(t) = 2000(1 + 0.2t)^{-3/2}$$

students per year t years from now. If the current student enrollment is 1000, find an expression giving the total student enrollment t years from now. What will the student enrollment be five years from now?

56. TV Viewers: Newsmagazine Shows   The number of viewers of a weekly TV newsmagazine show, introduced in the 1997 season, has been increasing at the rate of

$$3\left(2 + \frac{1}{2}t\right)^{-1/3} \qquad (1 \le t \le 6)$$

million viewers per year in its tth year on the air. The number of viewers of the program during its first year on the air is given by $9(5/2)^{2/3}$ million. Find how many viewers are expected in the 2002 season.

57. Demand: Ladies' Boots   The rate of change of the unit price p (in dollars) of Apex ladies' boots is given by

$$p'(x) = \frac{-250x}{(16 + x^2)^{3/2}}$$

where x is the quantity demanded daily in units of a hundred. Find the demand function for these boots if the quantity demanded daily is 300 pairs ($x = 3$) when the unit price is $50 per pair.

58. Supply: Ladies' Boots   The rate of change of the unit price p (in dollars) of Apex ladies' boots is given by

$$p'(x) = \frac{240x}{(5 - x)^2}$$

where x is the number of pairs that the supplier will make available in the market daily when the unit price is $p per pair. Find the supply equation for these boots if the quantity the supplier is willing to make available is 200 pairs daily ($x = 2$) when the unit price is $50 per pair.

59. Oil Spill   In calm waters, the oil spilling from the ruptured hull of a grounded tanker forms an oil slick that is circular in shape. If the radius r of the circle is increasing at the rate of

$$r'(t) = \frac{30}{\sqrt{2t + 4}}$$

feet per minute t minutes after the rupture occurs, find an expression for the radius at any time t. How large is the polluted area 16 minutes after the rupture occurred? [*Hint:* $r(0) = 0$.]

60. Life Expectancy of a Female   Suppose that in a certain country the life expectancy at birth of a female is chang-

ing at the rate of

$$g'(t) = \frac{5.45218}{(1 + 1.09t)^{0.9}}$$

years per year. Here t is measured in years, and $t = 0$ corresponds to the beginning of 1900. Find an expression $g(t)$ giving the life expectancy at birth (in years) of a female in that country if the life expectancy at the beginning of 1900 is 50.02 years. What is the life expectancy at birth of a female born in the year 2000 in that country?

61. Average Birth Height of Boys   Using data collected at Kaiser Hospital, pediatricians estimate that the average height of male children changes at the rate of

$$h'(t) = \frac{52.8706e^{-0.3277t}}{(1 + 2.449e^{-0.3277t})^2}$$

inches per year, where the child's height $h(t)$ is measured in inches and t, the child's age, is measured in years, with $t = 0$ corresponding to the age at birth. Find an expression $h(t)$ for the average height of a boy at age t if the height at birth of an average child is 19.4 inches. What is the height of an average eight-year-old boy?

62. Learning Curves   The average student enrolled in the 20-week Advanced Shorthand course at the American Institute of Stenography progresses according to the rule

$$N'(t) = 6e^{-0.05t} \qquad (0 \le t \le 20)$$

where $N'(t)$ measures the rate of change in the number of words per minute of dictation the student takes in shorthand after t weeks in the course. Assuming that the average student enrolled in this course begins with a dictation speed of 60 words per minute, find an expression $N(t)$ that gives the dictation speed of the student after t weeks in the course.

63. Sales: Loudspeakers   Two thousand pairs of Acrosonic model F loudspeaker systems were sold in the first year they appeared in the market. Since then, sales of these loudspeaker systems have been growing at the rate of

$$f'(t) = 2000(3 - 2e^{-t})$$

units per year, where t denotes the number of years these systems have been on the market. Determine the number of systems that were sold in the first five years after their introduction.

64. Amount of Glucose in the Bloodstream   Suppose that a patient is given a continuous intravenous infusion of glucose at a constant rate of r milligrams per minute. Then the rate at which the amount of glucose in the bloodstream is changing at time t due to this infusion is given by

$$A'(t) = re^{-at}$$

milligrams per minute, where a is a positive constant associated with the rate at which excess glucose is eliminated from the bloodstream and is dependent on the patient's metabolism rate. Derive an expression for the amount of glucose in the bloodstream at time t.
[*Hint:* $A(0) = 0$.]

65. **Concentration of a Drug in an Organ** A drug is carried into an organ of volume V cm^3 by a liquid that enters the organ at the rate of a cm^3/sec and leaves it at the rate of b cm^3/sec. The concentration of the drug in the liquid entering the organ is c g/cm^3. If the concentration of the drug in the organ at time t is increasing at the rate of

$$x'(t) = \frac{1}{V}(ac - bx_0)e^{-bt/V}$$

g/cm^3 per second, and the concentration of the drug in the organ initially is x_0 g/cm^3, show that the concentration of the drug in the organ at time t is given by

$$x(t) = \frac{ac}{b} + \left(x_0 - \frac{ac}{b}\right)e^{-bt/V}$$

SOLUTIONS TO SELF-CHECK EXERCISES 14.2

1. Let $u = 2x + 5$. Then $du = 2\,dx$, or $dx = \frac{1}{2}\,du$. Making the appropriate substitutions, we have

$$\int \sqrt{2x + 5}\,dx = \int \sqrt{u}\left(\frac{1}{2}\,du\right) = \frac{1}{2}\int u^{1/2}\,du$$

$$= \frac{1}{2}\left(\frac{2}{3}\right)u^{3/2} + C$$

$$= \frac{1}{3}(2x + 5)^{3/2} + C$$

2. Let $u = 2x^3 + 1$, so that $du = 6x^2\,dx$, or $x^2\,dx = \frac{1}{6}\,du$. Making the appropriate substitutions, we have

$$\int \frac{x^2}{(2x^3 + 1)^{3/2}}\,dx = \int \frac{\left(\frac{1}{6}\right)du}{u^{3/2}} = \frac{1}{6}\int u^{-3/2}\,du$$

$$= \left(\frac{1}{6}\right)(-2)u^{-1/2} + C$$

$$= -\frac{1}{3}(2x^3 + 1)^{-1/2} + C$$

$$= -\frac{1}{3\sqrt{2x^3 + 1}} + C$$

3. Let $u = 2x^2 - 1$, so that $du = 4x\,dx$, or $x\,dx = \frac{1}{4}\,du$. Then

$$\int xe^{2x^2-1}\,dx = \frac{1}{4}\int e^u\,du$$

$$= \frac{1}{4}e^u + C$$

$$= \frac{1}{4}e^{2x^2-1} + C$$

4. Let $C(t)$ denote the concentration of carbon monoxide in the air due to automobile exhaust t years from now. Then

$$C'(t) = f(t) = \frac{8(0.1t + 1)}{300(0.2t^2 + 4t + 64)^{1/3}}$$

$$= \frac{8}{300}(0.1t + 1)(0.2t^2 + 4t + 64)^{-1/3}$$

Integrating, we find

$$C(t) = \int \frac{8}{300}(0.1t + 1)(0.2t^2 + 4t + 64)^{-1/3}\, dt$$

$$= \frac{8}{300}\int (0.1t + 1)(0.2t^2 + 4t + 64)^{-1/3}\, dt$$

Let $u = 0.2t^2 + 4t + 64$, so that $du = (0.4t + 4)\, dt = 4(0.1t + 1)\, dt$, or

$$(0.1t + 1)\, dt = \frac{1}{4}\, du$$

Then

$$C(t) = \frac{8}{300}\left(\frac{1}{4}\right)\int u^{-1/3}\, du$$

$$= \frac{1}{150}\left(\frac{3}{2}u^{2/3}\right) + k$$

$$= 0.01(0.2t^2 + 4t + 64)^{2/3} + k$$

where k is an arbitrary constant. To determine the value of k, we use the condition $C(0) = 0.16$, obtaining

$$C(0) = 0.16 = 0.01(64)^{2/3} + k$$

or

$$0.16 = 0.16 + k$$

and $k = 0$. Therefore,

$$C(t) = 0.01(0.2t^2 + 4t + 64)^{2/3}$$

14.3 AREA AND THE DEFINITE INTEGRAL

An Intuitive Look

Suppose that a certain state's annual rate of petroleum consumption over a 4-year period is constant and is given by the function

$$f(t) = 1.2 \qquad (0 \le t \le 4)$$

where t is measured in years and $f(t)$ in millions of barrels per year. Then the state's total petroleum consumption over the period of time in question is

$$(1.2)(4 - 0) \qquad \text{(Rate of consumption} \cdot \text{Time elapsed)}$$

Figure 14.5
The total petroleum consumption is given by the area of the rectangular region.

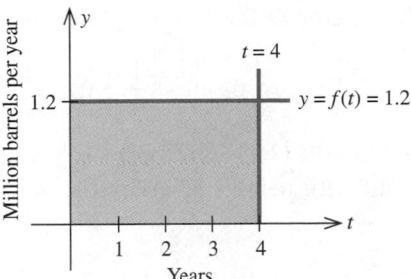

or 4.8 million barrels. If you examine the graph of *f* shown in Figure 14.5 you will see that this total is just the area of the rectangular region bounded above by the graph of *f*, below by the *t*-axis, and to the left and right by the vertical lines $t = 0$ (the *y*-axis) and $t = 4$, respectively.

Figure 14.6 shows the actual petroleum consumption of a certain New England state over a 4-year period from 1990 ($t = 0$) to 1994 ($t = 4$). Observe that the rate of consumption is not constant; that is, the function *f* is not a constant function. What is the state's total petroleum consumption over this 4-year period? It seems reasonable to conjecture that it is given by the "area" of the region bounded above by the graph of *f*, below by the *t*-axis, and to the left and right by the vertical lines $t = 0$ and $t = 4$, respectively.

This example raises two questions:

Figure 14.6
The daily petroleum consumption is given by the "area" of the shaded region.

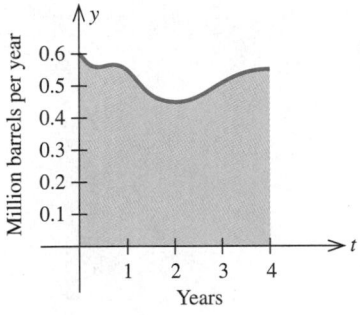

1. What is the "area" of the region shown in Figure 14.6?

2. How do we compute this area?

The Area Problem

The preceding example touches on the second fundamental problem in calculus: Calculate the area of the region bounded by the graph of a nonnegative function *f*, the *x*-axis, and the vertical lines $x = a$ and $x = b$ (Figure 14.7). This area is called the **area under the graph of** *f* on the interval $[a, b]$, or from *a* to *b*.

Figure 14.7
The area under the graph of f on $[a, b]$.

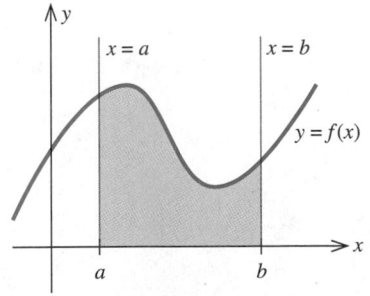

Defining Area—Two Examples

Just as we used the slopes of secant lines (quantities that we could compute) to help us define the slope of the tangent line to a point on the graph of a function, we now adopt a parallel approach and use the areas of rectangles (quantities that we can compute) to help us define the area under the graph of a function. We begin by looking at a specific example.

EXAMPLE I   Let $f(x) = x^2$ and consider the region *R* under the graph of *f* on the interval $[0, 1]$ (Figure 14.8a). In order to obtain an approximation of the area of *R*, let us construct four nonoverlapping rectangles as follows: Divide the interval [0, 1] into four subintervals

$$\left[0, \frac{1}{4}\right], \quad \left[\frac{1}{4}, \frac{1}{2}\right], \quad \left[\frac{1}{2}, \frac{3}{4}\right], \quad \text{and} \quad \left[\frac{3}{4}, 1\right]$$

of equal length 1/4. Next, construct four rectangles with these subintervals as bases and with heights given by the values of the function at the midpoints

$$\frac{1}{8}, \quad \frac{3}{8}, \quad \frac{5}{8}, \quad \text{and} \quad \frac{7}{8}$$

of each subinterval. Then each of these rectangles has width 1/4 and height

$$f\left(\frac{1}{8}\right), \quad f\left(\frac{3}{8}\right), \quad f\left(\frac{5}{8}\right), \quad \text{and} \quad f\left(\frac{7}{8}\right)$$

respectively (Figure 14.8b).

Figure 14.8
The area of the region under the graph of f on [0, 1] in (a) is approximated by the sum of the areas of the four rectangles in (b).

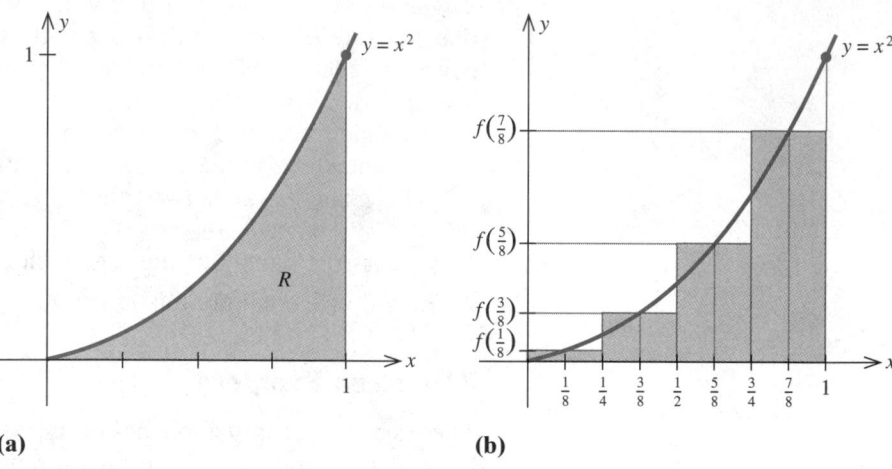

(a)   **(b)**

If we approximate the area A of S by the sum of the areas of the four rectangles, we obtain

$$A \approx \frac{1}{4} f\left(\frac{1}{8}\right) + \frac{1}{4} f\left(\frac{3}{8}\right) + \frac{1}{4} f\left(\frac{5}{8}\right) + \frac{1}{4} f\left(\frac{7}{8}\right)$$

$$= \frac{1}{4}\left[f\left(\frac{1}{8}\right) + f\left(\frac{3}{8}\right) + f\left(\frac{5}{8}\right) + f\left(\frac{7}{8}\right)\right]$$

$$= \frac{1}{4}\left[\left(\frac{1}{8}\right)^2 + \left(\frac{3}{8}\right)^2 + \left(\frac{5}{8}\right)^2 + \left(\frac{7}{8}\right)^2\right] \qquad \text{[Recall that } f(x) = x^2.\text{]}$$

$$= \frac{1}{4}\left(\frac{1}{64} + \frac{9}{64} + \frac{25}{64} + \frac{49}{64}\right) = \frac{21}{64}$$

or approximately 0.328125 square unit.

Following the procedure of Example 1, we can obtain approximations of the area of the region R using any number n of rectangles ($n = 4$ in Example 1). Figure 14.9a shows the approximation of the area A of R using 8 rectangles ($n = 8$), and Figure 14.9b shows the approximation of the area A of R using 16 rectangles.

Figure 14.9

As n increases, the number of rectangles increases and the approximation improves.

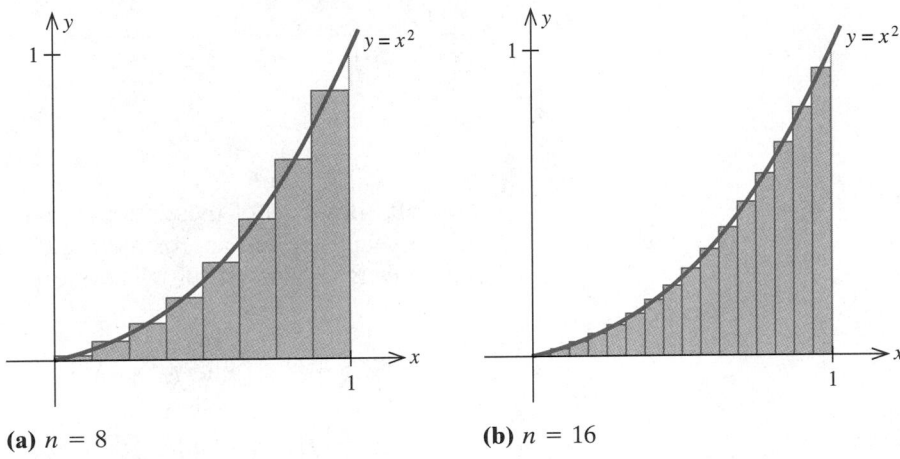

(a) $n = 8$ **(b)** $n = 16$

Table 14.1

Number of Rectangles n	Approximation of A
4	0.328125
8	0.332031
16	0.333008
32	0.333252
64	0.333313
100	0.333325
200	0.333331

These figures suggest that the approximations seem to get better as n increases. This is borne out by the results given in Table 14.1, which were obtained using a computer. Our computations seem to suggest that the approximations approach the number 1/3 as n gets larger and larger. This result suggests that we *define* the area of the region under the graph of $f(x) = x^2$ on the interval [0, 1] to be 1/3 square unit.

In Example 1 we chose the *midpoint* of each subinterval as the point at which to evaluate $f(x)$ to obtain the height of the approximating rectangle. Let us consider another example, this time choosing the *left endpoint* of each subinterval.

EXAMPLE 2 Let R be the region under the graph of $f(x) = 16 - x^2$ on the interval [1, 3]. Find an approximation of the area A of R using four subintervals of [1, 3] of equal length and picking the left endpoint of each subinterval to evaluate $f(x)$ to obtain the height of the approximating rectangle.

Solution The graph of f is sketched in Figure 14.10a. Since the length of [1, 3] is 2, we see that the length of each subinterval is 2/4, or 1/2. Therefore, the four subintervals are

$$\left[1, \frac{3}{2}\right], \quad \left[\frac{3}{2}, 2\right], \quad \left[2, \frac{5}{2}\right], \quad \text{and} \quad \left[\frac{5}{2}, 3\right]$$

Figure 14.10
The area of R in (a) is approximated by the sum of the areas of the four rectangles in (b).

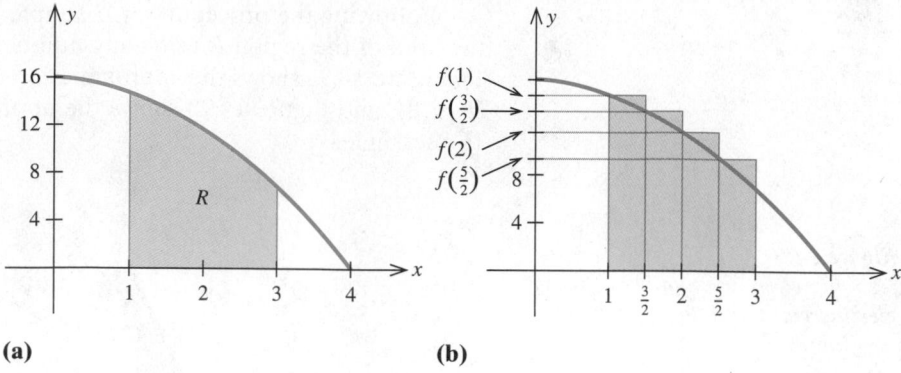

(a)

(b)

The left endpoints of these subintervals are 1, 3/2, 2, and 5/2, respectively, so the heights of the approximating rectangles are $f(1)$, $f(\frac{3}{2})$, $f(2)$, and $f(\frac{5}{2})$, respectively (Figure 14.10b). Therefore, the required approximation is

$$A \approx \frac{1}{2}f(1) + \frac{1}{2}f\left(\frac{3}{2}\right) + \frac{1}{2}f(2) + \frac{1}{2}f\left(\frac{5}{2}\right)$$

$$= \frac{1}{2}\left[f(1) + f\left(\frac{3}{2}\right) + f(2) + f\left(\frac{5}{2}\right)\right]$$

$$= \frac{1}{2}\left\{[16 - (1)^2] + \left[16 - \left(\frac{3}{2}\right)^2\right]\right.$$

$$\left. + [16 - (2)^2] + \left[16 - \left(\frac{5}{2}\right)^2\right]\right\} \qquad \text{[Recall that } f(x) = 16 - x^2.]$$

$$= \frac{1}{2}\left(15 + \frac{55}{4} + 12 + \frac{39}{4}\right) = \frac{101}{4}$$

or approximately 25.25 square units.   ○ ○ ○

Table 14.2

Number of Rectangles n	Approximation of A
4	25.2500
10	24.1200
100	23.4132
1000	23.3413
10,000	23.3341
50,000	23.3335
100,000	23.3334

Table 14.2 shows the approximations of the area A of the region R of Example 2 when n rectangles are used for the approximation and the heights of the approximating rectangles are found by evaluating $f(x)$ at the left endpoints. Once again we see that the approximations seem to approach a unique number as n gets larger and larger—this time the number is $23\frac{1}{3}$. This result suggests that we *define* the area of the region under the graph of $f(x) = 16 - x^2$ on the interval [1, 3] to be $23\frac{1}{3}$ square units.

Defining Area—The General Case

Examples 1 and 2 point the way to defining the area A under the graph of an arbitrary but continuous and nonnegative function f on an interval $[a, b]$ (Figure 14.11a).

Figure 14.11
The area of the region under the graph of f on [a, b] in (a) is approximated by the sum of the areas of the n rectangles shown in (b).

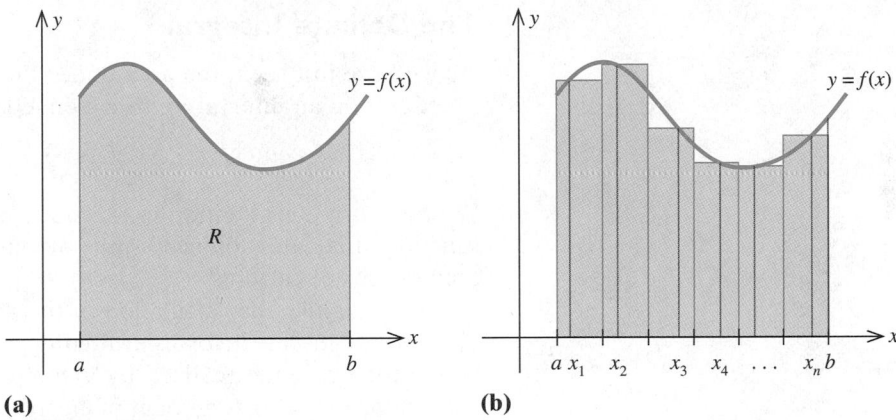

(a)   **(b)**

Divide the interval $[a, b]$ into n subintervals of equal length $\Delta x = (b - a)/n$. Next, pick n arbitrary points $x_1, x_2, \ldots, x_n$, called *representative points,* from the first, second, $\ldots$, nth subintervals, respectively (Figure 14.11b). Then, approximating the area A of the region R by the n rectangles of width Δx and heights $f(x_1), f(x_2), \ldots, f(x_n)$, so that the areas of the rectangles are $f(x_1)\,\Delta x, f(x_2)\,\Delta x, \ldots, f(x_n)\,\Delta x$, we have

$$A \approx f(x_1)\,\Delta x + f(x_2)\,\Delta x + \cdots + f(x_n)\,\Delta x$$

The sum on the right-hand side of this expression is called a **Riemann sum** in honor of the German mathematician Bernhard Riemann (1826–1866). Now, as the earlier examples seem to suggest, the Riemann sum will approach a unique number as n becomes arbitrarily large.* We define this number to be the area A of the region R.

THE AREA UNDER THE GRAPH OF A FUNCTION

Let f be a nonnegative continuous function on $[a, b]$. Then the area of the region under the graph of f is

$$A = \lim_{n \to \infty} [f(x_1) + f(x_2) + \cdots + f(x_n)]\,\Delta x \qquad (6)$$

where $x_1, x_2, \ldots, x_n$ are arbitrary points in the n subintervals of $[a, b]$ of equal width $\Delta x = (b - a)/n$.

◦ ◦ ◦

* Even though we chose the representative points to be the midpoints of the subintervals in Example 1 and the left endpoints in Example 2, it can be shown that each of the respective sums will always approach a unique number as n approaches infinity.

The Definite Integral

As we have just seen, the area under the graph of a continuous *nonnegative* function f on an interval $[a, b]$ is defined by the limit of the Riemann sum

$$\lim_{n \to \infty} [f(x_1) \, \Delta x + f(x_2) \, \Delta x + \cdots + f(x_n) \, \Delta x]$$

We now turn our attention to the study of limits of Riemann sums involving functions that are not necessarily nonnegative. Such limits arise in many applications of calculus.

For example, the calculation of the distance covered by a body traveling along a straight line involves evaluating a limit of this form. The computation of the total revenue realized by a company over a certain time period, the calculation of the total amount of electricity consumed in a typical home over a 24-hour period, the average concentration of a drug in a body over a certain interval of time, and the volume of a solid—all involve limits of this type.

We begin with the following definition.

DEFINITE INTEGRAL

Let f be defined on $[a, b]$. If

$$\lim_{n \to \infty} [f(x_1) \, \Delta x + f(x_2) \, \Delta x + \cdots + f(x_n) \, \Delta x]$$

exists for all choices of representative points $x_1, x_2, \ldots, x_n$ in the n subintervals of $[a, b]$ of equal width $\Delta x = (b - a)/n$, then this limit is called the **definite integral of f from a to b** and is denoted by $\int_a^b f(x) \, dx$. Thus,

$$\int_a^b f(x) \, dx = \lim_{n \to \infty} [f(x_1) \, \Delta x + f(x_2) \, \Delta x + \cdots + f(x_n) \, \Delta x] \qquad \textbf{(7)}$$

The number a is the **lower limit of integration**, and the number b is the **upper limit of integration**.

REMARKS

1. If f is nonnegative, then the limit in (7) is the same as the limit in (6) and, therefore, the definite integral gives the area under the graph of f on $[a, b]$.
2. The limit in (7) is denoted by the integral sign $\int$ because, as we will see later, the definite integral and the antiderivative of a function f are related.
3. It is important to realize that the definite integral $\int_a^b f(x) \, dx$ is a number, whereas the indefinite integral $\int f(x) \, dx$ represents a family of functions (the antiderivatives of f).
4. If the limit in (7) exists, we say that f is **integrable** on the interval $[a, b]$.

* * *

When Is a Function Integrable?

The following theorem, which we state without proof, guarantees that a continuous function is integrable.

| **INTEGRABILITY OF A FUNCTION** | Let f be continuous on $[a, b]$. Then f is integrable on $[a, b]$; that is, the definite integral $\int_a^b f(x)\,dx$ exists. |

Geometric Interpretation of the Definite Integral

If f is nonnegative and integrable on $[a, b]$, then we have the following geometric interpretation of the definite integral $\int_a^b f(x)\,dx$.

| **GEOMETRIC INTERPRETATION OF $\int_a^b f(x)\,dx$ FOR $f(x) \geq 0$ ON $[a, b]$** | If f is nonnegative and continuous on $[a, b]$, then $$\int_a^b f(x)\,dx \qquad (8)$$ is equal to the area of the region under the graph of f on $[a, b]$ (see Figure 14.12). |

Figure 14.12
If $f(x) \geq 0$ on $[a, b]$, then $\int_a^b f(x)\,dx =$ area under the graph of f on $[a, b]$.

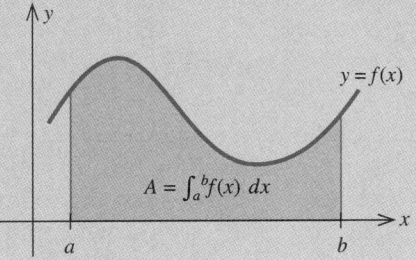

Next, let's extend our geometric interpretation of the definite integral to include the case where f assumes both positive and negative values on $[a, b]$. Consider a typical Riemann sum of the function f,

$$f(x_1)\,\Delta x + f(x_2)\,\Delta x + \cdots + f(x_n)\,\Delta x$$

corresponding to a partition of $[a, b]$ into n subintervals of equal width $(b - a)/n$, where $x_1, x_2, \ldots, x_n$ are representative points in the subintervals. The sum consists of n terms in which a positive term corresponds to the area of a rectangle of height $f(x_k)$ [for some positive integer k] lying above the x-axis, and a negative term corresponds to the area of a rectangle of height $-f(x_k)$ lying below the x-axis. (See Figure 14.13, which depicts a situation with $n = 6$.)

As n gets larger and larger, the sums of the areas of the rectangles lying above the x-axis seem to give a better and better approximation of the area of the region lying above the x-axis (see Figure 14.14). Similarly, the sums of

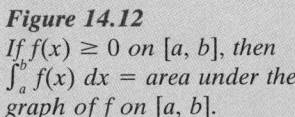 

Suppose f is nonpositive [that is, $f(x) \leq 0$] and continuous on $[a, b]$. Explain why the area of the region below the x-axis and above the graph of f is given by $-\int_a^b f(x)\,dx$.

Figure 14.13
The positive terms in the Riemann sum are associated with the areas of the rectangles that lie above the x-axis, and the negative terms are associated with the areas of those that lie below the x-axis.

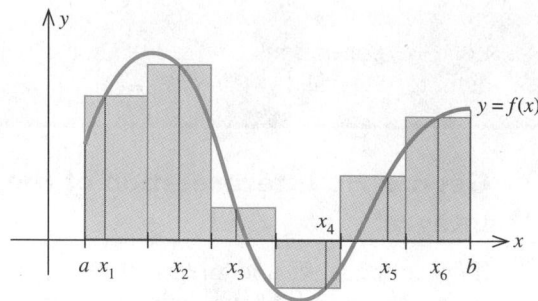

Figure 14.14
As n gets larger, the approximations get better. Here n = 12, and we are approximating with twice as many rectangles as in Figure 14.13.

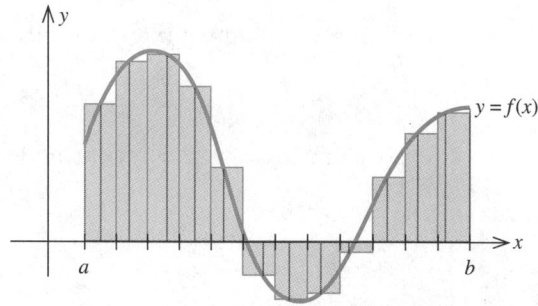

the areas of those rectangles lying below the *x*-axis seem to give a better and better approximation of the area of the region lying below the *x*-axis.

These observations suggest the following geometric interpretation of the definite integral for an arbitrary continuous function on an interval [*a*, *b*].

GEOMETRIC INTERPRETATION OF $\int_a^b f(x)\,dx$ **ON [a, b]**

If *f* is continuous on [*a*, *b*], then

$$\int_a^b f(x)\,dx$$

is equal to the area of the region above [*a*, *b*] minus the area of the region below [*a*, *b*] (see Figure 14.15).

Figure 14.15
$\int_a^b f(x)\,dx$ = Area of R_1 − Area of R_2 + Area of R_3.

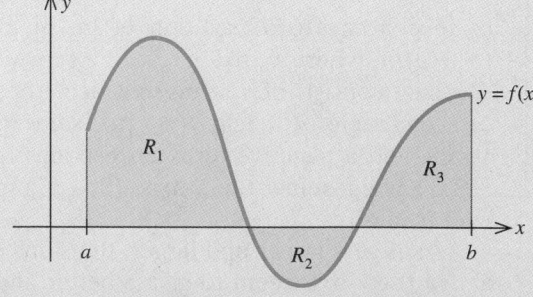

SELF-CHECK EXERCISE 14.3

Find an approximation of the area of the region R under the graph of $f(x) = 2x^2 + 1$ on the interval $[0, 3]$ using four subintervals of $[0, 3]$ of equal length and picking the midpoint of each subinterval as a representative point.

The Solution to Self-Check Exercise 14.3 can be found on page 967.

14.3 EXERCISES

In exercises 1 and 2, find an approximation of the area of the region R under the graph of f by computing the Riemann sum of f corresponding to the partition of the interval into the subintervals shown in the figures. In each case, use the midpoints of the subintervals as the representative points.

1.

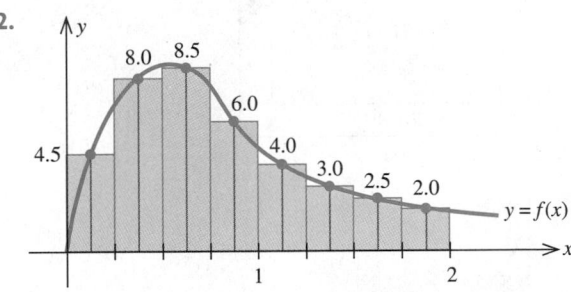

2.

3. Let $f(x) = 3x$.
   a. Sketch the region R under the graph of f on the interval $[0, 2]$ and find its exact area using geometry.
   b. Use a Riemann sum with four subintervals of equal length ($n = 4$) to approximate the area of R. Choose the representative points to be the left endpoints of the subintervals.

c. Repeat (b) with eight subintervals of equal length ($n = 8$).
   d. Compare the approximations obtained in (b) and (c) with the exact area found in (a). Do the approximations improve with larger n?

4. Repeat exercise 3 choosing the representative points to be the right endpoints of the subintervals.

5. Let $f(x) = 4 - 2x$.
   a. Sketch the region R under the graph of f on the interval $[0, 2]$ and find its exact area using geometry.
   b. Use a Riemann sum with five subintervals of equal length ($n = 5$) to approximate the area of R. Choose the representative points to be the left endpoints of the subintervals.
   c. Repeat (b) with ten subintervals of equal length ($n = 10$).
   d. Compare the approximations obtained in (b) and (c) with the exact area found in (a). Do the approximations improve with larger n?

6. Repeat exercise 5 choosing the representative points to be the right endpoints of the subintervals.

7. Let $f(x) = x^2$ and compute the Riemann sum of f over the interval $[2, 4]$ using
   a. Two subintervals of equal length ($n = 2$)
   b. Five subintervals of equal length ($n = 5$)
   c. Ten subintervals of equal length ($n = 10$)
   In each case, choose the representative points to be the midpoints of the subintervals.
   d. Can you guess at the area of the region under the graph of f on the interval $[2, 4]$?

8. Repeat exercise 7 choosing the representative points to be the left endpoints of the subintervals.

9. Repeat exercise 7 choosing the representative points to be the right endpoints of the subintervals.

10. Let $f(x) = x^3$ and compute the Riemann sum of f over the interval $[0, 1]$ using
   a. Two subintervals of equal length ($n = 2$)
   b. Five subintervals of equal length ($n = 5$)
   c. Ten subintervals of equal length ($n = 10$)
   In each case, choose the representative points to be the midpoints of the subintervals.
   d. Can you guess at the area of the region under the graph of f on the interval $[0, 1]$?

11. Repeat exercise 10 choosing the representative points to be the left endpoints of the subintervals.

12. Repeat exercise 10 choosing the representative points to be the right endpoints of the subintervals.

In exercises 13–16, find an approximation of the area of the region R under the graph of the function f on the interval [a, b]. In each case, use n subintervals and choose the representative points as indicated.

13. $f(x) = x^2 + 1$; $[0, 2]$; $n = 5$; midpoints

14. $f(x) = 4 - x^2$; $[-1, 2]$; $n = 6$; left endpoints

15. $f(x) = \dfrac{1}{x}$; $[1, 3]$; $n = 4$; right endpoints

16. $f(x) = e^x$; $[0, 3]$; $n = 5$; midpoints

17. **Real Estate**   Figure (a) shows a vacant lot with a 100-foot frontage in a development. To estimate its area, we introduce a coordinate system so that the *x*-axis coincides with the edge of the straight road forming the lower boundary of the property as shown in figure (b). Then, thinking of the upper boundary of the property as the graph of a continuous function *f* over the interval $[0, 100]$, we see that the problem is mathematically equivalent to finding the area under the graph of *f* on $[0, 100]$. To estimate the area of the lot using a Riemann sum, we divide the interval $[0, 100]$ into five equal subintervals of length 20 feet. Then, using surveyor's equipment, we measure the distance from the midpoint of each of these subintervals to the upper boundary of the property. These measurements give the values of $f(x)$ at $x = 10, 30, 50, 70,$ and 90. What is the approximate area of the lot?

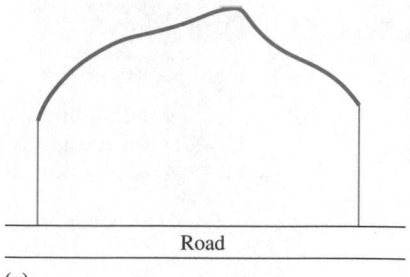

Road

(a)

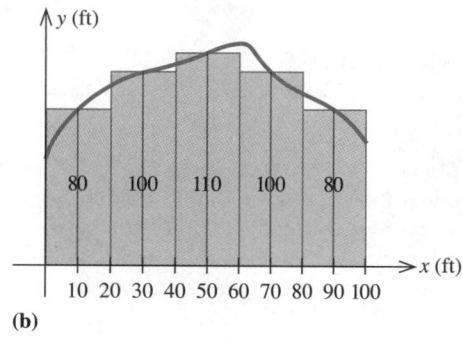

(b)

18. **Real Estate**   Use the technique of exercise 17 to obtain an estimate of the area of the vacant lot shown in the accompanying figures.

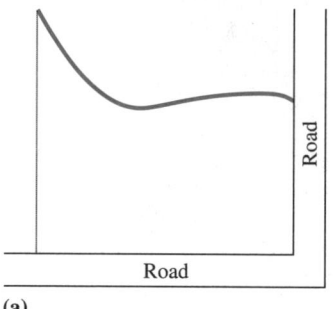

Road

(a)

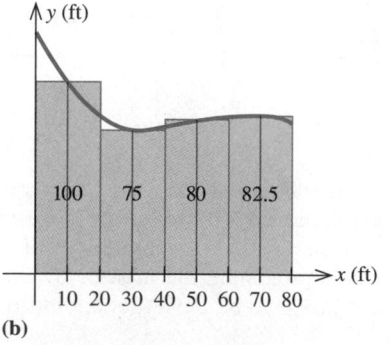

(b)

SOLUTION TO SELF-CHECK EXERCISE 14.3

The length of each subinterval is $\frac{3}{4}$. Therefore, the four subintervals are

$$\left[0, \frac{3}{4}\right], \quad \left[\frac{3}{4}, \frac{3}{2}\right], \quad \left[\frac{3}{2}, \frac{9}{4}\right], \quad \text{and} \quad \left[\frac{9}{4}, 3\right]$$

The representative points are $\frac{3}{8}$, $\frac{9}{8}$, $\frac{15}{8}$, and $\frac{21}{8}$, respectively. Therefore, the required approximation is

$$A = \frac{3}{4}f\left(\frac{3}{8}\right) + \frac{3}{4}f\left(\frac{9}{8}\right) + \frac{3}{4}f\left(\frac{15}{8}\right) + \frac{3}{4}f\left(\frac{21}{8}\right)$$

$$= \frac{3}{4}\left[f\left(\frac{3}{8}\right) + f\left(\frac{9}{8}\right) + f\left(\frac{15}{8}\right) + f\left(\frac{21}{8}\right)\right]$$

$$= \frac{3}{4}\left\{\left[2\left(\frac{3}{8}\right)^2 + 1\right] + \left[2\left(\frac{9}{8}\right)^2 + 1\right] + \left[2\left(\frac{15}{8}\right)^2 + 1\right] + \left[2\left(\frac{21}{8}\right)^2 + 1\right]\right\}$$

$$= \frac{3}{4}\left(\frac{41}{32} + \frac{113}{32} + \frac{257}{32} + \frac{473}{32}\right) = \frac{663}{32}$$

or approximately 20.72 square units.

14.4 THE FUNDAMENTAL THEOREM OF CALCULUS

The Fundamental Theorem of Calculus

In Section 14.3 we defined the definite integral of an arbitrary continuous function on an interval $[a, b]$ as a limit of Riemann sums. Calculating the value of a definite integral by actually taking the limit of such sums is tedious and in most cases impractical. It is important to realize that the numerical results we obtained in Examples 1 and 2 of Section 14.3 were *approximations* of the areas of the regions in question, even though these results enabled us to *conjecture* what the actual areas might be. Fortunately, there is a much better way of finding the exact value of a definite integral.

The following theorem shows how to evaluate the definite integral of a continuous function provided we can find an antiderivative of that function. Because of its importance in establishing the relationship between differentiation and integration, this theorem—discovered independently by Sir Isaac Newton (1642–1727) in England and Gottfried Wilhelm Leibniz (1646–1716) in Germany—is called the **Fundamental Theorem of Calculus.**

THE FUNDAMENTAL THEOREM OF CALCULUS

Let f be continuous on $[a, b]$. Then

$$\int_a^b f(x)\, dx = F(b) - F(a) \qquad (9)$$

where F is any antiderivative of f; that is $F'(x) = f(x)$.

We will explain why this theorem is true at the end of this section.

When we apply the Fundamental Theorem of Calculus, it is convenient to use the notation

$$F(x)\Big|_a^b = F(b) - F(a)$$

For example, using this notation, we write equation (9) as

$$\int_a^b f(x)\, dx = F(x)\Big|_a^b = F(b) - F(a)$$

EXAMPLE 1 Let R be the region under the graph of $f(x) = x$ on the interval $[1, 3]$. Use the Fundamental Theorem of Calculus to find the area A of R and verify your result by elementary means.

Solution The region R is shown in Figure 14.16a. Since f is nonnegative on $[1, 3]$, the area of R is given by the definite integral of f from 1 to 3; that is,

$$A = \int_1^3 x\, dx$$

Figure 14.16
The area of R can be computed in two different ways.

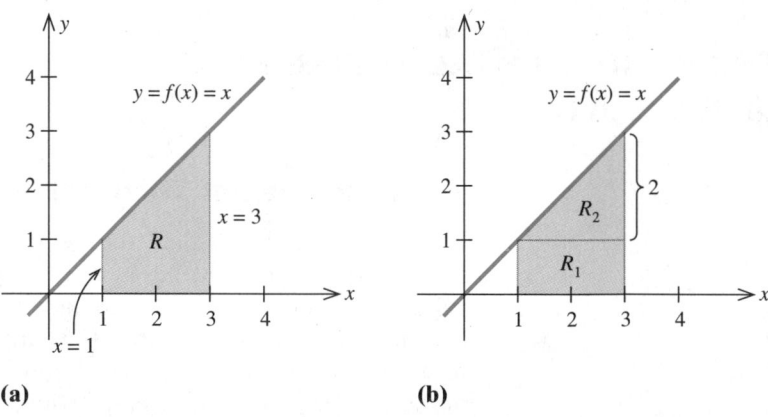

(a)   (b)

In order to evaluate the definite integral, observe that an antiderivative of $f(x) = x$ is $F(x) = \frac{1}{2}x^2 + C$, where C is an arbitrary constant. Therefore, by the Fundamental Theorem of Calculus, we have

$$A = \int_1^3 x\, dx = \frac{1}{2}x^2 + C\,\Big|_1^3$$

$$= \left(\frac{9}{2} + C\right) - \left(\frac{1}{2} + C\right) = 4 \text{ square units}$$

To verify this result by elementary means, observe that the area A is the area of the rectangle R_1 (width $\times$ height) plus the area of the triangle R_2 ($\frac{1}{2}$ base $\times$ height) (see Figure 14.16b); that is,

$$2(1) + \frac{1}{2}(2)(2) = 2 + 2 = 4$$

which agrees with the result obtained earlier.   ○ ○ ○

Observe that in the evaluation of the definite integral in Example 1, the constant of integration "dropped out." This is true in general, for if $F(x) + C$ denotes an antiderivative of some function f, then

$$F(x) + C \Big|_a^b = [F(b) + C] - [F(a) + C]$$
$$= F(b) + C - F(a) - C$$
$$= F(b) - F(a)$$

With this fact in mind, we may, in all future computations involving the evaluations of a definite integral, drop the constant of integration from our calculations.

Finding the Area Under a Curve

Having seen how effective the Fundamental Theorem of Calculus is in helping us find the area of simple regions, we now use it to find the area of more complicated regions.

EXAMPLE 2   In Section 14.3, we conjectured that the area of the region R under the graph of $f(x) = x^2$ on the interval $[0, 1]$ was $\frac{1}{3}$ square unit. Use the Fundamental Theorem of Calculus to verify this conjecture.

Solution   The region R is reproduced in Figure 14.17. Observe that f is non-negative on $[0, 1]$, so the area of R is given by $A = \int_0^1 x^2\, dx$. Since an antiderivative of $f(x) = x^2$ is $F(x) = \frac{1}{3}x^3$, we see, using the Fundamental Theorem

Figure 14.17
The area of R is $\int_0^1 x^2\, dx = \frac{1}{3}$.

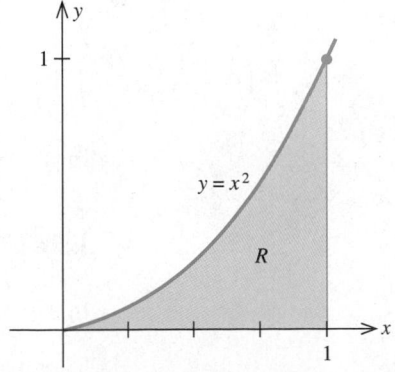

of Calculus, that

$$A = \int_0^1 x^2 \, dx = \frac{1}{3}x^3 \Big|_0^1 = \frac{1}{3}(1) - \frac{1}{3}(0) = \frac{1}{3} \text{ square unit}$$

as we wished to show.

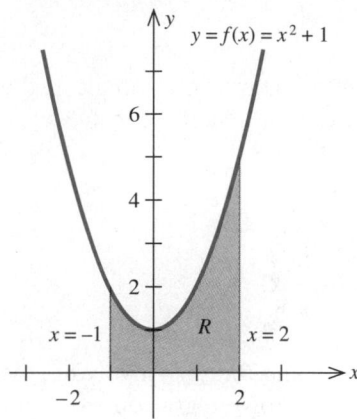

Figure 14.18
Area of $R = \int_{-1}^2 (x^2 + 1) \, dx$.

EXAMPLE 3   Find the area of the region R under the graph of $y = x^2 + 1$ from $x = -1$ to $x = 2$.

Solution   The region R under consideration is shown in Figure 14.18. Using the Fundamental Theorem of Calculus, we find that the required area is

$$\int_{-1}^2 (x^2 + 1) \, dx = \left(\frac{1}{3}x^3 + x \right) \Big|_{-1}^2$$

$$= \left[\frac{1}{3}(8) + 2 \right] - \left[\frac{1}{3}(-1)^3 + (-1) \right] = 6$$

or 6 square units.

Evaluating Definite Integrals

In Examples 4 and 5, we use the rules of integration of Section 14.1 to help us evaluate the definite integrals.

EXAMPLE 4   Evaluate $\int_1^3 (3x^2 + e^x) \, dx$.

Solution

$$\int_1^3 (3x^2 + e^x) \, dx = x^3 + e^x \Big|_1^3$$

$$= (27 + e^3) - (1 + e) = 26 + e^3 - e$$

EXAMPLE 5   Evaluate $\int_1^2 \left(\frac{1}{x} - \frac{1}{x^2} \right) dx$.

Solution

$$\int_1^2 \left(\frac{1}{x} - \frac{1}{x^2} \right) dx = \int_1^2 \left(\frac{1}{x} - x^{-2} \right) dx$$

$$= \ln |x| + \frac{1}{x} \Big|_1^2$$

$$= \left(\ln 2 + \frac{1}{2} \right) - (\ln 1 + 1)$$

$$= \ln 2 - \frac{1}{2} \qquad \text{(Recall } \ln 1 = 0.\text{)}$$

Consider the definite integral $\int_{-1}^{1} \frac{1}{x^2}\, dx$.

a. Show that a formal application of equation (9) leads to

$$\int_{-1}^{1} \frac{1}{x^2}\, dx = -\frac{1}{x}\Big|_{-1}^{1} = -1 - 1 = -2$$

b. Observe that $f(x) = 1/x^2$ is positive at each value of x in $[-1, 1]$ where it is defined. Therefore, one might expect that the definite integral with integrand f has a positive value, if it exists.

c. Explain this apparent contradiction in the result (a) and the observation (b).

Applications

EXAMPLE 6   The management of Staedtler Office Equipment has determined that the daily marginal cost function associated with producing battery-operated pencil sharpeners is given by

$$C'(x) = 0.000006x^2 - 0.006x + 4$$

where $C'(x)$ is measured in dollars per unit and x denotes the number of units produced. Management has also determined that the daily fixed cost incurred in producing these pencil sharpeners is $100. Find Staedtler's daily total cost for producing (a) the first 500 units and (b) the 201st through 400th units.

Solution

a. Since $C'(x)$ is the marginal cost function, its antiderivative $C(x)$ is the total cost function. The daily fixed cost incurred in producing the pencil sharpeners is $C(0)$ dollars. Since the daily fixed cost is given as $100, we have $C(0) = 100$. We are required to find $C(500)$. Let us compute $C(500) - C(0)$, the net change in the total cost function $C(x)$ over the interval $[0, 500]$. Using the Fundamental Theorem of Calculus, we find

$$C(500) - C(0) = \int_{0}^{500} C'(x)\, dx$$

$$= \int_{0}^{500} (0.000006x^2 - 0.006x + 4)\, dx$$

$$= 0.000002x^3 - 0.003x^2 + 4x \Big|_{0}^{500}$$

$$= [0.000002(500)^3 - 0.003(500)^2 + 4(500)]$$
$$\quad - [0.000002(0)^3 - 0.003(0)^2 + 4(0)]$$

$$= 1500$$

Therefore $C(500) = 1500 + C(0) = 1500 + 100 = 1600$, so the total cost incurred daily by Staedtler in producing 500 pencil sharpeners is $1600.

b. The daily total cost incurred by Staedtler in producing the 201st through 400th units of battery-operated pencil sharpeners is given by

$$C(400) - C(200) = \int_{200}^{400} C'(x)\, dx$$

$$= \int_{200}^{400} (0.000006x^2 - 0.006x + 4)\, dx$$

$$= 0.000002x^3 - 0.003x^2 + 4x \Big|_{200}^{400}$$

$$= 552$$

or $552.   ◦ ◦ ◦

Since $C'(x)$ is nonnegative for x in the interval $(0, \infty)$, we have the following geometrical interpretation of the two definite integrals in Example 6: $\int_0^{500} C'(x)\, dx$ is the area of the region under the graph of the function C' from $x = 0$ to $x = 500$, shown in Figure 14.19a, and $\int_{200}^{400} C'(x)\, dx$ is the area of the region shown in Figure 14.19b from $x = 200$ to $x = 400$.

Figure 14.19

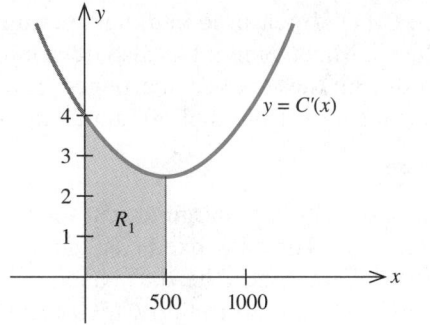

(a) Area of $R_1 = \int_0^{500} C'(x)\, dx$

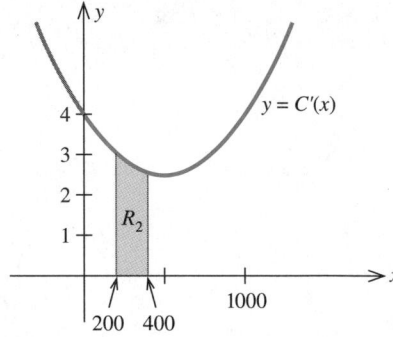

(b) Area of $R_2 = \int_{200}^{400} C'(x)\, dx$

EXAMPLE 7   An efficiency study conducted for the Elektra Electronics Company showed that the rate at which Space Commander walkie-talkies are assembled by the average worker t hours after starting work at 8 A.M. is given by the function

$$f(t) = -3t^2 + 12t + 15 \qquad (0 \le t \le 4)$$

Determine how many walkie-talkies can be assembled by the average worker in the first hour of the morning shift.

Solution   Let $N(t)$ denote the number of walkie-talkies assembled by the average worker t hours after starting work in the morning shift. Then we have

$$N'(t) = f(t) = -3t^2 + 12t + 15$$

Therefore, the number of units assembled by the average worker in the first hour of the morning shift is

$$N(1) - N(0) = \int_0^1 N'(t)\, dt = \int_0^1 (-3t^2 + 12t + 15)\, dt$$

$$= -t^3 + 6t^2 + 15t \Big|_0^1 = -1 + 6 + 15$$

$$= 20, \quad \text{or 20 units}$$

○ ○ ○

EXPLORING WITH TECHNOLOGY

You can demonstrate graphically that $\int_0^x t\, dt = \frac{1}{2}x^2$ as follows:

1. Plot the graphs of $y1 = \int_0^x t\, dt$ and $y2 = \frac{1}{2}x^2$ on the same set of axes in the viewing rectangle $[-5, 5] \times [0, 10]$.

2. Compare the graphs of $y1$ and $y2$ and draw the desired conclusion.

◐ ◐ ◐

EXAMPLE 8   A certain city's rate of electricity consumption is expected to grow exponentially with a growth constant of $k = 0.04$. If the present rate of consumption is 40 million kilowatt-hours (kwh) per year, what should the total production of electricity be over the next three years in order to meet the projected demand?

Solution   If $R(t)$ denotes the expected rate of consumption of electricity t years from now, then

$$R(t) = 40e^{0.04t}$$

million kwh per year. Next, if $C(t)$ denotes the expected total consumption of electricity over a period of t years, then

$$C'(t) = R(t)$$

Therefore, the total consumption of electricity expected over the next three years is given by

$$\int_0^3 C'(t)\,dt = \int_0^3 40e^{0.04t}\,dt$$

$$= \frac{40}{0.04}\,e^{0.04t}\,\Big|_0^3$$

$$= 1000(e^{0.12} - 1)$$

$$= 127.5, \quad \text{or } 127.5 \text{ million kwh}$$

the amount that must be produced over the next three years in order to meet the demand.   ◦ ◦ ◦

The definite integral $\int_{-3}^3 \sqrt{9 - x^2}\,dx$ cannot be evaluated using the Fundamental Theorem of Calculus because the method of this section does not enable us to find an antiderivative of the integrand. But the integral can be evaluated by interpreting it as the area of a certain plane region. What is the region? And what is the value of the integral?

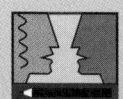

Validity of the Fundamental Theorem of Calculus

In order to demonstrate the plausibility of the Fundamental Theorem of Calculus for the case where f is nonnegative on an interval $[a, b]$, let us define an "area function" A as follows. Let $A(t)$ denote the area of the region R under the graph of $y = f(x)$ from $x = a$ to $x = t$, where $a \le t \le b$ (Figure 14.20).

If h is a small positive number, then $A(t + h)$ is the area of the region under the graph of $y = f(x)$ from $x = a$ to $x = t + h$. Therefore, the difference

$$A(t + h) - A(t)$$

is the area under the graph of $y = f(x)$ from $x = t$ to $x = t + h$ (Figure 14.21).

Figure 14.20
A(t) = Area under the graph of f from x = a to x = t.

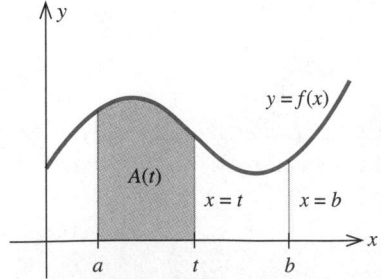

Figure 14.21
A(t + h) − A(t) = Area under the graph of f from x = t to x = t + h.

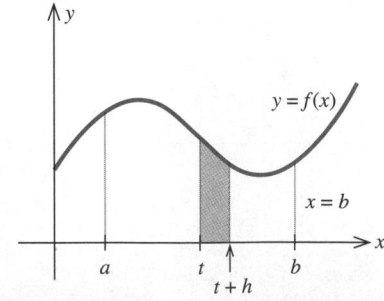

Figure 14.22
Area of rectangle $= h \cdot f(t)$.

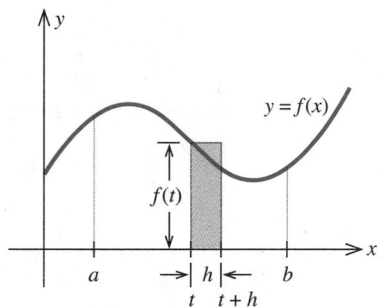

Now, the area of this last region can be approximated by the area of the rectangle of width h and height $f(t)$—that is, by the expression $h \cdot f(t)$ (Figure 14.22). Thus,

$$A(t + h) - A(t) \approx h \cdot f(t)$$

where the approximations improve as h is taken to be smaller and smaller.

Dividing both sides of the foregoing relationship by h, we obtain

$$\frac{A(t + h) - A(t)}{h} \approx f(t)$$

Taking the limit as h approaches zero, we find, by the definition of the derivative, that the left-hand side is

$$\lim_{h \to 0} \frac{A(t + h) - A(t)}{h} = A'(t)$$

The right-hand side, which is independent of h, remains constant throughout the limiting process. Because the approximation becomes exact as h approaches zero, we find that

$$A'(t) = f(t)$$

Since the foregoing equation holds for all values of t in the interval $[a, b]$, we have shown that the *area function A* is an antiderivative of the function $f(x)$. By Theorem 1 of Section 14.1, we conclude that $A(x)$ must have the form

$$A(x) = F(x) + C$$

where F is any antiderivative of f and C is an arbitrary constant. To determine the value of C, observe that $A(a) = 0$. This condition implies that

$$A(a) = F(a) + C = 0$$

Figure 14.23
The area of R is given by $A(b)$.

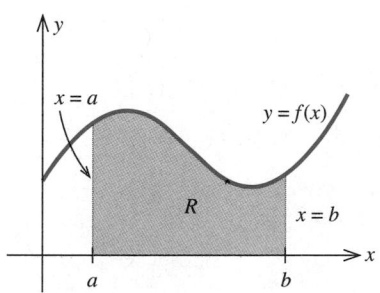

or $C = -F(a)$. Next, since the area of the region R is $A(b)$ (Figure 14.23), we see that the required area is

$$A(b) = F(b) + C$$
$$= F(b) - F(a)$$

Since the area of the region R is

$$\int_a^b f(x)\, dx$$

we have

$$\int_a^b f(x)\, dx = F(b) - F(a)$$

as we set out to show.

SELF-CHECK EXERCISES 14.4

1. Evaluate $\int_0^2 (x + e^x)\, dx$.

2. The daily marginal profit function associated with producing and selling Texa-Pep hot sauce is

$$P'(x) = -0.000006x^2 + 6$$

where x denotes the number of cases (each case contains 24 bottles) produced and sold daily and $P'(x)$ is measured in dollars per unit. The fixed cost is $400.

a. What is the total profit realizable from producing and selling 1000 cases of Texa-Pep per day?

b. What is the additional profit realizable if the production and sale of Texa-Pep are increased from 1000 to 1200 cases per day?

Solutions to Self-Check Exercises 14.4 can be found on page 980.

14.4 EXERCISES

In exercises 1–4, find the area of the region under the graph of the function f on the interval [a, b] using the Fundamental Theorem of Calculus. Then verify your result using geometry.

1. $f(x) = 2; [1, 4]$

2. $f(x) = 4; [-1, 2]$

3. $f(x) = 2x; [1, 3]$

4. $f(x) = -\frac{1}{4}x + 1; [1, 4]$

In exercises 5–16, find the area of the region under the graph of the function f on the interval [a, b].

5. $f(x) = 2x + 3; [-1, 2]$

6. $f(x) = 4x - 1; [2, 4]$

7. $f(x) = -x^2 + 4; [-1, 2]$

8. $f(x) = 4x - x^2; [0, 4]$

9. $f(x) = \frac{1}{x}; [1, 2]$

10. $f(x) = \frac{1}{x^2}; [2, 4]$

11. $f(x) = \sqrt{x}; [1, 9]$

12. $f(x) = x^3; [1, 3]$

13. $f(x) = 1 - \sqrt[3]{x}; [-8, -1]$

14. $f(x) = \frac{1}{\sqrt{x}}; [1, 9]$

15. $f(x) = e^x; [0, 2]$

16. $f(x) = e^x - x; [1, 2]$

In exercises 17–40, evaluate the definite integral.

17. $\int_2^4 3\, dx$

18. $\int_{-1}^2 -2\, dx$

19. $\int_1^3 (2x + 3)\, dx$

20. $\int_{-1}^0 (4 - x)\, dx$

21. $\int_{-1}^3 2x^2\, dx$

22. $\int_0^2 8x^3\, dx$

23. $\int_{-2}^2 (x^2 - 1)\, dx$

24. $\int_1^4 \sqrt{u}\, du$

25. $\int_1^8 4x^{1/3}\, dx$

26. $\int_1^4 2x^{-3/2}\, dx$

27. $\int_0^1 (x^3 - 2x^2 + 1)\, dx$

28. $\int_1^2 (t^5 - t^3 + 1)\, dt$

29. $\int_2^4 \frac{1}{x}\, dx$

30. $\int_1^3 \frac{2}{x}\, dx$

31. $\int_0^4 x(x^2 - 1)\, dx$

32. $\int_0^2 (x - 4)(x - 1)\, dx$

33. $\int_1^3 (t^2 - t)^2\, dt$

34. $\int_{-1}^1 (x^2 - 1)^2\, dx$

35. $\int_{-3}^{-1} \frac{1}{x^2}\, dx$

36. $\int_1^2 \frac{2}{x^3}\, dx$

37. $\int_1^4 \left(\sqrt{x} - \frac{1}{\sqrt{x}}\right) dx$

38. $\int_0^1 \sqrt{2x}(\sqrt{x} + \sqrt{2})\, dx$

39. $\int_1^4 \frac{3x^3 - 2x^2 + 4}{x^2}\, dx$

40. $\int_1^2 \left(1 + \frac{1}{u} + \frac{1}{u^2}\right) du$

41. Marginal Cost   A division of Ditton Industries manufactures a deluxe toaster oven. Management has determined that the daily marginal cost function associated with producing these toaster ovens is given by

$$C'(x) = 0.0003x^2 - 0.12x + 20$$

where $C'(x)$ is measured in dollars per unit and x denotes the number of units produced. Management has also determined that the daily fixed cost incurred in the production is $800.
a. Find the total cost incurred by Ditton in producing the first 300 units of these toaster ovens per day.
b. What is the total cost incurred by Ditton in producing the 201st through 300th units per day?

42. Marginal Revenue   The management of Ditton Industries has determined that the daily marginal revenue function associated with selling x units of their deluxe toaster ovens is given by

$$R'(x) = -0.1x + 40$$

where $R'(x)$ is measured in dollars per unit.
a. Find the daily total revenue realized from the sale of 200 units of the toaster oven.
b. Find the additional revenue realized when the production (and sales) level is increased from 200 to 300 units.

43. Marginal Profit   Refer to exercise 41. The daily marginal profit function associated with the production and sales of the deluxe toaster ovens is known to be

$$P'(x) = -0.0003x^2 + 0.02x + 20$$

where x denotes the number of units manufactured and sold daily and $P'(x)$ is measured in dollars per unit.
a. Find the total profit realizable from the manufacture and sale of 200 units of the toaster ovens per day.
[*Hint:* $P(200) - P(0) = \int_0^{200} P'(x) \, dx$, $P(0) = -800$.]

b. What is the additional daily profit realizable if the production and sale of the toaster ovens are increased from 200 to 220 units per day?

44. Efficiency Studies   Tempco Electronics, a division of Tempco Toys, Inc., manufactures an electronic football game. An efficiency study showed that the rate at which the games are assembled by the average worker t hours after starting work at 8 A.M. is

$$-\frac{3}{2}t^2 + 6t + 20 \qquad (0 \le t \le 4)$$

units per hour.
a. Find the total number of games the average worker can be expected to assemble in the 4-hour morning shift.
b. How many units can the average worker be expected to assemble in the first hour of the morning shift? In the second hour of the morning shift?

45. Speedboat Racing   In a recent pretrial run for the world water speed record, the velocity of the *Sea Falcon II* t seconds after firing the booster rocket was given by

$$v(t) = -t^2 + 20t + 440 \qquad (0 \le t \le 20)$$

feet per second. Find the distance covered by the boat over the 20-second period after the booster rocket was activated.
[*Hint:* The distance is given by $\int_0^{20} v(t) \, dt$.]

46. U.S. Census   According to the U.S. Census Bureau, the number of Americans aged 45 to 54 (which stood at 25 million at the beginning of 1990) will grow at the rate of

$$R(t) = 0.00933t^3 + 0.019t^2 - 0.10833t + 1.3467$$

million people per year, t years from the beginning of 1990. How many Americans aged 45 to 54 will be added to the population between 1990 and the year 2000?
Source: U.S. Census Bureau

47. Air Purification   To test air purifiers, the engineers run a purifier in a smoke-filled 10×20-foot room. While conducting a test for a certain brand of air purifier, they determined that the amount of smoke in the room was decreasing at the rate of

$$R(t) = 0.00032t^4 - 0.01872t^3 + 0.3948t^2$$
$$- 3.83t + 17.63 \qquad (0 \le t \le 20)$$

percent of the (original) amount of the smoke per minute, t minutes after the start of the test. How much smoke was left in the room 5 minutes after the start of the test? Ten minutes after the start of the test?
Source: Consumer Reports

USING TECHNOLOGY

FINDING DEFINITE INTEGRALS

Some graphing utilities have an operation for finding the definite integral of a function. If your graphing utility has this capability, use it to work through the example and exercises of this section.

EXAMPLE 1 Use the numerical integral operation of a graphing utility to evaluate

$$\int_{-1}^{2} \frac{2x + 4}{(x^2 + 1)^{3/2}} \, dx$$

Solution   Using the numerical integral operation of a graphing utility, we find

$$\int_{-1}^{2} \frac{2x + 4}{(x^2 + 1)^{3/2}} \, dx \approx 6.92592225992$$

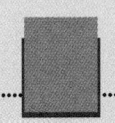 **EXERCISES**

In exercises 1–4, use a graphing utility to find the area of the region under the graph of f on the interval [a, b]. Express your answer to four decimal places.

1. $f(x) = 0.002x^5 + 0.032x^4 - 0.2x^2 + 2;\ [-1.1, 2.2]$

2. $f(x) = x\sqrt{x^3 + 1};\ [1, 2]$

3. $f(x) = \sqrt{x}e^{-x};\ [0, 3]$

4. $f(x) = \dfrac{\ln x}{\sqrt{1 + x^2}};\ [1, 2]$

In exercises 5–10, use a graphing utility to evaluate the definite integral.

5. $\displaystyle\int_{-1.2}^{2.3} (0.2x^4 - 0.32x^3 + 1.2x - 1)\ dx$

6. $\displaystyle\int_{1}^{3} x(x^4 - 1)^{3.2}\ dx$

7. $\displaystyle\int_{0}^{2} \dfrac{3x^3 + 2x^2 + 1}{2x^2 + 3}\ dx$

8. $\displaystyle\int_{1}^{2} \dfrac{\sqrt{x} + 1}{2x^2 + 1}\ dx$

9. $\displaystyle\int_{0}^{2} \dfrac{e^x}{\sqrt{x^2 + 1}}\ dx$

10. $\displaystyle\int_{1}^{3} e^{-x} \ln(x^2 + 1)\ dx$

11. Rework exercise 46, Exercise set 14.4, using your graphing utility.

12. Rework exercise 47, Exercise set 14.4, using your graphing utility.

SOLUTIONS TO SELF-CHECK EXERCISES 14.4

1. $\int_0^2 (x + e^x)\, dx = \dfrac{1}{2}x^2 + e^x \bigg|_0^2$

$$= \left[\dfrac{1}{2}(2)^2 + e^2\right] - \left[\dfrac{1}{2}(0) + e^0\right]$$

$$= 2 + e^2 - 1$$

$$= e^2 + 1$$

2. a. We want $P(1000)$. But

$$P(1000) - P(0) = \int_0^{1000} P'(x)\, dx = \int_0^{1000} (-0.000006x^2 + 6)\, dx$$

$$= -0.000002x^3 + 6x \bigg|_0^{1000}$$

$$= -0.000002(1000)^3 + 6(1000)$$

$$= 4000$$

So, $P(1000) = 4000 + P(0) = 4000 - 400$, or \$3600 per day $[P(0) = -C(0)]$.

b. The additional profit realizable is given by

$$\int_{1000}^{1200} P'(x)\, dx = -0.000002x^3 + 6x \bigg|_{1000}^{1200}$$

$$= [-0.000002(1200)^3 + 6(1200)]$$

$$\qquad - [-0.000002(1000)^3 + 6(1000)]$$

$$= 3744 - 4000$$

$$= -256$$

That is, the company sustains a loss of \$256 per day if production is increased to 1200 cases a day.

14.5 EVALUATING DEFINITE INTEGRALS

This section continues our discussion of the applications of the Fundamental Theorem of Calculus.

Properties of the Definite Integral

Before going on, we list the following useful properties of the definite integral, some of which parallel the rules of integration of Section 14.1.

PROPERTIES OF THE DEFINITE INTEGRAL

Let f and g be integrable functions; then

1. $\int_a^a f(x)\,dx = 0$

2. $\int_a^b f(x)\,dx = -\int_b^a f(x)\,dx$

3. $\int_a^b cf(x)\,dx = c\int_a^b f(x)\,dx \qquad$ (c, a constant)

4. $\int_a^b [f(x) \pm g(x)]\,dx = \int_a^b f(x)\,dx \pm \int_a^b g(x)\,dx$

5. $\int_a^b f(x)\,dx = \int_a^c f(x)\,dx + \int_c^b f(x)\,dx \qquad (a < c < b)$

Property 5 states that if c is a number lying between a and b so that the interval $[a, b]$ is divided into the intervals $[a, c]$ and $[c, b]$, then the integral of f over the interval $[a, b]$ may be expressed as the sum of the integral of f over the interval $[a, c]$ and the integral of f over the interval $[c, b]$.

Property 5 has the following geometrical interpretation when f is nonnegative. By definition

$$\int_a^b f(x)\,dx$$

is the area of the region under the graph of $y = f(x)$ from $x = a$ to $x = b$ (Figure 14.24). Similarly, we interpret the definite integrals

$$\int_a^c f(x)\,dx \quad \text{and} \quad \int_c^b f(x)\,dx$$

as the areas of the regions under the graph of $y = f(x)$ from $x = a$ to $x = c$ and from $x = c$ to $x = b$, respectively. Since the two regions do not overlap, we see that

$$\int_a^b f(x)\,dx = \int_a^c f(x)\,dx + \int_c^b f(x)\,dx$$

Figure 14.24

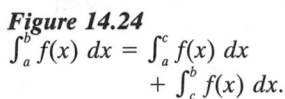
$\int_a^b f(x)\,dx = \int_a^c f(x)\,dx$
$\qquad + \int_c^b f(x)\,dx.$

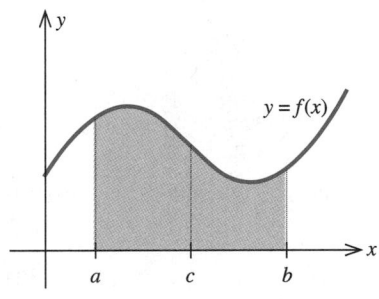

The Method of Substitution for Definite Integrals

Our first example shows two approaches generally used to evaluate a definite integral using the method of substitution.

EXAMPLE 1   Evaluate

$$\int_0^4 x\sqrt{9 + x^2}\,dx$$

Solution

Method 1   We first find the corresponding indefinite integral

$$I = \int x\sqrt{9 + x^2}\,dx$$

Make the substitution

$$u = 9 + x^2$$

so that

$$du = \frac{d}{dx}(9 + x^2)\, dx$$

$$= 2x\, dx$$

or

$$x\, dx = \frac{1}{2} du \qquad \text{(Dividing both sides by 2)}$$

Then

$$I = \int \frac{1}{2}\sqrt{u}\, du = \frac{1}{2}\int u^{1/2}\, du$$

$$= \frac{1}{3} u^{3/2} + C = \frac{1}{3}(9 + x^2)^{3/2} + C \qquad \begin{array}{l}\text{(Substituting}\\ 9 + x^2 \text{ for } u)\end{array}$$

Using this result, we now evaluate the given definite integral.

$$\int_0^4 x\sqrt{9 + x^2}\, dx = \frac{1}{3}(9 + x^2)^{3/2}\Big|_0^4$$

$$= \frac{1}{3}[(9 + 16)^{3/2} - 9^{3/2}]$$

$$= \frac{1}{3}(125 - 27) = \frac{98}{3} = 32\frac{2}{3}$$

Method 2 *Changing the limits of integration:* As before, we make the substitution

$$u = 9 + x^2 \qquad\qquad\qquad \textbf{(10)}$$

so that

$$du = 2x\, dx$$

or

$$x\, dx = \frac{1}{2} du$$

Next, observe that the given definite integral is evaluated *with respect to x* with the range of integration given by the interval [0, 4]. If we perform the integration *with respect to u* via the substitution (10), then we must adjust the range of integration to reflect the fact that the integration is being performed with respect to the new variable u. To determine the proper range of integration, note that when $x = 0$, equation (10) implies that

$$u = 9 + 0^2 = 9$$

which gives the required lower limit of integration with respect to u. Similarly, when $x = 4$,

$$u = 9 + 16 = 25$$

is the required upper limit of integration with respect to u. Thus, the range of integration when the integration is performed with respect to u is given by the interval $[9, 25]$. Therefore, we have

$$\int_0^4 x\sqrt{9 + x^2}\, dx = \int_9^{25} \frac{1}{2}\sqrt{u}\, du = \frac{1}{2}\int_9^{25} u^{1/2}\, du$$

$$= \frac{1}{3}u^{3/2}\Big|_9^{25} = \frac{1}{3}\left(25^{3/2} - 9^{3/2}\right)$$

$$= \frac{1}{3}(125 - 27) = \frac{98}{3} = 32\frac{2}{3}$$

which agrees with the result obtained using Method 1.   ◦ ◦ ◦

EXPLORING WITH TECHNOLOGY

Refer to Example 1. You can confirm the results obtained there by using a graphing utility as follows:

1. Use the numerical integration operation of the graphing utility to evaluate

$$\int_0^4 x\sqrt{9 + x^2}\, dx$$

2. Evaluate $\frac{1}{2}\int_9^{25}\sqrt{u}\, du$.

3. Conclude that $\int_0^4 x\sqrt{9 + x^2}\, dx = \frac{1}{2}\int_9^{25}\sqrt{u}\, du$.

◦   ◦   ◦

EXAMPLE 2   Evaluate

$$\int_0^2 xe^{2x^2}\, dx$$

Solution   Let $u = 2x^2$ so that $du = 4x\, dx$, or $x\, dx = \frac{1}{4}\, du$. When $x = 0$, $u = 0$, and when $x = 2$, $u = 8$. This gives the lower and upper limits of integration with respect to u. Making the indicated substitutions, we find

$$\int_0^2 xe^{2x^2}\, dx = \int_0^8 \frac{1}{4}e^u\, du = \frac{1}{4}e^u\Big|_0^8 = \frac{1}{4}\left(e^8 - 1\right)$$ ◦ ◦ ◦

EXAMPLE 3   Evaluate

$$\int_0^1 \frac{x^2}{x^3 + 1}\, dx$$

Solution   Let $u = x^3 + 1$ so that $du = 3x^2\, dx$, or $x^2\, dx = \frac{1}{3}\, du$. When $x = 0$, $u = 1$, and when $x = 1$, $u = 2$. This gives the lower and upper limits of integration with respect to u. Making the indicated substitutions, we find

$$\int_0^1 \frac{x^2}{x^3 + 1}\, dx = \frac{1}{3}\int_1^2 \frac{du}{u} = \frac{1}{3}\ln|u|\,\Big|_1^2$$

$$= \frac{1}{3}(\ln 2 - \ln 1) = \frac{1}{3}\ln 2$$

  ◑ ◑ ◑

Finding the Area Under a Curve

EXAMPLE 4   Find the area of the region R under the graph of $f(x) = e^{(1/2)x}$ from $x = -1$ to $x = 1$.

Solution   The region R is shown in Figure 14.25. Its area is given by

$$A = \int_{-1}^1 e^{(1/2)x}\, dx$$

In order to evaluate this integral, we make the substitution

$$u = \frac{1}{2}x$$

so that

$$du = \frac{1}{2}\, dx \qquad \text{or} \qquad dx = 2\, du$$

When $x = -1$, $u = -\frac{1}{2}$, and when $x = 1$, $u = \frac{1}{2}$. Making the indicated substitutions, we obtain

$$A = \int_{-1}^1 e^{(1/2)x}\, dx = 2\int_{-1/2}^{1/2} e^u\, du$$

$$= 2e^u\,\Big|_{-1/2}^{1/2}$$

$$= 2(e^{1/2} - e^{-1/2})$$

or approximately 2.08 square units.

  ◑ ◑ ◑

Figure 14.25
Area of $R = \int_{-1}^1 e^{(1/2)x}dx$.

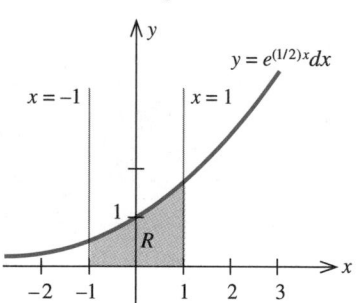

Let f be a function defined piecewise by the rule

$$f(x) = \begin{cases} \sqrt{x} & \text{if } 0 \le x \le 1 \\ \dfrac{1}{x} & \text{if } 1 < x \le 2 \end{cases}$$

How would you use property 5 of definite integrals to find the area of the region under the graph of f on $[0, 2]$? What is the area?

Average Value of a Function

The *average value* of a function over an interval provides us with an application of the definite integral. Recall that the average value of a set of n numbers is the number

$$\frac{y_1 + y_2 + \cdots + y_n}{n}$$

Now, suppose that f is a continuous function defined on $[a, b]$. Let us divide the interval $[a, b]$ into n subintervals of equal length $(b - a)/n$. Choose points $x_1, x_2, \ldots, x_n$ in the first, second, $\ldots$, nth subintervals, respectively. Then the average value of the numbers $f(x_1), f(x_2), \ldots, f(x_n)$, given by

$$\frac{f(x_1) + f(x_2) + \cdots + f(x_n)}{n}$$

is an approximation of the average of all the values of $f(x)$ on the interval $[a, b]$. This expression can be written in the form

$$\frac{(b - a)}{(b - a)} \left[f(x_1) \cdot \frac{1}{n} + f(x_2) \cdot \frac{1}{n} + \cdots + f(x_n) \cdot \frac{1}{n} \right]$$

$$= \frac{1}{b - a} \left[f(x_1) \cdot \frac{b - a}{n} + f(x_2) \cdot \frac{b - a}{n} + \cdots + f(x_n) \cdot \frac{b - a}{n} \right]$$

$$= \frac{1}{b - a} \left[f(x_1)\, \Delta x + f(x_2)\, \Delta x + \cdots + f(x_n)\, \Delta x \right] \qquad \textbf{(11)}$$

As n gets larger and larger, the expression (11) approximates the average value of $f(x)$ over $[a, b]$ with increasing accuracy. But the sum inside the brackets in (11) is a Riemann sum of the function f over $[a, b]$. In view of this, we have

$$\lim_{n \to \infty} \left[\frac{f(x_1) + f(x_2) + \cdots + f(x_n)}{n} \right]$$

$$= \frac{1}{b - a} \lim_{n \to \infty} \left[f(x_1)\Delta x + f(x_2)\Delta x + \cdots + f(x_n)\Delta x \right]$$

$$= \frac{1}{b - a} \int_a^b f(x)\, dx$$

This discussion motivates the following definition.

THE AVERAGE VALUE OF A FUNCTION

Suppose that f is integrable on $[a, b]$. Then the **average value of f** over $[a, b]$ is

$$\frac{1}{b - a} \int_a^b f(x)\, dx$$

EXAMPLE 5   Find the average value of the function $f(x) = \sqrt{x}$ over the interval $[0, 4]$.

Solution   The required average value is given by

$$\frac{1}{4-0}\int_0^4 \sqrt{x}\,dx = \frac{1}{4}\int_0^4 x^{1/2}\,dx$$

$$= \frac{1}{6}x^{3/2}\bigg|_0^4$$

$$= \frac{4}{3}$$

○ ○ ○

Applications

EXAMPLE 6   The interest rates charged by Madison Finance on auto loans for used cars over a certain 6-month period in 1999 are approximated by the function

$$r(t) = -\frac{1}{12}t^3 + \frac{7}{8}t^2 - 3t + 12 \qquad (0 \le t \le 6)$$

where t is measured in months and $r(t)$ is the annual percentage rate. What is the average rate on auto loans extended by Madison over the 6-month period?

Solution   The average rate over the 6-month period in question is given by

$$\frac{1}{6-0}\int_0^6 \left(-\frac{1}{12}t^3 + \frac{7}{8}t^2 - 3t + 12\right)dt$$

$$= \frac{1}{6}\left(-\frac{1}{48}t^4 + \frac{7}{24}t^3 - \frac{3}{2}t^2 + 12t\right)\bigg|_0^6$$

$$= \frac{1}{6}\left[-\frac{1}{48}(6^4) + \frac{7}{24}(6^3) - \frac{3}{2}(6^2) + 12(6)\right]$$

$$= 9$$

or 9% per year.

○ ○ ○

   EXAMPLE 7   The amount of a certain drug in a patient's body t days after it has been administered is

$$C(t) = 5e^{-0.2t}$$

units. Determine the average amount of the drug present in the patient's body for the first 4 days after the drug has been administered.

Solution The average amount of the drug present in the patient's body for the first 4 days after it has been administered is given by

$$\frac{1}{4-0}\int_0^4 5e^{-0.2t}\,dt = \frac{5}{4}\int_0^4 e^{-0.2t}\,dt$$

$$= \frac{5}{4}\left[\left(-\frac{1}{0.2}\right)e^{-0.2t}\Big|_0^4\right]$$

$$= \frac{5}{4}(-5e^{-0.8}+5)$$

$$\approx 3.44$$

or approximately 3.44 units.   ○ ○ ○

We now give a geometrical interpretation of the average value of a function f over an interval $[a, b]$. Suppose that $f(x)$ is nonnegative so that the definite integral

$$\int_a^b f(x)\,dx$$

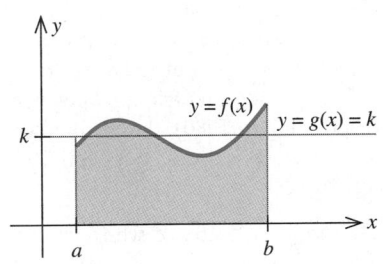

Figure 14.26
The average value of f over $[a, b]$ is k.

gives the area under the graph of f from $x = a$ to $x = b$ (Figure 14.26). Observe that, in general, the "height" $f(x)$ varies from point to point. Can we replace $f(x)$ by a constant function $g(x) = k$ (which has constant height) such that the areas under each of the two functions f and g are the same? If so, since the area under the graph of g from $x = a$ to $x = b$ is $k(b - a)$, we have

$$k(b - a) = \int_a^b f(x)\,dx$$

or

$$k = \frac{1}{b-a}\int_a^b f(x)\,dx$$

so that k is the average value of f over $[a, b]$. Thus, the average value of a function f over an interval $[a, b]$ is the height of a rectangle with base of length $(b - a)$ that has the same area as that of the region under the graph of f from $x = a$ to $x = b$.

SELF-CHECK EXERCISES 14.5

1. Evaluate $\int_0^2 \sqrt{2x + 5}\,dx$.

2. Find the average value of the function $f(x) = 1 - x^2$ over the interval $[-1, 2]$.

3. The median price of a house in a southwestern state between January 1, 1994, and January 1, 1999, is approximated by the function

$$f(t) = t^3 - 7t^2 + 17t + 140 \qquad (0 \le t \le 5)$$

where $f(t)$ is measured in thousands of dollars and t is expressed in years ($t = 0$ corresponds to the beginning of 1994). Determine the average median price of a house over that time interval.

Solutions to Self-Check Exercises 14.5 can be found on page 990.

14.5 EXERCISES

In exercises 1–28, evaluate the given definite integral.

1. $\int_0^2 x(x^2 - 1)^3 \, dx$

2. $\int_0^1 x^2(2x^3 - 1)^4 \, dx$

3. $\int_0^1 x\sqrt{5x^2 + 4} \, dx$

4. $\int_1^3 x\sqrt{3x^2 - 2} \, dx$

5. $\int_0^2 x^2(x^3 + 1)^{3/2} \, dx$

6. $\int_1^5 (2x - 1)^{5/2} \, dx$

7. $\int_0^1 \frac{1}{\sqrt{2x + 1}} \, dx$

8. $\int_0^2 \frac{x}{\sqrt{x^2 + 5}} \, dx$

9. $\int_1^2 (2x - 1)^4 \, dx$

10. $\int_1^2 (2x + 4)(x^2 + 4x - 8)^3 \, dx$

11. $\int_{-1}^1 x^2(x^3 + 1)^4 \, dx$

12. $\int_{-1}^1 (x^3 + \frac{3}{4})(x^4 + 3x)^{-2} \, dx$

13. $\int_1^5 x\sqrt{x - 1} \, dx$

14. $\int_1^4 x\sqrt{x + 1} \, dx$
    [*Hint:* Let $u = x + 1$.]

15. $\int_0^2 xe^{x^2} \, dx$

16. $\int_0^1 e^{-x} \, dx$

17. $\int_0^1 (e^{2x} + x^2 + 1) \, dx$

18. $\int_0^2 (e^t - e^{-t}) \, dt$

19. $\int_{-1}^1 xe^{x^2 + 1} \, dx$

20. $\int_0^4 \frac{e^{\sqrt{x}}}{\sqrt{x}} \, dx$

21. $\int_3^6 \frac{2}{x - 2} \, dx$

22. $\int_0^1 \frac{x}{1 + 2x^2} \, dx$

23. $\int_1^2 \frac{x^2 + 2x}{x^3 + 3x^2 - 1} \, dx$

24. $\int_0^1 \frac{e^x}{1 + e^x} \, dx$

25. $\int_1^2 \left(4e^{2u} - \frac{1}{u}\right) du$

26. $\int_1^2 \left(1 + \frac{1}{x} + e^x\right) dx$

27. $\int_1^2 \left(2e^{-4x} - \frac{1}{x^2}\right) dx$

28. $\int_1^2 \frac{\ln x}{x} \, dx$

In exercises 29–38, find the average value of the given function f over the indicated interval [a, b].

29. $f(x) = 2x + 3$; [0, 2]

30. $f(x) = 8 - x$; [1, 4]

31. $f(x) = 2x^2 - 3$; [1, 3]

32. $f(x) = 4 - x^2$; [−2, 3]

33. $f(x) = x^2 + 2x - 3$; [−1, 2]

34. $f(x) = x^3$; [−1, 1]

35. $f(x) = \sqrt{2x + 1}$; [0, 4]

36. $f(x) = e^{-x}$; [0, 4]

37. $f(x) = xe^{x^2}$; [0, 2]

38. $f(x) = \frac{1}{x + 1}$; [0, 2]

 39. **World Production of Coal**   A study proposed in 1980 by researchers from the major producers and consumers of the world's coal concluded that coal could and must play an important role in fueling global economic growth over the next 20 years. The world production of coal in 1980 was 3.5 billion metric tons. If output were to increase at the rate of $3.5e^{0.05t}$ billion metric tons per year in year t ($t = 0$ corresponding to 1980), determine how much coal will be produced worldwide between 1980 and the end of the century.

 40. **Newton's Law of Cooling**   A bottle of white wine at room temperature (68°F) is placed in a refrigerator at 4 P.M. Its temperature after t hours is changing at the rate of

$$-18e^{-0.6t}$$

degrees Fahrenheit per hour. By how many degrees will the temperature of the wine have dropped by 7 P.M.? What will the temperature of the wine be at 7 P.M.?

 41. **Net Investment Flow**   The net investment flow (rate of capital formation) of the giant conglomerate LTF Incorporated is projected to be

$$t\sqrt{\frac{1}{2}t^2 + 1}$$

million dollars per year in year t. Find the accruement on the company's capital stock in the second year.
Hint: The amount is given by

$$\int_1^2 t\sqrt{\frac{1}{2}t^2 + 1} \, dt$$

 42. **Oil Production**   Based on a preliminary report by a geological survey team, it is estimated that a newly discovered oil field can be expected to produce oil at the rate of

$$R(t) = \frac{600t^2}{t^3 + 32} + 5 \qquad (0 \le t \le 20)$$

thousand barrels per year, t years after production begins. Find the amount of oil that the field can be expected to yield during the first five years of production, assuming that the projection holds true.

43. **Depreciation: Double Declining-Balance Method** Suppose that a tractor purchased at a price of $60,000 is to be depreciated by the *double declining-balance method* over a period of ten years. It can be shown that the rate at which the book value will be decreasing is given by

$$R(t) = 13388.61e^{-0.22314t} \qquad (0 \le t \le 10)$$

dollars per year at year t. Find the amount by which the book value of the tractor will depreciate over the first five years of its life.

44. **Velocity of a Car** A car moves along a straight road in such a way that its velocity (in ft/sec) at any time t (in sec) is given by

$$v(t) = 3t\sqrt{16 - t^2} \qquad (0 \le t \le 4)$$

Find the distance traveled by the car in the 4 seconds from $t = 0$ to $t = 4$.

45. **Average Temperature** The temperature (in °F) in Boston over a 12-hour period on a certain December day was given by

$$T = -0.05t^3 + 0.4t^2 + 3.8t + 5.6 \qquad (0 \le t \le 12)$$

where t is measured in hours, with $t = 0$ corresponding to 6 A.M. Determine the average temperature on that day over the 12-hour period from 6 A.M. to 6 P.M.

46. **Whale Population** A group of marine biologists estimates that if certain conservation measures are implemented, the population of an endangered species of whale will be

$$N(t) = 3t^3 + 2t^2 - 10t + 600 \qquad (0 \le t \le 10)$$

where $N(t)$ denotes the population at the end of year t. Find the average population of the whales over the next ten years.

47. **Cable TV Subscribers** The manager of the Tele-Star Cable Television Service estimates that the total number of subscribers to the service in a certain city t years from now will be

$$N(t) = -\frac{40,000}{\sqrt{1 + 0.2t}} + 50,000$$

Find the average number of cable television subscribers over the next five years if this prediction holds true.

48. **Average Yearly Sales** The sales of the Universal Instruments Company in the first t years of its operation are approximated by the function

$$S(t) = t\sqrt{0.2t^2 + 4}$$

where $S(t)$ is measured in millions of dollars. What were Universal's average yearly sales over its first five years of operation?

49. **Concentration of a Drug in the Bloodstream** The concentration of a certain drug in a patient's bloodstream t hours after injection is

$$C(t) = \frac{0.2t}{t^2 + 1}$$

milligrams per cubic centimeter. Determine the average concentration of the drug in the patient's bloodstream over the first 4 hours after the drug is injected.

50. Refer to exercise 44. Find the average velocity of the car over the time interval $[0, 4]$.

51. **Flow of Blood in an Artery** According to a law discovered by nineteenth-century physician Jean Louis Marie Poiseuille, the velocity of blood (in cm/sec) r centimeters from the central axis of an artery is given by

$$v(r) = k(R^2 - r^2)$$

where k is a constant and R is the radius of the artery. Find the average velocity of blood along a radius of the artery (see accompanying figure).

[*Hint*: Evaluate $\dfrac{1}{R}\displaystyle\int_0^R v(r)\, dr$.]

Blood vessel

52. Prove property 1 of the definite integral.
[*Hint*: Let F be an antiderivative of f, and use the definition of the definite integral.]

53. Prove property 2 of the definite integral.
[*Hint*: See exercise 52.]

54. Verify by direct computation that

$$\int_1^3 x^2\, dx = -\int_3^1 x^2\, dx$$

55. Prove property 3 of the definite integral.

[*Hint:* See exercise 52.]

56. Verify by direct computation that

$$\int_1^9 2\sqrt{x}\,dx = 2\int_1^9 \sqrt{x}\,dx$$

57. Verify by direct computation that

$$\int_0^1 (1 + x - e^x)\,dx = \int_0^1 dx + \int_0^1 x\,dx - \int_0^1 e^x\,dx$$

What properties of the definite integral are demonstrated in this exercise?

58. Verify by direct computation that

$$\int_0^3 (1 + x^3)\,dx = \int_0^1 (1 + x^3)\,dx + \int_1^3 (1 + x^3)\,dx$$

What property of the definite integral is demonstrated here?

59. Verify by direct computation that

$$\int_0^3 (1 + x^3)\,dx = \int_0^1 (1 + x^3)\,dx$$
$$+ \int_1^2 (1 + x^3)\,dx + \int_2^3 (1 + x^3)\,dx$$

hence showing that property 5 may be extended.

SOLUTIONS TO SELF-CHECK EXERCISES 14.5

1. Let $u = 2x + 5$. Then $du = 2\,dx$, or $dx = \frac{1}{2}\,du$. Also, when $x = 0$, $u = 5$, and when $x = 2$, $u = 9$. Therefore,

$$\int_0^2 \sqrt{2x + 5}\,dx = \int_0^2 (2x + 5)^{1/2}\,dx$$
$$= \frac{1}{2}\int_5^9 u^{1/2}\,du$$
$$= \left(\frac{1}{2}\right)\left(\frac{2}{3}u^{3/2}\right)\Big|_5^9$$
$$= \frac{1}{3}[9^{3/2} - 5^{3/2}]$$
$$= \frac{1}{3}(27 - 5\sqrt{5})$$

2. The required average value is given by

$$\frac{1}{2 - (-1)}\int_{-1}^2 (1 - x^2)\,dx = \frac{1}{3}\int_{-1}^2 (1 - x^2)\,dx$$
$$= \frac{1}{3}\left(x - \frac{1}{3}x^3\right)\Big|_{-1}^2$$
$$= \frac{1}{3}\left[\left(2 - \frac{8}{3}\right) - \left(-1 + \frac{1}{3}\right)\right]$$
$$= 0$$

3. The average median price of a house over the stated time interval is given by

$$\frac{1}{5 - 0}\int_0^5 (t^3 - 7t^2 + 17t + 140)\,dt = \frac{1}{5}\left(\frac{1}{4}t^4 - \frac{7}{3}t^3 + \frac{17}{2}t^2 + 140t\right)\Big|_0^5$$
$$= \frac{1}{5}\left[\frac{1}{4}(5)^4 - \frac{7}{3}(5)^3 + \frac{17}{2}(5)^2 + 140(5)\right]$$
$$= 155.417$$

or $155,417.

14.6 AREA BETWEEN TWO CURVES

Suppose that a certain country's petroleum consumption is expected to grow at the rate of $f(t)$ million barrels per year, t years from now, for the next five years. Then the country's total petroleum consumption over the period of time in question is given by the area under the graph of f on the interval [0, 5] (Figure 14.27).

Next, suppose that because of the implementation of certain energy-conservation measures, the rate of growth of petroleum consumption is expected to be $g(t)$ million barrels per year instead. Then the country's projected total petroleum consumption over the 5-year period is given by the area under the graph of g on the interval [0, 5] (Figure 14.28).

Figure 14.27
At a rate of consumption $f(t)$ million barrels per year, the total petroleum consumption is given by the area of the region under the graph of f.

Figure 14.28
At a rate of consumption of $g(t)$ million barrels per year, the total petroleum consumption is given by the area of the region under the graph of g.

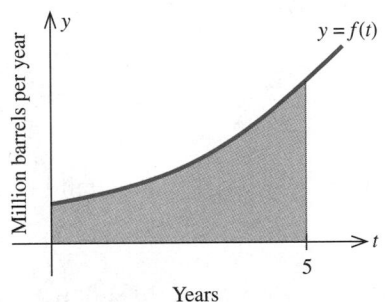

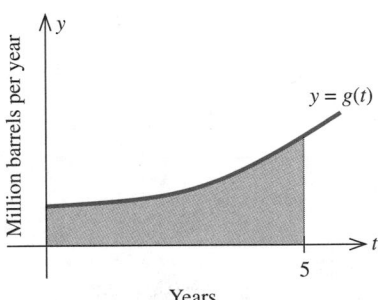

Therefore, the area of the shaded region S lying between the graphs of f and g on the interval [0, 5] (Figure 14.29) gives the amount of petroleum that would be saved over the 5-year period because of the conservation

Figure 14.29
The area of S gives the amount of petroleum that would be saved over the 5-year period.

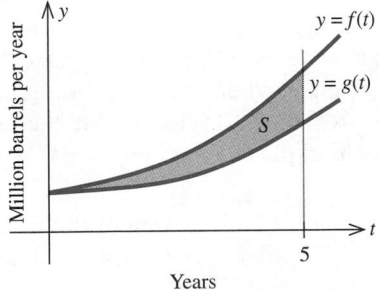

measures. But the area of S is given by

Area under the graph of f on $[a, b]$ − Area under the graph of g on $[a, b]$

$$= \int_0^5 f(t)\, dt - \int_0^5 g(t)\, dt$$

$$= \int_0^5 [f(t) - g(t)]\, dt \qquad \text{(By property 4, Section 14.5)}$$

This example shows that some practical problems can be solved by finding the area of a region between two curves, which, in turn, can be found by evaluating an appropriate definite integral.

Finding the Area Between Two Curves

We now turn our attention to the general problem of finding the area of a plane region bounded both above and below by the graphs of functions. First, consider the situation in which the graph of one function lies above that of another. More specifically, let R be the region in the xy-plane (Figure 14.30) that is bounded above by the graph of a continuous function f, below by a continuous function g where $f(x) \geq g(x)$ on $[a, b]$, and to the left and right by the vertical lines $x = a$ and $x = b$, respectively. From the figure, we see that

$$\text{Area of } R = \text{Area under } f(x) - \text{Area under } g(x)$$

$$= \int_a^b f(x)\, dx - \int_a^b g(x)\, dx$$

$$= \int_a^b [f(x) - g(x)]\, dx$$

upon using property 4 of the definite integral.

Figure 14.30
Area of $R = \int_a^b [f(x) - g(x)]\, dx$.

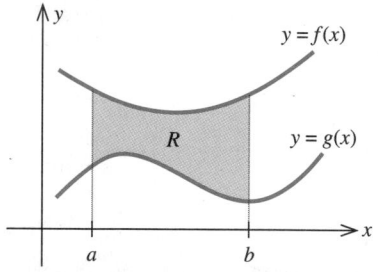

THE AREA BETWEEN TWO CURVES

Let f and g be continuous functions such that $f(x) \geq g(x)$ on the interval $[a, b]$. Then the area of the region bounded above by $y = f(x)$ and below by $y = g(x)$ on $[a, b]$ is given by

$$\int_a^b [f(x) - g(x)]\, dx \qquad (12)$$

Even though we assumed that both f and g were nonnegative in the derivation of (12), it may be shown that this equation is valid if f and g are not nonnegative (see exercise 48, page 1001). Also, observe that if $g(x)$ is 0 for all x—that is, when the lower boundary of the region R is the x-axis—(12) gives the area of the region under the curve $y = f(x)$ from $x = a$ to $x = b$, as we would expect.

EXAMPLE 1   Find the area of the region bounded by the x-axis, the graph of $y = -x^2 + 4x - 8$, and the lines $x = -1$ and $x = 4$.

Figure 14.31
Area of $R = -\int_{-1}^{4} g(x)\,dx$.

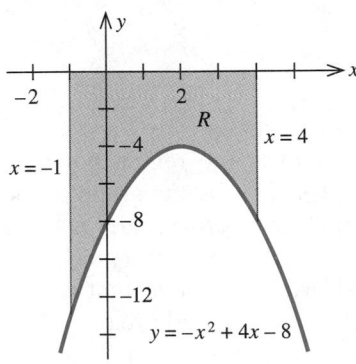

Figure 14.32
Area of $R = \int_{1}^{2} [f(x) - g(x)]\,dx$.

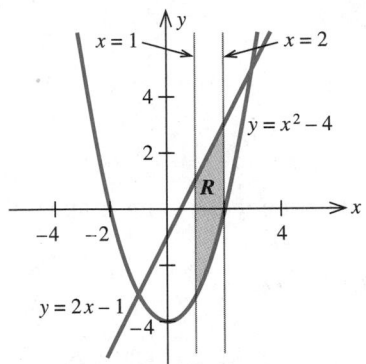

Figure 14.33
Area of $R = \int_{-1}^{3} [f(x) - g(x)]\,dx$.

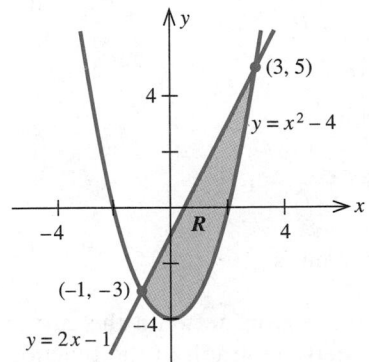

Solution   The region R under consideration is shown in Figure 14.31. We can view R as the region bounded above by the graph of $f(x) = 0$ (the x-axis) and below by the graph of $g(x) = -x^2 + 4x - 8$ on $[-1, 4]$. Therefore, the area of R is given by

$$\int_{a}^{b} [f(x) - g(x)]\,dx = \int_{-1}^{4} [0 - (-x^2 + 4x - 8)]\,dx$$

$$= \int_{-1}^{4} (x^2 - 4x + 8)\,dx$$

$$= \frac{1}{3}x^3 - 2x^2 + 8x \Big|_{-1}^{4}$$

$$= \left[\frac{1}{3}(64) - 2(16) + 8(4)\right] - \left[\frac{1}{3}(-1) - 2(1) + 8(-1)\right]$$

$$= 31\frac{2}{3}$$

or $31\frac{2}{3}$ square units.   ○ ○ ○

EXAMPLE 2   Find the area of the region R bounded by the graphs of

$$f(x) = 2x - 1 \quad \text{and} \quad g(x) = x^2 - 4$$

and the vertical lines $x = 1$ and $x = 2$.

Solution   We first sketch the graphs of the functions $f(x) = 2x - 1$ and $g(x) = x^2 - 4$, and the vertical lines $x = 1$ and $x = 2$, and identify the region R whose area is to be calculated (Figure 14.32). Since the graph of f always lies above that of g for x in the interval $[1, 2]$, we see from (12) that the required area is given by

$$\int_{1}^{2} [f(x) - g(x)]\,dx = \int_{1}^{2} [(2x - 1) - (x^2 - 4)]\,dx$$

$$= \int_{1}^{2} (-x^2 + 2x + 3)\,dx$$

$$= -\frac{1}{3}x^3 + x^2 + 3x \Big|_{1}^{2}$$

$$= \left(-\frac{8}{3} + 4 + 6\right) - \left(-\frac{1}{3} + 1 + 3\right) = \frac{11}{3}$$

or $3\frac{2}{3}$ square units.   ○ ○ ○

EXAMPLE 3   Find the area of the region R that is completely enclosed by the graphs of the functions

$$f(x) = 2x - 1 \quad \text{and} \quad g(x) = x^2 - 4$$

Solution   The region R is shown in Figure 14.33. First we find the points of intersection of the two curves. To do this, we solve the system that comprises the two equations $y = 2x - 1$ and $y = x^2 - 4$. Equating the two values of y gives

$$x^2 - 4 = 2x - 1$$
$$x^2 - 2x - 3 = 0$$
$$(x + 1)(x - 3) = 0$$

so $x = -1$ or $x = 3$. That is, the two curves intersect when $x = -1$ and $x = 3$.

Observe that we could also view the region R as the region bounded above by the graph of the function $f(x) = 2x - 1$, below by the graph of the function $g(x) = x^2 - 4$, and to the left and right by the vertical lines $x = -1$ and $x = 3$, respectively.

Next, since the graph of the function f always lies above that of the function g on $[-1, 3]$, we can use (12) to compute the desired area:

$$\int_a^b [f(x) - g(x)]\, dx = \int_{-1}^3 [(2x - 1) - (x^2 - 4)]\, dx$$

$$= \int_{-1}^3 (-x^2 + 2x + 3)\, dx$$

$$= -\frac{1}{3}x^3 + x^2 + 3x \Big|_{-1}^3$$

$$= (-9 + 9 + 9) - \left(\frac{1}{3} + 1 - 3\right) = \frac{32}{3}$$

or $10\frac{2}{3}$ square units.   ◦ ◦ ◦

EXAMPLE 4   Find the area of the region R bounded by the graphs of the functions

$$f(x) = x^2 - 2x - 1 \quad \text{and} \quad g(x) = -e^x - 1$$

and the vertical lines $x = -1$ and $x = 1$.

Figure 14.34
Area of $R = \int_{-1}^1 [f(x) - g(x)]\, dx$.

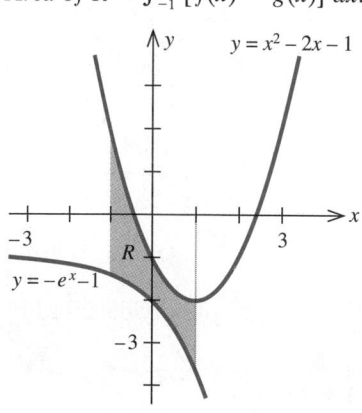

Solution   The region R is shown in Figure 14.34. Since the graph of the function f always lies above that of the function g, the area of the region R is given by

$$\int_a^b [f(x) - g(x)]\, dx = \int_{-1}^1 [(x^2 - 2x - 1) - (-e^x - 1)]\, dx$$

$$= \int_{-1}^1 (x^2 - 2x + e^x)\, dx$$

$$= \frac{1}{3}x^3 - x^2 + e^x \Big|_{-1}^1$$

$$= \left(\frac{1}{3} - 1 + e\right) - \left(-\frac{1}{3} - 1 + e^{-1}\right)$$

$$= \frac{2}{3} + e - \frac{1}{e} \text{ square units}$$ ◦ ◦ ◦

Expression (12), which gives the area of the region between the curves $y = f(x)$ and $y = g(x)$ for $a \leq x \leq b$, is valid when the graph of the function

f lies above that of the function g over the interval $[a, b]$. Example 5 shows how to use (12) to find the area of a region when the latter condition does not hold.

Figure 14.35
Area of R_1 = Area of R_2.

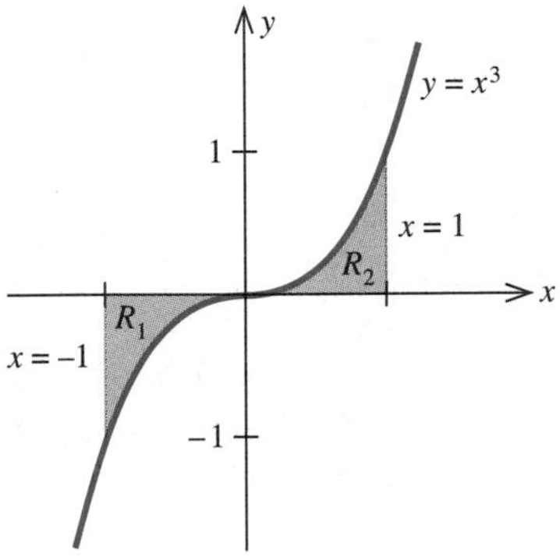

EXAMPLE 5 Find the area of the region bounded by the graph of the function $f(x) = x^3$, the x-axis, and the lines $x = -1$ and $x = 1$.

Solution The region R under consideration can be thought of as comprising the two subregions R_1 and R_2, as shown in Figure 14.35. Recall that the x-axis is represented by the function $g(x) = 0$. Since $g(x) \geq f(x)$ on $[-1, 0]$, we see that the area of R_1 is given by

$$\int_a^b [g(x) - f(x)]\, dx = \int_{-1}^0 (0 - x^3)\, dx = -\int_{-1}^0 x^3\, dx$$

$$= -\frac{1}{4}x^4 \Big|_{-1}^0 = 0 - \left(-\frac{1}{4}\right) = \frac{1}{4}$$

To find the area of R_2, we observe that $f(x) \geq g(x)$ on $[0, 1]$, so it is given by

$$\int_a^b [f(x) - g(x)]\, dx = \int_0^1 (x^3 - 0)\, dx = \int_0^1 x^3\, dx$$

$$= \frac{1}{4}x^4 \Big|_0^1 = \left(\frac{1}{4}\right) - 0 = \frac{1}{4}$$

Therefore, the area of R is $\frac{1}{4} + \frac{1}{4}$, or $\frac{1}{2}$, square unit.

By making use of symmetry, we could have obtained the same result by computing

$$-2\int_{-1}^0 x^3\, dx \quad \text{or} \quad 2\int_0^1 x^3\, dx$$

as you may verify.

○ ○ ○

A function is *even* if it satisfies the condition $f(-x) = f(x)$, and it is *odd* if it satisfies the condition $f(-x) = -f(x)$. Show that the graph of an even function is symmetric with respect to the y-axis, whereas the graph of an odd function is symmetric with respect to the origin. Explain why

$$\int_{-a}^a f(x)\, dx = 2\int_0^a f(x)\, dx \quad \text{if } f \text{ is even}$$

$$\int_{-a}^a f(x)\, dx = 0 \quad \quad \text{if } f \text{ is odd}$$

EXAMPLE 6 Find the area of the region completely enclosed by the graphs of the functions

$$f(x) = x^3 - 3x + 3 \quad \text{and} \quad g(x) = x + 3$$

Figure 14.36
Area of R_1 + Area of R_2 =
$\int_{-2}^{0} [f(x) - g(x)]\, dx$ +
$\int_{0}^{2} [g(x) - f(x)]\, dx.$

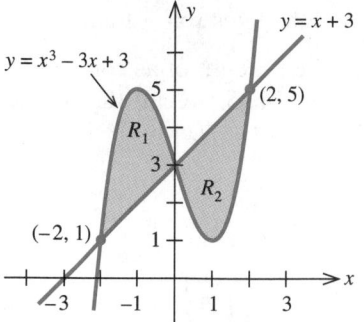

Solution   First sketch the graphs of $y = x^3 - 3x + 3$ and $y = x + 3$, and then identify the required region R. We can view the region R as being composed of the two subregions R_1 and R_2, as shown in Figure 14.36. By solving the equations $y = x + 3$ and $y = x^3 - 3x + 3$ simultaneously, we find the points of intersection of the two curves. Equating the two values of y, we have

$$x^3 - 3x + 3 = x + 3$$
$$x^3 - 4x = 0$$
$$x(x^2 - 4) = 0$$
$$x(x + 2)(x - 2) = 0 \quad \text{or} \quad x = 0, -2, \text{or } 2$$

Hence, the points of intersection of the two curves are $(-2, 1)$, $(0, 3)$, and $(2, 5)$.
For $-2 \le x \le 0$, we see that the graph of the function f lies above that of the function g, so the area of the region R_1 is, by virtue of (12),

$$\int_{-2}^{0} [(x^3 - 3x + 3) - (x + 3)]\, dx = \int_{-2}^{0} (x^3 - 4x)\, dx$$

$$= \frac{1}{4}x^4 - 2x^2 \Big|_{-2}^{0}$$

$$= -(4 - 8)$$

$$= 4 \quad \text{or} \quad 4 \text{ square units}$$

For $0 \le x \le 2$, the graph of the function g lies above that of the function f, and the area of R_2 is given by

$$\int_{0}^{2} [(x + 3) - (x^3 - 3x + 3)]\, dx = \int_{0}^{2} (-x^3 + 4x)\, dx$$

$$= -\frac{1}{4}x^4 + 2x^2 \Big|_{0}^{2}$$

$$= -4 + 8$$

$$= 4 \quad \text{or} \quad 4 \text{ square units}$$

Therefore, the required area is the sum of the area of the two regions $R_1 + R_2$—that is, $4 + 4$, or 8 square units.   ◦ ◦ ◦

Application

EXAMPLE 7   In a 1989 study for a developing country's Economic Development Board, government economists and energy experts concluded that if the Energy Conservation Bill were implemented in 1990, the country's oil consumption for the next five years would be expected to grow in accordance with the model

$$R(t) = 20e^{0.05t}$$

where t is measured in years ($t = 0$ corresponding to the year 1990) and $R(t)$ in millions of barrels per year. Without the government-imposed conservation measures, however, the expected rate of growth of oil consumption would be given by

$$R_1(t) = 20e^{0.08t}$$

millions of barrels per year. Using these models, determine how much oil would have been saved from 1990 through 1995 if the bill had been implemented.

Solution Under the Energy Conservation Bill, the total amount of oil that would have been consumed between 1990 and 1995 is given by

$$\int_0^5 R(t)\, dt = \int_0^5 20e^{0.05t}\, dt \tag{13}$$

Without the bill, the total amount of oil that would have been consumed between 1990 and 1995 is given by

$$\int_0^5 R_1(t)\, dt = \int_0^5 20e^{0.08t}\, dt \tag{14}$$

Equation (13) may be interpreted as the area of the region under the curve $y = R(t)$ from $t = 0$ to $t = 5$. Similarly, we interpret equation (14) as the area of the region under the curve $y = R_1(t)$ from $t = 0$ to $t = 5$. Furthermore, note that the graph of $y = R_1(t) = 20e^{0.08t}$ always lies on or above the graph of $y = R(t) = 20e^{0.05t}$ ($t \geq 0$). Thus, the area of the shaded region S in Figure 14.37 shows the amount of oil that would have been saved from 1990 to 1995 if the Energy Conservation Bill had been implemented. But the area of the region S is given by

$$\int_0^5 [R_1(t) - R(t)]\, dt = \int_0^5 [20e^{0.08t} - 20^{0.05t}]\, dt$$

$$= 20 \int_0^5 (e^{0.08t} - e^{0.05t})\, dt$$

$$= 20 \left(\frac{e^{0.08t}}{0.08} - \frac{e^{0.05t}}{0.05} \right) \Big|_0^5$$

$$= 20 \left[\left(\frac{e^{0.4}}{0.08} - \frac{e^{0.25}}{0.05} \right) - \left(\frac{1}{0.08} - \frac{1}{0.05} \right) \right]$$

$$\approx 9.3 \quad \text{or} \quad \text{approximately 9.3 square units}$$

Thus, the amount of oil that would have been saved is 9.3 million barrels.

Figure 14.37
Area of $S = \int_0^5 [R_1(t) - R(t)]\, dt$.

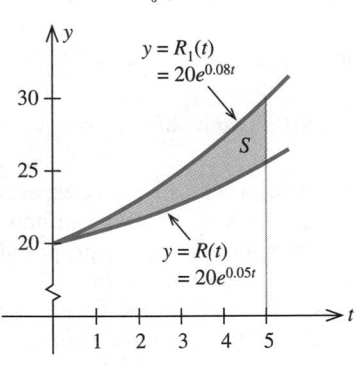

○ ○ ○

EXPLORING WITH TECHNOLOGY

Refer to Example 7. Suppose we want to construct a mathematical model giving the amount of oil saved from 1990 through the year 1990 + x, where $x \geq 0$. For example, in Example 7, $x = 5$.

1. Show that this model is given by

$$F(x) = \int_0^x [R_1(t) - R(t)]\, dt$$
$$= 250e^{0.08x} - 400e^{0.05x} + 150$$

[*Hint:* You may find it helpful to use some of the results of Example 7.]

2. Use a graphing utility to plot the graph of F in the viewing rectangle $[0, 10] \times [0, 50]$.

3. Find $F(5)$ and thus confirm the result of Example 7.

4. What is the main advantage of this model?

○ ○ ○

SELF–CHECK EXERCISES 14.6

1. Find the area of the region bounded by the graphs of $f(x) = x^2 + 2$ and $g(x) = 1 - x$ and the vertical lines $x = 0$ and $x = 1$.

2. Find the area of the region completely enclosed by the graphs of $f(x) = -x^2 + 6x + 5$ and $g(x) = x^2 + 5$.

 3. The management of the Kane Corporation, which operates a chain of hotels, expects its profits to grow at the rate of $1 + t^{2/3}$ million dollars per year t years from now. However, with renovations and improvements of existing hotels and the proposed acquisitions of new hotels, Kane's profits are expected to grow at the rate of $t - 2\sqrt{t} + 4$ million dollars per year in the next decade. What additional profits are expected over the next ten years if the group implements the proposed plans?

Solutions to Self-Check Exercises 14.6 can be found on page 1002.

14.6 EXERCISES

In exercises 1–8, find the area of the shaded region.

1.

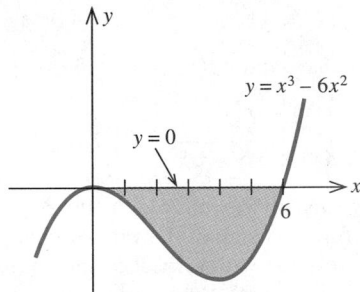

$y = x^3 - 6x^2$, $y = 0$

2.

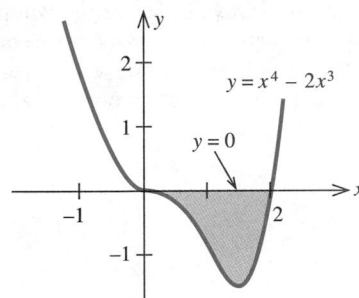

$y = x^4 - 2x^3$, $y = 0$

3.

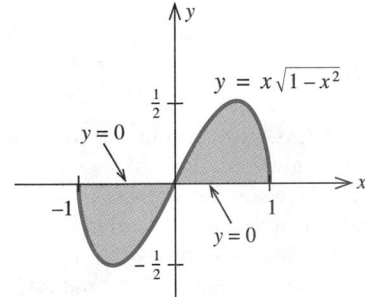

$y = x\sqrt{1-x^2}$, $y = 0$

4.

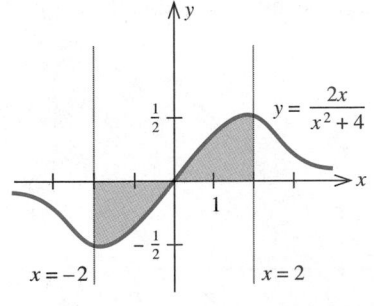

$y = \dfrac{2x}{x^2+4}$, $x = -2$, $x = 2$

5.

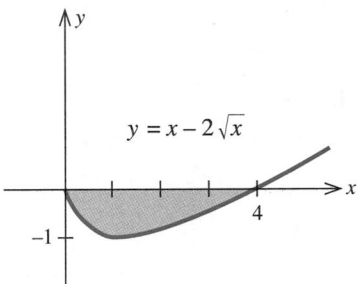

$y = x - 2\sqrt{x}$

6.

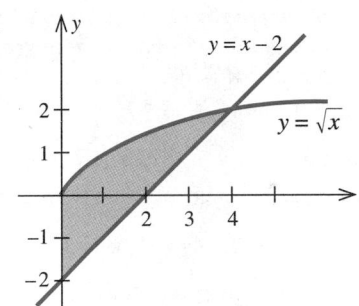

$y = x - 2$, $y = \sqrt{x}$

7.

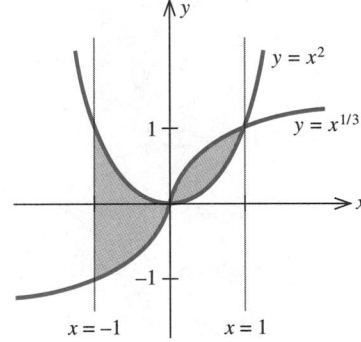

$y = x^2$, $y = x^{1/3}$, $x = -1$, $x = 1$

8.

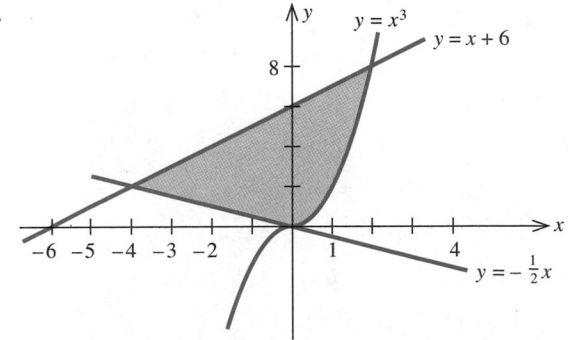

$y = x^3$, $y = x + 6$, $y = -\tfrac{1}{2}x$

In exercises 9–16, sketch the graph and find the area of the region bounded below by the graph of each of the given functions and above by the x-axis from x = a to x = b.

9. $f(x) = -x^2$; $a = -1, b = 2$

10. $f(x) = x^2 - 4$; $a = -2, b = 2$

11. $f(x) = x^2 - 5x + 4$; $a = 1, b = 3$

12. $f(x) = x^3$; $a = -1, b = 0$

13. $f(x) = -1 - \sqrt{x}$; $a = 0, b = 9$

14. $f(x) = \frac{1}{2}x - \sqrt{x}$; $a = 0, b = 4$

15. $f(x) = -e^{(1/2)x}$; $a = -2, b = 4$

16. $f(x) = -xe^{-x^2}$; $a = 0, b = 1$

In exercises 17–26, sketch the graphs of the functions f and g and find the area of the region enclosed by these graphs and the vertical lines x = a and x = b.

17. $f(x) = x^2 + 3, g(x) = 1$; $a = 1, b = 3$

18. $f(x) = x + 2, g(x) = x^2 - 4$; $a = -1, b = 2$

19. $f(x) = -x^2 + 2x + 3, g(x) = -x + 3$; $a = 0, b = 2$

20. $f(x) = 9 - x^2, g(x) = 2x + 3$; $a = -1, b = 1$

21. $f(x) = x^2 + 1, g(x) = \frac{1}{3}x^3$; $a = -1, b = 2$

22. $f(x) = \sqrt{x}, g(x) = -\frac{1}{2}x - 1$; $a = 1, b = 4$

23. $f(x) = \dfrac{1}{x}, g(x) = 2x - 1$; $a = 1, b = 4$

24. $f(x) = x^2, g(x) = \dfrac{1}{x^2}$; $a = 1, b = 3$

25. $f(x) = e^x, g(x) = \dfrac{1}{x}$; $a = 1, b = 2$

26. $f(x) = x, g(x) = e^{2x}$; $a = 1, b = 3$

In exercises 27–34, sketch the graph and find the area of the region bounded by the graph of the function f and the lines y = 0, x = a, and x = b.

27. $f(x) = x$; $a = -1, b = 2$

28. $f(x) = x^2 - 2x$; $a = -1, b = 1$

29. $f(x) = -x^2 + 4x - 3$; $a = -1, b = 2$

30. $f(x) = x^3 - x^2$; $a = -1, b = 1$

31. $f(x) = x^3 - 4x^2 + 3x$; $a = 0, b = 2$

32. $f(x) = 4x^{1/3} + x^{4/3}$; $a = -1, b = 8$

33. $f(x) = e^x - 1$; $a = -1, b = 3$

34. $f(x) = xe^{x^2}$; $a = 0, b = 2$

In exercises 35–40, sketch the graph and find the area of the region completely enclosed by the graphs of the given functions f and g.

35. $f(x) = x + 2$ and $g(x) = x^2 - 4$

36. $f(x) = -x^2 + 4x$ and $g(x) = 2x - 3$

37. $f(x) = x^2$ and $g(x) = x^3$

38. $f(x) = x^3 - 6x^2 + 9x$ and $g(x) = x^2 - 3x$

39. $f(x) = \sqrt{x}$ and $g(x) = x^2$

40. $f(x) = 2x$ and $g(x) = x\sqrt{x + 1}$

41. **Effect of Advertising on Revenue**   In the accompanying figure, the function f gives the rate of change of Odyssey Travel's revenue with respect to the amount x it spends on advertising with their current advertising agency. With the services of a different advertising agency, it is expected that Odyssey's revenue will grow at the rate given by the function g. Give an interpretation of the area A of the region S, and find an expression for A in terms of a definite integral involving f and g.

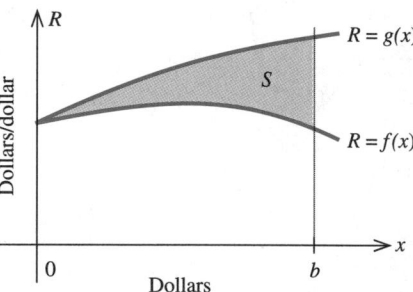

42. **Pulse Rate During Exercise**   In the accompanying figure, the function f gives the rate of increase of an individual's pulse rate when he walked a prescribed course on a treadmill 6 months ago. The function g gives the rate of increase of his pulse rate when he recently walked the same prescribed course. Give an interpretation of the area A of the region S, and find an expression for A in terms of a definite integral involving f and g.

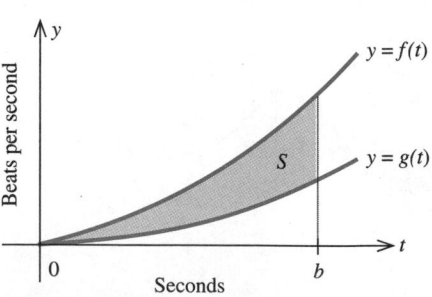

43. **Air Purification** In order to study the effectiveness of air purifiers in removing smoke, engineers run each purifier in a smoke-filled 10×20-foot room. In the accompanying figure, the function f gives the rate of change of the smoke level per minute, t minutes after the start of the test, when a brand A purifier is used. The function g gives the rate of change of the smoke level per minute when a brand B purifier is used. Give an interpretation of the area of the region S, and find an expression for the area of S in terms of a definite integral involving f and g.

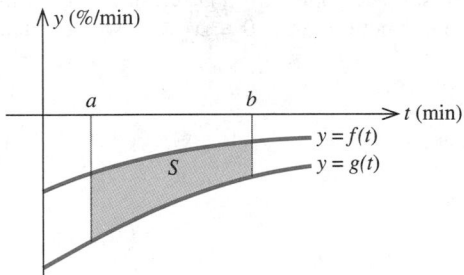

44. **Turbo-Charged Engine vs. Standard Engine** In tests conducted by *Auto Test Magazine* on two identical models of the Phoenix Elite—one equipped with a standard engine and the other with a turbo-charger—it was found that the acceleration of the former is given by

$$a = f(t) = 4 + 0.8t \qquad (0 \le t \le 12)$$

feet per second per second, t seconds after starting from rest at full throttle, whereas the acceleration of the latter is given by

$$a = g(t) = 4 + 1.2t + 0.03t^2 \qquad (0 \le t \le 12)$$

feet per second per second. How much faster is the turbo-charged model moving than the model with the standard engine, at the end of a 10-second test run at full throttle?

45. **Alternative Energy Sources** Because of the increasingly important role played by coal as a viable alternative energy source, the production of coal has been growing at the rate of

$$3.5e^{0.05t}$$

billion metric tons per year t years from 1980 (which corresponds to $t = 0$). Had it not been for the energy crisis, the rate of production of coal since 1980 might have been only

$$3.5e^{0.01t}$$

billion metric tons per year t years from 1980. Determine how much additional coal will be produced between 1980 and the end of the century as an alternative energy source.

46. **Effect of TV Advertising on Car Sales** Carl Williams, the new proprietor of Carl Williams Auto Sales, estimates that with extensive television advertising, car sales over the next several years could be increasing at the rate of

$$5e^{0.3t}$$

thousand cars per year t years from now, instead of at the current rate of

$$(5 + 0.5t^{3/2})$$

thousand cars per year t years from now. Find how many more cars Mr. Williams expects to sell over the next five years by implementing his advertising plans.

47. **Population Growth** In an endeavor to curb population growth in a Southeast Asian island state, the government has decided to launch an extensive propaganda campaign. Without curbs, the government expects the rate of population growth to have been

$$60e^{0.02t}$$

thousand people per year t years from now, over the next five years. However, successful implementation of the proposed campaign is expected to result in a population growth rate of

$$-t^2 + 60$$

thousand people per year t years from now, over the next five years. Assuming that the campaign is mounted, how many fewer people will there be in that country five years from now than there would have been if no curbs had been imposed?

48. Show that the area of a region R bounded above by the graph of a function f and below by the graph of a function g from $x = a$ to $x = b$ is given by

$$\int_a^b [f(x) - g(x)]\, dx$$

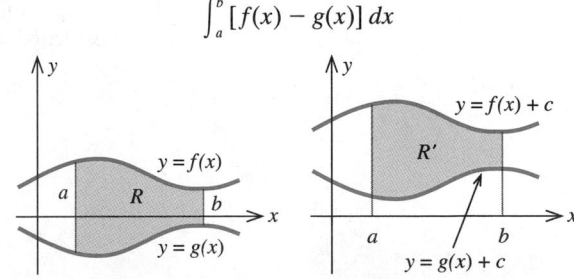

Hint: The validity of the formula was verified earlier for the case when both f and g were nonnegative. Now let f and g be two functions such that $f(x) \geq g(x)$ for $a \leq x \leq b$. Then there exists some nonnegative constant c such that the curves $y = f(x) + c$ and $y = g(x) + c$ are translated in the y-direction in such a way that the region R' has the same area as the region R (see the figures). Show that the area of R' is given by

$$\int_a^b \{[f(x) + c] - [g(x) + c]\}\, dx = \int_a^b [f(x) - g(x)]\, dx$$

SOLUTIONS TO SELF-CHECK EXERCISES 14.6

1. The region in question is shown in the accompanying figure. Since the graph of the function f lies above that of the function g for $0 \leq x \leq 1$, we see that the

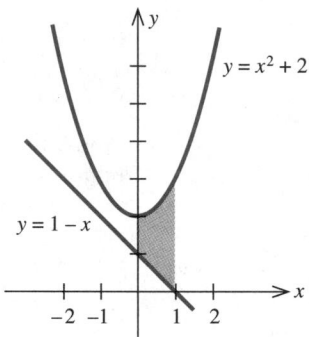

required area is given by

$$\int_0^1 [(x^2 + 2) - (1 - x)]\, dx = \int_0^1 (x^2 + x + 1)\, dx$$

$$= \frac{1}{3}x^3 + \frac{1}{2}x^2 + x \Big|_0^1$$

$$= \frac{1}{3} + \frac{1}{2} + 1$$

$$= \frac{11}{6}$$

or $\frac{11}{6}$ square units.

2. The region in question is shown in the accompanying figure. To find the points of intersection of the two curves, we solve the equation

$$-x^2 + 6x + 5 = x^2 + 5$$

$$2x^2 - 6x = 0$$

$$2x(x - 3) = 0$$

giving $x = 0$ or $x = 3$. Therefore, the points of intersection are $(0, 5)$ and $(3, 14)$.

Since the graph of f always lies above that of g for $0 \le x \le 3$, we see that the

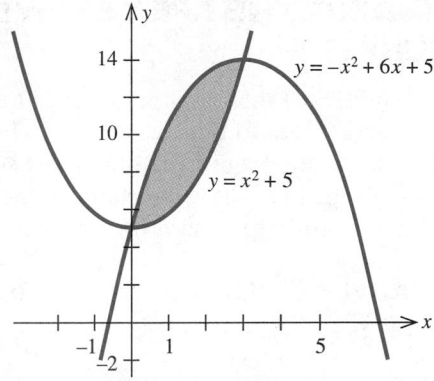

$y = -x^2 + 6x + 5$

$y = x^2 + 5$

required area is given by

$$\int_0^3 [(-x^2 + 6x + 5) - (x^2 + 5)]\, dx = \int_0^3 (-2x^2 + 6x)\, dx$$

$$= -\frac{2}{3}x^3 + 3x^2 \bigg|_0^3$$

$$= -18 + 27$$

$$= 9$$

or 9 square units.

3. The additional profits realizable over the next ten years are given by

$$\int_0^{10} [(t - 2\sqrt{t} + 4) - (1 + t^{2/3})]\, dt$$

$$= \int_0^{10} (t - 2t^{1/2} + 3 - t^{2/3})\, dt$$

$$= \frac{1}{2}t^2 - \frac{4}{3}t^{3/2} + 3t - \frac{3}{5}t^{5/3} \bigg|_0^{10}$$

$$= \frac{1}{2}(10)^2 - \frac{4}{3}(10)^{3/2} + 3(10) - \frac{3}{5}(10)^{5/3}$$

$$\approx 9.99$$

or approximately $10 million.

USING TECHNOLOGY

FINDING THE AREA BETWEEN TWO CURVES

The numerical integral operation can also be used to find the area between two curves. We do this by using the numerical integral operation to evaluate an appropriate definite integral or the sum (difference) of appropriate definite integrals. In the following example, the intersection operation is also used to advantage to help us find the limits of integration.

EXAMPLE 1   Use a graphing utility to find the area of the region R that is completely enclosed by the graphs of the functions

$$f(x) = 2x^3 - 8x^2 + 4x - 3 \quad \text{and} \quad g(x) = 3x^2 + 10x - 11$$

Solution   The graphs of f and g in the viewing rectangle $[-3, 4] \times [-20, 5]$ are shown in Figure T1. Using the intersection operation of a graphing utility,

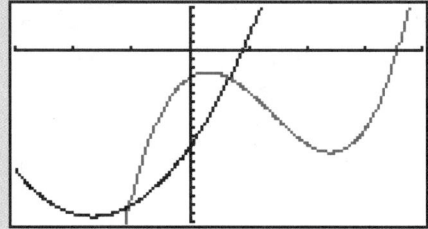

Figure T1
The region R is completely enclosed by the graphs of f and g.

we find the x-coordinates of the points of intersection of the two graphs to be approximately -1.04 and 0.65, respectively. Since the graph of f lies above that of g on the interval $[-1.04, 0.65]$, we see that the area of R is given by

$$A = \int_{-1.04}^{0.65} [(2x^3 - 8x^2 + 4x - 3) - (3x^2 + 10x - 11)] \, dx$$

$$= \int_{-1.04}^{0.65} (2x^3 - 11x^2 - 6x + 8) \, dx$$

Using the numerical integration function of a graphing utility, we find $A \approx 9.87$, and so the area of R is approximately 9.87 square units.   ◦ ◦ ◦

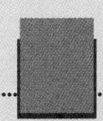

EXERCISES

In exercises 1–6, use a graphing utility to (a) plot the graphs of the functions f and g and (b) find the area of the region enclosed by these graphs and the vertical lines x = a and x = b. Express your answers accurate to four decimal places.

1. $f(x) = x^3(x - 5)^4$, $g(x) = 0$; $a = 1$, $b = 3$

2. $f(x) = x - \sqrt{1 - x^2}$, $g(x) = 0$; $a = -\frac{1}{2}$, $b = \frac{1}{2}$

3. $f(x) = x^{1/3}(x + 1)^{1/2}$, $g(x) = x^{-1}$; $a = 1.2$, $b = 2$

4. $f(x) = 2$, $g(x) = \ln(1 + x^2)$; $a = -1$, $b = 1$

5. $f(x) = \sqrt{x}$, $g(x) = \dfrac{x^2 - 3}{x^2 + 1}$; $a = 0$, $b = 3$

6. $f(x) = \dfrac{4}{x^2 + 1}$, $g(x) = x^4$; $a = -1$, $b = 1$

In exercises 7–12, use a graphing utility to (a) plot the graphs of the functions f and g and (b) find the area of the region totally enclosed by the graphs of these functions.

7. $f(x) = 2x^3 - 8x^2 + 4x - 3$ and $g(x) = -3x^2 + 10x - 10$

8. $f(x) = x^4 - 2x^2 + 2$ and $g(x) = 4 - 2x^2$

9. $f(x) = 2x^3 - 3x^2 + x + 5$ and $g(x) = e^{2x} - 3$

10. $f(x) = \frac{1}{2}x^2 - 3$ and $g(x) = \ln x$

11. $f(x) = xe^{-x}$ and $g(x) = x - 2\sqrt{x}$

12. $f(x) = e^{-x^2}$ and $g(x) = x^4$

14.7 APPLICATIONS OF THE DEFINITE INTEGRAL TO BUSINESS AND ECONOMICS

In this section, we consider several applications of the definite integral in the fields of business and economics.

Consumers' and Producers' Surplus

Figure 14.38
$D(x)$ is a demand function.

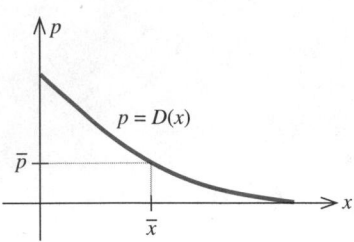

We begin by deriving a formula for computing the consumers' surplus. Suppose $p = D(x)$ is the demand function that relates the unit price p of a commodity to the quantity x demanded of it. Furthermore, suppose that a fixed unit market price has been established for the commodity and that corresponding to this unit price, the quantity demanded is $\bar{x}$ units (Figure 14.38). Then those consumers who would be willing to pay a unit price higher than $\bar{p}$ for the commodity would, in effect, experience a savings. This difference between what the consumers *would* be willing to pay for $\bar{x}$ units of the commodity and what they *actually* pay for them is called the **consumers' surplus.**

To derive a formula for computing the consumers' surplus, divide the interval $[0, \bar{x}]$ into n subintervals, each of length $\Delta x = \bar{x}/n$, and denote the right endpoints of these subintervals by $x_1, x_2, \ldots, x_n = \bar{x}$ (Figure 14.39).

We observe in Figure 14.39 that there are consumers who would pay a unit price of at least $D(x_1)$ dollars for the first Δx units of the commodity instead of the market price of $\bar{p}$ dollars per unit. The savings to these consumers is approximated by

$$D(x_1)\, \Delta x - \bar{p}\, \Delta x = [D(x_1) - \bar{p}]\, \Delta x$$

Figure 14.39
Approximating consumers' surplus by the sum of the rectangles r_1, $r_2, \ldots, r_n$.

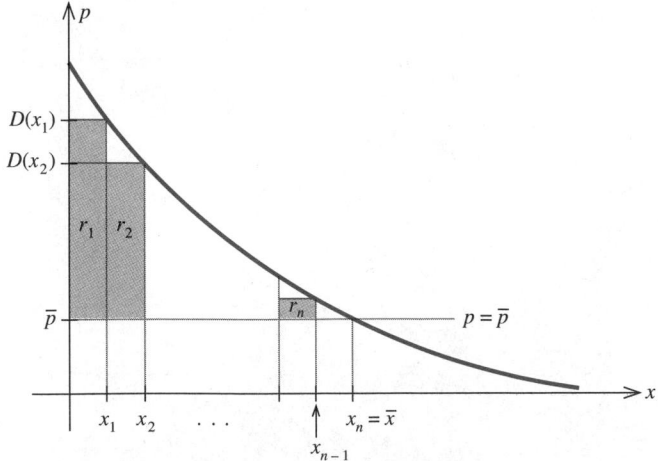

which is the area of the rectangle r_1. Pursuing the same line of reasoning, we find that the savings to the consumers who would be willing to pay a unit price of at least $D(x_2)$ dollars for the next Δx units (from x_1 through x_2) of the commodity, instead of the market price of $\overline{p}$ dollars per unit, is approximated by

$$D(x_2)\, \Delta x - \overline{p}\, \Delta x = [D(x_2) - \overline{p}]\, \Delta x$$

Continuing, we approximate the total savings to the consumers in purchasing $\overline{x}$ units of the commodity by the sum

$$[D(x_1) - \overline{p}]\, \Delta x + [D(x_2) - \overline{p}]\, \Delta x + \cdots + [D(x_n) - \overline{p}]\, \Delta x$$
$$= [D(x_1) + D(x_2) + \cdots + D(x_n)]\, \Delta x - \underbrace{[\overline{p}\, \Delta x + \overline{p}\, \Delta x + \cdots + \overline{p}\, \Delta x]}_{n \text{ terms}}$$
$$= [D(x_1) + D(x_2) + \cdots + D(x_n)]\, \Delta x - n\overline{p}\, \Delta x$$
$$= [D(x_1) + D(x_2) + \cdots + D(x_n)]\, \Delta x - \overline{p}\,\overline{x}$$

Now, the first term in the last expression is the Riemann sum of the demand function $p = D(x)$ over the interval $[0, \overline{x}]$ with representative points x_1, $x_2, \ldots, x_n$. Letting n approach infinity, we obtain the following formula for the consumers' surplus CS.

CONSUMERS' SURPLUS

Thus, the **consumers' surplus** is given by

$$CS = \int_0^{\overline{x}} D(x)\, dx - \overline{p}\,\overline{x} \tag{15}$$

where D is the demand function, $\overline{p}$ is the unit market price, and $\overline{x}$ is the quantity sold.

Figure 14.40
Consumers' surplus.

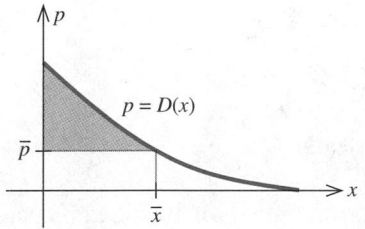

The consumers' surplus is given by the area of the region bounded above by the demand curve $p = D(x)$ and below by the straight line $p = \overline{p}$ from $x = 0$ to $x = \overline{x}$ (Figure 14.40). We can also see this if we rewrite equation (15) in the form

$$\int_0^{\overline{x}} [D(x) - \overline{p}]\, dx$$

and interpret the result geometrically.

Analogously, we can derive a formula for computing the producers' surplus. Suppose that $p = S(x)$ is the supply equation that relates the unit price p of a certain commodity to the quantity x that the supplier will make available in the market at that price.

Again, suppose that a fixed market price $\overline{p}$ has been established for the commodity and that, corresponding to this unit price, a quantity of $\overline{x}$ units will be made available in the market by the supplier (Figure 14.41). Then suppliers who would be willing to make the commodity available at a lower price stand to gain from the fact that the market price is set as such. The difference between what the suppliers actually receive and what they would be willing to receive is called the **producers' surplus.** Proceeding in a manner similar to the derivation of the equation for computing the consumers' surplus, we find that the producers' surplus *PS* is defined as follows:

THE PRODUCERS' SURPLUS

The **producers' surplus** is given by

$$PS = \bar{p}\bar{x} - \int_0^{\bar{x}} S(x)\, dx \qquad (16)$$

where $S(x)$ is the supply function, $\bar{p}$ is the unit market price, and $\bar{x}$ is the quantity supplied.

Geometrically, the producers' surplus is given by the area of the region bounded above by the straight line $p = \bar{p}$ and below by the supply curve $p = S(x)$ from $x = 0$ to $x = \bar{x}$ (Figure 14.42).

Figure 14.41
$S(x)$ is a supply function.

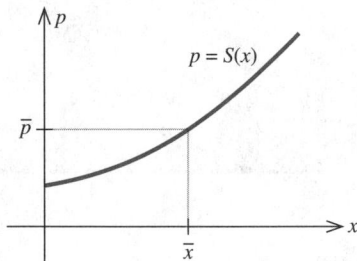

Figure 14.42
Producers' surplus.

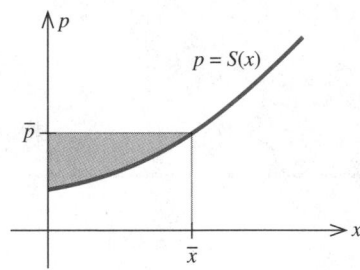

We can also show that the last statement is true by converting equation (16) to the form

$$\int_0^{\bar{x}} [\bar{p} - S(x)]\, dx$$

and interpreting the definite integral geometrically.

EXAMPLE 1   The demand function for a certain make of 10-speed bicycle is given by

$$p = D(x) = -0.001x^2 + 250$$

Figure 14.43
We wish to find consumers' surplus and producers' surplus when market price = equilibrium price.

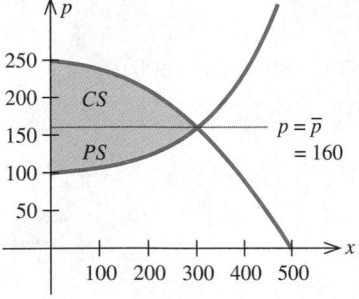

where p is the unit price in dollars and x is the quantity demanded in units of a thousand. The supply function for these bicycles is given by

$$p = S(x) = 0.0006x^2 + 0.02x + 100$$

where p stands for the unit price in dollars and x stands for the number of bicycles that the supplier will put on the market, in units of a thousand. Determine the consumers' surplus and the producers' surplus if the market price of a bicycle is set at the equilibrium price.

Solution   Recall that the equilibrium price is the unit price of the commodity when market equilibrium occurs. We determine the equilibrium price by solving for the point of intersection of the demand curve and the supply curve (Figure 14.43). To solve the system of equations

$$p = -0.001x^2 + 250$$
$$p = 0.0006x^2 + 0.02x + 100$$

we simply substitute the first equation into the second, obtaining

$$0.0006x^2 + 0.02x + 100 = -0.001x^2 + 250$$
$$0.0016x^2 + 0.02x - 150 = 0$$
$$16x^2 + 200x - 1,500,000 = 0$$
$$2x^2 + 25x - 187,500 = 0$$

Factoring this last equation, we obtain

$$(2x + 625)(x - 300) = 0$$

Thus, $x = -625/2$ or $x = 300$. The first number lies outside the interval of interest, so we are left with the solution $x = 300$, with a corresponding value of

$$p = -0.001(300)^2 + 250 = 160$$

Thus, the equilibrium point is $(300, 160)$; that is, the equilibrium quantity is 300,000 and the equilibrium price is $160. Setting the market price at $160 per unit and using formula (15) with $\bar{p} = 160$ and $\bar{x} = 300$, we find that the consumers' surplus is given by

$$CS = \int_0^{300} (-0.001x^2 + 250)\, dx - (160)(300)$$

$$= \left(-\frac{1}{3,000}x^3 + 250x \right)\Bigg|_0^{300} - 48,000$$

$$= -\frac{300^3}{3,000} + (250)(300) - 48,000$$

$$= 18,000$$

or $18,000,000. (Recall that x is measured in units of a thousand.)

Next, using formula (16), we find that the producers' surplus is given by

$$PS = (160)(300) - \int_0^{300} (0.0006x^2 + 0.02x + 100)\, dx$$

$$= 48{,}000 - (0.0002x^3 + 0.01x^2 + 100x)\Big|_0^{300}$$

$$= 48{,}000 - [(0.0002)(300)^3 + (0.01)(300)^2 + 100(300)]$$

$$= 11{,}700$$

or $11,700,000.

The Future and Present Value of an Income Stream

To introduce the notion of the future value and the present value of an income stream, suppose that a firm generates a stream of income over a period of time—the revenue generated by a large chain of retail stores over a 5-year period, for example. As the income is realized, it is reinvested and earns interest at a fixed rate. The *accumulated future income stream* over the 5-year period is the amount of money the firm ends up with at the end of that period.

The definite integral can be used to determine this **accumulated,** or **total, future income stream** over a period of time. The total future value of an income stream gives us a way to measure the value of such a stream. To find the total future value of an income stream, suppose that

$R(t) =$ Rate of income generation at any time t (Dollars per year)

$r =$ Interest rate compounded continuously

$T =$ Term   (Years)

Figure 14.44
The time interval $[0, T]$ is partitioned into n subintervals.

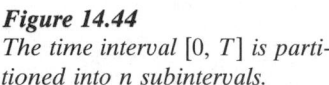
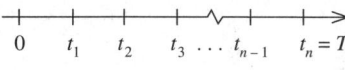

Let us divide the time interval $[0, T]$ into n subintervals of equal length $\Delta t = T/n$ and denote the right endpoints of these intervals by $t_1, t_2, \ldots, t_n = T$, as shown in Figure 14.44.

If R is a continuous function on $[0, T]$, then $R(t)$ will not differ by much from $R(t_1)$ in the subinterval $[0, t_1]$ provided that the subinterval is small (which is true if n is large). Therefore, the income generated over the time interval $[0, t_1]$ is approximately

$$R(t_1)\, \Delta t   \text{(Constant rate of income} \cdot \text{Length of time)}$$

dollars. The future value of this amount, T years from now, calculated as if it were earned at time t_1, is

$$[R(t_1)\, \Delta t] e^{r(T - t_1)}   \text{[Equation (3), Section 13.1]}$$

dollars. Similarly, the income generated over the time interval $[t_1, t_2]$ is approximately $P(t_2)\, \Delta t$ dollars and has a future value, T years from now, of approximately

$$[R(t_2)\, \Delta t] e^{r(T - t_2)}$$

dollars. Therefore, the sum of the future values of the income stream generated over the time interval $[0, T]$ is approximately

$$R(t_1)e^{r(T-t_1)}\,\Delta t + R(t_2)e^{r(T-t_2)}\,\Delta t + \cdots + R(t_n)e^{r(T-t_n)}\,\Delta t$$
$$= e^{rT}[R(t_1)e^{-rt_1}\,\Delta t + R(t_2)e^{-rt_2}\,\Delta t + \cdots + R(t_n)e^{-rt_n}\,\Delta t]$$

dollars. But this sum is just the Riemann sum of the function $e^{rT}R(t)e^{-rt}$ over the interval $[0, T]$ with representative points $t_1, t_2, \ldots, t_n$. Letting n approach infinity, we obtain the following result.

ACCUMULATED OR TOTAL FUTURE VALUE OF AN INCOME STREAM	The **accumulated,** or **total, future value** after T years of an income stream of $R(t)$ dollars per year, earning interest at the rate of r per year compounded continuously, is given by $$A = e^{rT}\int_0^T R(t)e^{-rt}\,dt \qquad \textbf{(17)}$$

EXAMPLE 2 Crystal Car Wash recently bought an automatic car-washing machine that is expected to generate \$40,000 in revenue per year, t years from now, for the next five years. If the income is reinvested in a business earning interest at the rate of 12% per year compounded continuously, find the total accumulated value of this income stream at the end of five years.

Solution We are required to find the total future value of the given income stream after five years. Using equation (17) with $R(t) = 40,000$, $r = 0.12$, and $T = 5$, we see that the required value is given by

$$e^{0.12(5)}\int_0^5 40{,}000e^{-0.12t}\,dt$$

$$= e^{0.6}\left[-\frac{40{,}000}{0.12}e^{-0.12t}\right]\Bigg|_0^5 \qquad \text{(Integrate using the substitution } u = -0.12t.)$$

$$= -\frac{40{,}000e^{0.6}}{0.12}(e^{-0.6} - 1) \approx 274{,}039.60$$

or approximately \$274,040. ◦ ◦ ◦

Another way of measuring the value of an income stream is by considering its present value. The **present value** of an income stream of $R(t)$ dollars per year over a term of T years, earning interest at the rate of r per year compounded continuously, is the principal P that will yield the same accumulated value as the income stream itself when P is invested today for a period of T years at the same rate of interest. In other words,

$$Pe^{rT} = e^{rT}\int_0^T R(t)e^{-rt}\,dt$$

Dividing both sides of the equation by e^{rT} gives the following result:

PRESENT VALUE OF AN INCOME STREAM

The **present value of an income stream** of $R(t)$ dollars per year, earning interest at the rate of r per year compounded continuously, is given by

$$PV = \int_0^T R(t)e^{-rt}\, dt \qquad \qquad \textbf{(18)}$$

EXAMPLE 3   The owner of a local cinema is considering two alternative plans for renovating and improving the theater. Plan A calls for an immediate cash outlay of $250,000, whereas plan B requires an immediate cash outlay of $180,000. It has been estimated that adopting plan A would result in a net income stream generated at the rate of

$$f(t) = 630,000$$

dollars per year, whereas adopting plan B would result in a net income stream generated at the rate of

$$g(t) = 580,000$$

dollars per year for the next three years. If the prevailing interest rate for the next five years were 10% per year, which plan would generate a higher net income by the end of three years?

Solution   Since the initial outlay is $250,000, we find, using equation (18) with $R(t) = 630,000$, $r = 0.1$, and $T = 3$, that the present value of the net income under plan A is given by

$$\int_0^3 630,000e^{-0.1t}\, dt - 250,000$$

$$= \frac{630,000}{-0.1}e^{-0.1t}\Big|_0^3 - 250,000 \qquad \text{(Integrate using the substitution } u = -0.1t.)$$

$$= -6,300,000e^{-0.3} + 6,300,000 - 250,000$$

$$\approx 1,382,845$$

or approximately $1,382,845.

To find the present value of the net income under plan B, we use equation (18) with $R(t) = 580,000$, $r = 0.1$, and $T = 3$, obtaining

$$\int_0^3 580,000e^{-0.1t}\, dt - 180,000$$

dollars. Proceeding as in the previous computation, we see that the required value is $1,323,254 (exercise 8).

Comparing the present value of each plan, we conclude that plan A would generate a higher net income by the end of three years.   ◦ ◦ ◦

REMARK   The function R in Example 3 is a constant function. If R is not a constant function, then we may need more sophisticated techniques of integration to evaluate the integral in equation (18). Exercise sets 15.1 and 15.3 in Chapter 15 contain problems of this type.   ◦ ◦ ◦

The Amount and Present Value of an Annuity

An annuity is a sequence of payments made at regular time intervals. The time period in which these payments are made is called the *term* of the annuity. Although the payments need not be equal in size, they are equal in many important applications, and we will assume that they are equal in our discussion. Examples of annuities are regular deposits to a savings account, monthly home mortgage payments, and monthly insurance payments.

The **amount of an annuity** is the sum of the payments plus the interest earned. A formula for computing the amount of an annuity A can be derived with the help of equation (17). Let

$$P = \text{Size of each payment in the annuity}$$
$$r = \text{Interest rate compounded continuously}$$
$$T = \text{Term of the annuity (in years)}$$
$$m = \text{Number of payments per year}$$

The payments into the annuity constitute a constant income stream of $R(t) = mP$ dollars per year. With this value of $R(t)$, equation (17) yields

$$A = e^{rT} \int_0^T R(t)e^{-rt}\, dt = e^{rT} \int_0^T mPe^{-rt}\, dt$$

$$= mPe^{rT} \left[-\frac{e^{-rt}}{r} \right]\Bigg|_0^T$$

$$= mPe^{rT} \left[-\frac{e^{-rT}}{r} + \frac{1}{r} \right] = \frac{mP}{r}(e^{rT} - 1)$$

This leads us to the following formula:

AMOUNT OF AN ANNUITY

The **amount of an annuity** is

$$A = \frac{mP}{r}(e^{rT} - 1) \tag{19}$$

where P, r, T, and m are as defined earlier.

 EXAMPLE 4   On January 1, 1985, Mr. Chapman deposited $2000 into an Individual Retirement Account (IRA) paying interest at the rate of 10% per year compounded continuously. Assuming that he deposits $2000 annually into the account, how much will he have in his IRA at the beginning of the year 2001?

Solution   We use equation (19), with $P = 2000$, $r = 0.1$, $T = 16$, and $m = 1$, obtaining

$$A = \frac{2000}{0.1}(e^{1.6} - 1)$$

$$\approx 79{,}060.65$$

Thus, Mr. Chapman will have approximately $79,061 in his account at the beginning of the year 2001.   ◦ ◦ ◦

EXPLORING WITH TECHNOLOGY

Refer to Example 4. Suppose Mr. Chapman wishes to know how much he will have in his IRA at any time in the future, not just at the beginning of 2001, as you were asked to compute in the example.

1. Using formula (17) and the relevant data from Example 4, show that the required amount at any time x (x measured in years, $x > 0$) is given by

$$A = f(x) = 20{,}000(e^{0.1x} - 1)$$

2. Use a graphing utility to plot the graph of f in the viewing rectangle $[0, 30] \times [2000, 400{,}000]$.

3. Using **ZOOM** and **TRACE**, or using the function evaluation capability of your graphing utility, use the result of part (2) to verify the result obtained in Example 4. Comment on the advantage of the mathematical model found in part (1).

◦   ◦   ◦

Using equation (18), we can derive the following formula for the *present value* of an annuity.

PRESENT VALUE OF AN ANNUITY

The **present value of an annuity** is given by

$$PV = \frac{mP}{r}(1 - e^{-rT}) \tag{20}$$

where P, r, T, and m are as defined earlier.

   EXAMPLE 5   Mr. Carson, the proprietor of a hardware store, wants to establish a fund from which he will withdraw $1000 per month for the next ten years. If the fund earns interest at the rate of 9% per year compounded continuously, how much money does he need to establish the fund?

Solution   We want to find the present value of an annuity with $P = 1000$, $r = 0.09$, $T = 10$, and $m = 12$. Using equation (20), we find

$$PV = \frac{12,000}{0.09}(1 - e^{-(0.09)(10)})$$

$$\approx 79,124.04$$

Thus, Mr. Carson needs approximately $79,124 to establish the fund.   ○ ○ ○

Lorentz Curves and Income Distributions

One method used by economists to study the distribution of income in a society is based on the **Lorentz curve,** named after American statistician M. D. Lorentz. To describe the Lorentz curve, let $f(x)$ denote the proportion of the total income received by the poorest $100x\%$ of the population for $0 \leq x \leq 1$. With this terminology, $f(0.3) = 0.1$ simply states that the lowest 30% of the income recipients receive 10% of the total income.

The function f has the following properties:

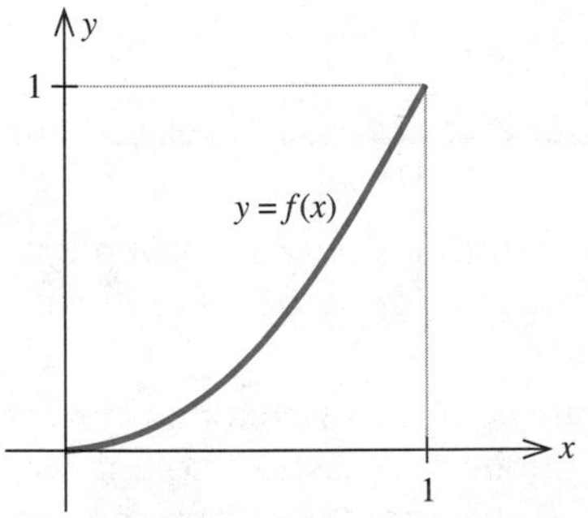

1. The domain of f is $[0, 1]$.

2. The range of f is $[0, 1]$.

3. $f(0) = 0$ and $f(1) = 1$.

4. $f(x) \leq x$ for every x in $[0, 1]$.

5. f is increasing on $[0, 1]$.

The first two properties follow from the fact that both x and $f(x)$ are fractions of a whole. Property 3 is a statement that 0% of the income recipients receive 0% of the total income and 100% of the income recipients receive 100% of the total income. Property 4 follows from the fact that the lowest $100x\%$ of the income recipients cannot receive more than $100x\%$ of the total income. A typical Lorentz curve is shown in Figure 14.45.

Figure 14.45
A Lorentz curve.

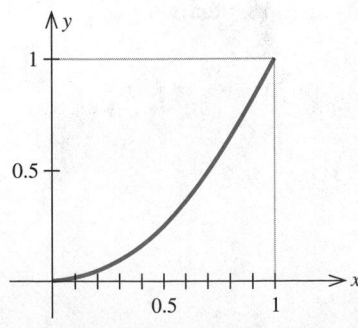

EXAMPLE 6   A developing country's income distribution is described by the function

$$f(x) = \frac{19}{20}x^2 + \frac{1}{20}x$$

a. Sketch the Lorentz curve for the given function.

b. Compute $f(0.2)$ and $f(0.8)$ and interpret your results.

Solution

a. The Lorentz curve is shown in Figure 14.46.

b.
$$f(0.2) = \frac{19}{20}(0.2)^2 + \frac{1}{20}(0.2) = 0.048$$

so that the lowest 20% of the people receive 4.8% of the total income.

$$f(0.8) = \frac{19}{20}(0.8)^2 + \frac{1}{20}(0.8) = 0.648$$

so that the lowest 80% of the people receive 64.8% of the total income.

Figure 14.46
The Lorentz curve $f(x) = \frac{19}{20}x^2 + \frac{1}{20}x$.

○ ○ ○

Figure 14.47

The closer the Lorentz curve is to the line, the more equitable the income distribution.

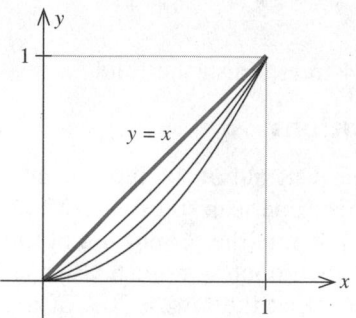

Next, let us consider the Lorentz curve described by the function $y = f(x) = x$. Since exactly $100x\%$ of the total income is received by the lowest $100x\%$ of income recipients, the line $y = x$ is called the **line of complete equality.** For example, 10% of the total income is received by the lowest 10% of income recipients, 20% of the total income is received by the lowest 20% of income recipients, and so on. Now, it is evident that the closer a Lorentz curve is to this line, the more equitable the income distribution is among the income recipients. But the proximity of a Lorentz curve to the line of complete equality is reflected by the area between the Lorentz curve and the line $y = x$ (Figure 14.47). The closer the curve is to the line, the smaller the enclosed area.

This observation suggests that we may define a number, called the **coefficient of inequality,** or **Gini Index,** of a Lorentz curve, as the ratio of the area between the line of complete equality and the Lorentz curve to the area under the line of complete equality. Since the area under the line of complete equality is $\frac{1}{2}$, we see that the coefficient of inequality is given by the following formula.

COEFFICIENT OF INEQUALITY OF A LORENTZ CURVE

The **coefficient of inequality** of a Lorentz curve is

$$L = 2 \int_0^1 [x - f(x)] \, dx \qquad (21)$$

The coefficient of inequality is a number between 0 and 1. For example, a coefficient of zero implies that the income distribution is perfectly uniform.

EXAMPLE 7   In a study conducted by a certain country's Economic Development Board with regard to the income distribution of certain segments of the country's work force, it was found that the Lorentz curves for the distribution of income of medical doctors and of movie actors and actresses are described by the functions

$$f(x) = \frac{14}{15}x^2 + \frac{1}{15}x \quad \text{and} \quad g(x) = \frac{5}{8}x^4 + \frac{3}{8}x$$

respectively. Compute the coefficient of inequality for each Lorentz curve. Which profession has a more equitable income distribution?

Solution   The required coefficients of inequality are, respectively,

$$L_1 = 2 \int_0^1 \left[x - \left(\frac{14}{15}x^2 + \frac{1}{15}x \right) \right] dx = 2 \int_0^1 \left(\frac{14}{15}x - \frac{14}{15}x^2 \right) dx$$

$$= \frac{28}{15} \int_0^1 (x - x^2) \, dx = \frac{28}{15} \left(\frac{1}{2}x^2 - \frac{1}{3}x^3 \right) \Big|_0^1$$

$$= \frac{14}{45} \approx 0.311$$

and
$$L_2 = 2\int_0^1 \left[x - \left(\frac{5}{8}x^4 + \frac{3}{8}x \right) \right] dx = 2\int_0^1 \left(\frac{5}{8}x - \frac{5}{8}x^4 \right) dx$$

$$= \frac{5}{4}\int_0^1 (x - x^4)\, dx = \frac{5}{4}\left(\frac{1}{2}x^2 - \frac{1}{5}x^5 \right)\Big|_0^1$$

$$= \frac{15}{40} \approx 0.375$$

We conclude that in this country the incomes of medical doctors are more evenly distributed than the incomes of movie actors and actresses.   ○ ○ ○

SELF–CHECK EXERCISE 14.7

The demand function for a certain make of exercise bicycle that is sold exclusively through cable television is

$$p = d(x) = \sqrt{9 - 0.02x}$$

where p is the unit price in hundreds of dollars and x is the quantity demanded per week. The corresponding supply function is given by

$$p = s(x) = \sqrt{1 + 0.02x}$$

where p has the same meaning as before and x is the number of exercise bicycles the supplier will make available at price p. Determine the consumers' surplus and the producers' surplus if the unit price is set at the equilibrium price.

The solution to Self-Check Exercise 14.7 can be found on page 1019.

14.7 EXERCISES

A calculator is recommended for this exercise set.

1. Consumers' Surplus   The demand function for a certain make of cartridge typewriter ribbon is given by

$$p = -0.01x^2 - 0.1x + 6$$

where p is the unit price in dollars and x is the quantity demanded each week, measured in units of a thousand. Determine the consumers' surplus if the market price is set at $4 per cartridge.

2. Consumers' Surplus   The demand function for a certain brand of compact disc is given by

$$p = -0.01x^2 - 0.2x + 8$$

where p is the wholesale unit price in dollars and x is the quantity demanded each week, measured in units

of a thousand. Determine the consumers' surplus if the wholesale market price is set at $5 per disc.

3. Consumers' Surplus   It is known that the quantity demanded of a certain make of portable hair dryer is x hundred units per week, and the corresponding wholesale unit price is

$$p = \sqrt{225 - 5x}$$

dollars. Determine the consumers' surplus if the wholesale market price is set at $10 per unit.

4. Producers' Surplus   The supplier of the portable hair dryers in exercise 3 will make x hundred units of hair dryers available in the market when the wholesale unit price is

$$p = \sqrt{36 + 1.8x}$$

dollars. Determine the producers' surplus if the whole-sale market price is set at $9 per unit.

5. **Producers' Surplus**   The supply function for the compact discs of exercise 2 is given by

$$p = 0.01x^2 + 0.1x + 3$$

where p is the unit wholesale price in dollars and x stands for the quantity that will be made available in the market by the supplier, measured in units of a thousand. Determine the producers' surplus if the wholesale market price is set at the equilibrium price.

6. **Consumers' and Producers' Surplus**   The management of the Titan Tire Company has determined that the quantity demanded x of their Super Titan tires per week is related to the unit price p by

$$p = 144 - x^2$$

where p is measured in dollars and x is measured in units of a thousand. Titan will make x units of the tires available in the market if the unit price is

$$p = 48 + \frac{1}{2}x^2$$

dollars. Determine the consumers' surplus and the producers' surplus when the market unit price is set at the equilibrium price.

7. **Consumers' and Producers' Surplus**   The quantity demanded x (in units of a hundred) of the Mikado miniature cameras per week is related to the unit price p (in dollars) by

$$p = -0.2x^2 + 80$$

and the quantity x (in units of a hundred) that the supplier is willing to make available in the market is related to the unit price p (in dollars) by

$$p = 0.1x^2 + x + 40$$

If the market price is set at the equilibrium price, find the consumers' surplus and the producers' surplus.

8. Refer to Example 3, page 1012. Verify that

$$\int_0^3 580,000e^{-0.1t}\, dt - 180,000 \approx 1,323,254$$

9. **Present Value of an Investment**   Suppose that an investment is expected to generate income at the rate of

$$R(t) = 200,000$$

dollars per year for the next five years. Find the present value of this investment if the prevailing interest rate is 8% per year compounded continuously.

10. **Franchises**   Ms. Gordon purchased a 15-year franchise for a computer outlet store that is expected to generate income at the rate of

$$R(t) = 400,000$$

dollars per year. If the prevailing interest rate is 10% per year compounded continuously, find the present value of the franchise.

11. **The Amount of an Annuity**   Find the amount of an annuity if $250 a month is paid into it for a period of 20 years earning interest at the rate of 8% per year compounded continuously.

12. **The Amount of an Annuity**   Find the amount of an annuity if $400 a month is paid into it for a period of 20 years earning interest at the rate of 8% per year compounded continuously.

13. **The Amount of an Annuity**   Mr. Jorgenson deposits $150 per month in a savings account paying 8% per year compounded continuously. Estimate the amount that will be in his account after 15 years.

14. **Custodial Accounts**   The Armstrongs wish to establish a custodial account to finance their children's education. If they deposit $200 monthly for ten years in a savings account paying 9% per year compounded continuously, find how much their savings account will be worth at the end of this period.

15. **IRA Accounts**   Refer to Example 4, page 1013. Suppose that Mr. Chapman makes his IRA payment on April 1, 1985, and annually thereafter. If interest is paid at the same initial rate, approximately how much will Mr. Chapman have in his account at the beginning of the year 2001?

16. **Present Value of an Annuity**   Estimate the present value of an annuity if payments are $800 monthly for 12 years and the account earns interest at the rate of 10% per year compounded continuously.

17. **Present Value of an Annuity**   Estimate the present value of an annuity if payments are $1200 monthly for 15 years and the account earns interest at the rate of 10% per year compounded continuously.

18. **Lottery Payments**   A state lottery commission pays the winner of the "Million Dollar" lottery 20 annual

installments of \$50,000 each. If the prevailing interest rate is 8% per year compounded continuously, find the present value of the winning ticket.

19. **Reverse Annuity Mortgages**   Mr. Sinclair wishes to supplement his retirement income by \$300 per month for the next ten years. He plans to obtain a reverse annuity mortgage (RAM) on his home to meet this need. Estimate the amount of the mortgage he will require if the prevailing interest rate is 12% per year compounded continuously.

20. **Reverse Annuity Mortgage**   Refer to exercise 19. Miss Santos wishes to supplement her retirement income by \$400 per month for the next 15 years by obtaining a RAM. Estimate the amount of the mortgage she will require if the prevailing interest rate is 9% per year compounded continuously.

21. **Lorentz Curves**   A certain country's income distribution is described by the function

$$f(x) = \frac{15}{16}x^2 + \frac{1}{16}x$$

a. Sketch the Lorentz curve for this function.
b. Compute $f(0.4)$ and $f(0.9)$ and interpret your results.

22. **Lorentz Curves**   In a study conducted by a certain country's Economic Development Board, it was found that the Lorentz curve for the distribution of income of college teachers was described by the function

$$f(x) = \frac{13}{14}x^2 + \frac{1}{14}x$$

and that of lawyers by the function

$$g(x) = \frac{9}{11}x^4 + \frac{2}{11}x$$

a. Compute the coefficient of inequality for each Lorentz curve.
b. Which profession has a more equitable income distribution?

23. **Lorentz Curves**   A certain country's income distribution is described by the function

$$f(x) = \frac{14}{15}x^2 + \frac{1}{15}x$$

a. Sketch the Lorentz curve for this function.
b. Compute $f(0.3)$ and $f(0.7)$.

24. **Lorentz Curves**   In a study conducted by a certain country's Economic Development Board, it was found that the Lorentz curve for the distribution of income of stockbrokers was described by the function

$$f(x) = \frac{11}{12}x^2 + \frac{1}{12}x$$

and that of high school teachers by the function

$$g(x) = \frac{5}{6}x^2 + \frac{1}{6}x$$

a. Compute the coefficient of inequality for each Lorentz curve.
b. Which profession has a more equitable income distribution?

SOLUTION TO SELF-CHECK EXERCISE 14.7

We find the equilibrium price and the equilibrium quantity by solving the system of equations

$$p = \sqrt{9 - 0.02x}$$
$$p = \sqrt{1 + 0.02x}$$

simultaneously. Substituting the first equation into the second, we have

$$\sqrt{9 - 0.02x} = \sqrt{1 + 0.02x}$$

Squaring both sides of the equation then leads to

$$9 - 0.02x = 1 + 0.02x$$
$$x = 200$$

Using Technology

CONSUMERS' SURPLUS AND PRODUCERS' SURPLUS

1. Re-solve Example 1, Section 14.7, using a graphing utility.

   [*Hint:* Use the intersection operation to find the equilibrium quantity and the equilibrium price. Use the numerical integration operation to evaluate the definite integral.]

2. Re-solve exercise 7, Section 14.7, using a graphing utility.

   [*Hint:* See exercise 1.]

3. The demand function for a certain brand of travel alarm clocks is given by

   $$p = -0.01x^2 - 0.3x + 10$$

   where p is the wholesale unit price in dollars and x is the quantity demanded each month, measured in units of a thousand. The supply function for this brand of clocks is given by

   $$p = -0.01x^2 + 0.2x + 4$$

   where p has the same meaning as before and x is the quantity, in thousands, the supplier will make available in the marketplace per month. Determine the consumers' surplus and the producers' surplus when the market unit price is set at the equilibrium price.

4. The quantity demanded of a certain make of compact disc organizer is x thousand units per week, and the corresponding wholesale unit price is

   $$p = \sqrt{400 - 8x}$$

   dollars. The supplier of the organizers will make x thousand units available in the market when the unit wholesale price is

   $$p = 0.02x^2 + 0.04x + 5$$

   dollars. Determine the consumers' surplus and the producers' surplus when the market unit price is set at the equilibrium price.

Therefore,
$$p = \sqrt{9 - 0.02(200)}$$
$$= \sqrt{5} \approx 2.24$$

and the equilibrium price is \$224 and the equilibrium quantity is 200. The consumers' surplus is given by

$$CS = \int_0^{200} \sqrt{9 - 0.02x} \, dx - (2.24)(200)$$

$$= \int_0^{200} (9 - 0.02x)^{1/2} \, dx - 448 \qquad \text{(Integrating by substitution)}$$

$$= -\frac{1}{0.02} \left(\frac{2}{3}\right) (9 - 0.02x)^{3/2} \Big|_0^{200} - 448$$

$$= -\frac{1}{0.03} (5^{3/2} - 9^{3/2}) - 448$$

$$\approx 79.32$$

or approximately \$7932. Next, the producers' surplus is given by

$$PS = (2.24)(200) - \int_0^{200} \sqrt{1 + 0.02x} \, dx$$

$$= 448 - \int_0^{200} (1 + 0.02x)^{1/2} \, dx$$

$$= 448 - \frac{1}{0.02} \left(\frac{2}{3}\right) (1 + 0.02x)^{3/2} \Big|_0^{200}$$

$$= 448 - \frac{1}{0.03} (5^{3/2} - 1)$$

$$\approx 108.66$$

or approximately \$10,866.

CHAPTER 14 SUMMARY OF PRINCIPAL FORMULAS AND TERMS

Formulas

1. Indefinite integral of a constant $\int k \, du = ku + C$

2. Power Rule $\int u^n \, du = \dfrac{u^{n+1}}{n + 1} + C$

3. Constant multiple rule $\int kf(u) \, du = k \int f(u) \, du$
$(k, \text{ a constant})$

4. Sum Rule $\int [f(u) \pm g(u)] \, du$
$$= \int f(u) \, du \pm \int g(u) \, du$$

5. Indefinite integral of the exponential function

$$\int e^u \, du = e^u + C$$

6. Indefinite integral of $f(u) = \dfrac{1}{u}$

$$\int \frac{du}{u} = \ln |u| + C$$

7. Method of substitution

$$\int f'(g(x))g'(x) \, dx = \int f'(u) \, du$$

8. Fundamental Theorem of Calculus

$$\int_a^b f(x) \, dx$$
$$= F(b) - F(a), F'(x) = f(x)$$

9. Area between two curves

$$\int_a^b [f(x) - g(x)] \, dx, f(x) \ge g(x)$$

10. Definite integral as the limit of a sum

$$\int_a^b f(x) \, dx = \lim_{n \to \infty} S_n,$$
where S_n is a Riemann sum

11. Average value of f over $[a, b]$

$$\frac{1}{b - a} \int_a^b f(x) \, dx$$

12. Consumers' surplus

$$CS = \int_0^{\bar{x}} D(x) \, dx - \bar{p}\bar{x}$$

13. Producers' surplus

$$PS = \bar{p}\bar{x} - \int_0^{\bar{x}} S(x) \, dx$$

14. Accumulated (future) value of an income stream

$$A = e^{rT} \int_0^T R(t)e^{-rt} \, dt$$

15. Present value of an income stream

$$PV = \int_0^T R(t)e^{-rt} \, dt$$

16. Present value of an annuity

$$PV = \frac{mP}{r}(1 - e^{-rT})$$

17. Amount of an annuity

$$A = \frac{mP}{r}(e^{rT} - 1)$$

18. Coefficient of inequality of a Lorentz curve

$$L = 2 \int_0^1 [x - f(x)] \, dx$$

Terms

Antiderivative

Integration

Indefinite integral

Integrand

Constant of integration

Riemann sum

Definite integral

Limits of integration

Lorentz curve

CHAPTER 14 REVIEW EXERCISES

In exercises 1–20, evaluate each indefinite integral.

1. $\int (x^3 + 2x^2 - x)\, dx$

2. $\int (\frac{1}{3}x^3 - 2x^2 + 8)\, dx$

3. $\int \left(x^4 - 2x^3 + \frac{1}{x^2}\right) dx$

4. $\int (x^{1/3} - \sqrt{x} + 4)\, dx$

5. $\int x(2x^2 + x^{1/2})\, dx$

6. $\int (x^2 + 1)(\sqrt{x} - 1)\, dx$

7. $\int \left(x^2 - x + \frac{2}{x} + 5\right) dx$

8. $\int \sqrt{2x + 1}\, dx$

9. $\int (3x - 1)(3x^2 - 2x + 1)^{1/3}\, dx$

10. $\int x^2(x^3 + 2)^{10}\, dx$

11. $\int \frac{x - 1}{x^2 - 2x + 5}\, dx$

12. $\int 2e^{-2x}\, dx$

13. $\int (x + \frac{1}{2})e^{x^2 + x + 1}\, dx$

14. $\int \frac{e^{-x} - 1}{(e^{-x} + x)^2}\, dx$

15. $\int \frac{(\ln x)^5}{x}\, dx$

16. $\int \frac{\ln x^2}{x}\, dx$

17. $\int x^3(x^2 + 1)^{10}\, dx$

18. $\int x\sqrt{x + 1}\, dx$

19. $\int \frac{x}{\sqrt{x - 2}}\, dx$

20. $\int \frac{3x}{\sqrt{x + 1}}\, dx$

In exercises 21–32, evaluate each definite integral.

21. $\int_0^1 (2x^3 - 3x^2 + 1)\, dx$

22. $\int_0^2 (4x^3 - 9x^2 + 2x - 1)\, dx$

23. $\int_1^4 (\sqrt{x} + x^{-3/2})\, dx$

24. $\int_0^1 20x(2x^2 + 1)^4\, dx$

25. $\int_{-1}^0 12(x^2 - 2x)(x^3 - 3x^2 + 1)^3\, dx$

26. $\int_4^7 x\sqrt{x - 3}\, dx$

27. $\int_0^2 \frac{x}{x^2 + 1}\, dx$

28. $\int_0^1 \frac{dx}{(5 - 2x)^2}$

29. $\int_0^2 \frac{4x}{\sqrt{1 + 2x^2}}\, dx$

30. $\int_0^2 xe^{(-1/2)x^2}\, dx$

31. $\int_{-1}^0 \frac{e^{-x}}{(1 + e^{-x})^2}\, dx$

32. $\int_1^e \frac{\ln x}{x}\, dx$

In exercises 33–36, find the function f given that the slope of the tangent line to the graph at any point (x, f(x)) is f′(x) and that the graph of f passes through the given point.

33. $f'(x) = 3x^2 - 4x + 1;\ (1, 1)$

34. $f'(x) = \frac{x}{\sqrt{x^2 + 1}};\ (0, 1)$

35. $f'(x) = 1 - e^{-x};\ (0, 2)$

36. $f'(x) = \frac{\ln x}{x};\ (1, -2)$

37. Let $f(x) = -2x^2 + 1$, and compute the Riemann sum of f over the interval $[1, 2]$ by partitioning the interval into five subintervals of the same length ($n = 5$), where the points p_i ($1 \leq i \leq 5$) are taken to be the *right* endpoints of the respective subintervals.

38. The management of the National Electric Corporation has determined that the daily marginal cost function associated with producing their automatic drip coffeemakers is given by

$$C'(x) = 0.00003x^2 - 0.03x + 20$$

where $C'(x)$ is measured in dollars per unit and x denotes the number of units produced. Management has also determined that the daily fixed cost incurred in producing these coffeemakers is $500. What is the total cost incurred by National in producing the first 400 coffeemakers per day?

39. Refer to exercise 38. Management has also determined that the daily marginal revenue function associated with producing and selling their coffeemakers is given by

$$R'(x) = -0.03x + 60$$

where x denotes the number of units produced and sold and $R'(x)$ is measured in dollars per unit.
a. Determine the revenue function $R(x)$ associated with producing and selling these coffeemakers.
b. What is the demand equation relating the wholesale unit price to the quantity of coffeemakers demanded?

40. The Franklin National Life Insurance Company purchased a new computer for $200,000. If the rate at which the computer's resale value changes is given by

the function

$$V'(t) = 3800(t - 10)$$

where t is the length of time since the purchase date and $V'(t)$ is measured in dollars per year, find an expression $V(t)$ that gives the resale value of the computer after t years. How much would the computer cost after six years?

41. The marketing department of the Vista Vision Corporation forecasts that sales of their new line of projection television systems will grow at the rate of

$$3000 - 2000e^{-0.04t} \quad (0 \le t \le 24)$$

units per month once they are introduced into the market. Find an expression giving the total number of projection television systems that Vista may expect to sell t months from the time they are put on the market. How many units of the television systems can Vista expect to sell during the first year?

42. Due to the increasing cost of fuel, the manager of the City Transit Authority estimates that the number of commuters using the city subway system will increase at the rate of

$$3000(1 + 0.4t)^{-1/2} \quad (0 \le t \le 36)$$

per month t months from now. If 100,000 commuters are currently using the system, find an expression giving the total number of commuters who will be using the subway t months from now. How many commuters will be using the subway 6 months from now?

43. The management of a division of Ditton Industries has determined that the daily marginal cost function associated with producing their hot-air corn poppers is given by

$$C'(x) = 0.00003x^2 - 0.03x + 10$$

where $C'(x)$ is measured in dollars per unit and x denotes the number of units manufactured. Management has also determined that the daily fixed cost incurred in producing these corn poppers is $600. Find the total cost incurred by Ditton in producing the first 500 corn poppers.

44. In 1980 the world produced 3.5 billion metric tons of coal. If output increased at the rate of

$$3.5e^{0.04t}$$

billion metric tons per year in year t ($t = 0$ corresponds to 1980), determine how much coal was produced worldwide between 1980 and the end of 1985.

45. Find the area of the region under the curve $y = 3x^2 + 2x + 1$ from $x = -1$ to $x = 2$.

46. Find the area of the region under the curve $y = e^{2x}$ from $x = 0$ to $x = 2$.

47. Find the area of the region bounded by the graph of the function $y = 1/x^2$, the x-axis, and the lines $x = 1$ and $x = 3$.

48. Find the area of the region bounded by the curve $y = -x^2 - x + 2$ and the x-axis.

49. Find the area of the region bounded by the graphs of the functions $f(x) = e^x$ and $g(x) = x$ and the vertical lines $x = 0$ and $x = 2$.

50. Find the area of the region that is completely enclosed by the graphs of $f(x) = x^4$ and $g(x) = x$.

51. Find the area of the region between the curve $y = x(x - 1)(x - 2)$ and the x-axis.

52. Based on current production techniques, the rate of oil production from a certain oil well t years from now is estimated to be

$$R_1(t) = 100e^{0.05t}$$

thousand barrels per year. Based on a new production technique, however, it is estimated that the rate of oil production from that oil well t years from now will be

$$R_2(t) = 100e^{0.08t}$$

thousand barrels per year. Determine how much additional oil will be produced over the next ten years if the new technique is adopted.

53. Find the average value of the function

$$f(x) = \frac{x}{\sqrt{x^2 + 16}}$$

over the interval $[0, 3]$.

54. The demand function for a brand of blank videocassettes is given by

$$p = -0.01x^2 - 0.2x + 23$$

where p is the wholesale unit price in dollars and x is the quantity demanded each week, measured in units of a thousand. Determine the consumers' surplus if the wholesale unit price is $8 per cassette.

55. The quantity demanded x (in units of a hundred) of the Sportsman 5×7 tents, per week, is related to the unit price p (in dollars) by

$$p = -0.1x^2 - x + 40$$

The quantity x (in units of a hundred) that the supplier is willing to make available in the market is related to the unit price by

$$p = 0.1x^2 + 2x + 20$$

If the market price is set at the equilibrium price, find the consumers' surplus and the producers' surplus.

56. Mr. Cunningham plans to deposit $4000 per year in his Keogh Retirement Account. If interest is compounded continuously at the rate of 8% per year, how much will he have in his retirement account after 20 years?

57. Miss Parker sold her house under an installment contract whereby the buyer gave her a down payment of $9000 and agreed to make monthly payments of $925 per month for 30 years. If the prevailing interest rate is 12% per year compounded continuously, find the present value of the purchase price of the house.

58. Ms. Carlson purchased a 10-year franchise for a health spa that is expected to generate income at the rate of

$$P(t) = 80,000$$

dollars per year. If the prevailing interest rate is 10% per year compounded continuously, find the present value of the franchise.

59. A certain country's income distribution is described by the function

$$f(x) = \frac{17}{18}x^2 + \frac{1}{18}x$$

a. Sketch the Lorentz curve for this function.
b. Compute $f(0.3)$ and $f(0.6)$ and interpret your results.
c. Compute the coefficient of inequality for this Lorentz curve.

60. The population of a certain Sunbelt city, currently 80,000, is expected to grow exponentially in the next five years with a growth constant of 0.05. If the prediction comes true, what will the average population of the city be over the next five years?

In addition to the basic rules of integration developed in Chapter 14, there are more sophisticated techniques for finding the antiderivatives of functions. We begin this chapter by looking at the method of integration by parts. We will also look at numerical methods of integration, which enable us to obtain approximate solutions to definite integrals, especially those whose exact value cannot be found otherwise. More specifically, we study the Trapezoidal Rule and Simpson's Rule. Numerical integration methods are especially useful when the integrand is known only at discrete points. Finally, we learn how to evaluate integrals in which the intervals of integration are unbounded. Such integrals, called *improper integrals*, play an important role in the study of probability, the last topic of this chapter.

What is the area of the oil spill caused by a grounded tanker? In Example 5, page 1044, you will see how to determine the area of the oil spill.

ADDITIONAL TOPICS IN INTEGRATION

15.1 INTEGRATION BY PARTS

The Method of Integration by Parts

Integration by parts is another technique of integration that, like the method of substitution discussed in Chapter 14, is based on a corresponding rule of differentiation. In this case, the rule of differentiation is the Product Rule, which asserts that if f and g are differentiable functions, then

$$\frac{d}{dx}[f(x)g(x)] = f(x)g'(x) + g(x)f'(x) \qquad \textbf{(1)}$$

If we integrate both sides of equation (1) with respect to x, we obtain

$$\int \frac{d}{dx} f(x)g(x)\, dx = \int f(x)g'(x)\, dx + \int g(x)f'(x)\, dx$$

or

$$f(x)g(x) = \int f(x)g'(x)\, dx + \int g(x)f'(x)\, dx$$

This last equation, which may be written in the form

$$\int f(x)g'(x)\, dx = f(x)g(x) - \int g(x)f'(x)\, dx \qquad \textbf{(2)}$$

is called the formula for **integration by parts.** This formula is useful since it enables us to express one indefinite integral in terms of another that may be easier to evaluate. Formula (2) may be simplified by letting

$$u = f(x) \qquad dv = g'(x)\, dx$$
$$du = f'(x)\, dx \qquad v = g(x)$$

giving the following version of the formula for integration by parts.

INTEGRATION BY PARTS FORMULA	$\int u\, dv = uv - \int v\, du \qquad \textbf{(3)}$

EXAMPLE 1   Evaluate $\int xe^x\, dx$.

Solution   No method of integration developed thus far enables us to evaluate the given indefinite integral in its present form. Therefore, we attempt to write it in terms of an indefinite integral that will be easier to evaluate. We use the integration by parts formula (3) by letting

$$u = x \quad \text{and} \quad dv = e^x\, dx$$

so that

$$du = dx \quad \text{and} \quad v = e^x$$

Then

$$\int x e^x \, dx = \int u \, dv$$

$$= uv - \int v \, du$$

$$= x e^x - \int e^x \, dx$$

$$= x e^x - e^x + C$$

$$= (x - 1) e^x + C \qquad \circ \; \circ \; \circ$$

The success of the method of integration by parts depends on the proper choice of u and dv. For example, if we had chosen

$$u = e^x \quad \text{and} \quad dv = x \, dx$$

in the last example, then

$$du = e^x \, dx \quad \text{and} \quad v = \frac{1}{2} x^2$$

so equation (3) would have yielded

$$\int x e^x \, dx = \int u \, dv$$

$$= uv - \int v \, du$$

$$= \frac{1}{2} x^2 e^x - \int \frac{1}{2} x^2 e^x \, dx$$

Since the indefinite integral on the right-hand side of this equation is not readily evaluated (it is, in fact, more complicated than the original integral!), choosing u and dv as shown has not helped us evaluate the given indefinite integral.

In general, we can use the following guidelines.

GUIDELINES FOR CHOOSING u AND dv

Choose u and dv so that

1. du is simpler than u.

2. dv is easy to integrate.

EXAMPLE 2 Evaluate $\int x \ln x \, dx$.

Solution Letting

$$u = \ln x \quad \text{and} \quad dv = x \, dx$$

we have

$$du = \frac{1}{x} \, dx \quad \text{and} \quad v = \frac{1}{2} x^2$$

Therefore
$$\int x \ln x \, dx = \int u \, dv = uv - \int v \, du$$

$$= \frac{1}{2}x^2 \ln x - \int \frac{1}{2}x^2 \cdot \left(\frac{1}{x}\right) dx$$

$$= \frac{1}{2}x^2 \ln x - \frac{1}{2}\int x \, dx$$

$$= \frac{1}{2}x^2 \ln x - \frac{1}{4}x^2 + C$$

$$= \frac{1}{4}x^2(2 \ln x - 1) + C$$

○ ○ ○

EXAMPLE 3   Evaluate $\int \dfrac{xe^x}{(x + 1)^2} \, dx$.

Solution   Let

$$u = xe^x \quad \text{and} \quad dv = \frac{1}{(x + 1)^2} dx$$

Then $du = (xe^x + e^x) \, dx = e^x(x + 1) \, dx$ and $v = -\dfrac{1}{x + 1}$. Therefore

$$\int \frac{xe^x}{(x + 1)^2} \, dx = \int u \, dv = uv - \int v \, du$$

$$= xe^x \left(\frac{-1}{x + 1}\right) - \int \left(-\frac{1}{x + 1}\right) e^x(x + 1) \, dx$$

$$= -\frac{xe^x}{x + 1} + \int e^x \, dx$$

$$= -\frac{xe^x}{x + 1} + e^x + C$$

$$= \frac{e^x}{x + 1} + C$$

○ ○ ○

The next example shows that repeated applications of the technique of integration by parts is sometimes required to evaluate an integral.

EXAMPLE 4   Evaluate $\int x^2 e^x \, dx$.

Solution   Let

$$u = x^2 \qquad \text{and} \quad dv = e^x \, dx,$$

so that
$$du = 2x \, dx \quad \text{and} \quad v = e^x$$

Then
$$\int x^2 e^x \, dx = \int u \, dv = uv - \int v \, du$$

$$= x^2 e^x - \int e^x(2x) \, dx = x^2 e^x - 2\int xe^x \, dx$$

To complete the solution of the problem, we need to evaluate the integral

$$\int xe^x \, dx$$

But this integral may be found using integration by parts. In fact, you will recognize that this integral is precisely that of Example 1. Using the results obtained there, we now find

$$\int x^2 e^x \, dx = x^2 e^x - 2[(x-1)e^x] + C = e^x(x^2 - 2x + 2) + C \quad \circ \ \circ \ \circ$$

---

a. Use the method of integration by parts to derive the formula

$$\int x^n e^{ax} \, dx = \frac{1}{a} x^n e^{ax} - \frac{n}{a} \int x^{n-1} e^{ax} \, dx$$

where n is a positive integer and a is a real number.

b. Use the formula of part (a) to evaluate

$$\int x^3 e^x \, dx$$

[*Hint:* You may find the results of Example 4 helpful.]

Application

EXAMPLE 5 The estimated rate at which oil will be produced from a certain oil well t years after production has begun is given by

$$R(t) = 100te^{-0.1t}$$

thousand barrels per year. Find an expression that describes the total production of oil at the end of year t.

Solution Let $T(t)$ denote the total production of oil from the well at the end of year t ($t \geq 0$). Then the rate of oil production will be given by $T'(t)$ thousand barrels per year. Thus,

$$T'(t) = R(t) = 100te^{-0.1t}$$

so

$$T(t) = \int 100te^{-0.1t} \, dt$$

$$= 100 \int te^{-0.1t} \, dt$$

We use the technique of integration by parts to evaluate this integral. Let

$$u = t \quad \text{and} \quad dv = e^{-0.1t} \, dt$$

so that

$$du = dt \quad \text{and} \quad v = -\frac{1}{0.1} e^{-0.1t} = -10e^{-0.1t}$$

Therefore
$$T(t) = 100 \left[-10te^{-0.1t} + 10 \int e^{-0.1t}\, dt \right]$$
$$= 100[-10te^{-0.1t} - 100e^{-0.1t}] + C$$
$$= -1000e^{-0.1t}(t + 10) + C$$

To determine the value of C, note that the total quantity of oil produced at the end of year 0 is nil, so $T(0) = 0$. This gives

$$T(0) = -1000(10) + C = 0$$

or
$$C = 10,000$$

Thus, the required production function is given by

$$T(t) = -1000e^{-0.1t}(t + 10) + 10,000$$

○ ○ ○

EXPLORING WITH TECHNOLOGY

Refer to Example 5.
1. Use a graphing utility to plot the graph of

$$T(t) = -1000e^{-0.1t}(t + 10) + 10,000$$

in the viewing rectangle $[0, 100] \times [0, 12,000]$. Use **TRACE** to see what happens when t is very large.
2. Verify the result of part (1) by evaluating $\lim_{t \to \infty} T(t)$. Interpret your results.

○　○　○

SELF–CHECK EXERCISES 15.1

1. Evaluate $\displaystyle\int x^2 \ln x\, dx$.

2. Since the inauguration of Ryan's Express at the beginning of 1996, the number of passengers (in millions) flying on this commuter airline has been growing at the rate of

$$R(t) = 0.1 + 0.2te^{-0.4t}$$

passengers per year ($t = 0$ corresponds to the beginning of 1996). Assuming that this trend continues through 2000, determine how many passengers will have flown on Ryan's Express by that time.

Solutions to Self-Check Exercises 15.1 can be found on page 1034.

15.1 EXERCISES

In exercises 1–26, evaluate each indefinite integral.

1. $\int xe^{2x}\,dx$

2. $\int xe^{-x}\,dx$

3. $\int xe^{x/4}\,dx$

4. $\int 6xe^{3x}\,dx$

5. $\int (e^x - x)^2\,dx$

6. $\int (e^{-x} + x)^2\,dx$

7. $\int (x + 1)e^x\,dx$

8. $\int (x - 3)e^{3x}\,dx$

9. $\int x(x + 1)^{-3/2}\,dx$

10. $\int x(x + 4)^{-2}\,dx$

11. $\int x\sqrt{x - 5}\,dx$

12. $\int \dfrac{x}{\sqrt{2x + 3}}\,dx$

13. $\int x \ln 2x\,dx$

14. $\int x^2 \ln 2x\,dx$

15. $\int x^3 \ln x\,dx$

16. $\int \sqrt{x} \ln x\,dx$

17. $\int \sqrt{x} \ln \sqrt{x}\,dx$

18. $\int \dfrac{\ln x}{\sqrt{x}}\,dx$

19. $\int \dfrac{\ln x}{x^2}\,dx$

20. $\int \dfrac{\ln x}{x^3}\,dx$

21. $\int \ln x\,dx$

[*Hint:* Let $u = \ln x$ and $dv = dx$.]

22. $\int \ln(x + 1)\,dx$

23. $\int x^2 e^{-x}\,dx$

[*Hint:* Integrate by parts twice.]

24. $\int e^{-\sqrt{x}}\,dx$

[*Hint:* First make the substitution $u = \sqrt{x}$; then integrate by parts.]

25. $\int x(\ln x)^2\,dx$

[*Hint:* Integrate by parts twice.]

26. $\int x \ln(x + 1)\,dx$

[*Hint:* First make the substitution $u = x + 1$; then integrate by parts.]

In exercises 27–32, evaluate each definite integral by using the method of integration by parts.

27. $\int_0^{\ln 2} xe^x\,dx$

28. $\int_0^2 xe^{-x}\,dx$

29. $\int_1^4 \ln x\,dx$

30. $\int_1^2 x \ln x\,dx$

31. $\int_0^2 xe^{2x}\,dx$

32. $\int_0^1 x^2 e^{-x}\,dx$

33. Find the function f given that the slope of the tangent line to the graph of f at any point $(x, f(x))$ is xe^{-2x} and that the graph passes through the point $(0, 3)$.

34. Find the function f given that the slope of the tangent line to the graph of f at any point $(x, f(x))$ is $x\sqrt{x + 1}$ and that the graph passes through the point $(3, 6)$.

35. Find the area of the region under the graph of $f(x) = \ln x$ from $x = 1$ to $x = 5$.

36. Find the area of the region under the graph of $f(x) = xe^{-x}$ from $x = 0$ to $x = 3$.

37. **Velocity of a Dragster** The velocity of a dragster t seconds after leaving the starting line is

$$100te^{-0.2t}$$

feet per second. What is the distance covered by the dragster in the first 10 seconds of its run?

38. **Production of Steam Coal** In keeping with the projected increase in worldwide demand for steam coal, the boiler-firing fuel used for generating electricity, the management of Consolidated Mining has decided to step up its mining operations. Plans call for increasing the yearly production of steam coal by

$$2te^{-0.05t}$$

million metric tons per year for the next 20 years. The current yearly production is 20 million metric tons. Find a function that describes Consolidated's total production of steam coal at the end of t years. How much coal will Consolidated have produced over the next 20 years if this plan is carried out?

39. **Concentration of a Drug in the Bloodstream** The concentration of a certain drug in a patient's bloodstream t hours after it has been administered is given by $C(t) = 3te^{-t/3}$ mg/ml. Find the average concentration of the drug in the patient's bloodstream over the first 12 hours after administration.

40. Alcohol-Related Traffic Accidents   As a result of increasingly stiff laws aimed at reducing the number of alcohol-related traffic accidents in a certain state, preliminary data indicate that the number of such accidents has been changing at the rate of

$$R(t) = -10 - te^{0.1t}$$

accidents per month t months after the laws took effect. There were 982 alcohol-related accidents for the year before the enactment of the laws. Determine how many alcohol-related accidents were expected during the first year the laws were in effect.

41. Record Sales   Sales of the latest recording by Brittania, a British rock group, are currently $2te^{-0.1t}$ units per week (each unit representing 10,000 albums), where t denotes the number of weeks since the recording's release. Find an expression that gives the total number of units of record albums sold as a function of t.

42. Rate of Return on an Investment   Suppose that an investment is expected to generate income at the rate of

$$P(t) = 30,000 + 800t$$

dollars per year for the next five years. Find the present value of this investment if the prevailing interest rate is 8% per year compounded continuously.

[*Hint:* Use formula (18), page 1012.]

43. Present Value of a Franchise   Ms. Hart purchased a 15-year franchise for a computer outlet store that is expected to generate income at the rate of

$$P(t) = 50,000 + 3000t$$

dollars per year. If the prevailing interest rate is 10% per year compounded continuously, find the present value of the franchise.

[*Hint:* Use formula (18), page 1012.]

44. Growth of HMOs   The membership of the Cambridge Community Health Plan, Inc. (a health maintenance organization), is projected to grow at the rate of $9\sqrt{t+1} \ln \sqrt{t+1}$ thousand people per year, t years from now. If the HMO's current membership is 50,000, what will the membership be five years from now?

SOLUTIONS TO SELF-CHECK EXERCISES 15.1

1. Let $u = \ln x$ and $dv = x^2 \, dx$, so that $du = \dfrac{1}{x} \, dx$ and $v = \dfrac{1}{3} x^3$. Then

$$
\begin{aligned}
\int x^2 \ln x \, dx &= \int u \, dv = uv - \int v \, du \\
&= \frac{1}{3} x^3 \ln x - \int \frac{1}{3} x^2 \, dx \\
&= \frac{1}{3} x^3 \ln x - \frac{1}{9} x^3 + C \\
&= \frac{1}{9} x^3 (3 \ln x - 1) + C
\end{aligned}
$$

2. Let $N(t)$ denote the total number of passengers who will have flown on Ryan's Express by the end of year t. Then $N'(t) = R(t)$, so that

$$
\begin{aligned}
N(t) &= \int R(t) \, dt \\
&= \int (0.1 + 0.2te^{-0.4t}) \, dt \\
&= \int 0.1 \, dt + 0.2 \int te^{-0.4t} \, dt
\end{aligned}
$$

We now use the technique of integration by parts on the second integral. Letting $u = t$ and $dv = e^{-0.4t}\, dt$, we have

$$du = dt \qquad \text{and} \qquad v = -\frac{1}{0.4}e^{-0.4t} = -2.5e^{-0.4t}$$

Therefore,

$$N(t) = 0.1t + 0.2\left[-2.5te^{-0.4t} + 2.5\int e^{-0.4t}\, dt\right]$$

$$= 0.1t - 0.5te^{-0.4t} - \frac{0.5}{0.4}e^{-0.4t} + C$$

$$= 0.1t - 0.5(t + 2.5)e^{-0.4t} + C$$

To determine the value of C, note that $N(0) = 0$, which gives

$$N(0) = -0.5(2.5) + C = 0$$

or
$$C = 1.25$$

Therefore,

$$N(t) = 0.1t - 0.5(t + 2.5)e^{-0.4t} + 1.25$$

The number of passengers who will have flown on Ryan's Express by the end of 2000 is given by

$$N(5) = 0.1(5) - 0.5(5 + 2.5)e^{-0.4(5)} + 1.25$$

$$= 1.242493$$

that is, 1,242,493 passengers.

15.2 NUMERICAL INTEGRATION

Approximating Definite Integrals

One method of measuring cardiac output is to inject 5–10 mg of a dye into a vein leading to the heart. After making its way through the lungs, the dye returns to the heart and is pumped into the aorta, where its concentration is measured at equal time intervals. The graph of the function c in Figure 15.1 shows the concentration of dye in a person's aorta, measured at 2-second intervals after 5 mg of dye have been injected. The person's cardiac output (measured in l/min) is computed using the formula

$$R = \frac{60D}{\displaystyle\int_0^{28} c(t)\, dt} \tag{4}$$

where D is the quantity of dye injected (see exercise 40, page 1051).

Now, in order to use formula (4), we need to evaluate the definite integral

$$\int_0^{28} c(t)\, dt$$

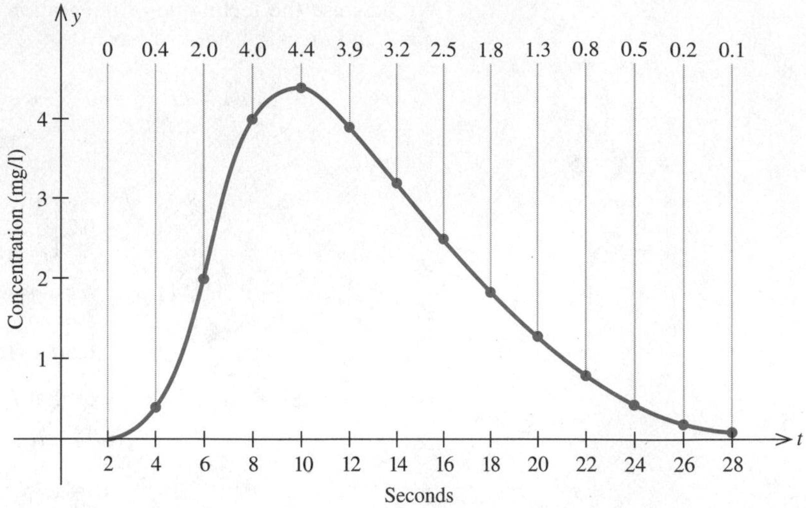

Figure 15.1
The function c gives the concentration of a dye measured at the aorta. The graph is constructed by drawing a smooth curve through a set of discrete points.

But we do not have the algebraic rule defining the integrand c for all values of t in $[0, 28]$. In fact, we are given its values only at a set of discrete points in that interval. In situations such as this, the Fundamental Theorem of Calculus proves useless because we cannot find an antiderivative of c. (We will complete the solution to this problem in Example 4.)

Other situations also arise in which an integrable function has an antiderivative that cannot be found in terms of elementary functions (functions that can be expressed as a finite combination of algebraic, exponential, logarithmic, and trigonometric functions). Examples of such functions are

$$f(x) = e^{x^2}, \quad g(x) = x^{-1/2}e^x, \quad \text{and} \quad h(x) = \frac{1}{\ln x}$$

Riemann sums provide us with a good approximation of a definite integral, provided the number of subintervals in the partitions is large enough. But there are better techniques and formulas, called *quadrature formulas,* that give a more efficient way of computing approximate values of definite integrals. In this section, we look at two rather simple but effective ways of approximating definite integrals.

The Trapezoidal Rule

We assume that $f(x) \geq 0$ on $[a, b]$ in order to simplify the derivation of the Trapezoidal Rule, but the result is valid without this restriction. We begin by subdividing the interval $[a, b]$ into n subintervals of equal length Δx by means of the $(n + 1)$ points $x_0 = a, x_1, x_2, \ldots, x_n = b$, where n is a positive integer (Figure 15.2).

Then the length of each subinterval is given by

$$\Delta x = \frac{b - a}{n}$$

Figure 15.2

The area under the curve is equal to the sum of the n subregions R_1, R_2, ..., R_n.

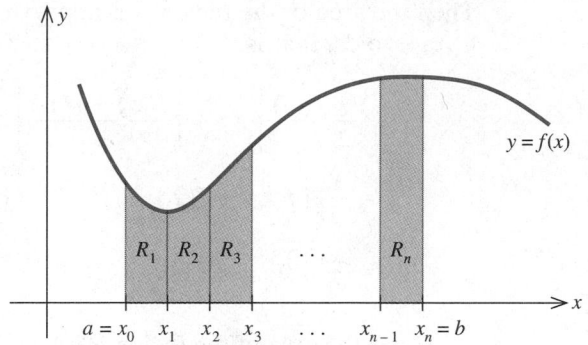

Furthermore, as we saw earlier, we may view the definite integral

$$\int_a^b f(x)\, dx$$

Figure 15.3

The area of R_1 is approximated by the area of the trapezoid.

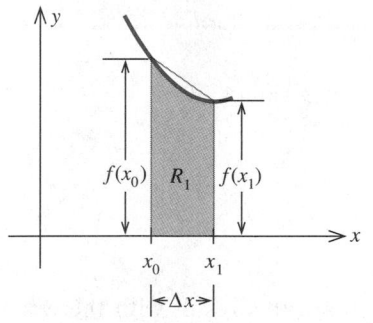

as the area of the region R under the curve $y = f(x)$ between $x = a$ and $x = b$. This area is given by the sum of the areas of the n nonoverlapping subregions R_1, R_2, ..., R_n, such that R_1 represents the region under the curve $y = f(x)$ from $x = x_0$ to $x = x_1$, and so on.

The basis for the Trapezoidal Rule lies in the approximation of each of the regions R_1, R_2, ..., R_n by a suitable trapezoid. This often leads to a much better approximation than one obtained by means of rectangles (a Riemann sum).

Let us consider the subregion R_1, shown magnified for the sake of clarity in Figure 15.3. Observe that the area of the region R_1 may be approximated by the trapezoid of width Δx whose parallel sides are of lengths $f(x_0)$ and $f(x_1)$. The area of the trapezoid is given by

$$\left[\frac{f(x_0) + f(x_1)}{2}\right] \Delta x \qquad \text{(Average of the lengths of the parallel sides times the width)}$$

square units. Similarly, the area of the region R_2 may be approximated by the trapezoid of width Δx and sides of lengths $f(x_1)$ and $f(x_2)$. The area of the trapezoid is given by

$$\left[\frac{f(x_1) + f(x_2)}{2}\right] \Delta x$$

Similarly, we see that the area of the last (nth) approximating trapezoid is given by

$$\left[\frac{f(x_{n-1}) + f(x_n)}{2}\right] \Delta x$$

Then the area of the region R is approximated by the sum of the areas of the n trapezoids, that is,

$$\left[\frac{f(x_0) + f(x_1)}{2}\right]\Delta x + \left[\frac{f(x_1) + f(x_2)}{2}\right]\Delta x + \cdots + \left[\frac{f(x_{n-1}) + f(x_n)}{2}\right]\Delta x$$

$$= \frac{\Delta x}{2}[f(x_0) + f(x_1) + f(x_1) + f(x_2) + \cdots + f(x_{n-1}) + f(x_n)]$$

$$= \frac{\Delta x}{2}[f(x_0) + 2f(x_1) + 2f(x_2) + \cdots + 2f(x_{n-1}) + f(x_n)]$$

Since the area of the region R is given by the value of the definite integral we wished to approximate, we are led to the following approximation formula, which is called the **Trapezoidal Rule.**

TRAPEZOIDAL RULE

$$\int_a^b f(x)\,dx \approx \frac{\Delta x}{2}[f(x_0) + 2f(x_1) + 2f(x_2)$$
$$+ \cdots + 2f(x_{n-1}) + f(x_n)]$$

(5)

where $\Delta x = \dfrac{b-a}{n}$.

The approximation generally improves with larger values of n.

 EXAMPLE 1   Approximate the value of

$$\int_1^2 \frac{1}{x}\,dx$$

using the Trapezoidal Rule with $n = 10$. Compare this result with the exact value of the integral.

Solution   Here $a = 1$, $b = 2$, and $n = 10$, so

$$\Delta x = \frac{b-a}{n} = \frac{1}{10} = 0.1$$

and

$$x_0 = 1, \quad x_1 = 1.1, \quad x_2 = 1.2, \quad x_3 = 1.3, \quad \ldots, \quad x_9 = 1.9, \quad x_{10} = 2$$

The Trapezoidal Rule yields

$$\int_1^2 \frac{1}{x}\,dx \approx \frac{0.1}{2}\left[1 + 2\left(\frac{1}{1.1}\right) + 2\left(\frac{1}{1.2}\right) + 2\left(\frac{1}{1.3}\right) + \cdots + 2\left(\frac{1}{1.9}\right) + \frac{1}{2}\right]$$

$$\approx 0.693771$$

In this case, we can easily compute the actual value of the definite integral under consideration. In fact,

$$\int_1^2 \frac{1}{x}\, dx = \ln x \Big|_1^2 = \ln 2 - \ln 1 = \ln 2$$
$$\approx 0.693147$$

Thus, the Trapezoidal Rule with $n = 10$ yields a result with an error of 0.000624 to six decimal places.   ○ ○ ○

 EXAMPLE 2 The demand function for a certain brand of perfume is given by

$$p = D(x) = \sqrt{10{,}000 - 0.01x^2}$$

where p is the unit price in dollars and x is the quantity demanded each week, measured in ounces. Find the consumers' surplus if the market price is set at $60 per ounce.

Solution When $p = 60$, we have

$$\sqrt{10{,}000 - 0.01x^2} = 60$$
$$10{,}000 - 0.01x^2 = 3600$$
$$x^2 = 640{,}000$$

or $x = 800$ since x must be nonnegative. Next, using the consumers' surplus formula [formula (15), page 1007] with $\bar{p} = 60$ and $\bar{x} = 800$, we find

$$CS = \int_0^{800} \sqrt{10{,}000 - 0.01x^2}\, dx - (60)(800)$$

It is not easy to evaluate this definite integral by finding an antiderivative of the integrand. Instead, let us use the Trapezoidal Rule with $n = 10$.

With $a = 0$ and $b = 800$, we find that

$$\Delta x = \frac{b - a}{n} = \frac{800}{10} = 80$$

and

$$x_0 = 0, \quad x_1 = 80, \quad x_2 = 160, \quad x_3 = 240, \quad \ldots, \quad x_9 = 720, \quad x_{10} = 800$$

so

$$\int_0^{800} \sqrt{10{,}000 - 0.01x^2}\, dx$$
$$\approx \frac{80}{2}\,[100 + 2\sqrt{10{,}000 - (0.01)(80)^2}$$
$$+ 2\sqrt{10{,}000 - (0.01)(160)^2} + \cdots + 2\sqrt{10{,}000 - (0.01)(720)^2}$$
$$+ \sqrt{10{,}000 - (0.01)(800)^2}\,]$$

$$= 40[100 + 199.3590 + 197.4234 + 194.1546 + 189.4835$$
$$+ 183.3030 + 175.4537 + 165.6985$$
$$+ 153.6750 + 138.7948 + 60]$$
$$\approx 70{,}293.82$$

Therefore, the consumers' surplus is approximately $70{,}294 - 48{,}000$, or $22,294.

Explain how you would approximate the value of $\int_0^2 f(x)\, dx$ using the Trapezoidal Rule with $n = 10$, where

$$f(x) = \begin{cases} \sqrt{1 + x^2} & \text{if } 0 \le x \le 1 \\[2mm] \dfrac{2}{\sqrt{1 + x^2}} & \text{if } 1 < x \le 2 \end{cases}$$

and find the value.

Simpson's Rule

Before stating Simpson's Rule, let us review the two rules we have used in approximating a definite integral. Let f be a continuous nonnegative function defined on the interval $[a, b]$. Suppose that the interval $[a, b]$ is partitioned by means of the $n + 1$ equally spaced points $x_0 = a, x_1, x_2, \ldots, x_n = b$, where n is a positive integer, so that the length of each subinterval is $\Delta x = (b - a)/n$ (Figure 15.4).

Figure 15.4
The area under the curve is equal to the sum of the n subregions R_1, R_2, $\ldots$, R_n.

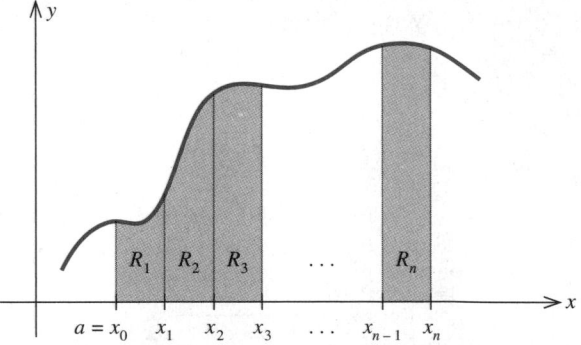

Let us concentrate on the portion of the graph of $y = f(x)$ defined on the interval $[x_0, x_2]$. In using a Riemann sum to approximate the definite integral, we are, in effect, approximating the function $f(x)$ on $[x_0, x_1]$ by the *constant* function $y = f(p_1)$, where p_1 is chosen to be a point in $[x_0, x_1]$; the function $f(x)$ on $[x_1, x_2]$ by the constant function $y = f(p_2)$, where p_2 lies in $[x_1, x_2]$; and so on. Using a Riemann sum, we see that the area of the region under the curve $y = f(x)$ between $x = a$ and $x = b$ is approximated by the area under the approximating "step" function (Figure 15.5a).

Figure 15.5

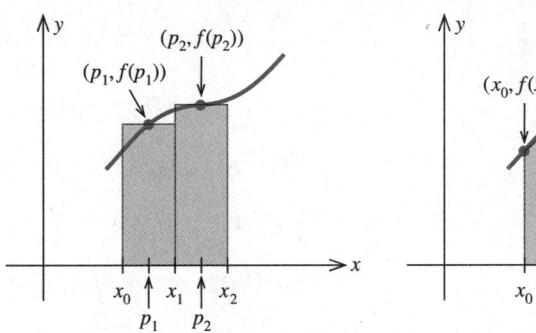
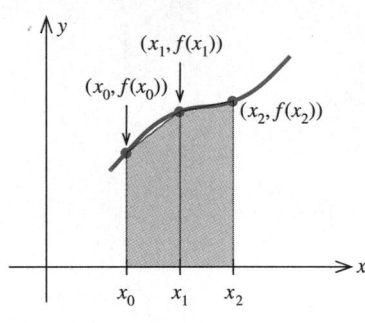

(a) The area under the curve is approximated by the area of the rectangles.

(b) The area under the curve is approximated by the area of the trapezoids.

When we use the Trapezoidal Rule, we are, in effect, approximating the function $f(x)$ on the interval $[x_0, x_1]$ by a *linear* function through the two points $(x_0, f(x_0))$ and $(x_1, f(x_1))$; the function $f(x)$ on $[x_1, x_2]$ by a *linear* function through the two points $(x_1, f(x_1))$ and $(x_2, f(x_2))$; and so on. Thus, the Trapezoidal Rule simply approximates the actual area of the region under the curve $y = f(x)$ from $x = a$ to $x = b$ by the area under the approximating polygonal curve (Figure 15.5b).

A natural extension of the preceding idea is to approximate portions of the graph of $y = f(x)$ by means of portions of the graphs of second-degree polynomials (parts of parabolas). It can be shown that given any three noncollinear points, there is a unique parabola that passes through the given points. Choose the points $(x_0, f(x_0))$, $(x_1, f(x_1))$, and $(x_2, f(x_2))$ corresponding to the first three points of the partition. Then we can approximate the function $f(x)$ on $[x_0, x_2]$ by means of a quadratic function whose graph contains these three points (Figure 15.6).

Although we will not do so here, it can be shown that the area under the parabola between $x = x_0$ and $x = x_2$ is given by

$$\frac{\Delta x}{3}[f(x_0) + 4f(x_1) + f(x_2)]$$

square units. Repeating this argument on the interval $[x_2, x_4]$, we see that the area under the curve between $x = x_2$ and $x = x_4$ is approximated by the area under the parabola between x_2 and x_4—that is, by

$$\frac{\Delta x}{3}[f(x_2) + 4f(x_3) + f(x_4)]$$

square units. Proceeding, we conclude that if n is even (Why?), then the area under the curve $y = f(x)$ from $x = a$ to $x = b$ may be approximated by the

Figure 15.6

Simpson's Rule approximates the area under the curve by the area under the parabola.

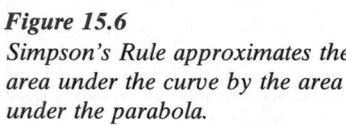
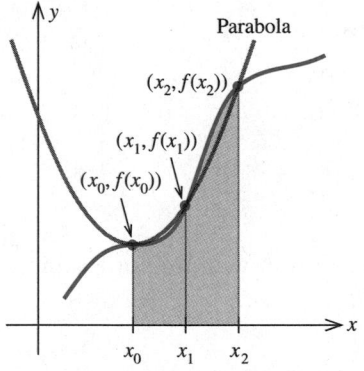

sum of the areas under the $n/2$ approximating parabolas—that is,

$$\frac{\Delta x}{3}[f(x_0) + 4f(x_1) + f(x_2)] + \frac{\Delta x}{3}[f(x_2) + 4f(x_3) + f(x_4)] + \cdots$$

$$+ \frac{\Delta x}{3}[f(x_{n-2}) + 4f(x_{n-1}) + f(x_n)]$$

$$= \frac{\Delta x}{3}[f(x_0) + 4f(x_1) + f(x_2) + f(x_2) + 4f(x_3) + f(x_4) + \cdots$$

$$+ f(x_{n-2}) + 4f(x_{n-1}) + f(x_n)]$$

$$= \frac{\Delta x}{3}[f(x_0) + 4f(x_1) + 2f(x_2) + 4f(x_3)$$

$$+ 2f(x_4) + \cdots + 4f(x_{n-1}) + f(x_n)]$$

The preceding is the derivation of the approximation formula known as **Simpson's Rule**.

SIMPSON'S RULE

$$\int_a^b f(x)\, dx \approx \frac{\Delta x}{3}[f(x_0) + 4f(x_1) + 2f(x_2) + 4f(x_3) + 2f(x_4)$$

$$+ \cdots + 4f(x_{n-1}) + f(x_n)] \tag{6}$$

where $\Delta x = \dfrac{b - a}{n}$ and n is even.

In using this rule, remember that n must be even.

 EXAMPLE 3   Find an approximation of

$$\int_1^2 \frac{1}{x}\, dx$$

using Simpson's Rule with $n = 10$. Compare this result with that of Example 1 and also with the exact value of the integral.

Solution   We have $a = 1$, $b = 2$, $f(x) = \dfrac{1}{x}$, and $n = 10$, so

$$\Delta x = \frac{b - a}{n} = \frac{1}{10} = 0.1$$

and

$$x_0 = 1, \quad x_1 = 1.1, \quad x_2 = 1.2, \quad x_3 = 1.3, \quad \ldots, \quad x_9 = 1.9, \quad x_{10} = 2$$

Simpson's Rule yields

$$\int_1^2 \frac{1}{x}\,dx \approx \frac{0.1}{3}[f(1) + 4f(1.1) + 2f(1.2) + \cdots + 4f(1.9) + f(2)]$$

$$= \frac{0.1}{3}\left[1 + 4\left(\frac{1}{1.1}\right) + 2\left(\frac{1}{1.2}\right) + 4\left(\frac{1}{1.3}\right) + 2\left(\frac{1}{1.4}\right) + 4\left(\frac{1}{1.5}\right)\right.$$

$$\left. + 2\left(\frac{1}{1.6}\right) + 4\left(\frac{1}{1.7}\right) + 2\left(\frac{1}{1.8}\right) + 4\left(\frac{1}{1.9}\right) + \frac{1}{2}\right]$$

$$\approx 0.693150$$

The Trapezoidal Rule with $n = 10$ yielded an approximation of 0.693771, which is 0.000624 off the value of $\ln 2 \approx 0.693147$ to six decimal places. Simpson's Rule yields an approximation with an error of 0.000003, a definite improvement over the Trapezoidal Rule.   ◦ ◦ ◦

EXAMPLE 4   Solve the problem posed at the beginning of this section. Recall that we wished to find a person's cardiac output by using the formula

$$R = \frac{60D}{\displaystyle\int_0^{28} c(t)\,dt}$$

where D (the quantity of dye injected) is equal to 5 mg and the function c has the graph shown in Figure 15.7. Use Simpson's Rule with $n = 14$ to estimate the value of the integral.

Figure 15.7
The function c gives the concentration of a dye measured at the aorta. The graph is constructed by drawing a smooth curve through a set of discrete points.

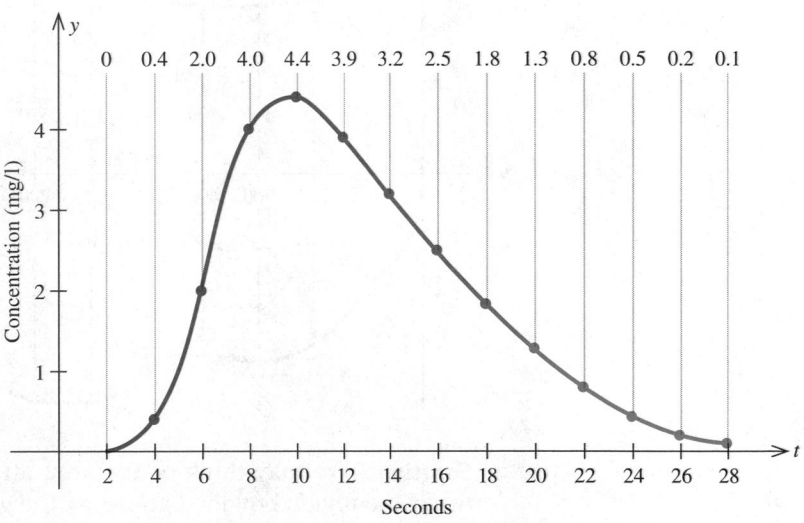

Solution   Using Simpson's Rule with $n = 14$ and $\Delta t = 2$ so that

$$t_0 = 0, \quad t_1 = 2, \quad t_2 = 4, \quad t_3 = 6, \quad \ldots, \quad t_{14} = 28$$

we obtain

$$\int_0^{28} c(t)\, dt \approx \frac{2}{3}[c(0) + 4c(2) + 2c(4) + 4c(6) + \cdots$$
$$+ 4c(26) + c(28)]$$
$$= \frac{2}{3}[0 + 4(0) + 2(0.4) + 4(2.0) + 2(4.0)$$
$$+ 4(4.4) + 2(3.9) + 4(3.2) + 2(2.5) + 4(1.8)$$
$$+ 2(1.3) + 4(0.8) + 2(0.5) + 4(0.2) + 0.1]$$
$$\approx 49.9$$

Therefore, the person's cardiac output is

$$R \approx \frac{60(5)}{49.9} \approx 6.0$$

or 6.0 l/min.

EXAMPLE 5   An oil spill off the coastline was caused by a ruptured tank in a grounded oil tanker. Using aerial photographs, the Coast Guard was able to obtain the dimensions of the oil spill (Figure 15.8). Use Simpson's Rule with $n = 10$ to estimate the area of the oil spill.

Figure 15.8
Simpson's Rule can be used to calculate the area of the oil spill.

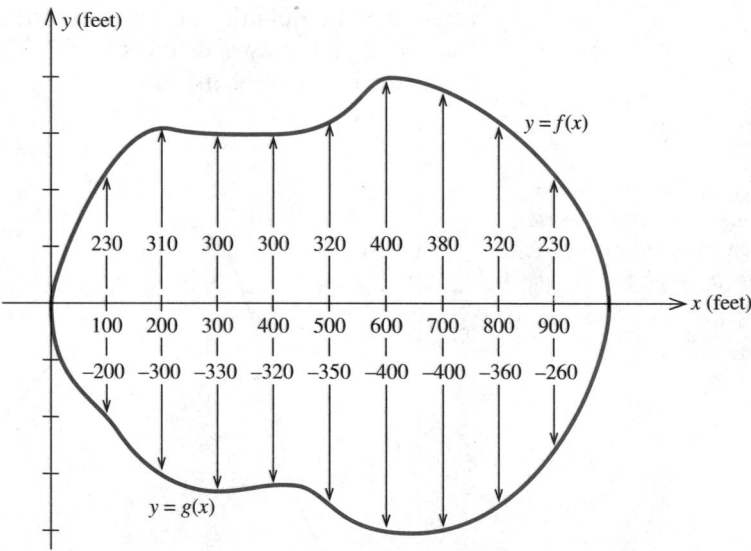

Solution   We may think of the area affected by the oil spill as the area of the plane region bounded above by the graph of the function $f(x)$ and below by the graph of the function $g(x)$ between $x = 0$ and $x = 1000$ (Figure 15.8). Then, the required area is given by

$$A = \int_0^{1000} [f(x) - g(x)]\, dx$$

Using Simpson's Rule with $n = 10$ and $\Delta x = 100$ so that

$$x_0 = 0, \quad x_1 = 100, \quad x_2 = 200, \quad \ldots, \quad x_{10} = 1000$$

we have

$$
\begin{aligned}
A &= \int_0^{1000} [f(x) - g(x)] \, dx \\
&\approx \frac{\Delta x}{3} \{ [f(x_0) - g(x_0)] + 4[f(x_1) - g(x_1)] + 2[f(x_2) - g(x_2)] \\
&\quad + \cdots + 4[f(x_9) - g(x_9)] + [f(x_{10}) - g(x_{10})] \} \\
&= \frac{100}{3} \{ [0 - 0] + 4[230 - (-200)] + 2[310 - (-300)] \\
&\quad + 4[300 - (-330)] + 2[300 - (-320)] + 4[320 - (-350)] \\
&\quad + 2[400 - (-400)] + 4[380 - (-400)] + 2[320 - (-360)] \\
&\quad + 4[230 - (-260)] + [0 - 0] \} \\
&= \frac{100}{3} [0 + 4(430) + 2(610) + 4(630) + 2(620) + 4(670) \\
&\quad + 2(800) + 4(780) + 2(680) + 4(490) + 0] \\
&= \frac{100}{3} (17{,}420) \\
&\approx 580{,}667
\end{aligned}
$$

or approximately 580,667 square feet.

Explain how you would approximate the value of $\int_0^2 f(x) \, dx$ using Simpson's Rule with $n = 10$, where

$$
f(x) = \begin{cases} \sqrt{1 + x^2} & \text{if } 0 \le x \le 1 \\ \dfrac{2}{\sqrt{1 + x^2}} & \text{if } 1 < x \le 2 \end{cases}
$$

and find the value.

Error Analysis

The following results give the bounds on the errors incurred when the Trapezoidal Rule and Simpson's Rule are used to approximate a definite integral (proof omitted).

ERRORS IN THE TRAPEZOIDAL AND SIMPSON APPROXIMATIONS

Suppose that the definite integral

$$\int_a^b f(x)\,dx$$

is approximated with n subintervals.

1. The *maximum* error incurred in using the Trapezoidal Rule is

$$\frac{M(b-a)^3}{12n^2} \qquad (7)$$

where M is a number such that $|f''(x)| \leq M$ for all x in $[a, b]$.

2. The *maximum* error incurred in using Simpson's Rule is

$$\frac{M(b-a)^5}{180n^4} \qquad (8)$$

where M is a number such that $|f^{(4)}(x)| \leq M$ for all x in $[a, b]$.

REMARK   In many instances, the actual error is less than the upper error bounds given.   ○ ○ ○

 EXAMPLE 6   Find bounds on the errors incurred when

$$\int_1^2 \frac{1}{x}\,dx$$

is approximated using (a) the Trapezoidal Rule and (b) Simpson's Rule with $n = 10$. Compare these with the actual errors found in Examples 1 and 3.

Solution

a. Here $a = 1$, $b = 2$, and $f(x) = 1/x$. To find a value for M, we compute

$$f'(x) = -\frac{1}{x^2} \quad \text{and} \quad f''(x) = \frac{2}{x^3}$$

Since $f''(x)$ is positive and decreasing on $(1, 2)$ (Why?), it attains its maximum value of 2 at $x = 1$, the left endpoint of the interval. Therefore, if we take $M = 2$, then $|f''(x)| \leq 2$. Using (7), we see that the maximum error incurred is

$$\frac{2(2-1)^3}{12(10)^2} = \frac{2}{1200} = 0.0016667$$

The actual error found in Example 1, 0.000624, is much less than the upper bound just found.

b. We compute

$$f'''(x) = \frac{-6}{x^4} \quad \text{and} \quad f^{(4)}(x) = \frac{24}{x^5}$$

Since $f^{(4)}(x)$ is positive and decreasing on $(1, 2)$ (just look at $f^{(5)}$ to verify this fact), it attains its maximum at the left endpoint of $[1, 2]$. Now,

$$f^{(4)}(1) = 24$$

and so we may take $M = 24$. Using (8), we obtain the maximum error of

$$\frac{24(2 - 1)^5}{180(10)^4} = 0.0000133$$

The actual error is 0.000003 (see Example 3). ○ ○ ○

SELF-CHECK EXERCISES 15.2

1. Use the Trapezoidal Rule and Simpson's Rule with $n = 8$ to approximate the value of the definite integral

$$\int_0^2 \frac{1}{\sqrt{1 + x^2}}\, dx$$

2. The graph in the accompanying figure shows the consumption of petroleum in the United States, in quadrillion BTU, from 1976 to 1990. Using Simpson's Rule with $n = 14$, estimate the average consumption during the 14-year period.

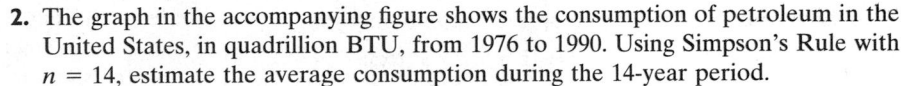
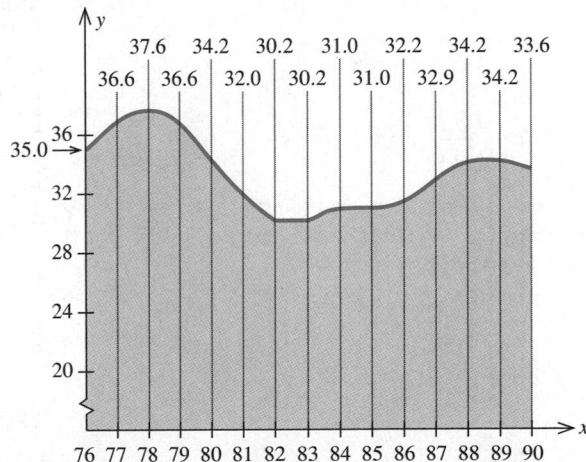

Source: The World Almanac

Solutions to Self-Check Exercises 15.2 can be found on page 1051.

15.2 EXERCISES

A calculator is recommended for this exercise set. In exercises 1–14, use the Trapezoidal Rule and Simpson's Rule to approximate the value of each definite integral. Compare your result with the exact value of the integral.

1. $\int_0^2 x^2 \, dx; n = 6$

2. $\int_1^3 (x^2 - 1) \, dx; n = 4$

3. $\int_0^1 x^3 \, dx; n = 4$

4. $\int_1^2 x^3 \, dx; n = 6$

5. $\int_1^2 \frac{1}{x} \, dx; n = 4$

6. $\int_1^2 \frac{1}{x} \, dx; n = 8$

7. $\int_1^2 \frac{1}{x^2} \, dx; n = 4$

8. $\int_0^1 \frac{1}{1+x} \, dx; n = 4$

9. $\int_0^4 \sqrt{x} \, dx; n = 8$

10. $\int_0^2 x\sqrt{2x^2 + 1} \, dx; n = 6$

11. $\int_0^1 e^{-x} \, dx; n = 6$

12. $\int_0^1 xe^{-x^2} \, dx; n = 6$

13. $\int_1^2 \ln x \, dx; n = 4$

14. $\int_0^1 x \ln (x^2 + 1) \, dx; n = 8$

In exercises 15–22, use the Trapezoidal Rule and Simpson's Rule to approximate the value of each definite integral.

15. $\int_0^1 \sqrt{1 + x^3} \, dx; n = 4$

16. $\int_0^2 x\sqrt{1 + x^3} \, dx; n = 4$

17. $\int_0^2 \frac{1}{\sqrt{x^3 + 1}} \, dx; n = 4$

18. $\int_0^1 \sqrt{1 - x^2} \, dx; n = 4$

19. $\int_0^2 e^{-x^2} \, dx; n = 4$

20. $\int_0^1 e^{x^2} \, dx; n = 6$

21. $\int_1^2 x^{-1/2}e^x \, dx; n = 4$

22. $\int_2^4 \frac{dx}{\ln x}; n = 6$

In exercises 23–28, find a bound on the error in approximating the given definite integral using (a) the Trapezoidal Rule and (b) Simpson's Rule with n intervals.

23. $\int_{-1}^1 x^5 \, dx; n = 10$

24. $\int_0^1 e^{-x} \, dx; n = 8$

25. $\int_1^3 \frac{1}{x} \, dx; n = 10$

26. $\int_1^3 \frac{1}{x^2} \, dx; n = 8$

27. $\int_0^2 \frac{1}{\sqrt{1 + x}} \, dx; n = 8$

28. $\int_1^3 \ln x \, dx; n = 10$

29. Trial Run of an Attack Submarine In a submerged trial run of an attack submarine, a reading of the sub's velocity was made every quarter hour, as shown in the accompanying table. Use the Trapezoidal Rule to esti-

mate the distance traveled by the submarine during the 2-hour period.

Time t (hours)	0	$\frac{1}{4}$	$\frac{1}{2}$	$\frac{3}{4}$
Velocity $V(t)$ (mph)	19.5	24.3	34.2	40.5

Time t (hours)	1	$\frac{5}{4}$	$\frac{3}{2}$	$\frac{7}{4}$	2
Velocity $V(t)$ (mph)	38.4	26.2	18	16	8

30. Real Estate Cooper Realty is considering development of a time-sharing condominium resort complex along the oceanfront property illustrated in the accompanying graph. In order to obtain an estimate of the area of this property, measurements of the distances from the edge of a straight road, which defines one boundary of the property, to the corresponding points on the shoreline are made at 100-foot intervals. Using Simpson's Rule with $n = 10$, estimate the area of the oceanfront property.

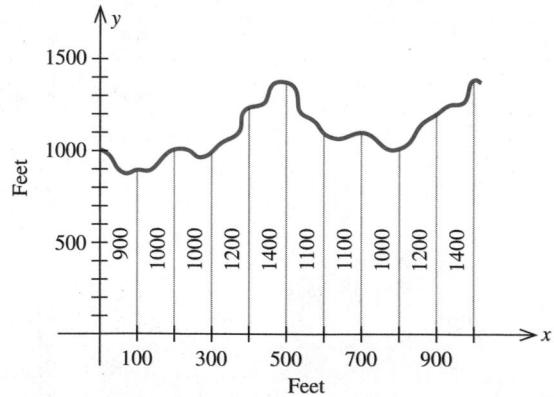

31. Fuel Consumption of Domestic Cars Thanks to smaller and more fuel-efficient models, American car makers have doubled their average fuel economy over a 13-year period from 1974 to 1987. The graph depicted in the figure gives the average fuel consumption in miles per gallon (mpg) of domestic-built cars over the period under consideration ($t = 0$ corresponds to the beginning of 1974). Use the Trapezoidal Rule to estimate the average fuel consumption of the domestic car built during this period.

[*Hint:* Approximate the integral $\frac{1}{13}\int_0^{13} f(t)\,dt$.]

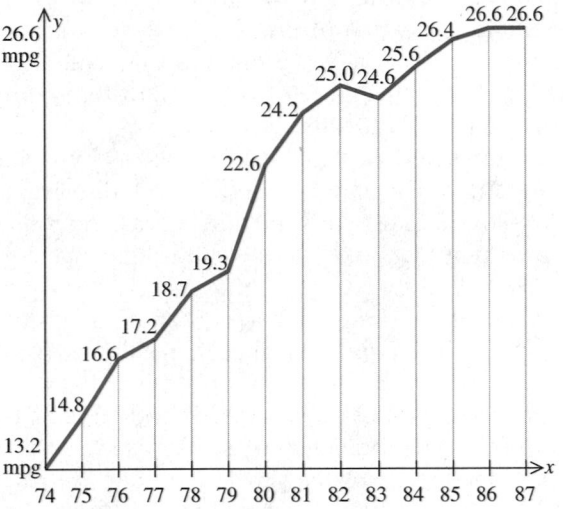

32. Average Temperature   The graph depicted in the accompanying figure shows the daily mean temperatures recorded during one September in Cameron Highlands. Using (a) the Trapezoidal Rule and (b) Simpson's Rule with $n = 10$, estimate the average temperature during that month.

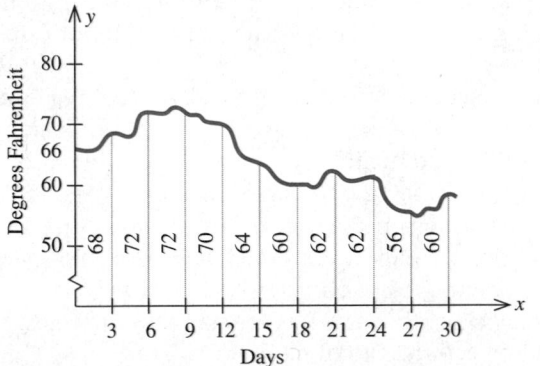

33. Consumers' Surplus   Refer to Section 14.7. The demand equation for the Sicard wristwatch is given by

$$p = \frac{50}{0.01x^2 + 1} \qquad (0 \le x \le 20)$$

where x (measured in units of a thousand) is the quantity demanded per week and p is the unit price in dollars. Use (a) the Trapezoidal Rule and (b) Simpson's Rule (take $n = 8$) to estimate the consumers' surplus if the market price is $25 per watch.

34. Producers' Surplus   Refer to Section 14.7. The supply function for the audio compact disc manufactured by the Herald Record Company is given by

$$p = S(x) = \sqrt{0.01x^2 + 0.11x + 38}$$

where p is the unit wholesale price in dollars and x stands for the quantity that will be made available in the market by the supplier, measured in units of a thousand. Use (a) the Trapezoidal Rule and (b) Simpson's Rule (take $n = 8$) to estimate the producers' surplus if the wholesale price is $8 per disc.

35. Air Pollution   The amount of nitrogen dioxide, a brown gas that impairs breathing, present in the atmosphere on a certain May day in the city of Long Beach has been approximated by

$$A(t) = \frac{136}{1 + 0.25(t - 4.5)^2} + 28 \qquad (0 \le t \le 11)$$

where $A(t)$ is measured in pollutant standard index (PSI), t is measured in hours, and $t = 0$ corresponds to 7 A.M. Use the Trapezoidal Rule with $n = 10$ to estimate the average PSI between 7 A.M. and noon.

[*Hint:* $\frac{1}{5}\int_0^5 A(t)\,dt$.]

36. Growth of Service Industries   It has been estimated that service industries, which currently make up 30% of the nonfarm work force in a certain country, will continue to grow at the rate of

$$R(t) = 5e^{1/(t+1)}$$

percent per decade t decades from now. Estimate the percentage of the nonfarm work force in the service industries one decade from now.

[*Hint:* (a) Show that the desired answer is given by $30 + \int_0^1 5e^{1/(t+1)}\,dt$, and (b) use Simpson's Rule with $n = 10$ to approximate the definite integral.]

37. Length of Infants at Birth   Medical records of infants delivered at the Kaiser Memorial Hospital show that the percentage of infants whose length at birth is between 19 and 21 inches is given by

$$P = 100 \int_{19}^{21} \frac{1}{2.6\sqrt{2\pi}} e^{-1/2[(x-20)/2.6]^2}\,dx$$

Use Simpson's Rule with $n = 10$ to estimate P.

In 1992 Chesebro was appointed professor of medicine at Harvard Medical School, after having conducted cardiovascular research at the Mayo Clinic for a number of years. In addition, he was made Associate Director for Research in the Cardiac Unit at Massachusetts General Hospital, one of Harvard's teaching affiliates.

James H. Chesebro, M.D.

Institution: Harvard Medical School, Massachusetts General Hospital

For over 20 years, James Chesebro has worked as an investigative cardiologist, diagnosing and treating patients suffering from heart and blood-vessel diseases. He specializes in researching ways to prevent blood clots from narrowing the heart's arteries. Even when patients have coronary bypass operations to correct problems, stresses Chesebro, "the piece of vein used to detour around blocked arteries is prone to plug up with clots."

Throughout his career, Chesebro has investigated the body's clotting mechanisms, as well as substances that may prevent, or even dissolve, clots. A study conducted in 1982 showed that bypass patients taking dipyridamole and aspirin were nearly three times less likely to develop clots in vein grafts than patients given placebos.

Recently, Chesebro has been working with a substance called hirudin derived from leech saliva. Hirudin has proved quite beneficial in blocking arterial blood-clot formations.

Determining the correct medication and dosage involves intense quantitative research. Colleagues from other branches of science such as nuclear medicine, molecular biology, and biochemistry play key roles in the research process. Chesebro and his colleagues have to understand how the heart's physiology and pharmacological variables such as distribution rates, concentration levels, elimination rates, and biological effects of particular substances dictate the choice and dosage of medication.

In considering whether a particular medication will bind effectively with specific body cells, researchers must determine bond tightness, speed, and length of time to reach optimal concentration. Researchers rely on complicated equations such as "integrating the area under a curve to find the amount of medication in the body at a given time," notes Chesebro. Without calculus, these equations can't be solved.

Whether physicians rely on medication or surgical procedures such as balloon angioplasty to open clogged passages, other factors play a major role in the outcome. Cholesterol and blood sugar levels, patient age, blood pressure, the geometry of the blockage, and the velocity and turbulence of blood flow all warrant consideration. According to Chesebro "linear modeling can be used to predict the contribution of patient variables and local blood-vessel variables in the outcome of opening blocked arteries."

The bottom line? Calculus has been a key contributor to Chesebro's success in preventing disabling arterial blood clots.

◦ ◦ ◦

38. Treadlives of Tires   Under normal driving conditions the percentage of Super Titan radial tires expected to have a useful treadlife of between 30,000 and 40,000 miles is given by

$$P = 100 \int_{30,000}^{40,000} \frac{1}{2000\sqrt{2\pi}} e^{-1/2[(x-40,000)/2000]^2} \, dx$$

Use Simpson's Rule with $n = 10$ to estimate P.

39. Measuring Cardiac Output   Eight milligrams of a dye are injected into a vein leading to an individual's heart. The concentration of the dye in the aorta measured at 2-second intervals is shown in the accompanying table. Use Simpson's Rule and the formula of Example 4 to estimate the person's cardiac output.

t	0	2	4	6	8	10	12
$C(t)$	0	0	2.8	6.1	9.7	7.6	4.8

t	14	16	18	20	22	24
$C(t)$	3.7	1.9	0.8	0.3	0.1	0

40. Derive the formula

$$R = \frac{60D}{\int_0^T C(t) \, dt}$$

for calculating the cardiac output of a person in 1/min. Here $C(t)$ is the concentration of dye in the aorta (in mg/l) at time t (in seconds) for t in $[0, T]$, and D is the amount of dye (in mg) injected into a vein leading to the heart.

Hint: Partition the interval $[0, T]$ into n subintervals of equal length Δt. The amount of dye that flows past the measuring point in the aorta during the time interval $[0, \Delta t]$ is approximately $C(t_1)(R \, \Delta t)/60$ (concentration times volume). Therefore, the total amount of dye measured at the aorta is

$$\frac{[C(t_1)R \, \Delta t + C(t_2)R \, \Delta t + \cdots + C(t_n)R \, \Delta t]}{60} = D$$

Take the limit of the Riemann sum to obtain

$$R = \frac{60D}{\int_0^T C(t) \, dt}$$

SOLUTIONS TO SELF-CHECK EXERCISES 15.2

1. We have $x = 0$, $b = 2$, and $n = 8$, so

$$\Delta x = \frac{b - a}{n} = \frac{2}{8} = 0.25$$

and $x_0 = 0, x_1 = 0.25, x_2 = 0.50, x_3 = 0.75, \ldots, x_7 = 1.75$, and $x_8 = 2$. The Trapezoidal Rule gives

$$\int_0^2 \frac{1}{\sqrt{1 + x^2}} \, dx$$

$$\approx \frac{0.25}{2}\left[1 + \frac{2}{\sqrt{1 + (0.25)^2}} + \frac{2}{\sqrt{1 + (0.5)^2}} + \cdots \right.$$

$$\left. + \frac{2}{\sqrt{1 + (1.75)^2}} + \frac{1}{\sqrt{5}}\right]$$

$$\approx 0.125(1 + 1.9403 + 1.7889 + 1.6000 + 1.4142 + 1.2494$$

$$+ 1.1094 + 0.9923 + 0.4472)$$

$$\approx 1.4427$$

Using Simpson's Rule with $n = 8$ gives

$$\int_0^2 \frac{1}{\sqrt{1+x^2}}\, dx$$

$$\approx \frac{0.25}{3}\left[1 + \frac{4}{\sqrt{1+(0.25)^2}} + \frac{2}{\sqrt{1+(0.5)^2}} + \frac{4}{\sqrt{1+(0.75)^2}}\right.$$

$$\left. + \cdots + \frac{4}{\sqrt{1+(1.75)^2}} + \frac{1}{\sqrt{5}}\right]$$

$$\approx \frac{0.25}{3}(1 + 3.8806 + 1.7889 + 3.2000 + 1.4142 + 2.4988 + 1.1094$$

$$+ 1.9846 + 0.4472)$$

$$\approx 1.4436$$

2. The average consumption of petroleum during the 14-year period is given by

$$\frac{1}{14}\int_0^{14} f(x)\, dx$$

where f is the function describing the given graph. Using Simpson's Rule with $a = 0$, $b = 14$, and $n = 14$ so that $\Delta x = 1$ and

$$x_0 = 0, \quad x_1 = 1, \quad x_2 = 2, \quad \ldots, \quad x_{14} = 14$$

we have

$$\frac{1}{14}\int_0^{14} f(x)\, dx$$

$$\approx \left(\frac{1}{14}\right)\left(\frac{1}{3}\right)[f(x_0) + 4f(x_1) + 2f(x_2) + 4f(x_3) + \cdots + 4f(x_{13}) + f(x_{14})]$$

$$= \frac{1}{42}[35 + 4(36.6) + 2(37.6) + 4(36.6) + 2(34.2)$$

$$+ 4(32.0) + 2(30.2) + 4(30.2) + 2(31.0) + 4(31.0)$$

$$+ 2(32.2) + 4(32.9) + 2(34.2) + 4(34.2) + 33.6]$$

$$\approx 33.4$$

or approximately 33.4 quadrillion BTU per year.

15.3 IMPROPER INTEGRALS

Improper Integrals

All of the definite integrals we have encountered have had finite intervals of integration. In many applications, however, we are concerned with integrals that have unbounded intervals of integration. Such integrals are called **improper integrals**.

Figure 15.9
The area of the unbounded region R can be approximated by a definite integral.

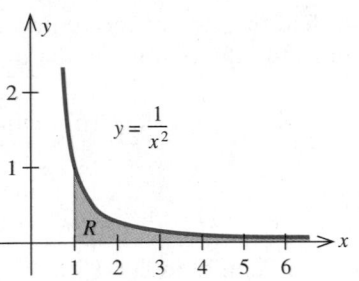

Figure 15.10
Area of shaded region $= \int_1^b \frac{1}{x^2}\,dx.$

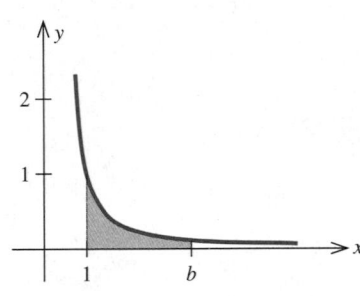

To lead us to the definition of an improper integral of a function f over an infinite interval, consider the problem of finding the area of the region R under the curve $y = f(x) = 1/x^2$ and to the right of the vertical line $x = 1$, as shown in Figure 15.9. Because the interval over which the integration must be performed is unbounded, the method of integration presented previously cannot be applied directly in solving this problem. However, we can approximate the region R by the definite integral

$$\int_1^b \frac{1}{x^2}\,dx \tag{9}$$

which gives the area of the region under the curve $y = f(x) = 1/x^2$ from $x = 1$ to $x = b$ (Figure 15.10). You can see that the approximation of the region R by the definite integral (9) improves as the upper limit of integration, b, becomes larger and larger. Figure 15.11 illustrates the situation for $b = 2, 3,$ and 4.

Figure 15.11
As b increases, the approximation of R by the definite integral improves.

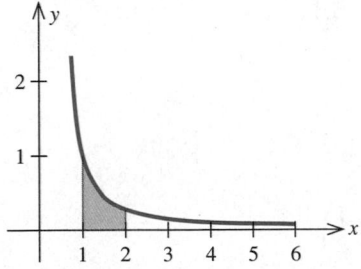

(a) Area of region under the graph of f on [1, 2].

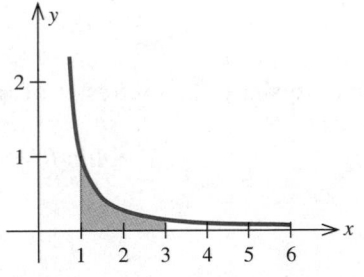

(b) Area of region under the graph of f on [1, 3].

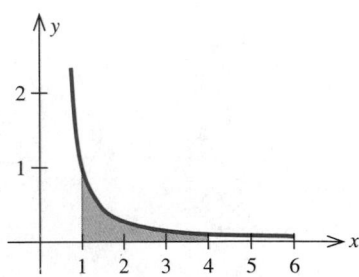

(c) Area of region under the graph of f on [1, 4].

This observation suggests that if we define a function $I(b)$ by

$$I(b) = \int_1^b \frac{1}{x^2}\,dx \tag{10}$$

then we can find the area of the required region R by evaluating the limit of $I(b)$ as b tends to infinity; that is, the area of R is given by

$$\lim_{b \to \infty} I(b) = \lim_{b \to \infty} \int_1^b \frac{1}{x^2}\,dx \tag{11}$$

EXAMPLE 1

a. Evaluate the definite integral $I(b)$ in equation (10).

b. Compute $I(b)$ for $b = 10, 100, 1000,$ and $10,000$.

c. Evaluate the limit in equation (11).

d. Interpret the results of (b) and (c).

Solution

a. $I(b) = \int_1^b \frac{1}{x^2}\,dx = -\frac{1}{x}\Big|_1^b = -\frac{1}{b} + 1$

b. From the result of (a),

$$I(b) = 1 - \frac{1}{b}$$

Therefore,

$$I(10) = 1 - \frac{1}{10} = 0.9$$

$$I(100) = 1 - \frac{1}{100} = 0.99$$

$$I(1000) = 1 - \frac{1}{1000} = 0.999$$

$$I(10,000) = 1 - \frac{1}{10,000} = 0.9999$$

c. Once again, using the result of (a), we find

$$\lim_{b \to \infty} I(b) = \lim_{b \to \infty} \int_1^b \frac{1}{x^2}\,dx$$

$$= \lim_{b \to \infty}\left(1 - \frac{1}{b}\right)$$

$$= 1$$

d. The result of (c) tells us that the area of the region R is 1 square unit. The results of the computations performed in (b) reinforce our expectation that $I(b)$ should approach 1, the area of the region R, as b approaches infinity.

⊙ ⊙ ⊙

The preceding discussion and the results of Example 1 suggest that we define the improper integral of a continuous function f over the unbounded interval $[a, \infty)$ as follows.

IMPROPER INTEGRAL OF f OVER $[a, \infty)$

Let f be a continuous function on the unbounded interval $[a, \infty)$. Then the improper integral of f over $[a, \infty)$ is defined by

$$\int_a^\infty f(x)\, dx = \lim_{b \to \infty} \int_a^b f(x)\, dx \tag{12}$$

if the limit exists.

If the limit exists, the improper integral is said to be **convergent.** An improper integral for which the limit in equation (12) fails to exist is said to be **divergent.**

EXAMPLE 2 Evaluate $\int_2^\infty \dfrac{1}{x}\, dx$ if it converges.

Solution
$$\int_2^\infty \frac{1}{x}\, dx = \lim_{b \to \infty} \int_2^b \frac{1}{x}\, dx$$
$$= \lim_{b \to \infty} \ln x \Big|_2^b$$
$$= \lim_{b \to \infty} (\ln b - \ln 2)$$

Since $\ln b \to \infty$ as $b \to \infty$, the limit does not exist and we conclude that the given improper integral is divergent.

⊙ ⊙ ⊙

a. Suppose f is continuous and nonnegative on $[0, \infty)$. Furthermore, suppose $\lim\limits_{x \to \infty} f(x) = L$, where L is a positive number. What can you say about the convergence of the improper integral $\int_0^\infty f(x)\, dx$? Explain your answer and illustrate with an example.

b. Suppose f is continuous and nonnegative on $[0, \infty)$ and satisfies the condition $\lim\limits_{x \to \infty} f(x) = 0$. What can you say about $\int_0^\infty f(x)\, dx$? Explain and illustrate your answer with examples.

EXAMPLE 3   Find the area of the region R under the curve $y = e^{-x/2}$ for $x \geq 0$.

Figure 15.12

Area of $R = \int_0^\infty e^{-x/2}\,dx$.

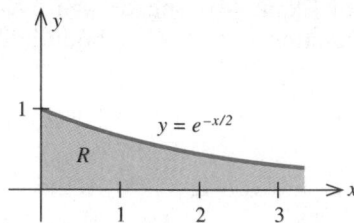

Solution   The region R is shown in Figure 15.12. Taking $b > 0$, we compute the area of the region under the curve $y = e^{-x/2}$ from $x = 0$ to $x = b$—namely,

$$I(b) = \int_0^b e^{-x/2}\,dx = -2e^{-x/2}\Big|_0^b = -2e^{-b/2} + 2$$

Then the area of the region R is given by

$$\lim_{b \to \infty} I(b) = \lim_{b \to \infty} (2 - 2e^{-b/2}) = 2 - 2\lim_{b \to \infty} \frac{1}{e^{b/2}}$$

$$= 2$$

or 2 square units.

EXPLORING WITH TECHNOLOGY

You can see how fast the improper integral in Example 3 converges, as follows:

1. Use a graphing utility to plot the graph of $I(b) = 2 - 2e^{-b/2}$ in the viewing window $[0, 50] \times [0, 3]$.

2. Use **TRACE** to follow the values of y for increasing values of x starting at the origin.

The improper integral defined in equation (12) has an interval of integration that is unbounded on the right. Improper integrals with intervals of integration that are unbounded on the left also arise in practice and are defined in a similar manner.

IMPROPER INTEGRAL OF f OVER $(-\infty, b]$

Let f be a continuous function on the unbounded interval $(-\infty, b]$. Then the improper integral of f over $(-\infty, b]$ is defined by

$$\int_{-\infty}^b f(x)\,dx = \lim_{a \to -\infty} \int_a^b f(x)\,dx \qquad (13)$$

if the limit exists.

In this case, the improper integral is said to be *convergent*. Otherwise, the improper integral is said to be *divergent*.

EXAMPLE 4 Find the area of the region R bounded above by the x-axis, below by the curve $y = -e^{2x}$, and on the right by the vertical line $x = 1$.

Figure 15.13

Area of $R = -\int_{-\infty}^{1} -e^{2x} \, dx.$

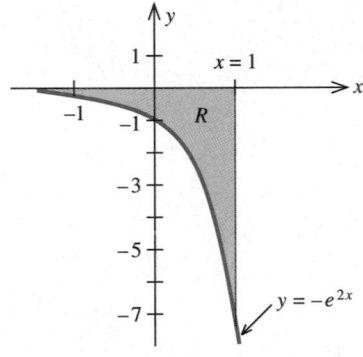

Solution The region R is shown in Figure 15.13. Taking $a < 1$, compute the area of the region bounded above by the x-axis ($y = 0$) and below by the curve $y = -e^{2x}$ from $x = a$ to $x = 1$—namely,

$$I(a) = \int_{a}^{1} [0 - (-e^{2x})] \, dx = \int_{a}^{1} e^{2x} \, dx$$

$$= \frac{1}{2} e^{2x} \Big|_{a}^{1} = \frac{1}{2} e^{2} - \frac{1}{2} e^{2a}$$

Then the area of the required region is given by

$$\lim_{a \to -\infty} I(a) = \lim_{a \to -\infty} \left(\frac{1}{2} e^{2} - \frac{1}{2} e^{2a} \right)$$

$$= \frac{1}{2} e^{2} - \frac{1}{2} \lim_{a \to -\infty} e^{2a}$$

$$= \frac{1}{2} e^{2}$$

○ ○ ○

Another improper integral found in practical applications involves the integration of a function f over the unbounded interval $(-\infty, \infty)$.

IMPROPER INTEGRAL OF f OVER $(-\infty, \infty)$

Let f be a continuous function over the unbounded interval $(-\infty, \infty)$. Let c be any real number, and suppose both the improper integrals

$$\int_{-\infty}^{c} f(x) \, dx \quad \text{and} \quad \int_{c}^{\infty} f(x) \, dx$$

are convergent. Then the improper integral of f over $(-\infty, \infty)$ is defined by

$$\int_{-\infty}^{\infty} f(x) \, dx = \int_{-\infty}^{c} f(x) \, dx + \int_{c}^{\infty} f(x) \, dx \qquad \textbf{(14)}$$

In this case, we say that the improper integral on the left in equation (14) is convergent. If either one of the two improper integrals on the right in (14) is divergent, then the improper integral on the left is not defined.

REMARK   Usually we choose $c = 0$.

○ ○ ○

EXAMPLE 5 Evaluate the improper integral

$$\int_{-\infty}^{\infty} x e^{-x^2} \, dx$$

and give a geometrical interpretation of the results.

Solution Take the point c in equation (14) to be $c = 0$. Let us first evaluate

$$\int_{-\infty}^{0} xe^{-x^2}\, dx = \lim_{a \to -\infty} \int_{a}^{0} xe^{-x^2}\, dx$$

$$= \lim_{a \to -\infty} -\frac{1}{2} e^{-x^2}\Big|_{a}^{0}$$

$$= \lim_{a \to -\infty} \left[-\frac{1}{2} + \frac{1}{2} e^{-a^2} \right] = -\frac{1}{2}$$

Next, we evaluate

$$\int_{0}^{\infty} xe^{-x^2}\, dx = \lim_{b \to \infty} \int_{0}^{b} xe^{-x^2}\, dx$$

$$= \lim_{b \to \infty} -\frac{1}{2} e^{-x^2}\Big|_{0}^{b}$$

$$= \lim_{b \to \infty} \left[-\frac{1}{2} e^{-b^2} + \frac{1}{2} \right] = \frac{1}{2}$$

Therefore,

$$\int_{-\infty}^{\infty} xe^{-x^2}\, dx = \int_{-\infty}^{0} xe^{-x^2}\, dx + \int_{0}^{\infty} xe^{-x^2}\, dx$$

$$= -\frac{1}{2} + \frac{1}{2}$$

$$= 0$$

The graph of $y = xe^{-x^2}$ is sketched in Figure 15.14. A glance at the figure tells us that the improper integral

$$\int_{-\infty}^{0} xe^{-x^2}\, dx$$

gives the negative of the area of the region R_1, bounded above by the x-axis, below by the curve $y = xe^{-x^2}$, and on the right by the y-axis ($x = 0$).

However, the improper integral

$$\int_{0}^{\infty} xe^{-x^2}\, dx$$

gives the area of the region R_2 under the curve $y = xe^{-x^2}$ for $x \geq 0$. Since the graph of f is symmetrical with respect to the origin, the area of R_1 is equal to the area of R_2. In other words,

$$\int_{-\infty}^{0} xe^{-x^2}\, dx = -\int_{0}^{\infty} xe^{-x^2}\, dx$$

Therefore,

$$\int_{-\infty}^{\infty} xe^{-x^2}\, dx = \int_{-\infty}^{0} xe^{-x^2}\, dx + \int_{0}^{\infty} xe^{-x^2}\, dx$$

$$= -\int_{0}^{\infty} xe^{-x^2}\, dx + \int_{0}^{\infty} xe^{-x^2}\, dx$$

$$= 0$$

as was shown earlier.

Figure 15.14

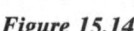

$$\int_{-\infty}^{\infty} xe^{-x^2}\, dx =$$
$$\int_{-\infty}^{0} xe^{-x^2}\, dx + \int_{0}^{\infty} xe^{-x^2}\, dx.$$

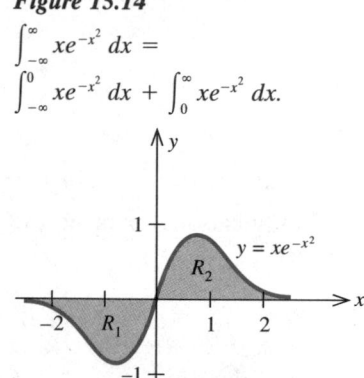

Perpetuities

Recall from Section 14.7 that the present value of an annuity is given by

$$PV \approx mP \int_0^T e^{-rt}\, dt = \frac{mP}{r}(1 - e^{-rT}) \tag{15}$$

Now, if the payments of an annuity are allowed to continue indefinitely, we have what is called a **perpetuity.** The present value of a perpetuity may be approximated by the improper integral

$$PV \approx mP \int_0^\infty e^{-rt}\, dt$$

obtained from formula (15) by allowing the term of the annuity, T, to approach infinity. Thus

$$mP \int_0^\infty e^{-rt}\, dt = \lim_{b \to \infty} mP \int_0^b e^{-rt}\, dt$$

$$= mP \lim_{b \to \infty} \int_0^b e^{-rt}\, dt$$

$$= mP \lim_{b \to \infty} \left[-\frac{1}{r} e^{-rt} \Big|_0^b \right]$$

$$= mP \lim_{b \to \infty} \left(-\frac{1}{r} e^{-rb} + \frac{1}{r} \right) = \frac{mP}{r}$$

THE PRESENT VALUE OF A PERPETUITY

The **present value PV of a perpetuity** is given by

$$PV = \frac{mP}{r} \tag{16}$$

where m is the number of payments per year, P is the size of each payment, and r is the interest rate (compounded continuously).

EXAMPLE 6   The Robinson family wishes to create a scholarship fund at a college. If a scholarship in the amount of $5000 is awarded annually beginning one year from now, find the amount of the endowment the family is required to make now. Assume that this fund will earn interest at a rate of 8% per year compounded continuously.

Solution   The amount of the endowment, A, is given by the present value of a perpetuity, with $m = 1$, $P = 5000$, and $r = 0.08$. Using formula (16), we find

$$A = \frac{(1)(5000)}{0.08}$$

$$= 62,500$$

or $62,500.

The improper integral also plays an important role in the study of probability theory, as we will see in Section 15.4.

SELF-CHECK EXERCISES 15.3

1. Evaluate $\int_{-\infty}^{\infty} \frac{x^3}{(1+x^4)^{3/2}} dx$.

2. Suppose that an income stream is expected to continue indefinitely. Then the present value of such a stream can be calculated from the formula for the present value of an income stream by letting T approach infinity. Thus, the required present value is given by

$$PV = \int_0^{\infty} P(t)e^{-rt} dt$$

Suppose that Marcia has an oil well in her backyard that generates a stream of income given by

$$P(t) = 20e^{-0.02t}$$

where $P(t)$ is expressed in thousands of dollars per year and t is the time in years from the present. Assuming that the prevailing interest rate in the foreseeable future is 10% per year compounded continuously, what is the present value of the income stream?

Solutions to Self-Check Exercises 15.3 can be found on page 1062.

15.3 EXERCISES

In exercises 1–10, find the area of the region under the given curve $y = f(x)$ over the indicated interval.

1. $f(x) = \dfrac{2}{x^2}; x \geq 3$ 2. $f(x) = \dfrac{2}{x^3}; x \geq 2$

3. $f(x) = \dfrac{1}{(x-2)^2}; x \geq 3$

4. $f(x) = \dfrac{2}{(x+1)^3}; x \geq 0$

5. $f(x) = \dfrac{1}{x^{3/2}}; x \geq 1$ 6. $f(x) = \dfrac{3}{x^{5/2}}; x \geq 4$

7. $f(x) = \dfrac{1}{(x+1)^{5/2}}; x \geq 0$

8. $f(x) = \dfrac{1}{(1-x)^{3/2}}; x \leq 0$

9. $f(x) = e^{2x}; x \leq 2$ 10. $f(x) = xe^{-x^2}; x \geq 0$

11. Find the area of the region bounded by the x-axis and the graph of the function

$$f(x) = \frac{x}{(1+x^2)^2}$$

12. Find the area of the region bounded by the x-axis and the graph of the function

$$f(x) = \frac{e^x}{(1+e^x)^2}$$

13. Consider the improper integral

$$\int_0^{\infty} \sqrt{x} \, dx$$

a. Evaluate $I(b) = \int_0^b \sqrt{x} \, dx$.

b. Show that

$$\lim_{b \to \infty} I(b) = \infty$$

thus proving that the given improper integral is divergent.

14. Consider the improper integral

$$\int_1^\infty x^{-2/3} \, dx$$

a. Evaluate $I(b) = \int_1^b x^{-2/3} \, dx$.

b. Show that

$$\lim_{b \to \infty} I(b) = \infty$$

thus proving that the given improper integral is divergent.

In exercises 15–40, evaluate each improper integral whenever it is convergent.

15. $\int_1^\infty \dfrac{3}{x^4} \, dx$

16. $\int_1^\infty \dfrac{1}{x^3} \, dx$

17. $\int_4^\infty \dfrac{2}{x^{3/2}} \, dx$

18. $\int_1^\infty \dfrac{1}{\sqrt{x}} \, dx$

19. $\int_1^\infty \dfrac{4}{x} \, dx$

20. $\int_2^\infty \dfrac{3}{x} \, dx$

21. $\int_{-\infty}^0 \dfrac{1}{(x-2)^3} \, dx$

22. $\int_2^\infty \dfrac{1}{(x+1)^2} \, dx$

23. $\int_1^\infty \dfrac{1}{(2x-1)^{3/2}} \, dx$

24. $\int_{-\infty}^0 \dfrac{1}{(4-x)^{3/2}} \, dx$

25. $\int_0^\infty e^{-x} \, dx$

26. $\int_0^\infty e^{-x/2} \, dx$

27. $\int_{-\infty}^0 e^{2x} \, dx$

28. $\int_{-\infty}^0 e^{3x} \, dx$

29. $\int_1^\infty \dfrac{e^{\sqrt{x}}}{\sqrt{x}} \, dx$

30. $\int_1^\infty \dfrac{e^{-\sqrt{x}}}{\sqrt{x}} \, dx$

31. $\int_{-\infty}^0 xe^x \, dx$

32. $\int_0^\infty xe^{-2x} \, dx$

33. $\int_{-\infty}^\infty x \, dx$

34. $\int_{-\infty}^\infty x^3 \, dx$

35. $\int_{-\infty}^\infty x^2(1+x^3)^{-2} \, dx$

36. $\int_{-\infty}^\infty x(x^2+4)^{-3/2} \, dx$

37. $\int_{-\infty}^\infty xe^{1-x^2} \, dx$

38. $\int_{-\infty}^\infty \left(x - \dfrac{1}{2}\right) e^{-x^2+x-1} \, dx$

39. $\int_{-\infty}^\infty \dfrac{e^{-x}}{1+e^{-x}} \, dx$

40. $\int_{-\infty}^\infty \dfrac{xe^{-x^2}}{1+e^{-x^2}} \, dx$

41. The Amount of an Endowment A university alumni group wishes to provide an annual scholarship in the amount of $1500 beginning next year. If the scholarship fund will earn interest at the rate of 8% per year compounded continuously, find the amount of the endowment the alumni are required to make now.

42. The Amount of an Endowment Mr. Thompson wishes to establish a fund to provide a university medical center with an annual research grant of $50,000 beginning next year. If the fund will earn interest at the rate of 9% per year compounded continuously, find the amount of the endowment he is required to make now.

43. Perpetual Net Income Streams The present value of a perpetual stream of income that flows continually at the rate of $P(t)$ dollars per year is given by the formula

$$PV \approx \int_0^\infty P(t)e^{-rt} \, dt$$

where r is the rate at which interest is compounded continuously. Using this formula, find the present value of a perpetual net income stream that is generated at the rate of

$$P(t) = 10,000 + 4000t$$

dollars per year.

$$\left[\text{Hint: } \lim_{b \to 0} \dfrac{b}{e^{rb}} = 0. \right]$$

44. Establishing a Trust Fund Mrs. Wilkinson wants to establish a trust fund that will provide her children and heirs with a perpetual annuity in the amount of

$$P(t) = (20 + t)$$

thousand dollars per year beginning next year. If the trust fund will earn interest at the rate of 10% per year compounded continuously, find the amount that she must place in the trust fund now.

[*Hint:* Use the formula given in exercise 43.]

45. Capital Value The capital value (present sale value) CV of property that can be rented on a perpetual basis for R dollars annually is approximated by the formula

$$CV \approx \int_0^\infty Re^{-it} \, dt$$

where i is the prevailing continuous interest rate.
a. Show that $CV = R/i$.
b. Find the capital value of property that can be rented at $10,000 annually when the prevailing continuous interest rate is 12% per year.

Solutions to Self-Check Exercises 15.3

1. Write

$$\int_{-\infty}^{\infty} \frac{x^3}{(1+x^4)^{3/2}} \, dx = \int_{-\infty}^{0} \frac{x^3}{(1+x^4)^{3/2}} \, dx + \int_{0}^{\infty} \frac{x^3}{(1+x^4)^{3/2}} \, dx$$

Now,

$$\int_{-\infty}^{0} \frac{x^3}{(1+x^4)^{3/2}} \, dx$$

$$= \lim_{a \to -\infty} \int_{a}^{0} x^3 (1+x^4)^{-3/2} \, dx$$

$$= \lim_{a \to -\infty} \frac{1}{4} (-2)(1+x^4)^{-1/2} \Big|_{a}^{0} \qquad \text{(Integrating by substitution)}$$

$$= -\frac{1}{2} \lim_{a \to -\infty} \left[1 - \frac{1}{(1+a^4)^{1/2}} \right]$$

$$= -\frac{1}{2}$$

Similarly, you can show that

$$\int_{0}^{\infty} \frac{x^3}{(1+x^4)^{3/2}} \, dx = \frac{1}{2}$$

Therefore,

$$\int_{-\infty}^{\infty} \frac{x^3}{(1+x^4)^{3/2}} \, dx = -\frac{1}{2} + \frac{1}{2}$$

$$= 0$$

2. The required present value is given by

$$PV = \int_{0}^{\infty} 20 e^{-0.02t} e^{-0.10t} \, dt$$

$$= 20 \int_{0}^{\infty} e^{-0.12t} \, dt$$

$$= 20 \lim_{b \to \infty} \int_{0}^{b} e^{-0.12t} \, dt$$

$$= -\frac{20}{0.12} \lim_{b \to \infty} e^{-0.12t} \Big|_{0}^{b}$$

$$= -\frac{500}{3} \lim_{b \to \infty} (e^{-0.12b} - 1)$$

$$= \frac{500}{3}$$

or approximately \$166,667.

15.4 APPLICATIONS OF PROBABILITY TO CALCULUS

Probability Density Functions

In this section, we will look at an application of the definite integral to the computation of probabilities associated with a continuous random variable. We begin by recalling the properties of a probability density function associated with a continuous random variable x.

PROBABILITY DENSITY FUNCTION

A **probability density function** of a random variable x in an interval I, where I may be bounded or unbounded, is a nonnegative function f having the following properties:

1. The total area of the region under the graph of f is equal to 1 (see Figure 15.15a).

2. The probability that an observed value of the random variable x lies in the interval $[a, b]$ is given by

$$P(a \le x \le b) = \int_a^b f(x)\,dx$$

(see Figure 15.15b).

Figure 15.15

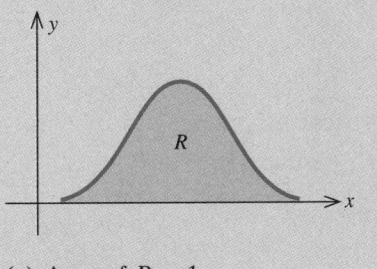

(a) Area of $R = 1$

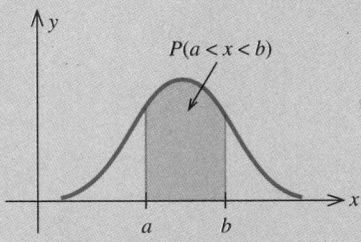

(b) $P(a < x < b)$ is the probability that an outcome of an experiment will lie between a and b.

A few comments are in order. First, a probability density function of a random variable x may be constructed using methods that range from theoretical considerations of the problem on the one extreme to an interpretation of data associated with the experiment on the other. Second, property 1 states that the probability that a continuous random variable takes on a value lying in its range is 1, a certainty, which is expected. Third, property 2 states that the probability that the random variable x assumes a value in an interval $a \le x \le b$ is given by the area of the region between the graph of f and the x-axis from $x = a$ to $x = b$. Because the area under one point of the graph of

f is equal to zero, we see immediately that $P(a \leq x \leq b) = P(a < x \leq b) = P(a \leq x < b) = P(a < x < b)$.

EXAMPLE 1 Show that each of the following functions satisfies the non-negativity condition and property 1 of probability density functions.

a. $f(x) = \dfrac{2}{27} x(x - 1)$ $(1 \leq x \leq 4)$

b. $f(x) = \dfrac{1}{3} e^{(-1/3)x}$ $(0 \leq x < \infty)$

Solution

a. Since the factors x and $(x - 1)$ are both nonnegative, we see that $f(x) \geq 0$ on $[1, 4]$. Next, we compute

$$\int_1^4 \frac{2}{27} x(x - 1)\, dx = \frac{2}{27} \int_1^4 (x^2 - x)\, dx$$

$$= \frac{2}{27} \left(\frac{1}{3} x^3 - \frac{1}{2} x^2 \right) \Big|_1^4$$

$$= \frac{2}{27} \left[\left(\frac{64}{3} - 8 \right) - \left(\frac{1}{3} - \frac{1}{2} \right) \right]$$

$$= \frac{2}{27} \left(\frac{27}{2} \right)$$

$$= 1$$

showing that property 1 of probability density functions holds as well.

b. First, $f(x) = \frac{1}{3} e^{(-1/3)x} \geq 0$ for all values of x in $[0, \infty)$. Next,

$$\int_0^\infty \frac{1}{3} e^{(-1/3)x}\, dx = \lim_{b \to \infty} \int_0^b \frac{1}{3} e^{(-1/3)x}\, dx$$

$$= \lim_{b \to \infty} -e^{(-1/3)x} \Big|_0^b$$

$$= \lim_{b \to \infty} \left(-e^{(-1/3)b} + 1 \right)$$

$$= 1$$

so the area under the graph of $f(x) = \frac{1}{3} e^{(-1/3)x}$ is equal to 1, as we set out to show.

◐ ◐ ◐

EXAMPLE 2

a. Determine the value of the constant k such that the function $f(x) = kx^2$ is a probability density function on the interval $[0, 5]$.

b. If x is a continuous random variable with the probability density function given in (a), compute the probability that x will assume a value between $x = 1$ and $x = 2$.

Solution

a. We compute

$$\int_0^5 kx^2\, dx = k \int_0^5 x^2\, dx$$

$$= \frac{k}{3} x^3 \Big|_0^5$$

$$= \frac{125}{3} k$$

Since this value must be equal to 1, we find that $k = 3/125$.

b. The required probability is given by

$$P(1 \le x \le 2) = \int_1^2 f(x)\, dx = \int_1^2 \frac{3}{125} x^2\, dx$$

$$= \frac{1}{125} x^3 \Big|_1^2 = \frac{1}{125}(8 - 1)$$

$$= \frac{7}{125}$$

The graph of the probability density function f and the area corresponding to the probability $P(1 \le x \le 2)$ are shown in Figure 15.16.

Figure 15.16
$P(1 \le x \le 2)$ *for the probability density function* $y = \dfrac{3}{125} x^2.$

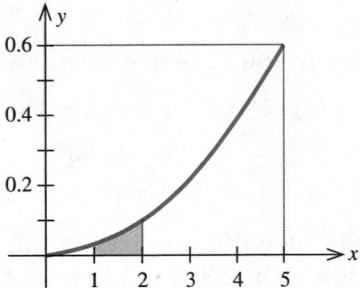

EXAMPLE 3 The TKK Products Corporation manufactures a 200-watt electric light bulb. Laboratory tests show that the life spans of these light bulbs have a distribution described by the probability density function

$$f(x) = 0.001 e^{-0.001x}$$

Determine the probability that a light bulb will have a life span of

a. 500 hours or less

b. more than 500 hours

c. more than 1000 hours but less than 1500 hours

Solution
Let x denote the life span of a light bulb.

a. The probability that a light bulb will have a life span of 500 hours or less is given by

$$P(0 \leq x \leq 500) = \int_0^{500} 0.001 e^{-0.001x}\, dx$$

$$= -e^{-0.001x}\Big|_0^{500} = -e^{-0.5} + 1$$

$$\approx 0.3935$$

b. The probability that a light bulb will have a life span of more than 500 hours is given by

$$P(x > 500) = \int_{500}^{\infty} 0.001 e^{-0.001x}\, dx$$

$$= \lim_{b \to \infty} \int_{500}^{b} 0.001 e^{-0.001x}\, dx$$

$$= \lim_{b \to \infty} -e^{-0.001x}\Big|_{500}^{b}$$

$$= \lim_{b \to \infty} (-e^{-0.001b} + e^{-0.5})$$

$$= e^{-0.5} \approx 0.6065$$

This result may also be obtained by observing that

$$P(x > 500) = 1 - P(x \leq 500)$$

$$= 1 - 0.3935 \qquad \text{[Using the result from (a)]}$$

$$\approx 0.6065$$

c. The probability that a light bulb will have a life span of more than 1000 hours but less than 1500 hours is given by

$$P(1000 < x < 1500) = \int_{1000}^{1500} 0.001 e^{-0.001x}\, dx$$

$$= -e^{-0.001x}\Big|_{1000}^{1500}$$

$$= -e^{-1.5} + e^{-1}$$

$$\approx -0.2231 + 0.3679$$

$$= 0.1448$$

Figure 15.17
The area under the graph of the exponential density function is equal to 1.

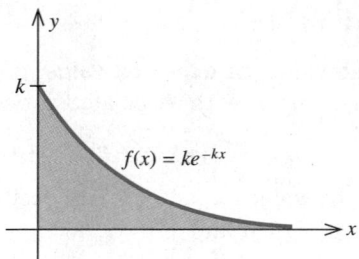

Figure 15.17
The area under the graph of the exponential density function is equal to 1.

The probability density function of Example 3 has the form

$$f(x) = ke^{-kx}$$

where $x \geq 0$ and k is a positive constant. Its graph is shown in Figure 15.17. Such a probability function is called an **exponential density function,** and a random variable associated with such a probability density function is said to be **exponentially distributed.** Exponential random variables are used to represent the life span of electronic components, the duration of telephone calls, the waiting time in a doctor's office, and the time between successive flight arrivals and departures in an airport, to mention but a few applications.

Another probability density function, and the one most widely used, is the **normal density function,** defined by

$$f(x) = \frac{1}{\sigma\sqrt{2\pi}} e^{-(1/2)[(x-\mu)/\sigma]^2}$$

where μ and σ are constants. The graph of the normal distribution is bell-shaped (Figure 15.18). Many phenomena, such as the heights of people in a given population, the weights of newborn infants, the IQs of college students, the actual weights of 16-ounce packages of cereals, and so on, have probability distributions that are normal.

Figure 15.18
The area under the bell-shaped normal distribution curve.

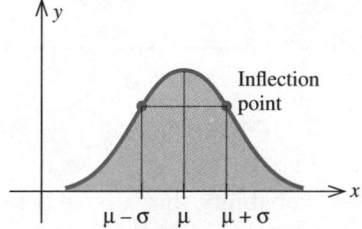

Areas under the standard normal curve (the normal curve with $\mu = 0$ and $\sigma = 1$) have been extensively computed and tabulated. Most problems involving the normal distribution can be solved with the aid of these tables.

Expected Value

The average value, or **expected value,** of a discrete variable X that takes on values $x_1, x_2, \ldots, x_n$ with associated probabilities $p_1, p_2, \ldots, p_n$ is defined by

$$E(X) = x_1 p_1 + x_2 p_2 + \cdots + x_n p_n$$

If each of the values $x_1, x_2, \ldots, x_n$ occurs with equal frequency, then $p_1 = p_2 = \cdots = p_n = 1/n$ and

$$E(X) = x_1 \left(\frac{1}{n}\right) + x_2 \left(\frac{1}{n}\right) + \cdots + x_n \left(\frac{1}{n}\right)$$

$$= \frac{x_1 + x_2 + \cdots + x_n}{n}$$

giving the familiar formula for computing the average value of the n numbers $x_1, x_2, \ldots, x_n$.

Now, suppose that x is a continuous random variable and f is the probability density function associated with it. For simplicity, let us first assume that $a \leq x \leq b$. We divide the interval $[a, b]$ into n subintervals of equal length $\Delta x = (b - a)/n$ by means of the $(n + 1)$ points $x_0 = a, x_1, x_2, \ldots, x_n = b$ (Figure 15.19). To find an approximation of the average value, or expected value, of x on the interval $[a, b]$, let us treat x as if it were a discrete random variable that takes on the values $x_1, x_2, \ldots, x_n$ with probabilities $p_1, p_2, \ldots, p_n$. Then

$$E(x) \approx x_1 p_1 + x_2 p_2 + \cdots + x_n p_n$$

Figure 15.19

Approximating the expected value of a random variable x on $[a, b]$ by a Riemann sum.

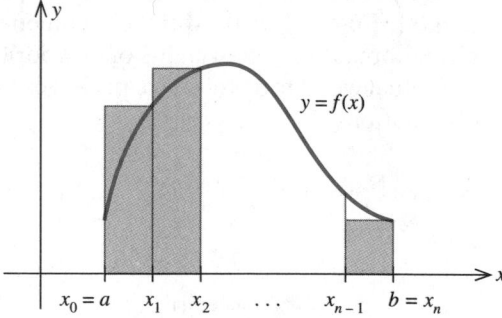

But p_1 is the probability that x is in the interval $[x_0, x_1]$, and this is just the area under the graph of f from $x = x_0$ to $x = x_1$, which may be approximated by $f(x_1) \Delta x$. The probabilities $p_2, \ldots, p_n$ may be approximated in a similar manner. Thus,

$$E(x) \approx x_1 f(x_1) \Delta x + x_2 f(x_2) \Delta x + \cdots + x_n f(x_n) \Delta x$$

which is seen to be the Riemann sum of the function $g(x) = xf(x)$ over the interval $[a, b]$. Letting n approach infinity, we obtain the following formula:

EXPECTED VALUE OF A CONTINUOUS RANDOM VARIABLE

Suppose that the function f defined on the interval $[a, b]$ is the probability density function associated with a continuous random variable x. Then the **expected value** of x is

$$E(x) = \int_a^b xf(x)\, dx \qquad (17)$$

If either $a = -\infty$ or $b = \infty$, then the integral in equation (17) becomes an improper integral.

The expected value of a random variable plays an important role in many practical applications. For example, if x represents the life span of a certain brand of electronic components, then the expected value of x gives the average life span of these components. If x measures the waiting time in a doctor's office, then $E(x)$ gives the average waiting time, and so on.

EXAMPLE 4 Show that if a continuous random variable x is exponentially distributed with the probability density function

$$f(x) = ke^{-kx}$$

then the expected value $E(x)$ is equal to $1/k$. Using this result, determine the average life span of a 200-watt electric light bulb manufactured by the TKK Products Corporation of Example 3.

Solution We compute

$$E(x) = \int_0^\infty xf(x)\, dx$$

$$= \int_0^\infty kxe^{-kx}\, dx$$

$$= k \lim_{b \to \infty} \int_0^b xe^{-kx}\, dx$$

Integrating by parts with

$$u = x \quad \text{and} \quad dv = e^{-kx}\, dx$$

so that

$$du = dx \quad \text{and} \quad v = -\frac{1}{k}e^{-kx}$$

we have

$$E(x) = k \lim_{b \to \infty} \left[-\frac{1}{k}xe^{-kx} \Big|_0^b + \frac{1}{k}\int_0^b e^{-kx}\, dx \right]$$

$$= k \lim_{b \to \infty} \left[-\left(\frac{1}{k}\right) be^{-kb} - \frac{1}{k^2}e^{-kx} \Big|_0^b \right]$$

$$= k \lim_{b \to \infty} \left[-\left(\frac{1}{k}\right) be^{-kb} - \frac{1}{k^2}e^{-kb} + \frac{1}{k^2} \right]$$

$$= -\lim_{b \to \infty} \frac{b}{e^{kb}} - \frac{1}{k}\lim_{b \to \infty} \frac{1}{e^{kb}} + \frac{1}{k}\lim_{b \to \infty} 1$$

Now, by taking a sequence of values of b that approaches infinity—for example, $b = 10, 100, 1000, 10,000, \ldots$—we see that, for a fixed k,

$$\lim_{b \to \infty} \frac{b}{e^{kb}} = 0$$

Therefore,

$$E(x) = \frac{1}{k}$$

as we set out to show. Next, since $k = 0.001$ in Example 3, we see that the average life span of the TKK light bulb is $1/(0.001) = 1000$ hours.   ◦ ◦ ◦

Before considering another example, let us summarize the important result obtained in Example 4.

THE AVERAGE VALUE OF AN EXPONENTIAL DENSITY FUNCTION	If a continuous random variable x is exponentially distributed with probability density function $$f(x) = ke^{-kx}$$ then the **expected value** of x is given by $$E = \frac{1}{k}$$

EXAMPLE 5   On a typical Monday morning, the time between successive arrivals of planes at Jackson International Airport is an exponentially distributed random variable x with expected value of 10 (minutes).

a. Find the probability density function associated with x.

b. What is the probability that between 6 and 8 minutes will elapse between successive arrivals of planes?

c. What is the probability that the time between successive arrivals of planes will be longer than 15 minutes?

Solution

a. Since x is exponentially distributed, the associated probability density function has the form $f(x) = ke^{-kx}$. Next, since the expected value of x is 10, we see that

$$E(x) = \frac{1}{k}$$

or

$$k = \frac{1}{10}$$

$$= 0.1$$

so the required probability density function is

$$f(x) = 0.1e^{-0.1x}$$

b. The probability that between 6 and 8 minutes will elapse between successive arrivals is given by

$$P(6 \leq x \leq 8) = \int_6^8 0.1e^{-0.1x}\, dx = -e^{-0.1x}\Big|_6^8$$

$$= -e^{-0.8} + e^{-0.6}$$

$$\approx 0.100$$

c. The probability that the time between successive arrivals will be longer than 15 minutes is given by

$$P(x > 15) = \int_{15}^{\infty} 0.1e^{-0.1x}\, dx$$

$$= \lim_{b \to \infty} \int_{15}^{b} 0.1e^{-0.1x}\, dx$$

$$= \lim_{b \to \infty} \left[-e^{-0.1x}\Big|_{15}^{b} \right]$$

$$= \lim_{b \to \infty} (-e^{-0.1b} + e^{-1.5}) = e^{-1.5}$$

$$\approx 0.22$$

ooo

SELF-CHECK EXERCISES 15.4

1. Determine the value of the constant k such that the function $f(x) = k(4x - x^2)$ is a probability density function on the interval $[0, 4]$.

2. Suppose that x is a continuous random variable with the probability density function of Self-Check exercise 1. Find the probability that x will assume a value between $x = 1$ and $x = 3$.

Solutions to Self-Check Exercises 15.4 can be found on page 1073.

15.4 EXERCISES

In exercises 1–8, show that the given function is a probability density function on the specified interval.

1. $f(x) = \dfrac{2}{32}x$; $(2 \leq x \leq 6)$

2. $f(x) = \dfrac{2}{9}(3x - x^2)$; $(0 \leq x \leq 3)$

3. $f(x) = \dfrac{3}{8}x^2$; $(0 \leq x \leq 2)$

4. $f(x) = \dfrac{3}{32}(x - 1)(5 - x)$; $(1 \leq x \leq 5)$

5. $f(x) = 20(x^3 - x^4)$; $(0 \leq x \leq 1)$

6. $f(x) = \dfrac{8}{7x^2}; \; (1 \le x \le 8)$

7. $f(x) = \dfrac{3}{14} \sqrt{x}; \; (1 \le x \le 4)$

8. $f(x) = \dfrac{12 - x}{72}; \; (0 \le x \le 12)$

9. a. Determine the value of the constant k such that the function $f(x) = k(4 - x)$ is a probability density function on the interval $[0, 4]$.
b. If x is a continuous random variable with the probability density function given in (a), compute the probability that x will assume a value between $x = 1$ and $x = 3$.

10. a. Determine the value of the constant k such that the function $f(x) = k/x^2$ is a probability density function on the interval $[1, 10]$.
b. If x is a continuous random variable with the probability density function of (a), compute the probability that x will assume a value between $x = 2$ and $x = 6$.

11. Life Span of a Plant Species   The life span of a certain plant species (in days) is described by the probability density function

$$f(x) = \frac{1}{100} e^{-x/100}$$

a. Find the probability that a plant of this species will live for 100 days or less.
b. Find the probability that a plant of this species will live more than 120 days.
c. Find the probability that a plant of this species will live more than 60 days but less than 140 days.

12. Average Waiting Time for Patients   The average waiting time for patients arriving at the Newtown Health Clinic between 1 P.M. and 4 P.M. on a weekday is an exponentially distributed random variable x with an expected value of 15 minutes.
a. Find the probability density function associated with x.
b. What is the probability that a patient arriving at the clinic between 1 P.M. and 4 P.M. will have to wait between 10 and 12 minutes?
c. What is the probability that a patient arriving at the clinic between 1 P.M. and 4 P.M. will have to wait longer than 15 minutes?

13. Expected Number of Chocolate Chips   The number of chocolate chips in each cookie of a certain brand has a distribution described by the probability density function

$$f(x) = \frac{1}{36}(6x - x^2) \qquad (0 \le x \le 6)$$

Find the expected number of chips in each cookie.

14. Reliability of Robots   The National Welding Company uses industrial robots in some of its assembly line operations. Management has determined that, on average, a robot breaks down after 1000 hours of use and that the lengths of time between breakdowns are exponentially distributed.
a. What is the probability that a robot selected at random will break down between 600 and 800 hours of use?
b. What is the probability that a robot will break down after 1200 hours of use?

15. Expressway Tollbooths   Suppose that the time intervals between arrivals of successive cars at an expressway tollbooth during rush hour are exponentially distributed and that the average time interval between arrivals is 8 seconds. Find the probability that the average time interval between arrivals of successive cars is more than 8 seconds.

16. Time Intervals Between Phone Calls   A study conducted by Uni-Mart, a mail-order department store, reveals that the time intervals between incoming telephone calls on its toll-free 800 line between 10 A.M. and 2 P.M. are exponentially distributed and that the average time interval is 30 seconds. What is the probability that the time interval between successive calls is more than 2 minutes?

17. Reliability of Microprocessors   The microprocessors manufactured by the United Motor Works Company, which are used in automobiles to regulate fuel consumption, are guaranteed against defects for 20,000 miles of use. Tests conducted in the laboratory under simulated driving conditions reveal that the distances driven before the microprocessors break down are exponentially distributed and that the average distance driven before the microprocessors fail is 100,000 miles. What is the probability that a microprocessor selected at random will fail during the warranty period?

SOLUTIONS TO SELF-CHECK EXERCISES 15.4

1. We compute

$$\int_0^4 k(4x - x^2)\, dx = k\left(2x^2 - \frac{1}{3}x^3\right)\Big|_0^4$$

$$= k\left(32 - \frac{64}{3}\right)$$

$$= \frac{32}{3}k$$

Since this value must be equal to 1, we see that $k = 3/32$.

2. The required probability is given by

$$P(1 \le x \le 3) = \int_1^3 f(x)\, dx$$

$$= \int_1^3 \frac{3}{32}(4x - x^2)\, dx$$

$$= \frac{3}{32}\left(2x^2 - \frac{1}{3}x^3\right)\Big|_1^3$$

$$= \frac{3}{32}\left[(18 - 9) - \left(2 - \frac{1}{3}\right)\right] = \frac{11}{16}$$

CHAPTER 15 SUMMARY OF PRINCIPAL FORMULAS AND TERMS

Formulas

1. Integration by parts $\displaystyle \int u\, dv = uv - \int v\, du$

2. Improper integral of f over $[a, \infty)$ $\displaystyle \int_a^\infty f(x)\, dx = \lim_{b \to \infty} \int_a^b f(x)\, dx$

3. Improper integral of f over $(-\infty, b]$ $\displaystyle \int_{-\infty}^b f(x)\, dx = \lim_{a \to -\infty} \int_a^b f(x)\, dx$

4. Improper integral of f over $(-\infty, \infty)$ $\displaystyle \int_{-\infty}^\infty f(x)\, dx = \int_{-\infty}^c f(x)\, dx$

$$+ \int_c^\infty f(x)\, dx$$

5. Present value of a perpetuity $\displaystyle PV = \frac{mP}{r}$

6. Trapezoidal Rule $\displaystyle \int_a^b f(x)\, dx \approx \frac{\Delta x}{2}[f(x_0) + 2f(x_1)$

$$+ 2f(x_2) + \cdots$$

$$+ 2f(x_{n-1}) + f(x_n)]$$

where $\Delta x = \dfrac{b - a}{n}$

7. Simpson's Rule

$$\int_a^b f(x)\,dx \approx \frac{\Delta x}{3}[f(x_0) + 4f(x_1)$$
$$+ 2f(x_2) + 4f(x_3)$$
$$+ 2f(x_4) + \cdots$$
$$+ 4f(x_{n-1}) + f(x_n)]$$

where $\Delta x = \dfrac{b-a}{n}$

8. Maximum error for Trapezoidal Rule

$\dfrac{M(b-a)^3}{12n^2}$, where $|f''(x)| \le M$

$(a \le x \le b)$

9. Maximum error for Simpson's Rule

$\dfrac{M(b-a)^5}{180n^4}$, where $|f^{(4)}(x)| \le M$

$(a \le x \le b)$

10. Probability an outcome of an experiment lies between a and b

$P(a \le x \le b) = \int_a^b f(x)\,dx$

11. Exponential density function $f(x) = ke^{-kx}$

12. Expected value $E(x) = x_1 p_1 + x_2 p_2 + \cdots + x_n p_n$

13. Expected value of a continuous random variable

$E(x) = \int_a^b x f(x)\,dx$

Terms

Improper integral

Convergent integral

Divergent integral

Perpetuity

Probability density function

Normal density function

Expected value

CHAPTER 15 REVIEW EXERCISES

In exercises 1–6, evaluate the integral.

1. $\displaystyle\int 2xe^{-x}\,dx$ **2.** $\displaystyle\int xe^{4x}\,dx$

3. $\displaystyle\int \ln 5x\,dx$ **4.** $\displaystyle\int_1^4 \ln 2x\,dx$

5. $\displaystyle\int_0^1 xe^{-2x}\,dx$ **6.** $\displaystyle\int_0^2 xe^{2x}\,dx$

7. Find the function f given that the slope of the tangent line to the graph of f at any point $(x, f(x))$ is

$$f'(x) = \frac{\ln x}{\sqrt{x}}$$

and that the graph of f passes through the point $(1, -2)$.

8. Find the function f given that the slope of the tangent line to the graph of f at any point $(x, f(x))$ is

$$f'(x) = \frac{e^x}{1 + e^x}$$

and that the graph of f passes through the point $(0, 0)$.

In exercises 9–14, evaluate each improper integral whenever it is convergent.

9. $\displaystyle\int_0^\infty e^{-2x}\, dx$

10. $\displaystyle\int_{-\infty}^0 e^{3x}\, dx$

11. $\displaystyle\int_3^\infty \frac{2}{x}\, dx$

12. $\displaystyle\int_2^\infty \frac{1}{(x+2)^{3/2}}\, dx$

13. $\displaystyle\int_2^\infty \frac{dx}{(1+2x)^2}$

14. $\displaystyle\int_1^\infty 3e^{1-x}\, dx$

In exercises 15–18, use the Trapezoidal Rule and Simpson's Rule to approximate the value of the definite integral.

15. $\displaystyle\int_1^3 \frac{dx}{1 + \sqrt{x}}; \, n = 4$

16. $\displaystyle\int_0^1 e^{x^2}\, dx; \, n = 4$

17. $\displaystyle\int_{-1}^1 \sqrt{1 + x^4}\, dx; \, n = 4$

18. $\displaystyle\int_1^3 \frac{e^x}{x}\, dx; \, n = 4$

19. Show that the function $f(x) = (3/128)(16 - x^2)$ is a probability density function on the interval $[0, 4]$.

20. Show that the function $f(x) = (1/9)x\sqrt{9 - x^2}$ is a probability density function on the interval $[0, 3]$.

21. a. Determine the value of the constant k such that the function $f(x) = kx\sqrt{4 - x^2}$ is a probability density function on the interval $[0, 2]$.
b. If x is a continuous random variable with the probability density function given in (a), compute the probability that x will assume a value between $x = 1$ and $x = 2$.

22. a. Determine the value of the constant k such that the function $f(x) = k/\sqrt{x}$ is a probability density function on the interval $[1, 4]$.
b. If x is a continuous random variable with the probability density function given in (a), compute the probability that x will assume a value between $x = 2$ and $x = 3$.

23. a. Determine the value of the constant k such that the function $f(x) = kx^2(3 - x)$ is a probability density function on the interval $[0, 3]$.
b. If x is a continuous random variable with the probability density function given in (a), compute the probability that x will assume a value between $x = 1$ and $x = 2$.

24. Records at the Centerville Hospital indicate that the length of time in days that a maternity patient stays in the hospital has a probability density function given by

$$P(t) = \frac{1}{4} e^{(-1/4)t}$$

a. What is the probability that a woman entering the maternity wing will be there more than 6 days?
b. What is the probability that a woman entering the maternity wing will be there less than 2 days?
c. What is the average length of time that a woman entering the maternity wing stays in the hospital?

25. The sales of the Starr Communication Company's newest video cartridge game, Laser Beams, are currently

$$te^{-0.05t}$$

units per month (each unit representing 1000 cartridges), where t denotes the number of months since the release of the cartridge. Find an expression that gives the total number of units of video cartridge games sold as a function of t. How many video games will be sold by the end of the first year?

26. The demand equation for a computer software program is given by

$$p = 2\sqrt{325 - x^2}$$

where p is the unit price in dollars and x is the quantity demanded per month in units of a thousand. Find the consumers' surplus if the market price is $30. Evaluate the definite integral using Simpson's Rule with $n = 10$.

27. Using aerial photographs, the Coast Guard was able to determine the dimensions of an oil spill along an embankment on a coastline, as shown in the accompanying figure. Using (a) the Trapezoidal Rule and (b) Simpson's Rule with $n = 10$, estimate the area of the oil spill.

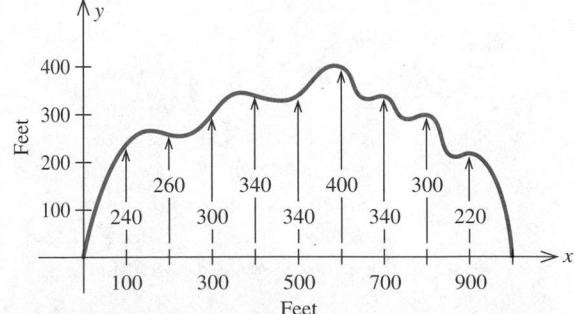

Up to now we have dealt with functions involving one variable. In many real-life situations, however, we encounter quantities that depend on two or more quantities. For example, the Consumer Price Index (CPI) compiled every month by the Bureau of Labor Statistics depends on the price of more than 95,000 consumer items from gas to groceries. In order to study such relationships, we need the notion of a function of several variables, and we study that first in this chapter. Next, generalizing the concept of the derivative of a function of one variable, we are led to the idea of the *partial derivatives* of a function of two or more variables. Partial derivatives enable us to study the rate of change of a function with respect to one variable while holding all the other variables constant. We next learn how to find the maximum and minimum values of a function of several variables. For example, we learn how a manufacturer can maximize her profits by producing the right amount of each of her products.

What should the dimensions of the swimming pool be? The operators of the Viking Princess, *a luxury cruise ship, are thinking about adding another swimming pool to the ship. The chief engineer has suggested that an area in the form of an ellipse, located in the rear of the promenade deck, would be suitable. Subject to this constraint, what are the dimensions of the largest pool that can be built? See Example 5, page 1124, for how to solve this problem.*

16

CALCULUS OF SEVERAL VARIABLES

1077

16.1 FUNCTIONS OF SEVERAL VARIABLES

Up to now, our study of calculus has been restricted to functions of one variable. In many practical situations, however, the formulation of a problem results in a mathematical model that involves a function of two or more variables. For example, suppose that the Ace Novelty Company determines that the profits are $6, $5, and $4 for three types of souvenirs it produces. Let x, y, and z denote the number of type-A, type-B, and type-C souvenirs to be made; then the company's profit is given by

$$P = 6x + 5y + 4z$$

and P is a function of the three variables x, y, and z.

Functions of Two Variables

Although this chapter deals with real-valued functions of several variables, most of our definitions and results are stated in terms of a function of two variables. One reason for adopting this approach, as you will soon see, is that there is a geometrical interpretation for this special case, which serves as an important visual aid. We can then draw upon the experience we gained from studying the two-variable case to help us understand the concepts and results connected with the more general case, which, by and large, is just a simple extension of the lower-dimensional case.

A FUNCTION OF TWO VARIABLES

A **real-valued function of two variables,** f, consists of

1. a set A of ordered pairs of real numbers (x, y) called the **domain** of the function and

2. a rule that associates with each ordered pair in the domain of f one and only one real number, denoted by $z = f(x, y)$.

The variables x and y are called **independent variables,** and the variable z, which is dependent on the values of x and y, is referred to as a **dependent variable.**

As in the case of a real-valued function of one real variable, the number $z = f(x, y)$ is called the **value of** f at the point (x, y). And, unless specified, the domain of the function f will be taken to be the largest possible set for which the rule defining f is meaningful.

EXAMPLE 1   Let f be the function defined by

$$f(x, y) = x + xy + y^2 + 2$$

Compute $f(0, 0)$, $f(1, 2)$, and $f(2, 1)$.

Solution   We have

$$f(0,0) = 0 + (0)(0) + 0^2 + 2 = 2$$
$$f(1,2) = 1 + (1)(2) + 2^2 + 2 = 9$$
$$f(2,1) = 2 + (2)(1) + 1^2 + 2 = 7$$

The domain of a function of two variables $f(x, y)$ is a set of ordered pairs of real numbers and may therefore be viewed as a subset of the xy-plane.

EXAMPLE 2   Find the domain of each of the following functions:

a. $f(x, y) = x^2 + y^2$ **b.** $g(x, y) = \dfrac{2}{x - y}$

c. $h(x, y) = \sqrt{1 - x^2 - y^2}$

Solution

a. $f(x, y)$ is defined for all real values of x and y, so the domain of the function f is the set of all points (x, y) in the xy-plane.

b. $g(x, y)$ is defined for all $x \neq y$, so the domain of the function g is the set of all points in the xy-plane except those lying on the line $y = x$ (Figure 16.1a).

c. We require that $1 - x^2 - y^2 \geq 0$ or $x^2 + y^2 \leq 1$, which is just the set of all points (x, y) lying on and inside the circle of radius 1 with center at the origin (Figure 16.1b).

Figure 16.1

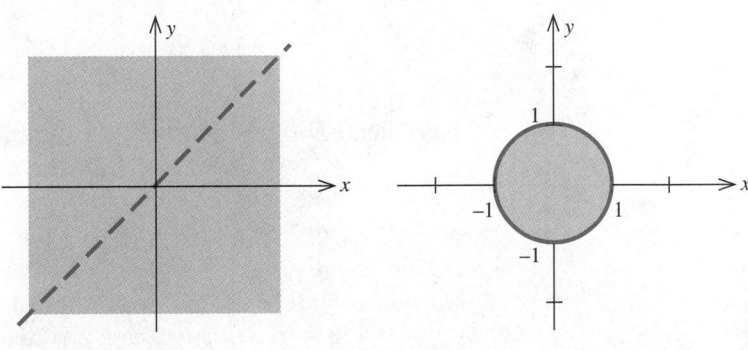

(a) Domain of g.   **(b)** Domain of h.

Applications

EXAMPLE 3   The Acrosonic Company manufactures a bookshelf loudspeaker system that may be bought fully assembled or in a kit. The demand equations that relate the unit prices, p and q, to the quantities demanded weekly, x and y, of the assembled and kit versions of the loudspeaker systems are given by

$$p = 300 - \frac{1}{4}x - \frac{1}{8}y \quad \text{and} \quad q = 240 - \frac{1}{8}x - \frac{3}{8}y$$

a. What is the weekly total revenue function $R(x, y)$?

b. What is the domain of the function R?

Solution

a. The weekly revenue realizable from the sale of x units of the assembled speaker systems at p dollars per unit is given by xp dollars. Similarly, the weekly revenue realizable from the sale of y units of the kits at q dollars per unit is given by yq dollars. Therefore, the weekly total revenue function R is given by

$$R(x, y) = xp + yq$$

$$= x\left(300 - \frac{1}{4}x - \frac{1}{8}y\right) + y\left(240 - \frac{1}{8}x - \frac{3}{8}y\right)$$

$$= -\frac{1}{4}x^2 - \frac{3}{8}y^2 - \frac{1}{4}xy + 300x + 240y$$

b. To find the domain of the function R, let us observe that the quantities x, y, p, and q must be nonnegative. This observation leads to the following system of linear inequalities:

$$300 - \frac{1}{4}x - \frac{1}{8}y \geq 0$$

$$240 - \frac{1}{8}x - \frac{3}{8}y \geq 0$$

$$x \geq 0$$

$$y \geq 0$$

The domain D of the function R is sketched in Figure 16.2.   ● ◐ ○

Figure 16.2
The domain of $R(x, y)$.

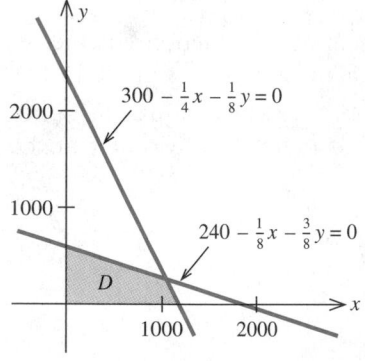

Suppose the total profit of a two-product company is given by $P(x, y)$, where x denotes the number of units of the first product produced and sold, and y denotes the number of units of the second product produced and sold. Fix $x = a$, where a is a positive number so that (a, y) is in the domain of P. Describe and give an economic interpretation of the function $f(y) = P(a, y)$. Next, fix $y = b$, where b is a positive number so that (x, b) is in the domain of P. Describe and give an economic interpretation of the function $g(x) = P(x, b)$.

EXAMPLE 4   The monthly payment that amortizes a loan of A dollars in t years when the interest rate is r per year is given by

$$P = f(A, r, t) = \frac{Ar}{12\left[1 - \left(1 + \dfrac{r}{12}\right)^{-12t}\right]}$$

Find the monthly payment for a home mortgage of $90,000 to be amortized over 30 years when the interest rate is 10% per year.

Solution   Letting $A = 90,000$, $r = 0.1$, and $t = 30$, we find the required monthly payment to be

$$P = f(90,000, 0.1, 30) = \frac{90,000(0.1)}{12\left[1 - \left(1 + \frac{0.1}{12}\right)^{-360}\right]}$$

$$\approx 789.81$$

or approximately $789.81.   ◐ ◐ ◐

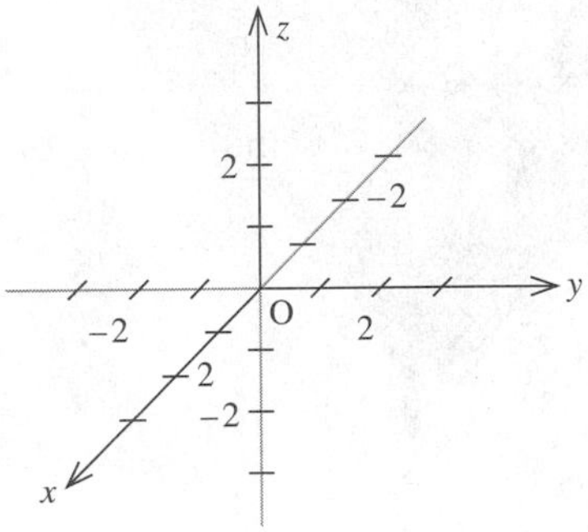

Graphs of Functions of Two Variables

In order to graph a function of two variables, we need a three-dimensional coordinate system. This is readily constructed by adding a third axis to the plane Cartesian coordinate system in such a way that the three resulting axes are mutually perpendicular and intersect at O. Observe that, by construction, the zeros of the three number scales coincide at the origin of the **three-dimensional Cartesian coordinate system** (Figure 16.3).

Figure 16.3
The three-dimensional Cartesian coordinate system.

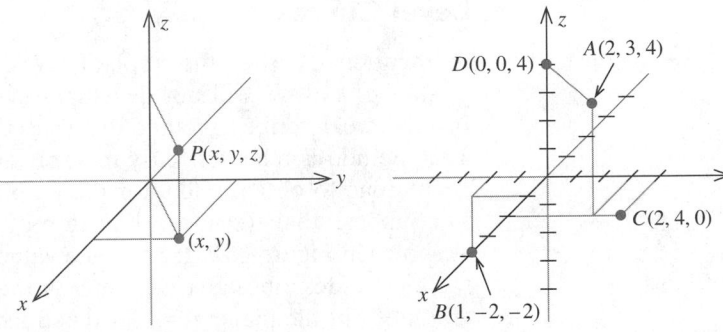

Figure 16.4

(a) A point in three-dimensional space.

(b) Some sample points in three-dimensional space.

A point in three-dimensional space can now be represented uniquely in this coordinate system by an **ordered triple** of numbers (x, y, z), and, conversely, every ordered triple of real numbers (x, y, z) represents a point in three-dimensional space (Figure 16.4a). For example, the points $A(2, 3, 4)$, $B(1, -2, -2)$, $C(2, 4, 0)$, and $D(0, 0, 4)$ are shown in Figure 16.4b.

Now, if $f(x, y)$ is a function of two variables x and y, the domain of f is a subset of the xy-plane. Let $z = f(x, y)$ so that there is one and only one point $(x, y, z) \equiv (x, y, f(x, y))$ associated with each point (x, y) in the domain of f. The totality of all such points makes up the **graph** of the function f and is, except for certain degenerate cases, a surface in three-dimensional space (Figure 16.5).

In interpreting the graph of a function $f(x, y)$, one often thinks of the value $z = f(x, y)$ of the function at the point (x, y) as the "height" of the point (x, y, z) on the graph of f. If $f(x, y) > 0$, then the point (x, y, z) is

Figure 16.5
The graph of a function in three-dimensional space.

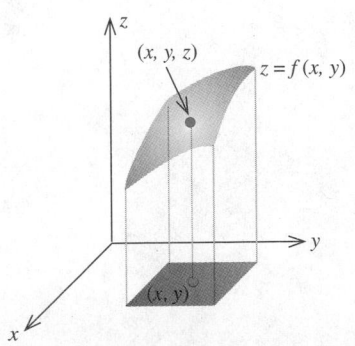

$f(x, y)$ units above the xy-plane; if $f(x, y) < 0$, then the point (x, y, z) is $|f(x, y)|$ units below the xy-plane.

In general, it is quite difficult to draw the graph of a function of two variables. But techniques have been developed that enable us to generate such graphs with minimum effort using a computer. Figure 16.6 shows the computer-generated graphs of two functions.

Figure 16.6
Two computer-generated graphs of functions of two variables.

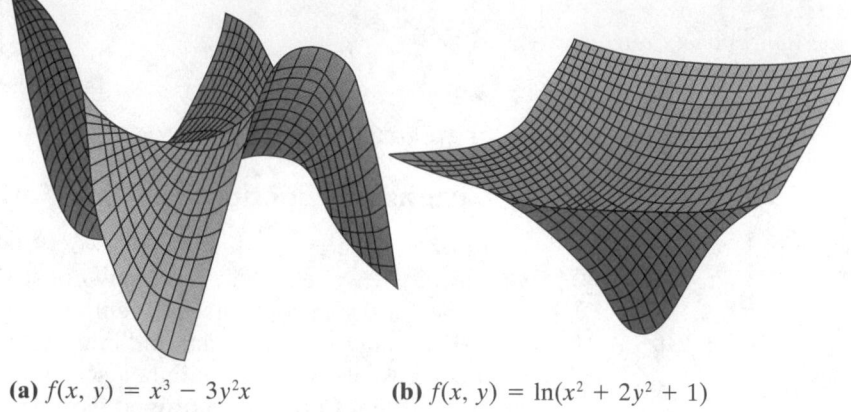

(a) $f(x, y) = x^3 - 3y^2x$ **(b)** $f(x, y) = \ln(x^2 + 2y^2 + 1)$

Level Curves

As mentioned earlier, the graph of a function of two variables is often difficult to sketch, and we will not develop a systematic procedure for sketching it. Instead, we describe a method that is used in constructing topographical maps. This method is relatively easy to apply and conveys sufficient information to enable one to obtain a feel for the graph of the function.

Suppose that $f(x, y)$ is a function of two variables x and y, with a graph as shown in Figure 16.7. If c is some value of the function f, then the equation $f(x, y) = c$ describes a curve lying on the plane $z = c$ called the **trace** of the graph of f in the plane $z = c$. If this trace is projected onto the xy-plane, the

Figure 16.7
The graph of the function $z = f(x, y)$ and its intersection with the plane $z = c$.

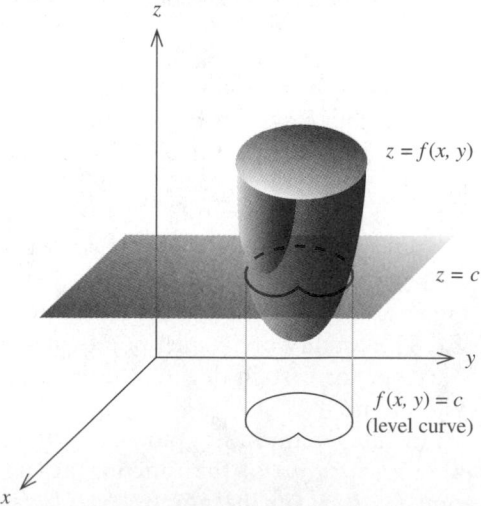

resulting curve in the xy-plane is called a **level curve.** By drawing the level curves corresponding to several admissible values of c, we obtain a **contour map.** Observe that, by construction, every point on a particular level curve corresponds to a point on the surface $z = f(x, y)$ that is a certain fixed distance from the xy-plane. Thus, by elevating or depressing the level curves that make up the contour map in one's mind, we can gct a feel for the general shape of the surface represented by the function f. Figure 16.8a shows a part of a mountain range with one peak; Figure 16.8b is the associated contour map.

Figure 16.8

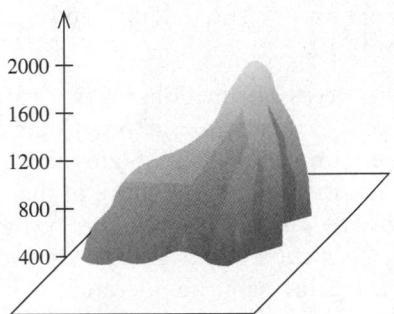

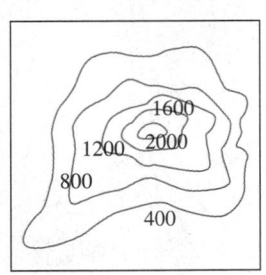

(a) A peak on a mountain range.

(b) A contour map for the mountain peak.

EXAMPLE 5   Sketch a contour map for the function $f(x, y) = x^2 + y^2$.

Solution   The level curves are the graphs of the equation $x^2 + y^2 = c$ for nonnegative numbers c. Taking $c = 0, 1, 4, 9,$ and 16, for example, we obtain

$$c = \ 0: x^2 + y^2 = 0$$
$$c = \ 1: x^2 + y^2 = 1$$
$$c = \ 4: x^2 + y^2 = 4 = 2^2$$
$$c = \ 9: x^2 + y^2 = 9 = 3^2$$
$$c = 16: x^2 + y^2 = 16 = 4^2$$

Figure 16.9

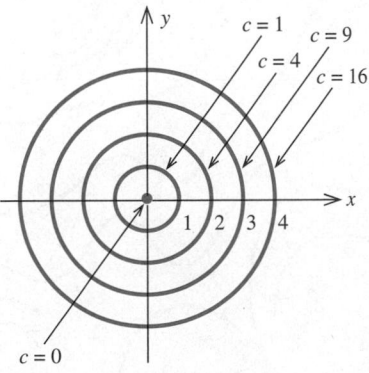

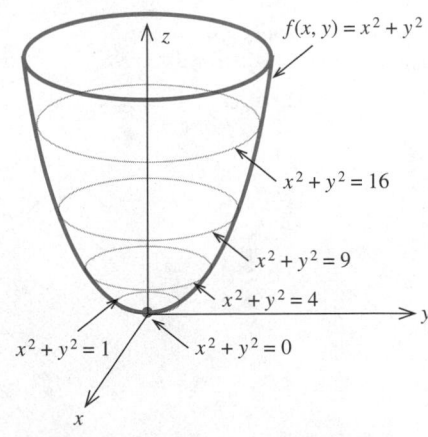

(a) Level curves of $f(x, y) = x^2 + y^2$.

(b) The graph of $f(x, y) = x^2 + y^2$.

The five level curves are concentric circles with center at the origin and radius given by $r = 0, 1, 2, 3,$ and 4, respectively (Figure 16.9a). A sketch of the graph of $f(x, y) = x^2 + y^2$ is included for your reference in Figure 16.9b.

● ● ●

Figure 16.10
Level curves for $f(x, y) = 2x^2 - y$.

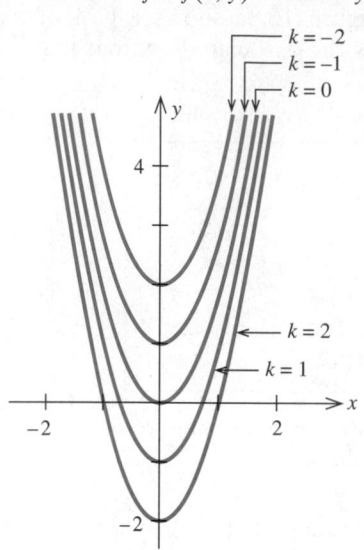

EXAMPLE 6   Sketch the level curves for the function $f(x, y) = 2x^2 - y$ corresponding to $z = -2, -1, 0, 1,$ and 2.

Solution   The level curves are the graphs of the equation $2x^2 - y = k$ or $y = 2x^2 - k$ for $k = -2, -1, 0, 1,$ and 2. The required level curves are shown in Figure 16.10.

● ● ●

Level curves of functions of two variables are found in many practical applications. For example, if $f(x, y)$ denotes the temperature at a location within the continental United States with longitude x and latitude y at a certain time of day, then the temperature at the point (x, y) is given by the "height" of the surface, represented by $z = f(x, y)$. In this situation, the level curve $f(x, y) = k$ is a curve superimposed on a map of the United States connecting points having the same temperature at a given time (Figure 16.11). These level curves are called **isotherms.**

Similarly, if $f(x, y)$ gives the barometric pressure at the location (x, y), then the level curves of the function f are called **isobars,** lines connecting points having the same barometric pressure at a given time.

As a final example, suppose that $P(x, y, z)$ is a function of three variables x, y, and z giving the profit realized when x, y, and z units of three products A, B, and C, respectively, are produced and sold. Then the equation $P(x, y, z) = k$, where k is a constant, represents a surface in three-dimensional space called a **level surface** of P. In this situation, the level surface represented by $P(x, y, z) = k$ represents the product mix that results in a profit of exactly k dollars. Such a level surface is called an **isoprofit surface.**

Figure 16.11
Isotherms: curves connecting points that have the same temperature.

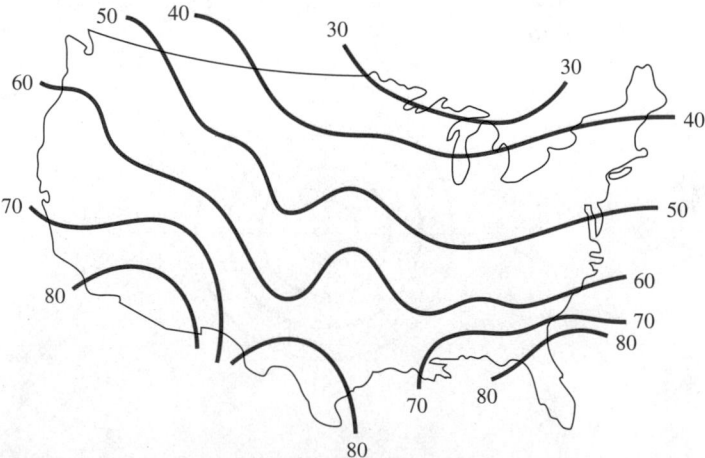

● ● ●

SELF-CHECK EXERCISES 16.1

1. Let $f(x, y) = x^2 - 3xy + \sqrt{x + y}$. Compute $f(1, 3)$ and $f(-1, 1)$. Is the point $(-1, 0)$ in the domain of f?

2. Find the domain of $f(x, y) = \dfrac{1}{x} + \dfrac{1}{x - y} - e^{x+y}$.

 3. The Odyssey Travel Agency has a monthly advertising budget of $20,000. Odyssey's management estimates that if they spend x dollars on newspaper advertising and y dollars on television advertising, then the monthly revenue will be

$$f(x, y) = 30x^{1/4}y^{3/4}$$

dollars. What will the monthly revenue be if Odyssey spends $5000 per month on newspaper ads and $15,000 per month on television ads? If Odyssey spends $4000 per month on newspaper ads and $16,000 per month on television ads?

Solutions to Self-Check Exercises 16.1 can be found on page 1088.

16.1 EXERCISES

1. Let $f(x, y) = 2x + 3y - 4$. Compute $f(0, 0)$, $f(1, 0)$, $f(0, 1)$, $f(1, 2)$, and $f(2, -1)$.

2. Let $g(x, y) = 2x^2 - y^2$. Compute $g(1, 2)$, $g(2, 1)$, $g(1, 1)$, $g(-1, 1)$, and $g(2, -1)$.

3. Let $f(x, y) = x^2 + 2xy - x + 3$. Compute $f(1, 2)$, $f(2, 1)$, $f(-1, 2)$, and $f(2, -1)$.

4. Let $h(x, y) = (x + y)/(x - y)$. Compute $h(0, 1)$, $h(-1, 1)$, $h(2, 1)$, and $h(\pi, -\pi)$.

5. Let $g(s, t) = 3s\sqrt{t} + t\sqrt{s} + 2$. Compute $g(1, 2)$, $g(2, 1)$, $g(0, 4)$, and $g(4, 9)$.

6. Let $f(x, y) = xye^{x^2+y^2}$. Compute $f(0, 0)$, $f(0, 1)$, $f(1, 1)$, and $f(-1, -1)$.

7. Let $h(s, t) = s \ln t - t \ln s$. Compute $h(1, e)$, $h(e, 1)$, and $h(e, e)$.

8. Let $f(u, v) = (u^2 + v^2)e^{uv^2}$. Compute $f(0, 1)$, $f(-1, -1)$, $f(a, b)$, and $f(b, a)$.

9. Let $g(r, s, t) = re^{s/t}$. Compute $g(1, 1, 1)$, $g(1, 0, 1)$, and $g(-1, -1, -1)$.

10. Let $g(u, v, w) = (ue^{vw} + ve^{uw} + we^{uv})/(u^2 + v^2 + w^2)$. Compute $g(1, 2, 3)$ and $g(3, 2, 1)$.

In exercises 11–18, find the domain of the function.

11. $f(x, y) = 2x + 3y$

12. $g(x, y, z) = x^2 + y^2 + z^2$

13. $h(u, v) = \dfrac{uv}{u - v}$ 14. $f(s, t) = \sqrt{s^2 + t^2}$

15. $g(r, s) = \sqrt{rs}$ 16. $f(x, y) = e^{-xy}$

17. $h(x, y) = \ln(x + y - 5)$

18. $h(u, v) = \sqrt{4 - u^2 - v^2}$

In exercises 19–24, sketch the level curves of the function corresponding to the given values of z.

19. $f(x, y) = 2x + 3y$; $z = -2, -1, 0, 1, 2$

20. $f(x, y) = -x^2 + y$; $z = -2, -1, 0, 1, 2$

21. $f(x, y) = 2x^2 + y$; $z = -2, -1, 0, 1, 2$

22. $f(x, y) = xy$; $z = -4, -2, 2, 4$

23. $f(x, y) = \sqrt{16 - x^2 - y^2}$; $z = 0, 1, 2, 3, 4$

24. $f(x, y) = e^x - y$; $z = -2, -1, 0, 1, 2$

25. The volume of a cylindrical tank of radius r and height h is given by

$$V = f(r, h) = \pi r^2 h$$

Find the volume of a cylindrical tank of radius 1.5 ft and height 4 ft.

26. IQs The IQ (intelligence quotient) of a person whose mental age is m years and whose chronological age is c years is defined as

$$f(m, c) = \frac{100m}{c}$$

What is the IQ of a nine-year-old child who has a mental age of 13.5 years?

27. Poiseuille's Law Poiseuille's Law states that the resistance R, measured in dynes, of blood flowing in a blood vessel of length l and radius r (both in centimeters) is given by

$$R = f(l, r) = \frac{kl}{r^4}$$

where k is the viscosity of blood (in dyne-sec/cm^2). What is the resistance, in terms of k, of blood flowing through an arteriole 4 cm long and of radius 0.1 cm?

28. Revenue Functions The Country Workshop manufactures both finished and unfinished furniture for the home. The estimated quantities demanded each week of its rolltop desks in the finished and unfinished versions are x and y units when the corresponding unit prices are

$$p = 200 - \frac{1}{5}x - \frac{1}{10}y$$

$$q = 160 - \frac{1}{10}x - \frac{1}{4}y$$

dollars, respectively.
a. What is the weekly total revenue function $R(x, y)$?
b. Find the domain of the function R.

29. For the total revenue function $R(x, y)$ of exercise 28, compute $R(100, 60)$ and $R(60, 100)$. Interpret your results.

30. Revenue Functions The Weston Publishing Company publishes a deluxe edition and a standard edition of

its English language dictionary. Weston's management estimates that the number of deluxe editions demanded is x copies per day and the number of standard editions demanded is y copies per day when the unit prices are

$$p = 20 - 0.005x - 0.001y$$

$$q = 15 - 0.001x - 0.003y$$

dollars, respectively.
a. Find the daily total revenue function $R(x, y)$.
b. Find the domain of the function R.

31. For the total revenue function $R(x, y)$ of exercise 30, compute $R(300, 200)$ and $R(200, 300)$. Interpret your results.

32. Volume of a Gas The volume of a certain mass of gas is related to its pressure and temperature by the formula

$$V = \frac{30.9T}{P}$$

where the volume V is measured in liters, the temperature T is measured in degrees Kelvin (obtained by adding 273° to the Celsius temperature), and the pressure P is measured in millimeters of mercury pressure.
a. Find the domain of the function V.
b. Calculate the volume of the gas at standard temperature and pressure—that is, when $T = 273°$K and $P = 760$ mm of mercury.

33. Surface Area of a Human Body An empirical formula by E. F. Dubois relates the surface area S of a human body (in square meters) to its weight W in kilograms and its height H in centimeters. The formula, given by

$$S = 0.007184W^{0.425}H^{0.725}$$

is used by physiologists in metabolism studies.
a. Find the domain of the function S.
b. What is the surface area of a human body that weighs 70 kilograms and has a height of 178 centimeters?

34. Arson for Profit A study of arson for profit was conducted by a team of paid civilian experts and police detectives appointed by the mayor of a large city. It was found that the number of suspicious fires in that city in 1992 was very closely related to the concentration of tenants in the city's public housing and to the level of reinvestment in the area in conventional mortgages by the ten largest banks. In fact, the number of fires was closely approximated by the formula

$$N(x, y) = \frac{100(1000 + 0.03x^2y)^{1/2}}{(5 + 0.2y)^2} \qquad (0 \le x \le 150; \\ 5 \le y \le 35)$$

where x denotes the number of persons per census tract and y denotes the level of reinvestment in the area in cents per dollar deposited. Using this formula, estimate the total number of suspicious fires in the districts of the city where the concentration of public housing tenants was 100 per census tract and the level of reinvestment was 20 cents per dollar deposited.

35. Continuously Compounded Interest If a principal of P dollars is deposited in an account earning interest at the rate of r per year compounded continuously, then the accumulated amount at the end of t years is given by

$$A = f(P, r, t) = Pe^{rt}$$

dollars. Find the accumulated amount at the end of three years if a sum of $10,000 is deposited in an account earning interest at the rate of 10% per year.

36. Home Mortgages The monthly payment that amortizes a loan of A dollars in t years when the interest rate is r per year is given by

$$P = f(A, r, t) = \frac{Ar}{12\left[1 - \left(1 + \frac{r}{12}\right)^{-12t}\right]}$$

a. Find the monthly payment for a home mortgage of $100,000 that will be amortized over 30 years with an interest rate of 8% per year. With an interest rate of 10% per year.
b. Find the monthly payment for a home mortgage of $100,000 that will be amortized over 20 years with an interest rate of 8% per year.

37. Home Mortgages Suppose that a home buyer secures a bank loan of A dollars to purchase a house. If the interest rate charged is r per year and the loan is to be amortized in t years, then the principal repayment at the end of i months is given by

$$B = f(A, r, t, i)$$

$$= A\left[\frac{\left(1 + \frac{r}{12}\right)^i - 1}{\left(1 + \frac{r}{12}\right)^{12t} - 1}\right] \qquad (0 \le i \le 12t)$$

Suppose that the Blakelys borrow a sum of $80,000 from a bank to help finance the purchase of a house and the bank charges interest at a rate of 9% per year. If the Blakelys agree to repay the loan in equal installments over 30 years, how much will they owe the bank after the 60th payment (5 years)? After the 240th payment (20 years)?

38. Force Generated by a Centrifuge A centrifuge is a machine designed for the specific purpose of subjecting materials to a sustained centrifugal force. The actual amount of centrifugal force, F, expressed in dynes (1 gram of force = 980 dynes) is given by

$$F = f(M, S, R) = \frac{\pi^2 S^2 MR}{900}$$

where S is in revolutions per minute (rpm), M is in grams, and R is in centimeters. Show that an object revolving at the rate of 600 rpm in a circle with a radius of 10 cm generates a centrifugal force that is approximately 40 times gravity.

39. Wilson Lot Size Formula The Wilson lot size formula in economics states that the optimal quantity Q of goods for a store to order is given by

$$Q = f(C, N, h) = \sqrt{\frac{2CN}{h}}$$

where C is the cost of placing an order, N is the number of items the store sells per week, and h is the weekly holding cost for each item. Find the most economical quantity of 10-speed bicycles to order if it costs the store $20 to place an order, and $5 to hold a bicycle for a week, and the store expects to sell 40 bicycles a week.

SOLUTIONS TO SELF-CHECK EXERCISES 16.1

1. $f(1, 3) = 1^2 - 3(1)(3) + \sqrt{1 + 3} = -6$

$f(-1, 1) = (-1)^2 - 3(-1)(1) + \sqrt{-1 + 1} = 4$

The point $(-1, 0)$ is not in the domain of f because the term $\sqrt{x + y}$ is not defined when $x = -1$ and $y = 0$. In fact, the domain of f consists of all real values of x and y that satisfy the inequality $x + y \geq 0$, the shaded half-plane shown in the accompanying figure.

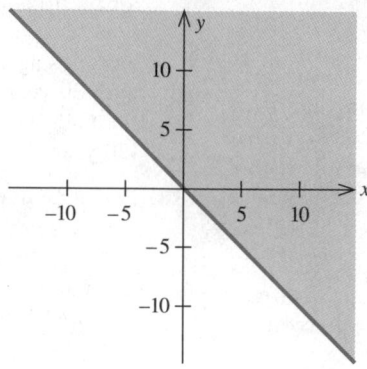

2. Since division by zero is not permitted, we see that $x \neq 0$ and $x - y \neq 0$. Therefore, the domain of f is the set of all points in the xy-plane not containing the y-axis ($x = 0$) and the straight line $x = y$.

3. If Odyssey spends $5000 per month on newspaper ads ($x = 5000$) and $15,000 per month on television ads ($y = 15,000$), then its monthly revenue will be given by

$$f(5000, 15,000) = 30(5000)^{1/4}(15,000)^{3/4}$$
$$\approx 341,926.06$$

or approximately $341,926. If the agency spends $4000 per month on newspaper ads and $16,000 per month on television ads, then its monthly revenue will be given by

$$f(4000, 16,000) = 30(4000)^{1/4}(16,000)^{3/4}$$
$$\approx 339,411.25$$

or approximately $339,411.

16.2 PARTIAL DERIVATIVES

Partial Derivatives

For a function $f(x)$ of one variable x, there is no ambiguity when we speak about the rate of change of $f(x)$ with respect to x since x must be constrained to move along the x-axis. The situation becomes more complicated, however,

Figure 16.12

We can approach a point in the plane from infinitely many directions.

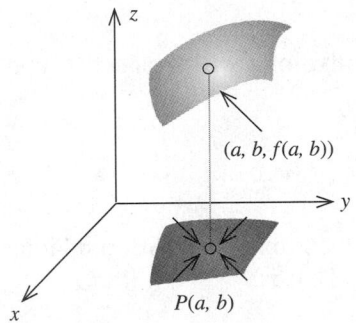

when we study the rate of change of a function of two or more variables. For example, the domain D of a function of two variables $f(x, y)$ is a subset of the plane (Figure 16.12), so if $P(a, b)$ is any point in the domain of f, there are infinitely many directions from which one can approach the point P. We may therefore ask for the rate of change of f at P along any of these directions.

However, we will not deal with this general problem. Instead, we will restrict ourselves to studying the rate of change of the function $f(x, y)$ at a point $P(a, b)$ in each of two *preferred directions*—namely, the direction parallel to the x-axis and the direction parallel to the y-axis. Let $y = b$, where b is a constant, so that $f(x, b)$ is a function of the one variable x. Since the equation $z = f(x, y)$ is the equation of a surface, the equation $z = f(x, b)$ is the equation of the curve C on the surface formed by the intersection of the surface and the plane $y = b$ (Figure 16.13).

Figure 16.13

The curve C is formed by the intersection of the plane $y = b$ with the surface $z = f(x, y)$.

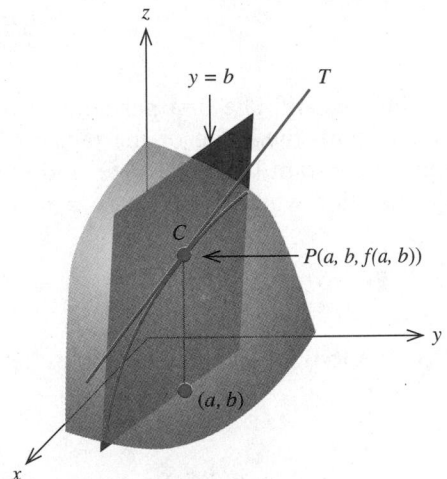

Because $f(x, b)$ is a function of one variable x, we may compute the derivative of f with respect to x at $x = a$. This derivative, obtained by keeping the variable y fixed and differentiating the resulting function $f(x, y)$ with respect to x, is called the **first partial derivative of f with respect to x at** (a, b), written

$$\frac{\partial z}{\partial x}(a, b) \quad \text{or} \quad \frac{\partial f}{\partial x}(a, b) \quad \text{or} \quad f_x(a, b)$$

Thus, $\dfrac{\partial z}{\partial x}(a, b) = \dfrac{\partial f}{\partial x}(a, b) = f_x(a, b) = \lim\limits_{h \to 0} \dfrac{f(a + h, b) - f(a, b)}{h}$

provided that the limit exists. The first partial derivative of f with respect to x at (a, b) measures both the slope of the tangent line T to the curve C and the rate of change of the function f in the x-direction when $x = a$ and $y = b$. We also write

$$\left.\frac{\partial f}{\partial x}\right|_{(a,b)} \equiv f_x(a, b)$$

Similarly, we define the **first partial derivative of f with respect to y** at (a, b), written

$$\frac{\partial z}{\partial y}(a, b) \quad \text{or} \quad \frac{\partial f}{\partial y}(a, b) \quad \text{or} \quad f_y(a, b)$$

as the derivative obtained by keeping the variable x fixed and differentiating the resulting function $f(x, y)$ with respect to y. That is,

$$\frac{\partial z}{\partial y}(a, b) = \frac{\partial f}{\partial y}(a, b) = f_y(a, b)$$

$$= \lim_{k \to 0} \frac{f(a, b + k) - f(a, b)}{k}$$

if the limit exists. The first partial derivative of f with respect to y at (a, b) measures both the slope of the tangent line T to the curve C, obtained by holding x constant (Figure 16.14), and the rate of change of the function f in the y-direction when $x = a$ and $y = b$. We write

$$\left.\frac{\partial f}{\partial y}\right|_{(a,b)} \equiv f_y(a, b)$$

Before looking at some examples, let us summarize these definitions.

Figure 16.14
The first partial derivative of f with respect to y at (a, b) measures the slope of the tangent line T to the curve C with x held constant.

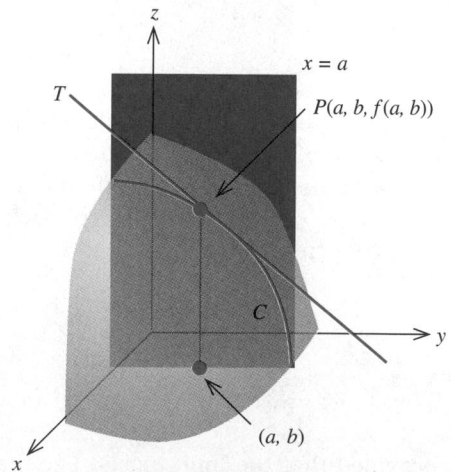

FIRST PARTIAL DERIVATIVES OF $f(x, y)$

Suppose that $f(x, y)$ is a function of the two variables x and y. Then the first partial derivative of f with respect to x at the point (x, y) is

$$\frac{\partial f}{\partial x} = \lim_{h \to 0} \frac{f(x + h, y) - f(x, y)}{h}$$

provided the limit exists. The first partial derivative of f with respect to y at the point (x, y) is

$$\frac{\partial f}{\partial y} = \lim_{k \to 0} \frac{f(x, y + k) - f(x, y)}{k}$$

provided the limit exists.

EXAMPLE I   Find the partial derivatives $\partial f/\partial x$ and $\partial f/\partial y$ of the function

$$f(x, y) = x^2 - xy^2 + y^3$$

What is the rate of change of the function f in the x-direction at the point $(1, 2)$? What is the rate of change of the function f in the y-direction at the point $(1, 2)$?

Solution   To compute $\partial f/\partial x$, think of the variable y as a constant and differentiate the resulting function of x with respect to x. Let us write

$$f(x, y) = x^2 - xy^2 + y^3$$

where the variable y to be treated as a constant is shown in color. Then,

$$\frac{\partial f}{\partial x} = 2x - y^2$$

To compute $\partial f/\partial y$, think of the variable x as being fixed—that is, as a constant—and differentiate the resulting function of y with respect to y. In this case,

$$f(x, y) = x^2 - xy^2 + y^3$$

so that

$$\frac{\partial f}{\partial y} = -2xy + 3y^2$$

The rate of change of the function f in the x-direction at the point $(1, 2)$ is given by

$$f_x(1, 2) = \frac{\partial f}{\partial x}\bigg|_{(1, 2)} = 2(1) - 2^2 = -2$$

That is, f decreases 2 units for each unit increase in the x-direction, y being kept constant ($y = 2$). The rate of change of the function f in the y-direction at the point (1, 2) is given by

$$f_y(1, 2) = \left.\frac{\partial f}{\partial y}\right|_{(1, 2)} = -2(1)(2) + 3(2)^2 = 8$$

That is, f increases 8 units for each unit increase in the y-direction, x being kept constant ($x = 1$).   ο ο ο

Refer to the Group Discussion question on page 1080. Suppose the management of the company has decided that the projected sales of the first product is a units. Describe how you might help management decide how many units of the second product the company should produce and sell in order to maximize the company's total profit. Justify your method to management. Suppose, however, management feels that b units of the second product can be manufactured and sold. How would you help management decide how many units of the first product to manufacture in order to maximize the company's total profit?

EXAMPLE 2   Compute the first partial derivatives of each of the following functions.

a. $f(x, y) = \dfrac{xy}{x^2 + y^2}$ **b.** $g(s, t) = (s^2 - st + t^2)^5$

c. $h(u, v) = e^{u^2 - v^2}$ **d.** $f(x, y) = \ln(x^2 + 2y^2)$

Solution

a. To compute $\partial f/\partial x$, think of the variable y as a constant. Thus,

$$f(x, y) = \frac{xy}{x^2 + y^2}$$

so that, upon using the Quotient Rule, we have

$$\frac{\partial f}{\partial x} = \frac{(x^2 + y^2)y - xy(2x)}{(x^2 + y^2)^2}$$

$$= \frac{y(y^2 - x^2)}{(x^2 + y^2)^2}$$

upon simplification and factorization. To compute $\partial f/\partial y$, think of the variable x as a constant. Thus,

$$f(x, y) = \frac{xy}{x^2 + y^2}$$

so that, upon using the Quotient Rule once again, we obtain

$$\frac{\partial f}{\partial y} = \frac{(x^2 + y^2)x - xy(2y)}{(x^2 + y^2)^2}$$

$$= \frac{x(x^2 - y^2)}{(x^2 + y^2)^2}$$

b. To compute $\partial g/\partial s$, we treat the variable t as if it were a constant. Thus,

$$g(s, t) = (s^2 - st + t^2)^5$$

Using the General Power Rule, we find

$$\frac{\partial g}{\partial s} = 5(s^2 - st + t^2)^4 \cdot (2s - t)$$

$$= 5(2s - t)(s^2 - st + t^2)^4$$

To compute $\partial g/\partial t$, we treat the variable s as if it were a constant. Thus,

$$g(s, t) = (s^2 - st + t^2)^5$$

$$\frac{\partial g}{\partial t} = 5(s^2 - st + t^2)^4(-s + 2t)$$

$$= 5(2t - s)(s^2 - st + t^2)^4$$

c. To compute $\partial h/\partial u$, think of the variable v as a constant. Thus,

$$h(u, v) = e^{u^2 - v^2}$$

Using the Chain Rule for exponential functions, we have

$$\frac{\partial h}{\partial u} = e^{u^2 - v^2} \cdot 2u$$

$$= 2ue^{u^2 - v^2}$$

Next, we treat the variable u as if it were a constant,

$$h(u, v) = e^{u^2 - v^2}$$

and we obtain

$$\frac{\partial h}{\partial v} = e^{u^2 - v^2} \cdot (-2v)$$

$$= -2ve^{u^2 - v^2}$$

d. To compute $\partial f/\partial x$, think of the variable y as a constant. Thus,

$$f(x, y) = \ln(x^2 + 2y^2)$$

so that the Chain Rule for logarithmic functions gives

$$\frac{\partial f}{\partial x} = \frac{2x}{x^2 + 2y^2}$$

Next, treating the variable x as if it were a constant, we find

$$f(x, y) = \ln(x^2 + 2y^2)$$

$$\frac{\partial f}{\partial y} = \frac{4y}{x^2 + 2y^2}$$

a. Let (a, b) be a point in the domain of $f(x, y)$. Put $g(x) = f(x, b)$ and suppose g is differentiable at $x = a$. Explain why you can find $f_x(a, b)$ by computing $g'(a)$. How would you go about calculating $f_y(a, b)$ using a similar technique? Give a geometric interpretation of these processes.

b. Let $f(x, y) = x^2y^3 - 3x^2y + 2$. Use the method of part (a) to find $f_x(1, 2)$ and $f_y(1, 2)$.

To compute the partial derivative of a function of several variables with respect to one variable, say x, we think of the other variables as if they were constants and differentiate the resulting function with respect to x.

EXAMPLE 3 Compute the first partial derivatives of the function

$$w = f(x, y, z) = xyz - xe^{yz} + x \ln y$$

Solution Here we have a function of three variables, x, y, and z, and we are required to compute

$$\frac{\partial f}{\partial x}, \quad \frac{\partial f}{\partial y}, \quad \text{and} \quad \frac{\partial f}{\partial z}$$

To compute f_x, we think of the other two variables, y and z, as fixed, and we differentiate the resulting function of x with respect to x, thereby obtaining

$$f_x = yz - e^{yz} + \ln y$$

To compute f_y, we think of the other two variables, x and z, as constants, and we differentiate the resulting function of y with respect to y. We then obtain

$$f_y = xz - xze^{yz} + \frac{x}{y}$$

Finally, to compute f_z, we treat the variables x and y as constants and differentiate the function f with respect to z, obtaining

$$f_z = xy - xye^{yz} \qquad \qquad \circ \ \circ \ \circ$$

EXPLORING WITH TECHNOLOGY

Refer to the preceding Group Discussion question. Let

$$f(x, y) = \frac{e^{\sqrt{xy}}}{(1 + xy^2)^{3/2}}$$

1. Compute $g(x) = f(x, 1)$ and use a graphing utility to plot the graph of g in the viewing rectangle $[0, 2] \times [0, 2]$.
2. Use the differentiation operation of your graphing utility to find $g'(1)$ and hence $f_x(1, 1)$.
3. Compute $h(y) = f(1, y)$ and use a graphing utility to plot the graph of h in the viewing rectangle $[0, 2] \times [0, 2]$.
4. Use the differentiation operation of your graphing utility to find $h'(1)$ and hence $f_y(1, 1)$.

$$\circ \quad \circ \quad \circ$$

The Cobb-Douglas Production Function

For an economic interpretation of the first partial derivatives of a function of two variables, let us turn our attention to the function

$$f(x, y) = ax^b y^{1-b} \tag{1}$$

where a and b are positive constants with $0 < b < 1$. This function is called the **Cobb-Douglas production function.** Here x stands for the amount of money expended for labor, y stands for the cost of capital equipment (buildings, machinery, and other tools of production), and the function f measures the output of the finished product (in suitable units) and is called, accordingly, the **production function.**

The partial derivative f_x is called the **marginal productivity of labor.** It measures the rate of change of production with respect to the amount of money expended for labor, with the level of capital expenditure held constant. Similarly, the partial derivative f_y, called the **marginal productivity of capital,** measures the rate of change of production with respect to the amount expended on capital, with the level of labor expenditure held fixed.

EXAMPLE 4   A certain country's production in the early years following World War II is described by the function

$$f(x, y) = 30x^{2/3} y^{1/3}$$

units, when x units of labor and y units of capital were used.

a. Compute f_x and f_y.

b. What are the marginal productivity of labor and the marginal productivity of capital when the amounts expended on labor and capital are 125 units and 27 units, respectively?

c. Should the government have encouraged capital investment rather than increasing expenditure on labor to increase the country's productivity?

Solution

a. $f_x = 30 \cdot \dfrac{2}{3} x^{-1/3} y^{1/3} = 20 \left(\dfrac{y}{x} \right)^{1/3}$

$f_y = 30x^{2/3} \cdot \dfrac{1}{3} y^{-2/3} = 10 \left(\dfrac{x}{y} \right)^{2/3}$

b. The required marginal productivity of labor is given by

$$f_x(125, 27) = 20 \left(\frac{27}{125} \right)^{1/3} = 20 \left(\frac{3}{5} \right)$$

or 12 units per unit increase in labor expenditure (capital expenditure is held constant at 27 units). The required marginal productivity of capital is given by

$$f_y(125, 27) = 10 \left(\frac{125}{27}\right)^{2/3} = 10 \left(\frac{25}{9}\right)$$

or $27\frac{7}{9}$ units per unit increase in capital expenditure (labor outlay is held constant at 125 units).

c. From the results of (b), we see that a unit increase in capital expenditure resulted in a much faster increase in productivity than a unit increase in labor expenditure would have. Therefore, the government should have encouraged increased spending on capital rather than on labor during the early years of reconstruction.

Substitute and Complementary Commodities

For another application of the first partial derivatives of a function of two variables in the field of economics, let us consider the relative demands of two commodities. We say that the two commodities are **substitute** (competitive) **commodities** if a decrease in the demand for one results in an increase in the demand for the other. Examples of competitive commodities are coffee and tea. Conversely, two commodities are referred to as **complementary commodities** if a decrease in the demand for one results in a decrease in the demand for the other as well. Examples of complementary commodities are automobiles and tires.

We now derive a criterion for determining whether two commodities A and B are substitute or complementary. Suppose that the demand equations that relate the quantities demanded, x and y, to the unit prices, p and q, of the two commodities are given by

$$x = f(p, q) \quad \text{and} \quad y = g(p, q)$$

Let us consider the partial derivative $\partial f/\partial p$. Since f is the demand function for commodity A, we see that, for fixed q, f is typically a decreasing function of p—that is, $\partial f/\partial p < 0$. Now, if the two commodities were substitute commodities, then the quantity demanded of commodity B would increase with respect to p—that is $\partial g/\partial p > 0$. A similar argument with p fixed shows that if A and B are substitute commodities, then $\partial f/\partial q > 0$. Thus, the two commodities A and B are substitute commodities if

$$\frac{\partial f}{\partial q} > 0 \quad \text{and} \quad \frac{\partial g}{\partial p} > 0$$

Similarly A and B are complementary commodities if

$$\frac{\partial f}{\partial q} < 0 \quad \text{and} \quad \frac{\partial g}{\partial p} < 0$$

SUBSTITUTE AND COMPLEMENTARY COMMODITIES	Two commodities A and B are **substitute commodities** if

$$\frac{\partial f}{\partial q} > 0 \quad \text{and} \quad \frac{\partial g}{\partial p} > 0 \tag{2}$$

Two commodities A and B are **complementary commodities** if

$$\frac{\partial f}{\partial q} < 0 \quad \text{and} \quad \frac{\partial g}{\partial p} < 0 \tag{3}$$

EXAMPLE 5 Suppose that the daily demand for butter is given by

$$x = f(p, q) = \frac{3q}{1 + p^2}$$

and the daily demand for margarine is given by

$$y = g(p, q) = \frac{2p}{1 + \sqrt{q}} \qquad (p > 0, q > 0)$$

where p and q denote the prices per pound (in dollars) of butter and margarine, respectively, and x and y are measured in millions of pounds. Determine whether these two commodities are substitute, complementary, or neither.

Solution We compute

$$\frac{\partial f}{\partial q} = \frac{3}{1 + p^2}$$

and

$$\frac{\partial g}{\partial p} = \frac{2}{1 + \sqrt{q}}$$

Since

$$\frac{\partial f}{\partial q} > 0 \quad \text{and} \quad \frac{\partial g}{\partial p} > 0$$

for all values of $p > 0$ and $q > 0$, we conclude that butter and margarine are substitute commodities.   ◐ ◐ ◐

Second-Order Partial Derivatives

The first partial derivatives $f_x(x, y)$ and $f_y(x, y)$ of a function $f(x, y)$ of the two variables x and y are also functions of x and y. As such, we may differentiate each of the functions f_x and f_y to obtain the **second-order partial derivatives of f.** (See Figure 16.15.) Thus, differentiating the function f_x with respect to x leads to the second partial derivative

$$f_{xx} \equiv \frac{\partial^2 f}{\partial x^2} = \frac{\partial}{\partial x}(f_x)$$

Figure 16.15
A schematic showing the four second-order partial derivatives of f.

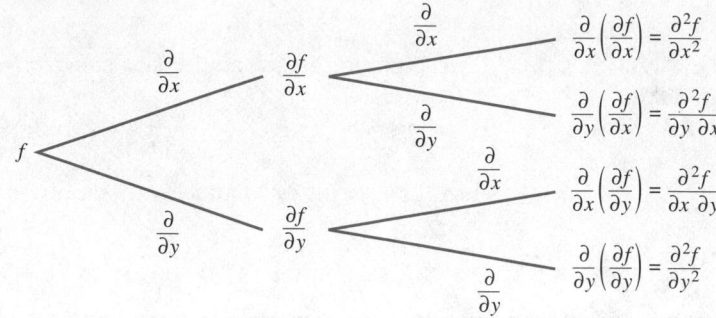

However, differentiation of f_x with respect to y leads to the second partial derivative

$$f_{xy} \equiv \frac{\partial^2 f}{\partial y\, \partial x} = \frac{\partial}{\partial y}(f_x)$$

Similarly, differentiation of the function f_y with respect to x and with respect to y leads to

$$f_{yx} \equiv \frac{\partial^2 f}{\partial x\, \partial y} = \frac{\partial}{\partial x}(f_y)$$

$$f_{yy} \equiv \frac{\partial^2 f}{\partial y^2} = \frac{\partial}{\partial y}(f_y)$$

respectively. Note that, in general, it is not true that $f_{xy} = f_{yx}$. However, in most practical applications f_{xy} and f_{yx} are equal.

EXAMPLE 6   Find the second-order partial derivatives of the function

$$f(x, y) = x^3 - 3x^2y + 3xy^2 + y^2$$

Solution   The first partial derivatives of f are

$$f_x = \frac{\partial}{\partial x}(x^3 - 3x^2y + 3xy^2 + y^2)$$

$$= 3x^2 - 6xy + 3y^2$$

$$f_y = \frac{\partial}{\partial y}(x^3 - 3x^2y + 3xy^2 + y^2)$$

$$= -3x^2 + 6xy + 2y$$

Therefore, $$f_{xx} = \frac{\partial}{\partial x}(f_x) = \frac{\partial}{\partial x}(3x^2 - 6xy + 3y^2)$$

$$= 6x - 6y = 6(x - y)$$

$$f_{xy} = \frac{\partial}{\partial y}(f_x) = \frac{\partial}{\partial y}(3x^2 - 6xy + 3y^2)$$
$$= -6x + 6y = 6(y - x)$$
$$f_{yx} = \frac{\partial}{\partial x}(f_y) = \frac{\partial}{\partial x}(-3x^2 + 6xy + 2y)$$
$$= -6x + 6y = 6(y - x)$$
$$f_{yy} = \frac{\partial}{\partial y}(f_y) = \frac{\partial}{\partial y}(-3x^2 + 6xy + 2y)$$
$$= 6x + 2$$

ooo

EXAMPLE 7 Find the second-order partial derivatives of the function

$$f(x, y) = e^{xy^2}$$

Solution   We have

$$f_x = \frac{\partial}{\partial x}(e^{xy^2})$$
$$= y^2 e^{xy^2}$$
$$f_y = \frac{\partial}{\partial y}(e^{xy^2})$$
$$= 2xy e^{xy^2}$$

so the required second-order partial derivatives of f are

$$f_{xx} = \frac{\partial}{\partial x}(f_x) = \frac{\partial}{\partial x}(y^2 e^{xy^2})$$
$$= y^4 e^{xy^2}$$
$$f_{xy} = \frac{\partial}{\partial y}(f_x) = \frac{\partial}{\partial y}(y^2 e^{xy^2})$$
$$= 2y e^{xy^2} + 2xy^3 e^{xy^2}$$
$$= 2y e^{xy^2}(1 + xy^2)$$
$$f_{yx} = \frac{\partial}{\partial x}(f_y) = \frac{\partial}{\partial x}(2xy e^{xy^2})$$
$$= 2y e^{xy^2} + 2xy^3 e^{xy^2}$$
$$= 2y e^{xy^2}(1 + xy^2)$$
$$f_{yy} = \frac{\partial}{\partial y}(f_y) = \frac{\partial}{\partial y}(2xy e^{xy^2})$$
$$= 2x e^{xy^2} + (2xy)(2xy)e^{xy^2}$$
$$= 2x e^{xy^2}(1 + 2xy^2)$$

ooo

SELF-CHECK EXERCISES 16.2

1. Compute the first partial derivatives of $f(x, y) = x^3 - 2xy^2 + y^2 - 8$.

2. Find the first partial derivatives of $f(x, y) = x \ln y + ye^x - x^2$ at $(0, 1)$ and interpret your results.

3. Find the second-order partial derivatives of the function of Self-Check exercise 1.

4. A certain country's production is described by the function

$$f(x, y) = 60x^{1/3}y^{2/3}$$

when x units of labor and y units of capital are used.

a. What are the marginal productivity of labor and the marginal productivity of capital when the amounts expended on labor and capital are 125 units and 8 units, respectively?

b. Should the government encourage capital investment rather than increased expenditure on labor at this time to increase the country's productivity?

Solutions to Self-Check Exercises 16.2 can be found on page 1104.

16.2 EXERCISES

In exercises 1–22, find the first partial derivatives of the function.

1. $f(x, y) = 2x + 3y + 5$

2. $f(x, y) = 2xy$

3. $g(x, y) = 2x^2 + 4y + 1$

4. $f(x, y) = 1 + x^2 + y^2$

5. $f(x, y) = \dfrac{2y}{x^2}$

6. $f(x, y) = \dfrac{x}{1 + y}$

7. $g(u, v) = \dfrac{u - v}{u + v}$

8. $f(x, y) = \dfrac{x^2 - y^2}{x^2 + y^2}$

9. $f(s, t) = (s^2 - st + t^2)^3$

10. $g(s, t) = s^2t + st^{-3}$

11. $f(x, y) = (x^2 + y^2)^{2/3}$

12. $f(x, y) = x\sqrt{1 + y^2}$

13. $f(x, y) = e^{xy+1}$

14. $f(x, y) = (e^x + e^y)^5$

15. $f(x, y) = x \ln y + y \ln x$

16. $f(x, y) = x^2e^{y^2}$

17. $g(u, v) = e^u \ln v$

18. $f(x, y) = \dfrac{e^{xy}}{x + y}$

19. $f(x, y, z) = xyz + xy^2 + yz^2 + zx^2$

20. $g(u, v, w) = \dfrac{2uvw}{u^2 + v^2 + w^2}$

21. $h(r, s, t) = e^{rst}$

22. $f(x, y, z) = xe^{y/z}$

In exercises 23–32, evaluate the first partial derivatives of the function at the given point.

23. $f(x, y) = x^2y + xy^2$; $(1, 2)$

24. $f(x, y) = x^2 + xy + y^2 + 2x - y$; $(-1, 2)$

25. $f(x, y) = x\sqrt{y} + y^2$; $(2, 1)$

26. $g(x, y) = \sqrt{x^2 + y^2}$; $(3, 4)$

27. $f(x, y) = \dfrac{x}{y}$; $(1, 2)$

28. $f(x, y) = \dfrac{x + y}{x - y}$; $(1, -2)$

29. $f(x, y) = e^{xy}$; $(1, 1)$

30. $f(x, y) = e^x \ln y$; $(0, e)$

31. $f(x, y, z) = x^2 yz^3$; $(1, 0, 2)$

32. $f(x, y, z) = x^2 y^2 + z^2$; $(1, 1, 2)$

In exercises 33–40, find the second-order partial derivatives of the given function. In each case, show that the mixed partial derivatives f_{xy} and f_{yx} are equal.

33. $f(x, y) = x^2 y + xy^3$

34. $f(x, y) = x^3 + x^2 y + x + 4$

35. $f(x, y) = x^2 - 2xy + 2y^2 + x - 2y$

36. $f(x, y) = x^3 + x^2 y^2 + y^3 + x + y$

37. $f(x, y) = \sqrt{x^2 + y^2}$

38. $f(x, y) = x\sqrt{y} + y\sqrt{x}$

39. $f(x, y) = e^{-x/y}$

40. $f(x, y) = \ln(1 + x^2 y^2)$

41. **Productivity of a Country** The productivity of a South American country is given by the function

$$f(x, y) = 20x^{3/4} y^{1/4}$$

when x units of labor and y units of capital are used.
a. What are the marginal productivity of labor and the marginal productivity of capital when the amounts expended on labor and capital are 256 units and 16 units, respectively?
b. Should the government encourage capital investment rather than increased expenditure on labor at this time in order to increase the country's productivity?

42. **Productivity of a Country** The productivity of a country in Western Europe is given by the function

$$f(x, y) = 40x^{4/5} y^{1/5}$$

when x units of labor and y units of capital are used.
a. What are the marginal productivity of labor and the marginal productivity of capital when the amounts ex-

pended on labor and capital are 32 units and 243 units, respectively?
b. Should the government encourage capital investment rather than increased expenditure on labor at this time in order to increase the country's productivity?

43. **Land Prices** The rectangular region R shown in the accompanying figure represents a city's financial district. The price of land in the district is approximated by the function

$$p(x, y) = 200 - 10\left(x - \frac{1}{2}\right)^2 - 15(y - 1)^2$$

where $p(x, y)$ is the price of land at the point (x, y) in dollars per square foot and x and y are measured in miles. Compute $\dfrac{\partial p}{\partial x}\,(0, 1)$ and $\dfrac{\partial p}{\partial y}\,(0, 1)$ and interpret your results.

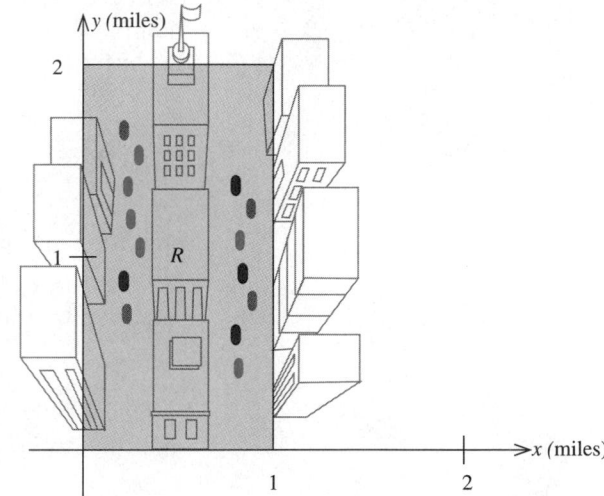

44. **Complementary and Substitute Commodities** In a survey conducted by *Home Entertainment* magazine it was determined that the demand equation for videocassette recorders (VCRs) is given by

$$x = f(p, q) = 10,000 - 10p + 0.2q^2$$

and the demand equation for videodisc players is given by

$$y = g(p, q) = 5000 + 0.8p^2 - 20q$$

where p and q denote the unit prices (in dollars) for the cassette recorders and disc players, respectively, and x

USING TECHNOLOGY

FINDING PARTIAL DERIVATIVES AT A GIVEN POINT

Suppose $f(x, y)$ is a function of two variables and we wish to compute

$$f_x(a, b) = \frac{\partial f}{\partial x}\bigg|_{(a, b)}$$

Recall that in computing $\partial f/\partial x$, we think of y as being fixed. But in this situation, we are evaluating $\partial f/\partial x$ at (a, b). Therefore, we set y equal to b. Doing this leads to the function g of one variable, x, defined by

$$g(x) = f(x, b)$$

It follows from the definition of the partial derivative that

$$f_x(a, b) = g'(a)$$

Thus, the value of the partial derivative $\partial f/\partial x$ at a given point (a, b) can be found by evaluating the derivative of a function of one variable. In particular, the latter can be found by using the numerical derivative operation of a graphing utility. We find $f_y(a, b)$ in a similar manner.

EXAMPLE 1 Let $f(x, y) = (1 + xy^2)^{3/2}e^{x^2y}$. Find (a) $f_x(1, 2)$ and (b) $f_y(1, 2)$.

Solution

a. Define $g(x) = f(x, 2) = (1 + 4x)^{3/2}e^{2x^2}$. Using the numerical derivative operation to find $g'(1)$, we obtain

$$f_x(1, 2) = g'(1) \approx 429.5832249$$

b. Define $h(y) = f(1, y) = (1 + y^2)^{3/2}e^y$. Using the numerical derivative operation to find $h'(2)$, we obtain

$$f_y(1, 2) = h'(2) \approx 181.74674899$$

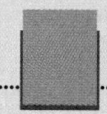

EXERCISES

In exercises 1–6, use a graphing utility to compute $\partial f/\partial x$ and $\partial f/\partial y$ at the given point.

1. $f(x, y) = \sqrt{x}(2 + xy^2)^{1/3}$; $(1, 2)$

2. $f(x, y) = \sqrt{xy}(1 + 2xy)^{2/3}$; $(1, 4)$

3. $f(x, y) = \dfrac{x + y^2}{1 + x^2y}$; $(1, 2)$

4. $f(x, y) = \dfrac{xy^2}{(\sqrt{x} + \sqrt{y})^2}$; $(4, 1)$

5. $f(x, y) = e^{-xy^2}(x + y)^{1/3}$; $(1, 1)$

6. $f(x, y) = \dfrac{\ln(\sqrt{x} + y^2)}{x^2 + y^2}$; $(4, 1)$

and y denote the number of cassette recorders and disc players demanded per week. Determine whether these two products are substitute, complementary, or neither.

45. Complementary and Substitute Commodities In the survey mentioned in exercise 44, it was determined that the demand equation for VCRs is given by

$$x = f(p, q) = 10,000 - 10p - e^{0.5q}$$

and the demand equation for blank tapes for the VCRs is given by

$$y = g(p, q) = 50,000 - 4000q - 10p$$

where p and q denote the unit prices, respectively, and x and y denote the number of VCRs and the number of blank VCR tapes demanded per week. Determine whether these two products are substitute, complementary, or neither.

46. Complementary and Substitute Commodities Refer to exercise 28, Exercises 16.1. Show that the finished and unfinished home furniture manufactured by the Country Workshop are substitute commodities.

[*Hint:* Solve the system of equations for x and y in terms of p and q.]

47. Revenue Functions The total weekly revenue (in dollars) of the Country Workshop associated with manufacturing and selling their rolltop desks is given by the function

$$R(x, y) = -0.2x^2 - 0.25y^2 - 0.2xy + 200x + 160y$$

where x denotes the number of finished units and y denotes the number of unfinished units manufactured and sold per week. Compute $\partial R/\partial x$ and $\partial R/\partial y$ when $x = 300$ and $y = 250$. Interpret your results.

48. Profit Functions The monthly profit (in dollars) of the Bond and Barker Department Store depends on the level of inventory x (in thousands of dollars) and the floor space y (in thousands of square feet) available for display of the merchandise, as given by the equation

$$P(x, y) = -0.02x^2 - 15y^2 + xy$$
$$+ 39x + 25y - 20,000$$

Compute $\partial P/\partial x$ and $\partial P/\partial y$ when $x = 4000$ and $y = 150$. Interpret your results. Repeat with $x = 5000$ and $y = 150$.

49. Volume of a Gas The volume V (in liters) of a certain mass of gas is related to its pressure P (in millimeters of mercury) and its temperature T (in degrees Kelvin) by the law

$$V = \frac{30.9T}{P}$$

Compute $\partial V/\partial T$ and $\partial V/\partial P$ when $T = 300$ and $P = 800$. Interpret your results.

50. Surface Area of a Human Body The formula

$$S = 0.007184W^{0.425}H^{0.725}$$

gives the surface area S of a human body (in square meters) in terms of its weight W in kilograms and its height H in centimeters. Compute $\partial S/\partial W$ and $\partial S/\partial H$ when $W = 70$ kg and $H = 180$ cm. Interpret your results.

SOLUTIONS TO SELF-CHECK EXERCISES 16.2

1.

$$f_x = \frac{\partial f}{\partial x} = 3x^2 - 2y^2$$

$$f_y = \frac{\partial f}{\partial y} = -2x(2y) + 2y$$

$$= 2y(1 - 2x)$$

2.

$$f_x = \ln y + ye^x - 2x \qquad f_y = \frac{x}{y} + e^x$$

In particular, $f_x(0, 1) = \ln 1 + 1e^0 - 2(0) = 1$

$$f_y(0, 1) = \frac{0}{1} + e^0 = 1$$

The results tell us that at the point $(0, 1)$, $f(x, y)$ increases 1 unit for each unit increase in the x-direction, y being kept constant; $f(x, y)$ also increases 1 unit for each unit increase in the y-direction, x being kept constant.

3. From the results of Self-Check exercise 1,

$$f_x = 3x^2 - 2y^2$$

Therefore,

$$f_{xx} = \frac{\partial}{\partial x}(3x^2 - 2y^2) = 6x$$

and

$$f_{xy} = \frac{\partial}{\partial y}(3x^2 - 2y^2) = -4y$$

Also, from the results of Self-Check exercise 1,

$$f_y = 2y(1 - 2x)$$

and so

$$f_{yx} = \frac{\partial}{\partial x}[2y(1 - 2x)] = -4y$$

and

$$f_{yy} = \frac{\partial}{\partial y}[2y(1 - 2x)] = 2(1 - 2x)$$

4. a. The marginal productivity of labor when the amounts expended on labor and capital are x and y units, respectively, is given by

$$f_x(x, y) = 60\left(\frac{1}{3}x^{-2/3}\right)y^{2/3} = 20\left(\frac{y}{x}\right)^{2/3}$$

In particular, the required marginal productivity of labor is given by

$$f_x(125, 8) = 20\left(\frac{8}{125}\right)^{2/3} = 20\left(\frac{4}{25}\right)$$

or 3.2 units per unit increase in labor expenditure, capital expenditure being held constant at 8 units. Next, we compute

$$f_y(x, y) = 60x^{1/3}\left(\frac{2}{3}y^{-1/3}\right) = 40\left(\frac{x}{y}\right)^{1/3}$$

and deduce that the required marginal productivity of capital is given by

$$f_y(125, 8) = 40\left(\frac{125}{8}\right)^{1/3} = 40\left(\frac{5}{2}\right)$$

or 100 units per unit increase in capital expenditure, labor expenditure being held constant at 125 units.

b. The results of (a) tell us that the government should encourage increased spending on capital rather than on labor.

16.3 MAXIMA AND MINIMA OF FUNCTIONS OF SEVERAL VARIABLES

Maxima and Minima

In Chapter 12 we saw how solving a practical problem formulated in terms of a function f of one variable often centers on determining an extreme value of f with respect to that variable. For example, we can solve the problem of finding a firm's production level x that will yield a maximum profit by finding the absolute maximum value of the profit function $P(x)$ with respect to x. Conversely, to find a manufacturer's production level x that will result in a minimal cost of operation, we find the absolute minimum value of the cost function $C(x)$ with respect to x.

The notion of an extreme value of a function plays an equally important role in the case of a function of several variables. As in the case of a function of one variable, it is important to distinguish between the concept of a relative maximum (relative minimum) of f and that of an absolute maximum (absolute minimum) of f. More specifically, a function $f(x, y)$ of two variables has a **relative maximum** at a point (a, b) in the domain of f if $f(x, y) \leq f(a, b)$ for all points (x, y) that are sufficiently close to the point (a, b). Similarly, f has a **relative minimum** at (a, b) if $f(x, y) \geq f(a, b)$ for all points (x, y) that are sufficiently close to the point (a, b). Geometrically, this means that no point on the graph of f corresponding to points (x, y) lying sufficiently close to the point (a, b) can be above (or below) the point $(a, b, f(a, b))$. Figure 16.16

Figure 16.16
f has a relative maximum at (a, b) and a relative minimum at (c, d).

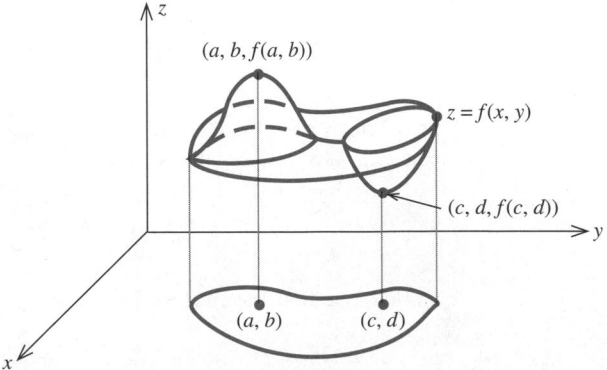

depicts the graph of a function f of two variables that has a relative maximum at the point (a, b) and a relative minimum at the point (c, d).

A function $f(x, y)$ of two variables has an **absolute maximum** at a point (a, b) if

$$f(x, y) \le f(a, b)$$

for *all* points (x, y) in the domain of f. Similarly, a function $f(x, y)$ has an **absolute minimum** at a point (a, b) if

$$f(x, y) \ge f(a, b)$$

for *all* points (x, y) in the domain of f. Geometrically, the absolute maximum and the absolute minimum of a function f are, respectively, the highest and the lowest points on the graph of f.

Just as in the case of a function of one variable, a relative extremum (relative maximum or relative minimum) may or may not be an absolute extremum. However, in order to simplify matters, we will assume that whenever an absolute extremum exists, it will occur at a point where f has a relative extremum.

Just as the first and second derivatives play an important role in determining the relative extrema of a function of one variable, the first and second partial derivatives are powerful tools for locating and classifying the relative extrema of functions of several variables.

Suppose now that a differentiable function $f(x, y)$ of two variables has a relative maximum (relative minimum) at a point (a, b) in the domain of f. From Figure 16.17 it is clear that at the point (a, b) the slope of the "tangent

Figure 16.17

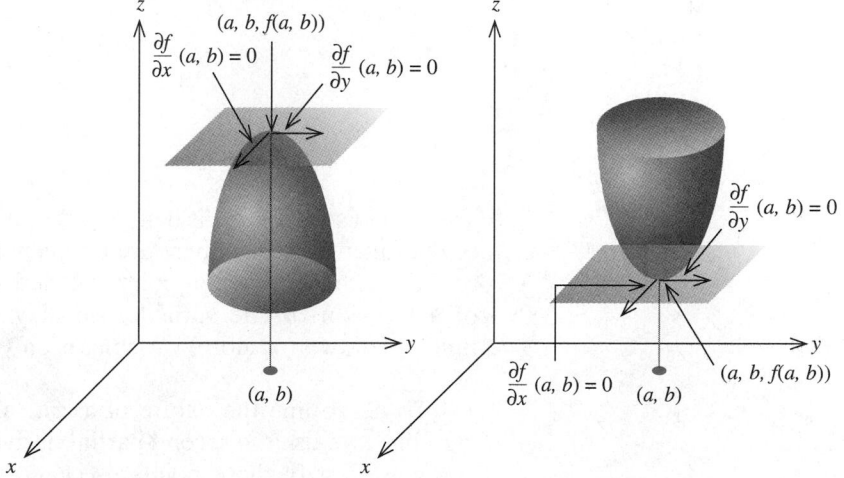

(a) f has a relative maximum at (a, b).

(b) f has a relative minimum at (a, b).

lines" to the surface in any direction must be zero. In particular, this implies that both

$$\frac{\partial f}{\partial x}(a, b) \quad \text{and} \quad \frac{\partial f}{\partial y}(a, b)$$

must be zero.

The point (a, b) is called a **critical point** of the function f. In case we are tempted to conclude that a critical point of a function f must automatically be a relative extremum of f, let us consider the graph of the function f in Figure 16.18. Here we have both

$$\frac{\partial f}{\partial x}(a, b) = 0 \quad \text{and} \quad \frac{\partial f}{\partial y}(a, b) = 0$$

Figure 16.18
The point $(a, b, f(a, b))$ is called a saddle point.

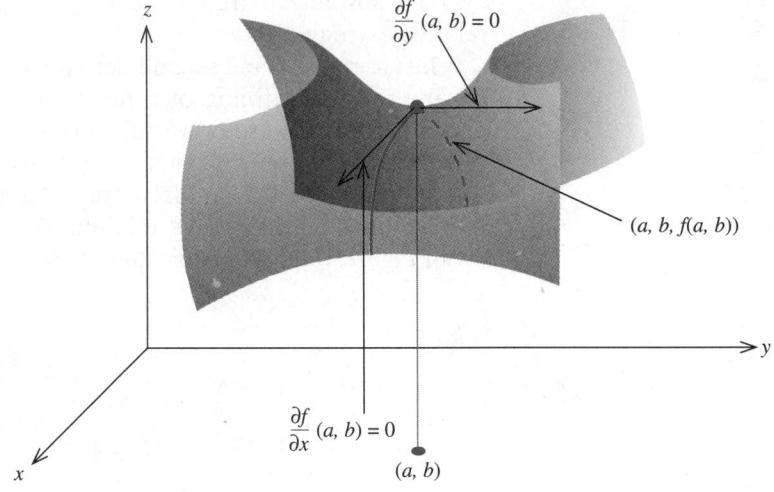

The point $(a, b, f(a, b))$ is neither a relative maximum nor a relative minimum of the function f since there are points nearby that are higher and others that are lower than it. Such a point is called a **saddle point.** Thus, as in the case of a function of one variable, we may conclude that a critical point of a function of two (or more) variables is only a candidate for a relative extremum of f.

To determine the nature of a critical point of a function $f(x, y)$ of two variables, we use the second partial derivatives of f. The resulting test, which helps us classify these points, is called the **Second Derivative Test** and is incorporated into the following procedure for finding and classifying the relative extrema of f.

**DETERMINING
RELATIVE EXTREMA**

1. Find the critical points of $f(x, y)$ by solving the system of simultaneous equations

$$f_x = 0$$
$$f_y = 0$$

2. The Second Derivative Test: Let

$$D(x, y) = f_{xx}f_{yy} - f_{xy}^2$$

Then

a. $D(a, b) > 0$ and $f_{xx}(a, b) < 0$ imply that $f(x, y)$ has a **relative maximum** at the point (a, b).

b. $D(a, b) > 0$ and $f_{xx}(a, b) > 0$ imply that $f(x, y)$ has a **relative minimum** at the point (a, b).

c. $D(a, b) < 0$ implies that $f(x, y)$ has neither a relative maximum nor a relative minimum at the point (a, b).

d. $D(a, b) = 0$ implies that the test is inconclusive, so some other technique must be used to solve the problem.

EXAMPLE 1   Find the relative extrema of the function

$$f(x, y) = x^2 + y^2$$

Solution   We have

$$f_x = 2x$$

and

$$f_y = 2y$$

To find the critical point(s) of f, we set $f_x = 0$ and $f_y = 0$ and solve the resulting system of simultaneous equations

$$2x = 0$$
$$2y = 0$$

obtaining $x = 0$, $y = 0$, or $(0, 0)$, as the sole critical point of f. Next, we apply the Second Derivative Test to determine the nature of the critical point $(0, 0)$. We compute

$$f_{xx} = 2, \quad f_{xy} = 0, \quad f_{yy} = 2$$

and

$$D(x, y) = f_{xx}f_{yy} - f_{xy}^2 = (2)(2) - 0 = 4$$

In particular, $D(0, 0) = 4$. Since $D(0, 0) > 0$ and $f_{xx}(0, 0) = 2 > 0$, we conclude that $f(x, y)$ has a relative minimum at the point $(0, 0)$. The relative minimum value, 0, also happens to be the absolute minimum of f. The graph of the function f, shown in Figure 16.19, confirms these results.   ◦ ◦ ◦

Figure 16.19
The graph of $f(x, y) = x^2 + y^2$.

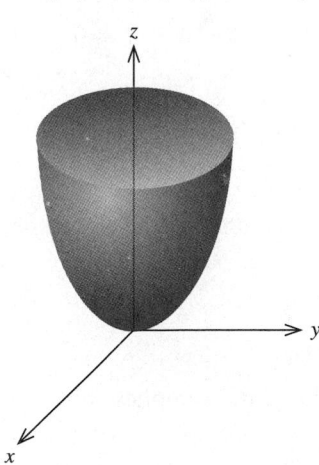

EXAMPLE 2 Find the relative extrema of the function.

$$f(x, y) = 3x^2 - 4xy + 4y^2 - 4x + 8y + 4$$

Solution We have

$$f_x = 6x - 4y - 4$$

and

$$f_y = -4x + 8y + 8$$

To find the critical points of f, we set $f_x = 0$ and $f_y = 0$ and solve the resulting system of simultaneous equations

$$6x - 4y = 4$$

$$-4x + 8y = -8$$

Multiplying the first equation by 2 and the second equation by 3, we obtain the equivalent system

$$12x - 8y = 8$$

$$-12x + 24y = -24$$

Adding the two equations gives $16y = -16$, or $y = -1$. We substitute this value for y into either equation in the system to get $x = 0$. Thus, the only critical point of f is $(0, -1)$. Next, we apply the Second Derivative Test to determine whether the point $(0, -1)$ gives rise to a relative extremum of f. We compute

$$f_{xx} = 6, \quad f_{xy} = -4, \quad f_{yy} = 8$$

and

$$D(x, y) = f_{xx}f_{yy} - f_{xy}^2 = (6)(8) - (-4)^2 = 32$$

Since $D(0, -1) = 32 > 0$ and $f_{xx}(0, -1) = 6 > 0$, we conclude that $f(x, y)$ has a relative minimum at the point $(0, -1)$. The value of $f(x, y)$ at the point $(0, -1)$ is given by

$$f(0, -1) = 3(0)^2 - 4(0)(-1) + 4(-1)^2 - 4(0) + 8(-1) + 4 = 0$$

○ ○ ○

Suppose $f(x, y)$ has a relative extremum (relative maximum or relative minimum) at a point (a, b). Let $g(x) = f(x, b)$ and $h(y) = f(a, y)$. Assuming that f and g are differentiable, explain why $g'(a) = 0$ and $h'(b) = 0$. Explain why these results are equivalent to the conditions $f_x(a, b) = 0$ and $f_y(a, b) = 0$.

EXAMPLE 3 Find the relative extrema of the function

$$f(x, y) = 4y^3 + x^2 - 12y^2 - 36y + 2$$

Solution To find the critical points of f, we set $f_x = 0$ and $f_y = 0$ simultaneously, obtaining

$$f_x = 2x = 0$$

$$f_y = 12y^2 - 24y - 36 = 0$$

The first equation implies that $x = 0$. The second equation implies that

$$y^2 - 2y - 3 = 0$$

$$(y + 1)(y - 3) = 0$$

a. Refer to the Second Derivative Test. Can the condition $f_{xx}(a, b) < 0$ in part 2a be replaced by the condition $f_{yy}(a, b) < 0$? Explain your answer. How about the condition $f_{xx}(a, b) > 0$ in part 2b?

b. Let $f(x, y) = x^4 + y^4$. Show that $(0, 0)$ is a critical point of f and that $D(0, 0) = 0$. Explain why f has a relative (in fact, an absolute) minimum at $(0, 0)$. Does this contradict the Second Derivative Test? Explain your answer.

that is, $y = -1$ or 3. Therefore, there are two critical points of the function f—namely, $(0, -1)$ and $(0, 3)$. Next, we apply the Second Derivative Test to determine the nature of each of the two critical points. We compute

$$f_{xx} = 2, \quad f_{xy} = 0, \quad f_{yy} = 24y - 24 = 24(y - 1)$$

Therefore,

$$D(x, y) = f_{xx}f_{yy} - f_{xy}^2 = 48(y - 1)$$

For the point $(0, -1)$,

$$D(0, -1) = 48(-1 - 1) = -96 < 0$$

Since $D(0, -1) < 0$, we conclude that the point $(0, -1)$ gives rise to a saddle point of f. For the point $(0, 3)$,

$$D(0, 3) = 48(3 - 1) = 96 > 0$$

Since $D(0, 3) > 0$ and $f_{xx}(0, 3) > 0$, we conclude that the function f has a relative minimum at the point $(0, 3)$. Furthermore, since

$$f(0, 3) = 4(3)^3 + (0)^2 - 12(3)^2 - 36(3) + 2$$
$$= -106$$

we see that the relative minimum value of f is -106.   ◦ ◦ ◦

Applications

As in the case of a practical optimization problem involving a function of one variable, the solution to an optimization problem involving a function of several variables calls for finding the *absolute* extremum of the function. Determining the absolute extremum of a function of several variables is more difficult than merely finding the relative extrema of the function. However, in many situations, the absolute extremum of a function actually coincides with the largest relative extremum of the function that occurs in the interior of its domain. We assume that the problems considered here belong to this category. Furthermore, the existence of the absolute extremum (solution) of a practical problem is often deduced from the geometric or physical nature of the problem.

EXAMPLE 4   The total weekly revenue (in dollars) that the Acrosonic Company realizes in producing and selling its bookshelf loud speaker systems is given by

$$R(x, y) = -\frac{1}{4}x^2 - \frac{3}{8}y^2 - \frac{1}{4}xy + 300x + 240y$$

where x denotes the number of fully assembled units and y denotes the number of kits produced and sold per week. The total weekly cost attributable to the production of these loudspeakers is

$$C(x, y) = 180x + 140y + 5000$$

dollars, where x and y have the same meaning as before. Determine how many assembled units and how many kits Acrosonic should produce per week to maximize its profit.

Solution   The contribution to Acrosonic's weekly profit stemming from the production and sale of the bookshelf loudspeaker systems is given by

$$P(x, y) = R(x, y) - C(x, y)$$

$$= \left(-\frac{1}{4}x^2 - \frac{3}{8}y^2 - \frac{1}{4}xy + 300x + 240y \right) - (180x + 140y + 5000)$$

$$= -\frac{1}{4}x^2 - \frac{3}{8}y^2 - \frac{1}{4}xy + 120x + 100y - 5000$$

To find the relative maximum of the profit function $P(x, y)$, we first locate the critical point(s) of P. Setting $P_x(x, y)$ and $P_y(x, y)$ equal to zero, we obtain

$$P_x = -\frac{1}{2}x - \frac{1}{4}y + 120 = 0$$

$$P_y = -\frac{3}{4}y - \frac{1}{4}x + 100 = 0$$

Solving the first of these equations for y yields

$$y = -2x + 480$$

which, upon substitution into the second equation, yields

$$-\frac{3}{4}(-2x + 480) - \frac{1}{4}x + 100 = 0$$

$$6x - 1440 - x + 400 = 0$$

$$x = 208$$

We substitute this value of x into the equation $y = -2x + 480$ to get

$$y = 64$$

Therefore, the function P has the sole critical point $(208, 64)$. To show that the point $(208, 64)$ is a solution to our problem, we use the Second Derivative Test. We compute

$$P_{xx} = -\frac{1}{2}, \quad P_{xy} = -\frac{1}{4}, \quad P_{yy} = -\frac{3}{4}$$

So

$$D(x, y) = \left(-\frac{1}{2} \right)\left(-\frac{3}{4} \right) - \left(-\frac{1}{4} \right)^2 = \frac{3}{8} - \frac{1}{16} = \frac{5}{16}$$

In particular, $D(208, 64) = 5/16 > 0$.

Since $D(208, 64) > 0$ and $P_{xx}(208, 64) < 0$, the point $(208, 64)$ yields a relative maximum of P. This relative maximum is also the absolute maximum of P. We conclude that Acrosonic can maximize its weekly profit by manufacturing 208 assembled units and 64 kits of their bookshelf loudspeaker systems. The maximum weekly profit realizable from the production and sale of these loudspeaker systems is given by

$$P(208, 64) = -\frac{1}{4}(208)^2 - \frac{3}{8}(64)^2 - \frac{1}{4}(208)(64)$$
$$+ 120(208) + 100(64) - 5000$$
$$= 10,680$$

or \$10,680. ○ ○ ○

EXAMPLE 5 A television relay station will serve towns A, B, and C, whose relative locations are shown in Figure 16.20. Determine a site for the location of the station if the sum of the squares of the distances from each town to the site is minimized.

Figure 16.20

Locating a site for a television relay station.

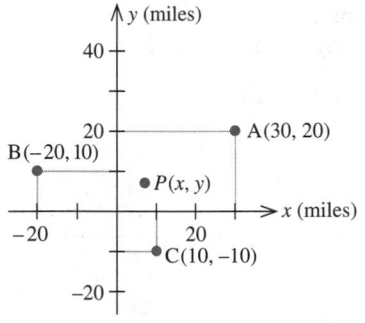

Solution Suppose the required site is located at the point $P(x, y)$. With the aid of the distance formula, we find that the square of the distance from town A to the site is

$$(x - 30)^2 + (y - 20)^2$$

The respective distances from towns B and C to the site are found in a similar manner, so that the sum of the squares of the distances from each town to the site is given by

$$f(x, y) = (x - 30)^2 + (y - 20)^2 + (x + 20)^2$$
$$+ (y - 10)^2 + (x - 10)^2 + (y + 10)^2$$

To find the relative minimum of $f(x, y)$, we first find the critical point(s) of f. Using the Chain Rule to find $f_x(x, y)$ and $f_y(x, y)$ and setting each equal to zero, we obtain

$$f_x = 2(x - 30) + 2(x + 20) + 2(x - 10) = 6x - 40 = 0$$
$$f_y = 2(y - 20) + 2(y - 10) + 2(y + 10) = 6y - 40 = 0$$

from which we deduce that $(20/3, 20/3)$ is the sole critical point of f. Since

$$f_{xx} = 6, \quad f_{xy} = 0, \quad \text{and} \quad f_{yy} = 6$$

we have

$$D(x, y) = f_{xx}f_{yy} - f_{xy}^2 = (6)(6) - 0 = 36$$

Since $D(20/3, 20/3) > 0$ and $f_{xx}(20/3, 20/3) > 0$, we conclude that the point $(20/3, 20/3)$ yields a relative minimum of f. Thus, the required site has coordinates $x = 20/3$ and $y = 20/3$. ○ ○ ○

SELF-CHECK EXERCISES 16.3

1. Let $f(x, y) = 2x^2 + 3y^2 - 4xy + 4x - 2y + 3$.
   a. Find the critical point of f.
   b. Use the Second Derivative Test to classify the nature of the critical point.
   c. Find the relative extremum of f if it exists.

2. The Robertson Controls Company manufactures two basic models of setback thermostats: a standard mechanical thermostat and a deluxe electronic thermostat. Robertson's monthly revenue (in hundreds of dollars) is

$$R(x, y) = -\frac{1}{8}x^2 - \frac{1}{2}y^2 - \frac{1}{4}xy + 20x + 60y$$

where x (in units of a hundred) denotes the number of mechanical thermostats manufactured and y (in units of a hundred) denotes the number of electronic thermostats manufactured per month. The total monthly cost incurred in producing these thermostats is

$$C(x, y) = 7x + 20y + 280$$

hundred dollars. Find how many thermostats of each model Robertson should manufacture per month in order to maximize its profits. What is the maximum profit?

Solutions to Self-Check Exercises 16.3 can be found on page 1116.

16.3 EXERCISES

In exercises 1–20, find the critical point(s) of the function. Then use the Second Derivative Test to classify the nature of each point, if possible. Finally, determine the relative extrema of the function.

1. $f(x, y) = 1 - 2x^2 - 3y^2$

2. $f(x, y) = x^2 - xy + y^2 + 1$

3. $f(x, y) = x^2 - y^2 - 2x + 4y + 1$

4. $f(x, y) = 2x^2 + y^2 - 4x + 6y + 3$

5. $f(x, y) = x^2 + 2xy + 2y^2 - 4x + 8y - 1$

6. $f(x, y) = x^2 - 4xy + 2y^2 + 4x + 8y - 1$

7. $f(x, y) = 2x^3 + y^2 - 9x^2 - 4y + 12x - 2$

8. $f(x, y) = 2x^3 + y^2 - 6x^2 - 4y + 12x - 2$

9. $f(x, y) = x^3 + y^2 - 2xy + 7x - 8y + 4$

10. $f(x, y) = 2y^3 - 3y^2 - 12y + 2x^2 - 6x + 2$

11. $f(x, y) = x^3 - 3xy + y^3 - 2$

12. $f(x, y) = x^3 - 2xy + y^2 + 5$

13. $f(x, y) = xy + \dfrac{4}{x} + \dfrac{2}{y}$

14. $f(x, y) = \dfrac{x}{y^2} + xy$

15. $f(x, y) = x^2 - e^{y^2}$

16. $f(x, y) = e^{x^2 - y^2}$

17. $f(x, y) = e^{x^2 + y^2}$

18. $f(x, y) = e^{xy}$

19. $f(x, y) = \ln(1 + x^2 + y^2)$

20. $f(x, y) = xy + \ln x + 2y^2$

21. Maximizing Profit   The total weekly revenue (in dollars) of the Country Workshop realized in manufacturing and selling its rolltop desks is given by

$$R(x, y) = -0.2x^2 - 0.25y^2 - 0.2xy + 200x + 160y$$

where x denotes the number of finished units and y denotes the number of unfinished units manufactured and sold per week. The total weekly cost attributable to the manufacture of these desks is given by

$$C(x, y) = 100x + 70y + 4000$$

dollars. Determine how many finished units and how many unfinished units the company should manufacture per week in order to maximize its profit. What is the maximum profit realizable?

22. **Maximizing Profit**   The total daily revenue (in dollars) that the Weston Publishing Company realizes in publishing and selling its English language dictionaries is given by

$$R(x, y) = -0.005x^2 - 0.003y^2 - 0.002xy + 20x + 15y$$

where x denotes the number of deluxe copies and y denotes the number of standard copies published and sold daily. The total daily cost of publishing these dictionaries is given by

$$C(x, y) = 6x + 3y + 200$$

dollars. Determine how many deluxe copies and how many standard copies Weston should publish per day to maximize its profits. What is the maximum profit realizable?

23. **Maximum Price**   The rectangular region R shown in the accompanying figure represents the financial district of a city. The price of land in the district is approximated by the function

$$p(x, y) = 200 - 10\left(x - \frac{1}{2}\right)^2 - 15(y - 1)^2$$

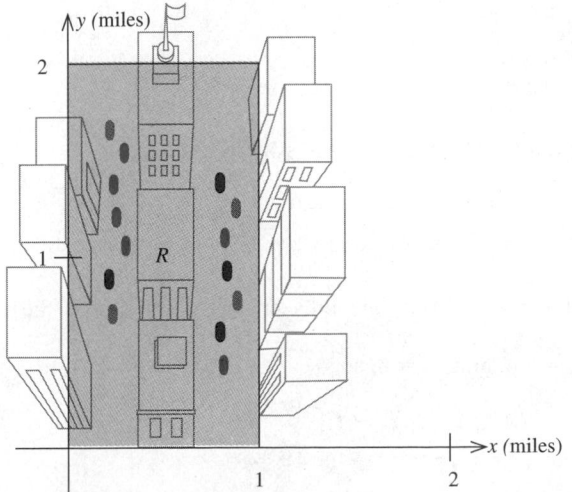

where $p(x, y)$ is the price of land at the point (x, y) in dollars per square foot and x and y are measured in miles. At what point in the financial district is the price of land highest?

24. **Maximizing Profit**   C & G Imports, Inc., imports two brands of white wine, one from Germany and the other from Italy. The German wine costs $4 a bottle, and the Italian wine can be obtained for $3 a bottle. It has been estimated that if the German wine retails at p dollars per bottle and the Italian wine is sold for q dollars per bottle, then

$$2000 - 150p + 100q$$

bottles of the German wine and

$$1000 + 80p - 120q$$

bottles of the Italian wine will be sold per week. Determine the unit price for each brand that will allow C & G to realize the largest possible weekly profit.

25. **Determining the Optimal Site**   An auxiliary electric power station will serve three communities, A, B, and C, whose relative locations are shown in the accompanying figure. Determine where the power station should be located if the sum of the squares of the distances from each community to the site is minimized.

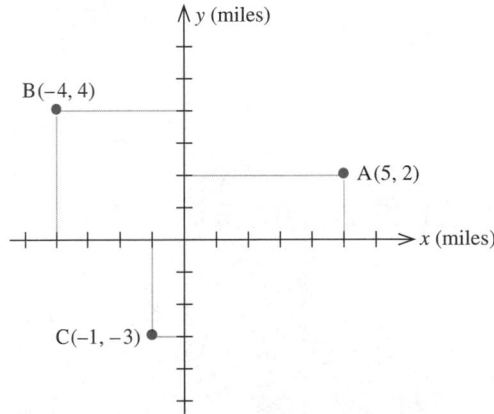

26. **Packaging**   An open rectangular box having a volume of 108 cubic inches is to be constructed from a tin sheet. Find the dimensions of such a box if the amount of material used in its construction is to be minimal. [*Hint:* Let the dimensions of the box be x'' by y'' by z''. Then

$$xyz = 108$$

and the amount of material used is given by

$$S = xy + 2yz + 2xz$$

Show that

$$S = f(x, y) = xy + \frac{216}{x} + \frac{216}{y}$$

Minimize $f(x, y)$.]

27. Packaging   Postal regulations specify that the combined length and girth of a parcel sent by parcel post may not exceed 108 inches. Find the dimensions of the rectangular package that would have the greatest possible volume under these regulations.

[*Hint:* Let the dimensions of the box be x'' by y'' by z'' (see the figure). Then

$$2x + 2z + y = 108$$

and the volume

$$V = xyz$$

Show that

$$V = f(x, z) = 108xz - 2x^2z - 2xz^2$$

Maximize $f(x, z)$.]

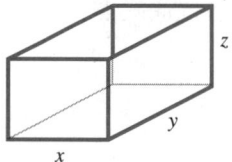

28. Minimizing Heating and Cooling Costs   A building in the shape of a rectangular box is to have a volume of 12,000 cubic feet (see figure). It is estimated that the annual heating and cooling costs will be \$2 per square foot for the top, \$4 per square foot for the front and back, and \$3 per square foot for the sides. Find the dimensions of the building that will result in a minimal annual heating and cooling cost. What is the minimal annual heating and cooling cost?

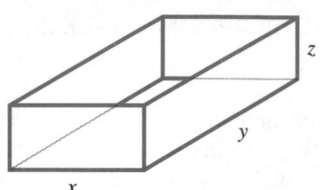

29. Packaging   An open box having a volume of 48 cubic inches is to be constructed. If the box is to include a partition that is parallel to a side of the box, as shown in the figure, and the amount of material used is to be minimal, what should the dimensions of the box be?

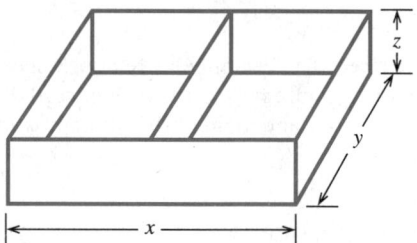

SOLUTIONS TO SELF–CHECK EXERCISES 16.3

1. a.   To find the critical point(s) of f, we solve the system of equations

$$f_x = 4x - 4y + 4 = 0$$
$$f_y = -4x + 6y - 2 = 0$$

obtaining $x = -2$ and $y = -1$. Thus, the only critical point of f is the point $(-2, -1)$.

b.   We have $f_{xx} = 4$, $f_{xy} = -4$, and $f_{yy} = 6$, so

$$D(x, y) = f_{xx}f_{yy} - f_{xy}^2$$
$$= (4)(6) - (-4)^2 = 8$$

Since $D(-2, -1) > 0$ and $f_{xx}(-2, -1) > 0$, we conclude that f has a relative minimum at the point $(-2, -1)$.

c. The relative minimum value of $f(x, y)$ at the point $(-2, -1)$ is

$$f(-2, -1) = 2(-2)^2 + 3(-1)^2 - 4(-2)(-1) + 4(-2) - 2(-1) + 3$$
$$= 0$$

2. Robertson's monthly profit is

$$P(x, y) = R(x, y) - C(x, y)$$
$$= \left(-\frac{1}{8}x^2 - \frac{1}{2}y^2 - \frac{1}{4}xy + 20x + 60y\right) - (7x + 20y + 280)$$
$$= -\frac{1}{8}x^2 - \frac{1}{2}y^2 - \frac{1}{4}xy + 13x + 40y - 280$$

The critical point of P is found by solving the system

$$P_x = -\frac{1}{4}x - \frac{1}{4}y + 13 = 0$$

$$P_y = -\frac{1}{4}x - y + 40 = 0$$

giving $x = 16$ and $y = 36$. Thus $(16, 36)$ is the critical point of P. Next,

$$P_{xx} = -\frac{1}{4}, \quad P_{xy} = -\frac{1}{4}, \quad P_{yy} = -1$$

and

$$D(x, y) = f_{xx}f_{yy} - f_{xy}^2$$
$$= \left(-\frac{1}{4}\right)(-1) - \left(-\frac{1}{4}\right)^2 = \frac{3}{16}$$

Since $D(16, 36) > 0$ and $P_{xx}(16, 36) < 0$, the point $(16, 36)$ yields a relative maximum of P. We conclude that the monthly profit is maximized by manufacturing 1600 mechanical and 3600 electronic setback thermostats per month. The maximum monthly profit realizable is

$$P(16, 36) = -\frac{1}{8}(16)^2 - \frac{1}{2}(36)^2 - \frac{1}{4}(16)(36) + 13(16) + 40(36) - 280$$

$$= 544$$

or $54,400.

16.4 CONSTRAINED MAXIMA AND MINIMA AND THE METHOD OF LAGRANGE MULTIPLIERS

Constrained Relative Extrema

In Section 16.3 we studied the problem of determining the relative extremum of a function $f(x, y)$ without placing any restrictions on the independent variables x and y except, of course, that the point (x, y) lie in the domain of f. Such a relative extremum of a function f is referred to as an **unconstrained relative extremum of f.** However, in many practical optimization problems, we must maximize or minimize a function in which the independent variables are subjected to certain further constraints.

In this section we discuss a powerful method for determining the relative extrema of a function $f(x, y)$ whose independent variables x and y are required to satisfy one or more constraints of the form $g(x, y) = 0$. Such a relative extremum of a function f is called a **constrained relative extremum of f.** We can see the difference between an unconstrained extremum of a function $f(x, y)$ of two variables and a constrained extremum of f, where the independent variables x and y are subjected to a constraint of the form $g(x, y) = 0$, by considering the geometry of the two cases. Figure 16.21a depicts the graph of a function $f(x, y)$ that has an unconstrained relative minimum at the point $(0, 0)$. However, when the independent variables x and y are subjected to an equality constraint of the form $g(x, y) = 0$, the points (x, y, z) that satisfy both $z = f(x, y)$ and the constraint equation $g(x, y) = 0$ lie on a curve C.

Figure 16.21

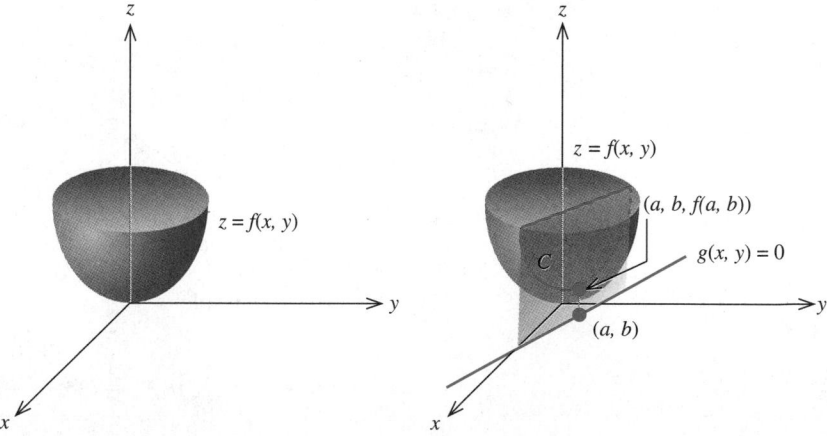

(a) $f(x, y)$ has an unconstrained relative extremum at $(0, 0)$

(b) $f(x, y)$ has a constrained relative extremum at $(a, b, f(a, b))$.

Therefore, the constrained relative minimum of f must also lie on C (Figure 16.21b).

Our first example involves an equality constraint $g(x, y) = 0$ in which we solve for the variable y explicitly in terms of x. In this case, we may apply the technique used in Chapter 12 to find the relative extrema of a function of one variable.

EXAMPLE 1 Find the relative minimum of the function

$$f(x, y) = 2x^2 + y^2$$

subject to the constraint $g(x, y) = x + y - 1 = 0$.

Solution Solving the constraint equation for y explicitly in terms of x, we obtain $y = -x + 1$. Substituting this value of y into the function $f(x, y) = 2x^2 + y^2$ results in a function of x,

$$h(x) = 2x^2 + (-x + 1)^2 = 3x^2 - 2x + 1$$

The function h describes the curve C lying on the graph of f on which the constrained relative minimum of f occurs. To find this point, use the technique developed in Chapter 12 to determine the relative extrema of a function of one variable:

$$h'(x) = 6x - 2 = 2(3x - 1)$$

Setting $h'(x) = 0$ gives $x = \frac{1}{3}$ as the sole critical point of the function h. Next we find

$$h''(x) = 6$$

and, in particular,

$$h''\left(\frac{1}{3}\right) = 6 > 0$$

Therefore, by the Second Derivative Test, the point $x = \frac{1}{3}$ gives rise to a relative minimum of h. We substitute this value of x into the constraint equation $x + y - 1 = 0$ to get $y = \frac{2}{3}$. Thus, the point $(\frac{1}{3}, \frac{2}{3})$ gives rise to the required constrained relative minimum of f. Since

$$f\left(\frac{1}{3}, \frac{2}{3}\right) = 2\left(\frac{1}{3}\right)^2 + \left(\frac{2}{3}\right)^2 = \frac{2}{3}$$

the required constrained relative minimum value of f is $\frac{2}{3}$ at the point $(\frac{1}{3}, \frac{2}{3})$. It may be shown that $\frac{2}{3}$ is, in fact, a constrained absolute minimum value of f (Figure 16.22).

Figure 16.22
f has a constrained absolute mini-mum of $\frac{2}{3}$ at $(\frac{1}{3}, \frac{2}{3})$.

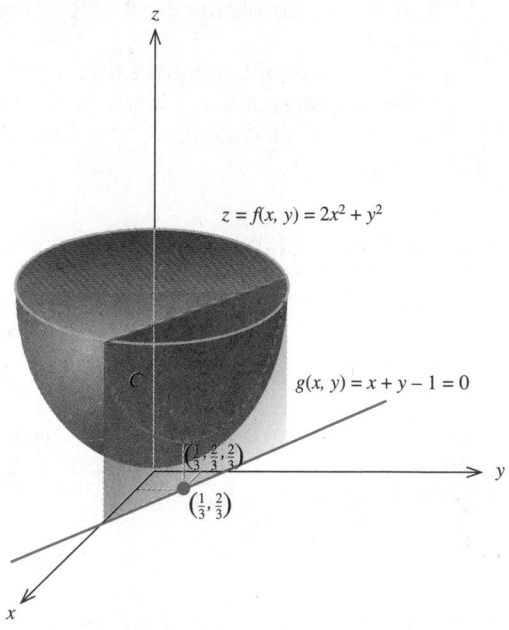

$z = f(x, y) = 2x^2 + y^2$

$g(x, y) = x + y - 1 = 0$

$(\frac{1}{3}, \frac{2}{3}, \frac{2}{3})$

$(\frac{1}{3}, \frac{2}{3})$

The Method of Lagrange Multipliers

The major drawback of the technique used in Example 1 is that it relies on our ability to solve the constraint equation $g(x, y) = 0$ for y explicitly in terms of x. This is not always an easy task. Moreover, even when we can solve the constraint equation $g(x, y) = 0$ for y explicitly in terms of x, the resulting function of one variable that is to be optimized may turn out to be unnecessarily complicated. Fortunately, an easier method exists. This method, called the **Method of Lagrange Multipliers** (Joseph Lagrange, 1736–1813), is as follows:

THE METHOD OF LAGRANGE MULTIPLIERS

To find the relative extremum of the function $f(x, y)$ subject to the constraint $g(x, y) = 0$ (assuming that these extreme values exist):

1. Form an auxiliary function

$$F(x, y, \lambda) = f(x, y) + \lambda g(x, y)$$

called the Lagrangian function (the variable λ is called the Lagrange multiplier).

2. Solve the system that comprises the equations

$$F_x = 0, \quad F_y = 0, \quad \text{and} \quad F_\lambda = 0$$

for all values of x, y, and λ.

3. Evaluate f at each of the points (x, y) found in step 2. The largest (smallest) of these values is the maximum (minimum) value of f.

Let us re-solve Example 1 using the Method of Lagrange Multipliers.

EXAMPLE 2   Using the Method of Lagrange Multipliers, find the relative minimum of the function

$$f(x, y) = 2x^2 + y^2$$

subject to the constraint $x + y = 1$.

Solution   Write the constraint equation $x + y = 1$ in the form $g(x, y) = x + y - 1 = 0$. Then form the Lagrangian function

$$\begin{aligned} F(x, y, \lambda) &= f(x, y) + \lambda g(x, y) \\ &= 2x^2 + y^2 + \lambda(x + y - 1) \end{aligned}$$

To find the critical point(s) of the function F, solve the system that comprises the equations

$$\begin{aligned} F_x &= 4x + \lambda = 0 \\ F_y &= 2y + \lambda = 0 \\ F_\lambda &= x + y - 1 = 0 \end{aligned}$$

Solving the first and second equations in this system for x and y in terms of λ, we obtain

$$x = -\frac{1}{4}\lambda \quad \text{and} \quad y = -\frac{1}{2}\lambda$$

which, upon substitution into the third equation, yields

$$-\frac{1}{4}\lambda - \frac{1}{2}\lambda - 1 = 0 \quad \text{or} \quad \lambda = -\frac{4}{3}$$

Therefore, $x = \frac{1}{3}$ and $y = \frac{2}{3}$, and $(\frac{1}{3}, \frac{2}{3})$ affords a constrained minimum of the function f, in agreement with the result obtained earlier.   ◦ ◦ ◦

The Method of Lagrange Multipliers may be used to solve a problem involving a function of three or more variables, as illustrated in the next example.

EXAMPLE 3   Use the Method of Lagrange Multipliers to find the minimum of the function

$$f(x, y, z) = 2xy + 6yz + 8xz$$

subject to the constraint

$$xyz = 12{,}000$$

(*Note:* The existence of the minimum is suggested by the geometry of the problem.)

Solution   Write the constraint equation $xyz = 12{,}000$ in the form $g(x, y, z) = xyz - 12{,}000$. Then the Lagrangian function is

$$F(x, y, z, \lambda) = f(x, y, z) + \lambda g(x, y, z)$$
$$= 2xy + 6yz + 8xz + \lambda(xyz - 12{,}000)$$

To find the critical point(s) of the function F, we solve the system that comprises the equations

$$F_x = 2y + 8z + \lambda yz = 0$$
$$F_y = 2x + 6z + \lambda xz = 0$$
$$F_z = 6y + 8x + \lambda xy = 0$$
$$F_\lambda = xyz - 12{,}000 = 0$$

Solving the first three equations of the system for λ in terms of x, y, and z, we have

$$\lambda = -\frac{2y + 8z}{yz}$$

$$\lambda = -\frac{2x + 6z}{xz}$$

$$\lambda = -\frac{6y + 8x}{xy}$$

Equating the first two expressions for λ leads to

$$\frac{2y + 8z}{yz} = \frac{2x + 6z}{xz}$$
$$2xy + 8xz = 2xy + 6yz$$
$$x = \frac{3}{4}y$$

Next, equating the second and third expressions for λ in the same system yields

$$\frac{2x + 6z}{xz} = \frac{6y + 8x}{xy}$$
$$2xy + 6yz = 6yz + 8xz$$
$$z = \frac{1}{4}y$$

Finally, substituting these values of x and z into the equation $xyz - 12{,}000 = 0$, the fourth equation of the first system of equations, we have

$$\left(\frac{3}{4}y\right)(y)\left(\frac{1}{4}y\right) - 12{,}000 = 0$$

$$y^3 = \frac{(12{,}000)(4)(4)}{3} = 64{,}000$$

or $y = 40$. The corresponding values of x and z are given by $x = \frac{3}{4}(40) = 30$ and $z = \frac{1}{4}(40) = 10$. Therefore, we see that the point $(30, 40, 10)$ gives the constrained minimum of f. The minimum value is

$$f(30, 40, 10) = 2(30)(40) + 6(40)(10) + 8(30)(10) = 7200$$

○ ○ ○

Applications

EXAMPLE 4 Refer to Example 3, Section 16.1. The total weekly profit (in dollars) that the Acrosonic Company realized in producing and selling its bookshelf loudspeaker systems is given by the profit function

$$P(x, y) = -\frac{1}{4}x^2 - \frac{3}{8}y^2 - \frac{1}{4}xy + 120x + 100y - 5000$$

where x denotes the number of fully assembled units and y denotes the number of kits produced and sold per week. Acrosonic's management decides that production of these loudspeaker systems should be restricted to a total of exactly 230 units per week. Under this condition, how many fully assembled units and how many kits should be produced per week to maximize Acrosonic's weekly profit?

Solution The problem is equivalent to maximizing the function

$$P(x, y) = -\frac{1}{4}x^2 - \frac{3}{8}y^2 - \frac{1}{4}xy + 120x + 100y - 5000$$

subject to the constraint

$$g(x, y) = x + y - 230 = 0$$

The Lagrangian function is

$$F(x, y, \lambda) = P(x, y) + \lambda g(x, y)$$
$$= -\frac{1}{4}x^2 - \frac{3}{8}y^2 - \frac{1}{4}xy + 120x + 100y$$
$$-5000 + \lambda(x + y - 230)$$

To find the critical point(s) of F, solve the following system of equations:

$$F_x = -\frac{1}{2}x - \frac{1}{4}y + 120 + \lambda = 0$$

$$F_y = -\frac{3}{4}y - \frac{1}{4}x + 100 + \lambda = 0$$

$$F_\lambda = x + y - 230 = 0$$

Solving the first equation of this system for λ, we obtain

$$\lambda = \frac{1}{2}x + \frac{1}{4}y - 120$$

which, upon substitution into the second equation, yields

$$-\frac{3}{4}y - \frac{1}{4}x + 100 + \frac{1}{2}x + \frac{1}{4}y - 120 = 0$$

$$-\frac{1}{2}y + \frac{1}{4}x - 20 = 0$$

Solving the last equation for y gives

$$y = \frac{1}{2}x - 40$$

When we substitute this value of y into the third equation of the system, we have

$$x + \frac{1}{2}x - 40 - 230 = 0$$

or

$$x = 180$$

The corresponding value of y is $(\frac{1}{2})(180) - 40$, or 50. Thus, the required constrained relative maximum of P occurs at the point (180, 50). Again, we can show that the point (180, 50) in fact yields a constrained absolute maximum for P. Thus, Acrosonic's profit is maximized by producing 180 assembled and 50 kit versions of their bookshelf loudspeaker systems. The maximum weekly profit realizable is given by

$$P(180, 50) = -\frac{1}{4}(180)^2 - \frac{3}{8}(50)^2 - \frac{1}{4}(180)(50)$$
$$+ 120(180) + 100(50) - 5000$$
$$= 10{,}312.5$$

or \$10,312.50.   ◦ ◦ ◦

Figure 16.23

A rectangular pool will be built in the elliptical poolside area.

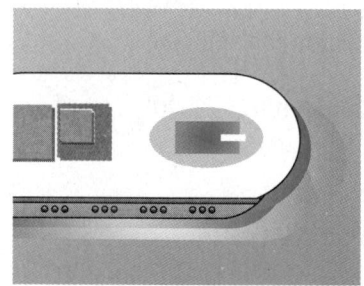

Figure 16.24

We want to find the largest rectangle that can be inscribed in the ellipse described by $x^2 + 4y^2 = 3600$.

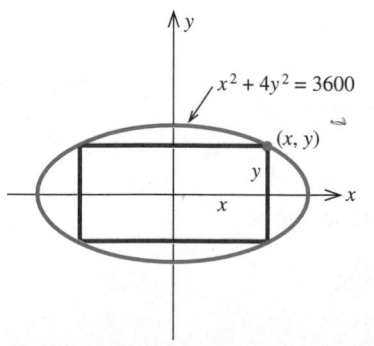

EXAMPLE 5   The operators of the *Viking Princess,* a luxury cruise liner, are contemplating the addition of another swimming pool to the ship. The chief engineer has suggested that an area in the form of an ellipse located in the rear of the promenade deck would be suitable. This location would provide a poolside area with sufficient space for passenger movement and placement of deck chairs (Figure 16.23). It has been determined that the shape of the ellipse may be described by the equation $x^2 + 4y^2 = 3600$, where x and y are measured in feet. *Viking's* operators would like to know the dimensions of the rectangular pool with the largest possible area that would meet these requirements.

Solution   In order to solve this problem, we need to find the rectangle inscribed in the ellipse with equation $x^2 + 4y^2 = 3600$ and having the largest area. Letting the sides of the rectangle be $2x$ and $2y$ feet, we see that the area of the rectangle is $A = 4xy$ (Figure 16.24). Furthermore, the point (x, y) must be constrained to move along the ellipse so that it satisfies the equation $x^2 + 4y^2 = 3600$. Thus, the problem is equivalent to maximizing the function

$$f(x, y) = 4xy$$

subject to the constraint $g(x, y) = x^2 + 4y^2 - 3600 = 0$. The Lagrangian function is

$$F(x, y, \lambda) = f(x, y) + \lambda g(x, y)$$
$$= 4xy + \lambda(x^2 + 4y^2 - 3600)$$

To find the critical point(s) of F, we solve the following system of equations:

$$F_x = 4y + 2\lambda x = 0$$
$$F_y = 4x + 8\lambda y = 0$$
$$F_\lambda = x^2 + 4y^2 - 3600 = 0$$

Solving the first equation of this system for λ, we obtain

$$\lambda = -\frac{2y}{x}$$

which, upon substitution into the second equation, yields

$$4x + 8\left(-\frac{2y}{x}\right)(y) = 0 \quad \text{or} \quad x^2 - 4y^2 = 0$$

that is, $x = \pm 2y$. Substituting these values of x into the third equation of the system, we have

$$4y^2 + 4y^2 - 3600 = 0$$

or, upon solving $y = \pm\sqrt{450} = \pm 15\sqrt{2}$. The corresponding values of x are $\pm 30\sqrt{2}$. Because both x and y must be nonnegative, we have $x = 30\sqrt{2}$ and $y = 15\sqrt{2}$. Thus, the dimensions of the pool with maximum area are $30\sqrt{2}$ ft by $60\sqrt{2}$ ft, or approximately 42 ft $\times$ 85 ft.   ○ ○ ○

EXAMPLE 6 Suppose that x units of labor and y units of capital are required to produce

$$f(x, y) = 100x^{3/4}y^{1/4}$$

units of a certain product (recall that this is a Cobb-Douglas production function). If each unit of labor costs \$200 and each unit of capital costs \$300, and a total of \$60,000 is available for production, determine how many units of labor and how many units of capital should be used in order to maximize production.

Solution The total cost of x units of labor at \$200 per unit and y units of capital at \$300 per unit is equal to $200x + 300y$ dollars. But \$60,000 is budgeted for production, so $200x + 300y = 60,000$, which we rewrite as

$$g(x, y) = 200x + 300y - 60,000 = 0$$

To maximize $f(x, y) = 100x^{3/4}y^{1/4}$ subject to the constraint $g(x, y) = 0$, we form the Lagrangian function

$$F(x, y, \lambda) = f(x, y) + \lambda g(x, y)$$
$$= 100x^{3/4}y^{1/4} + \lambda(200x + 300y - 60,000)$$

To find the critical point(s) of F, we solve the system that comprises the equations

$$F_x = 75x^{-1/4}y^{1/4} + 200\lambda = 0$$
$$F_y = 25x^{3/4}y^{-3/4} + 300\lambda = 0$$
$$F_\lambda = 200x + 300y - 60,000 = 0$$

Solving the first equation for λ, we have

$$\lambda = -\frac{75x^{-1/4}y^{1/4}}{200} = -\frac{3}{8}\left(\frac{y}{x}\right)^{1/4}$$

which, when substituted into the second equation, yields

$$25\left(\frac{x}{y}\right)^{3/4} + 300\left(-\frac{3}{8}\right)\left(\frac{y}{x}\right)^{1/4} = 0$$

Multiplying the last equation by $(x/y)^{1/4}$ then gives

$$25\left(\frac{x}{y}\right) - \frac{900}{8} = 0$$

or

$$x = \left(\frac{900}{8}\right)\left(\frac{1}{25}\right)y = \frac{9}{2}y$$

Substituting this value of x into the third equation of the first system of equations, we have

$$200\left(\frac{9}{2}y\right) + 300y - 60,000 = 0$$

from which we deduce that $y = 50$. Hence, $x = 225$. Thus, maximum production is achieved when 225 units of labor and 50 units of capital are used.   ○ ○ ○

When used in the context of Example 6, the negative of the Lagrange multiplier λ is called the **marginal productivity of money.** That is, if one additional dollar is available for production, then approximately $-\lambda$ units of a product can be produced. Here

$$\lambda = -\frac{3}{8}\left(\frac{y}{x}\right)^{1/4} = -\frac{3}{8}\left(\frac{50}{225}\right)^{1/4} \approx -0.257$$

so, in this case, the marginal productivity of money is 0.257. For example, if $65,000 is available for production instead of the originally budgeted figure of $60,000, then the maximum production may be boosted from the original

$$f(225, 50) = 100(225)^{3/4}(50)^{1/4}$$

or 15,448 units, to

$$15,448 + 5000(0.257)$$

or 16,733 units.

SELF-CHECK EXERCISES 16.4

1. Use the Method of Lagrange Multipliers to find the relative maximum of the function

$$f(x, y) = -2x^2 - y^2$$

subject to the constraint $3x + 4y = 12$.

2. The total monthly profit of the Robertson Controls Company in manufacturing and selling x hundred of its standard mechanical setback thermostats and y hundred of its deluxe electronic setback thermostats per month is given by the total profit function

$$P(x, y) = -\frac{1}{8}x^2 - \frac{1}{2}y^2 - \frac{1}{4}xy + 13x + 40y - 280$$

where P is in hundreds of dollars. If the production of setback thermostats is to be restricted to a total of exactly 4000 per month, how many of each model should Robertson manufacture in order to maximize its monthly profits? What is the maximum monthly profit?

Solutions to Self-Check Exercises 16.4 can be found on page 1129.

 EXERCISES

In exercises 1–16, use the Method of Lagrange Multipliers to optimize the given function subject to the given constraint.

1. Minimize the function $f(x, y) = x^2 + 3y^2$ subject to the constraint $x + y - 1 = 0$.

2. Minimize the function $f(x, y) = x^2 + y^2 - xy$ subject to the constraint $x + 2y - 14 = 0$.

3. Maximize the function $f(x, y) = 2x + 3y - x^2 - y^2$ subject to the constraint $x + 2y = 9$.

4. Maximize the function $f(x, y) = 16 - x^2 - y^2$ subject to the constraint $x + y - 6 = 0$.

5. Minimize the function $f(x, y) = x^2 + 4y^2$ subject to the constraint $xy = 1$.

6. Minimize the function $f(x, y) = xy$ subject to the constraint $x^2 + 4y^2 = 4$.

7. Maximize the function $f(x, y) = x + 5y - 2xy - x^2 - 2y^2$ subject to the constraint $2x + y = 4$.

8. Maximize the function $f(x, y) = xy$ subject to the constraint $2x + 3y - 6 = 0$.

9. Maximize the function $f(x, y) = xy^2$ subject to the constraint $9x^2 + y^2 = 9$.

10. Minimize the function $f(x, y) = \sqrt{y^2 - x^2}$ subject to the constraint $x + 2y - 5 = 0$.

11. Find the maximum and minimum values of the function $f(x, y) = xy$ subject to the constraint $x^2 + y^2 = 16$.

12. Find the maximum and minimum values of the function $f(x, y) = e^{xy}$ subject to the constraint $x^2 + y^2 = 8$.

13. Find the maximum and minimum values of the function $f(x, y) = xy^2$ subject to the constraint $x^2 + y^2 = 1$.

14. Maximize the function $f(x, y, z) = xyz$ subject to the constraint $2x + 2y + z = 84$.

15. Minimize the function $f(x, y, z) = x^2 + y^2 + z^2$ subject to the constraint $3x + 2y + z = 6$.

16. Find the maximum and minimum values of the function $f(x, y, z) = x + 2y - 3z$ subject to the constraint $z = 4x^2 + y^2$.

17. Maximizing Profit The total weekly profit (in dollars) realized by the Country Workshop in manufacturing and selling its rolltop desks is given by the profit function

$$P(x, y) = -0.2x^2 - 0.25y^2 - 0.2xy$$
$$+ 100x + 90y - 4000$$

where x stands for the number of finished units and y denotes the number of unfinished units manufactured and sold per week. The company's management decides to restrict the manufacture of these desks to a total of exactly 200 units per week. How many finished and how many unfinished units should be manufactured per week to maximize the company's weekly profit?

18. Maximizing Profit The total daily profit (in dollars) realized by the Weston Publishing Company in publishing and selling its dictionaries is given by the profit function

$$P(x, y) = -0.005x^2 - 0.003y^2 - 0.002xy$$
$$+ 14x + 12y - 200$$

where x stands for the number of deluxe editions and y denotes the number of standard editions sold daily. Weston's management decides that publication of these dictionaries should be restricted to a total of exactly 400 copies per day. How many deluxe copies and how many standard copies should be published per day to maximize Weston's daily profit?

19. Minimizing Construction Costs The management of UNICO Department Store decides to enclose an 800-sq-ft area outside their building to display potted plants. The enclosed area will be a rectangle, with one side provided by the external walls of the store. Two sides of the enclosure will be made of pine board, and the fourth side will be made of galvanized steel fencing material. If the pine board fencing costs $6 per running foot and the steel fencing costs $3 per running foot, determine the dimensions of the enclosure that will cost the least to erect.

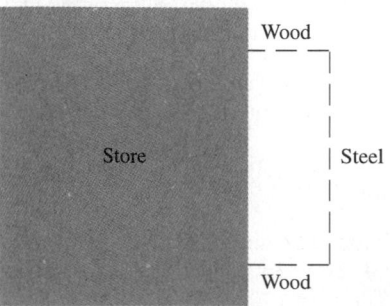

20. Minimizing Container Costs The Betty Moore Company requires that its corned beef hash containers have a capacity of 64 cubic inches, be right circular cylinders, and be made of a tin alloy. Find the radius and height of the least expensive container that can be made if the metal for the side and bottom costs 4 cents per square inch and the metal for the pull-off lid costs 2 cents per square inch.

Hint: Let the radius and height of the container be r and h inches, respectively. Then, the volume of the container is $\pi r^2 h = 64$, and the cost is given by

$$C(r, h) = 8\pi rh + 6\pi r^2$$

21. Minimizing Construction Costs An open rectangular box is to be constructed from material that costs $3 per square foot for the bottom and $1 per square foot for its sides. Find the dimensions of the box of greatest volume that can be constructed for $36.

22. Minimizing Construction Costs A closed rectangular box having a volume of 4 cubic feet is to be constructed. If the material for the sides costs $1 per square foot and the material for the top and bottom costs $1.50 per square foot, find the dimensions of the box that can be constructed with minimum cost.

23. Maximizing Sales The Ross-Simons Company has a monthly advertising budget of $60,000. Their marketing department estimates that if they spend x dollars on newspaper advertising and y dollars on television advertising, then the monthly sales will be given by

$$z = f(x, y) = 90x^{1/4}y^{3/4}$$

dollars. Determine how much money Ross-Simons should spend on newspaper ads and on television ads per month to maximize its monthly sales.

24. Maximizing Production John Mills, the proprietor of the Mills Engine Company, a manufacturer of model airplane engines, finds that it takes x units of labor and y units of capital to produce

$$f(x, y) = 100x^{3/4}y^{1/4}$$

units of the product. If a unit of labor costs $100 and a unit of capital costs $200, and $200,000 is budgeted for production, determine how many units should be expended on labor and how many units should be expended on capital in order to maximize production.

25. Use the Method of Lagrange Multipliers to solve exercise 28, Exercise Set 16.3.

SOLUTIONS TO SELF-CHECK EXERCISES 16.4

1. Write the constraint equation in the form $g(x, y) = 3x + 4y - 12 = 0$. Then the Lagrangian function is

$$F(x, y, \lambda) = -2x^2 - y^2 + \lambda(3x + 4y - 12)$$

To find the critical point(s) of F, we solve the system

$$F_x = -4x + 3\lambda = 0$$
$$F_y = -2y + 4\lambda = 0$$
$$F_\lambda = 3x + 4y - 12 = 0$$

Solving the first two equations for x and y in terms of λ, we find $x = \frac{3}{4}\lambda$ and $y = 2\lambda$. Substituting these values of x and y into the third equation of the system yields

$$3\left(\frac{3}{4}\lambda\right) + 4(2\lambda) - 12 = 0$$

or $\lambda = 48/41$. Therefore, $x = (3/4)(48/41) = 36/41$ and $y = 2(48/41) = 96/41$, and we see that the point $(36/41, 96/41)$ is the constrained maximum of f. The maximum value is

$$f\left(\frac{36}{41}, \frac{96}{41}\right) = -2\left(\frac{36}{41}\right)^2 - \left(\frac{96}{41}\right)^2$$
$$= -\frac{11{,}808}{1681} = -\frac{288}{41}$$

2. We want to maximize

$$P(x, y) = -\frac{1}{8}x^2 - \frac{1}{2}y^2 - \frac{1}{4}xy + 13x + 40y - 280$$

subject to the constraint

$$g(x, y) = x + y - 40 = 0$$

The Lagrangian function is

$$F(x, y, \lambda) = P(x, y) + \lambda g(x, y)$$
$$= -\frac{1}{8}x^2 - \frac{1}{2}y^2 - \frac{1}{4}xy + 13x$$
$$+ 40y - 280 + \lambda(x + y - 40)$$

To find the critical points of F, we solve the following system of equations:

$$F_x = -\frac{1}{4}x - \frac{1}{4}y + 13 + \lambda = 0$$

$$F_y = -\frac{1}{4}x - y + 40 + \lambda = 0$$

$$F_\lambda = x + y - 40 = 0$$

Subtracting the first equation from the second gives

$$-\frac{3}{4}y + 27 = 0 \quad \text{or} \quad y = 36$$

Substituting this value of y into the third equation yields $x = 4$. Therefore, in order to maximize its monthly profits, Robertson should manufacture 400 standard and 3600 deluxe thermostats. The maximum monthly profit is given by

$$P(4, 36) = -\frac{1}{8}(4)^2 - \frac{1}{2}(36)^2 - \frac{1}{4}(4)(36)$$
$$+ 13(4) + 40(36) - 280$$
$$= 526$$

or \$52,600.

16.5 DOUBLE INTEGRALS

A Geometric Interpretation of the Double Integral

In order to introduce the notion of the integral of a function of two variables, let us first recall the definition of the definite integral of a continuous function of one variable $y = f(x)$ over the interval $[a, b]$. We first divide the interval $[a, b]$ into n subintervals, each of equal length, by the points $x_0 = a < x_1 < x_2 < \cdots < x_n = b$ and define the Riemann sum by

$$S_n = f(p_1)h + f(p_2)h + \cdots + f(p_n)h$$

Figure 16.25
f(x, y) is a function defined over a rectangular region R.

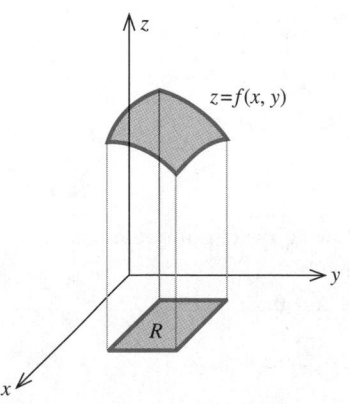

where $h = (b - a)/n$ and p_i is an arbitrary point in the interval $[x_{i-1}, x_i]$. The definite integral of f over $[a, b]$ is defined as the limit of the Riemann sum S_n as n tends to infinity, whenever it exists. Furthermore, recall that when f is a nonnegative continuous function on $[a, b]$, then the ith term of the Riemann sum, $f(p_i)h$, is an approximation (by the area of a rectangle) of the area under that part of the graph of $y = f(x)$ between $x = x_{i-1}$ and $x = x_i$, so that the Riemann sum S_n provides us with an approximation of the area under the curve $y = f(x)$ from $x = a$ to $x = b$. The integral

$$\int_a^b f(x)\, dx = \lim_{n \to \infty} S_n$$

gives the *actual* area under the curve from $x = a$ to $x = b$.

Now suppose that $f(x, y)$ is a continuous function of two variables defined over a region R. For simplicity, we assume for the moment that R is a rectangular region in the plane (Figure 16.25). Let us construct a Riemann sum for

Figure 16.26
Grid with m = 5 and n = 4.

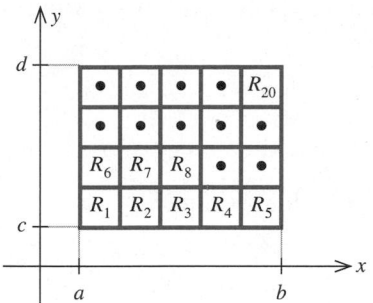

this function over the rectangle R by following a procedure that parallels the case for a function of one variable over an interval I. We begin by observing that the analogue of a *partition* in the two-dimensional case is a rectangular **grid** comprising mn rectangles, each of length h and width k, as a result of partitioning the side of the rectangle R of length $(b - a)$ into m segments and the side of length $(d - c)$ into n segments. By construction

$$h = \frac{b - a}{m} \quad \text{and} \quad k = \frac{d - c}{n}$$

A sample grid with $m = 5$ and $n = 4$ is shown in Figure 16.26.

Let us label the rectangles $R_1, R_2, R_3, \ldots, R_{mn}$. If (x_i, y_i) is *any* point in $R_i \, (1 \leq i \leq mn)$, then the **Riemann sum of $f(x, y)$ over the region R** is defined as

$$S(m, n) = f(x_1, y_1)hk + f(x_2, y_2)hk + \cdots + f(x_{mn}, y_{mn})hk$$

If the limit of $S(m, n)$ exists as both m and n tend to infinity, we call this limit the value of the **double integral of $f(x, y)$ over the region R** and denote it by

$$\iint\limits_R f(x, y)\, dA$$

If $f(x, y)$ is a nonnegative function, then the number $f(x_i, y_i)hk$ is the volume of a prism with base R_i and height $f(x_i, y_i)$ that provides us with an approximation of the volume of the solid under the surface $z = f(x, y)$ and bounded below by the rectangular region R_i (Figure 16.27).

Figure 16.27

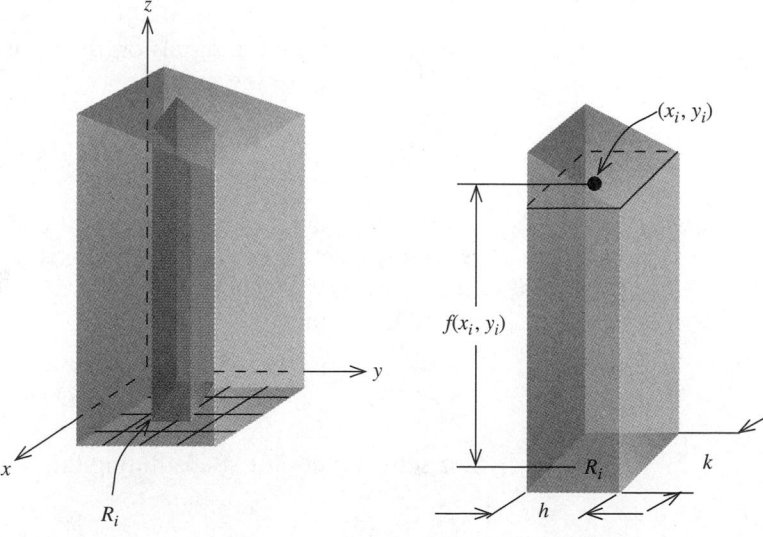

(a) Approximating the volume of a solid with rectangular prisms

(b) A prism with base R_i and height $f(x_i, y_i)$

Therefore, the Riemann sum $S(m, n)$ gives us an approximation of the volume of the solid bounded above by the surface $z = f(x, y)$ and below by the plane region R. As both m and n tend to infinity, the Riemann sum $S(m, n)$ approaches the *actual* volume under the solid.

Evaluating a Double Integral over a Rectangular Region

Let us turn our attention to the evaluation of the double integral

$$\iint_R f(x, y) \, dA$$

where R is the rectangular region shown in Figure 16.25. As in the case of the definite integral of a function of one variable, it turns out that the double integral can be evaluated without our having to first find an appropriate Riemann sum and then take the limit of that sum. Instead, as we will now see, the technique calls for evaluating two single integrals—the so-called **iterated integrals**—in succession, using a process that might be called "anti-partial differentiation." The technique is described in the following result, which we state without proof.

Let R be the rectangle defined by the inequalities $a \leq x \leq b$ and $c \leq y \leq d$ (Figure 16.26). Then

$$\iint_R f(x, y) \, dA = \int_c^d \left[\int_a^b f(x, y) \, dx \right] dy \tag{4}$$

where the iterated integrals on the right-hand side are evaluated as follows. We first compute the integral

$$\int_a^b f(x, y) \, dx$$

by treating y as if it were a constant and integrating the resulting function of x with respect to x (the "dx" reminds us that we are integrating with respect to x). In this manner we obtain a value for the integral that may contain the variable y. Thus,

$$\int_a^b f(x, y) \, dx = g(y)$$

for some function g. Substituting this value into equation (4) gives

$$\int_c^d g(y) \, dy$$

which may be integrated in the usual manner.

EXAMPLE 1   Evaluate $\iint_R f(x, y) \, dA$, where $f(x, y) = x + 2y$ and R is the rectangle defined by $1 \leq x \leq 4$ and $1 \leq y \leq 2$.

Solution   Using equation (4), we find

$$\iint_R f(x, y) \, dA = \int_1^2 \left[\int_1^4 (x + 2y) \, dx \right] dy$$

To compute

$$\int_1^4 (x + 2y) \, dx$$

we treat y as if it were a constant (remember that the "dx" reminds us that we are integrating with respect to x). We obtain

$$\int_1^4 (x + 2y) \, dx = \frac{1}{2}x^2 + 2xy \Big|_{x=1}^{x=4}$$

$$= \left[\frac{1}{2}(16) + 2(4)y \right] - \left[\frac{1}{2}(1) + 2(1)y \right]$$

$$= \frac{15}{2} + 6y$$

Thus,

$$\iint_R f(x, y) \, dA = \int_1^2 \left(\frac{15}{2} + 6y \right) dy = \left(\frac{15}{2}y + 3y^2 \right) \Big|_1^2$$

$$= (15 + 12) - \left(\frac{15}{2} + 3 \right) = 16\frac{1}{2}$$

● ● ●

Evaluating a Double Integral over a Plane Region

Up to now, we have assumed that the region over which a double integral is to be evaluated is rectangular. In fact, however, it is possible to compute the double integral of functions over rather arbitrary regions. The next theorem, which we state without proof, expands the number of types of regions over which we may integrate.

THEOREM 1

a. Suppose $g_1(x)$ and $g_2(x)$ are continuous functions on $[a, b]$ and the region R is defined by $R = \{(x, y) \mid g_1(x) \le y \le g_2(x); a \le x \le b\}$. Then

$$\iint_R f(x, y) \, dA = \int_a^b \left[\int_{g_1(x)}^{g_2(x)} f(x, y) \, dy \right] dx \tag{5}$$

(Figure 16.28a).

b. Suppose $h_1(y)$ and $h_2(y)$ are continuous functions on $[c, d]$ and the region R is defined by $R = \{(x, y) \mid h_1(y) \le x \le h_2(y); c \le y \le d\}$. Then

$$\iint_R f(x, y) \, dA = \int_c^d \left[\int_{h_1(y)}^{h_2(y)} f(x, y) \, dx \right] dy \tag{6}$$

(Figure 16.28b).

Figure 16.28

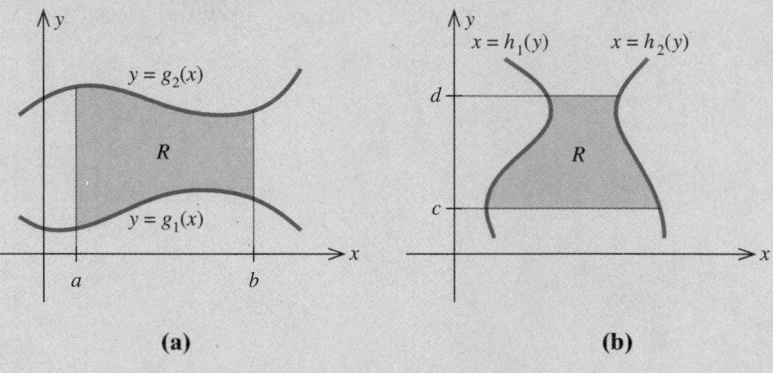

(a)   (b)

REMARKS

1. Observe that in (5) the lower and upper limits of integration with respect to y are given by $y = g_1(x)$ and $y = g_2(x)$. This is to be expected since, for a fixed value of x lying between $x = a$ and $x = b$, y runs between the lower curve defined by $y = g_1(x)$ and the upper curve defined by $y = g_2(x)$ [see Figure 16.28a]. Observe, too, that in the special case when $g_1(x) = c$ and $g_2(x) = d$, the region R is rectangular, and (5) reduces to (4).

2. For a fixed value of y, x runs between $x = h_1(y)$ and $x = h_2(y)$, giving the indicated limits of integration with respect to x in (6) [see Figure 16.28b].

3. Note that the two curves in Figure 16.28b are not graphs of functions of x (use the vertical-line test), but they are graphs of functions of y. It is this observation that justifies the approach leading to (6). ○ ○ ○

We now look at several examples.

Figure 16.29
R is the region bounded by $g_1(x) = x$ and $g_2(x) = 2x$ for $0 \leq x \leq 2$.

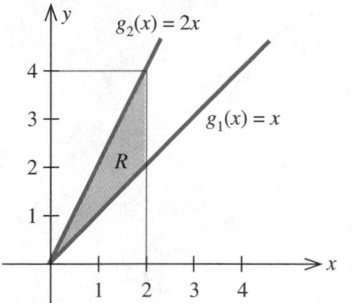

EXAMPLE 2   Evaluate $\iint_R f(x, y) \, dA$ given that $f(x, y) = x^2 + y^2$ and R is the region bounded by the graphs of $g_1(x) = x$ and $g_2(x) = 2x$ for $0 \leq x \leq 2$.

Solution   The region under consideration is shown in Figure 16.29. Using equation (5), we find

$$\iint_R f(x, y) \, dA = \int_0^2 \left[\int_x^{2x} (x^2 + y^2) \, dy \right] dx$$

$$= \int_0^2 \left[\left(x^2 y + \frac{1}{3} y^3 \right) \Big|_x^{2x} \right] dx$$

$$= \int_0^2 \left[\left(2x^3 + \frac{8}{3} x^3 \right) - \left(x^3 + \frac{1}{3} x^3 \right) \right] dx$$

$$= \int_0^2 \frac{10}{3} x^3 \, dx = \frac{5}{6} x^4 \Big|_0^2 = 13 \frac{1}{3}$$

● ● ●

EXAMPLE 3   Evaluate $\iint_R f(x, y) \, dA$, where $f(x, y) = xe^y$ and R is the plane region bounded by the graphs of $y = x^2$ and $y = x$.

Figure 16.30
R is the region bounded by $y = x^2$ and $y = x$.

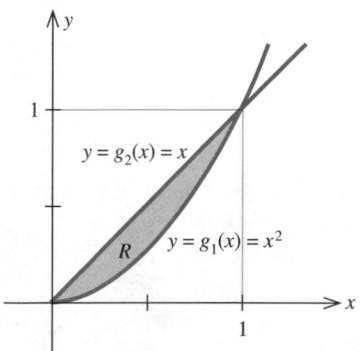

Solution   The region in question is shown in Figure 16.30. The point of intersection of the two curves is found by solving the equation $x^2 = x$, giving $x = 0$ and $x = 1$. Using equation (5), we find

$$\iint_R f(x, y) \, dA = \int_0^1 \left[\int_{x^2}^x xe^y \, dy \right] dx = \int_0^1 \left[xe^y \Big|_{x^2}^x \right] dx$$

$$= \int_0^1 (xe^x - xe^{x^2}) \, dx = \int_0^1 xe^x \, dx - \int_0^1 xe^{x^2} \, dx$$

and integrating the first integral on the right-hand side by parts,

$$= \left[(x - 1)e^x - \frac{1}{2} e^{x^2} \right] \Big|_0^1$$

$$= -\frac{1}{2} e - \left(-1 - \frac{1}{2} \right) = \frac{1}{2} (3 - e)$$

● ● ●

Refer to Example 3.

a. You can also view the region R as an example of the region shown in Figure 16.28(b). Doing so, find the functions h_1 and h_2 and the numbers c and d.

b. Find an expression for $\iint_R f(x, y) \, dA$ in terms of iterated integrals using formula (6).

c. Evaluate the iterated integrals of part (b) and hence verify the result of Example 3. [Hint: Integrate by parts twice.]

d. Does viewing the region R in two different ways make a difference?

Finding the Volume of a Solid by Double Integrals

As we saw in the last section, the double integral

$$\iint_R f(x, y) \, dA$$

gives the volume of the solid bounded by the graph of $f(x, y)$ over the region R.

THE VOLUME OF A SOLID UNDER A SURFACE	Let R be a region in the xy-plane and let f be continuous and nonnegative on R. Then the volume of the solid bounded above by the surface $z = f(x, y)$ and below by R is given by $$V = \iint_R f(x, y) \, dA$$

Figure 16.31
The plane region R defined by $y = \sqrt{1 - x^2}, 0 \le x \le 1$.

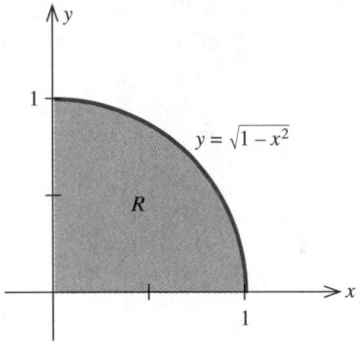

$y = \sqrt{1 - x^2}$

R

EXAMPLE 4   Find the volume of the solid bounded above by the plane $z = f(x, y) = y$ and below by the plane region R defined by $y = \sqrt{1 - x^2}, 0 \le x \le 1$.

Solution   The region R is sketched in Figure 16.31. Observe that $f(x, y) = y \ge 0$ for $y \in R$. Therefore, the required volume is given by

$$\iint_R y \, dA = \int_0^1 \left[\int_0^{\sqrt{1-x^2}} y \, dy \right] dx = \int_0^1 \left[\frac{1}{2} y^2 \Big|_0^{\sqrt{1-x^2}} \right] dx$$

$$= \int_0^1 \frac{1}{2} (1 - x^2) \, dx = \frac{1}{2} \left(x - \frac{1}{3} x^3 \right) \Big|_0^1 = \frac{1}{3}$$

or $\frac{1}{3}$ cubic unit. The solid is shown in Figure 16.32. Note that it is not necessary to make a sketch of the solid in order to compute its volume.

Figure 16.32
The solid bounded above by the plane $z = y$ and below by the plane region defined by $y = \sqrt{1 - x^2}, 0 \le x \le 1$.

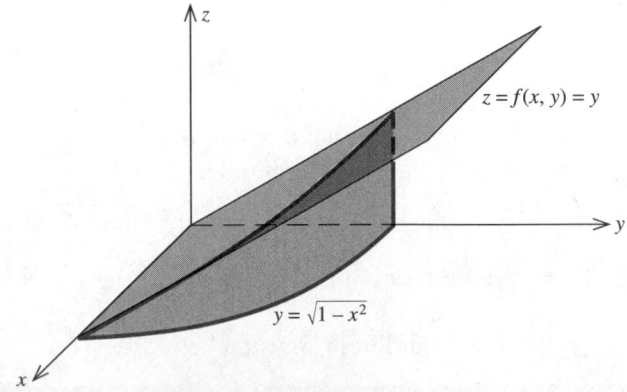

$z = f(x, y) = y$

$y = \sqrt{1 - x^2}$

Figure 16.33
The rectangular region R representing a certain district of a city is enclosed by a rectangular grid.

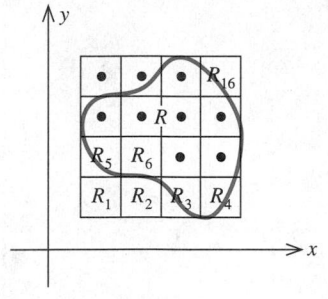

Population of a City

Suppose that the plane region R represents a certain district of a city and $f(x, y)$ gives the population density (the number of people per square mile) at any point (x, y) in R. Enclose the set R by a rectangle and construct a grid for it in the usual manner. In any rectangular region of the grid that has no point in common with R, set $f(x_i, y_i)hk = 0$ (see Figure 16.33). Then, corresponding to any grid covering the set R, the general term of the Riemann sum $f(x_i, y_i)hk$ (population density times area) gives the number of people living in that part of the city corresponding to the rectangular region R_i. Therefore, the Riemann sum gives an approximation of the number of people living in the district represented by R and, in the limit, the double integral

$$\iint_R f(x, y)\, dA$$

gives the actual number of people living in the district under consideration.

Figure 16.34
The rectangular region R represents a certain district of a city.

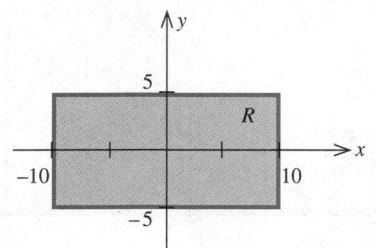

EXAMPLE 5 The population density of a certain city is described by the function

$$f(x, y) = 10{,}000e^{-0.2|x|-0.1|y|}$$

where the origin $(0, 0)$ gives the location of the city hall. What is the population inside the rectangular area described by

$$R = \{(x, y) \mid -10 \le x \le 10; \ -5 \le y \le 5\}$$

if x and y are in miles? (See Figure 16.34.)

Solution By symmetry, it suffices to compute the population in the first quadrant. (Why?) Then, upon observing that in this quadrant

$$f(x, y) = 10{,}000e^{-0.2x-0.1y} = 10{,}000e^{-0.2x}e^{-0.1y}$$

we see that the population in R is given by

$$\iint_R f(x, y)\, dA = 4\int_0^{10}\left[\int_0^5 10{,}000e^{-0.2x}\,e^{-0.1y}\, dy\right] dx$$

$$= 4\int_0^{10}\left[-100{,}000e^{-0.2x}e^{-0.1y}\Big|_0^5\right] dx$$

$$= 400{,}000(1 - e^{-0.5})\int_0^{10} e^{-0.2x}\, dx$$

$$= 2{,}000{,}000(1 - e^{-0.5})(1 - e^{-2})$$

or approximately 680,438 people.   ◦ ◦ ◦

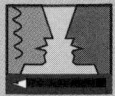

a. Consider the improper double integral $\iint\limits_{D} f(x, y)\, dA$ of the continuous function f of two variables defined over the plane region

$$D = \{(x, y)\mid 0 \le x < \infty \quad \text{and} \quad 0 \le y < \infty\}$$

Using the definition of improper integrals of functions of one variable (Section 15.3), explain why it makes sense to define

$$\iint\limits_{D} f(x, y)\, dA = \lim_{N \to \infty} \int_0^N \left[\lim_{M \to \infty} \int_0^M f(x, y)\, dx \right] dy$$

$$= \lim_{M \to \infty} \int_0^M \left[\lim_{N \to \infty} \int_0^N f(x, y)\, dy \right] dx$$

provided the limits exist.

b. Refer to Example 5. Assuming that the population density of the city is described by

$$f(x, y) = 10{,}000e^{-0.2|x|-0.1|y|}$$

for $-\infty < x < \infty$ and $-\infty < y < \infty$, show that the population outside the rectangular region

$$R = \{(x, y)\mid -10 < x < 10;\ -5 < y \le 5\}$$

of Example 5 is given by

$$4 \iint\limits_{D} f(x, y)\, dx\, dy - 680{,}438$$

(recall that 680,438 is the approximate population inside R).

c. Use the results of parts (a) and (b) to determine the population of the city outside the rectangular area R.

Average Value of a Function

In Section 14.5 we showed that the average value of a continuous function $f(x)$ over an interval $[a, b]$ is given by

$$\frac{1}{b - a} \int_a^b f(x)\, dx$$

That is, the average value of a function over $[a, b]$ is the integral of f over $[a, b]$ divided by the length of the interval. An analogous result holds for a function of two variables $f(x, y)$ over a plane region R. To see this, we enclose R by a rectangle and construct a rectangular grid. Let (x_i, y_i) be any point in the rectangle R_i of area hk. Now, the average value of the mn numbers $f(x_1, y_1), f(x_2, y_2), \ldots, f(x_{mn}, y_{mn})$ is given by

$$\frac{f(x_1, y_1) + f(x_2, y_2) + \cdots + f(x_{mn}, y_{mn})}{mn}$$

which can also be written as

$$\frac{hk}{hk}\left[\frac{f(x_1, y_1) + f(x_2, y_2) + \cdots + f(x_{mn}, y_{mn})}{mn}\right]$$

$$= \frac{1}{(mn)hk}[f(x_1, y_1) + f(x_2, y_2) + \cdots + f(x_{mn}, y_{mn})]hk$$

Now the area of R is approximated by the sum of the mn rectangles (*omitting* those having no points in common with R), each of area hk. Note that this is the denominator of the previous expression. Therefore, taking the limit as m and n both tend to infinity, we obtain the following formula for the *average value of $f(x, y)$ over R.*

AVERAGE VALUE OF $f(x, y)$ OVER THE REGION R

If f is integrable over the plane region R, then its average value over R is given by

$$\frac{\iint_R f(x, y)\, dA}{\text{Area of } R}$$

or

$$\frac{\iint_R f(x, y)\, dA}{\iint_R dA} \tag{7}$$

REMARK   If we let $f(x, y) = 1$ for all (x, y) in R, then

$$\iint_R f(x, y)\, dA = \iint_R dA = \text{Area of } R$$

○ ○ ○

EXAMPLE 6   Find the average value of the function $f(x, y) = xy$ over the plane region defined by $y = e^x$, $0 \le x \le 1$.

Figure 16.35
The plane region R defined by $y = e^x$, $0 \le x \le 1$.

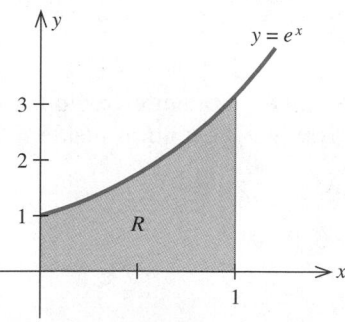

Solution   The region R is shown in Figure 16.35. The area of the region R is given by

$$\int_0^1 \left[\int_0^{e^x} dy\right] dx = \int_0^1 \left[y \Big|_0^{e^x}\right] dx$$

$$= \int_0^1 e^x\, dx$$

$$= e^x \Big|_0^1$$

$$= e - 1$$

square units. We would obtain the same result had we viewed the area of this region as the area of the region under the curve $y = e^x$ from $x = 0$ to $x = 1$.

Next, we compute

$$\iint\limits_{R} f(x, y)\, dA = \int_0^1 \left[\int_0^{e^x} xy\, dy \right] dx$$

$$= \int_0^1 \left[\frac{1}{2}xy^2 \Big|_0^{e^x} \right] dx$$

$$= \int_0^1 \frac{1}{2}xe^{2x}\, dx$$

$$= \frac{1}{4}xe^{2x} - \frac{1}{8}e^{2x} \Big|_0^1 \qquad \text{(Integrating by parts)}$$

$$= \left(\frac{1}{4}e^2 - \frac{1}{8}e^2 \right) + \frac{1}{8}$$

$$= \frac{1}{8}(e^2 + 1)$$

square units. Therefore, the required average value is given by

$$\frac{\iint\limits_{R} f(x, y)\, dA}{\iint\limits_{R} dA} = \frac{\frac{1}{8}(e^2 + 1)}{e - 1} = \frac{e^2 + 1}{8(e - 1)}$$

○ ○ ○

EXAMPLE 7 (Refer to Example 5.) The population density of a certain city (number of people per square mile) is described by the function

$$f(x, y) = 10{,}000e^{-0.2|x|-0.1|y|}$$

where the origin gives the location of the city hall. What is the average population density inside the rectangular area described by

$$R = \{(x, y) \mid -10 \le x \le 10;\ -5 \le y \le 5\}$$

where x and y are measured in miles?

Solution From the results of Example 5, we know that

$$\iint\limits_{R} f(x, y)\, dA \approx 680{,}438$$

From Figure 16.34, we see that the area of the plane rectangular region R is $(20)(10)$, or 200, square miles. Therefore, the average population inside R is

$$\frac{\iint\limits_{R} f(x, y)\, dA}{\iint\limits_{R} dA} = \frac{680{,}438}{200} = 3402.19$$

or approximately 3402 people per square mile.

○ ○ ○

SELF–CHECK EXERCISES 16.5

1. Evalute $\iint_R (x + y)\, dA$, where R is the region bounded by the graphs of $g_1(x) = x$ and $g_2(x) = x^{1/3}$.

2. The population density of a coastal town located on an island is described by the function

$$f(x, y) = \frac{5000xe^y}{1 + 2x^2} \qquad (0 \le x \le 4,\, -2 \le y \le 0)$$

where x and y are measured in miles (see the accompanying figure).

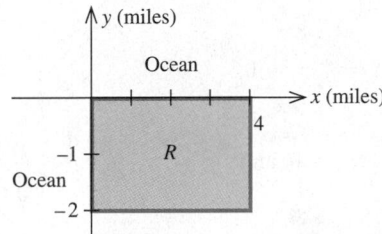

What is the population inside the rectangular area defined by $R = \{(x, y)\,|\,0 \le x \le 4,\, -2 \le y \le 0\}$? What is the average population density in the area?

Solutions to Self-Check Exercises 16.5 can be found on page 1144.

16.5 EXERCISES

In exercises 1–25, evaluate the double integral $\iint_R f(x, y)\, dA$ for the given function $f(x, y)$ and the region R.

1. $f(x, y) = y + 2x$; R is the rectangle defined by $1 \le x \le 2$ and $0 \le y \le 1$.

2. $f(x, y) = x + 2y$; R is the rectangle defined by $-1 \le x \le 2$ and $0 \le y \le 2$.

3. $f(x, y) = xy^2$; R is the rectangle defined by $-1 \le x \le 1$ and $0 \le y \le 1$.

4. $f(x, y) = 12xy^2 + 8y^3$; R is the rectangle defined by $0 \le x \le 1$ and $0 \le y \le 2$.

5. $f(x, y) = \dfrac{x}{y}$; R is the rectangle defined by $-1 \le x \le 2$ and $1 \le y \le e^3$.

6. $f(x, y) = \dfrac{xy}{1 + y^2}$; R is the rectangle defined by $-2 \le x \le 2$ and $0 \le y \le 1$.

7. $f(x, y) = 4xe^{2x^2 + y}$; R is the rectangle defined by $0 \le x \le 1$ and $-2 \le y \le 0$.

8. $f(x, y) = \dfrac{y}{x^2}e^{y/x}$; R is the rectangle defined by $1 \le x \le 2$ and $0 \le y \le 1$.

9. $f(x, y) = \ln y$; R is the rectangle defined by $0 \le x \le 1$ and $1 \le y \le e$.

10. $f(x, y) = \dfrac{\ln y}{x}$; R is the rectangle defined by $1 \le x \le e^2$ and $1 \le y \le e$.

11. $f(x, y) = x + 2y$; R is bounded by $x = 0$, $x = 1$, $y = 0$, and $y = x$.

12. $f(x, y) = xy$; R is bounded by $x = 0$, $x = 1$, $y = 0$, and $y = x$.

13. $f(x, y) = 2x + 4y$; R is bounded by $x = 1$, $x = 3$, $y = 0$, and $y = x + 1$.

14. $f(x, y) = 2 - y$; R is bounded by $x = -1$, $x = 1 - y$, $y = 0$, and $y = 2$.

15. $f(x, y) = x + y$; R is bounded by $x = 0$, $x = \sqrt{y}$, $y = 0$, and $y = 4$.

16. $f(x, y) = x^2y^2$; R is bounded by $x = 0$, $x = 1$, $y = x^2$, and $y = x^3$.

17. $f(x, y) = y$; R is bounded by $x = 0$, $x = \sqrt{4 - y^2}$, $y = 0$, and $y = 2$.

18. $f(x, y) = \dfrac{y}{x^3 + 2}$; R is bounded by $x = 0$, $x = 1$, $y = 0$, and $y = x$.

19. $f(x, y) = 2xe^y$; R is bounded by $x = 0$, $x = 1$, $y = 0$, and $y = x$.

20. $f(x, y) = 2x$; R is bounded by $x = e^{2y}$, $x = y$, $y = 0$, and $y = 1$.

21. $f(x, y) = ye^x$; R is bounded by $y = \sqrt{x}$ and $y = x$.

22. $f(x, y) = xe^{-y^2}$; R is bounded by $x = 0$, $y = x^2$, and $y = 4$.

23. $f(x, y) = e^{y^2}$; R is bounded by $x = 0$, $x = 1$, $y = 2x$, and $y = 2$.

24. $f(x, y) = y$; R is bounded by $x = 1$, $x = e$, $y = 0$, and $y = \ln x$.

25. $f(x, y) = ye^{x^3}$; R is bounded by $x = y/2$, $x = 1$, $y = 0$, and $y = 2$.

In exercises 26–33, use a double integral to find the volume of the solid shown in the figure.

26.

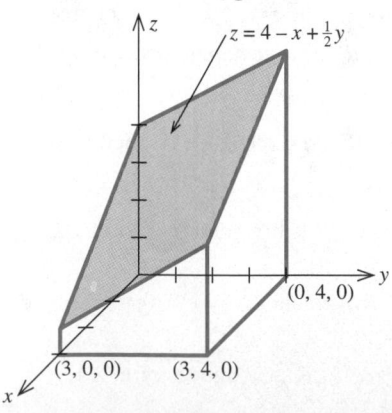

27.

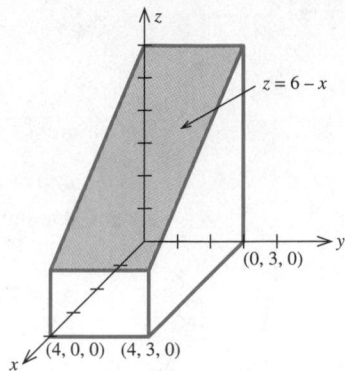

28.

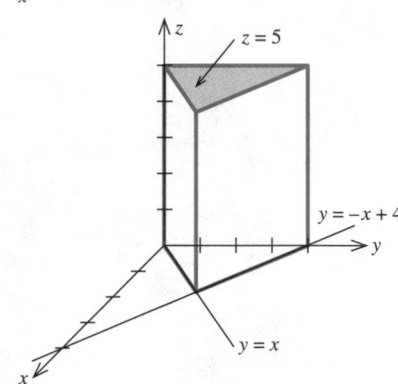

29.

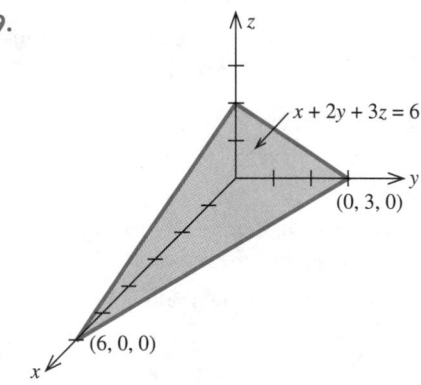

30.

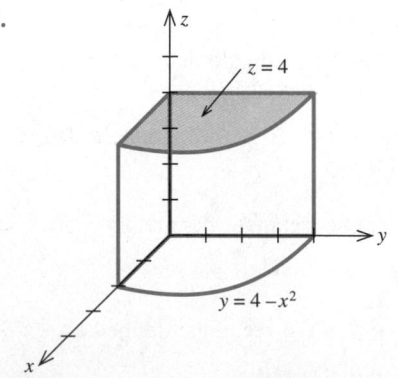

31.

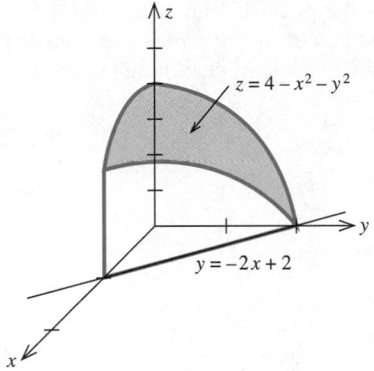

$z = 4 - x^2 - y^2$

$y = -2x + 2$

32.

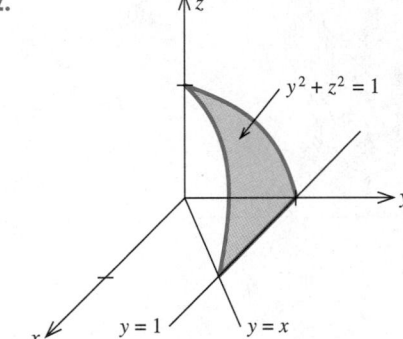

$y^2 + z^2 = 1$

$y = 1$ $y = x$

33.

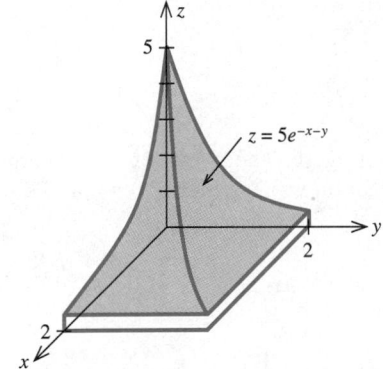

$z = 5e^{-x-y}$

In exercises 34–41, find the volume of the solid bounded above by the surface z = f(x, y) and below by the plane region R.

34. $f(x, y) = 4 - 2x - y$; $R = \{(x, y) \mid 0 \le x \le 1$ and $0 \le y \le 2\}$

35. $f(x, y) = 2x + y$; R is the triangle bounded by $y = 2x$, $y = 0$, and $x = 2$.

36. $f(x, y) = x^2 + y^2$; R is the rectangle with vertices $(0, 0)$, $(1, 0)$, $(1, 2)$, and $(0, 2)$.

37. $f(x, y) = e^{x+2y}$; R is the triangle with vertices $(0, 0)$, $(1, 0)$, and $(0, 1)$.

38. $f(x, y) = 2xe^y$; R is the triangle bounded by $y = x$, $y = 2$, and $x = 0$.

39. $f(x, y) = \dfrac{2y}{1 + x^2}$; R is the region bounded by $y = \sqrt{x}$, $y = 0$, and $x = 4$.

40. $f(x, y) = 2x^2y$; R is the region bounded by the graphs of $y = x$ and $y = x^2$.

41. $f(x, y) = x$; R is the region in the first quadrant bounded by the semicircle $y = \sqrt{16 - x^2}$, the x-axis, and the y-axis.

In exercises 42–47, find the average value of the given function f(x, y) over the plane region R.

42. $f(x, y) = 6x^2y^3$; $R = \{(x, y) \mid 0 \le x \le 2$ and $0 \le y \le 3\}$

43. $f(x, y) = x + 2y$; R is the triangle with vertices $(0, 0)$, $(1, 0)$, and $(1, 1)$.

44. $f(x, y) = xy$; R is the triangle bounded by $y = x$, $y = 2 - x$, and $y = 0$.

45. $f(x, y) = e^{-x^2}$; R is the triangle with vertices $(0, 0)$, $(1, 0)$, and $(1, 1)$.

46. $f(x, y) = xe^y$; R is the triangle with vertices $(0, 0)$, $(1, 0)$, and $(1, 1)$.

47. $f(x, y) = \ln x$; R is the region bounded by the graphs of $y = 2x$ and $y = 0$ from $x = 1$ to $x = 3$.

[*Hint:* Use integration by parts.]

48. Population Density The population density of a coastal town is described by the function

$$f(x, y) = \frac{10{,}000e^y}{1 + 0.5|x|} \qquad (-10 \le x \le 10, -4 \le y \le 0)$$

where x and y are measured in miles (see the accompanying figure). What is the population inside the rectangular area described by

$$R = \{(x, y) \mid -5 \le x \le 5 \text{ and } -2 \le y \le 0\}$$

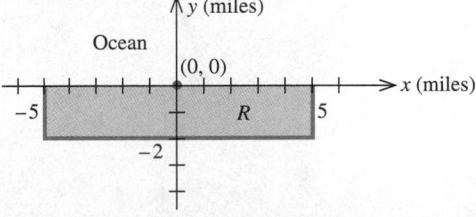

49. **Average Population Density** Refer to exercise 48. Find the average population density inside the rectangular area R.

50. **Average Profit** The Country Workshop's total weekly profit (in dollars) realized in manufacturing and selling its roll-top desks is given by the profit function

$$P(x, y) = -0.2x^2 - 0.25y^2 - 0.2xy$$
$$+ 100x + 90y - 4000$$

where x stands for the number of finished units and y stands for the number of unfinished units manufactured and sold per week. Find the average weekly profit if the number of finished units manufactured and sold varies between 180 and 200 and the number of unfinished units varies between 100 and 120 per week.

51. **Average Price of Land** The rectangular region R shown in the accompanying figure represents a city's financial district. The price of land in the district is approximated by the function

$$p(x, y) = 200 - 10\left(x - \frac{1}{2}\right)^2 - 15(y - 1)^2$$

where $p(x, y)$ is the price of land at the point (x, y) in dollars per square foot and x and y are measured in miles. What is the average price of land per square foot in the district?

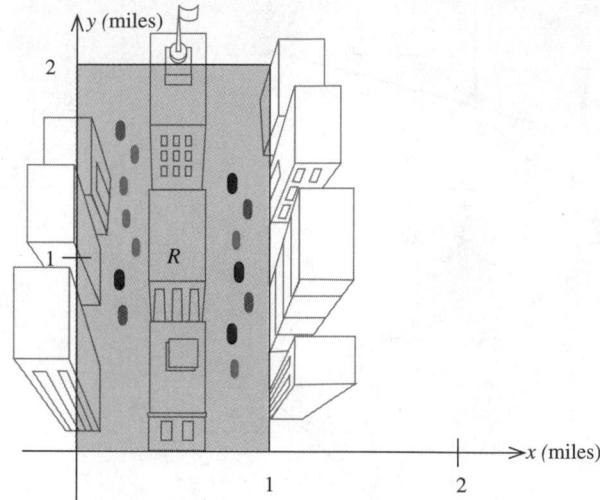

SOLUTIONS TO SELF-CHECK EXERCISES 16.5

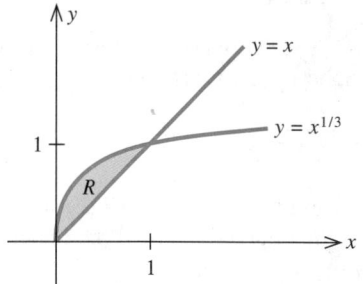

1. The region R is shown in the accompanying figure. The points of intersection of the two curves are found by solving the equation $x = x^{1/3}$, giving $x = 0$ and $x = 1$. Using equation (5), we find

$$\iint_R (x + y)\, dA = \int_0^1 \left[\int_x^{x^{1/3}} (x + y)\, dy\right] dx$$

$$= \int_0^1 \left[xy + \frac{1}{2}y^2\right]_x^{x^{1/3}} dx$$

$$= \int_0^1 \left[\left(x^{4/3} + \frac{1}{2}x^{2/3}\right) - \left(x^2 + \frac{1}{2}x^2\right)\right] dx$$

$$= \int_0^1 \left(x^{4/3} + \frac{1}{2}x^{2/3} - \frac{3}{2}x^2\right) dx$$

$$= \frac{3}{7}x^{7/3} + \frac{3}{10}x^{5/3} - \frac{1}{2}x^3 \Big|_0^1$$

$$= \frac{3}{7} + \frac{3}{10} - \frac{1}{2} = \frac{8}{35}$$

2. The population in R is given by

$$\iint_R f(x, y)\, dA = \int_0^4 \left[\int_{-2}^0 \frac{5000xe^y}{1 + 2x^2}\, dy \right] dx$$

$$= \int_0^4 \left[\frac{5000xe^y}{1 + 2x^2} \Big|_{-2}^0 \right] dx$$

$$= 5000(1 - e^{-2}) \int_0^4 \frac{x}{1 + 2x^2}\, dx$$

$$= 5000(1 - e^{-2}) \left[\frac{1}{4} \ln(1 + 2x^2) \Big|_0^4 \right]$$

$$= 5000(1 - e^{-2}) \left(\frac{1}{4} \right) \ln 33$$

or approximately 3779 people. The average population density inside R is

$$\frac{\displaystyle\iint_R f(x, y)\, dA}{\displaystyle\iint_R dA} = \frac{3779}{(2)(4)}$$

or approximately 472 people per square mile.

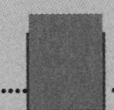

CHAPTER 16 SUMMARY OF PRINCIPAL TERMS

Terms

Three-dimensional Cartesian coordinate system

Level curve

Function of two variables:

  Domain

  First partial derivative

  Second-order partial derivative

  Relative maximum

  Relative minimum

  Absolute maximum

  Absolute minimum

  Critical point

Saddle point

Second Derivative Test

Riemann sum

Constrained relative extremum

Cobb-Douglas production function

Marginal productivity of labor

Marginal productivity of capital

Substitute commodities

Complementary commodities

Method of Lagrange Multipliers

Volume of a solid under a surface

CHAPTER 16 REVIEW EXERCISES

1. Let $f(x, y) = \dfrac{xy}{x^2 + y^2}$. Compute $f(0, 1)$, $f(1, 0)$, and $f(1, 1)$. Does $f(0, 0)$ exist?

2. Let $f(x, y) = \dfrac{xe^y}{1 + \ln xy}$. Compute $f(1, 1)$, $f(1, 2)$, and $f(2, 1)$. Does $f(1, 0)$ exist?

3. Let $h(x, y, z) = xye^z + \dfrac{x}{y}$. Compute $h(1, 1, 0)$, $h(-1, 1, 1)$, and $h(1, -1, 1)$.

4. Find the domain of the function $f(u, v) = \dfrac{\sqrt{u}}{u - v}$.

5. Find the domain of the function $f(x, y) = \dfrac{x - y}{x + y}$.

6. Find the domain of the function $f(x, y) = x\sqrt{y} + y\sqrt{1 - x}$.

7. Find the domain of the function $f(x, y, z) = \dfrac{xy\sqrt{z}}{(1 - x)(1 - y)(1 - z)}$.

In exercises 8–11, sketch the level curves of the function corresponding to the given values of z.

8. $z = f(x, y) = 2x + 3y$; $z = -2, -1, 0, 1, 2$

9. $z = f(x, y) = y - x^2$; $z = -2, -1, 0, 1, 2$

10. $z = f(x, y) = \sqrt{x^2 + y^2}$; $z = 0, 1, 2, 3, 4$

11. $z = f(x, y) = e^{xy}$; $z = 1, 2, 3$

In exercises 12–21, compute the first partial derivatives of the function.

12. $f(x, y) = x^2y^3 + 3xy^2 + \dfrac{x}{y}$

13. $f(x, y) = x\sqrt{y} + y\sqrt{x}$

14. $f(u, v) = \sqrt{uv^2 - 2u}$

15. $f(x, y) = \dfrac{x - y}{y + 2x}$ 16. $g(x, y) = \dfrac{xy}{x^2 + y^2}$

17. $h(x, y) = (2xy + 3y^2)^5$ 18. $f(x, y) = (xe^y + 1)^{1/2}$

19. $f(x, y) = (x^2 + y^2)e^{x^2 + y^2}$

20. $f(x, y) = \ln(1 + 2x^2 + 4y^4)$

21. $f(x, y) = \ln\left(1 + \dfrac{x^2}{y^2}\right)$

In exercises 22–27, compute the second-order partial derivatives of the function.

22. $f(x, y) = x^3 - 2x^2y + y^2 + x - 2y$

23. $f(x, y) = x^4 + 2x^2y^2 - y^4$

24. $f(x, y) = (2x^2 + 3y^2)^3$ 25. $g(x, y) = \dfrac{x}{x + y^2}$

26. $g(x, y) = e^{x^2 + y^2}$ 27. $h(s, t) = \ln\left(\dfrac{s}{t}\right)$

28. Let $f(x, y, z) = x^3y^2z + xy^2z + 3xy - 4z$. Compute $f_x(1, 1, 0)$, $f_y(1, 1, 0)$, and $f_z(1, 1, 0)$, and interpret your results.

In exercises 29–34, find the critical point(s) of the functions. Then use the Second Derivative Test to classify the nature of each of these points, if possible. Finally, determine the relative extrema of each function.

29. $f(x, y) = 2x^2 + y^2 - 8x - 6y + 4$

30. $f(x, y) = x^2 + 3xy + y^2 - 10x - 20y + 12$

31. $f(x, y) = x^3 - 3xy + y^2$

32. $f(x, y) = x^3 + y^2 - 4xy + 17x - 10y + 8$

33. $f(x, y) = e^{2x^2 + y^2}$

34. $f(x, y) = \ln(x^2 + y^2 - 2x - 2y + 4)$

In exercises 35–38, use the Method of Lagrange Multipliers to optimize the function subject to the given constraints.

35. Maximize the function $f(x, y) = -3x^2 - y^2 + 2xy$ subject to the constraint $2x + y = 4$.

36. Minimize the function $f(x, y) = 2x^2 + 3y^2 - 6xy + 4x - 9y + 10$ subject to the constraint $x + y = 1$.

37. Find the maximum and minimum values of the function $f(x, y) = 2x - 3y + 1$ subject to the constraint $2x^2 + 3y^2 - 125 = 0$.

38. Find the maximum and minimum values of the function $f(x, y) = e^{x-y}$ subject to the constraint $x^2 + y^2 = 1$.

In exercises 39–42, evaluate the double integrals.

39. $f(x, y) = 3x - 2y$; R is the rectangle defined by $2 \le x \le 4$ and $-1 \le y \le 2$.

40. $f(x, y) = e^{-x-2y}$; R is the rectangle defined by $0 \le x \le 2$ and $0 \le y \le 1$.

41. $f(x, y) = 2x^2y$; R is bounded by $x = 0$, $x = 1$, $y = x^2$, and $y = x^3$.

42. $f(x, y) = \dfrac{y}{x}$; R is bounded by $x = 1$, $x = 2$, $y = 1$, and $y = x$.

In exercises 43 and 44, find the volume of the solid bounded above by the surface z = f(x, y) and below by the plane region R.

43. $f(x, y) = 4x^2 + y^2$; $R = \{0 \le x \le 2 \text{ and } 0 \le y \le 1\}$

44. $f(x, y) = x + y$; R is the region bounded by $y = x^2$, $y = 4x$, and $y = 4$.

45. Find the average value of the function

$$f(x, y) = xy + 1$$

over the plane region R bounded by $y = x^2$ and $y = 2x$.

46. A division of Ditton Industries makes a 16-speed and a 10-speed electric blender. The company's management estimates that x units of the 16-speed model and y units of the 10-speed model are demanded daily when the unit prices are

$$p = 80 - 0.02x - 0.1y$$

and $\qquad q = 60 - 0.1x - 0.05y$

dollars, respectively.
  a. Find the daily total revenue function $R(x, y)$.
  b. Find the domain of the function R.
  c. Compute $R(100, 300)$ and interpret your result.

47. In a survey conducted by *Home Entertainment* magazine it was determined that the demand equation for compact disc (CD) players is given by

$$x = f(p, q) = 900 - 9p - e^{0.4q}$$

whereas the demand equation for compact audio discs is given by

$$y = g(p, q) = 20{,}000 - 3000q - 4p$$

where p and q denote the unit prices (in dollars) for the CD players and audio discs, respectively, and x and y denote the number of CD players and audio discs demanded per week. Determine whether these two products are substitute, complementary, or neither.

48. The Odyssey Travel Agency's monthly revenue depends on the amount of money x (in thousands of dollars) spent on advertising per month and the number of agents y in its employ in accordance with the rule

$$R(x, y) = -x^2 - 0.5y^2 + xy + 8x + 3y + 20$$

Determine the amount of money the agency should spend per month and the number of agents it should employ in order to maximize its monthly revenue.

49. The owner of the Rancho Grande wants to enclose a rectangular piece of grazing land along the straight portion of a river and then subdivide it using a fence running parallel to the sides. No fencing is required along the river. If the material for the sides costs $3 per running yard and the material for the divider costs $2 per running yard, what will the dimensions of a 303,750-yard pasture be if the cost of fencing material is kept to a minimum?

50. The production of Q units of a commodity is related to the amount of labor x and the amount of capital y (in suitable units) expended by the equation

$$Q = f(x, y) = x^{3/4}y^{1/4}$$

If an expenditure of 100 units is available for production, how should it be apportioned between labor and capital so that Q is maximized?
[*Hint:* Use the Method of Lagrange Multipliers to maximize the function Q subject to the constraint $x + y = 100$.]

The System of Real Numbers

In this appendix we briefly review the **system of real numbers.** This system comprises a set of objects called real numbers together with two operations, addition and multiplication, that enable us to combine two or more real numbers to obtain other real numbers. These operations are subjected to certain rules that we will state after first recalling the set of real numbers.

The set of real numbers may be constructed from the set of **natural** (also called **counting**) **numbers**

$$N = \{1, 2, 3, \ldots\}$$

by adjoining other objects (numbers) to it. Thus, the set

$$W = \{0, 1, 2, 3, \ldots\}$$

obtained by adjoining the single number 0 to N is called the set of **whole numbers.** By adjoining the *negatives* of the numbers 1, 2, 3, ... to the set W of whole numbers, we obtain the set of **integers**

$$I = \{\ldots, -3, -2, -1, 0, 1, 2, 3, \ldots\}$$

Next, consider the set

$$Q = \left\{ \frac{a}{b} \,\middle|\, a \text{ and } b \text{ are integers with } b \neq 0 \right\}$$

The set I of integers is contained in the set Q of **rational numbers.** To see this, observe that each integer may be written in the form a/b with $b = 1$, thus qualifying as a member of the set Q. The converse, however, is false, for the rational numbers (fractions) such as

$$\frac{1}{2}, \quad \frac{23}{25}, \quad \text{and so on}$$

are clearly not integers.

The sets N, W, I, and Q constructed thus far have

$$N \subset W \subset I \subset Q$$

That is, N is a proper subset of W, W is a proper subset of I, and so on.

Finally, consider the set Ir of all numbers that cannot be expressed in the form a/b, where a and b are integers ($b \neq 0$). The members of this set, called the **irrational numbers,** include $\sqrt{2}$, $\sqrt{3}$, π, and so on. The set

$$R = Q \cup Ir$$

That is, the set comprising all rational numbers as well as irrational numbers, is called the set of **real numbers** (Figure A.1).

Figure A.1

The set of all real numbers consists of the set of rational numbers plus the set of irrational numbers.

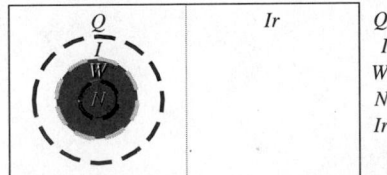

Q = Rationals
I = Integers
W = Whole numbers
N = Natural numbers
Ir = Irrationals

Note the following important representation of real numbers: Every real number has a decimal representation; a rational number has a representation in terms of a repeated decimal. For example,

$$\frac{1}{7} = 0.142857142857142857\ldots$$ (Note that the block of integers 142857 repeats.)

On the other hand, the irrational number $\sqrt{2}$ has a representation in terms of a nonrepeating decimal. Thus,

$$\sqrt{2} = 1.41421\ldots$$

As mentioned earlier, any two real numbers may be combined to obtain another real number. The operation of *addition,* written +, enables us to combine any two numbers a and b to obtain their sum, denoted by $a + b$. Another operation, called *multiplication* and written ·, enables us to combine any two real numbers a and b to form their product, the number $a \cdot b$ or, written more simply, ab. These two operations are subjected to the following rules of operation:

Given any three real numbers a, b, and c, we have

I. Under addition

1. $a + b = b + a$ (Commutative law of addition)

2. $a + (b + c) = (a + b) + c$ (Associative law of addition)

3. $a + 0 = a$ (Identity law of addition)

4. $a + (-a) = 0$ (Inverse law of addition)

II. Under multiplication

1. $ab = ba$ (Commutative law of multiplication)

2. $a(bc) = (ab)c$ (Associative law of multiplication)

3. $a \cdot 1 = a$ (Identity law of multiplication)

4. $a(1/a) = 1$ $(a \neq 0)$ (Inverse law of multiplication)

III. Under addition and multiplication

1. $a(b + c) = ab + ac$ (Distributive law for multiplication with respect to addition)

TABLES

Table I   *Compound amount, present value, and annuity*

	$i = \frac{1}{4}\%$			$i = \frac{1}{2}\%$		
n	$(1 + i)^n$	$s_{\overline{n}|i}$	$a_{\overline{n}|i}$	$(1 + i)^n$	$s_{\overline{n}|i}$	$a_{\overline{n}|i}$
1	1.0025 0000	1.0000 0000	0.9975 0623	1.0050 0000	1.0000 0000	0.9950 2488
2	1.0050 0625	2.0025 0000	1.9925 2492	1.0100 2500	2.0050 0000	1.9850 9938
3	1.0075 1877	3.0075 0625	2.9850 6227	1.0150 7513	3.0150 2500	2.9702 4814
4	1.0100 3756	4.0150 2502	3.9751 2446	1.0201 5050	4.0301 0013	3.9504 9566
5	1.0125 6266	5.0250 6258	4.9627 1766	1.0252 5125	5.0502 5063	4.9258 6633
6	1.0150 9406	6.0376 2523	5.9478 4804	1.0303 7751	6.0755 0188	5.8963 8441
7	1.0176 3180	7.0527 1930	6.9305 2174	1.0355 2940	7.1058 7939	6.8620 7404
8	1.0201 7588	8.0703 5110	7.9107 4487	1.0407 0704	8.1414 0879	7.8229 5924
9	1.0227 2632	9.0905 2697	8.8885 2357	1.0459 1058	9.1821 1583	8.7790 6392
10	1.0252 8313	10.1132 5329	9.8638 6391	1.0511 4013	10.2280 2641	9.7304 1186
11	1.0278 4634	11.1385 3642	10.8367 7198	1.0563 9583	11.2791 6654	10.6770 2673
12	1.0304 1596	12.1663 8277	11.8072 5384	1.0616 7781	12.3355 6237	11.6189 3207
13	1.0329 9200	13.1967 9872	12.7753 1555	1.0669 8620	13.3972 4018	12.5561 5131
14	1.0355 7448	14.2297 9072	13.7409 6314	1.0723 2113	14.4642 2639	13.4887 0777
15	1.0381 6341	15.2653 6520	14.7042 0264	1.0776 8274	15.5365 4752	14.4166 2465
16	1.0407 5882	16.3035 2861	15.6650 4004	1.0830 7115	16.6142 3026	15.3399 2502
17	1.0433 6072	17.3442 8743	16.6234 8133	1.0884 8651	17.6973 0141	16.2586 3186
18	1.0459 6912	18.3876 4815	17.5795 3250	1.0939 2894	18.7857 8791	17.1727 6802
19	1.0485 8404	19.4336 1727	18.5331 9950	1.0993 9858	19.8797 1685	18.0823 5624
20	1.0512 0550	20.4822 0131	19.4844 8828	1.1048 9558	20.9791 1544	18.9874 1915
21	1.0538 3352	21.5334 0682	20.4334 0477	1.1104 2006	22.0840 1101	19.8879 7925
22	1.0564 6810	22.5872 4033	21.3799 5488	1.1159 7216	23.1944 3107	20.7840 5896
23	1.0591 0927	23.6437 0843	22.3241 4452	1.1215 5202	24.3104 0322	21.6756 8055
24	1.0617 5704	24.7028 1770	23.2659 7957	1.1271 5978	25.4319 5524	22.5628 6622
25	1.0644 1144	25.7645 7475	24.2054 6591	1.1327 9558	26.5591 1502	23.4456 3803
26	1.0670 7247	26.8289 8619	25.1426 0939	1.1384 5955	27.6919 1059	24.3240 1794
27	1.0697 4015	27.8960 5865	26.0774 1585	1.1441 5185	28.8303 7015	25.1980 2780
28	1.0724 1450	28.9657 9880	27.0098 9112	1.1498 7261	29.9745 2200	26.0676 8936
29	1.0750 9553	30.0382 1330	27.9400 4102	1.1556 2197	31.1243 9461	26.9330 2423
30	1.0777 8327	31.1133 0883	28.8678 7134	1.1614 0008	32.2800 1658	27.7940 5397
31	1.0804 7773	32.1910 9210	29.7933 8787	1.1672 0708	33.4414 1666	28.6507 9997
32	1.0831 7892	33.2715 6983	30.7165 9638	1.1730 4312	34.6086 2375	29.5032 8355
33	1.0858 8687	34.3547 4876	31.6375 0262	1.1789 0833	35.7816 6686	30.3515 2592
34	1.0886 0159	35.4406 3563	32.5561 1234	1.1848 0288	36.9605 7520	31.1955 4818
35	1.0913 2309	36.5292 3722	33.4724 3126	1.1907 2689	38.1453 7807	32.0353 7132
36	1.0940 5140	37.6205 6031	34.3864 6510	1.1966 8052	39.3361 0496	32.8710 1624
37	1.0967 8653	38.7146 1171	35.2982 1955	1.2026 6393	40.5327 8549	33.7025 0372
38	1.0995 2850	39.8113 9824	36.2077 0030	1.2086 7725	41.7354 4942	34.5298 5445
39	1.1022 7732	40.9109 2673	37.1149 1302	1.2147 2063	42.9441 2666	35.3530 8900
40	1.1050 3301	42.0132 0405	38.0198 6336	1.2207 9424	44.1588 4730	36.1722 2786
41	1.1077 9559	43.1182 3706	38.9225 5697	1.2268 9821	45.3796 4153	36.9872 9141
42	1.1105 6508	44.2260 3265	39.8229 9947	1.2330 3270	46.6065 3974	37.7982 9991
43	1.1133 4149	45.3365 9774	40.7211 9648	1.2391 9786	47.8395 7244	38.6052 7354
44	1.1161 2485	46.4499 3923	41.6171 5359	1.2453 9385	49.0787 7030	39.4082 3238
45	1.1189 1516	47.5660 6408	42.5108 7640	1.2516 2082	50.3241 6415	40.2071 9640
46	1.1217 1245	48.6849 7924	43.4023 7047	1.2578 7892	51.5757 8497	41.0021 8547
47	1.1245 1673	49.8066 9169	44.2916 4137	1.2641 6832	52.8336 6390	41.7932 1937
48	1.1273 2802	50.9312 0842	45.1786 9463	1.2704 8916	54.0978 3222	42.5803 1778
49	1.1301 4634	52.0585 3644	46.0635 3580	1.2768 4161	55.3683 2138	43.3635 0028
50	1.1329 7171	53.1886 8278	46.9461 7037	1.2832 2581	56.6451 6299	44.1427 8635

	$i = \frac{1}{4}\%$			$i = \frac{1}{2}\%$		
n	$(1 + i)^n$	$s_{\overline{n}|i}$	$a_{\overline{n}|i}$	$(1 + i)^n$	$s_{\overline{n}|i}$	$a_{\overline{n}|i}$
51	1.1358 0414	54.3216 5449	47.8266 0386	1.2896 4194	57.9283 8880	44.9181 9537
52	1.1386 4365	55.4574 5862	48.7048 4176	1.2960 9015	59.2180 3075	45.6897 4664
53	1.1414 9026	56.5961 0227	49.5808 8953	1.3025 7060	60.5141 2090	46.4574 5934
54	1.1443 4398	57.7375 9252	50.4547 5265	1.3090 8346	61.8166 9150	47.2213 5258
55	1.1472 0484	58.8819 3650	51.3264 3656	1.3156 2887	63.1257 7496	47.9814 4535
56	1.1500 7285	60.0291 4135	52.1959 4669	1.3222 0702	64.4414 0384	48.7377 5657
57	1.1529 4804	61.1792 1420	53.0632 8847	1.3288 1805	65.7636 1086	49.4903 0505
58	1.1558 3041	62.3321 6223	53.9284 6730	1.3354 6214	67.0924 2891	50.2391 0950
59	1.1587 1998	63.4879 9264	54.7914 8858	1.3421 3946	68.4278 9105	50.9841 8855
60	1.1616 1678	64.6467 1262	55.6523 5769	1.3488 5015	69.7700 3051	51.7255 6075
61	1.1645 2082	65.8083 2940	56.5110 7999	1.3555 9440	71.1188 8066	52.4632 4453
62	1.1674 3213	66.9728 5023	57.3676 6083	1.3623 7238	72.4744 7507	53.1972 5824
63	1.1703 5071	68.1402 8235	58.2221 0557	1.3691 8424	73.8368 4744	53.9276 2014
64	1.1732 7658	69.3106 3306	59.0744 1952	1.3760 3016	75.2060 3168	54.6543 4839
65	1.1762 0977	70.4839 0964	59.9246 0800	1.3829 1031	76.5820 6184	55.3774 6109
66	1.1791 5030	71.6601 1942	60.7726 7631	1.3898 2486	77.9649 7215	56.0969 7621
67	1.1820 9817	72.8392 6971	61.6186 2974	1.3967 7399	79.3547 9701	56.8129 1165
68	1.1850 5342	74.0213 6789	62.4624 7355	1.4037 5785	80.7515 7099	57.5252 8522
69	1.1880 1605	75.2064 2131	63.3042 1302	1.4107 7664	82.1553 2885	58.2341 1465
70	1.1909 8609	76.3944 3736	64.1438 5339	1.4178 3053	83.5661 0549	58.9394 1756
71	1.1939 6356	77.5854 2345	64.9813 9989	1.4249 1968	84.9839 3602	59.6412 1151
72	1.1969 4847	78.7793 8701	65.8168 5774	1.4320 4428	86.4088 5570	60.3395 1394
73	1.1999 4084	79.9763 3548	66.6502 3216	1.4392 0450	87.8408 9998	61.0343 4222
74	1.2029 4069	81.1762 7632	67.4815 2834	1.4464 0052	89.2801 0448	61.7257 1366
75	1.2059 4804	82.3792 1701	68.3107 5146	1.4536 3252	90.7265 0500	62.4136 4543
76	1.2089 6291	83.5851 6505	69.1379 0670	1.4609 0069	92.1801 3752	63.0981 5466
77	1.2119 8532	84.7941 2797	69.9629 9920	1.4682 0519	93.6410 3821	63.7792 5836
78	1.2150 1528	86.0061 1329	70.7860 3411	1.4755 4622	95.1092 4340	64.4569 7350
79	1.2180 5282	87.2211 2857	71.6070 1657	1.4829 2395	96.5847 8962	65.1313 1691
80	1.2210 9795	88.4391 8139	72.4259 5169	1.4903 3857	98.0677 1357	65.8023 0538
81	1.2241 5070	89.6602 7934	73.2428 4458	1.4977 9026	99.5580 5214	66.4699 5561
82	1.2272 1108	90.8844 3004	74.0577 0033	1.5052 7921	101.0558 4240	67.1342 8419
83	1.2302 7910	92.1116 4112	74.8705 2402	1.5128 0561	102.5611 2161	67.7953 0765
84	1.2333 5480	93.3419 2022	75.6813 2072	1.5203 6964	104.0739 2722	68.4530 4244
85	1.2364 3819	94.5752 7502	76.4900 9548	1.5279 7148	105.5942 9685	69.1075 0491
86	1.2395 2928	95.8117 1321	77.2968 5335	1.5356 1134	107.1222 6834	69.7587 1135
87	1.2426 2811	97.0512 4249	78.1015 9935	1.5432 8940	108.6578 7968	70.4066 7796
88	1.2457 3468	98.2938 7060	78.9043 3850	1.5510 0585	110.2011 6908	71.0514 2086
89	1.2488 4901	99.5396 0527	79.7050 7581	1.5587 6087	111.7521 7492	71.6929 5608
90	1.2519 7114	100.7884 5429	80.5038 1627	1.5665 5468	113.3109 3580	72.3312 9958
91	1.2551 0106	102.0404 2542	81.3005 6486	1.5743 8745	114.8774 9048	72.9664 6725
92	1.2582 3882	103.2955 2649	82.0953 2654	1.5822 5939	116.4518 7793	73.5984 7487
93	1.2613 8441	104.5537 6530	82.8881 0628	1.5901 7069	118.0341 3732	74.2273 3818
94	1.2645 3787	105.8151 4972	83.6789 0900	1.5981 2154	119.6243 0800	74.8530 7282
95	1.2676 9922	107.0796 8759	84.4677 3966	1.6061 1215	121.2224 2954	75.4756 9434
96	1.2708 6847	108.3473 8681	85.2546 0315	1.6141 4271	122.8285 4169	76.0952 1825
97	1.2740 4564	109.6182 5528	86.0395 0439	1.6222 1342	124.4426 8440	76.7116 5995
98	1.2772 3075	110.8923 0091	86.8224 4827	1.6303 2449	126.0648 9782	77.3250 3478
99	1.2804 2383	112.1695 3167	87.6034 3967	1.6384 7611	127.6952 2231	77.9353 5799
100	1.2836 2489	113.4499 5550	88.3824 8346	1.6466 6849	129.3336 9842	78.5426 4477

	$i = \frac{3}{4}\%$			$i = 1\%$		
n	$(1 + i)^n$	$s_{\overline{n}|i}$	$a_{\overline{n}|i}$	$(1 + i)^n$	$s_{\overline{n}|i}$	$a_{\overline{n}|i}$
1	1.0075 0000	1.0000 0000	0.9925 5583	1.0100 0000	1.0000 0000	0.9900 9901
2	1.0150 5625	2.0075 0000	1.9777 2291	1.0201 0000	2.0100 0000	1.9703 9506
3	1.0226 6917	3.0225 5625	2.9555 5624	1.0303 0100	3.0301 0000	2.9409 8521
4	1.0303 3919	4.0452 2542	3.9261 1041	1.0406 0401	4.0604 0100	3.9019 6555
5	1.0380 6673	5.0755 6461	4.8894 3961	1.0510 1005	5.1010 0501	4.8534 3124
6	1.0458 5224	6.1136 3135	5.8455 9763	1.0615 2015	6.1520 1506	5.7954 7647
7	1.0536 9613	7.1594 8358	6.7946 3785	1.0721 3535	7.2135 3521	6.7281 9453
8	1.0615 9885	8.2131 7971	7.7366 1325	1.0828 5671	8.2856 7056	7.6516 7775
9	1.0695 6084	9.2747 7856	8.6715 7642	1.0936 8527	9.3685 2727	8.5660 1758
10	1.0775 8255	10.3443 3940	9.5995 7958	1.1046 2213	10.4622 1254	9.4713 0453
11	1.0856 6441	11.4219 2194	10.5206 7452	1.1156 6835	11.5668 3467	10.3676 2825
12	1.0938 0690	12.5075 8636	11.4349 1267	1.1268 2503	12.6825 0301	11.2550 7747
13	1.1020 1045	13.6013 9325	12.3423 4508	1.1380 9328	13.8093 2804	12.1337 4007
14	1.1102 7553	14.7034 0370	13.2430 2242	1.1494 7421	14.9474 2132	13.0037 0304
15	1.1186 0259	15.8136 7923	14.1369 9495	1.1609 6896	16.0968 9554	13.8650 5252
16	1.1269 9211	16.9322 8183	15.0243 1261	1.1725 7864	17.2578 6449	14.7178 7378
17	1.1354 4455	18.0592 7394	15.9050 2492	1.1843 0443	18.4304 4314	15.5622 5127
18	1.1439 6039	19.1947 1849	16.7791 8107	1.1961 4748	19.6147 4757	16.3982 6858
19	1.1525 4009	20.3386 7888	17.6468 2984	1.2081 0895	20.8108 9504	17.2260 0850
20	1.1611 8414	21.4912 1897	18.5080 1969	1.2201 9004	22.0190 0399	18.0455 5297
21	1.1698 9302	22.6524 0312	19.3627 9870	1.2323 9194	23.2391 9403	18.8569 8313
22	1.1786 6722	23.8222 9614	20.2112 1459	1.2447 1586	24.4715 8598	19.6603 7934
23	1.1875 0723	25.0009 6336	21.0533 1473	1.2571 6302	25.7163 0183	20.4558 2113
24	1.1964 1353	26.1884 7059	21.8891 4614	1.2697 3465	26.9734 6485	21.2433 8726
25	1.2053 8663	27.3848 8412	22.7187 5547	1.2824 3200	28.2431 9950	22.0231 5570
26	1.2144 2703	28.5902 7075	23.5421 8905	1.2952 5631	29.5256 3150	22.7952 0366
27	1.2235 3523	29.8046 9778	24.3594 9286	1.3082 0888	30.8208 8781	23.5596 0759
28	1.2327 1175	31.0282 3301	25.1707 1251	1.3212 9097	32.1290 9669	24.3164 4316
29	1.2419 5709	32.2609 4476	25.9758 9331	1.3345 0388	33.4503 8766	25.0657 8530
30	1.2512 7176	33.5029 0184	26.7750 8021	1.3478 4892	34.7848 9153	25.8077 0822
31	1.2606 5630	34.7541 7361	27.5683 1783	1.3613 2740	36.1327 4045	26.5422 8537
32	1.2701 1122	36.0148 2991	28.3556 5045	1.3749 4068	37.4940 6785	27.2695 8947
33	1.2796 3706	37.2849 4113	29.1371 2203	1.3886 9009	38.8690 0853	27.9896 9255
34	1.2892 3434	38.5645 7819	29.9127 7621	1.4025 7699	40.2576 9862	28.7026 6589
35	1.2989 0359	39.8538 1253	30.6826 5629	1.4166 0276	41.6602 7560	29.4085 8009
36	1.3086 4537	41.1527 1612	31.4468 0525	1.4307 6878	43.0768 7836	30.1075 0504
37	1.3184 6021	42.4613 6149	32.2052 6576	1.4450 7647	44.5076 4714	30.7995 0994
38	1.3283 4866	43.7798 2170	32.9580 8016	1.4595 2724	45.9527 2361	31.4846 6330
39	1.3383 1128	45.1081 7037	33.7052 9048	1.4741 2251	47.4122 5085	32.1630 3298
40	1.3483 4861	46.4464 8164	34.4469 3844	1.4888 6373	48.8863 7336	32.8346 8611
41	1.3584 6123	47.7948 3026	35.1830 6545	1.5037 5237	50.3752 3709	33.4996 8922
42	1.3686 4969	49.1532 9148	35.9137 1260	1.5187 8989	51.8789 8946	34.1581 0814
43	1.3789 1456	50.5219 4117	36.6389 2070	1.5339 7779	53.3977 7936	34.8100 0806
44	1.3892 5642	51.9008 5573	37.3587 3022	1.5493 1757	54.9317 5715	35.4554 5352
45	1.3996 7584	53.2901 1215	38.0731 8136	1.5648 1075	56.4810 7472	36.0945 0844
46	1.4101 7341	54.6897 8799	38.7823 1401	1.5804 5885	58.0458 8547	36.7272 3608
47	1.4207 4971	56.0999 6140	39.4861 6775	1.5962 6344	59.6263 4432	37.3536 9909
48	1.4314 0533	57.5207 1111	40.1847 8189	1.6122 2608	61.2226 0777	37.9739 5949
49	1.4421 4087	58.9521 1644	40.8781 9542	1.6283 4834	62.8348 3385	38.5880 7871
50	1.4529 5693	60.3942 5732	41.5664 4707	1.6446 3182	64.4631 8218	39.1961 1753

	$i = \frac{3}{4}\%$			$i = 1\%$		
n	$(1+i)^n$	$s_{\overline{n}|i}$	$a_{\overline{n}|i}$	$(1+i)^n$	$s_{\overline{n}|i}$	$a_{\overline{n}|i}$
51	1.4638 5411	61.8472 1424	42.2495 7525	1.6610 7814	66.1078 1401	39.7981 3617
52	1.4748 3301	63.3110 6835	42.9276 1812	1.6776 8892	67.7688 9215	40.3941 9423
53	1.4858 9426	64.7859 0136	43.6006 1351	1.6944 6581	69.4465 8107	40.9843 5072
54	1.4970 3847	66.2717 9562	44.2685 9902	1.7114 1047	71.1410 4688	41.5686 6408
55	1.5082 6626	67.7688 3409	44.9316 1193	1.7285 2457	72.8524 5735	42.1471 9216
56	1.5195 7825	69.2771 0035	45.5896 8926	1.7458 0982	74.5809 8192	42.7199 9224
57	1.5309 7509	70.7966 7860	46.2428 6776	1.7632 6792	76.3267 9174	43.2871 2102
58	1.5424 5740	72.3276 5369	46.8911 8388	1.7809 0060	78.0900 5966	43.8486 3468
59	1.5540 2583	73.8701 1109	47.5346 7382	1.7987 0960	79.8709 6025	44.4045 8879
60	1.5656 8103	75.4241 3693	48.1733 7352	1.8166 9670	81.6696 6986	44.9550 3841
61	1.5774 2363	76.9898 1795	48.8073 1863	1.8348 6367	83.4863 6655	45.5000 3803
62	1.5892 5431	78.5672 4159	49.4365 4455	1.8532 1230	85.3212 3022	46.0396 4161
63	1.6011 7372	80.1564 9590	50.0610 8640	1.8717 4443	87.1744 4252	46.5739 0258
64	1.6131 8252	81.7576 6962	50.6809 7906	1.8904 6187	89.0461 8695	47.1028 7385
65	1.6252 8139	83.3708 5214	51.2962 5713	1.9093 6649	90.9366 4882	47.6266 0777
66	1.6374 7100	84.9961 3353	51.9069 5497	1.9284 6015	92.8460 1531	48.1451 5621
67	1.6497 5203	86.6336 0453	52.5131 0667	1.9477 4475	94.7744 7546	48.6585 7050
68	1.6621 2517	88.2833 5657	53.1147 4607	1.9672 2220	96.7222 2021	49.1669 0149
69	1.6745 9111	89.9454 8174	53.7119 0677	1.9868 9442	98.6894 4242	49.6701 9949
70	1.6871 5055	91.6200 7285	54.3046 2210	2.0067 6337	100.6763 3684	50.1685 1435
71	1.6998 0418	93.3072 2340	54.8929 2516	2.0268 3100	102.6831 0021	50.6618 9539
72	1.7125 5271	95.0070 2758	55.4768 4880	2.0470 9931	104.7099 3121	51.1503 9148
73	1.7253 9685	96.7195 8028	56.0564 2561	2.0675 7031	106.7570 3052	51.6340 5097
74	1.7383 3733	98.4449 7714	56.6316 8795	2.0882 4601	108.8246 0083	52.1129 2175
75	1.7513 7486	100.1833 1446	57.2026 6794	2.1091 2847	110.9128 4684	52.5870 5124
76	1.7645 1017	101.9346 8932	57.7693 9746	2.1302 1975	113.0219 7530	53.0564 8638
77	1.7777 4400	103.6991 9949	58.3319 0815	2.1515 2195	115.1521 9506	53.5212 7364
78	1.7910 7708	105.4769 4349	58.8902 3141	2.1730 3717	117.3037 1701	53.9814 5905
79	1.8045 1015	107.2680 2056	59.4443 9842	2.1947 6754	119.4767 5418	54.4370 8817
80	1.8180 4398	109.0725 3072	59.9944 4012	2.2167 1522	121.6715 2172	54.8882 0611
81	1.8316 7931	110.8905 7470	60.5403 8722	2.2388 8237	123.8882 3694	55.3348 5753
82	1.8454 1691	112.7222 5401	61.0822 7019	2.2612 7119	126.1271 1931	55.7770 8666
83	1.8592 5753	114.5676 7091	61.6201 1930	2.2838 8390	128.3883 9050	56.2149 3729
84	1.8732 0196	116.4269 2845	62.1539 6456	2.3067 2274	130.6722 7440	56.6484 5276
85	1.8872 5098	118.3001 3041	62.6838 3579	2.3297 8997	132.9789 9715	57.0776 7600
86	1.9014 0536	120.1873 8139	63.2097 6257	2.3530 8787	135.3087 8712	57.5026 4951
87	1.9156 6590	122.0887 8675	63.7317 7427	2.3766 1875	137.6618 7499	57.9234 1535
88	1.9300 3339	124.0044 5265	64.2499 0002	2.4003 8494	140.0384 9374	58.3400 1520
89	1.9445 0865	125.9344 8604	64.7641 6875	2.4243 8879	142.4388 7868	58.7524 9030
90	1.9590 9246	127.8789 9469	65.2746 0918	2.4486 3267	144.8632 6746	59.1608 8148
91	1.9737 8565	129.8380 8715	65.7812 4981	2.4731 1900	147.3119 0014	59.5652 2919
92	1.9885 8905	131.8118 7280	66.2841 1892	2.4978 5019	149.7850 1914	59.9655 7346
93	2.0035 0346	133.8004 6185	66.7832 4458	2.5228 2869	152.2828 6933	60.3619 5392
94	2.0185 2974	135.8039 6531	67.2786 5467	2.5480 5698	154.8056 9803	60.7544 0982
95	2.0336 6871	137.8224 9505	67.7703 7685	2.5735 3755	157.3537 5501	61.1429 8002
96	2.0489 2123	139.8561 6377	68.2584 3856	2.5992 7293	159.9272 9256	61.5277 0299
97	2.0642 8814	141.9050 8499	68.7428 6705	2.6252 6565	162.5265 6548	61.9086 1682
98	2.0797 7030	143.9693 7313	69.2236 8938	2.6515 1831	165.1518 3114	62.2857 5923
99	2.0953 6858	146.0491 4343	69.7009 3239	2.6780 3349	167.8033 4945	62.6591 6755
100	2.1110 8384	148.1445 1201	70.1746 2272	2.7048 1383	170.4813 8294	63.0288 7877

	$i = 1\frac{1}{4}\%$			$i = 1\frac{1}{2}\%$		
n	$(1 + i)^n$	$s_{\overline{n}|i}$	$a_{\overline{n}|i}$	$(1 + i)^n$	$s_{\overline{n}|i}$	$a_{\overline{n}|i}$
1	1.0125 0000	1.0000 0000	0.9876 5432	1.0150 0000	1.0000 0000	0.9852 2167
2	1.0251 5625	2.0125 0000	1.9631 1538	1.0302 2500	2.0150 0000	1.9558 8342
3	1.0379 7070	3.0376 5625	2.9265 3371	1.0456 7838	3.0452 2500	2.9122 0042
4	1.0509 4534	4.0756 2695	3.8780 5798	1.0613 6355	4.0909 0338	3.8543 8465
5	1.0640 8215	5.1265 7229	4.8178 3504	1.0772 8400	5.1522 6693	4.7826 4497
6	1.0773 8318	6.1906 5444	5.7460 0992	1.0934 4326	6.2295 5093	5.6971 8717
7	1.0908 5047	7.2680 3762	6.6627 2585	1.1098 4491	7.3229 9419	6.5982 1396
8	1.1044 8610	8.3588 8809	7.5681 2429	1.1264 9259	8.4328 3911	7.4859 2508
9	1.1182 9218	9.4633 7420	8.4623 4498	1.1433 8998	9.5593 3169	8.3605 1732
10	1.1322 7083	10.5816 6637	9.3455 2591	1.1605 4083	10.7027 2167	9.2221 8455
11	1.1464 2422	11.7139 3720	10.2178 0337	1.1779 4894	11.8632 6249	10.0711 1779
12	1.1607 5452	12.8603 6142	11.0793 1197	1.1956 1817	13.0412 1143	10.9075 0521
13	1.1752 6395	14.0211 1594	11.9301 8466	1.2135 5244	14.2368 2960	11.7315 3222
14	1.1899 5475	15.1963 7988	12.7705 5275	1.2317 5573	15.4503 8205	12.5433 8150
15	1.2048 2918	16.3863 3463	13.6005 4592	1.2502 3207	16.6821 3778	13.3432 3301
16	1.2198 8955	17.5911 6382	14.4202 9227	1.2689 8555	17.9323 6984	14.1312 6405
17	1.2351 3817	18.8110 5336	15.2299 1829	1.2880 2033	19.2013 5539	14.9076 4931
18	1.2505 7739	20.0461 9153	16.0295 4893	1.3073 4064	20.4893 7572	15.6725 6089
19	1.2662 0961	21.2967 6893	16.8193 0759	1.3269 5075	21.7967 1636	16.4261 6837
20	1.2820 3723	22.5629 7854	17.5993 1613	1.3468 5501	23.1236 6710	17.1686 3879
21	1.2980 6270	23.8450 1577	18.3696 9495	1.3670 5783	24.4705 2211	17.9001 3673
22	1.3142 8848	25.1430 7847	19.1305 6291	1.3875 6370	25.8375 7994	18.6208 2437
23	1.3307 1709	26.4573 6695	19.8820 3744	1.4083 7715	27.2251 4364	19.3308 6145
24	1.3473 5105	27.7880 8403	20.6242 3451	1.4295 0281	28.6335 2080	20.0304 0537
25	1.3641 9294	29.1354 3508	21.3572 6865	1.4509 4535	30.0630 2361	20.7196 1120
26	1.3812 4535	30.4996 2802	22.0812 5299	1.4727 0953	31.5139 6896	21.3986 3172
27	1.3985 1092	31.8808 7337	22.7962 9925	1.4948 0018	32.9866 7850	22.0676 1746
28	1.4159 9230	33.2793 8429	23.5025 1778	1.5172 2218	34.4814 7867	22.7267 1671
29	1.4336 9221	34.6953 7659	24.2000 1756	1.5399 8051	35.9987 0085	23.3760 7558
30	1.4516 1336	36.1290 6880	24.8889 0623	1.5630 8022	37.5386 8137	24.0158 3801
31	1.4697 5853	37.5806 8216	25.5692 9010	1.5865 2642	39.1017 6159	24.6461 4582
32	1.4881 3051	39.0504 4069	26.2412 7418	1.6103 2432	40.6882 8801	25.2671 3874
33	1.5067 3214	40.5385 7120	26.9049 6215	1.6344 7918	42.2986 1233	25.8789 5442
34	1.5255 6629	42.0453 0334	27.5604 5644	1.6589 9637	43.9330 9152	26.4817 2849
35	1.5446 3587	43.5708 6963	28.2078 5822	1.6838 8132	45.5920 8789	27.0755 9458
36	1.5639 4382	45.1155 0550	28.8472 6737	1.7091 3954	47.2759 6921	27.6606 8431
37	1.5834 9312	46.6794 4932	29.4787 8259	1.7347 7663	48.9851 0874	28.2371 2740
38	1.6032 8678	48.2629 4243	30.1025 0133	1.7607 9828	50.7198 8538	28.8050 5163
39	1.6233 2787	49.8662 2921	30.7185 1983	1.7872 1025	52.4806 8366	29.3645 8288
40	1.6436 1946	51.4895 5708	31.3269 3316	1.8140 1841	54.2678 9391	29.9158 4520
41	1.6641 6471	53.1331 7654	31.9278 3522	1.8412 2868	56.0819 1232	30.4589 6079
42	1.6849 6677	54.7973 4125	32.5213 1874	1.8688 4712	57.9231 4100	30.9940 5004
43	1.7060 2885	56.4823 0801	33.1074 7530	1.8968 7982	59.7919 8812	31.5212 3157
44	1.7273 5421	58.1883 3687	33.6863 9536	1.9253 3302	61.6888 6794	32.0406 2223
45	1.7489 4614	59.9156 9108	34.2581 6825	1.9542 1301	63.6142 0096	32.5523 3718
46	1.7708 0797	61.6646 3721	34.8228 8222	1.9835 2621	65.5684 1398	33.0564 8983
47	1.7929 4306	63.4354 4518	35.3806 2442	2.0132 7910	67.5519 4018	33.5531 9195
48	1.8153 5485	65.2283 8824	35.9314 8091	2.0434 7829	69.5652 1929	34.0425 5365
49	1.8380 4679	67.0437 4310	36.4755 3670	2.0741 3046	71.6086 9758	34.5246 8339
50	1.8610 2237	68.8817 8989	37.0128 7575	2.1052 4242	73.6828 2804	34.9996 8807

Table I *(continued)*   **B** • TABLES   1157

	$i = 1\frac{1}{4}\%$			$i = 1\frac{1}{2}\%$		
n	$(1 + i)^n$	$s_{\overline{n}|i}$	$a_{\overline{n}|i}$	$(1 + i)^n$	$s_{\overline{n}|i}$	$a_{\overline{n}|i}$
51	1.8842 8515	70.7428 1226	37.5435 8099	2.1368 2106	75.7880 7046	35.4676 7298
52	1.9078 3872	72.6270 9741	38.0677 3431	2.1688 7337	77.9248 9152	35.9287 4185
53	1.9316 8670	74.5349 3613	38.5854 1660	2.2014 0647	80.0937 6489	36.3829 9690
54	1.9558 3279	76.4666 2283	39.0967 0776	2.2344 2757	82.2951 7136	36.8305 3882
55	1.9802 8070	78.4224 5562	39.6016 8667	2.2679 4398	84.5295 9893	37.2714 6681
56	2.0050 3420	80.4027 3631	40.1004 3128	2.3019 6314	86.7975 4292	37.7058 7863
57	2.0300 9713	82.4077 7052	40.5930 1855	2.3364 9259	89.0995 0606	38.1338 7058
58	2.0554 7335	84.4378 6765	41.0795 2449	2.3715 3998	91.4359 9865	38.5555 3751
59	2.0811 6676	86.4933 4099	41.5600 2419	2.4071 1308	93.8075 3863	38.9709 7292
60	2.1071 8135	88.5745 0776	42.0345 9179	2.4432 1978	96.2146 5171	39.3802 6889
61	2.1335 2111	90.6816 8910	42.5033 0054	2.4798 6807	98.6578 7149	39.7835 1614
62	2.1601 9013	92.8152 1022	42.9662 2275	2.5170 6609	101.1377 3956	40.1808 0408
63	2.1871 9250	94.9754 0034	43.4234 2988	2.5548 2208	103.6548 0565	40.5722 2077
64	2.2145 3241	97.1625 9285	43.8749 9247	2.5931 4442	106.2096 2774	40.9578 5298
65	2.2422 1407	99.3771 2526	44.3209 8022	2.6320 4158	108.8027 7215	41.3377 8618
66	2.2702 4174	101.6193 3933	44.7614 6195	2.6715 2221	111.4348 1374	41.7121 0461
67	2.2986 1976	103.8895 8107	45.1965 0563	2.7115 9504	114.1063 3594	42.0808 9125
68	2.3273 5251	106.1882 0083	45.6261 7840	2.7522 6896	116.8179 3098	42.4442 2783
69	2.3564 4442	108.5155 5334	46.0505 4656	2.7935 5300	119.5701 9995	42.8021 9490
70	2.3858 9997	110.8719 9776	46.4696 7562	2.8354 5629	122.3637 5295	43.1548 7183
71	2.4157 2372	113.2578 9773	46.8836 3024	2.8779 8814	125.1992 0924	43.5023 3678
72	2.4459 2027	115.6736 2145	47.2924 7431	2.9211 5796	128.0771 9738	43.8446 6677
73	2.4764 9427	118.1195 4172	47.6962 7093	2.9649 7533	130.9983 5534	44.1819 3771
74	2.5074 5045	120.5960 3599	48.0950 8240	3.0094 4996	133.9633 3067	44.5142 2434
75	2.5387 9358	123.1034 8644	48.4889 7027	3.0545 9171	136.9727 8063	44.8416 0034
76	2.5705 2850	125.6422 8002	48.8779 9533	3.1004 1059	140.0273 7234	45.1641 3826
77	2.6026 6011	128.2128 0852	49.2622 1761	3.1469 1674	143.1277 8292	45.4819 0962
78	2.6351 9336	130.8154 6863	49.6416 9640	3.1941 2050	146.2746 9967	45.7949 8485
79	2.6681 3327	133.4506 6199	50.0164 9027	3.2420 3230	149.4688 2016	46.1034 3335
80	2.7014 8494	136.1187 9526	50.3866 5706	3.2906 6279	152.7108 5247	46.4073 2349
81	2.7352 5350	138.8202 8020	50.7522 5389	3.3400 2273	156.0015 1525	46.7067 2265
82	2.7694 4417	141.5555 3370	51.1133 3717	3.3901 2307	159.3415 3798	47.0016 9720
83	2.8040 6222	144.3249 7787	51.4699 6264	3.4409 7492	162.7316 6105	47.2923 1251
84	2.8391 1300	147.1290 4010	51.8221 8532	3.4925 8954	166.1726 3597	47.5786 3301
85	2.8746 0191	149.9681 5310	52.1700 5958	3.5449 7838	169.6652 2551	47.8607 2218
86	2.9105 3444	152.8427 5501	52.5136 3909	3.5981 5306	173.2102 0389	48.1386 4254
87	2.9469 1612	155.7532 8945	52.8529 7688	3.6521 2535	176.8083 5695	48.4124 5571
88	2.9837 5257	158.7002 0557	53.1881 2531	3.7069 0723	180.4604 8230	48.6822 2237
89	3.0210 4948	161.6839 5814	53.5191 3611	3.7625 1084	184.1673 8954	48.9480 0234
90	3.0588 1260	164.7050 0762	53.8460 6036	3.8189 4851	187.9299 0038	49.2098 5452
91	3.0970 4775	167.7638 2021	54.1689 4850	3.8762 3273	191.7488 4889	49.4678 3696
92	3.1357 6085	170.8608 6796	54.4878 5037	3.9343 7622	195.6250 8162	49.7220 0686
93	3.1749 5786	173.9966 2881	54.8028 1518	3.9933 9187	199.5594 5784	49.9724 2055
94	3.2146 4483	177.1715 8667	55.1138 9154	4.0532 9275	203.5528 4971	50.2191 3355
95	3.2548 2789	180.3862 3151	55.4211 2744	4.1140 9214	207.6061 4246	50.4622 0054
96	3.2955 1324	183.6410 5940	55.7245 7031	4.1758 0352	211.7202 3459	50.7016 7541
97	3.3367 0716	186.9365 7264	56.0242 6698	4.2384 4057	215.8960 3811	50.9376 1124
98	3.3784 1600	190.2732 7980	56.3202 6368	4.3020 1718	220.1344 7868	51.1700 6034
99	3.4206 4620	193.6516 9580	56.6126 0610	4.3665 4744	224.4364 9586	51.3990 7422
100	3.4634 0427	197.0723 4200	56.9013 3936	4.4320 4565	228.8030 4330	51.6247 0367

	$i = 1\frac{3}{4}\%$			$i = 2\%$		
n	$(1 + i)^n$	$s_{\overline{n}|i}$	$a_{\overline{n}|i}$	$(1 + i)^n$	$s_{\overline{n}|i}$	$a_{\overline{n}|i}$
1	1.0175 0000	1.0000 0000	0.9828 0098	1.0200 0000	1.0000 0000	0.9803 9216
2	1.0353 0625	2.0175 0000	1.9486 9875	1.0404 0000	2.0200 0000	1.9415 6094
3	1.0534 2411	3.0528 0625	2.8979 8403	1.0612 0800	3.0604 0000	2.8838 8327
4	1.0718 5903	4.1062 3036	3.8309 4254	1.0824 3216	4.1216 0800	3.8077 2870
5	1.0906 1656	5.1780 8939	4.7478 5508	1.1040 8080	5.2040 4016	4.7134 5951
6	1.1097 0235	6.2687 0596	5.6489 9762	1.1261 6242	6.3081 2096	5.6014 3089
7	1.1291 2215	7.3784 0831	6.5346 4139	1.1486 8567	7.4342 8338	6.4719 9107
8	1.1488 8178	8.5075 3045	7.4050 5297	1.1716 5938	8.5829 6905	7.3254 8144
9	1.1689 8721	9.6564 1224	8.2604 9432	1.1950 9257	9.7546 2843	8.1622 3671
10	1.1894 4449	10.8253 9945	9.1012 2291	1.2189 9442	10.9497 2100	8.9825 8501
11	1.2102 5977	12.0148 4394	9.9274 9181	1.2433 7431	12.1687 1542	9.7868 4805
12	1.2314 3931	13.2251 0371	10.7395 4969	1.2682 4179	13.4120 8973	10.5753 4122
13	1.2529 8950	14.4565 4303	11.5376 4097	1.2936 0663	14.6803 3152	11.3483 7375
14	1.2749 1682	15.7095 3253	12.3220 0587	1.3194 7876	15.9739 3815	12.1062 4877
15	1.2972 2786	16.9844 4935	13.0928 8046	1.3458 6834	17.2934 1692	12.8492 6350
16	1.3199 2935	18.2816 7721	13.8504 9677	1.3727 8571	18.6392 8525	13.5777 0931
17	1.3430 2811	19.6016 0656	14.5950 8282	1.4002 4142	20.0120 7096	14.2918 7188
18	1.3665 3111	20.9446 3468	15.3268 6272	1.4282 4625	21.4123 1238	14.9920 3125
19	1.3904 4540	22.3111 6578	16.0460 5673	1.4568 1117	22.8405 5863	15.6784 6201
20	1.4147 7820	23.7016 1119	16.7528 8130	1.4859 4740	24.2973 6980	16.3514 3334
21	1.4395 3681	25.1163 8938	17.4475 4919	1.5156 6634	25.7833 1719	17.0112 0916
22	1.4647 2871	26.5559 2620	18.1302 6948	1.5459 7967	27.2989 8354	17.6580 4820
23	1.4903 6146	28.0206 5490	18.8012 4764	1.5768 9926	28.8449 6321	18.2922 0412
24	1.5164 4279	29.5110 1637	19.4606 8565	1.6084 3725	30.4218 6247	18.9139 2560
25	1.5429 8054	31.0274 5915	20.1087 8196	1.6406 0599	32.0302 9972	19.5234 5647
26	1.5699 8269	32.5704 3969	20.7457 3166	1.6734 1811	33.6709 0572	20.1210 3576
27	1.5974 5739	34.1404 2238	21.3717 2644	1.7068 8648	35.3443 2383	20.7068 9780
28	1.6254 1290	35.7378 7977	21.9869 5474	1.7410 2421	37.0512 1031	21.2812 7236
29	1.6538 5762	37.3632 9267	22.5916 0171	1.7758 4469	38.7922 3451	21.8443 8466
30	1.6828 0013	39.0171 5029	23.1858 4934	1.8113 6158	40.5680 7921	22.3964 5555
31	1.7122 4913	40.6999 5042	23.7698 7650	1.8475 8882	42.3794 4079	22.9377 0152
32	1.7422 1349	42.4121 9955	24.3438 5897	1.8845 4059	44.2270 2961	23.4683 3482
33	1.7727 0223	44.1544 1305	24.9079 6951	1.9222 3140	46.1115 7020	23.9885 6355
34	1.8037 2452	45.9271 1527	25.4623 7789	1.9606 7603	48.0338 0160	24.4985 9172
35	1.8352 8970	47.7308 3979	26.0072 5100	1.9998 8955	49.9944 7763	24.9986 1933
36	1.8674 0727	49.5661 2949	26.5427 5283	2.0398 8734	51.9943 6719	25.4888 4248
37	1.9000 8689	51.4335 3675	27.0690 4455	2.0806 8509	54.0342 5453	25.9694 5341
38	1.9333 3841	53.3336 2365	27.5862 8457	2.1222 9879	56.1149 3962	26.4406 4060
39	1.9671 7184	55.2669 6206	28.0946 2857	2.1647 4477	58.2372 3841	26.9025 8883
40	2.0015 9734	57.2341 3390	28.5942 2955	2.2080 3966	60.4019 8318	27.3554 7924
41	2.0366 2530	59.2357 3124	29.0852 3789	2.2522 0046	62.6100 2284	27.7994 8945
42	2.0722 6624	61.2723 5654	29.5678 0136	2.2972 4447	64.8622 2330	28.2347 9358
43	2.1085 3090	63.3446 2278	30.0420 6522	2.3431 8936	67.1594 6777	28.6615 6233
44	2.1454 3019	65.4531 5367	30.5081 7221	2.3900 5314	69.5026 5712	29.0799 6307
45	2.1829 7522	67.5985 8386	30.9662 6261	2.4378 5421	71.8927 1027	29.4901 5987
46	2.2211 7728	69.7815 5908	31.4164 7431	2.4866 1129	74.3305 6447	29.8923 1360
47	2.2600 4789	72.0027 3637	31.8589 4281	2.5363 4352	76.8171 7576	30.2865 8196
48	2.2995 9872	74.2627 8425	32.2938 0129	2.5870 7039	79.3535 1927	30.6731 1957
49	2.3398 4170	76.5623 8298	32.7211 8063	2.6388 1179	81.9405 8966	31.0520 7801
50	2.3807 8893	78.9022 2468	33.1412 0946	2.6915 8803	84.5794 0145	31.4236 0589

	$i = 1\frac{3}{4}\%$			$i = 2\%$		
n	$(1+i)^n$	$s_{\overline{n}|i}$	$a_{\overline{n}|i}$	$(1+i)^n$	$s_{\overline{n}|i}$	$a_{\overline{n}|i}$
51	2.4224 5274	81.2830 1361	33.5540 1421	2.7454 1979	87.2709 8948	31.7878 4892
52	2.4648 4566	83.7054 6635	33.9597 1913	2.8003 2819	90.0164 0927	32.1449 4992
53	2.5079 8046	86.1703 1201	34.3584 4632	2.8563 3475	92.8167 3746	32.4950 4894
54	2.5518 7012	88.6782 9247	34.7503 1579	2.9134 6144	95.6730 7221	32.8382 8327
55	2.5965 2785	91.2301 6259	35.1354 4550	2.9717 3067	98.5865 3365	33.1747 8752
56	2.6419 6708	93.8266 9043	35.5139 5135	3.0311 6529	101.5582 6432	33.5046 9365
57	2.6882 0151	96.4686 5752	35.8859 4727	3.0917 8859	104.5894 2961	33.8281 3103
58	2.7352 4503	99.1568 5902	36.2515 4523	3.1536 2436	107.6812 1820	34.1452 2650
59	2.7831 1182	101.8921 0405	36.6108 5526	3.2166 9685	110.8348 4257	34.4561 0441
60	2.8318 1628	104.6752 1588	36.9639 8552	3.2810 3079	114.0515 3942	34.7608 8668
61	2.8813 7306	107.5070 3215	37.3110 4228	3.3466 5140	117.3325 7021	35.0596 9282
62	2.9317 9709	110.3884 0522	37.6521 3000	3.4135 8443	120.6792 2161	35.3526 4002
63	2.9831 0354	113.3202 0231	37.9873 5135	3.4818 5612	124.0928 0604	35.6398 4316
64	3.0353 0785	116.3033 0585	38.3168 0723	3.5514 9324	127.5746 6216	35.9214 1486
65	3.0884 2574	119.3386 1370	38.6405 9678	3.6225 2311	131.1261 5541	36.1974 6555
66	3.1424 7319	122.4270 3944	38.9588 1748	3.6949 7357	134.7486 7852	36.4681 0348
67	3.1974 6647	125.5695 1263	39.2715 6509	3.7688 7304	138.4436 5209	36.7334 3478
68	3.2534 2213	128.7669 7910	39.5789 3375	3.8442 5050	142.2125 2513	36.9935 6351
69	3.3103 5702	132.0204 0124	39.8810 1597	3.9211 3551	146.0567 7563	37.2485 9168
70	3.3682 8827	135.3307 5826	40.1779 0267	3.9995 5822	149.9779 1114	37.4986 1929
71	3.4272 3331	138.6990 4653	40.4696 8321	4.0795 4939	153.9774 6937	37.7437 4441
72	3.4872 0990	142.1262 7984	40.7564 4542	4.1611 4038	158.0570 1875	37.9840 6314
73	3.5482 3607	145.6134 8974	41.0382 7560	4.2443 6318	162.2181 5913	38.2196 6975
74	3.6103 3020	149.1617 2581	41.3152 5857	4.3292 5045	166.4625 2231	38.4506 5662
75	3.6735 1098	152.7720 5601	41.5874 7771	4.4158 3546	170.7917 7276	38.6771 1433
76	3.7377 9742	156.4455 6699	41.8550 1495	4.5041 5216	175.2076 0821	38.8991 3170
77	3.8032 0888	160.1833 6441	42.1179 5081	4.5942 3521	179.7117 6038	39.1167 9578
78	3.8697 6503	163.9865 7329	42.3763 6443	4.6861 1991	184.3059 9558	39.3301 9194
79	3.9374 8592	167.8563 3832	42.6303 3359	4.7798 4231	188.9921 1549	39.5394 0386
80	4.0063 9192	171.7938 2424	42.8799 3474	4.8754 3916	193.7719 5780	39.7445 1359
81	4.0765 0378	175.8002 1617	43.1252 4298	4.9729 4794	198.6473 9696	39.9456 0156
82	4.1478 4260	179.8767 1995	43.3663 3217	5.0724 0690	203.6203 4490	40.1427 4663
83	4.2204 2984	184.0245 6255	43.6032 7486	5.1738 5504	208.6927 5180	40.3360 2611
84	4.2942 8737	188.2449 9239	43.8361 4237	5.2773 3214	213.8666 0683	40.5255 1579
85	4.3694 3740	192.5392 7976	44.0650 0479	5.3828 7878	219.1439 3897	40.7112 8999
86	4.4459 0255	196.9087 1716	44.2899 3099	5.4905 3636	224.5268 1775	40.8934 2156
87	4.5237 0584	201.3546 1971	44.5109 8869	5.6003 4708	230.0173 5411	41.0719 8192
88	4.6028 7070	205.8783 2555	44.7282 4441	5.7123 5402	235.6177 0119	41.2470 4110
89	4.6834 2093	210.4811 9625	44.9417 6355	5.8266 0110	241.3300 5521	41.4186 6774
90	4.7653 8080	215.1646 1718	45.1516 1037	5.9431 3313	247.1566 5632	41.5869 2916
91	4.8487 7496	219.9299 9798	45.3578 4803	6.0619 9579	253.0997 8944	41.7518 9133
92	4.9336 2853	224.7787 7295	45.5605 3860	6.1832 3570	259.1617 8523	41.9136 1895
93	5.0199 6703	229.7124 0148	45.7597 4310	6.3069 0042	265.3450 2094	42.0721 7545
94	5.1078 1645	234.7323 6850	45.9555 2147	6.4330 3843	271.6519 2135	42.2276 2299
95	5.1972 0324	239.8401 8495	46.1479 3265	6.5616 9920	278.0849 5978	42.3800 2254
96	5.2881 5429	245.0373 8819	46.3370 3455	6.6929 3318	284.6466 5898	42.5294 3386
97	5.3806 9699	250.3255 4248	46.5228 8408	6.8267 9184	291.3395 9216	42.6759 1555
98	5.4748 5919	255.7062 3947	46.7055 3718	6.9633 2768	298.1663 8400	42.8195 2505
99	5.5706 6923	261.1810 9866	46.8850 4882	7.1025 9423	305.1297 1168	42.9603 1867
100	5.6681 5594	266.7517 6789	47.0614 7304	7.2446 4612	312.2323 0591	43.0983 5164

	$i = 2\frac{1}{4}\%$			$i = 2\frac{1}{2}\%$		
n	$(1 + i)^n$	$s_{\overline{n}\|i}$	$a_{\overline{n}\|i}$	$(1 + i)^n$	$s_{\overline{n}\|i}$	$a_{\overline{n}\|i}$
1	1.0225 0000	1.0000 0000	0.9779 9511	1.0250 0000	1.0000 0000	0.9756 0976
2	1.0455 0625	2.0225 0000	1.9344 6955	1.0506 2500	2.0250 0000	1.9274 2415
3	1.0690 3014	3.0680 0625	2.8698 9687	1.0768 9063	3.0756 2500	2.8560 2356
4	1.0930 8332	4.1370 3639	3.7847 4021	1.1038 1289	4.1525 1563	3.7619 7421
5	1.1176 7769	5.2301 1971	4.6794 5253	1.1314 0821	5.2563 2852	4.6458 2850
6	1.1428 2544	6.3477 9740	5.5544 7680	1.1596 9342	6.3877 3673	5.5081 2536
7	1.1685 3901	7.4906 2284	6.4102 4626	1.1886 8575	7.5474 3015	6.3493 9060
8	1.1948 3114	8.6591 6186	7.2471 8461	1.2184 0290	8.7361 1590	7.1701 3717
9	1.2217 1484	9.8539 9300	8.0657 0622	1.2488 6297	9.9545 1880	7.9708 6553
10	1.2492 0343	11.0757 0784	8.8662 1635	1.2800 8454	11.2033 8177	8.7520 6393
11	1.2773 1050	12.3249 1127	9.6491 1134	1.3120 8666	12.4834 6631	9.5142 0871
12	1.3060 4999	13.6022 2177	10.4147 7882	1.3448 8882	13.7955 5297	10.2577 6460
13	1.3354 3611	14.9082 7176	11.1635 9787	1.3785 1104	15.1404 4179	10.9831 8497
14	1.3654 8343	16.2437 0788	11.8959 3924	1.4129 7382	16.5189 5284	11.6909 1217
15	1.3962 0680	17.6091 9130	12.6121 6551	1.4482 9817	17.9319 2666	12.3813 7773
16	1.4276 2146	19.0053 9811	13.3126 3131	1.4845 0562	19.3802 2483	13.0550 0266
17	1.4597 4294	20.4330 1957	13.9976 8343	1.5216 1826	20.8647 3045	13.7121 9772
18	1.4925 8716	21.8927 6251	14.6676 6106	1.5596 5872	22.3863 4871	14.3533 6363
19	1.5261 7037	23.3853 4966	15.3228 9590	1.5986 5019	23.9460 0743	14.9788 9134
20	1.5605 0920	24.9115 2003	15.9637 1237	1.6386 1644	25.5446 5761	15.5891 6229
21	1.5956 2066	26.4720 2923	16.5904 2775	1.6795 8185	27.1832 7405	16.1845 4857
22	1.6315 2212	28.0676 4989	17.2033 5232	1.7215 7140	28.8628 5590	16.7654 1324
23	1.6682 3137	29.6991 7201	17.8027 8955	1.7646 1068	30.5844 2730	17.3321 1048
24	1.7057 6658	31.3674 0338	18.3890 3624	1.8087 2595	32.3490 3798	17.8849 8583
25	1.7441 4632	33.0731 6996	18.9623 8263	1.8539 4410	34.1577 6393	18.4243 7642
26	1.7833 8962	34.8173 1628	19.5231 1260	1.9002 9270	36.0117 0803	18.9506 1114
27	1.8235 1588	36.6007 0590	20.0715 0376	1.9478 0002	37.9120 0073	19.4640 1087
28	1.8645 4499	38.4242 2178	20.6078 2764	1.9964 9502	39.8598 0075	19.9648 8866
29	1.9064 9725	40.2887 6677	21.1323 4977	2.0464 0739	41.8562 9577	20.4535 4991
30	1.9493 9344	42.1952 6402	21.6453 2985	2.0975 6758	43.9027 0316	20.9302 9259
31	1.9932 5479	44.1446 5746	22.1470 2186	2.1500 0677	46.0002 7074	21.3954 0741
32	2.0381 0303	46.1379 1226	22.6376 7419	2.2037 5694	48.1502 7751	21.8491 7796
33	2.0839 6034	48.1760 1528	23.1175 2977	2.2588 5086	50.3540 3445	22.2918 8094
34	2.1308 4945	50.2599 7563	23.5868 2618	2.3153 2213	52.6128 8531	22.7237 8628
35	2.1787 9356	52.3908 2508	24.0457 9577	2.3732 0519	54.9282 0744	23.1451 5734
36	2.2278 1642	54.5696 1864	24.4946 6579	2.4325 3532	57.3014 1263	23.5562 5107
37	2.2779 4229	56.7974 3506	24.9336 5848	2.4933 4870	59.7339 4794	23.9573 1812
38	2.3291 9599	59.0753 7735	25.3629 9118	2.5556 8242	62.2272 9664	24.3486 0304
39	2.3816 0290	61.4045 7334	25.7828 7646	2.6195 7448	64.7829 7906	24.7303 4443
40	2.4351 8897	63.7861 7624	26.1935 2221	2.6850 6384	67.4025 5354	25.1027 7505
41	2.4899 8072	66.2213 6521	26.5951 3174	2.7521 9043	70.0876 1737	25.4661 2200
42	2.5460 0528	68.7113 4592	26.9879 0390	2.8209 9520	72.8398 0781	25.8206 0683
43	2.6032 9040	71.2573 5121	27.3720 3316	2.8915 2008	75.6608 0300	26.1664 4569
44	2.6618 6444	73.8606 4161	27.7477 0969	2.9638 0808	78.5523 2308	26.5038 4945
45	2.7217 5639	76.5225 0605	28.1151 1950	3.0379 0328	81.5161 3116	26.8330 2386
46	2.7829 9590	79.2442 6243	28.4744 4450	3.1138 5086	84.5540 3443	27.1541 6962
47	2.8456 1331	82.0272 5834	28.8258 6259	3.1916 9713	87.6678 8530	27.4674 8255
48	2.9096 3961	84.8728 7165	29.1695 4777	3.2714 8956	90.8595 8243	27.7731 5371
49	2.9751 0650	87.7825 1126	29.5056 7019	3.3532 7680	94.1310 7199	28.0713 6947
50	3.0420 4640	90.7576 1776	29.8343 9627	3.4371 0872	97.4843 4879	28.3623 1168

Table I *(continued)*   **B** • TABLES   1161

	$i = 2\frac{1}{4}\%$			$i = 2\frac{1}{2}\%$		
n	$(1 + i)^n$	$s_{\overline{n}|i}$	$a_{\overline{n}|i}$	$(1 + i)^n$	$s_{\overline{n}|i}$	$a_{\overline{n}|i}$
51	3.1104 9244	93.7996 6416	30.1558 8877	3.5230 3644	100.9214 5751	28.6461 5774
52	3.1804 7852	96.9101 5661	30.4703 0687	3.6111 1235	104.4444 9395	28.9230 8072
53	3.2520 3929	100.0906 3513	30.7778 0623	3.7013 9016	108.0556 0629	29.1932 4948
54	3.3252 1017	103.3426 7442	31.0785 3910	3.7939 2491	111.7569 9645	29.4568 2876
55	3.4000 2740	106.6678 8460	31.3726 5438	3.8887 7303	115.5509 2136	29.7139 7928
56	3.4765 2802	110.0679 1200	31.6602 9768	3.9859 9236	119.4396 9440	29.9648 5784
57	3.5547 4990	113.5444 4002	31.9416 1142	4.0856 4217	123.4256 8676	30.2096 1740
58	3.6347 3177	117.0991 8992	32.2167 3489	4.1877 8322	127.5113 2893	30.4484 0722
59	3.7165 1324	120.7339 2169	32.4858 0429	4.2924 7780	131.6991 1215	30.6813 7290
60	3.8001 3479	124.4504 3493	32.7489 5285	4.3997 8975	135.9915 8995	30.9086 5649
61	3.8856 3782	128.2505 6972	33.0063 1086	4.5097 8449	140.3913 7970	31.1303 9657
62	3.9730 6467	132.1362 0754	33.2580 0573	4.6225 2910	144.9011 6419	31.3467 2836
63	4.0624 5862	136.1092 7221	33.5041 6208	4.7380 9233	149.5236 9330	31.5577 8377
64	4.1538 6394	140.1717 3083	33.7449 0179	4.8565 4464	154.2617 8563	31.7636 9148
65	4.2473 2588	144.3255 9477	33.9803 4405	4.9779 5826	159.1183 3027	31.9645 7705
66	4.3428 9071	148.5729 2066	34.2106 0543	5.1024 0721	164.0962 8853	32.1605 6298
67	4.4406 0576	152.9158 1137	34.4357 9993	5.2299 6739	169.1986 9574	32.3517 6876
68	4.5405 1939	157.3564 1713	34.6560 3905	5.3607 1658	174.4286 6314	32.5383 1099
69	4.6426 8107	161.8969 3651	34.8714 3183	5.4947 3449	179.7893 7971	32.7203 0340
70	4.7471 4140	166.5396 1758	35.0820 8492	5.6321 0286	185.2841 1421	32.8978 5698
71	4.8539 5208	171.2867 5898	35.2881 0261	5.7729 0543	190.9162 1706	33.0710 7998
72	4.9631 6600	176.1407 1106	35.4895 8691	5.9172 2806	196.6891 2249	33.2400 7803
73	5.0748 3723	181.1038 7705	35.6866 3756	6.0651 5876	202.6063 5055	33.4049 5417
74	5.1890 2107	186.1787 1429	35.8793 5214	6.2167 8773	208.6715 0931	33.5658 0895
75	5.3057 7405	191.3677 3536	36.0678 2605	6.3722 0743	214.8882 9705	33.7227 4044
76	5.4251 5396	196.6735 0941	36.2521 5262	6.5315 1261	221.2605 0447	33.8758 4433
77	5.5472 1993	202.0986 6337	36.4324 2310	6.6948 0043	227.7920 1709	34.0252 1398
78	5.6720 3237	207.6458 8329	36.6087 2675	6.8621 7044	234.4868 1751	34.1709 4047
79	5.7996 5310	213.3179 1567	36.7811 5085	7.0337 2470	241.3489 8795	34.3131 1265
80	5.9301 4530	219.1175 6877	36.9497 8079	7.2095 6782	248.3827 1265	34.4518 1722
81	6.0635 7357	225.0477 1407	37.1147 0004	7.3898 0701	255.5922 8047	34.5871 3875
82	6.2000 0397	231.1112 8763	37.2759 9026	7.5745 5219	262.9820 8748	34.7191 5976
83	6.3395 0406	237.3112 9160	37.4337 3130	7.7639 1599	270.5566 3966	34.8479 6074
84	6.4821 4290	243.6507 9567	37.5880 0127	7.9580 1389	278.3205 5566	34.9736 2023
85	6.6279 9112	250.1329 3857	37.7388 7655	8.1569 6424	286.2785 6955	35.0962 1486
86	6.7771 2092	256.7609 2969	37.8864 3183	8.3608 8834	294.4355 3379	35.2158 1938
87	6.9296 0614	263.5380 5060	38.0307 4018	8.5699 1055	302.7964 2213	35.3325 0671
88	7.0855 2228	270.4676 5674	38.1718 7304	8.7841 5832	311.3663 3268	35.4463 4801
89	7.2449 4653	277.5531 7902	38.3099 0028	9.0037 6228	320.1504 9100	35.5574 1269
90	7.4079 5782	284.7981 2555	38.4448 9025	9.2288 5633	329.1542 5328	35.6657 6848
91	7.5746 3688	292.2060 8337	38.5769 0978	9.4595 7774	338.3831 0961	35.7714 8144
92	7.7450 6621	299.7807 2025	38.7060 2423	9.6960 6718	347.8426 8735	35.8746 1604
93	7.9193 3020	307.5257 8645	38.8322 9754	9.9384 6886	357.5387 5453	35.9752 3516
94	8.0975 1512	315.4451 1665	38.9557 9221	10.1869 3058	367.4772 2339	36.0734 0016
95	8.2797 0921	323.5426 3177	39.0765 6940	10.4416 0385	377.6641 5398	36.1691 7089
96	8.4660 0267	331.8223 4099	39.1946 8890	10.7026 4395	388.1057 5783	36.2626 0574
97	8.6564 8773	340.2883 4366	39.3102 0920	10.9702 1004	398.8084 0177	36.3537 6170
98	8.8512 5871	348.9448 3139	39.4231 8748	11.2444 6530	409.7786 1182	36.4426 9434
99	9.0504 1203	357.7960 9010	39.5336 7968	11.5255 7693	421.0230 7711	36.5294 5790
100	9.2540 4630	366.8465 0213	39.6417 4052	11.8137 1635	432.5486 5404	36.6141 0526

	$i = 2\frac{3}{4}\%$			$i = 3\%$		
n	$(1 + i)^n$	$s_{\overline{n}\mid i}$	$a_{\overline{n}\mid i}$	$(1 + i)^n$	$s_{\overline{n}\mid i}$	$a_{\overline{n}\mid i}$
1	1.0275 0000	1.0000 0000	0.9732 3601	1.0300 0000	1.0000 0000	0.9708 7379
2	1.0557 5625	2.0275 0000	1.9204 2434	1.0609 0000	2.0300 0000	1.9134 6970
3	1.0847 8955	3.0832 5625	2.8422 6213	1.0927 2700	3.0909 0000	2.8286 1135
4	1.1146 2126	4.1680 4580	3.7394 2787	1.1255 0881	4.1836 2700	3.7170 9840
5	1.1452 7334	5.2826 6706	4.6125 8186	1.1592 7407	5.3091 3581	4.5797 0719
6	1.1767 6836	6.4279 4040	5.4623 6678	1.1940 5230	6.4684 0988	5.4171 9144
7	1.2091 2949	7.6047 0876	6.2894 0806	1.2298 7387	7.6624 6218	6.2302 8296
8	1.2423 8055	8.8138 3825	7.0943 1441	1.2667 7008	8.8923 3605	7.0196 9219
9	1.2765 4602	10.0562 1880	7.8776 7826	1.3047 7318	10.1591 0613	7.7861 0892
10	1.3116 5103	11.3327 6482	8.6400 7616	1.3439 1638	11.4638 7931	8.5302 0284
11	1.3477 2144	12.6444 1585	9.3820 6926	1.3842 3387	12.8077 9569	9.2526 2411
12	1.3847 8378	13.9921 3729	10.1042 0366	1.4257 6089	14.1920 2956	9.9540 0399
13	1.4228 6533	15.3769 2107	10.8070 1086	1.4685 3371	15.6177 9045	10.6349 5533
14	1.4619 9413	16.7997 8639	11.4910 0814	1.5125 8972	17.0863 2416	11.2960 7314
15	1.5021 9896	18.2617 8052	12.1566 9892	1.5579 6742	18.5989 1389	11.9379 3509
16	1.5435 0944	19.7639 7948	12.8045 7315	1.6047 0644	20.1568 8130	12.5611 0203
17	1.5859 5595	21.3074 8892	13.4351 0769	1.6528 4763	21.7615 8774	13.1661 1847
18	1.6295 6973	22.8934 4487	14.0487 6661	1.7024 3306	23.4144 3537	13.7535 1308
19	1.6743 8290	24.5230 1460	14.6460 0157	1.7535 0605	25.1168 6844	14.3237 9911
20	1.7204 2843	26.1973 9750	15.2272 5213	1.8061 1123	26.8703 7449	14.8774 7486
21	1.7677 4021	27.9178 2593	15.7929 4612	1.8602 9457	28.6764 8572	15.4150 2414
22	1.8163 5307	29.6855 6615	16.3434 9987	1.9161 0341	30.5367 8030	15.9369 1664
23	1.8663 0278	31.5019 1921	16.8793 1861	1.9735 8651	32.4528 8370	16.4436 0839
24	1.9176 2610	33.3682 2199	17.4007 9670	2.0327 9411	34.4264 7022	16.9355 4212
25	1.9703 6082	35.2858 4810	17.9083 1795	2.0937 7793	36.4592 6432	17.4131 4769
26	2.0245 4575	37.2562 0892	18.4022 5592	2.1565 9127	38.5530 4225	17.8768 4242
27	2.0802 2075	39.2807 5467	18.8829 7413	2.2212 8901	40.7096 3352	18.3270 3147
28	2.1374 2682	41.3609 7542	19.3508 2640	2.2879 2768	42.9309 2252	18.7641 0823
29	2.1962 0606	43.4984 0224	19.8061 5708	2.3565 6551	45.2188 5020	19.1884 5459
30	2.2566 0173	45.6946 0831	20.2493 0130	2.4272 6247	47.5754 1571	19.6004 4135
31	2.3186 5828	47.9512 1003	20.6805 8520	2.5000 8035	50.0026 7818	20.0004 2849
32	2.3824 2138	50.2698 6831	21.1003 2623	2.5750 8276	52.5027 5852	20.3887 6553
33	2.4479 3797	52.6522 8969	21.5088 3332	2.6523 3524	55.0778 4128	20.7657 9178
34	2.5152 5626	55.1002 2765	21.9064 0712	2.7319 0530	57.7301 7652	21.1318 3668
35	2.5844 2581	57.6154 8391	22.2933 4026	2.8138 6245	60.4620 8181	21.4872 2007
36	2.6554 9752	60.1999 0972	22.6699 1753	2.8982 7833	63.2759 4427	21.8322 5250
37	2.7285 2370	62.8554 0724	23.0364 1609	2.9852 2668	66.1742 2259	22.1672 3544
38	2.8035 5810	65.5839 3094	23.3931 0568	3.0747 8348	69.1594 4927	22.4924 6159
39	2.8806 5595	68.3874 8904	23.7402 4884	3.1670 2698	72.2342 3275	22.8082 1513
40	2.9598 7399	71.2681 4499	24.0781 0106	3.2620 3779	75.4012 5973	23.1147 7197
41	3.0412 7052	74.2280 1898	24.4069 1101	3.3598 9893	78.6632 9753	23.4123 9997
42	3.1249 0546	77.2692 8950	24.7269 2069	3.4606 9589	82.0231 9645	23.7013 5920
43	3.2108 4036	80.3941 9496	25.0383 6563	3.5645 1677	85.4838 9234	23.9819 0213
44	3.2991 3847	83.6050 3532	25.3414 7507	3.6714 5227	89.0484 0911	24.2542 7392
45	3.3898 6478	86.9041 7379	25.6364 7209	3.7815 9584	92.7198 6139	24.5187 1254
46	3.4830 8606	90.2940 3857	25.9235 7381	3.8950 4372	96.5014 5723	24.7754 4907
47	3.5788 7093	93.7771 2463	26.2029 9154	4.0118 9503	100.3965 0095	25.0247 0783
48	3.6772 8988	97.3559 9556	26.4749 3094	4.1322 5188	104.4083 9598	25.2667 0664
49	3.7784 1535	101.0332 8544	26.7395 9215	4.2562 1944	108.5406 4785	25.5016 5693
50	3.8823 2177	104.8117 0079	26.9971 6998	4.3839 0602	112.7968 6729	25.7297 6401

Table I (continued) **B** • TABLES 1163

	$i = 2\frac{3}{4}\%$			$i = 3\%$		
n	$(1 + i)^n$	$s_{\overline{n}|i}$	$a_{\overline{n}|i}$	$(1 + i)^n$	$s_{\overline{n}|i}$	$a_{\overline{n}|i}$
51	3.9890 8562	108.6940 2256	27.2478 5400	4.5154 2320	117.1807 7331	25.9512 2719
52	4.0987 8547	112.6831 0818	27.4918 2871	4.6508 8590	121.6961 9651	26.1662 3999
53	4.2115 0208	116.7818 9365	27.7292 7368	4.7904 1247	126.3470 8240	26.3749 9028
54	4.3273 1838	120.9933 9573	27.9603 6368	4.9341 2485	131.1374 9488	26.5776 6047
55	4.4463 1964	125.3207 1411	28.1852 6879	5.0821 4859	136.0716 1972	26.7744 2764
56	4.5685 9343	129.7670 3375	28.4041 5454	5.2346 1305	141.1537 6831	26.9654 6373
57	4.6942 2975	134.3356 2718	28.6171 8203	5.3916 5144	146.3883 8136	27.1509 3566
58	4.8233 2107	139.0298 5692	28.8245 0806	5.5534 0098	151.7800 3280	27.3310 0549
59	4.9559 6239	143.8531 7799	29.0262 8522	5.7200 0301	157.3334 3379	27.5058 3058
60	5.0922 5136	148.8091 4038	29.2226 6201	5.8916 0310	163.0534 3680	27.6755 6367
61	5.2322 8827	153.9013 9174	29.4137 8298	6.0683 5120	168.9450 3991	27.8403 5307
62	5.3761 7620	159.1336 8002	29.5997 8879	6.2504 0173	175.0133 9110	28.0003 4279
63	5.5240 2105	164.5098 5622	29.7808 1634	6.4379 1379	181.2637 9284	28.1556 7261
64	5.6759 3162	170.0338 7726	29.9569 9887	6.6310 5120	187.7017 0662	28.3064 7826
65	5.8320 1974	175.7098 0889	30.1284 6605	6.8299 8273	194.3327 5782	28.4528 9152
66	5.9924 0029	181.5418 2863	30.2953 4409	7.0348 8222	201.1627 4055	28.5950 4031
67	6.1571 9130	187.5342 2892	30.4577 5581	7.2459 2868	208.1976 2277	28.7330 4884
68	6.3265 1406	193.6914 2022	30.6158 2074	7.4633 0654	215.4435 5145	28.8670 3771
69	6.5004 9319	200.0179 3427	30.7696 5522	7.6872 0574	222.9068 5800	28.9971 2399
70	6.6792 5676	206.5184 2746	30.9193 7247	7.9178 2191	230.5940 6374	29.1234 2135
71	6.8629 3632	213.1976 8422	31.0650 8270	8.1553 5657	238.5118 8565	29.2460 4015
72	7.0516 6706	220.0606 2054	31.2068 9314	8.4000 1727	246.6672 4222	29.3650 8752
73	7.2455 8791	227.1122 8760	31.3449 0816	8.6520 1778	255.0672 5949	29.4806 6750
74	7.4448 4158	234.3578 7551	31.4792 2936	8.9115 7832	263.7192 7727	29.5928 8107
75	7.6495 7472	241.8027 1709	31.6099 5558	9.1789 2567	272.6308 5559	29.7018 2628
76	7.8599 3802	249.4522 9181	31.7371 8304	9.4542 9344	281.8097 8126	29.8075 9833
77	8.0760 8632	257.3122 2983	31.8610 0540	9.7379 2224	291.2640 7469	29.9102 8964
78	8.2981 7869	265.3883 1615	31.9815 1377	10.0300 5991	301.0019 9693	30.0099 8994
79	8.5263 7861	273.6864 9485	32.0987 9685	10.3309 6171	311.0320 5684	30.1067 8635
80	8.7608 5402	282.2128 7345	32.2129 4098	10.6408 9056	321.3630 1855	30.2007 6345
81	9.0017 7751	290.9737 2747	32.3240 3015	10.9601 1727	332.0039 0910	30.2920 0335
82	9.2493 2639	299.9755 0498	32.4321 4613	11.2889 2079	342.9640 2638	30.3805 8577
83	9.5036 8286	309.2248 3137	32.5373 6850	11.6275 8842	354.2529 4717	30.4665 8813
84	9.7650 3414	318.7285 1423	32.6397 7469	11.9764 1607	365.8805 3558	30.5500 8556
85	10.0335 7258	328.4935 4837	32.7394 4009	12.3357 0855	377.8569 5165	30.6311 5103
86	10.3094 9583	338.5271 2095	32.8364 3804	12.7057 7981	390.1926 6020	30.7098 5537
87	10.5930 0696	348.8366 1678	32.9308 3994	13.0869 5320	402.8984 4001	30.7862 6735
88	10.8843 1465	359.4296 2374	33.0227 1527	13.4795 6180	415.9853 9321	30.8604 5374
89	11.1836 3331	370.3139 3839	33.1121 3165	13.8839 4865	429.4649 5500	30.9324 7936
90	11.4911 8322	381.4975 7170	33.1991 5489	14.3004 6711	443.3489 0365	31.0024 0714
91	11.8071 9076	392.9887 5492	33.2838 4905	14.7294 8112	457.6493 7076	31.0702 9820
92	12.1318 8851	404.7959 4568	33.3662 7644	15.1713 6556	472.3788 5189	31.1362 1184
93	12.4655 1544	416.9278 3418	33.4464 9776	15.6265 0652	487.5502 1744	31.2002 0567
94	12.8083 1711	429.3933 4962	33.5245 7202	16.0953 0172	503.1767 2397	31.2623 3560
95	13.1605 4584	442.2016 6674	33.6005 5671	16.5781 6077	519.2720 2568	31.3226 5592
96	13.5224 6085	455.3622 1257	33.6745 0775	17.0755 0559	535.8501 8645	31.3812 1934
97	13.8943 2852	468.8846 7342	33.7464 7956	17.5877 7076	552.9256 9205	31.4380 7703
98	14.2764 2255	482.7790 0194	33.8165 2512	18.1154 0388	570.5134 6281	31.4932 7867
99	14.6690 2417	497.0554 2449	33.8846 9598	18.6588 6600	588.6288 6669	31.5468 7250
100	15.0724 2234	511.7244 4867	33.9510 4232	19.2186 3198	607.2877 3270	31.5989 0534

	$i = 3\frac{1}{2}\%$			$i = 4\%$		
n	$(1 + i)^n$	$s_{\overline{n}|i}$	$a_{\overline{n}|i}$	$(1 + i)^n$	$s_{\overline{n}|i}$	$a_{\overline{n}|i}$
1	1.0350 0000	1.0000 0000	0.9661 8357	1.0400 0000	1.0000 0000	0.9615 3846
2	1.0712 2500	2.0350 0000	1.8996 9428	1.0816 0000	2.0400 0000	1.8860 9467
3	1.1087 1788	3.1062 2500	2.8016 3698	1.1248 6400	3.1216 0000	2.7750 9103
4	1.1475 2300	4.2149 4288	3.6730 7921	1.1698 5856	4.2464 6400	3.6298 9522
5	1.1876 8631	5.3624 6588	4.5150 5238	1.2166 5290	5.4163 2256	4.4518 2233
6	1.2292 5533	6.5501 5218	5.3285 5302	1.2653 1902	6.6329 7546	5.2421 3686
7	1.2722 7926	7.7794 0751	6.1145 4398	1.3159 3178	7.8982 9448	6.0020 5467
8	1.3168 0904	9.0516 8677	6.8739 5554	1.3685 6905	9.2142 2626	6.7327 4487
9	1.3628 9735	10.3684 9581	7.6076 8651	1.4233 1181	10.5827 9531	7.4353 3161
10	1.4105 9876	11.7313 9316	8.3166 0532	1.4802 4428	12.0061 0712	8.1108 9578
11	1.4599 6972	13.1419 9192	9.0015 5104	1.5394 5406	13.4863 5141	8.7604 7671
12	1.5110 6866	14.6019 6164	9.6633 3433	1.6010 3222	15.0258 0546	9.3850 7376
13	1.5639 5606	16.1130 3030	10.3027 3849	1.6650 7351	16.6268 3768	9.9856 4785
14	1.6186 9452	17.6769 8636	10.9205 2028	1.7316 7645	18.2919 1119	10.5631 2293
15	1.6753 4883	19.2956 8088	11.5174 1090	1.8009 4351	20.0235 8764	11.1183 8743
16	1.7339 8604	20.9710 2971	12.0941 1681	1.8729 8125	21.8245 3114	11.6522 9561
17	1.7946 7555	22.7050 1575	12.6513 2059	1.9479 0050	23.6975 1239	12.1656 6885
18	1.8574 8920	24.4996 9130	13.1896 8173	2.0258 1652	25.6454 1288	12.6592 9697
19	1.9225 0132	26.3571 8050	13.7098 3742	2.1068 4918	27.6712 2940	13.1339 3940
20	1.9897 8886	28.2796 8181	14.2124 0330	2.1911 2314	29.7780 7858	13.5903 2634
21	2.0594 3147	30.2694 7068	14.6979 7420	2.2787 6807	31.9692 0172	14.0291 5995
22	2.1315 1158	32.3289 0215	15.1671 2484	2.3699 1879	34.2479 6979	14.4511 1533
23	2.2061 1448	34.4604 1373	15.6204 1047	2.4647 1554	36.6178 8858	14.8568 4167
24	2.2833 2849	36.6665 2821	16.0583 6760	2.5633 0416	39.0826 0412	15.2469 6314
25	2.3632 4498	38.9498 5669	16.4815 1459	2.6658 3633	41.6459 0829	15.6220 7994
26	2.4459 5856	41.3131 0168	16.8903 5226	2.7724 6978	44.3117 4462	15.9827 6918
27	2.5315 6711	43.7590 6024	17.2853 6451	2.8833 6858	47.0842 1440	16.3295 8575
28	2.6201 7196	46.2906 2734	17.6670 1885	2.9987 0332	49.9675 8298	16.6630 6322
29	2.7118 7798	48.9107 9930	18.0357 6700	3.1186 5145	52.9662 8630	16.9837 1463
30	2.8067 9370	51.6226 7728	18.3920 4541	3.2433 9751	56.0849 3775	17.2920 3330
31	2.9050 3148	54.4294 7098	18.7362 7576	3.3731 3341	59.3283 3526	17.5884 9356
32	3.0067 0759	57.3345 0247	19.0688 6547	3.5080 5875	62.7014 6867	17.8735 5150
33	3.1119 4235	60.3412 1005	19.3902 0818	3.6483 8110	66.2095 2742	18.1476 4567
34	3.2208 6033	63.4531 5240	19.7006 8423	3.7943 1634	69.8579 0851	18.4111 9776
35	3.3335 9045	66.6740 1274	20.0006 6110	3.9460 8899	73.6522 2486	18.6646 1323
36	3.4502 6611	70.0076 0318	20.2904 9381	4.1039 3255	77.5983 1385	18.9082 8195
37	3.5710 2543	73.4578 6930	20.5705 2542	4.2680 8986	81.7022 4640	19.1425 7880
38	3.6960 1132	77.0288 9472	20.8410 8736	4.4388 1345	85.9703 3626	19.3678 6423
39	3.8253 7171	80.7249 0604	21.1024 9987	4.6163 6599	90.4091 4971	19.5844 8484
40	3.9592 5972	84.5502 7775	21.3550 7234	4.8010 2063	95.0255 1570	19.7927 7388
41	4.0978 3381	88.5095 3747	21.5991 0371	4.9930 6145	99.8265 3633	19.9930 5181
42	4.2412 5799	92.6073 7128	21.8348 8281	5.1927 8391	104.8195 9778	20.1856 2674
43	4.3897 0202	96.8486 2928	22.0626 8870	5.4004 9527	110.0123 8169	20.3707 9494
44	4.5433 4160	101.2383 3130	22.2827 9102	5.6165 1508	115.4128 7696	20.5488 4129
45	4.7023 5855	105.7816 7290	22.4954 5026	5.8411 7568	121.0293 9204	20.7200 3970
46	4.8669 4110	110.4840 3145	22.7009 1813	6.0748 2271	126.8705 6772	20.8846 5356
47	5.0372 8404	115.3509 7255	22.8994 3780	6.3178 1562	132.9453 9043	21.0429 3612
48	5.2135 8898	120.3882 5659	23.0912 4425	6.5705 2824	139.2632 0604	21.1951 3088
49	5.3960 6459	125.6018 4557	23.2765 6450	6.8333 4937	145.8337 3429	21.3414 7200
50	5.5849 2686	130.9979 1016	23.4556 1787	7.1066 8335	152.6670 8366	21.4821 8462

	$i = 3\frac{1}{2}\%$			$i = 4\%$		
n	$(1 + i)^n$	$s_{\overline{n}|i}$	$a_{\overline{n}|i}$	$(1 + i)^n$	$s_{\overline{n}|i}$	$a_{\overline{n}|i}$
51	5.7803 9930	136.5828 3702	23.6286 1630	7.3909 5068	159.7737 6700	21.6174 8521
52	5.9827 1327	142.3632 3631	23.7957 6454	7.6865 8871	167.1647 1768	21.7475 8193
53	6.1921 0824	148.3459 4958	23.9572 6043	7.9940 5226	174.8513 0639	21.8726 7493
54	6.4088 3202	154.5380 5782	24.1132 9510	8.3138 1435	182.8453 5865	21.9929 5667
55	6.6331 4114	160.9468 8984	24.2640 5323	8.6463 6692	191.1591 7299	22.1086 1218
56	6.8653 0108	167.5800 3099	24.4097 1327	8.9922 2160	199.8055 3991	22.2198 1940
57	7.1055 8662	174.4453 3207	24.5504 4760	9.3519 1046	208.7977 6151	22.3267 4943
58	7.3542 8215	181.5509 1869	24.6864 2281	9.7259 8688	218.1496 7197	22.4295 6676
59	7.6116 8203	188.9052 0085	24.8177 9981	10.1150 2635	227.8756 5885	22.5284 2957
60	7.8780 9090	196.5168 8288	24.9447 3412	10.5196 2741	237.9906 8520	22.6234 8997
61	8.1538 2408	204.3949 7378	25.0673 7596	10.9404 1250	248.5103 1261	22.7148 9421
62	8.4392 0793	212.5487 9786	25.1858 7049	11.3780 2900	259.4507 2511	22.8027 8289
63	8.7345 8020	220.9880 0579	25.3003 5796	11.8331 5016	270.8287 5412	22.8872 9124
64	9.0402 9051	229.7225 8599	25.4109 7388	12.3064 7617	282.6619 0428	22.9685 4927
65	9.3567 0068	238.7628 7650	25.5178 4916	12.7987 3522	294.9683 8045	23.0466 8199
66	9.6841 8520	248.1195 7718	25.6211 1030	13.3106 8463	307.7671 1567	23.1218 0961
67	10.0231 3168	257.8037 6238	25.7208 7951	13.8431 1201	321.0778 0030	23.1940 4770
68	10.3739 4129	267.8268 9406	25.8172 7489	14.3968 3649	334.9209 1231	23.2635 0740
69	10.7370 2924	278.2008 3535	25.9104 1052	14.9727 0995	349.3177 4880	23.3302 9558
70	11.1128 2526	288.9378 6459	26.0003 9664	15.5716 1835	364.2904 5876	23.3945 1498
71	11.5017 7414	300.0506 8985	26.0873 3975	16.1944 8308	379.8620 7711	23.4562 6440
72	11.9043 3624	311.5524 6400	26.1713 4275	16.8422 6241	396.0565 6019	23.5156 3885
73	12.3209 8801	323.4568 0024	26.2525 0508	17.5159 5290	412.8988 2260	23.5727 2966
74	12.7522 2259	335.7777 8824	26.3309 2278	18.2165 9102	430.4147 7550	23.6276 2468
75	13.1985 5038	348.5300 1083	26.4066 8868	18.9452 5466	448.6313 6652	23.6804 0834
76	13.6604 9964	361.7285 6121	26.4798 9244	19.7030 6485	467.5766 2118	23.7311 6187
77	14.1386 1713	375.3890 6085	26.5506 2072	20.4911 8744	487.2796 8603	23.7799 6333
78	14.6334 6873	389.5276 7798	26.6189 5721	21.3108 3494	507.7708 7347	23.8268 8782
79	15.1456 4013	404.1611 4671	26.6849 8281	22.1632 6834	529.0817 0841	23.8720 0752
80	15.6757 3754	419.3067 8685	26.7487 7567	23.0497 9907	551.2449 7675	23.9153 9185
81	16.2243 8835	434.9825 2439	26.8104 1127	23.9717 9103	574.2947 7582	23.9571 0755
82	16.7922 4195	451.2069 1274	26.8699 6258	24.9306 6267	598.2665 6685	23.9972 1879
83	17.3799 7041	467.9991 5469	26.9275 0008	25.9278 8918	623.1972 2952	24.0357 8730
84	17.9882 6938	485.3791 2510	26.9830 9186	26.9650 0475	649.1251 1870	24.0728 7241
85	18.6178 5881	503.3673 9448	27.0368 0373	28.0436 0494	676.0901 2345	24.1085 3116
86	19.2694 8387	521.9852 5329	27.0886 9926	29.1653 4914	704.1337 2839	24.1428 1842
87	19.9439 1580	541.2547 3715	27.1388 3986	30.3319 6310	733.2990 7753	24.1757 8694
88	20.6419 5285	561.1986 5295	27.1872 8489	31.5452 4163	763.6310 4063	24.2074 8745
89	21.3644 2120	581.8406 0581	27.2340 9168	32.8070 5129	795.1762 8225	24.2379 6870
90	22.1121 7595	603.2050 2701	27.2793 1564	34.1193 3334	827.9833 3354	24.2672 7759
91	22.8861 0210	625.3172 0295	27.3230 ,1028	35.4841 0668	862.1026 6688	24.2954 5923
92	23.6871 1568	648.2033 0506	27.3652 2732	36.9034 7094	897.5867 7356	24.3225 5695
93	24.5161 6473	671.8904 2074	27.4060 1673	38.3796 0978	934.4902 4450	24.3486 1245
94	25.3742 3049	696.4065 8546	27.4454 2680	39.9147 9417	972.8698 5428	24.3736 6582
95	26.2623 2856	721.7808 1595	27.4835 0415	41.5113 8594	1012.7846 4845	24.3977 5559
96	27.1815 1006	748.0431 4451	27.5202 9387	43.1718 4138	1054.2960 3439	24.4209 1884
97	28.1328 6291	775.2246 5457	27.5558 3948	44.8987 1503	1097.4678 7577	24.4431 9119
98	29.1175 1311	803.3575 1748	27.5901 8308	46.6946 6363	1142.3665 9080	24.4646 0692
99	30.1366 2607	832.4750 3059	27.6233 6529	48.5624 5018	1189.0612 5443	24.4851 9896
100	31.1914 0798	862.6116 5666	27.6554 2540	50.5049 4818	1237.6237 0461	24.5049 9900

	i = 5%			i = 6%		
n	$(1 + i)^n$	$s_{\overline{n}|i}$	$a_{\overline{n}|i}$	$(1 + i)^n$	$s_{\overline{n}|i}$	$a_{\overline{n}|i}$
1	1.0500 0000	1.0000 0000	0.9523 8095	1.0600 0000	1.0000 0000	0.9433 9623
2	1.1025 0000	2.0500 0000	1.8594 1043	1.1236 0000	2.0600 0000	1.8333 9267
3	1.1576 2500	3.1525 0000	2.7232 4803	1.1910 1600	3.1836 0000	2.6730 1195
4	1.2155 0625	4.3101 2500	3.5459 5050	1.2624 7696	4.3746 1600	3.4651 0561
5	1.2762 8156	5.5256 3125	4.3294 7667	1.3382 2558	5.6370 9296	4.2123 6379
6	1.3400 9564	6.8019 1281	5.0756 9207	1.4185 1911	6.9753 1854	4.9173 2433
7	1.4071 0042	8.1420 0845	5.7863 7340	1.5036 3026	8.3938 3765	5.5823 8144
8	1.4774 5544	9.5491 0888	6.4632 1276	1.5938 4807	9.8974 6791	6.2097 9381
9	1.5513 2822	11.0265 6432	7.1078 2168	1.6894 7896	11.4913 1598	6.8016 9227
10	1.6288 9463	12.5778 9254	7.7217 3493	1.7908 4770	13.1807 9494	7.3600 8705
11	1.7103 3936	14.2067 8716	8.3064 1422	1.8982 9856	14.9716 4264	7.8868 7458
12	1.7958 5633	15.9171 2652	8.8632 5164	2.0121 9647	16.8699 4120	8.3838 4394
13	1.8856 4914	17.7129 8285	9.3935 7299	2.1329 2826	18.8821 3767	8.8526 8296
14	1.9799 3160	19.5986 3199	9.8986 4094	2.2609 0396	21.0150 6593	9.2949 8393
15	2.0789 2818	21.5785 6359	10.3796 5804	2.3965 5819	23.2759 6988	9.7122 4899
16	2.1828 7459	23.6574 9177	10.8377 6956	2.5403 5168	25.6725 2808	10.1058 9527
17	2.2920 1832	25.8403 6636	11.2740 6625	2.6927 7279	28.2128 7976	10.4772 5969
18	2.4066 1923	28.1323 8467	11.6895 8690	2.8543 3915	30.9056 5255	10.8276 0348
19	2.5269 5020	30.5390 0391	12.0853 2086	3.0255 9950	33.7599 9170	11.1581 1649
20	2.6532 9771	33.0659 5410	12.4622 1034	3.2071 3547	36.7855 9120	11.4699 2122
21	2.7859 6259	35.7192 5181	12.8211 5271	3.3995 6360	39.9927 2668	11.7640 7662
22	2.9252 6072	38.5052 1440	13.1630 0258	3.6035 3742	43.3922 9028	12.0415 8172
23	3.0715 2376	41.4304 7512	13.4885 7388	3.8197 4966	46.9958 2769	12.3033 7898
24	3.2250 9994	44.5019 9887	13.7986 4179	4.0489 3464	50.8155 7735	12.5503 5753
25	3.3863 5494	47.7270 9882	14.0939 4457	4.2918 7072	54.8645 1200	12.7833 5616
26	3.5556 7269	51.1134 5376	14.3751 8530	4.5493 8296	59.1563 8272	13.0031 6619
27	3.7334 5632	54.6691 2645	14.6430 3362	4.8223 4594	63.7057 6568	13.2105 3414
28	3.9201 2914	58.4025 8277	14.8981 2726	5.1116 8670	68.5281 1162	13.4061 6428
29	4.1161 3560	62.3227 1191	15.1410 7358	5.4183 8790	73.6397 9832	13.5907 2102
30	4.3219 4238	66.4388 4750	15.3724 5103	5.7434 9117	79.0581 8622	13.7648 3115
31	4.5380 3949	70.7607 8988	15.5928 1050	6.0881 0064	84.8016 7739	13.9290 8599
32	4.7649 4147	75.2988 2937	15.8026 7667	6.4533 8668	90.8897 7803	14.0840 4339
33	5.0031 8854	80.0637 7084	16.0025 4921	6.8405 8988	97.3431 6471	14.2302 2961
34	5.2533 4797	85.0669 5938	16.1929 0401	7.2510 2528	104.1837 5460	14.3681 4114
35	5.5160 1537	90.3203 0735	16.3741 9429	7.6860 8679	111.4347 7987	14.4982 4636
36	5.7918 1614	95.8363 2272	16.5468 5171	8.1472 5200	119.1208 6666	14.6209 8713
37	6.0814 0694	101.6281 3886	16.7112 8734	8.6360 8712	127.2681 1866	14.7367 8031
38	6.3854 7729	107.7095 4580	16.8678 9271	9.1542 5235	135.9042 0578	14.8460 1916
39	6.7047 5115	114.0950 2309	17.0170 4067	9.7035 0749	145.0584 5813	14.9490 7468
40	7.0399 8871	120.7997 7424	17.1590 8635	10.2857 1794	154.7619 6562	15.0462 9687
41	7.3919 8815	127.8397 6295	17.2943 6796	10.9028 6101	165.0476 8356	15.1380 1592
42	7.7615 8756	135.2317 5110	17.4232 0758	11.5570 3267	175.9505 4457	15.2245 4332
43	8.1496 6693	142.9933 3866	17.5459 1198	12.2504 5463	187.5075 7724	15.3061 7294
44	8.5571 5028	151.1430 0559	17.6627 7331	12.9854 8191	199.7580 3188	15.3831 8202
45	8.9850 0779	159.7001 5587	17.7740 6982	13.7646 1083	212.7435 1379	15.4558 3209
46	9.4342 5818	168.6851 6366	17.8800 6650	14.5904 8748	226.5081 2462	15.5243 6990
47	9.9059 7109	178.1194 2185	17.9810 1571	15.4659 1673	241.0986 1210	15.5890 2821
48	10.4012 6965	188.0253 9294	18.0771 5782	16.3938 7173	256.5645 2882	15.6500 2661
49	10.9213 3313	198.4266 6259	18.1687 2173	17.3775 0403	272.9584 0055	15.7075 7227
50	11.4673 9979	209.3479 9572	18.2559 2546	18.4201 5427	290.3359 0458	15.7618 6064

	$i = 7\%$			$i = 8\%$		
n	$(1+i)^n$	$s_{\overline{n}|i}$	$a_{\overline{n}|i}$	$(1+i)^n$	$s_{\overline{n}|i}$	$a_{\overline{n}|i}$
1	1.0700 0000	1.0000 0000	0.9345 7944	1.0800 0000	1.0000 0000	0.9259 2593
2	1.1449 0000	2.0700 0000	1.8080 1817	1.1664 0000	2.0800 0000	1.7832 6475
3	1.2250 4300	3.2149 0000	2.6243 1604	1.2597 1200	3.2464 0000	2.5770 9699
4	1.3107 9601	4.4399 4300	3.3872 1126	1.3604 8896	4.5061 1200	3.3121 2684
5	1.4025 5173	5.7507 3901	4.1001 9744	1.4693 2808	5.8666 0096	3.9927 1004
6	1.5007 3035	7.1532 9074	4.7665 3966	1.5868 7432	7.3359 2904	4.6228 7966
7	1.6057 8148	8.6540 2109	5.3892 8940	1.7138 2427	8.9228 0336	5.2063 7006
8	1.7181 8618	10.2598 0257	5.9712 9851	1.8509 3021	10.6366 2763	5.7466 3894
9	1.8384 5921	11.9779 8875	6.5152 3225	1.9990 0463	12.4875 5784	6.2468 8791
10	1.9671 5136	13.8164 4796	7.0235 8154	2.1589 2500	14.4865 6247	6.7100 8140
11	2.1048 5195	15.7835 9932	7.4986 7434	2.3316 3900	16.6454 8746	7.1389 6426
12	2.2521 9159	17.8884 5127	7.9426 8630	2.5181 7012	18.9771 2646	7.5360 7802
13	2.4098 4500	20.1406 4286	8.3576 5074	2.7196 2373	21.4952 9658	7.9037 7594
14	2.5785 3415	22.5504 8786	8.7454 6799	2.9371 9362	24.2149 2030	8.2442 3698
15	2.7590 3154	25.1290 2201	9.1079 1401	3.1721 6911	27.1521 1393	8.5594 7869
16	2.9521 6375	27.8880 5355	9.4466 4860	3.4259 4264	30.3242 8304	8.8513 6916
17	3.1588 1521	30.8402 1730	9.7632 2299	3.7000 1805	33.7502 2569	9.1216 3811
18	3.3799 3228	33.9990 3251	10.0590 8691	3.9960 1950	37.4502 4374	9.3718 8714
19	3.6165 2754	37.3789 6479	10.3355 9524	4.3157 0106	41.4462 6324	9.6035 9920
20	3.8696 8446	40.9954 9232	10.5940 1425	4.6609 5714	45.7619 6430	9.8181 4741
21	4.1405 6237	44.8651 7678	10.8355 2733	5.0338 3372	50.4229 2144	10.0168 0316
22	4.4304 0174	49.0057 3916	11.0612 4050	5.4365 4041	55.4567 5516	10.2007 4366
23	4.7405 2986	53.4361 4090	11.2721 8738	5.8714 6365	60.8932 9557	10.3710 5895
24	5.0723 6695	58.1766 7076	11.4693 3400	6.3411 8074	66.7647 5922	10.5287 5828
25	5.4274 3264	63.2490 3772	11.6535 8318	6.8484 7520	73.1059 3995	10.6747 7619
26	5.8073 5292	68.6764 7036	11.8257 7867	7.3963 5321	79.9544 1515	10.8099 7795
27	6.2138 6763	74.4838 2328	11.9867 0904	7.9880 6147	87.3507 6836	10.9351 6477
28	6.6488 3836	80.6976 9091	12.1371 1125	8.6271 0639	95.3388 2983	11.0510 7849
29	7.1142 5705	87.3465 2927	12.2776 7407	9.3172 7490	103.9659 3622	11.1584 0601
30	7.6122 5504	94.4607 8632	12.4090 4118	10.0626 5689	113.2832 1111	11.2577 8334
31	8.1451 1290	102.0730 4137	12.5318 1419	10.8676 6944	123.3458 6800	11.3497 9939
32	8.7152 7080	110.2181 5426	12.6465 5532	11.7370 8300	134.2135 3744	11.4349 9944
33	9.3253 3975	118.9334 2506	12.7537 9002	12.6760 4964	145.9506 2044	11.5138 8837
34	9.9781 1354	128.2587 6481	12.8540 0936	13.6901 3361	158.6266 7007	11.5869 3367
35	10.6765 8148	138.2368 7835	12.9476 7230	14.7853 4429	172.3168 0368	11.6545 6822
36	11.4239 4219	148.9134 5984	13.0352 0776	15.9681 7184	187.1021 4797	11.7171 9279
37	12.2236 1814	160.3374 0202	13.1170 1660	17.2456 2558	203.0703 1981	11.7751 7851
38	13.0792 7141	172.5610 2017	13.1934 7345	18.6252 7563	220.3159 4540	11.8288 6899
39	13.9948 2041	185.6402 9158	13.2649 2846	20.1152 9768	238.9412 2103	11.8785 8240
40	14.9744 5784	199.6351 1199	13.3317 0884	21.7245 2150	259.0565 1871	11.9246 1333
41	16.0226 6989	214.6095 6983	13.3941 2041	23.4624 8322	280.7810 4021	11.9672 3457
42	17.1442 5678	230.6322 3972	13.4524 4898	25.3394 8187	304.2435 2342	12.0066 9867
43	18.3443 5475	247.7764 9650	13.5069 6167	27.3666 4042	329.5830 0530	12.0432 3951
44	19.6284 5959	266.1208 5125	13.5579 0810	29.5559 7166	356.9496 4572	12.0770 7362
45	21.0024 5176	285.7493 1084	13.6055 2159	31.9204 4939	386.5056 1738	12.1084 0150
46	22.4726 2338	306.7517 6260	13.6500 2018	34.4740 8534	418.4260 6677	12.1374 0880
47	24.0457 0702	329.2243 8598	13.6916 0764	37.2320 1217	452.9001 5211	12.1642 6741
48	25.7289 0651	353.2700 9300	13.7304 7443	40.2105 7314	490.1321 6428	12.1891 3649
49	27.5299 2997	378.9989 9951	13.7667 9853	43.4274 1899	530.3427 3742	12.2121 6341
50	29.4570 2506	406.5289 2947	13.8007 4629	46.9016 1251	573.7701 5642	12.2334 8464

Table II   *Binomial probabilities*

							p					
n	*x*	**0.05**	**0.1**	**0.2**	**0.3**	**0.4**	**0.5**	**0.6**	**0.7**	**0.8**	**0.9**	**0.95**
2	0	0.902	0.810	0.640	0.490	0.360	0.250	0.160	0.090	0.040	0.010	0.002
	1	0.095	0.180	0.320	0.420	0.480	0.500	0.480	0.420	0.320	0.180	0.095
	2	0.002	0.010	0.040	0.090	0.160	0.250	0.360	0.490	0.640	0.810	0.902
3	0	0.857	0.729	0.512	0.343	0.216	0.125	0.064	0.027	0.008	0.001	
	1	0.135	0.243	0.384	0.441	0.432	0.375	0.288	0.189	0.096	0.027	0.007
	2	0.007	0.027	0.096	0.189	0.288	0.375	0.432	0.441	0.384	0.243	0.135
	3		0.001	0.008	0.027	0.064	0.125	0.216	0.343	0.512	0.729	0.857
4	0	0.815	0.656	0.410	0.240	0.130	0.062	0.026	0.008	0.002		
	1	0.171	0.292	0.410	0.412	0.346	0.250	0.154	0.076	0.026	0.004	
	2	0.014	0.049	0.154	0.265	0.346	0.375	0.346	0.265	0.154	0.049	0.014
	3		0.004	0.026	0.076	0.154	0.250	0.346	0.412	0.410	0.292	0.171
	4			0.002	0.008	0.026	0.062	0.130	0.240	0.410	0.656	0.815
5	0	0.774	0.590	0.328	0.168	0.078	0.031	0.010	0.002			
	1	0.204	0.328	0.410	0.360	0.259	0.156	0.077	0.028	0.006		
	2	0.021	0.073	0.205	0.309	0.346	0.312	0.230	0.132	0.051	0.008	0.001
	3	0.001	0.008	0.051	0.132	0.230	0.312	0.346	0.309	0.205	0.073	0.021
	4			0.006	0.028	0.077	0.156	0.259	0.360	0.410	0.328	0.204
	5				0.002	0.010	0.031	0.078	0.168	0.328	0.590	0.774
6	0	0.735	0.531	0.262	0.118	0.047	0.016	0.004	0.001			
	1	0.232	0.354	0.393	0.303	0.187	0.094	0.037	0.010	0.002		
	2	0.031	0.098	0.246	0.324	0.311	0.234	0.138	0.060	0.015	0.001	
	3	0.002	0.015	0.082	0.185	0.276	0.312	0.276	0.185	0.082	0.015	0.002
	4		0.001	0.015	0.060	0.138	0.234	0.311	0.324	0.246	0.098	0.031
	5			0.002	0.010	0.037	0.094	0.187	0.303	0.393	0.354	0.232
	6				0.001	0.004	0.016	0.047	0.118	0.262	0.531	0.735
7	0	0.698	0.478	0.210	0.082	0.028	0.008	0.002				
	1	0.257	0.372	0.367	0.247	0.131	0.055	0.017	0.004			
	2	0.041	0.124	0.275	0.318	0.261	0.164	0.077	0.025	0.004		
	3	0.004	0.023	0.115	0.227	0.290	0.273	0.194	0.097	0.029	0.003	
	4		0.003	0.029	0.097	0.194	0.273	0.290	0.227	0.115	0.023	0.004
	5			0.004	0.025	0.077	0.164	0.261	0.318	0.275	0.124	0.041
	6				0.004	0.017	0.055	0.131	0.247	0.367	0.372	0.257
	7					0.002	0.008	0.028	0.082	0.210	0.478	0.698
8	0	0.663	0.430	0.168	0.058	0.017	0.004	0.001				
	1	0.279	0.383	0.336	0.198	0.090	0.031	0.008	0.001			
	2	0.051	0.149	0.294	0.296	0.209	0.109	0.041	0.010	0.001		
	3	0.005	0.033	0.147	0.254	0.279	0.219	0.124	0.047	0.009		
	4		0.005	0.046	0.136	0.232	0.273	0.232	0.136	0.046	0.005	
	5			0.009	0.047	0.124	0.219	0.279	0.254	0.147	0.033	0.005
	6			0.001	0.010	0.041	0.109	0.209	0.296	0.294	0.149	0.051
	7				0.001	0.008	0.031	0.090	0.198	0.336	0.383	0.279
	8					0.001	0.004	0.017	0.058	0.168	0.430	0.663

Table II   *(continued)*

p

n	x	0.05	0.1	0.2	0.3	0.4	0.5	0.6	0.7	0.8	0.9	0.95
9	0	0.630	0.387	0.134	0.040	0.010	0.002					
	1	0.299	0.387	0.302	0.156	0.060	0.018	0.004				
	2	0.063	0.172	0.302	0.267	0.161	0.070	0.021	0.004			
	3	0.008	0.045	0.176	0.267	0.251	0.164	0.074	0.021	0.003		
	4	0.001	0.007	0.066	0.172	0.251	0.246	0.167	0.074	0.017	0.001	
	5		0.001	0.017	0.074	0.167	0.246	0.251	0.172	0.066	0.007	0.001
	6			0.003	0.021	0.074	0.164	0.251	0.267	0.176	0.045	0.008
	7				0.004	0.021	0.070	0.161	0.267	0.302	0.172	0.063
	8					0.004	0.018	0.060	0.156	0.302	0.387	0.299
	9						0.002	0.010	0.040	0.134	0.387	0.630
10	0	0.599	0.349	0.107	0.028	0.006	0.001					
	1	0.315	0.387	0.268	0.121	0.040	0.010	0.002				
	2	0.075	0.194	0.302	0.233	0.121	0.044	0.011	0.001			
	3	0.010	0.057	0.201	0.267	0.215	0.117	0.042	0.009	0.001		
	4	0.001	0.011	0.088	0.200	0.251	0.205	0.111	0.037	0.006		
	5		0.001	0.026	0.103	0.201	0.246	0.201	0.103	0.026	0.001	
	6			0.006	0.037	0.111	0.205	0.251	0.200	0.088	0.011	0.001
	7			0.001	0.009	0.042	0.117	0.215	0.267	0.201	0.057	0.010
	8				0.001	0.011	0.044	0.121	0.233	0.302	0.194	0.075
	9					0.002	0.010	0.040	0.121	0.268	0.387	0.315
	10						0.001	0.006	0.028	0.107	0.349	0.599
11	0	0.569	0.314	0.086	0.020	0.004						
	1	0.329	0.384	0.236	0.093	0.027	0.005	0.001				
	2	0.087	0.213	0.295	0.200	0.089	0.027	0.005	0.001			
	3	0.014	0.071	0.221	0.257	0.177	0.081	0.023	0.004			
	4	0.001	0.016	0.111	0.220	0.236	0.161	0.070	0.017	0.002		
	5		0.002	0.039	0.132	0.221	0.226	0.147	0.057	0.010		
	6			0.010	0.057	0.147	0.226	0.221	0.132	0.039	0.002	
	7			0.002	0.017	0.070	0.161	0.236	0.220	0.111	0.016	0.001
	8				0.004	0.023	0.081	0.177	0.257	0.221	0.071	0.014
	9				0.001	0.005	0.027	0.089	0.200	0.295	0.213	0.087
	10					0.001	0.005	0.027	0.093	0.236	0.384	0.329
	11							0.004	0.020	0.086	0.314	0.569
12	0	0.540	0.282	0.069	0.014	0.002						
	1	0.341	0.377	0.206	0.071	0.017	0.003					
	2	0.099	0.230	0.283	0.168	0.064	0.016	0.002				
	3	0.017	0.085	0.236	0.240	0.142	0.054	0.012	0.001			
	4	0.002	0.021	0.133	0.231	0.213	0.121	0.042	0.008	0.001		
	5		0.004	0.053	0.158	0.227	0.193	0.101	0.029	0.003		
	6			0.016	0.079	0.177	0.226	0.177	0.079	0.016		
	7			0.003	0.029	0.101	0.193	0.227	0.158	0.053	0.004	
	8			0.001	0.008	0.042	0.121	0.213	0.231	0.133	0.021	0.002
	9				0.001	0.012	0.054	0.142	0.240	0.236	0.085	0.017
	10					0.002	0.016	0.064	0.168	0.283	0.230	0.099
	11						0.003	0.017	0.071	0.206	0.377	0.341
	12							0.002	0.014	0.069	0.282	0.540

Table II   *(continued)*

n	x	0.05	0.1	0.2	0.3	0.4	0.5	0.6	0.7	0.8	0.9	0.95
13	0	0.513	0.254	0.055	0.010	0.001						
	1	0.351	0.367	0.179	0.054	0.011	0.002					
	2	0.111	0.245	0.268	0.139	0.045	0.010	0.001				
	3	0.021	0.100	0.246	0.218	0.111	0.035	0.006	0.001			
	4	0.003	0.028	0.154	0.234	0.184	0.087	0.024	0.003			
	5		0.006	0.069	0.180	0.221	0.157	0.066	0.014	0.001		
	6		0.001	0.023	0.103	0.197	0.209	0.131	0.044	0.006		
	7			0.006	0.044	0.131	0.209	0.197	0.103	0.023	0.001	
	8			0.001	0.014	0.066	0.157	0.221	0.180	0.069	0.006	
	9				0.003	0.024	0.087	0.184	0.234	0.154	0.028	0.003
	10				0.001	0.006	0.035	0.111	0.218	0.246	0.100	0.021
	11					0.001	0.010	0.045	0.139	0.268	0.245	0.111
	12						0.002	0.011	0.054	0.179	0.367	0.351
	13							0.001	0.010	0.055	0.254	0.513
14	0	0.488	0.229	0.044	0.007	0.001						
	1	0.359	0.356	0.154	0.041	0.007	0.001					
	2	0.123	0.257	0.250	0.113	0.032	0.006	0.001				
	3	0.026	0.114	0.250	0.194	0.085	0.022	0.003				
	4	0.004	0.035	0.172	0.229	0.155	0.061	0.014	0.001			
	5		0.008	0.086	0.196	0.207	0.122	0.041	0.007			
	6		0.001	0.032	0.126	0.207	0.183	0.092	0.023	0.002		
	7			0.009	0.062	0.157	0.209	0.157	0.062	0.009		
	8			0.002	0.023	0.092	0.183	0.207	0.126	0.032	0.001	
	9				0.007	0.041	0.122	0.207	0.196	0.086	0.008	
	10				0.001	0.014	0.061	0.155	0.229	0.172	0.035	0.004
	11					0.003	0.022	0.085	0.194	0.250	0.114	0.026
	12					0.001	0.006	0.032	0.113	0.250	0.257	0.123
	13						0.001	0.007	0.041	0.154	0.356	0.359
	14							0.001	0.007	0.044	0.229	0.488
15	0	0.463	0.206	0.035	0.005							
	1	0.366	0.343	0.132	0.031	0.005						
	2	0.135	0.267	0.231	0.092	0.022	0.003					
	3	0.031	0.129	0.250	0.170	0.063	0.014	0.002				
	4	0.005	0.043	0.188	0.219	0.127	0.042	0.007	0.001			
	5	0.001	0.010	0.103	0.206	0.186	0.092	0.024	0.003			
	6		0.002	0.043	0.147	0.207	0.153	0.061	0.012	0.001		
	7			0.014	0.081	0.177	0.196	0.118	0.035	0.003		
	8			0.003	0.035	0.118	0.196	0.177	0.081	0.014		
	9			0.001	0.012	0.061	0.153	0.207	0.147	0.043	0.002	
	10				0.003	0.024	0.092	0.186	0.206	0.103	0.010	0.001
	11				0.001	0.007	0.042	0.127	0.219	0.188	0.043	0.005
	12					0.002	0.014	0.063	0.170	0.250	0.129	0.031
	13						0.003	0.022	0.092	0.231	0.267	0.135
	14							0.005	0.031	0.132	0.343	0.366
	15								0.005	0.035	0.206	0.463

Table III   *The standard normal distribution*

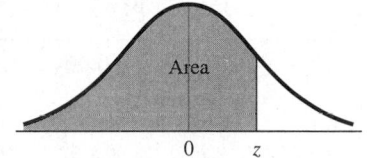

$$F_z(z) = P[Z \le z]$$

z	0.00	0.01	0.02	0.03	0.04	0.05	0.06	0.07	0.08	0.09
−3.4	0.0003	0.0003	0.0003	0.0003	0.0003	0.0003	0.0003	0.0003	0.0003	0.0002
−3.3	0.0005	0.0005	0.0005	0.0004	0.0004	0.0004	0.0004	0.0004	0.0004	0.0003
−3.2	0.0007	0.0007	0.0006	0.0006	0.0006	0.0006	0.0006	0.0005	0.0005	0.0005
−3.1	0.0010	0.0009	0.0009	0.0009	0.0008	0.0008	0.0008	0.0008	0.0007	0.0007
−3.0	0.0013	0.0013	0.0013	0.0012	0.0012	0.0011	0.0011	0.0011	0.0010	0.0010
−2.9	0.0019	0.0018	0.0017	0.0017	0.0016	0.0016	0.0015	0.0015	0.0014	0.0014
−2.8	0.0026	0.0025	0.0024	0.0023	0.0023	0.0022	0.0021	0.0021	0.0020	0.0019
−2.7	0.0035	0.0034	0.0033	0.0032	0.0031	0.0030	0.0029	0.0028	0.0027	0.0026
−2.6	0.0047	0.0045	0.0044	0.0043	0.0041	0.0040	0.0039	0.0038	0.0037	0.0036
−2.5	0.0062	0.0060	0.0059	0.0057	0.0055	0.0054	0.0052	0.0051	0.0049	0.0048
−2.4	0.0082	0.0080	0.0078	0.0075	0.0073	0.0071	0.0069	0.0068	0.0066	0.0064
−2.3	0.0107	0.0104	0.0102	0.0099	0.0096	0.0094	0.0091	0.0089	0.0087	0.0084
−2.2	0.0139	0.0136	0.0132	0.0129	0.0125	0.0122	0.0119	0.0116	0.0113	0.0110
−2.1	0.0179	0.0174	0.0170	0.0166	0.0162	0.0158	0.0154	0.0150	0.0146	0.0143
−2.0	0.0228	0.0222	0.0217	0.0212	0.0207	0.0202	0.0197	0.0192	0.0188	0.0183
−1.9	0.0287	0.0281	0.0274	0.0268	0.0262	0.0256	0.0250	0.0244	0.0239	0.0233
−1.8	0.0359	0.0352	0.0344	0.0336	0.0329	0.0322	0.0314	0.0307	0.0301	0.0294
−1.7	0.0446	0.0436	0.0427	0.0418	0.0409	0.0401	0.0392	0.0384	0.0375	0.0367
−1.6	0.0548	0.0537	0.0526	0.0516	0.0505	0.0495	0.0485	0.0475	0.0465	0.0455
−1.5	0.0668	0.0655	0.0643	0.0630	0.0618	0.0606	0.0594	0.0582	0.0571	0.0559
−1.4	0.0808	0.0793	0.0778	0.0764	0.0749	0.0735	0.0722	0.0708	0.0694	0.0681
−1.3	0.0968	0.0951	0.0934	0.0918	0.0901	0.0885	0.0869	0.0853	0.0838	0.0823
−1.2	0.1151	0.1131	0.1112	0.1093	0.1075	0.1056	0.1038	0.1020	0.1003	0.0985
−1.1	0.1357	0.1335	0.1314	0.1292	0.1271	0.1251	0.1230	0.1210	0.1190	0.1170
−1.0	0.1587	0.1562	0.1539	0.1515	0.1492	0.1469	0.1446	0.1423	0.1401	0.1379
−0.9	0.1841	0.1814	0.1788	0.1762	0.1736	0.1711	0.1685	0.1660	0.1635	0.1611
−0.8	0.2119	0.2090	0.2061	0.2033	0.2005	0.1977	0.1949	0.1922	0.1894	0.1867
−0.7	0.2420	0.2389	0.2358	0.2327	0.2296	0.2266	0.2236	0.2206	0.2177	0.2148
−0.6	0.2743	0.2709	0.2676	0.2643	0.2611	0.2578	0.2546	0.2514	0.2483	0.2451
−0.5	0.3085	0.3050	0.3015	0.2981	0.2946	0.2912	0.2877	0.2843	0.2810	0.2776

Table III   (continued)

$$F_z(z) = P[Z \le z]$$

z	0.00	0.01	0.02	0.03	0.04	0.05	0.06	0.07	0.08	0.09
−0.4	0.3446	0.3409	0.3372	0.3336	0.3300	0.3264	0.3228	0.3192	0.3156	0.3121
−0.3	0.3821	0.3783	0.3745	0.3707	0.3669	0.3632	0.3594	0.3557	0.3520	0.3483
−0.2	0.4207	0.4168	0.4129	0.4090	0.4052	0.4013	0.3974	0.3936	0.3897	0.3859
−0.1	0.4602	0.4562	0.4522	0.4483	0.4443	0.4404	0.4364	0.4325	0.4286	0.4247
−0.0	0.5000	0.4960	0.4920	0.4880	0.4840	0.4801	0.4761	0.4721	0.4681	0.4641
0.0	0.5000	0.5040	0.5080	0.5120	0.5160	0.5199	0.5239	0.5279	0.5319	0.5359
0.1	0.5398	0.5438	0.5478	0.5517	0.5557	0.5596	0.5636	0.5675	0.5714	0.5753
0.2	0.5793	0.5832	0.5871	0.5910	0.5948	0.5987	0.6026	0.6064	0.6103	0.6141
0.3	0.6179	0.6217	0.6255	0.6293	0.6331	0.6368	0.6406	0.6443	0.6480	0.6517
0.4	0.6554	0.6591	0.6628	0.6664	0.6700	0.6736	0.6772	0.6808	0.6844	0.6879
0.5	0.6915	0.6950	0.6985	0.7019	0.7054	0.7088	0.7123	0.7157	0.7190	0.7224
0.6	0.7257	0.7291	0.7324	0.7357	0.7389	0.7422	0.7454	0.7486	0.7517	0.7549
0.7	0.7580	0.7611	0.7642	0.7673	0.7704	0.7734	0.7764	0.7794	0.7823	0.7852
0.8	0.7881	0.7910	0.7939	0.7967	0.7995	0.8023	0.8051	0.8078	0.8106	0.8133
0.9	0.8159	0.8186	0.8212	0.8238	0.8264	0.8289	0.8315	0.8340	0.8365	0.8389
1.0	0.8413	0.8438	0.8461	0.8485	0.8508	0.8531	0.8554	0.8577	0.8599	0.8621
1.1	0.8643	0.8665	0.8686	0.8708	0.8729	0.8749	0.8770	0.8790	0.8810	0.8830
1.2	0.8849	0.8869	0.8888	0.8907	0.8925	0.8944	0.8962	0.8980	0.8997	0.9015
1.3	0.9032	0.9049	0.9066	0.9082	0.9099	0.9115	0.9131	0.9147	0.9162	0.9177
1.4	0.9192	0.9207	0.9222	0.9236	0.9251	0.9265	0.9278	0.9292	0.9306	0.9319
1.5	0.9332	0.9345	0.9357	0.9370	0.9382	0.9394	0.9406	0.9418	0.9429	0.9441
1.6	0.9452	0.9463	0.9474	0.9484	0.9495	0.9505	0.9515	0.9525	0.9535	0.9545
1.7	0.9554	0.9564	0.9573	0.9582	0.9591	0.9599	0.9608	0.9616	0.9625	0.9633
1.8	0.9641	0.9649	0.9656	0.9664	0.9671	0.9678	0.9686	0.9693	0.9699	0.9706
1.9	0.9713	0.9719	0.9726	0.9732	0.9738	0.9744	0.9750	0.9756	0.9761	0.9767
2.0	0.9772	0.9778	0.9783	0.9788	0.9793	0.9798	0.9803	0.9808	0.9812	0.9817
2.1	0.9821	0.9826	0.9830	0.9834	0.9838	0.9842	0.9846	0.9850	0.9854	0.9857
2.2	0.9861	0.9864	0.9868	0.9871	0.9875	0.9878	0.9881	0.9884	0.9887	0.9890
2.3	0.9893	0.9896	0.9898	0.9901	0.9904	0.9906	0.9909	0.9911	0.9913	0.9916
2.4	0.9918	0.9920	0.9922	0.9925	0.9927	0.9929	0.9931	0.9932	0.9934	0.9936
2.5	0.9938	0.9940	0.9951	0.9943	0.9945	0.9946	0.9948	0.9949	0.9951	0.9952
2.6	0.9953	0.9955	0.9956	0.9957	0.9959	0.9960	0.9961	0.9962	0.9963	0.9964
2.7	0.9965	0.9966	0.9967	0.9968	0.9969	0.9970	0.9971	0.9972	0.9973	0.9974
2.8	0.9974	0.9975	0.9976	0.9977	0.9977	0.9978	0.9979	0.9979	0.9980	0.9981
2.9	0.9981	0.9982	0.9982	0.9983	0.9984	0.9984	0.9985	0.9985	0.9986	0.9986
3.0	0.9987	0.9987	0.9987	0.9988	0.9988	0.9989	0.9989	0.9989	0.9990	0.9990
3.1	0.9990	0.9991	0.9991	0.9991	0.9992	0.9992	0.9992	0.9992	0.9993	0.9993
3.2	0.9993	0.9993	0.9994	0.9994	0.9994	0.9994	0.9994	0.9995	0.9995	0.9995
3.3	0.9995	0.9995	0.9995	0.9996	0.9996	0.9996	0.9996	0.9996	0.9996	0.9997
3.4	0.9997	0.9997	0.9997	0.9997	0.9997	0.9997	0.9997	0.9997	0.9997	0.9998

CHAPTER I

Exercises 1.1, page 8

1. (3, 3); Quadrant I   **3.** (2, −2); Quadrant IV

5. (−4, −6); Quadrant III   **7.** A **9.** E, F, and G

11. F **13.–19.** See the following figure.

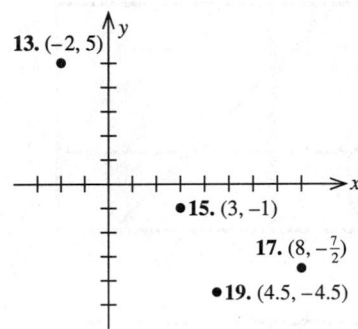

13. (−2, 5)
15. (3, −1)
17. $(8, -\frac{7}{2})$
19. (4.5, −4.5)

21. 5   **23.** $\sqrt{61}$ **25.** (−8, −6) and (8, −6)

29. 3400 miles   **31.** Route 1   **33.** Model C

Exercises 1.2, page 22

1. 1/2   **3.** Not defined   **5.** 5   **7.** 5/6

9. $\dfrac{d - b}{c - a}\,(a \neq c)$ **11. a.** 4   **b.** −8   **13.** Parallel

15. Perpendicular   **17.** $a = -5$ **19.** $y = -3$

21. e   **23.** a   **25.** f   **27.** $y = 2x - 10$

29. $y = 2$ **31.** $y = 3x - 2$ **33.** $y = x + 1$

35. $y = 3x + 4$ **37.** $y = 5$

39. $y = \frac{1}{2}x$; $m = 1/2$; $b = 0$

41. $y = \frac{2}{3}x - 3$; $m = 2/3$; $b = -3$

43. $y = -\frac{1}{2}x + \frac{7}{2}$; $m = -1/2$; $b = \frac{7}{2}$

45. $y = \frac{1}{2}x + 3$ **47.** $y = -6$ **49.** $y = b$

51. $y = \frac{2}{3}x - \frac{2}{3}$ **53.** $k = 8$

55.
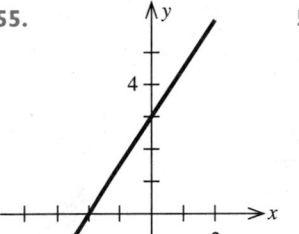

57.
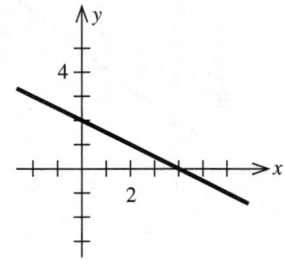

59.

63. $y = -2x - 4$ **65.** $y = \frac{1}{8}x - \frac{1}{2}$ **67.** Yes

69. a. $y = 0.55x$ **b.** 2000

71. a. and **b.**

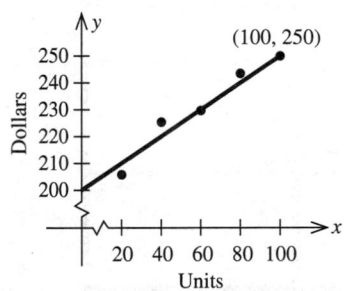

c. $y = \frac{1}{2}x + 200$ **d.** $227

73. a. and b.

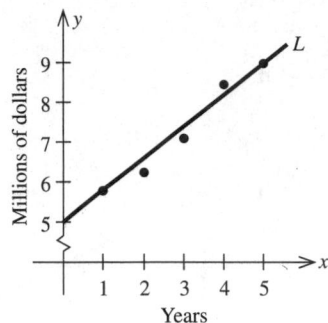

c. $y = 0.8x + 5$ **d.** \$12.2 million

75. a. a family of parallel lines having slope m
  b. a family of straight lines that pass through the point $(0, b)$

Using Technology Exercises 1.2, page 29

1.

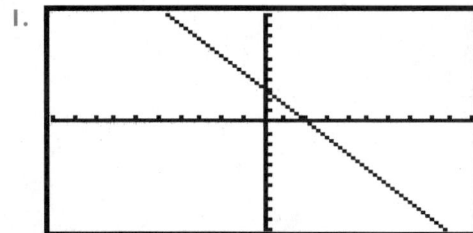

3.

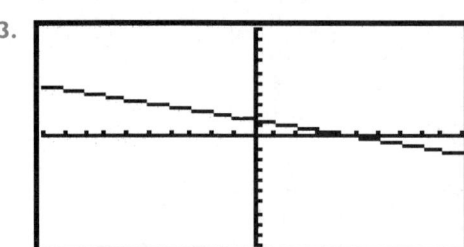

5.

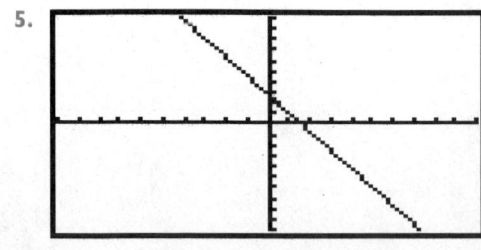

7. a.

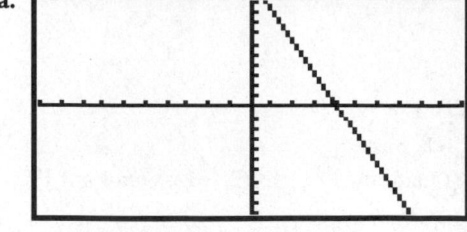

b.

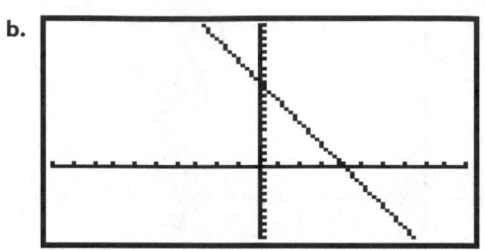

9. a.

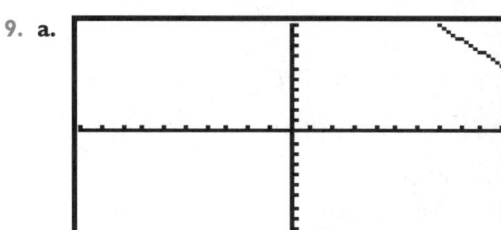

b.

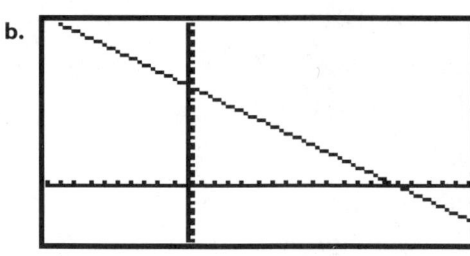

11.

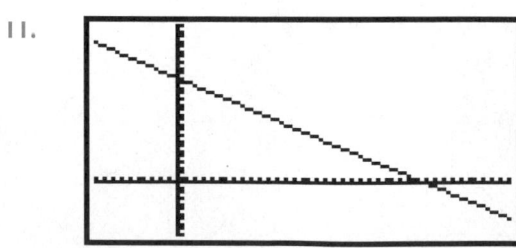

13.

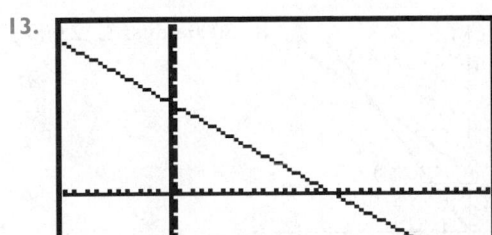

15.

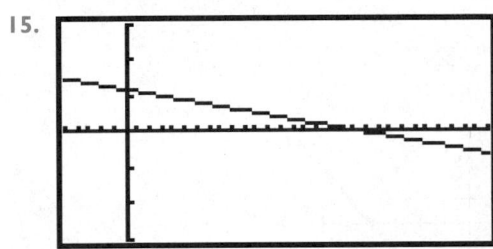

17.

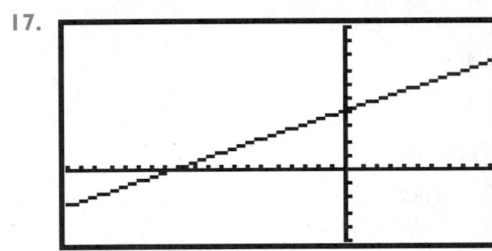

Exercises 1.3, page 38

1. Yes; $y = -\frac{2}{3}x + 2$ 3. Yes; $y = \frac{1}{2}x + 2$

5. Yes; $y = \frac{1}{2}x + \frac{9}{4}$ 7. No   9. No

11. **a.** $C(x) = 8x + 40{,}000$ **b.** $R(x) = 12x$
    c. $P(x) = 4x - 40{,}000$
    d. Loss of $8000; profit of $8000

13. $m = -1, b = 2$ 15. $900{,}000; $800,000

17. $6 billion; $43.5 billion; $81 billion

19. **a.** $y = 1.053x$ **b.** $652.86

21. $C(x) = 0.6x + 12{,}100; R(x) = 1.15x;$
    $P(x) = 0.55x - 12{,}100$

23. **a.** $12,000/year
    b. $V = 60{,}000 - 12{,}000t$

c.

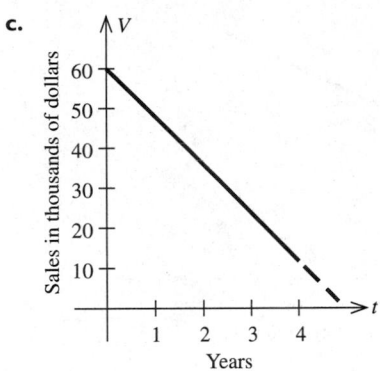

d. $24,000

25. $900,000; $800,000

27. **a.** $m = a/1.7; b = 0$ **b.** 117.65 mg

29. **a.** $F = \frac{9}{5}C + 32$ **b.** 68°F   **c.** 21.1°C   31. L_2

33. **a.**

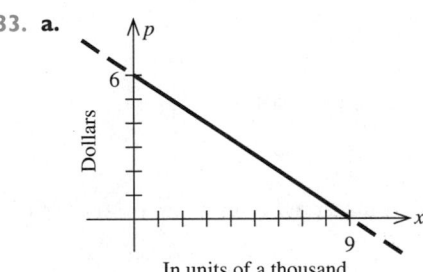

In units of a thousand

b. 3000

35. **a.**

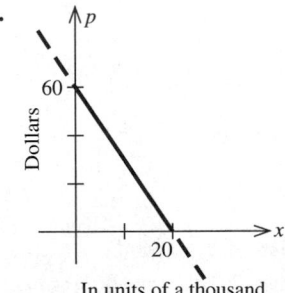

In units of a thousand

b. 10,000

37. $p = -\frac{3}{40}x + 130;$ $130; 1733

39. 2500 units

41. a.

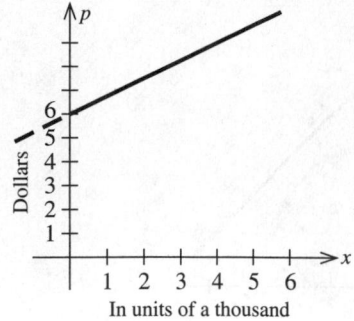

b. 2667 units

43. a.

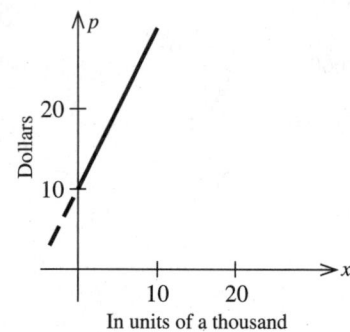

b. 2000 units

45. $p = \frac{1}{2}x + 40$ (x is measured in units of a thousand)

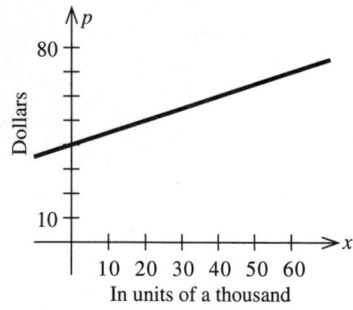

60,000 units

Using Technology Exercises 1.3, page 43

1. 2.2875   **3.** 2.880952381   **5.** 7.2851648352

7. 2.4680851064

Exercises 1.4, page 52

1. (2, 10)   **3.** (4, 2/3)   **5.** (−4, −6)

7. 1000 units; $15,000   **9.** 600 units; $240

11. a.

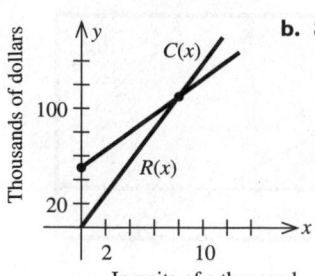

b. 8000 units; $112,000

c.

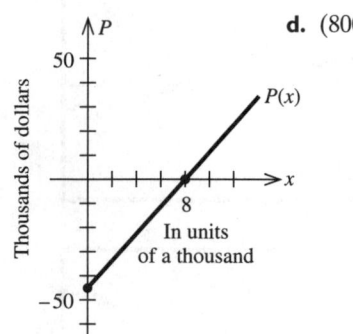

d. (8000, 0)

13. 9259 units; $83,331

15. a. $C_1(x) = 18,000 + 15x$
$C_2(x) = 15,000 + 20x$

b.

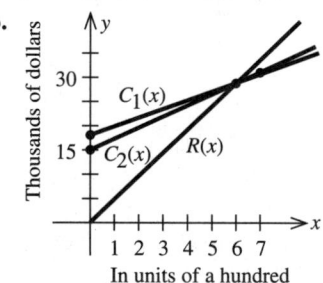

c. Machine II; machine II; machine I

d. ($1500); $1500; $4750

17. 8000 units; $9   **19.** 2000 units; $18

21. a. $p = -0.08x + 725$ **b.** $p = 0.09x + 300$
c. 2500 VCRs; $525

23. 300 fax machines; $600

25. a. $\dfrac{b-d}{c-a}; \dfrac{bc-ad}{c-a}$

b. If c is increased, x gets smaller and p gets larger.
c. If b is decreased, x decreases and p decreases.

27. a. $m_1 = m_2$ and $b_2 \neq b_1$

    b. $m_1 \neq m_2$

    c. $m_1 = m_2$ and $b_1 = b_2$

Using Technology Exercises 1.4, page 55

1. (0.6, 6.2)   **3.** (3.8261, 0.1304)

5. (386.9091, 145.3939)

Exercises 1.5, page 61

  1. a. $y = 2.3x + 1.5$

    b.

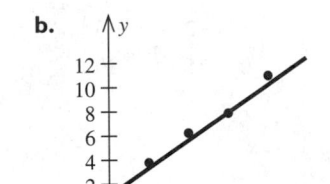

  3. a. $y = -0.77x + 5.74$

    b.

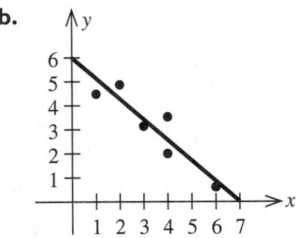

  5. a. $y = 1.2x + 2$

    b.

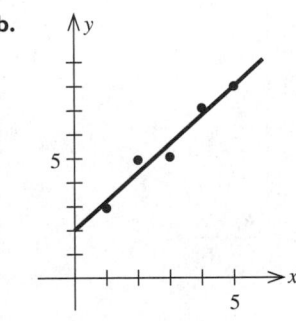

  7. a. $y = 0.34x - 0.9$

    b.

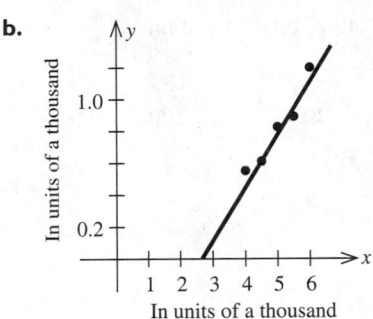

    c. 1276 applications

  9. a. $y = -2.8x + 440$

    b.

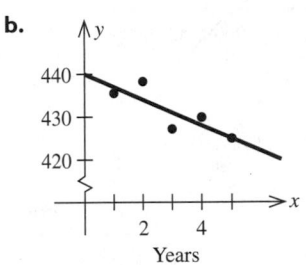

    c. 420

11. a. $y = 6.23x + 167.6$ **b.** 541 acres

13. a. $y = 2.8x + 17.6$ **b.** \$40,000,000

15. a. $y = 4.842x + 11.842$ **b.** 103.8 billion cans

Using Technology Exercises 1.5, page 65

1. $y = 3.8639 + 2.3596x$ **3.** $y = 3.5525 - 1.1948x$

5. a. $y = 13.321x + 72.571$ **b.** 192 million tons

Chapter 1 Review Exercises, page 68

1. 5   **3.** 5   **5.** $x = -2$ **7.** $x + 10y - 38 = 0$

9. $5x - 2y + 18 = 0$ **11.** $y = -\frac{1}{2}x - 3$

13. $3x + 4y - 18 = 0$ **15.** $3x + 2y + 14 = 0$

17.

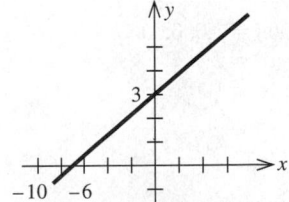

19. a. $f(x) = x + 2.4$ **b.** \$5.4 million

21. b. ≈ 117 mg

23. a. \$22,500/year   **b.** $V = -22,500t + 300,000$

25. $p = -0.05x + 200$

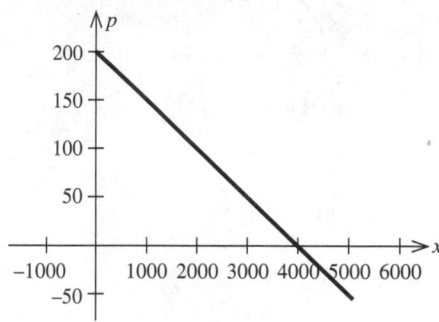

27. $(2, -3)$ **29.** $(2500, 50,000)$

31. a. $y = 0.25x$ **b.** 1600

CHAPTER 2

Exercises 2.1, page 78

1. Unique solution; $(2, 1)$ **3.** No solution

5. Unique solution; $(3, 2)$

7. Infinitely many solutions; $(t, \frac{2}{5}t - 2)$, t a parameter

9. Unique solution; $(1, -2)$

11. No solution   **13.** $k = -2$

15. $\begin{aligned} x + \quad y &= \quad 500 \\ 42x + 30y &= 18,600 \end{aligned}$ **17.** $\begin{aligned} x + \quad y &= 100 \\ 2.5x + 3y &= 280 \end{aligned}$

19. $\begin{aligned} x + \quad y &= 1000 \\ .25x + .75y &= 650 \end{aligned}$

21. $\begin{aligned} .06x + .08y + .12z &= 21,600 \\ z &= 2x \\ .12z &= .08y \end{aligned}$

23. $\begin{aligned} 8000x + 12,000y + 16,000z &= 1,000,000 \\ x &= 2y \\ x + \quad y + \quad z &= 100 \end{aligned}$

25. $\begin{aligned} 10x + \quad 6y + \quad 8z &= 100 \\ 10x + 12y + \quad 6z &= 100 \\ 5x + \quad 4y + 12z &= 100 \end{aligned}$

Exercises 2.2, page 93

1. $\begin{bmatrix} 2 & -3 & | & 7 \\ 3 & 1 & | & 4 \end{bmatrix}$

3. $\begin{bmatrix} 0 & -1 & 2 & | & 6 \\ 2 & 2 & -8 & | & 7 \\ 0 & 3 & 4 & | & 0 \end{bmatrix}$

5. $\begin{aligned} 3x + 2y &= -4 \\ x - \quad y &= \quad 5 \end{aligned}$ **7.** $\begin{aligned} x + 3y + 2z &= 4 \\ 2x \quad\quad &= 5 \\ 3x - 3y + 2z &= 6 \end{aligned}$

9. Yes   **11.** No   **13.** Yes   **15.** No   **17.** No

19. $\begin{bmatrix} 1 & 2 & | & 4 \\ 0 & -5 & | & -10 \end{bmatrix}$ **21.** $\begin{bmatrix} 1 & -2 & | & -3 \\ 0 & 16 & | & 20 \end{bmatrix}$

23. $\begin{bmatrix} 1 & 2 & 3 & | & 6 \\ 0 & -1 & -5 & | & -7 \\ 0 & -7 & -7 & | & -14 \end{bmatrix}$

25. $\begin{bmatrix} -6 & -11 & 0 & | & -5 \\ 2 & 4 & 1 & | & 3 \\ 1 & -2 & 0 & | & -10 \end{bmatrix}$

27. $\begin{bmatrix} 3 & 9 & | & 6 \\ 2 & 1 & | & 4 \end{bmatrix} \xrightarrow{\frac{1}{3}R_1} \begin{bmatrix} 1 & 3 & | & 2 \\ 2 & 1 & | & 4 \end{bmatrix}$

$\xrightarrow{R_2 - 2R_1} \begin{bmatrix} 1 & 3 & | & 2 \\ 0 & -5 & | & 0 \end{bmatrix} \xrightarrow{-\frac{1}{5}R_2}$

$\begin{bmatrix} 1 & 3 & | & 2 \\ 0 & 1 & | & 0 \end{bmatrix} \xrightarrow{R_1 - 3R_2} \begin{bmatrix} 1 & 0 & | & 2 \\ 0 & 1 & | & 0 \end{bmatrix}$

29. $\begin{bmatrix} 1 & 3 & 1 & | & 3 \\ 3 & 8 & 3 & | & 7 \\ 2 & -3 & 1 & | & -10 \end{bmatrix} \xrightarrow[R_3 - 2R_1]{R_2 - 3R_1}$

$\begin{bmatrix} 1 & 3 & 1 & | & 3 \\ 0 & -1 & 0 & | & -2 \\ 0 & -9 & -1 & | & -16 \end{bmatrix} \xrightarrow{-R_2}$

$\begin{bmatrix} 1 & 3 & 1 & | & 3 \\ 0 & 1 & 0 & | & 2 \\ 0 & -9 & -1 & | & -16 \end{bmatrix} \xrightarrow[R_3 + 9R_2]{R_1 - 3R_2}$

$\begin{bmatrix} 1 & 0 & 1 & | & -3 \\ 0 & 1 & 0 & | & 2 \\ 0 & 0 & -1 & | & 2 \end{bmatrix} \xrightarrow[-R_3]{R_1 + R_3}$

$\begin{bmatrix} 1 & 0 & 0 & | & -1 \\ 0 & 1 & 0 & | & 2 \\ 0 & 0 & 1 & | & -2 \end{bmatrix}$

31. $(2, 0)$

33. $(-1, 2, -2)$

35. $(4, -2)$

37. $(-1, 2)$

39. $(\frac{7}{5}, -\frac{1}{9}, -\frac{2}{3})$

41. $(19, -7, -15)$

43. $(3, 0, 2)$

45. $(1, -2, 1)$

47. $(-20, -28, 13)$

49. $(4, -1, 3)$

51. 300 acres of corn, 200 acres of wheat

53. In 100 lb of blended coffee, use 40 lb of the $2.50/lb coffee and 60 lb of the $3/lb coffee.

55. 200 children and 800 adults

57. $40,000 in a savings account, $120,000 in mutual funds, $80,000 in money-market certificates

59. 60 compact, 30 intermediate, and 10 full-size cars

61. 4 oz of food I, 2 oz of food II, 6 oz of food III

63. 240 front orchestra seats, 560 rear orchestra seats, 200 front balcony seats

Using Technology Exercises 2.2, page 98

1. $x_1 = 3, x_2 = 1, x_3 = -1$, and $x_4 = 2$

3. $x_1 = 5, x_2 = 4, x_3 = -3$, and $x_4 = -4$

5. $x_1 = 1, x_2 = -1, x_3 = -2, x_4 = 0$, and $x_5 = 3$

Exercises 2.3, page 108

1. **a.** One solution   **b.** $(3, -1, 2)$

3. **a.** One solution   **b.** $(2, 4)$

5. **a.** Infinitely many solutions
   b. $(4 - t, -2, t)$; t, a parameter

7. **a.** No solution

9. **a.** Infinitely many solutions
   b. $(2, -1, 2 - t, t)$; t, a parameter

11. **a.** Infinitely many solutions
    b. $(2 - 3s, 1 + s, s, t)$; s, t, parameters

13. $(2, 1)$ 15. No solution   17. $(1, -1)$

19. $(2 + 2t, t)$; t, a parameter

21. $(\frac{4}{3} - \frac{2}{3}t, t)$; t, a parameter

23. $(-2 + \frac{1}{2}s - \frac{1}{2}t, s, t)$; s, t, parameters

25. $(-1, \frac{17}{7}, \frac{23}{7})$

27. $(1 - \frac{1}{4}s + \frac{1}{4}t, s, t)$; s, t, parameters

29. No solution   31. $(2, -1, 4)$

33. $x = 20 + z, y = 40 - 2z$; 25 compact cars, 30 mid-sized cars, and 5 full-sized cars; 30 compact cars, 20 mid-sized cars, and 10 full-sized cars

37. **a.**
$$\begin{aligned} x_1 - x_2 &= 200 \\ x_1 \qquad\qquad - x_5 &= 100 \\ -x_2 + x_3 \qquad\quad + x_6 &= 600 \\ -x_3 + x_4 \qquad &= 200 \\ x_4 - x_5 + x_6 &= 700 \end{aligned}$$

b. $x_1 = s + 100$
$x_2 = s - 100$
$x_3 = s - t + 500$
$x_4 = s - t + 700$
$x_5 = s$
$x_6 = t$
$(250, 50, 600, 800, 150, 50)$
$(300, 100, 600, 800, 200, 100)$
c. $(300, 100, 700, 900, 200, 0)$

39. $k = 6$; $(-2, 2)$

Using Technology Exercises 2.3, page 111

1. $(1 + t, 2 + t, t)$; t, a parameter

3. $(-\frac{17}{7} + \frac{6}{7}t, 3 - t, -\frac{18}{7} + \frac{4}{7}t, t)$; t, a parameter

5. No solution

Exercises 2.4, page 120

1. 4×4; 4×3; 1×5; 4×1 3. 2; 3; 8

5. D; $D^T = [1 \quad 3 \quad -2 \quad 0]$

7. 3×2; 3×2; 3×3; 3×3

9. $\begin{bmatrix} 1 & 6 \\ 6 & -1 \\ 2 & 2 \end{bmatrix}$ 11. $\begin{bmatrix} 1 & 1 & -4 \\ -1 & -8 & 1 \\ 6 & 3 & 1 \end{bmatrix}$

13. $\begin{bmatrix} 3 & 5 & 9 \\ 4 & 10 & 13 \end{bmatrix}$ 15. $\begin{bmatrix} 3 & -4 & -16 \\ 17 & -4 & 16 \end{bmatrix}$

17. $\begin{bmatrix} -1.9 & 3.0 & -0.6 \\ 6.0 & 9.6 & 1.2 \end{bmatrix}$

19. $\begin{bmatrix} \frac{7}{2} & 3 & -1 & \frac{10}{3} \\ -\frac{19}{6} & \frac{2}{3} & -\frac{17}{2} & \frac{23}{3} \\ \frac{29}{3} & \frac{17}{6} & -1 & -2 \end{bmatrix}$

21. $x = \frac{5}{2}, y = 7, z = 2$, and $u = 3$

23. $x = 2, y = 2, z = -\frac{7}{3}$, and $u = 15$

31. $\begin{bmatrix} 3 \\ 2 \\ -1 \\ 5 \end{bmatrix}$ **33.** $\begin{bmatrix} 1 & 3 & 0 \\ -1 & 4 & 1 \\ 2 & 2 & 0 \end{bmatrix}$

35. $\begin{bmatrix} 220 & 215 & 210 & 205 \\ 220 & 210 & 200 & 195 \\ 215 & 205 & 195 & 190 \end{bmatrix}$

37. a. $D = \begin{bmatrix} 2960 & 1510 & 1150 \\ 1100 & 550 & 490 \\ 1230 & 590 & 470 \end{bmatrix}$

b. $E = \begin{bmatrix} 3256 & 1661 & 1265 \\ 1210 & 605 & 539 \\ 1353 & 649 & 517 \end{bmatrix}$

Using Technology Exercises 2.4, page 123

1. $\begin{bmatrix} 15 & 38.75 & -67.5 & 33.75 \\ 51.25 & 40 & 52.5 & -38.75 \\ 21.25 & 35 & -65 & 105 \end{bmatrix}$

3. $\begin{bmatrix} -5 & 6.3 & -6.8 & 3.9 \\ 1 & 0.5 & 5.4 & -4.8 \\ 0.5 & 4.2 & -3.5 & 5.6 \end{bmatrix}$

5. $\begin{bmatrix} 16.44 & -3.65 & -3.66 & 0.63 \\ 12.77 & 10.64 & 2.58 & 0.05 \\ 5.09 & 0.28 & -10.84 & 17.64 \end{bmatrix}$

7. $\begin{bmatrix} 7.4 & 7.2 & 2.9 \\ -0.1 & 5.9 & 1.4 \\ -4 & 3 & -6.9 \\ 1.5 & -1.4 & 11.2 \end{bmatrix}$

Exercises 2.5, page 133

1. 2×5; not defined   **3.** 1×1; 7×7

5. $n = s$ and $m = t$ **7.** $\begin{bmatrix} -1 \\ 3 \end{bmatrix}$ **9.** $\begin{bmatrix} 9 \\ -10 \end{bmatrix}$

11. $\begin{bmatrix} 4 & -2 \\ 9 & 13 \end{bmatrix}$ **13.** $\begin{bmatrix} 2 & 9 \\ 5 & 16 \end{bmatrix}$ **15.** $\begin{bmatrix} 0.57 & 1.93 \\ 0.64 & 1.76 \end{bmatrix}$

17. $\begin{bmatrix} 6 & -3 & 0 \\ -2 & 1 & -8 \\ 4 & -4 & 9 \end{bmatrix}$ **19.** $\begin{bmatrix} 5 & 1 & -3 \\ 1 & 7 & -3 \end{bmatrix}$

21. $\begin{bmatrix} -4 & -20 & 4 \\ 4 & 12 & 0 \\ 12 & 32 & 20 \end{bmatrix}$ **23.** $\begin{bmatrix} 4 & -3 & 2 \\ 7 & 1 & -5 \end{bmatrix}$

27. $AB = \begin{bmatrix} 10 & 7 \\ 22 & 15 \end{bmatrix}$, $BA = \begin{bmatrix} 5 & 8 \\ 13 & 20 \end{bmatrix}$

31. $A = \begin{bmatrix} -2 & -1 \\ 5 & 2 \end{bmatrix}$ **33. a.** $A^T = \begin{bmatrix} 2 & 5 \\ 4 & -6 \end{bmatrix}$

35. $AX = B$, where $A = \begin{bmatrix} 2 & -3 \\ 3 & -4 \end{bmatrix}$, $X = \begin{bmatrix} x \\ y \end{bmatrix}$,

and $B = \begin{bmatrix} 7 \\ 8 \end{bmatrix}$

37. $AX = B$, where $A = \begin{bmatrix} 2 & -3 & 4 \\ 0 & 2 & -3 \\ 1 & -1 & 2 \end{bmatrix}$, $X = \begin{bmatrix} x \\ y \\ z \end{bmatrix}$,

and $B = \begin{bmatrix} 6 \\ 7 \\ 4 \end{bmatrix}$

39. $AX = B$, where $A = \begin{bmatrix} -1 & 1 & 1 \\ 2 & -1 & -1 \\ -3 & 2 & 4 \end{bmatrix}$, $X = \begin{bmatrix} x_1 \\ x_2 \\ x_3 \end{bmatrix}$,

and $B = \begin{bmatrix} 0 \\ 2 \\ 4 \end{bmatrix}$

41. a. $AB = \begin{bmatrix} 51,400 \\ 54,200 \end{bmatrix}$

b. The first entry shows that William's total stock-holdings are $51,400; the second shows that Michael's stockholdings are $54,200.

43. a. $B = \begin{bmatrix} 20,000 \\ 22,000 \\ 25,000 \\ 30,000 \end{bmatrix}$

b. $7,160,000 in New York, $2,860,000 in Connecticut, and $1,430,000 in Massachusetts; the total profit is $11,450,000.

45. $AB = \begin{bmatrix} 1575 & 1590 & 1560 & 975 \\ 410 & 405 & 415 & 270 \\ 215 & 205 & 225 & 155 \end{bmatrix}$

47. a. $AC = \begin{bmatrix} 348,400 \\ 273,200 \\ 632,400 \end{bmatrix}$ **b.** $AD = \begin{bmatrix} 478,200 \\ 369,800 \\ 877,400 \end{bmatrix}$

c. $BC = \begin{bmatrix} 233,400 \\ 239,000 \\ 517,600 \end{bmatrix}$ **d.** $BD = \begin{bmatrix} 320,100 \\ 324,500 \\ 718,200 \end{bmatrix}$

e. $(A+B)C = \begin{bmatrix} 581,800 \\ 512,200 \\ 1,150,000 \end{bmatrix}$

f. $(A+B)D = \begin{bmatrix} 798,300 \\ 694,300 \\ 1,595,600 \end{bmatrix}$

g. $A(D-C) = \begin{bmatrix} 129,800 \\ 96,600 \\ 245,000 \end{bmatrix}$

h. $B(D-C) = \begin{bmatrix} 86,700 \\ 85,500 \\ 200,600 \end{bmatrix}$

i. $(A+B)(D-C) = \begin{bmatrix} 216,500 \\ 182,100 \\ 445,600 \end{bmatrix}$

Using Technology Exercises 2.5, page 137

1. $\begin{bmatrix} 18.66 & 15.2 & -12 \\ 24.48 & 41.88 & 89.82 \\ 15.39 & 7.16 & -1.25 \end{bmatrix}$

3. $\begin{bmatrix} 20.09 & 20.61 & -1.3 \\ 44.42 & 71.6 & 64.89 \\ 20.97 & 7.17 & -60.65 \end{bmatrix}$

5. $\begin{bmatrix} 32.89 & 13.63 & -57.17 \\ -12.85 & -8.37 & 256.92 \\ 13.48 & 14.29 & 181.64 \end{bmatrix}$

7. $\begin{bmatrix} 18.66 & 24.48 & 15.39 \\ 15.2 & 41.88 & 7.16 \\ -12 & 89.82 & -1.25 \end{bmatrix}$

9. $\begin{bmatrix} 87 & 68 & 110 & 82 \\ 119 & 176 & 221 & 143 \\ 51 & 128 & 142 & 94 \\ 28 & 174 & 174 & 112 \end{bmatrix}$;

$\begin{bmatrix} 113 & 117 & 72 & 101 & 90 \\ 72 & 85 & 36 & 72 & 76 \\ 81 & 69 & 76 & 87 & 30 \\ 133 & 157 & 56 & 121 & 146 \\ 154 & 157 & 94 & 127 & 122 \end{bmatrix}$

11. $\begin{bmatrix} 170 & 18.1 & 133.1 & -106.3 & 341.3 \\ 349 & 226.5 & 324.1 & 164 & 506.4 \\ 245.2 & 157.7 & 231.5 & 125.5 & 312.9 \\ 310 & 245.2 & 291 & 274.3 & 354.2 \end{bmatrix}$

Exercises 2.6, page 149

5. $\begin{bmatrix} 3 & -5 \\ -1 & 2 \end{bmatrix}$ **7.** Does not exist

9. $\begin{bmatrix} 2 & -11 & -3 \\ 1 & -6 & -2 \\ 0 & -1 & 0 \end{bmatrix}$ **11.** Does not exist

13. $\begin{bmatrix} -\frac{13}{10} & \frac{7}{5} & \frac{1}{2} \\ \frac{2}{5} & -\frac{1}{5} & 0 \\ -\frac{7}{10} & \frac{3}{5} & \frac{1}{2} \end{bmatrix}$

15. $\begin{bmatrix} 3 & 4 & -6 & 1 \\ -2 & -3 & 5 & -1 \\ -4 & -4 & 7 & -1 \\ -4 & -5 & 8 & -1 \end{bmatrix}$

17. a. $A = \begin{bmatrix} 2 & 5 \\ 1 & 3 \end{bmatrix}$, $X = \begin{bmatrix} x \\ y \end{bmatrix}$, $B = \begin{bmatrix} 3 \\ 2 \end{bmatrix}$

b. $x = -1, y = 1$

19. a. $A = \begin{bmatrix} 2 & -3 & -4 \\ 0 & 0 & -1 \\ 1 & -2 & 1 \end{bmatrix}$, $X = \begin{bmatrix} x \\ y \\ z \end{bmatrix}$, $B = \begin{bmatrix} 4 \\ 3 \\ -8 \end{bmatrix}$

b. $x = -1, y = 2,$ and $z = -3$

21. a. $A = \begin{bmatrix} 1 & 4 & -1 \\ 2 & 3 & -2 \\ -1 & 2 & 3 \end{bmatrix}$, $X = \begin{bmatrix} x \\ y \\ z \end{bmatrix}$, $B = \begin{bmatrix} 3 \\ 1 \\ 7 \end{bmatrix}$

b. $x = 1, y = 1,$ and $z = 2$

23. **a.** $A = \begin{bmatrix} 1 & 1 & -1 & 1 \\ 2 & 1 & 1 & 0 \\ 2 & 1 & 0 & 1 \\ 2 & -1 & -1 & 3 \end{bmatrix}$, $X = \begin{bmatrix} x_1 \\ x_2 \\ x_3 \\ x_4 \end{bmatrix}$,

$B = \begin{bmatrix} 6 \\ 4 \\ 7 \\ 9 \end{bmatrix}$

b. $x_1 = 1$, $x_2 = 2$, $x_3 = 0$, and $x_4 = 3$

25. **a.** $x = 24/5$, $y = 23/5$ **b.** $x = 2/5$, $y = 9/5$

27. **a.** $x = -1$, $y = 3$, $z = 2$
    b. $x = 1$, $y = 8$, $z = -12$

29. **a.** $x = -2/17$, $y = -10/17$, $z = -60/17$
    b. $x = 1$, $y = 0$, $z = -5$

31. **a.** $x_1 = 1$, $x_2 = -4$, $x_3 = 5$, $x_4 = -1$
    b. $x_1 = 12$, $x_2 = -24$, $x_3 = 21$, $x_4 = -7$

33. **a.** $A^{-1} = \begin{bmatrix} -\frac{5}{2} & -\frac{3}{2} \\ 2 & 1 \end{bmatrix}$

35. **a.** $ABC = \begin{bmatrix} 4 & 10 \\ 2 & 3 \end{bmatrix}$, $A^{-1} = \begin{bmatrix} 3 & -5 \\ 1 & -2 \end{bmatrix}$,

$B^{-1} = \begin{bmatrix} 1 & -3 \\ -1 & 4 \end{bmatrix}$, $C^{-1} = \begin{bmatrix} \frac{1}{8} & -\frac{3}{8} \\ \frac{1}{4} & \frac{1}{4} \end{bmatrix}$

37. **a.** 3214; 3929   **b.** 4286; 3571   **c.** 3929; 5357

39. **a.** \$6 million to Organization I, \$10 million to Organization II, and \$8 million to Organization III
    b. \$8 million to Organization I, \$6 million to Organization II, and \$5 million to Organization III

Using Technology Exercises 2.6, page 153

1. $\begin{bmatrix} 0.36 & 0.04 & -0.36 \\ 0.06 & 0.05 & 0.20 \\ -0.19 & 0.10 & 0.09 \end{bmatrix}$

3. $\begin{bmatrix} 0.01 & -0.09 & 0.31 & -0.11 \\ -0.25 & 0.58 & -0.15 & -0.02 \\ 0.86 & -0.42 & 0.07 & -0.37 \\ -0.27 & 0.01 & -0.05 & 0.31 \end{bmatrix}$

5. $\begin{bmatrix} 0.30 & 0.85 & -0.10 & -0.77 & -0.11 \\ -0.21 & 0.10 & 0.01 & -0.26 & 0.21 \\ 0.03 & -0.16 & 0.12 & -0.01 & 0.03 \\ -0.14 & -0.46 & 0.13 & 0.71 & -0.05 \\ 0.10 & -0.05 & -0.10 & -0.03 & 0.11 \end{bmatrix}$

Exercises 2.7, page 163

1. **a.** \$10 million   **b.** \$160 million
   c. Agricultural; manufacturing and transportation

3. $x = 23.75$ and $y = 21.25$

5. $x = 42.86$ and $y = 57.14$

9. **a.** \$318.2 million worth of agricultural goods and \$336.4 million worth of manufactured products
   b. \$198.2 million worth of agricultural products and \$196.4 million worth of manufactured goods

11. **a.** \$443.75 million, \$381.25 million, and \$281.25 million worth of agricultural products, manufactured goods, and transportation, respectively
    b. \$243.75 million, \$281.25 million, and \$221.25 million worth of agricultural products, manufactured goods, and transportation, respectively

13. \$45 million and \$75 million

15. \$34.4 million, \$33 million, and \$21.6 million

Using Technology Exercises 2.7, page 165

1. The final outputs of the first, second, third, and fourth industries are 602.62, 502.30, 572.57, and 523.46 units, respectively.

3. The final outputs of the first, second, third, and fourth industries are 143.06, 132.98, 188.59, and 125.53 units, respectively.

Chapter 2 Review Exercises, page 169

1. $\begin{bmatrix} 2 & 2 \\ -1 & 4 \\ 3 & 3 \end{bmatrix}$ 3. $[-6 \quad -2]$

5. $x = 2$, $y = 3$, $z = 1$, and $w = 3$

7. $a = 3$, $b = 4$, $c = -2$, $d = 2$, and $e = -3$

9. $\begin{bmatrix} 8 & 9 & 11 \\ -10 & -1 & 3 \\ 11 & 12 & 10 \end{bmatrix}$ 11. $\begin{bmatrix} 6 & 18 & 6 \\ -12 & 6 & 18 \\ 24 & 0 & 12 \end{bmatrix}$

13. $\begin{bmatrix} -11 & -16 & -15 \\ -4 & -2 & -10 \\ -6 & 14 & 2 \end{bmatrix}$ **15.** $\begin{bmatrix} -3 & 17 & 8 \\ -2 & 56 & 27 \\ 74 & 78 & 116 \end{bmatrix}$

17. $x = 1, y = -1$ **19.** $x = 1, y = 2,$ and $z = 3$

21. No solution

23. $x = 1, y = 0,$ and $z = 1$

25. $\begin{bmatrix} \frac{2}{5} & -\frac{1}{5} \\ -\frac{1}{5} & \frac{3}{5} \end{bmatrix}$ **27.** $\begin{bmatrix} -1 & 2 \\ 1 & -\frac{3}{2} \end{bmatrix}$

29. $\begin{bmatrix} \frac{5}{4} & \frac{1}{4} & -\frac{7}{4} \\ -\frac{1}{4} & -\frac{1}{4} & \frac{3}{4} \\ -\frac{3}{4} & \frac{1}{4} & \frac{5}{4} \end{bmatrix}$ **31.** $\begin{bmatrix} -\frac{1}{5} & \frac{2}{5} & 0 \\ \frac{2}{3} & -\frac{1}{3} & \frac{1}{3} \\ -\frac{1}{30} & \frac{1}{15} & -\frac{1}{6} \end{bmatrix}$

33. $\begin{bmatrix} \frac{3}{2} & 1 \\ -\frac{7}{2} & -1 \end{bmatrix}$ **35.** $\begin{bmatrix} \frac{2}{5} & -\frac{3}{5} \\ \frac{1}{5} & \frac{1}{5} \end{bmatrix}$

37. $A^{-1} = \begin{bmatrix} \frac{2}{7} & \frac{3}{7} \\ \frac{1}{7} & -\frac{2}{7} \end{bmatrix}; x = -1, y = -2$

39. $A^{-1} = \begin{bmatrix} 1 & -\frac{2}{5} & \frac{4}{5} \\ -1 & 1 & -1 \\ -\frac{1}{2} & \frac{3}{5} & -\frac{7}{10} \end{bmatrix}; x = 1, y = 2,$ and $z = 4$

41. $5180, $4540, and $5550

43. 30 of each type

CHAPTER 3

Exercises 3.1, page 180

1.
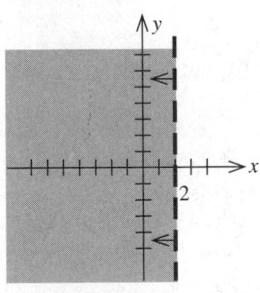

3.
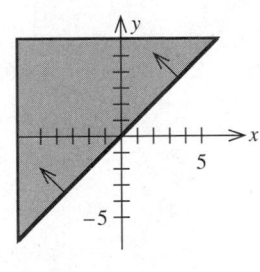

5.
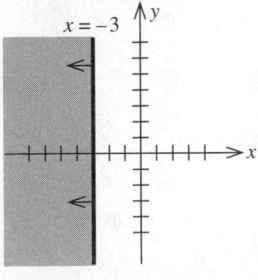

7.
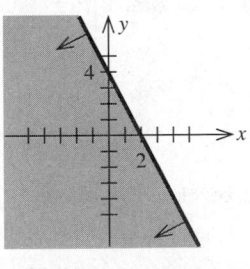

9.
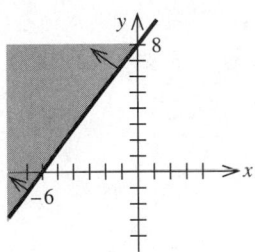

11. $x \geq 1,\quad x \leq 5,\quad y \geq 2,\quad$ and $\quad y \leq 4$

13. $2x - y \geq 2,\quad 5x + 7y \geq 35,\quad$ and $\quad x \leq 4$

15. $x - y \geq -10,\quad 7x + 4y \leq 140,\quad$ and $\quad x + 3y \geq 30$

17. $x + y \geq 7,\quad x \geq 2,\quad y \geq 3,\quad$ and $\quad y \leq 7$

19.
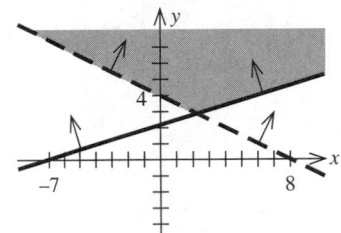
Unbounded

21.
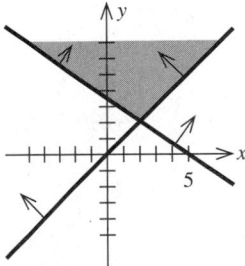
Unbounded

23.
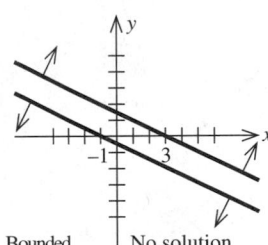
Bounded   No solution

25.
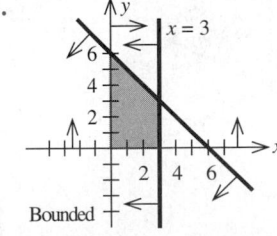
Bounded

27.

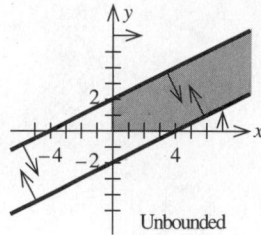

Unbounded

29.

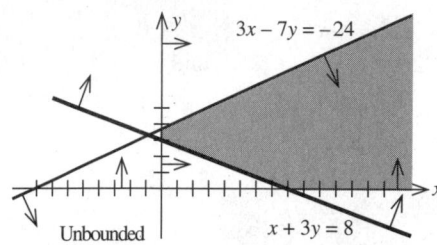

$3x - 7y = -24$

$x + 3y = 8$

Unbounded

31.

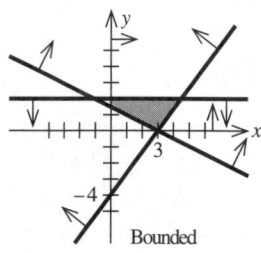

Bounded

33.

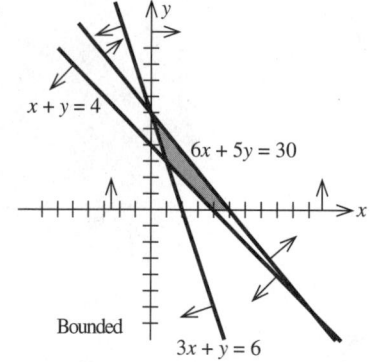

$x + y = 4$

$6x + 5y = 30$

$3x + y = 6$

Bounded

35.

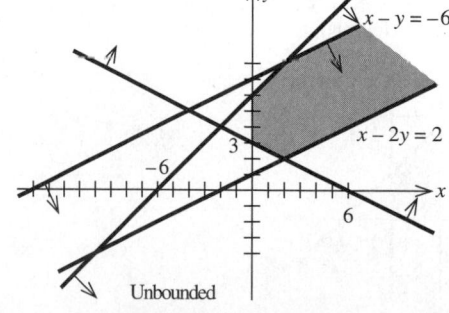

$x - y = -6$

$x - 2y = 2$

Unbounded

Exercises 3.2, page 188

1. Maximize $P = 3x + 4y$ subject to
$$6x + 9y \leq 300$$
$$5x + 4y \leq 180$$
$$x \geq 0, y \geq 0$$

3. Maximize $P = 2x + (3/2)y$ subject to
$$3x + 4y \leq 1000$$
$$6x + 3y \leq 1200$$
$$x \geq 150, y \geq 0$$

5. Minimize $C = 2x + 5y$ subject to
$$30x + 25y \geq 400$$
$$x + 0.5y \geq 10$$
$$2x + 5y \geq 40$$
$$x \geq 0, y \geq 0$$

7. Minimize $C = 32{,}000 - x - 3y$ subject to
$$x + y \leq 6000$$
$$x + y \geq 2000$$
$$x \leq 3000$$
$$y \leq 4000$$
$$x \geq 0, y \geq 0$$

9. Maximize $P = 18x + 12y + 15z$ subject to
$$2x + y + 2z \leq 900$$
$$3x + y + 2z \leq 1080$$
$$2x + 2y + z \leq 840$$
$$x \geq 0, y \geq 0, z \geq 0$$

11. Maximize $P = 26x + 28y + 24z$ subject to
$$\tfrac{5}{4}x + \tfrac{3}{2}y + \tfrac{3}{2}z \leq 310$$
$$x + y + \tfrac{3}{4}z \leq 205$$
$$x + y + \tfrac{1}{2}z \leq 190$$
$$x \geq 0, y \geq 0, z \geq 0$$

13. Minimize $C = 60x_1 + 60x_2 + 80x_3 + 80x_4$
$$+ 70x_5 + 50x_6 \text{ subject to}$$
$$x_1 + x_2 + x_3 \leq 300$$
$$x_4 + x_5 + x_6 \leq 250$$
$$x_1 + x_4 \geq 200$$
$$x_2 + x_5 \geq 150$$
$$x_3 + x_6 \geq 200$$
$$x_1 \geq 0, x_2 \geq 0, \ldots, x_6 \geq 0$$

15. Maximize $P = 180x + 200y + 300z$ subject to
$$2.5x + 3y + 4z \leq 70$$
$$x \leq 9$$
$$y \leq 12$$
$$z \leq 6$$
$$x \geq 0, y \geq 0, z \geq 0$$

Exercises 3.3, page 201

1. Max: 35; min: 5   3. No max. value; min: 27

5. Max: 44; min: 15   7. $x = 0, y = 6, P = 18$

9. $x = 14, y = 3, C = 58$

11. Any point (x, y) lying on the line segment joining $(5, 20)$ to $(12, 6)$, $C = 90$

13. $x = 3, y = 3, C = 75$

15. $x = 15, y = 17.5, P = 115$

17. $x = 10, y = 38, P = 134$

19. Max: $x = 6, y = 33/2, P = 258$
    Min: $x = 15, y = 3, P = 186$

21. Max: $x = 5, y = 15, P = 70$
    Min: $x = 0, y = 5, P = 20$

23. No model A, 2000 model B; $P = \$80,000$

25. 120 model A, 160 model B; $P = \$480$

27. $16 million in homeowner loans, $4 million in auto loans; $P = \$2.08$ million

29. 65 acres of crop A, 80 acres of crop B; $P = \$25,750$

31. 2000 from I to A, 4000 from I to B, 1000 from II to A, 0 from II to B; $C = \$18,000$

33. $22,500 in growth stocks and $7500 in speculative stocks; maximum return: $5250

35. 750 urban and 750 suburban families; $P = \$2737.50$

37. **a.** True   **b.** True

Chapter 3 Review Exercises, page 208

1. Max: 18. Any point (x, y) lying on the line segment joining $(0, 6)$ to $(3, 4)$; min: 0

3. $x = 0, y = 4, P = 20$ 5. $x = 0, y = 0, C = 0$

7. $x = 3, y = 10, P = 29$

9. $x = 20, y = 0, C = 40$

11. Max: $x = 22, y = 0, Q = 22$
    Min: $x = 3, y = 5/2, Q = 11/2$

13. $40,000 in each company; $P = \$13,600$

15. 93 model A, 180 model B

CHAPTER 4

Exercises 4.1, page 227

1. In final form; $x = \frac{30}{7}, y = \frac{20}{7}, u = 0, v = 0, P = \frac{220}{7}$

3. Not in final form; pivot element is $\frac{1}{2}$, lying in the first row, second column

5. In final form: $x = \frac{1}{3}, y = 0, z = \frac{13}{3}, u = 0, v = 6, w = 0, P = 17$

7. Not in final form; pivot element is 1, lying in the third row, second column

9. In final form; $x = 30, y = 0, z = 0, u = 10, v = 0, P = 60$ and $x = 0, y = 30, z = 0, u = 10, v = 0, P = 60$

11. $x = 6, y = 3, u = 0, v = 0, P = 96$

13. $x = 6, y = 6, u = 0, v = 0, w = 0, P = 60$

15. $x = 0, y = 4, z = 4, u = 0, v = 0, P = 36$

17. $x = 0, y = 3, z = 0, u = 90, v = 0, w = 75, P = 12$

19. $x = 15, y = 3, z = 0, u = 2, v = 0, w = 0, P = 78$

21. $x = \frac{5}{4}, y = \frac{15}{2}, z = 0, u = 0, v = \frac{15}{4}, w = 0, P = 90$

23. $x = \frac{15}{7}, y = 0, z = \frac{39}{7}, s = 0, t = \frac{1}{7}, u = 0, v = \frac{2}{7}, P = 36\frac{3}{7}$

27. No model A, 2000 model B; $P = \$80,000$

29. 65 acres of crop A, 80 acres of crop B; $P = \$25,750$

31. 22 minutes of morning advertising time, 3 minutes of evening advertising time; maximum exposure: 6,200,000 viewers

33. 80 units of model A, 80 units of model B, and 60 units of model C; maximum profit: $5760

35. 9000 bottles of formula I, 7833 bottles of formula II, 6000 bottles of formula III; maximum profit: $4986.67

Using Technology Exercises 4.1, page 233

1. $x = 1.2, y = 0, z = 1.6, w = 0$, and $P = 8.8$

3. $x = 1.6, y = 0, z = 0, w = 3.6$, and $P = 12.4$

Exercises 4.2, page 246

1. $x = 4, y = 0, C = -8$

3. $x = 4, y = 3, C = -18$

5. $x = 0, y = 13, z = 18, w = 14, C = -111$

7. $x = \frac{5}{4}, y = \frac{1}{4}, u = 2, v = 3$, and $C = P = 13$

9. $x = 5, y = 10, z = 0, u = 1, v = 2$, and $C = P = 80$

11. Maximize $P = 4u + 6v$ subject to
$$u + 3v \le 2$$
$$2u + 2v \le 5; \quad x = 4, y = 0, C = 8$$
$$u \ge 0, v \ge 0$$

13. Maximize $P = 60u + 40v + 30w$ subject to
$$6u + 2v + w \leq 6$$
$$u + v + w \leq 4; \quad x = 10, y = 20, C = 140$$
$$u \geq 0, v \geq 0, w \geq 0$$

15. Maximize $P = 10u + 20v$ subject to
$$20u + v \leq 200$$
$$10u + v \leq 150; \quad x = 0, y = 0, z = 10, C = 1200$$
$$u + 2v \leq 120$$
$$u \geq 0, v \geq 0$$

17. Maximize $P = 10u + 24v + 16w$ subject to
$$u + 2v + w \leq 6$$
$$2u + v + w \leq 8; \quad x = 8, y = 0, z = 8, C = 80$$
$$2u + v + w \leq 4$$
$$u \geq 0, v \geq 0, w \geq 0$$

19. Maximize $P = 6u + 2v + 4w$ subject to
$$2u + 6v \qquad \leq 30$$
$$4u \qquad + 6w \leq 12; \quad x = \tfrac{1}{3}, y = \tfrac{4}{3}, z = 0, C = 26$$
$$3u + v + 2w \leq 20$$
$$u \geq 0, v \geq 0, w \geq 0$$

21. Loc. I: 500 to warehouse A, 200 to warehouse B; Loc. II: 200 to warehouse B, 400 to warehouse C; shipping costs: $20,800

23. 8 oz. orange juice; 6 oz. pink grapefruit juice; 178 calories

Using Technology Exercises 4.2, page 252

1. $x = \tfrac{4}{3}, y = \tfrac{10}{3}, z = 0, C = \tfrac{14}{3}$

3. $x = 0.9524, y = 4.2857, z = 0, C = 6.0952$

Exercises 4.3, page 265

1. Maximize $C = -P = -2x + 3y$
subject to $-3x - 5y \leq -20$
$$3x + y \leq 16$$
$$-2x + y \leq 1$$
$$x \geq 0, y \geq 0$$

3. Maximize $P = -C = -5x - 10y - z$
subject to $-2x - y - z \leq -4$
$$-x - 2y - 2z \leq -2$$
$$2x + 4y + 3z \leq 12$$
$$x \geq 0, y \geq 0, z \geq 0$$

5. $x = 5, y = 2, P = 9$

7. $x = 4, y = 0, C = -8$

9. $x = 4, y = \tfrac{2}{3}, P = \tfrac{20}{3}$

11. $x = 3, y = 2, P = 7$

13. $x = 24, y = 0, z = 0, P = 120$

15. $x = 0, y = 17, z = 1, C = -33$

17. $x = \tfrac{46}{7}, y = 0, z = \tfrac{50}{7}, P = \tfrac{142}{7}$

19. $x = 0, y = 0, z = 10, P = 30$

21. 80 acres of crop A, 68 acres of crop B; $P = \$25{,}600$

23. $50 million worth of home loans, $10 million worth of commercial-development loans; maximum return: $4.6 million

25. 0 units of product A, 280 units of product B, 280 units of product C; $P = \$7560$

27. 10 oz. food A, 4 oz. food B, 40 mg cholesterol; there are infinitely many solutions

Chapter 4 Review Exercises, page 270

1. $x = 3, y = 4, u = 0, v = 0, P = 25$

3. $x = 56/5, y = 2/5, z = 0, u = 0, v = 0, P = 23\tfrac{3}{8}$

5. $x = 3/2, y = 1, C = 13/2$

7. $x = 3/4, y = 0, z = 7/4, C = 60$

9. $x = 45, y = 0, P = 135$

11. $x = 5, y = 2, P = 16$

13. 30 units product B; $P = \$180$

15. $50,000 in stocks, $100,000 in bonds, $50,000 in money-market funds; maximum return: $21,500/year

CHAPTER 5

Exercises 5.1, page 283

1. $80; $580   **3.** $836   **5.** $1000   **7.** 146 days

9. 10% per year   **11.** $1718.19   **13.** $4974.47

15. $27,566.93   **17.** $261,751.04   **19.** $214,986.69

21. $10\tfrac{1}{4}$% per year   **23.** 8.3% per year   **25.** $29,277.61

27. $30,255.95   **29.** 24% per year   **31.** $558.34

33. $182,326   **35.** $3.8 million   **37.** $26,267.49

39. a. $34,626.88   **b.** $33,886.16   **c.** $33,506.76

41. $23,329.48   **43.** 4.2%   **45.** 8.5%

47. a. A family of straight lines with varying slope and P-intercept
b. A family of straight lines emanating from the point $(0, P)$ with varying slope

Using Technology Exercises 5.1, page 287

1. $5872.78   **3.** $475.49   **5.** 8.95% per year

7. 10.20% per year

9. :PROGRAM: PREVAL   **11.** $94,038.74
:Disp "A"
:Input A
:Disp "r"
:Input r
:Disp "t"
:Input t
:Disp "m"
:Input m
:A(1+r/m)^(−m*t)→P
:Disp "PRESENT VALUE IS"
:Disp P

13. $62,244.96

Exercises 5.2, page 295

1. $15,937.42   **3.** $54,759.35   **5.** $37,965.57

7. $28,733.19   **9.** $15,558.61   **11.** $15,011.29

13. $109,658.91   **15.** $455.70   **17.** $44,526.45

19. $9850.12   **21.** $608.54

23. Between $120,141 and $143,927

25. Between $103,875 and $123,593

Using Technology Exercises 5.2, page 297

1. $59,622.15   **3.** $8453.59

5. :PROGRAM: PVAN   **7.** $45,983.53
:Disp "R"
:Input R
:Disp "i"
:Input i
:Disp "N"
:Input N
:(R/i)(1−(1+i)^(−N))→P
:Disp "AMOUNT IS"
:Disp P

9. $18,344.08

Exercises 5.3, page 305

1. $14,902.95   **3.** $444.24   **5.** $622.13

7. $731.79   **9.** $1491.19   **11.** $516.76

13. $172.95   **15.** $16,274.54

17. a. $212.27   **b.** $1316.36; $438.79

19. a. $387.21; $304.35   **b.** $1939.56; $2608.80

21. $1174.02; $27,887.98; $39,639.86; $83,233.30

23. $3135.48   **25.** $1449.74

27. $2090.41; $4280.21   **29.** $33,835.20

Using Technology Exercises 5.3, page 309

1. $3645.40   **3.** $18,443.75

5. :PROGRAM: SINKFD   **7.** $916.26
:Disp "S"
:Input S
:Disp "i"
:Input i
:Disp "N"
:Input N
:S*i/(1+i)^N−1)→R
:Disp "R is"
:Disp R

9. $809.31   **11.** $45,069.31

Exercises 5.4, page 316

1. 30   **3.** −4.5   **5.** −3, 8, 19, 30, 41   **7.** $x + 6y$

9. 795   **11.** 792   **13.** 550

15. a. 275   **b.** −280   **17.** 37 weeks   **19.** $7.90

21. b. $800   **23.** G.P.; 256; 508   **25.** Not a G.P.

27. G.P.; 1/3; $364\frac{1}{3}$ **29.** 3; 0   **31.** 293,866

33. $41,149.12   **35.** Annual raise of 8% per year

37. a. $20,113.57   **b.** $87,537.38   **39.** $25,165.82

41. $39,321.60; $110,678.40

Chapter 5 Review Exercises, page 320

1. a. $7320.50   **b.** $7387.28   **c.** $7422.53
d. $7446.77

3. a. 12%   **b.** 12.36%   **c.** 12.5509%
d. 12.6825%

5. $30,000.29   **7.** $5557.68   **9.** $7861.70

11. $694.49   **13.** $332.73   **15.** 7.179%

17. $2,592,702; $8,612,002   **19.** $15,000

21. $218.64   **23.** $13,026.89

25. a. $965.55   **b.** $227,598   **c.** $42,684

27. $19,573.56   **29.** $205.09

CHAPTER 6

Exercises 6.1, page 332

1. $\{x \mid x$ is a gold medalist in the 1996 Summer Olympic Games$\}$

3. $\{x \mid x$ is an integer greater than 2 and less than 8$\}$

5. $\{2, 3, 4, 5, 6\}$ 7. $\{-2\}$

9. **a.** T   **b.** F   11. **a.** F   **b.** F

13. T   15. **a.** T   **b.** F

17. **a.** and **b.**

19. **a.** $\varnothing, \{1\}, \{2\}, \{1, 2\}$
   b. $\varnothing, \{1\}, \{2\}, \{3\}, \{1, 2\}, \{1, 3\}, \{2, 3\}, \{1, 2, 3\}$
   c. $\varnothing, \{1\}, \{2\}, \{3\}, \{4\}, \{1, 2\}, \{1, 3\}, \{1, 4\}, \{2, 3\}, \{2, 4\},$
   $\{3, 4\}, \{1, 2, 3\}, \{1, 2, 4\}, \{1, 3, 4\}, \{2, 3, 4\}, \{1, 2, 3, 4\}$

21. $\{1, 2, 3, 4, 6, 8, 10\}$

23. $\{$Jill, John, Jack, Susan, Sharon$\}$

25. **a.**

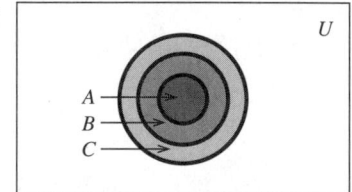

b.

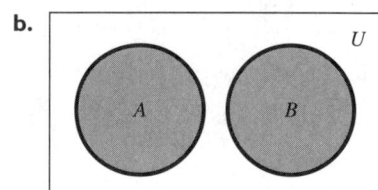

c.

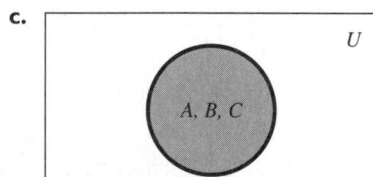

27. **a.**

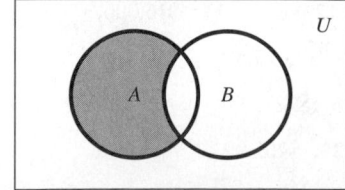

b.

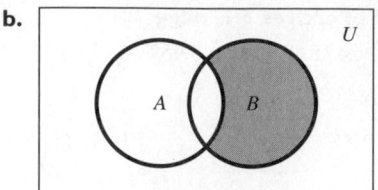

29. **a.**

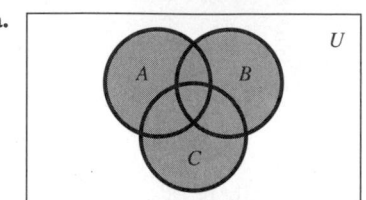

b.

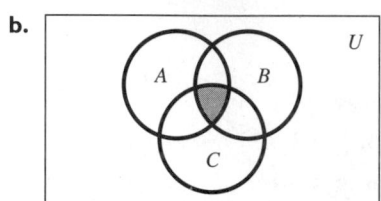

31. **a.**

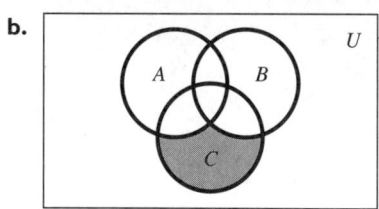

b.

33. **a.** $\{2, 4, 6, 8, 10\}$ **b.** $\{1, 2, 4, 5, 6, 8, 9, 10\}$
   c. U

35. **a.** $C = \{1, 2, 4, 5, 8, 9\}$ **b.** $\varnothing$ **c.** U

37. **a.** Not disjoint   **b.** Disjoint

39. **a.** The set of all employees at the Universal Life Insurance Company who do not drink tea
   b. The set of all employees at the Universal Life Insurance Company who do not drink coffee

41. **a.** The set of all employees at the Universal Life Insurance Company who drink tea but not coffee

b. The set of all employees at the Universal Life Insurance Company who drink coffee but not tea

43. a. The set of all employees in a hospital who are not doctors
b. The set of all employees in a hospital who are not nurses

45. a. The set of all employees in a hospital who are female doctors
b. The set of all employees in a hospital who are both doctors and administrators

47. a. $D \cap F$ **b.** $R \cap F^C \cap L^C$

49. a. B^C **b.** $A \cap B$ **c.** $A \cap B \cap C^C$

51. a. $A \cap B \cap C$; the set of tourists who have taken the underground, a cab, and a bus over a 1-week period in London
b. $A \cap C$; the set of tourists who have taken the underground and a bus over a 1-week period in London
c. B^c; the set of tourists who have not taken a cab over a 1-week period in London

53. a.

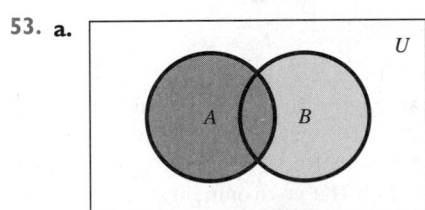

b.

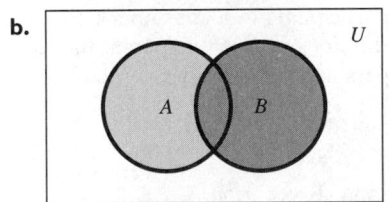

55.

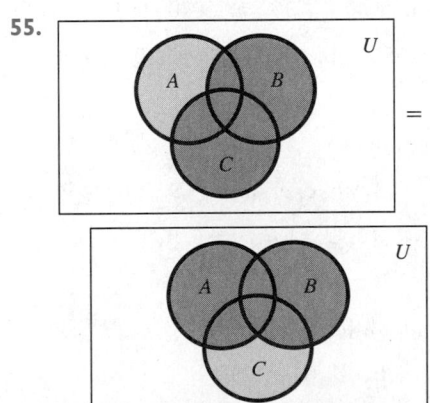

57.

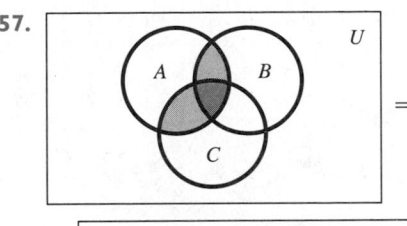

=

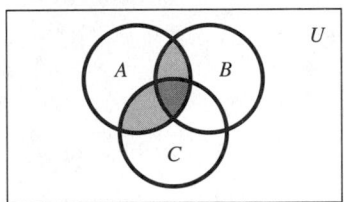

61. a. x, y, v, r, w, u **b.** v, r
63. a. s, t, y **b.** t, z, w, x, s **65.** $A \subset C$

Exercises 6.2, page 339

3. a. 4   **b.** 5   **c.** 7   **d.** 2
7. 3
9. a. 40   **b.** 120   **c.** 60
11. 11   **13.** 2   **15.** 600; 100
17. 61
19. a. 106   **b.** 64   **c.** 38   **d.** 14
21. a. 182   **b.** 118   **c.** 56   **d.** 18
23. a. 13   **b.** 15   **25. a.** 38   **b.** 64
27. 5
29. a. 62   **b.** 33   **c.** 25   **d.** 38
31. a. 108   **b.** 15   **c.** 45   **d.** 12

Exercises 6.3, page 347

1. 12   **3.** 64   **5.** 24
7. 24   **9.** 60   **11.** 5^{50}
13. 30   **15.** 9990
17. a. 36
b.

```
F ── D ── NW
         W
         C
         S
         E
         NE
  ── I ── NW
         W
         C
         S
         E
         NE
  ── R ── NW
         W
         C
         S
         E
         NE
```

19. 1024; 59,049 **21.** 2730

23. 217

Exercises 6.4, page 360

1. 360 **3.** 10 **5.** 120

7. 20 **9.** n **11.** 1

13. 35 **15.** 1 **17.** 84

19. $\dfrac{n(n-1)}{2}$ **21.** $\dfrac{n!}{2}$

23. Permutation **25.** Combination

27. Permutation **29.** Combination

31. $P(4, 4) = 24$ **33.** $P(4, 4) = 24$

35. $P(9, 9) = 362,880$ **37.** $C(12, 3) = 220$

39. 151,200 **41.** $C(12, 3) = 220$

43. $C(100, 3) = 161,700$ **45.** $P(6, 6) = 720$

47. $P(12, 6) = 665,280$

49. a. $P(10, 10) = 3,628,800$
b. $P(3, 3) \cdot P(4, 4) \cdot P(3, 3) \cdot P(3, 3) = 5184$

51. a. $P(20, 20) = 20!$
b. $P(5, 5) \cdot P[(4, 4)]^5 = 5!(4!)^5 = 955,514,880$

53. $P(2, 1) \cdot P(3, 1) = 6$

55. $C(3, 3) \cdot [C(8, 6) + C(8, 7) + C(8, 8)] = 37$

57. a. $C(12, 3) = 220$ **b.** $C(11, 2) = 55$
c. $C(5, 1) \cdot C(7, 2) + C(5, 2) \cdot C(7, 1) + C(5, 3) = 185$

59. $P(7, 3) + C(7, 2) \cdot P(3, 2) = 336$

61. $[C(5, 1) \cdot C(3, 1) \cdot C(6, 2)][C(4, 1) + C(3, 1)] = 1575$

63. $C(10, 8) + C(10, 9) + C(10, 10) = 56$

65. $10 \cdot C(4, 1) = 40$

67. $4 \cdot C(13, 5) - 40 = 5108$

69. $13C(4, 3) \cdot 12C(4, 2) = 3744$

71. $C(6, 2) = 15$

73. $C(12, 6) + C(12, 7) + C(12, 8) + C(12, 9) + C(12, 10) + C(12, 11) + C(12, 12) = 2510$

75. $4! = 24$

Using Technology Exercises 6.4, page 365

1. $1.307674368 \times 10^{12}$ **3.** $2.56094948229 \times 10^{16}$

5. 674,274,182,400 **7.** 133,784,560 **9.** 4,656,960

11. 658,337,004,000

Chapter 6 Review Exercises, page 367

1. {3} **3.** {4, 6, 8, 10} **5.** Yes **7.** Yes

9.

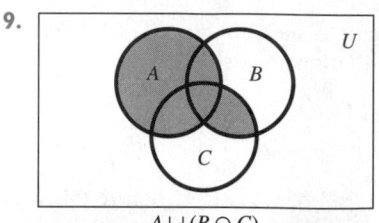

$A \cup (B \cap C)$

11.

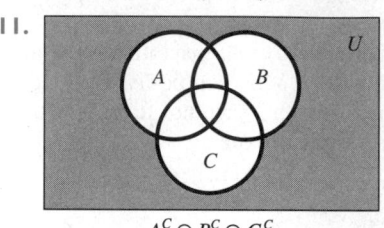

$A^C \cap B^C \cap C^C$

17. The set of all participants in a consumer behavior survey who both avoided buying a product because it is not recyclable and boycotted a company's products because of its record on the environment

19. The set of all participants in a consumer behavior survey who both did not use cloth diapers rather than disposable diapers and voluntarily recycled their garbage

21. 150 **23.** 270 **25.** 70 **27.** 190 **29.** 120

31. None **33.** 720 **35.** 720

37. a. 50,400 **b.** 5040 **39. a.** 5040 **b.** 3600

41. a. $C(15, 4) = 1365$ **b.** $C(15, 4) - C(10, 4) = 1155$

CHAPTER 7

Exercises 7.1, page 379

1. {a, b, d, f}; {a} **3.** {b, c, e}; {a} **5.** No **7.** S

9. ∅ **11.** Yes **13.** $E \cup F$ **15.** G^C

17. $(E \cup F \cup G)^C$

19. ∅, {a}, {b}, {c}, {a, b}, {a, c}, {b, c}, S

21. a. $S = \{B, R\}$ **b.** ∅, {B}, {R}, S

23. **a.** $S = \{(H, 1), (H, 2), (H, 3), (H, 4), (H, 5), (H, 6),$
$(T, 1), (T, 2), (T, 3), (T, 4), (T, 5), (T, 6)\}$
b. $\{(H, 2), (H, 4), (H, 6)\}$

25. $S = \{(d, d, d), (d, d, n), (d, n, d), (n, d, d), (d, n, n),$
$(n, d, n), (n, n, d), (n, n, n)\}$

27. **a.** {ABC, ABD, ABE, ACD, ACE, ADE, BCD,
BCE, BDE, CDE}
b. 6   **c.** 3   **d.** 6

29. **a.** E^c **b.** $E^c \cap F^c$ **c.** $E \cup F$
d. $(E \cap F^c) \cup (E^c \cap F)$

31. **a.** $\{x \mid x > 0\}$ **b.** $\{x \mid 0 < x \le 2\}$
c. $\{x \mid x > 2\}$

33. **a.** $S = \{0, 1, 2, 3, \ldots, 10\}$ **b.** $E = \{0, 1, 2, 3\}$
c. $F = \{5, 6, 7, 8, 9, 10\}$

35. **a.** $S = \{0, 1, 2, \ldots, 20\}$
b. $E = \{0, 1, 2, \ldots, 9\}$ **c.** $F = \{20\}$

39. 2^n

Exercises 7.2, page 388

1. $\{(H, H)\}, \{(H, T)\}, \{(T, H)\}, \{(T, T)\}$

3. $\{(D, m)\}, \{(D, f)\}, \{(R, m)\}, \{(R, f)\}, \{(I, m)\}, \{(I, f)\}$

5. $\{(1, i)\}, \{(1, d)\}, \{(1, s)\}, \{(2, i\}, \{(2, d)\}, \{(2, s)\}, \ldots,$
$\{(5, i)\}, \{(5, d)\}, \{(5, s)\}$

7. $\{(A, Rh^+)\}, \{(A, Rh^-)\}, \{(B, Rh^+)\}, \{(B, Rh^-)\},$
$\{(AB, Rh^+)\}, \{(AB, Rh^-)\}, \{(O, Rh^+)\}, \{(O, Rh^-)\}$

9.

Grade	A	B	C	D	F
Probability	.10	.25	.45	.15	.05

11. **a.** $S = \{(0 < x \le 200), (200 < x \le 400),$
$(400 < x \le 600), (600 < x \le 800),$
$(800 < x \le 1000), (x > 1000)\}$
b.

Number of Cars (x)	Probability
$0 < x \le 200$	.075
$200 < x \le 400$	.1
$400 < x \le 600$	.175
$600 < x \le 800$	.35
$800 < x \le 1000$	.225
$x > 1000$	.075

13.

Event	A	B	C	D	E
Probability of an Event	.026	.199	.570	.193	.012

15.

Number of Figures Produced (in Dozens)	30	31	32
Probability	.125	0	.1875

Number of Figures Produced (in Dozens)	33	34	35	36
Probability	.25	.1875	.125	.125

17. .469   19. **a.** .856   **b.** .144   21. .46

23. **a.** $\frac{1}{4}$ **b.** $\frac{1}{2}$ **c.** $\frac{1}{13}$ 25. $\frac{3}{8}$ 27. .95

29. **a.** .633   **b.** .276

31. There are two ways of obtaining a sum of 7.

33. No   35. No   37. **a.** $\frac{1}{6}$ **b.** $\frac{5}{6}$ **c.** 1

39. **a.** $\frac{3}{8}$ **b.** $\frac{1}{2}$ **c.** $\frac{1}{4}$

Exercises 7.3, page 398

1. 1/2   3. 1/36   5. 1/9   7. 1/52

9. 3/13   11. 12/13   13. .002; .998

15. $P(a) + P(b) + P(c) \ne 1$

17. Since the five events are not mutually exclusive, Property 3 cannot be used; that is, he could win more than one purse.

19. The two events are not mutually exclusive; hence, the probability of the given event is $\frac{1}{6} + \frac{1}{6} - \frac{1}{36} = \frac{11}{36}$.

21. $E^c \cap F^c = \{e\} \ne \varnothing$

23. $P(G \cup C)^c \ne 1 - P(G) - P(C)$; he has not considered the case in which a customer buys both glasses and contact lenses.

25. **a.** 0   **b.** .7   **c.** .8   **d.** .3

27. **a.** $\frac{1}{2}, \frac{3}{8}$ **b.** $\frac{1}{2}, \frac{5}{8}$ **c.** $\frac{1}{8}$ **d.** $\frac{3}{4}$ 29. .33

31. **a.** .16   **b.** .38   **c.** .22

33. **a.** .90   **b.** .40   **c.** .40

35. **a.** .6   **b.** .332   **c.** .232   **d.** .6

37. .32   39. True

Exercises 7.4, page 407

1. 1/32   **3.** 31/32

5. $P(A) = 13 \cdot C(4, 2)/C(52, 2) = .0588$

7. $C(26, 2)/C(52, 2) = .245$

9. $[C(3, 2) \cdot C(5, 2)]/C(8, 4) = 3/7$

11. $[C(5, 3) \cdot C(3, 1)]/C(8, 4) = 3/7$

13. $[C(3, 2) \cdot C(1, 1)]/8 = 3/8$

15. 1/8   **17.** $C(10, 6)/2^{10} = .205$

19. a. $C(4, 2)/C(24, 2) \approx .022$
  b. $1 - C(20, 2)/C(24, 2) \approx .312$

21. a. $C(6, 2)/C(80, 2) \approx .005$
  b. $1 - C(74, 2)/C(80, 2) \approx .145$

23. a. .12; $C(98, 10)/C(100, 12) \approx .013$
  b. .15; .015

25. $[C(12, 8) \cdot C(8, 2) + C(12, 9) \cdot C(8, 1) +$
  $C(12, 10)]/C(20, 10) \approx .085$

27. a. 3/5   **b.** $C(3, 1)/C(5, 3) = .3$
  c. $1 - C(3, 3)/C(5, 3) = .9$

29. 1/729   **31.** .0001   **33.** .10

35. $40/C(52, 5) \approx .0000154$

37. $[4C(13, 5) - 40]/C(52, 5) \approx .00197$

39. $[13C(4, 3) \cdot 12C(4, 2)]/C(52, 5) \approx .00144$

41. a. .618   **b.** .059   **43.** .03

Exercises 7.5, page 422

1. a. .4   **b.** .33   **3.** .3   **5.** Independent

7. Independent   **9. a.** .24   **b.** .76

11. a. .5   **b.** .4   **c.** .2   **d.** .35
  e. No   **f.** No

13. a. .4   **b.** .3   **c.** .12   **d.** .30
  e. Yes   **f.** Yes

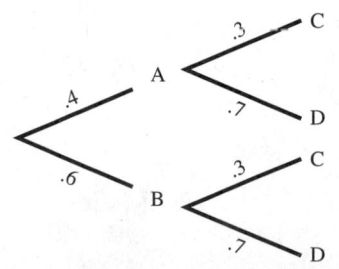

15. 4/11   **17. a.** 1/4   **b.** 13/51   **c.** 12/51

19. .012   **21. a.** .064   **b.** .414   **c.** .57   **d.** .43

23. a. 1/2   **b.** 1/2   **c.** 1/2   **25.** .014

27. a.

L: 30-year loans
M: 15-year loans
N: 20-year loans
T: 10-year loans
P: loans of 5 years or less
  b. .15   **c.** .29

29. 1/15

31. a. .757; .569; .393; .520; .720   **b.** No

33. $P(A \cap B) \neq P(A) \cdot P(B)$

35. a. .81   **b.** .729

37. a. .0000001   **b.** .927   **c.** .071

39. a. No   **b.** Yes

41. $\dfrac{P(A)}{P(A) + P(B)}$

Exercises 7.6, page 431

1.

3. a. .45   **b.** .22   **5. a.** .48   **b.** .33

7. a. .08   **b.** .15   **c.** .35

9. a. 1/12   **b.** 1/4   **c.** 1/18   **d.** 3/14

11. 4/17   **13.** .0784

15.

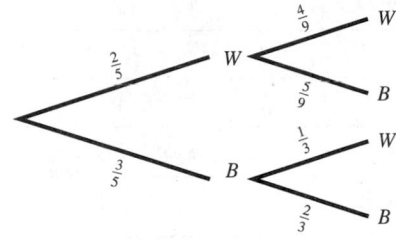

17. .53   **19.** .422

21. a. 3/4   **b.** 2/9   **23.** .856

25. a. .03   **b.** .29   **27.** .35

29. a. .30   **b.** .10

31. .62   **33.** .65   **35.** .93

Exercises 7.7, page 449

1. Yes   **3.** Yes   **5.** No

7. Yes   **9.** No

11. a. Given that the outcome state 1 has occurred, the conditional probability that the outcome state 1 will occur is .3.

 b. .7   **c.** [.48   .52]

13. $X_0 T = [.4 \quad .6]$;

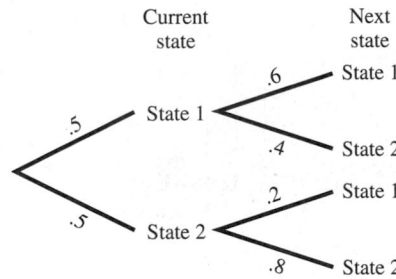

15. $X_2 = [.576 \quad .424]$ **17.** $X_2 = [\frac{5}{16} \quad \frac{27}{64} \quad \frac{17}{64}]$

19. Regular   **21.** Not regular

23. Regular   **25.** Not regular

27. $[\frac{3}{11} \quad \frac{8}{11}]$ **29.** $[\frac{2}{7} \quad \frac{5}{7}]$

31. $[\frac{3}{13} \quad \frac{8}{13} \quad \frac{2}{13}]$ **33.** $[\frac{3}{19} \quad \frac{8}{19} \quad \frac{8}{19}]$

35. a.

	Current state		Next state

(tree diagram)

State 1 →(.8) State 1
State 1 →(.2) State 2
State 2 →(.9) State 1
State 2 →(.1) State 2
with .5 and .5 to State 1 and State 2

 b.
$$T = \begin{array}{c} L \\ R \end{array} \begin{bmatrix} .8 & .2 \\ .9 & .1 \end{bmatrix}$$
with column labels $L \quad R$

 c. $X_0 = [.5 \quad .5]$ (columns $L \quad R$)   **d.** .85

37. a.
$$\begin{array}{c} A \\ U \\ N \end{array} \begin{bmatrix} .85 & .10 & .05 \\ 0 & .95 & .05 \\ .10 & .05 & .85 \end{bmatrix}$$
with column labels $A \quad U \quad N$

 b. $A \quad U \quad N$
[.50   .15   .35]

 c. $A \quad U \quad N$
[.424   .262   .314]

39. After one year A has 47%, B has 37%, and C has 16%; after two years A has 38%, B has 42%, and C has 20%.

41. Business: 36%; humanities: 23.8%; education: 15%; natural sciences and others: 25.1%.

43. a. 40.8% one wage earner; 59.2% two wage earners
 b. 30% one wage earner; 70% two wage earners

45. a. 72.5% in single-family homes; 27.5% in condominiums
 b. 70% in single-family homes; 30% in condominiums

47. $33\frac{1}{3}$% each

49. Manufacturer A 16.7% of the market, manufacturer B 54.2% of the market, and manufacturer C 29.2% of the market.

Using Technology Exercises 7.7, page 456

1. $X_5 = [.204489 \quad .131869 \quad .261028 \quad .186814 \quad .2158]$

3. Manufacturer A will have 23.95% of the market, manufacturer B will have 49.71% of the market share, and manufacturer C will have 26.34% of the market share.

Chapter 7 Review Exercises, page 458

1. a. 0   **b.** .6   **c.** .6   **d.** .4   **e.** 1

3. a. .49   **b.** .39   **c.** .48

5. a. .019   **b.** .981

7. .18   **9.** .06   **11.** .49

13. a. .284   **b.** .984   **15.** 2/15

17. .01   **19.** .510   **21.** Not regular

23. Regular   **25.** [.3675   .36   .2725]

27. $\begin{bmatrix} \frac{3}{7} & \frac{4}{7} \\ \frac{3}{7} & \frac{4}{7} \end{bmatrix}$ **29.** $\begin{bmatrix} .457 & .200 & .343 \\ .457 & .200 & .343 \\ .457 & .200 & .343 \end{bmatrix}$

31. .457   **33. a.** .68   **b.** .0529

CHAPTER 8

Exercises 8.1, page 468

1. a. See (b)

b.

Outcome	GGG	GGR	GRG	RGG
Value	3	2	2	2

Outcome	GRR	RGR	RRG	RRR
Value	1	1	1	0

c. {GGG}

3. Any positive integer   **5.** 1/6

7. Any positive integer; infinite discrete

9. $0 \le x < \infty$, continuous

11. Any positive integer; infinite discrete

13. a. .20   **b.** .60   **c.** .30   **d.** 1

15.

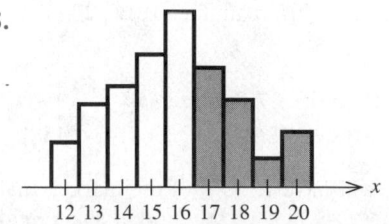

17. a.

x	1	2	3	4	5	6
$P(X = x)$	$\frac{1}{6}$	$\frac{1}{6}$	$\frac{1}{6}$	$\frac{1}{6}$	$\frac{1}{6}$	$\frac{1}{6}$

y	1	2	3	4	5	6
$P(Y = y)$	$\frac{1}{6}$	$\frac{1}{6}$	$\frac{1}{6}$	$\frac{1}{6}$	$\frac{1}{6}$	$\frac{1}{6}$

b.

$x + y$	2	3	4	5	6	7
$P(X + Y = x + y)$	$\frac{1}{36}$	$\frac{2}{36}$	$\frac{3}{36}$	$\frac{4}{36}$	$\frac{5}{36}$	$\frac{6}{36}$

$x + y$	8	9	10	11	12
$P(X + Y = x + y)$	$\frac{5}{36}$	$\frac{4}{36}$	$\frac{3}{36}$	$\frac{2}{36}$	$\frac{1}{36}$

19. a.

x	0	1	2	3	4
$P(X = x)$	.017	.067	.033	.117	.233

x	5	6	7	8	9	10
$P(X = x)$	.133	.167	.1	.05	.067	.017

b.

21.

x	1	2	3	4	5
$P(X = x)$	.007	.029	.021	.079	.164

x	6	7	8	9	10
$P(X = x)$	.15	.20	.207	.114	.029

Using Technology Exercises 8.1, page 471

1.

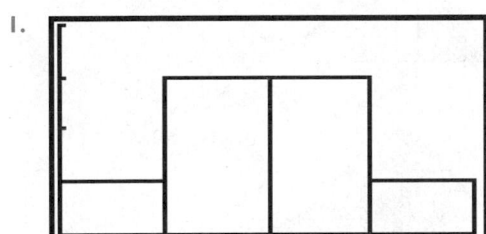

3.

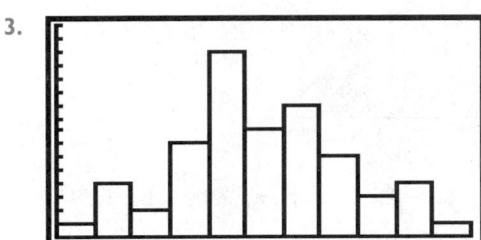

Exercises 8.2, page 483

1. a. 2.6

b.

x	0	1	2	3	4	
$P(X = x)$	0	.1	.4	.3	.2	; 2.6

3. 0.86   **5.** $78.50   **7.** 0.91

9. 0.12   **11.** 1.73   **13.** $800

15. $162.50   **17.** First project

19. a. Dahl: 8.52; Farthington: 7.25
b. Farthington Auto Sales

21. a. 3 to 7   **b.** 7 to 3

23. −10.5¢   **25.** 4 to 1; 1 to 4   **27.** .5625

29. .6154   **31.** .5714   **33. b.** −$15.79

35. Mean: $10.94; mode: $11.00; median: $10.95

Exercises 8.3, page 494

1. $\mu = 2$, $\text{Var}(X) = 1$, $\sigma = 1$

3. $\mu = 0$, $\text{Var}(X) = 1$, $\sigma = 1$

5. $\mu = 518$, $\text{Var}(X) = 1891$, $\sigma = 43.5$

7. a   **9.** 1.56

11. $\mu = 4.5$, $\text{Var}(X) = 5.25$

13. a. Let X = The annual birth rate during the years
1981–1990

b.

x	15.5	15.6	15.7	15.9	16.2	16.7
$P(X = x)$	.2	.1	.3	.2	.1	.1

c. $\mu = 15.84$, $\text{Var}(X) = 0.1224$, $\sigma = 0.350$

15. a. Mutual fund A: $\mu = \$620$, $\text{Var}(X) = \$267,600$;
Mutual fund B: $\mu = \$520$, $\text{Var}(X) = \$137,600$
b. Mutual fund A   **c.** Mutual fund B

17. 1

19. $\mu = \$139,600$; $\text{Var}(X) = \$1,443,840,000$; $\sigma = \$37,998$

21. a. At least .75   **b.** At least .96

23. At least 75%   **25.** At least 7/16   **27.** $c \geq 0.1$

Using Technology Exercises 8.3, page 499

1. a.

b. $\mu = 4$, $\sigma = 1.40$

3. a.

b. $\mu = 17.34$, $\sigma = 1.11$

5. a. Let X denote the random variable that gives the
weight of a carton of sugar.

b.

x	4.96	4.97	4.98	4.99	5.00	5.01
$P(X = x)$	$\frac{3}{30}$	$\frac{4}{30}$	$\frac{4}{30}$	$\frac{1}{30}$	$\frac{1}{30}$	$\frac{5}{30}$

x	5.02	5.03	5.04	5.05	5.06
$P(X = x)$	$\frac{3}{30}$	$\frac{3}{30}$	$\frac{4}{30}$	$\frac{1}{30}$	$\frac{1}{30}$

$\mu \approx 5.00$; $\text{Var}(X) \approx 0.0009$; $\sigma \approx 0.03$

Exercises 8.4, page 509

1. Yes

3. No. There are more than two outcomes to the experiment.

5. No. The probability of an accident on a clear day is not the same as the probability of an accident on a rainy day.

7. .296   9. .0512   11. .132

13. 21/32   15. .0041   17. ≈ .116

19. a. $P(X = 0) \approx .08$, $P(X = 1) \approx .26$,
    $P(X = 2) \approx .35$, $P(X = 3) \approx .23$,
    $P(X = 4) \approx .08$, $P(X = 5) \approx .01$

b.

x	0	1	2	3	4	5
$P(X = x)$	.08	.26	.35	.23	.08	.01

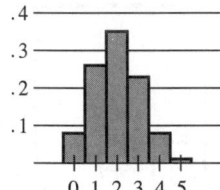

c. $\mu = 2$, $\sigma \approx 1.1$

21. No. The probability that at most 1 is defective is $P(X = 0) + P(X = 1) = .74$.

23. ≈ .0002

25. a. ≈ .633   b. ≈ .367

27. a. ≈ .273   b. ≈ .650

29. .3125   31. .9133

33. a. ≈ .003988   b. .000006   c. ≈ 3.997 × 10⁻⁹

35. ≈ .392   37. 7 times

39. a. 1200   b. ≈ 21.91

Exercises 8.5, page 520

1. .9265   3. .0401   5. .8657

7. a.

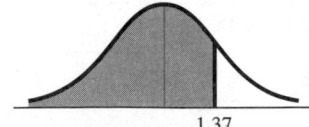

b. .9147

9. a.

b. .2578

−0.65

11. a.

b. .8944

−1.25

13. a.

b. .2266

0.68 2.02

15. a. 1.23   b. −0.81   17. a. 1.9   b. −1.9

19. a. .9772   b. .9192   c. .7333

Exercises 8.6, page 530

1. a. .2206   b. .2206   c. .3034

3. a. .0228   b. .0228   c. .4772   d. .7258

5. a. .0038   b. .0918   c. .4082   d. .2514

7. .6247   9. 0.62%

11. A:80; B:73; C:62; D:54

13. a. .4207   b. .4254   c. .0122

15. a. .2877   b. .0008   c. .7287

17. .9265

19. a. .0037   b. Drug is very effective

21. 2142

Chapter 8 Review Exercises, page 534

1. a. {WWW, BWW, WBW, WWB, BBW, BWB, WBB, BBB}

b.

Outcome	WWW	BWW	WBW	WWB
Value of X	0	1	1	1

Outcome	BBW	BWB	WBB	BBB
Value of X	2	2	2	3

c.

x	0	1	2	3
$P(X = x)$	$\frac{1}{35}$	$\frac{12}{35}$	$\frac{18}{35}$	$\frac{4}{35}$

d.
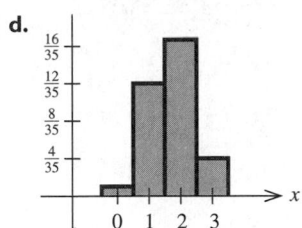

3. a. .8   **b.** $\mu = 2.7$; $\sigma = 1.42$

5. .6915
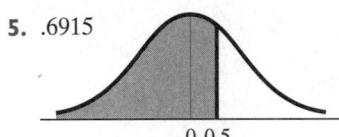

7. .4649
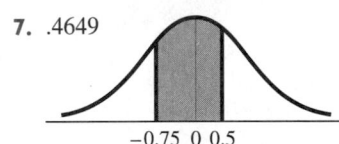

9. 2.42   **11.** −2.03   **13.** .6915   **15.** .2417

17. .2646; .9163   **19.** At least .75

21. $\mu = 120$, $\sigma = 10.1$ **23.** 0.9738

CHAPTER 9

Exercises 9.1, page 541
1. 9   **3.** 1   **5.** 4

7. 7   **9.** $\frac{1}{5}$ **11.** 2

13. 2   **15.** 1   **17.** True

19. False   **21.** False   **23.** False

25. False   **27.** $\dfrac{1}{(xy)^2}$ **29.** $\dfrac{1}{x^{5/6}}$

31. $\dfrac{1}{(s+t)^3}$ **33.** $x^{13/3}$ **35.** $\dfrac{1}{x^3}$

37. x **39.** $\dfrac{9}{x^2 y^4}$ **41.** $\dfrac{y^8}{x^{10}}$

43. $2x^{11/6}$ **45.** $-2xy^2$ **47.** $2x^{4/3}y^{1/2}$

49. 2.828   **51.** 5.196   **53.** 31.62

55. 316.2   **57.** $\dfrac{3\sqrt{x}}{2x}$ **59.** $\dfrac{2\sqrt{3y}}{3}$

61. $\dfrac{\sqrt[3]{x^2}}{x}$ **63.** $\dfrac{2x}{3\sqrt{x}}$ **65.** $\dfrac{2y}{\sqrt{2xy}}$

67. $\dfrac{xy}{y\sqrt[3]{xz^2}}$

Exercises 9.2, page 550
1. $9x^2 + 3x + 1$ **3.** $4y^2 + y + 8$

5. $-x - 1$ **7.** $\frac{2}{3} + e - e^{-1}$

9. $6\sqrt{2} + 8 + \frac{1}{2}\sqrt{x} - \frac{11}{4}\sqrt{y}$

11. $x^2 + 6x - 16$ **13.** $a^2 + 10a + 25$

15. $x^2 + 4xy + 4y^2$ **17.** $4x^2 - y^2$

19. $-2x$ **21.** $2t(2\sqrt{t} + 1)$

23. $2x^3(2x^2 - 6x - 3)$ **25.** $7a^2(a^2 + 7ab - 6b^2)$

27. $e^{-x}(1 - x)$ **29.** $\frac{1}{2}x^{-5/2}(4 - 3x)$

31. $(2a + b)(3c - 2d)$ **33.** $(2a + b)(2a - b)$

35. $-2(3x + 5)(2x - 1)$ **37.** $3(x - 4)(x + 2)$

39. $2(3x - 5)(2x + 3)$ **41.** $(3x - 4y)(3x + 4y)$

43. $(x^2 + 5)(x^4 - 5x^2 + 25)$

45. $x^3 - xy^2$

47. $4(x - 1)(3x - 1)(2x + 2)^3$

49. $4(x - 1)(3x - 1)(2x + 2)^3$

51. $2x(x^2 + 2)^2(5x^4 + 20x^2 + 17)$

53. −4 and 3   **55.** −1 and $\frac{1}{2}$

57. 2 and 2   **59.** −2 and $\frac{3}{4}$

61. $\frac{1}{2} + \frac{1}{4}\sqrt{10}$ and $\frac{1}{2} - \frac{1}{4}\sqrt{10}$

63. $-1 + \frac{1}{2}\sqrt{10}$ and $-1 - \frac{1}{2}\sqrt{10}$

Exercises 9.3, page 558
1. $\dfrac{x - 1}{x - 2}$ **3.** $\dfrac{3(2t + 1)}{2t - 1}$

5. $-\dfrac{7}{(4x - 1)^2}$ **7.** $\dfrac{1}{(2x + 3)^2}$

9. $\dfrac{e^x(1 - 2e^x)}{(e^x + 1)^3}$ **11.** −8

13. $\dfrac{3x - 1}{2}$ **15.** $\dfrac{t + 20}{3t + 2}$

17. $-\dfrac{x(2x - 13)}{(2x - 1)(2x + 5)}$ **19.** $\dfrac{x^4 - 1}{x}$

21. $-\dfrac{x + 27}{(x - 3)^2(x + 3)}$ **23.** $-\dfrac{3a + 10}{(a + 2)(a - 2)}$

25. $\dfrac{x + 1}{x - 1}$ **27.** $\dfrac{y - x}{x^2 y^2}$

29. $\dfrac{4x^2 + 7}{\sqrt{2x^2 + 7}}$ **31.** $\dfrac{x - 1}{x^2 \sqrt{x + 1}}$

33. $\dfrac{x - 1}{(2x + 1)^{3/2}}$ **39.** $\dfrac{\sqrt{3} + 1}{2}$

41. $\dfrac{\sqrt{x} + \sqrt{y}}{x - y}$ **43.** $\dfrac{(\sqrt{a} + \sqrt{b})^2}{a - b}$

45. $\dfrac{x}{3\sqrt{x}}$ **47.** $-\dfrac{2}{3(1 + \sqrt{3})}$

49. $-\dfrac{x + 1}{\sqrt{x + 2}\,(1 - \sqrt{x + 2})}$

Exercises 9.4, page 564

1. False   **3.** False

5. ![number line with open interval between 3 and 6] $\to x$
    0   3   6

7. ![number line from -1 to 4] $\to x$
    -1   0   4

9. ![number line open at 0 going left] $\to x$
    0

11. $(-\infty, 2)$ **13.** $(-\infty, -5]$ **15.** $(-4, 6)$

17. $(-\infty, -3) \cup (3, \infty)$ **19.** $(-2, 3)$

21. $[-3, 5]$ **23.** $(-\infty, 1] \cup [\tfrac{3}{2}, \infty)$

25. $(-\infty, -3] \cup (2, \infty)$ **27.** $[0, 1) \cup (1, \infty)$

29. 4   **31.** 2   **33.** $5\sqrt{3}$

35. $\pi + 1$ **37.** 2   **39.** False

41. False   **43.** True   **45.** False

47. True   **49.** False   **51.** $[362, 488.7]$

53. \$12,300   **55.** \$18,666.67   **57.** $|x - 0.5| < 0.01$

59. Between 1000 and 4000 units

Chapter 9 Review Exercises, page 566

1. $\frac{27}{8}$ **3.** $\frac{1}{144}$ **5.** $\frac{1}{4}$

7. $4(x^2 + y)^2$ **9.** $\dfrac{2x}{3z}$ **11.** $6xy^7$

13. $9x^2 y^4$ **15.** $-2\pi r^2(\pi r - 50)$

17. $(4 - x)(4 + x)$ **19.** $-2(x - 1)(x + 3)$

21. $\dfrac{180}{(t + 6)^2}$ **23.** $\dfrac{78x^2 - 8x - 27}{3(2x^2 - 1)(3x - 1)}$

25. $\frac{1}{2}; -\frac{3}{4}$ **27.** $0; 1; -3$

29. $(-2, \infty)$ **31.** $(-\infty, -4) \cup (5, \infty)$

33. 4   **35.** $\pi - 6$ **37.** $[-2, \frac{1}{2}]$

39. $(-1, 4)$ **41.** $\dfrac{1}{\sqrt{x} + 1}$

43. $1 + \sqrt{6}, 1 - \sqrt{6}$ **45.** \$100

CHAPTER 10

Exercises 10.1, page 578

1. $21, -9, 5a + 6, -5a + 6, 5a + 21$

3. $-3, 6, 3a^2 - 6a - 3, 3a^2 + 6a - 3, 3x^2 - 6$

5. $\dfrac{8}{15}, 0, \dfrac{2a}{a^2 - 1}, \dfrac{2(2 + a)}{a^2 + 4a + 3}, \dfrac{2(t + 1)}{t(t + 2)}$

7. $8, \dfrac{2a^2}{\sqrt{a - 1}}, \dfrac{2(x + 1)^2}{\sqrt{x}}, \dfrac{2(x - 1)^2}{\sqrt{x - 2}}$

9. $5, 1, 1$ **11.** $\frac{5}{2}, 3, 3, 9$

13. Yes   **15.** Yes

17. $(-\infty, \infty)$ **19.** $(-\infty, 0) \cup (0, \infty)$

21. $(-\infty, \infty)$ **23.** $(-\infty, 5]$

25. $(-\infty, -1) \cup (-1, 1) \cup (1, \infty)$

27. $[-3, \infty)$ **29.** $(-\infty, -2) \cup (-2, 1]$

31. **a.** $(-\infty, \infty)$
    b. $6, 0, -4, -6, -\frac{25}{4}, -6, -4, 0$
    c.

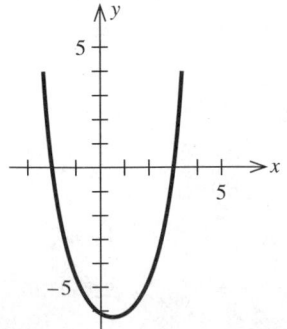

33.

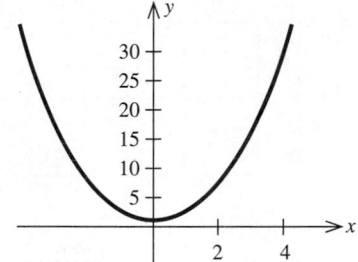

35.

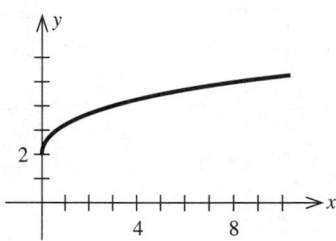

37.

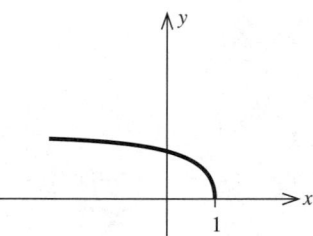

39.

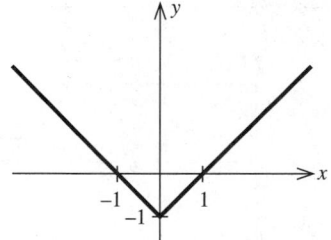

41.

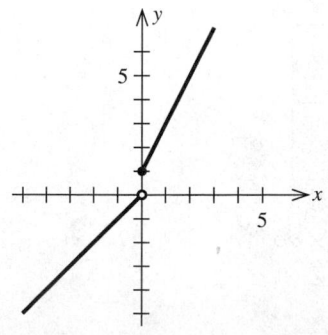

43.

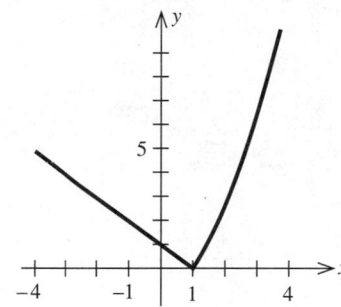

45. Yes   **47.** No   **49.** Yes   **51.** Yes

53. 10π in.   **55. a.** From 1985 to 1990
b. From 1990 on
c. 1990; $3.5 billion

57. a.
$$f(t) = \begin{cases} 0.0185t + 0.58 & \text{if } 0 \le t \le 20 \\ 0.015t + 0.65 & \text{if } 20 < t \le 30 \end{cases}$$
b. 0.0185/yr from 1960 through 1980; 0.015/yr from 1980 through 1990
c. 1980

59. a. $0.06x$ **b.** $12.00; $0.34   **61.** 160 mg

63. a. $30 + 0.15x$; $25 + 0.20x$
b.

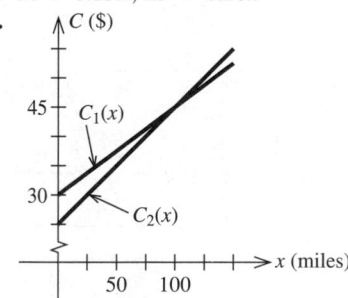

c. Acme

65. $620,000; $540,000

67. a. $0 \le r \le 0.2$
b. 40; 30; 0; as the distance r increases, the velocity of the blood decreases.

69. 20; 26

71. 0.77. When the proportion of popular votes won by the Democratic presidential candidate is 0.60, the proportion of seats in the House of Representatives won by Democratic candidates is 0.77.

73. a. (0, 12]

b.

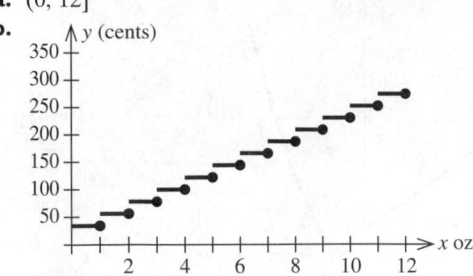

Using Technology Exercises 10.1, page 587

1.

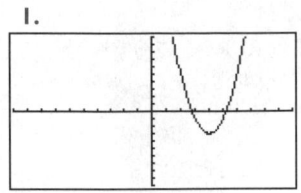

3.

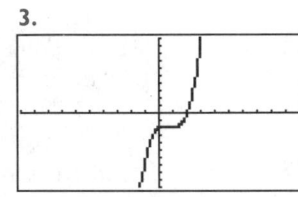

5.

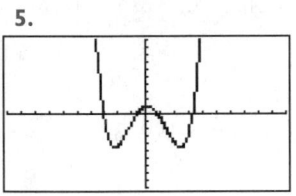

7.

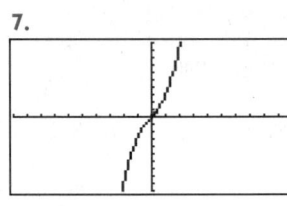

9. a.

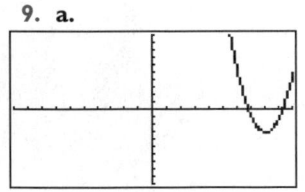

b.

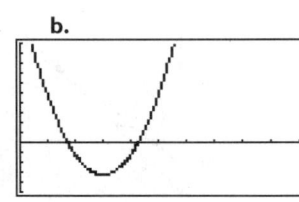

11. a.

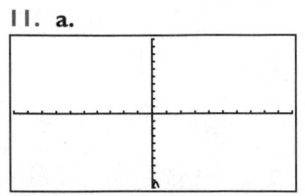

b.

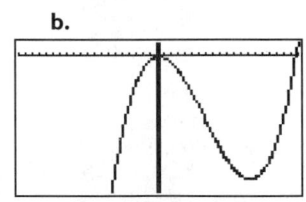

13. a.

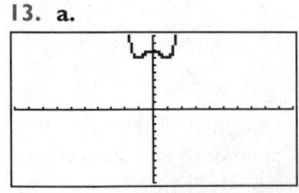

b.

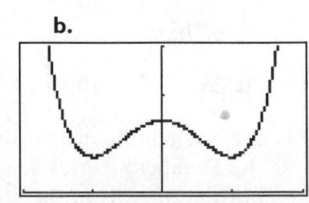

15. a.

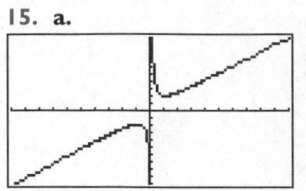

b.

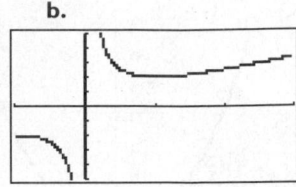

17. a.

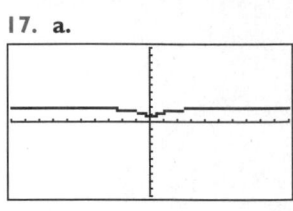

b.

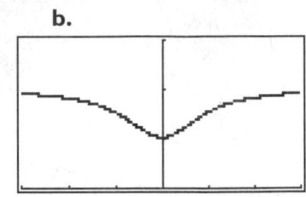

19. a.

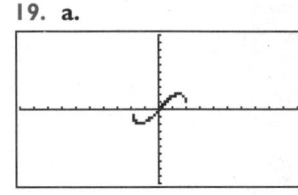

b.

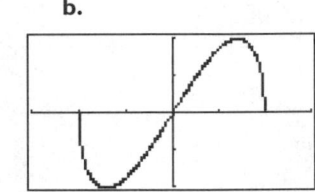

21.

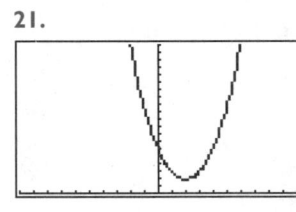

23.

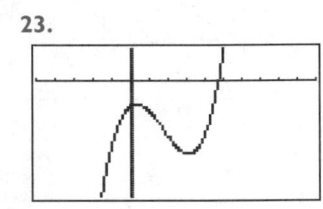

25.

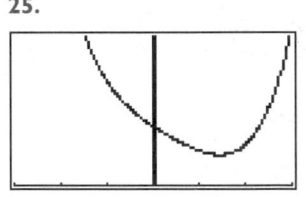

27.

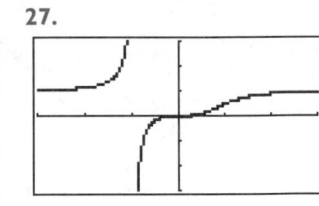

29.

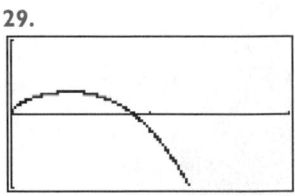

31. 18   **33.** 2   **35.** 18.5505   **37.** 17.3850

39. 4.1616   **41.** 1.7214

43. a.

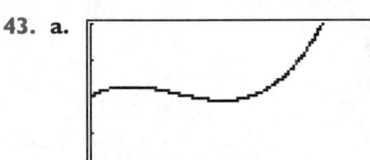

b. 2.1762%; 1.9095%

45. a.

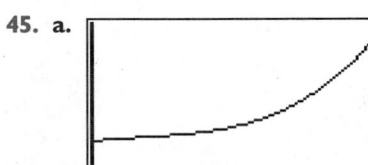

b. 44.7; 52.7; 129.2

Exercises 10.2, page 595

1. $f(x) + g(x) = x^3 + x^2 + 3$

3. $f(x)g(x) = x^5 - 2x^3 + 5x^2 - 10$

5. $\dfrac{f(x)}{g(x)} = \dfrac{x^3 + 5}{x^2 - 2}$

7. $\dfrac{f(x)g(x)}{h(x)} = \dfrac{x^5 - 2x^3 + 5x^2 - 10}{2x + 4}$

9. $f(x) + g(x) = x - 1 + \sqrt{x + 1}$

11. $f(x)g(x) = (x - 1)\sqrt{x + 1}$

13. $\dfrac{g(x)}{h(x)} = \dfrac{\sqrt{x + 1}}{2x^3 - 1}$

15. $\dfrac{f(x)g(x)}{h(x)} = \dfrac{(x - 1)\sqrt{x + 1}}{2x^3 - 1}$

17. $\dfrac{f(x) - h(x)}{g(x)} = \dfrac{x - 2x^3}{\sqrt{x + 1}}$

19. $f(x) + g(x) = x^2 + \sqrt{x} + 3$;
$f(x) - g(x) = x^2 - \sqrt{x} + 7$;
$f(x)g(x) = (x^2 + 5)(\sqrt{x} - 2)$; $\dfrac{f(x)}{g(x)} = \dfrac{x^2 + 5}{\sqrt{x} - 2}$

21. $f(x) + g(x) = \dfrac{(x - 1)\sqrt{x + 3} + 1}{x - 1}$;
$f(x) - g(x) = \dfrac{(x - 1)\sqrt{x + 3} - 1}{x - 1}$;
$f(x)g(x) = \dfrac{\sqrt{x + 3}}{x - 1}$; $\dfrac{f(x)}{g(x)} = (x - 1)\sqrt{x + 3}$

23. $f(x) + g(x) = \dfrac{2(x^2 - 2)}{(x - 1)(x - 2)}$;
$f(x) - g(x) = \dfrac{-2x}{(x - 1)(x - 2)}$;
$f(x)g(x) = \dfrac{(x + 1)(x + 2)}{(x - 1)(x - 2)}$; $\dfrac{f(x)}{g(x)} = \dfrac{(x + 1)(x - 2)}{(x - 1)(x + 2)}$

25. $f(g(x)) = x^4 + x^2 + 1$; $g(f(x)) = (x^2 + x + 1)^2$

27. $f(g(x)) = \sqrt{x^2 - 1} + 1$; $g(f(x)) = x + 2\sqrt{x}$

29. $f(g(x)) = \dfrac{x}{x^2 + 1}$; $g(f(x)) = \dfrac{x^2 + 1}{x}$

31. 49   **33.** $\dfrac{\sqrt{5}}{5}$

35. $f(x) = 2x^3 + x^2 + 1$ and $g(x) = x^5$

37. $f(x) = x^2 - 1$ and $g(x) = \sqrt{x}$

39. $f(x) = x^2 - 1$ and $g(x) = \dfrac{1}{x}$

41. $f(x) = 3x^2 + 2$ and $g(x) = \dfrac{1}{x^{3/2}}$

43. $3h$ **45.** $-h(2a + h)$ **47.** $2a + h$

49. $C(x) = 0.6x + 12,100$

51. a. $P(x) = -0.000003x^3 - 0.07x^2 + 300x - 100,000$
b. \$182,375

53. a. $N(t) = \dfrac{7}{1 + 0.02\left[\dfrac{10t + 150}{t + 10}\right]^2}$
b. 1.27 million units; 1.74 million units; 1.85 million units

55. $N(x(t)) = 9.94\left[\dfrac{(t + 10)^2}{(t + 10)^2 + 2(t + 15)^2}\right]$; 2.24 million jobs; 2.48 million jobs

Exercises 10.3, page 605

1. Polynomial function; degree 6

3. Polynomial function; degree 6

5. Some other function

7. $28,800   **9.** 104 mg   **11.** $400,000

13. a. $R(x) = \dfrac{100x}{40 + x}$ **b.** 60%

15. $72,000

17. $\dfrac{110}{\frac{1}{2}t + 1} - 26(\frac{1}{4}t^2 - 1)^2 - 52$; $32, $6.71, $3; the gap was closing.

19. a.

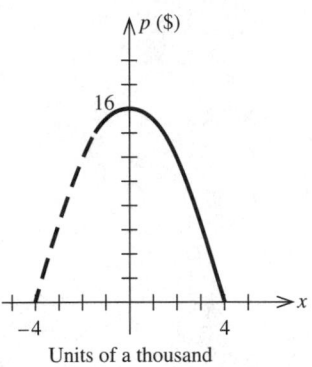

Units of a thousand

b. 3000 units

21. a.

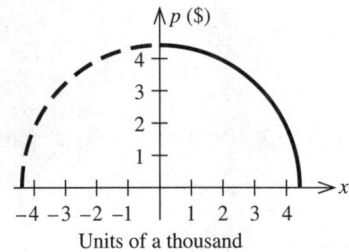

Units of a thousand

b. 3000

23. $p = \sqrt{-x^2 + 100}$; 6614 units

25. a.

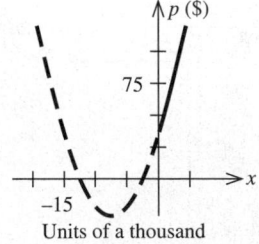

Units of a thousand

b. $76

27. a.

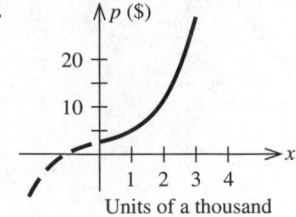

Units of a thousand

b. $15

29. $p = \frac{1}{10}\sqrt{x} + 10$; $30

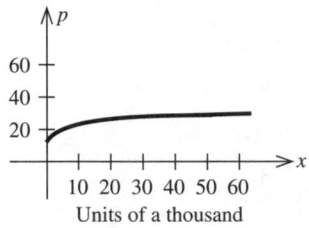

Units of a thousand

31. Equilibrium quantity is 5000; equilibrium price is $65.

33. 2000 units; $52   **35.** 8000 units; $80

Exercises 10.4, page 626

1. $\lim\limits_{x \to -2} f(x) = 3$ **3.** $\lim\limits_{x \to 3} f(x) = 3$ **5.** $\lim\limits_{x \to -2} f(x) = 3$

7. The limit does not exist.

9.

x	1.9	1.99	1.999
$f(x)$	4.61	4.9601	4.9960

x	2.001	2.01	2.1
$f(x)$	5.004	5.0401	5.41

$\lim\limits_{x \to 2} (x^2 + 1) = 5$

11.

x	−0.1	−0.01	−0.001
$f(x)$	−1	−1	−1

x	0.001	0.01	0.1
$f(x)$	1	1	1

The limit does not exist.

13.

x	0.9	0.99	0.999
f(x)	100	10,000	1,000,000

x	1.001	1.01	1.1
f(x)	1,000,000	10,000	100

The limit does not exist.

15.

x	0.9	0.99	0.999	1.001	1.01	1.1
f(x)	2.9	2.99	2.999	3.001	3.01	3.1

$$\lim_{x \to 1} \frac{x^2 + x - 2}{x - 1} = 3$$

17.

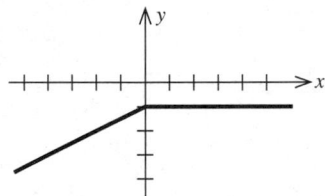

$$\lim_{x \to 0} f(x) = -1$$

19.

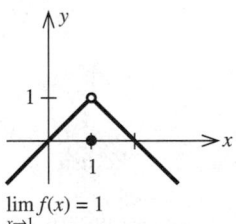

$$\lim_{x \to 1} f(x) = 1$$

21.

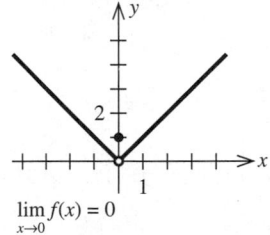

$$\lim_{x \to 0} f(x) = 0$$

23. 3

25. 3   **27.** −1   **29.** 2   **31.** −4   **33.** $\frac{5}{4}$

35. 2   **37.** $\sqrt{171} = 3\sqrt{19}$ **39.** $\frac{3}{2}$ **41.** −1

43. −6   **45.** 2   **47.** $\frac{1}{6}$ **49.** 2   **51.** −1

53. −10   **55.** The limit does not exist.   **57.** $\frac{5}{3}$

59. $\frac{1}{2}$ **61.** $\frac{1}{3}$ **63.** $\lim_{x \to \infty} f(x) = \infty$; $\lim_{x \to -\infty} f(x) = \infty$

65. 0; 0   **67.** $\lim_{x \to \infty} f(x) = -\infty$; $\lim_{x \to -\infty} f(x) = -\infty$

69.

x	1	10	100	1000
f(x)	0.5	0.009901	0.0001	0.000001

x	−1	−10	−100	−1000
f(x)	0.5	0.009901	0.0001	0.000001

$\lim_{x \to \infty} f(x) = 0$ and $\lim_{x \to -\infty} f(x) = 0$

71.

x	1	5	10	100
f(x)	12	360	2910	2.99×10^6

x	1000	−1	−5
f(x)	2.999×10^9	6	−390

x	−10	−100	−1000
f(x)	−3090	-3.01×10^6	-3.0×10^9

$\lim_{x \to \infty} f(x) = \infty$ and $\lim_{x \to -\infty} f(x) = -\infty$

73. 3   **75.** 3   **77.** $\lim_{x \to -\infty} f(x) = -\infty$ **79.** 0

81. a. $0.5 million; $0.75 million; $1,166,667; $2 million; $4.5 million; $9.5 million
b. The limit does not exist; as the percentage of pollutant to be removed approaches 100, the cost becomes astronomical.

83. $2.20; the average cost of producing x video discs will approach $2.20/disc in the long run.

85. a. $24 million; $60 million; $83.1 million
b. $120 million

87. a. 76.1 cents/mile; 30.5 cents/mile; 23 cents/mile; 20.6 cents/mile; 19.5 cents/mile

b.

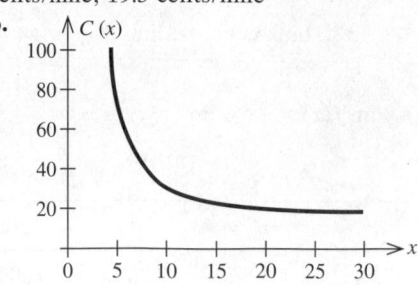

c. The average cost will approach 17.8 cents/mile.

89. No

Using Technology Exercises 10.4, page 632

1. 5   **3.** 3   **5.** $\frac{2}{3}$ **7.** $\frac{1}{2}$ **9.** e^2

13. a.

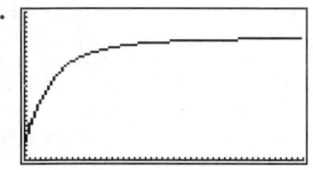

b. 25,000

Exercises 10.5, page 644

1. 3; 2; the limit does not exist.

3. The limit does not exist; 2; the limit does not exist.

5. 0; 2; the limit does not exist.

7. −2; 2; the limit does not exist.

9. True   **11.** True   **13.** False   **15.** True

17. False   **19.** True   **21.** 6   **23.** $-\frac{1}{4}$

25. The limit does not exist.   **27.** −1   **29.** 0

31. −4   **33.** The limit does not exist.   **35.** 4

37. 0   **39.** 0; 0   **41.** 2; 3

43. $x = 0$; conditions 2 and 3

45. Continuous everywhere   **47.** $x = 0$; condition 3

49. $x = 0$; condition 3   **51.** $(-\infty, \infty)$ **53.** $(-\infty, \infty)$

55. $(-\infty, \frac{1}{2}) \cup (\frac{1}{2}, \infty)$ **57.** $(-\infty, -2) \cup (-2, 1) \cup (1, \infty)$

59. $(-\infty, \infty)$ **61.** $(-\infty, \infty)$ **63.** $(-\infty, \infty)$

65. $(-\infty, \infty)$ **67.** −1 and 1   **69.** 1 and 2

71. f is discontinuous at $x = 1, 2, \ldots, 11$

73. Michael makes progress toward solving the problem until $x = x_1$. Between $x = x_1$ and $x = x_2$, he makes no further progress. But at $x = x_2$ he suddenly achieves a breakthrough, and at $x = x_3$ he proceeds to complete the problem.

75. Conditions 2 and 3 are not satisfied at each of these points.

77.

79.

f is discontinuous at $x = \frac{1}{2}, 1, 1\frac{1}{2}, \ldots, 4$.

81. a. ∞; if the speed of the fish is very close to that of the current, the energy expended by the fish is enormous.
b. ∞; if the speed of the fish increases greatly, so does the amount of energy required to swim L feet.

83. $k = -4$ **85. a.** No   **b.** No

87. a. f is a polynomial of degree 3.
b. $f(-1) = -4$ and $f(1) = 4$

89. a. f is continuous on [14, 16]   **b.** $f(14) \approx -6.06$ and $f(16) \approx 1.60$

91. $x = 2$ **93.** ≈ 1.34 **95.** False   **97. c.** No

Using Technology Exercises 10.5, page 652

1. $x = 0, 1$ **3.** $x = 2$ **5.** $x = 0, \frac{1}{2}$

7. $x = -\frac{1}{2}, 2$ **9.** $x = -2, 1$

11.

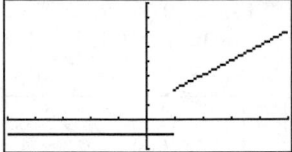

13.

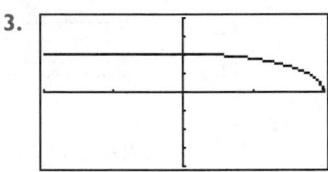

15.

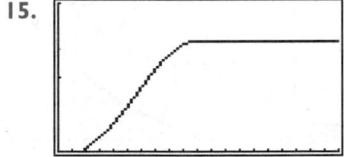

Exercises 10.6, page 669

1. 1.5; 0.5833; 1.25   **3.** 3.075; −21.15

5. a. Car A
b. They are traveling at the same speed.
c. Car B
d. Both cars covered the same distance.

7. a. P_2 **b.** P_1 **c.** Bactericide B; bactericide A

9. 0   **11.** 2   **13.** $6x$ **15.** $-2x + 3$

17. 2; $y = 2x + 7$ **19.** 6; $y = 6x - 3$

21. 1/9; $y = \frac{1}{9}x - \frac{2}{3}$

23. a. $4x$
b. $y = 4x - 1$
c.

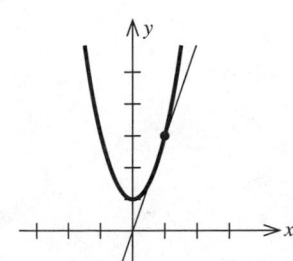

25. a. $2x - 2$
b. $(1, 0)$
c.

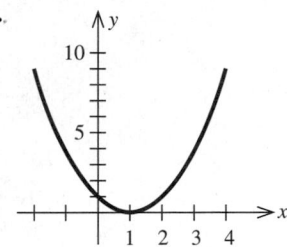

d. 0

27. a. 6; 5.5; 5.1
b. 5
c. The computations in (a) show that as h approaches zero, the average velocity approaches the instantaneous velocity.

29. a. 130 ft/sec; 128.2 ft/sec; 128.02 ft/sec
b. 128 ft/sec
c. The computations in (a) show that as the time intervals over which the average velocity is computed become smaller and smaller, the average velocity approaches the instantaneous velocity of the car at $t = 20$.

31. a. 5 sec   **b.** 80 ft/sec   **c.** 160 ft/sec

33. a. $-\frac{1}{6}$ liter/atmosphere   **b.** $-\frac{1}{4}$ liter/atmosphere

35. a. $-\frac{2}{3}x + 7$ **b.** $333 per quarter; −$13,000 per quarter

37. $6 billion/yr; $10 billion/yr

39. Average rate of change of the seal population over $[a, a + h]$; instantaneous rate of change of the seal population at $x = a$

41. Average rate of change of the country's industrial production over $[a, a + h]$; instantaneous rate of change of the country's industrial production at $x = a$

43. Average rate of change of atmospheric pressure over $[a, a + h]$; instantaneous rate of change of atmospheric pressure at $x = a$

45. a. Yes   **b.** No   **c.** No

47. a. Yes   **b.** Yes   **c.** No

49. a. No   **b.** No   **c.** No

51. 5.06060; 5.06006; 5.060006; 5.0600006; 5.06000006; $5.06/case

53.

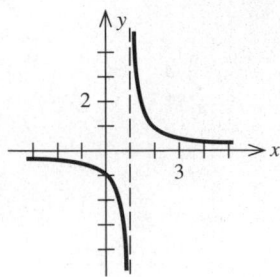

55.

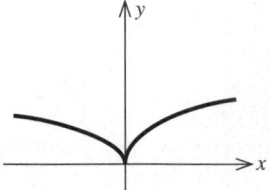

Yes; no; the graph of f has a kink at $x = 0$.

Using Technology Exercises 10.6, page 677

1. a. $y = 4x - 3$
b.

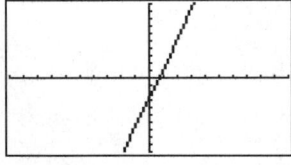

3. a. $y = -7x - 8$
b.

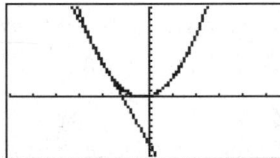

5. a. $y = 9x - 11$
b.

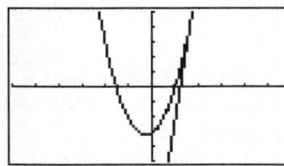

7. a. $y = 2$
b.

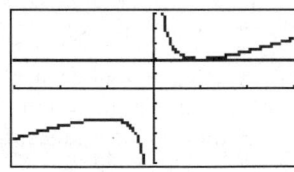

9. a. $y = \frac{1}{4}x + 1$
b.

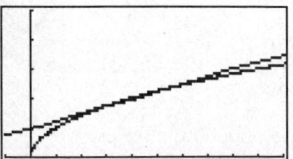

11. a. 4
b. $y = 4x - 1$
c.

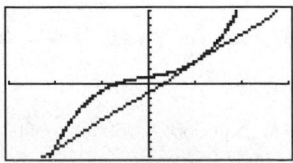

13. a. 20
b. $y = 20x - 35$
c.

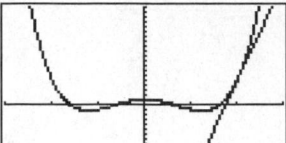

15. a. 0.75
b. $y = 0.75x - 1$
c.

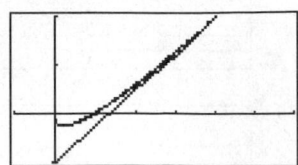

17. a. -0.25
b. $y = -0.25x + 0.75$
c.

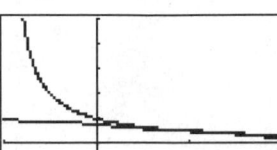

19. a. 4.02
b. $y = 4.02x - 3.57$
c.

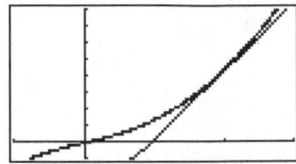

Chapter 10 Review Exercises, page 679

1. a. $(-\infty, 9]$
b. $(-\infty, -1) \cup (-1, \frac{3}{2}) \cup (\frac{3}{2}, \infty)$

3. a.

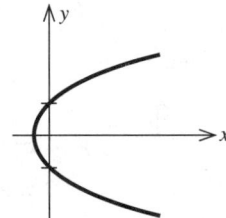

b. No
c. Yes

5. a. $\dfrac{2x + 3}{x}$ **b.** $\dfrac{1}{x(2x + 3)}$ **c.** $\dfrac{1}{2x + 3}$ **d.** $\dfrac{2}{x} + 3$

7. 2   **9.** 0   **11.** The limit does not exist.

13. $\frac{9}{2}$ **15.** $\frac{1}{2}$ **17.** 1   **19.** The limit does not exist.

21.

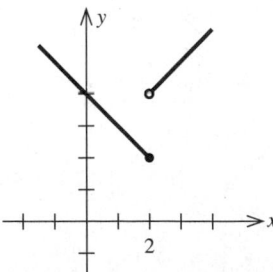

4; 2; the limit does not exist.

23. $x = -\frac{1}{2}, 1$ **25.** $x = 0$ **27.** 3

29. $\frac{3}{2}$; $y = \frac{3}{2}x + 5$ **31. a.** Yes **b.** No

33. a. $S(t) = t + 2.4$ **b.** $5.4 million

35. $(6, \frac{21}{2})$ **37.** 6000; $22 **39.** $45,000

41.

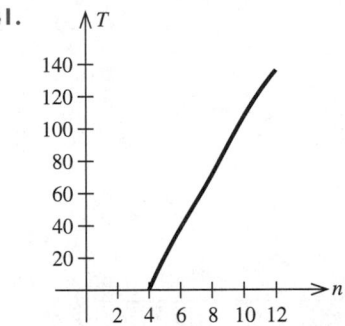

As the length of the list increases, the time taken to learn the list increases by a very large amount.

43. $C(x) = \begin{cases} 5 & \text{if} & 1 \le x \le 100 \\ 9 & \text{if} & 100 < x \le 200 \\ 12.50 & \text{if} & 200 < x \le 300 \\ 15.00 & \text{if} & 300 < x \le 400 \\ 7 + 0.02x & \text{if} & x > 400 \end{cases}$

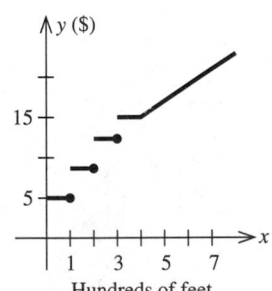

The function is discontinuous at $x = 100, 200,$ and 300.

CHAPTER 11

Exercises 11.1, page 691

1. 0 **3.** $5x^4$ **5.** $2.1x^{1.1}$ **7.** $6x$ **9.** $2\pi r$

11. $\frac{3}{x^{2/3}}$ **13.** $\frac{3}{2\sqrt{x}}$ **15.** $-84x^{-13}$ **17.** $10x - 3$

19. $-3x^2 + 4x$ **21.** $0.06x - 0.4$ **23.** $2x - 4 - \frac{3}{x^2}$

25. $16x^3 - 7.5x^{3/2}$ **27.** $-\frac{3}{x^2} - \frac{8}{x^3}$ **29.** $-\frac{16}{t^5} + \frac{9}{t^4} - \frac{2}{t^2}$

31. $2 - \frac{5}{2\sqrt{x}}$ **33.** $-\frac{4}{x^3} + \frac{1}{x^{4/3}}$

35. a. 20 **b.** -4 **c.** 20 **37.** 3 **39.** 11

41. $m = 5$; $y = 5x - 4$ **43.** $m = -2$; $y = -2x + 2$

45. a. $(0, 0)$ **b.**

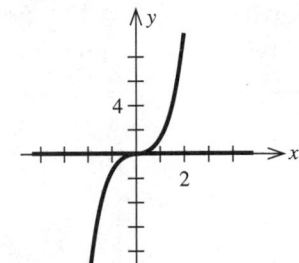

47. a. $(-2, -7), (2, 9)$
  b. $y = 12x + 17$ and $y = 12x - 15$
  c.

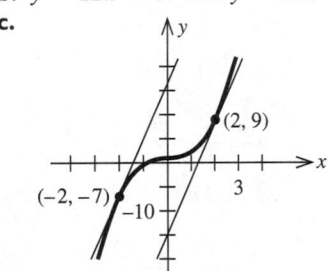

49. a. $(0, 0)$; $(1, -\frac{13}{12})$
  b. $(0, 0)$; $(2, -\frac{8}{3})$; $(-1, -\frac{5}{12})$
  c. $(0, 0)$; $(4, \frac{80}{3})$; $(-3, \frac{81}{4})$

51. a. $\frac{16\pi}{9}$ cm³/cm **b.** $\frac{25\pi}{4}$ cm³/cm

53. -89.01; -4.36; if you make 0.25 stop per mile, your average speed will decrease at the rate of approximately 89.01 mph/stop/mile. If you make 2 stops per mile, your average speed will decrease at the rate of approximately 4.36 mph/stop/mile.

55. a. 15 pts/yr; 12.6 pts/yr; 0 pts/yr **b.** 10 pts/yr

57. 155/month; 200/month

59. 32 turtles/yr; 428 turtles/yr; 3260 turtles

61. a. $120 - 30t$ **b.** 120 ft/sec **c.** 240 ft

63. a. 2.495 million **b.** 405,000/year

65. a. $20\left(1 - \dfrac{1}{\sqrt{t}}\right)$

b. 50 mph; 30 mph; 33.43 mph

c. −8.28; 0; 5.86; at 6:30 A.M., the average velocity is decreasing at the rate of 8.28 mph/hr; at 7 A.M., it is unchanged; and at 8 A.M., it is increasing at the rate of 5.86 mph.

Using Technology Exercises 11.1, page 693

1. 1   **3.** 0.4226   **5.** 0.1613

7. a.

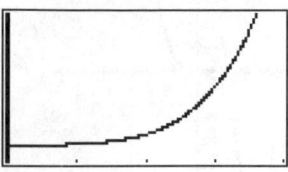

b. 3.4295 parts/million; 105.4332 parts/million

9. a.

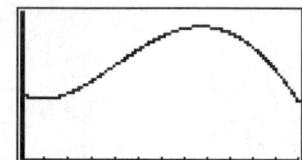

b. Decreasing at the rate of 9 days/yr; increasing at the rate of 13 days/yr

Exercises 11.2, page 705

1. $6x^2 + 2$ **3.** $4t - 1$ **5.** $9x^2 + 2x - 6$

7. $4x^3 + 3x^2 - 1$ **9.** $5w^4 - 4w^3 + 9w^2 - 6w + 2$

11. $\dfrac{25x^2 - 10x^{3/2} + 1}{x^{1/2}}$ **13.** $\dfrac{3x^4 - 10x^3 + 4}{x^2}$

15. $\dfrac{-1}{(x - 2)^2}$ **17.** $\dfrac{3}{(2x + 1)^2}$ **19.** $-\dfrac{2x}{(x^2 + 1)^2}$

21. $\dfrac{s^2 + 2s + 4}{(s + 1)^2}$ **23.** $\dfrac{1 - 3x^2}{2\sqrt{x}(x^2 + 1)^2}$ **25.** $\dfrac{x^2 - 2x - 2}{(x^2 + x + 1)^2}$

27. $\dfrac{2x^3 - 5x^2 - 4x - 3}{(x - 2)^2}$

29. $\dfrac{-2(x^5 + 12x^4 - 8x^3 + 16x + 64)}{(x^4 - 16)^2}$ **31.** 8   **33.** −9

35. $2(3x^2 - x + 3)$; 10   **37.** $\dfrac{-3x^4 + 2x^2 - 1}{(x^4 - 2x^2 - 1)^2}$; $-\dfrac{1}{2}$

39. 60; $y = 60x - 102$ **41.** $-\dfrac{1}{2}$; $y = -\dfrac{1}{2}x + \dfrac{3}{2}$

43. $y = 7x - 5$ **45.** $(\frac{1}{3}, \frac{50}{27})$; (1, 2)

47. $(\frac{4}{3}, -\frac{770}{27})$; (2, −30)   **49.** $y = -\frac{1}{2}x + 1$; $y = 2x - \frac{3}{2}$

51. 0.125; 0.5; 2; 50

53. −5000/min; −1600/min; 7000; 4000

55. a. $\dfrac{180}{(t + 6)^2}$ **b.** 3.7; 2.2; 1.8; 1.1

c.

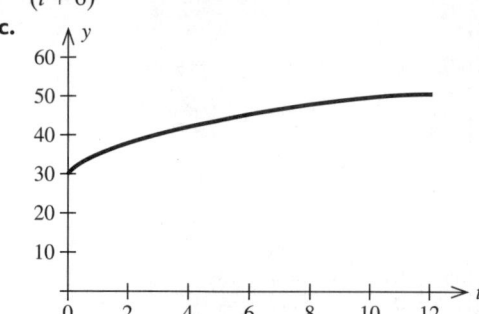

; yes

d. 50 words/min

57. Dropping at the rate of 0.0375 parts/million/year; dropping at the rate of 0.006 parts/million/year

Using Technology Exercises 11.2, page 709

1. 0.8750   **3.** 0.0774   **5.** −0.5000

7. 87,322 per year

Exercises 11.3, page 720

1. $8(2x - 1)^3$ **3.** $10x(x^2 + 2)^4$

5. $6x^2(1 - x)(2 - x)^2$ **7.** $\dfrac{-4}{(2x + 1)^3}$ **9.** $3x\sqrt{x^2 - 4}$

11. $\dfrac{3}{2\sqrt{3x - 2}}$ **13.** $\dfrac{-2x}{3(1 - x^2)^{2/3}}$ **15.** $-\dfrac{6}{(2x + 3)^4}$

17. $\dfrac{-1}{(2t - 3)^{3/2}}$ **19.** $-\dfrac{3(16x^3 + 1)}{2(4x^4 + x)^{5/2}}$

21. $\dfrac{-4(3x + 1)}{(3x^2 + 2x + 1)^3}$ **23.** $6x(2x^2 - x + 1)$

25. $3(t^{-1} - t^{-2})^2(-t^{-2} + 2t^{-3})$ **27.** $\dfrac{1}{2\sqrt{x} - 1} + \dfrac{1}{2\sqrt{x} + 1}$

29. $-12x(4x - 1)(3 - 4x)^3$

31. $6(x - 1)(2x - 1)(2x + 1)^3$ **33.** $-\dfrac{15(x + 3)^2}{(x - 2)^4}$

35. $\dfrac{3\sqrt{t}}{2(2t+1)^{5/2}}$ **37.** $\dfrac{-1}{2\sqrt{u+1}\,(3u+2)^{3/2}}$

39. $-\dfrac{2x(3x^2+1)}{(x^2-1)^5}$ **41.** $-\dfrac{2x(3x^2+1)^2(3x^2+13)}{(x^2-1)^5}$

43. $-\dfrac{3x^2+2x+1}{\sqrt{2x+1}\,(x^2-1)^2}$ **45.** $-\dfrac{t^2+2t-1}{2\sqrt{t+1}\,(t^2+1)^{3/2}}$

47. $\frac{4}{3}u^{1/3}$; $6x$; $8x(3x^2-1)^{1/3}$

49. $-\dfrac{2}{3u^{5/3}}$; $6x^2-1$; $-\dfrac{2(6x^2-1)}{3(2x^3-x+1)^{5/3}}$

51. $\frac{1}{2}u^{-1/2}-\frac{1}{2}u^{-3/2}$; $3x^2-1$; $\dfrac{(3x^2-1)(x^3-x-1)}{2(x^3-x)^{3/2}}$

53. $y=-33x+57$ **55.** $y=\frac{43}{5}x-\frac{54}{5}$

57. 0.333 million/wk; 0.304 million/wk; 16 million; 22.7 million

59. a. $0.027(0.2t^2+4t+64)^{-1/3}(0.1t+1)$
    b. 0.009 parts per million

61. a. $0.21t^2(t-3)(t-7)^3$
    b. 90.72; 0; −90.72; at 8 A.M., the level of nitrogen dioxide is increasing; at 10 A.M., the level stops increasing; and at 11 A.M., the level is decreasing.

63. $\dfrac{3450t}{(t+25)^2\sqrt{\frac{1}{2}t^2+2t+25}}$; 2.9 beats/min²; 0.7 beat/min²; 0.2 beat/min²; 179 beats/min

65. 160π ft²/sec   **67.** −27 mph/decade; 19 mph

69. $\dfrac{1.42(140t^2+3500t+21,000)}{(3t^2+80t+550)^2}$; 87,322 jobs/year

71. Decreasing at the rate of $5.86/passenger/year

1. 0.5774   **3.** 0.9390   **5.** −4.9498

7. 10,146,200/decade; 7,810,520/decade

Exercises 11.4, page 736

1. a. $C(x)$ is always increasing because as the number of units x produced increases, the greater the amount of money that must be spent on production.
    b. 4000

3. a. $114; $120.16; $138.12   **b.** $114; $120; $138

5. a. $\dfrac{5000}{x}+2$ **b.** $-\dfrac{5000}{x^2}$

7. $0.0002x^2-0.06x+120+\dfrac{5000}{x}$; $0.0004x-0.06-\dfrac{5000}{x^2}$

9. a. $-0.04x^2+800x$ **b.** $-0.08x+800$
    c. 400; when the level of production is 5000 units, the production of the next speaker system will bring in additional revenue of $400.

11. a. $-0.04x^2+600x-300,000$ **b.** $-0.08x+600$
    c. 200; −40
    d. The profit increases as production increases, peaking at 7500 units; beyond this level, profit falls.

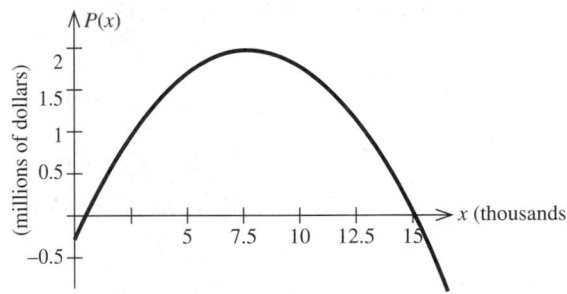

13. a. $-0.006x^2+180x$; $-0.000002x^3+0.014x^2+60x-60,000$
    b. $0.000006x^2-0.04x+120$; $-0.012x+180$; $-0.000006x^2+0.028x+60$
    c. 64; 156; 92   **d.**

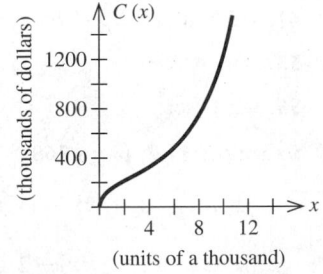

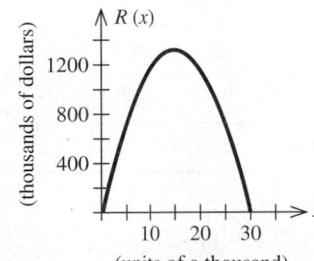

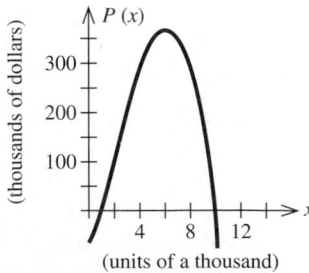

15. a. $0.000004x - 0.02 - \dfrac{60,000}{x^2}$

b. -0.0024; 0.0194; the average cost is decreasing when 5000 TV sets are produced and increasing when 10,000 units are produced.

17. 0.712 billion/billion dollars

21. \$0.209 billion/billion dollars

23. $\frac{5}{3}$; elastic   **25.** 1; unitary   **27.** 0.104; inelastic

29. a. Inelastic; elastic   **b.** When $p = 8.66$
c. Increase   **d.** Increase

31. a. Inelastic   **b.** Increase

33. $\dfrac{25}{50 - p}$; for $p > 25$, demand is elastic; for $p = 25$, demand is unitary; and for $p < 25$, demand is inelastic.

Exercises 11.5, page 745

1. $8x - 2$; 8   **3.** $6x^2 - 6x$; $6(2x - 1)$

5. $4t^3 - 6t^2 + 12t - 3$; $12(t^2 - t + 1)$

7. $10x(x^2 + 2)^4$; $10(x^2 + 2)^3(9x^2 + 2)$

9. $6t(2t^2 - 1)(6t^2 - 1)$; $6(60t^4 - 24t^2 + 1)$

11. $14x(2x^2 + 2)^{5/2}$; $28(2x^2 + 2)^{3/2}(6x^2 + 1)$

13. $(x^2 + 1)(5x^2 + 1)$; $4x(5x^2 + 3)$

15. $\dfrac{1}{(2x + 1)^2}$; $-\dfrac{4}{(2x + 1)^3}$

17. $\dfrac{2}{(s + 1)^2}$; $-\dfrac{4}{(s + 1)^3}$

19. $-\dfrac{3}{2(4 - 3u)^{1/2}}$; $-\dfrac{9}{4(4 - 3u)^{3/2}}$

21. $72x - 24$ **23.** $-\dfrac{6}{x^4}$

25. $\frac{81}{8}(3s - 2)^{-5/2}$ **27.** $192(2x - 3)$

29. 128 ft/sec; 32 ft/sec^2

31. a. and **b.**

t	0	1	2	3	4	5	6	7
$N'(t)$	0	2.7	4.8	6.3	7.2	7.5	7.2	6.3
$N''(t)$				0.6	0		-0.6	-1.2

33. a. $\frac{1}{4}t(t^2 - 12t + 32)$ **b.** 0 ft/sec; 0 ft/sec; 0 ft/sec
c. $\frac{3}{4}t^2 - 6t + 8$ **d.** 8 ft/sec^2; -4 ft/sec^2; 8 ft/sec^2
e. 0 ft; 16 ft; 0 ft

35. -3.09; 0.35; ten minutes after the test began, the smoke was being removed at the rate of 3%/min and the rate of purification was also increasing at the rate of 0.35/min^2.

37. $f(x) = x^{n+1/2}$

Using Technology Exercises 11.5, page 747

1. -18 **3.** 15.2762   **5.** -0.6255 **7.** 0.1973

9. -68.46214; at the beginning of 1988, the rate of the rate of the rate at which banks were failing was 68 banks/yr/yr/yr.

Exercises 11.6, page 757

1. a. $-\frac{1}{2}$ **b.** $-\frac{1}{2}$ **3. a.** $-\dfrac{1}{x^2}$ **b.** $-\dfrac{y}{x}$

5. a. $2x - 1 + \dfrac{4}{x^2}$ **b.** $3x - 2 - \dfrac{y}{x}$

7. a. $\dfrac{1 - x^2}{(1 + x^2)^2}$ **b.** $-2y^2 + \dfrac{y}{x}$

9. $-\dfrac{x}{y}$ **11.** $\dfrac{x}{2y}$ **13.** $1 - \dfrac{y}{x}$ **15.** $-\dfrac{y}{x}$

17. $-\dfrac{\sqrt{y}}{\sqrt{x}}$ **19.** $2\sqrt{x + y} - 1$ **21.** $-\dfrac{y^3}{x^3}$

23. $\dfrac{2\sqrt{xy} - y}{x - 2\sqrt{xy}}$ **25.** $\dfrac{6x - 3y - 1}{3x + 1}$ **27.** $\dfrac{2(2x - y^{3/2})}{3x\sqrt{y} - 4y}$

29. $-\dfrac{2x^2 + 2xy + y^2}{x^2 + 2xy + 2y^2}$ **31.** $y = 2$ **33.** $y = -\frac{3}{2}x + \frac{5}{2}$

35. $\dfrac{2y}{x^2}$ **37.** $\dfrac{2y(y - x)}{(2y - x)^3}$

39. a. $\dfrac{dV}{dt} = \pi r\left(r\dfrac{dh}{dt} + 2h\dfrac{dr}{dt}\right)$ **b.** 3.6π cu in./sec

41. Dropping at the rate of 111 tires per week

43. Increasing at the rate of 44 10-packs per week

45. Dropping at the rate of 3.7 cents/carton/week

47. 0.37; inelastic   **49.** 160π ft^2/sec   **51.** 17 ft/sec

53. 7.69 ft/sec

55. 1.5 ft/sec

Exercises 11.7, page 768

1. $4x\,dx$ **3.** $(3x^2 - 1)\,dx$ **5.** $\dfrac{dx}{2\sqrt{x + 1}}$

7. $\dfrac{6x + 1}{2\sqrt{x}}\,dx$ **9.** $\dfrac{x^2 - 2}{x^2}\,dx$

11. $\dfrac{-x^2 + 2x + 1}{(x^2 + 1)^2}\, dx$ 13. $\dfrac{6x - 1}{2\sqrt{3x^2 - x}}\, dx$

15. **a.** $2x\, dx$ **b.** 0.04   **c.** 0.0404

17. **a.** $-\dfrac{dx}{x^2}$ **b.** -0.05 **c.** -0.05263

19. 3.167   21. 7.0358   23. 1.983   25. 0.298

27. 2.50146   29. ± 8.64 cm^3 31. ± 0.076 cm^3

33. $7.9 billion   35. $2000   37. 75 cents

39. 0.133%   41. $\pm 1.91\%$

43. **a.** $\dfrac{10{,}000\left\{\left[1 - \left(1 + \dfrac{r}{12}\right)^{-360}\right] - 30r\left(1 + \dfrac{r}{12}\right)^{-361}\right\}}{\left[1 - \left(1 + \dfrac{r}{12}\right)^{-360}\right]^2}\, dr$

b. $17.27; $25.91; $34.54; $43.17

Using Technology Exercises 11.7, page 771

1. 7.5787   3. 0.031220185778   5. -0.0198761598

7. $26.60/month; $35.47/month; $44.34/month

9. 625

Chapter 11 Review Exercises, page 774

1. $15x^4 - 8x^3 + 6x - 2$ 3. $\dfrac{6}{x^4} - \dfrac{3}{x^2}$

5. $-\dfrac{1}{t^{3/2}} - \dfrac{6}{t^{5/2}}$ 7. $1 - \dfrac{2}{t^2} - \dfrac{6}{t^3}$

9. $2x + \dfrac{3}{x^{5/2}}$ 11. $\dfrac{2t}{(2t^2 + 1)^2}$

13. $\dfrac{1}{\sqrt{x}(\sqrt{x} + 1)^2}$ 15. $\dfrac{2x(x^4 - 2x^2 - 1)}{(x^2 - 1)^2}$

17. $72x^2(3x^3 - 2)^7$ 19. $\dfrac{2t}{\sqrt{2t^2 + 1}}$

21. $\dfrac{-4(3t - 1)}{(3t^2 - 2t + 5)^3}$ 23. $\dfrac{2(x^2 + 1)(x^2 - 1)}{x^3}$

25. $4t^2(5t + 3)(t^2 + t)^3$, or $4t^5(5t + 3)(t + 1)^3$

27. $\dfrac{1}{2\sqrt{x}}(x^2 - 1)^2(13x^2 - 1)$

29. $-\dfrac{12x + 25}{2\sqrt{3x + 2}\,(4x - 3)^2}$

31. $2(12x^2 - 9x + 2)$ 33. $\dfrac{2t(t^2 - 12)}{(t^2 + 4)^3}$

35. $\dfrac{2}{(2x^2 + 1)^{3/2}}$ 37. $\dfrac{2x}{y}$

39. $-\dfrac{2x}{y^2 - 1}$ 41. $\dfrac{x - 2y}{2x + y}$

43. **a.** $(2, -25)$ and $(-1, 14)$
    b. $y = -4x - 17; y = -4x + 10$

45. $y = -\dfrac{\sqrt{3}}{3}x + \dfrac{4}{3}\sqrt{3}$

47. $-\dfrac{48}{(2x - 1)^4}; (-\infty; \tfrac{1}{2}) \cup (\tfrac{1}{2}, \infty)$

49. **a.** $2.20; $2.20   **b.** $\dfrac{2500}{x} + 2.2; -\dfrac{2500}{x^2}$

c. $\lim\limits_{x \to \infty}\left(\dfrac{2500}{x} + 2.2\right) = 2.2$

51. **a.** $2000x - 0.04x^2$;
    $-0.000002x^3 - 0.02x^2 + 1000x - 120{,}000$;
    $0.000002x^2 - 0.02x + 1000 + \dfrac{120{,}000}{x}$

b. $0.000006x^2 - 0.04x + 1000; 2000 - 0.08x$;
    $-0.000006x^2 - 0.04x + 1000$;
    $0.000004x - 0.02 - \dfrac{120{,}000}{x^2}$

c. 934; 1760; 826

d. $-0.0048; 0.010125$; at a production level of 5000, the average cost is decreasing by 0.48 cents/unit. At a production level of 8000, the average cost is increasing by 1.0125 cents/unit.

CHAPTER 12

Exercises 12.1, page 791

1. Decreasing on $(-\infty, 0)$ and increasing on $(0, \infty)$

3. Increasing on $(-\infty, -1) \cup (1, \infty)$ and decreasing on $(-1, 1)$

5. Decreasing on $(-\infty, 0) \cup (2, \infty)$ and increasing on $(0, 2)$

7. Decreasing on $(-\infty, -1) \cup (1, \infty)$ and increasing on $(-1, 1)$

9. Increasing on $(20.2, 20.6) \cup (21.7, 21.8)$, constant on $(19.6, 20.2) \cup (20.6, 21.1)$, and decreasing on $(21.1, 21.7) \cup (21.8, 22.7)$

11. Increasing on $(-\infty, \infty)$

13. Decreasing on $(-\infty, \tfrac{3}{2})$ and increasing on $(\tfrac{3}{2}, \infty)$

15. Decreasing on $\left(-\infty, -\frac{\sqrt{3}}{3}\right) \cup \left(\frac{\sqrt{3}}{3}, \infty\right)$ and increasing on $\left(-\frac{\sqrt{3}}{3}, \frac{\sqrt{3}}{3}\right)$

17. Increasing on $(-\infty, -2) \cup (0, \infty)$ and decreasing on $(-2, 0)$

19. Increasing on $(-\infty, 3) \cup (3, \infty)$

21. Decreasing on $(-\infty, 0) \cup (0, 3)$ and increasing on $(3, \infty)$

23. Decreasing on $(-\infty, 2) \cup (2, \infty)$

25. Decreasing on $(-\infty, 1) \cup (1, \infty)$

27. Increasing on $(-\infty, 0) \cup (0, \infty)$

29. Increasing on $(-1, \infty)$

31. Increasing on $(-4, 0)$ and decreasing on $(0, 4)$

33. Increasing on $(-\infty, 0) \cup (0, \infty)$

35. Increasing on $(-\infty, 1)$ and decreasing on $(1, \infty)$

37. Relative maximum: $f(0) = 1$; relative minima: $f(-1) = 0$ and $f(1) = 0$

39. Relative maximum: $f(-1) = 2$; relative minimum: $f(1) = -2$

41. Relative maximum: $f(1) = 3$; relative minimum: $f(2) = 2$

43. Relative minimum: $f(0) = 2$

45. a   47. d

49. Relative minimum: $f(2) = -4$

51. Relative minimum: $f(2) = 2$

53. Relative minimum: $f(0) = 2$

55. Relative maximum: $g(0) = 4$; relative minimum: $g(2) = 0$

57. Relative minimum: $F(3) = -5$; relative maximum: $F(-1) = \frac{17}{3}$

59. Relative minimum: $g(3) = -19$

61. Relative minimum: $f(\frac{1}{2}) = \frac{63}{16}$

63. None

65. Relative maximum: $f(-3) = -4$; relative minimum: $f(3) = 8$

67. Relative maximum: $f(1) = \frac{1}{2}$; relative minimum: $f(-1) = -\frac{1}{2}$

69. Relative maximum: $f(0) = 0$

71. Relative minimum: $f(1) = 0$

73. Increasing on $(0, 4000)$ and decreasing on $(4000, \infty)$

75. Decreasing on $(0, 5)$ and increasing on $(5, 10)$; after declining from 1984 through 1989, the index began to increase after 1989.

79. Increasing on $(0, 4)$ and decreasing on $(4, 5)$; the cash in the Central Provident Fund will be increasing from 1995 to 2015 and decreasing from 2015 to 2025.

81. Increasing on $(0, 1)$ and decreasing on $(1, 4)$

85. If $a > 0$, f is decreasing on $(-\infty, -b/2a)$ and increasing on $(-b/2a, \infty)$. If $a < 0$, f is increasing on $(-\infty, -b/2a)$ and decreasing on $(-b/2a, \infty)$.

87. **a.** $f'(x) = -2x$ if $x \neq 0$ and is not defined if $x = 0$.
    b. No

91. $f'(x) = 0$ for all x, so f is a constant function.

Using Technology Exercises 12.1, page 798

1. **a.** f is decreasing on $(-\infty, -0.2934)$ and increasing on $(-0.2934, \infty)$
   b. Relative minimum: $f(-0.2934) = -2.5435$

3. **a.** f is increasing on $(-\infty, -1.6144) \cup (0.2390, \infty)$ and decreasing on $(-1.6144, 0.2390)$
   b. Relative maximum: $f(-1.6144) = 26.7991$; relative minimum: $f(0.2390) = 1.6733$

5. **a.** f is decreasing on $(-\infty, -1) \cup (0.33, \infty)$ and increasing on $(-1, 0.33)$
   b. Relative maximum: $f(0.33) = 1.11$; relative minimum: $f(-1) = -0.63$

7. **a.** f is decreasing on $(-1, -0.71)$ and increasing on $(-0.71, 1)$
   b. Relative minimum: $f(-0.71) = -1.41$

9. **a.**

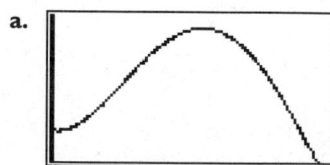

b. f is decreasing on $(0, 0.2398) \cup (6.8758, 12)$ and increasing on $(0.2398, 6.8758)$
c. $(6.8758, 200.14)$

11. a.

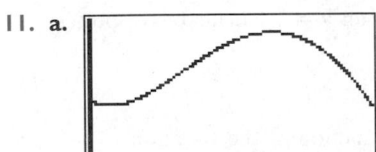

b. f is decreasing on $(0, 0.8343) \cup (7.6726, 12)$ and increasing on $(0.8343, 7.6726)$

13. Increasing from 7 A.M. to 11:30 A.M. and decreasing from 11:30 A.M. to 6 P.M.; highest at 11:30 A.M. when it reaches 164 PSI

Exercises 12.2, page 811

1. Concave downward on $(-\infty, 0)$ and concave upward on $(0, \infty)$; inflection point: $(0, 0)$

3. Concave downward on $(-\infty, 0) \cup (0, \infty)$

5. Concave upward on $(-\infty, 0) \cup (1, \infty)$ and concave downward on $(0, 1)$; inflection points: $(0, 0)$ and $(1, -1)$

7. Concave downward on $(-\infty, -2) \cup (-2, 2) \cup (2, \infty)$

9. a   **11.** b

13. a. $D_1'(t) > 0$, $D_2'(t) > 0$, $D_1''(t) > 0$, and $D_2''(t) < 0$ on $(0, 12)$
b. With or without the proposed promotional campaign, the deposits will increase, but with the promotion, the deposits will increase at an increasing rate whereas without the promotion, the deposits will increase at a decreasing rate.

15. At the time t_0, corresponding to its t-coordinate, the restoration process is working at its peak.

23. Concave upward on $(-\infty, \infty)$

25. Concave downward on $(-\infty, 0)$; concave upward on $(0, \infty)$

27. Concave upward on $(-\infty, 0) \cup (3, \infty)$; concave downward on $(0, 3)$

29. Concave downward on $(-\infty, 0) \cup (0, \infty)$

31. Concave downward on $(-\infty, 4)$

33. Concave downward on $(-\infty, 2)$; concave upward on $(2, \infty)$

35. Concave upward on $(-\infty, -\sqrt{6}/3) \cup (\sqrt{6}/3, \infty)$; concave downward on $(-\sqrt{6}/3, \sqrt{6}/3)$

37. Concave downward on $(-\infty, 1)$; concave upward on $(1, \infty)$

39. Concave upward on $(-\infty, 0) \cup (0, \infty)$

41. Concave upward on $(-\infty, 2)$; concave downward on $(2, \infty)$

43. Concave upward on $(-\infty, -1) \cup (1, \infty)$; concave downward on $(-1, 1)$

45. $(0, -2)$ **47.** $(1, -15)$ **49.** $(0, 1)$ and $(\frac{2}{3}, \frac{11}{27})$

51. $(0, 0)$ **53.** $(1, 2)$ **55.** $(-\sqrt{3}/3, 3/2)$ and $(\sqrt{3}/3, 3/2)$

57. Relative maximum: $f(1) = 5$

59. None

61. Relative maximum: $f(-1) = -\frac{22}{3}$; relative minimum: $f(5) = -\frac{130}{3}$

63. Relative maximum: $f(-3) = -6$; relative minimum: $f(3) = 6$

65. None

67. Relative minimum: $f(-2) = 12$

69. Relative maximum: $g(1) = \frac{1}{2}$; relative minimum: $g(-1) = -\frac{1}{2}$

71. Relative maximum: $f(0) = 0$; relative minimum: $f(\frac{4}{3}) = \frac{256}{27}$

73. Relative minimum: $g(4) = -\frac{1}{108}$

75. a. N is increasing on $(0, 12)$.
b. $N''(t) < 0$ on $(0, 6)$ and $N''(t) > 0$ on $(6, 12)$
c. The rate of growth of the number of help-wanted advertisements was decreasing over the first 6 months of the year and increasing over the last 6 months.

77. $(100, 4600)$; the sales increase rapidly until $100,000 is spent on advertising; after that, any additional expenditure results in increased sales but at a slower rate of increase.

79. 10 A.M.   **81.** $(18, 20{,}310)$; 1452 ft/sec

85. $\left(-\dfrac{b}{3a}, f\left(-\dfrac{b}{3a}\right)\right)$

Using Technology Exercises 12.2, page 817

1. a. f is concave upward on $(-\infty, 0) \cup (1.1667, \infty)$ and concave downward on $(0, 1.1667)$
b. $(1.1667, 1.1153)$; $(0, 2)$

3. a. f is concave downward on $(-\infty, 0)$ and concave upward on $(0, \infty)$
b. $(0, 2)$

5. a. f is concave downward on $(-\infty, 0)$ and concave upward on $(0, \infty)$
b. $(0, 0)$

7. a. f is concave downward on $(-\infty, -2.4495) \cup (0, 2.4495)$ and concave upward on $(-2.4495, 0) \cup (2.4495, \infty)$
b. $(2.4495, 0.3402)$

9. a.

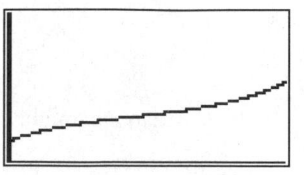

11. a.

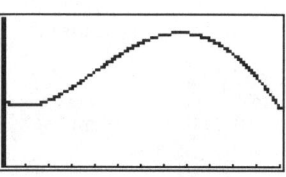

b. $(5.5318, 35.9483)$
c. $t = 5.5318$

b. $(3.9024, 77.0919)$

Exercises 12.3, page 829

1. Horizontal asymptote: $y = 0$

3. Horizontal asymptote: $y = 0$; vertical asymptote: $x = 0$

5. Horizontal asymptote: $y = 0$; vertical asymptotes: $x = -1$ and $x = 1$

7. Horizontal asymptote: $y = 3$; vertical asymptote: $x = 0$

9. Horizontal asymptotes: $y = 1$ and $y = -1$

11. Horizontal asymptote: $y = 0$; vertical asymptote: $x = 0$

13. Horizontal asymptote: $y = 0$; vertical asymptote: $x = 0$

15. Horizontal asymptote: $y = 1$; vertical asymptote: $x = -1$

17. None

19. Horizontal asymptote: $y = 1$: vertical asymptotes: $t = -3$ and $t = 3$

21. Horizontal asymptote: $y = 0$; vertical asymptotes: $x = -2$ and $x = 3$

23. Horizontal asymptote: $y = 2$; vertical asymptote: $t = 2$

25. Horizontal asymptote: $y = 1$; vertical asymptotes: $x = -2$ and $x = 2$

27. None

29. f is the derivative function of the function g

31.

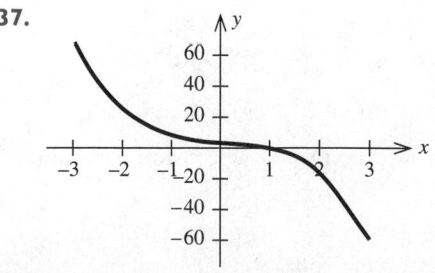

33.

35.

37.

39.

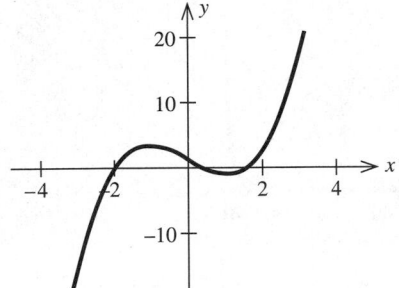

41.

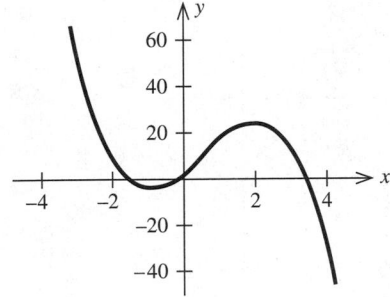

43.

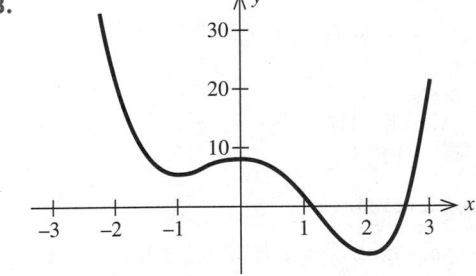

45.

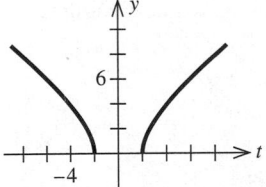

47.

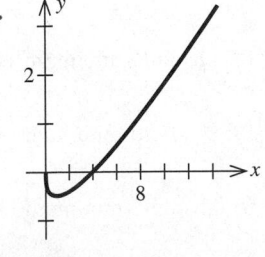

49.

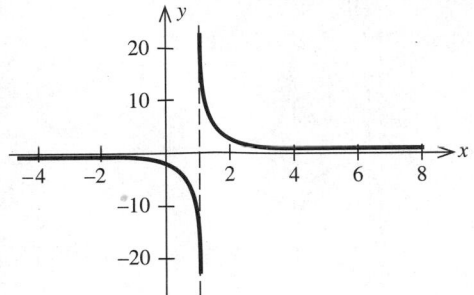

51.

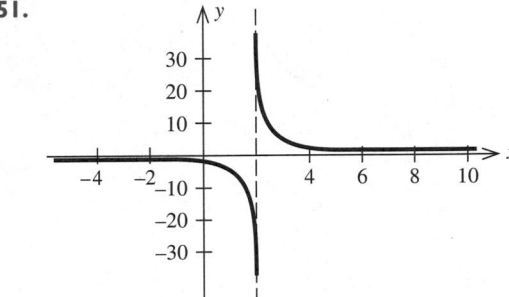

53.

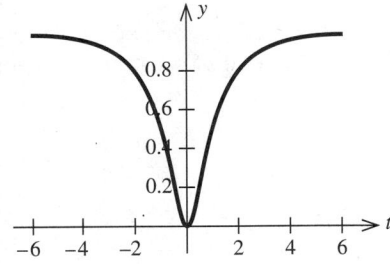

55.

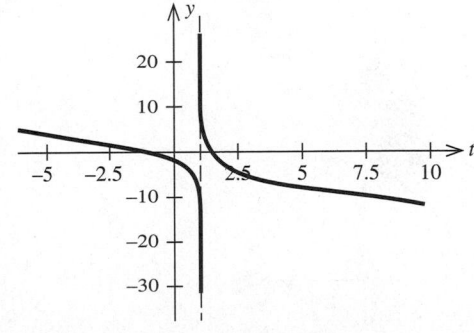

57.

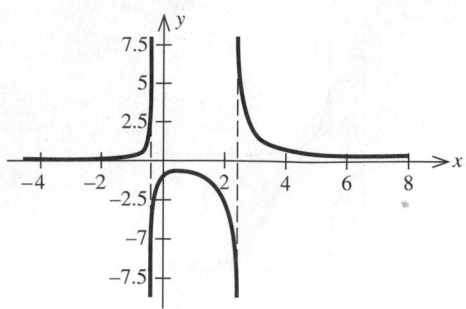

59.

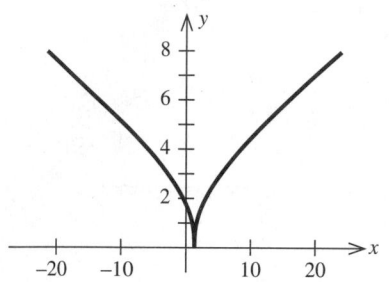

61. a. $y = 2.2$ **b.** $2.20/disc

63. a. a

b. The initial speed of the reaction approaches a moles/liter/sec as the amount of substrate becomes arbitrarily large.

65.

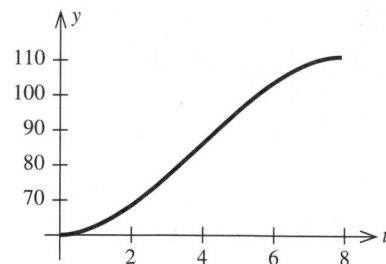

67.

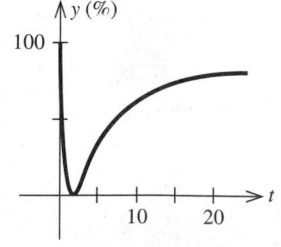

69.

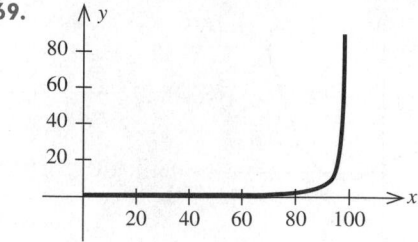

Using Technology Exercises 12.3, page 835

1.

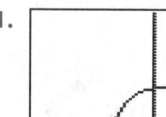

3.

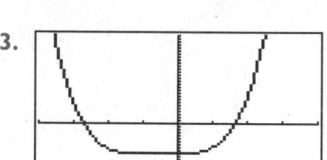

5. $-0.9733, 2.3165, 4.6569$ **7.** $-1.1301, 2.9267$

9. 1.5142

Exercises 12.4, page 848

1. None

3. Absolute minimum value: 0

5. Absolute maximum value: 3; absolute minimum value: -2

7. Absolute maximum value: 3; absolute minimum value: $-\frac{27}{16}$

9. Absolute minimum value: $-\frac{41}{8}$

11. No absolute extrema

13. Absolute maximum value: 1

15. Absolute maximum value: 5; absolute minimum value: -4

17. Absolute maximum value: 10; absolute minimum value: 1

19. Absolute maximum value: 19; absolute minimum value: -1

21. Absolute maximum value: 16; absolute minimum value: -1

23. Absolute maximum value: 3; absolute minimum value: $\frac{5}{3}$

25. Absolute maximum value: $\frac{37}{3}$; absolute minimum value: 5

27. Absolute maximum value: ~1.04; absolute minimum value: −1.5

29. No absolute extrema

31. Absolute maximum value: 1; absolute minimum value: 0

33. Absolute maximum value: 0; absolute minimum value: −3

35. Absolute maximum value: $\sqrt{2}/4 \approx 0.35$; absolute minimum value: $-\frac{1}{3}$

37. Absolute maximum value: $\sqrt{2}/2$; absolute minimum value: $-\sqrt{2}/2$

39. 144 ft   **41.** 3000

43. $f(6) = 3.60$, $f(0.5) = 1.13$; the number of nonfarm, full-time, self-employed women over the time interval from 1963 to 1993 reached its highest level, 3.6 million, in 1993.

45. 5000   **47.** 3333

49. a. $0.0025x + 80 + \dfrac{10,000}{x}$ **b.** 2000

c. 2000   **d.** Same

51. 533

53. a. 2 days after the organic waste was dumped into the pond
b. 3.5 days after the organic waste was dumped into the pond

63. c.

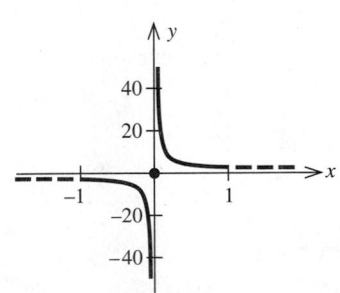

Using Technology Exercises 12.4, page 853

1. Absolute maximum value: 145.8985; absolute minimum value: −4.3834

3. Absolute maximum value: 16; absolute minimum value: −0.1257

5. Absolute maximum value: 2.8889; absolute minimum value: 0

7. a.

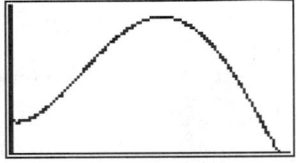

b. 200.1410 banks per year

9. a.

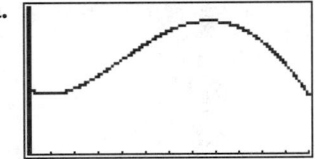

b. Absolute maximum value: 108.8756; absolute minimum value: 49.7773

Exercises 12.5, page 863

1. 750 yd × 1500 yd; 1,125,000 sq yd

3. $10\sqrt{2}$ ft × $40\sqrt{2}$ ft

5. $\frac{16}{3}$ inches × $\frac{16}{3}$ inches × $\frac{4}{3}$ inches

7. 5.04 inches × 5.04 inches × 5.04 inches

9. 18 inches × 18 inches × 36 inches; 11,664 cu in.

11. $r = \frac{36}{\pi}$ inches; $l = 36$ inches; $\frac{46,656}{\pi}$ cu in.

13. $\frac{2}{3}\sqrt[3]{9}$ ft × $\sqrt[3]{9}$ ft × $\frac{2}{5}\sqrt[3]{9}$ ft   **15.** 250; $62,500; $250

17. $w \approx 13.86$ inches; $h \approx 19.60$ inches

19. $x = 750\sqrt{2}$ ft   **21.** 56.57 mph   **23.** 45; 44,721

Chapter 12 Review Exercises, page 868

1. a. f is increasing on $(-\infty, 1) \cup (1, \infty)$
b. No relative extrema
c. Concave down on $(-\infty, 1)$; concave up on $(1, \infty)$
d. $(1, -\frac{17}{3})$

3. a. f is increasing on $(-1, 0) \cup (1, \infty)$; decreasing on $(-\infty, -1) \cup (0, 1)$
b. Relative maximum: 0; relative minimum: −1

c. Concave up on $(-\infty, -\sqrt{3}/3) \cup (\sqrt{3}/3, \infty)$; concave down on $(-\sqrt{3}/3, \sqrt{3}/3)$
d. $(-\sqrt{3}/3, -5/9); (\sqrt{3}/3, -5/9)$

5. a. f is increasing on $(-\infty, 0) \cup (2, \infty)$; decreasing on $(0, 1) \cup (1, 2)$
b. Relative maximum: 0; relative minimum: 4
c. Concave up on $(1, \infty)$; concave down on $(-\infty, 1)$
d. None

7. a. f is decreasing on $(-\infty, 1) \cup (1, \infty)$
b. No relative extrema
c. Concave down on $(-\infty, 1)$; concave up on $(1, \infty)$
d. $(1, 0)$

9. a. f is increasing on $(-\infty, -1) \cup (-1, \infty)$
b. No relative extrema
c. Concave down on $(-1, \infty)$; concave up on $(-\infty, -1)$
d. No inflection point

11.

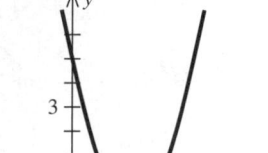

13.

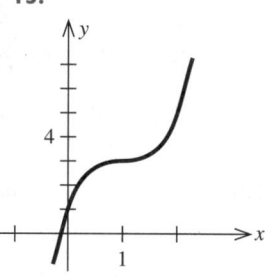

15.

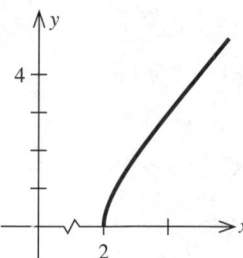

17.

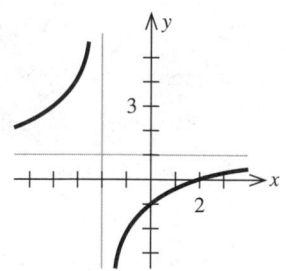

19. Vertical asymptote: $x = -\frac{3}{2}$; horizontal asymptote: $y = 0$

21. Vertical asymptotes: $x = -2$, $x = 4$; horizontal asymptote: $y = 0$

23. Absolute minimum value: $-\frac{25}{8}$

25. Absolute maximum value: 5; absolute minimum value: 0

27. Absolute maximum value: -16; absolute minimum value: -32

29. Absolute maximum value: $\frac{8}{3}$; absolute minimum value: 0

31. Absolute maximum value: $\frac{1}{2}$; absolute minimum value: $-\frac{1}{2}$

33. $4000   **35.** 168   **37.** 10 A.M.

39. 20,000 cases

CHAPTER 13

Exercises 13.1, page 880

1. a. 16   **b.** 27   **3. a.** 3   **b.** $\sqrt{5}$

5. a. -3 **b.** 8

7. a. $4x^3$ **b.** $5xy^2\sqrt{x}$ **9. a.** $2/a^2$ **b.** $\frac{1}{3}b^2$

11. a. $8x^9y^6$ **b.** $16x^4y^4z^6$

13. 2   **15.** 3   **17.** 3   **19.** $\frac{5}{4}$

21.

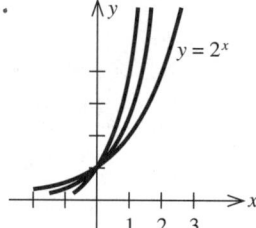

23.

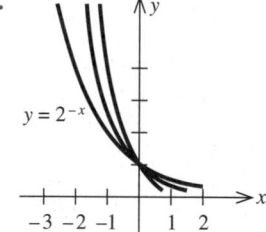

25.

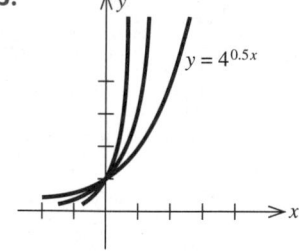

27.

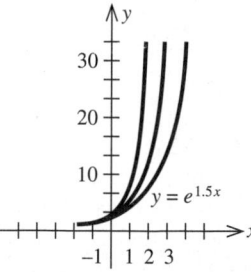

29.

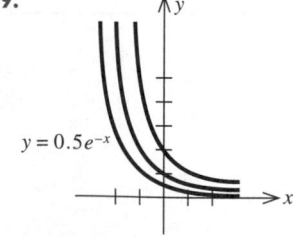

31. Investment A

33. a. $33,885.14   **b.** $33,565.38

35. a. $16,262.79   **b.** $12,047.77   **c.** $6,611.96

39. Bank B

Using Technology Exercises 13.1, page 883

1.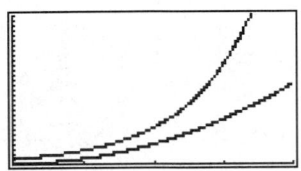

3.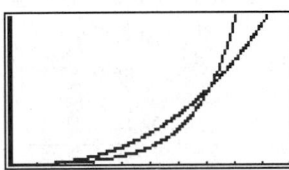

5.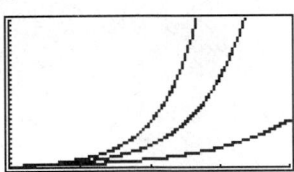

7.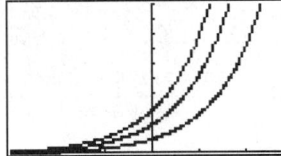

9.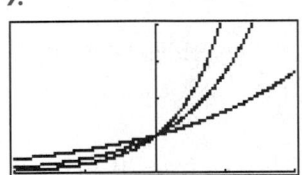

Exercises 13.2, page 892

1. $\log_2 64 = 6$ **3.** $\log_3 \frac{1}{9} = -2$ **5.** $\log_{1/3} \frac{1}{3} = 1$

7. $\log_{32} 8 = \frac{3}{5}$ **9.** $\log_{10} 0.001 = -3$ **11.** 1.0792

13. 1.2042 **15.** 1.6813 **17.** $\log x + 4 \log (x + 1)$

19. $\frac{1}{2} \log (x + 1) - \log (x^2 + 1)$ **21.** $\ln x - x^2$

23. $-\frac{3}{2} \ln x - \frac{1}{2} \ln (1 + x^2)$ **25.** $x \ln x$

27.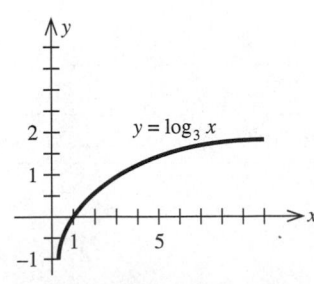

29.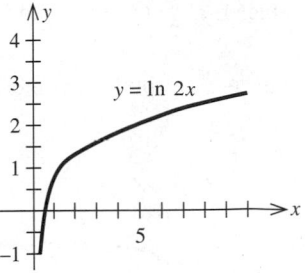

31.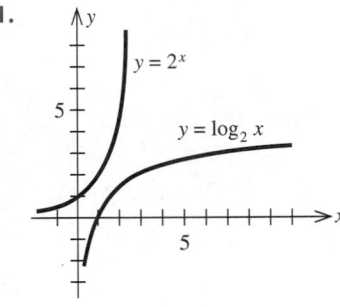

33. 5.1986 **35.** -0.0912 **37.** -8.0472

39. -4.9041 **41.** $-2 \ln \left(\dfrac{A}{B} \right)$ **43.** 105.7 mm

45. a. $10^3 I_0$
    b. $100,000$ times greater
    c. $10,000,000$ times greater

47. $34\frac{1}{2}$ hr earlier, at 1:30 P.M.

Exercises 13.3, page 901

1. $3e^{3x}$ **3.** $-e^{-t}$ **5.** $e^x + 1$ **7.** $x^2 e^x (x + 3)$

9. $\dfrac{2e^x (x - 1)}{x^2}$ **11.** $3(e^x - e^{-x})$ **13.** $-\dfrac{1}{e^w}$

15. $6e^{3x-1}$ **17.** $-2xe^{-x^2}$ **19.** $\dfrac{3e^{-1/x}}{x^2}$

21. $25e^x (e^x + 1)^{24}$ **23.** $\dfrac{e^{\sqrt{x}}}{2\sqrt{x}}$ **25.** $e^{3x+2}(3x - 2)$

27. $\dfrac{2e^x}{(e^x + 1)^2}$ **29.** $2(8e^{-4x} + 9e^{3x})$ **31.** $6e^{3x}(3x + 2)$

33. $y = 2x - 2$

35. f is increasing on $(-\infty, 0)$ and decreasing on $(0, \infty)$

37. Concave downward on $(-\infty, 0)$; concave upward on $(0, \infty)$

39. $(1, e^{-2})$

41. Absolute maximum value: 1; absolute minimum value: e^{-1}

43. Absolute minimum value: -1; absolute maximum value: $2e^{-3/2}$

45.

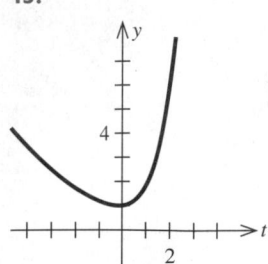

47.

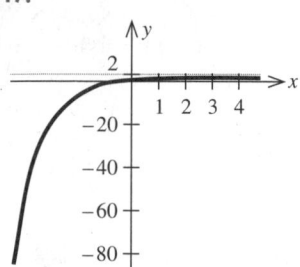

49. $-\$6065/\text{day}$; $-\$3679/\text{day}$; $-\$2231/\text{day}$; $-\$1353/\text{day}$

51. b. 4505 per year; 273 cases per year

53. a. -1.68 cents/case   **b.** $\$40.36$ per case

55. Tenth year

57. 1.8; -0.1084; -0.2286; -0.1279; initially, the number of barrels of oil needed to fuel productivity ($\$1000$ worth of economic output) was increasing, but after one decade the number of barrels required was decreasing.

Using Technology Exercises 13.3, page 904

1. 5.4366   **3.** 12.3929   **5.** 0.1861

7. a. 50   **c.**

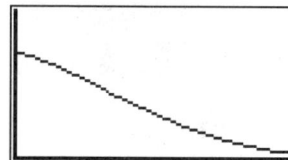

9. a.

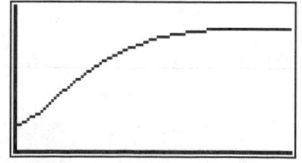

b. 4.2720 billion/half century

11. a. 196 aphids/stem; 180 aphids/stem
    b. Increasing at the rate of 226 aphids/month; decreasing at the rate of 238 aphids/month

Exercises 13.4, page 912

1. $\dfrac{5}{x}$ **3.** $\dfrac{1}{x+1}$ **5.** $\dfrac{8}{x}$ **7.** $\dfrac{1}{2x}$ **9.** $\dfrac{-2}{x}$

11. $\dfrac{2(4x-3)}{4x^2-6x+3}$ **13.** $\dfrac{1}{x(x+1)}$ **15.** $x(1+2\ln x)$

17. $\dfrac{2(1-\ln x)}{x^2}$ **19.** $\dfrac{3}{u-2}$ **21.** $\dfrac{1}{2x\sqrt{\ln x}}$

23. $\dfrac{3(\ln x)^2}{x}$ **25.** $\dfrac{3x^2}{x^3+1}$ **27.** $\dfrac{(x\ln x+1)e^x}{x}$

29. $\dfrac{e^{2t}[2(t+1)\ln(t+1)+1]}{t+1}$ **31.** $\dfrac{1-\ln x}{x^2}$ **33.** $-\dfrac{1}{x^2}$

35. $\dfrac{2(2-x^2)}{(x^2+2)^2}$ **37.** $(x+1)(5x+7)(x+2)^2$

39. $(x-1)(x+1)^2(x+3)^3(9x^2+14x-7)$

41. $\dfrac{(2x^2-1)^4(38x^2+40x+1)}{2(x+1)^{3/2}}$ **43.** $3^x \ln 3$

45. $(x^2+1)^{x-1}[2x^2+(x^2+1)\ln(x^2+1)]$

47. $y = x - 1$

49. f is decreasing on $(-\infty, 0)$ and increasing on $(0, \infty)$

51. Concave upward: $(-\infty, -1) \cup (1, \infty)$; concave downward: $(-1, 0) \cup (0, 1)$

53. $(-1, \ln 2)$ and $(1, \ln 2)$

55. Absolute minimum value: 1; absolute maximum value: $3 - \ln 3$

57.

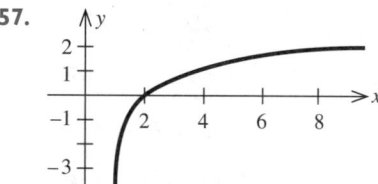

Exercises 13.5, page 922

1. a. 0.05
   b. 400
   c.

t	0	10	20	100	1000
Q	400	660	1087	59,365	2.07×10^{24}

3. a. $Q(t) = 100e^{0.035t}$
   b. 266 minutes
   c. $Q(t) = 1000e^{0.035t}$

5. a. 55.5 yr   **b.** 14.25 billion   **7.** 8.7 lb/sq in.

9. $Q(t) = 100e^{-0.049t}$; 70.6 grams **11.** 13,412 years ago

13.

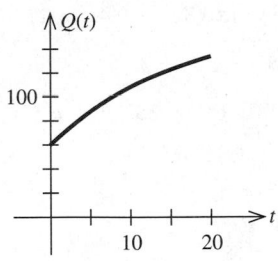

a. 60 words per minute
b. 107 words per minute
c. 136 words per minute

15.

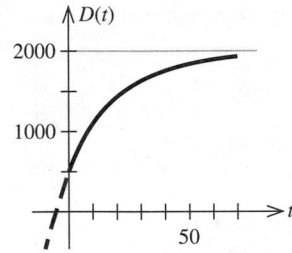

a. 573 computers; 1177 computers; 1548 computers; 1925 computers
b. 2000 computers
c. 46 computers/month

17. a. 11 **b.** 937 **c.** 1000 **19.** 3%; 65%

21. 1080 **23. b.** $\dfrac{1}{k}\ln A$

Chapter 13 Review Exercises, page 926

1. a.–b.

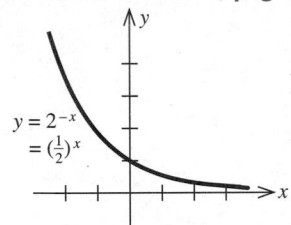

$y = 2^{-x}$
$= \left(\tfrac{1}{2}\right)^x$

3. $\log_{16} 0.125 = -\tfrac{3}{4}$ **5.** 2 **7.** $x + 2y - z$

9.

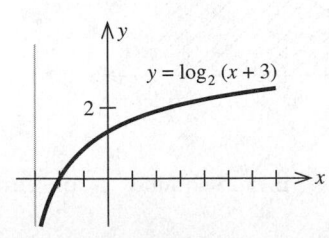

$y = \log_2 (x + 3)$

11. $(2x + 1)e^{2x}$ **13.** $\dfrac{(1 - 4t)}{2\sqrt{t}e^{2t}}$ **15.** $\dfrac{2(e^{2x} + 2)}{(1 + e^{-2x})^2}$

17. $(1 - 2x^2)e^{-x^2}$ **19.** $(x + 1)^2 e^x$ **21.** $\dfrac{2xe^{x^2}}{e^{x^2} + 1}$

23. $\dfrac{x + 1 - x \ln x}{x(x + 1)^2}$ **25.** $\dfrac{4e^{4x}}{e^{4x} + 3}$ **27.** $\dfrac{1 + e^x(1 - x \ln x)}{x(1 + e^x)^2}$

29. $-\dfrac{9}{(3x + 1)^2}$ **31.** 0

33. $6x(x^2 + 2)^2(3x^3 + 2x + 1)$ **35.** $y = -(2x - 3)e^{-2}$

37.

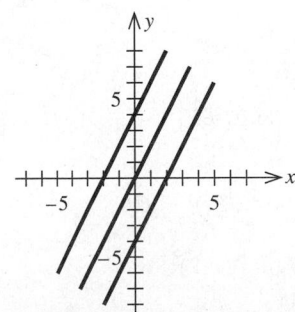

$y = xe^{-2x}$

39. Absolute maximum value: $1/e$ **41.** 12% per year

43. a. $Q(t) = 2000e^{0.01831t}$ **b.** 161,992

45.

$D(t)$ graph with values 4000, 2000 marked and t-axis at 50

a. 1175; 2540; 3289 **b.** 4000

CHAPTER 14

Exercises 14.1, page 940

5. b. $y = 2x + C$

graph of parallel lines

7. b. $y = \frac{1}{3}x^3 + C$

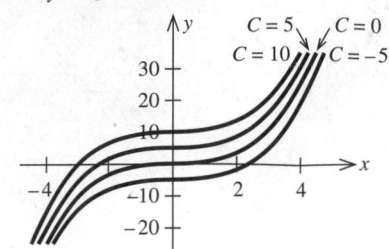

9. $6x + C$ **11.** $\frac{1}{4}x^4 + C$ **13.** $-\frac{1}{3x^3} + C$

15. $\frac{3}{5}x^{5/3} + C$ **17.** $-\frac{4}{x^{1/4}} + C$ **19.** $-\frac{2}{x} + C$

21. $\frac{2}{3}\pi t^{3/2} + C$ **23.** $3x - x^2 + C$

25. $\frac{1}{3}x^3 + \frac{1}{2}x^2 - \frac{1}{2x^2} + C$ **27.** $4e^x + C$

29. $x + \frac{1}{2}x^2 + e^x + C$ **31.** $x^4 + \frac{2}{x} - x + C$

33. $\frac{2}{7}x^{7/2} + \frac{4}{5}x^{5/2} - \frac{1}{2}x^2 + C$ **35.** $\frac{2}{3}x^{3/2} + 6\sqrt{x} + C$

37. $\frac{1}{9}u^3 + \frac{1}{4}u^2 - \frac{1}{3}u + C$ **39.** $\frac{2}{3}t^3 - \frac{3}{2}t^2 - 2t + C$

41. $\frac{1}{3}x^3 - 2x - \frac{1}{x} + C$ **43.** $\frac{1}{3}s^3 + s^2 + s + C$

45. $e^t + \frac{t^{e+1}}{e+1} + C$ **47.** $\frac{1}{2}x^2 + x - \ln|x| - \frac{1}{x} + C$

49. $\ln|x| + \frac{4}{\sqrt{x}} - \frac{1}{x} + C$ **51.** $x^2 + x + 1$

53. $x^3 + 2x^2 - x - 5$ **55.** $x - \frac{1}{x} + 2$ **57.** $x + \ln|x|$

59. $\sqrt{x}$ **61.** $e^x + \frac{1}{2}x^2 + 2$ **63.** $f(t) = \frac{4}{3}t^{3/2}$

65. \$3370   **67.** 5000 units; \$34,000

69. a. $-t^3 + 6t^2 + 45t$ **b.** 212   **71.** 21,960

73. 120,000

75. 1.9424 sq m   **77.** $\frac{1}{2}k(R^2 - r^2)$

79. $9\frac{7}{9}$ ft/sec²; 396 ft   **81.** 36 ft/sec²

Exercises 14.2, page 953

1. $\frac{1}{5}(4x + 3)^5 + C$ **3.** $\frac{1}{3}(x^3 - 2x)^3 + C$

5. $-\frac{1}{2(2x^2 + 3)^2} + C$ **7.** $\frac{2}{3}(t^3 + 2)^{3/2} + C$

9. $\frac{1}{20}(x^2 - 1)^{10} + C$ **11.** $-\frac{1}{5}\ln|1 - x^5| + C$

13. $\ln(x - 2)^2 + C$ **15.** $\frac{1}{2}\ln(0.3x^2 - 0.4x + 2) + C$

17. $\frac{1}{6}\ln|3x^2 - 1| + C$ **19.** $-\frac{1}{2}e^{-2x} + C$

21. $-e^{2-x} + C$ **23.** $-\frac{1}{2}e^{-x^2} + C$ **25.** $e^x + e^{-x} + C$

27. $\ln(1 + e^x) + C$ **29.** $2e^{\sqrt{x}} + C$

31. $-\frac{1}{6(e^{3x} + x^3)^2} + C$ **33.** $\frac{1}{8}(e^{2x} + 1)^4 + C$

35. $\frac{1}{2}(\ln 5x)^2 + C$ **37.** $\ln|\ln x| + C$

39. $\frac{2}{3}(\ln x)^{3/2} + C$ **41.** $\frac{1}{2}e^{x^2} - \frac{1}{2}\ln(x^2 + 2) + C$

43. $\frac{2}{3}(\sqrt{x} - 1)^3 + 3(\sqrt{x} - 1)^2 + 8(\sqrt{x} - 1)$
$+ 4\ln|\sqrt{x} - 1| + C$

45. $\frac{(6x + 1)(x - 1)^6}{42} + C$

47. $5 + 4\sqrt{x} - x - 4\ln(1 + \sqrt{x}) + C$

49. $-\frac{1}{252}(1 - v)^7(28v^2 + 7v + 1) + C$

51. $\frac{1}{2}[(2x - 1)^5 + 5]$ **53.** $e^{-x^2+1} - 1$

55. $21,000 - \frac{20,000}{\sqrt{1 + 0.2t}}; 6858$ **57.** $p(x) = \frac{250}{\sqrt{16 + x^2}}$

59. $30(\sqrt{2t + 4} - 2); 14,400\pi$ sq ft

61. $\frac{71.86887}{1 + 2.449e^{-0.3277t}} - 1.43760; 59.6$ inches

63. 24,555 pairs

Exercises 14.3, page 965

1. 4.27 sq units

3. a. 6 sq units

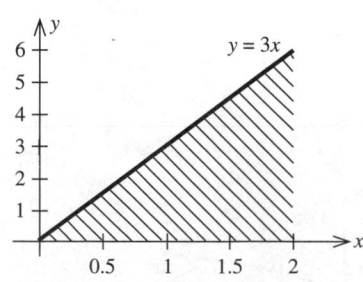

b. 4.5 sq units   **c.** 5.25 sq units   **d.** Yes

5. a. 4 sq units

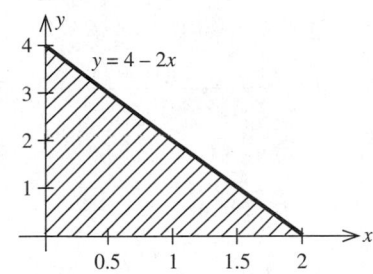

b. 4.8 sq units   **c.** 4.4 sq units   **d.** Yes

7. a. 18.5 sq units   **b.** 18.64 sq units
c. 18.66 sq units   **d.** ≈18.7 sq units

9. a. 25 sq units   **b.** 21.12 sq units
c. 19.88 sq units   **d.** ≈19.9 sq units

11. a. 0.0625 sq unit   **b.** 0.16 sq unit
c. 0.2025 sq unit   **d.** ≈0.2 sq unit

13. 4.64 sq units   **15.** 0.95 sq unit   **17.** 9400 sq ft

Exercises 14.4, page 976

1. 6 sq units   **3.** 8 sq units   **5.** 12 sq units

7. 9 sq units   **9.** ln 2 sq units   **11.** $17\frac{1}{3}$ sq units

13. $18\frac{1}{4}$ sq units   **15.** $(e^2 - 1)$ sq units   **17.** 6

19. 14   **21.** $18\frac{2}{3}$ **23.** $\frac{4}{3}$ **25.** 45   **27.** $\frac{7}{12}$

29. ln 2   **31.** 56   **33.** $\frac{256}{15}$ **35.** $\frac{2}{3}$ **37.** $2\frac{2}{3}$

39. $19\frac{1}{2}$ **41. a.** $4100   **b.** $900

43. a. $2800   **b.** $219.20   **45.** $10,133\frac{1}{3}$ ft

47. 46% of the smoke; 24% of the smoke

Using Technology Exercises 14.4, page 979

1. 6.1787   **3.** 0.7873   **5.** −0.5888   **7.** 2.7044

9. 3.9973   **11.** 37.7 million

Exercises 14.5, page 993

1. 10   **3.** $\frac{19}{15}$ **5.** $32\frac{4}{15}$ **7.** $\sqrt{3} - 1$ **9.** $24\frac{1}{5}$

11. $\frac{32}{15}$ **13.** $18\frac{2}{15}$ **15.** $\frac{1}{2}(e^4 - 1)$ **17.** $\frac{1}{2}e^2 + \frac{5}{6}$

19. 0   **21.** 2 ln 4   **23.** $\frac{1}{3}(\ln 19 - \ln 3)$

25. $2e^4 - 2e^2 - \ln 2$ **27.** $\frac{1}{2}(e^{-4} - e^{-8} - 1)$ **29.** 5

31. $\frac{17}{3}$ **33.** −1   **35.** $\frac{13}{6}$ **37.** $\frac{1}{4}(e^4 - 1)$

39. 120.3 billion metric tons   **41.** ≈$2.24 million

43. $40,339   **45.** 26°F   **47.** 16,863

49. 0.071 mg/cm³/hr   **51.** $\dfrac{2k}{3} R^2$ cm/sec

Exercises 14.6, page 999

1. 108 sq units   **3.** $\frac{2}{3}$ sq unit   **5.** $2\frac{2}{3}$ sq units

7. $1\frac{1}{2}$ sq units   **9.** 3   **11.** $3\frac{1}{3}$ **13.** 27

15. $2(e^2 - e^{-1})$ **17.** $12\frac{2}{3}$ **19.** $3\frac{1}{3}$ **21.** $4\frac{3}{4}$

23. 12 − ln 4   **25.** $e^2 - e - \ln 2$ **27.** $2\frac{1}{2}$ **29.** $7\frac{1}{3}$

31. $\frac{3}{2}$ **33.** $e^3 - 4 + \dfrac{1}{e}$ **35.** $20\frac{5}{6}$ **37.** $\frac{1}{12}$ **39.** $\frac{1}{3}$

41. a. S is the additional revenue that Odyssey Travel could realize by switching to the new agency

b. $S = \displaystyle\int_0^b [g(x) - f(x)]\,dx$

43. a. S is the additional amount of smoke that brand B will remove over brand A in the time interval $[a, b]$

b. $S = \displaystyle\int_a^b [f(t) - g(t)]\,dt$

45. 42.79 million metric tons   **47.** 57,179 people

Using Technology Exercises 14.6, page 1005

1. a.   **3. a.**

b. 1074.2857   **b.** 0.9961

5. a.   **7. a.**

b. 5.4603   **b.** 25.8549

9. a.

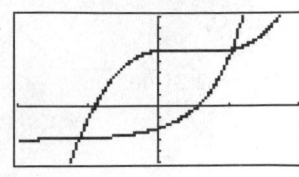

11. a.

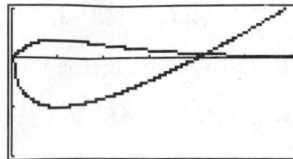

b. 10.5144

b. 3.5799

Exercises 14.7, page 1017

1. $11,667   **3.** $6667   **5.** $11,667

7. Consumers' surplus: $13,333; producers' surplus: $11,667

9. $824,200   **11.** $148,239   **13.** $52,203

15. $76,615   **17.** $111,869   **19.** $20,964

21. a.

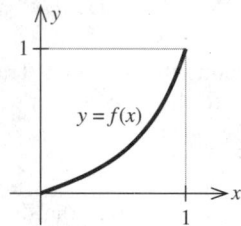

b. 0.175; 0.816

23. a.

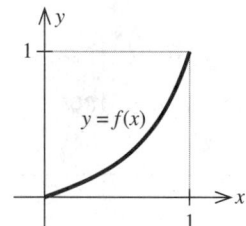

b. 0.104; 0.504

Using Technology Exercises 14.7, page 1020

1. Consumers' surplus: $18,000,000; producers' surplus: $11,700,000

3. Consumers' surplus: $33,120; producers' surplus: $2,880

Chapter 14 Review Exercises, page 1023

1. $\frac{1}{4}x^4 + \frac{2}{3}x^3 - \frac{1}{2}x^2 + C$ **3.** $\frac{1}{5}x^5 - \frac{1}{2}x^4 - \frac{1}{x} + C$

5. $\frac{1}{2}x^4 + \frac{2}{5}x^{5/2} + C$ **7.** $\frac{1}{3}x^3 - \frac{1}{2}x^2 + 2\ln|x| + 5x + C$

9. $\frac{3}{8}(3x^2 - 2x + 1)^{4/3} + C$

11. $\frac{1}{2}\ln(x^2 - 2x + 5) + C$ **13.** $\frac{1}{2}e^{x^2+x+1} + C$

15. $\frac{1}{6}(\ln x)^6 + C$ **17.** $\dfrac{(11x^2 - 1)(x^2 + 1)^{11}}{264} + C$

19. $\frac{2}{3}(x + 4)\sqrt{x - 2} + C$ **21.** $\frac{1}{2}$ **23.** $5\frac{2}{3}$

25. -80 **27.** $\frac{1}{2}\ln 5$ **29.** 4   **31.** $\dfrac{e - 1}{2(1 + e)}$

33. $f(x) = x^3 - 2x^2 + x + 1$ **35.** $f(x) = x + e^{-x} + 1$

37. -4.28

39. a. $-0.015x^2 + 60x$ **b.** $p = -0.015x + 60$

41. $3000t - 50,000(1 - e^{-0.04t})$; 16,939

43. $3100   **45.** 15 sq units   **47.** $\frac{2}{3}$ sq unit

49. $e^2 - 3$ sq units   **51.** $\frac{1}{2}$ sq unit   **53.** $\frac{1}{3}$

55. $2083; $3333   **57.** $98,973

59. a.

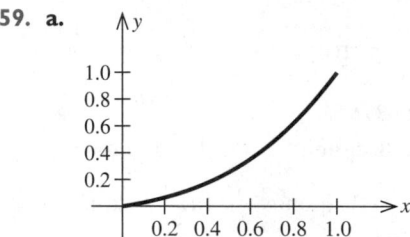

b. 0.1017; 0.3733   **c.** 0.315

CHAPTER 15

Exercises 15.1, page 1033

1. $\frac{1}{4}e^{2x}(2x - 1) + C$ **3.** $4(x - 4)e^{x/4} + C$

5. $\frac{1}{2}e^{2x} - 2(x - 1)e^x + \frac{1}{3}x^3 + C$ **7.** $xe^x + C$

9. $\dfrac{2(x + 2)}{\sqrt{x + 1}} + C$ **11.** $\frac{2}{3}x(x - 5)^{3/2} - \frac{4}{15}(x - 5)^{5/2} + C$

13. $\dfrac{x^2}{4}(2\ln 2x - 1) + C$ **15.** $\dfrac{x^4}{16}(4\ln x - 1) + C$

17. $\frac{2}{9}x^{3/2}(3\ln\sqrt{x} - 1) + C$ **19.** $-\dfrac{1}{x}(\ln x + 1) + C$

21. $x(\ln x - 1) + C$ **23.** $-(x^2 + 2x + 2)e^{-x} + C$

25. $\frac{1}{4}x^2[2(\ln x)^2 - 2\ln x + 1] + C$ **27.** $2\ln 2 - 1$

29. $4\ln 4 - 3$ **31.** $\frac{1}{4}(3e^4 + 1)$

33. $-\frac{1}{2}xe^{-2x} - \frac{1}{4}e^{-2x} + \frac{13}{4}$ **35.** $5\ln 5 - 4$

37. 1485 ft   **39.** 2.04 mg/ml

41. $-20e^{-0.1t}(t + 10) + 200$ **43.** $521,087

Exercises 15.2, page 1048

1. 2.7037; 2.6667; $2\frac{2}{3}$ **3.** 0.2656; 0.2500; $\frac{1}{4}$

5. 0.6970; 0.6933; $\approx$0.6931   **7.** 0.5090; 0.5004; $\frac{1}{2}$

9. 5.2650; 5.3046; $\frac{16}{3}$ **11.** 0.6336; 0.6321; $\approx$0.6321

13. 0.3837; 0.3863; $\approx$0.3863   **15.** 1.1170; 1.1114

17. 1.3973; 1.4052   **19.** 0.8806; 0.8818

21. 3.7757; 3.7625   **23. a.** 3.6   **b.** 0.0324

25. a. 0.013   **b.** 0.00043

27. a. 0.0078125   **b.** 0.0002848

29. 52.84 miles   **31.** 21.65 mpg

33. a. $142,374   **b.** $142,698

35. 103.9 PSI   **37.** $\approx$30%   **39.** $\approx$6.42 1/min

Exercises 15.3, page 1060

1. $\frac{2}{3}$ sq unit   **3.** 1 sq unit   **5.** 2 sq units

7. $\frac{2}{3}$ sq unit   **9.** $\frac{1}{2}e^4$ sq units   **11.** 1 sq unit

13. a. $\frac{2}{3}b^{3/2}$ **15.** 1   **17.** 2   **19.** Divergent

21. $-\frac{1}{8}$ **23.** 1   **25.** 1   **27.** $\frac{1}{2}$ **29.** Divergent

31. -1 **33.** Divergent   **35.** 0   **37.** 0

39. Divergent   **41.** $18,750

43. $\dfrac{10{,}000r + 4000}{r^2}$ dollars

45. b. $83,333

Exercises 15.4, page 1071

9. a. $k = \frac{1}{8}$ **b.** $\frac{1}{2}$

11. a. 0.63   **b.** 0.30   **c.** 0.30

13. 3   **15.** 0.37   **17.** 0.18

Chapter 15 Review Exercises, page 1074

1. $-2(1 + x)e^{-x} + C$ **3.** $x(\ln 5x - 1) + C$

5. $\frac{1}{4}(1 - 3e^{-2})$ **7.** $2\sqrt{x}(\ln x - 2) + 2$

9. $\frac{1}{2}$ **11.** Divergent   **13.** $\frac{1}{10}$

15. 0.8421; 0.8404   **17.** 2.2379; 2.1791

21. a. $k = \frac{3}{8}$ **b.** 0.6495

23. a. $k = \frac{4}{27}$ **b.** 0.4815

25. $-20e^{-0.05t}(t + 20) + 400$; 48,761

27. a. 274,000 sq ft   **b.** 278,667 sq ft

CHAPTER 16

Exercises 16.1, page 1085

1. $f(0, 0) = -4$; $f(1, 0) = -2$; $f(0, 1) = -1$; $f(1, 2) = 4$; $f(2, -1) = -3$

3. $f(1, 2) = 7$; $f(2, 1) = 9$; $f(-1, 2) = 1$; $f(2, -1) = 1$

5. $g(1, 2) = 4 + 3\sqrt{2}$; $g(2, 1) = 8 + \sqrt{2}$; $g(0, 4) = 2$; $g(4, 9) = 56$

7. $h(1, e) = 1$; $h(e, 1) = -1$; $h(e, e) = 0$

9. $g(1, 1, 1) = e$; $g(1, 0, 1) = 1$; $g(-1, -1, -1) = -e$

11. All real values of x and y

13. All real values of u and v except those satisfying the equation $u = v$

15. All real values of r and s satisfying $rs \geq 0$

17. All real values of x and y satisfying $x + y > 5$

19.   **21.**

23.

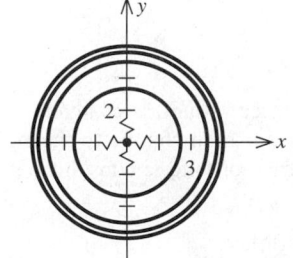

25. 9π cu ft   **27.** 40,000k dynes

29. $25,500; $23,580   **31.** $8310; $7910

33. a. The set of all nonnegative values of W and H
   b. 1.871 sq m

35. $13,498.59   **37.** $76,704.11; $50,814.62   **39.** 18

Exercises 16.2, page 1100

1. 2; 3   **3.** $4x$; 4   **5.** $-\dfrac{4y}{x^3};\dfrac{2}{x^2}$

7. $\dfrac{2v}{(u+v)^2}$; $-\dfrac{2u}{(u+v)^2}$

9. $3(2s-t)(s^2-st+t^2)^2$; $3(2t-s)(s^2-st+t^2)^2$

11. $\dfrac{4x}{3(x^2+y^2)^{1/3}};\dfrac{4y}{3(x^2+y^2)^{1/3}}$ **13.** ye^{xy+1}; xe^{xy+1}

15. $\ln y + \dfrac{y}{x};\dfrac{x}{y}+\ln x$ **17.** $e^u \ln v;\dfrac{e^u}{v}$

19. $yz + y^2 + 2xz$, $xz + 2xy + z^2$; $xy + 2yz + x^2$

21. ste^{rst}; rte^{rst}; rse^{rst} **23.** $f_x(1, 2) = 8; f_y(1, 2) = 5$

25. $f_x(2, 1) = 1; f_y(2, 1) = 3$

27. $f_x(1, 2) = \frac{1}{2}; f_y(1, 2) = -\frac{1}{4}$

29. $f_x(1, 1) = e; f_y(1, 1) = e$

31. $f_x(1, 0, 2) = 0; f_y(1, 0, 2) = 8; f_z(1, 0, 2) = 0$

33. $f_{xx} = 2y; f_{xy} = 2x + 3y^2 = f_{yx}; f_{yy} = 6xy$

35. $f_{xx} = 2; f_{xy} = f_{yx} = -2; f_{yy} = 4$

37. $f_{xx} = \dfrac{y^2}{(x^2+y^2)^{3/2}}; f_{xy} = f_{yx} = -\dfrac{xy}{(x^2+y^2)^{3/2}};$
    $f_{yy} = \dfrac{x^2}{(x^2+y^2)^{3/2}}$

39. $f_{xx} = \dfrac{1}{y^2}e^{-x/y}; f_{xy} = \dfrac{y-x}{y^3}e^{-x/y} = f_{yx}; f_{yy} =$
    $\dfrac{x}{y^3}\left(\dfrac{x}{y}-2\right)e^{-x/y}$

41. a. $f_x = 7.5; f_y = 40$ **b.** Yes

43. $P_x = 10$—at $(0, 1)$, the price of land is changing at the rate of $10 per sq ft per mile change to the right; $P_y = 0$—at $(0, 1)$, the price of land is constant per mile change upward.

45. Complementary commodities

47. $30/unit change in finished desks; $-$25/unit change in unfinished desks. The weekly revenue increases by $30/unit for each additional finished desk produced

(beyond 300) when the level of production of unfinished desks remains fixed at 250; the revenue decreases by $25/unit when each additional unfinished desk (beyond 250) is produced and the level of production of finished desks remains fixed at 300.

49. 0.039 l/degree; -0.015 l/mm of mercury. The volume increases by 0.039 l when the temperature increases by 1 degree (beyond 300 K) and the pressure is fixed at 800 mm of mercury. The volume decreases by 0.15 l when the pressure increases by 1 mm of mercury (beyond 800 mm) and the temperature is fixed at 300 K.

Using Technology Exercises 16.2, page 1103

1. 1.3124; 0.4038   **3.** -1.8889; 0.7778

5. -0.3863; -0.8497

Exercises 16.3, page 1114

1. $(0, 0)$; relative maximum value: $f(0, 0) = 1$

3. $(1, 2)$; saddle point: $f(1, 2) = 4$

5. $(8, -6)$; relative minimum value: $f(8, -6) = -41$

7. $(1, 2)$ and $(2, 2)$; saddle point: $f(1, 2) = -1$; relative minimum value: $f(2, 2), = -2$

9. $(-\frac{1}{3}, \frac{11}{3})$ and $(1, 5)$; saddle point: $f(-\frac{1}{3}, \frac{11}{3}) = -\frac{319}{27}$; relative minimum value: $f(1, 5) = -13$

11. $(0, 0)$ and $(1, 1)$; saddle point: $f(0, 0) = -2$; relative minimum value: $f(1, 1) = -3$

13. $(2, 1)$; relative minimum value: $f(2, 1) = 6$

15. $(0, 0)$; saddle point: $f(0, 0) = -1$

17. $(0, 0)$; relative minimum value: $f(0, 0) = 1$

19. $(0, 0)$; relative minimum value: $f(0, 0) = 0$

21. 200 finished units and 100 unfinished units; $10,500

23. Price of land ($200/sq ft) is highest at $(\frac{1}{2}, 1)$

25. $(0, 1)$ gives desired location

27. 18 inches $\times$ 36 inches $\times$ 18 inches

29. 6 inches $\times$ 4 inches $\times$ 2 inches

Exercises 16.4, page 1127

1. Min of $\frac{3}{4}$ at $(\frac{3}{4}, \frac{1}{4})$ **3.** Max of $-\frac{7}{4}$ at $(2, \frac{7}{2})$

5. Min of 4 at $\left(\sqrt{2}, \dfrac{\sqrt{2}}{2}\right)$ and $\left(-\sqrt{2}, -\dfrac{\sqrt{2}}{2}\right)$

7. Max of $-\frac{3}{4}$ at $(\frac{3}{2}, 1)$

9. Max of $2\sqrt{3}$ at $\left(\dfrac{\sqrt{3}}{3}, -\sqrt{6}\right)$ and $\left(\dfrac{\sqrt{3}}{3}, \sqrt{6}\right)$

11. Max of 8 at $(2\sqrt{2}, 2\sqrt{2})$ and $(-2\sqrt{2}, -2\sqrt{2})$;
    min of -8 at $(2\sqrt{2}, -2\sqrt{2})$ and $(-2\sqrt{2}, 2\sqrt{2})$

13. Max: $\dfrac{2\sqrt{3}}{9}$; min: $-\dfrac{2\sqrt{3}}{9}$

15. Min of $\frac{18}{7}$ at $(\frac{9}{7}, \frac{6}{7}, \frac{3}{7})$

17. 140 finished and 60 unfinished units

19. $10\sqrt{2}$ ft $\times$ $40\sqrt{2}$ ft   21. 2 ft $\times$ 2 ft $\times$ 3 ft

23. $15,000 on newspaper and $45,000 on TV

25. 30 ft $\times$ 40 ft $\times$ 10 ft; $7200

Exercises 16.5, page 1141

1. $\frac{7}{2}$ 3. 0   5. $4\frac{1}{2}$

7. $(e^2 - 1)(1 - e^{-2})$ 9. 1

11. $\frac{2}{3}$ 13. $\frac{188}{3}$ 15. $\frac{84}{5}$

17. $2\frac{2}{3}$ 19. 1   21. $\frac{1}{2}(3 - e)$

23. $\frac{1}{4}(e^4 - 1)$ 25. $\frac{2}{3}(e - 1)$ 27. 48

29. 6   31. $\frac{19}{6}$

33. $5(1 - 2e^{-2} + e^{-4})$ 35. 16

37. $\frac{1}{2}(e - 1)^2$ 39. $\frac{1}{2} \ln 17$

41. $\frac{64}{3}$ 43. $\frac{4}{3}$

45. $1 - \dfrac{1}{e}$ 47. $\frac{1}{8}(9 \ln 3 - 4)$

49. ≈ 2166 people/sq mi   51. $194/sq ft

Chapter 16 Review Exercises, page 1146

1. $0, 0, \frac{1}{2}$; no   3. $2, -(e + 1), -(e + 1)$

5. The set of all ordered pairs (x, y) such that $y \ne -x$

7. The set of all (x, y, z) such that $z \ge 0$ and $x \ne 1$, $y \ne 1$, and $z \ne 1$

9.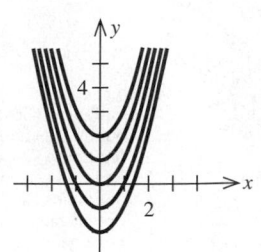

11.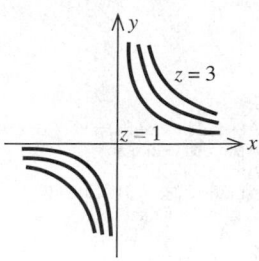

13. $f_x = \sqrt{y} + \dfrac{y}{2\sqrt{x}}$; $f_y = \dfrac{x}{2\sqrt{y}} + \sqrt{x}$

15. $f_x = \dfrac{3y}{(y + 2x)^2}$; $f_y = -\dfrac{3x}{(y + 2x)^2}$

17. $10y(2xy + 3y^2)^4$; $10(x + 3y)(2xy + 3y^2)^4$

19. $2x(1 + x^2 + y^2)e^{x^2 + y^2}$; $2y(1 + x^2 + y^2)e^{x^2 + y^2}$

21. $\dfrac{2x}{x^2 + y^2}$; $-\dfrac{2x^2}{y(x^2 + y^2)}$

23. $f_{xx} = 12x^2 + 4y^2$; $f_{xy} = 8xy = f_{yx}$; $f_{yy} = 4x^2 - 12y^2$

25. $g_{xx} = \dfrac{-2y^2}{(x + y^2)^3}$; $g_{yy} = \dfrac{2x(3y^2 - x)}{(x + y^2)^3}$;

    $g_{xy} = \dfrac{2y(x - y^2)}{(x + y^2)^3} = g_{yx}$

27. $h_{ss} = -\dfrac{1}{s^2}$; $h_{st} = h_{ts} = 0$; $h_{tt} = \dfrac{1}{t^2}$

29. $(2, 3)$; relative minimum value: $f(2, 3) = -13$

31. $(0, 0)$ and $(\frac{3}{2}, \frac{9}{4})$; saddle point at $f(0, 0) = 0$; relative minimum value: $f(\frac{3}{2}, \frac{9}{4}) = -\frac{27}{16}$

33. $(0, 0)$; relative minimum value: $f(0, 0) = 1$

35. $f(\frac{12}{11}, \frac{20}{11}) = -\frac{32}{11}$

37. Relative maximum value: $f(5, -5) = 26$; relative minimum value: $f(-5, 5) = -24$

39. 48   41. $\frac{2}{63}$ 43. $11\frac{1}{3}$ cu units

45. 3   47. Complementary

49. 337.5 yd $\times$ 900 yd

BASIC RULES OF DIFFERENTIATION

1. $\dfrac{d}{dx}(c) = 0$, $\quad c$ a constant

2. $\dfrac{d}{dx}(u^n) = nu^{n-1}\dfrac{du}{dx}$

3. $\dfrac{d}{dx}(u \pm v) = \dfrac{du}{dx} \pm \dfrac{dv}{dx}$

4. $\dfrac{d}{dx}(cu) = c\dfrac{du}{dx}$, $\quad c$ a constant

5. $\dfrac{d}{dx}(uv) = u\dfrac{dv}{dx} + v\dfrac{du}{dx}$

6. $\dfrac{d}{dx}\left(\dfrac{u}{v}\right) = \dfrac{v\dfrac{du}{dx} - u\dfrac{dv}{dx}}{v^2}$

7. $\dfrac{d}{dx}(e^u) = e^u\dfrac{du}{dx}$

8. $\dfrac{d}{dx}(\ln u) = \dfrac{1}{u} \cdot \dfrac{du}{dx}$

BASIC RULES OF INTEGRATION

1. $\displaystyle\int du = u + C$

2. $\displaystyle\int kf(u)\,du$
$= k\displaystyle\int f(u)\,du$, $\quad k$ a constant

3. $\displaystyle\int [f(u) \pm g(u)]\,du$
$= \displaystyle\int f(u)\,du \pm \int g(u)\,du$

4. $\displaystyle\int u^n\,du = \dfrac{u^{n+1}}{n+1} + C$, $\quad n \neq -1$

5. $\displaystyle\int e^u\,du = e^u + C$

6. $\displaystyle\int \dfrac{du}{u} = \ln|u| + C$

FORMULAS

EQUATION OF A STRAIGHT LINE

a. point-slope form $\quad y - y_1 = m(x - x_1)$
b. slope-intercept form $\quad y = mx + b$
c. general form $\quad Ax + By + C = 0$

EQUATION OF THE LEAST-SQUARES LINE

$$y = mx + b$$

where m and b satisfy the **normal equations**

$$nb + (x_1 + x_2 + \cdots + x_n)m = y_1 + y_2 + \cdots + y_n$$

$$(x_1 + x_2 + \cdots + x_n)b + (x_1^2 + x_2^2 + \cdots + x_n^2)m = x_1y_1 + x_2y_2 + \cdots + x_ny_n$$

COMPOUND INTEREST

$$A = P(1 + i)^n \qquad (i = r/m, n = mt)$$

where A is the accumulated amount at the end of n conversion periods, P is the principal, r is the interest rate per year, m is the number of conversion periods per year, and t is the number of years.

EFFECTIVE RATE OF INTEREST

$$r_{\text{eff}} = \left(1 + \frac{r}{m}\right)^m - 1$$

where r_{eff} is the effective rate of interest, r is the nominal interest rate per year, and m is the number of conversion periods per year.

FUTURE VALUE OF AN ANNUITY

$$S = R\left[\frac{(1 + i)^n - 1}{i}\right]$$

PRESENT VALUE OF AN ANNUITY

$$P = R\left[\frac{1 - (1 + i)^{-n}}{i}\right]$$

AMORTIZATION FORMULA

$$R = \frac{Pi}{1 - (1 + i)^{-n}}$$